PHYSIOLOGY

PHYSIOLOGY

Fifth Edition

Edited by Ewald E. Selkurt, Ph.D.

Distinguished Professor, Department of Physiology and Biophysics,
Indiana University School of Medicine,
Indianapolis

LITTLE, BROWN AND COMPANY   BOSTON/TORONTO

Library of Congress Catalog Card No. 83-80301

ISBN 0-316-78038-3

Printed in the United States of America

DON

CONTENTS

CONTRIBUTING AUTHORS

The contributors to *Physiology* are affiliated with the Department of Physiology and Biophysics, Indiana University School of Medicine, Indianapolis.

EWALD E. SELKURT, Ph.D., *Distinguished Professor; Editor*

WILLIAM M. ARMSTRONG, Ph.D., *Professor*

JULIUS J. FRIEDMAN, Ph.D., *Professor*

KALMAN GREENSPAN,* Ph.D., *Professor*

WILLIAM V. JUDY,† Ph.D., *Adjunct Associate Professor*

LEON K. KNOEBEL, Ph.D., *Professor*

RICHARD A. MEISS,‡ Ph.D., *Associate Professor*

WARD W. MOORE,§ Ph.D., *Associate Dean*

SIDNEY OCHS, Ph.D., *Professor*

RODNEY A. RHOADES, Ph.D., *Professor and Chairman*

CARL F. ROTHE, Ph.D., *Professor*

*Professor of Physiology and Medicine, Terre Haute (Indiana) Center for Medical Education.
†Clinical Research Division, Methodist Hospital, Indianapolis.
‡Joint Appointment in Department of Obstetrics and Gynecology.
§Director of Medical Science Program, Bloomington, Indiana.

PREFACE

The basic science of physiology has advanced markedly since the publication eight years ago of the fourth edition of this text. Yet in the Fifth Edition we have tried to maintain a compact style suitable for a good foundation in mammalian physiology for professional courses. To do this, revision has involved dropping out considerable material and substituting updated information. Several chapters have been entirely rewritten (e.g., Chapter 1, on the cell membrane and biological transport, and Chapters 19 through 21, on respiration). Most of the other chapters have also undergone appreciable revision. The text has been redesigned in format for greater simplicity.

Established facts are emphasized, and complementary subject matter that is included in other basic science disciplines—for example, anatomy and biochemistry—has been limited to review of essentials. Although basic principles are stressed, sufficient clinical application is provided to lay a strong foundation for subsequent development of the subject matter in a clinical setting.

The text is profusely illustrated, many of the figures having been updated. Particularly noteworthy is the clarity of the most complex figures. For this we are particularly indebted to our artists, Phil Wilson, Rebecca Wilson, and Deborah Eads. We are also grateful for the secretarial help of Marsha Hunt and Ann Hollingsworth.

E. E. S.

PHYSIOLOGY

THE CELL MEMBRANE AND BIOLOGICAL TRANSPORT

1

William M. Armstrong

Physiology is concerned with the overall functioning of the tissues and organs of which the body is composed. Essentially, these tissues and organs are *organized* assemblies of large numbers of cells, often of several different types. Through the operation of various *control mechanisms* (e.g., neural, hormonal), these complex assemblies respond to specific stimuli in an integrated fashion. One sees, then, that the cell is, in a very real sense, a fundamental functional unit in relation to the body as a whole, and that the behavior of single cells is a logical takeoff point for the study of more complex physiological systems.

Living cells are themselves complex and highly organized. In addition to a fluid phase (intracellular fluid or cytoplasm), the interior of a typical cell contains a number of *inclusions* or *organelles* (e.g., nucleus, mitochondria, lysosomes). Modern research has shown that these subcellular particles are again highly organized structures that play a vital role in the overall activity of the cell. For didactic purposes the detailed study of these structures is, nowadays, to a large extent, included in the formal disciplines of

biochemistry and cell biology. Physiology proper is often considered to begin at the point where the cell as a whole interacts with its external environment or with neighboring cells, that is, with the exchanges of matter and energy that take place across its outer limiting membrane, or *plasma membrane*. In keeping with this practice this chapter will focus on these exchanges. The student should remember, however, that this division of different aspects of cell function between various academic disciplines is purely a matter of convenience. It has no real basis in the domain of biology or medicine and is, quite properly, often ignored in the classroom and completely disregarded in the clinic or research laboratory.

Cytoplasm, or intracellular fluid, differs markedly in composition from the external medium (blood or interstitial fluid) bathing the cell. In intact, normally functioning cells, some of these differences in composition remain virtually constant over long periods of time. At the same time there are continuous exchanges of material between the cell interior and the external environment, with some substances entering the cell and others leaving it. Evidently, cells must possess a mechanism or mechanisms by which the entry and exit of dissolved solutes are regulated. Two alternative kinds of mechanism have been postulated. One invokes the idea that the cytoplasm as a whole is capable of selectively accumulating certain substances while excluding others. The other assigns the control of the transfer of materials between the cell and its environment to a special region at or near the periphery of the cell, the *cell membrane, or plasma membrane.* At present, although the idea of functional selectivity by the cytoplasm as a whole is vigorously championed by some workers (Ling, 1962), most physiologists accept the special role of the membrane in the regulation of cellular composition as the most satisfactory interpretation of the available experimental evidence. Therefore, in this chapter, cellular transport mechanisms will be discussed from this point of view.

THE CELL MEMBRANE

STRUCTURE. At present, the existence of the plasma membrane as a discrete entity is not seriously challenged, even by the few who dispute its importance in controlling the exchange of material between the cell interior and the external environment. Moreover, during recent years, there has been a rapidly growing awareness of the importance of internal membranous structures (e.g., the *sarcoplasmic reticulum* of muscle) as regulators of overall cellular function. It is a truism of contemporary biology that a complete understanding of the function of any system requires a detailed knowledge of its molecular architecture. The problem of the molecular structure of cell membranes, particularly as it relates to their functional properties, therefore is of crucial importance to the effective progress of physiology and medicine (Andreoli et al., 1978).

Unfortunately, we are still a long way from a complete understanding of the fine details of the structure of the cell membrane. Nevertheless, recent theoretical and experimental research has reached a stage where the arrangement of its major components can be inferred with some confidence. In this section, one current model of membrane structure, the fluid mosaic model of Singer (Singer and Nicolson, 1972), will be discussed, since it appears to represent the most convincing synthesis of theoretical concepts and experimental data presently available. However, some cautionary comments are appropriate.

First, there is by no means unanimous agreement on even the grosser aspects of membrane structure. A number of models of varying ingenuity, complexity, and plausibility have been proposed (Hendler, 1971; see also the chapter by

Robertson in Andreoli et al., 1978). Second, as a result of vigorous research utilizing a variety of physical, chemical, and biological techniques, our insights into the molecular structure of membranes are expanding so rapidly that any model of the membrane that can be proposed at present must be regarded as no more than a temporary expedient. Third, in considering any generalization about membrane structure, one must bear in mind that cell membranes exhibit considerable diversity in composition. Thus, there may be—and often are—significant differences in the structural details of the membranes of different cells and between different membranous structures within the same cell. This phenomenon becomes especially apparent when one considers highly specialized membranous organelles such as the *myelin sheath* of nerve or the *inner mitochondrial membrane.* Finally, the inertia inherent in human affairs has led to the retention, in otherwise unexceptionable textbooks, of earlier models for membrane structure (e.g., that of Davson, 1970).* One supposes that time will remedy this.

CHEMICAL COMPOSITION OF CELL MEMBRANES. The principal structural components of cell membranes are lipids (mainly cholesterol and phospholipids), proteins, and oligosaccharides. In addition, membranes contain water (some of which is probably in a more or less highly *ordered* state because of its close association with the ionized and polar groups of phospholipids and proteins) and small amounts of low-molecular-weight species such as inorganic ions. By far the most abundant components are proteins and lipids. Together, these account for approximately 95 percent, by weight, of the nonaqueous components of the membrane. By contrast, oligosaccharides normally account for approximately 3 to 4 percent and are often found as integral components of membrane glycoproteins. Their importance, however, far outweighs their relative scarcity, since specific polysaccharides are frequently key elements in the cell surface *receptors* that are implicated in the recognition and binding of ligand molecules. Important interactions of this type are those involved in immunological responses.

In general, the plasma membranes of eukaryotic cells have a protein/lipid ratio, by weight, of approximately 1/1. In specialized membranous structures the ratio may be different. For example, the myelin sheath of nerve (see Chap. 2) is a membranous structure that appears to function principally as a barrier to the movement of ions (i.e., the passage of electrical current). Apart from this, the myelin is

---

*For a brief obituary of this long-lived model, the reader is referred to the fourth edition of this text.

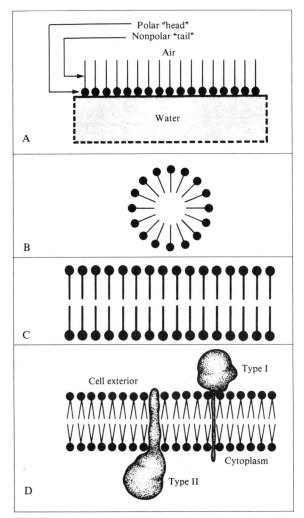

Fig. 1-1. (A) Orientation of a monolayer of amphiphilic lipid at the interface between a polar and a nonpolar phase. **(B)** Phospholipid micelle in a polar liquid. **(C)** Phospholipid bilayer. **(D)** Insertion of intrinsic proteins in the cell membrane.

relatively inert. The protein/lipid weight ratio in myelin is about 1/4. By contrast, the inner mitochondrial membrane, which is the site of a variety of enzymatic reactions, has a protein/lipid weight ratio of about 3/1.

ORGANIZATION OF MEMBRANE LIPIDS AND PROTEINS. In considering the problem of cell membrane ultrastructure, the basic question is the way in which the lipid and protein moieties of the membrane are arranged. These substances contain both polar and nonpolar groups and are therefore

amphiphilic, that is, display both *hydrophilic* and *lipophilic* or *hydrophobic* properties. Their polar groups tend to interact strongly with water or other polar solvents (hydrophilic reactions), and at the same time their nonpolar residues tend to interact with nonpolar solvents or other nonpolar entities (hydrophobic reactions). Thus, if proteins or lipids are spread as a layer one molecule thick (*monolayer*) at the interface between water and a nonpolar phase (e.g., oil or air), they tend to orient themselves so that their polar groups are in contact with the polar phase while their nonpolar portions are in contact with the nonpolar phase (Fig. 1-1A). In aqueous solutions, polar-nonpolar lipids are likely to clump together to form *micelles* (Fig. 1-1B), in which the polar "heads" are oriented outward toward the water, while the nonpolar "tails" are tucked inside the micelle, where their interactions with the polar water molecules are minimized and their interactions with each other are maximized. Similarly, proteins in aqueous solution are oriented so that as many as possible of their ionic residues are in contact with water (mostly at the outside of the protein molecule) and as many as possible of their hydrophobic residues are folded inside the molecule, away from the external aqueous environment.

This tendency of amphiphilic substances to orient themselves so that, as far as possible, both hydrophilic and hydrophobic (van der Waals) interactions are maximized is a reflection of a very general principle embodied in the second law of thermodynamics: The state of maximum stability for any system is that state in which the free energy of the system is at a minimum. Looked at in another way, the state of minimum free energy is the most probable state for the system to achieve as a result of spontaneous changes; or, conversely, one can say that spontaneous changes in any system always proceed in the direction of decreasing free energy. Applying this principle to the problem of the orientation of lipids and proteins in cell membranes, which may be regarded essentially as a thin layer, about 75 to 100 Å in thickness, bounded on both sides by an aqueous solution, one can predict that, within the constraints imposed by the overall geometry of the membrane, both these molecular species will dispose themselves in such a way that their capacity for hydrophilic and hydrophobic interaction with neighboring molecules or groups is maximized. A conformation that permits phospholipids to achieve this condition within the cell membrane is the *bilayer* (Fig. 1-1C). In this structure, the polar head groups of the phospholipids make contact with the aqueous media on both sides of the membrane, while the predominantly hydrocarbon tails are sequestered inside the bilayer.

The suggestion that the lipid moiety of the cell membrane

is organized as a bilayer was initially put forward about 50 years ago on the basis of studies with lipids extracted from red blood cell "ghosts," i.e., red blood cells from which the internal contents have escaped following *hemolysis* (see Appendix). Since then, this concept has not only remained viable but has been strongly reinforced by recent experimental work (Singer and Nicolson, 1972). Thus, despite the fact that a number of models for cell membrane structure have been proposed in which most membrane lipids are supposed to be organized as micelles, either alone or as mixed lipoprotein micelles (Lucy, 1968; Hendler, 1971), the existence of a lipid bilayer as an essential structural component of the cell membrane now seems well established. In particular, artificially produced lipid bilayers have been shown to possess many properties that are similar to those of naturally occurring cell membranes (Christensen, 1975). One may therefore conclude (Singer and Nicolson, 1972) that the greater part (at least 70 percent) of the membrane lipid is in the bilayer form, though some fraction (about 30 percent) may be organized in a different form, either alone or with proteins (i.e., as lipoproteins).

In the model proposed by Singer and Nicolson (1972), membrane proteins are divided into two classes. Singer and Nicolson originally called these classes *integral* and *peripheral* proteins, but the preferred terminology at present seems to be *intrinsic* and *extrinsic* proteins. (Such abrupt changes in fashion are somewhat baffling, but one must keep up to date!) Intrinsic proteins account for about 70 percent of the total amount of protein associated with the membrane. They are rather tightly bound to the membrane and can only be removed from it by relatively harsh treatment (e.g., extraction with detergents). Thus, they are considered to be integral components of the membrane matrix. In Singer and Nicolson's model, intrinsic proteins are considered to be inserted into the lipid bilayer as discrete entities at somewhat irregular intervals. Looked at from above, the membrane would present (on a microscale, of course) something of the aspect of a mosaic structure— hence the use of the term *mosaic* by Singer and Nicolson.

The insertion of intrinsic proteins into the lipid bilayer is illustrated in simple schematic form in Figure 1-1D. Two points about this schema may be noted. First, although it was not a part of Singer and Nicolson's original proposal, it is now believed that intrinsic proteins nearly always pass completely through the bilayer and are partly exposed to the predominantly aqueous media on both sides of it. Second, the disposition of these proteins with respect to the bilayer is not symmetrical. *Type I* proteins have most of their mass in the aqueous medium outside the membrane. *Type II* proteins reside mainly on the inner (cytoplasmic) side. As dis-

cussed later, this is not a trivial difference. A third important point (not shown in Fig. 1-1D) is that many intrinsic proteins, especially those involved in membrane transport (see p. 10), are not *monomers*. They are *oligomers*, i.e., they are composed of several subunits. For example, the ubiquitous membrane-bound Na^+-K^+ adenosine triphosphatase (ATPase) has two kinds of subunits, an alpha unit (90,000 daltons) and a beta unit (40,000 daltons), the latter being a glycoprotein. The complete enzyme has a total of four subunits and has the composition $\alpha_2\beta_2$. In the acetylcholine receptor protein that is present in the postsynaptic membrane of the neuromuscular junction (see Chap. 3), five different kinds of protein subunits have been identified. (The debilitating disease myasthenia gravis is characterized by a deficiency in this receptor protein—a good example of a pathological condition associated with a specific membrane defect.)

Within the lipid bilayer the polypeptide chains of intrinsic proteins are so arranged that the greatest possible number of hydrophobic (lipophilic) groups is externalized. Thus, van der Waals interactions between the "submerged" portions of the intrinsic proteins and the hydrophobic "tails" of the lipid bilayer are mainly responsible for "anchoring" the protein components in the complex shown schematically in Figure 1-1D.

The second class of proteins associated with the membrane are the so-called extrinsic proteins. These lie on the surface of the membrane, outside the matrix structure illustrated in Figure 1-1D and are not shown in this figure. They are attached to the membrane proper mainly by polar interactions between their polar groups and the polar "heads" of the membrane lipids and can be detached by relatively gentle methods, e.g., hypotonic solutions, strong salt solutions, or removal of divalent cations during membrane fractionation. In principle, one would expect to find extrinsic proteins associated with both the outer and the cytoplasmic surface of the membrane. In practice, they are nearly always found on the cytoplasmic side. This may, however, be an artifact, since extrinsic proteins at the outer surface could easily become detached and "lost" during early stages of membrane fractionation.

MEMBRANE DYNAMICS. In recent years a number of powerful techniques have become available for the investigation of molecular motion within the membrane. These include electron spin resonance, nuclear magnetic resonance, Raman spectroscopy, and the use of fluorescent probes. As a result, much is now known about the freedom of individual molecules to move within the membrane and about the constraints that restrict such movement.

Turning first to lipids, the study of intramolecular motion (the relative movement of different parts of a single molecule) and of the rotational motion of whole molecules indicates that the interior of the bilayer is a fluid structure (the basis of Singer and Nicolson's "fluid mosaic").* Investigation of the translational movement of lipids in membranes has revealed the important fact that whereas translational movement within the plane of the bilayer, that is, within one-half of the bilayer (monolayer), occurs readily (at least above the transition temperature*), migration across the bilayer (from one monolayer to the other—the so-called flip-flop) is difficult and probably occurs very rarely. This is easily understood on theoretical grounds. Lateral movement within one monolayer does not necessitate any change in the relative orientations of polar and nonpolar portions of the lipid molecule. On the other hand, flip-flop requires that the polar head group of the moving molecule be dragged through the nonpolar interior of the bilayer—an inherently improbable occurrence on energetic grounds.

It is easily seen that the same arguments apply, with even more force, to the large intrinsic protein molecules. These may have some freedom of movement in the plane of the bilayer; indeed, they are often compared to icebergs floating in a "sea" of lipid. Without wishing to push the analogy too far, one might add that, like real icebergs, they may drift but are very unlikely to turn cartwheels!

An interesting consequence of the relative absence of flip-flop movements on the part of membrane lipids is as follows: By suitable techniques, e.g., freeze fracture, it is possible to "peel" apart the two halves of the bilayer. The lipid composition of each half can then be determined. When this is done, it is usually found that individual lipids are asymmetrically distributed between the two halves. For example, with human red blood cells, it was found that, although both halves of the bilayer had identical amounts (per unit weight) of total phospholipid, the outside monolayer was relatively rich in sphingomyelin and phosphatidyl-

*This fluidity is very dependent on the lipid composition of the membrane. Certain lipids, notably cholesterol, exert a "condensing" effect and reduce the bilayer fluidity. Furthermore, highly interesting "phase transitions" have been reported both for artificial bilayers and natural membranes. These are characterized by a sudden marked transition from a highly fluid to a viscous, gel-like state (and vice versa) over a very narrow range of temperature. The transition temperatures for these changes are, again, highly composition dependent (e.g., cholesterol can markedly increase the transition temperature of a mixed lipid bilayer). Membrane fluidity and phase transitions may well turn out to be of great significance in physiology and medicine. It has been suggested, for example, that altered membrane fluidity may be an important factor in the action of certain anesthetics and that its composition dependence might be a significant variable in temperature adaptation. However, such suggestions are still largely speculative.

choline. The inner monolayer had relatively small proportions of these substances but had high proportions of phosphatidylethanolamine and phosphatidylserine.

FUNCTIONS OF MEMBRANE PROTEINS AND LIPIDS. Clearly (Fig. 1-1C), the lipid bilayer forms the structural core or matrix of the membrane. In addition, it forms an important barrier to the passage of charged particles, or *ions,* across the membrane and effectively restricts their diffusive movement to a limited number of channels that are associated with specific membrane proteins. Since ions are the charge carriers involved in the passage of *electrical currents* across cell membranes, the lipid bilayer is responsible for the high electrical resistance possessed by these structures. Measured membrane resistances are of the order of 1000 to 2000 ohm·cm^{-2}. A film of extracellular fluid or cytoplasm with the same dimensions (1 cm^2 × 100 Å thick) would have an electrical resistance of approximately 1×10^{-4} ohm, i.e., about 10 million times less.

The high electrical resistance of the cell membrane is important for the following reason: The normal function of cells requires that they maintain an electrical potential difference (the *membrane potential,* or *resting potential*) across their membranes. This is an essential part of the property of excitability in nerve and muscle cells (see Chaps. 2 and 3) and plays a key role in many other cell and tissue functions, e.g., absorption and secretion. As will be discussed later, the existence of cell membrane potentials is associated with the asymmetrical distribution of ions (notably K$^+$ and Na$^+$ ions) between the cell interior and its environment. The high electrical resistance of the cell membrane permits these ionic concentration differences to be preserved without prohibitive demands on the cell's energy resources. Finally, the lipid bilayer plays a role in the *permeability* of the membrane to certain substances. Molecules that have a certain degree of lipid solubility (see Factors Influencing Diffusive Transport) appear to penetrate cells by solution in, and passage through, the lipid bilayer. The significance of this fact to the practicing physician is that many drugs belong to this class of substance.

Type I intrinsic proteins have all their functional properties (as well as most of their mass) in the aqueous environment outside the lipid bilayer. The *intramembranous* part of these proteins is usually a relatively short amino acid sequence (about 20 or so amino acid residues) sufficient to span the bilayer in a single strand as an alpha-helix and "anchor" the protein to it. Proteins of this type are frequently involved in immunological reactions. Examples are the major sialoglycoprotein of the red blood cell membrane and the histocompatibility antigens of human cells. The glycoprotein nature of type I intrinsic proteins is noteworthy; e.g., in the most intensively studied example, the major sialoglycoprotein of the red blood cell, two-thirds of the total molecular mass consists of carbohydrate residues.

Type II intrinsic proteins (Fig. 1-1D) are characterized by the fact that most of their mass lies in the cytoplasm and that only a very small fraction of the molecule (10 percent or less) is exposed to the external environment on the noncytoplasmic side of the membrane. They appear to fall into two major categories. The first category consists of proteins in which all *functional properties* are located on the cytoplasmic side of the membrane. These proteins appear to be, in the main, membrane-bound enzymes (e.g., cytochrome b_5, cytochrome b_5 reductase). The second category includes proteins that subserve *transmembrane* functions, e.g., transport processes. Examples are the ubiquitous Na^+-K^+ ATPase, associated with Na^+ "pumping" in many cells; Ca^{2+} ATPase; the acetylcholine receptor protein; and the so-called (because of the position in which it appears during sodium dodecyl sulfate [SDS] gel electrophoresis) band 3 protein of the human red cell. Band 3 protein is involved in anion transport across the red cell membrane. All these proteins are involved in the transfer of specific substances across the cell membrane. It seems reasonable to suppose that proteins engaged in various kinds of transmembrane signaling (e.g., hormone receptors) will also be found to belong to this category. These proteins are, again, glycoproteins. They are also oligomers, and their oligomeric structure extends to the intramembranous moiety. It seems highly probable that the transport sites in these proteins (channels or pores) are located at the interfaces between the oligomeric subunits.

The precise function of many extrinsic proteins remains to be clarified. In some cases they may be components of enzyme complexes; e.g., the band 6 protein of the human red blood cell is a subunit of the enzyme glyceraldehyde phosphate dehydrogenase. Spectrin, a large and complex extrinsic protein, has been identified in the red blood cell membrane. It has been suggested that spectrin provides a protein network (cytoskeleton) that stabilizes the lipid bilayer. So far, this idea remains speculative. However, it is noteworthy that large polypeptides, comparable in size to spectrin, have been found in other kinds of cells, e.g., smooth muscle cells and macrophages.

THE TRANSPORT OF SMALL MOLECULES AND IONS ACROSS THE CELL MEMBRANE

MECHANISMS OF MEMBRANE TRANSPORT: GENERAL PRINCIPLES. The term *membrane transport*, as generally employed in physiology, has come to have a rather arbitrarily restricted meaning. Ordinarily, any general discussion, such as that presented here, is largely confined to transmembrane movements of water and low-molecular-weight solutes, the latter being assumed to be in the monomeric state. However, many cells of the body subserve important transport functions that are not subject to the rules to be discussed in detail subsequently. These transport functions may be absorptive, secretory, or protective. For example, the epithelial cells lining the lumen of the small intestine absorb fats. Multimolecular aggregates such as micelles (Fig. 1-1B) play an important role in this process. Again, the transmembrane transport of large molecules such as proteins is of vital importance to the normal functioning of the body. One may cite as examples the intestinal absorption of immunoglobulins by the newborn and the secretion of salivary and pancreatic enzymes from the cells in which they are synthesized to the lumen of the alimentary canal. Scavenging by the cells of the reticuloendothelial system of macromolecular or particulate matter of exogenous (e.g., bacterial) origin is an important element in the body's defense against infection or other toxic invasion.

It is generally believed that the process of *exocytosis* and its converse, *endocytosis,* are notable factors in the transport of relatively large molecules and aggregates. In this process a segment of the cell membrane folds around the transported species to form a vesicle that then separates either externally (exocytosis) or internally (endocytosis) from the rest of the membrane. The membranous "wall" of the vesicle is then degraded enzymatically, thus releasing the transported species to the external or internal medium. These processes, though they are of great physiological importance and will in some instances be discussed elsewhere in this book, will not be considered further in this chapter.

Even with the restrictions that have been noted, the characterization of membrane transport processes is complex, particularly because certain substances (e.g., water, Na^+ and K^+ ions) may be transported across the cell membrane at the same time by at least two different mechanisms. However, most attempts to categorize the transmembrane transport of simple solutes are structured around two basic considerations. These are the *kinetics* and the *energetics* of membrane transport. By a happy coincidence, the "rule of twos" applies in turn to each of these considerations. Thus, *kinetically,* membrane transport processes can be categorized as either *diffusive* or *saturable.* *Energetically,* they can be classified as *passive* (downhill) or *active* (uphill). Passive (downhill) transport processes are those in which net solute transfer ceases when the solute is at equilibrium across the cell membrane; i.e., when the *driv-*

ing force or gradient that effects a net transmembrane flow or flux of solute becomes zero. Thus, passive transport tends to dissipate those forces or gradients that can give rise to net flows of energy or matter across the cell membrane.

Clearly, systems in which passive flows alone occurred would be inconsistent with life. Therefore, in living cell membranes, one finds, as one would expect, a complex interplay and balance between passive and active transport processes. In the latter, the end point (zero net flow) is a steady state that differs from equilibrium in that finite driving forces (and the gradients that give rise to them) are conserved. It is clear, on thermodynamic grounds, that such processes, if they are to operate continuously, require an energy input from an external source. In the final analysis, this input can only be achieved at the expense of the metabolic energy of the cell. Indeed, a large part of the total metabolism is coupled to active transport in certain cells, such as those of the renal cortex, where the transport of large amounts of water and solutes is a major cellular activity.

DIFFUSIVE TRANSPORT. Diffusive membrane transport processes are so called because they bear a striking resemblance to diffusion in homogeneous physical systems, i.e., gases and liquids. Diffusion results from the spontaneous tendency for any substance (e.g., a gas or a solute in aqueous solution) to distribute itself uniformly throughout the whole space available to it. For example, if a solute is present initially at a higher concentration in one region of a solution than in another, with the passage of time the concentration difference disappears, and the solute concentration becomes uniform throughout the solution. As long as a solute concentration difference, or *concentration gradient,* exists between two different regions of the solution, there will be a spontaneous net movement of solute from the region of higher to the region of lower concentration. When all concentration gradients have been dissipated, *net* solute movement between different regions of the solution will cease. The solution is then said to be in a state of equilibrium with respect to the solute.

The rate of net solute diffusion between two regions of a solution containing unequal concentrations of solute is obviously equal to $-dS/dt$, the amount of solute passing from the region of higher to that of lower solute concentration per unit time. At constant temperature, this quantity is given by Fick's equation, as follows:

$$J = -\frac{dS}{dt} = DA\,(S_1 - S_2) = DA\,\Delta S \qquad (1)$$

In this equation, S_1 and S_2 are the solute concentrations in the regions of higher and lower concentration respectively;

A is the cross-sectional area of the boundary between these regions (the area across which diffusion is taking place); and D is the *diffusion coefficient,* or *diffusivity,* of the solute. It is apparent from equation 1 that D is numerically equal to the rate of diffusion across the unit area (1 cm²) when the concentration gradient, ΔS, across the boundary layer is unity. Thus, D is a measure of the inherent ability of the solute molecules to move through the solution. It is also clear from equation 1 that the net rate of diffusion at constant temperature depends on the concentration gradient, ΔS, and on the magnitude of D; D has been found to depend on temperature, becoming larger as the temperature is increased, and also varies with the molecular weight of the dissolved substance, becoming smaller as the molecular weight increases. It likewise depends to some extent on molecular shape as well as on molecular size. With large molecules in particular, the extent and strength of their interactions with other molecules, and hence the overall resistance to their diffusion in aqueous solution, are governed by shape as well as size. In these circumstances the relationship between D and molecular weight can be complex.

In thermodynamic terms the tendency of a solution to move spontaneously in the direction of equilibrium can be viewed as a reflection of the general tendency of natural systems to strive to attain the condition of minimum free energy, or maximum *entropy.* It can be shown by the methods of statistical thermodynamics that this condition corresponds to the most probable state of the system, which is, in turn, the state of *maximum disorder,* or *randomness.*

When a concentration gradient, ΔS, for a solute exists across a cell membrane, the rate of solute diffusion across the membrane is given by

$$J = \frac{PA}{x} \cdot \Delta S \qquad (2)$$

Equation 2 is analogous to equation 1 except for the term x, which represents the thickness of the membrane, and the term P, the *permeability coefficient* for the solute, which replaces the diffusion coefficient (D) in equation 1. The substitution of P for D merely reflects the fact that in these conditions it is the rate of diffusion of the solute within the membrane, rather than its diffusivity in free solution, that is the rate-determining step in the penetration process.

Equation 2 states that, for a diffusive process, the net rate of penetration, J, is a linear function of ΔS. This is illustrated graphically in Figure 1-2A. It is evident that the slope of the line relating J to ΔS in the graph is equal to PA/x. Thus, if one knows A and x, P is readily determined. In practice, A, the area of the membrane, and x may be difficult to determine with accuracy. However, if these vari-

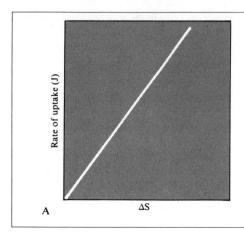

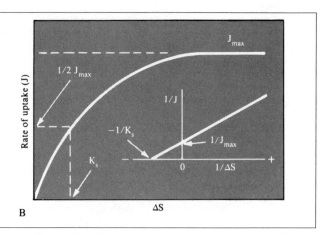

Fig. 1-2. Kinetic characteristics of transport processes. (A) Diffusion type. (B) Carrier-mediated transport. (From H. Stern and D. L. Nancy. *The Biology of Cells*. New York: Wiley, 1965. P. 326.)

ables are assumed to remain constant, the *relative* permeabilities of a given cell species for a number of solutes can be derived from uptake studies such as the one illustrated in Figure 1-2A.

It is important to note that in the present discussion and elsewhere in this chapter, *net* rates of transport are considered. If the concentration of a permeable solute is finite on both sides of a membrane, there will be, simultaneously, an *influx* of solute into the cell and an *efflux* of solute out of the cell. The *net flux* is obviously the difference between these two *unidirectional fluxes*. Now, in general, a flow or flux of material in any system implies a conjugate *driving force*. In thermodynamic terms the concentration gradient (ΔS of equation 2) represents the force driving the flow, J, in a diffusive process. The term PA/x is an example of a coupling coefficient that expresses the relationship between ΔS and J. A familiar example of an equation describing a linear relationship between a flow and its conjugate driving force is Ohm's law, $E = IR$, which relates the current or flow of electricity (I) between two points in a conductor to the driving force, E, the difference in electrical potential between these two points. The electrical resistance, R, is an expression of the coupling coefficient between E and I. Systems of linear equations of this kind form the basis of Onsager's analysis of the thermodynamics of nonequilibrium systems. In recent years, Onsager's theory has been extensively applied to the theoretical analysis of membrane transport (Katchalsky and Curran, 1965).

It is plain from equation 2 that the net flow in diffusive transport is always in the direction of the concentration gradient; i.e., net flow is *downhill* in an energetic sense. Hence, diffusion is an example of what is frequently called *passive* transport. When ΔS becomes zero, net flux is also

zero. Clearly, this is the condition for *equilibrium* (zero net driving force) and occurs when the concentrations of the solute inside and outside the cell are equal. Note, however, that this is always true only for uncharged solutes. With charged species (i.e., ions), the condition that equilibrium corresponds to equal concentrations inside and outside the cell holds only if there is no electrical potential difference across the cell membrane, a situation that virtually never exists with living cells (except very transiently as in the *action potentials* of nerve and muscle cells—see Chaps. 2 and 3). The implications of this fact are explored in detail under Ion Transport and Bioelectric Potentials.

FACTORS INFLUENCING DIFFUSIVE TRANSPORT. As a result of extensive studies on the penetration of cell membranes by a variety of compounds, the German physiologist E. Overton proposed two generalizations that apply to substances that cross the membrane by diffusive mechanisms. Although they were formulated in the latter part of the nineteenth century, these generalizations (which have become familiar to generations of physiology students as "Overton's rules") have stood the test of time remarkably well.

Overton's first rule is that for predominantly nonpolar molecules *permeability is directly proportional to lipid solubility*. *Lipid solubility* is defined as the *partition* or *distribution coefficient* (K) for a given substance between water and a lipid or fat solvent (e.g., chloroform, benzene) that is immiscible with water. That is, $K = C_1/C_2$, where C_1 and C_2

are the equilibrium concentrations of the solute in question in the nonaqueous medium and in water, respectively. This relationship implies that, for predominantly nonpolar molecules, high lipid solubility is associated with high permeability, and vice versa.*

The role of lipid solubility in cell penetration is further underlined by the behavior of weak electrolytes, i.e., electrolytes that undergo *reversible dissociation* of the type

$$HA \rightleftharpoons H^+ + A^- \tag{3}$$

or

$$BH^+ \rightleftharpoons H^+ + B \tag{4}$$

within the physiological pH range. It is found that, although weak electrolytes (e.g., acetic acid) penetrate cells readily in the nonionized state, they fail to do so when *ionized*. This finding is in keeping with the fact that K values for these compounds are much greater in the nonionized than in the ionized state.

The implications for drug therapy are of considerable practical importance, since many drugs are weak electrolytes. Consider the absorption from the alimentary canal into the bloodstream of two orally administered drugs, one of which (A) dissociates according to equation 3, while the other (B) dissociates in conformity with equation 4. In any given situation the degree of ionization of both these drugs will be determined by the relationship between the pH of the surrounding medium and the pK (pH of half-dissociation) of the drug. For drug A the degree of ionization will decrease as the ambient pH decreases; the opposite will be true for drug B. Let us suppose that both our hypothetical drugs have a pK of 6 (and that neither is chemically altered by the enzymes of gastrointestinal juice). In the stomach, where the pH of the gastric juice is ordinarily about 2 to 3, virtually all of drug A will be in the nonionized form and thus will pass freely through the membranes of the gastric mucosa, whereas drug B will be almost completely ionized and will not be absorbed to any great extent during its passage through the stomach. On the other hand, further down the intestine, where the alkaline secretions of the pancreas increase the luminal pH to 7 or 8, drug B will be largely converted to the nonionized form and readily absorbed. In consequence, if both A and B are administered

together, one would expect to find an effective level of A in the bloodstream before B appeared in quantity. Such considerations (in addition to intrinsic therapeutic potency, toxicity, etc.) underlie the many chemical variants of a basic therapeutic agent that are frequently available to the physician.

Overton's second rule is concerned with predominantly polar or hydrophilic molecules. It states that, for these molecules, permeability is *inversely related to molecular size;* i.e., the smaller the molecule, the more readily it penetrates the cell, and vice versa. A much fuller discussion of the role of lipid solubility, molecular size, and dissociation in the permeability of nonelectrolytes and weak electrolytes is given by Davson (1970).

The model of the cell membrane shown in Figure 1-1D is easily reconciled with Overton's rules. Since, in this model, the fluid matrix of the membrane is lipid in nature, one can easily see how lipid solubility can be the major determinant of permeability for nonpolar molecules. Similarly, the model readily permits the existence of discrete aqueous pores through which small, water-soluble molecules and ions (in addition to water itself) might be expected to move. As already mentioned, the most likely location of these pores is between the subunits of oligomeric intrinsic proteins. Current evidence indicates that the apparent average pore diameter in the membranes of a number of cell species is about 8 Å.

CARRIER-MEDIATED TRANSPORT. With many substances of physiological importance (e.g., sugars, amino acids), the experimental curve relating transport rate to the transmembrane concentration difference (ΔS) is frequently of the type shown in Figure 1-2B. At first, as ΔS is increased, the transport rate also increases. Finally, however, a value of ΔS is reached at which the transport rate becomes maximal and independent of further increases in ΔS (J_{max} in Fig. 1-2B). Nonlinear kinetic behavior of this type is referred to as *saturation kinetics* and is often exhibited by systems in which the kinetic process involves the reversible combination of the substrate with a receptor site (e.g., enzyme reactions).

In the case of membrane transport it is postulated that the transported solute or *substrate* enters the cell by combining reversibly with a specific membrane component or *carrier* at the outer surface of the membrane. The carrier-substrate complex so formed moves across to the inner surface of the membrane, where it dissociates, releasing free substrate to the cell interior. The carrier then moves back across the membrane to the outer boundary, where it combines with a second molecule of substrate, and the cycle begins again.

---

*Some discussion of the meaning of the term *high lipid solubility* in the context of cell physiology seems appropriate. Clearly, it must mean that K exceeds a certain critical value. A survey of the pertinent literature shows that Overton and his successors set this value at about 0.02. Hence, the word *high* in the present context is a very relative one indeed! Conversely, we may note that the term *low lipid solubility* is used by membrane physiologists to denote K values that are two to three orders of magnitude less than this.

Thus, a relatively small number of carrier molecules operating cyclically can transport large amounts of substrate. On this basis the occurrence of a maximum in the rate of transport with increasing substrate concentration is readily explained as being due to saturation of the available carrier sites when the external substrate concentration reaches a sufficiently high level. Clearly, an analogous explanation applies to net outward transport across the cell membrane.

With one simplifying assumption, the kinetic behavior of carrier-mediated transport systems can be quantitatively described by an equation identical to the well-known Michaelis-Menten equation of enzyme kinetics, i.e.,

$$J/J_{max} = \frac{[S]}{[S] + K_s} \tag{5}$$

In this equation, J is the transport rate at a given substrate concentration, [S], and K_s is analogous to the Michaelis-Menten constant, K_m; i.e., K_s is numerically equal to that substrate concentration for which $J = \frac{1}{2} J_{max}$.

The assumption previously mentioned is that, of the three steps involved in carrier-mediated transport, i.e., an *association* between substrate and carrier at one side of the membrane, a *translocation* of the carrier-substrate complex across the membrane, and a *dissociation* of this complex at the other side of the membrane, the *translocation* step is by far the slowest. With this assumption, $1/K_s$ gives a measure of the *affinity* of the carrier for the transported substance. Note that this affinity is *reciprocally* related to the magnitude of K_s.

In practice, this assumption is well justified for many membrane transport processes.* This has proved useful because, as is well known from enzyme studies, equation 5 can be used as the basis for a kinetic analysis of the action of specific inhibitors. Unlike diffusive processes, carrier-mediated transport is highly sensitive to these inhibitors. In terms of this analysis the action of competitive inhibitors is characterized by an increase in the apparent value of K_s with no change in J_{max}. Noncompetitive inhibitors lower J_{max} without affecting K_s. As shown in the insert of Figure 1-2B, K_s and J_{max} are easily determined from a plot of $1/J$ against $1/[S]$ for any given set of conditions.

A further application of this kind of kinetic analysis has been to show that substances that are closely related in chemical structure (e.g., hexose sugars, some amino acids, alkali metal cations) often compete for a common carrier in

*There are, of course, exceptions. Many of the complexities associated with enzyme kinetics and receptor binding studies have been noted also in transport kinetics. A number of these complexities are discussed in the article by Wilbrandt and Rosenberg (1961) cited at the end of this chapter.

the cell membrane. This approach has been extremely useful in analyzing the different carrier-mediated transport pathways that exist in various cell types and in constructing *affinity* or *selectivity sequences* for the transport of chemically related species by specific membrane carriers. These are simply obtained by ranking, in descending order, the reciprocals of the K_s values for the transported species involved.†

TRANSPORT ENERGETICS: THE COUPLING OF TRANSPORT PROCESSES. In terms of energetics, we may note that while diffusive transport is *always* passive, carrier-mediated transport processes may be passive or active. The corollary to this is that *active transport always involves a membrane carrier*. As previously noted for diffusive transport, net passive (downhill) carrier-mediated transport occurs in the direction of the substrate concentration gradient and ceases when the substrate is at equilibrium across the membrane. For uncharged solutes this corresponds to the condition $\Delta S = 0$. At this point the substrate is transported at the same rate in both directions across the membrane (*exchange diffusion*). Equilibrating processes of this kind are often referred to as *facilitated diffusion*. They do not depend on metabolism and can, in fact, be studied in systems such as isolated membrane vesicles in which metabolic energy production is entirely absent.

Uphill carrier-mediated transport processes are of two kinds: those that directly utilize metabolic energy to effect solute transfer (primary active transport) and those in which uphill transfer of one solute species is achieved at the expense of the transmembrane energy gradient for another (secondary active transport).

Primary active transport processes ("pumps") appear to be less common than was formerly thought. In general, they appear to be restricted to systems that operate through specific membrane-bound ATPase, e.g., the Na^+-K^+ ATPase (the Na^+ pump) or the Ca^{2+} ATPase that is found in

†It is important to note that such characteristics as the existence of competition between structurally related species and of sensitivity to specific inhibitors may, in some cases, be the only criteria whereby carrier-mediated transport can be distinguished from diffusive movement. This occurs when one is constrained to work with concentrations of the transported solute that are far below its K_s. It will be clear from an inspection of Figure 1-2 that, under these conditions, the kinetics of carrier-mediated transport become virtually linear. A slightly embarrassing example of this is the transport of urea across the membrane of the human red blood cell. In the lecture hall and in the laboratory, this has been exhibited to generations of admiring students as the example, par excellence, of diffusive transport. There is now increasing evidence (Macey, 1979) that it is a carrier-mediated process! Charity compels one to assume that the reason why this escaped the notice of so many perspicacious professors for so many years lies in the enormous (in physiological terms) K_s involved (about 2 *M*).

the membranes of many cell species. Though many of the detailed molecular mechanisms involved are still unknown, these transport systems can directly utilize the energy liberated by the hydrolysis of the terminal phosphate bond of ATP to transport ions (Na^+, K^+, Ca^{2+}) against a net energy gradient.

The Na^+-K^+–ATPase pump is an example of a *coupled transport* process. To operate successfully, it requires that transmembrane transport of Na^+ and K^+ occur at the same time. Normally, Na^+ is moved (pumped) outward across the cell membrane in an energetically uphill direction (see p. 12). The movement of K^+ is inward and may or may not be uphill, depending on the cell species involved and the specific experimental situation. Coupled movement of two dissolved solutes in opposite directions is called *antiport*. Analogous movement, but in the same direction, is called *symport*.

A further point of interest about the Na^+-K^+ pump is the stoichiometry or *coupling ratio* between Na^+ and K^+. This is normally not unity and in most cases appears to be 3 Na^+/ 2 K^+. Since the moving species are electrically charged, it is clear that the operation of this pump results in *net charge transfer,* in an outward direction, across the cell membrane. This will tend to generate a transmembrane electrical potential difference, oriented so that the exterior of the membrane is positive with respect to the inside. Transport processes of this kind are called *electrogenic,* and there is indeed evidence that the Na^+-K^+ pump contributes to the resting membrane potential of some cell species. Alternatively, since a continuous transfer of electrical charge across a membrane is equivalent to a flow of current, transport processes of this kind are sometimes called *rheogenic*.

Secondary active transport is another example of *coupled transport*. In processes of this kind, *uphill* transport of one species is obligatorily coupled to the simultaneous downhill transport of another. The transmembrane energy gradient for the downhill part of the overall process serves as the energy source for its uphill component. Coupled processes of this kind are perhaps most easily understood by considering some specific examples, such as the four diagrammed in Figure 1-3. These are chosen from the wide variety of such processes that are known, or suspected, to occur in certain epithelial tissues, e.g., the renal proximal tubules, gallbladder, and small intestine. More specifically, they represent coupled transport systems that have been identified in the *apical* membranes of the absorptive cells in these tissues.

The first transapical coupled transport process shown in Figure 1-3 (reading from top to bottom on the left-hand side) is Na^+-Cl^-–coupled symport. This is the major entry route across the apical cell membrane for these ions in a number

of epithelia. There is compelling evidence that the stoichiometry of this process is 1/1; i.e., no net charge is transported across the membrane by this mechanism, which is therefore *electroneutral*. It would not be expected to have any effect on the apical membrane potential. This prediction has been confirmed by direct experimentation (Garcia-Diaz and Armstrong, 1980).

The second process shown, Na^+-glucose symport, is an example of Na^+-coupled transport of a nonelectrolyte. Processes like this are of key importance in the reabsorption (or absorption) of organic metabolites by the kidney and intestine (see Chaps. 22 and 26) and are also found in other tissues (Schultz and Curran, 1970). Since one of the transported solutes is electrically charged and the other is not, it will be evident, regardless of the Na^+/glucose coupling ratio, that this transport process must be rheogenic. In fact, since the cell membrane potential is normally oriented so that the inside is negative with respect to the outside, one would predict that the inward flow of Na^+ ions induced by Na^+-coupled glucose entry should *depolarize* the cell membrane (i.e., reduce the magnitude of the electrical potential difference across it). This prediction was experimentally confirmed for the small intestine by White and Armstrong (1971).

The third process illustrated in Figure 1-3 is a Na^+-H^+ antiport or exchange. This has been shown to be electroneutral (i.e., one Na^+ ion exchanges for one H^+ ion). However, because H^+ ions produced as a result of cell metabolism can be "extracted" from the cell interior by Na^+-H^+ exchange, this mechanism is important, in some cells at least, in the regulation of intracellular pH (Roos and Boron, 1981).

Finally in Figure 1-3, a Cl^--HCO_3^- antiport is indicated. This again is usually considered to be a 1/1 electroneutral process. Since HCO_3^- ions play an important regulatory role with respect to intracellular pH (see Chap. 24), Cl^--HCO_3^- exchange is important in this context. This exchange, by removing HCO_3^- ions from the cell interior, would tend to acidify this region. Clearly, the Na^+-H^+ exchange already discussed would have the opposite effect.

Specific membrane inhibitors have been found for all the four coupled processes just discussed. The loop diuretics furosemide and bumetanide (so called because their primary target in the kidney is the loop of Henle) are potent inhibitors of Na^+-Cl^- symport. The diuretic amiloride inhibits Na^+-H^+ exchange (though this is *not* the primary physiological basis of its diuretic action). The glycoside phlorhizin combines avidly with the sugar-binding site of the Na^+-glucose symport but is not transported across the cell membrane. Finally, Cl^--HCO_3^- exchange is specifically in-

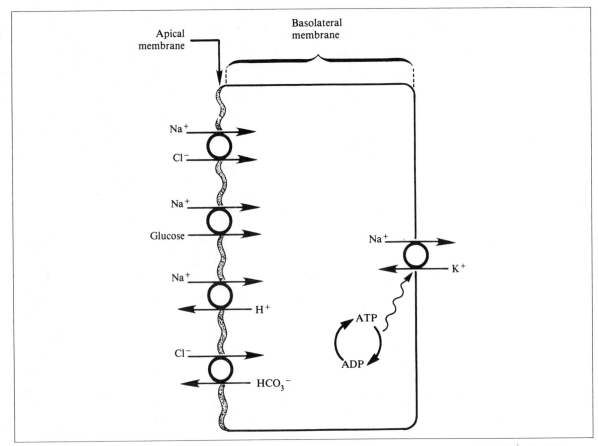

Fig. 1-3. Coupled transport processes in epithelial cells. *ATP* = adenosine triphosphate; *ADP* = adenosine diphosphate.

hibited by a series of substituted stilbene derivatives (known "in the trade" by such evocative acronyms as SITS and DIDS). Such inhibitors have proved extremely useful in identifying and characterizing the mechanisms involved in coupled transport.

In principle, none of these transport processes depends directly on metabolism. This is elegantly demonstrated by the fact that they can be shown to occur in isolated vesicles prepared from the apical membranes of cells from the kidney and intestine, among others. However, in vesicular systems, after their initial induction, the transport processes tend to run down; i.e., the net rate of transport declines with time. This is because the energy gradients that drive them are progressively dissipated. In living cells, metabolic processes serve to *conserve* these gradients and thus contribute indirectly to the continuous operation of gradient-dependent transport. As a reminder of this, the basolateral Na^+-K^+ pump is included in the schematic representation of an epithelial cell shown in Figure 1-3. By removing Na^+ continuously from the cell interior, this pump conserves the transapical Na^+ gradient, that is, the energy source for three of the coupled transfer processes shown.

The examples in Figure 1-3 are all representative of highly specific coupled transport mechanisms; that is, they involve membrane carriers that possess special combining sites for both transported solutes. Normally, these sites are highly selective. For example, in coupled processes that are sodium dependent, other cations, including alkali metal cations, cannot, as a rule, substitute effectively for Na^+. Similarly, the Na^+-glucose co-transport system illustrated in Figure 1-3 will accept only other hexose sugars (e.g., galactose) that share certain structural features with D-glucose (i.e., a pyranose ring with a carbon attached to C_5 and the OH on C_2 in the same stereochemical orientation).

In addition to these highly selective coupled transport mechanisms, there are other forms of coupling that are much less specific, well exemplified by the net osmotic movement of water that is induced by a transmembrane solute gradient (see p. 14). Such movement can occur even when water is, in chemical terms, at equilibrium across the membrane and is clearly driven by the solute gradient. In a purely mathematical and phenomenological context, this kind of coupling has been successfully analyzed by the methods of nonequilibrium thermodynamics (Katchalsky and Curran, 1965). In terms of specific molecular models it remains something of a puzzle. When, as frequently happens, water is one of the species transported, this kind of coupling is often called *frictional coupling,* the general idea being that energy transfer between the different molecular species that are moving across the membrane occurs by mechanisms analogous to mechanical friction. Although this has proved to be a useful approach for mathematical analysis, it is, for descriptive purposes, not entirely satisfactory.

WATER MOVEMENT: OSMOSIS AND OSMOTIC PRESSURE.
Virtually all cells are freely permeable to water. The regulation of the water content of cells and tissues is an important factor in the maintenance of the overall *fluid and electrolyte balance* of the body (see Chap. 23), which is crucial in human health and disease. Indeed, one of the most striking contributions of physiology to modern clinical medicine lies in the sophisticated system of fluid and electrolyte therapy that has been and is still being developed to deal with imbalances in body water and electrolytes resulting from pathological conditions or from accidental or surgically induced trauma. It is therefore not surprising that the movement of water across cell membranes and the forces governing this movement have been of prime interest to physiologists for many years.

Osmotic Phenomena in Dilute Solutions.
Consider a vessel divided into two parts by a membrane that is permeable to water but impermeable to dissolved solutes (a *semipermeable* membrane). If the compartment on one side of the membrane contains pure water and the other compartment contains an aqueous solution, water will flow spontaneously from the side containing pure water to the side containing the solution. This spontaneous net flow of water is called *osmosis,* or *osmotic flow.* Its origin may be explained as follows: As noted at the beginning of the chapter, all natural processes tend to proceed spontaneously in the direction of equilibrium. In the system under consideration, there are, initially, two concentration gradients across the membrane: (1) a solute concentration gradient from the side containing the solution to the side containing pure water and (2) because of the diluting effect of the solute on the water in the side containing the solution, a concentration gradient for water in the opposite direction. In the absence of external restraints, both water and solute would diffuse freely in the direction of their respective concentration gradients until mixing was complete. Because of the restraint imposed on the system by the semipermeable membrane, the solute cannot diffuse into the side containing pure water. The only net movement of material that can take place across the membrane is a flow of water into the side containing the solution. If, instead of pure water and a solution, two solutions containing unequal concentrations of solutes were used, a similar osmotic flow of water would take place. In this case, osmosis would occur from the more dilute to the more concentrated solution.

Clearly, unless some additional restraint is imposed, water will continue to move in the direction of its concentration gradient* as long as that gradient exists; i.e., until the concentrations of water on both sides of the membrane become equal. In the case of two solutions, equilibrium would be achieved when the solute concentrations on both sides of the membrane became equal. In a system containing pure water and an aqueous solution, osmotic equilibrium cannot, in principle at least, be realized in this way at any finite solute concentration. However, it is evident that if a force or pressure equal and opposite to the force generated by the concentration gradient for water across the membrane could be applied to the side containing the solution, osmotic flow could be prevented. The hydrostatic pressure (ΔP) required to prevent osmotic flow of water into a given solution is called the *osmotic pressure* ($\Delta \pi$) of that solution.

There are many ways in which osmotic flow can be prevented by the application of an external force. One of the simplest—and it also permits the osmotic pressure to be determined directly—is utilization of the force of gravity. The solution whose osmotic pressure is to be measured is placed inside a thin, semipermeable bag, which is then immersed in a vessel containing water. One end of a fine capillary tube is inserted into the bag, and the apparatus (called an *osmometer*) is adjusted so that the capillary tube is vertical. Initially, the apparatus is adjusted so that the liquid levels in the capillary and in the outer vessel, apart from the slight rise due to capillary action, are the same. As water

---

*This concentration gradient can then be said to be the driving force for osmotic water flow. More exactly, this force is related to the chemical potential gradient for water.

enters the bag by osmosis, the level of the liquid in the capillary rises until the hydrostatic pressure developed is just sufficient to balance the osmotic driving force across the wall of the bag. If the volume of the column of solution in the capillary is very small compared with the total volume enclosed in the bag (so that the total amount of water that enters the bag and its diluting effect on the solution within the bag are negligible), the osmotic pressure of the original solution is given by

$$\Delta\pi = \Delta P = h\rho g \tag{6}$$

where $\Delta\pi$ is the osmotic pressure difference across the membrane, h is the height of the column of solution in the capillary necessary to balance the osmotic driving force, ρ is the density of the solution, and g is the acceleration due to gravity. For moderately dilute solutions, ρ does not differ significantly from the density of water; hence, the height in centimeters of the column in the capillary will give the osmotic pressure directly in centimeters of water. Since 1034 cm of water is equivalent to 76 cm of mercury or 1 standard atmosphere, $\Delta\pi$ is readily obtained in either of these units.

Osmotic Pressure and Solute Concentration: Units of Osmotic Concentration. It is clear from the preceding discussion that the magnitude of the osmotic pressure in any given solution depends only on the difference between the concentration of water in that solution and its concentration in the pure liquid. This dependence, however, is of little practical use in physiology. Useful and meaningful relationships can be obtained if osmotic pressure is expressed in terms of solute concentration. Fortunately, despite the fact that the solute, aside from its diluting effect on the concentration of water in the solution, is not a fundamental factor in the generation of osmotic pressure, a simple relationsip between osmotic pressure and solute concentration does exist. The reason is that, in moderately dilute solutions, there is a simple complementary relationship between the concentration of water and that of solute.

Consider a solution containing n_1 moles of water and n_2 moles of solute per unit volume. The *molar fractions* of water and solute are respectively $X_1 = n_1/(n_1 + n_2)$ and $X_2 = n_2/(n_1 + n_2)$. Hence, $(X_1 + X_2) = 1$, and $X_2 = (1 - X_1)$. Evidently, it is the factor $(1 - X_1)$ that determines the osmotic pressure of a solution in contact with pure water. Therefore, the osmotic pressure is directly proportional to X_2 or solute concentration. For a dilute solution the quantitative relationship between osmotic pressure and solute concentration is given by the van't Hoff equation

$$\Delta\pi = CRT \tag{7}$$

where C is the solute concentration, R is the gas constant, and T is the absolute temperature.

It is apparent from equation 7 that, if the concentration of solute in a given solution is known, its osmotic pressure can readily be calculated. Conversely, given the osmotic pressure, the solute concentration can be obtained from this equation. It is important to realize that there is a difference between concentrations as they relate to osmotic activity and ordinary chemical concentrations. Since, as has already been pointed out, the solute has no intrinsic effect on osmotic pressure, the osmotic pressure of a solution is independent of the nature of the solute particles and depends only on their number per unit volume. In other words, in a given volume of solution, equal numbers of dissolved particles will contribute equally to osmotic pressure whether the particles are large molecules, small molecules, or ions. For this reason, the osmotic activities of solutions containing equal *chemical* concentrations of different solutes will not necessarily be identical. Consider, for example, two solutions, one containing 0.1 M sucrose and the other containing 0.1 M NaCl. Although the concentrations of these two solutions are equal in chemical terms (in moles/L), their osmotic pressures are not the same, because NaCl exists in solutions as Na^+ and Cl^- ions. Consequently, the osmotic pressure of a 0.1 M solution of NaCl is approximately twice that of a 0.1 M sucrose solution. Thus, it is apparent that, with electrolyte solutions, if one wishes to relate the osmotic pressure or effective osmotic concentration of solute to its chemical concentration, one must multiply the term C in equation 7 by n, where n is the number of ions produced by one molecule of electrolyte and is sometimes referred to as the van't Hoff coefficient.

The situation is further complicated by the fact that this simple relationship between the number of ions formed by an electrolyte and its osmotic activity holds only for very dilute solutions. Because of the attractive forces between ions of opposite charge and between individual ions and water molecules, the apparent value of n for a given electrolyte varies with concentration. Also, the concentration dependence of n is different for different electrolytes. At physiological concentrations the divergence of n from its limiting value in dilute solution is sufficiently great to affect appreciably the accuracy of results calculated on the basis of equation 7. Further, the physiologist is frequently confronted with solutions, such as blood or urine, that contain complex mixtures of solutes, both electrolytes and nonelectrolytes, in widely different concentrations. In this situation the need for a practical unit of osmotic concentrations that is independent of the apparent variation of n with concentration for different individual solutes will readily be appreciated. The *osmol* (Osm) is such a unit. A solution having an osmotic pressure of 22.4 standard atmospheres is said to have an effective osmotic concentration of 1 Osm per liter or to be an *osmolar* solution, regardless of its chemical composition. For any individual substance the osmol is defined as the weight in grams that gives rise to an osmotic pressure of 22.4 standard atmospheres when dissolved in 1 liter of solution. For osmotic purposes, the concentration of any given solution can be expressed directly in terms of *osmols per liter,* or *osmolarity.* In physiological work the milliosmol (mOsm) is usually employed as the unit of osmotic concentration (1 Osm = 1000 mOsm). In the case of dilute solutions one may write for nonelectrolytes (e.g., sucrose, urea) *milliosmols = millimols* and for electrolytes *milliosmols = n × millimols,* where n has its ideal value, i.e., the number of ions formed by each molecule of the electrolyte. For solutions containing a single solute, these "ideal" relationships may be used for approximate purposes at physiological concentrations.

Determination of Osmotic Concentration. In principle, the osmolarities of physiological solutions can be determined by directly measuring their osmotic pressure and converting it to osmols or milliosmols. Because of technical complications, however, this method is unsuitable for routine physiological or clinical investigations, in which rapid determination of the osmolarities of large numbers of samples is usually required. An alternative method of determining osmolarities, which is at once simpler in practice, more rapid, and more accurate than all but the most elaborate instruments for the direct measurement of osmotic pressure, is based on the relative freezing points of solutions.

Osmotic pressure is one of the so-called *colligative* properties of dilute solutions; i.e., its magnitude is related to the concentration (number/unit volume) of dissolved particles and is not affected (in very dilute solutions at least) by such factors as their size, shape, or chemical composition. Other colligative properties of dilute solutions are vapor-pressure lowering, depression of the freezing point, and elevation of the boiling point. These four properties of solutions are very simply related, so that if one of them is known for a given set of circumstances, the others may be calculated readily.

In the case of freezing-point depression, it can be shown that, in very dilute solutions, the amount by which a nondissociating solute like sucrose lowers the freezing point of pure water is 1.86°C per mole. This factor, the *cryoscopic constant,* is the same for all nondissociating solutes, provided they can be considered to behave ideally. Under similar conditions, one would expect an electrolyte like NaCl,

which gives rise to 2 ions per molecule, to lower the freezing point of water by 2 × 1.86, or 3.72°C per mole, and so on. In practice, it is found that the factor 1.86°C per mole is subject to the same kind of concentration dependence as the van't Hoff coefficient n. Therefore, for practical purposes, an *osmolar* (or osmolal) solution may be *defined* as one that lowers the freezing point of water by 1.86°C regardless of the precise conditions or of its exact chemical composition. Since, as already pointed out, such a solution has an osmotic pressure of 22.4 atmospheres, the following relations between osmolarity, freezing-point lowering, and osmotic pressure are at once apparent:

$$\text{Milliosmols} = \frac{\Delta T_f \times 1000}{1.86}$$

and

$$\text{Osmotic pressure (atm)} = \frac{\Delta T_f \times 22.4}{1.86}$$

where ΔT_f is the difference between the freezing point of the solution and that of pure water (0°C).

Nowadays, a number of instruments (osmometers) are available with which the freezing points of solutions can be measured rapidly and accurately and which give a direct readout in milliosmols of the results obtained. Thus, lowering the freezing-point (cryoscopy) is at present the usual method of choice for the determination of osmolarities in physiological and clinical investigations.

Osmotic Behavior of Cells: The Solute Reflection Coefficient. To understand the mechanisms of osmotic water movement across cell membranes and the nature of the driving forces involved, it is necessary to bear in mind that under physiological conditions cell membranes are not semipermeable (i.e., impermeable to all solutes present in the system). Rather, they are selectively permeable, or *permselective* with respect to different solutes. In other words, some solutes do not penetrate the cell membranes to any significant extent, while others cross the cell membrane but do so much more slowly than water. Still others can move across the membrane as fast, or almost as fast, as water. We can express these differences in rates of penetration in another way by saying that the cell membrane *discriminates,* in varying degrees, between water and various solutes. Total discrimination occurs when the solute is completely excluded ("reflected") by the membrane. Partial discrimination means that the solute can penetrate, but less readily than water (i.e., it is partially reflected, relative to water, by the cell membrane). Finally, if the solute and water penetrate the cell membrane with equal ease, the membrane cannot discriminate between

Table 1-1. Some values of the solute reflection coefficient (σ)

Cell or tissue	Sucrose	Glycerol	Urea
Red blood cell (human)	—	0.88	0.62
Axolemma (squid)	—	0.96	0.7
Schwann cells	—	0.09	—
Rat intestine (luminal surface)	0.99	—	0.81
Frog skeletal muscle fiber	1.0	0.86	0.82
Corneal epithelium (rabbit)	1.0	—	1.0
Kidney slices (*Necturus*)	1.0	0.77	0.52
Capillary wall	0.058	—	—

them at all. There is no "reflection" of the solute relative to water. It is important to note that in this last case the membrane simply will not sense the presence of the solute in aqueous solution.

The quantitative development of this idea (Katchalsky and Curran, 1965) has led to a very useful index of permeability. This is the *solute reflection coefficient* (σ). It is a measure of the degree of discrimination or selectivity for water, compared with solute, possessed by the membrane. If the solute is completely impermeant, $\sigma = 1$. If the solute and water penetrate with equal ease, $\sigma = 0$. When the solute penetrates less readily than water, $0 < \sigma < 1$.

The utility of σ as a measure of permeability lies in the fact that its theoretical derivation involves no assumptions concerning specific mechanisms of penetration (Katchalsky and Curran, 1965). Unlike the classic permeability coefficient P (equation 2) that is based on the assumption that permeation is diffusive, σ can be used to compare rates of penetration by solutes that cross the membrane by different mechanisms (e.g., diffusion, carrier-mediated transport). In addition, the experimental determination of σ is straightforward.*

It should be noted that σ depends not only on the specific solute, but also on the type of membrane under consideration. That is, σ values for the same solute differ in different cell species (Table 1-1).

The fact that cell membranes have widely different permeabilities, relative to water, for various solutes (Table 1-1) is of crucial importance to the osmotic behavior of living tissues. To illustrate this point, let us consider a simple experiment with red blood cells.

The total osmotic concentration of mammalian blood

plasma is about 300 mOsm. Present estimates indicate that in most cases, including red cells, the intracellular osmolarity in vivo is essentially the same as that of plasma. Intracellular osmolarity is principally due to the presence inside the cell of K^+, Na^+, and Cl^- ions, together with smaller amounts of HCO_3^-, phosphates, amino acids, and so on. In red cells, as in many other cells, the sum of Na^+ and K^+ ions greatly exceeds the total amount of small anions present. To preserve electroneutrality it is necessary to postulate the existence within the cell of large, nondiffusible or "fixed" anions. These are partly the ionized anionic groups (e.g., $-COO^-$ groups) of cellular proteins and also include smaller anionic molecules (e.g., ATP) that are "nondiffusible" in the sense that they do not readily move across the cell membrane.† If human red blood cells are suspended in a solution containing 300 mOsm (approximately 0.15 M) NaCl, i.e., a NaCl solution that has the same osmolarity as the cytoplasm or is *isosmotic* with the latter, the volume of the cells will not change. This is because the cell membrane senses the full osmotic effect of the solutes. We may then write

$$(\Delta\pi i)_{out} = (\Delta\pi i)_{in} \qquad (8)$$

where $\Delta\pi i$ symbolizes the osmotic pressure difference generated across the cell membrane by a solute for which $\sigma = 1$ (an impermeant solute). Equation 8 means that, under the conditions of our experiment, there is no osmotic gradient for water across the cell membrane and therefore no driving force that would induce a net transmembrane water flow in either direction.

A solution such as 300 mOsm NaCl (physiological saline) that causes no change in cell volume is called an *isotonic* solution. Similarly, solutions that cause cells to swell (e.g., NaCl in less than 300-mOsm concentration) are *hypotonic*, and solutions that cause shrinkage of the cell are *hypertonic*. It will be apparent that, when $\sigma \sim 1$, *tonicity* and *osmolarity* are interchangeable terms. It will also be apparent that under these circumstances the only way in which one can induce a net osmotic flow of water across the cell membrane is to create directly a concentration or chemical potential gradient for water across it. To do this, one must adjust the concentration of the external solute so that the resulting solution is either hyposmotic or hyperosmotic with respect to the cytoplasm. Note that when this is done, the net

---

*Theoretically, *negative* values (i.e., values less than zero) of σ are possible. This means that the solutes in question cross the membrane more rapidly than water. However, at present, this does not seem to be of much practical importance in physiological systems.

†Although Na^+, K^+, and Cl^- can exchange fairly freely across the red blood cell membrane, *net* transmembrane movements of salt are slow. Therefore, the osmotically active cell contents can be regarded as a "nonpenetrating solute" ($\sigma \sim 1$) in osmotic experiments of relatively short duration. The NaCl in the external medium can be similarly regarded.

volume flow across the cell membrane is a flow of water only.

Returning to our imaginary experiment, if we now suspend human red blood cells in an isosmotic (300-mOsm) urea solution, we will find that the cells swell rapidly. In fact, they will *hemolyze* (see Appendix) within a very short time. Under these conditions there is a rapid osmotic volume flow into the cells, despite the fact that there is no intrinsic transmembrane gradient for water—the external and internal osmolarities are exactly the same. The explanation for this is as follows. Because urea penetrates the red blood cell membrane, i.e., it is not completely reflected (Table 1-1), the *effective* osmotic pressure, across this membrane, of a urea solution ($\Delta\pi_p$) is less than its true osmotic pressure ($\Delta\pi_i$) as determined, for example, from the freezing point depression. For any combination of membrane and external solute, the relationship between the effective osmotic pressure and the true osmotic pressure is given by the following equation.

$$\Delta\pi_p = \sigma \cdot \Delta\pi_i$$

In our example, $\Delta\pi_i$ is 300 mOsm, and σ is 0.62 (Table 1-1). Therefore, $\Delta\pi_p$ is (300 × 0.62), or 186 mOsm. This means that as soon as the cells are exposed to 300 mOsm urea, an effective osmotic gradient of 114 mOsm is generated across the cell membrane. This causes fluid to enter the cells.

Pictorial representations of the transmembrane osmotic relationships immediately following immersion of human red blood cells in 300 mOsm NaCl or 300 mOsm urea appear in Figure 1-4A and B respectively. Figure 1-4C underlines some further important differences between the osmotic response of cells to solutes for which $\sigma = 1$ and those for which σ is less than 1. First, as indicated by the horizontal line at the top of Figure 1-4C, $\Delta\pi_i$ for an isotonic solution does not change with time. By contrast, with a penetrating solute like urea, the initial osmotic pressure across the membrane ($\Delta\pi_p$) declines with time and may eventually approach zero (curve in Fig. 1-4C). This reflects the fact that with a penetrating solute in the external solution, the fluid that enters the cell across the membrane is not pure water but a mixture of water and solute. The composition of this moving fluid relative to the medium from which it originates depends on σ. As has already been said, if $\sigma = 1$, no solute enters the cell, and the moving fluid is pure water. If $\sigma = 0$, none of the solute is "reflected" by the membrane, and the moving fluid is identical in composition to the external medium. If $0 < \sigma < 1$, the moving fluid contains a lower concentration of solute than the external

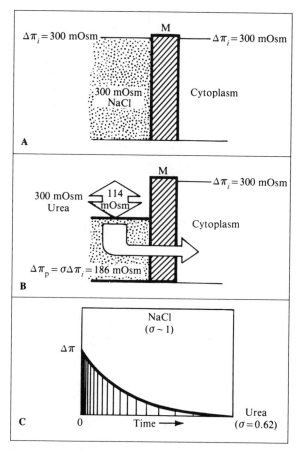

Fig. 1-4. (A and B) Initial osmotic pressure relationships across the membrane (M) of a human red blood cell suspended in 300 mOsm NaCl (A) and 300 mOsm urea (B). The arrow in B indicates the direction of osmotic volume flow. (C) Time dependence of the effective transmembrane osmotic pressures of NaCl and urea.

medium from which it is derived. This is because a fraction of the solute is reflected or "filtered out" by the membrane, a phenomenon sometimes referred to as "molecular sieving."

It is instructive to calculate, for the specific example under consideration, the composition of the fluid that crosses the red blood cell membrane. Since σ represents the fraction of urea relative to water that is "reflected" by the membrane, "$1 - \sigma$" clearly represents the relative fraction of urea that is admitted into the cell. The concentration of urea in the moving fluid is thus 300 × (1 − 0.62), or 114 mOsm. Note that this is the same as the initial transmembrane osmotic pressure difference already calculated (see also Fig. 1-4B). Indeed, it is the failure of the membrane to sense or "see" this fraction of the dissolved urea that is the cause of the initial osmotic pressure difference shown schematically in Figure 1-4B.

Finally, it is clear from the curve in Figure 1-4C that when σ is less than 1, there is no functional relationship between osmolarity and tonicity. In principle, *all solutions containing a penetrating solute are hypotonic,* whatever their osmolarity may be.

Mechanisms of Water Movement Across Cell Membranes. In terms of the Singer-Nicolson model of membrane structure presented earlier (Fig. 1-1D), water or an aqueous solution is considered to move through pores or channels in the membrane. Two mechanisms can be postulated for this movement. One is diffusion and the other is bulk flow. Bulk flow is considered in more detail in Chapter 11. It is the movement en masse of a given volume of fluid that occurs, for example, when water flows through a pipe or conduit. Recent studies have revealed that when water moves across porous membranes, it can do so by either diffusion or bulk flow. The degree to which one or the other will predominate is determined by the size of the membrane pores. If the diameter of the pores is not much greater than that of the water molecule itself (~ 1.5 Å), water movement will occur almost exclusively by diffusion. If pore diameter is relatively large (e.g., 30 to 40 Å), as in the walls of blood capillaries, water will penetrate almost entirely by bulk flow. In cell membranes, most studies to date indicate an average pore diameter of about 8 to 10 Å. In these circumstances, one would predict that water movement should be about 50 percent diffusive, the other 50 percent occurring by bulk flow. This prediction is supported by experimental evidence (Dick, 1966).

The existence of bulk flow can generate yet another transport mechanism for dissolved solutes called *solvent drag.* Simply stated, it means that solutes can be literally swept along in the moving stream of solvent. Solvent drag can significantly affect overall rates of solute movement in some physiological systems (e.g., the walls of blood capillaries and certain epithelial systems).

ION TRANSPORT AND BIOELECTRIC POTENTIALS

THE RESTING MEMBRANE POTENTIAL. If a microelectrode (tip diameter 1 μm or less) is inserted into a cell (Fig. 1-5), a *potential difference* is observed between it and a reference electrode immersed in the external medium. This potential difference, which in resting cells is usually such that the intracellular electrode is *negative* with respect to the external reference electrode, is the *resting potential* (E_R), or

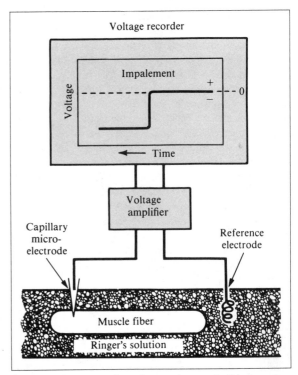

Fig. 1-5. Measurement of a resting membrane potential. An open-tip capillary microelectrode filled with KCl solution is inserted into the cell, and the potential difference between it and a reversible half-cell (e.g., calomel or silver–silver chloride) immersed in the bathing medium is recorded. At the moment of impalement there is a sharp deflection in the voltage trace.

membrane potential, of the cell. By convention, the potential of the external reference electrode is taken as zero at all times. Representative values for the resting potential are -70 to -90 mv for excitable cells (nerve, muscle), -30 to -70 mv for epithelial cells (e.g., small intestine, renal tubular cells), and -10 mv for red blood cells. Recalling that ions, which are electrically charged entities, can cross cell membranes either by simple diffusive transport through aqueous channels or via carrier-mediated processes, it will be evident that ion transport must frequently involve net charge movement across the membrane. This will, in fact, always hold for net diffusive ionic transfer, and, indeed, the pathways for such transfer are often referred to as conductive pathways. In addition, carrier-mediated ionic transfer, except for the coupled electroneutral processes already discussed may also be, in effect, conductive; i.e., their operation will result in the passage of an electrical current across the membrane. Under these circumstances, the membrane potential is an electrical force that may either assist or impede a given ion's passage across the membrane, depending on the direction of movement and the sign of its charge. In other words, in calculating the transmembrane driving force for any ion, the membrane potential must be taken into account. We may note that the membrane potential is normally so oriented that it assists the movement of cations into the cell and opposes their exit from the cell interior. The opposite is true for anions.

TRANSMEMBRANE DRIVING FORCES FOR IONS: EQUILIBRIUM POTENTIALS. Let us begin this section by recalling that, *for any permeant species,* the transmembrane equilibrium condition is that condition in which the net transmembrane driving force for the given species is zero. Let us also recall that the total driving force for an ion has two components—the membrane potential and any difference existing between the extracellular and intracellular concentration—and we at once reach the following conclusion: As long as the membrane potential has a finite (not zero) value, an ion cannot be at equilibrium across the membrane when its intracellular and extracellular concentrations are the same. The equilibrium state must, rather, correspond to an *unequal* distribution of the ion across the cell membrane in which the chemical (concentration) and electrical components of the total driving force are equal and opposed. Turning to living cells, we find that they are indeed characterized by unequal distribution of ions between the cell interior and the external environment. We will discuss this asymmetrical distribution of ions in some detail for K^+, Na^+, and Cl^-, since these are, quantitatively, the most important ions in the cell and its environment. It should be remembered, however, that similar asymmetries exist for other ions, e.g., Ca^{2+}, H^+, and HCO_3^-.

The interior of a typical mammalian cell contains about 20 to 30 times as much K^+ as Na^+. In blood plasma or interstitial fluid, this ratio is reversed. Chloride is also unequally distributed between the cell and its environment, being present in lower concentration in the cell interior than in the external fluid. Table 1-2 shows these relationships in detail for two cell types, frog skeletal muscle fibers and human red blood cells.

The following questions now arise: Could some or all of these ions be at equilibrium across the muscle fiber or red blood cell membrane, despite the marked differences in their extracellular and intracellular concentrations? Furthermore, if they are not, how can one determine the net driving forces and the mechanisms required to maintain their steady-state distribution across the cell membrane? In principle, there are a number of ways to approach these questions. One of the simplest and most convenient is to utilize the concept of equilibrium potentials. The *equilibrium potential,* E_j, for an ion, j, is the electrical potential difference required to neutralize exactly the net chemical driving force across the cell membrane for j under any given set of conditions. E_j can be calculated as follows: The total driving force for j is its *electrochemical potential difference,* $\Delta\bar{\mu}_j$, across the membrane. As already stated, $\Delta\bar{\mu}_j$ is the *algebraic* sum of a chemical and an electrical component, i.e.,

$$\Delta\bar{\mu}_j = 2.3 \, RT \log [j_i] / [j_o] + zE_RF \qquad (9)$$

The first term on the right-hand side of equation 9 represents the chemical component of $\Delta\bar{\mu}_j$. In this term, R is the gas constant, T is the absolute temperature, and $[j_i]$ and $[j_o]$ are the extracellular and intracellular concentrations, respectively, of j. The second term on the right is the electrical component of $\Delta\bar{\mu}_j$. In this term, z is the valence, E_R is the

Table 1-2. Ionic concentrations, measured resting potentials, and calculated equilibrium potentials in frog skeletal muscle and human red blood cells

Cell type	Concentration (mEq)[a]			Equilibrium potential (mv)[b]			
	Na^+	K^+	Cl^-	E_R	E_{Na}	E_K	E_{Cl}
Frog skeletal muscle				−90	+56	−105	−86
Fibers	13	140	~3				
Plasma	110	2.5	90				
Human blood cells				−7 to −14	+55	−86	−9
Cells	19	136	78				
Plasma	155	5	112				

[a]Concentrations for cells and fibers in milliequivalents per liter of cell water. Plasma concentrations in millequivalents per liter.
[b]Minus sign indicates inside negative.

membrane potential,* and F is Faraday's constant (96,500 coulombs). It is clear that by measuring $[j_i]$, $[j_o]$, and E_R under any given set of conditions and inserting the values obtained in equation 9, one can calculate the magnitude and direction of $\Delta\bar{\mu}_j$ for those conditions. If $[j_i]$ and $[j_o]$ are known, equation 9 can be used to calculate E_j. To do this, one sets $\Delta\bar{\mu}_j = 0$ (the equilibrium condition for j). Then, substituting E_j and E_R and rearranging equation 9, one obtains

$$E_j = \frac{-2.3 \, RT}{zF} \log [j_i]/[j_o] \tag{10}$$

Equation 10 is usually called the Nernst equation. Note that E_j in this equation is a *calculated,* not a measured, parameter. Note also that, with E_j in millivolts, the factor 2.3 RT/zF has the approximate value (60/z) at body temperature.

When $[j_i]$, $[j_o]$, and E_R are known, equations 9 and 10 provide a very simple and straightforward estimate of the transmembrane driving force for an ion. This is simply $E_R - E_j$.

In Table 1-2 this type of analysis is applied to frog skeletal muscle and human red blood cells. The Na^+, K^+, and Cl^- concentrations listed are average figures from the published literature (Harris, 1960). The data shown for skeletal muscle make it clear that, within the limits of experimental error, $E_R = E_{Cl}$. Therefore, one may conclude that in this tissue, Cl^- ions are in electrochemical equilibrium ("passively" distributed) across the fiber membrane and that no direct energy expenditure is required to maintain a steady state with respect to Cl^- in the living muscle fiber. In other words, the inwardly directed concentration gradient for Cl^- is balanced by an equal and opposite outwardly directed electrical gradient.

With regard to potassium, it is seen that in skeletal muscle fibers under normal physiological conditions, E_K is significantly greater than E_R, though both are oriented in the same direction. This means that an E_R value some 10 to 15 mv greater than that observed would be required to balance the outwardly directed concentration gradient for this ion. In other words, K^+ is present within the fibers in excess of the amount corresponding to electrochemical equilibrium, although the excess is relatively small. Because of this net outwardly directed electrochemical potential gradient, K^+ must leak continuously out of the fibers, and the loss must continuously be made good. Thus, in skeletal muscle, there is a need for an inwardly directed, energy-requiring potas-

sium pump that can transport K^+ against an electrochemical gradient. This, as already discussed, is one of the functions of the Na^+-K^+-ATPase pump in the membrane.

As for the distribution of K^+ and Cl^- between red blood cells and plasma, it is seen from Table 1-2 that the situation is qualitatively similar to, but quantitatively different from, that in muscle. In recent years, several workers have succeeded in penetrating red blood cells with microelectrodes and have recorded potentials in the range -7 to -14 mv. Once again, these figures are in essential agreement with the calculated value of E_{Cl} (Table 1-2), indicating an equilibrium distribution of Cl^- across the red blood cell membrane. By contrast, the calculated value of E_K (Table 1-2) shows that the concentration of K^+ ions inside the red blood cell is far in excess of the amount required for electrochemical equilibrium. An inwardly directed, active potassium pump must therefore be a major component of the potassium-transporting machinery in these cells.

Inspection of Table 1-2 shows clearly that the situation with respect to intracellular Na^+ is not only quantitatively but also qualitatively different from that of K^+ and Cl^-. In both muscle and red blood cells, E_{Na} is different from E_R, both in magnitude and in orientation; i.e., E_{Na} requires that the inside surface of the membrane be *positive* with respect to the outside, whereas in reality the opposite is true.† That is, the intracellular Na^+ concentration in muscle and red blood cells is much *lower* than that required for electrochemical equilibrium, and both the chemical and electrical gradients tend to "push" Na^+ into these cells. Furthermore, the same appears to be true for virtually all cells, not only those of humans and other animals, but also for higher plant cells and microorganisms. Since cells are permeable to Na^+, and since many cell species exist in an environment rich in Na^+ ions, there is an almost universal requirement for a mechanism that removes Na^+ from the cell interior as fast as it enters. Removal of Na^+ is usually accomplished against an electrochemical potential gradient. Consequently, the mechanism involved must be an active process. In this context, the ubiquity of the Na^+-K^+-ATPase pump is readily understood.

THE NATURE OF THE RESTING MEMBRANE POTENTIAL. For many years the origin of the resting potential has been thought to lie in the asymmetrical distribution of ions across the cell membrane. There is now compelling evidence that this is indeed the case. In principle, any ion that is unequally

---

*In this chapter, E_R is used to denote the *measured* potential difference across a cell membrane. E_j (specifically, E_K, E_{Na}, etc.) is the *calculated* equilibrium potential for a given ion.

†This "inversion" of the sign of E_{Na} is of fundamental importance in relation to the phenomenon of "overshoot" in the action potentials of nerve and muscle (see Chaps. 2, 3, and 14).

distributed across the cell membrane can contribute to the membrane potential. However, because of their relative abundance, the only ions that need to be considered in practice are K^+, Na^+, and Cl^-. To examine in detail the genesis of the membrane potential in terms of ionic concentration differences, we will focus, as an example, on the skeletal muscle fiber, specifically the frog sartorius fiber, for which the relevant data are shown in Table 1-2. It should be borne in mind, however, that with appropriate quantitative adjustments, the same considerations apply to most cell types.

Summarizing the situation illustrated in Table 1-2, we have, for the skeletal muscle fiber in the steady state, the following net flows of ions: There is an outward diffusive movement (leak) of K^+ and an inward leak of Na^+ across the fiber membrane. The resulting tendency for the cell to lose K^+ and gain Na^+ continuously is counterbalanced by the operation of the Na^+-K^+ pump so that, overall, the inward and outward flows of Na^+ and of K^+ are the same. By contrast, Cl^- appears to be in essential equilibrium across the fiber membrane, so that there is no net transmembrane movement of Cl^-. Thus, in normal physiological situations, where $[Cl_i]$, $[Cl_o]$, and E_R are virtually constant, there should not be any significant contribution by Cl^- to the membrane potential. This parameter must be determined by the net movements of K^+ and Na^+ that occur.

Let us first examine the possible role of the Na^+ pump in the generation of E_R. Since the stoichiometry of this coupled transport process is 3 Na^+/2K^+, it does indeed give rise to a net outward flow of positive charge across the membrane. Thus, at first glance, one might suppose that the (outside-positive) membrane potential could be accounted for in terms of this electrogenic mechanism alone. Careful experiments have shown, however, that in the resting state the Na^+-pump rate in muscle is far too low to generate a membrane potential of the order of -90 mv (Table 1-2). In fact, its contribution to the overall membrane potential under these conditions is too small to evaluate with accuracy. As an approximation we may say that it probably lies somewhere between 0 and 5 mv.*

We therefore conclude that the resting membrane potential in muscle is largely due to the passive leaks of K^+ and Na^+ across the fiber membrane that exist under steady-state conditions. In analyzing the effects of these leaks in detail, we note first that they represent flows of electrical current across the membrane. Hence, we may characterize

the passive permeability properties of membranes to ions in electrical terms; i.e., the total current, I_t, traversing the membrane under various conditions, can be separated into its individual components, I_K, I_{Na}, I_{Cl}, and so on. Similarly, total membrane conductance, G_t (which is the reciprocal of the membrane resistance) can be characterized in terms of individual *ionic conductances,* G_K, G_{Na}, G_{Cl}, and so on. This approach has led to the use of *equivalent electrical circuits* to represent these aspects of membrane function, and it has proved highly important in analyzing transient phenomena, such as action potentials in muscle and nerve (see Chap. 2).

Returning to our immediate concern, the origin of the muscle fiber membrane potential, we note that the passive K^+ current, I_K, in the steady state, represents an outward movement of positive charge. Any increase in I_K would therefore tend to make the outer surface of the membrane more positive with respect to the inner surface; i.e., it would tend to increase the magnitude of E_R, or *hyperpolarize* the membrane; I_K is therefore a *hyperpolarizing* current. Conversely, I_{Na} tends to increase the relative positivity of the inner membrane surface. It is a *depolarizing* current that reflects a net *inward* movement of positive charge. Clearly, I_K and I_{Na} have opposing effects on E_R, the first tending to make it larger (more negative inside), and the second tending to make it smaller. The final result, i.e., the actual value of E_R, must represent some kind of balance between these opposed tendencies. A detailed examination of this "balance" and of how it is brought about should lead us to a reasonable understanding of the resting membrane potential.

Returning to Table 1-2, we see that E_R in skeletal muscle (-90 mv) is much closer to E_K (-105 mv) than it is to E_{Na} (56 mv). This immediately suggests to us that K^+ rather than Na^+ is the major determinant of E_R. This seems, at first glance, a little paradoxical in view of the fact that the net driving force for Na^+, i.e., $E_R - E_{Na}$, is much larger than the corresponding driving force for K^+ (Table 1-2), and one would predict that I_K and I_{Na} respectively should depend, among other things, on the magnitude of the appropriate driving forces for K^+ and Na^+. The apparent paradox disappears if one recalls that the flow of an ion across the membrane, and consequently the effect of that ion on E_R, depends, according to Ohm's law, on the product of its driving force and its conductance, i.e., $I_j = (E_R - E_j) \cdot G_j$. To explain why E_R in muscle is much more dependent on K^+ than on Na^+, we merely have to assume that, for the muscle fiber, $G_K \gg G_{Na}$. Since ionic conductance and ionic permeability are directly related, we can state our assumption in the form $P_K \gg P_{Na}$.

---

*Under certain laboratory conditions, the Na^+ pump in muscle can be induced to generate potential differences of about 25 to 30 mv across the fiber membrane. Under normal physiological conditions, the electrogenic Na^+ pump in some cell types may contribute more substantially to the membrane potential than appears to be the case in muscle.

These simple ideas lie at the heart of a brilliant quantitative analysis of cell membrane potentials by D. E. Goldman (1943) that is still one of the cornerstones of membrane theory. His conclusions can be summarized in terms of the celebrated Goldman or constant field equation, which, for the specific case under consideration (E_R in the skeletal muscle fiber), can be written (Hodgkin and Horowicz, 1959)

$$E_R = -60 \log \frac{[K^+_i] + \alpha [Na^+_i]}{[K^+_o] + \alpha [Na^+_o]} \tag{11}$$

where α is the *ratio* P_{Na}/P_K.

The predictions that equation 11 makes concerning the behavior of E_R with respect to the ratios $[K^+_i]/[K^+_o]$ and $[Na^+_i]/[Na^+_o]$ respectively can be summarized as follows: If α is extremely small (i.e., approaches zero), equation 11 reduces to the Nernst equation for K^+ (equation 10), E_R depends only on $[K^+_i]/[K^+_o]$ and becomes equal to E_K. Under these conditions, the membrane will behave as an *ideal, or perfect, K^+ electrode*; that is, if one varies $[K^+_i]/[K^+_o]$ and plots E_R as a function of this ratio, the result will be a straight line with a slope of 60 mv per *decade* change in $[K^+_i]/[K^+_o]$ (see Fig. 1-5). Conversely, if α were extremely large, E_R would approach E_{Na} and the membrane would behave as a *perfect Na^+ electrode*.* Of more immediate interest to us is the situation where α is small but of sufficient magnitude to exert some influence on E_R. In this situation, equation 11 predicts that E_R should be close to E_K but somewhat smaller than it, due to the effect of Na^+. Furthermore, the effect of Na^+ on E_R and the deviation of E_R from E_K should increase as $[K^+_o]$ is reduced.

The applicability of appropriate forms of the Goldman equation has been tested experimentally with many kinds of cells. The results show that this equation does indeed provide a highly satisfactory description of the behavior of membrane potentials under a variety of conditions. Figure 1-6 illustrates one such investigation of E_R in frog skeletal muscle by Hodgkin and Horowicz (1959). The straight line AB in this figure represents the results to be expected if the fiber membrane behaved as a perfect K^+ electrode (equation 10) throughout the experiment. It is seen that when $[K^+_o]$ exceeds 10 mM, the muscle fiber closely approximates this behavior, indicating that under these conditions E_R is completely dependent on K^+. As $[K^+_o]$ is decreased below 10 mM, the effect on E_R of a small but finite P_{Na} becomes apparent (open circles). The curve AC in Figure 1-

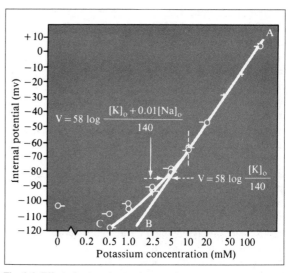

Fig. 1-6. Effect of external potassium on the membrane potential of a single fiber taken from the frog sartorius muscle. (From A. L. Hodgkin and P. Horowicz. Influence of potassium and chloride ions on the membrane potential of single muscle fibres. *J. Physiol.* [Lond.] 148:135, 1959.)

6 is calculated from equation 11, assuming $\alpha = 0.01$. One sees that these calculated values of E_R agree rather well with its measured values (circles), indicating that in frog skeletal muscle α lies in the range 0.01 to 0.02.

In summary, then, we may conclude that in skeletal muscle (and in nerve) E_R is, to a large extent, a composite of the *diffusion potentials* generated by an outward leak of K^+ and an inward leak of Na^+ under steady-state conditions. Because $P_K >> P_{Na}$ (or $G_K >> G_{Na}$), E_R, under normal conditions, is primarily dependent on K^+. A similar conclusion may be drawn for a variety of cell types. In some cells (e.g., smooth muscle fibers, epithelial cells), the ratio P_{Na}/P_K may be significantly larger than in skeletal muscle and nerve. Consequently, E_R may be smaller in these cells. Finally, a relatively small contribution, which may vary from one cell type to another, from an electrogenic Na^+-K^+ pump to the total magnitude of E_R can be postulated.

SOME FURTHER COMMENTS ON IONIC ASYMMETRIES: THE GIBBS-DONNAN EQUILIBRIUM. As already indicated, the asymmetrical distribution of Na^+ across the membrane of many cell species can be entirely accounted for by the operation of the Na^+ pump. Also, as we have seen, this mechanism plays an important role in the regulation of intracellular K^+ concentrations. In the case of an ion that is passively distributed across the cell membrane (e.g., Cl^- in

---

*As first postulated by A. L. Hodgkin and A. F. Huxley (see Hodgkin, 1958), large, rapid, and transient changes in α are the mechanism by which action potentials are generated and propagated in excitable tissues such as nerve and muscle (see Chap. 2).

skeletal muscle) one must look for mechanisms other than specific ion pumps by which an asymmetrical distribution can be achieved and maintained.

Cells contain considerable amounts of "fixed" or "nondiffusible" anions, that is, anionic groups that cannot leave the cell interior. These include the excess anionic groups of cellular proteins with a preponderance of negative charges at normal intracellular pH, as well as a number of low-molecular-weight anions (e.g., ATP) to which the cell membrane is impermeable. This suggested to physiologists that a purely passive thermodynamic mechanism, the Gibbs-Donnan equilibrium (often called simply the Donnan equilibrium), long since known to chemists as a process whereby unequal distribution of ions across nonliving membranes could be accomplished, might be an important determinant of the distribution of some permeable ions across the living cell membrane. We will close this chapter with a brief consideration of this mechanism (see also Davson, 1970).

For illustrative purposes, let us consider an artificial system in which two solutions, one containing KCl only and the other containing KCl together with KPr (the potassium salt of a large anion, Pr^-), are separated by a membrane that is permeable to water, K^+, and Cl^- but not to Pr^-. If the membrane were permeable to Pr^- as well as to K^+ and Cl^-, all three ions would diffuse freely until they were equally distributed throughout the whole solution on both sides of the membrane. By making the membrane impermeable to one ionic species (Pr^-), one imposes a restriction on free diffusion that gives rise to a number of important consequences. Of particular interest in the present context is that the impermeability of the membrane to Pr^- has a marked effect on the equilibrium distribution of the diffusible ions K^+ and Cl^- across the membrane. This distribution is given by the Gibbs-Donnan relationship, which states that *at equilibrium the products of the concentrations of the diffusible ions on each side of the membrane are equal.* For the conditions under consideration (if we designate the solution containing Pr^- by the subscript 1 and the contralateral solution by the subscript 2), this may be written

$$[K^+_1] \times [Cl^-_1] = [K^+_2] \times [Cl^-_2] \tag{12}$$

or

$$[K^+_1]/[K^+_2] = [Cl^-_2]/[Cl^-_1] \tag{13}$$

Certain consequences of this relationship will be immediately apparent. Clearly, since the solution on each side of the membrane must be, as a whole, electrically neutral, $[K^+_2]$ must be equal to $[Cl^-_2]$. Also, since in the solution containing the nondiffusible ion Pr^- the condition of overall electroneutrality requires that there be sufficient K^+ ions

present to balance the negative charges on the Pr^- ions as well as those on the Cl^- ions, $[K^+_1]$ must be greater than $[Cl^-_1]$. Again, equation 12 requires that the sum of $[K^+_1]$ and $[Cl^-_1]$ be greater than the sum of $[K^+_2]$ and $[Cl^-_2]$.* Thus, $[K^+_1] > [K^+_2]$ and $[Cl^-_1] < [Cl^-_2]$.

These considerations lead to the important generalization that in a Gibbs-Donnan system at equilibrium there is a concentration gradient across the membrane for each species of diffusible ion present. This gradient (the Gibbs-Donnan ratio) is identical in magnitude for each diffusible ionic species but is opposite in direction for cations and anions. Thus, the equilibrium condition cannot be specified in terms of chemical concentrations alone, since the individual concentration gradients would, if unopposed, result in an equal distribution of diffusible ions across the membrane. In the Gibbs-Donnan situation, the concentration gradients for diffusible ions that exist at equilibrium must therefore be neutralized by an equal and opposite driving force. This force is an equilibrium potential, generated by the unequal distribution of diffusible ions across the membrane and given by the Nernst equation in the following form:

$$E_K = -60 \log \frac{[K^+_1]}{[K^+_2]} = 60 \log \frac{[Cl^-_2]}{[Cl^-_1]} = E_{Cl}$$

Note that the side of the membrane in contact with the solution containing the nondiffusible anion is *negative* with respect to the opposite side. The opposite would hold true if the impermeant species were a cation.

The Gibbs-Donnan equilibrium has a long (and, to generations of students, agonizing!) history in physiology. Its debut in this field was most probably its incorporation into the celebrated Starling hypothesis (see Chap. 12) for the regulation of fluid movement across the blood capillary wall. Obviously, the term *equilibrium* in the Gibbs-Donnan context applies only to the diffusible ions present and not to the system as a whole. For example, in the model system just considered, Pr^-, which is physically constrained to remain on one side of the membrane, cannot reach equilibrium. Because of this, and because there is an excess of *total diffusible ions* on the same side, water also is not at equilibrium. There is, in fact, an osmotic pressure difference across the membrane such that water will move into the solution containing Pr^-. This excess osmotic pressure arising from the presence of Pr^- on only one side of the mem-

---

*This follows from the algebraic fact that, if $ab = c^2$, where a and b are unequal, then $(a + b) > 2c$. A geometrical illustration of this is the fact that the sum of the lengths of two adjacent sides of a rectangle is greater than twice the length of the side of a square having the same area as the rectangle (Davson, 1970).

brane is called the *colloid osmotic pressure,* or *oncotic pressure.* In the blood capillary wall, the σ values for the (negatively charged) plasma proteins are much higher than the corresponding values for small ions. Consequently, extracellular fluid contains considerably less protein than does blood plasma. As a result, there are significant, if small, differences in the ionic content of plasma and extracellular fluid, and the oncotic pressure across the capillary wall is an important regulatory factor in the distribution of fluid between intravascular and extravascular compartments. Damage to the capillary wall can increase its permeability to proteins and induce the condition of *edema,* in which there is an excess of fluid in the extravascular region. In similar fashion, so-called Donnan forces play a role in the composition of the *glomerular filtrate* produced in nephrons (see Chap. 22).

In the area of cell and membrane physiology, the history of the Gibbs-Donnan equilibrium has been more episodic. At one time, it was believed that E_R in muscle and nerve approximated E_K and E_{Cl}, and it was seriously contended that E_R was, in fact, generated by Donnan forces operating on K^+ and Cl^-. Indeed, many of today's senior physiologists and clinicians first encountered the subject of membrane potentials in this context. In view of the discussion of the origin of membrane potentials that has been presented, it is hoped that the reader realizes that we no longer adhere to this view. Nevertheless, the idea that the Gibbs-Donnan effect plays a significant role in the overall regulation of cell Cl^- (and, to some extent, cell K^+) is still a viable component of contemporary membrane theory.

APPENDIX

HEMOLYSIS OF RED BLOOD CELLS. In some ways, the topic of red blood cell hemolysis relates more directly to Chapter 10 than to the present chapter. However, for many years, human red blood cells have been a popular experimental model in laboratory exercises designed to illustrate basic principles of membrane permeation by water and dissolved solutes. There are several reasons for this, which include the following: their ready availability in quantity (1 mm³ of human blood contains about 5 million red cells); their ease of handling; and the fact that, since they are free cells whose whole surface is exposed to the external medium, the volume changes they undergo in response to changes in their osmotic environment and the rates of entry of various substances into them are easily determined. Although there are undoubtedly differences in detail in the osmotic behavior of different cells, the red cell may be considered a fairly

typical representative, in this respect, of animal cells in general.

OSMOTIC (HYPOTONIC) HEMOLYSIS: OSMOTIC FRAGILITY. In the following discussion, two experimental conditions will be assumed. The first is that the cells are suspended in a volume of external solution that is very large compared with the cytoplasmic volume. Therefore, any movement of water or solute from the external medium to the cell interior will have no measurable effect on the composition of the former. Second, we will assume that the response of the cells to different external *milieux* will be followed for a short time (1–2 min) only. Under these conditions we may legitimately conclude that, provided the cell membranes remain intact, no significant net movement of osmotically active material from the cell interior to the external medium will occur; i.e., σ = 1 for cytoplasmic solutes. The same conclusion applies to "impermeant" external solutes, e.g., NaCl.

If red blood cells are placed in a hypotonic solution, e.g., a NaCl solution with a concentration below 300 mOsm or any solution containing a permeable solute (e.g., urea), they swell because of osmotic water entry. Swelling in red blood cells is accompanied by a change in shape. Normally, the cells are biconcave disks with an average major diameter of about 7.2 μm. When water enters them as the result of an osmotic gradient, the cells become progressively spherical. In this way they can increase their volume by a maximum of about 67 percent without any appreciable change in surface area. According to present ideas, the red blood cell membrane possesses very little tensile strength. Thus, if the cells cannot, by becoming spherical, neutralize the osmotic gradient imposed on them, they cannot further increase their volume without damage to the membrane. Entry of water beyond the point at which the cells become fully spherical results in damage to the membrane and escape of the characteristic red protein *hemoglobin* from the cell interior. This phenomenon is called *hemolysis.* The residual cells left following hemolysis are called *ghosts,* or *stroma.* The cell volume beyond which hemolysis occurs is the *critical hemolytic volume.* Within the limits imposed by hemolysis, swelling and shrinking of red cells are reversible.

OSMOTIC FRAGILITY. The resistance of red blood cells to hypotonic hemolysis is often characterized in terms of their *osmotic fragility.* This is determined by the following method. Aliquots of the cells are suspended in NaCl solutions that have a range of concentrations (e.g., from 400 mOsm to about 100 mOsm or less). The percentage of cells

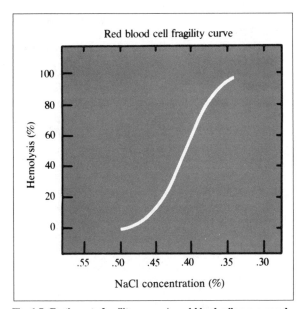

Red blood cell fragility curve

Fig. 1-7. Erythrocyte fragility curve. As red blood cells are exposed to a decreasing extracellular concentration of sodium chloride and therefore an increasing osmotic stress, the percentage of cells that hemolyze increases. The sigmoid shape of the curve is assumed to represent the hemolytic behavior of cells of varying age. Older cells hemolyze with lower osmotic stress, while younger cells are more resistant to osmotic stress.

that hemolyze in each of these solutions is measured and plotted as a function of external NaCl concentration. The resulting *fragility curve* (Fig. 1-7) is S shaped because, even in a normal population of red blood cells, all the cells do not hemolyze at exactly the same transmembrane osmotic gradient; e.g., young cells that have been in the circulation for only a short time are less susceptible to osmotic hemolysis than older cells. From the fragility curve, the concentration of NaCl that causes 50 percent hemolysis is determined. This gives an estimate of the average fragility for the given sample of cells.

Osmotic fragility is of interest in a wide variety of clinical situations. For example, certain diseases are associated with significant changes, from the normal level, in the fragility of the patient's red blood cells. Microcytic anemias (e.g., pernicious anemia), in which the cells are smaller than normal, are characterized by decreased osmotic fragility (i.e., the cells have an increased resistance to osmotic hemolysis). Conversely, increased fragility is characteristic of macrocytic anemias. In hemolytic jaundice (as the name implies) the osmotic fragility of the patient's red blood cells is markedly increased. This is probably due to the presence of

excessive amounts of bile salts in the bloodstream. Increased osmotic fragility may also be found in the red blood cells of patients suffering from bacterial infections, snakebite, or chronic heavy metal (e.g., mercury) poisoning (see the next section).

NONOSMOTIC HEMOLYSIS. The term *nonosmotic hemolysis* is used to describe the action of a variety of agents (hemolysins) that can impair or destroy the selective permeability of the red blood cell membrane by interacting specifically with one or more of its components. Note that relatively minute quantities of these substances in the bloodstream of patients or in a solution bathing red blood cells can cause massive hemolysis. No significant alteration of the cell's osmotic *milieu* is required.

Some examples of hemolytic agents and of their mode of action are as follows:

1. Detergents and bile salts: These interact with cholesterol in the bilayer and thus perturb the structural and functional integrity of the cell membrane.
2. Heavy metals (e.g., mercury): These combine with specific groups (e.g., $-$ SH groups) in membrane proteins that contribute to the maintenance of selective membrane permeability.
3. Phospholipases: These enzymes occur in hemolytic venoms and bacterial toxins. A number of them have been well studied. For example, phospholipase A_2 removes one fatty acid residue from lecithin to form lysolecithin. Sphingomyelinase specifically removes the polar head group from sphingomyelin. Thus, these enzymes disrupt the phospholipids of the bilayer by highly selective mechanisms.

BIBLIOGRAPHY

Andreoli, T. E., Hoffman, J. F., and Fanestil, D. D. (eds.). *Membrane Physiology*. New York: Plenum, 1978.

Christensen, H. N. *Biological Transport* (2nd ed.). London: Benjamin, 1975.

Davson, H. *A Textbook of General Physiology* (4th ed). Baltimore: Williams & Wilkins, 1970.

Dick, D. A. T. *Cell Water*. Washington, D.C.: Butterworth, 1966.

Garcia-Diaz, J. F., and Armstrong, W. M. The steady-state relationship between sodium and chloride transmembrane electrochemical potential differences in *Necturus* gallbladder. *J. Membrane Biol.* 55:213–222, 1980.

Goldman, D. E. Potential, impedance and rectification in membranes. *J. Gen. Physiol.* 27:37–60, 1943.

Harris, E. J. *Transport and Accumulation in Biological Systems* (2nd ed.). New York: Academic, 1960.

Hendler, R. W. Biological membrane ultrastructure. *Physiol. Rev.* 51:66–97, 1971.

Hodgkin, A. L. The Croonian Lecture. Ionic movements and electrical activity in giant nerve fibers. *Proc. R. Soc. [Biol.]* 148:1–37, 1958.

Hodgkin, A. L., and Horowicz, P. The influence of potassium and chloride ions on the membrane potential of single muscle fibres. *J. Physiol.* (Lond.) 148:127–160, 1959.

Katchalsky, A., and Curran, P. F. *Nonequilibrium Thermodynamics in Biophysics*. Cambridge, Mass.: Harvard University Press, 1965.

Ling, G. N. *A Physical Theory of the Living State*. Boston: Blaisdell, 1962.

Lucy, J. A. Ultrastructure of membranes: Bimolecular organization. *Br. Med. Bull.* 24:127–134, 1968.

Macey, R. I. Transport of Water and Nonelectrolytes Across Red Cell Membranes. In G. Giebisch, D. C. Tosteson, and H. H. Ussing (eds.), *Membrane Transport in Biology*. New York: Springer, 1979. Vol. 2, pp. 1–57.

Roos, A., and Boron, W. F. Intracellular pH. *Physiol. Rev.* 61:296–434, 1981.

Schultz, S. G., and Curran, P. F. Coupled transport of sodium and organic solutes. *Physiol. Rev.* 50:637–718, 1970.

Singer, S. J., and Nicolson, G. L. The fluid mosaic model of the structure of cell membranes. *Science* 175:720–731, 1972.

White, J. F., and Armstrong, W. M. Effect of transported solutes on membrane potentials in bullfrog small intestine. *Am. J. Physiol.* 221:194–201, 1971.

Wilbrandt, W., and Rosenberg, T. The concept of carrier transport and its corollaries in pharmacology. *Pharmacol. Rev.* 13:109–183, 1961.

GENERAL PROPERTIES OF NERVE

2

Sidney Ochs

One of the fundamental properties of cells is excitability, which may be defined as the capacity of cells to respond to changes in the external environment with an electrical change and in some cells an accompanying movement. Excitation is demonstrated even in a unicellular organism such as the ameba. When this cell is subjected to either mechanical or chemical stimulation on part of its surface, the response is the formation of pseudopods and movement of the ameba. Much evidence suggests that these reactions are initiated at the surface membrane where special physicochemical properties relate ions to membrane potential, as has been discussed in Chapter 1. Here, we will discuss those special electrical changes present in the membrane that, while found in the single-cell organism, become more highly developed in the multicellular organism. In the course of evolutionary specialization, cells whose chief function is concerned with excitation and conduction arose, namely, the *neurons*. Other cells that also show excitation and conduction, the *effectors,* became specialized for contraction and movement of the organism. In higher species these form the muscles (see Chap. 3).

Several representative neurons found in the vertebrate are shown in Figure 2-1. The sensory neuron with its cell body in the dorsal root ganglion has one long fiber passing outward to innervate the skin, muscle, and other parts of the body. At its termination in the periphery the afferent fiber ending is often combined with other cells to form a sensory receptor organ. Impulses pass in the usual (orthograde) direction from the receptor ending to the nerve cell body in the dorsal root ganglion, where another fiber process of the cell carries the impulses into the central nervous system (CNS). Its endings terminate on synapses made on nerve cells within the spinal cord or other parts of the CNS. A motoneuron, such as is typically found in the central portion of the gray matter of the spinal cord, may have a large, wide, ramifying system of the treelike branches, the *dendrites*. These arise from the cell body of the neuron, also referred to as the *soma,* or *perikaryon.* Many synapses take place on the dendrites as well as on the soma. As can be observed in electron microscope studies, there is no merging of the membrane of one cell with another at the synapse. The presynaptic cell excites the postsynaptic cell by the

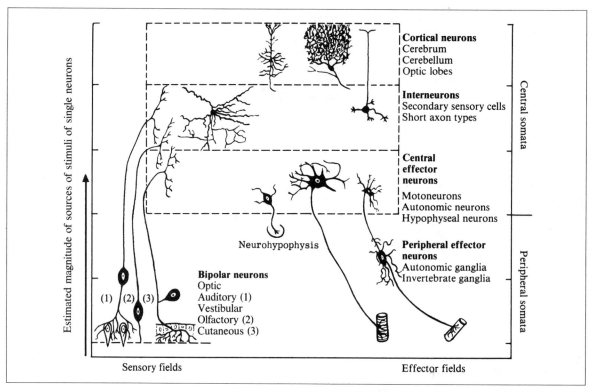

Fig. 2-1. Varieties of neurons. Sensory neurons in the higher animal forms are typically bipolar cells, shown to the left. Both the fiber portion receiving sensory input and the portion synapsing centrally are long. Effector neurons often have a long terminal axon, except for those of the autonomic nervous system. Shown in the dashed box are cells with very great arborizations of dendrites of pyramidal and Golgi cells. (From D. Bodian. Introductory survey of neurons. *Col. Spring Harbor Symp. Quant. Biol.* **17**:3, 1952.)

release of a special neurotransmitter substance (see Chaps. 3 and 5).

Most neurons have all their portions within the CNS. The cerebral cortex (see Chaps. 8 and 9), the thin layer of gray matter over the surface of the brain, contains the *pyramidal cell* as one of its characteristic cells (Fig. 2-1). This cell takes its name from the pyramidal form of the cell body. The apex of the pyramid is pointed radially outward to the surface, from which a large dendrite, the *apical dendrite,* takes its origin. The apical dendrite passes upward for some distance before dividing into smaller and smaller branches, thus making available a relatively large area of dendritic surface receiving synapses from other axons terminating on the branches. Thus, the neuron may be influenced by many other cells of the cortex or of other regions in the brain. Its axon passes to other parts of the cortex or to other regions of the CNS. Another type of cortical neuron is the *stellate* cell. It has a short axon and performs an integrative function, synapsing with many other neurons. Regions of the CNS where cell bodies and dendrites and their connections predominate are *gray* in color. Regions where axon tracts

are compact have a whiter appearance and are referred to as *white* areas.

Most of what is known of nerve excitability has been obtained from studies made on peripheral nerve. As shown in Figure 2-2, a nerve trunk is composed of a large number of fibers of varying size, with diameters ranging from a fraction of 1 μm up to approximately 20 μm. Fibers above 2 μm in diameter have a myelin sheath around them and are known as *myelinated nerve fibers.* Myelin has a large content of lipid, apparent when it is stained with dyes having an affinity for it (e.g., osmium tetroxide). These show the fiber to be covered by a sheath of myelin (Fig. 2-2). Fibers below 2 μm in diameter are the *unmyelinated nerve fibers.* Elec-

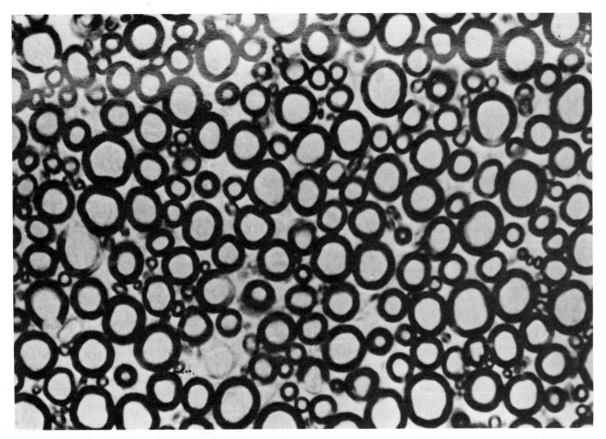

Fig. 2-2. Cross section of myelinated nerve fibers. The dense bands around the fibers are myelin sheaths. The smallest groups of fibers are the unmyelinated fibers, which typically are found in bundles, and are not visible here. ×425.

tron microscope studies of the myelin sheath of the myelinated fibers have revealed that the myelin is composed of layers that are regularly arranged in the form of lamellae (Fig. 2-3). The myelin layers are laid down by the Schwann cells positioned at intervals of 1 to 2 mm along the length of the fibers. The axon is thus left uncovered at the *nodes,* where, as will be discussed later, excitation takes place. The myelinated portions between the nodes are referred to as the *internodes*. Electron microscope studies of growing nerves during embryogenesis have revealed that the myelin is laid down by a jelly-roll wrapping process of membrane around the axon, to produce the typical lamellated appearance (Fig. 2-3). The nerve fibers in the CNS are similarly myelinated by their analogous satellite cells—the *glial cells*. A single glial cell may myelinate several CNS fibers, whereas one Schwann cell lays down myelin over the internodal length of a peripheral nerve fiber. The thinner unmyelinated fibers have been shown in electron micrographs to have one layer of myelin around them. In peripheral nerve, a Schwann cell engulfs a number of unmyelinated fibers—up to 12 or more. The bundles of fibers with Schwann cells at intervals along their length are known as the fibers of Remak.

Under the electron microscope the giant fibers of squid and cuttlefish are seen to have several irregular myelinlike layers around them, supplied by satellite cells analogous to Schwann cells. The giant nerve fibers attain diameters of 0.5 to 1.0 mm. The use of such large-diameter fibers has been of inestimable service in the development of the modern theory of nerve excitation. Their use in studies of excitation will be discussed later in this chapter. First, we will take up the most basic properties of excitation using the frog nerve as a model.

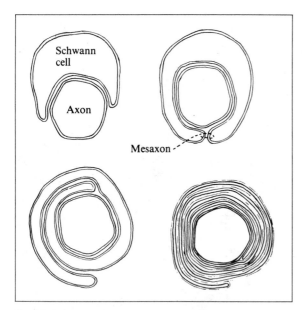

Fig. 2-3. Embryological development of the myelin sheath. Early in development the Schwann cell becomes approximated to the axon. The membranes of the Schwann cell meet to form the mesaxon. One surface pushes forward and in so doing wraps a double membrane around the nerve. Myelin is thus laid down in a jelly-roll fashion. (Adapted from B. B. Geren. The formation from the Schwann cell surface of myelin in the peripheral nerves of chick embryos. *Exp. Cell Res.* 7:560, 1954.)

EXCITATION AND RECORDING OF THE ACTION POTENTIAL

The frog sciatic nerve has long been favored for the study of the basic properties of excitation and conduction because it can survive for many hours after it has been removed from the animal. The isolated sciatic nerve is placed in a closed chamber containing moist air where electrical contact can be made with it through electrodes at various points along its length. As indicated in Figure 2-4, two stimulating electrodes are used to convey a brief pulse of current through a portion of the nerve at one end. When a current of adequate strength is used to excite the fibers, a nerve impulse, an *action potential* (which will be described subsequently), passes along the nerve. If the nerve is innervating a muscle, a brief contraction (twitch) of the muscle innervated by the motor fibers results. Some simple observations can be made using the muscle twitch as an index of effective nerve excitation. Stimulation below a strength effective to excite is called a *subthreshold stimulus.*

The frog nerve's sciatic nerve trunk is composed of thousands of fibers. As we will note later on, adequate excitation of a fiber causes it to give rise to a propagated action potential that travels down the fiber to activate a group of muscles; such excitation, once initiated in the nerve fibers, is not increased by increasing the strength of the stimulation, and the nerve is said to be excited in an *all-or-none* fashion.

A strength of stimulation causing stimulation of a few fibers gives rise to a just discernible motor response, referred to as a *threshold stimulation.* A further increase of stimulation causes more fibers to respond, and the twitch is increased in size. When all nerve fibers are excited, the result is a *maximal response.* An additional increase in strength, a *supramaximal stimulation,* does not further increase the twitch size.

The action potential is excited by a brief electrical current led into the nerve through a pair of stimulating electrodes, and the electrical change that constitutes the nerve impulse is recorded by placing electrodes along the sciatic nerve and connecting them through an amplifier to an oscilloscope, as shown in Figure 2-4. The nerve impulse is a small, brief negative voltage change that is picked up by these electrodes and must be considerably amplified before it can be displayed upon the screen of an oscilloscope. A brief description of this instrument will be necessary to explain the display of the response. A small beam of electrons is focused on the inner face of the oscilloscope tube and visualized by the fluorescence it produces. Voltages applied to the horizontal and vertical plates arranged along the path of the beam cause it to be displaced horizontally and vertically on the face of the screen. A special sweep circuit is used to apply appropriate voltages on the horizontal plates to make the beam move uniformly in the horizontal direction from the left to the right side of the tube screen. This constitutes the *time axis.* After a horizontal sweep is completed, the beam is shut off, and it returns to the starting position until a trigger pulse starts the movement once again. With the beam moving in the horizontal (X) direction, voltages applied to the vertical (Y) plates cause the beam to rise or fall, giving rise to a time display of fast-changing voltages.

The brief electrical signal seen on the oscilloscope trace at the time of the brief stimulating pulse applied to the nerve is an *artifact* caused by an adventitious electrical coupling to the recording electrodes. It serves as a convenient indication of the time when the stimulus is applied. After the stimulus a *latency* period is seen as the oscilloscope beam moves along with no further voltage change until the beginning of the action potential is recorded. The latency depends on the speed at which the nerve impulse is propagated along the nerve from the stimulating electrodes to

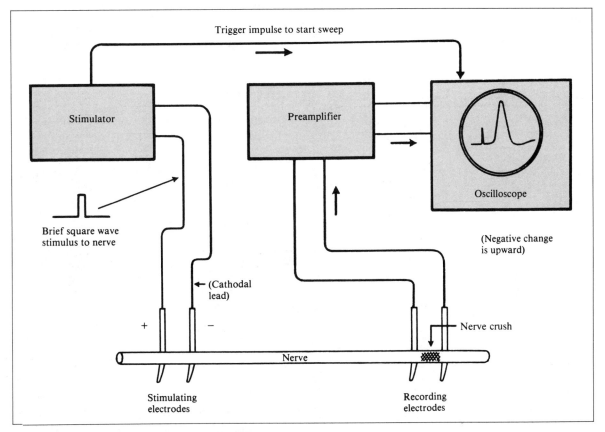

Fig. 2-4. Excitation and recording of action potentials. The stimulator starts the horizontal sweep of the oscilloscope beam with a trigger pulse, and at some set time thereafter a brief pulse of current is delivered through the stimulating electrodes on which the nerve is placed. Farther along the nerve, electrodes pick up the propagated action potential after a conduction delay. After amplification, the action potential is displayed on the vertical plates of the oscilloscope. The combined time movement in the horizontal direction and voltage changes in the vertical direction give an X-Y graphic display.

the recording electrodes and the distance from the cathode of the stimulating electrode to the first recording electrode (Fig. 2-4). There is some variation in the site of origin of excitation under the cathode lead of the stimulating electrode. The excitation may, because of the spread of current in the perineural sheath, be initiated a millimeter or more away, but this usually is not too large an error. As the action potential passes under the first electrode, it becomes negative relative to the other electrode, and the beam is seen to rise vertically on the oscilloscope (Fig. 2-5A). The action potential continues past the first electrode until it reaches an intermediate position between the recording electrodes so that no difference in potential exists and the beam returns to the baseline level. Then, as the action potential passes under the second recording electrode, it becomes negative with respect to the first electrode, and the beam is directed downward. As the action potential moves on, the beam returns to the baseline, thus displaying the *diphasic action potential.*

To convert the diphasic action potential to the *monophasic action potential,* the nerve is crushed between the two recording electrodes. The nerve impulse can then pass under the first electrode as before but not under the second one, leaving only the first negative deflection to be recorded on the oscilloscope (Fig. 2-5B).

As noted, the action potential is *all-or-none.* Conduction of the action potential does not decrease in amplitude in its passage down the length of the fibers. If its amplitude is decreased by application of a narcotic that only partially decreases excitability over a short length of nerve, the ac-

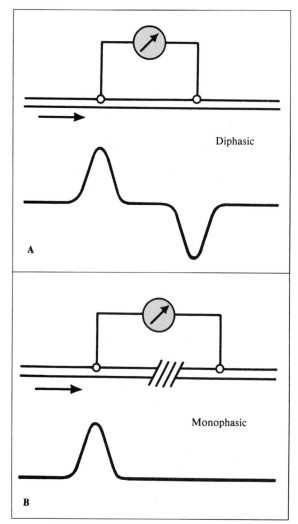

A

B

Fig. 2-5. Diphasic (A) and monophasic (B) action potentials. The arrows show the direction of an action potential as it moves along a nerve in (A) past the first and then the second recording electrode. As it passes each electrode, that electrode becomes relatively negative with respect to the other, accounting for the deflection first one way, then the other, giving rise to the diphasic action potential. In (B), monophasic action potential is seen because the crush of the nerve prevents the action potentional from passing to the second recording electrode. (From S. Ochs. *Elements of Physiology*. New York: Wiley, 1965. P. 27.)

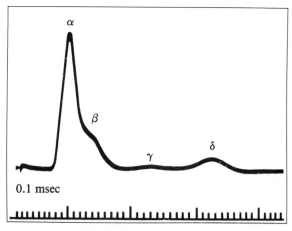

Fig. 2-6. Compound action potential. The A fibers are subdivided into subgroups labeled with Greek letters having successively higher stimulation thresholds and slower conduction velocities. (Note that a large γ group is present only in motor nerves.) (From H. S. Gasser. Pain-producing impulses in peripheral nerves. *Association for Research in Nervous and Mental Disease, Proceedings* [1942] 23:48, 1943.)

tion potential can recover its full amplitude after passing through the narcotized region.

The monophasic action potential is followed by a smaller, longer-lasting potential, the *negative afterpotential,* merging with a much-longer-lasting *positive afterpotential.* These are discussed with respect to the giant nerve axon, where their analysis is easier.

The fibers of a mixed nerve such as the frog sciatic nerve have different diameters (see Fig. 2-2). The action potential excited at the lower stimulus strengths comes from the most excitable fibers, from the group of fibers with the largest diameters. These fibers are classified as the *A fiber group,* which is divided into subgroups of successively smaller sizes, designated by Greek letters: *alpha, beta, gamma,* and *delta.* The alpha fibers, ranging in diameter from about 15 to 18 μm in the frog sciatic nerve, are the most excitable subgroup of the A group. With just threshold stimulus strengths, fibers of the alpha group are excited. As the stimulation strength is increased, more fibers of the alpha group are added to the population responding until the action potential response reaches a maximum. A further increase in stimulation strength does not increase the size of the alpha group response. With greater strengths a hump appears on the descending portion of the action potential. This potential is due to the excitation of the beta fibers (Fig. 2-6).

Still greater stimulation strengths similarly excite later-appearing responses. The smaller-fiber groups have a lower conduction velocity besides having higher thresholds. Advantage is taken of this property to separate the action potentials of the different subgroups by increasing the distance between the stimulating and and recording electrodes. Just as the faster and slower runners in a race are lined up at the start but in the course of the race spread out, so in nerve the different action potentials representing the subgroups traveling at their various speeds spread out. The spread is best seen when the distance between stimulating and recording electrodes is increased, as is shown for the alpha and beta groups in Figure 2-7.

If a second stimulus is initiated too soon after a first response, it does not excite an action potential and the nerve is said to show *absolute refractoriness*. It is inexcitable no matter how strong a stimulus is used. In the alpha group of A fibers of the frog sciatic nerve, absolute refractoriness lasts a little longer than 1 msec and is followed by a period of *relative refractoriness*. In this state some response is possible if the strength of stimulus is increased, and as the fibers recover their excitability, the amplitude of response increases along an S-shaped curve until the action potential equals that of the control response (Fig. 2-8). The duration of the relative refractory period for the alpha group of the frog sciatic nerve is approximately 3 to 5 msec.

STIMULATING CURRENTS AND EXCITABILITY

Electrical conduction in electrolyte solutions takes place by the movement of ions. As discussed in Chapter 1, the major ions of extracellular fluids in mammalian tissue are Na^+ and Cl^-. The positively charged Na^+ moves toward the cathode, and the negative Cl^- moves toward the anode when a current is impressed on tissues. Tissues in general can be considered as electrolyte solutions of various ionic compositions, with mainly K^+ and the larger molecular anions present within the nerve and muscle fibers. Ions moving as a result of an imposed potential constitute a current flow; the number of ions flowing is directly related to the

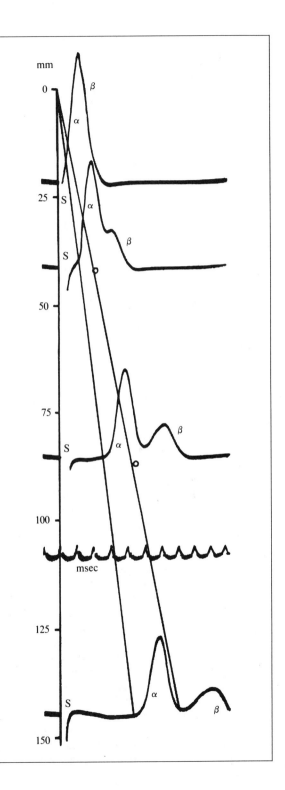

Fig. 2-7. Separation of alpha (α) and beta (β) groups. If the distance between stimulated and recording sites along a nerve is successively increased, the α and β groups become easier to distinguish. The velocities of the two groups are indicated by the differing slopes of the two lines drawn through the beginning (the foot) of their action potentials. At a greater distance the foot of the β group is clearly separated from that of the α group. (From J. Erlanger and H. S. Gasser. *Electrical Signs of Nervous Activity.* [2nd ed.]. Philadelphia: University of Pennsylvania Press, 1968. P. 13.)

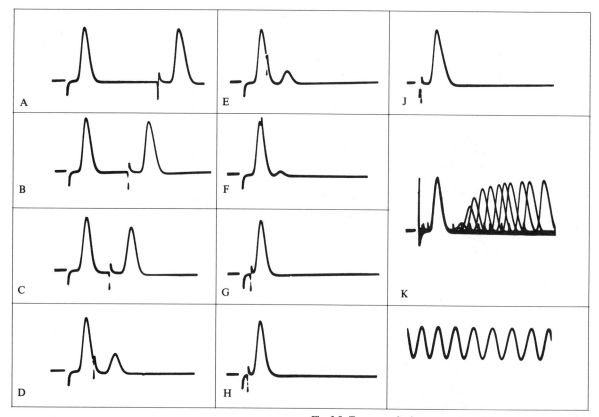

Fig. 2-8. Two monophasic responses at decreasing time intervals from (A) to (J) give rise to a decrease of the second test response in (C) to (F) and its absence in (G) to (J) (relative and absolute refractoriness). In (K) a series of double-shock traces is overlayed to show in a more compact form the decreasing amplitude of the test response during refractoriness. Callibration in the lowermost trace is 1000 Hz. (From S. Ochs. *Elements of Neurophysiology.* New York: Wiley, 1965. P. 42.)

quantity of current. The proportion of the current carried by any given ion species also depends on its charge and mobility. Lines drawn between the electrodes signify the direction, and the number of lines the density, of the current flow (Fig. 2-9A). By convention, current flows from the anode of the stimulating electrode, through the nerve trunk, and then back to the cathode.

As a general rule, nerve cells are excited in the region close to or under the cathodal electrode. The current flowing between the nerve fibers is ineffective as regards excitation; only that part of the current flowing out of the fibers at the cathode constitutes the effective portion. The lines of current pictured in Figure 2-9 show only the form of current flow with no particular designation of the ion species. The ion species that carries current varies in the cells and in the extracellular fluid. This is an important principle. The cations, mainly Na^+ and K^+, move toward the cathode, while the anions, mainly Cl^-, move toward the anode. The ions actually carrying the current through the nerve are thus moving in both directions. The current carried does not depend on the particular ion but on its charge, mobility, and

amount. Potassium is a major ion carrying current within the nerve fiber. Sodium is a major ion in the extracellular fluid. Both ions are included in the lines of current flow from the anode to the cathode of stimulating electrodes.

The membranes of nerve fibers (and other excitable cells such as muscle) offer a high resistance to the passage of electrical current, the resistance pictured by a suitable resistance element in an electrical model of the nerve (Fig. 2-10). Excitation occurs when the current flowing outward across the membrane reduces its *resting membrane potential,* i.e., causes a *depolarization.* When the depolarization reaches a critical level, the impedance decreases, a greater flow of ion currents occurs, and an action potential is excited. In a later

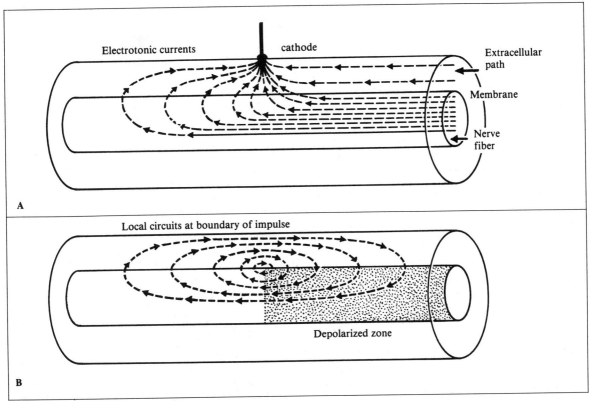

Fig. 2-9. Paths of applied current and local circuits resulting from an action potential. In (A), the lines of current from an applied source are shown passing from the nerve to the cathode. Part of the current passes between the fibers, and part inside. The latter current is the effective portion, and excitation occurs in the region of the cathode as current passes outward across the nerve membranes and depolarizes the membranes. In (B), a similar direction of outward current in the region in front of an active part of the fibers depolarizes and can excite that region of nerve. (From A. L. Hodgkin. Evidence for electrical transmission in nerve. *J. Physiol.* 90:183, 1937.)

section of this chapter, the ionic events related to excitation and the production of an action potential will be discussed in more detail.

Because of the high resistance of the membranes and their electrical *capacitance,* an application of electrical currents gives rise to a distribution of potential along the length of the fiber to which the term *electrotonus* is applied. We can define *capacitance* as the ability to hold a charge, i.e., the storage of ions in the capacitance giving rise to a potential difference. Electrotonus refers not only to distributed voltages along the nerve resulting from applied currents, but also to the potentials produced by the current flows resulting from action potentials in the nerve, in part to stored charge in the capacitance of the membrane. Distributed currents and resulting voltage changes can affect the nearby membrane so that its excitability changes. This will be described later in a discussion of local currents in connection with the flow of electrotonic currents noted in Figure 2-9B.

The duration of a pulse of exciting current needed to excite an action potential is given by the *strength-duration curve* (Fig. 2-11). This curve is determined by the strength of stimulating current required to reach threshold with different durations of stimulating current. As the duration of the pulse is shortened, the strength required to reach threshold must be raised. With long stimulus durations the stimulating current required is called the *rheobase.* *Chronaxie* is defined as that duration of stimulation required to excite at a strength twice the rheobase (Fig. 2-11). The more excitable tissues have shorter chronaxies. Nerve, for example, has a shorter chronaxie than does muscle. Stimulation at certain points over the surface of muscles is more effective in eliciting motor contractions than stimulation at other points. These *motor points* are places where there is a

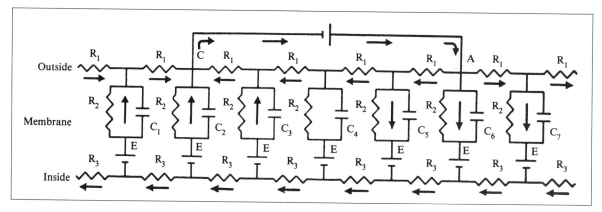

confluence of motor nerves entering the muscle. The strength-duration curve over the motor point shows the short chronaxie and strength-duration curve typical of nerve excitation. This property is made use of in testing for nerve degeneration in cases of nerve lesions and for the recovery of function after nerve regeneration takes place (see below).

IONIC BASIS OF MEMBRANE POTENTIALS

RESTING POTENTIALS. An electrical potential exists across the membrane of nerve and muscle. It was demonstrated by the negative potential seen at a crushed or injured site on one part of the nerve or muscle relative to the intact portion. This *injury potential* or *demarcation potential* is due to current passing from the *polarized,* intact part of the fiber to the injured, *depolarized* portion. The source of current is the asymmetry of ions across the normal membrane, giving rise to the resting membrane potential in the manner described in Chapter 1. In the Nernst equation

$$E = \frac{RT}{F} \ln \frac{[K^+]_i}{[K^+]_o}$$

where E represents the electrical driving force in equilibrium with the chemical driving force due to the ratio of $[K^+]$ across the membrane (see Chap. 1). An important point to note is that ion concentration differences of only several picomols (10^{-12} moles) across the membrane can produce a voltage of some -90 mv. The sign refers to the inside of the cell, with the outside taken as zero.

The giant nerve axons obtained from squid and cuttlefish have diameters up to 0.5 mm or more, and in these fibers the *transmembrane* potential can be readily measured by inserting a microelectrode into the fiber from one end with another electrode placed in the outside medium. Potentials of approximately 50 to 60 mv are found, with the inside nega-

Fig. 2-10. Electrical model of the membrane. The arrows in this model show the direction of current through an electrical network representing the nerve membrane. The model is composed of resistance and capacitance elements with a battery representing resting membrane potential. (From J. F. Fulton. *A Textbook of Physiology* [17th ed.] Philadelphia: Saunders, 1955.)

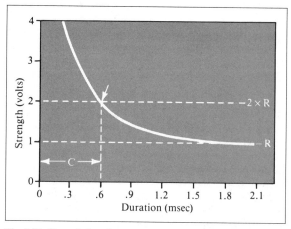

Fig. 2-11. Strength-duration curve. The curve is derived from the strength required to attain a given level of response when various durations of stimulation are used. The longer pulse durations approach the rheobase (*R*). Chronaxie (*C*) is the duration of the stimulus required for excitation at a strength of stimulation twice rheobase (*arrow*). (From S. Ochs. *Elements of Neurophysiology.* New York: Wiley, 1965. P. 20.)

tive to the outside. The transmembrane potential or *resting membrane potential* (E_M) is accounted for by the Nernst equation. Additionally, microelectrodes with tips measuring 0.5 μm or less can be inserted through the membrane of smaller nerve fibers and nerve cell bodies as well as into muscle cells. By this means, resting membrane potentials of

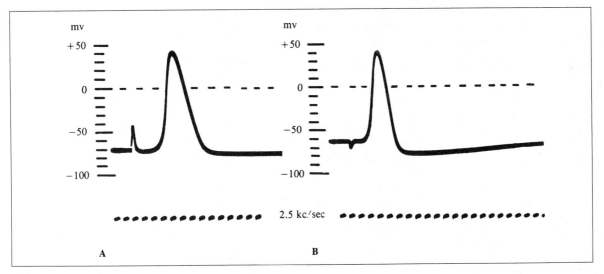

Fig. 2-12. Action potentials from giant nerve fibers. In (A), the action potential is recorded from a giant nerve fiber in situ with a microelectrode. The resting membrane potential is 70 mv, inside negative. An action potential shows an overshoot of approximately 35 mv. In (B), an internal capillary electrode inserted from one end measures a smaller resting membrane potential of approximately 60 mv. A hyperpolarization follows the action potential—the positive afterpotential. (From A. L. Hodgkin. Ionic movements and electric activity in giant nerve fibres. *Proc. R. Soc.* [Biol.] 148:5, 1958.)

the order of 50 to 90 mv have been found in a wide variety of tissues in various species, including humans.

On changing the concentration of K^+ in the external medium, the potential changes are seen to conform to the Nernst relationship, except that at the lower concentrations of K^+ in the outside medium a voltage somewhat lower than that calculated from nerve potential is found. This is due to the contribution of Na^+ to the resting membrane potential. In muscle fibers there is also a large contribution of Cl^- to the resting membrane potential. It acts as a "battery" in parallel with the K^+ "battery" with the same orientation of potential as that of K^+.

It is possible to squeeze the axoplasm out of the giant axons and replace it with fluids of known composition. In such *perfused axons,* when K^+ was present at its usual concentration, the usual resting potential was seen. When K^+ was diluted with a nonelectrolyte such as sucrose to lower its internal concentration, the membrane potential was reduced according to the expectations of the Nernst equation. With equal concentration of K^+ across the membrane, the resting potential was close to zero, as expected from the Nernst equation.

ACTION POTENTIALS. With a micropipette inserted into a giant fiber and a resting membrane potential of 60 mv, stimulation of the fiber gives rise to action potentials having amplitudes of 90 mv (Fig. 2-12). An action potential larger in amplitude than the resting membrane potential was a new and unexpected finding. The excess potential, approximately 30 mv beyond that of a simple depolarization, was called the *overshoot.* The explanation of the overshoot resulted in a new theory of nerve action, the *sodium hypothesis,* now generally accepted as the basis of the action potential. Sodium ion is high in concentration outside the nerve fiber and low inside. In effect it constitutes a "battery" in opposition to the resting membrane potential insofar as the Na^+ concentration difference across the membrane is so directed that it tends to make Na^+ enter the fiber. The negative charge inside the cell attracting the positively charged Na^+ ion adds an electrical driving force. Radioactive tracer studies have shown that Na^+ in actuality continually enters the cells. This process would soon result in the entrance of sufficient Na^+ into the cells to bring the resting membrane potential to zero. It is counteracted by the operation of the *sodium pump* (see Chap. 1), whereby Na^+ is pumped out of the cell interior and in effect reduces the content and the effect of Na^+ on the resting membrane potential.

The action potential is explained by the sodium hypothesis. When the resting potential across the membrane of the nerve is reduced (depolarized) to a *critical level,* Na^+ permeability rapidly increases, allowing Na^+ to move into the cell within a fraction of the millisecond. The entry of Na^+ accounts for the 30-mv overshoot of the action potential. The voltage across the membrane moves toward the *equilibrium potential for Na⁺* (E_{Na^+}), computed from the Nernst equation and the concentration ratio of Na⁺ across the membrane. At E_{Na^+} further Na⁺ entry is prevented.

The increase of Na⁺ permeability does not continue indefinitely because of a process called *inactivation,* which shuts off further entry of Na⁺. Additionally, after a brief lag, K⁺ permeability increases, and K⁺ leaves the interior of the fiber, bringing its positive charge out of the cell. Thus, the membrane voltage returns to the resting membrane potential to cause the termination of the action potential. The increase in K⁺ permeability continues for some time, causing the *afterhyperpolarization* or *positive afterpotential* (Fig. 2-12). The size of the afterpotential is related to the level of the resting membrane potential and the deviation existing from the *equilibrium potential for K⁺* (E_{k^+}). This potential is computed from the resting membrane potential difference (V_M) and the Nernst potential for K⁺. The hyperpolarization is larger when the nerve has been depolarized below its normal resting membrane potential.

It must be emphasized that only a few picomols of Na⁺ and K⁺ are exchanged across the membrane to give rise to the action potential. The net effect of these ion shifts is that the fiber has gained a tiny bit of Na⁺ (3–6 picomols) and lost an equivalent amount of K⁺ for each action potential. The sodium pump will later redress the imbalance of the gain of Na⁺ and loss of K⁺ after a number of action potentials to restore the original ionic asymmetry. The small ionic change occurring as a result of each discharge allows a great many action potentials to be generated before any significant effect on ionic concentration and resting membrane potential is seen.

The analysis of the sodium hypothesis was accomplished by Hodgkin, Huxley, and Katz using electrodes placed inside the length of giant nerve axons. By means of a control circuit completed through the outside electrodes, the nerve could be "clamped" at a fixed level of potential (Fig. 2-13). The entire axon membrane could thus be depolarized (to excite it) and the relatively large flow of current through the whole of the membrane measured without the complication of the conduction of an action potential. On depolarization, an early brief inflow of current appeared that was related to the entry of Na⁺. Later, the current became reversed as a result of the outward flow of K⁺. This sequence was shown in experiments in which the nerve was immersed in solutions from which Na⁺ was removed—for example, by replacing Na⁺ with sucrose or choline. The early inward portion of the current was eliminated while the outward current—i.e., that due to K⁺ outflow—remained (Fig. 2-14).

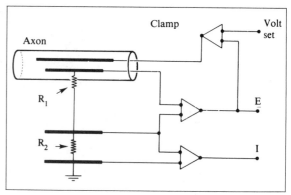

Fig. 2-13. Electrodes are shown inserted lengthwise into a giant axon for voltage clamping. The clamping voltage is set by the upper clamp amplifier, and the voltage change desired, led into the axon through the longer electrode, is measured by the amplifier (*E*) across the membrane resistance (*R₁*). The resulting membrane current (*I*) is measured across the external resistance (*R₂*) through the lower amplifier.

The permeability changes to Na⁺ and K⁺ underlying these ionic flows are conceived of as the successive opening of selective *channels* in the membrane, first to Na⁺ and then to K⁺, allowing the Na⁺ to flow in and the K⁺ to flow out through the channels. The behavior of nerve in response to various other ions shows in general that these channels are apparently separate, though not perfectly selective to these ion species.

The electrical resistivity of the nerve membrane is high. As shown in electron micrographs, the membrane thickness is of the order of 75 Å. Physicochemical studies show it to be composed of a bilayer of lipoproteins with globular proteins embedded in the lipid, where some act as the channels for ion movement. Just at the onset of the action potential, the electrical resistivity of the membrane falls drastically, to about one-fortieth of its original value. As shown in Figure 2-15, there is a *Na⁺ conductance* (G_{Na^+}) increase, which is soon followed by a *K⁺ conductance* (G_{K^+}) increase. Conductance is the inverse of resistance. Conductance increases may be viewed as being due to a changed conformation of globular proteins embedded in the membrane, opening the channel through which the ions move, each ion having a specific channel. To describe the conductance changes related to permeability, Hodgkin (1958) considers K⁺ conductance to be controlled by a variable raised to the fourth power. Intuitively, we can consider that the increased conductance comes about when four associated ions are grouped at some critical site within the K⁺ channel. A more complicated situation controls Na⁺ permeability with three ions considered to occupy a critical site to bring

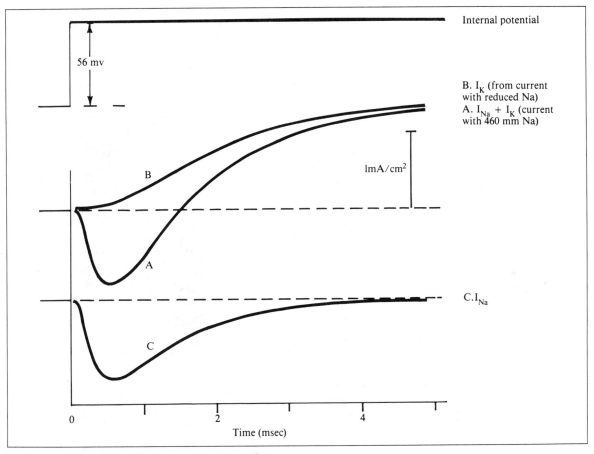

Fig. 2-14. Curve *A* shows the inward component of current flow (*downward direction*) followed by an outward current in response to a depolarization of −56 mv. In a sodium-deficient medium (curve *B*) the inward component is absent, while the potassium current (I_K) remains. Subtraction of curve *B* from curve *A* gives, in curve *C*, the sodium current (I_{Na}). (After A. L. Hodgkin and A. F. Huxley. Currents carried by sodium and potassium ions through the membrane of the giant axon of *Loligo*. *J. Physiol.* [Lond.] 116:459, 1952.)

about an increase in Na^+ conductance. An opposite process controlled by another variable, *inactivation*, brings about a closing of Na^+ channels.

The underlying idea is that the control of Na^+ and K^+ is brought about by the voltage difference existing across the membrane. The crucial event is a quick depolarization to reach the threshold or *critical level* for the initiation of increased Na^+ conductance. If the voltage change is applied too slowly, inactivation occurs, preventing excitation.

Support for the sodium hypothesis has been obtained in

several ways. Radioactive isotopes of Na^+ and K^+ show an extra uptake of Na^+ with a corresponding loss of K^+ in nerve or muscle fibers following a repetitive series of action potentials. The amounts conformed to the expected relation of charge moved across the membrane, which was on the order of several picomols/cm^2 per action potential.

Specific toxins acting on giant axon membranes have helped to reveal the properties of the active Na^+ channel and the permeability increase on depolarization. *Tetrodotoxin*, a toxin obtained from the puffer fish, acts in very low concentrations to reduce Na^+ permeability and to block the activation of the Na^+ permeability increase during excitation. It does so by entering the Na^+ channels at or just within the outer face, thus blocking Na^+ entry. *Batrachotoxin*, another toxin that blocks in very small concentrations, acts to keep the Na^+ channels open and thus to make the nerve inexcitable. Local anesthetics such as procaine block excitability by blocking the permeability of both

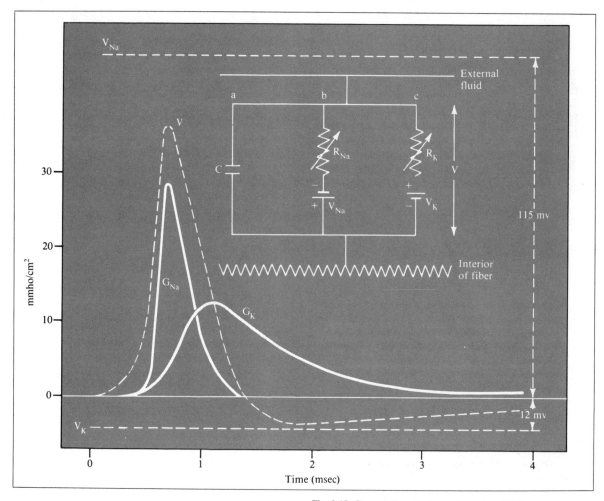

Fig. 2-15. Computed conductance change in millimhos per square centimeter to Na (G_{Na}) and to K (G_K), accounting for the action potential (V). Notice that the result of Na$^+$ entry is to drive the membrane potential to the Na equilibrium (V_{Na}). The membrane at rest is close to the equilibrium potential for K$^+$ (V_K). An electrical representation of the Na-channel resistance (R_{Na}) and K-channel resistance (R_K) are shown as variables and in series with their respective voltages (V_{Na}) and (V_K), giving rise to the total voltage (V) across the membrane. Also shown is the capacitance (C) across the membrane. (From A. L. Hodgkin. The Croonian lecture. Ionic movements and electric activity in giant nerve fibers. *Proc. R. Soc.* [Biol.] 148:12, 1958.)

Na$^+$ and K$^+$ channels. *Tetraethylammonium* blocks K$^+$ channels at their inner surface.

Calcium is associated in some important way with excitability, as is shown with regard to the property of rhythmicity. Nerve axons do not usually discharge rhythmically, but do so in low-Ca^{2+} media. In *hypocalcemia*, abnormal excitability of nerve and muscle is observed, often with a period of sustained rhythmic nerve discharge following an excitation, leading to a *tetanic* muscular discharge.

LOCAL CURRENTS AND CONDUCTION: CONTINUOUS AND SALTATORY

The conduction of the action potential is explained on the basis of the electrical properties of the membrane: its high resistance and electrical capacity distributed all along the length of the fiber (see Fig. 2-10). At the region where an action potential has been excited, either by application of a pulse of electric current or naturally at sensory nerve endings (see Chap. 3), the voltage change across the membrane that constitutes the action potential can be considered as a "battery." This voltage causes current to spread out into the adjoining membrane, depolarizing it and bringing it to the critical level for excitation and initiation of an action potential. The process is repeated again and again, constituting the propagated action potential (see Fig. 2-9B).

The ionic difference between the local potential changes due to local currents and the currents involved in the action potential cannot be stressed too much. The local currents are passive, with no specificity of the species of ions carrying the current. In the nerve fiber, K^+ and anions carry the major part of the current, while Na^+ and Cl^- carry the major part of the current outside the fiber. These current flows follow the general law of electrical conductance in an electrolyte medium. In contrast, the active action potential is, as previously described, the result of specific ionic processes, first Na^+ and then K^+ moving through their respective specific channels in the membrane when excitation occurs.

Conduction velocity is determined by the resistances inside the fiber and across the membrane, neglecting for the time being the electrical capacitance of the membrane (see Fig. 2-10). Considering the voltage change due to an action potential across the membrane as a battery, the current passively flowing in the nearby membrane falls off with distance from that point. The fall-off is expressed by the length constant (λ) given by the resulting potential recorded along the length of the fiber at the point where the voltage falls to 1/e of its original level,

$$\frac{V}{V_o} = e^{-(x/\lambda)}$$

where V = measured voltage at different distances, x, from the source voltage, V_o. When $x = \lambda$, $V = 1/e \cdot V_o$. This formula is used to determine the length constant, λ. It can be seen intuitively that the greater the λ, the farther the electrotonic voltage reaches out in the membrane. In effect, excitation of the action potential will then occur at a more distant site and thus give rise to a faster conduction velocity.

A much faster conduction velocity is seen in myelinated than in unmyelinated nerve fibers. Single myelinated nerve fibers can be isolated for a study of their physiological properties, which requires care and patience. The fiber is shown to be more excitable at the nodes, where there is no myelin-

ation, than at the internodes (Fig. 2-16). This finding follows from the theory of *saltatory conduction,* which takes into account the fact that the myelin acts as an electrical insulator. Electrical currents can enter or leave only at the nodes. Excitation takes place at the node when the membrane is depolarized to the critical level, as was described for the giant axon membrane. Na^+ permeability then rapidly increases, and Na^+ enters, causing a *local current* to pass down inside the axon and out the neighboring nodes. The outward current across the nodal membrane causes excitation when the critical level of depolarization is reached. Sodium ions enter that node, and the same process is repeated. Excitation, therefore, instead of occurring continuously along the membrane, as is the case in the giant axon fiber and in the small unmyelinated axons, occurs at the nodes and is thus discontinuous, or *saltatory.* As a result, the velocity is higher than for a correspondingly sized unmyelinated fiber. Velocities of up to 120 meters per second are attained in the large, 20- to 22-μm myelinated fibers of animals and humans.

The current flowing from node to node within a fiber does not normally excite the nerve fibers lying alongside it, but it is not wholly without effect. Under special conditions, small alterations in the excitability of nearby fibers can be shown. In some diseases in which the myelin is lost, abnormal excitation of neighboring fibers may take place.

METABOLISM RELATED TO EXCITABILITY

After generating an action potential, the nerve has gained several picomols/cm² of Na^+ and lost an equivalent amount of K^+. Complete restitution of ionic asymmetry after nerve activity requires the pumping out of the extra Na^+ gained and the recapture of the lost K^+. Otherwise, the ionic composition of the fiber would eventually change after a long time, with a loss of resting potential and excitability. Sodium pumping requires the operation of an enzyme located in the membrane, *sodium-potassium–activated* adenosine triphosphate (Na^+-K^+ ATPase). This enzyme is polarized so that its high Na^+-binding side faces internally and its K^+-binding portion faces externally. *Adenosine triphosphate* is required by the Na^+ pump, the ATP supplying energy through its energy-rich phosphate bond ($\sim$P). Oxygen is utilized by the nerve to generate ATP through oxidative phosphorylation in the mitochondria found all along the inside of the axons.

If metabolic blocking agents, such as cyanide or dinitrophenol, are applied to the giant nerve fiber or injected into it, the resulting block of oxidative phosphorylation causes a fall in ATP and a consequent cessation of the outward flux

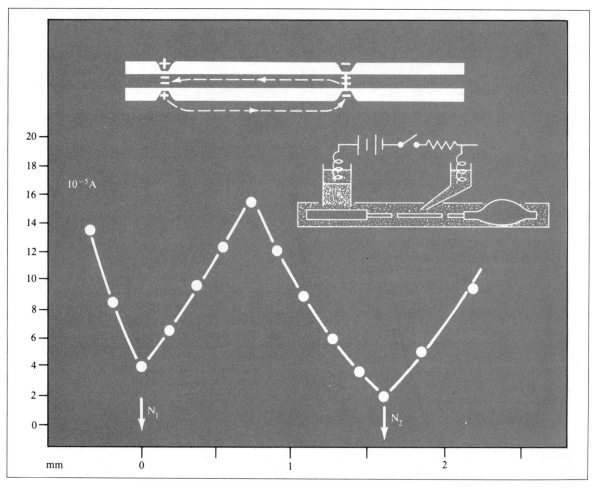

Fig. 2-16. Transmission in a myelinated nerve occurs via currents passing down along the inside of the fiber and out the adjacent nodes, as shown in the insert above. The internodal portion covered with myelin is less excitable than the nodes. Threshold to electrical currents for a microelectrode placed at various points along a single fiber, as shown in the insert above, is lowest at the two nodes, N_1 and N_2, and is high in the internodal region. (From I. Tasaki. *Nervous Transmission*, 1953. P. 6. Courtesy of Charles C Thomas, Publisher, Springfield, Ill.)

of Na^+. Na^+ then accumulates inside the fiber. When the Na^+ pump of the giant nerve fiber is poisoned by these agents, the resting potential and action potentials can continue for a relatively long time (90 minutes or more) before a change in the concentrations of Na^+ and K^+ sufficient to block excitation takes place. The reason is the large volume of axoplasm and the small ionic changes that result from the relatively small inward flux of Na^+. In comparison, much faster concentration changes occur in the smaller fibers of myelinated nerves, where potentials fail within a matter of 15 minutes.

When ATP is injected into giant axons poisoned with cyanide or dinitrophenol, Na^+ pumping resumes. In the giant axon, *arginine phosphate* can also cause a resumption of pumping because $\sim P$ is transferred from arginine phosphate to ATP to restore its normal level. In mammalian

tissues, this metabolic storage function is performed by *creatine phosphate*.

The nerve, like other tissues in the body, utilizes glucose to supply the energy needed to perform its function. The glucose undergoes *glycolysis* to give rise to pyruvate, and *acetyl coenzyme A (CoA)*, which then enters the *citric acid cycle* in the mitochondrion. In the operation of the citric acid cycle, the oxygen taken up by the terminal end of the *electron transfer* chain is utilized, and ATP is generated. Various toxic agents are known to block *glycolysis,* the citric acid cycle, and the electron transfer chain of mitochondria and thus produce a block of metabolism. Such diseases as diabetes or thiamine deficiency, which alter energy metabolism, can cause neurological disturbances through a defect of ATP supply, resulting in ionic changes and changed excitability. As will be described, defects in metabolism also adversely affect the transport of materials within the fibers.

AXOPLASMIC TRANSPORT

Electron microscope and biochemical studies have revealed that the nerve cell body has a high level of protein synthesis. The small bodies within the cytoplasm of the neuron soma, the *Nissl bodies,* when stained a deep blue with aniline dyes, are seen in electron microscope studies to be made up of a system of channels, the *endoplasmic reticulum,* along which small particulate structures, the *ribosomes,* are bound. Protein chain formation is known to occur on the ribosomes. In recent years, proteins, polypeptides, and other materials have been shown to move from the cells and then outward in the fibers by a process called *axoplasmic transport.*

The rate and characteristics of downflow are directly measured by injecting labeled amino acid (^{3}H-leucine) into the seventh lumbar (L7) dorsal root ganglia or into the L7 motoneuron region of the spinal cord of an animal such as the cat. The precursor is taken up by the nerve cell bodies and is rapidly incorporated into proteins and polypeptides, which then move down the sciatic nerve fibers at a rate of 410 mm per day (Fig. 2-17). The rate of the advancing crest of radioactivity is linear and independent of nerve fiber size or function, either sensory or motor. The latter is shown by the similar characteristic pattern and rate of outflow in motor fibers when ^{3}H-leucine is injected into the spinal cord near the motoneurons of the L7 lumbar segment. The same rate is seen in a variety of animal species and it is probably the same in humans. Specific substances, such as norepinephrine in sympathetic nerve fibers and acetylcholinesterase in cholinergic nerves, are also transported at a fast rate, as shown by accumulation of these substances at nerve crushes or ligations.

A hypothesis proposed to account for axoplasmic transport is that a *transport filament* binds the various materials transported, which move down the microtubules found in the nerve fibers by side arms. This model is based on the sliding-filament theory of muscle (see Chap. 3). The side arms require an energy supply, and fast axoplasmic trans-

port has been shown to be closely dependent on oxidative metabolism. Nitrogen, anoxia, azide, cyanide, and dinitrophenol, all of which block oxidative phosphorylation, also block axoplasmic transport. When transport is blocked in mammalian nerves, membrane excitability is also lost; both functions are gone in about 15 minutes. This loss indicates a common supply of ATP to both the Na$^+$ pump and the transport mechanism in the nerve fibers (Fig. 2-18).

The anoxic block of nerve action potentials and axoplasmic transport is reversible. After periods of anoxia in vivo produced by the application of high pressures in cuffs around the hind limb of cats to block blood flow, electrical responses and fast axoplasmic transport can recover after *compression ischemia* lasting up to 5 or 6 hours. With ischemic compression for times longer than 6 or 7 hours, irreversible block and Wallerian degeneration of the nerve fibers occur (see Degeneration and Regeneration).

Microtubules, considered to be the stationary elements in the nerve along which transport of material occurs, are found in various other cells and are probably present in all cells. They are associated with the translocation of materials and the maintenance of cellular form. Microtubules are sensitive to *mitotic blocking agents* such as *colchicine* and *Vinca* alkaloids (e.g., *vinblastine*), which arrest and block cell division. These tubulin-binding agents act on the microtubules in the spindles of dividing cells, causing them to disassemble and thus block cell division. These agents can also block axoplasmic transport. Electron microscopy of nerve fibers exposed to colchicine or vinblastine shows that these agents cause a disassembly of microtubules.

A more subtle action of these agents on microtubules or the transport mechanism must be considered. The neuropathy resulting from the action of these agents on microtubules in nerve is the chief limitation in the use of vinblastine and other *Vinca* alkaloids for the treatment of leukemia and other neoplasms. With the use of a desheathed nerve preparation, it was recently found that Ca^{2+} is required to maintain axoplasmic transport. There are a number of Ca^{2+} regulatory mechanisms present in the nerve fibers that act to keep the level of Ca^{2+} at approximately 10^{-7} to 10^{-6} M. At these levels the activator protein *calmodulin* enables the Ca^{2+}-Mg^{2+}–activated ATPase (Ca^{2+}-Mg^{2+} ATPase) to utilize energy for movement. In the transport filament model, the Ca^{2+}-Mg^{2+} ATPase is considered to be present on the side arms (Fig. 2-18). The calmodulin-activated Ca^{2+}-Mg^{2+} ATPase is able to utilize the ~P of ATP and, by its attachment to the transport filament and the resulting conformational movement of the side arm, cause the transport filament and the materials bound to it to move axially in the fiber.

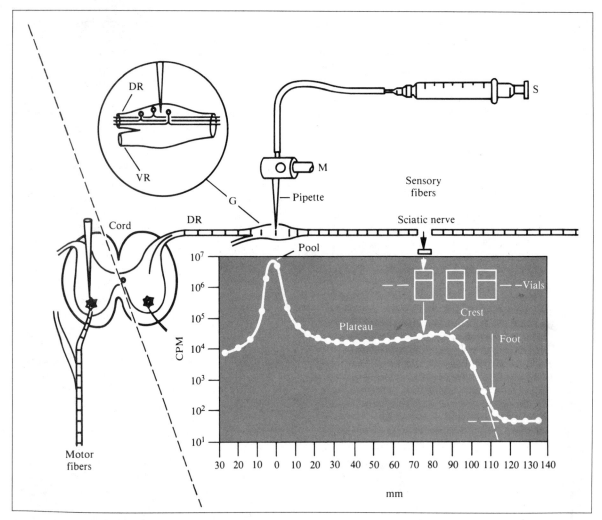

Fig. 2-17. Technique of isotope labeling and outflow pattern of fast transport. The L7 dorsal root ganglion (*G*) is shown with a pipette inserted for injection of ³H-leucine. Insert shows ganglion with T-shaped nerve cells. The dorsal root (*DR*), ganglion, and sciatic nerve are cut into 5-mm segments. Each segment is placed in a vial and solubilized; scintillation fluid is added, and a count is made. The display of activity in counts per minute (*CPM*) is shown using a logarithmic scale on the ordinate and the distance in millimeters on the abscissa, taking zero as the ganglion. A typical pattern for a 6.5-hr downflow is shown. A high level of activity remains in the ganglion and falls distally to a plateau before rising to a crest. A sharp drop is seen at the front of the crest down to the baseline level at its foot. At the left of the dashed line, an injection of the L7 ventral horn is shown for uptake of precursor by the motoneurons. The ventral root and sciatic nerve are cut and measured for outflow as in the case of ganglion injection. (From S. Ochs, and R. M. Worth. Axoplasmic Transport in Normal and Pathological Systems. In S. G. Waxman [ed.], *Physiology and Pathobiology of Axons*. New York: Raven, 1978. P. 252.)

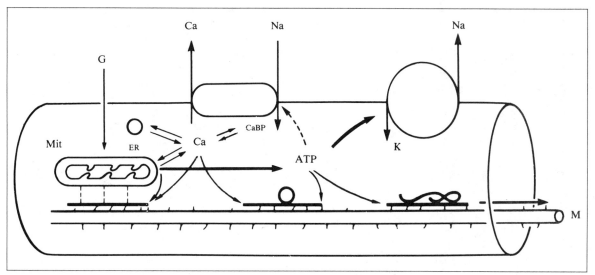

Fig. 2-18. Transport filament hypothesis. Glucose (*G*) enters the fiber, and, after oxidative phosphorylation in the mitochondrion (*Mit*), ATP is produced. A pool of ATP is shown supplying energy to the sodium pump, which controls the level of Na^+ and K^+ in the fiber and movement of the "transport filaments," shown as black bars. The various components bound to it and carried down the fiber include: (1) the mitochondria, attaching as indicated by dashed lines to the transport filament intermittently for fast forward movement; (2) particulates or vesicles; (3) soluble proteins shown in folded configuration, and other polypeptides. Thus, a wide range of component types and sizes are carried down the fiber at the same fast rate. Cross-bridges between the transport filament and the microtubules effect the movement when supplied by ATP. Also shown is calcium (*Ca*), which is regulated in the axon by sequestration in mitochondria and endoplasmic reticulum (*ER*) and binding to calcium-binding proteins (*CaBP*). It is removed from the fiber by a Ca^{2+}-Na^+ exchange mechanism or a Ca^{2+} pump. The calcium at 10^{-6} to $10^{-5}M$ activates calmodulin, and it in turn activates the Ca^{2+}-Mg^{2+} ATPase to provide energy for the movement of the transport filaments and the components bound to them. (From S. Ochs. Axoplasmic Transport. In G. J. Siegel et al. [eds.], *Basic Neurochemistry*. Boston: Little, Brown, 1981. P. 432.)

Completing the picture of the movement of materials in nerve fibers, there is, in addition to the anterograde transport that has been described, a *retrograde transport* of material. An agent that shows retrograde movement is horseradish peroxidase. It is picked up by nerve terminals, carried in the retrograde direction, and accumulates in the cell bodies. Its use has been of signal importance in tracing nerve fibers in the CNS. In the same way, viruses and tetanus toxin pass up within the nerve fiber to the central nervous system, there to produce neuropathological changes.

DEGENERATION AND REGENERATION

When a peripheral nerve is crushed or severed, the part distal to the crush shows the process known as *Wallerian degeneration*. After a few days, the myelin of the fibers shows undulation in its form and a failure to conduct. Over a period of weeks, it shows *beading,* then *ovoid formation,* with a change to *spheres.* These are later absorbed, leaving only Schwann cells and fibrous tissue (Fig. 2-19). Lubińska (1977) has shown that the earliest changes occur in myelinated fibers in the middle of the internode, just under the Schwann cell body. Here, an indentation is seen with, soon after, unduloid changes in the nearby region in the fiber. These changes spread out over the fiber. Using this sign of the earliest changes seen during degeneration, a fairly rapid proximodistal progression of degeneration at a rate of up to several hundred millimeters per day is measured.

A day or so after nerve transection, yet another process is seen. The central ends of fibers above an interruption sprout buds and a growth of fine fibers that grow down into the amputated part of the nerve. This process of regeneration of these neurites is similar to the embryonic growth of nerve fibers in that some unknown directive force, a *neurotropic stimulus,* acts on the growing tips of the neurites to direct them. The Schwann cells in the degenerated sheaths of the distal parts of the transected nerve may have such a directive influence on this growth, but little is known about the subject. The surgeon repairing nerve lesions assists regeneration by bringing the cut ends into close apposition and stitching the epineurial sheaths together. Fibrous growth is to be prevented in order not to impede new fiber growth

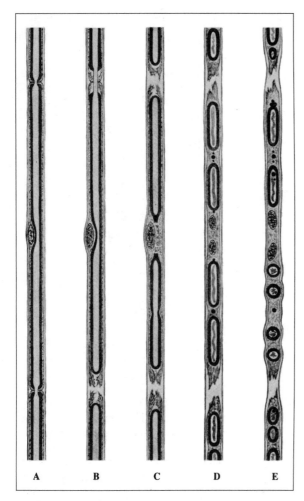

Fig. 2-19. Stages of Wallerian degeneration of an amputated part of myelinated nerve fiber shown after periods of days and weeks. Normal fiber (A). Early in degeneration (B) there is a retraction of myelin from the nodes. Later (C), long ovoids are formed that still later (D and E) become smaller and round up into vesicular structures. (From J. Z. Young. Factors influencing the regeneration of nerve. *Adv. Surg.* 1:178, 1949.)

teristic changes known as *chromatolysis*. Deep blue staining of the Nissl bodies is decreased or lost and then slowly returns over a period of months. The chromatolytic reaction is due to the dispersion of the Nissl bodies within the soma. Along with this change, water uptake increases, and swelling of the cell body occurs, with a displacement of the nucleus. The nucleic acid content and protein of the cell are increased later on in the course of chromatolysis.

Various hypotheses have been advanced to account for the phenomenon of chromatolysis. The most likely explanation is that some "signal" substance that normally moves up the fiber by retrograde transport acts to control the level of protein synthesis in the cell. On transection the signal is lost, and the cell undergoes the series of changes characteristic of chromatolysis with an increase in protein synthesis related to the regeneration process.

BIBLIOGRAPHY

Albers, R. W., Siegel, G. J., Katzman, R., and Agranoff, B. W. *Basic Neurochemistry* (3rd ed.). Boston: Little, Brown, 1981.

Bodian, D. Introductory survey of neurons. *Cold Spring Harbor Symp. Quant. Biol.* 17:1–13, 1952.

Drachman, D. B. (ed.). Trophic functions of the neuron. *Ann. N.Y. Acad. Sci.* 228:1–423, 1974.

Erlanger, J., and Gasser, H. S. *Electrical Signs of Nervous Activity.* Philadelphia: University of Pennsylvania Press, 1937 (1968 reprinting).

Gasser, H. S. Pain-producing impulses in peripheral nerves. *Res. Nerv. Ment. Dis. Proc.* 23:44–62, 1943.

Geren, B. B. The formation from the Schwann cell surface of myelin in the peripheral nerves of chick embryos. *Exp. Cell Res.* 7:558–562, 1954.

Hodgkin, A. L. The Croonian Lecture. Ionic movements and electric activity in giant nerve fibres. *Proc. R. Soc.* [*Biol.*] 148:1–37, 1958.

Hubbard, J. I. *The Peripheral Nervous System.* New York: Plenum, 1974.

Jacobson, M. *Developmental Neurobiology* (2nd ed.). New York: Plenum, 1970.

Katz, B. *Nerve, Muscle and Synapse.* New York: McGraw-Hill, 1966.

Landon, D. N. (ed.). *The Peripheral Nerve.* New York: Wiley, 1976.

Lubińska, L. Early course of Wallerian degeneration in myelinated fibres of the rat phrenic nerve. *Brain Res.* 130:47–63, 1977.

Ochs, S. *Elements of Neurophysiology.* New York: Wiley, 1965.

Ochs, S. Physiology of Nerves. In P. J. Vinken and G. W. Bruyn (eds.), *Diseases of Nerves* (*Handbook of Clinical Neurology*). New York: American Elsevier, 1970. Vol. 7, Chap. 3.

Ochs, S. *Axoplasmic Transport and Its Relation to Other Nerve Functions.* New York: Wiley-Interscience, 1982.

Sumner, A. J. *The Physiology of Peripheral Nerve Disease.* Philadelphia: Saunders, 1980.

Young, J. Z. Factors influencing the regeneration of nerves. *Adv. Surg.* 1:165–220, 1949.

down into the amputated nerve stump. Scars may cause the outgrowing fibers to turn around, forming a whorl known as a *neuroma,* which may be painful (see Chap. 3). The problem of successful regeneration is to direct the regenerating fibers to their right "addresses," namely, the target cells they would normally innervate. This is a goal that is yet to be achieved.

During the several weeks and months of regeneration of new fibers, the nerve cell body undergoes a series of charac-

RECEPTORS AND EFFECTORS

3

SENSATION AND
NEUROMUSCULAR TRANSMISSION

Sidney Ochs

SENSATION

Sensations are defined as feelings or impressions produced by stimulation of a sensory organ or afferent nerves. This definition is extended to include the action of sensory nerves that signal information not consciously registered—for example, the rate of the heartbeat, or impulses from vascular pressure receptors. Sensations may be categorized as *exteroceptive* (those relating to stimuli from the external world), *interoceptive* (those from viscera), or *proprioceptive* (those from somatic structures within the organism).

Sensations are also categorized by the quality of sensa-tion, as *epicritic* and *protopathic*. Epicritic sensations are those readily localized on the surface of the skin. One clinical test for epicritic sensibility is *two-point localization*. The points of a pair of dividers set at various distances are placed simultaneously on the skin, and the subject is asked whether one or two points are perceived. A closer separation can be discriminated over protruding portions of the body (nose, lips, ears, and fingertips). Protopathic sensations are poorly localized. They may be experienced as aches that seem to be present over a wide area or to be poorly localized.

According to Müller's *law of specific nerve energies,* when particular nerve fibers are excited in any way, the sensations aroused are those that would be experienced by a normal excitation of the receptors of those fibers. For example, if optic nerve fibers are electrically stimulated, the sensation experienced is one of light. Mechanical stimulation, such as that produced by pressure applied to the side of the eyeball, will also bring about a visual sensation. Similarly, fibers subserving heat perception will, when adequately excited by any means, give rise to a sensation of heat.

Particular sensations can be elicited at discrete points over the skin. The points appear to be associated anatomically with specific receptors for cold, heat, pain, and so on. Critical studies of the presumed relation of the various receptor organs identified histologically to the various sensory points within the skin do not show this association in all cases. Over large areas, basketlike endings sensitive to touch that surround the hair follicles in hairy portions and free nerve terminals appear to be the main type of receptor organ. According to Weddell (1961), the pattern of the termination of the free nerve endings can affect the quality of sensation; this depends in part on the overlapping pattern of the nerve fiber terminations in the receptive area. Stimulation of the cornea where only free nerve endings are present gives rise not only to pain, but also to sensations of touch, cold, and heat. The free nerve afferent endings in the skin and other structures are no longer considered to educe only pain. In experiments involving the skin of the cat, cold, heat, and pressure stimulation were effective stimuli in exciting action potential responses from C fibers. The receptors for these stimuli are categorized as noncorpuscular or free nerve endings. The cutaneous *corpuscular* receptors have been categorized as *encapsulated* and *nonencapsulated* receptors. They have classically been associated with specific sensations. Of the corpuscular encapsulated endings we include the *Pacinian corpuscle* (touch), *Ruffini's cell* (warmth), *Meissner's corpuscle* (touch), *Golgi-Mazzoni corpuscle* (pressure), *Krause's corpuscle* (cold). Merkel's touch receptor is an example of a nonencapsulated receptor. The significance of these structures in modifying sensory reception will be discussed later.

THE SENSORY CODE. Receptors do not usually respond with a single action potential. When sensory terminations are adequately excited, a repetitive discharge is generally elicited, with the rate of discharge related to the intensity of the stimulation applied.

Rhythmicity of discharge of sensory nerves on stimulation is seen when recording from sensory fibers known to take origin from muscle (see Chap. 5). Using small groups of relatively isolated muscles, recordings can be made from only one or a few afferent nerve fibers. A regular rhythmic discharge is recorded from individual sensory fibers on stretch of the muscle, the rate depending on the degree of stretch. The same behavior is shown for a carotid stretch receptor when subjected to different pressures (Fig. 3-1A). A similar regular repetitive discharge in afferent fibers is

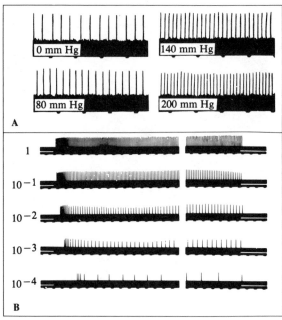

Fig. 3-1. Rate of discharge at different stimulus strengths. (A) A single fiber from a carotid sinus pressure receptor is shown with discharge rates at different levels of intracarotid pressure. As the intrasinus pressure (mm Hg) is increased, the discharge rate increases. (From D. W. Bronk and G. Stella. The response to steady pressures of single end organs in the isolated carotid sinus. *Am. J. Physiol.* 110:71, 1935.) (B) The discharge from a single optic receptor of the *Limulus* eye is shown at different intensities of illumination. The filled bar underneath signifies the time of stimulation. As the illumination strength is decreased by factors of 10, the discharge rate is seen to decrease. (From E. F. MacNichol, Jr. Visual Receptors as Biological Transducers. In R. G. Grenell and L. J. Mullins [eds.], *Molecular Structure and Functional Activity of Nerve Cells*. Arlington, Va.: American Institute of Biological Sciences, 1956. P. 36.)

also shown for discharges in *Limulus* optic nerves at different levels of illumination (Fig. 3-1B).

The rate of sensory discharge increases approximately with the logarithm of the strength of the stimulus. A similar logarithmic relationship had been found by Fechner and Weber in psychophysical studies (Granit, 1956). When blindfolded subjects were asked to compare weights placed in the palm of the hand, the ability to estimate small increments in weight depended on the weight already present. The relationship of stimulus to sensation is shown by the expression

$$\Delta S = \frac{\Delta W}{W}$$

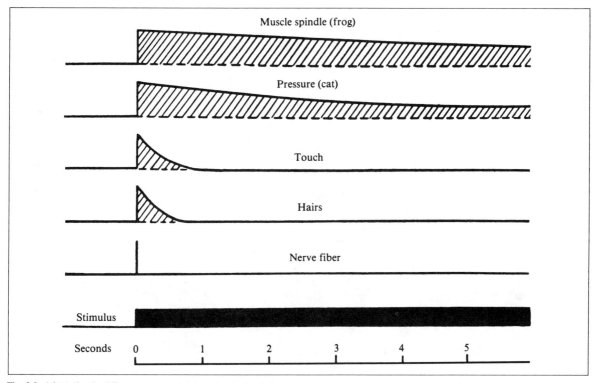

Fig. 3-2. Adaptation in different receptors. Adaptation is shown for various sensory receptors to a constant stimulus represented by a solid black bar in the lower part of the figure. The hatched parts of the various receptors show the rate at which the single units of the receptor are discharging. Adaptation is least for the muscle spindle. For the pressure receptor, adaptation is moderate, and for touch and hair receptors, adaptation is rapid. The nerve fiber itself responds with a very brief discharge. (From E. D. Adrian. *The Basis of Sensation.* New York: Norton, 1928.)

where ΔS = increment of sensation, ΔW = small weight added, and W = weight present. By integrating,

$$S = K_1 \log W + K_2$$

where S refers to the sensation to the total weight, W, and K_1 and K_2 are constants. That the rate of discharge in a sensory nerve bears a logarithmic relationship to the degree of stimulation suggests that it is the factor controlling the psychological appreciation of the stimulus. There are, however, deviations from this relation at the extreme ranges. A more general representation of the relationship between stimulation and sensation that holds over a wider range of stimulation is the *power function;* as described by Stevens, the relationship is

$$\psi = k\phi^n$$

where the sensation (psychological) magnitude, ψ, is related to the (stimulus) physical magnitude, ϕ, with a power exponent, n. The factor n has values ranging from 0.33 to 3.5, depending on the sensory modality; k is a constant determined by the choice of units employed.

Adaptation is a decrease of sensation on continued excitation of sensory receptors and occurs as a result of processes within the receptors. When action potentials are recorded from the afferent nerves of various receptors during a continued application of a stimulus, the frequency of discharge is seen to decline (Fig. 3-2). The rate of decline varies with the type of receptor involved. Adaptation is moderately slow for stretch receptors in muscle, whereas it is rapid for touch. On the other hand, no adaptation is seen in carotid stretch receptors, as would be expected from its role in the regulation of blood pressure (see Chap. 17).

Central mechanisms also play a part in adaptation to sensory input. Stimuli continually present are often ignored. For example, the weight of clothes on the body goes unnoticed unless attention is drawn to it. Mechanisms that

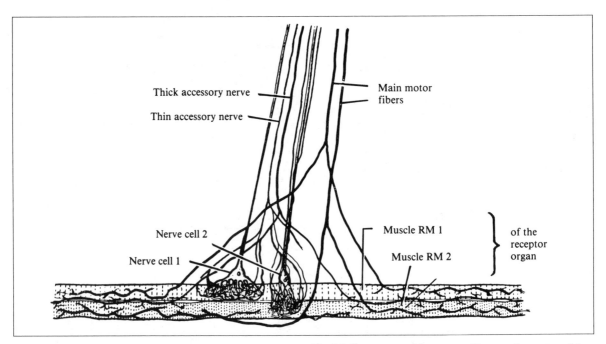

have to do with the attention to stimuli are likewise involved in the adaptive process. There are some central mechanisms that can inhibit sensory influx; by this means the field of sensory attention becomes "cleared" and remains open for new sensory information (see Chap. 9).

TRANSDUCTION. The mechanisms underlying excitation can be directly studied in the single stretch receptor cells of a crustacean (Fig. 3-3). The afferent nerve of the stretch receptor has its branched dendritic portions wrapped around the central region of the receptor's muscular portion. When the muscular portion is pulled on and is elongated, a deformation of the dendritic branches of the sensory stretch receptor occurs, resulting in a repetitive neural discharge. Because the cell body is close to the dendrites, the underlying events can readily be recorded by means of a microelectrode inserted in the cell body. The discharge rate depends on the degree to which the receptor is stretched and the depolarization produced by stretching, the *generator potential* (Fig. 3-4). This depolarization is produced within the deformed dendrites. The generator potential is brought about by ionic change subsequent to deformation of the membrane, the mechanical change opening ionic channels so that both Na^+ and K^+ can move through the membrane. This tends to bring the resting membrane potential toward zero. The depolarization spreads to the rest of the cell (cf. Chap. 2), reaching the initial segment of the axon

Fig. 3-3. Crustacean stretch receptors. Two stretch receptors of the crayfish are shown with the dendritic portions of the cells wrapped around the muscular parts of two different receptors. Nerve cell 1 is from the slow-adapting receptor, and nerve cell 2 is from the fast-adapting receptor. Movements of the muscular portions of the receptor distort the membrane of the dendritic portion of the receptors and cause repetitive discharge in those sensory afferents. Also shown are motor fibers that act in a feedback system to control the level of sensitivity of the stretch receptors. (*RM* = the two types of receptor muscles.) (From J. S. Alexandrowicz. Muscle receptor organs in the abdomen of *Homarus vulgaris* and *Palinurus vulgaris*. *Q. J. Micr. Sci.* 92:190, 1951.)

adjacent to the cell body, the site at which the fiber fires repetitively. The amplitude of the generator potential here determines the rate of discharge.

It has been mentioned that repetitive activity is not usually excited in peripheral axons by a maintained depolarization unless the axons are subjected to special conditions such as a lower Ca^{2+} level (see Chap. 2). The initial segment differs from axons in this respect insofar as a repetitive response to a maintained depolarization is its normal mode of behavior.

Two types of stretch receptors are found: *slow-adapting* and *fast-adapting* receptors. Fast-adapting receptors show a more rapid adaptation and decrease of rate of discharge with maintained stretch than the slow-adapting receptors, which maintain their firing rates. This difference in behavior

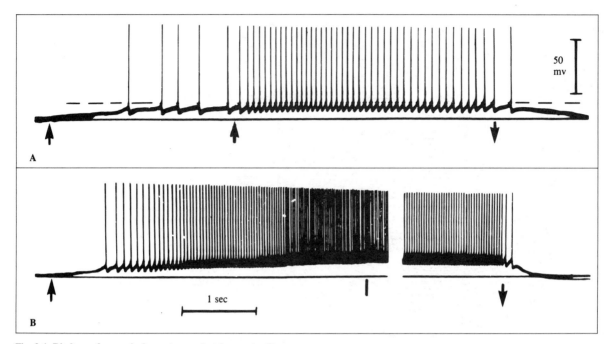

Fig. 3-4. Discharge from a single crustacean stretch receptor fiber.
In (A), the discharge of a single slow-adapting stretch receptor recorded from inside the soma with a microelectrode is shown. A
stretch is gradually applied, resulting in depolarization, indicated
by the shift in the baseline. The rate of discharge from the receptor
increases with the level of depolarization. With a maintained stretch
a fairly constant level of discharge is maintained until the stretch is
released; then depolarization decreases, and the rate of discharge of
the receptor falls off. In (B), a greater stretch gives rise to a higher
level of depolarization and a more rapid rate of discharge. (From
C. Eyzaguirre and S. W. Kuffler. Processes of excitation in the
dendrites and in the soma of single isolated sensory nerve cells of
the lobster and crayfish. *J. Gen. Physiol.* **39**:87–119, 1955.)

is related to the generator potential. If depolarization is held
constant by a current introduced into the receptor cell body
through an electrode, the rate of the repetitive discharge
continues undiminished for as long as the depolarization is
maintained.

The pacinian corpuscle, the mechanoreceptor subserving
touch in the mammal, has a number of onionlike layers
around the terminal sensory nerve in the receptor (Fig. 3-5).
When these tissues are removed by microdissection, and
the naked central afferent nerve fiber is delicately deformed
by a stylus, a rapid repetitive discharge can be excited in its
sensory fiber. The action potential discharges are preceded
by a generator potential that spreads from the deformed part
of the fiber termination to initiate a series of propagated

action potentials from the first node of the afferent fiber.
The generator potential is brief in duration in this receptor
and gives rise to a correspondingly brief repetitive discharge.

The onionlike layers of membrane around the receptor
terminal serve to modify or limit its range of responses.
They may also protect the nerve ending from excessive
stimulation, but for the most part they act as a filter, modifying the way in which deformation leads to excitation. This
phenomenon is seen in other sensory receptors in which
secondary structures enable the sensory fibers to respond to
selected aspects of the environment (e.g., the auditory hair
cells with which the nerve terminal is associated in the ear
(see Chap. 4). This is also seen with the static and dynamic
hair cell receptors of the semicircular canals, where the
utricle and saccule are excited by fluid movement or gravity, enabling the sensory organs to signal position and
movement of the head in certain directions in space (see
Chap. 4). In the eye (see Chap. 4) a number of complex
structures have evolved to enable the optic nerve cells of
the retina to respond selectively to light over a wide range of
illumination and to colors.

In the specialized sensory organs, secondary cells themselves may give rise to electrical potentials in response to
physical or chemical stimulation. These *receptor potentials*
act on sensory afferent nerve terminals closely apposed to
the cells in the receptors to excite repetitive discharges in

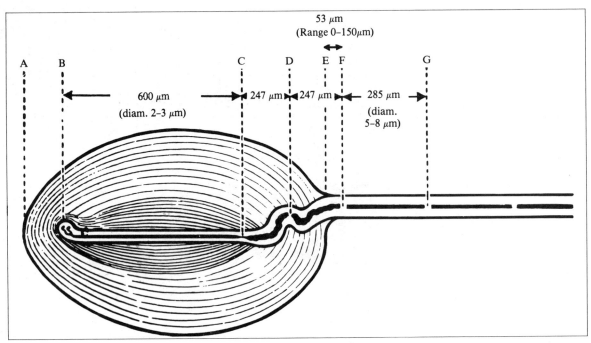

Fig. 3-5. The Pacinian corpuscle. The letters *C* and *D* indicate the nodes of the sensory fiber within the receptor, *F* and *G* nodes those outside. The sensory fiber between *B* and *C* is surrounded by a lamellated structure extended to the corpuscle from *A* and *E*. (From T. A. Quilliam and M. Sato. The distribution of myelin on nerve fibers from Pacinian corpuscles. *J. Physiol.* [Lond.] 129:167, 1955.)

them. The receptor cells are specialized so that transduction of physical inputs approach such high sensitivities that movements of the membrane in the ear of atomic dimensions and, in the case of the eye, only a few photons can be perceived.

Unlike the mechanoreceptors or special sense organs, the properties of the free nerve terminals of the skin are less amenable to direct study and must be inferred from the discharge of the afferent nerve fibers in response to an appropriate stimulation. It is possible to describe them by their adaptive properties; the rapidly adapting velocity or acceleration type of receptor in contrast to the slow-adapting type of receptor. The Golgi-Mazzoni corpuscle is, like the Pacinian corpuscle, a rapidly adapting type of mechanoreceptor. On the other hand, slow-adapting responses are elicited from special domelike pressure receptors seen in certain regions of the hairy skin, i.e., those innervated by Merkel's cells, a class of touch receptor. These cells continue to discharge on applications of long-maintained pressure, falling off in rate somewhat with time. Another slowly adapting receptor is Ruffini's cell, situated intradermally.

Both rapid- and slow-adapting mechanical receptors are found in receptors of the smooth or *glabrous* skin of the palms, soles, and lips.

PAIN. Pain does not result from an excessive stimulation of all the various sensory nerves present, but from stimulation of specific fibers for pain. Pain may range from a mildly disagreeable sensation to one so overpowering that all else is driven from the mind. Some researchers consider itch and tickle a milder stimulation of the same afferent fibers subserving pain. An argument against this view is that severe itching, or *pruritus,* may appear as a separate symptom in certain diseases without being accompanied by the sensation of pain.

Pain is often described by its quality. It may be called "dull," "burning," "sharp," and so on—the terms make a lengthy list. The difference between the epicritic or sharp localizable pain and the duller, less easily localized protopathic pain can be demonstrated by a simple experiment. Touching the toe with the ember of a hot match elicits a sharp *first pain,* followed after a few moments by a duller *second pain.* This *double-pain* sensation has been interpreted as epicritic pain carried in faster-conducting pain fibers, followed by responses in the more slowly conducted protopathic pain fibers. If the skin of the lower leg is similarly excited, the time between the two pain sensations is decreased, and when the knee and then the thigh are stimulated, the time is even less. The results are in accord with more slowly conducting C fibers carrying pain sensation: these are the protopathic impulses, or widespread pains, and the delta group of A fibers, those of the faster, more localizable pain sensations.

The technique of studying pain in animals is similar to that for other sensory modalities, i.e., recording from single fibers while applying an adequate stimulation. In the case of pain the adequate stimulus is one that is damaging, or *noxious*. By this means it has been appreciated that the pain receptor or *nociceptor* has a relatively high threshold and other characteristics that distinguish it as a specific sensory modality.

The nociceptive fibers appear to be *chemoreceptive,* i.e., they respond to certain chemical substances released from damaged cells. This may be histamine, *H substance,* or P factors. *Bradykinin* and other substances that elicit responses in pain fibers in low concentration have been implicated. In experiments devised to test putative pain substances, the agents are applied to the base of exposed blisters or injected into blisters or into the skin. In such experiments, serotonin, acetylcholine, and prostaglandin, among other agents, have been found effective in eliciting pain and have been considered as possible pain-mediating agents.

Experimental attempts to determine pain thresholds (e.g., by application of graded radiant heat to a blackened spot on the forehead or fingernail of the human subject) have been used in a search for effective analgesic agents. In some cases, analgesic agents have been reported to elevate the pain threshold. Pain-threshold experiments are generally unsuccessful, however, because subjective factors are of overriding importance in human responses. The central mechanisms responsible for the alleviation of pain by analgesics or for the marked variation in the perception of pain experienced by a given patient are as yet little understood.

An important characteristic of pain relative to central nervous processes is its tendency to *irradiate* and to give rise to *referred pain.* For example, pain due to heart damage may be experienced by the patient as pain in the left upper arm or as pain passing down the left arm into the hand. Pain caused by kidney disease may be felt as back pain, and the ureteral dilation and trauma incident to the passing of a kidney stone may be experienced as a severe pain in the flank. Pain from a stomach ulcer may be referred to the right shoulder. These misdirections of pain sensation appear to be due to the excitation of a common pool of neurons within the spinal cord, brainstem, or cortex acted on by different afferent sources. According to this theory, pain is a functional "spillover" or irradiation of activity in a set of neurons acted on by two different sets of afferent inputs. In the *gate* theory of pain, intermediary cells in the substantia gelatinosa of the cord are considered to receive several incoming sensory afferent inputs, the activity of the larger-fiber inputs acting to suppress the pain impulses arising from the smaller pain afferents. Such cells have not, however, been identified, and it is possible that some cells acting as "gates" are present at higher levels, in the thalamus or cortex. In 1978, a reevaluation of the theory was presented by Wall, who retained the main concept but rejected the cellular mechanism for gating proposed earlier.

Painful states may be long enduring. For example, a painful procedure such as the drilling of teeth may give rise to a feeling of pain long after the drilling ceases. The functional change in neurons producing a continued painful state may exist at various levels within the central nervous system (CNS).

The neuroma produced at the end of a cut nerve (see Chap. 2) is similarly a source of pain and may also give the sensation that a missing member is still present, a phenomenon known as a *phantom limb.* In one of the striking examples given by Livingston (1947), the patient experienced an amputated arm as present, with the fist clenched so tightly that it constantly ached. Temporary relief may be experienced by the injection of procaine into the neuroma or by its excision. However, procaine is usually not effective for long. Spinal cord section of the afferent tracts may also afford temporary relief, but another intervention may be required. A frontal or prefrontal lobotomy is effective in some cases and may have to be resorted to for intractable pain.

The patient's alertness and attitude at the time of injury are important in determining the degree of pain experienced. It has long been noted in wartime and in catastrophes that seriously wounded patients, as a result of stress or emotion, may experience no pain at all at the time of injury. Descending neural impulses from the brain acting on gate cells in the spinal cord may determine which types of ascending impulses will reach the brain and consciousness.

NEUROMUSCULAR TRANSMISSION

Transmission between a motor nerve and the muscle fibers it innervates takes place at the *end-plate* (Fig. 3-6). Trans-

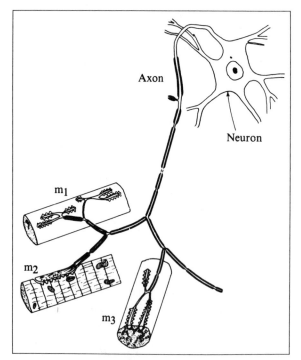

Fig. 3-6. The motor unit. The motorneuron is shown with branches terminating on muscle fibers (m_1, m_2, and m_3), at the end-plates. These are seen as indentations of the surface with folds in the membrane. (Modified from R. Miledi. From nerve to muscle. *Discovery* Oct. 1962.)

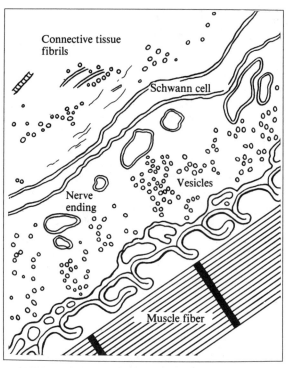

Fig. 3-7. Neuromuscular junction. A longitudinal section of a nerve fiber lying within a channel in the muscle surface is diagrammed from an electron micrograph. The nerve fiber is approximately 1.5 μm in diameter. Within the nerve terminal, two types of particulate structure are shown, mitochondria and small vesicles believed to contain the transmitter substance acetylcholine. (From R. Birks, H. E. Huxley, and B. Katz. The fine structure of the neuromuscular junction of the frog. *J. Physiol.* [Lond.] 150:130, 1960.)

mission at the end-plate is effected by the release of a special chemical agent from the nerve terminal, a *neurotransmitter,* which in the vertebrate is *acetylcholine* (ACh). When the motor nerve is fired, ACh from the nerve endings moves across the gap between the nerve and muscle membrane in the end-plate to bind to a *receptor protein.* The result is a special potential change in the muscle membrane, which then leads to a propagated action potential in the muscle fiber and to muscular contraction. This process will be described in detail in the following sections.

The proof that a special neurotransmitter agent was involved was first given by Loewi in 1921. Using an isolated beating frog heart, he found that a substance was released into the fluid medium on stimulation of the vagus nerve of the heart to cause it to slow in rate and stop (see Chap. 17). This *vagus stuff* in turn caused a slowing in the rate of another similarly isolated frog heart. The vagus stuff was later identified as ACh. Acetylcholine is labile, and if blood is present, the enzyme *cholinesterase* in the blood hydro-

lyzes the ACh to acetate and choline, a much less active form. If *anticholinesterase agents,* such as *eserine* or *neostigmine,* are present, the enzymatic action of cholinesterase is blocked, and ACh can then be detected in the presence of blood which also contains an esterase.

Electron microscopy of the end-plate region shows that there is no merging or continuity of the nerve membrane with the muscle membrane. The nerve fiber occupies a cleft or channel within the muscle membrane, the space between the two membranes being approximately 300 to 500 Å (Fig. 3-7). Vesicular bodies approximately 500 Å in diameter are found within the terminal portion of the motor nerve, which is believed to contain the transmitter substance ACh. Mitochondria are also seen within the nerve endings. They are present wherever metabolic activity occurs and a source of energy is required (see Chap. 2). Their presence in the terminals is related to the resynthesis of ACh from the choline taken up by the terminal to participate again in transmission, as will be discussed in the following sections.

THE END-PLATE POTENTIAL. The analysis of neuromuscular transmission was markedly advanced by Claude Bernard's discovery that *curare* blocks transmission between nerve and muscle. The agent has no effect on the excitability of the nerve or muscle. With curare present, *direct* stimulation produces propagated action potentials from either the nerve or muscle membrane, the latter accompanied by normal muscle contractions. However, with curare present, stimulation of the motor nerve to excite the muscle *indirectly* through the nerve does not elicit muscle contractions.

The study of transmission may be carried out in vitro with a convenient preparation, the frog sartorius muscle. In a chamber the individual fibers of the isolated muscle can be readily seen, as can the approximate region of the nerve terminals ending on the muscle fibers. The actual end-plates can be seen with a microscope provided with Nomarski optics. With a microelectrode placed inside a muscle fiber close to an end-plate, an action potential with the usual overshoot is seen, but with a preceding hump as well (Fig. 3-8A). This extra potential is not seen when recording outside the end-plate region as indicated in Figure 3-8B. Another phenomenon shown in Figure 3-8A is the occurrence of small potentials recorded in the region of the end-plate— the miniature end-plate potentials that will be described later.

The step of voltage at the onset of the action potential and the hump following were suspected of having a different origin than the ordinary action potential because their size depended on the recording side; i.e., whether the recording microelectrode was close to or far from the end-plate. This

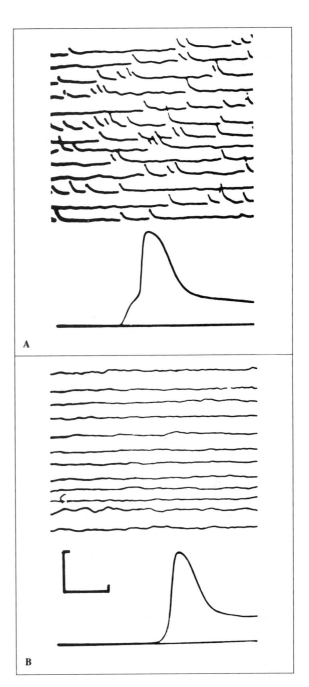

Fig. 3-8. Spontaneous miniature end-plate potentials at the myoneural junction. The records were obtained with intracellular recording. (A) A microelectrode was placed inside a frog muscle fiber at the nerve-muscle junction. (B) An electrode was placed 2 mm away into the same muscle fiber. The uppermost portions of (A) and (B) were recorded with slow speed and high amplification (calibrations: 3–6 mv and 47 msec); the lower portions of (A) and (B) show the response to a nerve impulse with fast speed and low-gain recording (calibrations: 50 mv and 2 msec). The stimulus was applied to the nerve at the beginning of the sweep; response (A) (at the end-plate) shows the steplike initial EPP, and the response (B) shows the propagated action potential wave, delayed by conduction over a distance of 2 mm. The figure shows that the spontaneous activity is restricted to the myoneural junction. (From J. del Castillo and B. Katz. Biophysical aspects of neuro-muscular transmission. *Prog. Biophys. Mol. Biol.* 6:141, 1956.)

inference was verified when a small amount of the purified form of D-tubocurarine was added to the bath in an amount sufficient to block indirect excitation of muscle contraction and the action potential but not the humplike early potential (Fig. 3-9). This depolarization rises about 1 msec after nerve stimulation, reaches a peak, and then declines over a longer time period of some 15 to 20 msec. The potential recorded from the end-plate region is referred to as the *end-plate potential* (EPP). With increased amounts of D-tubocurarine added to the bath, the EPP is diminished in amplitude and may be completely blocked. The responses in Figure 3-9 were recorded with an extracellular electrode placed close to the end-plate. Except for the amplitude of the overshoot, the figure shows the same features recorded internally from the muscle with microelectrodes shown in Figure 3-8A. As the amount of D-tubocurarine in the bath was successively increased, the EPP became diminished in amplitude, and the propagated action potential was initiated with increasingly greater delays until the amplitude of the EPP fell below threshold for the initiation of an action potential. D-Tubocurarine, then, effectively blocks an indirectly excited action potential in the muscle.

The EPP differs from the action potential in several important respects. First, it is not an all-or-none response; it diminishes in amplitude with distance from the region of the end-plate (Fig. 3-10). Also, unlike the propagated action potential, the EPP shows the property of *summation*. When EPPs are too small to excite an action potential, their summation may bring the membrane to threshold and fire a propagated action potential. The propagated action potentials are brought about by a brief increased permeability, first to Na^+ and then, after an interval, to K^+, as in the case of the giant nerve axon (see Chap. 2). The EPP represents a depolarization brought about by a simultaneous increase in the permeability of both Na^+ and K^+. The extracellular fluid is mainly composed of Na^+ and Cl^-, with small amounts of K^+ and other ions present. When, after ACh binds to the receptor, the permeability to Na^+ and K^+ increases, these cations move according to their concentration and electrical gradients. The result is a reduction in the membrane potential (-12 mv) toward the *equilibrium potential* for the EPP (E_{EPP}). This represents the electrochemical equilibrium for both Na^+ and K^+ together.

The blocking agent D-tubocurarine acts by its binding to the receptor protein; consequently, the ACh released from the nerve ending cannot bind to all the sites available. With increasing curarization the EPP is reduced in size, the extent of reduction depending on the amount of D-tubocurarine bound to the receptors, either rendering them unavailable to the transmitter ACh or leaving them open to

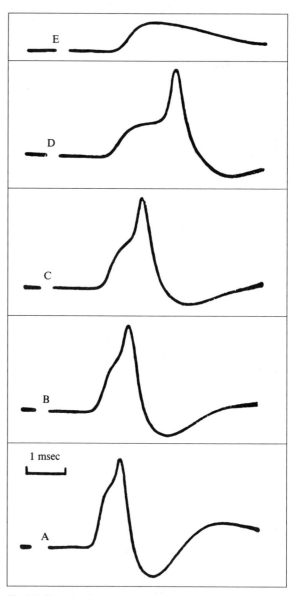

Fig. 3-9. Records taken at the end-plate region. (A) Before application of D-tubocurarine. (B, C, and D). During progressive curarization, the diminution of the initial end-plate potential (EPP) and the progressive lengthening of the spike latent period. (E) Pure EPP; no spike is set up. (From S. W. Kuffler. Electric potential changes at an isolated nerve-muscle junction. *J. Neurophysiol.* 5:23, 1940.)

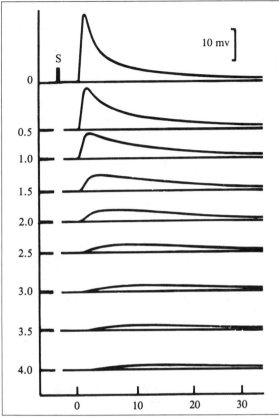

Fig. 3-10. The end-plate potential is recorded with a microelectrode from just under the muscle membrane in the end-plate region. It is a depolarization that occurs after a latency period following the stimulus at S. The EPP is relatively long lasting, and the height decreases with distance, as shown by the numbers indicating in millimeters the distance of the recording microelectrode from the end-plate. (From P. Fatt and B. Katz. An analysis of the end-plate potential recorded with an intracellular electrode. *J. Physiol.* [Lond.] 115:326, 1951.)

brane, in the manner previously discussed with respect to local currents in the nerve membrane (see Chap. 2). The amplitude of these currents and the decrease in size of the EPP with distance from the end-plate (see Fig. 3-9) can be accounted for by the electrical characteristics of the muscle, i.e., its internal resistivity and membrane resistance.

The evidence that ACh is in fact the neurotransmitter was substantiated by a study of the effect of ACh applied directly to the end-plate region through a micropipette, using the technique referred to as *iontophoresis*. The ACh in a microelectrode is ejected from the tip by passing an electrical pulse of current through it. The released ACh causes an increased permeability to Na^+ and K^+ and a depolarization in the muscle membrane with an appearance similar to that of an EPP (Fig. 3-11). This *ACh potential* can, if it reaches a high enough level, set off a propagated action potential. When ACh is released from a microelectrode inserted just under the membrane of the end-plate, it is ineffective. Only the outer surface of the receptor substance in the membrane of the end-plate is responsive to ACh. The specific cholinesterase acetylcholinesterase (AChE) present at the end-plate acts to terminate the depolarization produced by ACh. It is located adjacent to the receptor protein between the nerve and muscle membrane. When an anticholinesterase agent such as eserine is added to a muscle bath, the EPP is much larger than usual and is prolonged in duration. Normally, a portion of the ACh released from the nerve terminal is hydrolyzed by the AChE and does not reach the receptor on the end-plate. In the main, the ACh that does reach the receptor is terminated in its action by AChE.

Choline, succinylcholine, and decamethonium show some similarity to ACh in chemical structure. They produce a much longer-lasting depolarization at the end-plate, which acts to block neurotransmission. The mode of action of these agents, referred to as *depolarizing blocking agents,* differs radically from that of curare, which does not cause a depolarization.

Besides the curariform and depolarizing groups of neuromuscular blocking agents, other types of neuromuscular block are produced by agents acting on the membrane of the motor endings to interfere with the mechanism of release of the transmitter substance. The blocking agent *Clostridium botulinum* toxin, an extremely potent poison, is selectively taken up from the circulation by motor nerve endings, where it causes a block in the release of ACh. *Black widow spider venom* blocks by causing a complete release of all the transmitter, with disappearance of vesicles from the terminal, also resulting in a block. Death following exposure to these agents comes about from a paralysis of the respiratory muscles.

its action. Normally, the EPP is relatively large, with an amplitude greater than 50 to 60 mv, and more than able to depolarize the adjoining muscle membrane to excite a propagated action potential in the muscle membrane. When the EPP has been reduced in amplitude below a critical level, it cannot sufficiently depolarize the muscle membrane to give rise to a propagated action potential.

The increased permeability to Na^+ and K^+ brought about by the action of ACh is relatively brief—on the order of 1 or 2 msec. The relatively long duration of the EPP is due to the resultant flow of local currents in the adjoining muscle mem-

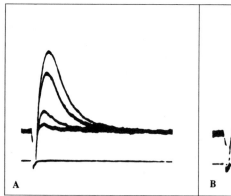 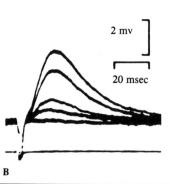 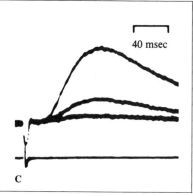

Fig. 3-11. Responses of a junctional sensitive spot to ACh pulses of variable strength delivered at different positions of a pipette. In (A), the pulse duration varied from 1 to 2 msec; in (B), from 2 to 3 msec; and in (C), from 3 to 5 msec. The distance from source to receptors was increased from (A) to (C). In (A), the receptors were located within the source radius; in (B) and (C), the calculated distances were 14 μm and 21 μm respectively. Notice that the time to peak depends on the position of the pipette but not on the pulse strength. Same time calibration for (A) and (B). (From A. Feltz and A. Mallart. An analysis of acetylcholine responses of junctional and extrajunctional receptors of frog muscle fibres. *J. Physiol.* [Lond.] 218:91, 1971.)

MINIATURE END-PLATE POTENTIALS AND THE RELEASE MECHANISM. A continual and irregular discharge of small potentials several millivolts in amplitude has been noted in the vicinity of the end-plate (see Fig. 3-8). They have a duration and other properties similar to those of the EPP except for their very much smaller size. On this basis they have been termed *miniature end-plate potentials* (MEPPs). The MEPPs, like EPPs, are blocked by D-tubocurarine, and their duration is lengthened by the presence of the anti-AChE agents eserine and neostigmine.

Each MEPP is considered to represent the discharge of ACh contained in a vesicle. Each unit or quantum of the discharge consists of some 10,000 to 20,000 molecules of ACh. This number was determined by using iontophoresis and releasing a known amount of ACh. Comparison of the depolarization produced by the ACh released (Fig. 3-11) with the depolarization produced by the MEPPs made it possible to estimate the number of molecules per quantum.

It is widely held that the simultaneous release of a number of quanta from the nerve terminal brings about an EPP. In support of this concept, it has been shown that the rate of discharge of the MEPP is increased greatly by the depolarization of the motor nerve terminals. On this account, when the motor fiber is excited, an action potential invades the nerve terminal to depolarize it and thus allow the release of a large number of MEPPs. These are sufficiently synchronized to summate and give rise to the EPP.

When ACh is applied iontophoretically to the end-plate, there is less than a 100-μsec latency for the resulting depolarization. This is a small fraction of the usual latency, which ranges from approximately 0.5 to 2.5 msec on nerve stimulation. Most of that time is taken up by the process of the release of ACh from the nerve terminal. The release of the neurotransmitter requires that Ca^{2+} be present in the medium. The most likely hypothesis is that when the nerve

action potential depolarizes the nerve membrane, Ca^{2+} enters the terminal and acts on the vesicle, allowing it to move to and merge with the inside face of the nerve terminal membrane. With Ca^{2+} present, an *exocytosis* takes place whereby the vesicle opens to the outer surface and its contents are emptied into the synaptic cleft to reach the postsynaptic membrane. The need for Ca^{2+} to initiate transmitter release has been shown by the use of tetrodotoxin-treated preparations to block action potentials in the terminal. A microelectrode placed on the nerve terminal to depolarize it can initiate a release of transmitter if the medium contains Ca^{2+}. This result was shown by timing a pulse of depolarizing current immediately after a pulse to a Ca^{2+}-containing micropipette placed on the nerve terminal for iontophoretic release of Ca^{2+} (Fig. 3-12).

The movement of vesicles to the inner wall of the terminal and the discharge of their contents into the synaptic cleft appear to involve fairly complex mechanisms. Some series of internal events are required to account for the temporal changes seen when the nerve is stimulated at high rates. Repetitive stimulation decreases the size of the EPPs. Alternatively, periods of high-frequency stimulation of the motor nerve may be followed by an increase in the size of the EPP for a short time, or a *posttetanic potentiation* (PTP). This

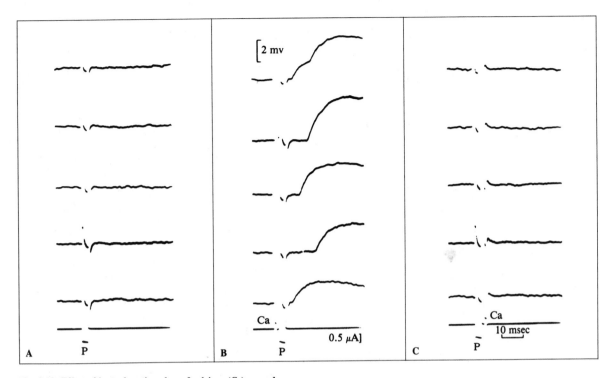

Fig. 3-12. Effect of iontophoretic pulses of calcium (Ca) on end-plate response. Depolarizing pulses (P), and calcium were applied from a twin-barrel micropipette to a small part of the nerve-muscle junction. Intracellular recording from the end-plate region of a muscle fiber. Bottom traces show current pulses through the pipette. (A) Depolarizing pulse alone. (B) Calcium pulse precedes depolarizing pulse. (C) Depolarization precedes calcium pulse. Temperature 4°C. (From B. Katz and R. Miledi. Timing of calcium action during neuromuscular transmission. *J. Physiol.* **[Lond.] 189:537, 1967.)**

phenomenon appears to be due to an increased *mobilization* of the transmitter-containing vesicles. There seem to be both short- and longer-lasting changes related to the mobilization of vesicles. There appears to be a storage pool of vesicles that takes a longer time for mobilization. These longer-period changes merge with still longer-lasting processes of transmitter replacement provided by axoplasmic transport (see Chap. 2).

Additionally, there are local processes that involve the resynthesis of ACh from choline taken up into nerve terminals by a carrier present in the membrane, the acetate provided as *acetyl coenzyme A* through metabolism (Fig. 3-13). By this means, ACh is resynthesized and repackaged into the vesicles to be used later on in transmission. In the process of endocytosis whereby a vesicle is formed inside the terminal, the protein *clathrin* is laid down on the inner surface of the terminal membrane. It undergoes a folding, forming the vesicle from the surface membrane enclosed within the basketlike clathrin structure.

Neuromuscular transmission is poor in the muscle disease *myasthenia gravis*, apparently because of a deficiency in the number of receptors that is due to a circulating immune protein. Transmission fails in such muscles because the resulting EPPs are too low in amplitude to excite propagated action potentials in the muscle membrane. Anticholinesterase agents, such as eserine, neostigmine, and similar agents, can improve neuromuscular transmission by augmenting the EPP, allowing it to attain the critical level of depolarization for muscle action potential activation and contraction.

TROPHIC INFLUENCE OF NERVE ON MUSCLE. When the nerve to a skeletal muscle is transected, the denervated muscle undergoes a series of profound biochemical and functional changes. These take place over a period of some days or weeks. Increased excitability and sensitivity to injected ACh is seen early, with the muscle showing *denervation hypersensitivity*. In this state, *fibrillation,* large irregular discharges, and abnormal sensitivity to mechanical stimulation and ACh are found. Metabolism is abnormal, thus con-

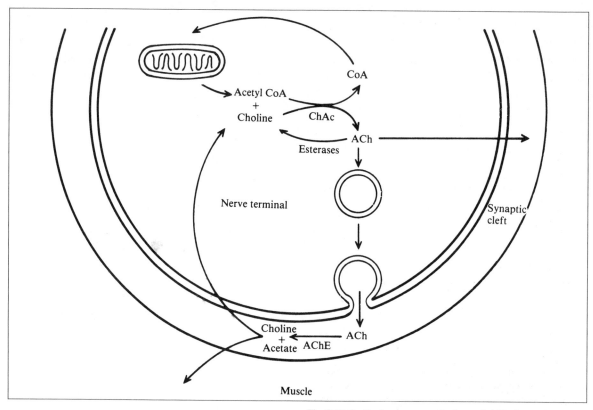

Fig. 3-13. Synthesis, storage, and release of ACh. Arrows indicate various metabolic cycles that interlock in the synthesis of ACh. Releases of ACh into the synaptic cleft is indicated in quantal form from a vesicle and in nonquantal form (*arrow*). Choline uptake from the synaptic cleft by nerve terminals and muscles is indicated by arrows crossing the nerve terminal and muscle boundaries. Choline acetylase (ChAc) is used to resynthesize ACh, which is then packaged into the vesicles shown by arrows, carried to the membrane, and discharged. (From J. H. Hubbard [ed.]. *The Peripheral Nervous System.* New York: Plenum, 1974.)

tributing to the slow shrinkage in muscle size over the weeks or months that is known as *denervation atrophy.* Furthermore, whereas only the end-plate region is normally highly sensitive to ACh, the area of sensitivity to ACh spreads slowly outward from the end-plate region of a denervated muscle, so that in a matter of days or weeks, depending on the species, most or all of the membrane becomes sensitive to ACh. This change in property was shown by iontophoretic application of ACh at graduated steps all along the length of the fibers. If the muscle becomes reinnervated, the increased sensitivity along the fiber diminishes until only the end-plate as usual shows a high sensitivity to ACh. An interpretation of this phenomenon is that a trophic substance carried down inside the nerve fibers by axoplasmic transport (see Chap. 2) passes into the muscle fiber to control the synthesis of the receptor protein within it. In its absence there is a *derepression* of receptor protein synthesis, and new receptor protein becomes inserted into the muscle membrane outside the end-plate region.

Following nerve interruption or when as a result of *Clos-*

tridium botulinum poisoning the release of transmitter is blocked, the muscle also develops hypersensitization. Thus, one possibility is that ACh is the trophic material. However, the failure of the application of ACh to denervated muscle to prevent hypersensitization in vitro and other evidence suggest that some as yet unknown trophic substance controls the synthesis of receptor protein in the muscle.

A number of different trophic or control agents appear to be carried from the nerve into the muscle. In the mammal, the white muscles are known to have a faster twitch duration than the red muscles. The more slowly contracting red muscles are tonic in function (see Chap. 5), and their longer twitch duration is related to the prolonged contractions required for the maintenance of posture. If the motor nerves to the fast and slow muscles are cut and the nerves united so as to regenerate into the different muscles, the muscles are then innervated by the different nerves. In such *cross inner-vations* the fast muscles become slowly contracting, and the slow muscles tend to become fast contracting. This result is explained by the concept that trophic substances carried down within the two types of nerve fiber determine the functional properties of muscles they innervate. Another possible explanation is that the muscles change as a result of different patterns of excitation. In any event, it appears that certain diseases, such as muscular dystrophy and other ill-defined muscle diseases, may well have their basis in an alteration of the proper supply of trophic substances by the nerves innervating them.

EFFECTORS: STRIATED, SMOOTH, AND CARDIAC MUSCLE

Richard A. Meiss

The task of muscle is to perform some mechanical action in response to a neural or hormonal command. In the more familiar instances, muscle serves as the final link between the voluntary activity of the CNS and the external environment. Obvious examples include muscles used for locomotion and speech. Equally important, however, are those muscles under hormonal or involuntary nervous control that assist in maintaining a stable internal environment. The heart, viscera, and peripheral circulatory system, for example, contain muscle that is not controlled voluntarily.

A consideration of the role of muscle in the total integrated function of the whole body or of a specific organ or system must rest on an understanding of muscle as a contractile system. In this sense, muscle is a type of motor, one that consumes fuel, produces both desired mechanical effects and unwanted thermal and chemical waste products, and is subject to a variety of controls. Muscle may also be considered a form of *transducer,* since it changes electrical (nerve) signals into mechanical motion.

In spite of the diversity of muscle types found in the human body, some properties are common to all; the adaptation of a specific muscle to a specific task involves modification and refinement of one or more of these basic properties until it becomes a dominant characteristic. On the basis of their predominant functional and anatomical characteristics, muscles are usually grouped into three major categories. The most familiar group is *skeletal muscle,* so called because it is almost always found attached in some way to skeletal structures. This category is also referred to as *striated* (striped) muscle, a reference to its microscopic structure; it is also called *voluntary* muscle, since it is usually (but not always) under the voluntary control of the somatic nervous system. A second category is that of *visceral muscle*. As the name implies, this type is found largely associated with the viscera, or internal organs. It is also categorized structurally as *smooth* muscle because it lacks the microscopic striations of skeletal muscle. Since it is usually (but not always) under involuntary control of the autonomic nervous system, it is also called *involuntary* muscle. The third category is *heart (cardiac) muscle,* with an obvious anatomical location. While this muscle type has many microscopic similarities to skeletal muscle and some functional similarities to visceral muscle, its mode of control, its unique location, and its vital importance merit putting it into a separate category.

The aim of this chapter is to provide a basic understanding of the shared properties common to muscle in general and then to examine how special modifications of the common properties are used to physiological advantage.

MUSCLE STRUCTURE

SKELETAL MUSCLE. Because muscles are able only to pull and not to push, they are all anatomically located so that some sort of force can be applied to relengthen them after contraction. This universal necessity for providing a lengthening force imposes an important restriction on the possible anatomical arrangements of all muscle types. Hence, skeletal limb muscles are usually arranged around a joint in antagonistic pairs, so that when one muscle shortens, the other is lengthened (Fig. 3-14). The muscles of the heart, having shortened to pump out blood, are relengthened by the force of new blood entering the pumping chambers of the heart. Similarly, shortened muscle in the walls of the intestine is relengthened when that segment of gut is actively filled with material moved from another region. In some cases, muscles are arranged to work against elastic tissue structures that cause the relaxed muscle to return to its resting position; the muscles responsible for facial expression are an example.

Skeletal muscles are usually distinct and relatively large structures, and their gross anatomy is straightforward. A whole muscle and the main structures of the muscle cell are shown in Figure 3-15. The individual muscle cells (called *fibers*) are from 20 to 90 microns (μm) in diameter and may be many centimeters long, depending on the length of the whole muscle (Fig. 3-15C). The fibers are multinucleate, with the nuclei located at intervals directly beneath the *sarcolemma,* which is a very thin connective tissue sheath enclosing each individual fiber. The sarcolemma also contains the usual *plasma membrane* as one of its components. The cytoplasmic region is largely filled with the contractile substance, the *myofibrils* (Fig. 3-15D). These bear the characteristic cross-striated pattern, about which more will be said later. The muscle fibers are grouped into bundles called *fasciculi* (Fig. 3-15B), arranged in an approximately parallel fashion and fastened together side by side with connective tissue called the *perimysium.* Additionally, each fiber is surrounded by a delicate sheet of connective tissue, the *endomysium,* which is separate and distinct from the sarcolemma. The large fiber bundles are further bound together to form a complete muscle with its connective tissue cover-

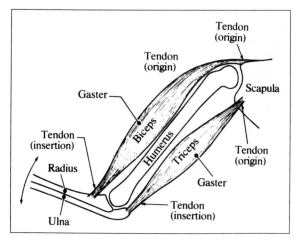

Fig. 3-14. A highly simplified representation of two major muscles of the upper arm. They are arranged antagonistically, so that contraction of the biceps will stretch out the triceps and contraction of the triceps will relengthen the biceps. (In actuality, the biceps has a double origin, as its name implies, and the triceps has a triple origin and a more complex insertion than is shown here.) (From R. A. Meiss. Muscle: Striated, Smooth, Cardiac. In E. E. Selkurt [ed.], *Basic Physiology for the Health Sciences* [2nd ed.]. Boston: Little, Brown, 1982.)

ing, the *epimysium.* At both ends, connective structures, the *tendons,* serve to connect the bundled fibers to the skeleton. The end of the muscle attached to the more stationary skeletal part is its *origin;* the end attached to the part to be moved is the *insertion;* and the thickened middle portion is the *belly,* or *gaster.*

Many variations on the basic arrangement are found. The *biceps,* for example, is a relatively long muscle with a double origin and a single insertion. Its fibers are arranged to give it a roughly circular cross section. The *diaphragm,* on the other hand, is a flat sheet of roughly parallel fibers that originate on the skeleton but insert on a central tendon. Some striated muscles in the tongue and in the upper esophagus neither originate nor insert on the skeleton. The gross structures and connections of other skeletal muscles are similarly tailored to their location and major function.

SMOOTH MUSCLE. Smooth muscles existing as separate organs are relatively rare. Smooth muscle usually forms an intimate part of the wall of a hollow tubular or saclike organ; often it forms the largest portion of the wall thickness. The individual smooth muscle cells (Fig. 3-16) are small, tapered structures 5 to 10 μm in diameter and from 200 to 500 μm long. The nucleus (one per cell) is large and centrally

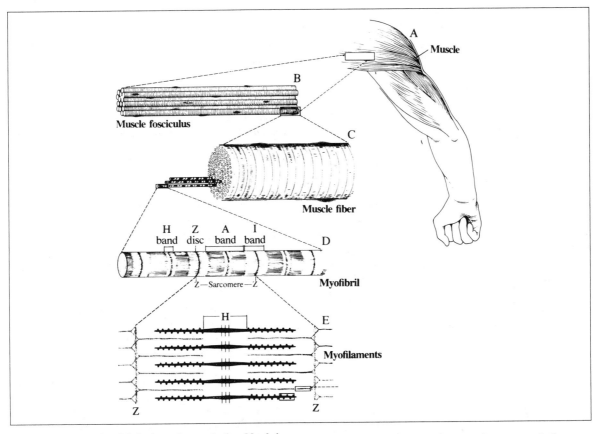

Fig. 3-15. The hierarchy of skeletal muscle organization. Muscle is organized into anatomical units that can be subdivided several times into increasingly smaller functional units. A whole muscle (A) can be subdivided into fiber bundles called fasciculi (B). These can be further broken down into individual muscle fibers (C), which are single muscle cells. The cells contain fibrous structures, the myofibrils (D). Each myofibril is composed of a series arrangement of repeating structures, the sarcomeres. The sarcomere is the smallest level of organization capable of contraction. Every sarcomere is composed of complex protein filaments, the myofilaments (E) (see Fig. 3-18), which are in turn composed of large molecules of the fibrous protein myosin (thick filaments) or the smaller globular protein actin (thin filaments), along with other protein components mentioned later in the text. Although muscles throughout the body exhibit a wide range of shapes and sizes, at the ultimate functional level their structures are remarkably similar. (From W. Bloom and D. W. Fawcett. *A Textbook of Histology* [10th ed.]. Philadelphia: Saunders, 1975.)

located; the myofibrillar substance shows no periodic cross striations. The cells are grouped by connective tissue into a relatively parallel array with a definite directional orientation. In many organs, notably the small intestine (Fig. 3-17), the smooth muscle is arranged into a circular layer (running around the circumference of the tube) and a longitudinal layer (running along the length of the tube). In other organs, such as the stomach and uterus, there are three or more layers, with various distinct orientations. Very small blood vessels, on the other hand, may have a single muscle cell completely encircling the structure. A specialized smooth muscle structure, one found at numerous points in the circulatory system and the viscera, is the *sphincter*. This is a region of extra-thick, circularly oriented smooth muscle encircling a tube. Contraction of the sphincter smooth muscle closes off that region of the tube and prevents flow of its contents. Since a sphincter possesses a natural geometrical mechanical advantage, high fluid pressure (such as may be found in the urinary bladder, for example) may be contained with little muscular effort. In the skin, tiny individual smooth muscles are responsible for causing hairs to stand

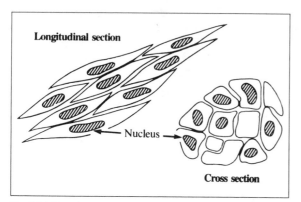

Fig. 3-16. Smooth muscle cells in longitudinal and cross sections. The cells are long and slender, tapering at the ends. In the cross section, some cells are cut, so their nuclei do not appear in the illustration. At many regions throughout the population of cells, adjacent cells are in close contact; this very close proximity aids in cell-to-cell electrical communication. Connective tissue fibers form a network in which the cells are embedded and to which they may also be attached. There are no tendons as such. (Modified from R. A. Meiss. Muscle: Striated, Smooth, Cardiac. In E. E. Selkurt [ed.], *Basic Physiology for the Health Sciences* [2nd ed.]. Boston: Little, Brown, 1982. P. 216.)

erect and for raising goose pimples. Thus, as was the case for skeletal muscle, the anatomy of smooth muscle has many variations to suit a variety of physiological needs.

CARDIAC MUSCLE. The gross anatomy of cardiac muscle is treated in Chapter 13. It will suffice here to say that the cardiac muscle cells are much shorter (approximately 100 μm) and smaller in diameter (9 to 20 μm) than skeletal muscle fibers (Fig. 3-18). The cells are arranged in a branching network with a predominantly longitudinal organization; the outer portions of the cells are covered by a sarcolemma, and the end-to-end connections of the cells are made at the *intercalated disk,* which also functions in the electrical interconnection of the cells. The nuclei are centrally located, and the myofibrillar substance has the cross-striated pattern found in skeletal muscle.

The electron microscope has permitted extensive investigation of the fine structure of muscle in recent years, and a great deal of progress has been made in understanding how the contractile process operates. Here again, studies of skeletal muscle have yielded the clearest picture.

FINE MUSCLE STRUCTURE. The smallest organized unit of the contractile mechanism of skeletal muscle is the *sarcomere,* as illustrated in Figures 3-18C, 3-19, and 3-20. A single *myofibril* is composed of many sarcomeres arranged both end to end and side by side. The cross striations visible in the light microscope can be associated with the dark and light regions of the sarcomere; the thin dark bands that define the ends of the sarcomere are the *Z lines.* On either side of a given Z line is a light area, the *I band,* and midway between any two Z lines is a dark area, the *A band.* Down the center of the A band is a darker stripe, the *M line,* surrounded on both sides by a lighter *H zone.* Close inspection reveals that the A and I bands are made up of the myofilaments, which are oriented along the longitudinal axis of the muscle, that is, perpendicular to the Z line. The thin filaments of the I band, which attach at one end to the Z line, are seen to penetrate among the thicker filaments of the A band; thus, the A band and the I band overlap, and the degree of overlap can vary. The lightness of the H zone is due to the absence of I filaments in this region of the A band; if there is less overlap, the H zone will be wider and the Z lines will be farther apart.

Biochemical and biophysical studies have shown the fibrous protein *myosin* to be the principal constituent of the thick filaments. The individual myosin molecules possess a very long (about 1200 Å), straight tail and a double globular head portion. When packed together to form a thick myofilament, the tails of the myosin molecules lie side by side and form the "backbone" of the myofilament; the globular head portions protrude from the structure at regular intervals to form the cross-bridges seen in Figures 3-19 and 3-22. The globular head of the myosin molecules contains the special biochemical and enzymatic properties (to be discussed later) that enable the thick filaments to participate in the contraction process. The myosin tails appear to serve a structural function. Because the myosin molecules pack together in a compact way, the protruding heads extend in all directions from the thick filaments in a spiral fashion that undergoes one complete "turn" every 426 Å; this is the distance between successive cross-bridges as "seen" by any single thin filament. The center region of the thick filament (the pseudo H zone) is devoid of any cross-bridges because this region contains myosin tails only; the myosin molecules are arranged in a "head-out" fashion, starting at the center of the filament. Thus, the thick filament is a symmetrical structure, with identical halves extending in either direction from the M line (see Fig. 3-19B). This built-in directionality is important in the contraction process.

The thin filaments are composed of three separate proteins. The largest contributor to the structure is *actin.* This is a globular protein whose relatively small molecules (55 Å) are approximately spherical; they are joined together to

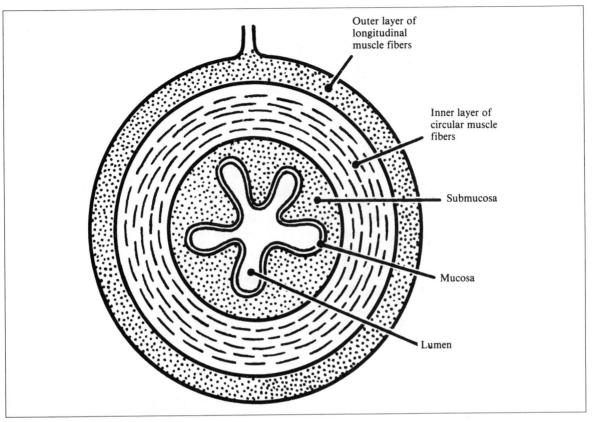

Fig. 3-17. A cross-sectional representation of the small intestine. The smooth muscle in this organ is arranged into two layers whose fibers are oriented at right angles to each other. The outer muscle layer is made up of longitudinal fibers; its fibers and fiber bundles run in the direction of the length of the intestine. The fibers of the inner, circular layer run around the intestine. Because the intestine is filled with incompressible fluid, contraction of the longitudinal layer will cause the intestine to become shorter and fatter, while contraction of the circular layer will cause the diameter to decrease and the total length of the intestine to increase, thus stretching the longitudinal muscle fibers. Compare this type of antagonistic relationship with the more direct one shown in Figure 3-14. (From R. A. Meiss. Muscle: Striated, Smooth, Cardiac. In E. E. Selkurt [ed.], *Basic Physiology for the Health Sciences*. Boston: Little, Brown, 1975. P. 197.)

form two long bead-chain structures, which in turn wrap about each other to form a helix, which completes a half-turn every seven actin subunits (see Fig. 3-19C). This type of structure comprises the whole 1-μ length of the thin filament; at one end the thin filament is attached to the Z line protein and then extends through to be a thin filament (1 μm long) of the adjacent sarcomere. Because the two faces of the individual actin molecules are not chemically identical, actin filaments also show a directionality important to the contraction process. That is, the portions of an actin filament on either side of a Z line are oppositely directed along their entire lengths. Stated in another way, the actin filaments interdigitating with the thick filaments from either end of a single sarcomere have opposite directionalities that are matched to the bidirectional character of the thick filaments.

The two other proteins found on the thin filaments are *troponin* and *tropomyosin*. A pair of the small troponin molecules lie on opposite sides of the actin helix every seven-actin half-turn (see Figs. 3-19C and 3-22). From each troponin site a long, slender tropomyosin molecule extends

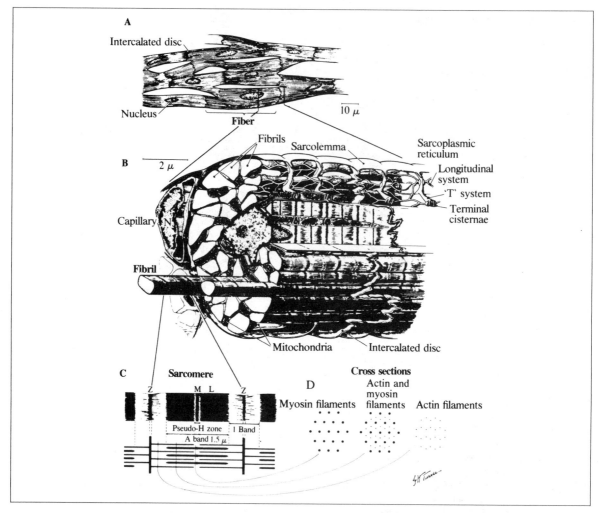

A

Intercalated disc

Nucleus

Fiber

10 µ

B   2 µ

Fibrils

Sarcolemma

Sarcoplasmic
reticulum

Longitudinal
system

'T' system

Terminal
cisternae

Capillary N

Fibril

Mitochondria   Intercalated disc

C   **Sarcomere**   **D**   **Cross sections**

Z   M L   Z   Actin and
myosin
filaments

Myosin filaments   filaments   Actin filaments

Pseudo-H zone   I Band

A band 1.5 µ

Fig. 3-18. The anatomical organization of heart muscle. As shown in (A), the fibers of heart muscle are short and branched and interconnect with one another. Each fiber (cell) has one nucleus. The individual fibers (B) are supplied with many mitochondria and are internally organized into fibrils (or myofibrils). Each fibril in turn is composed of a chain of sarcomeres (C), which are made up of two sets of myofilaments (C, D). Thus, heart muscle shares some of the anatomical features of both skeletal muscle (Figs. 3-15, 3-19) and smooth muscle (Fig. 3-16). (From E. Braunwald, J. Ross, and E. H. Sonnenblick. *Mechanism of Contraction of the Normal and Failing Heart.* Boston: Little, Brown, 1968. P. 4.)

for seven actin units along the groove formed by the two strands of the helix. It is the presence of the troponin and the tropomyosin molecules that is responsible for controlling the interaction between the thick and thin filaments during the contraction process.

The precise regularity of the sarcomere in skeletal muscle is also found in cardiac muscle (see Fig. 3-18C), and the previous description is again applicable. In smooth muscle, however, no such order is apparent. Smooth muscle myofilaments, both thick and thin, are plentiful, but they are not arranged in a precisely interdigitating array. The muscle contains both myosin and actin, which are identified with the thick and thin filaments respectively. No Z lines or sarcomeres as such are present, although scattered plentifully

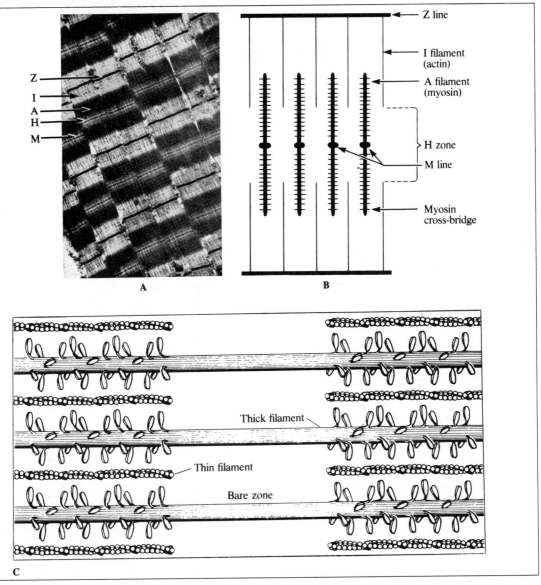

Fig. 3-19. The ultrastructural organization of skeletal muscle. (A) The complex pattern of bands produced by the overlap of the two sets of myofilaments is shown in this electron micrograph. The lightest areas represent the I bands, and the wide dark areas, the A bands. The lighter region at the center of the A band is the H zone; and the center of the H zone is made lighter still by the absence here of cross-bridges on the myosin filaments. (From H. E. Huxley. Muscle Cells. In T. Brachet and E. Mirsky [eds.], *The Cell*. New York and London: Academic, 1960. Vol. 4, p. 371.) (B) Diagram of a single sarcomere, the fundamental organized unit of contractile apparatus. The lengths of the filaments are drawn to scale, but the lateral spacings have been exaggerated for clarity. The width of the bridge-free portion of the H zone is constant; the total width of the H zone is measured between the tips of I filaments extending from the opposite Z lines, and this dimension will vary, depending on the degree of myofilament overlap. By convention the sarcomere is considered to run from Z line to Z line. (From R. A. Meiss. Muscle: Striated, Smooth, Cardiac. In E. E. Selkurt [ed.], *Basic Physiology for the Health Sciences* [2nd ed.]. Boston: Little, Brown, 1982. P. 218.) (C) The assembly of the various protein components of the myofilaments into a portion of a sarcomere. The helical "bead-chain" structure of the actin polymer is shown in the thin filaments, as is the presence of the molecules of troponin and tropomyosin in the "groove" region of the actin helix (this may be seen more clearly in Fig. 3-22). The heads of the myosin molecules protrude from the thick filaments, the major portion of which is composed of the myosin tail portions. No heads protrude from the center portion of the thick filaments because this region includes tail portions only; no cross-bridges can be formed in this region of the sarcomere. The M line protein shown in (B) is not shown here. (From J. M. Murray and A. Weber. The cooperative action of muscle proteins. *Sci. Am.* 230(2):58, 1974. Copyright © 1974 by Scientific American, Inc. All rights reserved.)

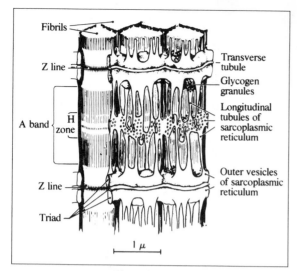

Fig. 3-20. The internal membrane system of skeletal muscle. Contraction is initiated by a release of stored calcium ions from the longitudinal tubules of the sarcoplasmic reticulum. This "trigger" calcium has only a short distance to diffuse to reach the myofilaments in the A band, where the ultimate reactions leading to contraction take place. It is by means of this internal membrane system that an action potential on the surface membrane of the fiber rapidly makes its effect felt at the center of the fiber. (After L. D. Peachy. The sarcoplasmic reticulum and transverse tubules of the frog's sartorius. *J. Cell. Biol.* 25:222, 1965.)

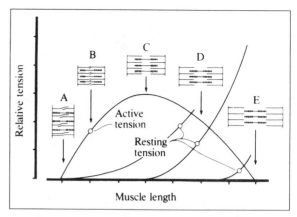

Fig. 3-21. The structural basis for the length-force relationship in skeletal muscle. When the muscle is so highly stretched as to pull the thin filaments completely out from among the thick filaments (E), no force is developed. If partial overlap is allowed (D), less than maximum force is generated. At C, there is optimal overlap, and maximum force is developed. In the region between C and A, the force is proportional to the degree of thick and thin filament overlap. As the muscle becomes very short (B), thin filaments from opposite sides of the sarcomere interfere with each other, and thick filaments are forced against the Z lines, leading to a rapid fall in force. When there is no overlap (E), no force is generated. If the curve shown in this figure were drawn for a single sarcomere, the transitions between the different regions would be abrupt. Because many sarcomeres participate in a contraction and are not all in exact register with each other, the length-tension curve for a whole muscle would be smoothed out as shown. The behavior illustrated is the structural basis for the length-force relationship in skeletal muscle and probably for heart muscle also. It is likely that in the future a similar mechanism will be demonstrated in smooth muscle as well. (From R. A. Meiss. Muscle: Striated, Smooth, Cardiac. In E. E. Selkurt [ed.], *Physiology for the Health Sciences* [2nd ed.]. Boston: Little, Brown, 1982. P. 221.)

throughout the cytoplasm (myoplasm) are the *dense bodies,* small oblong structures to which thin myofilaments attach. Thus, it appears that while nonstriated (smooth) muscle does not have the precise and esthetically pleasing regularity of skeletal muscle, it nevertheless contains the same essential component parts.

<hr>

CONTRACTION

THE SLIDING FILAMENT THEORY OF CONTRACTION. Skeletal Muscle. The function of muscle is closely related to its observable ultramicroscopic structure. It is not surprising, then, that skeletal muscle, with its beautifully organized ultrastructure, should have provided the major portion of our understanding of the contractile process.

The sarcomere is the fundamental contractile unit of skeletal muscle. A series of interrelated anatomical observations and experimental results have led to the *sliding filament theory* of muscle contraction, based on the properties of the sarcomere. Briefly, the points of evidence for this theory are these: when muscles that have been either mechanically stretched out or allowed to contract were

fixed and examined under the electron microscope, the distance between Z lines was found to vary directly with the overall muscle length. However, the dimensions of the A bands remained constant, while those of the I bands and the H zone varied. The lengths of both the actin and the myosin myofilaments remained constant. Therefore, a change in overall muscle length was directly related to the degree of thick and thin myofilament overlap, with muscle length changes being accompanied by a "sliding" of the array of myofilaments of the I band into or out of the array of A band filaments. Precise studies on single fibers of living muscle showed that the force the muscle could exert was closely related to the degree of myofilament overlap and hence to the number of myosin-filament cross-bridges that could attach to nearby actin filaments (see Fig. 3-21). When the muscle was stretched so far that there was no myofilament overlap, it could develop no force. On the other hand, when the muscle had contracted to so short a length that the I band filaments coming in from opposite sides of the A band interfered with each other, force was greatly reduced. By these and other means it has been established that the amount of force that skeletal muscle can exert is directly proportional to the degree of thick and thin myofilament overlap, and that the function of the myosin cross-bridges is to slide the thin filaments past the thick filaments in a hand-over-hand sort of way and thereby to produce an overall shortening of the muscle. This, in essence, is the sliding filament theory; to date, it has survived all experimental tests and appears to approach the truth more closely than do other theories of contraction.

Cardiac Muscle. Because of the structural similarities between cardiac and skeletal muscle, it is reasonable to expect a similarity of contractile mechanisms. Although experimental tests in cardiac muscle have not been so precise, there appears at present to be no reason to expect any fundamentally different contractile mechanism. At present, all evidence indicates that the sliding filament theory is adequate to account for the function of cardiac muscle.

Smooth Muscle. The adequacy of the sliding filament theory is less easy to settle in the case of smooth muscle, because its structure is much less well organized than that of skeletal muscle. However, actin and myosin filaments are present in close proximity to each other, and other biophysical and biochemical criteria are met, so it is entirely possible (and also highly likely) that smooth muscle also contracts by a sliding filament mechanism.

INITIATION AND CONTROL OF CONTRACTION. Control by Muscle Cell Membrane. The greatest degree of control of contractile activity is exerted at the level of the *muscle cell membrane*. It is here that the signal from the CNS is translated into a form to which the contractile apparatus can respond. Of the three major types of muscle, skeletal muscle is the one most completely controlled via its cell membrane.

The plasma membrane of a typical skeletal muscle cell (of the *fast,* or *twitch,* type) has functional properties similar to those of an unmyelinated nerve axon (see Chap. 2). A *resting potential* of −90 mv (inside negative) is typical, and excitation of the membrane leads to a rapid, overshooting, propagated *action potential* of about 2-msec duration, which is very brief compared with the duration of the contraction it causes. The distribution of Na^+ and K^+ ions

across the membrane and the conductance changes that accompany the action potential are similar to those of nerve. A *refractory period* follows the action potential, and the ionic asymmetry of the resting state is maintained by an active sodium pump.

In the case of nerve, the function of membrane activity is to transmit a message down the length of the fiber only; in muscle, there are two functions. The first is like that of nerve; the message to contract must be delivered to the whole length of muscle as nearly simultaneously as possible to ensure a coordinated contraction. In addition, there must be some communication from the outer membrane of the fiber to its interior in order to activate the contractile apparatus rapidly. Calculations show that a chemical substance released from the periphery of a skeletal muscle fiber cannot diffuse inward nearly fast enough to account for the rapid onset of contractile activity following an action potential. Muscle must therefore possess a special inwardly conducting system as well as an axially conducting system. The inwardly conducting function has been identified with an anatomical component of the muscle, the *transverse-tubule system* (commonly called the *T-tubule system*). This is made up of a network of fine tubes penetrating from the surface membrane deep into the muscle fiber at the level of the Z line of each sarcomere (see Fig. 3-20). The interior of this network of tubes is continuous with the extracellular space; hence, the T tubules may be regarded as an inward extension of the outer cell membrane. As such, they do not disrupt the membrane barrier between the cell interior (cytoplasm) and the extracellular space. It is the function of the T tubule system to conduct the electrical impulse inward. If chemical means are used to destroy the T system while leaving the rest of the muscle intact, the electrical activity of the surface membrane does not penetrate the fiber, and no contraction results in response to an action potential.

As the T tubules penetrate the fiber, they make close contact with yet another membranous component of the muscle interior, the *sarcoplasmic reticulum* (SR). In contrast to the transverse tubules, the SR, a system of *longitudinal tubules* terminating with *outer vesicles* in each sarcomere, runs parallel to the myofilaments from Z line (where the T tubules are) to Z line. It is at the level of the SR that the final electrically controlled event in the *excitation-contraction coupling* sequence takes place. The SR functions as a storage, release, and uptake site for *calcium ions*. When an impulse from the surface action potential arrives down the T tubules, the SR releases stored calcium ions. These ions diffuse across the short distance from the SR to the cross-bridge region of myofilament overlap, where they provide the signal for the contractile process to begin.

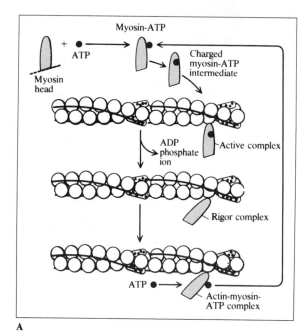

A

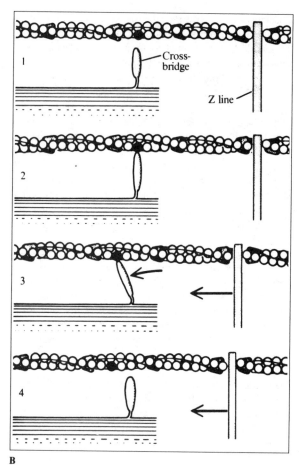

B

Extensive biochemical studies of the interactions between calcium, adenosine triphosphate (ATP), actin, and myosin have led to a good understanding of the means whereby calcium exercises its control over the contractile process. The key to calcium control has been found to be the presence on the thin filaments of the two additional proteins troponin and tropomyosin.

Figure 3-22A outlines the basic biochemical steps in the contraction process. The sequence of reactions shown was deduced from studies in which the tails of the myosin molecules had been chemically removed; only the enzymatically active head portions (which are identified with the cross-bridges of intact muscle) were used. The actin filaments used were intact (but separated from the Z lines) and contained their normal complement of troponin and tropomyosin. An excess amount of calcium, sufficient to activate the process fully, was present. This partly disassembled system was used for experimental convenience and mimics very well the actual process in intact muscle. Figure 3-22B shows the mechanical counterpart (in intact muscle) of the biochemical reactions.

The reaction sequence begins when a myosin head combines with a molecule of ATP, forming a "charged" myosin-ATP intermediate (step 1). This intermediate species is permitted to interact with an actin molecule (step 2); the actin-myosin combination forms an ATPase system that quickly splits ATP into adenosine diphosphate and an inor-

Fig. 3-22. (A) The chemical events of the contraction cycle as outlined in the text. In the drawing the troponin molecule (the oblong structure lying on the actin filament) is fully occupied by calcium ions (small dots), and hence all of its inhibitory activity is suppressed. Thus, as soon as the charged myosin-ATP intermediate is formed, it combines with an actin molecule, and the ATP is split. Sufficient ATP is present to allow the rigor complex to be broken up and the myosin recycled through the process. **(B)** The events of the contraction cycle as they occur in intact muscle. Here, the myosin head is permanently attached to the thick filament. When ATP is split, the myosin head converts the chemical energy into mechanical energy and undergoes the change in position shown; this action moves the thin filament parallel to the thick filament, and a slight shortening of the muscle results. When the myosin head lets go of the actin molecule, it returns to its original position and is free to repeat the whole cycle. Many such cycles, when added together, can produce a large movement of the muscle. (From J. M. Murray and A. Weber. The cooperative action of muscle proteins. *Sci. Am.* 230(2):58, 1974. Copyright © 1974 by Scientific American, Inc. All rights reserved.)

ganic phosphate ion (step 3). This is the energy-liberating and motion-producing step in the contraction of intact muscle. Following the splitting of the ATP molecule, the myosin head is not free to leave the actin to which it is bound (this association is the *rigor complex*) until it is combined with a new ATP molecule (step 1 again). The myosin head then becomes detached from the actin (step 4) and becomes a new "charged" myosin-ATP intermediate (as in step 1) and may go on to repeat the whole sequence as many times as the calcium concentration (see below) and the ATP supply allow.

Control over this process is exercised by calcium ions; in the absence of calcium the troponin-tropomyosin complex will not allow the myosin-ATP "charged" intermediate to combine with any actin molecule, and the sequence is halted at the end of step 1. The muscle is now ready to contract but is inhibited from further action; this is the normal state of resting muscle. When calcium ions (released from the sarcoplasmic reticulum in intact muscle) combine with troponin, it releases its inhibition of the reaction, and steps 2,3,4, and 1 are free to proceed through as many cycles as possible. The function of the long tropomyosin molecule is to transmit the troponin "inhibition release" message to those actin molecules not directly covered by the troponin molecule itself. Removal of calcium ions from the troponin molecule (as by the action of the sarcoplasmic reticulum in intact muscle) reestablishes the troponin inhibition of the contraction process; no further ATP is split, and the muscle relaxes. Agents such as caffeine cause the sarcoplasmic reticulum to take up calcium ions less readily, and thus the contraction is prolonged. If the ATP supply for the muscle becomes exhausted for some reason, the reaction sequence must stop at the end of step 3. The myosin remains bound to the actin (the rigor complex), and the whole muscle becomes stiff and rigid; this is the cause of the phenomenon of *rigor mortis* that takes place following death, as ATP supply diminishes.

Cardiac muscle also contains a T system and a well-developed SR. In addition, the sarcolemma is able to act as a calcium storage and release site. Because of the smaller size of the cardiac muscle cell, calcium released at the surface membrane can diffuse to a significant proportion of the myofilaments. The essential features of excitation-contraction coupling in cardiac muscle appear not to differ significantly from the scheme that has been worked out for skeletal muscle.

Smooth muscle contractile activity is also under the control of calcium ions, but the mechanism by which the control is exercised is quite different. In skeletal muscle the proteins associated with the thin filaments (*actin-linked reg-*

ulation) control the contraction; in smooth muscle the contractile process is controlled by proteins associated with the thick filaments (*myosin-linked regulation*). Each of the paired enzymatically active head portions of the smooth muscle myosin molecules contains a so-called *light chain*, a firmly-bound protein molecule whose molecular weight is much less than that of myosin itself. The adenosine triphosphatase (ATPase) activity of the actin-myosin association is low until the terminal phosphate of an ATP molecule is bound to each of the light chains. This process of *phosphorylation* activates the enzymatic activity of the actin-myosin complex, which then is free to hydrolyze additional ATP molecules to provide the energy for contraction. An enzyme called *myosin light-chain kinase* is responsible for the crucial myosin phosphorylation; for the kinase to phosphorylate the myosin light chains, the presence of calcium ions is necessary. At rest, the internal free calcium concentration of the smooth muscle cell is very low (as in skeletal muscle at rest), the myosin is not phosphorylated, and no contractile activity occurs. During contraction, when the internal calcium concentration rises sharply, phosphorylation takes place, and contractile activity is believed to proceed by essentially the same type of cyclic mechanism as is found in skeletal muscle.

The light-chain kinase enzyme itself can exist in active and inactive forms. The transition to the active (calcium-sensitive) form depends on the presence of yet another calcium-dependent regulatory protein, *calmodulin* (also called calcium-dependent regulator). When the internal calcium concentration is lowered by mechanisms that function to cause relaxation, the light-chain phosphorylation reaction ceases. Another enzyme in the system, a *phosphatase,* then *dephosphorylates* the myosin light chains, inactivating actin-myosin ATPase. This phosphatase is believed to be continuously active in the cell; during contraction its activity is overwhelmed by the light-chain kinase activity. For this reason, the process of relaxation in smooth muscle appears to be different from a mere reversal of the contraction process. These interrelationships are summarized in Figure 3-23.

The contrast with the situation in skeletal muscle is important. Resting skeletal muscle exists in a state of "inhibited readiness" that calcium ions can release; resting smooth muscle actin-myosin cannot begin ATP hydrolysis until it is placed in a state of readiness by calcium-controlled myosin light-chain phosphorylation. While the net effects of a varying intracellular calcium concentration are outwardly similar in many respects, the different underlying mechanisms, although not fully understood (especially in the case of smooth muscle), provide a basis for the understanding of

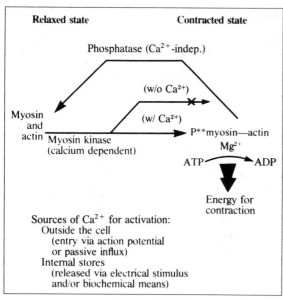

Relaxed state   **Contracted state**

Phosphatase (Ca^{2+}-indep.)

(w/o Ca^{2+})

Myosin and actin

(w/ Ca^{2+})

Myosin kinase (calcium dependent)

P^{**}myosin—actin

Mg^{2+}

ATP → ADP

Energy for contraction

Sources of Ca^{2+} for activation:
  Outside the cell
    (entry via action potential
    or passive influx)
  Internal stores
    (released via electrical stimulus
    and/or biochemical means)

Fig. 3-23. A proposed basic regulatory scheme for vertebrate smooth muscle. In this mechanism, the phosphorylation of the myosin light chains by myosin light-chain kinase (complexed with calmodulin) is the principal event that allows myosin and actin interaction. Various other pathways are responsible for controlling the intracellular calcium concentration and activating the myosin light-chain kinase enzyme. Relaxation is achieved by the action of a light-chain phosphatase enzyme whose activity appears not to be regulated by other cellular functions. While many further details of this mechanism are unclear, this proposed schema is likely to account for many of the properties of smooth muscle activation. (From R. A. Meiss. Muscle: Striated, Smooth, Cardiac. In E. E. Selkurt [ed.], *Basic Physiology for the Health Sciences* [2nd ed.]. Boston: Little, Brown, 1982. P. 229.)

the function of smooth muscle under a variety of physiological and pharmacological conditions.

Types of Excitation of the Muscle Membrane. The details of the transmission of the nerve signal to the muscle membrane have been given in this chapter. The accounts given there and here treat principally the *fast* or *twitch* type of muscle fibers, which make up most of the human limb musculature. Another type of specialized skeletal muscle fiber, the *slow* or *tonic* type, is associated with some muscles of the trunk and elsewhere (e.g., some of the intercostal muscles). This fiber type, which will not be discussed further, is specialized for slower, more sustained contractions and is characterized by (among other things) a less regularly organized contractile protein arrangement, a greatly reduced internal membrane system (T tubules and SR) and a graded,

incremental membrane response to nerve stimulation. Mechanical responses may also be graded by the extent of membrane depolarization. While all skeletal muscles appear to share most of their biochemical and physiological characteristics to a large extent, because of the comparative simplicity, physiological importance, and regularity of structure and behavior of fast skeletal muscle, general discussions (including this one) of skeletal muscle are likely to focus on this type. The reader is referred to the literature (e.g., Close, 1972) for consideration of the other skeletal muscle types.

The basic mechanical response of skeletal muscle, the *twitch,* is a single, brief contraction produced by the arrival of a single nerve action potential (Fig. 3-24). For a given muscle fiber under constant conditions of temperature, fiber length, and so on, the magnitude of the twitch is a constant, all-or-none property. The duration of a twitch, usually a fraction of a second, is too short to be useful for most tasks. However, if the fiber is restimulated before it has relaxed completely, the second twitch will add its mechanical effect to the first, a process called *mechanical summation.* Repeated stimulation will produce a state of continued contraction, or *tetanus;* if the interval between stimuli is short enough, the separate twitches will merge into a smooth contraction, a *fused tetanus.* Because of mechanical summation a tetanus produces more force than does a twitch; the usual physiological pattern of activation of a muscle fiber is a tetanus (not necessarily fused) of the appropriate duration.

The motor axons that innervate a muscle undergo varying degrees of branching before they terminate in a number of end-plates. Thus, a single axon may have terminals on many separate muscle fibers. A single axon, together with all the muscle fibers it innervates, is a *motor unit.* Since activation of a muscle fiber via the motor end-plate produces an all-or-none contraction (though the maximum force may be increased by tetanic activation), the size of a typical motor unit is related to the fineness of control required of the whole muscle. In muscles in which a high degree of control is exercised over fine movements, the motor axon will control only a few muscle fibers, and hence an action potential from that axon will produce contraction in only a small portion of the whole muscle. Other muscles are specialized for large, rapid movements with little fine control and typically have large motor units, so that a single motor axon will control a relatively large portion of the whole muscle. This variable ratio of muscle fibers to motor axons is the *innervation ratio.*

It is possible for experimental or diagnostic purposes to observe the electrical activity of skeletal muscle in the intact body by applying recording electrodes to the skin above the

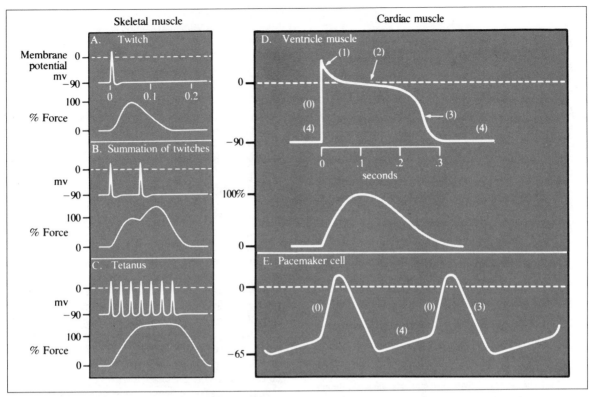

Fig. 3-24. Membrane and contractile events in skeletal muscle. (A) The upper trace displays the muscle membrane potential in millivolts. The single action potential shown leads to the single contractile event, a twitch, shown in the lower trace. (The action potential trace time for all the traces is marked off in tenths of a second, and the amplitude of all the contraction tracing is with reference to the peak twitch tension.) It is important to note that the duration of the twitch is much longer than that of the action potential that triggered it. (B) Because the refractory period of the muscle cell membrane is over long before the mechanical activity ceases, the muscle may be restimulated while some contractile force is still present. This restimulation leads to mechanical summation of the two twitches, and more force is produced than in a single twitch. (C) If action potentials follow each other quickly enough (but not too quickly—why?), no relaxation occurs between successive twitches, and a fused tetanus results. (Modified from R. A. Meiss. Muscle: Striated, Smooth, Cardiac. In E. E. Selkurt [ed.], *Basic Physiology for the Health Sciences* [2nd ed.]. Boston: Little, Brown, 1982. P. 231.)

muscle in question. The pattern produced by the combined action potentials of many motor units is called an *electromyogram* (EMG) (Fig. 3-25). It may be extended to the observation of a single motor unit by inserting a fine needle electrode, insulated everywhere but at the tip, into the muscle.

Central Nervous System Control. The most highly refined control over muscular movement is exercised by the coordinated activity of the CNS. Sensory receptors located in the muscle and its antagonist and in the joints and tendons, together with visual and aural observation, provide the CNS with a feedback of information concerning the speed, force, position, and so on, of the muscle or limb. This feedback allows the motor signals to the muscle to be modified to produce the desired actions. (See Chap. 5 for the details of this process.)

Smooth muscle is also controlled by the CNS; the modes of regulation may be divided into two broad categories. *Multiunit* smooth muscles include the *iris* (which controls the diameter of the pupil of the eye), the *ciliary muscle* (which controls the focal length of the lens of the eye), the *pilomotor muscles* (which erect the hairs of the skin), and

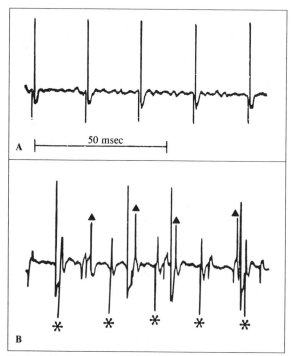

Fig. 3-25. Electromyographs of human triceps muscle. These recordings were made with a fine needle electrode inserted into the muscle and represent the activity of single motor units. (A) Record made during a low level of voluntary activity. Action potentials are occurring at a rate of approximately 45/sec, and the overall muscle force is low. (B) During moderate voluntary activity, at least two other motor units (marked by separate symbols) in the vicinity of the recording electrode are active, and the overall muscle force is higher. (These recordings were kindly supplied by Dr. John Kincaid of the Indiana University School of Medicine.)

many *blood vessel muscles.* This type of smooth muscle is controlled much as skeletal muscle is, having functional or separate motor units and responding to a single nerve impulse with a single twitch. Fused tetanes, associated with a series of action potentials, usually occur. The transfer of the information from nerve to muscle is via a chemical mediator (ACh in some instances), although highly structured motor end-plates are not found.

The visceral smooth muscle, on the other hand, behaves as if all the cells of an extensive region were connected into one large motor unit. Hence, this muscle type is termed *unitary.* In a given region, changes in the membrane potential of any one cell are shared by all the adjacent cells, so the tissue behaves electrically as if all the cell interiors were in direct contact with one another. Such an arrangement is a *functional syncytium;* this type of muscle tends to have pe-

riodic spontaneous variations in membrane potential. When depolarization is sufficient, action potentials and contraction may occur. Myoneural junctions are not present as such, and the transmitter substances (which may have either a stimulating or an inhibiting effect) are released into the extracellular space among the cells. In such muscles the function of nerves is largely to modify and modulate, rather than to initiate, the independent spontaneous and rhythmic control inherent in the muscle cell membranes.

In many smooth muscle tissues, notably the stomach and intestine, periodic fluctuations in the membrane potential ("slow waves") provide the basic timing of the rhythmic contractile activity of the gut musculature. The slow waves do not cause contractions by themselves, but during that portion of the slow-wave cycle in which the membrane is most depolarized, action potentials ("spikes") can arise and trigger muscle contractions. In this way the periodic muscular contractions are controlled and coordinated by the intrinsic electrical activity of the tissue. Such nerve-independent contractile activity is called *myogenic,* in contrast to the *neurogenic* activity of skeletal muscles. Additionally, there are smooth muscles in the circulatory and respiratory systems whose contraction may be independent of their own cell membrane electrical activity and may instead be controlled by factors such as pH, oxygen tension, and circulating hormones.

While the electrical activity of smooth muscle cells is often complex and varies greatly among the tissue types, some features are held in common with nerve and skeletal and cardiac muscle. The balance between extracellular Na^+ and intracellular K^+ is a primary determinant of the resting potential, which in smooth muscles tends to be smaller in magnitude (less negative; i.e., from -35 to -60 mv) than in other excitable tissues. The resting potential may also be unsteady or undergo rhythmic fluctuations (slow waves) or show repeated periods of gradual depolarization terminating in action potentials much like the pacemaker activity of some cardiac muscle cells (see Chap. 14). Other ions, notably Ca^{2+}, may influence the magnitude or stability of the membrane potential. Action potentials in smooth muscles likewise have features in common with those from other excitable tissues. While the spikes are often longer in duration and may not show an overshoot, membrane conductance changes to Na^+ and K^+ (and also Ca^{2+}) do appear to be involved. In addition, some of the periodic fluctuations in smooth muscle membrane potentials may be due to the action of spontaneously rhythmic electrogenic pumps (see Chap. 1) and not associated with ion conductance changes.

The electrical properties of cardiac muscle lie somewhere between those of striated and smooth muscle. Although the contraction of cardiac muscle fibers is closely controlled by their cell membranes, the electrical activity of the cell membranes is only slightly influenced by the nervous system. Mammalian hearts, under proper conditions, can continue to beat even after they are removed from the body; the beat is myogenic, and its origin lies within the heart itself. It is the special membrane properties of various types of cardiac muscle cells that account for many of the remarkable automatic and rhythmic properties of heart muscle function (see Chap. 14).

THE MECHANICAL FUNCTION OF MUSCLE

ISOTONIC AND ISOMETRIC CONTRACTION. One important function of skeletal muscle is to do work in the physical sense; that is, to exert a force that can cause motion of some object (Fig. 3-26). When the force is constant, as when an object is lifted upward, the contraction is *isotonic* (meaning "same force"). On the other hand, if the object is too heavy to be moved, force will be generated without any change in the muscle length, and the contraction is *isometric* ("same length"). Practically speaking, most contractions are both isotonic and isometric. When we lift a load, it does not move at first (isometric contraction); when sufficient force has been developed, the load is lifted. Since it becomes no heavier as it is lifted, force stays constant (isotonic contraction). Sometimes a contraction occurs in which force continually increases and motion continuously occurs; the drawing of a bow and arrow is an example of this *auxotonic* type of contraction. Finally, in many cases, force may be constant with the muscle lengthening, not shortening; lowering an object or descending stairs involves this type of activity.

Length-Force Relationship. In the laboratory, two fundamental mechanical properties of muscle have been revealed by carefully controlling the conditions of contraction. The first of these is the *length-force relationship* (see Fig. 3-27, A and B; see also Fig. 3-21). If a number of isometric contractions are made with the muscle adjusted to a different length before each contraction, it will be found that there is a length at which the maximum force can be produced. When the muscle is made to be longer or shorter than the optimum length, it will develop less than its maximum force. In muscles with skeletal attachments at both ends, this property is largely obscured because the range of length changes the skeleton allows is not very great. Also, the lever action of the skeleton provides a constantly changing mechanical advantage; this can make the force vary with limb position as well as muscle length. For heart muscle,

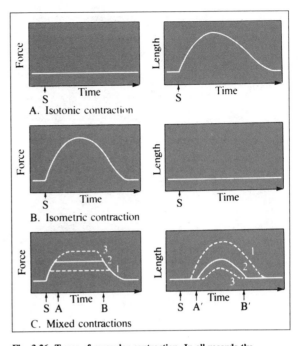

Fig. 3-26. Types of muscular contraction. In all records the stimulus is delivered at the point marked *S* on the time axis; increasing force and decreasing length (shortening) are plotted as upward movements. (A) Isotonic contraction. Here, the stimulated muscle exerts no force; all of its contractile activity is expressed as shortening. (B) Isometric contraction. Here, the muscle was prevented from shortening, and contractile activity resulted in the development of force only. (C) Mixed contractions. In this case, the muscle was presented with a load it could lift. The isometric record is shown at the left. Force increases (following stimulation) until, at point *A*, the isometric force becomes equal to the load (*solid line, trace 2*). Between points *A* and *B* (while the load is being lifted) the force is constant (isotonic conditions). The length (shortening) record shown at the right shows no movement between *S* and *A'* (isometric conditions), and shortening followed by relaxation between *A'* and *B'*. After *B'* the load is no longer being lifted; movement has ceased, and relaxation is now isometric. The dotted traces represent different loads on the muscle. If the load is small (*trace 1, left*), the isometric portion of the contraction is short, because the muscle can soon exert enough force to lift it. Correspondingly, the isometric portion of the contraction is long (*right panel*), and the load is lifted quickly and much shortening occurs. If the load is large (*trace 3, left*), the isometric portion is longer, and the isotonic portion is shorter. Less shortening occurs, and the muscle moves at a lower rate of speed. (From R. A. Meiss. Muscle: Striated, Smooth, Cardiac. In E. E. Selkurt [ed.], *Basic Physiology for the Health Sciences* [2nd ed.]. Boston: Little, Brown, 1982. P. 232.)

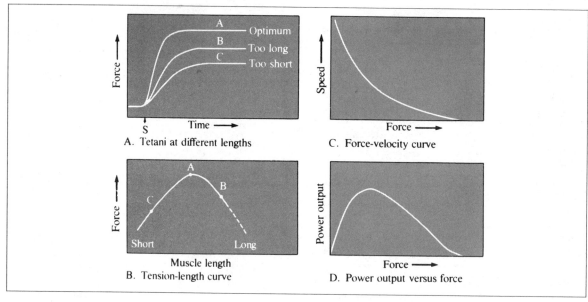

A. Tetani at different lengths

C. Force-velocity curve

B. Tension-length curve

D. Power output versus force

however, the length-force property can be of great importance, since it allows the heart to contract with more force when it has been filled with an increased amount of blood. This property, which is a basis for Starling's law of the heart, will be treated more fully in Chapter 15. As was discussed earlier, the functional basis for the length-force relationship is the change in amount of myofilament overlap at different muscle lengths.

Force-Velocity Relationship. A second fundamental property of muscle is the *force-velocity relationship* (Fig. 3-27, C). It is well known from everyday experience that the heavier a load, the less quickly it can be lifted. Stated another way: the greater the force, the lower the velocity (speed of shortening). When the load (force) is so large that it cannot be lifted, the contraction is isometric (speed = 0). When there is no load at all, the speed of movement is the greatest. Between these two extremes, speed varies with load in a way that is characteristic of the individual muscle; the general form of the curve is similar to that in Figure 3-27, C. The shape of this curve is of some theoretical interest, because it provides information on the way the muscle liberates energy. A further extension of the curve, however, is of practical interest. From elementary physics we know that the term *work* can be defined as a force acting through some distance (work = force × distance). Also, we find that the term *power* is a measure of the rate of doing work (power = work/time). A little more algebraic manipulation

Fig. 3-27. Consequences of changes in experimental conditions. (A) Three fused tetanes at different resting muscle lengths. At *A* the length was optimal for the production of force, while at *B* and *C* it was too long or too short, respectively. (B) The force values from *A* are plotted against muscle length to form the *length-tension* (force) *curve*. Compare this with the behavior of muscle shown in Figure 3-21. The dotted line shows an area of possible damage to the muscle resulting from extreme stretch. (C) In Figure 3-26 it was shown that light loads could be more quickly moved than heavy ones. If the speeds (velocities) of shortening for a large number of different loads (forces) are plotted against the various forces, a characteristic *force-velocity curve* is obtained. This curve is similar in shape for a wide variety of muscles and expresses the general notion that the greater the load, the more slowly it can be lifted. (D) If power (force × velocity) is plotted against the force exerted, the curve shows a distinct optimum force for the production of power. Operation of a muscle at this region of the curve is not an automatic property of the muscle but involves decision making in the CNS. (From R. A. Meiss. Muscle: Striated, Smooth, Cardiac. In E. E. Selkurt [ed.], *Basic Physiology for the Health Sciences* [2nd ed.]. Boston: Little, Brown, 1982. P. 233.)

leads to the conclusion that the amount of power produced is the product of force and velocity (power = force × velocity). To relate this to the performance of muscle, the force-velocity relationship may be reexpressed as in Figure 3-27, D: force × velocity (= power) is plotted against force. When the force is maximum, power output is zero (since velocity is zero, no work is done), and when the velocity is at its maximum, power output is again very low, since force is very small. Between these two extremes, the curve relating power output to force shows a definite maximum; i.e., there is a combination of force and velocity at which maximum power output occurs. By way of analogy, an automobile engine also has a certain speed of revolution at which its power output is optimum.

The realization that muscles have some of the characteristics of engines can aid in the rational design of efficient muscle-powered machines. A bicycle equipped with a gear shift is a case in point. Let us assume that on flat and level ground the cyclist is moving the leg muscles at such a force and speed that the power output is at its optimum. When the cyclist encounters a hill, he or she is required to exert more force; the speed of leg movement will thus fall, and the cyclist will no longer be working at the optimal point on the power curve. For this reason the bicycle maker has provided a selection of gear ratios. The cyclist shifts to low gear, and the speed of the legs can again increase; the force required of the leg muscles is reduced, and he or she is again working at the optimal point. (The bicycle is now moving more slowly, however; you can't get something for nothing.) In going down the hill, the situation is reversed; the speed of the leg muscles is now too high, the force too low, and again the optimal conditions are not met. By shifting into high gear, the cyclist will increase the required force; the leg muscle speed will fall, and conditions can return to optimum. From arguments of this sort, it can be appreciated that there is, for example, an optimal slant to a pedestrian ramp, an optimal weight of boxes for a worker to lift and stack. Successful athletes, whether knowingly or not, employ many such physiologically sound techniques to produce the most from their muscles.

It should be kept in mind, however, that a situation may call for muscular activity to be performed under force-velocity conditions far from the optimum; in such cases the CNS can select the appropriate number of motor units to be activated and determine the duration of activity of each to produce the desired motion. Mechanically optimal performance is not obligatory, but it is useful at times. It should also be noted that under isometric conditions, no external work is performed; nevertheless, the contractile apparatus of the muscle is activated and is consuming energy in order to maintain force. In this case, the efficiency of the muscle as a producer of motion is zero. The maximum possible efficiency of the isotonically contracting human muscle is about 25 percent; that is, of the energy consumed by muscle, 25 percent can appear as external work, and 75 percent is wasted as heat. The "waste" heat from inefficient muscular contraction is not always physiologically wasted, however, since muscle plays a large role in the thermal economy of the body (see Chap. 29). The efficiency of human muscle

is about the same as that of an automobile engine, better than that of a steam engine, and much less than that of an electric motor.

The length-force and force-velocity properties of skeletal muscle are also found in smooth and cardiac muscle. The contraction cycle of cardiac muscle is mechanically complex (see Chaps. 13 and 15), involving a mixture of conditions ranging from nearly isometric to a quasi-isotonic phase during which the afterload is continuously changing. Under controlled conditions, both isolated cardiac and smooth muscle in vitro can be made to perform in much the same way as skeletal muscle, lending support to the notion that the basic contractile mechanisms have much in common. The wide range in the functional capability of the intact heart (see Chap. 15) may be more readily appreciated if it is seen largely as the expression of the complex interaction between the force-velocity and length-force properties basic to muscle in general.

ENERGY SOURCES. The preceding discussion implies that muscles are motors and that they consume fuel in order to produce work. This is true, but in some respects muscles are more versatile than automobile engines because the body can produce the proper muscle fuel from a variety of foods. The fuel actually utilized in the process of contraction is the ubiquitous biological high-energy compound ATP. During the course of contraction, the ATP molecule is degraded to *adenosine diphosphate* (ADP) and *inorganic phosphate* (P_i), and the energy thus liberated is used by the contractile apparatus to produce force and motion. Several biochemical pathways supply the muscle stores of ATP. The most available "ready reserve" source of ATP is the compound *creatine phosphate* (CP). It exists in equilibrium with ATP according to the following reaction series:

1. $ATP \rightarrow ADP + P_i$ (liberation of energy for contraction)
2. $ADP + CP \rightarrow ATP + creatine$ (replenishment of ATP)
3. $Creatine + ATP \rightarrow ADP + CP$ (storage of ATP energy in the CP molecule)

Reaction 1 expresses the liberation of energy for contraction by ATP breakdown. In reaction 2, the ADP from 1 is rephosphorylated to ATP, using the energy stored in CP. In reaction 3, ATP produced elsewhere in the muscle is converted to its reserve storage form, CP. The equilibrium constants of reactions 2 and 3 are such that there is approximately 30 times as much CP as ATP. Thus, ATP can be quickly supplied to the contractile apparatus for some time, even when all non-CP sources are shut off.

The ATP supplied to the CP "reservoir" is produced

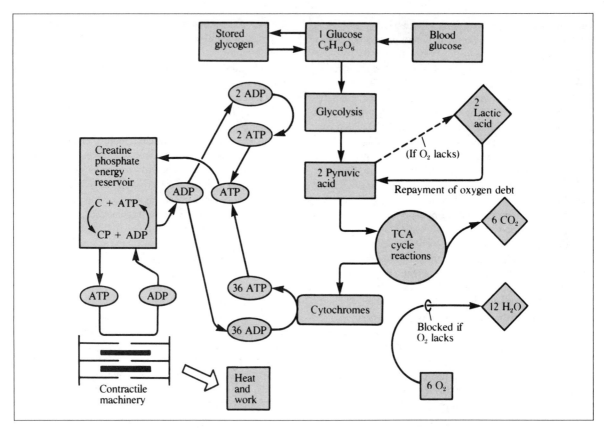

Fig. 3-28. The major metabolic sources of energy for muscular contraction. The number before each of the chemical species represents the net number of molecules of that substance arising from the breakdown of one molecule of glucose. Note that complete aerobic breakdown of glucose produces 18 times as much ATP as does glycolytic breakdown alone. The diamond-shaped boxes represent waste products as far as the immediate energy needs of the muscle are concerned. (From R. A. Meiss. Muscle: Striated, Smooth, Cardiac. In E. E. Selkurt [ed.], *Basic Physiology for the Health Sciences* [2nd ed.]. Boston: Little, Brown, 1982. P. 236.)

primarily by two well-known biochemical pathways, one *aerobic* (oxygen requiring) and one *anaerobic* (oxygen independent) (Fig. 3-28). The anaerobic pathway is the sequence of cytoplasmic reactions called *glycolysis*. The aerobic pathway is the mitochondrial reaction sequence variously called the *Krebs cycle, citric acid cycle,* or *tricarboxylic acid* (TCA) *cycle.* The less efficient of these pathways is glycolysis. This series of reactions takes *ADP* and *glucose* (which is stored in muscle largely as its polymer *glycogen*) and from them produces *pyruvic acid* and a small amount of ATP. Anaerobic glycolysis will soon come to a halt if the pyruvic acid is allowed to accumulate. By far the more efficient pathway (in terms of ATP production) is the TCA cycle. In the presence of an adequate supply of oxygen, the TCA cycle reactions utilize the pyruvic acid produced in glycolysis to rephosphorylate large amounts of ADP to ATP. Water and carbon dioxide are produced as waste products. If, as in the case of heavy exercise, the oxygen supply is inadequate to permit the TCA cycle to operate, the accumulating pyruvic acid may be temporarily

disposed of by reducing it to *lactic acid.* At the expense of accumulating lactic acid, then, glycolysis can continue to operate and produce ATP for contraction. The muscle is now said to have acquired an *oxygen debt.* When the oxygen supply is again adequate, some of it must be used to oxidize the accumulated lactic acid back to pyruvic acid and then to carbon dioxide and water by the TCA cycle. The muscular soreness sometimes experienced after heavy exercise is partially due to the accumulation of lactic acid.

While most skeletal muscles can sustain some degree of oxygen debt, cardiac muscle can sustain very little. The chest pain of *angina pectoris* results when the blood supply of the heart has become insufficient to adjust to the extra oxygen demands of exercise. Any sudden interruption of the normal cardiac blood supply, such as the blockage of a coronary artery by a blood clot (coronary thrombosis), leads rapidly to a depletion of the energy supply of the affected region and may also lead rapidly to death.

FUNCTIONAL ADAPTATIONS OF MUSCLE

The combined mechanical attributes of a muscle at any given time (as described by its length-force characteristic, its force-velocity relationship, and so on) are referred to as its *contractility*. Over a short span of time, the contractility of skeletal muscle is constant except for the property of fatigue, which is a reversible depletion of the metabolic fuel supply brought about by overuse. Certain skeletal muscles, such as those in the arms, become fatigued quickly but are well adapted for performing rapid and forceful movements. Other muscles, notably those of the trunk and the legs, can maintain a force for long periods of time with little fatigue. These *postural* muscles are not so well adapted, however, for rapid movement. Skeletal muscles vary in their ability to sustain an oxygen debt. An adequate supply of oxygen is assured during periods of heavy exercise (see Chap. 29), when the oxygen-carrying blood flow to the muscles is greatly increased through several physiological mechanisms, one of which is the relaxation of the smooth muscle sphincters that constrict the muscle blood vessels.

Certain long-term adaptive changes are found in skeletal muscle. One of these is *hypertrophy,* or an increase in muscle size and force capability. Increased use of a muscle, especially isometric exercise, results in an increase in the cytoplasm and myofilament content, although no new muscle cells are formed. Lack of use of a muscle, such as occurs during long confinement to bed or in a cast or as a result of nerve paralysis (e.g., poliomyelitis), can result in the opposite process, muscle *atrophy*. Connections with the CNS and some amount of exercise appear necessary to prevent atrophy, which results in an actual loss of contractile protein.

Yet another long-term change is *muscular dystrophy,* a complex of conditions characterized by progressive muscular weakness and a loss of contractile material, with partial replacement by noncontractile tissue. Other muscular disorders include *myotonia,* which is a failure of muscles to relax after a normal contraction because of continued spontaneous muscle membrane electrical activity, and *contracture,* which is a maintained force in the contractile apparatus independent of cell membrane activity. Many disorders of the CNS produce severe muscular symptoms (convulsions, trembling, tetanus) when there is nothing wrong with the muscles themselves.

In contrast to the situation in skeletal muscle, cardiac muscle is subject to many influences that can greatly modify its contractility. This feature is important in enabling the heart to make a wide range of adjustments in its mechanical properties in order to cope with changing physiological demands. Factors that modify the performance of cardiac muscle are called *inotropic agents*. Not all inotropic factors function solely by changing contractility, however. For example, the increase in contractile force associated with an increase in resting fiber length (see the preceding section) represents both the geometric effects due to changing length and the effects of length on the level of internal activation. An increase in heart rate, however, increases the force capabilities of the muscle without any change in fiber length. The increased contractility caused by increased heart rate is described by the *force-frequency relationship*. The muscle is said to have been *potentiated* by the rate increase and will now have a force-velocity curve and a force-length property distinctly different from those at the lower rate. A related phenomenon is that of *postextrasystolic potentiation*. If for some reason the heart produces a beat sooner than it should (*extrasystole*), this beat will be followed by a *compensatory pause*. The contraction following the longer-than-normal interval will show increased force; also, if for some reason the heart drops a beat, the next beat will be potentiated. The value of these mechanisms will become apparent in Chapter 15.

Certain drugs and hormones can also modify cardiac contractility. For example, *epinephrine* (Adrenalin) (a hormone released from the adrenal medulla in times of stress) will cause an increase in contractility over and above that due to its rate-increasing effect. For this reason, epinephrine is used as a therapeutic agent. Increasing the extracellular concentration of *calcium ions* has a similar type of effect. The drug *digitalis,* which is administered to failing hearts, produces a strengthened heartbeat without affecting the heart rate.

Finally, as in skeletal muscle, an increased load on the heart muscle for an extended period (such as that produced by chronic high blood pressure) can lead to *cardiac hypertrophy* (enlarged heart). Although cardiac hypertrophy represents a useful adaptation to increased stress, the enlarged heart may be subject to disturbances in impulse conduction.

Smooth muscle possesses two mechanical adaptations that suit it well to its physiological roles. First, it is capable of maintaining a force (called *tonus*) for very long periods without the consumption of large amounts of energy and consequent fatigue. This property is vital, for example, in keeping sphincters closed over extended periods. The other notable property of smooth muscle is its ability to adjust its resting length over wide ranges without large changes in resting tension. Thus, the urinary bladder is able to contain variable volumes of urine without distress, the stomach can expand to accommodate a full meal, and the uterus is able to adjust to the growing fetus.

The hypertrophy that smooth muscle undergoes differs from that of other muscle types in terms of its basic cause. Smooth muscle forms large portions of the target organs of several of the reproductive hormones. At the time of puberty the rising concentration of *estrogens,* for example, induces a marked hypertrophy of the uterine muscle and associated structures. As long as satisfactory hormone levels are maintained, the hypertrophied muscle remains in that state. Withdrawal of hormonal support, as by removal of the ovaries, will result in atrophy of the muscle.

BIBLIOGRAPHY

SENSATION AND NEUROMUSCULAR TRANSMISSION

Aidley, D. J. *The Physiology of Excitable Cells.* (2nd ed.). London: Cambridge University Press, 1978.

Birks, R., Huxley, H. E., and Katz, B. The fine structure of the neuromuscular junction of the frog. *J. Physiol.* (Lond.) 150:134–144, 1960.

Carlson, F. D., and Wilkie, D. R. Effectors: Striated, Smooth, and Cardiac Muscle. *Muscle Physiology.* Englewood Cliffs, N.J.: Prentice-Hall, 1974.

Close, R. I. Dynamic properties of mammalian skeletal muscles. *Physiol. Rev.* 52:129, 1972.

Davis, H. Some principles of sensory receptor action. *Physiol. Rev.* 41:391–416, 1961.

del Castillo, J., and Katz, B. Biophysical aspects of neuro-muscular transmission. *Prog. Biophys. Mol. Biol.* 6:121–170, 1956.

Drachman, D. B. (ed.). Trophic functions of the neuron. *Ann. N.Y. Acad. Sci.* 228:1–423, 1974.

Eccles, J. C. *The Understanding of the Brain.* New York: McGraw-Hill, 1973.

Feltz, A., and Mallart, A. An analysis of acetylcholine responses of junctional and extrajunctional receptors of frog muscle fibres. *J. Physiol.* (Lond.) 218:85–100, 1971.

Granit, R. *The Basis of Motor Control.* New York: Academic, 1970.

Granit, R. *Receptors and Sensory Perceptors.* New Haven: Yale University Press, 1956.

Guth, L. "Trophic" influence of nerve on muscle. *Physiol. Rev.* 48:645–687, 1969.

Hoyle, G. How is muscle turned on and off? *Sci. Am.* 222(4):84, 1970.

Hubbard, J. I. (ed.). *The Peripheral Nervous System.* New York: Plenum, 1974.

Hubbard, J. I., Llinas, R., and Quastel, D. M. J. *Electrophysiological Analysis of Synaptic Transmission.* Baltimore: Williams & Wilkins, 1969.

Huxley, A. F. Review lecture: Muscle contraction. *J. Physiol.* (Lond.) 243:1, 1974.

Huxley, H. E. The contraction of muscle. *Sci. Am.* 199(5):66, 1958.

Huxley, H. E. The mechanism of muscular contraction. *Sci. Am.* 213(6):18, 1965.

Iggo, A. Cutaneous Receptors. In J. I. Hubbard (ed.), *The Peripheral Nervous System.* New York: Plenum, 1974.

Katz, B. *Nerve, Muscle, and Synapse.* New York: McGraw-Hill, 1966.

Katz, B., and Miledi, R. Timing of calcium action during neuromuscular transmission. *J. Physiol.* (Lond.) 189:535–544, 1967.

Krnjević, K. Chemical nature of synaptic transmission in vertebrates. *Physiol. Rev.* 54:418–540, 1974.

Kuffler, S. W. Electric potential changes at an isolated nerve–muscle junction. *J. Neurophysiol.* 5:18–26, 1940.

Livingston, W. K. *Pain Mechanisms: A Physiologic Interpretation of Causalgia and Its Related States.* New York: Macmillan, 1947.

Loewi, O. Über humorale Übertragbarkeit der Herznervenwirkung. *Pflügers Arch.* 189:239–242, 1921.

Melzack, R., and Wall, P. D. Pain mechanisms: A new theory. *Science* 150:971–979, 1965.

Merton, P. A. How we control the contraction of our muscles. *Sci. Am.* 226(5):30, 1972.

Murray, J. M., and Weber, A. The cooperative action of muscle proteins. *Sci. Am.* 230(2):58, 1974.

Phillis, J. W. *The Pharmacology of Synapses.* Elmsford, N.Y.: Pergamon, 1970.

Rosenblith, W. A. (ed.). *Sensory Communication.* New York: Wiley, 1961.

Stevens, S. S. The Psychophysics of Sensory Function. In W. A. Rosenblith (ed.), *Sensory Communication.* New York: Wiley, 1961.

Sweet, W. H. Pain. In J. Field (ed.), *Handbook of Physiology.* Washington, D.C.: American Physiological Society, 1959. Section 1: Neurophysiology. Vol. 1.

Wall, P. D. The gate control theory of pain mechanisms. *Brain* 101:1–18, 1978.

SPECIAL SENSES

THE EYE

William M. Armstrong

The eye is a complex and intricate physiological device. Since it subserves the all-important function of vision, it has been investigated in great detail. Space considerations preclude all but a general survey of the basic principles of the physiology of vision in a textbook of this size. The interested student is referred to the general references listed at the end of the chapter for more detailed accounts of specialized aspects of the subject. In particular, the book by Campbell et al. (1974) gives a readable and relatively uncomplicated account of physiological optics and its clinical applications. Davson's (1980) monograph is a scholarly summary, within the confines of a single volume, of current knowledge and research in the physiology of vision.

The physiological role of the eye is twofold. First, it is an optical instrument that collects light from objects in the external environment and projects images of them on a light-sensitive organ, the *retina*. Second, it is a sensory receptor that translates these optical images into information that is transmitted to the visual areas of the brain. For convenience, these two functions of the eye will be considered separately. However, it should be remembered that, in the physiological functioning of the eye, they are closely related.

GENERAL STRUCTURE OF THE EYE

Figure 4-1 represents a horizontal cross section of the right eye. The body, or globe, of the eye has three coats. The outer coat is a layer of connective tissue called the *sclera*. Part of this is seen as the "white" of the eye. The anterior one-sixth of the sclera is a specialized transparent structure, the *cornea,* which forms part of the *dioptric,* or light-transmitting, apparatus of the eye. The front of the eye,

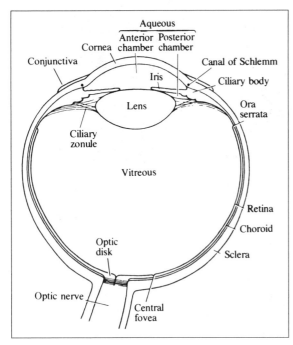

Fig. 4-1. General structure of the eye.

except for the cornea, is covered by a delicate membranous tissue, the *conjunctiva,* which also lines the eyelids.

The middle coat of the eyeball is the *choroid.* This is essentially a layer of highly vascular tissue that is important for the nourishment of the outer layers of the inner coat, or *retina.* Anteriorly, the middle or vascular coat is continued as the *ciliary body* and *iris* (Fig. 4-1). The choroid, ciliary body, and iris are known collectively as the *uvea.* The degree of pigmentation of the iris determines the color of the eye.

The major portion of the inner coat of the eye is the *retina.* Anteriorly, the retina terminates at the *ora serrata* (Fig. 4-1). From this, the inner coat continues forward as the *ciliary epithelium,* which forms part of the ciliary body.

The lens, which together with the cornea forms the dioptric apparatus, is supported by the *ciliary zonule;* this in turn is attached to the *ciliary body.* Between the cornea and the iris lies the *anterior chamber* of the eye. It and the smaller *posterior chamber,* which is bounded anteriorly by the iris and posteriorly by the ciliary zonule, are filled with a transparent aqueous medium, the *aqueous humor.* The main body of the eyeball contains a transparent jelly-like substance, the *vitreous humor.* This essentially aqueous solution has a high viscosity because it contains appreciable

quantities of the mucopolysaccharide *hyaluronic acid.* In addition, its structure is maintained by a branching network of fibers, consisting largely of a protein, *vitrosin.*

It is appropriate at this point to consider some of the structural elements in more detail. The cornea is made up of three main regions: an outer region, five or six cells thick, the *epithelium;* a central region, the *stroma,* which accounts for some 90 percent of the total corneal thickness; and finally, a single layer of cells, the *endothelium.* Both the epithelium and the endothelium are typical epithelial sheets (see Chap. 1). The stroma is a highly ordered arrangement of parallel collagenous fibers embedded in a relatively structureless ground substance. These fibers tend to have remarkably uniform diameters. In the opaque sclera, they are arranged in a more random fashion and have variable diameters. When the cornea swells because of uptake of excess water, it becomes opaque. The opacity seems to be the result of disarrangement of the ordered array of stromal fibers. In the isolated cornea, swelling may be induced by cold, metabolic inhibitors or an increase in *intraocular pressure.* Both the epithelium and the endothelium are permeable to salts, and it appears that, in vivo, K^+, Na^+, and Cl^- ions tend spontaneously to enter the stroma both from the fluid bathing the outer surface (the tears) and from that bathing the inner surface (the aqueous humor). The entry of salt, which is partly, at least, due to Donnan forces (see Chap. 1), is accompanied by an osmotic entry of water; if uncompensated, this will lead to corneal swelling and opacity. Thus, the maintenance of corneal transparency requires an outward active transport of ions (with a concomitant osmotic movement of water) across the epithelium or endothelium, or both. In the mammalian cornea, the exact nature of the active processes involved is still not completely known (Davson, 1980; Zadunaisky, 1971). In the amphibian cornea an active transport of chloride from stroma to tears is a major factor in the regulation of stromal hydration, and a similar process seems to be a significant component of the mammalian regulatory system (Zadunaisky, 1971).

A peculiarity of the cornea is that it can be readily transplanted from one individual to another of the same species (*homografted*), perhaps because there is a relative lack of antigenic responses in the cornea (Davson, 1980). Whatever the reason, the importance of corneal transplants in the treatment of visual defects due to corneal injury or dysfunction is, of course, well known.

The cornea and conjunctiva are protected by a fluid film of tears that is secreted continuously by the lacrimal glands, together with mucous and oily secretions from other secretory organs in the conjunctiva and eyelids. Without this protection the cornea would rapidly dehydrate through loss of water by evaporation. There appears to be no secretion of tears during sleep, and secretion decreases markedly with advancing age. Reflex secretion of tears occurs in response to a variety of stimuli, e.g., irritation of the cornea, conjunctiva, or nasal mucosa; thermal stimuli (including "hot," peppery foods) applied to the mouth and tongue; excessively bright lights; and vomiting, coughing, and yawning. *Psychic weeping* occurs as a result of emotional upsets. A drainage system consisting of the *canaliculi,* the *lacrimal sac,* and the *nasolacrimal* duct drains off secretions that exceed the loss due to evaporation. The formation of tears is called *lacrimation.* Tears serve to wash away irritating particles and fumes and also contain a bactericidal enzyme, *lysozyme.*

Blinking, a rapid temporary closure of the eyes, occurs reflexly in response to actual or potential mechanical insults. It is effected by contraction of muscles in the upper and lower eyelids. In addition to protecting the eyes from specific hazards, blinking spreads the protective film of lacrimal secretions over the cornea and conjunctiva. During waking hours, spontaneous blinking occurs every few seconds. Defective closure of the eyelids or defective lacrimal secretion can give rise to the pathological condition known as *keratitis sicca.* Reflex blinking can be as fast as 0.1 second.

The aqueous humor has an ionic composition basically similar to that of the blood, but it has a much lower protein content (about 5–15 mg/dl in humans). Although the details of the process are unknown, it seems well established that aqueous humor is secreted continuously by the cells of the ciliary epithelium into the posterior chamber, whence it flows into the anterior chamber. Because it is produced continuously, the aqueous humor must be drained away continuously if the *intraocular pressure* is to remain normal. The *canal of Schlemm* (Fig. 4-1) is an important element in the drainage system of the eye; through this canal, fluid drains into veins leaving the eye.

Normal intraocular pressure in the human is about 15 mm Hg (Davson, 1980), with no significant difference between the sexes and no appreciable correlation with age. Chronically elevated intraocular pressure (resulting, for example, from a defect in intraocular drainage) can give rise to *glaucoma.* Glaucoma can cause irreversible damage to certain structures of the eye, including the retina, and is a major cause of blindness. For this reason the measurement of intraocular pressure is of great importance in ophthalmological practice. It is done with an instrument called a *tonometer.* Clinical tonometers are of two types, *impression tonometers* and *applanation tonometers.* Impression tonometers measure the depth to which a weighted plunger depresses the cornea. Applanation tonometers measure the area of flattening when a metal surface is applied to the cornea with a known force (see Davson, 1980, and Campbell et al., 1974, for further details).

The lens is a transparent biconvex entity consisting of an outer elastic envelope, the *capsule,* which encloses the *lens substance.* The latter is composed of fibers and interstitial cements. In addition, the *epithelium,* a single layer of cells, covers the anterior surface of the lens substance. The tension of the capsule tends to make the lens assume as spherical a conformation as possible. This tendency is counteracted by the pull of the fibers of the ciliary zonules around the lens equator, which help give the lens a more flattened form. With respect to its internal composition the lens behaves as if it were a single cell; i.e., the lens substance has a high K^+ and a low Na^+ and Cl^- content. The maintenance of the internal ionic composition of the lens depends on metabolism; cooling or metabolic inhibitors, for instance, decrease internal K^+ and increase internal Na^+ and Cl^-. It has been suggested that the lens behaves like other epithelial systems such as the amphibian skin (see Chap. 1) in that the active transport of solutes from the interior to the external medium is largely controlled by the epithelium.

The transparency of the lens, like that of the cornea, seems to depend on the maintenance of a very precise arrangement of the fibers within the lens substance, and this in turn probably depends on the maintenance of a normal salt and water content. Lack of transparency in the lens is the condition called *cataract.* It may be localized in spots or may involve the whole of the lens substance. Cataract may result from metabolic or nutritional defects, from trauma (including radiation damage), or simply from age. The last condition (senile cataract) is, unfortunately, rather frequent in occurrence. Abnormalities in calcium metabolism, such as may arise from parathyroidectomy or rickets, can lead to cataract, suggesting a role of calcium in the maintenance of lens transparency.

The structure of the retina is considered under The Retina.

PHYSICAL PROPERTIES OF LIGHT

The visible light spectrum ranges from 400 to 800 *nanometers* or *millimicra* (1 nm = 1 mμ = 1×10^{-9} m), with blue or violet light having the shortest wavelength and red the longest. Immediately adjacent to the visible spectrum are the areas of ultraviolet and infrared radiation. Although not perceived by the eye as such, radiation in these areas is biologically important; for example, ultraviolet radiation is the cause of the tanning and burning effects of the sun on the skin. Infrared radiation has heating effects on the body.

From the time of Newton, scientists have debated whether light is a stream of corpuscles having the properties of matter, or a form of wave reflected or emitted from material objects. Careful experimentation has shown that light actually has properties of both matter and waves and consists of electrical and magnetic energy that travels in packets, or *photons,* at a speed of 186,000 miles per second.

The wavelike properties of light are of particular significance in considering the optical function of the eye. They enable light to be focused, reflected, and refracted like other waves. One important property of wave phenomena is *Huygens's principle,* which states that *any point on a wave front may be regarded as a new source of light.* For example, if light passes through a narrow slit, the rays fan out when they leave the slit, as if that slit were a new source of light. If a screen is placed beyond the slit to intercept the rays, they form a central image on the screen with less intense secondary images of the slit on either side of the central image (Fig. 4-2A). This formation of a *diffraction pattern* is a consequence of the wave nature of light.

Diffraction is of practical importance because it limits the sharpness of the image and consequently the *degree of optical resolution.* For example, if two objects are very close together, their diffraction patterns may overlap and it may be impossible to separate them optically, no matter how much the image is magnified. Ideally, the *limit of resolution* or *resolving power* for any optical instrument that is diffraction-limited is about one-half the wavelength of the light it uses.* In practice, resolving power may be considerably less.

REFRACTION AND IMAGE FORMATION

When light rays strike an object, they may be reflected, absorbed, or transmitted through it. Even rays that are transmitted do not escape some alteration. The entering light rays are bent to an extent, depending on the angle at which they strike the surface and on the *refractive indices* of the surrounding medium and of the object. The refractive index (n) of any optical medium is formally defined as the *ratio of the velocity of light in a vacuum to its velocity in the*

---

*This is one reason why the electron microscope is such an important tool in cellular biology. The waves associated with a high-energy electron beam are much shorter than those of visible light. Thus, one can visualize subcellular structures that are far too small to be separated from their surroundings by the ordinary light microscope.

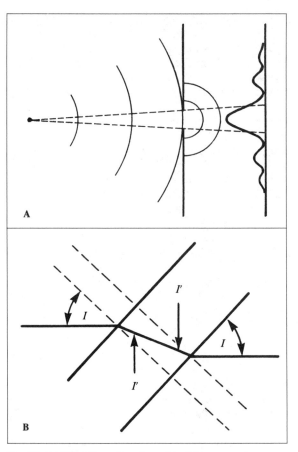

Fig. 4-2. (A) Diffraction pattern formed by light rays passing through a narrow slit. Dashed lines indicate the path that light rays would follow if diffraction did not occur. (B) Bending of light rays at the points of entrance and emergence from a medium with a higher refractive index. I and I' are angles of incidence and of refraction.

given medium. In practice, air rather than a vacuum may be taken for reference purposes, since the refractive index of air (1.0003) is close to unity.

As light rays pass from one medium into a second medium with a higher refractive index (i.e., is more optically dense), the rays are bent toward a perpendicular drawn through the surface between the media. Conversely, as the rays emerge from the more to the less optically dense medium, they are bent away from the perpendicular (Fig. 4-2B). The degree of bending, or *refraction,* depends on the refractive indices of the two media and the angle at which

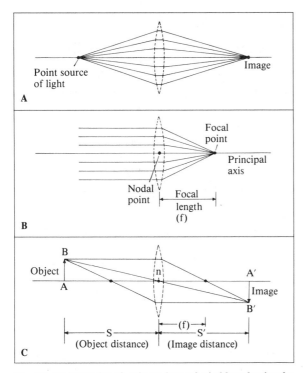

Fig. 4-3. (A) Formation of an image by a spherical lens showing the greater deviation of peripheral rays. (B) Illustration of focal length as the distance from the lens at which parallel rays are brought to focus. (C) Construction of an image when lens strength, object size, and object distance are known.

the light rays strike the surface between the media. Quantitatively, this is expressed in *Snell's law of refraction*,

$$n \sin I = n' \sin I' \tag{1}$$

where n and n′ are the refractive indices of the two media, I is the *angle of incidence,* and I′ is the *angle of refraction* (Fig. 4-2B). One notes from equation 1 that, given constant values of n and n′, it is the ratio of the sines of these angles, not the ratio of the angles themselves, that remains constant. Hence, the degree of refraction decreases as I becomes larger. When I is maximal—i.e., 90 degrees—there is no refraction.

The property of refraction is basic to the operation of lenses, as may be seen in Figure 4-3A. The central rays from the point source of light strike almost perpendicularly to the lens surface and are bent only to a small degree, whereas the peripheral rays strike the surface at an acute angle and bend considerably more. If the surface of the lens is spher-

ical, the transmitted rays intersect at a point on the other side of the lens to form an image of the point source. The lens illustrated is a converging, convex spherical lens. The surface of such a lens is a portion of a sphere; i.e., it has the same curvature in all planes. One or both surfaces may be spherical. A lens that causes light to diverge is constructed by making the spherical surfaces concave rather than convex.

Since refraction depends on the angle of incidence, the light rays passing through the peripheral portion of a spherical convex lens do not focus at exactly the same point as those passing through the center. This phenomenon is called *spherical aberration.* An additional limitation is the variation of the refractive index with the wavelength of the light used, blue light being bent more than red. Thus, if white light is used, the rays of different wavelengths form separate images, and an expanded image is produced. This characteristic is termed *chromatic aberration.** The optical system of the eye possesses both spherical and chromatic aberration to some degree.

GEOMETRICAL OPTICS

Geometrical optics is the study of the properties of lenses and the formation of images. Two important applications of this branch of optics are the determinations of the strength of a lens and of the size and position of the image it forms. When parallel rays of light fall on a lens, they come to focus at a point called the *focal point,* as in Figure 4-3B. The *focal length* is the distance from the focal point to the lens. The focal length is a convenient measure of the strength of a lens. In physiological optics the lens strength is usually expressed in *diopters,* the reciprocal of the focal length expressed in meters.

Diopters (D) = 1/focal length (meters)

For example, if a lens has a focal length of 0.1 meter, its power is 10 D.

IMAGE FORMATION AND SIZE

The position at which an image is formed by a lens can be determined geometrically by tracing the path of several rays through the lens, as shown in Figure 4-3C. The determination requires only that object size, object distance, and focal

*In high-quality optical instruments, complex combinations of lenses are employed to reduce both spherical and chromatic aberration.

length of the lens be known. There are three rays whose paths can be readily traced. First, a ray that leaves a point source (the end of the arrow) and passes through the center of the lens is bent as it enters the lens and again as it leaves. Because the two surfaces of the lens are parallel at the center, there is little deviation, and the ray passes on, its direction essentially unchanged. Second, a ray that proceeds parallel to the principal axis is bent sufficiently to pass through the focal point on the principal axis on the image side of the lens. As this ray proceeds further, it intersects the ray passing through the center of the lens, and an image is formed at the intersection. Third, a ray that passes through the focal point on the object side of the lens is bent just enough to meet the other two rays at their point of intersection.*

If the lens is made stronger or the object is moved farther from the lens, the image is formed closer to the lens. The object distance, image distance, and lens strength are therefore interrelated. This relation is expressed by the *lens formula*

$$\frac{1}{S} + \frac{1}{S'} = \frac{1}{f} \tag{2}$$

where S is the object distance, S' is the image distance, and f is the focal length.

Example 1. If an object 20 cm from the lens forms an image 20 mm from the lens, what is the lens strength? First, all values should be expressed in meters. The lens strength can then be obtained in diopters, as follows:

$$\frac{1}{0.2} + \frac{1}{0.02} = \frac{1}{f}$$

$$\frac{1}{f} = 55 \text{ D}$$

Example 2. If an object is placed 40 cm from a lens of 50-D strength, what is the image distance?

$$\frac{1}{0.4} + \frac{1}{S'} = 50$$

Multiplying by 0.4S' gives

$$\frac{0.4S'}{0.4} + \frac{0.4S'}{S'} = 50 \times 0.4S'$$

$$S' = 0.021 \text{ meter}$$

If a point source of light is placed at the focal point of a lens, the rays are parallel after passing through the lens and do not come to a focus (or focus only at an infinite distance from the lens). If the source is moved away from the focal

*Note that the image formed by a simple spherical lens is inverted.

point, an image is first formed at a great distance from the lens, gradually moving closer as the source retreats from the lens. When the source is an infinite distance from the lens, the light rays falling on the lens are parallel, and the image forms at the focal point. This can be shown by the lens equation. Since S is infinite, the equation becomes

$$\frac{1}{S'} = \frac{1}{f}$$

For any optical system there is a point beyond which the object can be considered for all practical purposes to be at infinity. For the eye this distance is 6 meters (20 ft). Any further movement beyond that point has little effect on the image distance.

The size of the image is directly proportional to the size of the object and the ratio of the image and the object distances. For example:

if

AB = object length

and

A'B' = image length

then

$$\frac{A'B'}{AB} = \frac{S'}{S}$$

These relations can be seen in Figure 4-3C and follow from the fact that ABn and A'B'n are similar triangles.

OPTICAL FUNCTION OF THE EYE

The basic optical principles that have been outlined can be applied to the function of the eye. The structure of the eye is shown in Figure 4-1. Light rays striking the eye first enter the *cornea*, which has a high degree of curvature and a refractive index of 1.33. These two factors combine to cause a considerable bending of the light rays as they enter the cornea. The resting eye has a power of 67 D. Most of this optical power (45 D) is in the cornea. When one swims underwater, most of the refractive power of the cornea is lost, since water has a refractive index of 1.33. After transversing the cornea, light passes through the *anterior chamber* (refractive index 1.33) without further alteration but is bent when it enters the *crystalline lens* of the eye. The lens has a graded refractive index that varies from 1.41 at the center to 1.39 at the periphery. This graded index decreases the power of the lens, making it equivalent to one having a uniform index of 1.39. The power of the lens is approximately 20 D at rest but may be increased by as much as 12 D by accommodation. *Accommodation* is the process by which the refractive power of the eye is modified for viewing nearby objects. It is produced by a thickening of the center of the lens, mostly at the anterior surface.

As already mentioned, the lens has an elastic capsule that tends to make it assume a spherical shape. However, this tendency is opposed by the tension within the *sclera,* or outer coat of the eyeball. The tension is due simply to the intraocular pressure, normally about 15 mm Hg, and is transmitted to the lens by the ring-shaped *ciliary muscle,* which encircles the lens, and by the *suspensory ligaments,* which are attached to the periphery of the lens. The tension applied to the lens by the ligaments flattens it by a stretching process. However, when the ciliary muscle contracts, its diameter decreases, thereby decreasing the tension on the suspensory ligaments and allowing the lens to assume a more spherical shape. Thus, accommodation depends on the elastic properties of the lens capsule. It is illustrated in Figure 4-4.

Associated with accommodation of the lens is a constriction of a sphincterlike *pupillary muscle* that eliminates rays passing through the peripheral portions of the lens. This action decreases spherical aberration, which might otherwise be a limiting factor in near vision. When one observes a nearby object (accommodation required), the axis of the eye is shifted by the *extrinsic muscles* to train both eyes on the object (see Binocular Vision).

For most purposes the optical properties of the compound lens system of the eye may be considered equivalent to those of a single refractive surface, the *optical center* or *nodal point* of which is situated 5 mm behind the anterior corneal surface and 15 mm in front of the *retina.* This simplified model, called the *reduced eye of Listing,* makes it possible to apply the lens equation to the eye. The normal power of the eye and the power of accommodation may be readily determined. For example, in the normal (*emmetropic*) eye at rest, distant objects are brought to focus on the retina. If the object distance in this case is considered to be infinite, equation 2 becomes

$$\frac{1}{\infty} + \frac{1}{S'} = \frac{1}{f} \text{ or } \frac{1}{S'} = \frac{1}{f} \tag{3a}$$

Since the image is focused on the retina, $S' = 15$ mm; therefore,

$$\frac{1}{.015} = \frac{1}{f}$$

giving 67 D as the optical strength of the resting eye.

The maximal optical strength of the eye can be determined by measuring the shortest distance at which an object may be seen distinctly. This is called the *near point* of vision. In the young adult the distance is about 10 cm. The maximal strength is then

$$\frac{1}{0.10} + \frac{1}{.015} = \frac{1}{f} \tag{3b}$$

$$\frac{1}{f} = 77 \text{ D}$$

The difference in strength between the resting and the maximally accommodated eye is the *power of accommodation.* This is about 12 D in children and 10 D in young adults. With increasing age the elasticity of the lens decreases, thereby reducing the power of accommodation and causing the near point to recede (Table 4-1). The decrease in accom-

Fig. 4-4. Contraction of ciliary muscles on viewing near object. This causes the lens to become more spherical (accommodation) and the iris to constrict, decreasing the area of its central opening (the pupil).

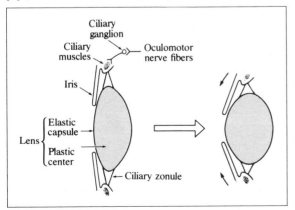

Table 4-1. Effect of Age on Amplitude of Accommodation and Near Point

Age (years)	Amplitude (D)	Near point (cm)
10	11.3	8.8
20	9.6	10.4
30	7.8	12.8
40	5.4	18.5
50	1.9	53.6
60	1.2	83.3
70	1.0	100.0

modation with age is known as *presbyopia*. It is an inevitable result of the aging process. Between ages 40 and 50 the near point normally recedes beyond arm's length, making reading glasses necessary.

OPTICAL ABNORMALITIES

Two of the most common optical defects result from an abnormal size of the eyeball: hyperopia, or farsightedness, and myopia, or nearsightedness. *Hyperopia* (or *hypermetropia*) is characterized by an abnormally short eyeball with a decreased distance from lens to retina. Much less frequently it is due to insufficient refractive power of the optical system. This defect causes the image of a distant object to be formed behind the retina in the resting eye (Fig. 4-5). The hyperopic person can focus the image on the retina by partially accommodating the lens. This, of course, is not a normal situation and leads to eye fatigue. Because some accommodation is required even for distant viewing, the near point is more distant than normal, and near vision is deficient—hence the descriptive term *farsightedness*. The hyperopic condition can be remedied by placing the appropriate convex spherical lens before the eye (Fig. 4-5).

Myopia is characterized by an abnormally long eyeball with an increased lens-to-retina distance. Occasionally, this condition is produced by an abnormally great curvature of the cornea or lens. The defect causes the image of a distant object to be formed before the retina in the resting eye (Fig. 4-5). There is no simple physiological way by which the defect can be compensated, since the normal process of accommodation only aggravates the deficiency. Vision of near objects is not impaired, however, and the near point is closer than normal. Myopia can be remedied by placing the appropriate concave spherical lens before the eye (Fig. 4-5).

Astigmatism is a common optical defect that is most often due to an abnormal curvature of the cornea. Normally, the corneal surface is spherical. In astigmatism the surface is ellipsoid, or egg-shaped (Fig. 4-6), so that the rays traveling in one plane are bent more drastically than those in another. As a result, the rays in one plane may focus on the retina while those in another do not. Astigmatism can be corrected by a *cylindrical* lens. Such a lens may be thought of as a portion of a cylinder that is cut longitudinally. If the lens is viewed from above, the curvature is seen in cross section (Fig. 4-6). Rays traveling in the horizontal plane also "see" the lens as a curved surface and are bent accordingly. Rays traveling in the vertical plane "see" the lens as a rectangu-

Fig. 4-5. Optical dimensions of the normal and abnormal eye. In the emmetropic eye the nodal point of the lens system is 15 mm before the retina. In the hyperopic eye the nodal point is less than 15 mm before the retina, causing the focal point to fall behind the retina. This defect is corrected by a spherical converging lens placed before the eye. In the myopic eye, rays focus before the retina. This defect is corrected by a spherical diverging lens.

Fig. 4-6. Optic features of astigmatism. In the uncorrected eye, light rays (shown by solid lines) focus behind the retina in one plane and on the retina in another. The defect is corrected by a cylindrical lens so placed that the curvature is only in the plane exhibiting the defect.

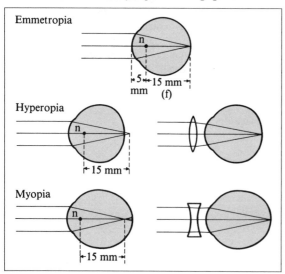

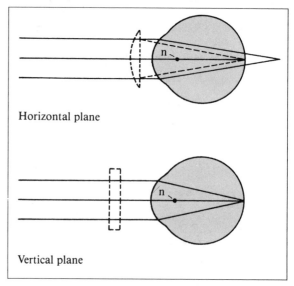

lar object and are not bent as they pass through it. By use of a cylindrical lens, the rays not focused on the retina may be brought to focus at that point. Thus, it is necessary to place the curvature of the correcting lens in the same plane as the rays having an abnormal focal point. The longitudinal axis of the correcting lens then is perpendicular to the plane of the astigmatism.

THE OPHTHALMOSCOPE

An ophthalmoscope is a device used for examination of the optical and physical properties of the eye. It consists essentially of a light source and a mirror or prism that reflects light into the eye and onto the retina. Some of the light is reflected from the retina, which becomes, in effect, a new light source. The light passes out of the eye through the lens and cornea. If the eye is normal and relaxed, the focal point of the lens and cornea coincides with the position of the retina. Consequently, light rays reflected from the retina are bent just enough to be rendered parallel as they pass through the lens and leave the eye; if the eye is myopic, the rays are convergent.

The degree of abnormality of the eye may be readily determined by a variety of means with the ophthalmoscope. For example, if the parallel rays emerging from the emmetropic eye pass through a +1 diopter lens, an image of the retina is formed 1 meter from the lens. An additional converging lens for the hyperopic eye or diverging lens for the myopic eye must be used to bring the image to this same point. The strength of the additional lenses required is a measure of the abnormality that exists. This method illustrates the principle underlying the use of the ophthalmoscope. At present, more indirect methods permitting rapid precise measurement of the degree of abnormality are utilized (see Campbell et al., 1974).

THE RETINA

The retina (see Fig. 4-1) is an extremely complex organ consisting largely of nervous tissue. It is in fact an outgrowth of the central nervous system. Two special areas of its posterior aspect should be noted: the *optic disk* and the *central fovea* (see Fig. 4-1). The optic disk is the region where the fibers of the *optic nerve* leave the retina through the *optic foramen* (a canal in the bony socket of the eye) and where the *central artery of the retina* (a branch of the ophthalmic artery) and *retinal veins* enter and leave the retinal region. The optic disk contains no visual receptor and is therefore a *physiological blind spot*. The *fovea* is a small depression in the retinal surface, important in relation to visual acuity (p. 119). The fovea and its immediate surroundings form the *yellow spot* or *macula lutea,* so called because of its yellow pigmentation.

The general structure of the nervous elements of the retina is illustrated diagrammatically in Figure 4-7. Three main layers of cells can be distinguished. These are (going from the inside of the retina toward the outer or vitreous humor side) the *receptor layer,* consisting of *rod cells* (rods) and *cone cells* (cones); a second layer, the *bipolar cells* or *inner nuclear layer;* and a layer of *ganglion cells.* Note that light entering the eyeball must pass through the ganglion layer and bipolar layer before impinging on the photoreceptor layer. In addition to the nerve cells just enumerated, the retina contains numerous *neuroglial* cells that act as supporting and insulating elements.

The rods and cones are so called because of their characteristic shapes. Both have the same general structure, but the rods are usually much thinner than the cones. The light-sensitive pigment is contained in the *outer segment* (OS, Fig. 4-7). At the other end of these cells are the *synaptic bodies,* which cause the photoreceptor cells to synapse with the bipolar cells. The synaptic regions contain many *synaptic vesicles* (S); these are thought to release chemical transmitters. From the bipolar cells, impulses are transmitted to the ganglion cells and hence to the higher centers, the *lateral geniculate body* and the occipital lobe of the cerebral cortex. The axons of the ganglion cells form the fibers of the optic nerve.

The neural pathways in the retina are complex, and their detailed analysis is far from complete. The general pattern may be outlined as follows. The bipolar cells are of two main types, *midget bipolars* (MB, Fig. 4-7), each of which synapses with one cone, and *diffuse bipolars*. The latter are divided into *rod bipolars* and *flat bipolars* (RB and FB, Fig. 4-7). Rod bipolars synapse with up to 50 rod receptors. Flat bipolars connect with about seven cones. Bipolars that synapse with both rods and cones do not seem to occur. Similarly, one finds two types of ganglion cells, *midget ganglion* cells (MG, Fig. 4-7), which synapse with a single midget bipolar, thus providing an exclusive pathway from a single cone to the higher neural centers, and *diffuse ganglion* cells (DG, Fig. 4-7). The latter connect with a number of bipolar cells and may therefore respond to stimuli originating from both rods and cones.

Thus, the vertical pathways through the retina display both *divergence* (stimulation of several bipolar cells by a single receptor and, in turn, stimulation of several ganglion cells by a single bipolar) and *convergence;* i.e., a single ganglion cell may respond to a number of receptors. The lateral spread of nervous impulses through the body of the

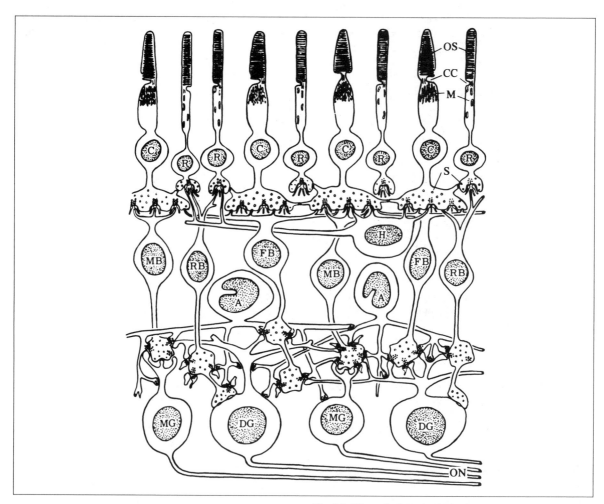

Fig. 4-7. Schematic diagram of the retina showing rod cells (*R*), cone cells (*C*) and their specialized structures, the outer segment (*OS*), connecting cilium (*CC*), mitochondria (*M*), and synaptic vesicles (*S*). Other structures shown are midget (*MB*), rod (*RB*), and flat (*FB*) bipolar cells, together with midget (*MG*) and diffuse (*DG*) ganglion cells. The axons of the ganglion cells are the fibers of the optic nerve (*ON*). Horizontal cells (*H*) and amacrine cells (*A*) are interneurons.

retina is thus assured. This spread is greatly assisted by the action of two kinds of interneurons, *horizontal cells* and *amacrine cells*. Overall, there would appear to be a great deal of convergence in the human retina, since the number of rods (about 125 million) and cones (about 6 million) is much greater than the number of ganglion cells (about 1 million).

SCOTOPIC VISION AND PHOTOPIC VISION

The rods contain a reddish purple pigment called *rhodopsin,* or *visual purple,* which is bleached on exposure to light. The bleaching process causes excitation of the receptor by

means not yet understood. The bleaching and regenerative processes involve changes in the rhodopsin molecule that can be studied in vivo or in vitro. The first stage in the bleaching process is the formation of orange pigments (*lumirhodopsin* followed quickly by *metarhodopsin*), which then split into *retinene* and *opsin* (visual yellow) (Fig. 4-8). At first, the retinene is in the *trans* isomer form. Some of this retinene is converted to the *cis* form by a photochemical

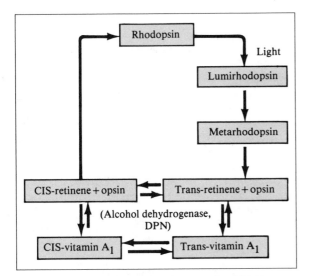

Fig. 4-8. Degradation of rhodopsin when exposed to light. Resynthesis from cis-retinene and opsin to rhodopsin occurs in darkness.

process. Cis-retinene in a solution containing opsin is converted in the dark to rhodopsin. Only one of the several cis isomers (11-cis) is the precursor of rhodopsin. The trans-retinene that is not isomerized to the cis form is reduced to vitamin A_1. Before the regeneration of rhodopsin can occur, vitamin A_1 must be isomerized from the trans form to the cis form. This action occurs elsewhere in the body by a process not yet understood. Meanwhile, the active form of vitamin A_1 is withdrawn from the blood, to be used in regeneration of visual purple.

Rhodopsin is most readily bleached by light having a wavelength of 500 nm. Its sensitivity falls off on either side of 500 nm; it is about 40 percent as sensitive to light at the blue end of the spectrum. The sensitivity to light in the red region is very low. For this reason, wearing red-tinted glasses allows full adaptation of the rods to night vision even while general illumination is normal. Since some cones are stimulated by red radiation, they do not adapt under these circumstances.

Photopic vision is the term for the visual function performed by the cones. The cones, as a group, are most sensitive to light in the region of 550 nm. The sensitivity of the cones extends throughout the entire visible spectrum but is much less at the two ends of the spectrum than at the center. This is the reason yellow seems much brighter than blue or red. This spectral variation in sensitivity appears to be a summated effect of several different types of cones, each of which responds to a limited portion of the spectrum.

VISUAL ACUITY

The ability of the eye to detect a separation between two adjacent objects depends on the retina's ability to perceive a separation between the images that fall on it. The separation of two adjacent images on the retina depends on the adequacy of illumination, the fidelity with which the light rays are transmitted by the optical system, and the diffraction pattern of the image on the retina. In addition, the density of the packing of the retinal receptors is a determining factor in the resolution of images. If the separation between images on the retina is to be perceived, a receptor must be present in the intervening space. Since the diffraction pattern tends to diffuse the edges of the image, the separation may not be clear-cut. There may be, in fact, a "gray" zone between two images, which the retina must recognize as a separation.

In viewing nearby objects, the iris of the eye constricts to block peripheral rays and minimize spherical aberration. The visual area of the cortex is apparently able to compensate in some fashion for chromatic aberration. Thus, the resolving power of the eye is determined primarily by retinal "grain" or receptor density and the unavoidable diffraction of rays.

The minimal separation between images that can be detected may be conveniently expressed in terms of the visual angle. This is the angle the separation subtends in the visual field. The minimal visual angle for the normal eye is approximately 1 minute. The images of objects undergoing close scrutiny are projected onto the *macula lutea* or yellow spot, the portion of the retina where the *cones,* the receptors subserving detail vision, are most densely packed. In other portions of the retina the visual angle is greater because the cone population is less dense—and for other reasons that will be discussed shortly.

In clinical studies, visual acuity is determined by use of test charts or letters. The patient is situated 20 feet from the chart, so that the emmetropic eye will not be accommodated. The Snellen test chart, which is most often used, is composed of block letters so constructed that the width of the trace and the separation between limbs of the letter are standardized. The letter E, for example, is so constructed that the width of the horizontal bars is equal to the separation between them. The chart contains lines of various sizes of type. The test is carried out by having the patient read the smallest type possible from the 20-foot distance, each eye being tested separately. For example, if the patient can read at 20 feet what the normal person can read at that distance, his or her acuity is expressed as 20/20 and corresponds to a visual angle of 1 minute. If the smallest type the patient can discern at 20 feet corresponds to what the normal person can read at 40 feet, his or her acuity is expressed as 20/40 and the visual angle is 2 minutes.

Visual acuity is not uniform over the entire retina. It is greatest in the central region, and particularly in the *fovea,* where the photoreceptors are most closely packed. The central area of the fovea contains only cones, which are the receptors concerned with fine detail and color vision. In addition, the synapses of the cones with the bipolar cells and the connections of the latter with their ganglion cells are of the one-to-one midget type. Other factors contribute to sharp definition of the image in the foveal region: The bipolar and ganglion cell layers are displaced laterally, and the retinal blood vessels largely bypass the area. Thus, entering light can impinge more directly on the receptor cells. Furthermore, the yellow pigment of the macula lutea helps to reduce loss of light by reflection. As a result, visual acuity in the central fovea is about twice that just outside the macula lutea and about 40 times that at the retinal border.

As one moves away in a peripheral direction from the fovea, the cones become less abundant and the rods more abundant. Convergence also increases. Vision in the peripheral region of the retina is therefore indistinct, but overall light sensitivity is increased, since a number of receptors can combine to stimulate a single ganglion cell.

Thus, only those images that are projected onto the fovea are perceived clearly and sharply by the brain. However, the peripheral parts of the retina play an important role in *detecting* objects or movement within the visual field. Once an object is detected peripherally, the eyes, head, or both are rapidly turned to bring the image onto the fovea and into sharp focus.

Visual acuity and the contrast between an object and its background may be intensified by mechanisms within the retina. When a point source of light is focused on a portion of the retina, some fibers from this area that were previously silent begin to fire. These are called *on* fibers. Other fibers, previously discharging, become silent. These are termed *off* fibers. Still other fibers rapidly adapt to existing light, discharging for a brief period at the beginning and after the end of a light stimulus. These are the *on-off* fibers. Movement of an object across the retinal field leaves a trail of on-off fiber activity in its wake. The fact that the on-off fibers are more numerous than the other two types may account for the great ability of the eye to detect small movements. Normally, the eye does not fix on an object persistently but scans an area by movements that are rapid and small in amplitude. Apparently, this scanning action helps maintain visual acuity, since the image begins to fade if the movements are blocked. The on-off response may be important

in this respect, for the scanning tends to create a halo of on-off impulses at the border of the image. There are also lateral neuronal connections in the retina that may be important in visual acuity. When a small area is illuminated, inhibitory impulses sent to surrounding nonilluminated areas tend to intensify the image.

Although visual acuity is lower in the peripheral regions of the retina than in the macula lutea, night vision is better because of the increased population of rods, which are specifically adapted for night vision. The rods are much more sensitive to low levels of illumination than the cones but have a lower level of visual acuity. (Rod vision is about the same throughout the retina and is one-twentieth as acute as the cone vision at the fovea centralis.) The lower visual acuity is due, at least in part, to the convergence of a number of rods on a single line to the cortex.

ADAPTATION TO ILLUMINATION

If a bright light is shone into the eye, the pupil immediately constricts. This is the *light reflex,* which is initiated at the retina and passes by way of the pretectal region and the ciliary ganglion to cause contraction of the sphincter pupillae muscles. If only one eye is stimulated, the reaction occurs in the pupil illuminated—the *direct light reflex*—and in the opposite pupil as well—the *consensual light reflex.* In tabes dorsalis the light reflex may be lost, while the pupillary reaction during accommodation remains unimpaired. This condition is called the *Argyll Robertson pupil.* The reaction of the pupil to light is usually a temporary phenomenon whose purpose is to protect the retina from too intense illumination. As time passes, the retina adapts to the new level of illumination, and the pupil returns to its original size. The maximal diameter of the pupil is 8 mm, and the minimal diameter is about 1.5 mm. Thus, the area, and therefore the amount of light entering, can be changed 30-fold. However, the eye is able to adapt to changes in light intensity of about 10-billion-fold because of changes in the retina.

The reactions of the retina to changes in light intensity may be more readily understood by considering the adaptation to darkness. If an individual enters a dark room, there is a period during which he perceives nothing followed by one in which the objects around him gradually become more discernible. If, during this time, periodic measurements of the light threshold or the just perceptible light stimulus are made, a gradual decrease in this threshold is found, as shown in Figure 4-9. Both the rods and the cones begin to adapt immediately. The cones initially adapt more rapidly, so the threshold measured in the early stages of adaptation

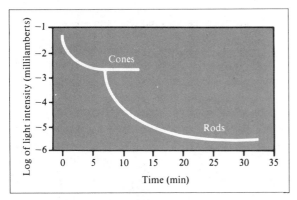

Fig. 4-9. Adaptation of rods and cones to darkness. The change in light threshold is shown on a logarithmic scale.

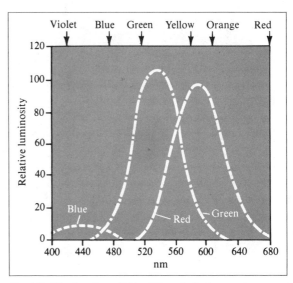

Fig. 4-10. Wavelength sensitivity of the retinal color receptors according to the Young-Helmholtz theory.

is essentially that of the cones alone. The cones increase their sensitivity 20-fold to 50-fold in the first 5 minutes, after which little further adaptation of these receptors occurs. The rods, on the other hand, adapt more slowly but to a much greater degree. After the cones have reached their maximal sensitivity, the threshold of the eye as a whole continues to decrease, because of the adaptation of the rods. Since these adaptations proceed at different rates, the dark adaptation curve for the eye breaks sharply at the point where the rods become more sensitive than the cones. The adaptation of the rods is essentially complete in 25 to 30 minutes, although it may continue slowly for several hours thereafter. In passing from a dark area to a light one, the process is reversed. The two-step change in sensitivity is not seen; the photosensitive material in the rods is immediately bleached, and the cones then determine the light threshold.

The process of adaptation is a manifestation of the properties of the photosensitive chemical substance in the light receptor. The nature of this substance has been studied in much more detail in the rods than in the cones.

COLOR VISION

The most widely accepted theory of color vision is the Young-Helmholtz theory. It is based on the assumption that three receptors, red, green, and blue (violet), subserve color vision. The sensitivity of these receptors, as deduced from indirect evidence, is shown in Figure 4-10. It can be seen that, although each has its maximum sensitivity in a certain portion of the spectrum, there is considerable overlap between the receptors. The sensation of blue is due to stimulation of the blue receptor alone. The sensation of blue-green

is due to stimulation of both the blue and the green receptors. On the other end of the spectrum, the sensation of red is due only to stimulation of the red receptor. However, shifting toward the green portion of the spectrum results in the perception of the sensations of orange and then yellow because of stimulation of both the red and green receptors. Thus, the various hues of the spectrum, plus the extraspectral color purple, may be differentiated on the basis of the three color receptors. If all the receptors are stimulated simultaneously in the correct proportions, the sensation of white is perceived.

For 150 years the Young-Helmholtz theory, first advanced in 1803 by Stephen Young, was without direct substantiation. Indirect evidence from studies of color perception had made it possible to formulate the relationship between the three hypothetical receptors shown in Figure 4-10. However, in 1964, experiments were performed that showed conclusively that such color receptors do exist. Brown and Wald succeeded in measuring the absorption spectrum of single rods and cones in an excised portion of human retina. They found that the rods display an absorption peak at about 505 nm (Fig. 4-11). They also recorded the absorption of three types of cones: a blue-sensitive cone with an absorption maximum at 450 nm, two green-sensitive cones having absorption maxima at 525 nm, and a red-sensitive cone with its absorption maximum at 555 nm. The results of these remarkable experiments are shown in Figure 4-12. It should be noted that these absorption spectra

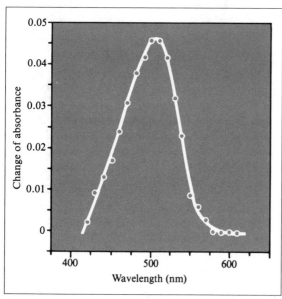

Fig. 4-11. The difference spectrum of a single rod in the human retina (parafoveal region). The spectrum was recorded in a darkened room using very low light intensities to determine the absorption. The experiment was repeated after bleaching with a flash of yellow light. The curve represents the difference between the two absorption spectra. (From P. K. Brown and G. Wald. *Science* 144:45–52, 1964. Copyright © 1964 by the American Association for the Advancement of Science.)

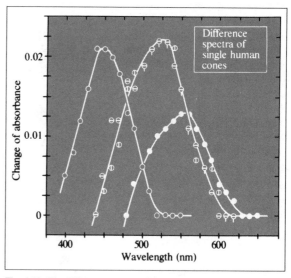

Fig. 4-12. The difference spectra of four cones in the parafoveal region of the retina. Difference spectra represent reduction in light absorption after bleaching with flash of yellow light. The three types of receptors have maximal absorbance at 450 nm (blue sensitive), 525 nm (green sensitive), and 555 nm (red sensitive). (From P. K. Brown and G. Wald. *Science* 144:45–52, 1964. Copyright © 1964 by the American Association for the Advancement of Science.)

agree well with the results obtained in color perception studies.

Equally convincing are the results obtained by Mac-Nichol (1964) on the goldfish retina. Using photometric techniques also, he was able to measure the absorption spectra of more than 100 individual cones and demonstrated that they fell into just three groups, having maxima at 455, 530, and 625 nm. Since this species is known to be color perceptive, the evidence strongly favors the trichromatic theory of color vision. MacNichol's findings also indicate that the mode of excitation of cones strongly resembles that of rods, with a pigment being bleached in the receptor when it is presented with an adequate stimulus.

COLOR BLINDNESS

Color blindness is the inability to perceive a portion of the spectrum or to distinguish between colors recognized by the normal person as being different. The confusion arises between colors that are adjacent in the spectrum. According to the Young-Helmholtz theory, color blindness is due to absence or reduced sensitivity of one or more of the three color receptors. It is present to some degree in 9 percent of males and 2 percent of females.

For example, an person may be red-blind. This condition is termed *protanopia* and is due to the absence or deficiency of the red receptor. Without red-sensitive cones the red portion of the spectrum is not detected. Apparently, the spectral range of the eye is also reduced, since the protanope does not perceive radiation in the red range, particularly beyond 640 nm. The classic example of lack of red perception is the protanope who appeared at a funeral wearing a bright red tie. The protanope has difficulty distinguishing between adjacent colors in the yellow, green, and orange spectral regions, but his or her perception of blue is normal.

Instead of outright lack of a color receptor, some persons may show a color weakness. For example, if this occurs in the red region, the person requires an abnormal amount of red and green to match an intermediate color such as yellow. This condition is known as *protanomaly*. Protanopia and protanomaly are the most common forms of color blindness.

A less common form of color blindness is *deuteranopia*,

which is due to lack of the green receptor. The entire spectrum appears to the deuteranope to be composed of yellows and blues. In this instance, the green receptor shifts to the red spectral range and gives the sensation of yellow in the yellow, green, orange, and red ranges. Thus, the deuteranope is unable to distinguish between adjacent colors in the region. A weakness of the green receptor is termed *deuteranomaly*. The deuteranope makes many of the same mistakes as the protanope in matching colors. However, the two conditions can be distinguished by the protanope's lack of perception of red.

Tritanopia is a rare form of color blindness that is due to lack of the blue receptor. The spectrum is shortened at the blue end for those with this defect.

The types of color blindness just described are due to lack of a single type of receptor. Persons having them are termed *dichromats*, since their entire spectrum is subjectively composed of two colors or combinations thereof. Normal persons are *trichromats*. In rare instances, two of the receptors are absent; persons with this defect are *monochromats*. They have no perception of color as such. Apparently, they detect only one color and gradations thereof. Total color blindness has also been observed. In this case the cones seemingly do not function at all. The spectral sensitivity curve for persons so afflicted is the same as for persons with scotopic vision (maximum at 500 nm).

VISUAL FIELD

The visual field is the portion of the external environment that is represented on the retina. The extent of this field is limited at most points by anatomical factors, such as the supraorbital ridges above and the nose medially. The temporal field is limited only by the orientation of the eye and the sensitivity of the peripheral portions of the retina.

The visual field is also limited to a small extent by the structure of the retina. Since there are no receptors at the exit of the optic nerve from the retina, images falling on this area are not perceived. This is the *blind spot* of the retina. It is situated 10 degrees from the fovea centralis on the nasal side of the retina and encompasses about 3 degrees of visual field.

The extent of the visual field projected onto the retina may be determined by use of the *perimeter*. The perimeter is a metal band shaped like a half-circle, and the subject is situated so that one eye is at its center. If the eye being tested is directed straight ahead at the middle of the band, the two ends of the perimeter are 90 degrees removed from the visual axis. If a small target is moved in from the periphery along the perimeter, the image moves from the periph-

eral to the central regions of the retina. The subject then indicates when the target is in view. By repeating this test along the horizontal, vertical, and several intervening meridians, one can determine the extent of the visual field. Normally, the visual field for the eye extends 50 degrees upward, 80 degrees downward, 60 degrees nasally, and more than 90 degrees temporally. The field is most extensive for a white target and becomes successively smaller when blue, red, and green targets are used. The extent of the color fields is said to be a relative matter, depending on the brightness of the target.

VISUAL PATHWAYS

In humans, each optic nerve contains about 1 million fibers. In the *optic chiasm*, fibers from the nasal region of each retina cross over; those from the temporal regions do not (Fig. 4-13). The nerve trunks that emerge proximally from the optic chiasm are called the *optic tracts*. The visual pathways that carry information from the retina to the cortex are most easily understood by considering the effects on vision of lesions in various parts of them (Fig. 4-13). First, however, one should note that images formed on the retina are inverted and reversed from left to right (the cortex compensates for this physical inversion). Thus, for example, an object in the lower-left-hand quadrant of the visual field will project an image onto the upper-right-hand quadrant of the retina. Also, images formed on the portions of the visual field of each eye that overlap are projected onto the same area of the cortex.

A lesion of the optic nerve produces blindness in the corresponding eye. A lesion at the optic chiasm involves the fibers from the nasal portion of each retina that cross to join contralateral temporal fibers at this point. The result is *hemianopia*, or blindness in one-half of the visual field of each eye. The blindness is referred to the *visual field* rather than the *retinal field*. Thus, this defect would be termed *bitemporal hemianopia*.

If the lesion occurs in the optic tracts, it interrupts pathways serving the same visual fields in each retina and produces *contralateral homonymous hemianopia*. Since the lesion is on the side opposite the visual field it affects, it is termed *contralateral*. In the instance shown in Figure 4-13 the lesion is on the right and the impaired portion of the field is on the left. *Homonymous* refers to the fact that the same side of the visual field is affected in each eye.

The fibers of the optic tract synapse in the *lateral geniculate body* of the thalamus. The latter sends fibers via the *geniculocalcarine tract* to the *visual cortex*, which is located in the occipital lobe of the cerebrum on the same side. A

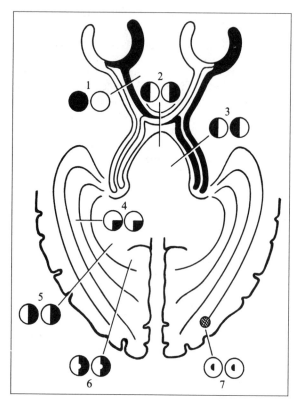

Fig. 4-13. The visual pathways showing sites of interruption of pathways and the resultant abnormalities in visual fields. (1) Optic nerve: blindness on the side of lesion, with normal contralateral field. (2) Optic chiasm: bitemporal hemianopia. (3) Optic tract: contralateral homonymous hemianopia. (4) Medial fibers of the optic radiation: contralateral inferior homonymous quadrantanopia. (5) Optic radiation in the parietal lobe: contralateral homonymous hemianopia. (6) Optic radiation in the posterior parietal lobe and occipital lobe: contralateral homonymous hemianopia with macular sparing. (7) Tip of the occipital lobe: contralateral homonymous hemianopic scotoma. (From D. O. Harrington. *The Visual Fields* [2nd ed.]. St. Louis: Mosby, 1964.)

lesion in the geniculocalcarine tract, as in the optic tract, may cause contralateral homonymous hemianopia. If the lesion is less extensive, only one quadrant is obliterated and *contralateral homonymous quadrantanopia* results. If the lesion occurs in the posterior parietal lobe and occipital lobe, a *contralateral homonymous hemianopia with macular sparing* results. The sparing of macular vision occurs because macular fibers terminate before the calcarine cortex. A lesion of the macular field on one side causes *hemianopic scotoma. Scotoma* is a generalized term referring to a blindness or weakness in a portion of the visual field.

BINOCULAR VISION

The extent of the visual field for a single eye has been noted previously. Most of the visual field is the same for both the right and the left eye. If both eyes fix straight ahead on an object, the visual axes of the eyes are directed toward the object, and the centers of the two visual fields coincide. All points within approximately 60 degrees of this center are seen by both eyes.

Each point on the visual cortex within the binocular field receives impulses from a point on each retina. These retinal points are called *corresponding points*. Each gives rise to the same sensation in the cortex. The retina-to-cortex connections are so arranged that when the foveas are trained on the same point, so also are the corresponding points on the two retinas. Thus, the same information is sent to the cortex from all corresponding points on both retinas.*

If an object is moved toward an observer, the orientation of the eyes must be shifted so that the foveas remain trained on the object and the corresponding points of the retinas remain matched. This is *convergence,* a reflex rather than a voluntary act. It is accomplished by contraction of an extrinsic striated muscle, the *medial rectus,* which draws the visual axis medially as the object approaches.

Each eye has six extrinsic muscles; they rotate the eyeball on the horizontal, vertical, transverse, and oblique axes. When the eyes are following a moving object, these muscles act in concert to keep the object trained on corresponding points of the retinas. They are capable of moving the eye approximately 50 degrees in any direction from the normal position. If, for some reason, the coordination between these muscles is lost, the images formed in the two eyes no longer fall on corresponding points. A condition known as *diplopia* then exists, in which a double image is projected onto the cortex. If this condition is chronic, one of the images is suppressed and the corresponding eye suffers deterioration of its performance, a condition known as *amblyopia.*

When both eyes fail to focus on the same object, a condition known as *strabismus,* one eye looks directly at the object of interest and the other (the deviating eye) is turned elsewhere. Strabismus can arise from a number of causes: dysfunction of the extrinsic muscles, errors of refraction, and anatomical abnormalities, for example. Depending on the causative agent, strabismus can be corrected by eyeglasses, muscle training exercises, or surgical treatment.

*Because of the separation of the eyes there are small differences in the shape and pattern of objects in the two visual fields. The closer an object is to the observer, the greater these differences. The observer associates the differences with distance between himself or herself and the object; hence, depth perception depends importantly on these small differences.

THE EAR

Carl F. Rothe

The ear is an organ of exquisite sensitivity and superb design. It is stimulated by energies infinitesimally small, since at threshold it will respond to an energy flux of a millionth of one billionth of a watt per square centimeter. As the function of the ear is described, note the relative ease by which pathological or traumatic changes may occur, and imagine the deleterious effects of calcification of various parts, reduction in compliance, or changes in motion due to inflammation.

Sound is produced and transmitted by the vibratory motion of bodies or air molecules that are displaced from a position of equilibrium in a direction parallel to the direction of propagation, are rapidly restored to this position, overshoot, and again are restored to continue the cycle. Thus, sound waves produced by a tuning fork are waves of alternating compression (increased pressure) and expansion (reduced pressure) of air as the tines of the fork swing back and forth. Physiologically, hearing is the subjective interpretation of the sensations produced by vibrations of a frequency and energy adequate to stimulate the auditory apparatus.

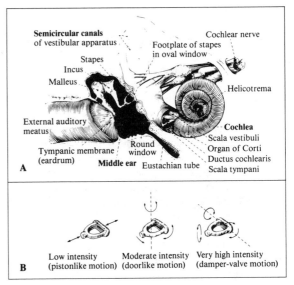

Fig. 4-14. (A) Diagram of the human hearing apparatus. (Redrawn from B. Melloni. In *Dorland's Illustrated Medical Dictionary* [25th ed.]. Philadelphia: Saunders, 1974.) (B) Modes of vibration of the stapes. At high intensities, near the threshold of feeling, the primary mode of rotation is shifted to a horizontal axis, so that as one edge of the stapes goes in, the other comes out, protecting the inner ear. (After G. von Békésy. *Experiments in Hearing.* New York: McGraw-Hill, 1960. Copyright © 1960 by McGraw-Hill Book Company. Used with the permission of McGraw-Hill Book Company.)

OUTLINE OF THE HEARING PROCESS

Sound enters the external auditory meatus (Fig. 4-14) and impinges on the tympanic membrane (eardrum), putting it into motion. The tympanic membrane in turn is coupled to the auditory ossicles, or bones of the middle ear, which transmit the sound to the inner ear, the cochlea. The cochlea of the living animal is curled in the form of a snail; hence the name. The movement of the auditory ossicles sets into motion the oval window, which separates the aqueous perilymph of the inner ear from the air in the middle ear. The motion of the fluid in the scala vestibuli of the inner ear (the vestibule) causes the basilar membrane to move in a pattern determined by the frequency and intensity of the sound. Movement of fluid in the scala tympani in turn moves the round window. The motion of the basilar membrane stimulates sensory elements in the inner ear (the organ of Corti), so that nerve action potentials are transmitted by the auditory nerve to the auditory cortex. We then perceive these impulses as sound. The vestibular apparatus or labyrinth is part of the inner ear and functions to provide

sensory inputs for the maintenance of the postural equilibrium (see The Ear: Vestibular Apparatus).

The *frequency* of sound is given in hertz (Hz) (1 Hz = 1 cycle/sec). The psychophysiological appreciation of frequency is pitch. The resonant frequency of a system forced to vibrate depends inversely on the mass in motion (inertia) and directly on the restoring force (elasticity) acting to bring the body back to equilibrium. The *quality* of a sound refers to the sensation perceived when one hears a mixture of related frequencies. Although musical instruments such as a flute produce a relatively pure sound of only one frequency, most musical sounds are composed of a fundamental frequency plus various amounts of harmonics or overtones that are integral multiples of the fundamental frequency. Nonintegral multiples are also present in some sounds, as from bells. Such a mixture of frequencies, regularly repeated, determines the quality of a musical sound and its distinctiveness. *Noise,* on the other hand, is composed of a random mixture of unrelated frequencies. The *duration* of a sound also helps to make it distinctive. Whereas a plucked violin string continues to produce sound for several seconds, the soft tissue of the body effectively dampens the vibrations of such sound-producing organs as the heart.

The *intensity* of a sound is expressed in physical terms as the *amount of energy transmitted per second* through a unit *area* perpendicular to the direction of travel of the wave, i.e., power flux. The usual units are watts/cm^2. This intensity is dependent on the fluctuation in air pressure. The psychophysiological appreciation of intensity is proportional to the 0.6 power (i.e., approximately the square root) of the sound intensity. It is also related to the frequency. Our ears are most sensitive to sounds with frequencies between 500 and 5000 Hz. If the frequency is increased above or decreased below this level, our ears are less sensitive. This difference in sensitivity is marked in that it requires about 10,000 times more sound *power* to hear a 100- or a 15,000-Hz sound than it does to hear a 2000-Hz sound (Fig. 4-15A). The *pressure* fluctuations must be 100 times as great at 15,000 Hz to be heard as well as a 2000-Hz sound, that is, to sound equally *loud*.

To represent the extreme range of the intensity of sound, the *decibel* (dB) is used. The decibel is a ratio that is somewhat related to perception of intensity. The just noticeable difference of loudness is about 1 dB but ranges between 0.3 and 5.0 dB, depending on the frequency and absolute sound level. A decibel is defined as 10 times the logarithm of the ratio of the intensity of the sound in question to the intensity of a reference sound. It is not an absolute unit but always is related to some reference level. In equation form

$$N_{(dB)} = 10 \log_{10} \frac{I_1}{I_2} = 20 \log_{10} \frac{P_1}{P_2}$$

in which I is the intensity in watts/cm^2 and P is pressure fluctuation in dynes/cm^2. If intensity is in terms of amplitude or pressure change instead of power flux, the factor is $2 \times 10 = 20$ instead of 10, because for a given system $W = P^2/R$. Here, W is power in watts, P is pressure fluctuation in dynes/cm^2, and R is acoustical resistance. (Note the analogous equation for voltage, E, and current, I, in an electrical circuit: $W = EI = E^2/R$.) Thus, a unit change in power is related to the square (2 times the logarithm) of the potential or pressure fluctuation, assuming a constant resistance. The reference level generally used for intensity is a power flux of 10^{-16} watts/cm^2 and for sound pressure a fluctuation of 0.0002 dynes/cm^2. (Normal atmospheric pressure is about 1 million dynes/cm^2 but is relatively constant, allowing pressure equilibrium throughout the body. However, explosive decompression at high altitude or outer space produces extreme stress on the ear, as well as on other structures of the body.) The reference levels, at about 2000 Hz, are about at the threshold of hearing under ideal conditions. Since a dyne of force is roughly equivalent to 1 mg of weight, the usable range of sensitivity to the pressure fluctuations varies between about 0.2 μg/cm^2 at threshold and 2 gm/cm^2 (1.5 mm Hg), the latter as a result, for example, of a sonic boom. This is a ratio of $1:10^7$ or 140 dB (20×7), since pressure, not power, is the unit considered. The human voice, during normal conversation, develops about 50 microwatts of sound. The sound intensity at the ear of the listener is about 10^{-10} watts/cm^2 (60 dB).

Audiometers are instruments for measuring hearing ability. There are two basic types: the pure-tone audiometer, producing tones of various frequencies and intensity levels, and the speech audiometer, providing for the presentation of speech-testing material by live or recorded voice at different intensities. These instruments are used clinically to measure two basic dimensions of hearing: (1) the threshold of hearing various tones and test words and (2) how clearly speech can be understood when it is presented at a comfortably loud level. Hearing may be impaired in varying degrees for either one or both of these dimensions.

In the pure-tone audiometer, the test frequencies usually are at octave (doubling the frequency) or half-octave intervals, ranging from 125 to 8000 Hz. This frequency range is somewhat less than the capability of the young adult ear, which may perceive frequencies as low as about 10 Hz and as high as about 23,000 Hz without pain. Dogs and some very young children may hear frequencies as high as 40,000 Hz. The thresholds of hearing at the extremes of the frequency range require a high energy flux (Fig. 4-15A). The intensities on an audiometer are scaled in decibels of hearing threshold level (decibels of *hearing loss* on older audiometers). The zero reference is related to the performance of normal ears rather than to the physicist's standard zero sound pressure level of 0.0002 dyne/cm^2. This threshold reference is more convenient, since the ear is not equally responsive at all frequencies. Moreover, the typical normal ear even at the most easily heard frequencies is not quite able to hear sounds as faint as the physicist's zero. The audiometric zero is based on extensive hearing surveys of otologically normal ears of well-motivated young people under ideal conditions. A hearing threshold 25 dB above the ISO 1964 reference level in the range of 500 to 2000 Hz is considered to be a slight hearing impairment.

The ability to hear high-frequency sounds suffers with age, as does the ability to focus for close vision. This poor hearing of the elderly, called *presbycusis*, is particularly obvious in men (Fig. 4-15B). A man 50 years of age normally has a hearing loss of about 25 dB at 4000 Hz com-

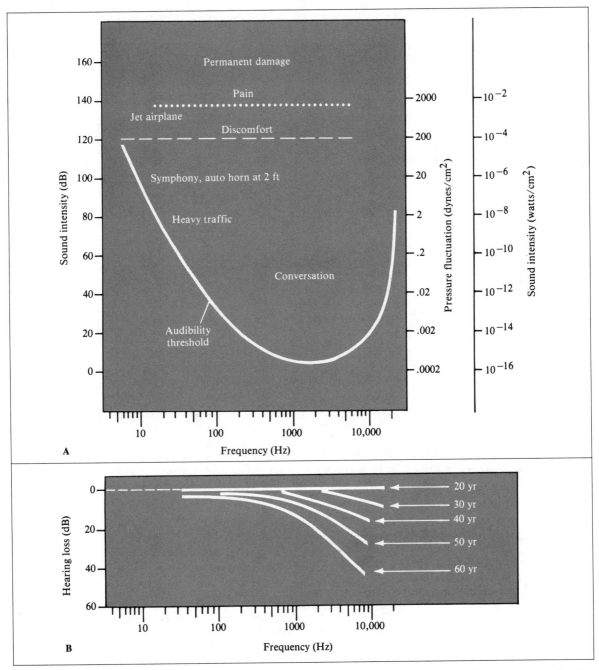

Fig. 4-15. (A) Sound intensity as related to audibility at various frequencies, the relative pressure, and energy fluxes. Sound intensity that is uncomfortable or damaging is relatively independent of frequency. **(B)** Presbycusis. Progressive loss of hearing ability for high frequencies in men at various ages compared with hearing at 20 years. (Data for [B] obtained from C. C. Bunch. Age variations in auditory acuity. *Arch. Otolaryngol.* 9:625–636, 1929; J. C. Webster et al. San Diego County Fair hearing survey. *J. Acoust. Soc. Am.* 22:473–483, 1950.)

pared with a young adult. This is a ratio of $10^{-25/10}$, which is $10^{-2.5}$, or a loss of $1:300$. Presbycusis is complex, including hair cell degeneration, loss of neuron population, and middle ear impairment (Davis and Silverman, 1970).

THE EXTERNAL EAR

The external ear consists of the pinna (auricle), the external auditory meatus, and the tympanic membrane. The *meatus* provides a passage for sound to enter the middle and inner ear. These delicate structures are surrounded and protected by the bone of the skull. The meatus prevents the entrance of large insects and objects that might damage the paper-thin (0.1-mm) eardrum. Furthermore, it acts to keep the air moist and near body temperature ($\pm$ 0.2°C), essential conditions if the eardrum is to function adequately. The *tympanic membrane*, or eardrum, is roughly conical in shape, so that a degree of rigidity is provided for coupling to the auditory bones.

THE MIDDLE EAR

The middle ear is air filled and contains the *auditory ossicles*. These bones (see Fig. 4-14A) weigh, in all, about 55 mg. The *malleus,* or hammer, is fastened to the tympanic membrane; the *incus,* or anvil, is firmly attached to the malleus and then acts on the *stapes,* or stirrup, which is attached to the oval window of the inner ear.

The primary function of the middle ear is to couple efficiently the movements of the low-density air to the high-density aqueous medium of the inner ear. If the energy flux is to be transferred efficiently, there must be impedance matching. That is, the relatively high amplitude but low force of movement of the air must be efficiently coupled to the high resistance to movement (inertia) of the fluid of the inner ear, so that the maximum amount of power (force × velocity) is transmitted. The human tympanic membrane has an area of about 0.7 cm². It is coupled through the bones to a much smaller (0.03 cm²) oval window. The coupling acts to convert the movements of easily compressible air to the higher forces necessary to put into motion the aqueous perilymph of the inner ear. This arrangement is a pressure transformer. It is analogous to holding an automobile door open with one finger while the car is moving; for although there may be a relatively small wind force on each unit area of the door, when the forces are transferred to the one small fingertip, a high force is experienced. In addition, there is a small lever action of the ossicular chain of bones (1.3/1.0). By these mechanisms the force per unit area is increased about 15 times, while the amplitude of vibration is little changed. The minute amounts of energy available are thus efficiently transferred to the inner ear with relatively small loss.

A second important function of the middle ear is to protect the exquisite structure of the inner ear from excessive movements. With low-intensity sound, the motion of the stapes is probably pistonlike (see Fig. 4-14B). With a sound of moderate intensity, the axis of rotation of the malleus and incus is such that the stapes rocks about a vertical axis at one edge of the oval window. This motion, though minute in magnitude, is analogous to that of a door opening and closing. Loud, low-pitched sounds cause the axis of rotation to shift so that the major axis of rotation is horizontal, across the oval window. Under these conditions one edge goes in and the other comes out, as diagrammed in Figure 4-14B. Excessive movements of the fluid within the cochlea are thus prevented by this short-circuiting procedure—a crucial protective mechanism in that the mode of motion can change instantly. It provides a means of protecting the inner ear from transient sounds such as explosions, which occur much too rapidly for any reflex mechanism.

Another protective device of the middle ear is the reflex action of the muscles, which functions in a manner similar to, and fully as fast as, a blink of the eye. The *tensor tympani* acts to pull the malleus and tympanic membrane into the middle ear, as the name implies. The *stapedius* tends to pull the stapes out of the oval window. The two muscles, acting together, snub low-frequency vibrations and so help protect the inner ear. However, above about 2000 Hz, this mechanism provides scant protection. Furthermore, the response time of this reflex is at least 50 msec, and about one-sixth of a second of a loud, reflex-stimulating tone is required for maximal protection. It is therefore of little value in shielding the inner ear from explosions or blows on the ear. Indeed, protection is not fully adequate, for continued exposure to loud noise is a well-documented cause of hearing loss.

These two protective mechanisms, in conjunction with the characteristics of the inner ear, provide the tremendous dynamic range of the ear, so that we can perceive, without damage, pressures 10 million times that at threshold.

The *eustachian tubes* act to equalize pressures. The body fluids absorb gases because the total gas pressure in tissue is about 60 mm Hg below atmospheric pressure. Consequently, air must periodically enter the middle ear if a partial vacuum and impairment of hearing are not to develop. Air enters or leaves by way of the pharyngeal slits of the eustachian tubes when a person yawns or swallows.

Chronic *otitis media* from infection of the middle ear may so damage the ossicles and their supporting structures that surgical replacement of the ossicles by a single columella between the tympanic membrane and the oval window is required. Since air conduction is impaired, this disorder is an example of *conduction deafness*. A common cause of hearing disability is *otosclerosis*. This disease causes a significant hearing loss in about 1 percent of whites (twice as frequent in females as in males) but is rare among blacks. There is destruction and regrowth of bone about the otic capsule. In most instances otosclerosis produces a primary middle ear lesion. The otosclerotic foci invade the stapes footplate, gradually reducing its mobility. In the early stages of the disease, only the low frequencies are affected. However, as the invasion of the footplate continues, the high frequencies are also involved. With complete stapes ankylosis (immobility), all frequencies are about equally affected, and a hearing loss of about 60 to 65 dB will be present—the maximum loss that can be imposed by a conductive lesion. That part of the impairment from reduced mobility of the stapes may be relieved by surgery, but surgery is not effective at all in reducing impairment attributable to sensorineural involvement. The surgical procedure of choice used to be fenestration, in which a new oval window was created in the horizontal semicircular canal. Without an impedance-matching transformer system even the best surgical results left a mild hearing deficit of about 25 dB. Later techniques were developed that retained the transformer action of the ossicular chain. At the present time the procedure of choice is stapedectomy with the substitution of a prosthesis.

THE INNER EAR

Movement of the stapes causes movement of the perilymph fluid, basilar membrane, and organ of Corti, which in turn triggers neural impulses. Many theories have been proposed to explain frequency analysis by the ear and the triggering of the auditory nerves. The *volley* and *place* theories of frequency discrimination, described next, are generally accepted.

Very-low-frequency pressure waves cause the perilymph to move back and forth through the helicotrema, a minute opening connecting the scala vestibuli to the scala tympani (see Fig. 4-14A). They have little effect on the basilar membrane. At somewhat higher frequencies, for example 30 Hz, the pressure waves tend to short-circuit through the basilar membrane because of the inertia of the fluids; thus they cause it to move back and forth. The movement of the basilar membrane causes distortion of the hair cells, which in turn initiates *volleys* of neural impulses. Under these conditions, frequency discrimination is performed by the cere-

bral cortex. At these low frequencies the volleys of neural impulses correspond to the fluctuations in pressure of the incoming sound.

At high frequencies, perception of sound frequency is based on the *place* where the maximal movement of the basilar membrane occurs. A pattern of the sound pressure changes may still be seen in the action potential pattern to about 3000 Hz, but above about 120 Hz the place discrimination process becomes important. The basilar membrane is relatively massive at the distal or apical end (about 0.5 mm wide); here, it has a relatively low stiffness, and, because of its distance from the stapes, a relatively large mass of perilymph is involved in moving the distal end. Thus, low frequencies tend to act here. On the other hand, just inside the oval window and stapes the supporting structure is lighter, narrower (0.04 mm), and more rigid. Most important, there is less fluid to move. High frequencies tend to produce the maximal effect here; that is, the amplitude of the oscillation is greater at this place than farther on in the cochlea (Fig. 4-16).

Von Békésy (1960) provided much information concerning the dynamic activity of the inner ear. Using a microscope and microtechniques with extreme care, he bored holes in the walls of the cochlear channels, applied minute specks of silver, and with a stroboscope actually measured the displacement at various places along the membrane when sounds of various frequencies were applied to the ears of cadavers (Fig. 4-16C). The basilar membrane undulates maximally at a certain place along the membrane when stimulated by a specific frequency. The nature of the undulation is shown in Figure 4-16. This kind of wave form is called a *traveling wave*. Because of the relatively large mass of fluid and low tension on the basilar membrane, the resonance theory of Helmholtz, which pictured the inner ear as being tuned like a harp, is not adequate.

Studies of inner ears damaged by high-intensity sounds of a specific frequency support the theory that sound frequencies act maximally on specific sites along the basilar membrane. Frequencies above about 10,000 Hz will be discriminated rather poorly because of the closeness of the frequency scale on the basilar membrane (Fig. 4-16A). Thus, there is mechanical analysis of frequency by localization of the place of maximal displacement of the basilar membrane. Low frequencies are localized at the far end, and high frequencies act next to the inner ear windows.

The *structure* of the inner ear is complex. A cross-sectional drawing of the cochlea is presented in Figure 4-17. The total diameter is about 3 mm, and its volume is about 100 μl (the volume of two drops of water!). Pressure waves come in along the top, the scala vestibuli, and cause the

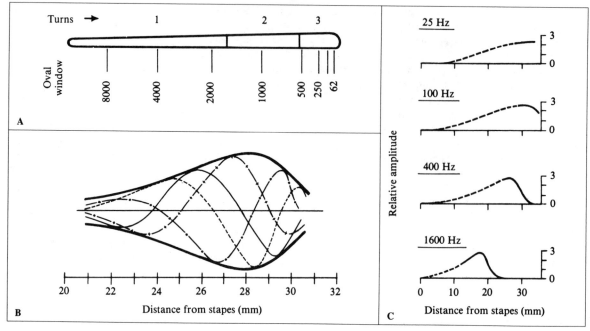

Fig. 4-16. (A) Localization of pitch discrimination along the basilar membrane of humans deduced from experiments with guinea pigs. Lesions were produced in the basilar membrane, and electrical audiograms were then made. In the human cochlea there are three turns, with high frequencies acting near the stapes and oval window. (After S. S. Stevens et al. The localization of pitch perception on the basilar membrane. *J. Gen. Psychol.* 13:297–315, 1935.) (B) Pattern of motion of a traveling wave with maximal amplitude of the envelope of motion at about 28 mm from the stapes. (C) Relative amplitude of motion of cochlear partition of a cadaver specimen. (After G. von Békésy. *Experiments in Hearing.* New York: McGraw-Hill, 1960. P. 462. Copyright © 1960 by McGraw-Hill Book Company. Used with the permission of McGraw-Hill Book Company.)

basilar membrane to move up and down. The lower drawing is at higher magnification than the upper and shows the hair cells and auditory nerves. The *tectorial membrane* is a rather rigid and massive structure that is in contact with the hair cells.

Because of the geometry of the attachments of the tectorial membrane and the organ of Corti, sound-induced vibrations produce a shearing action that causes the hairs to distort the cuticular plates of the hair cells. The movement of these hairs somehow excites the auditory nerves. The amount of movement is exceedingly small. A 3000-Hz sound of just threshold intensity moves the tympanic membrane back and forth about 10^{-10} cm. The amplitude of movement is much less than the diameter of a hydrogen molecule (2×10^{-8} cm). The basilar membrane moves only a small fraction as much as this. The auditory apparatus is indeed extremely sensitive. The reason we do not hear blood flowing through the vessels of the tympanic membrane is probably that the otherwise audible higher frequencies are damped so that the flow is steady. It is of significance that there are no blood vessels in the organ of Corti. Its nutrient supply is dependent upon the secretory epithelium of the ductus cochlearis.

INITIATION OF NEURAL IMPULSE. The most recent evidence indicates that neural impulses are triggered in some

manner by the development of *receptor potentials* resulting from relative movements of parts of the organ of Corti. These potentials are detected outside the ear as *cochlear microphonics.*

A unique feature of the inner ear is the steady potential difference across the hair cell membrane available for the development of the receptor potential. The *perilymph,* an ultrafiltrate of plasma, fills the scala vestibuli and the scala tympani and surrounds the otic labyrinth; it is in direct communication with the cerebrospinal fluid and is at the same potential as the rest of the body (Figs. 4-14 and 4-17). The fluid filling the organ of Corti (the *cortilymph*) is similar in composition and potential to the perilymph. The *endolymph* fills the interior of the ductus cochlearis (the scala media),

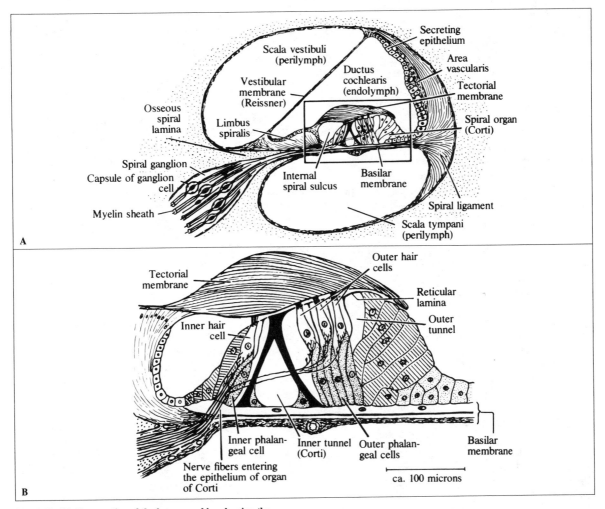

Fig. 4-17. (A) Cross section of the human cochlea showing the spiral organ of Corti and the three ducts of the inner ear. (B) Basilar membrane with organ of Corti at higher magnification. The scala vestibuli and tympani are filled with perilymph, and the ductus cochlearis is filled with endolymph. (Modified from A. T. Rasmussen. *Outlines of Neuro-Anatomy* [3rd ed.]. Dubuque, Iowa: Brown, 1943. Pp. 45, 47.)

located between the basilar membrane and the vestibular membrane. It is probably produced by the secreting epithelium (the stria vascularis) of the ductus cochlearis (Fig. 4-17). This fluid has, in the ductus cochlearis (scala media), an electrical potential of about 70 mv positive (not negative) with respect to the rest of the body. The potassium concentration is high, and the sodium concentration is low. The mechanism maintaining this potential and the electrolyte concentration gradients is not clearly understood. It is highly dependent on oxidative metabolism and is not particularly sensitive to changes in sodium or other ionic concentrations within the duct. With a few minutes of severe hypoxia the potential drops to near zero, and if the hypoxia is prolonged, it attains a large negative value. On return of blood flow carrying adequate amounts of oxygen, it returns to normal in a matter of seconds. Destruction of the secreting epithelium abolishes the potential. Thus the positive endolymph potential is apparently a secretory potential rather than a diffusion potential. With this positive 50 to 80 mv outside the hair cells and the usual negative 20 to 80 mv inside, there is a uniquely high (up to 160 mv) potential across the membranes of the hair cells.

Slight movement of the hairs by the relative motion of the

tectorial membrane and the basilar membrane probably distorts the hair cell membrane and opens pores to allow a flow of ions to initiate the partial depolarization of the hair cell membrane—the receptor, cochlear, or trigger potential. Present evidence suggests that this potential "acts directly on the unmyelinated dendrites of the afferent neurons found at the sides and bases of the hair cells" to excite the cochlear nerves (Gulick, 1971).

The receptor potential and cochlear microphonics are not nerve action potentials, for there is no distinct threshold, no refractory period, no all-or-none response, but a potential that is proportional to the displacement of the basilar membrane at moderate sound pressure levels. As a pressure front impinges on the tympanic membrane, there is a delay of 0.1 msec or less before the initiation of a cochlear microphonic, then a delay of about 0.7 msec before the nerve spikes are seen. The hair cells next to the stapes respond to all frequencies, but in the third turn of the cochlea (see Fig. 4-16), receptor potentials are produced by relatively low frequencies only, supporting the place theory of frequency analysis.

Although sound of a particular frequency sets into motion a relatively large part of the basilar membrane, some mechanism, probably mediated through the auditory cortex, sharpens the sensation so that frequencies differing by 2 to 3 Hz can be distinguished in the range of 60 to 1000 Hz when presented separately. Above 1000 Hz the ability to discriminate between different frequencies is about 2 per 1000. Thus, 8000- and 8016-Hz tones, if heard alternately, will be perceived as just different in frequency. If two tones of similar loudness and nearly the same frequency are heard simultaneously, a phenomenon called *beats* will be heard. There will be a variation in loudness that occurs at a frequency equal to the *difference* of the frequencies of the two tones. Beats between two tones can be detected up to a beat frequency of about six per second. Listening for these beats is the technique used by musicians to tune their instruments. If the beat frequency is greater than about six per second, a sensation of dissonance occurs.

Loudness discrimination is possible because sounds of higher intensity cause a greater movement over a wider area of the basilar membrane than do those of low intensity. By moving more hairs to a greater degree, more hair cells are stimulated to excite more auditory nerves and also to increase the nerve impulse frequency to give the sensation of greater loudness. In addition, the inner hair cells (Fig. 4-17) have a higher threshold and, when stimulated, may therefore add to the sensation of loudness.

Sensorineural deafness, involving damage to the inner ear or the auditory nerve, may occur at any age as a result of infections or trauma. Some antibiotics of the streptomycin group are ototoxic and thus cause degeneration of the organ of Corti.

Prolonged exposure over a period of years to high-intensity noise can result in *noise-induced hearing impairment.* Thus, hearing conservation measures are recommended if the noise level exceeds 85 dB over the range of 300 to 2400 Hz. At this level of noise, conversation is not possible unless the voice is raised and the people are within a few inches of each other. The availability of high-power audio amplifiers for popular music has made permanent hearing impairment by such "noise pollution" a significant problem for many young people. *Acoustic trauma* to the organ of Corti may be caused by a single high-intensity noise greater than about 140 dB. When the sound is a pure tone, histologically detectable damage to a localized spot along the basilar membrane occurs. The maximum loss in hearing ability occurs at a frequency about a half-octave above the frequency of the damaging tone, a site nearer the oval window. Hearing impairment by noise is dependent on not only the intensity of the noise but also its frequency and duration.

Sound intensity of 160 dB, a level attained, for example, close to a jet engine, has a power flux of 1 watt/cm^2. Since the soft tissues of the body absorb sound, particularly high-frequency sound, and convert it to heat, there is a significant and dangerous increase in body temperature under these conditions. Furthermore, the pressure fluctuations are of the order of 20 gm/cm^2 (about 15 mm Hg), and delicate tissues such as those of the ear and brain can be literally torn apart by the vibrations. Ultrasound intensities of over 1 watt/cm^2 may cause serious biological damage from excessive temperature increases, pressure variations, or cavitation; levels of 100 milliwatts/cm^2 or less are currently considered safe for diagnostic studies. However, continuing research may reveal serious dangers from even this low level in some tissues of certain individuals.

High-intensity sound can be used to destroy pathological structures in the brain. Ultrasonic frequencies of about 1 million per second, well beyond the audible range, may be focused to impinge on a small area within the body. When such beams of high-intensity sound are focused on a desired spot in the brain, tissue can be selectively destroyed by the heat generated without serious damage to intervening tissue. The echoes from pulses of ultrasound of *low* intensity are being used to visualize internal structures of the body, such as heart valves and the cerebral ventricles, as well as fetuses.

OLIVOCOCHLEAR EFFERENT NERVES. In addition to the afferent cochlear nerves, there are efferent nerves—the olivocochlear bundles—going to the hair cells. Their function is not clear. They have vesiculated, densely granulated endings. They appear to be inhibitory, raising the threshold of the auditory nerve fibers, especially to noise.

LOCALIZATION OF SOURCE OF SOUND

The process by which we localize the source of a sound is complex in that several mechanisms are involved. The *time of arrival* of the pressure front (e.g., from a click) provides one mechanism. If the source is to the person's right, the sound first stimulates the right ear and then the left ear. By orienting sound sources and having a blindfolded subject indicate the direction of the sound source, or by using two earphones and an electronically produced time-delay for one of the ears, it has been found that the normal person can distinguish differences in the apparent location of the source when the arrival times of two clicks are spaced as closely as about 50 μsec. A separation of about 1 msec gives a sense of localization to the side of the earliest sounds. If the clicks are more than about 10 msec apart, the sensation is not that of localization but of hearing two separate sounds. In the interpretation of heart sounds, for example, if the aortic and pulmonary valves close within an interval less than 20 msec, the sound will be perceived as a single sound and not as a split sound (see Chap. 13).

With low frequencies there is localization based on the *phase* relationship. This is closely similar to the time-of-arrival mechanism but involves a smoothly changing type of sound (sine wave) rather than sharp waves such as are associated with clicks.

Intensity of sound at each ear provides another important mechanism, particularly at high frequencies. The head tends to shadow the sound, and since we can distinguish a difference in intensity of about 1 dB, this ability furnishes a mechanism for localization of sound. At high frequencies the difference in intensity, because of the shadowing effect, is as high as 30 dB; thus, rather fine localization (± 10 degrees) is obtainable. This difference in intensity of the high frequencies is of prime importance in the appreciation of stereophonic sound. Interestingly enough, at about 3000 Hz, the frequency of maximal sensitivity of the human ear, both the time of arrival and the intensity mechanism are less than optimal; hence, these frequencies are difficult to localize. The human voice, having transient, clicklike sounds and lower frequencies, is easily localized.

THE EAR: VESTIBULAR APPARATUS

Julius J. Friedman

Spatial orientation of the body during both rest and movement is maintained, in part, by reflex activity that originates from the vestibular apparatus (labyrinth), the nonacoustical part of the inner ear. The vestibular apparatus is part of a multimodal system (Fig. 4-18), including visual and somatic receptors that go to the vestibular nuclei directly or through the vestibular cerebellum or the reticular formation. The ingoing signals are integrated in the vestibular nuclei, giving rise to outgoing responses that modify oculomotor and spinal motor control. The vestibular apparatus consists of the *semicircular canals,* which are dynamic receptors concerned with rotational movement, and the otolith organs, the *utricle and saccule,* which are static receptors stimulated by movements of slow tilting and linear acceleration.

STRUCTURE OF THE LABYRINTH

The horizontal anterior and posterior membranous semicircular canals (Fig. 4-19) are enclosed in the *bony labyrinth* of the temporal bone and are surrounded by *perilymph,* a fluid similar to cerebrospinal fluid. They are aligned in three planes almost perpendicular to each other, with the anterior canal on one side in a plane parallel to the opposite posterior canal. The canals are continuous with the utricle and are filled with *endolymph,* a fluid similar to intracellular fluid.

Near the point of junction with the utricle, each canal is equipped with an enlarged region called the *ampulla.* The ampulla contains the receptor organ, the *crista ampullaris* (Fig. 4-19), a ridge of columnar epithelial hair cells possessing hair filaments that are embedded in a gelatinous mass called the *cupula.* The cupula projects from the crista to the opposite wall of the ampulla and represents a moveable, leakproof partition that can be distorted by movements of endolymph within the canals. The hair filaments protruding from each sensory hair cell are composed of 50 to 100 *stereocilia* and 1 *kinocilium.* The kinocilia of the cells of the horizontal crista are always located at the peripheral margin of the cells on the side toward the utricle, whereas the kinocilia of the cells of the posterior and anterior cristae are on the canal side.

The utricle and saccule are situated anterior to the semicircular canals and communicate with each other

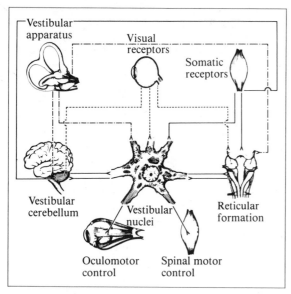

Fig. 4-18. **Multimodal nature of the system concerned with spatial orientation of the body.**

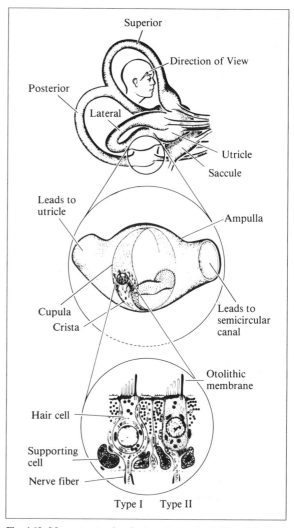

Fig. 4-19. **Macrostructural and microstructural relations of the human vestibular apparatus.**

through the endolymphatic ducts. The receptor organs of the utricle and saccule, called *maculae,* are structurally similar to the cristae. The stereocilia and the kinocilia of the utricular macula are oriented in a more complex pattern. The kinocilia of this sensory organ are pointed in both directions.

Embedded in the gelatinous layer of the maculae are numerous small particles of calcium carbonate called *otoliths,* which represent an inertial mass that tends to remain stationary as the head moves, thereby distorting the hair filaments. The macula of the utricle is oriented horizontally when the head is in the erect position, whereas the macula of the saccule is oriented vertically.

The nerve endings in the cristae and maculae communicate with the central nervous system via the vestibular branch of the eighth nerve.

FUNCTION OF THE VESTIBULAR APPARATUS

STIMULATION OF THE SEMICIRCULAR CANALS. Under resting conditions there is a balanced basal discharge of impulses from the ampullae on both sides of the head. Angular acceleration of as little as 0.5 degree/sec^2 is sufficient to displace the cupula and bend the cilia. Displacement of the stereocilia toward the kinocilia results in hair cell depolarization and excitation, whereas the opposite displacement produces hyperpolarization and inhibition.

Because of the inertia of endolymph, movement of this fluid lags behind the movement of the canal during rotational acceleration. This lag is equivalent to a reverse flow of endolymph.

As the speed of rotation is held constant and acceleration becomes zero, the endolymph lag disappears. Because of its elasticity, the cupula returns to the resting position within 10 to 20 seconds, and the discharge of impulses from the ampullae is restored to the basal frequency. With deceleration, the inertia of the endolymph causes it to be displaced in the direction of rotation, producing a postrotatory sensa-

Rotation	Endolymph displacement (postrotary)	POST-ROTATIONAL		
		Vertigo	Nystagmus	Falling
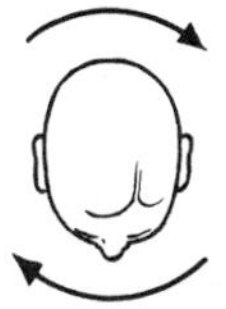 Right horizontal	Posterior 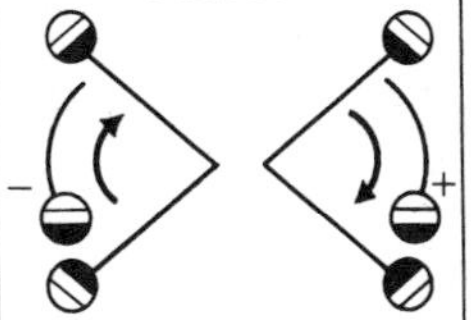 / Anterior	Spinning left	Horizontal left	Turning right
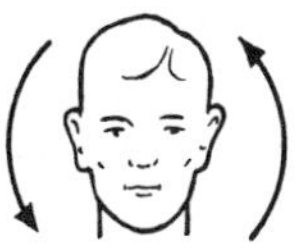 Right transverse	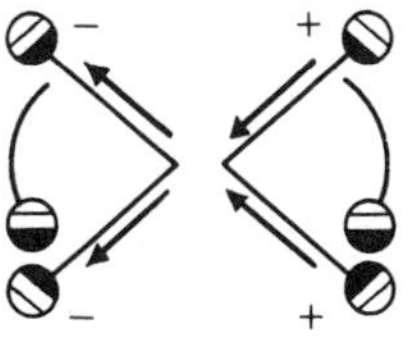	Falling left	Rotary left	Falling right
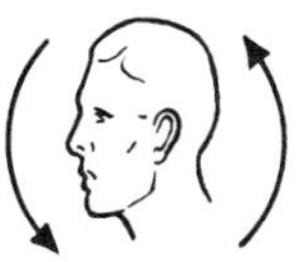 Forward medial	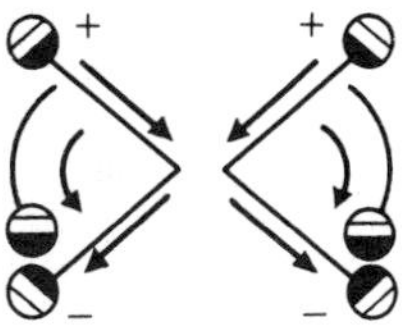	Falling backward	Vertical upward	Forward

Fig. 4-20. Effects of rotation in the three main axes on the post-rotatory endolymph displacement. The dark segment and thin line of each crista denote respectively the relative position of the kinocilia and stereocilia. Rotations in the opposite directions will produce reverse endolymph displacement and signs.

tion of rotation in the opposite direction (Fig. 4-20). The canals are oriented so that various combinations of ampullae may be excited or inhibited, depending on the axis of rotation (Fig. 4-20).

EFFECTS OF SEMICIRCULAR CANAL STIMULATION. Stimulation of the semicircular canals produces the subjective manifestations of vertigo, nausea, and other autonomic nervous system responses, in addition to the objective manifestations associated with changes in tonus of eye muscles (nystagmus) and of the antigravity muscles (falling reaction).

Vertigo. Vertigo is a sensation of spinning that may cause a person to lose balance and fall. The direction of the sensation of spinning depends on which semicircular canals are stimulated. In every case the vertigo is oriented in the direction opposite to the endolymph displacement.

Therefore, the vertigo is in the direction of rotation during the period of rotation, whereas the sensation experienced during the postrotatory state is in the direction opposite to the original rotation.

Autonomic responses resulting from vertigo include nausea and vomiting, pallor, perspiration, and, with intense stimulation, cardiac, vasomotor, and respiratory influences that can lead to hypotension and hyperpnea.

Alterations in Muscle Tone. Stimulation of the semicircular canals causes a change in muscle tone that results in nystagmus, past pointing, and falling reactions. *Nystagmus* is an oscillation of the eye consisting of a slow and a fast movement. The slow movement, always in the direction of the flow of the endolymph displacement, is a result of a reflex from the cristae, which travels via the vestibular nuclei in the brainstem to the eye muscles. The fast movement, the recovery phase of eye rotation, determines the direction of a nystagmus; it is due to the activity of an efferent central nervous system component of the vestibular reflex from the reticular formation in the brainstem. Rotation in the horizontal plane produces a horizontal nystagmus, medial rotation produces a vertical nystagmus, and transverse rotation produces a rotatory nystagmus. *Past pointing* and *falling reactions* are the results of a change in tonus of the antigravity muscles in which the tone of the muscles increases in the side toward which the endolymph is displaced and decreases on the opposite side. Thus, if the ground gives way under a person's right foot, the person and his or her head lurch to the right, shifting the endolymph to the left. This reflex action causes an immediate extension of the right arm and leg and flexion of the left arm and leg accompanied by deviation of the eyes to the left. These movements are protective righting reflexes.

A sensation of rotation can be caused by an imbalance in the bilateral basal discharge of impulses resulting from loss of function on one side. Fortunately, because of the multimodal nature of the system, a loss of one component can be compensated for by adaptation of the other components.

CALORIC STIMULATION. Douching the ear with warm or cool water produces convection currents in the endolymph that cause stimulation of the crista. This method has an advantage over stimulation by rotation because the vestibular system of each ear can be tested separately. It is used in the diagnosis of vestibular disorders.

When the head is held backward at 60 degrees, the horizontal canals are brought into vertical position. The douche causes a greater change in the temperature of the endolymph in the part of the canal lying near the external auditory meatus than in the more deeply situated parts. With a cold douche the currents created move away from the ampulla; with a warm douche they move toward it. These convection currents stimulate the crista, and *horizontal nystagmus* and vertigo result. With the head upright, both sets of vertical canals are stimulated, and *rotary nystagmus* results, away from the ear douched with cold water and toward the ear douched with warm water.

STIMULATION OF THE UTRICLE AND SACCULE. The utricle is the most important gravitationally influenced organ in the labyrinth. Its macula responds to slow tilting, linear acceleration, and centrifugal force. With the head in the erect position, the utricular macula is in the horizontal plane, and the otoliths exert uniform pressure down on the hair cells. In this position there is no distortion of the hair filaments, and the discharges from both right and left utricles are in balance. Linear acceleration or a tilting of the head causes the otoliths to be displaced because of inertia and the force of gravity. As a result of the multidirectional orientation of the cilia, this displacement produces distortion of the hair filaments and elicits a complex pattern of change in the frequency of discharge of the hair cells. Specifically, if the

left utricle is tilted to the left, the discharge rate will increase, whereas a tilt to the right decreases the discharge of the left utricle. Complementary changes occur on the opposite side; these heighten the imbalance between the right and the left utricular discharge. In humans, these movements are believed to be involved with vibration, but the mechanism remains obscure.

BIBLIOGRAPHY

THE EYE

Brindley, G. S. *Physiology of the Retina and Visual Pathway* (2nd ed.). Baltimore: Williams & Wilkins, 1970.

Brown, P. K., and Wald, G. Visual pigments in single rods and cones of the human retina. *Science* 144:45, 1964.

Campbell, C. J., Koester, C. J., Rittler, M. C., and Takaberry, R. B. *Physiological Optics*. New York: Harper & Row, 1974.

Davson, H. *The Physiology of the Eye* (4th ed.). New York: Academic, 1980.

MacNichol, E. F., Jr. Retinal mechanisms of color vision. *Vision Res.* 4:119, 1964.

Moses, R. M. *Adler's Physiology of the Eye* (7th ed.). St. Louis: Mosby, 1981.

Zadunaisky, J. A. Electrophysiology and Transparency of the Cornea. In G. Giebisch (ed.), *Electrophysiology of Epithelial Cells*. New York: Schattauer, 1971. Pp. 225–255.

THE EAR

Beranek, L. L. Noise. *Sci. Am.* 215:66–76. December, 1966.

Davis, H., and Silverman, S. R. (eds.). *Hearing and Deafness* (3rd ed.). New York: Holt, Rinehart & Winston, 1970.

Gelfand, S. A. *Hearing. An Introduction to Psychological and Physiological Acoustics*. New York: Marcel Dekker, 1981.

Gulick, W. L. *Hearing: Physiology and Psychophysics*. New York: Oxford University Press, 1971.

Jerger, J. (ed.). *Modern Developments in Audiology* (2nd ed.). New York: Academic, 1973.

Siebert, W. M. Hearing and the Ear. In J. H. U. Brown and D. S. Gann (eds.), *Engineering Principles in Physiology*. New York: Academic, 1973. Vol. 1, Chap. 7.

von Békésy, G. *Experiments in Hearing*. New York: McGraw-Hill, 1960.

THE EAR: VESTIBULAR APPARATUS

Fischer, J. *The Labyrinth: Physiology and Functional Tests*. New York: Grune & Stratton, 1956.

Gernandt, B. E. Vestibular Mechanism. In J. Field (ed.), *Handbook of Physiology*. Washington, D.C.: American Physiological Society, 1959. Section 1: Neurophysiology. Vol. 1, pp. 549–564.

Howard, I. P., and Templeton, W. B. *Human Spatial Orientation*. New York: Wiley, 1966.

Roberts, T. D. M. *Neurophysiology of Postural Mechanisms*. New York: Plenum, 1967.

REFLEXES AND REFLEX MECHANISMS

5

Sidney Ochs

Descartes, in the seventeenth century, first clearly defined the basic behavior pattern of the reflex. He gave an example of a foot placed near a fire, which, when painfully stimulated, is quickly withdrawn. (In modern terms this is called a *flexor withdrawal reflex*.) Involved in reflex activity are excitation of sensory receptors, conduction over afferent nerve fibers, and finally, after specific central nervous system (CNS) activity, excitation of the motor nerves innervating the musculature to give an appropriate movement.

Reflex responses can be described as *machinelike* in character, implying a reproducibility of response. They have also been called *purposeful,* since reflexes are generally of use to the organism. For example, shining a beam of light into the eye causes a constriction of the pupil, the *pupillary reflex*. This reflex response helps to prevent the retina from being subjected to intense illumination. Similarly, reflex control of the muscles of the eardrum acts to prevent excessive sounds from being transmitted to the inner ear. Another example is the *pinna reflex,* found in the cat and the dog. When a foreign object enters the outer ear,

a series of reflex ear twitches act to dislodge the object. The purposive nature of reflexes is further demonstrated in the spinal frog. Pithing this animal's brain results in a short period of depressed reflexes (*spinal shock*) (see the next section). After recovery, when an irritant acid solution is applied to one flank, the leg on that side is brought up, and wiping movements are performed. If the leg is held, the other leg performs the wiping movements. The direction of the limb toward the spot of excitation on the flank is an example of *local sign*. This implies that stimuli entering the nervous system have sufficient localizing information for the resulting reflex actions to be directed to the appropriate site.

In the last century, the purposiveness of reflexes was taken to indicate that there was some kind of primitive consciousness in the spinal cord that directs reflex activity. Sherrington (1906), however, pointed out that the apparently purposive nature of reflex responses represents a selection, during evolution, of responses that have survival value. Certain reflexes damaging to the organism, rather than protective, may be elicited, further indicating the lack of conscious direction in reflex actions.

Some actions are partially voluntary and partially reflex. For example, a mixture of volition and reflex activity is involved in the act of swallowing. Food in the mouth may be voluntarily rejected, but once it has passed beyond the fauces, reflexes coordinated by a swallowing center in the medulla are set in motion.

A practical use made of reflexes is illustrated in anesthesiology. Touching the eyelids causes a reflex closure, the *eyelid reflex,* which is lost with moderately deep anesthesia. Touching the cornea of the eye causes a *corneal reflex;* the lid blinks to cover and protect the cornea. This reflex is diminished or lost when the brainstem has been depressed to a perilous degree. The *pupillary light reflex,* constriction of the pupils to light shining on the eye, is one of the last reflexes to disappear in deep anesthesia, along with respiratory and cardiovascular reflex control mechanisms whose centers are in the medulla. Another reflex of value in judging whether the depth of anesthesia is satisfactory for operative manipulation is the response produced by pinching the skin (one example of *nociceptive,* i.e., injury-provoking, stimulation). This evokes a *flexor withdrawal reflex* in which the limb flexes away from the site where the noxious stimulus is applied. When this reflex is absent, anesthesia is generally considered sufficiently deep to permit operative procedures.

As will be discussed in more detail in a later section, the *stretch reflexes* of skeletal muscle are of great importance for normal posture and locomotion and form our basic concepts of the reflex mechanism. A stretch reflex occurs as a result of a brief extension of the muscle. This can be elicited by tapping on a tendon so that a quick reflex contraction of the muscle, a *phasic* or tendon reflex, occurs. A maintained stretch produces a maintained tonic reflex tension in the muscle. Muscles are termed *flexors* or *extensors* depending on whether they flex or extend a limb at a joint. As will be discussed more fully later on, there is a special relationship of flexor and extensor muscles acting around a joint in a stretch reflex response: When one muscle is excited reflexly, the other is inhibited, and vice versa.

The various reflexes are so interrelated in the animal that the result is one continuous, smooth, well-directed behavior pattern, each reflex succeeding and merging with the next in rapid sequence. The interrelation of one reflex with another is demonstrated in locomotion. An animal in which the spinal cord has been cut some weeks previously to allow it to recover its reflex excitability is suspended in a harness. When the foot is gently but quickly pressed upward, the slight spreading of the toe pads suddenly results in a powerful downward thrust of the leg, the *extensor-thrust* reflex. During locomotion the limb is flexed and brought up from the ground. After the body of the animal has carried it forward, the limb is extended and comes into contact with the surface. With that contact the extensor thrust is excited, and the limb is converted into a rigid column to give a pole-vaulting effect to the body, carrying it forward over the extended limb. The reflex is then quickly inhibited, permitting the leg to flex, and the cycle of flexion and extension is repeated.

The reflex relationship of the two hind limbs is shown in the *crossed reflexes.* When the spinal animal is suspended in a body harness with its legs pendant and a noxious stimulus is applied to one limb, a flexor withdrawal is produced. The contralateral hind limb then may show a *crossed extension.* Forceful flexion of one leg may also cause a reflex extension of the contralateral limb. Some animals do not walk or run but hop or gallop, with both hind legs flexing and extending together. Rabbits are hopping animals, and in this species reflex flexing of one hind limb usually causes a *crossed flexion* of the contralateral hind limb rather than the crossed extension seen in the cat and the dog. Another type of response associated with locomotion, the *walking reflex,* may be observed when reflex excitability is high. Noxious stimulation of one leg of a spinal dog to cause flexion not only may produce a crossed extension of the opposite limb but may also be followed by an extension of the originally reflexly flexed leg. This action proceeds in alternating fashion with a fairly constant rhythm, so that there is a cyclic pattern of flexion and extension of the two hind limbs that may persist for several minutes.

Walking and running are activities of the quadruped requiring all four legs. When one leg is stimulated, not only does the leg of the contralateral side show crossed extension, but the foreleg may also extend on the side on which the hind limb is flexed. The opposite foreleg may become extended. Such patterns are referred to as *reflex figures.* The reflexes elicited by these stimulations are part of patterns within the spinal cord controlling normal locomotor behavior of the four-legged animal.

LOWER AND UPPER MOTOR CONTROL

The basic machinery for the most fundamental types of reflex mechanisms is found within the spinal cord and the brainstem. In *local reflexes,* some of which are described in the preceding section, a complex interplay of control is present. Sherrington used such terms as *prepotency* and *dominance* to describe the interaction and ordering of reflexes. Noxious stimulation is usually prepotent, as demonstrated in competition with the *scratch reflex*. To excite the latter, a stimulus is used to imitate an insect crawling on the skin. This gives rise to a rhythmic scratching movement of the limb, directed toward the stimulus. If the scratch reflex is induced and then followed by nociceptive stimulation to the skin, the scratch reflex ceases. The prepotent flexor withdrawal response to the noxious stimulation takes command of the neural channels. Only a relatively few lower motoneurons are available to respond to reflex action, and some mechanisms as yet unknown act to switch command from one reflex to the other. Sherrington applied the term *final common path* to the motoneurons innervating the muscles. All the varieties of reflex activities and the resulting complex behavior displayed by animals are eventually funneled through the relatively few motoneurons of the final common path.

If, as a result of compression, asphyxiation, or neurotropic diseases such as poliomyelitis, the final common path motoneurons are destroyed, there is a total loss of reflex excitability. It is followed later by an atrophic change of the muscles innervated by these neurons (see Chap. 3), part of the condition termed a *lower motor lesion*. If the spinal cord is transected above the level of the motoneurons, so that the descending connections from higher motor centers are interrupted, a *higher motor lesion* develops. Immediately on or after the making of the section, a diminution or elimination of motor reflexes occurs with a lessened muscle tone, a state known as *spinal shock*. A gradual recovery of tone and reflexes then can take place over a period of minutes, months, or longer, depending on the species. Spinal shock is not due to the immediate effects of the lesion—i.e., a possible stimulation produced in cutting across nerve fibers; if, after recovery from spinal shock, a second cut is made below the level of the first, no further onset of shock ensues. The degree of spinal shock is more profound and the recovery time more prolonged, the higher the animal is in the phylogenetic series. In the frog, much of the reflex function returns within a few minutes. In the dog and cat, reflexes begin to return within several hours, full recovery taking several weeks. In the monkey, months are required for recovery, and in humans it may take many months, with only a limited recovery of reflexes. These differences in the time course of recovery from spinal shock in the different species are due to *encephalization,* the greater dependence of the lower motor centers on the higher motor control mechanisms that have developed in the brains of higher species.

Not only are the somatic reflexes depressed during spinal shock, but one also finds a similar depression of the *visceral reflexes*. Some of the higher visceral control mechanisms will be outlined later in the chapter, but because it is an important aspect of reflex loss, the effect of spinal shock on urinary bladder function is mentioned here. A reflex emptying of the bladder normally occurs when it is filled with urine to a certain level and stretch receptors in the musculature of the wall are activated. The reflex mechanisms are localized within the lower spinal cord but can also be controlled in part by higher centers. Voluntary inhibition of the reflex is thus possible. Spinal cord transection interrupts the upper influences, and the threshold for the local spinal reflex becomes greatly elevated. In such an event the bladder may attain a considerable size before the threshold is reached for reflex bladder emptying, a condition known as *automatic* or *reflex bladder*. In the clinical management of spinal cases, the bladder must be drained by a catheter or the urine manually expressed at intervals. The procedure is continued for some days or weeks, until the threshold falls and reflex emptying occurs. Other visceral reflex mechanisms, those of vasomotion and sweating, are also depressed and gradually recover after spinal cord section.

Spinal shock appears to be due to the removal of upper influences acting on lower motor mechanisms in a mainly excitatory fashion. Removal of the excitation results, therefore, in a reduced level of a local excitatory state in the lower motor centers and a depression of spinal reflexes and tone. Possibly spinal shock is also due to the removal of inhibitory control over local inhibition of reflexes.

Later in the course of recovery from spinal shock, the excitability of somatic and visceral reflexes may increase far beyond normal levels, so that a mild stimulation can excite a very great response, a *mass discharge*. Profuse sweating, flushing, urination, defecation, and flexor or extensor motor spasms of the limbs are seen during a mass discharge in spinal humans. It is possible that sprouting of fibers from the entering afferent fibers and interneurons synapsing on motoneurons is responsible for these later manifestations of exaggerated neural activity. Or the membrane properties of the cells themselves may have changed, so that their reflex excitation thresholds are abnormally low.

THE FLEXOR REFLEX AND SOME GENERAL PROPERTIES OF REFLEX POOLS

The flexor reflex acts to protect against a potential trauma. Painful, i.e., *nociceptive,* stimulation of the skin of a limb causes a contraction of the limb muscles with a movement of the limb—or even, if strong enough, the entire body— away from the source of stimulation. Flexor reflexes are widespread, engaging many muscles of a limb and, if strong, they may spread to other limbs. For example, moderate nociceptive stimulation of the foot may cause only the foot to be withdrawn. If stronger stimulation is used, the lower leg is also flexed. With still stronger stimulation, the upper leg is flexed as well. Furthermore, other limbs may enter into the reflex when very strong excitation is used. The *Babinski reflex,* routinely looked for during clinical examination, is, in adults, an abnormally augmented flexor withdrawal to a relatively mild stimulation. The sole of the foot is stroked along the outer edge. Normally, the response is a downward movement of the big toe (plantar flexion). In the Babinski reflex an upward movement of the big toe takes place, often accompanied by a fanning of the other toes. If reflex excitability is high, the foot may be everted from the site of stimulation and the whole limb withdrawn. The presence of the Babinski reflex has been associated with an upper motor lesion.

A brief electrical stimulus applied to the central end of a cut flexor muscle nerve causes a brief reflex contraction in other flexor muscles of that limb. The extensive distribution of the response of flexor muscles to stimulation of the nerve of one muscle is brought about by the widespread synaptic connections within the spinal cord. Two classes of afferent fiber synaptic termination and distribution are present in the cord. In one type, afferent fibers synapse directly on the motoneurons in *circumscribed* fashion (Fig. 5-1A). In the other, afferent fibers synapse on interneurons, which in turn have widespread or diffuse spread of synaptic contact with other interneurons that eventually synapse with motoneurons (Fig. 5-1B). The stretch reflexes are subserved by the circumscribed type of connections, and the flexor withdrawal reflex by the diffuse type.

The entering collateral fibers make synaptic contact either on interneurons or on motoneurons via enlarged endings, the *synaptic boutons.* These are present on nearly all neurons, being particularly dense over the surface of the motoneurons. One class of boutons terminating on the motoneurons acts to excite them, while another set inhibits the discharge of the motoneuron. This result can be inferred from interaction studies made by stimulating two afferent nerves selected from the various nerves innervating the limb muscles. Each afferent nerve eventually activates a number of motoneurons in the pool innervating a part of the peripheral limb musculature to give rise to a flexor reflex. The type of effect observed depends on whether a weak or strong stimulus is used to excite the afferent nerves. With weak stimuli, only a few cells of the motoneuron pool are excited. The cells that do not fire may be *subliminally* excited. When two afferent nerves are stimulated simultaneously, each producing subliminal excitation in a group of neurons common to them, a summation of subliminal excitability occurs, and a large response is obtained (Fig. 5-2,

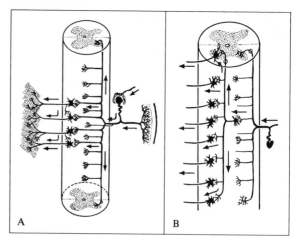

Fig. 5-1. Diffuse and circumscribed distribution of afferent fibers terminating in the spinal cord. In (A), a circumscribed monosynaptic type of termination is shown. The fibers synapse directly on the motor horn cells, usually within the spinal cord segment of entry of those fibers. In (B), a diffuse transmission within the CNS is shown. Sensory fibers terminate on interneurons within the spinal cord that, by multiple branches, engage a large number of motoneurons that in turn are distributed to many muscles of the peripheral musculature. (From S. Ramón y Cajal. *Histologie du Système Nerveux.* Madrid: Consejo Superior de Investigaciones. Científicas, 1952. Vol. 1, pp. 531, 532.)

top). This is called *facilitation.* If the shocks to the afferent nerve inputs are made so strong that a group of cells is excited by each input, the response to stimulation of the two afferent inputs together will be less than the sum of each of the responses elicited individually (Fig. 5-2, bottom). This effect demonstrates the phenomenon of *occlusion.*

The presence of multiple synapses over the surface of the motoneuron accounts for the variety of complex interactions seen when inputs from different sources are presented at different times. The spatiotemporal interactions observed with stimuli led into the same afferent nerve at different times are those of *temporal interaction.* If the strength of the first shock is low in magnitude, so that only relatively few cells are fired, with other cells subliminally excited, the later test shocks may show an increased excitatory effect in the motoneuron membrane, *temporal facilitation.* If stronger conditioning shocks are used, the later stimulus shows an inhibition brought about by the action of inhibiting neurons, which, as we will see, may persist for up to several hundred milliseconds. The long-lasting inhibitions are brought about in part by an effect on the postsynaptic neuronal membrane and in part by interneurons acting on the motoneurons. The complexities of interaction with this

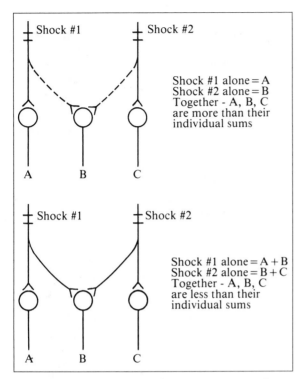

Fig. 5-2. *(Top)* Facilitation. Excitation of one sensory nerve, *#1*, excites a group of neurons, *A*, and a collateral branch, shown by the dashed line, subliminally excites a group represented by *B*. Similarly, excitation of sensory nerve *#2* excites neuron group *C* and subliminally group *B*. The sum of the two shocks presented separately is *A* + *C*. If *#1* and *#2* are excited together, the two subliminal excitations of group B summate, and the resulting discharge of *A*, *B*, and *C* together is more than the sum of the individual shocks. *(Bottom)* Occlusion. A shock to *#1* excites neuron groups *A* and *B*. A shock to *#2* excites groups *B* and *C*. A shock to *#1* and *#2* together, because of their common termination on *B*, is less than the sum of their individual responses. (From S. Ochs. *Elements of Neurophysiology.* New York: Wiley, 1965. P. 286.)

type of reflex activation are simplified by the use of the monosynaptic reflex system described in the following section.

STRETCH MYOTATIC REFLEXES: THE MONOSYNAPTIC REFLEX SYSTEM. A sudden stretch of a limb muscle gives rise to a reflex contraction of that same muscle. The classic example of a *stretch*, or *myotatic*, *reflex* is the *patellar reflex*, or *knee jerk*, produced when the tendon at the knee is given a light tap. The quadriceps muscles are abruptly stretched, thereby exciting stretch *spindle receptors* in the muscle. A volley passes via its afferents to the cord to excite the motoneurons, producing an extension of the lower leg. The reflex nature of the response is shown by cutting the dorsal roots through which the afferent fibers from that muscle enter the spinal cord. In this case, stretch of the muscle no longer elicits a reflex contraction. The reflex nature of the response to a maintained stretch on the muscle is shown by the increase of tension in the stretched muscle and the drop of tension in it when the dorsal roots are cut to interrupt the reflex arc.

Muscles usually operate in pairs for the movement of a limb, so that when one set of muscles, the *agonistic* muscles, is contracting, the opposing *antagonistic* muscles are inhibited and relaxed. This coordinated opposed action of agonist and antagonist muscles is called *reciprocal innervation*. To demonstrate the reflex properties of a reciprocally innervated muscle group (e.g., of the anterior tibialis and gastrocnemius muscles), the cut ends of the tendons are stretched, and the tension reflexly produced is recorded. A stretch of the muscle elicits a reflex contraction in the same muscle and an inhibition of the antagonistic muscle. Reflex inhibition is mediated by inhibitory discharges that act on the motoneurons innervating the antagonistic muscle, as will be described in the following section.

Both the stretch reflexes excited by a quick stretch of the muscle and the slow myotatic reflexes produced by a maintained stretch are excited by the *spindle receptors* distributed among the skeletal muscle fibers. A more detailed description of these muscle receptors will be given in a later section of this chapter. The afferent fibers of the spindle receptors enter the cord to synapse directly on the motoneurons with the circumscribed type of connectivity shown in Figure 5-1A. Thus, only one synapse between the afferents and motoneurons is involved.

The dorsal root fibers can be arranged so that they may be stimulated with a brief impulse. The electrical reflex discharge is recorded in the ventral roots (Fig. 5-3). An artifact signifies the time of stimulation, and a latency of approximately 1.5 msec occurs before the *monosynaptic response* is recorded in the ventral root (Fig. 5-3). This response is then followed by an irregular discharge lasting approximately 15 msec, the *multisynaptic*, or *polysynaptic*, *response*. As indicated in Figure 5-3, the monosynaptic response is so termed because it is excited by fibers synapsing directly on the motoneurons; the multisynaptic response occurs when a number of intervening neurons are activated that terminate on motoneurons.

Evidence that the monosynaptic response involves only a single synapse has been obtained from a knowledge of the conduction time of impulses in afferent and efferent fibers

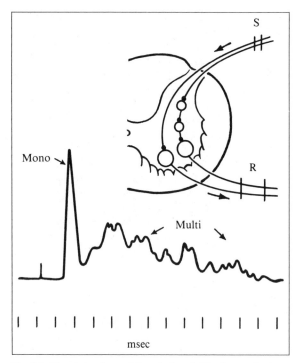

Fig. 5-3. Electrical reflexes of the spinal cord. The diagram in the upper portion of this figure shows a stimulus (*S*) of afferent fibers in the dorsal root that terminate directly on the motoneurons to give the monosynaptic (*mono*) responses (*R*) recorded in the ventral root. Terminations on interneurons give rise to multisynaptic (*multi*) discharges. Electrical reflex response patterns are shown in the lower part of the figure. (From S. Ochs. *Elements of Neurophysiology.* New York: Wiley, 1965. P. 302.)

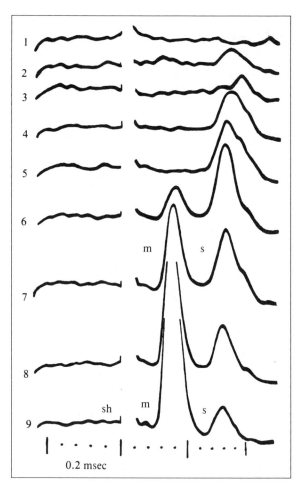

Fig. 5-4. Focal stimulation within the CNS. With a stimulating needle electrode in the oculomotor nucleus, the motoneurons are excited, as are afferent branches terminating on another group of motoneurons. As the strength of the shock is successively increased (shown in the traces from above, downward), a small longer latency response is seen, corresponding to a monosynaptic response with synaptic (*s*) delay. As the strength is further increased, an earlier membrane (*m*) response is seen that is due to a direct excitation of the motoneurons. Notice that as the size of the direct response increases, it occludes and reduces the synaptic response. Even at the strongest shock levels, synaptic delay is no shorter than 0.5 msec. *Sh* = shock artifact. (After R. Lorente de Nó. Transmission of impulses through cranial motor nuclei. *J. Neurophysiol.* 2:409, 1939.)

and the time required for synaptic transmission at the motoneurons. The delay is similar to the synaptic delay at the neuromuscular junction (see Chap. 3). To show this, a brief electrical stimulus is delivered through the tips of needle electrodes inserted into the ventral horn so as to excite a population of motoneurons (Fig. 5-3). A double electrical response is recorded in the ventral root (Fig. 5-4). The first spike (m) is accounted for by the direct excitation of the motoneurons or of their axons. Little latency is seen other than the time of conduction from the directly stimulated motoneurons to the recording electrodes on the ventral root. The m response is followed after an additional latency of 0.5 to 1.0 msec by an s response, which is due to excitation of afferent fibers synapsing on the motoneurons, the *reflexomotor collaterals of Cajal.* The time between the m and the s responses cannot be reduced below 0.5 msec no matter how strong the stimulation. This represents the synaptic delay time, which ranges from 0.5 to 1.0 msec.

As the strength of a dorsal root stimulus is gradually increased, more dorsal root fibers and more synaptic endings on the motoneuron cells are excited. The size of the resulting monosynaptic response in the ventral root increases in sigmoidal fashion with increasing stimulus strength. When the stimulus is weak, there is a subliminal activation of a large number of fringe cells. When the stimulus strength is increased, a large number of neurons are fired, and the fringe group becomes smaller. However, even with a maximal response, a significant fringe group exists, as is shown by the phenomenon of *posttetanic potentiation* (PTP). A PTP is demonstrated by a tetanic stimulation of the dorsal root for several minutes, resulting in augmented monosynaptic responses. This phenomenon persists for some minutes before there is a gradual return of responses to control levels. The mechanism responsible for PTP is explained by a change of synaptic transmission, discussed later in this chapter. To anticipate that discussion: The chemical mediator that is considered to underlie synaptic excitation appears to be mobilized by the tetanic activation with a temporarily increased probability of release of transmitter from the boutons, as shown by the time course of PTP.

The more prolonged electrical discharge appearing after the monosynaptic response, the *multisynaptic response,* is believed to be due to a repetitive activity in chains of interneurons terminating on motoneurons (see Fig. 5-3). The motoneurons that give rise to multisynaptic discharges are distributed mainly to flexor muscles, while the monosynaptic response is restricted to the same muscles from which the spindle afferents originate, i.e., from extensor or flexor muscles. The multisynaptic response is present not only in the ventral root of the spinal cord segment of its stimulated afferent root but also in the ventral roots of a number of segments of the spinal cord, as would be expected from the diffuse type of connectivity (see Fig. 5-1B).

A variable factor determining the proportion of monosynaptic to multisynaptic responses is the strength of stimulation. The most excitable afferent fibers, the largest-diameter subgroup of A fibers (IA), take origin from the spindle stretch receptors within the muscles. On activation, these fibers give rise to the monosynaptic response. Fibers of smaller diameter give rise to multisynaptic discharges. By reducing the activity of the spinal cord, either by the use of barbiturate anesthesia or by means of weak shocks to the afferent nerve fibers, one can obtain monosynaptic responses in relative isolation. If the nerve selected for afferent stimulation is purely a skin sensory nerve (e.g., sural nerve), the reflex response in the ventral root is typically only a multisynaptic discharge with little evidence of a monosynaptic discharge.

FACILITATION AND INHIBITION OF MONOSYNAPTIC RESPONSES. Muscles acting in the same manner at a joint (e.g., the lateral and medial gastrocnemius muscles) are *synergists*. Facilitation may be clearly demonstrated in the motoneurons of such muscles by using monosynaptic responses obtained by stimulating synergistic nerves. For example, the medial gastrocnemius nerve is excited by a *conditioning stimulus* and concurrently or at various intervals thereafter by a *test stimulus* delivered to the central end of the lateral gastrocnemius nerve. The result is that the test response is augmented (Fig. 5-5A). The conditioning stimulus is usually adjusted so that only a few motoneurons fire, giving rise to a just liminal excitation. The test stimulus given concurrently is much augmented, and as the time between the two volleys is increased over a period of approximately 15 msec, facilitation gradually falls to zero (Fig. 5-5A).

Because facilitation is seen when the two nerves are stimulated concurrently with no time for an intervening synapse, this is termed a *direct facilitation*. It is brought about by afferent fibers from each set of afferents from the two muscles synapsing directly on their synergistic motoneurons. The concurrent excitation in the synergistic nerve causes an excitatory activity in the membranes of the fringe group motoneurons acted on in common by the afferents of the synergists, bringing them to excitation (cf. Fig. 5-2). Reversing the order of test and conditioning stimulations presented to the synergistic pair of nerves gives rise to the same pattern of facilitation; i.e., they show a *reciprocal facilitation*.

Because of the simplicity of the monosynaptic response, consisting of one presynaptic fiber ending on one postsynaptic neuron, a similar direct system for testing *inhibition* from reciprocally paired inputs of antagonistic muscles can be carried out. In this case, however, because a special inhibitory interneuron is interposed (to be more fully discussed in the next section), there is a difference in the pattern seen. Using a reciprocal pair of muscle nerves (e.g., anterior tibial and gastrocnemius muscles), a test stimulus delivered to one afferent input is preceded by a conditioning shock to the nerve of the antagonistic muscle, and the test response shows a diminution; i.e., it is inhibited (Fig. 5-5B). When the conditioning and test stimuli are delivered concurrently, very little inhibitory effect may be seen, but with a small increase in the conditioning-test interval, the inhibition increases to a peak at an interval of 0.5 msec between the volleys (Fig. 5-5B). As the time between the two stimuli is increased further, the test response gradually regains its

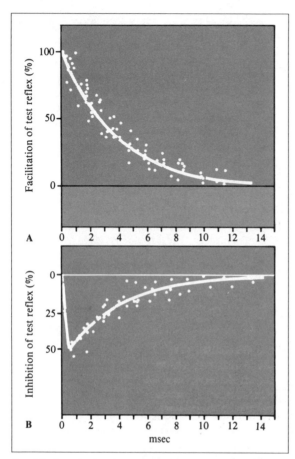

Fig. 5-5. Direct facilitation and "direct" inhibition. The monosynaptic response is used as a test response and is preceded at the time intervals shown on the abscissa by a conditioning shock. In (A), the conditioning shock is a volley in a synergistic nerve. In (B), the conditioning volley is to an antagonistic nerve. Both facilitatory and inhibitory effects decline gradually to control levels in approximately 15 msec. (From D. P. C. Lloyd. Facilitation and inhibition of spinal motoneurons. *J. Neurophysiol.* 9:425, 429, 1946.)

normal size over a period of approximately 15 msec. An explanation of this difference between facilitation and inhibition requires an understanding of their synaptic mechanisms, described in the next section. But we note here that the extra synaptic delay due to the inhibitory interneuron is responsible for the different pattern seen with inhibition.

MECHANISMS UNDERLYING SYNAPTIC TRANSMISSION

The preceding discussion dealt with populations of motoneurons involved in the monosynaptic response. A deeper analysis requires a knowledge of the events occurring in individual motoneurons. Microelectrodes were first successfully used for this purpose by Eccles and his colleagues. These were inserted into the substance of the spinal cord until the entry of the tip of the electrode into the soma of the motoneuron was indicated by the sudden appearance of a negative potential of approximately 70 mv, the resting membrane potential. With a microelectrode inside the motoneuron cell body, the stimulation of its axon in the ventral root gave rise to a large action potential with overshoot (Fig. 5-6A). The latency of the antidromic response was very short, as expected from the short conduction distance between the ventral root and the cell. When the dorsal root was stimulated to produce an orthodromic activation, a depolarizing potential appeared with a latency of 0.3 to 0.5 msec (Fig. 5-6B). This response is recorded in the cell after it is excited synaptically and thus is referred to as an *excitatory postsynaptic potential* (EPSP). The EPSP lasts approximately 15 msec and has a form similar to that of the end-plate potential (EPP) recorded in the muscle (see Chap. 3). It increases in size as the volley to the afferent fibers is increased in strength and as more afferent fibers add their synaptic excitation onto the motoneuron. When the EPSP becomes large enough, and a critical level of depolarization is reached, an action potential is excited in the motoneuron (Fig. 5-6C). The action potential propagates down the axon and excites the muscles innervated by that motoneuron.

The EPSP size is related to the number of presynaptic fibers activating bouton endings distributed over the membrane of the motoneuron. Each synapse gives rise to a small current, adding to the EPSP. Their addition accounts for the facilitation found in the interaction studies discussed in the preceding section in which the EPSP produced by a conditioning volley summates with a second EPSP produced by a volley in a synergistic nerve. The temporal facilitation is brought about by the summation of EPSPs if these occur in a period of approximately 15 msec, the time course of the EPSP (Fig. 5-7).

When an antagonistic nerve to the motoneuron is stimulated, a smaller hyperpolarizing type of response, an *inhibitory postsynaptic potential* (IPSP), is seen in the motoneuron (Fig. 5-8). The effect of the inhibitory excitation is to reduce membrane potential below the critical level for firing and so to block excitation. As a result of the inhibitory action, an EPSP cannot depolarize the cell to the critical level (Fig. 5-8). The IPSP also increases the ionic conductance of the membrane, which further acts to reduce the effectiveness of an EPSP to excite an action potential.

The delay found for the IPSP is greater than that for an EPSP because of the additional interneuron on the *inhibi-*

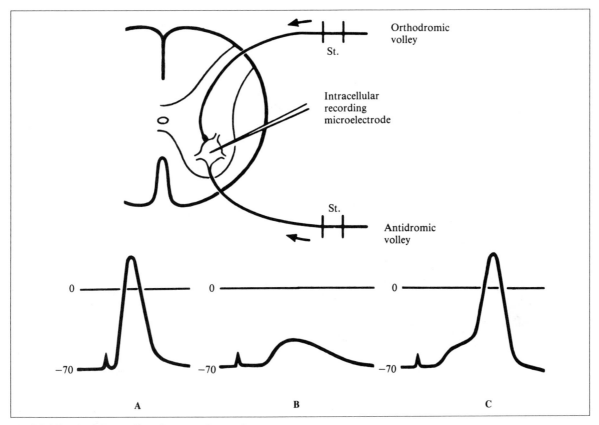

Fig. 5-6. Microelectrode recording of responses from motoneurons. A recording microelectrode is shown inside a motoneuron in the top diagram. Stimulation (*St.*) of the cell's fiber in the ventral root gives rise to an antidromic response (A). Stimulation orthodromically with a weak shock gives rise, after a synaptic delay, to an excitatory postsynaptic potential (EPSP) (B). Stimulation orthodromically with a stronger shock gives rise to EPSP and an action potential (C). The response recorded in the soma on excitation of the ventral root results from conduction in the reverse direction, and it is therefore called an *antidromic response* (Greek, *dromos*, or "running"). This is distinguished from propagation in the usual direction, i.e., *orthodromic* excitation via afferent nerve excitation.

tory line (Fig. 5-9). The afferent fibers have collaterals that synapse onto the inhibitory cells. They have short axons that synapse onto the antagonist motoneurons to give rise to an IPSP in them. The synaptic delay caused by transmission through these intervening cells accounts for the extra time needed for the inhibition to reach a maximum in the interaction studies mentioned in the preceding section (see Fig. 5-5B).

The action potential of the motoneuron does not normally arise from the cell body membrane. Evidence from a variety of cells indicates that the propagated action potential actually originates from the axon just distal to the cell body, in the unmyelinated *initial segment*. The soma membrane is, in fact, less excitable than the initial segment, as shown by the occasional failure of antidromic invasion into the soma. Evidence for this position was obtained from the directly visualized stretch receptor neurons of crustaceans (see Chap. 3). With recording electrodes placed at the soma and at the initial segment, responses were seen to be initiated first at the initial segment. The sequence of events during normal orthodromic excitation is therefore as follows: (1) An EPSP is excited in the soma by synaptic activity, resulting in a flow of current outward through the initial segment of the cell; (2) excitation of an action potential follows at the initial segment, with propagation down the axon to the muscle; (3) at the same time, an action potential is propagated back into the cell body to give rise to a *soma-dendrite action potential*.

The initial segment is the common site on the cell where

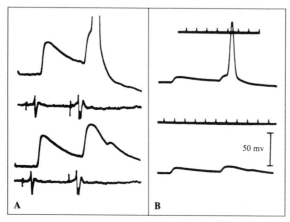

Fig. 5-7. Synaptic potentials of motoneurons. Two EPSPs recorded from inside a motoneuron in response to two successive weak shocks are shown in (A) at high and in (B) at lower amplification. The EPSPs are smaller and longer-lasting depolarizing potentials. The first EPSP of the upper traces is ineffective, but the second, occurring a short time afterward, excites an action potential. In the lower trace, EPSPs only are seen. A small shift in excitability caused the second EPSP to be ineffective. (From J. C. Eccles. *Neurophysiological Basis of Mind*. London: Oxford University Press, 1953. P. 133.)

the depolarizations of the EPSP are tallied up. If a large number of excitatory synapses are fired, a larger flow of outward current passes through the membrane of the initial segment. If at the same time, or just previously, an inhibitory input is excited and IPSPs are elicited, the resulting hyperpolarization of the initial segment also adds up at the initial segment, reducing the depolarization below the critical level required for the firing of propagated action potentials (see Fig. 5-8, C–E).

SYNAPTIC TRANSMITTERS. The irreducible latency between the activation of the afferent nerve terminals and the beginning of an EPSP is taken as evidence that a transmitter substance is released from the terminals to cause an excitation of the postsynaptic membrane. However, the transmitter substances giving rise to the EPSP and IPSP are as yet unknown. The transmitters for EPSP and IPSP are known to differ because of the opposite electrical actions they produce on the postsynaptic membrane of the motoneurons, depolarization and hyperpolarization respectively.

The ionic permeability increases mainly to Na^+ and K^+ in response to excitation, causing the membrane to shift toward a zero equilibrium potential. The inhibitory transmitter substance, on the other hand, produces a more selec-

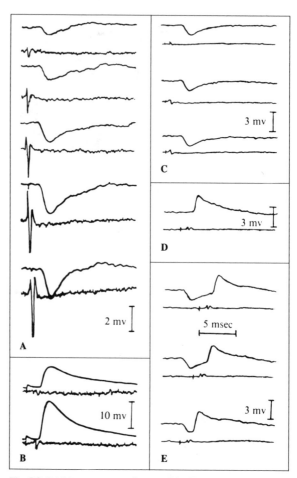

Fig. 5-8. Inhibitory postsynaptic potentials. The hyperpolarizing IPSPs recorded with a microelectrode from within motoneurons are shown at increasing strengths of stimulation in (A). The action potential of the afferents in the dorsal root volley is shown on the lower traces. As the strength is increased, the size of the IPSP increases. The IPSPs are to be compared with EPSPs excited in these same motoneurons (B). In (C), inhibitory volleys are set up and then (D) an excitatory volley. Finally, in (E), an excitatory volley follows soon after an inhibitory volley. The IPSP decreases the resultant depolarization brought about by the EPSP to below the critical level for firing. (J. C. Eccles. *Neurophysiological Basis of Mind*. London: Oxford University Press, 1953. P. 155.)

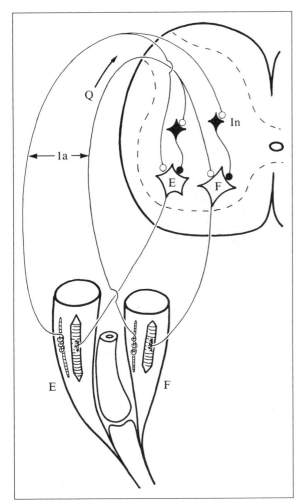

Fig. 5-9. Afferent fibers from spindle stretch receptors of extensor muscle (*E*) are shown terminating on motoneurons, their efferent axons innervating that same muscle. Collateral branches from the afferent fibers branch to terminate on cells in an intermediate group of cord neurons. These are inhibitory cells terminating on flexor (*F*) motoneurons to inhibit their discharge. Conversely, excitation fibers from flexor muscles terminate onto flexor motoneurons and extensor motoneurons. (Modified from J. C. Eccles. *The Understanding of the Brain* [2nd ed.]. New York: McGraw-Hill, 1977.)

tive permeability increase to K$^+$ and to Cl$^-$, with a resultant shift to an equilibrium potential at a higher membrane potential (i.e., from -70 to -80 mv). These different ionic permeability changes produced by EPSP and IPSP activation were revealed by the use of double-barreled microelectrodes inserted into motoneurons. One barrel recorded the EPSP or IPSP; the other was used to pass current and different ions into the cell. By this means it was shown that the inhibitory transmitter permits only ions of smaller overall size to pass through the membrane, apparently because a smaller channel opening is produced in the membrane by the inhibitory transmitter, as compared to the wider channels open by the excitatory transmitter (Fig. 5-10).

Another interesting difference between the action of the two transmitters is that the convulsant agent strychnine blocks the effect of the inhibitory transmitter without affecting excitatory activation. Some recent candidates for transmitter substances are believed to be amino acids. Glutamic acid has been considered as possibly the excitatory transmitter, and gamma-aminobutyric acid and glycine the inhibitory transmitter, with most evidence favoring glycine as the transmitter in the cord.

RENSHAW CELL INHIBITION. A special type of inhibitory interneuron, the *Renshaw cell,* has been found related to the motoneurons within the ventral horn portion of the spinal cord (Fig. 5-11). Collateral fibers from the motoneurons synapse on the Renshaw cells, their axons in turn synapsing onto a number of motoneurons in their immediate vicinity to give rise to a long-lasting inhibition in them. The repetitive discharge of the Renshaw cell is responsible for the long-lasting inhibitory effect.

The functional significance of the inhibition brought about by Renshaw cell activity is as yet not completely determined. It may act to sharpen the effect of those motoneurons that have fired by inhibiting other motoneurons in the pool that have not fired. Apparently, Renshaw cell effects are greater on the *tonically* active motoneurons that supply muscles important in maintaining posture for long periods of time than on responses that are rapid (*phasic*), those responsible for movement. Phasic changes take place when posture is interrupted, to be followed by another tonic position or a series of movements.

The Renshaw cells have a further significance. The motoneuron collateral fiber ending on the Renshaw cell releases acetylcholine (ACh) as its transmitter. This action was inferred by Eccles from what he termed *Dale's principle,* namely, that the same transmitter is released at all branches of a neuron. Since the motoneuron releases ACh from its terminals at the neuromuscular junction, on this principle it

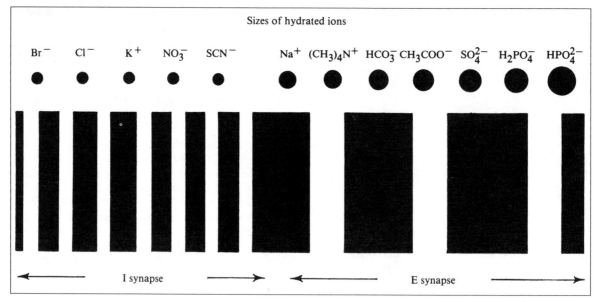

Fig. 5-10. The results with ion substitutions using multiple-barrel microelectrodes indicate that the smaller-sized ions Br^- to SCN^- can pass through the inhibitory (*I*) activated synapses, while after activation by the excitatory (*E*) transmitter substance, larger ions Na^+ to HPO^- can penetrate. The difference in the resulting equilibrium potentials will determine whether a depolarization or a hyperpolarization PSP will occur. (From J. C. Eccles. *Physiology of Nerve Cells.* Baltimore: Johns Hopkins Press, 1957. P. 220.)

should also release ACh at the motoneuron collaterals ending in the cord on the Renshaw cells. Injection of substances into the bloodstream that are known to excite or block ACh transmission at the neuromuscular junction should likewise excite or block synaptic activation of the Renshaw cells. The substance dihydro-β-erythroidine, which, unlike D-tubocurarine, can pass the blood-brain barrier, was found to block Renshaw cell activity (Fig. 5-11). Conversely, ACh and cholinomimetic substances were able to prolong the discharge of the Renshaw cell, as expected from the analogous actions postulated. While the evidence supports the inference that the collaterals release ACh as the neurotransmitter exciting Renshaw cells, a number of examples are known of neurons containing several transmitters, and it has not been excluded that different transmitters may be released from different branches of the same neuron. It should be emphasized, however, that as yet no such evidence for this is in hand, and the standard position, Dale's principle, is that the same neurotransmitter is released from all the branches of a given neuron.

THE GAMMA LOOP SYSTEM. As has been noted in Chapter 3, muscle spindle receptors are excited by stretch. The large-diameter afferent fibers arise from the central portion of the spindle receptor as anulospiral endings. Elongation of the spindle deforms the sensory terminations of the fibers, thereby exciting discharges with a rate proportional to the degree of stretch (see Chap. 3). Because the spindles are

anatomically in parallel with the contracting muscles, as shown in Figure 5-12A, a twitch that contracts the *extrafusal muscles* (muscle fibers other than the spindle) will cause a relaxation of the stretch receptor, *unloading* the tension on it. The result is a diminution or cessation of the discharge in the spindle afferent fibers known as the *pause.* This is seen by recording from single fibers isolated from the dorsal roots. A given fiber is identified as coming from a spindle stretch receptor by the pause (Fig. 5-12A).

The spindle receptor has a complex structure (Fig. 5-13). Small motor fibers synapse on the muscular parts of the spindle receptor present on either side of the central receptor portion. The small motor fibers fall into the gamma range of fiber sizes and are thus referred to as *gamma motor* or *fusimotor fibers.* When the gamma motor fibers are fired, the muscular portions of the spindle receptor contract and stretch out the central portion of the receptor. This has the same effect as a pull on the muscle, elongating the spindle receptor and causing a discharge of the receptor. With an increased excitation of gamma discharge, a smaller stretch of the spindle is required to elicit an afferent discharge.

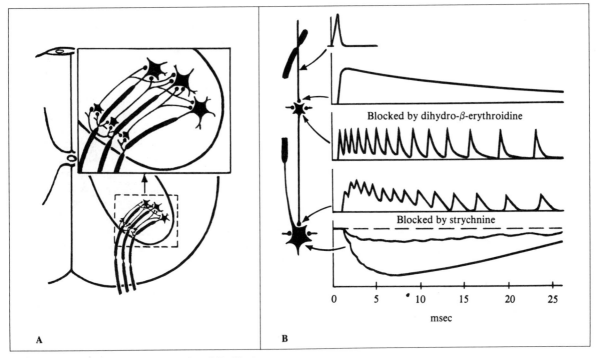

A

B

Fig. 5-11. Renshaw cell activity of the spinal cord. In (A), the ventral horn of the spinal cord is shown and, in the inset, an enlarged view. From the axons of three motoneurons, collaterals can be seen taking origin from the first node. These synapse on smaller cells (Renshaw cells), which in turn send axons to the surrounding motoneurons and branches back to the same motoneuron. In (B), the collateral from the motor horn cell axon is shown at the top synapsing on a Renshaw cell. The Renshaw cell, when it discharges, does so at a very rapid rate and gives rise to a series of discharges that can be blocked by dihydro-β-erythroidine. The Renshaw cell in turn synapses on motoneurons, and at this point, inhibitory mediator substance is released in successive summated IPSPs, which can be blocked by strychnine. (From J. C. Eccles et al. Cholinergic and inhibitory synapses in a pathway from motor-axon collaterals to motoneurons. *J. Physiol.* [Lond.] 126:542, 557, 1954.)

When the gamma system is less active or stopped altogether, a greater stretch must be placed on the spindle to cause the same afferent discharge. If great enough, gamma activity can overcome the pause, and the stretch receptor can discharge at a high rate even during a twitch response.

Activity in the CNS, by exciting or inhibiting the gamma motor fiber discharge, can in turn alter the sensitivity of the spindle receptor. An excessive discharge of the gamma neurons appears to underlie the pathological state of *spasticity* resulting from an increased discharge from higher brain centers augmenting gamma motoneuron activity. This mecha-

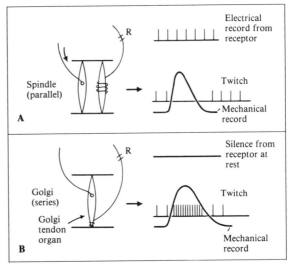

Fig. 5-12. Muscle receptor organs. Two main types of muscle receptor organs are pictured, the spindle stretch receptor (A) and the Golgi tendon organ (B). The spindle is placed in parallel with the muscle fibers. When the muscle is caused to contract, the electrical discharge recorded from a spindle receptor shows a decrease, the "pause." The Golgi tendon organ is in series with skeletal muscles, and contraction of the muscle during a twitch causes a speeding up of the discharge rate during the time of the muscle contraction.

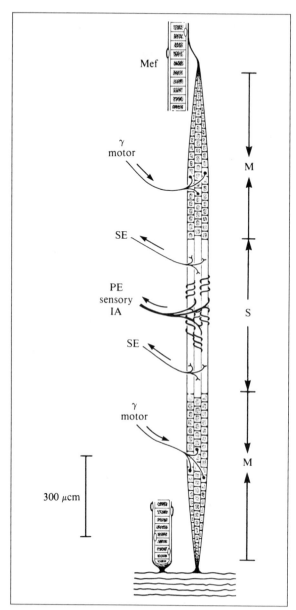

Fig. 5-13. The primary sensory (*PE*) nerve endings of the mammalian spindle terminate in an annulospiral configuration around the center of the receptor. These are type 1A sensory fibers. Secondary receptor (*SE*) fibers terminate at the sides in the receptor (*S*) area. In addition, a muscular (*M*) portion of the receptor is shown containing striped fibers. These are innervated by gamma (γ) motor fibers. Portions of extrafusal muscle fibers (*Mef*) are shown at the side. (Modified from D. Barker. The innervation of the muscle spindles. *Q. J. Micros. Sci.* 89:143–186, 1948.)

nism is responsible for the condition of exaggerated hypertonus and increased reflex excitability known as *decerebrate rigidity*. It is produced by transecting the brainstem of a dog or cat at the level of the midbrain between the colliculi. The position of the animal in decerebrate rigidity is one of extreme hypertension of the limbs, with the head arched back and the tail elevated, in the posture known as *opisthotonos*. This state has been called a caricature of standing. If, in such a decerebrate preparation, the dorsal roots are cut, the limb immediately becomes flaccid, proving that the decerebrate state is produced by reflex activity of stretch reflexes. An augmented gamma activity was shown by the early appearance of increased discharges in recordings taken from thin slips from the ventral roots and identified as gamma motor fiber discharge; discharges increase during the onset of decerebrate rigidity.

If the leg of an animal showing decerebrate rigidity is forcibly flexed, it resists the applied flexion until at a certain point the resistance suddenly melts away. This phenomenon is known as the *knife-clasp reflex* or the *lengthening reaction*. At some high level of stretch, sensory afferents from the tendon regions of muscles, the *Golgi tendon organs,* reach threshold, and they have an inhibitory effect on the motoneurons. This may be a protective reflex. The Golgi tendon organs, unlike the spindles, are placed in a series with respect to the skeletal muscle (see Fig. 5-12B). Such afferents may be demonstrated in the following way: a single sensory nerve fiber isolated in the dorsal root is identified as coming from a Golgi tendon organ if it shows an increased repetitive activity during a twitch of the muscle rather than the pause, as appears in the case of a spindle afferent (see Fig. 5-12B). Another identifying characteristic is the lack of effect of increased gamma motor activity on the discharge rate of the Golgi receptor.

Synaptic boutons are found terminating on afferent terminals within the spinal cord. These *presynaptic endings* can inhibit the discharge of the afferent terminal and the cell on which they synapse. The precise relation of presynaptic inhibition to the usual postsynaptic inhibition is a matter of present investigation but seems to act as another means of selective control of afferent input to the nervous system.

BIBLIOGRAPHY

Barker, D. The innervation of the muscle spindles. *Q. J. Micros. Sci.* 89:143–186, 1948.

Barker, D. (ed.). *Symposium on Muscle Receptors.* Hong Kong: Hong Kong University Press, 1962.

Boyd, J. A. The isolated mammalian muscle spindle. *Trends Neurosci.* 3:258–264, 1980.

Boyd, J. A. Eyzaguirre, C., Mathews, P. B. C., and Rushworth, G. *The Role of the Gamma System in Movement and Posture*. New York: Association for the Aid of Crippled Children, 1968.

Creed, R. S., Denny-Brown, D., Eccles, J. C., Liddell, E. G. T, and Sherrington, C. S. *Reflex Activity of the Spinal Cord*. London: Oxford University Press, 1932. (Reprinted, 1972.)

DeMyer, W. D. *Technique of the Neurologic Examination* (2nd ed.). New York: McGraw-Hill, 1974.

Desmedt, J. E. (ed.). Human Reflexes. Pathophysiology of Motor Systems. Methodology of Human Reflexes. In *New Developments in Electromyography and Clinical Neurophysiology*. New York: Karger, 1973. Vol. 3.

Eccles, J. C. *Neurophysiological Basis of Mind*. London: Oxford University Press, 1953.

Eccles, J. C. *Physiology of Nerve Cells*. Baltimore: Johns Hopkins University Press, 1957.

Eccles, J. C. *The Physiology of Synapses*. New York: Academic, 1964.

Eccles, J. C. The Dynamic Loop Hypothesis of Movement Control. In K. N. Leibovic (ed.), *Information Processing in the Nervous System*. New York: Springer-Verlag, 1969.

Eccles, J. C. *The Understanding of the Brain* (2nd ed.). New York: McGraw-Hill, 1977.

Mathews, P. B. C. *Mammalian Muscle Receptors and Their Central Actions*. Baltimore: Williams & Wilkins, 1972.

Ochs, S. *Elements of Neurophysiology*. New York: Wiley, 1965.

Ottoson, D. Morphology and physiology of muscle spindles. In R. Llinas and W. Precht (eds.), *Frog Neurobiology*. Heidelberg: Springer-Verlag, 1976. Pp. 643–675.

Phillis, J. W. *The Pharmacology of Synapses*. Elmsford, N.Y.: Pergamon, 1970.

Sherrington, C. *The Integrative Action of the Nervous System*. Cambridge, Engl.: Cambridge University Press, 1906. (Rev. reprint, New Haven: Yale University Press, 1947.)

Willis, W. D., and Coggeshall, R. E. *Sensory Mechanisms of the Spinal Cord*. New York: Plenum, 1978.

PROPERTIES OF THE CEREBRUM AND HIGHER SENSORY FUNCTION

6

Sidney Ochs

The term *encephalization* refers to the increased power of command that the more headward part of the central nervous system exerts on the lower reflex mechanisms found in the higher phyla. The cerebral cortex makes its appearance late in phylogenetic development and has thus been taken to be the seat of intelligence. But, as will be seen, subcortical centers are also important for the exercise of higher functions.

Clinical observations indicating that certain functions could be localized within the cerebral cortex were confirmed when Fritsch and Hitzig in 1870 showed that electrical stimulation of certain regions of the cortex gave rise to *motor responses* (see Chap. 8). Later, various other areas were found to receive *sensory inputs*. Still other areas in the cortex, designated as *associational*, were relegated to the control of higher functions (see Chap. 9). However, the concept of strictly separable motor and sensory centers in the cortex has been modified with the realization that complex interrelations exist between them as indicated by the term *sensorimotor center*. We will, for didactic reasons, treat them separately. In this chapter the electrical activity of the cerebrum with relation to the changes seen during sleep and wakefulness, sensory reception in the cortex, and synaptic transmission and its neural connectivity will be discussed. A discussion of motor control from the cerebrum will be reserved for Chapter 8.

THE ELECTROENCEPHALOGRAM

In 1875, Caton first recorded the continuous electrical activity in the brain by placing electrodes on the exposed cerebrum of animals. In 1929, Berger, using the string galvanometer, showed that similar electrical activity could be recorded through the intact human skull. Of great importance was his finding that this electrical activity changed with sleep or altered attention. The availability of improved electronic instrumentation since then has allowed the *electroencephalogram* (EEG) to become a valuable clinical tool in the study of epilepsy, brain tumor localization, and the

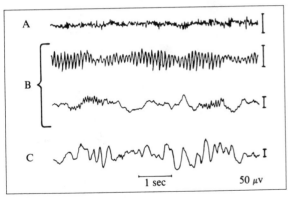

A

B

C

1 sec   50 μv

Fig. 6-1. The EEG during different states of sleep and wakefulness. The EEG of the alert state (A) shows a low-amplitude, fast *beta*-wave type of activity. In the drowsy or light sleep state (B), *alpha* activity is seen. There may be periods in the drowsy state when a mixture of alpha- and slower-wave activity is seen. In deep sleep (C), large-amplitude, slow *delta* waves are predominant. (From W. Penfield and T. C. Erickson. *Epilepsy and Cerebral Localization*, 1941. P. 401. Courtesy of Charles C Thomas, Publisher, Springfield, Illinois.)

diagnosis of other neurological conditions. It is also of value in relating brain changes to higher function (see Chap. 9).

To record the EEG, electrodes are placed over the surface of the skull in a regular pattern using *monopolar* or *bipolar* recording. Monopolar recording is accomplished with an *active electrode* placed on the scalp and an *indifferent electrode* placed on inactive tissue at a distance. The earlobes are usually used as the site for the indifferent electrode. Bipolar recording is accomplished with two electrodes placed over active sites on the scalp. The potential recorded between them at any instant is the algebraic resultant of voltage appearing under each electrode. Figure 6-1 shows an EEG taken from a human subject. A fairly regular 8 to 12 per second series of waves may be readily distinguished in portions of the records. These are known as the *alpha waves,* although the term *Berger rhythm* is sometimes used in honor of their discoverer. The alpha rhythm is usually present in a subject who is relaxed, with eyes closed, in a quiet room. If the subject is asked to open his eyes, the alpha rhythm is abruptly blocked and replaced with the smaller-amplitude and faster-frequency *beta* waves. The same blocking of the alpha rhythm can occur if the subject is asked to visualize a scene with his eyes closed and a sudden sound is made or if the subject is otherwise alerted to some unexpected sensory stimulus.

If the blood supply to the brain is suddenly interrupted, the EEG activity disappears within 15 to 20 seconds, with

loss of consciousness. If the blood supply to the brain is returned within approximately 5 minutes, recovery is generally complete (see Metabolism). The correlation of EEG activity to consciousness is clear.

The various rhythms in the EEG may be grouped according to the dominant frequencies found. The alpha waves range from 8 to 12 Hz, the beta waves from 18 to 30 Hz, and the *theta* waves from 4 to 7 Hz. Those of still slower frequencies, below 4 Hz, are referred to as *delta* waves. They are usually larger in amplitude than the alpha waves and are often seen during sleep (see below). The wave form of the EEG may be difficult to analyze into dominant frequencies by simple inspection. Automatic frequency analyzers have been used to measure the EEG spectrum, usually based on Fourier analysis. Frequency analyzers assess the relative amount of activity in selected divisions of the frequency spectrum. The relative amount of activity present in the various frequency bands within a short sampling time is then displayed so that the proportion of alpha, beta, theta, and delta activity can be determined numerically.

The EEG waves are believed to reflect activity within the dendrites of pyramidal cell neurons of the cerebral cortex (Fig. 6-2). A large number of the apical dendrites of the pyramidal cells are arranged vertically in the cortex, their finer branches most numerous in the upper (molecular) layer of the cortex. Many such neuronal elements must be synchronously activated to produce the EEG voltages recorded. The voltages at the surface of the cerebral cortex result from the current flow between the active region of the apical dendrites and the rest of the neuron and are explainable on the basis of general electrical principles. Voltage differences occurring between the upper and lower shafts of the apical dendrites or the cell bodies result in the flow of current in the extracellular spaces from polarized to depolarized parts of the cell. Current flow in the brain occurs in a three-dimensional conducting volume, giving rise to the complex electrical potentials that can be recorded from the surface of the brain. For purposes of this discussion it will suffice to consider only the simplest case. Areas of depolarization on the neuron membrane—for example, the dendrites—are called *sinks*. Current flows into them from polarized parts of the membrane, referred to as *sources*. An external electrode records the voltage drop, the size of the potential recorded being a function of the number of neurons synchronously discharging. The depolarized regions must also have similar spatial orientations so that their currents can add together.

A propensity for rhythmic response appears to be a fundamental property of cells in the cerebral cortex. To study the origin of rhythmicity, isolated regions or *islands* of cere-

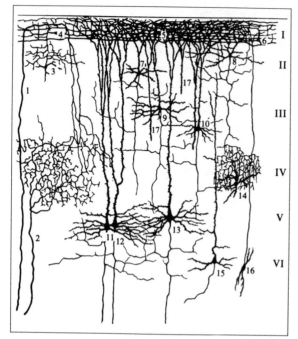

Fig. 6-2. Section through the cerebral cortex with characteristic cells through all the layers I–VI. Pyramidal cells 7, 10, 11, 12, and 13 have apical dendrites extending up to the molecular (first) layer, where they arborize profusely. Other cells include short axon cell 3 and Golgi type II cell 14. (From H.-T. Chang. Dendritic potential of cortical neurons produced by direct electrical stimulation of the cerebral cortex. *J. Neurophysiol.* 14[3]:1, 1951.)

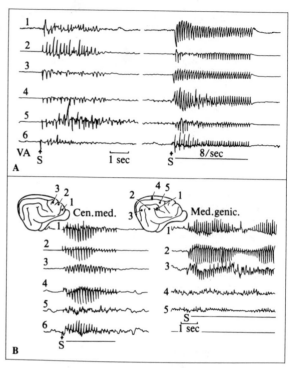

Fig. 6-3. Recruiting waves excited by stimulation of various nonspecific thalamic nuclei. In (A), to the right, a train of eight stimuli per second was initiated at *S*. The evoked responses are large and follow the rate of stimulation. To the left, a single stimulus was delivered, and a train of repetitive responses was excited. In (B), nuclei of the nonspecific group are excited with single shocks, as in (A), with trains of stimulation. Recruiting waves are apparent in the various regions of the cat's cortex. (From J. Hanberry and H. H. Jasper. Independence of diffuse thalamo-cortical projection system shown by specific nuclear destructions. *J. Neurophysiol.* 16:256, 1953.)

bral cortex have been prepared. These are made by means of cortical cuts produced some days or weeks previously, the blood supply to the island left intact but with all neuronal connections severed. When an island is electrically stimulated, rhythmic responses are excited. Without such stimulation the islands remain electrically silent. An external source of activation, therefore, is considered to *trigger* rhythmic behavior from the cortex. A trigger or *pacemaker* for the EEG waves located in the thalamus was shown by using electrodes insulated except at their tips and inserted into various nonspecific and interlaminar regions of the thalamus. Stimuli delivered at a rate close to that of the naturally occurring EEG *spindle waves* evoked similar rhythmic waves in the cortex (Fig. 6-3A). Spindle waves normally show *recruitment*, in that the successive waves in the brief train increase in amplitude as more cells add their discharge to the group. Then the cells fall out of the re-

sponding group, thus accounting for the waxing and waning of the amplitude of the waves in the spindle. Spindles are seen in sleep and in barbiturate anesthesia. A single shock delivered to the thalamus may excite a long train of spindle waves in the cerebral cortex (Fig. 6-3B).

Single-cell studies of thalamic nuclei have shown that a long-lasting hyperpolarization of approximately 100 msec occurs in their neurons following an action potential. This long period of positivity is close to that of waves seen in the EEG and is considered to be responsible for the pacemaker activity in the thalamus. The long-lasting inhibition appears to be produced by the same mechanisms in the cord as when Renshaw cells are activated (see Chap. 5).

ALERTING AND THE SLEEP-WAKING CYCLE

During sleep a shift to lower-frequency waves is seen in the EEG. In general, the EEG from alert subjects is largely composed of *fast-wave (FW) activity,* whereas sleeping individuals show a large proportion of delta waves, or *slow-wave (SW) activity.*

By transecting the brainstem of cats between the colliculi of the midbrain, a *cerveau isolé* preparation is produced (cut B in Fig. 6-4). The EEG taken from the brain of this preparation shows long runs of slow-wave activity, as is typical of a sleeping animal. This cut interrupts the upward influences coming both from the specific sensory afferents and from the *reticular formation.* It was pointed out by comparative anatomists many years ago that the reticular formation, composed of cells with interspersed fibers, might well have an integrative function. Collaterals branching off from the various sensory tracts passing up to the cerebral cortex relay impulses into the reticular formation on their passage through the brainstem. These collateral inputs were demonstrated by recording *evoked responses,* i.e., a grouped discharge of neurons in the reticular formation elicited by peripheral sensory stimulation. Evoked responses are elicited in response to sound, skin touch, or a light flash to the eye. The evoked responses recorded in the reticular formation to all these different sensory sources showed *occlusion;* that is, a second evoked response was reduced in amplitude if it followed too soon after a first such response. Occlusion was also shown between responses to the different sensory modalities. Occlusion indicates that the different excitatory inputs on entering the reticular formation synapse onto a common group of cells, reflect the general level of sensory excitation. The fibers of the reticular formation leading to the cerebral cortex signal the general state of the organism, whether alert or sleeping, as will be described in the next section.

The reticular formation can be selectively destroyed by making lesions in the medial portion of the medulla, the lateral part of the brainstem containing the specific afferent pathways passing upward to the thalamus. Lesions made in the middle part of the medulla containing the reticular formation gave rise to a *comatose* state. The EEG recorded from the animals showed the slow, large-amplitude EEG pattern characteristic of deep sleep. When lesions were placed laterally in the brainstem, in the tract containing the ascending specific sensory pathways, the animals remained behaviorally in a waking state and had an EEG pattern typical of an alert animal.

Alerting or *arousal* refers to the change seen in the EEG

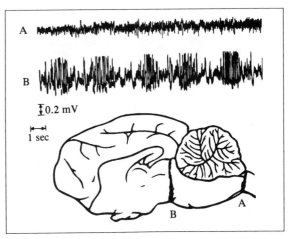

Fig. 6-4. EEG patterns after brainstem transections. The medial surface of the brain is represented in the diagram. Cut *A* produces an *encéphale isolé* preparation. The EEG record taken from this preparation and shown in line *A* has the low-amplitude and fast-wave activity characteristic of the alert animal. A transection made through the midcollicular region of the brainstem (cut *B*) produces a *cerveau isolé* preparation. The EEG pattern (line *B*) in this case is of the sleep type, with larger-amplitude slow waves. (From F. Bremer. L'activité cérébrale au cours du sommeil et de la narcose. Contribution à l'étude du mécanisme du sommeil. *Bull. Acad. R. Med. Belg.* 2:68, 1937. Modified from M. A. B. Brazier. *The Electrical Activity of the Nervous System* [3rd ed.]. New York: Macmillan, 1968. P. 244.)

of a drowsing animal when it is suddenly presented with a brief sensory stimulation, such as a click of sound or pinching the skin. In this event the slow-wave activity is abruptly replaced by a low-amplitude pattern of fast EEG activity that may persist for some time after the stimulus (Fig. 6-5A). The pattern of arousal or *desynchronization* of the EEG is found to be widespread in the cortex. A similar period of alerting of the EEG could be obtained following electrical stimulation of the reticular formation by means of electrodes inserted into its substance (Fig. 6-5B).

When a complete transection is made below the brainstem, an *encéphale isolé* preparation is produced (see cut A in Fig. 6-4). The animals may be studied for several hours thereafter. The EEG pattern (see EEG record A in Fig. 6-4) is fast; i.e., it is one representative of a state of wakefulness. The difference between this section and the cerveau isolé is that sensory inputs coming into the brainstem from the trigeminal nerve innervating the head are retained in the encéphale isolé. These inputs into the reticular formation activate it and produce the wakeful EEG state in the cortex.

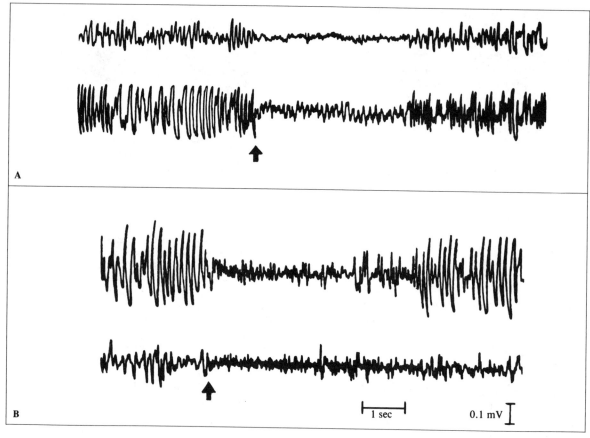

Fig. 6-5. Alerting responses in the brain. In (A), the EEG obtained from the motor cortex and visual cortex is shown. After a sudden sensory stimulation (*arrow*), which may be a noise, a touch, or a flash of light, the EEG pattern changes to the alert type. Alerting or desynchronization of the EEG lasts for a brief time before return of the regular alpha or spindle type of record characteristic of the drowsy animal. In (B), a brief electrical stimulation is delivered to the reticular formation, and it has a similar alerting effect on the EEG of the cortex. (From F. Bremer. Neurophysiological Mechanisms in Cerebral Arousal. In G. E. W. Wolstenholme and M. O'Connor [eds.], *Ciba Foundation Symposium on the Nature of Sleep*. Boston: Little, Brown, 1960. Pp. 32, 38.)

Destruction of the trigeminal nerve afferents in an encéphale isolé preparation, a major sensory input, promptly brings about the slow-wave EEG pattern typical of sleep.

The state of sleep, while generally thought to be due to a diminished corticopetal sensory influence, may also be actively induced by electrical stimulation of thalamic nuclei. When electrodes are chronically implanted in this region in an animal, this is shown by behavioral changes in the animal on stimulation of thalamic regions at low frequencies. The animal exhibits diminished motor activity, seeks out a resting place, curls up, and appears to fall into a natural state of sleep. It may be wakened by stimulation at another site in the brainstem or by a faster rate of stimulation at the same sleep-producing site. The systems involved in sleep-waking states will be discussed in the following section.

THE MONOAMINE THEORY OF SLEEP-WAKING

A group of monoamine-containing neurons have recently been shown to be present in the midbrain and diencephalon.

These fiber systems are present in the cord, brainstem, and cortex. In the brainstem they have a close relation to the control centers of visceral functions (see Chap. 8), and they appear to be involved in control of the sleep-waking state. The cells and their fibers are visualized by freeze-drying a section of the brain and treating it with formalin. A characteristic fluorescence of fibers may then be seen microscopically.

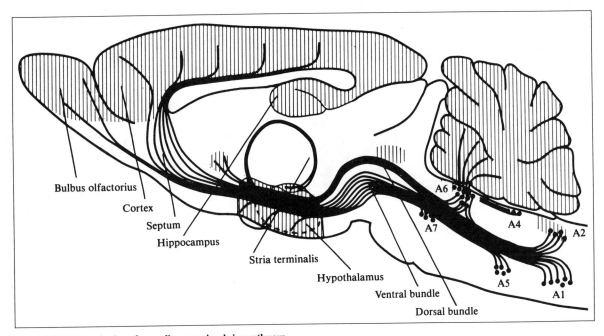

Fig. 6-6. Sagittal projection of ascending norepinephrine pathways. The striped regions indicate fiber termination areas. Descending norepinephrine pathways are not shown. (From U. Ungerstedt. Stereotoxic mapping of the monoamine pathways in the rat brain. *Acta Physiol. Scand.* [Suppl.] 367:27, 1971.)

Figure 6-6 shows the cell groups of origin and the terminations of three main groups of monoaminergic fibers in the brainstem. The *norepinephrine* group of fibers terminating in the cortex seems to be involved in the sleep-waking cycle as an activator (see below). The *dopaminergic* system extending through the brainstem (Fig. 6-7) and its dopamine-containing fibers have another pattern of distribution, with the fibers terminating in the striatum. This region of the brain is involved in motor control (see Chap. 8). Some of the other connections of the dopaminergic system are to the cerebellum and hypothalamus. In these two groups, as in the peripheral autonomic nervous system and the chromaffin tissue of the adrenal medulla, a set of related enzymes is found that leads from the amino acid tyrosine to the transmitters dopamine and norepinephrine. In the third group, phenylalanine, through related biochemical reactions, becomes yet another transmitter, 5-hydroxytryptamine (5-HT), or *serotonin*. The distribution of the *serotonergic* fibers is shown in Figure 6-8. This system is apparently also involved in the sleep-waking cycle.

An uptake system for their transmitters is present in the terminals of these neurons, a mechanism that may act to terminate transmitter action. A local synthesis of the transmitters also occurs in the fibers and in the terminals, and it is of interest that the enzyme required for such synthesis, dopamine-β-hydroxylase, is transported down to the terminals within the dense-core vesicles containing the transmitter norepinephrine by axoplasmic transport (see Chap. 2). Catabolic enzymes are also involved in the regulation of the level of the monoamine transmitters within the terminals through the enzymes *monoamine oxidase* and *catechol-O-methyltransferase*. It is possible to build up higher levels of monoamines in these fibers by treating animals with anti-monoamine oxidase agents such as isoniazid. These usually give rise to signs of increased behavioral excitability. Conversely, the monoamines may be depleted by the use of *reserpine* or other agents that can interfere with the uptake of nonepinephrine or 5-HT to cause in turn behavioral changes in the direction of sedation.

PARADOXICAL (REM) SLEEP

The study of sleep was revolutionized when it was recognized that *rapid eyeball movements* (REM) occur periodically during sleep. REM sleep is associated with an EEG pattern that paradoxically consists of fast low-amplitude waves looking very much like an awake or an alert EEG. This is termed *paradoxical sleep* or *fast-wave* (FW) sleep. If the subject is awakened during a REM period, he or she

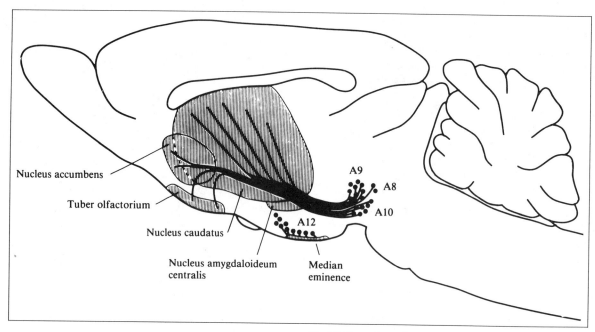

Fig. 6-7. Dopamine pathways shown in sagittal projection. The striped regions show fiber termination areas. (From U. Ungerstedt. Stereotoxic mapping of the monoamine pathways in the rat brain. *Acta Physiol. Scand.* [Suppl.] 367:27, 1971.)

reports a dream. REM periods were found to occur five or six times during a 7-hour sleep session and to occupy approximately 20 percent of the total sleep period in the adult.

REM periods with the same alertlike phases of the fast-wave EEG have been found during sleep in many species. In the cat they are associated with a marked decrease in the tonus of the neck muscles recorded electromyographically. Using the decreased muscle tone as an index of the paradoxical phase or REM period of sleep, it was shown that the decorticate cat still had such episodes, indicating a lower brainstem origin for the paradoxical-sleep periodicity. When lesions were made in the upper pons, these periodic changes were blocked. The rhombencephalon therefore appears to be the site of origin of these cycles of the alertlike EEG activity during sleep; hence, FW sleep is at times referred to as *rhombencephalic sleep*. Such active periods are initiated in the pons to cause alerting of the cortex. Judging from loss of neck muscle tonus, a downward discharge of the inhibitory region of the medullary reticular formation to the lower motor center (see Chap. 8) also takes place.

The periodic changes occurring during sleep are of great significance. As already mentioned, dreaming is associated with the FW periods, and subjects repeatedly wakened during this phase of sleep and prevented from fulfilling the usual time spent in dreaming may suffer hallucinations and other behavioral changes. In animals, too, such deprivation

leads to abnormal behavior. The implications for psychiatry are apparent. On the other hand, there are well-authenticated cases of people who require only a few hours or less of sleep and yet appear to be normal.

Jouvet (1972) has proposed that the serotonin system of neurons is involved in SW sleep. Depletion of serotonin pharmacologically, by injection of *p*-chlorophenylalanine or by the destruction of the cell bodies of the 5-HT fibers in the raphe nucleus, leads to a constantly wakeful state, an "insomniac" animal. This finding suggests that the system initiates or brings about sleep. Activation of the catecholamine systems is probably related to FW sleep, an activation of the cortex.

Destruction of the dorsal norepinephrine bundle or the groups originating from the A8 cell body of the norepinephrine fibers (see Fig. 6-7) leads to a decrease of low-voltage FW activity and sleeplike behavior; i.e, *hypersomnia* results.

The ontogeny of sleep patterns throws some light on the relation of 5-HT and norepinephrine to sleep and wakefulness. Newborn animals show mainly REM sleep, while the SW pattern develops later. The dual system of control of the sleep-waking cycle is shown by the relation of 5-HT and

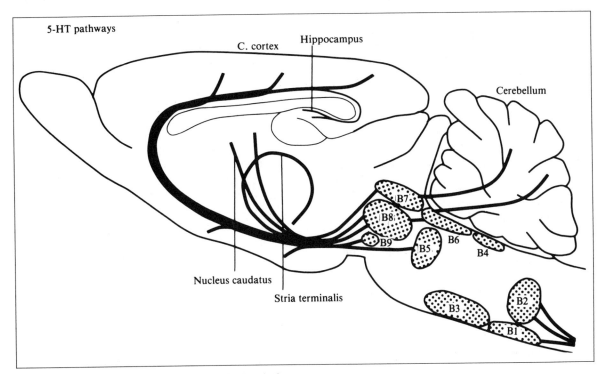

Fig. 6-8. Schematic diagram of the central 5-HT cell groups in the lower brainstem. The ascending, descending, and cerebellar 5-HT projections are shown in sagittal section. (From K. Fuxe and G. Jonsson. Further mapping of central 5-hydroxytryptamine neurons: Studies with the neurotoxic d. hydroxytryptamines. *Adv. Biochem. Psychopharmacol.* 10:2, 1974. Courtesy of Raven Press, New York.)

norepinephrine levels in neonates and in animals subjected to a variety of brainstem lesions. Slow-wave activity is associated with increased 5-HT levels and FW activity, with relatively more norepinephrine.

We have two fundamental problems to contend with in understanding the cyclic shift from sleep to wakefulness and back again. One pertains to the neuronal mechanisms that regulate cyclic changes. No doubt these involve interactions between the monoaminergic neuronal systems described. Another system has also been implicated, namely, the *acetylcholine* neurons present within the brainstem and cortex. These may have to do with the initiation of sleep, but it is uncertain how they and the ascending fiber systems in the cerebral cortex act to regulate consciousness. For an understanding of those mechanisms, a knowledge of the action of synaptic networks in the cortex is required.

SYNAPTIC CONNECTION AND TRANSMITTER CONTENT

New techniques have shown the full extent of the dendritic arborization of pyramidal and stellate cells in the cortex, only partially seen in Golgi preparations. Injection of the cell with procion yellow or with ^{3}H-glycine, which, after incorporation into protein, is spread by axoplasmic transport, reveals the great extent of the very fine branches in radioautographs. Synapses on the dendrites have features resembling the presynaptic terminals at the neuromuscular junction (see Chap. 3). The terminal bouton endings typically contain both vesicular structures (approximately 500 Å in diameter) and mitochondria. In the cerebral cortex these presynaptic terminals may end on special spinelike projections arising from the shafts of the apical dendrites of pyramidal cells to give rise to *axodendritic junctions*. Additionally, synaptic junctions similar to those on the somas of the spinal cord motoneurons are found on the cell bodies (see Chap. 5) at the *axosomatic junctions*.

A further advance in the study of the synaptic terminals was made possible through the development of the technique of *subcellular fractionation*. The brain is homogenized in an isotonic sucrose solution and subjected to centrifugation at different speeds. One fraction can be selected that contains presynaptic terminals separated from their fibers and sealed in a spherical structure. Vesicles, mitochondria, and, in some cases, part of the postsynaptic membranes torn off in the course of homogenization can be identified in these *synaptosomes*. On hyposmotic treatment of the synaptosomes, their membranes are ruptured and the vesicles are released. Pharmacological and biological studies of the vesicles have shown that acetylcholine, 5-HT, and norepinephrine are present in synaptic terminals.

A very high concentration of *glutamic acid* and *gamma-aminobutyric acid* (GABA) is characteristic of CNS tissue. By means of microiontophoresis, very small amounts of glutamic acid or GABA can be released in the immediate vicinity of neurons in the cortex. As a result, their discharge rate is greatly changed. Glutamic acid excites a rapid increase in the rate of discharge of most neurons, whereas GABA markedly inhibits such discharge. The fact that these effects are found with extremely low concentrations of these substances suggests that they may be excitatory and inhibitory transmitters respectively; or they may act as *modulators*, i.e., agents that change the excitatory state of cortical neurons in a more general fashion and over a longer period of time than does a neurotransmitter. A series of other putative neurotransmitters has been investigated.

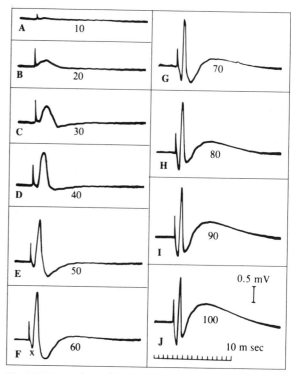

Fig. 6-9. Direct cortical responses (dendritic potentials). The responses elicited by direct stimulation of the surface in recording a few millimeters away are shown at successively increasing strengths of stimulation. The response is a slow-wave response, broader at the weaker stimulus strength, and at the higher strengths of stimulation it is followed by a longer-lasting second negative wave. (From H.-T. Chang. Dendritic potential of cortical neurons produced by direct electrical stimulation of the cerebral cortex. *J. Neurophysiol.* 14:1, 1951.)

DIRECTLY EVOKED RESPONSES

The EEG represents the outcome of the complex temporal and spatial interplay of neuronal activity in the cortex. To further the analysis of the electrical responses of the cortex and the neurons responsible, the discharge of a relatively large population of cortical neurons is synchronized by directly stimulating the exposed surface of the cortex with a single, brief pulse of current. This results in a characteristic surface-negative response recorded a few millimeters away, a *superficial response*, or a *direct cortical response* (DCR), which is seen as a surface-negative wave. Its amplitude is related to the strength of stimulation (Fig. 6-9). The response is due to an excitation of axons in the molecular (outermost) layer extending tangentially in the cortex and synapsing on the apical dendrites of pyramidal cells, where the response is generated (Fig. 6-10).

The nature of the negative-appearing DCR is not as yet completely known. It appears from interaction studies that it represents a graded response in the apical dendrites but does not have the properties expected of an excitatory postsynaptic potential (see Chap. 5). Some portions of the evoked responses of the cortex, to be described in the following section, appear to be composed of similar responses.

SENSORY AREAS AND EVOKED RESPONSES

Sensory inputs of various modalities are localized to specified regions in the cerebral cortex: visual input to a part of the occipital cortex; auditory response to part of the temporal lobe; and sensations (e.g., touch, pressure) from the skin to the postcentral cortex of the parietal lobe of the brain.

Nerve fiber pathways from peripheral sensory organs of the eye, ear, and skin have a relay on cells within the various specific thalamic nuclei, and the fibers of these nuclei in turn terminate within the separate sensory regions of the cerebral cortex. Histological studies of the cortex show that these *primary sensory areas* have distinctive cellular features, appearing as a six-layered structure. The visual area

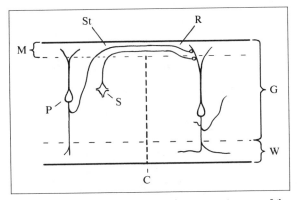

Fig. 6-10. Neuronal connections underlying the negative wave of the DCR. Stimulation (*St*) of axons in the molecular layer (*M*) inferred by passage over a cut (*C*) of all other layers. They may arise from pyramidal cell (*P*) collaterals or stellate cells (*S*). In the recording site (*R*), synapse of these axons on apical dendrites is shown, the latter generating the response. (From S. Ochs and H. Suzuki. Transmission of direct cortical responses. *Electroencephalogr. Clin. Neurophysiol.* 19:234, 1965.)

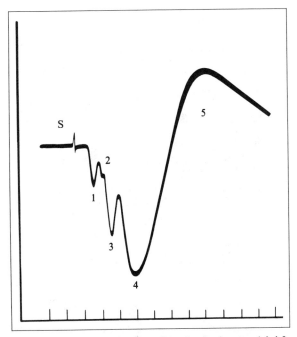

Fig. 6-11. Primary cortical response from the visual cortex. A brief electrical stimulation of the afferent pathway leading to a primary sensory area of the cerebral cortex (visual cortex in this case) gives rise to a series of fast, spikelike waves (1, 2, 3) inscribed on a slow positive wave (4), which is succeeded by a slow negative wave (5). Spike wave 1— and possibly 2— is due to the activity of the specific afferent endings within the visual cortex. Waves from 3 on are due to intracortical activity. (From L. I. Malis and L. Kruger. Multiple response and excitability of cat's visual cortex. *J. Neurophysiol.* 19:174, 1956.)

of the cortex has a striate appearance that is due to the position of the large number of terminating afferent sensory fibers and small *granular cells* in the fourth layer. These small cells predominate in sensory regions, and sensory cortical regions are therefore often called granular cortex. Because of the variation of fibers and the different cell types found in the different layers of the different cortical areas, these regions have been characterized and given separate letter or number designations, with the aim of relating these differences to different cortical functions. Such *cytoarchitectonic* maps, however, have in the past been carried to the point where too fine an areal division of the cortex was made. Differences in cell shapes are due in part to mechanical distortions produced by the folds of the surface. For these reasons only, the distinctly different cytoarchitectural regions are currently considered as having a possible relation to function.

Electrical studies have confirmed the general features of localization of the visual, auditory, and somesthetic receptor areas in the cortex. If the receptor organs are briefly stimulated by a light flash to the eye, a click to the ear, or a touch to the skin, a characteristic electrical response can be detected in the primary sensory area of the cortex to which the fibers of each sense modality project. A more consistent response is seen when a brief single electrical shock is delivered to a sensory projection tract leading to the primary receptor area. This *primary cortical response* consists of a series of small fast waves that continue on to a slower 10- to

20-msec positive wave, followed in turn by a slow negative wave lasting 10 to 20 msec (Fig. 6-11).

The same complex response is seen in each of the corresponding primary sensory areas when its specific sensory relay nuclear group (somesthetic, visual, or auditory) is stimulated. The first of the brief spikelike waves is the sign of activity in the specific afferent fibers projecting into the cortex.

The specific afferent fibers synapse on granular cells in the fourth layer of the cortex. In turn, the granular cells synapse on other neurons and eventually on pyramidal cells, to give rise to the slower-wave components of the response. The positive portion of the response is due to excitation of elements activated in deeper layers of the cortex, probably on the soma and lower dendrites of pyramidal cells. The negative portion of the sensorily evoked response is due to a later activation and depolarization of the apical

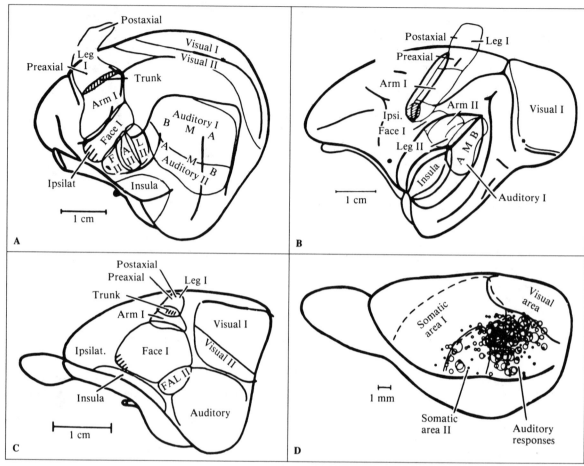

Fig. 6-12. Sensory maps of several animal species. (A) is a sensory map for the cat. Two visual areas are shown, and the somesthetic area I is represented by leg, arm, and face areas. A second somesthetic area, II, is also shown, and two auditory areas as well are found. (B) is a corresponding map for the monkey, (C) for the rabbit, and (D) for the rat. Notice that in general the relative positions of the main sensory areas have a similar pattern over the surface of different brains. (P. Bard. *Medical Physiology* [11th ed.]. St. Louis: Mosby, 1961. P. 1217.)

dendrites near the surface. It is likely that the same electrogenesis is involved as that described for the negative-wave DCR; the negative phase of the sensorily evoked response shows occlusive interaction with the DCR, and, as previously mentioned, occlusive interaction indicates that a common element (in this case the apical dendrites) underlies both responses.

Electrical stimulation of the primary sensory cortex of humans under local anesthesia gives rise to sensations experienced as either visual, auditory, or somesthetic, depending on the sensory area stimulated: for the visual area, optical; for the auditory area, sound; and for the somesthetic cortex, sensations referred to the skin of the body. Sensory areas relating to olfactory, gustatory, and alimentary sensations are also known. Tumors of the uncus, for example, may be associated with disturbing sensations of smell.

Within each of the specific sensory regions of the cortex

there is an internal spatial organization—a topographical localization of parts of the receptor field to parts of the sensory area, as shown for several species in Figure 6-12. The correspondence of a point within the sensory field to a point of the cortex is referred to as *point-to-point representation*. This correspondence was shown experimentally by the use of a point source of light flashed at a given locus in the visual field of an animal while the extent of the visual cortex was explored with a monopolar electrode. A point on

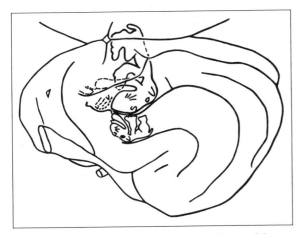

Fig. 6-13. Figurine representation of the somesthetic area of the dog. The parts of the schematic dog lie over their corresponding tactile areas as mapped by evoked-potential techniques. A second somesthetic area is also found that is a smaller duplication of the first, somesthetic area II. Note the greater disproportionate size of the snout region in this species. (P. Bard. *Medical Physiology* [11th ed.]. St. Louis: Mosby, 1961. P. 1216.)

point source of light falls on the retina, a relatively large number of cells may be discharged. Also, a number of cells in the surrounding area are inhibited. Retinal cells therefore show complex spreads of excitatory and inhibitory interaction fields. The same sort of surrounding *inhibition* (and excitation) occurs at relay stations as the impulses ascend to the cortex. Within the cortex the specific afferent fibers branch profusely and have multiple synaptic terminations on a large number of neurons within the sensory cortex. These afferents thus engage a relatively large volume of cortex, and the "grain" of the cortical region excited by a single afferent fiber would be much too coarse to provide an appreciation of fine detail in the sensory field if no other mechanism of discrimination between points in the visual field were operating. Some additional process must come into play to help "sharpen" the representation of points in the visual field. The sharpening could come about by a surrounding inhibitory effect, such as that described for the Renshaw system in the spinal cord (see Chap. 5). Sharpening could also occur by the overlapping of afferent terminations ending with a cortical region so as to bring about the excitation of a much smaller number of cortical cells through the process of occlusion.

the visual cortex was found where the evoked response had the greatest response amplitude. Reducing the strength of the light and using a small point source allowed the area of the cortex representing the point stimulated in the visual field to be localized to within less than a millimeter.

The topographical relationship within a sensory area may be shown by a figurine, as in Figure 6-13. Such representation more readily shows the extent of the area given over in the cortex to a sensory modality that has a relation to a species' behavior. For example, in the primate, the thumb area in the somesthetic region is large, as is the face region, especially the mouth and tongue. This is also the case in humans, and the size of these areas conforms with their importance. Another example is the large extent of the cortical somesthetic representation of the snout in the pig.

Sensory afferent projections to other regions than the primary sensory areas have been found in the cortex. These *secondary sensory areas* (Fig. 6-13) also have a topographical organization with a point-to-point representation of the sensory field. The functional relation of the secondary receptor areas to the first sensory areas is at present unknown.

Point-to-point localization does not mean that fibers are activated in a direct line leading from a sensory point within the receptor field to a point in the sensory region of the cortex. Studies of the retina have shown that when a pin-

SINGLE CELL SENSORY RESPONSES: COLUMNAR ARRANGEMENT. New principles of organization of neurons in primary sensory regions have been found by recording the activity of single cells within the sensory regions of the cortex. With the tip of a microelectrode close to an active cell, there is a rapid decrement of potential with distance from the discharging surface. Its activity can therefore be recorded in isolation, the more distant cells contributing too small a voltage to be recorded. Unit cell studies have the disadvantage that a statistical survey must be made to determine what a given population of neurons is doing in any small area. For example, during an evoked response to a light flash, some neuronal units increase their rate of firing, others show a decrease, and still others are not affected at all. These varied types of unit response indicate that much of the function of the primary receptor areas underlying vision is complex. Recording from the cortex reveals that some cortical cells in the visual cortex give rise to a discharge at the onset of light stimulation, an *on response,* and others at the termination of a light stimulation, an *off response.* Still other cells show a combination, the *on-off* cells. Excitation (on response) and inhibition (off response) were discovered to be functionally related, so that in a region of the cortex surrounding excitatory on units, an inhibitory discharge of off units may be found, and vice versa (Fig. 6-14). In Figure 6-14, action potentials occur at the

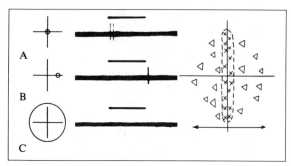

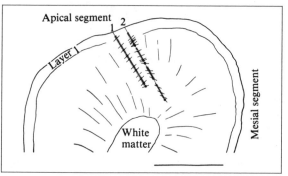

Fig. 6-14. Responses of a unit to a stimulation with circular spots of light in the visual field. The receptive area is activated by (A) a 1-degree spot in the center; (B) the same spot displaced 3 degrees to the right; (C) an 8-degree spot covering a large part of the receptive field. X = excitatory; Δ = inhibitory. (From D. H. Hubel and T. N. Wiesel. Receptive fields of single neurons in the cat's striate cortex. *J. Physiol.* [Lond.] 148: 578, 1959.)

Fig. 6-15. Two recording penetrations of microelectrodes (1 and 2) into the visual cortex are shown, along with responses from neurons recorded at different distances from the surface. The lines intersecting the tracks represent the orientation of the light slits used for stimulation of the visual field that gave maximal neural responses. Notice the similarity of orientations for the cells along track 1, which is nearly perpendicular to the surface, compared to the changes in orientation of responsiveness of the cells along track 2. Horizontal line = 1 mm. (From D. H. Hubel and T. N. Wiesel. Shape and arrangement of columns in cat's striate cortex. *J. Physiol.* [Lond.] 165: 561, 1963.)

start of the light, and these are excitatory; the response after its termination is inhibitory, as in trace B in the figure.

Unit studies of neurons in the sensory cortex have revealed an extraordinary *columnar* organization. It was first discovered in the somesthetic sensory area, where individual neurons responded either to movement of hairs of the body surface or to rotation of limb and digit joints. On recording in the cortex downward from the surface, a series of neurons was found to respond successively in the same way to a given stimulus type. The neurons responding to a given type of sensory input seem to be arranged vertically in columns. The column pattern was seen only when the electrode was passed perpendicularly down from the surface, the tip encountering cell after cell giving rise to the same response, as, for example, to a limb rotation or a touch to a particular place on the skin surface. The regions of the skin surface from which the columns of responding cells could be elicited were generally discrete and located on the body contralateral to the somesthetic cortex. In the secondary sensory areas of the cortex, the peripheral body surface area from which an excitation could elicit responses was more widespread, and, in addition, the same neurons could be excited from bilateral areas of the body surface.

A similar system of neurons having a columnar organization was found in the visual cortex. Units in the visual cortex responded to slits of light that were oriented at a certain specific angle in the visual field (Fig. 6-15). If the orientation of the slit was varied, the light stimulus brought no response from those units. Recording from cells at successive levels below the surface of the cortex showed that the cells responded only to the slit at the same orientation.

Vertical columns of cells nearby responded to slits of other orientations or to other types of visual stimuli, for example, a movement of a slit of light in one certain direction, or different-shaped contours of light stimulation; in the latter case the response is referred to as a *complex type* of cell response.

Cells in the primary visual areas are found connected for binocular vision response: A binocular neuron in one visual cortex responding to a given type of slit orientation responds to the same stimulus presented in the other visual field.

Use of the simple sensory systems of lower organisms has shown some basic mechanisms involved in vision that may help reveal some principles for study of the complex system of higher forms. The *ommatidium* of the arthropod *Limulus* consists of a small cluster of light receptor cells. A number of these form the compound eye, with an *eccentric cell* found in close apposition to them. Discharge in the axon of the eccentric cell has been recorded, the rate of its discharge being determined by the degree of light shining into the ommatidium. There are collateral interactions between the axons of the eccentric cells of the eye such that discharges in one will inhibit the response in those around it. This *lateral inhibition* has some relation to the effect of Renshaw cell activity (see Chap. 5) in the cord, though there is no intervening inhibitory cell in *Limulus ommatidia*.

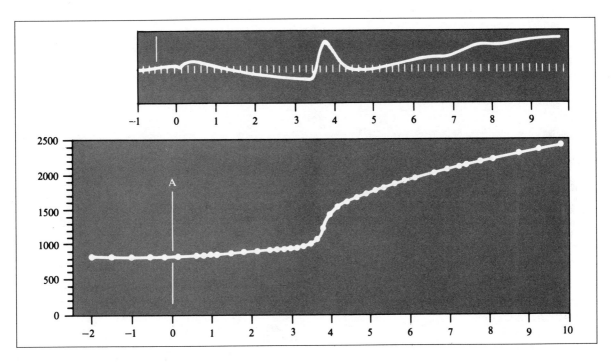

Fig. 6-16. Asphyxial changes in the cerebral cortex. In the upper curve, the steady potential is recorded; in the lower curve, the electrical resistivity of the cortex. At *A*, the brain is asphyxiated by sudden deprivation of its blood supply. A latent period of several minutes ensues before a rapid increase in resistivity is seen. At the same time a negative depolarization of the brain takes place. (From A. van Harreveld and S. Ochs. Cerebral impedance changes after circulatory arrest. *Am. J. Physiol.* 187:184, 1956.)

STEADY-POTENTIAL CHANGES AND SPREADING DEPRESSION

The apical dendrites of pyramidal cells are not only related to the production of EEG waves, as discussed in a preceding section, but they also give rise to a *steady potential*. Steady potentials are recorded from the surface of the cerebrum by means of a DC coupled amplifier and for this reason are also called *DC potentials*. Changes in the steady potential have been related to differences in the polarization of the upper and lower portions of the pyramidal cells. The relation of the steady potential to activity is shown by the negative shift in the steady potential when stimulating the nonspecific thalamic nuclei. The change in voltage is due to the termination of axons from the thalamus onto cortical dendrites that act to depolarize them.

The steady potential is also related to metabolism. If the blood supply to the brain is clamped off or nitrous oxide is breathed to initiate anoxia, little change is seen during a latent period, lasting approximately 2 to 5 minutes; then there is a rapid negative shift of the steady potential (Fig. 6-16). At the time of the negative shift, referred to as the *asphyxial potential*, the electrical resistance (impedance) of the cortex increases markedly. Both phenomena, the negative change in steady potential and the increased electrical impedance of the cortex, are related to the loss of normal ion permeability of the dendrites, with an increased entry of Na^+, Cl^-, and water from the extracellular space into the dendrites. Evidence for this event is a measured swelling of the apical dendrites occurring with the use of freeze-substitution to prepare the tissue for electron microscopy. The asphyxial potential and ionic changes herald an irreversible damage of cortical neurons. Consequently, it is imperative that oxygenated blood be supplied to the brain before this change takes place. In humans, a permanent loss of consciousness can result if anoxia lasts more than about 5 minutes at 38°C. The time can be considerably lengthened if the brain temperature is lowered. Advantage is taken of this fact in surgery in which brain circulation must be stopped

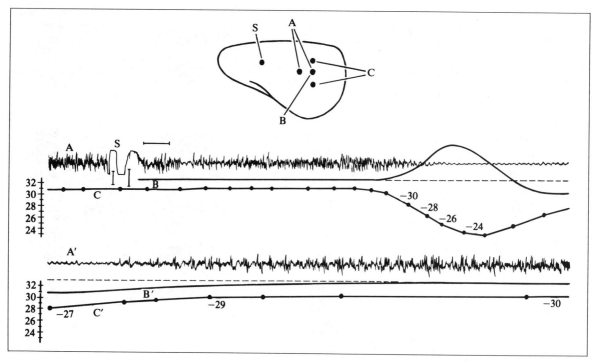

Fig. 6-17. Spreading depression of the cerebral cortex. The EEG is shown in line *A,* the steady potential of the surface of the brain in line *B,* and the electrical conductivity (inverse of resistivity) of the cerebral cortex on line *C.* These records are continued below *A',* *B',* and *C'.* The electrode positions on the surface of the brain are shown in the diagram. A brief, intense stimulation of the brain is employed at *S* to start spreading depression. As depression spreads out and occupies the region of the recording electrodes, the EEG shows a diminution, and at the same time a negative variation is seen in the steady-potential level of the brain. A decrease in conductivity also occurs at this time. Recovery occurs some minutes later. (From A. van Harreveld and S. Ochs. Electrical and vascular concomitants of spreading depression. *Am. J. Physiol.* 189:159, 1957.)

for longer periods where techniques to lower body temperature artificially are employed.

Some aspects of the changes seen after asphyxiation are also found to occur in the phenomenon known as *spreading depression.* Spreading depression is elicited by mechanical, chemical, or electrical stimulation of the surface of the brain. It is characterized by a depression of the EEG and of evoked responses that spreads through the cortex without regard to the functional area involved—sensory, motor, or associational—at the rate of 2 to 5 mm per minute. In an area occupied by spreading depression, increased permeability and ionic changes occur in the neurons, mainly in the apical dendrites of the pyramidal neurons, with a negative shift of the steady potential and an increased electrical resistivity seen (Fig. 6-17). Apparently, both K^+ and glutamate are released from the involved cells to act on nearby cells to excite spreading depression in them. The process of ionic shifts is repeated in nearby cortical regions to give rise to the slow spread. The apical dendrites of the involved neurons show evidence of swelling, with an entry of Na^+, Cl^-, and water into them, similar to the events occurring during the asphyxial change. However, unlike the asphyxial change, a recovery takes place within a few minutes, the Na^+ pump in the dendrites and cells acting to restore nor-

mal ionic conditions. The ionic shifts—increase of K^+ in the extracellular spaces and decrease of Na^+—have been shown using ion-selective microelectrodes. Spreading depression occurs readily in the rat, rabbit, and guinea pig. In higher species it is more readily seen after cortical exposure or an application of Ringer's solution with an augmented amount of K^+ present. It may also occur pathologically in humans following a concussion or in some convulsive states.

The term *spreading depression* may be somewhat misleading because a small amount of convulsive spike activity is commonly seen under certain conditions in which a slow spread of a convulsive discharge occurs rather than depression; this is referred to as a *spreading convulsion*. It is due to a "release" of activity of cells in lower cortical layers that may not always be invaded by spreading depression.

METABOLISM

A close dependence of brain function on oxidative metabolism is shown by the rapid changes in function with loss of blood supply. As noted, consciousness is lost in approximately 10 to 15 seconds. Concomitantly, the EEG changes to an isoelectric pattern; i.e., a loss of rhythmicity is seen. If blood circulation is returned within approximately 5 minutes, consciousness returns with a return of EEG activity. Similar loss of cortical function due to metabolic interference can come about through a hypoglycemia such as that produced by an overdose of insulin. If hypoglycemia is prolonged, irreversible neuronal damage may occur, with loss of higher function as in the case of prolonged anoxia.

The maintenance of adequate circulation to the brain is accomplished by a number of important reflex mechanisms outside the brain (see Chap. 17). The fine control of circulation within the brain apparently depends not on local neural control but on metabolic mechanisms, primarily by variation in P_{CO_2} level, which in turn is determined by local metabolism. Increase in P_{CO_2} increases cerebral blood flow, and vice versa (Fig. 6-18). The use of a Fick technique (see Chap. 11) for estimation of cerebral blood flow gives values of approximately 60 ml/100 gm/min, and this flow does not show much variation with changing conditions such as sleep or waking. Adequate blood flow is maintained in fairly constant fashion as systemic blood pressure varies from approximately 70 to 150 mm Hg and more; i.e., it shows autoregulation. Low blood pressures are dangerous, however, for with decreases to or below a level of approximately 40 to 50 mm Hg, as may occur in shock (see Chap. 18), cerebral blood flow becomes inadequate for proper oxygen uptake and metabolism. Coma and permanent loss of cerebral function may occur if this condition is not soon reversed.

A more subtle alteration of metabolism is possible. As noted earlier, neurons have a high rate of protein synthesis that supplies materials carried outward in the fibers by axoplasmic transport to maintain normal nerve function and synaptic transmission. At present, not enough is known about the effects of various diseases on this transport function. There is, however, a growing awareness of its role. For example, brain damage can occur as a result of protein deficiencies in the diets of infants, with a resultant subnormal intelligence later in life.

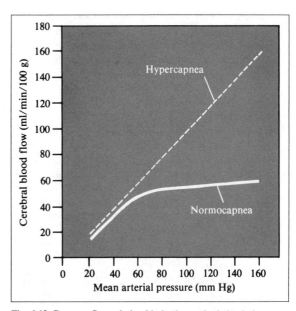

Fig. 6-18. Pressure-flow relationship in the cerebral circulation. Autoregulation of blood flow is present at normal arterial blood levels of CO_2 (normocapnea) but is abolished by high arterial blood levels of CO_2 (hypercapnea). Then cerebral blood flow varies as a linear function of mean arterial blood pressure. (From R. M. Berne and M. N. Levy. *Cardiovascular Physiology* [3rd ed.]. St. Louis: Mosby, 1977. P. 246.)

A large portion of the cells in the CNS and in the cortex are neuroglial (glial) cells. The exact function of the glial cells is not understood. According to one theory, the glial cells control the transport of materials from the blood capillaries to the neurons. This role has classically been that of the *blood-brain barrier,* usually ascribed to capillaries, which controls entry of substances to the extracellular space and thence to cells. Part of the evidence for such an intermediating role for glia comes from electron microscopic studies showing that there is very little extracellular space in central nervous tissue, possibly too little for the diffusion of the materials from the blood capillaries to the neurons. However, the size of the extracellular space is likely to be similar to that found in other tissues—about 20 percent. This has been determined from electrical impedance studies and the increase in impedance on asphyxiation when Na^+, Cl^-, and water enter the cells, with a corresponding reduction of the extracellular space. The seeming lack of an extracellular space noted in the usual electron micrographs (and in turn an anatomical evaluation of the size of the extracellular compartment) is due to the previously discussed translocation of Na^+, Cl^-, and water into cells during the time the tissue is removed and prepared for electron microscopic examination. An extracellular space with a value closer to that obtained from measurement of cortical impedance has been shown in electron micrographs using freeze-substitution.

BIBLIOGRAPHY

Anderson, P., and Andersson, S. S. *Physiological Basis of the Alpha Rhythm*. New York: Appleton-Century-Crofts, 1968.

Berne, R. M., and Levy, M. N. *Cardiovascular Physiology* (3rd ed.). St. Louis: Mosby, 1977. P. 246.

Bureš, J., Burešová, O., and Krivanel, J. *The Mechanism and Applications of Leão's Spreading Depression of Electroencephalographic Activity*. New York: Academic, 1974.

Chang, H.-T. Dendritic potential of cortical neurons produced by direct electrical stimulation of the cerebral cortex. *J. Neurophysiol.* 14:1–21, 1951.

Davson, H. The Blood-Brain Barrier. In G. H. Bourne (ed.), *The Structure and Function of Nervous Tissue*. New York: Academic, 1972. Vol. 4.

Fuxe, K., and Jonsson, G. Further mapping of central 5-hydroxytryptamine neurons: Studies of the neurotoxic dihydroxytryptamines. *Adv. Biochem. Psychopharmacol.* 10:1–12, 1974.

Hubel, D. H., and Wiesel, T. N. Receptive fields of single neurons in the cat's striate cortex. *J. Physiol.* (Lond.) 148:574–591, 1959.

Hubel, D. H., and Wiesel, T. N. Receptive fields of single neurons in the cat's striate cortex. *J. Physiol.* (Lond.) 160:106–154, 1962.

Hubel, D. H., and Wiesel, T. N. Shape and arrangement of columns in cat's striate cortex. *J. Physiol.* (Lond.) 165:559–568, 1963.

Jasper, H. H., Ward, A. A., Jr., and Pope, A. (eds.). *Basic Mechanisms of the Epilepsies*. Boston: Little, Brown, 1969.

Jouvet, M. The role of monoamines of acetylcholine containing neurons in the regulation of the sleep-waking cycle. *Ergeb. Physiol.* 64:166–307, 1972.

Kandel, E. R., and Schwartz, J. H. *Principles of Neural Science*. New York: Elsevier/North-Holland, 1981.

Kety, S. The Cerebral Circulation. In J. Field, H. W. Magoun, and V. E. Hall (eds.), *Handbook of Physiology*. Washington, D. C.: American Physiological Society, 1960. Section 1: Neurophysiology. Vol. 3, chap. 61.

Krnjević, K. Iontophoretic studies on cortical neurons. *Int. Rev. Neurobiol.* 7:41–98, 1964.

Magoun, H. W. *The Waking Brain* (2nd ed.). Springfield, Ill.: Thomas, 1963.

McGeer, P. L., Eccles, J. C., and McGeer, E. G. *Molecular Neurology of the Mammalian Brain*. New York: Plenum, 1978.

McIlwain, H. and Bachelard, H. S. *Biochemistry and the Nervous System* (4th ed.). London: Churchill Livingstone, 1971.

Moruzzi, G. The sleep-waking cycle. *Ergeb. Physiol.* 64:1–165, 1972.

Penfield, W. G., and Jasper, H. H. *Epilepsy and the Functional Anatomy of the Human Brain*. Boston: Little, Brown, 1954.

Phillis, J. *The Pharmacology of Synapses*. Elmsford, N.Y.: Pergamon, 1970.

Resnick, O. The Role of Biogenic Amines in Sleep. In C. D. Clemente, D. P. Purpura, and F. E. Mayer (eds.), *Sleep and the Maturing Nervous System*. New York and London: Academic, 1972. Chap. 6.

Ross Conference. *Brain Damage in the Fetus and Newborn from Hypoxia or Asphyxia*. Columbus, Ohio: Ross Laboratories, 1967.

Rossi, G. F., and Zanchetti, A. The brain stem reticular formation. *Arch. Ital. Biol.* 95:195–435, 1957.

Shepherd, G. M. *The Synaptic Organization of the Brain* (2nd ed.). New York: Oxford University Press, 1979.

Sholl, D. A. *The Organization of the Cerebral Cortex*. London: Methuen, 1956.

Siesjo, B. K. *Brain Energy Metabolism*. New York: Wiley, 1981.

Tower, D. B. (ed.-in-chief). *The Nervous System*. 3 vols. New York: Raven, 1975.

van Harreveld, A. *Brain Tissue Electrolytes*. Washington, D.C.: Butterworth, 1966.

van Harreveld, A. The Extracellular Space in the Vertebrate Central Nervous System. In G. H. Bourne (ed.), *The Structure and Function of Nervous Tissue*. New York: Academic, 1972. Vol. 4.

von Bonin, G. *Some Papers on the Cerebral Cortex*. Springfield, Ill.: Thomas, 1960.

REGULATION OF VISCERAL FUNCTION

Carl F. Rothe

THE AUTONOMIC NERVOUS SYSTEM

Homeostasis, the constant and optimal internal environment of the body, is maintained in large part by the actions of the autonomic nervous system (ANS), for it is this part of the nervous system that provides a fine control of the visceral or internal functions of the body. It is generally involuntary and acts on the internal effectors such as (1) non-striated (smooth) muscle, as in the intestine; (2) cardiac muscle; (3) exocrine glands, e.g., the sweat and salivary glands; and (4) some endocrine glands. The endocrine system also participates in the control of the visceral function of the body, but it is generally slower and acts through the release of internal secretions (hormones), which are transported by the cardiovascular system. The ANS, along with the endocrine system, thus aids in keeping the internal environment of the body at such a composition and temperature that cellular life may proceed optimally.

CHARACTERISTICS OF THE AUTONOMIC NERVOUS SYSTEM

Anatomically, the ANS is the efferent pathway linking the control centers in the brain and the effector organs, such as smooth muscle and secretory cells. However, physiologically, control of visceral function must include sensors, afferent pathways, and central control centers as well. The sensory afferents from the viscera in the vagus and splanchnic nerves, for instance, serve both the somatic and the autonomic nervous system. Other sensors, such as those of blood plasma osmolarity and carbon dioxide partial pressure, are located within the cells of the central nervous system itself. The ANS differs from the somatic nervous system in that the motoneurons that come into immediate functional relationship with the effector cells lie completely outside the central nervous system. These motoneurons are called postganglionic, since they synapse with the preganglionic fibers coming from the spinal cord in ganglia (clumps of neuron cell bodies) located in chains beside the spinal

141

cord or in the organ innervated. The adrenal medullae are an exception, for they have no postganglionic neurons. The adrenal medullary tissue is of the same embryological origin as postganglionic tissue, however, and the gland functions in a manner analogous to that of postganglionic fibers by releasing catecholamines (see Mediators).

The ANS may be divided, both functionally and structurally, into the *sympathetic division,* with neurons leaving the spinal cord from the thoracic and lumbar segments (Fig. 7-1), and the *parasympathetic division,* with neurons leaving from the cranial and sacral segments (Fig. 7-2).

THE SYMPATHETIC DIVISION. The sympathetic division tends to act in a widespread manner to prepare the body for emergencies and vigorous muscular activity. Life is generally possible without the sympathetic division; but the animal must be sheltered, for the sympathectomized animal is much less resistant to extremes of environmental temperature, hypoxia, and other forms of stress. The cardiovascular system is no longer finely adjusted to provide for prolonged or severe exertion. The animal tends to be weak and apathetic.

The synapses of the sympathetic division are characteristically located either in the sympathetic paravertebral chain on each side of the spinal cord throughout its length or in special ganglia such as the celiac ganglion (*solar plexus*) (see Fig. 7-1). The interconnections between the paravertebral ganglia facilitate the diffuse action of the sympathetic division. In addition, the adrenal medullae release the sympathetic mediators epinephrine and norepinephrine into the bloodstream for even more complete and diffuse distribution.

Homeostatic needs are anticipated because of interconnections with the cortex of the brain. For instance, the mere thought of a fight activates the sympathetic division, which in turn increases the activity of the cardiovascular respiratory systems, curtails gastrointestinal activity, causes the release of glucose from glycogen stores, channels more blood to muscles, and starts sweating for removal of excess heat. The pupils of the eyes dilate, the eyes tend to protrude, and the eyelids widen. In many animals the hairs of the back and tail bristle. All these reactions are in anticipation of vigorous muscular activity. The sympathetic division thus acts in emergencies to adjust the organism to an adverse environment or a rapid change in internal requirements.

Besides participating in massive discharge for emergency situations, the various parts of the sympathetic division function in everyday life in independent and discrete ways. The connections of the sympathetic division with various

organs and its actions on them are summarized in Figure 7-1 and bear careful study.

THE PARASYMPATHETIC DIVISION. The parasympathetic division acts more discretely on individual organs or regions than does the sympathetic division. The abdominal viscera are innervated by preganglionic neurons that leave the cranial part of the cord and travel via the vagus nerves. The synapses are generally located *within* the innervated organ. The vagi contain most of the parasympathetic efferent fibers, although they consist mainly of sensory afferents that participate in visceral reflexes, such as those from the abdomen and the pressure and chemical receptors located in the aortic arch and the heart. There are also a few somatic efferent fibers. Dogs, in addition, have in the cervical vagus trunks some cervical postganglionic sympathetic nerve fibers to the head. The sacral part of the parasympathetic division is primarily concerned with the emptying of the pelvic organs and with reproductive functions.

Whereas the sympathetic division is mainly involved with homeostasis during voluntary muscular activity, the parasympathetic division is involved with restorative vegetative functions such as digestion and rest. Imagine an old man sleeping after eating. His heart rate is low, his breathing is noisy because of bronchial constriction, and his pupils are small. Saliva may run from the corners of his mouth, and rumbles from his abdomen reveal much intestinal activity.

The mediator that is released at the parasympathetic nerve endings is the same as in the somatic nervous system and ganglia—acetylcholine. The actions on various organs are presented in Figure 7-2 and should be studied.

THE HYPOTHALAMUS. The hypothalamus (see Chap. 8) is the center for interrelating the visceral and somatic functions of the body. Not only are motor actions accompanied by widespread and complex visceral responses, but visceral activity also modifies somatic reactions. For instance, during digestion the increased blood flow through the intestine tends to reduce the muscular capacity for work. Such interdependence of visceral and somatic functions is implied in the concept that central control of both functions is exerted through neural mechanisms that are located at common levels in the spinal cord, brainstem, diencephalon, and cerebral cortex; and that both have a common sensory inflow. Many levels of the central nervous system interact in the control of the circulatory system.

If the spinal cord is sectioned at about the first thoracic vertebra, there is an immediate depression not only of somatic but also of autonomic reflexes. Blood pressure drops as a result of loss of peripheral vasoconstrictor tone,

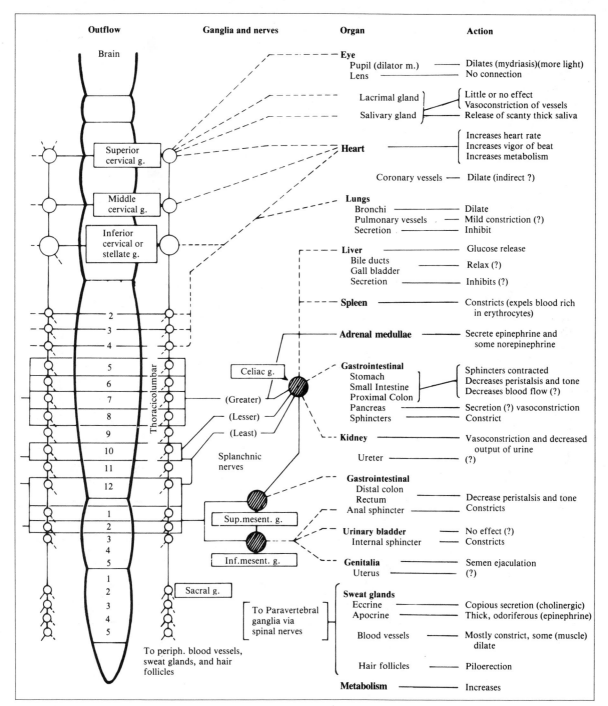

Outflow	Ganglia and nerves	Organ	Action

Eye
Pupil (dilator m.) —— Dilates (mydriasis)(more light)
Lens —— No connection

Lacrimal gland ⎱ —— Little or no effect
—— Vasoconstriction of vessels
Salivary gland ⎰ —— Release of scanty thick saliva

Heart —— ⎱ Increases heart rate
Increases vigor of beat
Increases metabolism ⎰

Coronary vessels —— Dilate (indirect ?)

Lungs
Bronchi —— Dilate
Pulmonary vessels —— Mild constriction (?)
Secretion —— Inhibit

Liver —— Glucose release
Bile ducts —— Relax (?)
Gall bladder
Secretion —— Inhibits (?)

Spleen —— Constricts (expels blood rich in erythrocytes)

Adrenal medullae —— Secrete epinephrine and some norepinephrine

Gastrointestinal —— ⎱ Sphincters contracted
Stomach Decreases peristalsis and tone
Small Intestine Decreases blood flow (?) ⎰
Proximal Colon
Pancreas —— Secretion (?) vasoconstriction
Sphincters —— Constrict

Kidney —— Vasoconstriction and decreased output of urine

Ureter —— (?)

Gastrointestinal
Distal colon —— Decrease peristalsis and tone
Rectum
Anal sphincter —— Constricts

Urinary bladder —— No effect (?)
Internal sphincter —— Constricts

Genitalia —— Semen ejaculation
Uterus —— (?)

Sweat glands
Eccrine —— Copious secretion (cholinergic)
Apocrine —— Thick, odoriferous (epinephrine)

Blood vessels —— Mostly constrict, some (muscle) dilate

Hair follicles —— Piloerection

Metabolism —— Increases

Superior cervical g.
Middle cervical g.
Inferior cervical or stellate g.

Thoracicolumbar

2 3 4 5 6 7 8 9 10 11 12
1 2 3 4 5
1 2 3 4 5

Celiac g.
(Greater)
(Lesser)
(Least)
Splanchnic nerves
Sup.mesent. g.
Inf.mesent. g.
Sacral g.
To Paravertebral ganglia via spinal nerves
To periph. blood vessels, sweat glands, and hair follicles

Fig. 7-1. Sympathetic division of the autonomic nervous system. The celiac, superior, and inferior mesenteric ganglia are called pre-vertebral (or paravertebral) ganglia, since the synapses (ganglia) are not in the sympathetic trunk along the spinal column. Solid lines = cholinergic fibers; dashed lines = adrenergic (postganglionic) fibers.

143

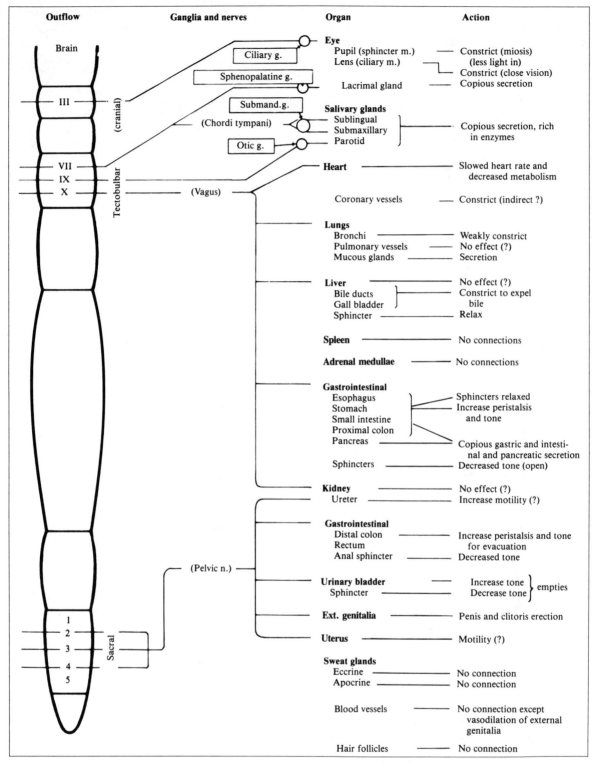

Fig. 7-2. Parasympathetic division of the autonomic nervous system (all nerve fibers are cholinergic).

sweating is absent, and the body temperature is poorly regulated and tends to approach the environmental temperature. The evacuation reflexes for the bladder and bowel are greatly impaired, as are the sexual reflexes. With the passage of time, many of these reflexes return, in part, with the further development of the spinal segmental reflexes; but regulation of blood pressure and of body temperature is severely limited, and micturition, defecation, and sexual reflexes are incomplete.

The decerebrate preparation responds in a similar fashion except that many cardiovascular centers located in the medulla are intact, and cardiovascular regulation is therefore more nearly normal.

If the hypothalamus is left intact by sectioning the brain above the diencephalon, removing only the cerebral cortex, the animal has normal vegetative regulation of homeostasis, but the adjustments appropriate for the anticipation of muscular activity are missing.

As discussed in Chapters 8 and 9 and other appropriate chapters, the hypothalamus is concerned with the regulation of (1) the cardiovascular system, (2) respiration, (3) body temperature, (4) body water and electrolyte concentrations, (5) food intake and gastric and pancreatic secretion, (6) hypophyseal secretion and sexual and maternal behavior, and (7) emotional states.

RECIPROCAL ACTION. Most of the visceral organs have a dual, antagonistic innervation in that the actions of the two divisions are diametrically opposed. For instance, the sympathetic division increases cardiac activity, and the parasympathetic division decreases it. The parasympathetic division acts to enhance and accelerate visceral functions such as gastrointestinal motility, while the sympathetic division generally acts to constrict visceral blood vessels and to inhibit visceral functions such as digestion and elimination.

An exception to the generalization of reciprocal action is the control of secretory function of the salivary glands and pancreas; both divisions seem to stimulate secretion. In addition, not all organs have dual innervation. The sympathetic division has exclusive innervation of the adrenal medullae, spleen, pilomotor muscles, sweat glands (although the postganglionic fibers are cholinergic), and probably the vasomotor muscles of the blood vessels of the viscera, skin, and skeletal muscle, if not all blood vessels. The parasympathetic division innervates the ciliary and the sphincter muscles of the eye. Functionally, the pupil of the eye has a dual, antagonistic innervation; structurally, however, the dilator muscles have sympathetic innervation, whereas the constrictor muscles have parasympathetic innervation.

TONE. The ANS generally maintains a "tone," a basal level of activity, which then may be either increased or decreased by central control. For example, the flow of blood through an innervated muscle is about one-third to one-half that observed following interruption of sympathetic nerve supply to the muscle, indicating that a sympathetic tone is acting under normal conditions to constrict the vessels. Because of this vasomotor tone, the blood vessels can be either dilated by a decrease in sympathetic division activity or further constricted by an increase in activity (see Chap. 17).

The parasympathetic division, by way of the vagus, provides a "vagal tone" to the heart. If the vagi of an animal under basal conditions are sectioned, the heart rate increases as a result of removal of this inhibiting effect. Parasympathetic tone tends to increase the intrinsic motility of the intestinal tract. If the lower vagi are sectioned, there is serious and prolonged, but not permanent, gastric and intestinal dysfunction. Generalized sympathetic division activity tends to inhibit intestinal motility.

The presence of dual innervation and the possibility of either increasing or decreasing the ANS tone permit a wide range of control.

MEDIATORS

The chemical mediators released at the nerve endings of the autonomic nerve fibers act on *receptors* to produce, in turn, an effector response. A receptor may be conceived of as that molecular structure with which a single molecule of mediator or drug interacts. The site may be either an enzyme or the structural configuration of the cell membrane.

The mediator released by the preganglionic fibers of both divisions is acetylcholine (ACh). On the other hand, the postganglionic mediator is not the same for all fibers (Fig. 7-3). The parasympathetic postganglionic fibers are all cholinergic; that is, they release ACh as the transmitter agent. The mediator substance released at most of the postganglionic sympathetic nerve endings is *norepinephrine*. Such fibers are *adrenergic*. *Epinephrine* and some norepinephrine are released into the blood by the chromaffin cells of the adrenal medullae. Norepinephrine is also called levarterenol, arterenol, or noradrenalin. Epinephrine is norepinephrine plus an N methyl group and is also called methylarterenol or adrenalin. *Isoproterenol,* a synthetic catecholamine, is norepinephrine with an N isopropyl group. These compounds as a class are called catecholamines. In addition, some postganglionic sympathetic fibers, such as those going to the sweat glands, are cholinergic, releasing ACh. The best evidence for this conclusion is that their activity is blocked by

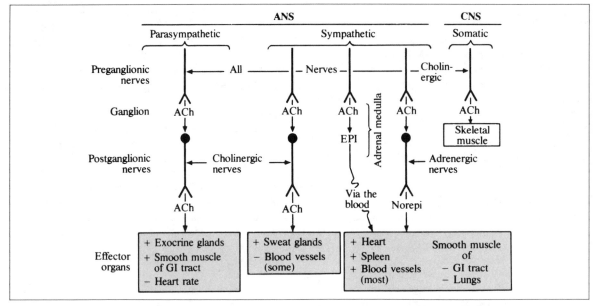

Fig. 7-3. Classification of nerves in terms of transmitter or mediator released and location of cell body and synapse. Plus sign = stimulatory; minus sign = inhibitory.

atropine, a drug that blocks the action of ACh on these cells.

The *mode of action* of ACh and the catecholamines on excitable membranes, such as those of the intestinal smooth muscle and the heart, is apparently that of modifying the rate of polarization and depolarization, thus changing the frequency of spontaneous spike discharge that is the basis of the autorhythmicity characteristic of these tissues. Acetylcholine increases and epinephrine decreases the frequency of spontaneous spike discharge of intestinal smooth muscle. The fact that each spike is followed by a slight contraction explains the effect of these mediators on this tissue. On the other hand, ACh slows or stops the heartbeat by hyperpolarizing the pacemaker cell membrane and slowing depolarization, both actions being probably due to an increased permeability of the resting membrane to potassium. The net effect is opposite to that of ACh on intestinal smooth muscle or on the neuromuscular junction, for here depolarization is enhanced. Epinephrine accelerates the heart by increasing the rate of depolarization of the pacemaker during diastole. Thus, the response of a particular end-organ to ANS activity depends on the individual makeup of that organ.

Epinephrine and norepinephrine generally act in the same manner in that both tend to constrict the blood vessels, stimulate the heart, and inhibit the viscera. However, the blood vessels of skeletal muscle are constricted by norepinephrine and, under some conditions, dilated by epineph-

rine. Because epinephrine and norepinephrine show different relative degrees of influence between tissues, two adrenergic receptors are hypothesized: *alpha* and *beta*. Blocking agents such as phentolamine seem to block only the alpha receptors, while drugs such as propranolol block only the beta receptors. The biochemical characteristics of the receptors are still unknown. The *alpha receptor,* when stimulated, generally excites smooth muscle contraction (except in the intestine) and may act via changes in permeability of the cell membrane to change its excitability. The *beta receptor,* when stimulated, tends to be associated with increased metabolism, glycogenolysis, and lipolysis. (The vigor of cardiac muscle activity is also increased.) Vascular or intestinal smooth muscle relaxation following beta-receptor stimulation may be associated with an increased sodium pump activity that stabilizes the membrane by increasing its polarization. The beta receptor seems to be related to cyclic adenosine monophosphate and adenyl cyclase. The complex effect of epinephrine on skeletal muscle blood flow (Fig. 7-4) seems to be the net result of both alpha-receptor and beta-receptor stimulation. In the intestine the alpha-receptor response is weak, and so relaxation is usually seen. Norepinephrine causes much less beta response than does epinephrine. On the other hand, isoproter-

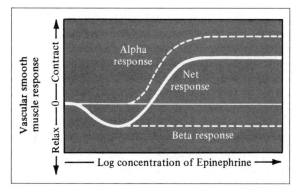

Fig. 7-4. Net effect of various concentrations of epinephrine on the vasculature. (After a suggestion by D. O. Allen.)

enol induces a beta response (increased metabolism, vasodilation, and increased vigor of cardiac contraction).

The mediators must be removed from the site of action if they are to be effective under dynamic conditions. *Acetyl-cholinesterases* destroy ACh by hydrolysis—normally within less than a minute if ACh is introduced into the bloodstream. The enzymes are also found in high concentration around nerve endings. The catecholamines are relatively stable in blood but are rapidly inactivated in body tissues. The major mechanism for catecholamine inactivation is the active uptake of the catecholamine by the postganglionic sympathetic nerve endings; other mechanisms involve degradation (see, for example, Goodman, 1980).

THE ADRENAL MEDULLAE

The release of epinephrine into the blood provides a mechanism for spreading the effects of generalized sympathetic division activity to all cells of the body. These effects are primarily on metabolic processes. Adrenal medullary secretions are important factors in the homeostatic control of blood glucose concentration and release of free fatty acids from adipose tissue during conditions of stress, such as violent muscular activity, hypotension, asphyxia, and hypoxia. Epinephrine secreted by the adrenals facilitates synaptic transmission, increases the heart rate and strength of contraction, and increases metabolism. Although the secretions stimulate the cardiovascular system, the direct adrenergic innervation can produce still higher maximal levels of activity.

The adrenal medullae, like the rest of the sympathetic division, are not essential for life if emergency demands are minimized. In fact, it is difficult to show deficiencies under nonstressful conditions following removal of these organs. However, a reduction in glycogenolysis and a reduced stress tolerance are seen. Functionally, the adrenal medullary cells are equivalent to sympathetic postganglionic nerve cells. Their cholinergic preganglionic innervation and their embryological origins are similar. The cells are stimulated to secrete under the same conditions of stress as those affecting the adrenergic postganglionic neurons. In both, the catecholamines are held in minute granules and so are protected from enzymatic degradation. The neuroadrenomedullary and ganglionic junctions respond similarly to ganglionic blocking or facilitatory agents. The one clear difference is the secretion of epinephrine by the medullae and, on the other hand, the secretion of norepinephrine by the adrenergic fibers. The endocrine secretion of the adrenal medullae supplements the neural action of the sympathetic division activity. In addition to the adrenal medullae, similar chromaffin cells secreting small amounts of epinephrine are located adjacent to the paravertebral chain of sympathetic ganglia, near the carotid sinus, heart, and liver.

Adrenal medullary secretions are controlled by centers located in the posterior hypothalamus. Under normal resting conditions the catecholamine concentration is 0.1 to 0.5 μg per liter of blood. The secretion is about 75 percent epinephrine, although there are marked species differences. At birth, most of the secretion is norepinephrine. With pain, excitement, anxiety, hypoglycemia from insulin injection, cold, hemorrhage, anoxia, or similar stresses, the systemic concentrations increase to as much as 100 μg per liter. The maximal rate of release is about 2 μg min per kilogram of body weight. There is no clear evidence of separate control of norepinephrine and epinephrine secretion.

Because of the metabolic stimulation and glycogenolytic actions, epinephrine from the adrenal medullae has a particular value in homeostasis in response to decreases in body temperature. Cardiac stimulation and facilitation of synaptic transmission by epinephrine are remarkable during hypothermia.

EFFECTORS: SMOOTH MUSCLE

Smooth muscle is the effector organ of much of the ANS. *Multiunit* smooth muscle has innervation similar to that of skeletal muscle. Little or no spontaneous or rhythmic activity is present. The nerve supply is excitatory. Examples include the muscle of the iris, the piloerector and nictitating membrane muscles of animals such as the cat, and probably vascular smooth muscle. *Visceral* smooth muscle, on the other hand, contracts spontaneously and rhythmically in response to the local chemical environment or to stretch.

This activity is modified by the innervation from the ANS, which either inhibits or excites further activity.

In most, if not all, types of smooth muscle under normal conditions, electrical activity of the muscle membrane precedes contraction (see Chap. 3). A prolonged contraction comes about as a result of repetitive action potentials. The evidence is not certain for all organs because of the minute size of the smooth muscle cells, most of which have diameters of about 5μm or less. In the case of visceral smooth muscle and cardiac muscle, the spontaneous contraction results from an unstable resting membrane potential of the pacemaker cell, which, on reaching threshold, initiates an all-or-none action potential that is transmitted from cell to cell. Stretching visceral smooth muscle in many cases depolarizes the membrane to initiate an action potential. Sodium and potassium are important in determining the resting membrane potential, while sodium and calcium are involved in the action potential (see Chap. 3). The smooth muscle cell membrane is apparently much more permeable to sodium at rest than is that of neurons or skeletal muscle, because cellular metabolism and sodium-pump activity are important factors in determining visceral smooth muscle tension.

EFFECTS OF THE AUTONOMIC NERVOUS SYSTEM ON SPECIFIC ORGANS

THE EYES. The adjustment of the eyes to varying light conditions and to focusing at various distances is controlled by the ANS. Sympathetic transmission to the radially oriented muscle of the iris causes it to dilate and to allow more light to enter the eye. Neural impulses of the parasympathetic division make the sphincter muscles of the iris contract and thereby cause constriction of the pupil. During excitement the pupils of the eye are characteristically dilated (*mydriasis*), whereas poisoning with anticholinesterase compounds (Drug Enhancing or Depressing ANS Functions, p. 150) causes a characteristic pinpoint pupil (*miosis*).

Focusing of the lens of the eye is controlled almost exclusively by the parasympathetic division. The lens at rest is focused for distant vision and is relatively flat, but excitation by the parasympathetic division causes the ciliary muscles to contract and make the lenses more spherical for accommodation for near vision.

THE HEART. As will be discussed in the chapters on circulation, vigorous muscular activity is accompanied by an increased sympathetic division activity, causing the heart to beat faster and to contract more vigorously, so that more blood is pumped. At some later time the parasympathetic

division, by slowing the heart rate, serves a restorative function.

SMOOTH MUSCLE OF THE SYSTEMIC BLOOD VESSELS. The blood vessels of the abdominal viscera and skin are generally constricted by generalized sympathetic activity. On the other hand, the vessels of the heart and active muscles dilate during general sympathetic activity in response to the increased metabolism (see Chap. 17). Sympathetic cholinergic nerves for active vasodilation of skeletal muscle act to increase the flow of blood even before contraction starts. The blood vessels of the brain are little influenced by the ANS unless the blood pressure becomes abnormally high. Then, sympathetic vasoconstriction provides some protection. Dilation of the arteries and veins of the skin leads to increased heat loss from the body. It follows a decrease in the sympathetic constrictor tone. Blood vessels other than those to the external genitalia have no parasympathetic innervation. Increased blood flow of salivary glands following stimulation via the parasympathetic division is due to the increase in metabolism of the organs and the release of *bradykinin,* a polypeptide that is a potent vasodilator on the one hand and an intestinal and uterine smooth muscle stimulant on the other. The evidence for a *direct* vasodilating action by the parasympathetic or sympathetic division has been uncertain. Much, but not all, of the data can be explained either by a reduction in sympathetic vasoconstrictor tone or by the release of vasodilator materials resulting from the enhanced metabolism brought about by increased activity of the tissues, e.g., heart, muscle, or glands (see Chap. 17).

If the flow of blood to the skin, intestinal tract, and kidneys is reduced (e.g., during exercise and hemorrhage), blood flow is available for diversion to active muscle, the heart, and the brain. On the other hand, muscle blood flow is decreased during asphyxia.

SPLEEN. The contraction of the spleen and subsequent discharge of blood rich in erythrocytes is another effective mechanism for maintenance of homeostasis during physiological stress, such as exercise, hemorrhage, and anoxia. The erythrocyte storage function of the spleen is of minor importance in man.

RESPIRATORY TRACT. The autonomic system apparently has a minor effect on the functioning of the lungs. Consistent with the concept of the importance of the sympathetic division for voluntary muscular activity, sympathetic division stimulation causes the bronchi to dilate, thus allowing air to enter more easily. The parasympathetic division acts in an opposite manner, and also stimulates the secretion of mucus.

EXOCRINE GLANDS. Most of the exocrine glands of the body are stimulated to secrete as a result of parasympathetic division activity. An exception is the *mammary gland.* Prolactin is the primary hormone controlling milk production and its secretion into the alveoli of the breasts. Stimulation of the myoepithelial cells surrounding the alveoli for milk ejection or "let-down" (the exocrine function) is by oxytocin released from the posterior pituitary (the neurohypophysis). Emotional disturbances leading to generalized sympathetic stimulation inhibit the release of oxytocin, depress milk ejection, and so lead to inadequate nursing.

The *nasal* and *lacrimal glands* secrete mucus and tears, thereby providing a protective function. The *salivary glands,* when stimulated by the parasympathetic division, supply a copious secretion for mastication and digestion. The *glands of the stomach* and *pancreas* are stimulated by parasympathetic division impulses and provide an increased secretion of pancreatic and digestive juices for digestion. Not all control of these secretions is neural, however.

The sympathetic division has little or no effect on exocrine secretion except that its activity generally causes vasoconstriction, which tends to reduce secretion. Exceptions are the effects on the salivary glands and pancreas. Sympathetic stimulation causes a small but definite increase in the flow of saliva, which is thicker and more viscous than that following parasympathetic division activity. Splanchnic nerve stimulation (sympathetic) causes some pancreatic secretion and changes in the zymogen granules of the acinar cells of the pancreas.

The *eccrine* type of *sweat gland,* producing copious, watery sweat (*hidrosis*), is innervated by sympathetic nerves with no connections from the parasympathetic division. Innervation is by cholinergic fibers, an exception to the general rule that sympathetic postganglionic fibers are adrenergic.

The *apocrine glands,* located primarily in the axillary and pubic regions, secrete a thick material that is decomposed by skin flora to produce characteristic body odors. Although closely related to the eccrine glands, they respond to adrenergic stimulation. In fact, there is evidence that they are not innervated, but respond to circulatory epinephrine only. Whereas the eccrine glands are associated with heat loss, the apocrine glands are associated with fear, anxiety, and vigorous muscular or sexual activity.

REPRODUCTIVE SYSTEM. The role of the ANS in normal sexual activity is complex. Engorgement of *erectile tissue* of both sexes (penis and clitoris) is by parasympathetic dilation of the arterioles of the tissue. With restricted venous outflow, the increased blood flow engorges the venous sinusoids. Without sexual excitement a sympathetic tone appears to keep blood flow low and the tissue flaccid. In addition, there is parasympathetic stimulation of exocrine glands of the *vagina,* which supply mucus for lubrication and protection. Whereas the ANS is clearly involved in psychogenic erection, engorgement of the erectile tissue can also be elicited by physical stimuli involving a simple spinal reflex. In the male the contraction of the vas deferens and seminal vesicles and *ejaculation* of semen are innervated by the sympathetic division. However, spinal cord transection above the third lumbar segment does not abolish ejaculation. It is not clear whether the muscular contractions of the female *orgasm* are also mediated by the sympathetics. The role of the ANS in *parturition* is minimal, although the release of oxytocin from the neurohypophysis is crucial. The sympathetic division is apparently not essential for gestation.

SMOOTH MUSCLE OF THE GASTROINTESTINAL TRACT. Although the gastrointestinal system has its own intrinsic neural control, both divisions of the ANS affect gastrointestinal activity. Parasympathetic division activity generally increases gastrointestinal motility. The role of the sympathetic division is primarily an inhibiting one, seen only in some diseases and during stress.

The esophagus is supplied by parasympathetic nerves. The involuntary parts of the process of swallowing are mediated by the parasympathetic division, which initiates a wave of constriction down the musculature of the esophagus.

SMOOTH MUSCLE OF THE PELVIC VISCERA. The parasympathetic system is involved in contraction of the bladder and lower colon for emptying of these organs. A spinal reflex arc, including parasympathetic fibers, activates micturition and defecation. Excitement and generalized sympathetic activity will tend to inhibit those functions. Furthermore, since the external sphincters are under somatic control, elimination can be stopped voluntarily.

RESPONSE TO DENERVATION

Immediately following sectioning of sympathetic or parasympathetic nerves, the denervated organ loses most of its tone and activity, but not permanently. Compensation develops that is primarily an increased sensitivity to circulating chemical agents. Denervation supersensitivity is most pronounced if the postganglionic nerves are severed. This was first studied in connection with *Horner's syndrome,* which results from interruption of the sympathetic division supply to the face. As a result of lack of sympathetic tone the pupils are constricted (miosis), the eyelids droop (ptosis), and the face is flushed. With the passage of time, however, these symptoms tend to disappear. Denervation supersensitivity also follows parasympathetic nerve sectioning. Even skeletal muscle shows increased sensitivity to the mediator after denervation, for sectioning of the motor nerve causes the motor end-plates to become hypersensitive to ACh.

Since preganglionic denervation causes less sensitization than does postganglionic denervation, it provides a surgical tool. For example, *Raynaud's disease* involves painful paroxysmal cutaneous vasospasms (usually in the fingers and toes) following exposure to cold or emotional stress. The vasospasm is so intense that gangrene sometimes results from the lack of circulation. If the postganglionic fibers are sectioned, it is only a matter of time before the blood vessels respond excessively to circulating epinephrine. Preganglionic denervation, however, eliminates the neural constrictor influence without the development of marked hypersensitivity to the circulating epinephrine.

TRAINING THE AUTONOMIC NERVOUS SYSTEM

With operant conditioning, it is clear that ANS responses to various stresses and stimuli can be modified. Such visceral learning would be of great therapeutic significance and is the subject of continuing research.

DRUGS ENHANCING OR DEPRESSING ANS FUNCTIONS

Drugs that mimic or block sympathetic or parasympathetic division activity provide a tool for a better understanding of normal physiological processes and may be used to reverse or counteract some diseases. Pharmacology texts should be consulted for details.

PREGANGLIONIC SIMULATING DRUGS. Preganglionic simulating agents (e.g., ACh in very high concentrations and nicotine) mimic the activity of the preganglionic neurons and thereby stimulate postganglionic neurons of both divisions. Since nicotine and "nicotinic" drugs such as carbachol stimulate not only sympathetic and parasympathetic postganglionic fibers but also skeletal muscle, it is thought that the receptor system at the postganglionic neurons is similar to the system at the neuromuscular junction.

PARASYMPATHOMIMETIC OR CHOLINERGIC DRUGS. Because ACh is so rapidly destroyed in blood after injection, it does not have the same effects throughout the body as parasympathetic division activity does. However, drugs such as methacholine, muscarine, and pilocarpine are less rapidly inactivated and have effects in the body similar to those of the ACh released from parasympathetic endings. These drugs also simulate the activity of sympathetic *cholinergic* fibers; e.g., profuse sweating is seen following administration.

The parasympathetic effects may be potentiated if ACh destruction is inhibited. An agent such as eserine or prostigmine inhibits ACh hydrolysis and thus potentiates parasympathetic activity. Because of this anticholinesterase activity, transmission at sympathetic ganglia and neuromuscular junctions is also enhanced. The poisoning action of most organic phosphorus pesticides and nerve gases is by way of anticholinesterase activity.

The action of the alkaloid muscarine on viscera such as cardiac muscle, exocrine glands, and smooth muscle is similar to that of ACh. The effects are termed the *muscarinic actions* of ACh. Muscarine has little effect on skeletal muscle or ganglionic transmission. *Nicotine,* on the other hand, affects autonomic ganglia—both parasympathetic and sympathetic—and skeletal muscle much as ACh does. Thus, the nicotinic actions of ACh may be differentiated from the muscarinic actions.

SYMPATHOMIMETIC OR ADRENERGIC DRUGS. Sympathomimetic or adrenergic drugs mimic the effects of sympathetic division discharge. Drugs such as ephedrine, amphetamine, and isoproterenol have effects similar to the action of epinephrine and norepinephrine, which are natural hormones. They differ, however, in potency at various effector sites, in mode of action, and in the duration of their activity.

BLOCKING AGENTS. Two general types of blocking agents are used for research and therapy. *Depolarizing agents,* such as nicotine in high concentration, block by maintaining depolarization of the excitable membrane. At first, the effector organ is stimulated, but then blockage of transmission occurs. *Antidepolarizers* generally block by competing with ACh for the receptor sites on the postganglionic or postjunction membrane. Hexamethonium and tetraethylammonium act in this manner at the ganglia. *d*-Tubocurarine has a similar blocking action at the skeletal muscle neuromuscular junction, and atropine blocks at the visceral sites.

At the ganglia the nicotinic effects of ACh are blocked by agents such as tetraethylammonium and trimethaphan camphorsulfonate, but atropine or curare is relatively ineffective. Ganglionic blocking agents are used primarily to block sympathetic activity, as in reducing hypertension, but also act on the parasympathetic division ganglia.

At the neuromuscular junction of skeletal muscle the receptor agent for ACh is different from that in smooth muscle, for here atropine is not effective, while *d*-tubocurarine is. Succinylcholine is an example of a depolarizing blocking agent, whereas curare reversibly binds the receptor.

At the visceral effector organs, the muscarinic effects of ACh are blocked by atropine and scopolamine. Atropine blocks not only the excitatory effects of ACh, such as those on the intestine or exocrine glands, but also the inhibiting effects, such as in the heart.

The direct excitatory (alpha) responses of sympathetic effector cells to agents that cause vasoconstriction are blocked by certain drugs, for example, phentolamine or phenoxybenzamine. After administration of these drugs, epinephrine dilates certain blood vessels; the constricting effects are blocked. The drugs do not block the beta receptors, for catecholamines continue to stimulate metabolism and inhibit the gastrointestinal tract.

The beta response to epinephrine is blocked by propranolol. The use of such agents aids in studying the action of catecholamines.

Reserpine appears to block the sympathetic system, but in fact it acts to reduce the amount of catecholamines in the brain and adrenal medullae available for subsequent release after neural stimulation. Indeed, the catecholamines in the heart and blood vessels approach zero in animals treated with high doses of reserpine.

HOMEOSTASIS AND CONTROL OF PHYSIOLOGICAL SYSTEMS

Each cell of the body requires an environment that supplies nutrients and removes metabolic wastes. The concept of *a constant and optimal internal environment* as a necessity for normal function was first formulated by Claude Bernard a century ago. In 1929, Cannon further developed the concept of this condition, which he called *homeostasis,* and emphasized the role of the ANS. One of the cardinal principles of physiology is that homeostatic mechanisms operate to counteract changes in the internal environment that are induced either by changes in the external environment or by activity of the individual. Thus, disturbances of the internal environment as the result of exercise, nutritional imbalance, trauma, and disease are minimized.

An example of homeostasis is the control of body temperature. If the internal temperature drops, homeostatic mechanisms act to reduce heat loss from the body and to increase heat production. Consequently, these mechanisms limit a decrease in body temperature and so maintain this variable relatively constant. On the other hand, a cold-blooded animal does not possess homeostatic temperature control systems; its body temperature thus tends to be similar to that of its environment.

Homeostatic mechanisms act to minimize the difference between the actual and optimal responses of a system and are therefore biological examples of *negative feedback control*. The level of the controlled variable is sensed, and action is taken that opposes any change from the desired level. If the response increases, a signal is fed back to an effector mechanism in a negative or inhibitory manner, so that the subsequent response is reduced. Conversely, a decrease in response elicits a subsequent increase. A familiar example of a negative feedback control system is the thermostatic control of room temperature. Room temperature is measured by a temperature-sensing element in a thermostat and compared with a reference (desired or optimal temperature) in such a manner that an *error signal* develops when discrepancies exist. If the room is too hot, the furnace is turned off by the action of the thermostat; if it is too cold, the furnace is turned on. The error signal from the thermostat is used to adjust the system automatically so as to minimize any deviation between the measured temperature and the desired temperature. If the system is properly designed, room temperature can be held nearly constant despite wide fluctuations in outside temperature.

Homeostatic feedback mechanisms of mammals are exceedingly complex and interrelated but are generally amenable to analysis using the approach of engineers. These workers, utilizing the principle of negative feedback control, have made great advances in the design of control systems used in the operation of devices such as automatic airplane pilots, missile guidance systems, and servomechanisms for industrial automation. The primary purpose of using negative feedback in these mechanisms is to provide high accuracy and stability of function in spite of changes in the external environment or changes in the systems themselves. The same general principle acts in mammals to keep constant and optimal such diverse variables as body temperature, blood pressure, blood sugar concentration, electrolyte concentrations, muscle tone, and blood carbon dioxide levels, to mention only a few. The ANS is an important part of most homeostatic mechanisms.

SYSTEMS

The concept of a system—a set of components that act together and that can be treated as a whole—has been found to be highly useful for the study of dynamic, complex biomedical situations. With a system, *inputs* are those stimuli, forcings, or disturbances that act on it, while the *output* or response is the consequence of such inputs. Furthermore, the relationship between output and input is the "law" describing the system, usually written as a mathematical equation. In some cases, such as that describing the relationship between the input forcing of a transducer and the resulting pen deflection of a recorder, the equation is simple; in others, such as those predicting changes in cardiac function in response to changes in pressure at the pressoreceptors, the relationship is so involved that complex differential equations are required to provide even an approximation. If the purpose of the analysis is to understand and predict the behavior of complex systems, the input-to-output relationship written in mathematical form with carefully defined terms is an effective form for the theory describing the system.

SYSTEMS ANALYSIS, SIMULATION, AND MODELS

As our understanding of biological systems becomes more extensive, it becomes more difficult to organize the information, to check it for consistency, and to predict the consequences of various stimuli or changes of parameters on a system. *Mathematical biology, systems analysis,* and *simulation* are different words for the general approach of (1) describing our understanding through models written in the form of mathematical equations and (2) studying the models with computers to check for internal consistency and for agreement with available experimental data. A mathematical model may provide a concise and accurate summary of thousands of experiments and may permit accurate predictions under a wide variety of conditions.

A *model* is a representation of reality, be it a stereotype, archetype, caricature, syndrome, diagram, prejudice, equation, or verbal description. In comparison to verbal models, mathematical models are more precise, less ambiguous, and more readily permit quantitative verification. The use of computers facilitates the study of multifactor interaction and dynamic (i.e., transient or moment-by-moment) responses. In the process of developing and testing the model—called *simulation*—new experiments are suggested, for gaps in knowledge often become obvious. Not only do models provide a concise summary of pertinent observations, but for many students they also provide a more satisfying description of the system than catalogues of cases, tables of data, or graphs.

CHARACTERISTICS OF HOMEOSTATIC MECHANISMS

The operation of a negative feedback control system may be partitioned into several separate functions (Fig. 7-5). First, information about the magnitude of the response must be sensed (the *sensor*). This is fed to a *controller*. In some systems the response signal is compared to a *reference* signal by subtraction to generate an *error signal*. The error signal is *fed back* to the *effector mechanism* in such a manner (negative) as to *minimize the error*.

Regulation of the amount of food eaten is an example of a complex biological control system. Hunger is the error signal. The level of intake needed to satisfy the hunger is widely variable, depending in part on the individual's physical activity, psychological status, and general health. If the food intake is inadequate, the hunger increases and the individual eats more and thereby reduces the hunger. The arterial baroreceptor reflex, illustrated in Figure 17-2, is another example of a negative feedback control system.

At the molecular level, negative feedback control systems also operate by controlling enzymatic reactions. Here, all facets may be combined into a system in which an increase in the concentration of the product of a chemical reaction inhibits the reaction itself, so that the product concentration is held relatively constant.

The response of a system may be controlled by two general approaches. (1) *Open loop,* or control based on *compensation* for the effect of each source of disturbance: The quantitative effects of all the important disturbances on the

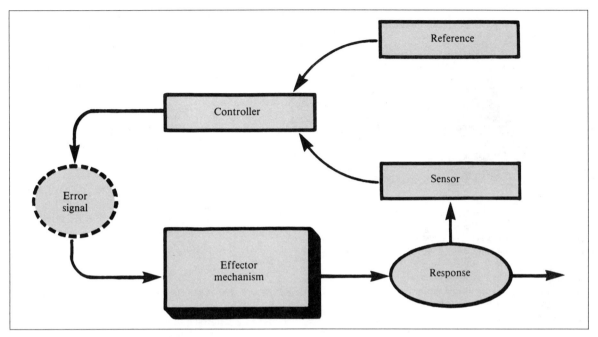

Fig. 7-5. Components of a negative feedback control system.

system must be determined. The magnitudes of the disturbances are sensed or measured, and the system is then adjusted accordingly. Aiming a gun, after taking into account wind and distance, is an example of this approach. Its effectiveness depends upon precise knowledge of all disturbing variables, both as to their influence and as to their magnitude. In addition, the mechanism itself must be precise and stable, features not characteristic of biological systems. (2) *Closed loop:* In this approach, control is attained by measuring or sensing the response and *feeding back* the deviation from optimal to the effector mechanism. Knowledge of the source and magnitude of the disturbance is immaterial—providing they are not beyond the range of control. The effector mechanism may be unstable and imprecise in constructional details—a hallmark of living organisms. With a closed loop, close and reliable control is attainable. A control system in which the output or response is held constant is often called a *regulator,* while one in which the response is made to follow the pattern of a varying input signal is called a *servomechanism.* Control (in the sense of negative-feedback control) implies that the response of the total system will be held near certain desired values even though disturbing inputs to the system fluctuate, or there are changes in the effector system itself.

REFERENCE LEVEL. The *reference level* (or set point) is the desired condition. In the control of room temperature the reference is the temperature setting of the thermostat. In the healthy individual, it is usually the optimal level of the variable and aids in defining what is meant by normal. With a control system the regulated variable may be changed by changing the reference level. Many pathological conditions can be traced to such a change, although the details of how and why are far from being solved. For example, fever is primarily a change in the reference level for the control of body temperature. Hypertension is possibly a change in the reference level for the control of blood pressure. When the reference is changed, the controlled function fluctuates about the new level.

Comparing the response of the system to the input or reference by subtraction is one way to obtain the error signal. A different approach is to divide the reference signal by a signal representing the response. As applied to an amplifier, this type of control provides an automatic gain control. Although potentially of wide importance in studying biological systems, especially those involving enzymatic actions, this form of description has not been widely exploited as yet.

OPERATING POINT. Since in most biological systems there is no clearly defined arithmetic comparator for the genera-

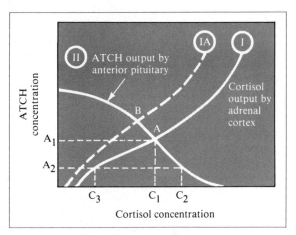

Fig. 7-6. Concept of *operating point* in control of blood cortisol concentration. Curve *I* shows the effect of ACTH on blood cortisol concentration and curve *II*, the effect of cortisol concentration on blood ACTH secreted by the pituitary. Curve *IA*, shows the relationship after damage to the adrenal cortex. Point *A* is the normal operating point. Point *B* is the operating point with a damaged adrenal cortex.

tion of an error signal by subtracting a value representing the response from a reference value, most biological systems function around an operating point determined by the characteristics of the chemical and physical structures of the system. The *operating point* of two interacting subsystems is that level of responses that leads to a stable value of the mutually influenced variable. In effect, the values of the mutually influenced variables are obtainable from the solution of the simultaneous equations describing the input-output response of each of the interacting subsystems. An example may help.

Cortisol synthesis and secretion into the blood are stimulated by the adrenocorticotrophic hormone (ACTH), curve I, Figure 7-6. An increase in ACTH causes an increase in cortisol concentration. (Here, the dependent variable, cortisol, is plotted on the horizontal axis, contrary to convention.) At the same time, as the blood cortisol concentration increases, the cortisol inhibits the anterior pituitary, so that ACTH synthesis and thus blood concentration *decrease* (curve II, Fig. 7-6). Only at ACTH concentrations of A_1 and cortisol concentrations of C_1 will the system be stable. If, for example, exogenous cortisol is administered to produce a transient increase in cortisol concentration of C_2, this will act to inhibit ACTH secretion (curve II, Fig. 7-6), and so ACTH concentration will drop to A_2. This would stimulate only enough cortisol to tend to give a cortisol concentration of C_3. At the same time, the increased cortisol will be elimi-

nated faster than is normal by the liver and other tissues of the body, because the concentration is higher than usual. Because cortisol destruction then exceeds production, the concentration will decrease toward C_1. If it undershoots slightly (lower-than-normal cortisol), ACTH production will increase above normal, because the lower-than-normal cortisol concentration will cause less inhibition.

The operating point may change. If the adrenal cortex becomes damaged, or cortisol is metabolized more rapidly than normal, so that the cortisol output at various ACTH concentrations becomes that shown by curve IA, then the operating point will equilibrate at point B—a somewhat lower-than-normal blood cortisol concentration in conjunction with a somewhat higher-than-normal ACTH concentration. This is of diagnostic value. A low ACTH in conjunction with a lower-than-normal cortisol concentration suggests malfunction at the pituitary rather than at the adrenal level.

FEEDBACK FACTOR. The feedback factor has been variously defined, but it generally means the ratio of the response to a disturbing influence of an uncontrolled system to that of the controlled system.

$$\text{Feedback factor} = \frac{\text{uncontrolled response}}{\text{controlled response}}$$

It is a measure of the effectiveness of the control system. As an example, the control of body temperature in mammals is excellent. If the body temperature of a group of subjects in a room with the temperature at 40°C was 37.8°C, while that of the same persons under similar conditions but at 2° C was 37.2°C, the feedback factor for these people was

$$\frac{40°C - 2°C \text{ (uncontrolled response to environment)}}{37.8°C - 37.2°C \text{ (controlled response)}} = \frac{38}{0.6} = 63$$

The blood pressure is regulated with a feedback factor of about 5. This means that an influence that would cause only a 12 mm Hg change in the normal animal would cause a 60 mm Hg change in the animal without control. Hypertension might be explained by some factor causing an increase in blood pressure in an animal that has a feedback factor too low to provide adequate compensation or that has a control system that unfortunately adapts over a long period to the pathological pressures.

It is generally difficult to obtain a completely uncontrolled response from a biological system, since it is hard—and in most cases probably impossible—to remove all the regulatory features without damaging the normal, basic biological mechanisms. The concept of feedback factors is of value, however, in assessing the importance of parts of a biological control system or in testing the adequacy of a homeostatic mechanism. For example, if before anesthetization the change in a dog's blood pressure in response to a temporary loss of blood volume is -2 mm Hg, and after anesthetization the response to a similar loss is -20 mm Hg, it can be said that the efficiency of the control of blood pressure is reduced by a factor of 20/2, or 10, by the anesthetic agent. For use as a diagnostic approach to disease, standardized stresses or test situations must be developed.

MODES OF CONTROL. Several modes of control are found in biological systems: (1) In *proportional* control the signal (E) driving the effector system is directly proportional to the error (ε) signal ($E = k_1\varepsilon$, where k_1 is a gain or amplification factor). With simple proportional control, the error is relatively large if the disturbance is large. (2) In *integral* control, to reduce the steady-state error, the error is integrated (added up as time goes by) to develop an additional signal giving proportional-plus-integral control ($E = k_1\varepsilon + k_2\int\varepsilon \, dt$). (3) In *derivative* control, to anticipate, in effect, the degree of corrective action needed and so to reduce "hunting" (overshooting and undershooting around the desired value), the rate of change of error can also be used with proportional control ($E = k_1\varepsilon + k_3 \, d\varepsilon/dt$). For example, control of blood pressure and body temperature is in part rate dependent. A rapid change has a greater effect than a slow change. Because all physical systems have energy-storing features (e.g., inertia and compliance), the response never exactly follows the stimulus, and so perfect control during transient changes, even with these more complex modes, is impossible. (4) *On-off* control is the familiar type of control used with automatic appliances such as home furnaces, refrigerators, and pumps. When the controlled variable exceeds set limits, the unit is turned on or off. Though it is discontinuous, this type of control can be highly effective, especially if used as a supplement to open-loop control.

LIMITS OF CONTROL. A control system will perform accurately only within certain limits of change in components or energy supply. If the energy supply is markedly reduced or some component is seriously damaged, the effector mechanism may operate at maximal, yet inadequate, levels. With a serious continuing loss of blood, the blood pressure may be maintained for a while by compensatory mechanisms, but a point will be reached at which the limits of compensation have been exceeded. Then, the pressure will fall, often in a fulminating manner.

When a biological feedback system is driven to its limits because of malfunction or external stress, the ideal form of treatment is to find and correct the cause of the change rather than attempting to augment compensatory changes that might lead to overloading of the system and possible fulminating failure. Defective function of any link in the homeostatic system is ordinarily reflected in some degree of hyperfunction or hypofunction of other elements in that system. This is the reason symptoms are commonly clustered into characteristic *syndromes.* Because the various parts of a homeostatic system are connected functionally into a closed loop, the chain of cause and effect among symptoms is often difficult to discover unless studied in the light of overall system operation.

POSITIVE FEEDBACK, RESPONSE TIME, AND STABILITY. If the effect of the feedback is to increase, rather than decrease, the deviation from the desired level, the feedback is said to be positive. Positive feedback may be defined as a situation in which an *increase* in response, when fed back to the control mechanism, causes a still *further increase* in response. Positive feedback occurs if the feedback factor is less than 1; i.e., if the "controlled" response to a disturbing influence is greater than the "uncontrolled" response with no feedback. Positive feedback, if not limited, causes a progressive change in response, so that eventually the system operates at a maximal or minimal level—a vicious cycle.

If the thermostat for a furnace were connected in reverse, so that an increase in temperature turned the furnace *on,* the result would be an example of unlimited positive feedback. Biological examples of unlimited positive feedback are relatively rare, since the condition results in extremes of response. The response of a nerve to an adequate stimulus is a good example of positive feedback, however. If the stimulus to nerve action is inadequate, there is no response; but if the stimulus is adequate, once the response of the nerve starts, the process reinforces itself (positive feedback), so that the nerve fiber responds maximally. Only later is the process reversed.

Positive feedback can lead to death. For example, if the heart starts to fail significantly, less blood is pumped, the blood pressure falls, and there is consequently a lesser supply of blood to the heart itself. This weakens it *further,* leading to a further decrease in blood pumped and eventually to death.

The *response time* of a system is a measure of the time required for it to respond to a change in conditions and is defined, for the simplest system, as 63.2 percent ($1 - 1/e$) of the time taken to achieve the final total change in response. Following a sudden change in conditions, the error signal, even in an efficient control system, is relatively large at first, but then it declines toward zero as the desired level is again approached.

Because of the interaction between energy-storing devices, oscillations may occur in response to a sudden change in input or to disturbances. The transport time of materials through the vascular system and the diffusion of products from sites of synthesis are conducive to oscillations. In most control systems, oscillations are a sign of malfunctioning or poor control. The blood pressure sometimes oscillates in an animal with a failing cardiovascular system. Predicting the conditions for system stability (i.e., lack of sustained oscillations) is a major part of the complex science of control-system design.

For accurate control, a high feedback factor is essential, so that the effects of changes in the metabolic energy supply, parts of the system, or environment will be reduced toward zero. Such systems, however, tend to be unstable and to oscillate, especially if the feedback system has a relatively long response time (i.e., is sluggish). The response of the somatic nervous system is speedy; therefore, accurate control of posture by rapidly responding muscles is possible. In some disorders, however, the feedback system is sluggish or the feedback factor (gain) is too high, so there is tremor on attempts to make fine movements.

BIOLOGICAL CONTROL SYSTEMS. Many biological variables are controlled directly. For example, the amount of carbon dioxide in the arterial blood is controlled by homeostatic mechanisms that regulate the degree of lung ventilation so that the proper amount of carbon dioxide is exhaled. Other variables, such as the amount of oxygen in the arterial blood, tend to be regulated in an indirect manner, for the arterial oxygen tension is dependent largely on the amount of oxygen in the air and the degree of lung ventilation; this in turn is controlled directly by the amount of carbon dioxide in the blood. Finding the primary variables and the interrelationships of these homeostatic mechanisms is one of the challenges of modern physiology.

It must be emphasized that a negative feedback control system is not highly dependent on the accuracy and stability of its components. A heart may be severely damaged by disease, yet the blood pressure may be normal if the person avoids strenuous exertion. There are marked differences in the dimensions of the organs of people and in their functional reserves between birth and old age, yet these differences do not preclude normal function.

Finally, biological homeostatic mechanisms often possess several feedback systems that act on the same variable, providing constancy and reliability to the organism (though complexity to the physiologist and student). As an example, there are several types of sensors and effector mechanisms for the control of arterial blood pressure.

The ANS is an essential part of most mammalian control systems. As better understanding of the homeostatic mechanisms and sites of failure is obtained, more effective therapy for various pathological conditions may be devised.

In conclusion: "The essence of physiology is regulation. It is this concern with 'purposeful' system responses which distinguishes physiology from biophysics and biochemistry. Thus, physiologists study the regulation of breathing, of cardiac output, of blood pressure, of water balance, of body temperature and of a host of other biological phenomena" (Grodins et al., 1954).

BIBLIOGRAPHY

The Autonomic Nervous System

Altman, P. L., and Dittmer, D. S. (eds.). *Biology Data Book* (2nd ed.). Bethesda, Md.: Federation of American Societies for Experimental Biology, 1973. Vol. 2, pp. 1150–1180.

Bevan, J. A., Bevan, R. D., and Duckles, S. P. Adrenergic Regulation of Vascular Smooth Muscle. In D. Bohr et al. (eds.), *Handbook of Physiology*. Washington: American Physiological Society, 1980. Section 2: The Cardiovascular System. Vol. 2, Vascular Smooth Muscle, pp. 515–566.

Burn, J. H. *Autonomic Nervous System* (4th ed.). Oxford, Engl.: Blackwell, 1971.

Goodman, L. S., and Gilman, A. (eds.). *The Pharmacological Basis of Therapeutics* (6th ed.). New York: Macmillan, 1980.

Korner, P. I. Central Nervous Control of Autonomic Cardiovascular Function. In R. M. Berne and N. Sperelakis (eds.), *Handbook of Physiology*. Bethesda, Md.: American Physiological Society, 1979. Section 2: The Cardiovascular System. Vol. 1, The Heart, pp. 691–739.

Malmejac, J. Activity of adrenal medulla and its regulation. *Physiol. Rev.* 44:186–218, 1964.

Homeostasis and Control
of Physiological Systems

Cannon, W. B. Organization for physiological homeostasis. *Physiol. Rev.* 9:399–431, 1929.

Gold, H. J. *Mathematical Modeling of Biological Systems—An Introductory Guidebook*. New York: Wiley, 1977.

Grodins, F. S., Gray, J. S., Schroeder, K. R., Norins, A. L., and Jones, R. W. Respiratory responses to CO_2 inhalation. A theoretical study of a nonlinear biological regulator. *J. Appl. Physiol.* 7:283–308, 1954.

Jones, R. W. *Principles of Biological Regulation: An Introduction to Feedback Systems*. New York: Academic, 1973.

Riggs, D. S. *Control Theory and Physiological Feedback Mechanisms*. Baltimore: Williams & Wilkins, 1970.

Yamamoto, W. S., and Brobeck, J. R. *Physiological Controls and Regulations*. Philadelphia: Saunders, 1965.

Yamamoto, W. S., and Walton, E. S. On the evolution of the physiological model. *Annu. Rev. Biophys. Bioeng.* 4:81–102, 1975.

HIGHER SOMATIC AND VISCERAL CONTROL

Sidney Ochs

MOTOR AREAS OF THE CEREBRAL CORTEX

In 1870, Fritsch and Hitzig first showed that electrical stimulation of certain parts of the cortex produced movements of specific portions of the peripheral musculature. Since then, others have shown, in an increasingly refined way, the correspondence of points within such *motor areas* to the peripheral musculature. In Figure 8-1, the motor area of the human brain is labeled with the body musculature excited from the various cortical sites. A general similarity or homology of motor areas and their muscle control patterns within the cerebral cortex has been found for the different species. A relationship of motor function to the underlying cellular structure was indicated for the area of the brain numbered 4 by Brodmann (Fig. 8-1), which contains large *pyramidal tract* (PT) *cells* also known as *Betz cells*. The large PT cells are stimulated by applied electrical currents to give responses in the different limb muscles. However, there is not an exact correspondence of motor control of a given musculature to a specific PT cell popula-

tion; the motor area is known to extend beyond area 4. Also, axons from a given part of the motor cortex are widely distributed to many motor nuclei of the lower motor centers—although with a predominant convergence onto one group of motoneurons innervating a particular muscle.

With the use of a constant level of anesthesia and other measures to control the stability of the experimental preparation, a fairly consistent and detailed representation of the peripheral musculature in the motor region may be found. The motor area determined in this way is called the *precentral motor area*. Figure 8-2 shows the result obtained from a *Macaca mulatta* brain, in which a figurative map was constructed from those points of the cortex having the lowest threshold for responses in the various muscles represented. As was indicated, the correspondence of motor areas with the various cytoarchitectonic regions of the cortex is only approximate. Portions of the motor cortex are seen to fall outside area 4 into area 6 of Brodmann, where larger muscle groups of the back and the upper limb muscles are represented. Area 4 controls finer muscles, namely, those of the

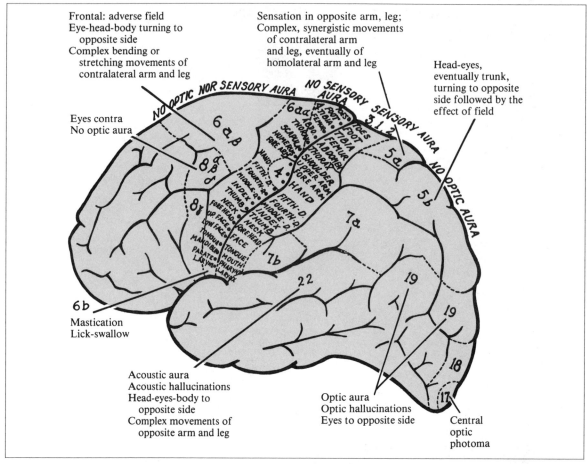

Frontal: adverse field
Eye-head-body turning to
    opposite side
Complex bending or
    stretching movements of
    contralateral arm and leg

Sensation in opposite arm, leg;
Complex, synergistic movements
    of contralateral arm
    and leg, eventually of
    homolateral arm and leg

Head-eyes,
eventually trunk,
turning to opposite
side followed by the
effect of field

Eyes contra
No optic aura

Mastication
Lick-swallow

Acoustic aura
Acoustic hallucinations
Head-eyes-body to
    opposite side
Complex movements of
    opposite arm and leg

Optic aura
Optic hallucinations
Eyes to opposite side

Central
optic
photoma

Fig. 8-1. Motor areas of the human brain. The motor areas of the brain are in front of the central sulcus. The various parts of the peripheral musculature excited by stimulation in that region are labeled. Various other regions in different areas of the brain also can give motor responses as indicated. Notice the correspondence of the motor area to the sensory regions posterior to the sulcus. (From P. Bard. *Medical Physiology* [11th ed.]. St. Louis: Mosby, 1961. P. 1208.)

lower arms, digits, tongue, and face. These areas of finer muscle control are disproportionately large with respect to the control areas of other parts of the musculature. Therefore, lesions of the motor cortex are more destructive to the finer and complex manipulative movements of the digits, while the grosser movements of the axial musculature or of muscles of the upper limbs remain relatively less affected. In humans and primates in general, PT fibers controlling fine movements end directly on spinal motoneurons. However, a large share of the control is apparently also subserved in tracts other than the pyramidal, i.e., in the *extrapyramidal tracts,* which, after interneuronal discharge, also influence the motoneurons.

A greater susceptibility to disruption of fine control is characteristically seen in patients who have had strokes with lesions in the internal capsule interrupting a large part

of the fiber downflow from the motor cortex. A patient so affected has a disturbance of posture and movement. The position of the limbs is said to show a reversion to primitive patterns of motor control—flexion of the upper arms and hyperextension of the lower legs with increased reflexes, i.e., *spasticity.* The lower motor centers and brainstem have been *released* by the capsular lesion from upper cortical control. A positive Babinski sign is present with upper motoneuron lesions. But it is in connection with fine complex movements that the disturbance is most apparent. The

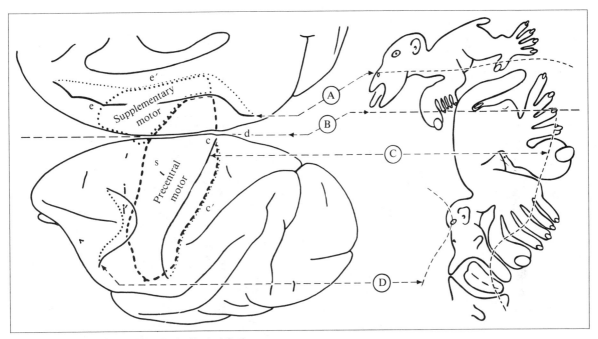

Fig. 8-2. Motor areas of the monkey brain. To the left, the precentral motor area is shown with dashed lines; to the right is a map of the body musculature represented within the area. In addition to this primary motor area, a smaller supplementary motor area can be found. Much of it is hidden from view on the medial aspect of the brain. For both the primary and secondary motor areas the thumb area is relatively very large, as are the areas for the digits of the hand and foot. The face is quite large, with a large tongue area. The posterior band of the central sulcus, the anterior bank of the inferior precentral, and the lower bank of the sulcus cinguli have been cut away; c = central sulcus; c' = bottom of central sulcus; i = inferior precentral sulcus; i' = bottom of inferior precentral sulcus; d = medial edge of hemisphere; e = sulcus cinguli; e' = bottom of sulcus cinguli. (From C. N. Woolsey et al. Patterns of localization in precentral and "supplementary" motor areas and their relation to the concept of a premotor area. *Association for Research in Nervous and Mental Disease, Proceedings* [1950] 30:244, 252, 1952.)

attempt to move the fingers—opening and closing them or attempting to approximate the fingers—is difficult or not possible.

In the lower limb the defect gives rise to the spastic gait typical of a stroke patient. The leg is held stiffly extended, and during walking it is swung outward and forward in a sticklike fashion. The rhythmic flexions and extensions of the limbs at the hip, knee, and ankle joints that are part of the normal walking pattern are lost. The earlier assignation of a coarse control function to area 6 was due in part to an

unrecognized involvement with a distinct second motor area, the *supplementary motor area,* which can also be represented by a figurative map. Part of this area is turned around and hidden from view on the medial surface of the hemisphere (Fig. 8-2). The responses to stimulation of the supplementary motor area give rise to the more generalized movements affecting large muscular groups. The efferent pathway from the supplementary area projects separately into the PT and does not relay through the precentral motor area of the cortex.

The region from which motor responses are obtained represents a complex interplay of neurons having facilitatory and excitatory effects on the eventual motor outflow that then passes down in the PT. Anesthesia, which might be expected to have an effect on the complex pattern of excitation, causes a decrease in the size of the area from which a motor response of a given muscle is obtained by monopolar anodal excitation. Studies indicate that the organization of the motor cortex consists in widely overlapping areas of muscle "representations" instead of pointlike areas or discrete regions controlling particular muscle groups. The shrinkage of the area of excitability of a given muscle group with increased anesthesia is probably due to the loss of facilitation from surrounding neurons ending on the pyramidal cells. When local anesthesia is used, or chronically implanted electrodes are employed for stimulation without any

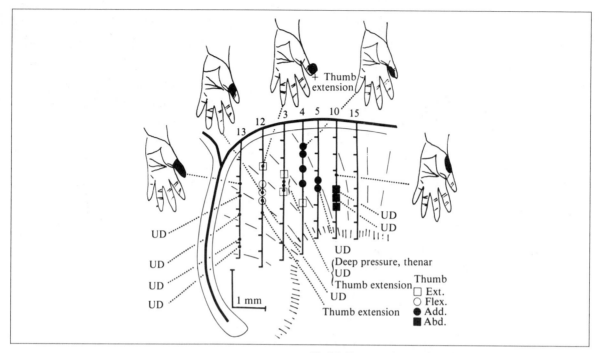

Fig. 8-3. Reconstruction of electrode tracks and cell locations. Several electrode penetrations (*solid lines, identified by numbers*) passed through efferent zones projecting to various thumb muscles. *UD* represents cells undriven by peripheral stimulation. The peripheral motor effects are indicated by symbols explained in the figure. Cortical spots stimulated without evoking motor effects are shown by short solid lines perpendicular to the lines indicating tracks. Positions of cells encountered are indicated by dots and connected with dotted lines to descriptions of receptive fields and adequate stimuli. (From J. Rosen and H. Asanuma. Peripheral afferent inputs to the forelimb area of the monkey motor cortex: Input-output relations. *Exp. Brain Res.* 14:261, 1972.)

anesthesia, motor responses are generally seen to be elicited over large areas of the cortex.

A technique used to restrict excitation to a small group of pyramidal cells is the insertion of semimicroelectrodes into the cortex to effect a stimulation at various depths in the cortex. The effective current is a cathodal pulse, which probably depolarizes cell bodies or, rather, the initial segments of the pyramidal cells. With this technique, excitation can be seen at much lower thresholds than with surface monopolar anodal shocks, and small muscle fields can be excited from groups of neurons aligned in columnar fashion (Fig. 8-3). As the microelectrode tip is moved down through the depth of the cortex, it excites cells having the same motor effect. The threshold is lowest at the fifth layer, where the larger pyramidal cells contributing to the corticospinal outflow are found.

Fairly large groups of cells are involved by stimulation of even the small region around such electrode tips. This is part of a pattern of coordination of excitation via interconnections between a great many neurons. The resultant inputs acting on single PT cells have been shown by recording intracellularly from neurons identified as PT cells by antidromic excitation. In this technique the axons in the PT are stimulated while a recording is made from a PT cell in the

cortex. The method is similar to that used to identify the motoneurons of the spinal cord by an antidromic volley (see Chap. 5). While recording is being carried out from an identified PT cell, a stimulation of a primary sensory input to the cortex gives rise to either a discharge or a change in the ongoing discharge rate of that PT cell.

Study of the properties of PT cells by means of such microelectrode recording has revealed an inhibitory system in the cortex similar to that of the Renshaw cell system in the spinal cord. The PT neuron has collaterals synapsing onto inhibitory cells, giving rise to long-lasting hyperpolarizations and a decreased excitability in other PT neurons when those cells are activated.

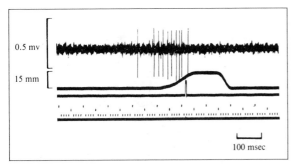

Fig. 8-4. Discharge of a precentral neuron in association with one phase of a repetitive-movement task. A monkey grasped a horizontal lever and pulled it forward through 15 mm with its right hand. The movement of the lever is indicated in the second trace. The top trace shows the burst of nerve impulses produced by a single neuron in the left precentral gyrus discharging in relation to one performance of this movement task. The cell in the motor cortex began firing about 100 msec before the beginning of the movement. The bursts always began before the movement and always preceded the beginning of electromyographic activity in flexor muscles. Brackets indicate 0.5 mv for cell spikes and 15 mm for movement trace. (From R. Porter. Functions of the mammalian cerebral cortex in movement. *Prog. Neurobiol.* [Pt. 1] 1:27, 1973.)

The motor cortex is the site of voluntary control of motor behavior. The mechanism of such control is at present unknown, but recent microelectrode studies have shown some of the early events occurring during a voluntary movement. Microelectrodes were chronically implanted into the motor cortex of monkeys that were trained to make a movement, e.g., depress a lever. A discharge of neurons was found to precede the motor response (Fig. 8-4). This presumably represents some part of the integration of command, which then eventually results in the discharge of the PT cells controlling the peripheral musculature.

Interconnections between associational areas and motor areas permit other parts of the brain to affect motor control. The sensory receptor areas within the cortex have close connections to the motor areas, and electrical stimulation of parts of the cortex outside the "true" motor area may be compounded of sensory and associational cortex excitation that then acts on the motor neurons. A correspondence of the peripheral representation present in the somesthetic sensory cortical areas 1, 2, and 3 with the motor control regions of area 4 can be seen in Figure 8-1. In some species there is more overlap of sensory and musculature representations, and the term *sensorimotor cortex* rather than separate motor and sensory regions more aptly expresses this close integration.

RETICULAR FORMATION

The sensory activation exerted by the reticular formation on the cerebral cortex was referred to in Chapter 6. Historically, the motor control from the reticular formation acting on lower motor centers was discovered first. Stimulation within certain areas of the reticular formation results in a *facilitatory* effect on lower motor centers, whereas stimulation in other reticular formation areas gives rise to *inhibitory* effects, as is shown in Figure 8-5. While a reflex, the knee jerk, was being elicited continuously at regular intervals, stimulation through electrodes stereotaxically placed in the reticular formation of the lower medial part of the medulla gave rise to a rapid inhibition. On cessation of stimulation the reflexes recovered within a few seconds. A similar effect of reticular formation stimulation on cortically evoked limb movements was also found from this region. If stimulation was administered to the upper lateral portion of the reticular formation, a facilitation of knee jerk reflexes was found (Fig. 8-6). A similar effect was seen with cortically evoked limb movements. By this means, the presence of *facilitatory and inhibitory regions* within the reticular formation could be plotted (Fig. 8-6; see also Fig. 8-5).

There is also evidence of cortical and subcortical areas controlling these reticular formation regions, mainly by exertion of an inhibitory effect. Loss of the normal inhibitory control descending from the cortex and subcortical centers onto the facilitatory regions of the reticular formation centers could result in a *release* of excitatory influences in them. In turn, their descending influence on the brainstem and spinal cord is seen as exaggerated reflex activity, as *spasticity*.

A loss of discharge from the inhibitory reticular formation could also help produce the exaggerated reflex activity associated with spasticity. An important control of the spinal cord gamma motoneurons has been located in the reticular formation, and an increased gamma discharge has been associated with the spasticlike condition of decerebrate rigidity. Making a cut between the colliculi of the brainstem produces a *decerebrate rigidity* preparation. The legs are stiffly extended, and the head is arched back in a caricature of standing. An explanation of this phenomenon is that the inhibition from the reticular formation is reduced along with an increased excitation of vestibulospinal downflow and an excitatory downflow from the reticular formation.

Not all of this is due only to an augmentation of gamma motor activity. Stimulation of the reticular formation can still cause facilitation or inhibition of the monosynaptic response. When a monosynaptic reflex is elicited by stimula-

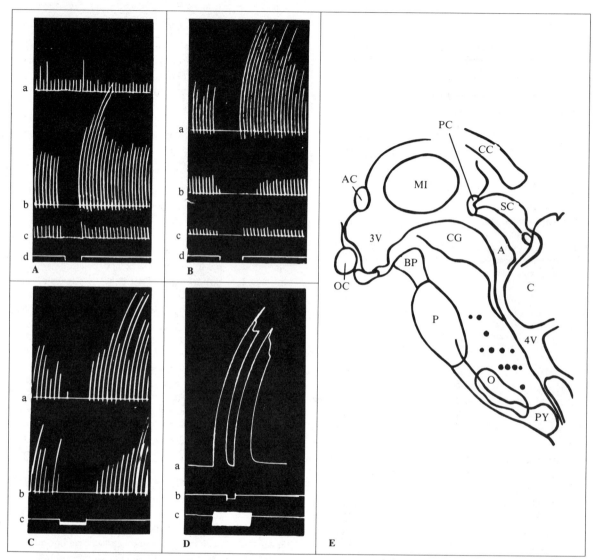

Fig. 8-5. Inhibitory regions in the reticular formation. To the right, in (E), is a sagittal view of the brainstem, with dots in the lower portion indicating regions effective for causing inhibitory effects. These effects are shown in (A) and (B) as an inhibition of knee jerks at the time of reticular formation stimulation (see lower segment lines). In (C) and (D), stimulation to the inhibitory regions is seen to be effective for inhibiting cortically evoked movements. *A* = aqueduct; *AC* = anterior commissure; *BP* = basis pedunculi; *C* = cerebellum; *CG* = periaqueductal gray; *F* = facial nucleus and nerve; *H* = hypoglossal nucleus and nerve; *MI* = massa intermedia; *MLF* = median longitudinal fasciculus; *OC* = optic chiasma; *P* = pons; *PC* = posterior commissure; *PN* = posterior column nuclei; *PY* = pyramidal tract; *R* = inferior reticular nucleus; *SC* = superior colliculus; *T* = nucleus of the spinal fifth tract; *TS* = tractus solitarius; *V* = vestibular nucleus; *3V* = third ventricle; *4V* = fourth ventricle. (From R. Rhines and H. W. Magoun. Brainstem facilitation of cortical motor response. *J. Neurophysiol.* **9:**221, 1946.)

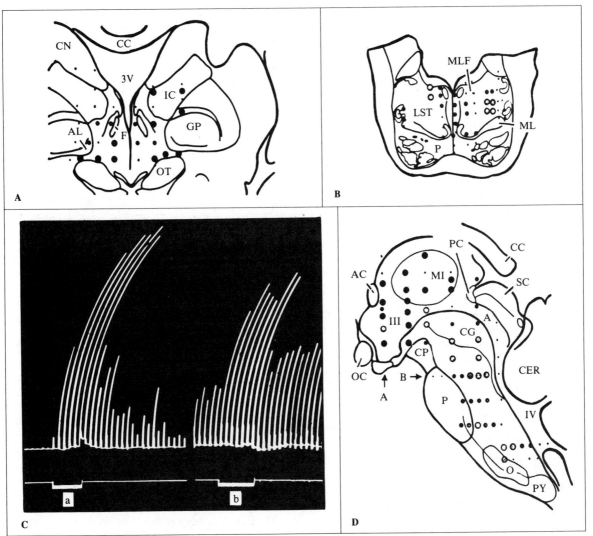

Fig. 8-6. Facilitatory regions of the reticular formation. (D) is a sagittal view of the brainstem. (A) and (B) are cross sections of the reticular formation, with dots representing the areas from which facilitatory effects are obtained. In general, these facilitatory regions are higher in the brainstem and more lateral than is the inhibitory region. In (C), cortically evoked responses are facilitated during the period of stimulation of the facilitatory region of the reticular formation (signal *a* on lower line). At the right, a knee jerk reflex is similarly facilitated during the period of stimulation of the reticular formation (signal *b* on lower line). Anatomical structures: *A* = aqueduct; *AC* = anterior commissure; *AL* = ansa lenticularis; *BC* = brachium conjunctivum; *BIC* = brachium of inferior colliculus; *BP* = brachium pontis; *C* = nucleus centralis; *CC* = corpus callosum; *CER* = cerebellum; *CG* = central gray; *CL* = nucleus centralis lateralis; *CN* = caudate nucleus; *CP* = cerebral peduncle; *F* = fornix; *GP* = globus pallidus; *IC* = internal capsule; *L* = lateral nucleus of thalamus; *LST* = lateral spinothalamic tract; *M* = medial nucleus of thalamus; *MB* = mammillary body; *MI* = massa intermedia; *ML* = medial lemniscus; *MLF* = medial longitudinal fasciculus; *MT* = mammillothalamic tract; *O* = inferior olive; *OC* = optic chiasma; *OT* = optic tract; *P* = pons; *PC* = posterior commissure; *PU* = putamen; *PY* = pyramid; *RN* = red nucleus; *SC* = superior colliculus; *SN* = subthalamic nucleus; *VL* = nucleus ventralis lateralis; *VPL* = nucleus ventralis posterolateralis; *VPM* = nucleus ventralis posteromedialis; *III* or *3V* = third ventricle; *IV* = fourth ventricle. (From R. Rhines and H. W. Magoun. Brainstem facilitation of cortical motor response. *J. Neurophysiol.* 9:220, 222, 1946.)

tion of the central end of cut dorsal roots, the gamma moto-neurons are excluded from the response. Thus, there is some control over the motoneurons that does not primarily involve the gamma system. The proportion of such motoneuron control varies in different species. In the cat and dog, most of the reticulospinal influence on the cord is through interneurons that eventually terminate on the motoneurons. In primates and humans there is a greater synaptic influence from the brain that acts directly on the motoneurons. The synaptic terminations are found on the soma and on the dendrites of the motoneurons.

The earlier view that the functional control exerted from the reticular formation is "global" in its effect on lower motor centers had to be modified as a result of further study. Mixed facilitatory and inhibitory effects have been found. Under certain experimental conditions—for example, when anesthesia is sufficiently deepened—a simplification of the complex control of reflex responses from the reticular formation takes place. This was shown by stimulation via electrodes chronically implanted in the brainstem reticular formation. By this means, anesthesia could be lessened or eliminated, and a coordinated movement of the limbs occurred rather than generalized facilitation from the "facilitatory" areas or inhibition from the "inhibitory" areas. Part of the descending motor control exerted by the reticular formation region appears to derive from the *monoamine neurons* discussed in Chapter 6.

CEREBELLUM

The cerebellum, situated behind the cerebrum, is involved in those mechanisms relating to motor coordination and body equilibrium. An animal with its cerebellum ablated shows an inability to adjust the position of its limbs with respect to a goal. It overshoots or undershoots its goal when, for example, reaching for food: a condition called *dysmetria*. Movements, instead of occurring smoothly, are disjointed or jerky, showing *asynergia*. A cerebellar tremor is also seen during a voluntary movement. The tremor is coarse and becomes greater as the goal is being approached—an *intention* tremor. Decerebellated animals appear weaker (*asthenic*); the weakness of the muscle is described as a diminution of tone, or *hypotonia*.

Removal of different parts of the cerebellum has different effects on motor behavior. The phylogenetically newer posterior and lateral portions (the *neocerebellum*) are joined with the cortex by extensive afferent and efferent fiber connections. This part of the cerebellum is associated with cortically controlled movements. The phylogenetically older portion of the cerebellum (the *paleocerebellum*) is located anteriorly and medially in the cerebellum and is concerned with tonic postural mechanisms as well as with locomotion, i.e., with processes that are more automatic in nature. The flocculonodular portions of the cerebellum (part of the *archicerebellum*) are found laterally and are concerned with body equilibrium. Nerve impulses from the vestibular sensory receptor organs (semicircular canals) give information as to the position of the head with respect to spatial orientation (see Chap. 4). The afferent fibers of these receptors terminate in the flocculonodular portion of the cerebellum. Ablation of either the vestibular organs or the flocculonodular lobes causes impairment of orientation and loss of equilibrium. A human so damaged falls if asked to close the eyes, because he or she thereby loses the remaining visual cues for equilibrium. Sense modalities such as vision and sound are represented in topographical fashion on the cerebellum, as are sensory inputs from the skin and from muscle receptors (Fig. 8-7).

Analysis of the function of the cerebellum shows that it acts as a feedback control mechanism that smoothes and integrates motor activity initiated in the cerebrum. How the integration of various sensory inputs within the cerebellum comes about is suggested by the studies of Eccles (1974), Ito (1978), Llinás (1975), and their colleagues. The cells of the cerebellar cortex are arranged in three layers. In the outermost (molecular) layer are the large cell bodies of the *Purkinje cells*. These have large ramified dendrites that spread fanwise at right angles to the folia of the cerebellum. Between the dendritic branches are the parallel fibers, which course through the dendrites in the direction of the folia and make synaptic contact with them (Fig. 8-8). The parallel fibers originate from granular cells found in the middle layer of the cerebellum. The deepest layer is the white matter, representing fibers entering and leaving the cerebellar cortex (Fig. 8-9).

There are two sensory inputs to the cerebellar cortex. One, originating from the inferior olivary nucleus of the brainstem, ascends into the molecular layer to intertwine with and synapse on the dendrites of the Purkinje cells as *climbing fiber* afferents. The other afferent input is made up of the *mossy fibers,* representing sensory inputs from the spinal cord, brainstem, and upper motor areas of the brain. These end in bulbous *glomeruli*—synapses on the granule cells, which are excitatory, while an inhibitory control is exerted on them by the Golgi cells (Fig. 8-9). The axons of the granule cells ascend to the molecular layer, where their fibers divide to become the parallel fibers, synapsing on the apical dendrites of the Purkinje cells through which they pass. In addition to this excitatory input to the Purkinje cells, there are inhibitory interneurons, the *stellate* and *bas-*

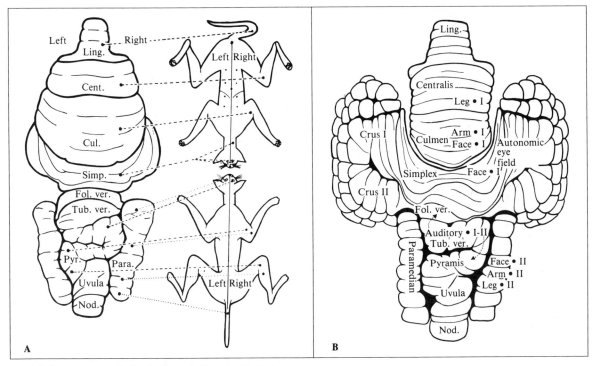

Fig. 8-7. Somatotopic localization in the cerebellum. In (A), the localizations of motor functions found by stimulating areas over the cerebellar surface are shown. In (B), the localization within the cerebellum of various sensory inputs using evoked-response techniques is shown. (From J. L. Hampson, C. R. Harrison, and C. N. Woolsey. *Res. Publ. Assoc. Nerv. Ment. Dis.* 30:304, 1952.)

ket cells, synapsing on the Purkinje cell dendrites and cell body respectively to contribute an inhibitory influence on its outflow (Fig. 8-9). The Purkinje cell represents the "final common path" from the cerebellum, its activity being a resultant of the excitatory and inhibitory influences acting on it. The Purkinje axons are inhibitory to the deeper relay nuclei of the cerebellum (fastigial, interpositus, and dentate). The neurons of these nuclei in turn synapse on other brainstem groups (vestibular, reticular formation, and the red nucleus), which form tracts descending to the lower motor spinal reflex centers (Fig. 8-10). Another major outflow from the dentate nucleus of the cerebellum goes to the ventrolateral nucleus of the thalamus and, after a relay there, to the sensorimotor cortex of the cerebrum to regulate its motor control functions and the pattern of outflow of PT neurons.

The interconnections of the neurons of the cerebellum with their fiber outputs passing to the cortex and spinal cord through the deep cerebellar nuclei are part of an interconnected loop of neurons (Fig. 8-11). Exactly how this network operates as a whole to smooth ongoing movement is not yet understood, but the function of such loop connectivities is likely to yield to a systems approach.

BASAL GANGLIA

An important but little understood group of nuclei of the brainstem are the phylogenetically older motor centers, the *basal ganglia*. These include the caudate nucleus, putamen, globus pallidus (the striate body), subthalamic nucleus of Luys, substantia nigra, and red nucleus. Their stimulation may produce a series of complex stereotyped movements—for example, a turning of the body or a sudden arrest of ongoing movement. Inhibitory effects from the basal ganglia on lower reflexes have been obtained resembling the effect seen on stimulating the inhibitory regions of the reticular formation. *Parkinson's disease* appears to be a disorder of basal ganglia function. It is characterized by an alternating activation of extensor and flexor muscle groups seen as a characteristic tremor. Destruction of the efferent outflow of the globus pallidus by electrocoagulation or by focused high-frequency sound waves has in some cases resulted in

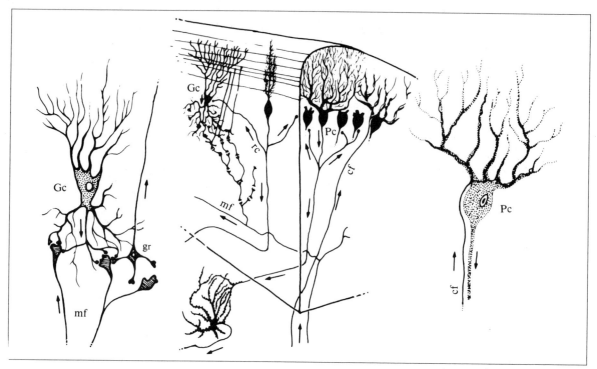

Fig. 8-8. Details of cerebellar neurons. (*Left*) Afferent mossy fibers (*mf*) terminate on granule cells (*gr*) of the granular layer and on Golgi cells (*Gc*). (*Right*) Afferent climbing fibers (*cf*) terminate on dendrites of Purkinje cells (*Pc*). (*Middle*) Purkinje cells have an extensive dendritic arborization across the folium in the molecular layer, where parallel fibers thread through and synapse on them. Efferent outflow is from the Purkinje axon. (After C. A. Fox. In E. C. Crosby, T. H. Humphrey, and E. W. Lauer [eds.]. *Correlative Anatomy of the Nervous System.* New York: Macmillan, 1962; and from J. C. Eccles, Functional meaning of the patterns of synaptic connections in the cerebellum. *Perspect. Biol. Med.* 8:289, 1965. Copyright © 1965, The University of Chicago Press.)

dramatic relief. Dopaminergic fibers from the globus pallidus terminating in the striatum are also important. A deficiency of dopamine was found in the striatum of patients with parkinsonism. L-Dopa is synthesized into the transmitter dopamine in the dopamine neurons (see Chap. 6), and on that basis L-dopa was given in large amounts to such patients and benefited a substantial number.

HIGHER CENTERS OF VISCERAL CONTROL

In later chapters the physiology of various visceral systems—cardiovascular, respiratory, genitourinary, digestive, and others—is thoroughly dealt with. Here, we discuss only briefly the mechanisms of upper visceral control located in the central nervous system and their relation to those visceral systems and endocrine systems of general homeostatic regulation.

The peripheral part of the autonomic nervous system, with its cell bodies located in the intermediary portions of the gray matter of the spinal cord, is discussed in Chapter 7. This system, like the somatic nervous system, has afferent fibers entering the spinal cord via the dorsal roots that synapse eventually on cell bodies within the intermediary

parts of the gray matter of the spinal cord. These neurons give rise to efferent fibers emerging from the ventral roots. The autonomic motor nerves are distributed in a more or less diffuse fashion to the heart, various smooth muscles, and glandular structures throughout the body. As in the somatic nervous system, the local cord reflex centers controlling the autonomic visceral functions are in turn controlled by higher brain centers. In spinal shock, transection of the spinal cord gives rise not only to the profound somatic changes characteristic of this state, but also to vascular, thermoregulatory, and urinary bladder and other urogenital changes that parallel the reflex changes described for the somatic nervous system.

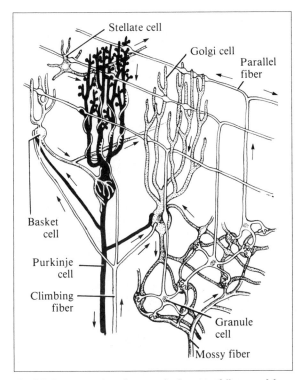

Stellate cell

Golgi cell

Parallel fiber

Basket cell

Purkinje cell

Climbing fiber

Granule cell

Mossy fiber

Fig. 8-9. Interconnection of neurons in the cortex follows an elaborate but stereotyped pattern. Each Purkinje cell is associated with a single climbing fiber and forms many synaptic junctions with it. The climbing fiber also branches to the basket cells and Golgi cells. Mossy fibers come in contact with the terminal "claws" of granule cell dendrites in a structure called a cerebellar glomerulus. The axons of the granule cells ascend to the molecular layer, where they bifurcate to form parallel fibers. Each parallel fiber comes in contact with many Purkinje cells, but usually it forms only one synapse with each cell. The stellate cells connect the parallel fibers with the dendrites of the Purkinje cell, the basket cells mainly with the Purkinje cell soma. Most Golgi cell dendrites form junctions with the parallel fibers, but some join the mossy fibers; Golgi cell axons terminate at the cerebellar glomeruli. Arrows indicate direction of nerve conduction. (From R. R. Llinás. The cortex of the cerebellum. *Sci. Am.* 232:58, 1975. Copyright © 1975 by Scientific American, Inc. All rights reserved.)

Another relationship of higher central nervous system structure to visceral functions is found in the brain regions connected to and regulating the pituitary gland (hypophysis). The relation of the hypothalamus to the hypophysis was suggested by Frölich's study of a syndrome in adolescent boys consisting of obesity, dwarfing, and sexual infantilism (see Chap. 30). The syndrome was finally traced to certain portions of the hypothalamus. It is of interest here to describe the control this area of the brain has on visceral functions connected with food and water intake, body fluid and ion regulation, cardiovascular and respiratory control, gastrointestinal regulation, genitourinary function, and sexual and maternal behavior. These visceral functions all relate to the internal economy of the organism or to broader events in the organism's life, such as sex and the struggle for food, and they generally have an attendant emotional involvement. Emotion is a predominant factor in human behavior and, by inference, in the behavior of animals as well. Emotional reactions will be discussed elsewhere (see Chap. 9).

The hypothalamus may, for present purposes, be considered a controlling station for most of the visceral functions that have been mentioned, serving as a "high-level" reflex control system. As will be noted in the next chapter, other regions of the brain have important interconnections with the hypothalamus. The limbic system and the neocortex, among other regions, exert, at a higher level, a regulation of visceral functions by their effects on the hypothalamus. The relation of this region to behavioral and emotional mechanisms has in recent years been brought to attention by the discovery of endogenous opioid neurons in the hypothalamus and elsewhere in the brain and their relation to mechanisms controlling emotion (see Chap. 9).

ROLE OF THE HYPOTHALAMUS IN VISCERAL CONTROL

The importance of the hypothalamus as a higher regulating center for the autonomic nervous system lies in its influence on cardiovascular control, respiration, temperature regulation, water and electrolyte balance, control of food intake, and sexual and maternal function. We will briefly take note of these functions in a general way.

CARDIOVASCULAR CONTROL. An animal in a state of spinal shock shows various degrees of loss of vasomotor activity in the parts of the nervous system below the transected level. The peripheral vessels are dilated, blood pressure is reduced, and response to reflex vasomotor activity is lessened. Vasodilation follows from the loss of normal reflex vasomotor tone acting on the vessels. If the transection of the brain is made anterior to the hypothalamus, control of the nervous system is relatively normal insofar as vasomotor activity is concerned, and blood pressure is not affected.

Anesthesia, if very deep, causes a similar state because of its central action with a loss of reflex adjustments. The anesthetic agent chloroform sensitizes the heart to adrenergic action, leading to irregularities of rhythm, extrasystoles,

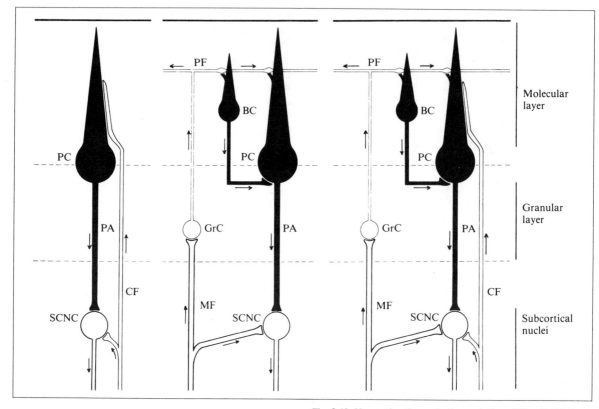

Fig. 8-10. Neuronal pathways in the cerebellum. *PC* = Purkinje cell; *BC* = basket cell; *PA* = Purkinje cell axon; *GrC* = granule cell; *MF* = mossy fiber; CF = climbing fiber; *SCNC* = subcortical nuclear cell. The directions of impulse flow are shown by arrows. (From J. C. Eccles. The Dynamic Loop Hypothesis of Movement Control. In K. N. Leibovic [ed.], *Information Processing in the Nervous System.* New York: Springer-Verlag, 1969. P. 248.)

and fibrillation. Cardiovascular control originates from the hypothalamus with a pathway traced from the hypothalamus and descending in the brainstem via sympathetic nerves that exit from the spinal cord at the thoracic levels to innervate the heart. The release of adrenergic transmitter may lead to fibrillation with chloroform present.

Parts of the brain higher than the hypothalamus also have been shown to have vascular effects. Stimulation of the cingulate cortex and the neocortex can give rise to localized vasodilation and constrictions of skin, muscle, and body organs.

RESPIRATION. As indicated in Chapter 21, there is much evidence that a fundamental pacemaker activity underlies the rhythm of respiration. This rhythm is also controlled by higher nervous centers, so that a voluntary control of respiration can occur. Higher centers in the brain, afferent discharges from the lungs and body, and the oxygen, carbon dioxide content, and pH of the blood all affect the rhythm of respiration. But the nature of the rhythmic changes in cells that lead to periodic discharge is at present little understood. One would like to know what series of molecular events causes the rhythmic changes in those cells. A similar problem relates to the pacemaker cells of the heart (see Chap. 14) and electroencephalographic activity. The generator potential of sensory nerve endings related to rhythmic discharge (see Chap. 3) is an analogous system whose study may eventually lead to a better understanding of pacemaker activity.

Evidence exists (see Chap. 21) that the hypothalamus has an excitatory effect on the inspiratory neurons in the medulla. Higher control by the cerebral cortex is indicated

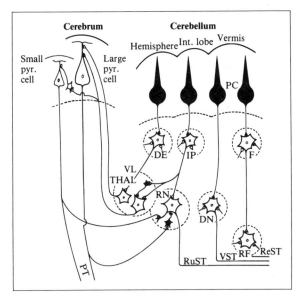

Fig. 8-11. Various efferent pathways of Purkinje cells (*PC*) in the vermis, intermediate lobe, and hemisphere of the cerebellum. All inhibitory cells are shown in black. *F* = fastigial nucleus; *IP* = interpositus nucleus; *DE* = dentate nucleus; *RF* = reticular formation; *DN* = Deiters' nucleus; *RN* = red nucleus; *VL THAL* = ventrolateral nucleus of thalamus; *PYR* = pyramidal cell; *ReST* = reticulospinal tract; *VST* = vestibulospinal tract; *RuST* = rubrospinal tract. The pathways are shown from small and large pyramidal cells to the VL thalamus and red nucleus. (From J. C. Eccles. The Dynamic Loop Hypothesis of Movement Control. In K. N. Leibovic [ed.], *Information Processing in the Nervous System.* New York: Springer-Verlag, 1969. P. 255.)

by the modifications of respiration that occur with speech, chewing, swallowing, and smelling. Stimulation of the cingulate cortex and other parts of the limbic system has also been found to affect respiration.

TEMPERATURE REGULATION. The relation of the hypothalamus to temperature regulation is shown when, on local heating of small regions within the hypothalamus, peripheral changes are found in the body acting in the direction of removal or loss of body heat from the animal (see Chap. 28). Conversely, cooling of the blood passing to the hypothalamus will cause changes in the direction of conserving body heat or increasing heat production by shivering. Shivering results in an increase in body warmth, and electrical stimulation of the posterior hypothalamus is effective in producing shivering. Electrodes placed in the hypothalamus

show potential changes indicative of neural activity during heating or cooling of the blood or of the cells in this region. If the outflow from the hypothalamus is cut or the hypothalamus destroyed, heat-regulatory mechanisms are interfered with. In such cases an animal may be changed from its *homeothermic* to a *poikilothermic* state. In a cold environment the poikilothermic animal will show a reduction in body temperature, with a fall toward that of the external temperature. Conversely, if the animal is placed in a warmer-than-usual environment, body temperature will rise. In other words, the homeostatic reflex adjustments mediated by the hypothalamus in the homeotherm are drastically interfered with or lost.

Sweating is part of the hypothalamic regulation of body temperature in some species, including humans, and operates in conjunction with the aforementioned control of the peripheral vascular bed. Temperature control mechanisms provide appropriate vasoconstriction to aid the body's heat retention, or vasodilation to assist its heat loss.

WATER AND ELECTROLYTE BALANCE. Receptors sensitive to changes in osmotic pressure of the blood, called *osmoreceptors*, have been localized within the anterobasal part of the hypothalamus. Injection of hypertonic fluid into the arterial supply of the hypothalamus causes the release of antidiuretic hormone (ADH) from the posterior pituitary (neurohypophysis) by the reflex excitation of the osmoreceptors of the hypothalamus (see Chap. 30). An increased discharge of neural units within the supraoptic nuclei of the hypothalamus is also found following injection of small amounts of hypertonic fluids into the carotid arteries. Electrical stimulation within the hypothalamus of goats has been shown to lead to excessive thirst and excessive intake of water to the point where death can ensue. Axons from cells in the anterobasal region of the hypothalamus (supraoptic and paraventricular nuclei) pass down into the neurohypophysis. The ADH formed in the cell bodies moves down inside the fibers by axoplasmic transport (see Chap. 2) to be stored and later released as required from the pituicytes of the neurohypophysis. This hormone is required for the normal return of water from the kidney tubules to the blood (see Chap. 22). In the absence of ADH, polyuria results.

CONTROL OF FOOD INTAKE. Within the hypothalamus there are areas that on stimulation cause an animal to seek food and other regions that stop the satiated animal from continuing to eat. Lesions made in the medial part of the hypothalamus will make an animal eat excessively; i.e., it shows *hyperphagia*. Such an animal may more than double its normal weight. On the other hand, lesions in the lateral part of the hypothalamus will cause a reduction of food intake, or *hypophagia*. If the lesions are complete, *aphagia* may result. Electrical stimulation of the medial nuclei makes an animal eat less, and stimulation of the lateral hypothalamic nuclei causes increased or continued eating. The medial nuclei are therefore regarded as a *satiety center,* the lateral nuclei as a *hunger center*. These two centers within the hypothalamus receive appropriate sensory inputs when hunger is present or after sufficient food is taken, and together they regulate the intake of food with relation to body needs. Receptors responding to glucose levels seem to be one important factor in stopping eating by exciting the satiety center. An increase in the level of food intake over a long enough time causes *obesity*. The deposition of fat in excess amounts in obesity is related to later degenerative changes in the cardiovascular and other systems and is associated with a shorter life span. Obesity may also be brought about through influences from the higher centers (limbic, brain, and cortex) acting on the food intake mechanisms of the hypothalamus. Some pharmacological agents, such as Dexedrine, appear to act on the hypothalamus. The adrenergic drug apparently depresses the sensitivity of the hunger center. Higher centers may also be involved. In the cephalic phase of digestion (see Chap. 26), the sight and smell of food stimulate the intake of food, and conditioned reflexes associated with food and eating are of great importance in the level of food intake or the desire to eat.

REGULATION OF SEXUAL AND MATERNAL BEHAVIOR. The implantation of minute amounts of estrogen within the hypothalamus causes changed behavior patterns concerned with mating. Other limbic areas controlling overt behavior of this function will be discussed in Chapter 9.

The secretion of gonadotropic hormones by the adenohypophysis in both males and females is under the control of the hypothalamus. Lesions placed in the anterior hypothalamus prevent the release of luteinizing hormone, and a condition of constant estrus results. Lesions placed in the posterior tuberal area cause gonadal atrophy. When the adenohypophysis is removed from its normal site in the sella turcica and transplanted to the kidney capsule or anterior chamber of the eye, it is unable to secrete gonadotropins in amounts adequate to maintain gonadal activity. Hormonal release from the adenohypophysis requires *releasing factors* that pass from the hypothalamus to the anterior pituitary via the pituitary portal circulation. These mechanisms of control will be discussed in Chapter 30.

BIBLIOGRAPHY

Anand, B. K. Nervous regulation of food intake. *Physiol. Rev.* 41:677–708, 1961.

Asanuma, H. Recent developments in the study of the columnar arrangement of neurons within the motor cortex. *Physiol. Rev.* 55:143–156, 1975.

Asanuma, H., and Rosén, I. Topographical organization of cortical efferent zones projecting to distal forelimb muscles in the monkey. *Exp. Brain Res.* 14:243–256, 1972.

Brooks, C. M., Gilbert, J. L., Levey, H. A., and Curtis, D. R. *Humors, Hormones, and Neurosecretion.* New York: State University of New York, 1962.

Brooks, V. B. Information Processing in the Motorsensory Cortex. In K. N. Leibovic (ed.), *Information Processing in the Central Nervous System.* New York: Springer-Verlag, 1969.

Chan-Palay, V. *Cerebellar Dentate Nucleus. Organization, Cytology and Transmitters.* Heidelberg: Springer-Verlag, 1977.

Denny-Brown, D. *The Cerebral Control of Movement.* Springfield, Ill.: Thomas, 1966.

Desmedt, J. E. Patterns of motor commands during various types of voluntary movement in man. *Trends in Neurosci.* 3(11):265–268.

Eccles, J. C. The Dynamic Loop Hypothesis of Movement Control. In K. N. Leibovic (ed.), *Information Processing in the Nervous System.* New York: Springer-Verlag, 1969.

Eccles, J. C. The cerebellum as a computer: Patterns in space and time. *J. Physiol.* (Lond.) 228:1–32, 1973.

Eccles, J. C. *The Understanding of the Brain* (2nd ed.). New York: McGraw-Hill, 1977.

Eccles, J. C., Ito, M., and Szentagothai, J. *The Cerebellum as a Neuronal Machine.* Berlin: Springer-Verlag, 1967.

Evarts, E. V. Brain mechanisms of movement. *Sci. Am.* 241:146–156, 1979.

Granit, R. *The Basis of Motor Control.* New York: Academic, 1970.

Ito, M. Recent Advances in Cerebellar Physiology and Pathology. In R. A. P. Kark, R. N. Rosenberg, and L. J. Schut (eds.), *Advances in Neurology.* New York: Raven, 1978. Vol. 21, pp. 59–84.

Lassek, A. M. *The Pyramidal Tract.* Springfield, Ill.: Thomas, 1954.

Liddell, E. G. T., and Phillips, E. G. Overlapping areas in the motor cortex of the baboon. *J. Physiol.* (Lond.) 122:392–399, 1951.

Llinás, R. The Cerebellar Cortex. In D. B. Tower (ed.), *The Nervous System.* New York: Raven, 1975. Vol. 1, pp. 235–244.

Llinás, R. R. The cortex of the cerebellum. *Sci. Am.* 232:58, 1975.

Llinás, R. R. (ed.). *Neurobiology of Cerebellar Evolution and Development.* Chicago: American Medical Association, 1969.

McGeer, P. L., Eccles, J. C., and McGeer, E. G. *Molecular Neurobiology of the Mammalian Brain.* New York: Plenum, 1978.

Marks, J. *The Treatment of Parkinsonism with L-DOPA.* New York: American Elsevier, 1974.

Penfield, W. *The Excitable Cortex in Conscious Man.* Springfield, Ill.: Thomas, 1958.

Porter, R. Functions of the mammalian cerebral cortex in movement. *Prog. Neurobiol.* (Pt. 1) 1:1–51, 1973.

Rosén, J., and Asanuma, H. Peripheral afferent inputs to the forelimb area of the monkey motor cortex: Input-output relations. *Exp. Brain Res*. 14:257–273, 1972.

Shepherd, G. M. *The Synaptic Organization of the Brain*. New York: Oxford University Press, 1979.

Szentagothai, J. Local Neuron Circuits of the Neocortex. In F. O. Schmitt and F. G. Worden (eds.), *The Neurosciences. Fourth Study Program*. Cambridge: MIT Press, 1979. Pp. 399–415.

Wiesendanger, M. The pyramidal tract. *Ergeb. Physiol*. 61:72–136, 1969.

Woolsey, C. N., Settlage, P. G., Meyer, D. R., Sencer, W., Hamuy, T. P., and Travis, A. M. Patterns of localization in precentral and "supplementary" motor areas and their relation to the concept of a premotor area. *Association for Research in Nervous and Mental Disease, Proceedings* (1950) 30:238–264, 1952.

Sidney Ochs

Higher nervous functions involve those complex behaviors resulting from interactions with the environment whereby an animal learns to adapt and to attain its goals. The field of study of the higher nervous functions is a difficult one. Inferences from our own subjective experience conditions the objective observations made in the laboratory. At issue is the view taken of the mind and its relations to the nervous system. The subject will be divided into *emotional* and *intellectual* behavior, with full recognition of the difficulty of a sharp separation between these realms of discourse.

EMOTION AND THE LIMBIC SYSTEM

Emotion has two aspects: internal awareness and external display. The release of an emotional display appears at times to be almost reflexlike. For example, we may have a feeling of sorrow or depression, but a stronger sorrow may lead to uncontrollable weeping and sobbing. We may be amused or give way to sudden, uncontrolled laughter. We may feel pleasure or enter a state of ecstasy. Displeasure

and anger in the extreme can turn to rage. These human experiences have some parallels in experimental investigation of the emotions in animals. We ascribe emotional states to those displays in the animal corresponding to the behavior seen in humans, usually the sudden reflexlike and explosive changes.

The similarity of brain structures in the human and the animal suggests generalizations for the localization of higher nervous function in brain regions with similar form. For the most part the hypothalamus and the *limbic system* are involved. The limbic system is a group of nuclear regions with interconnections found around the brainstem on its medial and ventral aspect. Although the hypothalamus is sometimes included in the limbic system because of important connections made by the fornix carrying impulses from the limbic structures to the hypothalamus, it will be treated separately.

An extreme state of angry behavior known as *sham rage* can be produced in animals by amputation of the brain anterior to the hypothalamus. Such an animal shows periodic displays of rage, with claws extended, jaws open, and muscles tense. On provocation, or even gentle restraint, the animal growls, thrashes about, lunges, and snaps. The rage display is not directed specifically to the source of provocation because the higher centers necessary to direct such an attack are lost. Sympathetic discharge from the hypothalamus is shown by the increased heart rate and blood pressure, pupillary dilation, hair bristling, and salivation (see Chap. 7). An augmented sympathetic discharge is not seen if the hypothalamus has been destroyed. If, in the monkey, a cut is made in the base of the brain just anterior to the hypothalamus, a rage state is produced, but in this case the animal can focus its attacks. The extreme ferocity of this preparation is noteworthy. Some components of a rage reaction can be elicited from still lower brainstem structures. A chronic decerebrate preparation shows some aspects of ragelike behavior referred to as *pseudoaffective*. However, while some of the components of rage display are evoked from the lower brain structures, the more organized response requires the hypothalamus to be present.

Localized electrical stimulation of the hypothalamus by means of chronically implanted electrodes can clearly bring about a display of ragelike behavior, along with signs of sympathetic discharge. Some experiments indicate that sham rage elicited by such electrical stimulation of the hypothalamus does not have a subjective component; an animal may purr and even respond to petting between displays and during the stimulation. Stimulation directly in the hypothalamus and conditioning, however (see the following section), would suggest a subjective aspect to the electrically elicited rage. While all of the display may not be conditioned, animals can be conditioned to avoid hypothalamus stimulation by techniques to be described in a later section. The hypothalamus appears to act as a control center for the outward expression of rage responses without necessarily being the place where emotions as such are experienced.

Papez (1937) related limbic brain structures (Fig. 9-1) to emotions by theorizing that afferent activity passes from the thalamus to the hypothalamus, then via mamillary bodies to the anterior thalamic nucleus, and from there to the cingulate cortex. The cingulate cortex was supposed to be the site of the subjective appreciation of the emotional states, not only of rage but also of sexual and feeding behaviors.

The ablation studies of temporal lobes of monkey brain by Klüver and Bucy further directed attention to the relation of limbic brain structures to behavior. In their work, a number of limbic structures were removed, including the amygdaloid nuclei. Monkeys with both temporal lobes ablated showed bizarre behavior patterns. They lost their nor-mal fear of objects such as snakes and other similar genetically determined "fear" objects. They seemed to "explore" their environment orally, putting edible and inedible objects alike into their mouths, no matter how abnormal or repulsive these objects might be with respect to the usual behavior for the animal. They would not necessarily swallow them, spitting out inedible objects and eating food. They appeared to show lack of recent memory by returning soon after to an object previously spat out and again putting it into their mouths. The picking up of any and all objects for oral exploration might suggest a lack of visual discrimination. However, animals with ablated temporal lobes can learn to discriminate visually between symbols (e.g., a square versus a circle) when presented simultaneously in a standard visual discrimination experiment. Apparently, the character of the objects is not recognized, or the animal cannot inhibit oral exploration of them. Animals with lesions in the amygdaloid nuclei of the temporal lobes also showed a steady gain in weight. Their level of food intake seemed to have been raised by the lesion, or they become less active and therefore accumulated fat.

Temporal lobe lesions give rise to peculiar sexual behavior. The animals take other species or even inanimate objects as sex partners and in general show an increase in sexual activity. Other limbic structures are related to sexual activity. The retrosplenial cortex in the rat is homologous to the cingulate cortex, a limbic brain structure. If it is destroyed, mother rats do not retrieve their pups from danger. Female rats so damaged show an abnormally increased responsiveness to the male. In the male rat, electrical stimulation of cingulate cortical regions can give rise to erection.

The two constellations of behavior (feeding and sex) have been referred to by MacLean (1955) as "preservation of the self and preservation of the species." Obviously, these two biological goals are required for carrying on animal life. Aggressive or placid emotive states connected with hunger and satiation are of biological significance with regard to the organism's drive toward or inattention to those goals.

It has been discovered that electrical stimulation of certain regions in the brain may produce a "pleasurable" experience in the animal or be *positively rewarding* for the animal. Electrodes were chronically implanted in several subcortical structures of the rat: the hypothalamus and the septum, hippocampus, and other limbic brain regions. Connections from the electrodes were made to an electrical stimulating apparatus that could be controlled by a switch placed inside the animal's box. The animal, when pressing the switch, caused a brief electrical stimulation to be delivered to its own brain. Once the animal learned this, it would return to the switch and administer *self-stimulation* by re-

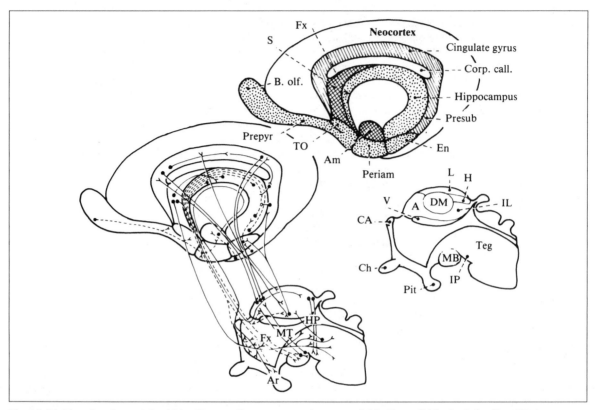

Fig. 9-1. Limbic system shown at the right, with upper diagram representing the medial aspect of the brain, lower diagram representing the medial aspect of the brainstem. The corresponding pair of diagrams at the left show their regional interconnections. Three types of region are distinguished: (1) the *paleocortex,* represented by stippling, including as its chief structure the hippocampus and the olfactory structures, olfactory bulb (*B. olf.*) and prepyriformis (*Prepyr*); (2) the *juxtallocortex,* shown by slanted lines, which includes the cingulate cortex and presubiculum (*Presub*); and (3) the subcortical structures considered in relation to the limbic system (cross-hatched in the upper diagram), the septum (*S*) and *amygdaloid complex (Am)*. These regions are related to one another and to brainstem regions as shown at the left. Important efferent connections are made from the hippocampus to the mamillary bodies in the hypothalamus via the fornix (*Fx*) and from the mamillary bodies to the anterior thalamic nucleus via the mamillothalamic tract (*MT*). Connections from the septum to lower brainstem centers are carried by the medial forebrain bundle, which is not labeled. A = anterior nucleus of the thalamus; Am = amygdaloid complex; Ar = arcuate nucleus; *B. olf.* = olfactory bulb; CA = anterior commissure; Ch = optic chiasm; *Corp. call.* = corpus callosum; DM = dorsomedial nucleus of the thalamus; En = entorhinal area; Fx = fornix; H = habenular complex; HP = habenulointerpeduncular tract; IL = intralaminar thalamic nuclei; IP = interpeduncular nucleus; L = lateral thalamic nucleus; MB = mamillary bodies; MT = mamillothalamic tract; *Periam* = periamygdaloid-cortex; *Pit* = pituitary; *Prepyr* = prepyriform; *Presub* = presubiculum; S = septal region; *Teg* = mid-brain tegmentum; *TO* = olfactory tubercle; V = ventral nucleus of the thalamus. (From J. V. Brady, The Paleocortex and Behavioral Motivation. In H. F. Harlow and C. N. Woolsey (eds.), *Biological and Biochemical Bases of Behavior*. Madison: The University of Wisconsin Press. Plate 23. Copyright 1958 by the Regents of the University of Wisconsin.)

peatedly pressing the switch. Apparently, such self-administered stimulation is so *positively reinforcing* (pleasurable) that the animal may incessantly press the switch for hours on end without stopping. Having learned to self-stimulate, it will endure painful electrical shocks applied through a floor grid in order to get to the lever and administer self-stimulation. It will prefer self-stimulation to other biological goals such as food or mating. Such self-stimulation has been demonstrated in a wide variety of animal species.

Along with evidence of some special distribution within the limbic structures of positively reinforcing sites, other regions close to these same structures presumably give rise to painful or unpleasant sensations when electrically excited, as shown by the *aversive* responses of the animal. When the electrodes are inserted into such *negatively reinforcing* sites, an animal will stimulate itself once and not press the switch again. Furthermore, it will exert considerable effort to prevent excitation of those negatively reinforcing regions.

The positively reinforcing sites are related to the medial forebrain bundle found in the lateral hypothalamus. These fibers contain catecholamines, and it is likely that the effect of self-stimulation is on these fibers to release transmitters from their terminals. Catecholamines have been implicated in emotional excitatory states, possibly producing pleasurable experiences for the animal. Besides the catecholamines, norepinephrine and dopamine, other amines, acetylcholine, and serotonin have been implicated as possible neurotransmitters acting in systems related to mood alteration. In recent years it has been discovered that small peptides are present in the brain that have opiumlike properties. These *opioids,* known as *enkephalins* and *endorphins,* were recognized after receptors binding opium were isolated from brain, and in the course of such studies the presence of opioid substances was predicated. It was discovered that proopiomelanocortin (POMC), a 31,000-mw peptide, contains within its sequence ACTH and β-lipotropin next to one another in the chain. The opioid β-endorphin is derived from β-lipotropin, and β-endorphin in turn gives rise to metenkephalin. These opioids are present in the brain as well as in the pituitary, where the role of ACTH has long been appreciated. What is surprising is the occurrence of mood-altering substances in these sites.

The use of a radioimmune method for localization studies has shown enkephalin to be high in the globus pallidus and to be present in the caudate, hypothalamus, periaqueductal gray, amygdala, and spinal cord and low in the cortex and cerebellum. On the other hand, β-endorphin is high in the hypothalamus and low in the other regions. The evidence indicates that the two peptides represent separate systems, with enkephalin having the properties of a neurotransmitter or a neuromodulator. Both enkephalin and β-endorphin appear to have morphinelike properties, as judged from behavioral studies. They act as positive reinforcers of bar pressing and produce catalepsy, and animals become dependent to them following long-term administration. They also are analgesic agents and produce euphoric states in human subjects. Such evidence makes it likely that the opioid systems are responsible for the feelings of pleasure or satisfaction that accompany positive reinforcement.

LOCALIZATION OF LEARNING IN THE BRAIN

In the previous section the relation of parts of the brain to fundamental biological goals that are emotionally charged was emphasized. The alteration of innate behavior with respect to experiences in the environment (i.e., learning) will now be discussed. In their well-known studies, Pavlov and his colleagues investigated what is now known as *classical conditioning.* An animal is conditioned according to the Pavlovian technique by first receiving a sensory stimulus to which it will respond in an unlearned manner—the *unconditioned stimulus.* Food juices placed in the mouth of a hungry dog by means of an implanted cannula represent such a stimulus. This unconditioned food stimulus gives rise to an *unconditioned response,* in this case salivation. The measure of the response is the amount of saliva produced. If, at the time that the unconditioned food stimulus is being presented (or a short time beforehand) a bell is rung, the animal after a number of such pairings will salivate in response to the ringing of the bell alone.

The sound of the bell is termed the *conditioned stimulus,* and salivation in response to it is a *conditioned response.* According to Pavlov, some new neural connections are made in the cerebral cortex after conditioning has occurred, so that the conditioned stimulus can take the place of the unconditioned stimulus with which it has been paired. The basis of Pavlov's belief was a report of an inability to condition decorticate dogs. Subsequent investigations have shown that simple types of classical conditioning can occur in the decorticate animal, and this finding directed attention to subcortical regions that may be involved in learning. Stimulation of the centrencephalic system localized in the nonspecific thalamic nuclei causes a petit mal type of widespread convulsive discharge synchronized in both hemispheres with a loss of consciousness (see Chap. 6). On this basis the centrencephalic system was proposed as the ''highest'' level in Jackson's sense—namely, where the more complex higher functions are controlled or organized.

That some "higher" level of learning is accomplished in subcortical sites is suggested by the use of direct stimulation of subcortical structures as an unconditioned stimulus and peripheral stimulation or a sensory stimulus such as a sound as the conditioned stimulus to cause conditioning. A *deviational response* (a turning of the head) could be obtained from stimulation of limbic structures and conditioned by sound. From the septum, sexual responses, as the unconditioned responses, could be conditioned to a tone. A more stereotyped behavior such as the *display of rage* obtained from stimulating some subcortical structures could not be conditioned.

In contrast to the simple type of classical conditioning, a greater involvement of the cortex in learning is shown by use of *instrumental, or operant, conditioning*. In instrumental conditioning, an animal makes some response that at first may be accidental or part of its normal repertoire of activity. The experimenter arranges conditions so that when the desired response is made, the animal is immediately rewarded with food or some other reward. It soon learns to discriminate between rewarded responses and other kinds of behavior and will, if sufficiently motivated, consistently perform the responses that have been so *reinforced*. In Skinner's type of experimental arrangement, an animal is placed in a small box with a small bar projecting from one wall. In moving about inside the box, the animal will occasionally strike the bar. If, when a hungry rat strikes the bar, a food pellet is delivered into a food hopper within the box, the animal learns to strike the bar repetitively for food. If that reinforcement is discontinued, the animal gradually decreases its rate of responding until no more responses are made, a phenomenon known as *extinction*.

Various schedules of reinforcement have been described. An animal may be required to press the bar a fixed number of times to obtain reinforcement. Or it may learn that reinforcement will take place only when the bar is pressed after a fixed interval of time. These and other schedules are of interest insofar as they have been shown to be differentially affected by drugs. Cortical ablation or spreading depression, when present in both cortices, will prevent the learning or execution of an instrumental response.

While recent studies have shown the possibility of subcortical conditioning, one should not lose site of the importance of the cortex for higher functions. It is possible to stimulate the cortex directly for use as a conditioned stimulus in order to elicit a conditioned response. The cortex appears to be critically involved in more complex types of learning tasks. Attention has long been directed to the frontal lobes, an associational region thought to be connected with learning. Animals with frontal lobes ablated show a defect of *delayed response* performance. In a delayed response trial, an animal is allowed to see food bait placed under one of two specially marked covers. Then, after a delay, the animal is permitted to choose one or the other of the covers to obtain the bait. Successful performance is measured by the proper choice of the cover over the food bait. Delayed response behavior is greatly interfered with by frontal lobe lesions. Intense electrical stimulation or convulsive activity induced in the frontal cortex is also effective in interrupting delayed responding. Similar disruptive activity induced in other cortical areas is ineffective. Animals with the frontal lobes ablated do not show a loss of ability to discriminate one object from another if both objects are presented simultaneously. An interpretation of this result is that the frontal cortex is related to recent memory—or more likely that animals with ablated frontal lobes are more distractible when they must wait before choosing. They may not have a sufficiently prolonged interest in the test object during the time it is out of sight. Once learning has been achieved, disruption of the activity of the frontal lobes is ineffective in interrupting delayed responding. This finding demonstrates a fundamental difference in the mechanisms responsible for the *acquisition* of learned behaviors and for their *retention*. Retention (i.e., memory mechanisms) will be discussed in a later section.

ELECTROENCEPHALOGRAPHIC CORRELATES OF CONDITIONING

Changes in the electroencephalogram (EEG) have been correlated with a classical conditioning procedure using the alerting reaction (see Chap. 6). Shining a light in the eye gives rise to alerting, seen as a period of low-amplitude fast waves (Fig. 9-2B). To show EEG conditioning, a tone signal that does not by itself produce alerting is used. A tone that does not cause alerting is arranged for by *habituation* to the sound signal. The first presentation of a sound stimulus produces alerting; with repeated presentations the alerting response is diminished, and eventually the sound no longer produces alerting. If the sound is changed to another tone close in frequency to the first, it will produce alerting. In the example in Figure 9-2A, a sound stimulus to which the subject had been habituated no longer caused alerting. Alerting was caused in Figure 9-2B by presentation of a light signal. In Figure 9-2C, early in the course of conditioning, the light signal was preceded by a tone as a conditioning stimulus. After a number of such pairings, conditioning occurred in response to the sound stimulus, as shown in Figure 9-2D.

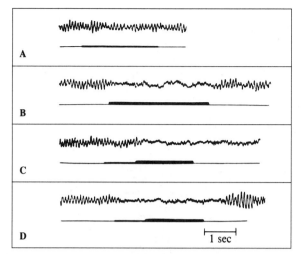

Fig. 9-2. Conditioning of the EEG. In (A), an EEG of a resting human is shown, with no alerting response to a tone signal (time of presentation of tone shown by thickening of the line underneath). In (B), alerting to a light stimulus is seen (light on at time shown by heavier line underneath). In (C), the tone signal precedes the light, and no conditioning has yet occurred. After a number of such pairings, conditioning is seen in (D) when alerting occurs at the onset of the *tone* signal. (From F. Morrell. Interseizure disturbances in focal epilepsy. *Neurology* 6:329, 1956.)

Notice that alerting took place in response to the conditioning signal that preceded the onset of the light stimulus.

An interesting type of EEG conditioning is shown by the use of a light flashed at the rate of 7.5 times per second as the unconditioned stimulus, the flashing light exciting a series of evoked waves in the cortex. A tone signal to which the animal was habituated was selected as the conditioned stimulus and was presented a short time before the light stimulus. After a number of such pairings the tone signal elicited conditioned responses at the same frequency as that of the light flashes used as the unconditioned stimulus. This conditioned response was seen to be widespread in a number of cortical regions. *Discrimination* was shown when, in a later stage of conditioning, the conditioned response became localized to the visual cortex (the unconditioned primary areas).

The limbic cortex—in particular, the hippocampus—has a connection with the learning processes. When the cortex shows the faster, smaller-amplitude waves of alerting, the hippocampus gives rise to large waves at the rate of five to seven per second. During conditioning, EEG changes registered from the hippocampus via chronically implanted electrodes show a characteristic shift to a lower frequency during learning. The electrical correlate of learning suggests that the hippocampus is involved in an early stage of learning.

MEMORY MECHANISMS

Intellectual ability is dependent on memories, to which the individual is continually adding. It must be admitted that little is known of the neuronal basis underlying memory, although some general statements may be made. Memories can exist for years, perhaps a lifetime, and thus would appear to depend on some permanent cellular changes. Memories persist after normal cerebral activity has been temporarily stopped, either by a degree of anesthetization sufficient to decrease all EEG activity of the cerebrum or by cooling to a low enough body temperature to cause a cessation of EEG activity. All this supports the conclusion that memory stores are due to some permanent structural change of the neurons rather than being the result of some pattern of neuronal activity.

Another general observation concerning memory is that recent memories are less stable than older ones. Clinical experience indicates that a blow on the head may produce a *retrograde amnesia,* i.e., a loss of memory extending back to include a period of several hours or more before the blow. The susceptibility of recent memories to interruption of neural activity was experimentally shown by application of electrical shock to the heads of rats at different intervals after a conditioning stimulus was presented. Animals given electroshock a short time after a learning session do not learn conditioned responses, even in numerous conditioning and shock sessions. If the interval between the conditioning experience and the administration of the electroshock is lengthened to an hour or more, shocked animals can be conditioned as well as unshocked animals. The destructive influence of electroshocks on recent memory shows a curve of decreasing effect: It is greatest when shocks are given just after the training session, and the effect diminishes in regular fashion as more time between training sessions and the shock is allowed. A similar destructive effect on conditioned learning was found by using a quickly induced short period of anesthesia or a rapid cooling of the brain after conditioning sessions.

Two types of memory are thus distinguishable: (1) neuronal activity responsible for recent memories and (2) molecular changes in neurons responsible for long-term memories.

In an attempt to find the place in the brain where long-term memories are stored, investigators have studied the effects of various cortical lesions on the retention of a learned response. Lashley (1950), from his work on the rat, concluded that memories, or *engrams,* could not be specifically located and that the amount of cortex damaged was related to the degree of memory loss. On the other hand, the importance of the temporal lobe with respect to storage and retrieval of memories in humans has been noted. Electrical stimulation of the temporal cortex of conscious patients under local anesthesia has been shown to give rise to vivid memories. Stimulation of cortical areas outside the temporal lobe did not elicit memories.

The memory responses obtained by electrical stimulation of the temporal region might be a misinterpretation or altered interpretation of present experience with respect to past experiences. Thus, a mother electrically stimulated in this region appeared to be seeing her child present in the operating room with all the accompanying sensations and sounds of an actual experience that had occurred previously. Similar hallucinatory effects may also occur in patients as a result of the excitation produced by a tumor in the temporal region, these giving rise to *auras,* i.e., sensory alterations or hallucinations preceding a generalized epileptic attack.

The fact that electrical stimulation of the temporal cortex can excite a past memory suggests that neuronal activity in the normal temporal lobe cortex is somehow connected with a memory. The memory does not exist in this part of the cortex per se, but it can trigger other regions, e.g., the centrencephalic region in the thalamus. A subcortical focus is indicated by the fact that, after an ablation of the temporal cortex, stimulation of the underlying fibers is also effective in eliciting memories. The temporal lobe cortex may have an interpretative role to play with respect to such deeper memory stores. Two commonly observed memory defects associated with pathological changes in the brain have been recognized: an inability to retrieve information (the loss of recent memory usually being most marked) and the loss of memory required to retain new information. Usually, if there is a recovery of memory, both types recover together, although there may be a greater loss of past memories than of those related to new acquisitions. Also, memories often recover in a spotty fashion, with "holes" remaining, which are more or less filled in with time.

The study of acquisition of memories and their utilization was further advanced by use of the *split-brain* preparation (Fig. 9-3). To produce a split-brain cat or monkey, the optic chiasm and the corpus callosum tract connecting the two

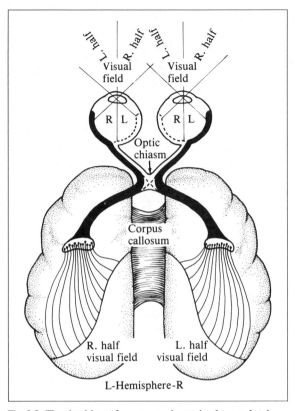

Fig. 9-3. The visual input from one eye is restricted to one hemisphere by cutting the optic chiasm in the midline. The corpus callosum is also sectioned to produce a split-brain animal. (From R. W. Sperry. Cerebral organization and behavior. *Science* 133:1750, 1961. Copyright © 1961 by the American Association for the Advancement of Science.)

hemispheres, as well as the crossing fibers of the anterior commissure, are cut in the anteroposterior plane. The visual information input from each eye can therefore pass only to the hemisphere of the corresponding side. With one eye covered and the other eye open, the animal is taught a visual discrimination in the hemisphere on the side on which the eye is open. After training is completed on that side, the eye of the opposite side is uncovered and the eye on the trained side covered. The animal is then unable to perform the visual discrimination to which it had previously been trained. The habit remains localized to the brain on the side on which the eye was open during training. The use of this technique makes it possible for two habits to be laid down, one on each side; even opposing habits may be trained into the two sides. On one side the animal may learn to respond to a square and not to a circle; on the other side, to a circle and not to a square. With both eyes open the animal responds positively to one or the other symbol. At times, one habit dominates, at times the other; the animal does not respond partially to the habits laid down on each side nor is there evidence of a conflict between the sides.

If the optic chiasm is cut in the midline as usual, with the corpus callosum remaining uncut, sensory information can be channeled to the other side at the cortical level. When, after an animal was trained with one eye open and the other covered, and then later the previously open eye was closed and the other opened, it easily recognized the visual stimulus and responded to it, providing that a callosal transfer had occurred.

By means of the phenomenon of spreading depression (see Chap. 6), one or both cerebral cortices may be functionally incapacitated for up to several hours. During such *temporary decortication* of both hemispheres, animals may learn only a simple type of classical conditioned response, while the acquisition of a complex response is prevented. With spreading depression present in one hemisphere, an animal will learn with the cortex not occupied by spreading depression. The localization of the learned response to the cortex not depressed during training sessions was shown when later the trained cortex was depressed and the animal then behaved as an untrained or naive animal. Such *lateralization* of the learned response to one hemisphere may remain for days, weeks, or months even though the interhemispheric connections in the corpus callosum might be expected to transfer the habit from the trained side to the other side. Thus, the transfer can occur at the time the animal receives *reinforcement*. The transfer of lateralized conditioned responses was used to great effectiveness by Bureš. Spreading depression was used to lateralize a conditioned response in one hemisphere of rats. Another conditioned response related to the first response was lateralized in the other cortex. After conditioning the animals could bring these two conditioned responses together to make a chained conditioned response.

Attempts have been made to show the sites where memories are stored by means of conditioned responses and placing lesions in various limbic and other subcortical structures. Lesions in the septum and in the pathways connecting the hypothalamus and hippocampus to other regions will produce defects of learning and loss of conditioned responses. A number of subcortical regions have been implicated, but further analysis is required to determine whether memory or motivation is involved in such lesions.

Much is still to be learned about memory. If all memories are present subcortically, does the cortex play a role in retrieving them? Is part of the memory laid down in one site and part in another, and, if so, how are they gathered together to give rise to a whole memory sequence? Finally, much evidence suggests a molecular basis for permanent memories. Changes in the synthesis of specific proteins may be responsible for long-term memories. But how is neuronal activity transformed into specific proteins? How do such specific proteins later effect a neural discharge of a given pattern? Too little is known as yet to give a satisfactory answer to these questions.

HIGHER FUNCTIONS IN THE HUMAN CEREBRUM

A further advance in the study of brain function was made possible through the split-brain studies carried out by Sperry and his associates on human patients who required surgery for various reasons, including severe epilepsy. With the corpus callosum cut and techniques of sensory stimulation arranged so that visual and tactile stimuli could be presented to one eye or hand, a clear difference in function of the two hemispheres was shown. With a word flashed briefly (less than 0.1 sec) in the left visual field, the normal crossed connections of the optic nerve give rise to an activation of the right visual cortex. In this cortex (right) the split-brain human cannot verbalize what he has seen, but through the control exerted by the right hemisphere over his left hand he can pick out from the group of objects hidden from his sight the object he is told to select (Fig. 9-4). In spite of his correct manual selection, the subject cannot describe what he is doing with his left hand, which is under the control of the right or *minor* hemisphere. The subject may even be able to write out the correct word with the left hand without being subjectively aware of the term. If, similarly, a word is flashed in his right visual field, the object can be named by the subject. This is the *major* hemisphere or the hemisphere by which verbal behavior is controlled (Fig. 9-5). The most interesting result found in split-brain patients is that the one hemisphere does not know what is going on in the other hemisphere. There is, however, some degree of subconscious awareness. Certain words or objects presented to the right hemisphere having embarrassing or emotional value may give rise to an emotional reaction, but without the subject's being able consciously to identify its source. Consciousness in the right hemisphere falls far short of that in the dominant (left) hemisphere. Studies of brain lesions have shown that the conscious awareness of speech is perceived in Wernicke's area in the left hemisphere. Le-

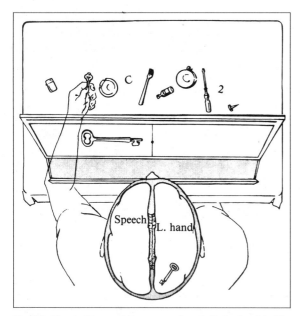

Fig. 9-4. Visuotactile associations are made only if visual and tactile stimuli both project to the same hemisphere. The subject here denies seeing anything but a flash of light on left. However, if the left hand is allowed to "make a guess," the correct object is selected or even an object "used with" the projected stimulus if the subject is so instructed. (R. W. Sperry, M. S. Gazzaniga, and J. E. Bogen. Interhemispheric Relationships. The Neocortical Commissures; Syndromes of Hemispheric Disconnection. In P. J. Jinken and G. W. Bruyn (eds.), *Handbook of Clinical Neurology*. Amsterdam: North-Holland, 1969. Vol. 1, p. 280.)

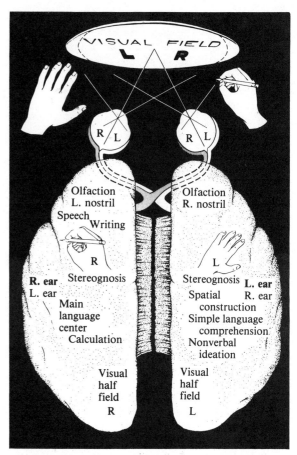

Fig. 9-5. Functional lateralization in left and right hemispheres shown after cutting corpus callosum. Language is located in the left hemisphere; spatial construction and nonverbal ideation are located mainly in the right hemisphere. (From R. W. Sperry. Perception in the absence of the neocortical commisure. *Res. Publ. Assoc. Res. Nerv. Ment. Dis.* 48:129, 1970.)

sions in the right or minor hemisphere cause defects related to a disturbance of body-sense awareness and spatial relationships in general (Fig. 9-5).

The studies of callosal section can be related to observations of patients who have lesions in portions of the brain that cause *agnosia* (failure to interpret or recognize objects or speech or parts of one's own body) or *apraxia* (failure to execute learned movements). As has been noted, Wernicke's area in the left hemisphere (Fig. 9-6) appears to be needed for the comprehension of spoken language, destruction of this area leading to language agnosia. On recognition of a spoken command, neural pathways are excited that lead to the left premotor area. From this region, control over the left motor area is possible and, through it, movement of the *right* hand, normally controlled by the left cortex. To move the *left* hand, excitation of nerve fibers pass-

ing via the corpus callosum to the right precentral motor cortex allows for a control of movement through this region (Fig. 9-7). If the corpus callosum is cut, a spoken command to move the left hand cannot be executed, and an *apraxia* is thereby revealed. This is termed the *disconnection syndrome*.

When a second language is learned at a later age (e.g., at 10 years, as compared with 5 years), damage to the language areas of the brain usually results in the loss of both the native and the second language. However, there are

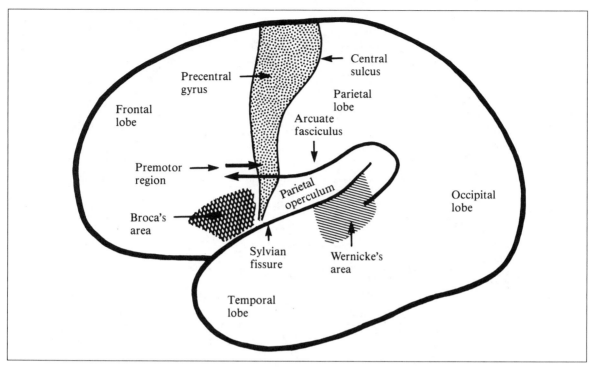

Fig. 9-6. Left hemisphere of human showing the pathway used in the performance of motor acts by the right limbs and the cranial muscles in response to a verbal command. The precentral gyrus (*stippled*) contains the classic motor cortex. At the lower end, near the Sylvian fissure, is the face motor area—the region controlling movements of the muscles of the face, jaw, tongue, palate, and vocal cords. The path from Wernicke's area, where an auditory command is received and interpreted, leads to the premotor region as shown by the curved line. A shorter line gives the path from the precentral gyrus to the pyramidal motor system descending to the brainstem and spinal cord for control of movement on the opposite (right) side. (From N. Geschwind. The apraxias: Neural mechanisms of disorders of learned movement. *Am. Sci.* 63:189, 1975. Reprinted by permission of *American Scientist*, Journal of Sigma Xi, The Scientific Research Society.)

cases of *aphasia*—a loss of speech ability—in which one language is retained better than another, suggesting a differential localization of at least some aspects of language acquisition or utilization. There is evidence that the right cortex participates in language with respect to synthesis, while the left hemisphere performs better in the process of analysis.

A number of voluntary control movements are bilaterally represented. These include use of the muscles of the lips, mouth, larynx, and axial musculature. For example, a patient with callosal section can move the facial muscles and walk properly. The difference between unilateral and bilateral control lies in the fact that fine motor control, such as that in the fingers in humans and primates, is a province of the pyramidal tract, a control unilaterally represented in the motor cortex for the most part. The extrapyramidal control in the precentral motor areas subserves the large muscle groups, primarily axial muscles, and these are generally bilaterally represented in the cortex (cf. Chap. 8 and Fig. 8-2).

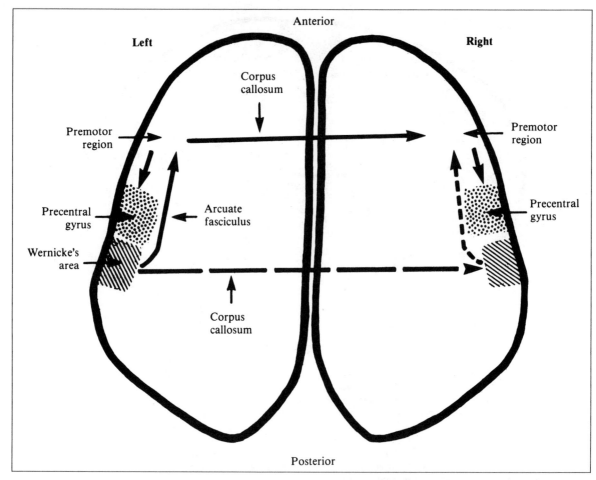

Fig. 9-7. Human brain viewed from above showing the intrahemispheric callosal pathways used in carrying out movements with the right or left limbs in response to a verbal command. Solid lines show path from Wernicke's area on the left to the premotor region and from the premotor region to the precentral gyrus for motor outflow controlling the right side. Another solid line from the premotor region on the left to the premotor area is shown on the right side for control of the right precentral gyrus and movement on the left side. Destruction of the anterior part of the corpus callosum results in inability to respond with the left arm to a spoken command. Dashed lines show possible alternative paths. (From N. Geschwind. The apraxias: Neural mechanisms of disorders of learned movement. *Am. Sci.* 63:189, 1975. Reprinted by permission of *American Scientist,* Journal of Sigma Xi, The Scientific Research Society.)

BIBLIOGRAPHY

Bureš, J., Burešová, O., and Krivanek, J. *The Mechanism and Applications of Lēao's Spreading Depression of Electroencephalographic Activity*. New York: Academic, 1974.

Fulton, J. F. *Frontal Lobotomy and Affective Behavior: A Neurophysiological Analysis*. New York: Norton, 1951.

Gaito, J. (ed.). *Macromolecules and Behavior* (2nd ed.). New York: Appleton-Century-Crofts, 1971.

Gazzaniga, M. S. *The Bisected Brain*. New York: Appleton-Century-Crofts, 1970.

Geschwind, N. The apraxias: Neural mechanisms of disorders of learned movement. *Am. Sci.* 63:188–195, 1975.

Herrick, C. J. Memorial symposium: Neurophysiology of learning and behavior. *Fed. Proc.* 20:601–631, 1961.

Horn, G., and Hinde, R. A. *Short-Term Changes in Neural Activity and Behavior*. Cambridge, Engl.: Cambridge University Press, 1970.

Klüver, H., and Bucy, P. C. Preliminary analysis of functions of the temporal lobes in monkeys. *Arch. Neurol. Psychiatry* 42:979–1000, 1939.

Landauer, T. K. (ed.). *Readings in Physiological Psychology*. New York: McGraw-Hill, 1967.

Lashley, K. S. In search of the engram. *Symp. Soc. Exp. Biol.* 4:454–482, 1950.

MacLean, P. D. Psychosomatic disease and visceral brain. *Psychosom. Med.* 17:355–366, 1955.

Miller, N. E. Learning of visceral and glandular responses. *Science* 163:434–445, 1969.

Papez, J. W. A proposed mechanism of emotion. *Arch. Neurol. Psychiatry* 38:725–745, 1937.

Pavlov, I. P. *Conditioned Reflexes*. New York: Dover, 1960.

Penfield, W. Functional localization in temporal and deep sylvian areas. *Res. Publ. Assoc. Res. Nerv. Ment. Dis.* 36:210–226, 1958.

Snyder, S. H., and Childers, S. R. Opiate receptors and opioid peptides. *Annu. Rev. Neurosci.* 2:35–64, 1979.

Sperry, R. W. Perception in the absence of the neocortical commissures. *Res. Publ. Assoc. Res. Nerv. Ment. Dis.* 48:123–138, 1970.

Sperry, R. W., Gazzaniga, M. S., and Bogen, J. E. Interhemispheric Relationships: The Neocortical Commissures; Syndromes of Hemisphere Disconnection. In P. J. Vinken and G. W. Bruyn (eds.), *Handbook of Clinical Neurology*. Amsterdam: North-Holland, 1969. Vol. 4, chap. 14.

Vaid, J., and Lambert, W. E. Differential cerebral involvement in the cognitive functioning of bilinguals. *Brain Lang.* 8:92–110, 1979.

FUNCTIONAL PROPERTIES OF BLOOD

10

Julius J. Friedman

Blood is a tissue consisting of a fluid plasma in which are suspended a number of formed elements (erythrocytes, leukocytes, and thrombocytes). It serves to maintain a constant cellular environment by circulating continuously through every tissue, thereby providing a medium of exchange among the tissues. Because of its circulation throughout the body, blood plays an integral part in almost every functional activity. These activities include the following:

1. Respiration: Blood transports oxygen and carbon dioxide between the lungs and the tissues.
2. Nutrition: Blood transports amino acids, glucose, fatty acids, and other nutrients from the gastrointestinal tract to the various sites of utilization.
3. Excretion: Blood transports waste products, such as urea, uric acid, and creatinine, from the tissues to the kidneys.
4. Acid-base regulation: Hemoglobin and plasma proteins act as buffers.

5. Body fluid regulation: Blood redistributes fluid between tissues.
6. Temperature regulation: High specific heat and thermal conductivity enable blood to accommodate and dissipate heat readily.
7. Hormonal regulation: Blood transports hormones from the site of release to sites of action.
8. Immune reactions: Antibodies and phagocytes protect against foreign substances and organisms.

PLASMA

Plasma is a pale amber fluid that contains numerous dissolved materials, including proteins, carbohydrates, lipids, electrolytes, pigments, and hormones, among others.

The plasma proteins—albumin, globulins and fibrinogen—can be separated and quantified by means of electrophoresis. A normal electrophoretic pattern (Fig. 10-1A) illustrates the differences in mobility of the proteins in an

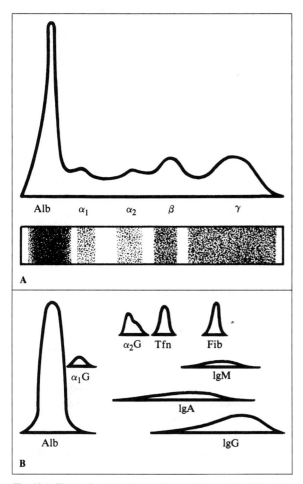

Fig. 10-1. Electrophoretic patterns of normal plasma identifying the plasma protein fractions obtained by electrophoresis (A) and agarose gel electrophoresis (B). Alb = albumin; Tfn = transferrin; Fib = fibrinogen. α₁G and α₂G are globulins, and IgM, IgA, and IgG are immunoglobulins.

ticuloendothelial and lymphatic systems (β_1-, β_2-, and α-globulins). Thus, marked changes in the electrophoretic pattern occur in disorders such as infectious diseases, liver dysfunction, and multiple myeloma. In some renal diseases, large quantities of albumin are lost to the urine.

Many of the physical and chemical properties of plasma are influenced by the plasma proteins. They determine the specific gravity of plasma, which is 1.028, as well as the plasma viscosity. Albumin and the globulins substantially determine the colloidal osmotic pressure of plasma. Globulins and fibrinogen determine the suspension stability of the red blood cells in plasma. Fibrinogen is essential for coagulation. The globulins are important in antibody formation and as hormones (e.g., thyroglobulin) and cellular enzymes. The plasma proteins also act as substrates for the release of peptides such as angiotensin and bradykinin and as binding vehicles for the transport of such substances as metal ions, fatty acids, steroids, hormones, and drugs. All the proteins contribute to the buffering capacity of blood.

FORMED ELEMENTS

The formed elements of the blood include erythrocytes, leukocytes, and thrombocytes (platelets).

ERYTHROCYTES. Erythrocytes, or red blood cells, are anucleated, biconcave disks that are formed in bone marrow by means of *erythropoiesis*. They are about 2 μm thick and 8 μm in diameter (with a surface area of about 120 μm^2 and a volume of about 85 μm^3). By virtue of its biconcave shape, the red cell can undergo marked changes in shape without changes in volume, as in the distortions that take place during passage through the capillaries. It can also endure increases in volume without marked changes in membrane tension during periods of osmotic stress. The major function of the red cell is to transport oxygen and carbon dioxide. The biconcave shape of the cell is ideally suited to this function, providing the maximal surface area and minimal diffusion distance in proportion to its volume.

The number of red blood cells in the circulatory system can be estimated by counting those in a representative sample of blood with the aid of the microscope and a graduated slide called a *hemacytometer* or with an electronic cell counter. There are 5.5 to 6.0 million red blood cells/mm^3 of blood in the normal adult male and 4.5 to 5.0 million/mm^3 of blood in the normal adult female. The number varies with age, activity, climate, and altitude, as well as with disease states.

The relative red blood cell content of blood is described by the *hematocrit ratio*. This ratio is derived by first render-

electric field. Figure 10-1B shows some of the major plasma protein fractions obtained by means of gel electrophoresis. The degree of separation of these components is determined by molecular size, configuration, and charge. Figure 10-2 indicates that the total plasma protein concentration is 5 to 7 gm per deciliter of plasma, with albumin at a concentration of 3 to 4 gm per deciliter, the globulins at about 1.5 to 2.0 gm per deciliter, and fibrinogen at a concentration of 0.2 to 0.4 gm per deciliter.

These plasma proteins are formed in part in the liver (albumin, fibrinogen, α_1- and α_2-globulins) and in the re-

Shape	Molecular weight	Plasma concentration gm/dl	% Total protein
Albumin	68,000	3.5	55
α_1 Globulin	80,000	.84	
α_2 Globulin	90,000	.76	38
γ Globulin	200,000	.6	
Fibrinogen	400,000	.3	7
Total		6.0	100

Fig. 10-2. The shapes, molecular weights, concentrations, and relative amounts of the major plasma proteins.

ing a blood sample incoagulable and then centrifuging it. Since the specific gravity of red cells is about 1.088, compared with 1.028 for plasma, the cells become packed at the bottom of the tube, and the less dense plasma lies above them. The white blood cells and platelets possess a specific gravity intermediate between that of plasma and red cells, forming the buff-colored layer above the red cell column. The hematocrit ratio equals the height of the red cell column divided by the height of the total blood column. The white blood cells are not included in the hematocrit ratio. The venous hematocrit of a normal adult is about 45 percent. Because the cells are disk-shaped, the small space between adjacent cells is filled with plasma. Estimates are that approximately 4 to 5 percent of the volume of packed cells is plasma. Thus, the corrected venous hematocrit reading is approximately 43 percent.

Care must be taken in evaluating changes, since the hematocrit varies with changes in sampling site as well as in many clinical conditions. Furthermore, the hematocrit is not uniform throughout the body. That of venous blood is slightly greater than that of arterial blood because of fluid and electrolyte shifts. In addition, the hematocrit varies from tissue to tissue.

The most reliable estimate of body hematocrit is derived from independent estimations of total plasma and total red cell volumes. Body hematocrit calculated as the ratio of total red cell volume to total blood volume is about 35 percent. The ratio of the total body hematocrit to the venous hematocrit, referred to as the *F cells ratio*, remains fairly constant at 0.91 under a variety of conditions of cardiovascular stress. However, it increases during pregnancy and is susceptible to change in severe hemorrhage and transfusion.

Red Blood Cell Production. Red cells are produced in the bone marrow at a normal rate of approximately 400 to 500 ml of packed red cells per month. The rate of red cell production (erythropoiesis) is probably determined by the oxygen tension of the blood, although the precise mechanism is not clear. Erythropoiesis appears to be controlled primarily by the action of a renal erythropoietic factor, released during renal hypoxia, on erythropoietin, a glycoprotein (mw 30,000).

Overstimulation of red cell production can cause the red blood cell population to increase excessively, resulting in a condition called *polycythemia*. Although the oxygen-carrying capacity of blood is enhanced in polycythemia, the viscosity is also increased, and the heart must perform additional work to pump the more viscous blood through the circulatory system. To accommodate the added work load, the heart hypertrophies. Polycythemia and cardiac hypertrophy are most prevalent among people living at higher elevations, who are continuously exposed to low oxygen tensions.

Red Cell Destruction. The red cell travels about 700 miles in its lifetime of about 120 days and during this period is subjected to considerable mechanical and chemical stress, which eventually leads to its destruction. Approximately 1 percent of the red cells is replaced daily, and the total red cell volume is replaced every 4 months. Red cells are destroyed primarily in the spleen. Continuous friction between red cells and the vessel walls, as well as the action of chemical substances in the circulation, apparently alters the integrity of the red cell membrane, thereby rendering it more susceptible to osmotic stresses.

When red blood cells are subjected to progressively increasing osmotic stress by a progressive reduction of the extracellular sodium chloride concentration from the isotonic level, they begin to hemolyze—few at first, then substantially more, and finally the few remaining. The sigmoid character of the degree of hemolysis represents the hemolytic behavior of cells of different ages. Thus, older cells hemolyze with lower osmotic stress, whereas the younger cells are most resistant to osmotic influence. These characteristics are taken into account in fragility tests, which are used to distinguish between hereditary and acquired red blood cell disorders.

The number of red cells in the circulation at any one time is determined by the balance between production and destruction. When production is less than destruction, red cell volume and hematocrit fall, and *anemia* results.

Anemia. Anemia is a condition in which the number of red blood cells per cubic millimeter, the amount of hemoglobin per deciliter of blood, or the red cell fraction of blood (hematocrit) is below normal levels. It may result from a number of causes, including hemorrhage (acute and chronic), maturation deficiency, aplastic bone marrow, and enhanced fragility. Each situation influences both the size and the hemoglobin content of the red cell in a characteristic manner; these two variables are used to classify anemia. The cell of normal size is *normocytic*. A cell larger than normal is *macrocytic,* while one smaller than normal is *microcytic.* The extent of microcytosis or macrocytosis is reflected in the mean corpuscular volume (MCV), calculated as follows:

$$MCV = \frac{\text{hematocrit ratio} \times 1000}{\text{red cell count in millions/mm}^3}$$

The normal range of red cell volume is 75 to 95 μm^3. A volume below 75 μm^3 is microcytic, whereas one above 95 μm^3 is macrocytic.

A cell possessing a normal concentration of hemoglobin is said to be *normochromic*. Red cells with a subnormal hemoglobin concentration are *hypochromic,* and those with an above-normal concentration are *hyperchromic.*

The degree of hemoglobin alteration is reflected in the mean corpuscular hemoglobin (MCH) determination, calculated as follows:

$$MCH = \frac{\text{hemoglobin (gm/1000 ml)}}{\text{red cell count in millions/mm}^3}$$

The normal range of hemoglobin per red blood cell is 24 to 34 picograms (pg), with an average hemoglobin content of 30 pg. This figure, together with an average cell volume of 85 μm^3, provides an average red cell hemoglobin concentration of 35 gm per deciliter of red cells. With an average hematocrit of 0.45, the blood hemoglobin concentration is about 16 gm per deciliter.

NORMOCYTIC, NORMOCHROMIC ANEMIA. Although red blood cells can be of normal size and contain the normal concentration of hemoglobin, anemia may be present if the number of red cells in the circulation is below normal, a condition that arises from acute hemorrhage. As a result, both plasma and red cells are lost from the circulation. The plasma volume is restored in a short time by compensatory responses (see Chap. 12). The process of red blood cell production is slow, however, and the circulation remains deficient in red cells for some time. Since the bulk of the cells present in the circulation during this period are the same ones that were present prior to the hemorrhage, the cells are essentially normal. Consequently, the anemia resulting from acute hemorrhage is normocytic and normochromic.

MICROCYTIC, HYPOCHROMIC ANEMIA (IRON DEFICIENCY ANEMIA). Iron deficiency anemia often occurs in women of childbearing age, in cases of chronic hemorrhage, and in infants.

During chronic hemorrhage, red cell loss from the circulation continues over a period of time, and the replacement of these cells must be maintained throughout the period. The persistent and extended stimulus to erythropoiesis imposes a continuous demand on the iron stores of the body and leads to a condition of iron deficiency. Consequently, the hemoglobin content of newly formed red cells is lower than normal (hypochromic), and the cells are smaller than normal (microcytic).

The iron requirement of infants is so great that nutritional iron deficiency often develops, producing a microcytic, hypochromic anemia.

MACROCYTIC, HYPERCHROMIC ANEMIA. In 1929, Castle discovered that a particular anemia could be alleviated by administration of the product formed by incubating meat with gastric juice. This principle was called the *antianemic, or hematinic, factor.* The meat component was labeled the *extrinsic factor* and was later identified as vitamin B_{12}. The gastric juice component was called the *intrinsic factor.* Vitamin B_{12} contained in ingested meat combines with the intrinsic factor to form the hematinic factor, which is absorbed through the intestinal wall and stored in the liver. In the absence of either of these factors, the maturation of red cells is altered, and excessively large red cells (macrocytes) are formed in reduced number. They contain a greater-than-normal amount of hemoglobin because of their increased size, and therefore are hyperchromic. The clinical condition that characterizes these changes is called *pernicious, or addisonian, anemia.*

Pernicious anemia may occur under any circumstances that interfere with any phase of the incorporation or utilization of the antianemic factor. Consequently, a reduction in extrinsic factor due to dietary deficiency, reduced gastric juice secretion, reduced intestinal absorption, or decreased liver function may lead to pernicious anemia. Because the liver is the storage site of the hematinic factor, either the addition of liver to the diet or administration of liver extract is generally sufficient to alleviate the condition.

Other megaloblastic anemias, which resemble pernicious anemia but have an obscure pathogenesis, occur in tropical sprue and pregnancy, as well as in conditions of folic acid and vitamin C deficiencies; these are known as *nutritional anemias.*

APLASTIC ANEMIA. Occasionally, one encounters a person with malfunctioning bone marrow and loss of the capacity to produce normal cells. In acute conditions, such as those due to excessive exposure to radiation, the cells in circulation are essentially those that were present prior to exposure and that are therefore normocytic and normochromic. However, since destruction proceeds independently of production, the red cell count falls progressively. Aplastic bone marrow can also occur spontaneously.

HEMOLYTIC ANEMIA. Anemia may result when enhanced red cell hemolysis takes place because of chemical agents, hereditary defects of cell formation, enhanced reticuloendothelial destruction, and hypersplenism. Venoms, bacterial toxins, and some drugs produce hemolysis by means of lysis of the cell membrane. *Familial hemolytic anemia* is a hereditary condition characterized by the production of small, spherical red cells called microcytes, or spherocytes. These cells are normochromic but hemolyze readily because their shape imparts great tension to the membrane during the deformations encountered by the cell as it passes through minute vessels. *Sickle cell anemia,* particularly prevalent among blacks, is another hereditary abnormality in which red cells shaped like sickles or crescents are formed during the sickling crisis that occurs as a result of abnormal hemoglobin and therefore are highly fragile. In addition, changes in the nature of the hemoglobin in these cells modify their binding of oxygen and impair gaseous exchange.

LEUKOCYTES. Leukocytes, or white blood cells, are primarily neutrophilic granulocytes about 12 μ in diameter with multilobular nuclei. Their major function is to protect against the invasion of bacterial infection. These cells are capable of passing through the vascular endothelium and entering the tissue spaces (diapedesis) in accordance with local needs. They are apparently attracted by chemical substances (chemotaxis) released by the bacteria and can engulf and digest the foreign substances by means of ameboid movement (phagocytosis). Chemotaxis appears to involve antigen-antibody complexes and complement. Normally, there are approximately 5000 to 7000 white blood cells/mm³ of blood.

The concentration of leukocytes in the circulation varies in a cyclical manner, with about a 20-day period. This represents the result of a regulatory function that involves rapid release of stored cells and slow activation of granulopoiesis. Excessive activation of granulopoiesis can lead to leukemia.

The leukocytes form part of a broader system that provides protection against infection, namely, the *reticuloendothelial system.* It includes all phagocytic cells, fixed and mobile, which are primarily situated in the liver, spleen, lymph nodes, lung, and gastrointestinal tract.

THROMBOCYTES. Thrombocytes, or platelets, are small, colorless bodies ranging in size from 2 to 4 μm in diameter with a lifespan of 7 to 10 days. There are 150,000 to 300,000 platelets/mm³ of blood. Thus, approximately 40,000 platelets are turned over in the spleen and throughout the vascular system. The number of platelets remains fairly constant for extended periods, suggesting the existence of some regulatory process. Platelets possess dense granules that are believed to be the location of the coagulation factor, platelet factor 3. These cells contain, or have absorbed on them, serotonin, histamine, catecholamines, adenosine, and ribonucleic materials. They also contain a retractile protein, *thrombosthenin,* which is responsible for clot retraction.

Platelets are of great importance in hemostasis. They support endothelial integrity, contribute serotonin and catecholamines for vasoconstriction, undergo transformation to form a platelet plug, establish a matrix for more effective coagulation, and retract to draw margins of wounds closer together.

BLOOD VOLUME

The total blood volume in a subject is best determined by estimating both the plasma and the red cell volumes independently and adding them together. These volumes are estimated by means of a dilution principle wherein a known amount (A) of indicator, which is essentially restricted to the vascular system, is introduced into an unknown volume (V). Sufficient time is allowed for complete mixing, and the concentration of indicator (C) in a representative sample is determined. The volume of dilution is then calculated as

$$V = A/C$$

PLASMA VOLUME. Plasma volume (PV) is estimated by injecting into the circulatory system about 5 microcuries of radioactively iodinated (^{125}I) serum albumin. Unfortunately, albumin leaves the circulation slowly (about 10 percent/hr) so that it is necessary to obtain at least two samples—10 and 20 minutes after injection—and extrapolate from these two points to zero time. This procedure provides an estimate of the mixed indicator concentration in the plasma before any extravascular loss occurred.

An average plasma concentration of indicator in a 70-kg male would be about 0.0015 microcurie per milliliter of plasma. Dividing this concentration into the dose administered provides an estimate of about 3200 ml of plasma.

RED BLOOD CELL VOLUME. Red blood cell volume (RCV) is generally estimated with red cells labeled with radioactive chromium (^{51}Cr). The procedure employed is as follows: Approximately 20 to 30 ml of blood is withdrawn from the subject and mixed with 5 ml of an anticoagulant acid citrate dextrose (ACD) solution. A dose of 15 to 20 microcuries of

sodium chromate (^{51}Cr) is added to the blood, which is gently agitated at room temperature for 30 to 45 minutes. This provides about 90-percent binding of ^{51}Cr. Tagging is terminated by adding 50 to 100 mg of ascorbic acid, which reduces the chromate.

The labeled blood is injected intravenously and allowed to mix in the circulation for about 10 minutes. A sample is obtained, corrected for unlabeled ^{51}Cr, and the concentration is determined. An average concentration is about 0.011 microcurie per milliliter of red cells. Dividing the corrected concentration into the corrected dose provides an estimate of 1800 ml of red cells for a 70-kg adult male.

The times allowed for vascular mixing apply to a subject under reasonably normal cardiovascular conditions. In cases of cardiovascular stress in which tissue blood flow is severely altered, longer mixing times will be necessary.

TOTAL BLOOD VOLUME. Total blood volume (BV) is calculated by the expression

$$BV = PV + RCV$$

The total blood volume of a normal 70-kg adult male is about 5000 ml. The blood volume of a normal adult female is about 4500 ml. It is possible to obtain an estimate of total blood volume from a single determination of plasma or red cell volume and a corrected venous hematocrit according to the expressions

$$BV = PV/(100 - hct \times F \text{ cells})$$

$$BV = RCV/(hct \times F \text{ cells})$$

Total blood volume varies under circumstances involving the integrity of the vascular system, the state of the individual's water balance, and the state of physiological activity. These will be discussed in detail in Chapters 18 and 23.

Changing posture from a lying to a standing position will result in a fall in blood volume due to a reduction in plasma volume. This difference reflects the change in transcapillary dynamics on standing, which produces enhanced filtration (see Chap. 12). Exercise produces similar variations in blood volume. Adaptation to low tissue oxygen tensions occurring at high altitude and with physical training increases red cell and blood volume. In pregnancy, both plasma and red cell volumes increase.

The blood volume is not uniformly distributed throughout the body. Approximately 80 percent of the total blood volume is contained within the low-pressure system (veins, right heart, and lungs). Because of the high distensibility of the pulmonary vasculature and the large mass of muscle, these two tissues are important reservoirs of blood that participate in large changes in volume during periods of circulatory stress. About 50 percent of a 400- to 1000-ml hemorrhage comes from the intrathoracic compartment. During orthostatism, approximately 700 ml is translocated from the thorax to the legs.

Changes in blood volume distribution become apparent as a result of passive or active adjustment of the capacitance vasculature. Passive behavior depends on the elastic recoil of the vessel system, whereas active behavior is primarily related to sympathetic activity and venomotor tone.

BLOOD GROUPS AND TRANSFUSION

In conditions of anemia or hemorrhage it is often necessary to transfuse additional volumes of blood into the circulation in order to improve the hemodynamics of the system. Care must be used in selecting the blood to be transfused, since not all blood is compatible. Incompatibilities reflect variations in principles present in the cells and in the plasma of different individuals.

Blood has been classified, according to its antigenic activity, as types O, A, B, and AB. An individual's blood type is determined genetically. Present in the γ-globulin fraction of plasma are anti-A and anti-B antibodies. When red blood cells are mixed with their antagonistic antigens, agglutination of the donor red cells results.

Fortunately, when an antigen is present in the blood, the antagonistic antibody is absent. Type O blood does not usually contain any antigens, and both alpha and beta antibodies are present, whereas type AB blood possesses both antigen A and antigen B; consequently, no antibodies are present. The blood groups, their constituents, and their percentage distribution are listed in Table 10-1.

Since type O blood possesses no antigens, type O cells cannot be agglutinated by any of the antibodies. Consequently, type O blood can be transfused into any other type without reaction, and type O persons are referred to as "universal donors." Conversely, type AB blood is without

Table 10-1. Blood group constituents

Group	Antigen	Antibody	Percent of population
O	—	α,β	46
A	A	β	42
B	B	α	9
AB	A + B	—	3

antibodies and will not agglutinate any type of donor blood. Persons with type AB blood are referred to as "universal recipients."

There are a number of other factors in the blood that can lead to low-level incompatibilities or sensitization. The most prominent of these, the *Rh factor,* is present in approximately 85 percent of the population. When it is present, the blood is Rh-positive. Antibodies to the Rh factor are not normally present in the blood but arise following the exposure of individuals with Rh-negative blood to the Rh antigen. An Rh-negative mother bearing an Rh-positive child may become sensitized and develop Rh antibodies. These antibodies can then diffuse into the fetal circulation, to produce an Rh reaction and death (erythroblastosis fetalis). Usually, a number of exposures are required to establish an antibody titer sufficiently high to produce fetal death.

Other factors, such as M, N, and P, are occasionally considered in blood typing, particularly in circumstances of disputed parenthood.

HEMOSTASIS

The rupturing of a blood vessel results in blood loss that, if unchecked, would eventually lead to a hemorrhagic shock and death. During normal activity, minor accidents frequently occur in which minute blood vessels are ruptured, yet blood loss is minimal. When a blood vessel is ruptured, hemorrhage is limited by a reduction of blood flow to the site of injury by means of local vasoconstriction, as well as by platelet aggregation, coagulation, and clot retraction.

VASOCONSTRICTION. Numerous influences can produce vasoconstriction (see Chap. 17). With vascular injury, platelets release serotonin, which may provide the stimulus for constriction of small arteries and arterioles. By reducing blood flow from the severed vessel, which allows the local concentration of the platelet and coagulation factors to increase, vasoconstriction enables platelet aggregation and coagulation to proceed with greater effectiveness.

PLATELET AGGREGATION. The role of platelets in hemostasis is a very important one. In addition to releasing serotonin, the platelets, under the influence of adenosine diphosphate released from damaged cells and from the platelets themselves, undergo transformation from a disk to a sphere and develop increased adhesiveness, which leads to formation of the platelet plug. The sphere-shaped platelets grow pseudopodia, which attach to collagen at the site of injury and thereby establish a framework on which effective coagulation can proceed. It is thought that prosta-

Table 10-2. Common synonyms of blood-clotting factors

Factor	Synonym
I	Fibrinogen
II	Prothrombin
III	Thromboplastin (tissue)
IV	Calcium ion
V	Proaccelerin, labile factor, plasma accelerator globulin (AcG)
VI	Discontinued
VII	Proconvertin, stable factor, precursor of serum prothrombin conversion accelerator (pro-SPCA)
VIII	Antihemophilic factor (AHF), antihemophilic globulin (AHG), thromboplastinogen, plasma thromboplastic factor (PTF), platelet cofactor I
IX	Plasma thromboplastic component (PTC), Christmas factor, platelet cofactor II
X	Stuart-Prower factor
XI	Plasma thromboplastin antecedent
XII	Hageman factor
XIII	Fibrin-stabilizing factor

glandin E_2 may be involved in platelet transformation, since aspirin, which blocks prostaglandin E_2, also blocks the transformation.

COAGULATION. Coagulation represents the complex sequence of interactions involving platelets, plasma, and tissue factors and culminating in the formation of a fibrin meshwork at the site of vessel rupture. Numerous theories have been introduced to account for this phenomenon, but uncertainty about it still exists because of the limited availability of purified factors. There are numerous synonyms for the coagulation factors, most of which are included in Table 10-2.

Coagulation proceeds in three stages: (I) formation of prothrombin-converting enzyme; (II) conversion of prothrombin to thrombin by this enzyme; and (III) conversion of fibrinogen to the fibrin clot through the action of thrombin. A scheme of coagulation is presented in Figure 10-3.

Stage I. Formation of Prothrombin-Converting Enzyme. Coagulation may occur in response to tissue injury (extrinsic) or take place without apparent tissue damage, such as intravascularly or in a test tube (intrinsic).

INTRINSIC COAGULATION. Blood obtained free of tissue factors can clot readily, indicating that plasma itself possesses the factors necessary to form prothrombin-converting enzyme. Coagulation begins by the contact of blood with

negatively charged surfaces, collagen, or phospholipids. This process then activates the Hageman factor (XII), which in turn activates plasma thromboplastic antecedent (PTA, XI). In the presence of Ca^{2+} and phospholipid, PTA activates the plasma thromboplastic component (PTC, IX). The PTC forms a complex with the antihemophilic factor (AHF, VIII), Ca^{2+}, and phospholipid to activate the Stuart-Prower factor (X). The complex formation is relatively slow and may be a rate-limiting step in intrinsic coagulation.

EXTRINSIC COAGULATION. The active principle in tissue is thromboplastin (III), which is continuously available and is released or activated on damage of tissue. Tissue thromboplastin in the presence of calcium ions and phospholipid apparently acts on or with the precursor substance of serum prothrombin conversion accelerator (pro-SPCA, VII) to activate the Stuart-Prower factor. The activated Stuart-Prower factor acts with activated proaccelerin (V), Ca^{2+}, and phospholipid to form what may be regarded as the prothrombin-converting enzyme. These coagulation factors remain in the plasma for a relatively short period because of their dilution, inactivation, degradation, and removal by the liver.

Stage II. Conversion of Prothrombin to Thrombin. Since blood clots rapidly in the presence of thrombin, significant amounts, as such, cannot be present in the circulation. Instead, thrombin is represented by a precursor, *prothrombin,* which is a plasma globulin fraction with a molecular weight of about 63,000, formed in the liver. In the presence of the converting enzyme, prothrombin is converted to thrombin. Thrombin, by activating factor V, accelerates its own formation.

Stage III. Conversion of Fibrinogen to Fibrin. Fibrinogen is converted to fibrin by thrombin, which partially proteolyzes fibrinogen by splitting off two peptides. The resulting fibrin monomer is unstable and dissolves easily in urea. However, in the presence of calcium ions and fibrin-stabilizing factor (factor XIII) derived from plasma transglutaminase, fibrin becomes cross-linked to form an insoluble clot.

Clot Retraction. After the clot forms, it retracts, extruding fluid identical to plasma except for the absence of fibrinogen and factors VII and IX. This fluid is *serum.* Platelets are necessary for clot retraction, since the process does not occur in conditions of platelet deficiency. The platelets apparently contain a contractile protein, *thrombosthenin,* which contracts when acted on by adenosine triphosphate. The function of clot retraction is thought to be the drawing

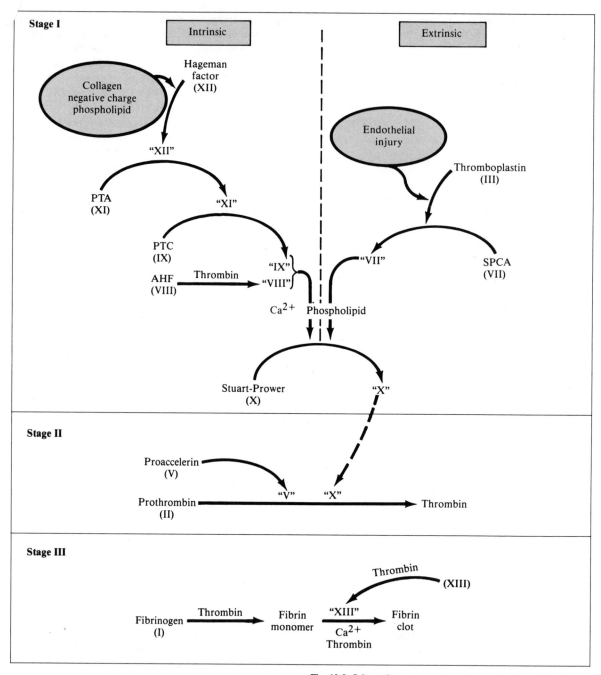

Fig. 10-3. Schematic representation of the mechanism of blood coagulation, viewed as a three-stage process.

of the wound surfaces together, recanalizing thrombosed vessels, or promoting more effective thrombosis.

FIBRINOLYSIS. Blood clots are not permanent. After some time the clot is dissolved, or lysed. The process of lysis involves the conversion of an inactive precursor, *profibrinolysin* (plasminogen), to the active form, *fibrinolysin* (plasmin). Apparently, the proteolytic fragments from activated Hageman factor convert an activator from its inactive form. This activator then converts profibrinolysin to fibrinolysin. Once formed, the fibrinolysin feeds back on the Hageman factor to increase the formation of the proteolytic fragments. Fibrinolysin then degrades the fibrin meshwork, disrupting the clot.

Evidence suggests that there may be a balance between fibrin clot formation and lysis. Interruption of coagulation by heparin administration results in an elevation of profibrinolysin concentration. This process has also been thought responsible for the maintenance of normal capillary flow by preventing microthrombus formation.

NATURAL INHIBITORS OF COAGULATION. Since tissue breakdown and platelet destruction probably occur under normal circumstances, and since some platelets are no doubt damaged and probably cause thromboplastin to enter the circulation to form thrombin, it would seem possible that coagulation in vivo might occur. Fortunately, inherent inhibitors, namely, antithromboplastin, heparin, and antithrombin, control clotting at every stage. Consequently, when small quantities of thromboplastin are released from tissue, inactivation by antithromboplastin occurs. However, even in situations of low-level release, the concentration of thromboplastin in the plasma may become sufficiently great to activate Stuart-Prower factor. Once again, coagulation is prevented, this time by virtue of the presence of heparin.

Heparin is a polysaccharide produced primarily in the basophilic mast cells distributed throughout the pericapillary tissue. Heparin prevents coagulation by inhibiting prothrombin conversion and possibly thrombin itself. Coagulation is also inhibited by the presence of antithrombin, a complex of substances that prevents the conversion of fibrinogen to fibrin and is thought to clear thrombin from the blood once clotting is completed. Otherwise, it is conceivable that thrombin would accumulate progressively and produce widespread coagulation.

COAGULATION DEFICIENCIES. A decreased rate of coagulation and a tendency toward bleeding may result from an insufficiency of any of the principles shown in Figure 10-3 except the Hageman factor. Deficiencies arise in several ways. Dietary or absorption inadequacy of vitamin K will result in a low blood prothrombin level (prothrombinemia). A reduced platelet count (thrombocytopenia) will lead to a deficiency of cofactors. Since fibrinogen, prothrombin, and probably other factors in coagulation are formed in the liver, liver disease that alters prothrombin or fibrinogen synthesis will result in coagulation deficiencies. The classic hereditary condition of hemophilia results from the absence of antihemophilic factor, which leads to coagulation deficiency and a bleeding tendency. This condition is an inherited, sex-linked abnormality of males, due to an abnormal gene on the X chromosome. In each case the replacement of the missing factor is usually sufficient to restore coagulation. Coagulation may also be impeded by excessive concentrations of fibrinolysin such as are seen in fibrinolytic purpura.

COAGULATION EXCESSES. When the concentration of any of the inherent anticoagulants is low, when endothelial damage is great, or when blood is stagnant, coagulation in vivo may occur. A clot that remains fixed in a vessel is called a *thrombus,* and the condition is called *thrombosis.* As blood flows by a thrombus, fragments (*emboli*) may break away and circulate throughout the vascular system. An embolus continues to circulate until it reaches a vessel with a cross-sectional area smaller than its own. It then becomes lodged in the vessel, effectively occluding the vessel and depriving the tissue supplied by the vessel of adequate nutrition. When the area supplied is extensive or is situated in a vital organ, death can result. Usually, the occluded region, if restricted in size, becomes invaded by a collateral circulation, which may restore an adequate blood supply. When collateral circulation is inadequate, the occluded area degenerates to form an *infarct.*

ANTICOAGULANT AGENTS. The most prominent anticoagulant agent administered clinically in circumstances of excessive coagulation in vivo is heparin. *Dicumarol* and *warfarin* (Coumadin) are drugs that act in the liver to prevent prothrombin formation by antagonizing the action of vitamin K. Because calcium ion is intimately associated with both the formation of the converting enzyme and the conversion of prothrombin, effective removal of calcium ion from the plasma prevents coagulation. However, the use of calcium precipitants or complexing agents is restricted to in vitro anticoagulation because a reduction of the calcium ion concentration can lead to cardiac arrest. *Sodium citrate* is a compound that removes calcium ion from solution by forming a poorly dissociating calcium salt. This substance is used routinely in blood banks as an anticoagulant.

COAGULATION TESTS. Numerous tests are in use to assess the effectiveness of the coagulation system: clotting time, prothrombin consumption time, thromboplastin generation, thrombin generation time, and others. The clotting time is the most comprehensive assay, since it is sensitive to deficiencies of most of the coagulating principles. The other tests are used to identify the coagulation deficiency more specifically. Details of the tests are to be found in any standard hematology text.

BIBLIOGRAPHY

Biggs, R. *Human Blood Coagulation, Hemostasis, and Thrombosis* (2nd ed.). Oxford, England: Blackwell, 1976.

MacFarlane, R. G., and Robb-Smith, A. H. T. (eds.). *The Functions of Blood*. New York: Academic, 1961.

Putnam, F. W. (ed.). *The Plasma Proteins* (2nd ed.). New York: Academic, 1977.

Ratnoff, O. D. *Bleeding Syndromes*. Springfield, Ill.: Thomas, 1960.

Ratnoff, O. D. Some recent advances in the study of hemostasis. *Circ. Res.* 35:1–14, 1974.

Wintrobe, M. M. *Clinical Hematology* (5th ed.). Philadelphia: Lea & Febiger, 1971.

FLUID DYNAMICS

Carl F. Rothe

The rate of blood flow through tissues is a vital determinant of the adequacy of cellular nutrition and waste removal. The physical factors governing this flow are the same as those modifying fluid flow through any living or nonliving system. A thorough understanding of cardiovascular function therefore requires the application of these principles of fluid dynamics. Once the concepts are understood, the effects of disease on the function of the cardiovascular system can be appreciated more fully, for one may reasonably predict the consequence of various changes. Most of the principles are summarized by the equations given in this chapter. The same equations, with differences in parameters or added terms, are applicable to flow in any system (e.g., airflow in the lungs).

FACTORS INFLUENCING FLOW

The rate of flow of a fluid is proportional to the driving pressure gradient. Expressed mathematically, this is

$$\dot{Q} = \Delta P/R = G \cdot \Delta P \qquad (1)$$

The flow, $\dot{Q}$, is expressed in units of volume per unit of time (e.g., ml/min). The *perfusion pressure,* or *effective pressure gradient,* along the vessel in question, $P_1 - P_2$ or ΔP (Fig. 11-1A), is expressed in millimeters of mercury (mm Hg or torr), or newtons per square meter (N/m^2). The unit of pressure in the System International, the pascal, or N/m^2, is rather small, since 1 mm Hg = 133.3 N/m^2. Therefore, the kilopascal (kPa) is commonly used. A pressure of 1 kPa equals 7.5 mm Hg. The *conductance,* G, is the proportionality constant of equation 1 and is expressed in flow units divided by pressure units (e.g., ml/min per mm Hg). The reciprocal of conductance is *resistance,* R. The resistance opposing the flow of fluid cannot be measured directly but is calculated from simultaneous measurements of the flow and pressure gradient using equation 1. It is expressed as the units of pressure divided by the units of flow. The equation $\dot{Q} = P/R$ is fundamentally the same as Ohm's law of electrical circuits: I = E/R. There is no commonly ac-

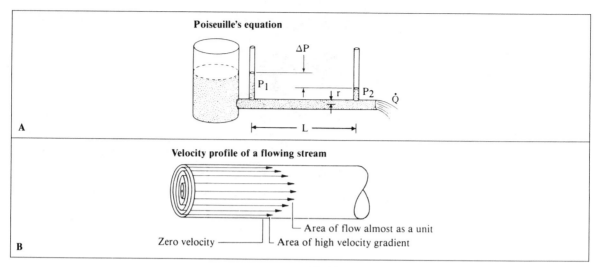

Poiseuille's equation

Velocity profile of a flowing stream

Area of flow almost as a unit
Zero velocity — Area of high velocity gradient

Fig. 11-1. A. Factors in Poiseuille's equation. B. The velocity profile of the flowing stream is parabolic, with the highest velocity in the center and no flow at the wall.

cepted resistance unit analogous to the ohm. Flow is directly proportional to the pressure gradient, which in the case of a tissue vascular bed is the arterial minus the venous pressure. Note that if the pressure gradient is increased, the flow increases, whereas if the resistance is increased, the flow decreases.

POISEUILLE'S EQUATION. Poiseuille's equation describes in more detail the factors determining the resistance to flow (Fig. 11-1A). Poiseuille's equation is

$$\dot{Q} = \Delta P \times \frac{\pi r^4}{8L} \times \frac{1}{\eta} \qquad (2)$$

Here, r is the radius of the vessel, L is its length, and η is the coefficient of viscosity. Note that flow will increase if the pressure gradient or radius is increased or if the length or viscosity is decreased.

There are several important consequences from the mathematical derivation of Poiseuille's equation. The primary assumption, attributed to Sir Isaac Newton, is that fluid flows as infinitely thin layers or laminae sliding past each other, giving *laminar* flow. The force per unit area (*shear stress*) tending to produce this sliding is directly proportional to the relative velocity between two adjacent layers (*shear rate*). The shear rate is a velocity gradient and is the longitudinal displacement per second divided by the distance between layers. Hence, the dimensions are: cm/sec per centimeter. (This simplifies to $\sec^{-1}$ because the units of length cancel.) The ratio of shear stress to shear rate is called the viscosity of the fluid. Like friction, there is dissipation of

energy to heat because of the relative motion of the molecules past each other. If the viscosity is independent of shear rate, the fluid is said to be *newtonian*. Blood and many colloidal suspensions may not be newtonian, for at low shear rates (less than $100 \sec^{-1}$, i.e., 10 cm/sec per millimeter of thickness) blood viscosity becomes sensitive to shear rate, as will be described later.

From the theoretical derivation and from actual measurement, the fluid at the wall of a blood vessel is stationary; the velocity is zero (Fig. 11-1B). However, the velocity gradient (shear rate) at this point is maximum. At the center of the tube the velocity is maximum, whereas the velocity gradient is zero.

There are two types of use for an equation such as equation 2: (1) The equation *describes* the characteristic of a physical *system* (see Chap. 7). It is the law relating input to output of the system; i.e., it describes how the variable $\dot{Q}$ is dependent on the various independent variables ΔP, r, η, and L. The equation is considered valid and the assumptions used in arriving at it are not disproved if, after each of the independent variables and the flow have been measured, the predicted flow equals the measured flow. Statistical analyses are used to estimate whether the discrepancies between measurement and prediction are likely to be real or a result of random variation. (2) The equation may also be used to *estimate* one of the independent variables. Assuming that the equation is valid under the conditions used, if flow and all but one other variable are measured, the value of this other variable may be calculated. One way to determine viscosity is to measure the rate of flow of the fluid through a tube of known dimensions using a known pressure gradient.

Note that the flow is inversely proportional to the viscosity (η) of the fluid. Syrup, for example, is a highly viscous fluid, whereas water has a relatively low viscosity. The viscosity of air is much less than that of water but is not negligible; it is about one-thirtieth of that of water at body temperature.

Since flow through a resistance vessel is directly proportional to the fourth power of the radius of that vessel, the resistance is markedly influenced by the radius of minute blood vessels such as the arterioles. For example, at a constant pressure gradient, a 19 percent increase in arteriolar radius permits a 100 percent increase in flow. Physiologically, resistance to the flow of blood is determined primarily by the caliber of the small arteries, arterioles, and precapillary sphincters, which in turn is dependent on changes in the vasomotor tone. This vasomotor tone is the state of tonic contraction of the smooth muscles of these resistance vessels (see Fig. 11-3 and Chaps. 12 and 17).

The *total peripheral resistance* (TPR) is the resistance to flow of blood in the systemic vasculature as a whole. It is the mean aortic pressure minus the mean central venous pressure divided by the total flow through the system (the cardiac output). Because the venous pressure is much less than the arterial pressure in the systemic circuit, the venous pressure can often be neglected in calculating overall vascular resistance. However, to compute the resistance to flow through a low-pressure circuit, such as the pulmonary circulation, requires a knowledge of the pressure in the pulmonary veins as well as in the pulmonary artery for an accurate estimate. The TPR is decreased in exercise and is increased in essential hypertension. Because the circulation consists of many parallel branches, the calculated value of the total peripheral resistance is less than that of any resistance through a particular organ or region. When using total peripheral resistance as a diagnostic tool, one should remember that pathological conditions causing an increase in resistance to flow in one organ may be compensated by opposite changes in other organs, so the total peripheral resistance may remain constant.

VISCOSITY OF BLOOD. Plasma is about 1.5 times as viscous as water. Whole blood has a variable viscosity of between 2 and 15 times that of water. The causes of this range of blood viscosity are physiologically important.

The *hematocrit ratio* has a marked effect upon the viscosity and flow of whole blood, as may be seen in Figure 11-2A and B. If the hematocrit were zero, the viscosity would be that of plasma, but as the proportion of blood that is cells increases, the viscosity of the blood increases. Beyond the normal hematocrit ratio of about 0.45, the blood becomes much more viscous. Thus, if the concentration of erythrocytes is abnormally large, as in polycythemia, the flow tends to be diminished because of the increased viscosity of the blood. In contrast, in anemia the proportion of cells is so reduced that the flow increases, if the perfusion pressure is the same, because of a decrease in viscosity of the blood.

Temperature has an inverse effect on viscosity. The viscosity of water and plasma increases about 2.6 percent per degree Celsius as the temperature is lowered from normal body temperature. This rise in viscosity of cold blood is an important factor in determining blood flow when extremities are cooled by immersion in cold water or when hypothermia is used in surgery.

Fluid velocity may influence blood viscosity, for although blood behaves as a newtonian fluid under normal physiological conditions in all the vasculature, under some pathological conditions of nearly stagnant flow the viscosity of blood increases. Since flow is then not linearly proportional to the driving pressure—that is, the velocity of each layer is not proportional to the force acting on it—the blood is *nonnewtonian*. Because the calculated viscosity of blood varies with shear rate or the diameter of the flow channel (Fig. 11-2A and below), the ratios of shear stress to shear rate are called the *apparent viscosities* under the various conditions. Because the geometry of the minute blood vessel network is so complex that calculation of viscosity from pressure-flow data and the Poiseuille equation is impossible, the viscosity of blood is sometimes compared with that of a saline solution or plasma—fluids of known rheological characteristics. Unfortunately, such a comparison to give relative viscosity may be erroneous, because perfusion with a solution other than blood may lead to changes in blood vessel dimensions, especially at the microvascular level.

A major cause of the increase in viscosity of blood at abnormally low shear rates is a tendency for erythrocytes to aggregate into rouleaux (i.e., they resemble a stack of coins). Platelets may aggregate or leukocytes may adhere to the vessel walls. Indeed, at very low shear stresses, blood does not flow. Since force is required to break up the aggregates so that flow can continue through the capillaries, less driving force is available to overcome the remaining viscous friction. Although normal blood does not undergo rapid aggregation, the tendency toward cell clumping may be important under conditions such as traumatic or burn shock, when the blood tends to *sludge*.

FLOW IN MINUTE VESSELS. The apparent viscosity of blood flowing through arterioles and capillaries is less than that of

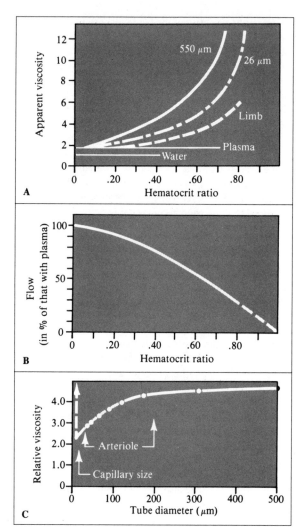

Fig. 11-2. A. Effect of the concentration of erythrocytes (hematocrit ratio) on the apparent viscosity of blood perfused through tubing of various diameters and an isolated dog hind limb. There is less effect of hematocrit ratio on viscosity if the blood flows through small, 26-μm diameter tubes than if the flow is through larger, 550-μm glass tubes. (Data from L. E. Bayliss. Rheology of blood and lymph. In Frey-Wyssling [ed.], *Deformation and Flow in Biological Systems.* New York: Interscience, 1952. Pp. 355–418.) There is even less effect when blood flows through a hind limb. (Data from S. R. F. Whittaker and F. R. Winton. The apparent viscosity of blood flowing in the isolated hind limb of the dog, and its variation with corpuscular concentration. *J. Physiol.* 78:339–369, 1933.) B. As the hematocrit ratio of blood is increased, the flow of blood through isolated tissue decreases, other factors being constant. (Derived from data of S. R. F. Whittaker and F. R. Winton. The apparent viscosity of blood flowing in the isolated hind limb of the dog, and

its variation with corpuscular concentration. *J. Physiol.* 78:339–369, 1933.) C. The effect of tube diameter on the relative viscosity of blood. (Data from R. Fahraeus and T. Lindqvist. The viscosity of the blood in narrow capillary tubes. *Am. J. Physiol.* 96:562–568, 1931.)

blood flowing in large-diameter tubes (Fig. 11-2C). This *Fåhraeus-Lindqvist* effect seems to be a result of a tendency of the erythrocytes to accumulate in the center of the flow stream. As a result of this *axial accumulation* of the erythrocytes, a "lubricating layer" of plasma lines the vessel wall. This reduces the viscous energy dissipation. Because the velocity of the cells along the axis of the flow stream is faster than the average velocity of the blood, the residence time of the cells in the capillaries is relatively less than that of the plasma. The hematocrit of the blood in the capillaries may be as little as 20 percent of that found in the large arteries and veins. Even though the relative viscosity in 10-μm diameter vessels may be as little as 50 percent of that seen in large (>200-μm internal diameter) vessels, if the vessel diameter is decreased even more, the apparent viscosity increases. The 8-μm diameter eyrthrocytes are very readily distorted, so that the cells can easily be perfused through the tissue and 5- to 7-μm diameter capillaries, but some force is required to distort the cells. In vessels less than about 4μm in diameter, the erythrocytes tend to occlude the resistance vessels, especially at precapillary sphincters, very narrow arterioles, or at the site of leukocyte adhesion. The apparent viscosity then becomes infinite, and flow ceases. Factors influencing the Fåhraeus-Lingqvist effect and axial accumulation are not all fully understood, but include shear rate (flow), cell concentration, and cell deformability (see Schmid Schönbein, 1976; Lipowsky et al., 1980; Charm and Kurland, 1974). The reduction in apparent viscosity, in small-diameter tubes in comparison to large ones, is particularly noticeable with abnormally high hematocrit ratios (Fig. 11-2A). Axial accumulation, by permitting plasma skimming into side branches, is a factor in determining the distribution of erythrocytes and plasma in the capillary bed (see Chap. 12).

CONTINUITY EQUATION. Because the mass of material in a closed system is constant (conservation of mass principle), flow of material leaving a section of blood vessel must equal that entering, or else volume will be gained (stored) or lost at that point. Furthermore, mean flow equals cross-sectional area (A) times mean velocity (v). Because the cross-sectional area of the vascular system increases as blood progresses from the heart toward the periphery, the mean blood velocity decreases (see Fig. 16-3). Thus, in a human aorta about 2 cm^2 in area, the mean velocity is about 50 cm per second, but in the 4-μm diameter capillaries it is only about 0.1 cm per second because the *total* cross-sectional area in humans is over 1000 cm^2 (1 square foot). This principle is expressed in the continuity equation

$$v_1 \times A_1 = \text{flow} = v_2 \times A_2 \tag{3}$$

where A_1 is the cross-sectional area at one point, and A_2 is the area at another point.

EQUATION OF MOTION. When a pressure or force is applied to a system having mass, it does not start moving instantly. Inertia must be considered. Newton's laws of motion may be applied to a flowing stream to obtain an estimate of the distribution of the force, since the algebraic sum of the forces is zero (d'Alembert's principle). The basic equation applied to a mechanical system, such as a weight suspended on a spring and moving in a viscous liquid, is

$$M a + R v + S x = F \tag{4}$$

Here, F is the applied force; x is the displacement of the spring from the neutral or zero force position; S is the parameter defining the stiffness of the spring (force/unit displacement); v is the velocity or rate of change of position (dx/dt); R is the resistance or coefficient of friction related to the viscosity of the fluid and the geometry of the moving object; a is the acceleration or rate of change of velocity; and M is the mass. The first term (M a) represents the inertial part of the applied force; the second, the frictional or viscous part; and the third, the elastic part.

When applied to a flowing fluid, Newton's equation may be transformed by dividing equation 4 by the cross-sectional area of the vessel, giving

$$M\ddot{Q} + R\dot{Q} + SQ = P \tag{5}$$

where $\ddot{Q}$ is the rate of change of flow or acceleration of the bolus of fluid, $\dot{Q}$ is the flow, Q is the distending volume, and P is the applied pressure at that point (force per unit area). (Note: The units for M, R, and S are different in equation 5 from those in equation 4.) Thus, the pressure developed in the ventricle during systole is partitioned into three parts acting to (1) accelerate the bolus of blood, $M\ddot{Q}$; (2) overcome the resistance to flow through the aortic valve and aorta, $R\dot{Q}$; and (3) distend the aorta, SQ. At the peak of ventricular pressure, the inertial and viscous terms each normally represent less than about 5 percent of the total force. Most of the developed pressure is acting to distend the aorta and so to provide the pressure to move the blood through the periphery.

The preceding analysis is only an approximation, for it considers the parameters of mass, viscosity, and stiffness as lumped at discrete points rather than as continuously distributed throughout the system. Nonetheless, consideration of inertia, for example, is crucial in understanding the dynamics of ventricular ejection, closing of the valves, and the resonant behavior of fluid systems. A further consequence of the physics of the system is that energy is dissipated (i.e., lost to the environment) *only* by the frictional term. The energy stored as momentum on acceleration and the energy used to distend the aorta are available later to drive the blood through the periphery.

BERNOULLI'S EQUATION. Rather than considering a balance of forces, as in equations 4 and 5, one may obtain another useful description of a system by considering a balance of energy. Bernoulli's equation is based on the principle of conservation of energy and states that the total energy, E, of a laminar (streamline) flow stream *without resistance* is constant and equals the sum of the potential energy of pressure, P V (here, V is volume of fluid); the potential energy due to gravity above a reference plane, M g h (here, M is mass; g, the acceleration by gravity; and h, the height); and the kinetic energy due to the velocity, v, of the mass that is flowing, ½ M v^2. Thus,

$$E = P V + M g h + \tfrac{1}{2} M v^2 \tag{6}$$

The driving force or *effective pressure gradient* for the propulsion of blood is the total energy gradient, i.e., the difference between the total energy, as expressed in equation 6, at the upstream point of measurement and that at the downstream end. Thus, the effective pressure gradient includes not only the usual *pressure* type of potential energy but also hydrostatic or *gravitational* potential energy and the *inertial* or kinetic energy of the moving mass. The hydrostatic factor is important in understanding the factors determining the flow to and from the legs (see Chap. 16). The inertial factor accounts for over half of the driving force in the pulmonary artery or great veins during vigorous exercise. During one part of the ejection phase of the cardiac cycle, the inertial term is great enough so that flow continues even though the pressure gradient is reversed (see Chap. 13). When the rate of flow changes, as in the arteries, the resistance to flow, more correctly termed *impedance,* includes this inertial term. With inertance or compliance in the system, the flow pattern does not coincide with that of the pressure but lags behind the pressure if the inertial component predominates, or leads if the arterial compliance (see Capacitance and Compliance) is dominant. The mathematics describing the relationship between the driving arterial pulse and the resulting flow becomes complex (see Noordergraaf, 1969; Bergel, 1972; McDonald, 1974).

A consequence of the Bernoulli principle is that, if the velocity of flow increases because of a constriction in a tube (see equation 3) or because blood is being rapidly propelled from the heart, more of the total energy is in the form of kinetic energy (velocity) and less is in the form of potential energy (pressure). Under most physiological conditions this factor is minor. However, it does help to explain the closure of the valves of the heart, for blood rapidly flowing past partially closed valve leaflets tends to suck them together.

The orientation of a catheter to measure the lateral pressure of a flowing stream is important, for if it is directed upstream against the direction of flow, it will give a somewhat higher value than if it is directed at right angles to the flow stream. If the opening is oriented toward the flow stream, as the flow impinges on the cannula, some kinetic energy due to the inertia of the blood is converted to potential energy and is added to the lateral pressure. In the larger systemic arteries this "end-pressure" or kinetic factor is minor, but in the pulmonary vessels, with high flow velocities and low lateral pressures, it becomes significant, for an open-ended catheter pointing into a stream flowing at 100 cm per second will have a pressure 4 mm Hg greater than that measured with a catheter with openings in the side. At 200 cm per second (a value not uncommon in the aorta or pulmonary artery at the peak of outflow) the difference is 15.9 mm Hg.*

---

*In working with equations such as these, recall that force = mass times acceleration, and weight = mass times the gravitational constant. Thus, the M g H in the second term of equation 6 is a force (in newtons) due to gravity (g = 9.806 m/sec^2) acting on the mass (in kg). Multiplying by height (in m) gives work in newton-meters [kg · (9.81 m/sec^2) · m = N · m], i.e., the potential energy available due to height. In the third term ($\frac{1}{2}$ M v^2), since a newton is kg · m/sec^2, one obtains the appropriate units of work.

$$\frac{1}{2} \text{ kg} \cdot (\text{m/sec})^2 = \frac{1}{2} (\text{kg} \cdot \text{m/sec}^2) \text{ m} = \frac{1}{2} \text{ N} \cdot \text{m}$$

Note: A newton-meter is a joule of work, and a newton-meter per second is a watt: the rate of doing work or power.

In solving the catheter orientation problem, total work and the height are assumed to be constant, the kinetic energy and potential energy of pressure are set equal to each other (P V = $\frac{1}{2}$ M v^2), and the resulting equation is solved for P.

The mass (M) moving is density times volume (ρV). By assuming that this volume is equal to the pressure volume, one obtains

$$P = \frac{\frac{1}{2} Mv^2}{V} = \frac{\frac{1}{2} \rho Vv^2}{V} = \frac{1}{2} \rho v^2$$

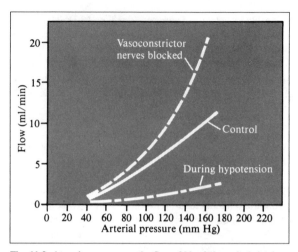

Fig. 11-3. At a given pressure, the flow of blood through skeletal muscle is influenced by the vasomotor tone. If impulse conduction along vasoconstrictor nerves is blocked, the flow is increased. The arteriolar diameters become larger (i.e., dilate). Conversely, during severe hypotension as a result of hemorrhage, there is an intense, neurally induced vasoconstriction of the blood vessels in vascularly isolated muscles of dogs, so that the flow is reduced even though the perfusion pressure may be normal.

TRANSMURAL PRESSURE. The *transmural pressure*, P_T, is the difference between the pressure inside a vessel, P_i, and the pressure outside the vessel, P_o.

$$P_T = P_i - P_o \tag{7}$$

Often, the outside pressure is atmospheric and so may be considered zero.

The pressure gradient *along* a resistance vessel is a major determinant of flow (equation 1). The transmural pressure *across* a resistance vessel is also a factor determining the resistance to flow, for as the transmural pressure is increased, the vessels are passively distended and the resistance to flow is decreased, thereby allowing an even higher flow than would be found in a rigid vessel of constant internal diameter (Figs. 11-3 and 11-4). Contraction of vascular smooth muscle in conjunction with the elastic elements counteracts the distending force and so regulates the flow of blood to the tissue. This phenomenon explains in part the

---

Here, P is the pressure (N/m^2, or pascal) that results from the conversion of kinetic to potential energy; ρ is the density (kg/m^3); v is velocity (m/sec); and a newton (N) is equivalent to 1 kg · m/sec^2. Note that 7.5 mm Hg = 1 kPa = 1000 pascal of pressure.

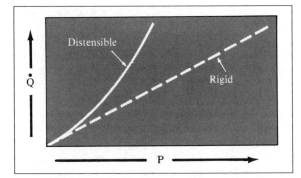

Fig. 11-4. Effect of vessel distensibility on flow. As the driving pressure in a distensible vessel is increased, the flow increases more rapidly than that in a rigid vessel of the same resting diameter. This is because both the pressure gradient and the radius are increased by increases in the driving pressure.

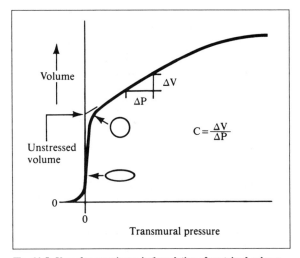

Fig. 11-5. *Vascular capacitance* **is the relation of contained volume (V) to transmural pressure (P). Compliance (C) is a slope of curve estimated as change in volume (ΔV) from a small change in transmural pressure (ΔP). Unstressed volume is the virtual volume of a vessel estimated by extrapolating the pressure-volume curve to zero transmural pressure. At abnormally high pressures, vessel walls are stiff and the compliance is so small that little change in volume occurs as pressure is increased. (From C. F. Rothe. Measurement of Circulatory Capacitance and Resistance. In R. J. Linden [ed.], *Techniques in the Life Sciences, Cardiovascular Physiology.* Ireland: Elsevier, 1983. Vol. 3.)**

nonlinear pressure-flow characteristics of blood through a living tissue (see Fig. 11-3).

CAPACITANCE AND COMPLIANCE. Hollow organs and tubes show an increase in volume with an increase in transmural pressure. This property is called *distensibility*. Compliance (C) is a general term describing the change in dimension (strain) resulting from a change in transmural pressure (stress). It is usually defined as

$$C = \frac{\Delta V}{\Delta P} \qquad (8)$$

where ΔV is the change in volume in response to a change in pressure, ΔP. The coefficient of stiffness, S, of the third term in equation 5 represents the reciprocal of compliance. The histological structure of most biological blood vessels—and indeed most tissues—is such that with distention, especially beyond the normal operating range, the compliance decreases (i.e., they become stiffer) (Fig. 11-5). This limits excessive pooling of blood in the lower limbs during standing. Blood vessels contain a volume at zero transmural pressure. By extrapolating a linear part of the pressure-volume relationship to zero pressure, an *unstressed volume* is obtained (Fig. 11-5). The added volume required to distend the vessel to the pressure under consideration is the *stressed volume,* and is the product of compliance (C) and distending pressure (P). Venoconstriction, resulting from a change in vascular smooth muscle activity, may cause a change in transmural pressure without a change in vascular volume by a change in compliance (Fig. 11-6, curve A), or a change in unstressed volume (Fig. 11-6, curve B), or both. The term

vascular capacitance is used to relate volume to transmural distending pressure over the full range from zero to maximum volume. Vascular capacitance cannot be quantified with a single number. Large veins, or those in the skin, collapse to an elliptical shape at near zero pressure (Fig. 11-5). Here large changes in volume can be induced by small changes in pressure. Veins within an organ are tethered by surrounding tissue and so will not collapse even with moderately negative transmural pressures.

Because a large organ will require a greater volume for a given pressure change than a small organ, compliance is usually normalized to the volume per 100 gm or per kilogram of tissue. The total body vascular compliance is about 3 ml/mm Hg per kilogram body weight. The compliance of the lesser circulation (heart and lungs) is about 1 ml/mm Hg per kilogram body weight. Of a total blood volume of about 75 ml per kilogram of body weight, only about 25 ml per kilogram is required to distend the vasculature, leaving an unstressed volume of about 50 ml per kilogram.

The combination of resistance, compliance, and inertance of the blood in the arterial system determines the arterial

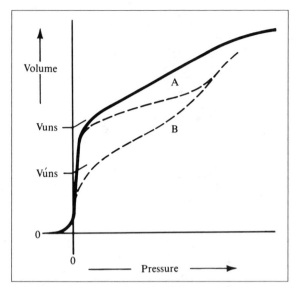

Fig. 11-6. Increased activity of smooth muscle in walls of veins may reduce contained volume with no change in transmural pressure, either by a change in compliance with no change in unstressed volume (Vuns, curve A) or active venoconstriction may induce a change in unstressed volume (V'uns) with no change in compliance in the normal range (curve B), or both compliance and unstressed volume may change. (From C. F. Rothe. Measurement of Circulatory Capacitance and Resistance. In R. J. Linden [ed.], *Techniques in the Life Sciences, Cardiovascular Physiology*. Ireland: Elsevier, 1983. Vol. 3.)

input impedance, the ratio of pressure gradient to flow at a specified frequency. Because of its compliance, the aortic input impedance for pulsatile frequencies between 1 and 15 Hz is only about 10 percent of that for the steady flows and so acts to reduce the load on the heart, as well as to aid in continuing flow to the peripheral tissues during cardiac diastole.

LAPLACE EQUATION. Because blood flow is proportional to the fourth power of the radius of the arterioles and other resistance vessels, small changes in vessel dimensions may have profound effects on blood flow. Therefore, the factors influencing vascular dimensions, such as transmural pressure and wall tension, are important contributors to the state of vascular resistance. The Laplace equation states that the tension (T) in the wall of a vessel required to maintain a given radius is proportional to the product of the *transmural pressure* (P) and the *radius* (r).

$$T = Pr \qquad (9)$$

A derivation of the Laplace equation provides a valuable insight into simple mathematical reasoning. If the wall of the vessel is to be stationary, the *distending force* pushing the vessel wall out must be balanced by a *restraining force* produced by the tensions of the fibers in the wall. The distending force is equal to the area of a longitudinal sectional plane times the transmural pressure. In the case of a cylinder the distending force equals $P \times 2rL$, where P is the pressure in gm/cm^2, r is the radius, and L is the length of the cylindrical vessel (Fig. 11-7). The restraining force is equal to $T \times 2L$, where T is the tension per unit length and L is the length. The factor 2 is used because both edges of the vessel are acting to hold it together. (In this simple derivation the walls of the vessel are assumed to be infinitely thin, an assumption that is not valid for calculating wall tensions of small arteries, for example.) Equating these two expressions gives the relationship $T = Pr$. (For a sphere the equation becomes $T = Pr/2$.) In a larger vessel the forces in the wall required to keep it from distending at a given pressure must be higher than in a smaller vessel. For the heart to eject blood at a given pressure, the myocardial tension must be greater in a large heart than in a small heart (see Chap. 15). In aortic aneurysm the walls of the aorta are damaged, so the diameter becomes much larger than normal. This development is exceedingly serious; the enlarged vessel is much more likely to rupture because of the high tensions in its walls, even at normal arterial pressures. The capillaries, because of their minute size, have wall tensions of less than 10 mg per centimeter of vessel length and so remain intact in spite of wall thicknesses of about 1 μm.

TURBULENCE. Blood flow through vascular channels is normally streamline in nature. Under special conditions, turbulent fluid flow, with swirls or eddies, may develop. In this event, flow is approximately proportional to the square root of the pressure drop, so a given increase in pressure results in less than a proportionate increase in flow.

$$\dot{Q} \approx k \sqrt{\Delta P} \qquad (10)$$

The reason is that the potential (driving) energy is dissipated in the swirls and eddies as heat. If turbulence is of sufficient magnitude, audible sound like the hiss of air from a compressed-air outlet is produced.

If the ratio of inertial to viscous forces of a flowing stream exceeds a given value, a slight disturbance in a streamline flow pattern will lead to the development of a randomly oriented turbulent pattern. Turbulence is likely when the ratio, called Reynolds number (Re), exceeds a value of about 1000 for blood and 2000 for fluids (liquid or gas) with-

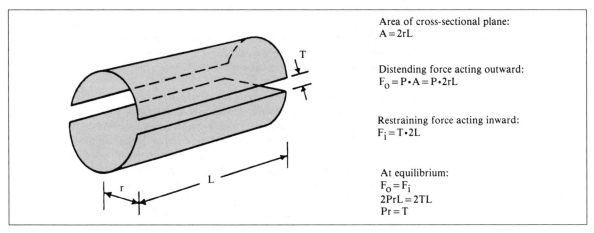

Area of cross-sectional plane:
$A = 2rL$

Distending force acting outward:
$F_O = P \cdot A = P \cdot 2rL$

Restraining force acting inward:
$F_i = T \cdot 2L$

At equilibrium:
$F_O = F_i$
$2PrL = 2TL$
$Pr = T$

Fig. 11-7. Derivation of Laplace equation.

out suspended particulates (e.g., erythrocytes). Reynolds number may be computed for a cylindrical conduit with the following equation:

$$Re = \frac{\rho \, v \, D}{\eta} \tag{11}$$

where ρ is the density of the fluid in gm/cm$^3 \cdot$ g (here, g is the gravitational constant, 980 cm/sec^2), v is the average velocity in cm/sec, D is the diameter of the vessel in centimeters, and η is the viscosity in gm $\cdot$ sec/cm^2. Other factors being constant, an increase in velocity is likely to result in turbulence. Writing the equation in terms of flow, $\dot{Q}$, rather than velocity gives

$$Re = \frac{4 \, \rho \, \dot{Q}}{\pi \, D \, \eta} \tag{12}$$

Because blood density is reasonably constant, the factors tending to cause turbulence are an increase in the rate of blood flow, such as occurs in exercise; a decrease in blood viscosity, as seen in anemia; and a decrease in the radius of the blood vessel, as occurs with stenotic valves or coarctation of a blood vessel such as the aorta.

The streamline flow of fluid may also break into organized swirls (called *vortex shedding*) and produce sounds of a rather definite pitch at flow velocities much *lower* than those causing fully developed random turbulence. As a fluid flows past an obstacle in the flow stream, a slight disturbance causes the flow wake to fluctuate at first toward one side. Inertial forces continue the stream into a higher-velocity area, which then forces it back in the opposite direction,

where it overshoots and then is forced to swing back to repeat the cycle. The fluttering of a flag in a stiff breeze is produced by the same mechanism. Described another way, the fluid tends to follow the curvature of the obstacle and so curves in a swirl around it to form a vortex. This causes an area of low pressure; hence, the stream tends to flow into that area from the opposite direction. The cycle is repeated indefinitely, with swirls of opposite direction being set up, and vortices shed and carried downstream. Each time the wake of the flow stream swings back and forth, it changes the lateral pressure on the wall of the vessels.

The frequency and location of cardiovascular murmurs under physiological conditions are explainable by this process. Note that the fluctuating wake has a maximal component of fluctuations at right angles to the flow stream. If lateral fluctuations of pressure from the oscillating flow stream are of sufficient magnitude and frequency, audible sound is produced. In this kind of turbulence, relatively more sound is produced with a given driving pressure than in the randomly oriented flow pattern of classic Reynolds turbulence. It can occur under conditions in which Reynolds number is low.

In summary:

1. Factors regulating flow are the pressure gradient along the vessel, the viscosity of the fluid, and the physical dimensions of the resistance vessel.
2. An increase in transmural pressure, by distending the resistance blood vessels, tends to increase flow irrespective of its concomitant effect on the driving pressure.
3. With a constant pressure gradient the smooth muscle

tension (vasomotor tone) acts to reduce the caliber of the resistance vessels and so reduces flow.

4. The apparent viscosity of blood is increased (hence the flow decreases if other factors remain constant) as a result of increased hematocrit ratio, decreased temperature, turbulence, an increase in minute vessel size from approximately 10 to 300 μm, or erythrocyte aggregation if flow rate is reduced toward zero.

MEASUREMENT OF BLOOD FLOW

Understanding of the performance of the circulatory systems requires a measurement of the flow of blood. In measuring flow, two sources of error must always be considered: errors in the measurement itself, and deleterious effects of the measurement process on the functioning of the organism. In cardiovascular research, if highly accurate results are to be obtained, measurements must be made without disrupting the organism by concomitant surgery, anesthesia, pain, impeding the normal blood flow, inducing blood clotting, or causing uncompensated loss of blood. Much has been learned about the cardiovascular system, however, without the necessity of placing such stringent requirements on the measuring process.

The simplest and fundamentally the most accurate way to measure blood flow is to collect a known volume of vascular outflow over a measured time interval by means of a graduate and a stopwatch. This method, although simple, is obviously of no value for measuring cardiac output, because the removal of the blood would seriously affect the animal. However, it is of value in measuring the flow through small amounts of tissue.

The flow of blood from small segments of tissue may be measured by cannulating a vein and counting the number of drops of blood per minute leaving the cannula. If the volume of the drops is known, the flow may be determined. The maximum flow measurable is limited to about 10 ml per minute. Furthermore, the size of the drops is dependent on a multitude of variables; thus, reliable measurement of complex fluids such as blood is difficult.

To measure the pattern of pulsatile blood flow, electromagnetic and ultrasonic flowmeters are available (Geddes and Baker, 1975; Cobbold, 1974). The probes of these instruments can, if desired, be implanted in an experimental animal so that painless measurements can be made at a later date without anesthetics. The artery does not have to be cannulated, since the devices detect the flow through the walls of intact blood vessels.

The *electromagnetic flowmeter* is based on the principle of an electric generator. If a conductor is moved through a magnetic field, an electrical potential is developed in the conductor that is proportional to the length and velocity of the conductor. Blood going through a magnetic field forms a single conductor; consequently, electrodes can be placed on both sides of the blood vessel to pick up the potential developed.

The *ultrasonic flowmeter* is based on the principle that the velocity of sound along a flowing stream in relation to a stationary point is higher going downstream than going upstream. One-half the difference between the two velocities is the stream velocity. By the use of techniques similar to those used for radar and sonar, a pulse of sound is produced, and the time required for it to travel downstream is compared with the time for a similar pulse of sound to be transmitted upstream. The Doppler principle of change in frequency following reflection from moving particles (red blood cells) is currently the principle most widely used. Ultrasonic flowmeters are available to measure blood flow without penetrating the skin.

Indicator dilution provides a way to measure blood flow. An indicator is injected into the bloodstream, samples are taken after mixing, and the flow is computed with respect to the degree of dilution of the indicator. The method is widely used for determining cardiac output and is fully as accurate as more direct methods if proper conditions are met.

The *material conservation principle* is used to account for the indicator injected. If a known amount of a substance is injected into a stream flowing into a particular region, it must either (1) leave the region, (2) accumulate in the region, (3) be converted to some other material, or (4) be excreted by some route other than the flow stream.

The *indicator dilution principle* is based on the fact that an amount of a substance (A) mixed uniformly in a volume (V) has a concentration (C) such that by definition of concentration

$$C = \frac{A}{V} \tag{13}$$

This *dilution method* (see also Chaps. 10 and 23) is used to estimate the volumes of various fluid compartments in the body by rearranging the equation to $V = A/C$. A material is used that equilibrates within the entire fluid compartment in question. Any losses of indicator from the fluid compartment must be accounted for by the material conservation principle.

Indicator dilution can also be used to measure flow. Mean flow ($\bar{Q}$) is volume (V) moved per unit time (t).

$$\bar{Q} = \frac{V}{t} \tag{14}$$

This volume of fluid passing a given point in a unit of time may be estimated by the dilution principle, by substituting $A/C = V$ into equation 14. If the amount of indicator added or removed by an organ is known, and is divided by the change in concentration of a substance as it passes through the organ, the flow can be calculated, as indicated below. The material conservation principle is used to account for the indicator used.

FICK METHOD (CONSTANT INFUSION OR REMOVAL). The blood flow through an organ can be determined by measuring the amount of a substance removed by the organ per minute and dividing by the change in concentration of the substance in the blood as it goes through the organ. For *cardiac output,* for example, measurement of the flow of blood through the heart and thus through the lungs is desired. The transport of oxygen by the lungs provides a convenient indicator agent. The volume of oxygen consumed per minute, A/t, may be measured by means of a spirometer. Oxygen consumption is measured over an interval of 5 to 10 minutes. The concentration of oxygen in the venous blood coming to the right heart and lungs (C_v) and the concentration of oxygen leaving the lungs and left heart in the arterial blood (C_a) are measured to give the change in concentration (ΔC). This change in concentration is the dilution. The *Fick equation* is derived as follows:

$$\dot{A}_a = \dot{A}_v + \dot{A}_r$$

where $\dot{A}_a$ is the amount of indicator (oxygen) leaving via the arteries per unit time, $\dot{A}_v$ is the amount entering via the veins per unit time, and $\dot{A}_r$ is the amount entering the bloodstream via the respiratory system. (It is the rate of oxygen consumption [ml/min] in this case.) Since flow ($\dot{Q}$) times concentration (C) equals amount transported per unit time ($\dot{A}$),

$$\dot{Q}_a C_a = \dot{Q}_v C_v + \dot{A}_r$$

If we can assume that the venous return ($\dot{Q}_v$) equals the cardiac output ($\dot{Q}_a$) and that the arterial and venous indicator (oxygen) concentrations are constant though different, then

$$\dot{Q}_a(C_a - C_v) = \dot{A}_r$$
$$\dot{Q} = \frac{\dot{A}_r}{C_a - C_v} = \frac{A/t}{\Delta C} \tag{15}$$

The technique and calculations are shown schematically in Figure 11-8. The problem to be solved is: If each 100 ml of blood going through the lungs takes up 5 ml oxygen, how much blood must flow per minute to take up 200 ml of oxygen?

Arterial blood samples are obtained by arterial puncture. The major problem with this method is obtaining a valid sample of mixed venous blood. This blood must be representative of all the venous blood. Since the kidneys usually extract only about 2 to 3 volumes of oxygen per 100 volumes of blood, the brain about 6, and the heart and active muscle 10 to 15, the sample must be obtained after mixing. Mixing is adequate only at sites beyond the right ventricle. Cardiac catheterization is thus routinely used to obtain these samples. The catheter is inserted into an arm vein and advanced through the veins into the right heart and on through the pulmonary valve into the pulmonary artery. The procedure is safe, but to the patient it is a rather heroic one, so the apprehension tends to give abnormally high values.

The direct Fick is the standard method of determining the cardiac output in humans but is subject to error, particularly during severe exertion. It is a *determination* of mean flow, since the oxygen consumption is averaged over a 5- to 10-minute period. The venous sample should be taken continuously during this period if accurate results are desired. It is not adequate for measuring rapid changes in cardiac output, since it is impossible to measure the true oxygen consumption over short intervals of time because of the variable amount of air held in the lungs from breath to breath during such changes. Repeated determinations using the Fick and dye dilution methods (to be discussed) agree within a range of ± 15 percent. Much of this variation is probably due to true minute-by-minute variation in the cardiac output.

The indicator dilution method, using the Fick equation, provides a convenient tool for the measurement of mean *organ blood flow* is the indicators used are removed from the blood by these organs. The liver removes Bromsulphalein, and the kidneys remove para-aminohippurate. The brain and heart absorb nitrous oxide to provide arteriovenous differences that enable flow to be measured, assuming that the amount of material accumulated is related to the solubility of the gas in the tissue. The pattern of "washout" of radioactive xenon or krypton may also be used to calculate flow.

STEWART-HAMILTON METHOD (SINGLE INJECTION). The Stewart-Hamilton method is based on the rapid injection of a definite, known amount of indicator. If the amount injected and the mean concentration over a given time (the minute-concentration) are measured, the flow can be calculated. It is in many ways more convenient than, and is fully as accurate as, the direct Fick method. In practice, for

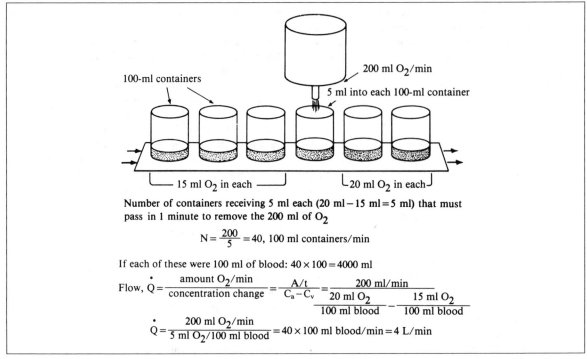

Number of containers receiving 5 ml each (20 ml − 15 ml = 5 ml) that must pass in 1 minute to remove the 200 ml of O_2

$$N = \frac{200}{5} = 40, \; 100 \text{ ml containers/min}$$

If each of these were 100 ml of blood: $40 \times 100 = 4000$ ml

$$\text{Flow, } \dot{Q} = \frac{\text{amount } O_2/\text{min}}{\text{concentration change}} = \frac{A/t}{C_a - C_v} = \frac{200 \text{ ml/min}}{\dfrac{20 \text{ ml } O_2}{100 \text{ ml blood}} - \dfrac{15 \text{ ml } O_2}{100 \text{ ml blood}}}$$

$$\dot{Q} = \frac{200 \text{ ml } O_2/\text{min}}{5 \text{ ml } O_2/100 \text{ ml blood}} = 40 \times 100 \text{ ml blood/min} = 4 \text{ L/min}$$

Fig. 11-8. Derivation of the Fick equation for the measurement of cardiac output or blood flow.

cardiac outputs, a known amount of an indicator such as indocyanine green or albumin labeled with ^{131}I is injected into a vein. It is diluted and mixed with the blood flowing through the heart and lungs. The resulting concentration of indicator in the arterial blood is then measured and the cardiac output calculated.

To understand this procedure, first imagine a flowing stream (Fig. 11-9A). If the total flow containing an added amount of dye (A) is collected over a definite period of time—for example, 1 minute—and the dye is then mixed throughout the total collected volume to give a concentration (C), the volume can be calculated from the basic dilution equation V = A/C. This is the volume per minute, i.e., the rate of flow.

Next, imagine that the dye is uniformly mixed across the flow stream and that a definite proportion (assume one-tenth) of the flow is collected as in Figure 11-9B. Starting before the dye appears and ending after the dye has cleared the sampling site, a volume containing dye is collected for 1 minute and is thoroughly mixed so that the mean concentration may again be determined. If, in the sample, the proportion of the total amount of dye is exactly equal to the pro-

portion of the total flow collected over 1 minute (i.e., one-tenth of each), the total flow can again be calculated. Since the proportionality constants for the dye and the volume can be canceled (Fig. 11-9B), it is not even necessary to know what proportion of the flow is collected. This proportion must, however, be constant throughout the collection, and the dye must be uniformly mixed across the flow stream.

Finally, instead of collecting a sample over a given time interval and mixing, one may analyze individual samples or make a continuous measure of the indicator concentration with a densitometer (Fig. 11-9C). The mean concentration is mathematically calculated (Fig. 11-9D) over the time interval, usually 1 minute. The *Stewart-Hamilton indicator dilution equation* is

$$\dot{Q} = \frac{A}{\int_0^\infty C \, dt} = \frac{A}{\Sigma C \Delta t} \tag{16}$$

where Δt is the fraction of the unit time interval.

The concentrations at various fractions of a minute are added together to give $\Sigma C \Delta t$, the *minute concentration*, a

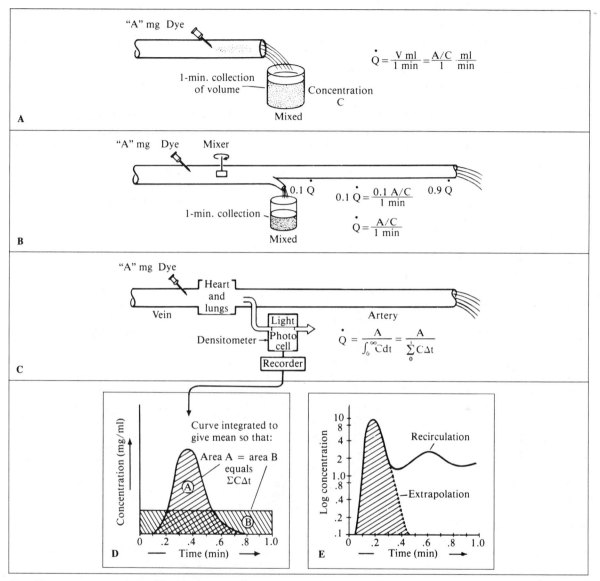

Fig. 11-9. Derivation of the Stewart-Hamilton indicator dilution
equation for the measurement of cardiac output by the dye dilution
method. A. Total flow stream collection. B. Collection of a fraction
of the flow stream with no recirculation. C. Measuring dynamic
concentration changes with a densitometer. D. Record of dye concentration and the derivation of the concentration-time factor,
which is usually the mean concentration during 1 minute (i.e., the
minute-concentration). E. Effect of recirculation and the method of
extrapolation using a semilogarithmic plot of dye concentration.

term somewhat analogous to *respiratory minute volume*. Usually, after less than a minute the concentration has declined to so near zero that in actuality no further additions are made to the sum, called the integral. Because fractional units of time are used, this is a mean concentration averaged over one unit of time—ordinarily a minute.

In practice, a known amount of dye is injected into the bloodstream. It must mix across the total flow stream and, for cardiac output determinations, does so in the right or the left ventricle. A representative fraction of the cardiac output containing changing concentrations of dye is obtained by taking a continuous sample of arterial blood, as shown schematically in Figure 11-9C. The concentration of dye is analyzed and plotted (solid curve, Fig. 11-9E) over an interval long enough for a complete passage of dye through the circulatory system. In calculating the mean concentration of dye over a certain period (the minute concentration), one must calculate the area under the curve either by measuring with an instrument called a planimeter or by summation of instantaneous values. The latter consists in finding the concentration during the first second and adding it to that at the second second, adding to the third, and so summating throughout the whole minute. Dividing this sum by the number of samples measured during the minute, 60, gives the minute concentration. During the first few samples after the indicator injection, the concentration will be zero. If there were no recirculation of dye, there would be no dye in the last samples and so no additions to be made to the sum. This situation is analogous to the calculation of the daily rainfall over 1 year. Some days are without rain, and, for these, a zero amount of rain is added in the calculation.

Recirculation is a potentially serious problem associated with this method. Dye in the arterial side rapidly returns through the venous system to recirculate and be diluted a second time. Recirculation must be eliminated in the calculation of the mean concentration. Since the circulation time through the coronary vessels of the heart is about 15 seconds, recirculation usually distorts the curve. The problem is solved by making the assumption that the tail of the curve follows an exponential decay; that is, dye is washed out at a rate proportional to the amount that is present. This assumption is usually valid, so that plotting a logarithm of the concentration will give a straight downslope if plotted against time. Thus, the downslope of the curve, as shown in Figure 11-9E, may be extrapolated to some value that can be considered to be of no significance and thus called zero in the summing process. (A logarithmic curve approaches but does not reach zero; hence, an arbitrary baseline of about 1 percent of the peak concentration must be established.) Fi-

nally, note that an increase in cardiac ouput, by diluting the dye to a greater extent, results in a decrease in the concentration-time factor. Special-purpose computers are now available to speed the computation.

In the determination of flow by this technique, several major assumptions are made. Conditions must be such that the assumptions are valid, or appropriate corrections are made. (1) The indicator must be adequately *mixed* within the system. The flow at the point of mixing is the flow that is calculated. This condition usually presents no problem in measuring cardiac output, since the ventricles mix the blood vigorously. It does preclude, however, the possibility of injecting dye in the aortic arch and sampling at the femoral artery to obtain accurate information about aortic blood flow. (2) There must be no *loss* (or gain) of indicator. The dye must not be lost in tubing, absorbed on artery walls, or lost by passing through the capillary walls. If indicator is lost, there will be an apparently higher diluting volume and thus a higher indicated flow. (3) In flow determinations, corrections must be made for *recirculation*. There must be enough data on the downward side of the curve to obtain an accurate estimate of the downslope for the extrapolation. Injection of dye near the heart usually solves the problem, but in some clinical conditions, such as heart failure, it may be serious. (4) A *representative* sample must be obtained. That is, the sampling rate should be proportional to the flow at all moments during the curve for the determination. Since the dilution is made over a period of 10 to 20 heartbeats, this source of error is not particularly serious. Fluctuations in cardiac output from respiratory efforts do cause variations, however. The same problem of proportionate sampling exists in obtaining a mixed venous sample for the direct Fick method. (5) An *accurate determination* of concentration must be made. If a single calibration factor is used, the calibration curve must be linear and pass through zero. The detector must respond to the indicator in question and not to extraneous indicator—a difficult feat with external counting when radioisotopes are used. Good practice dictates that duplicate or triplicate determinations be made.

Both the dye dilution and Fick methods measure the mean flow over a period of many seconds to several minutes. There are, therefore, variations from one determination to the next because of variations in cardiac output from moment to moment. The respiratory cycle is a particularly important source of this type of error, since the cardiac output often fluctuates in response to changes in intrathoracic pressure and respiratory center activity.

A discussion of the factors influencing cardiac output in man is presented in Chapter 15.

BIBLIOGRAPHY

Bergel, D. H. (ed.). *Cardiovascular Fluid Dynamics.* New York: Academic, 1972. Vols. 1 and 2.

Bloomfield, D. A. (ed.). *Dye Curves: The Theory and Practice of Indicator Dilution.* Baltimore: University Park Press, 1974.

Charm, S. E., and Kurland, G. S. *Blood Flow and Microcirculation.* New York: Wiley, 1974.

Cobbold, R. S. C. *Transducers for Biomedical Measurements: Principles and Applications.* New York: Wiley, 1974.

Geddes, L. A., and Baker, L. E. *Principles of Applied Biomedical Instrumentation* (2nd ed.). New York: Wiley, 1975.

Gow, B. S. Circulatory Correlates: Vascular Impedance, Resistance, and Capacity. In D. F. Bohr, et al (eds.), *Handbook of Physiology.* Bethesda, Md.: American Physiological Society, 1980. Section 2: The Cardiovascular System, vol. II, Vascular Smooth Muscle. Pp. 353–408.

Lassen, N. A., and Perl, W. *Tracer Kinetic Methods in Medical Physiology.* New York: Raven, 1979.

Lipowsky, H. H., Usami, S., and Chien, S. In vivo measurements of "apparent viscosity" and microvessel hematocrit in the mesentery of the cat. *Microvasc. Res.* 19:297–319, 1980.

McDonald, D. A. *Blood Flow in Arteries* (2nd ed.). Baltimore: Williams & Wilkins, 1974.

Milnor, W. R. *Hemodynamics.* Baltimore: Williams and Wilkins, 1982.

Noordergraaf, A. Hemodynamics. In H. P. Schwann (ed.), *Biological Engineering.* New York: McGraw-Hill, 1969. Pp. 391–545.

Replogle, R. L., Meiselman, H. J., and Merrill, E. W. Clinical implications of blood rheology studies. *Circulation* 36:148–160, 1967.

Schmid-Schönbein, H. Microrheology of erythrocytes, blood viscosity and distribution of blood flow in the microcirculation. *Int. Rev. Physiol.* (Cardiovascular Physiology II) 9:1–62, 1976.

MICROCIRCULATION

12

Julius J. Friedman

The cardiovascular system is organized for the maintenance of a homeostatic environment for the tissues of an organism. Cardiac and peripheral vascular functions are coordinated to transport blood to and from the capillary-venular networks where exchange of nutrients and cell products takes place between blood and tissue fluids.

ANATOMICAL ORGANIZATION

A microcirculatory unit consists of blood vessels in serial and parallel arrangements: arterioles, metarterioles, precapillary sphincters, capillaries, arteriovenous anastomoses, venules, and collecting venules.

In the mesoappendix of the rat these vessels are arranged as diagrammed in Figure 12-1. Blood flow enters the microcirculatory unit through the arteriole, which terminates as terminal arterioles or metarterioles. The metarteriole differs from the arteriole in that the smooth muscle cells in its vessel wall are not continuous.

Numerous capillaries originate from metarterioles and terminal arterioles, and they branch and anastomose repeatedly to produce a great increase in surface area. At the entrance of many capillaries are the precapillary sphincters, which regulate the number of open capillaries and thus the effective surface area of the capillary-venular network. Capillaries merge to form venules, which then form collecting venules to drain the microcirculatory unit. While it is functionally correct to regard the microcirculation as consisting of units, extensive arteriolar and venular anastomoses provide extensive collateral circuits.

FUNCTIONAL ORGANIZATION

Functionally, the series and parallel coupled components of the microcirculation may be classified as *resistance vessels, exchange vessels, shunt vessels,* and *capacitance vessels.*

RESISTANCE VESSELS. Resistance to flow is present in all blood vessels, precapillary as well as postcapillary. The *precapillary resistance elements* are the small arteries, ar-

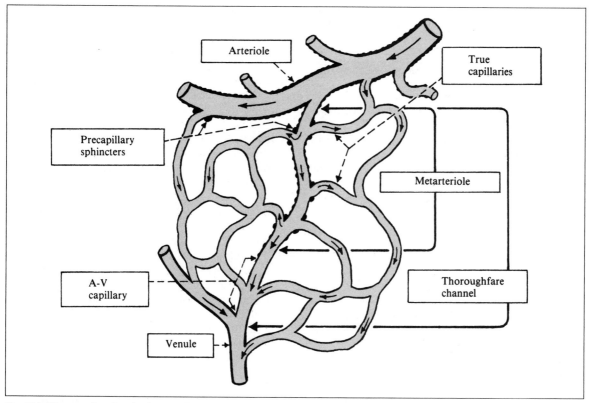

Fig. 12-1. Diagram of a microcirculatory unit in the rat. A-V = arteriovenous. (After B. W. Zweifach. *Factors Regulating Blood Pressure*. New York: Josiah Macy, Jr. Foundation, 1949.)

terioles, and precapillary sphincters. The small arteries and arterioles represent the major variable resistance elements that determine the extent of total tissue blood flow. The precapillary sphincters, on the other hand, by adjusting the number of capillaries open, determine the distribution of capillary blood flow, the extent of capillary flow velocity, capillary surface area, and the mean extravascular diffusion distance. In short, precapillary sphincter activity exerts a great influence on the nature of transcapillary exchange.

Postcapillary resistance vessels include the muscular venules and small veins. Although they show only small changes in resistance, their strategic position, immediately postcapillary, enables them to influence capillary pressure markedly. The ratio of precapillary to postcapillary resistance primarily determines capillary hydrostatic pressure.

EXCHANGE VESSELS. Effective exchange between the vascular and extravascular compartments occurs across capil-

laries and venules. These vessels are uniquely suited for exchange by virtue of their high surface area/volume ratio and their thin walls. Moreover, since exchange vessels are usually located within 20 to 50 μm of tissue cells, diffusion distances are minimal. Greater exchange occurs across the venous end of the exchange vasculatures, because it is more permeable to water and solutes than is the arterial end. In different tissues the capillary walls possess varying relative permeabilities.

SHUNT VESSELS. All elements that bypass the effective exchange circulation of a tissue serve as shunt vessels. They include arteriovenous anastomoses and preferential (thoroughfare) channels. Except for those in the skin that subserve temperature regulation, their precise functional significance is unclear. When effective shunts are present, some fraction of the total flow to an organ or region will not participate in exchange.

CAPACITANCE VESSELS. The small veins, because of their high compliance and large number, are the major capacitance elements of the vascular system. According to estimates the venous system contains about 70 percent of the total blood volume.

FUNCTIONAL ACTIVITY OF THE MICROCIRCULATION

NORMAL BLOOD FLOW. In contrast to the laminar flow seen in small arterial and venous vessels, laminar flow characteristics in the microcirculation are less evident. Red cells flow through capillaries in single file at varying rates and are usually distorted because of a restricted vessel lumen. Arteriolar flow velocity has been estimated at 4.6 mm per second, and venular flow velocity, at 2.6 mm per second. Red blood cell velocity in the capillary is estimated to be about 1 mm per second.

INTERMITTENCY OF BLOOD FLOW. Direct examination of the microcirculation reveals that flow is very intermittent.

Flow in any single capillary depends primarily on the state of the precapillary sphincters and arterioles. These elements exhibit alternate phases of contraction and relaxation with a period of about 30 seconds to several minutes. Such phasic contractile activity has been termed *vasomotion.* Vasomotion determines the distribution of blood flow between various regions of the capillary bed; vasomotion is the response of vascular smooth muscle to local chemical, myogenic, and neurogenic influences.

Phasic contractions similar to vasomotion are also seen in arterioles. However, arteriolar vasomotion is probably influenced more by central nervous impulses traveling through the sympathetic outflow to the arterioles than by local humoral influences.

TRANSCAPILLARY FLUID AND MOLECULAR TRANSPORT

Normal tissue function requires that water and solutes pass between blood and tissue. This transfer of material across the capillary-venular wall occurs by means of filtration, diffusion, and cytopempsis. The rate and extent of transcapillary movement depend on the nature of the molecule and of the capillary structure.

CAPILLARY STRUCTURE. Examination of capillaries by electron microscopy shows them to consist of a single layer of *endothelial cells.* At the outer surface of the endothelial cells is an amorphous mucopolysaccharide matrix called the *basement membrane,* which forms a layer about 500 Å thick, contributing to the large molecular permeability characteristics of the capillary. Capillaries have been classified according to the nature of their endothelial perforations (fenestrations) and the extent of their basement membrane and perivascular investment. Capillaries of muscle, skin, heart, and lung (Fig. 12-2A) have an unperforated endothelium, a prominent continuous basement membrane, and substantial pericapillary investment. Liver capillaries (Fig. 12-2B) possess large endothelial perforations, a faint discontinuous basement membrane, and no significant pericapillary investment. Intestinal capillaries (Fig. 12-2C)

Fig. 12-2. Diagrammatic representation of the electron microscopic characteristics of capillaries from different tissues. A. Muscle. B. Liver. C. Intestine. EC = endothelial cell; IC = intercellular gap; L = vessel lumen; V = vesicle; F = fenestration; BM = basement membrane; P = pericyte; R = red cell; S = extravascular space. (After H. S. Bennett, J. H. Luft, and J. C. Hampton. Morphological characteristics of vertebrate blood capillaries. *Am. J. Physiol.* 196:381–390, 1959.)

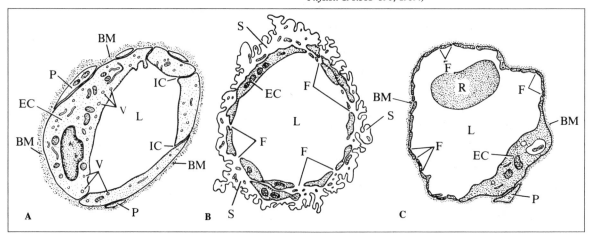

possess thin, fenestrated endothelial walls with a continuous but indistinct basement membrane and only a slight perivascular investment.

CAPILLARY PORES. Transcapillary exchange of lipid-insoluble molecules, such as electrolytes, glucose, and amino acids, takes place in part through aqueous-filled "pores," which are associated with the intercellular spaces. These spaces are, on the average, about 200 Å wide, with one or two foci of narrowing of the intercellular space where the adjacent endothelial cell membranes become closely approximated and appear to fuse. However, fusion is incomplete, and effective openings of about 40 Å in radius are present. These small openings constitute less than 0.02 percent of the capillary surface. Despite the limited area of their openings, they provide sufficient capacity to permit very rapid exchange of small molecules (mw < 10,000), primarily by diffusion. In addition to these small pores, large pores (radius 250 Å) also exist that are associated primarily with the venous end of the exchange system. It is estimated that for every large pore there are approximately 30,000 small pores.

Fenestrations (500–1000 Å), which may be considered as giant pores, when present, must certainly be prominent elements in capillary permeability. Capillaries of the liver, with numerous large fenestrations, possess a relatively high permeability. Capillaries of the intestine and kidney, which have intermediate-sized fenestrations, also have intermediate permeabilities. Capillaries of muscle, skin, and heart, which have no fenestrations, have low permeabilities.

CYTOPEMPSIS (VESICULAR TRANSPORT). Electron microscopy reveals the presence of abundant vesicles 500 to 800 Å in diameter in the endothelial cytoplasm and at the endothelial membrane. There are also numerous invaginations at each surface of the endothelial membranes. These vesicles presumably participate in transcapillary transport by engulfing plasma at the vascular border of the membrane, migrating through the cell cytoplasm by means of Brownian movement, and discharging their contents at the interstitial border of the endothelial cell. Vesicles are not very efficient at transporting small molecules, since these are rapidly exchanged across the capillary wall by diffusion. They may be important in the transcapillary transport of large molecules such as the plasma proteins.

TRANSCAPILLARY FLUID MOVEMENT. Net transcapillary fluid movement (J_v) is determined by the relationship between capillary and interstitial hydrostatic and oncotic forces acting across the capillary wall, as first described by Starling. This movement may be formulated as

$$J_v = K_f [(P_c - P_i) - \sigma(\pi_c - \pi_i)]$$

The difference between capillary (P_c) and interstitial fluid (P_i) hydrostatic pressures represents the transcapillary hydrostatic pressure gradient (ΔP). The difference between plasma (π_c) and interstitial fluid (π_i) oncotic pressures, as modified by the albumin osmotic reflection coefficient (σ), represents the effective transcapillary oncotic gradient ($\sigma\Delta\pi$). K_f is the capillary filtration coefficient.

CAPILLARY FILTRATION COEFFICIENT. The capillary filtration coefficient (K_f) is defined as the volume of fluid filtered in 1 minute by 100 gm of tissue for a change of 1 mm Hg in capillary pressure. K_f reflects the status of hydraulic conductivity and of surface area of the capillary wall.

The filtration of water through the capillary wall of the vessels of the human forearm is approximately 0.0057 ml per minute × 100 gm of tissue × capillary pressure (cm H_2O). Fluid filters from the kidney, intestine, and liver at rates considerably higher than those from muscle or skin. However, because of the great mass of muscle and skin, the total body filtration rate resulting from elevations of central venous pressure is approximately 0.0061 ml per minute × 100 gm of tissue × venous pressure (cm H_2O). In an adult, an increase in central venous pressure of 10 cm H_2O will create a loss of 250 ml of fluid from plasma in 10 minutes.

CAPILLARY FORCES. Capillary Hydrostatic Pressure. Under normal circumstances, capillary hydrostatic pressure probably exerts the major part of the control over transcapillary fluid movement. Capillary pressure (P_c) is determined by arterial pressure (P_a), peripheral venous pressure (P_v), precapillary resistance (r_a), and postcapillary resistance (r_v) and may be formulated as

$$P_c = [(r_v/r_a) P_a + P_v] / [1 + (r_v/r_a)]$$

Thus, elevations of P_a, P_v, and r_v or reductions of r_a will produce elevations of P_c. Since r_a is usually four to five times greater than r_v, about an 80 to 85 percent change in P_v is reflected in the capillaries compared with about a 15 to 20 percent change in P_a. This apportionment will be modified by changes in either r_a or r_v.

Plasma Colloidal Osmotic (Oncotic) Pressure. The major contributors to plasma oncotic pressure are the plasma proteins, albumin, and the globulins (Table 12-1). Albumin, the

Table 12-1. Osmotic pressure of plasma proteins

Protein	Molecular weight	Concentration (gm/dl)	Osmotic pressure (mm Hg)
Albumin	68,000	5.0	20
Globulins	200,000	1.5	5
Fibrinogen	500,000	0.5	1

smallest and most concentrated of the plasma proteins, makes the largest contribution. In addition, approximately 30 percent of the total effective osmotic pressure results from the unequal distribution of electrolytes in an ionized colloidal system caused by the Gibbs-Donnan phenomenon (see Chap. 1).

INTERSTITIAL FORCES. The interstitial space is connective tissue that consists of a collagen fiber framework with a ground substance between the fibers formed from water, salts, proteins, and glycosaminoglycans (mucopolysaccharides). The collagen fibers resist changes in tissue configuration and volume and also absorb macromolecules. The most important constituent of the glycosaminoglycans is hyaluronic acid. Because of its coiled configuration, hyaluronic acid occupies a space 1000 times greater than its own unhydrated size. At the concentration existing in tissues, the hyaluronic acid acts as a gel.

Interstitial Fluid Hydrostatic Pressure. The hydrostatic pressure of the interstitial fluid (P_i) has been estimated to range from −7 to +1 mm Hg, depending on the technique used. This pressure is determined by the volume of fluid in, and the compliance of, the interstitial space. On the average, interstitial fluid volume is about 15 to 20 percent of total body weight. Variations in this volume depend on the relationship between inflow (capillary filtration) and outflow (lymph flow). The compliance of the interstitial space is determined by the presence of collagen and the state of hydration. At low states of hydration the interstitial matrix behaves as a gel, and compliance is low. Thus, small changes in interstitial fluid volume will result in relatively large changes in interstitial fluid pressure. With increased hydration the gel continuity is disrupted and free fluid spaces result, imparting a relatively high compliance to the interstitial space. In this situation, large changes in interstitial fluid volume will produce relatively small changes in interstitial fluid pressure.

Interstitial Fluid Oncotic Pressure. Interstitial fluid oncotic pressure is established by the plasma proteins in the interstitial space. The average interstitial protein concentration is about 40 percent that of plasma. Thus, at a plasma protein concentration of 6 gm per deciliter, interstitial protein concentration would be about 2.4 gm per deciliter. This concentration will vary with changes in transcapillary fluid and protein transport. If capillary filtration is increased, interstitial protein concentration will be reduced as some of the protein is washed out of the interstitial space. If, on the other hand, capillary permeability to protein increases, the concentration of interstitial protein will increase as well.

A protein concentration of 1 gm per deciliter develops an oncotic pressure of about 3 mm Hg; thus, tissue oncotic pressure should be about 7 to 8 mm Hg. However, the process of estimating interstitial oncotic pressure based on protein concentration alone is difficult, because interstitial protein is not uniformly distributed. The presence of hyaluronic acid creates islands of protein-free fluid. One gram of hyaluronic acid can hold as much as a deciliter of water exclusive of albumin. Thus, effective interstitial oncotic pressure is probably higher than 7 to 8 mm Hg.

NORMAL TRANSCAPILLARY FLUID MOVEMENT. Transcapillary exchange of fluid under normal conditions is determined primarily by the balance between capillary pressure and plasma oncotic pressure. However, this balance can be modified significantly by changes in interstitial fluid hydrostatic and oncotic pressures. When capillary hydrostatic pressure exceeds plasma oncotic pressure, filtration results. When capillary hydrostatic pressure is less than plasma oncotic pressure, absorption occurs. Filtration takes place at the arterial end of the capillary, where the hydrostatic pressure exceeds oncotic pressure. As blood moves along the exchange vessels, hydrostatic pressure decreases, while plasma oncotic pressure increases until filtration stops, and absorption from the interstitial space into the vascular lumen occurs.

An alternative concept of transcapillary fluid movement considers the effect of vasomotion. During the dilator phase of vasomotion, capillary hydrostatic pressure is high, and filtration occurs along the entire capillary, whereas the constrictor phase of vasomotion produces a reduction in pressure and permits absorption to take place in the same vessel.

VARIATIONS IN TRANSCAPILLARY MOVEMENT

The net exchange of fluid across the capillary membrane may be influenced by variations in either capillary or tissue hydrostatic pressure, plasma or tissue protein concentra-

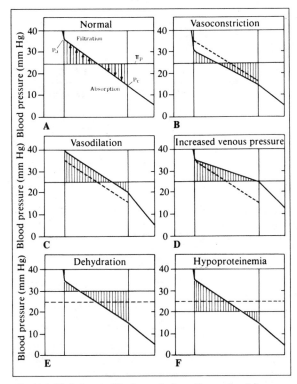

Fig. 12-3. Variations in filtration and absorption produced by changes in arterial and venous pressures and resistances and in plasma colloidal osmotic pressure. The broken lines represent normal levels. (See text for further reference.)

tion, lymphatic drainage, capillary permeability, or capillary surface area.

ALTERATIONS IN CAPILLARY HYDROSTATIC PRESSURE. The normal balance between filtration and absorption is illustrated in Figure 12-3A. For simplicity, changes in the interstitial forces are not considered in the illustration. A reduction in capillary pressure (Fig. 12-3B) results in reduced filtration and increased absorption. Increasing capillary pressure (Fig. 12-3C) increases filtration and reduces absorption. Marked increases in filtration and reductions in absorption follow elevations of venous pressure or postcapillary resistance (Fig. 12-3D).

Changes in Plasma Colloidal Osmotic (Oncotic) Pressure. A reduction of plasma albumin concentration reduces the colloidal osmotic pressure of plasma. A reduction in the absorption force throughout the length of the capillary follows, so that filtration increases and absorption decreases

(Fig. 12-3F). In conditions of hyperproteinemia or dehydration, the plasma protein concentration is elevated, and consequently plasma oncotic pressure is elevated (Fig. 12-3E). This change reduces the filtration force at the arterial end of the capillary and increases the absorption force at the venous end, so that a net movement of fluid from the interstitial compartment into the circulation takes place, and circulating plasma volume tends to rise.

Changes in Tissue Pressure. Either enhanced filtration or reduced lymphatic drainage leads to an increase in interstitial fluid volume and tissue pressure. An elevation in tissue pressure increases the force of absorption throughout the capillary. Thus, filtration at the arterial end of the capillary declines, while absorption at the venous end increases, and further filtration of fluid from the circulatory system is impeded. In this respect, excessive filtration of fluid from the circulation is somewhat self-limiting. However, before tissue pressure achieves such a high level, filtration must exceed lymphatic drainage for a sufficient length of time or by a sufficient magnitude. In such a situation, interstitial fluid accumulation becomes excessive, and edema develops.

CHANGES IN CAPILLARY PERMEABILITY. The oncotic pressure exerted by the plasma proteins is effective only as long as the transport of the proteins through the capillary wall is adequately restricted. Under circumstances of increased capillary permeability, protein molecules are able to diffuse through the capillary endothelium at a more rapid rate than normal. This increased rate of diffusion results in a reduction of the plasma oncotic pressure and an increase in tissue oncotic pressure. Consequently, the absorptive force in the circulation is reduced and the relative filtration force is enhanced. In conditions of increased capillary permeability, then, filtration of fluid from the circulation into the extravascular spaces increases markedly, and the volume of circulating plasma declines.

TRANSCAPILLARY SOLUTE TRANSPORT

The net movement of molecules in solution occurs by means of passive diffusion as described by the Fick relationship (see Chap. 1).

Diffusion of molecules across a porous barrier such as the capillary wall is somewhat restricted; this diffusion is described by the expression

$$J_s = PS\Delta C$$

Table 12-2. Permeability of mammalian muscle capillaries to lipid-insoluble molecules

Substance	Molecular weight (gm)	Diffusion (D) (cm²/sec)	Molecular radius (approx.) (cm)	Permeability (P) cm/sec / 100 gm
H_2O	18	3.4×10^{-5}	1.5×10^{-8}	28×10^{-5}
NaCl	58	2.0	2.3	15
Urea	60	1.95	2.6	14
Glucose	180	0.90	3.7	6
Sucrose	342	0.70	4.8	4
Raffinose	504	0.64	5.7	3
Inulin	5,500	0.24	13.0	0.3
Myoglobin	17,000	0.17	19.0	0.1
Serum albumin	67,000	0.085	36.0	0.001

Source: Modified from E. M. Landis and J. R. Pappenheimer. Exchange of Substances Through the Capillary Walls. In W. F. Hamilton (ed.), *Handbook of Physiology*. Washington, D.C.: American Physiological Society, 1963. Section 2: Circulation. Vol. 2, pp. 1035–1073.

where J_s is the solute transport, P the permeability, S the surface area, and ΔC is the concentration gradient across the barrier.

Capillary permeability depends in part on the relationship between the physical, chemical, and electrical properties of the molecule compared with those of the barrier. Because the capillary membrane is composed substantially of lipid, lipid-soluble molecules utilize the entire exchange surface area in accordance with their oil/water partition coefficients. Thus, their permeability is high, and the extent of transport is limited primarily by the rate of delivery to the exchange surface (flow limited). Lipid-insoluble molecules presumably travel through the barrier by a number of pathways: intercellular junctions, fenestrae, leaks, vesicles, and vesicular channels. For small molecules, these pathways provide little resistance, and exchange is flow limited. For larger molecules the barrier is progressively restrictive as molecular weights increase, and the extent of exchange is limited primarily by diffusion (diffusion limited). The relationship between molecular size and capillary permeability in skeletal muscle is presented in Table 12-2.

Transcapillary solute transport also takes place by means of convection (filtration) due to "solvent drag." As plasma fluid is filtered, the smaller molecules move along with the solvent at their plasma concentration (bulk flow). Large molecules are partially restricted because of their geometric shape and are separated to some degree from the filtered fluid (molecular sieving). Thus, the filtered concentration of larger molecules is less than that of plasma.

The relationship for J_s that includes both convective and diffusive components, as derived by Kedem and Katchalsky, is

$$J_s = (1 - \sigma) J_v \bar{C}_p + PS\Delta C$$

where σ is the solvent drag reflection coefficient, J_v the filtration rate, and $\bar{C}_p$ the average solute concentration within the transport pathway (channel or pore).

TRANSCAPILLARY OXYGEN EXCHANGE

Effective nutrition of tissue depends on the mutual participation of adequate tissue blood flow, blood flow distribution, and transcapillary exchange. The capillaries are distributed throughout the tissue so that the average diffusion distance from capillary to cell is probably about 10 μm. This short diffusion distance is critical to effective tissue nutrition, since the time for transport by diffusion varies inversely with distance. At a distance of 10 μm, approximately 0.01 second is required for oxygen to reach 90 percent of its equilibrium value, whereas at a distance of 1 mm, about 100 seconds are necessary.

Figure 12-4A illustrates the influence of tissue blood flow, oxygen consumption, and diffusion distance on the capillary and interstitial PO_2. Blood PO_2 is greatest at the arterial end of the capillary and decreases in an exponential manner as the blood flows through the capillary, reaching its lowest level at the venous end. The PO_2 at any point along the capillary depends on the relationship between blood flow rate and oxygen extraction by the tissue. Interstitial PO_2 is

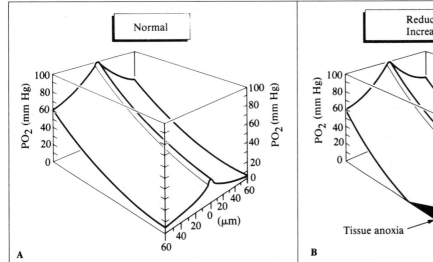
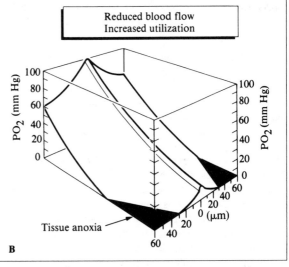

Fig. 12-4. The longitudinal and radial concentration gradients for oxygen in and around the capillary during (A) normal and (B) reduced blood flow or increased utilization. (After G. Thews. Gaseous Diffusion in the Lungs and Tissues. In E. B. Reeve and A. C. Guyton [eds.], *Physical Bases of Circulatory Transport: Regulation and Exchange*. Philadelphia: Saunders, 1967, p. 335.)

determined by the relationship between supply and utilization. Since supply is by way of diffusion, and since the effectiveness of diffusive transport diminishes rapidly with distance, the PO_2 at any point in the interstitial space is defined by the level of PO_2 in the adjacent capillary segment and its distance from that capillary.

At normal flow, consumption, and microvascular tone, the PO_2 profile in and around the capillary is as illustrated in Figure 12-4A, which reveals that all of the tissue is oxygenated to an adequate extent. As blood flow decreases, as consumption increases, or as diffusion distance increases because of precapillary vasoconstriction, PO_2 in the tissue can decline to hypoxic and anoxic levels that are incompatible with tissue integrity (Fig. 12-4B). In such a situation, local factors elicit some degree of precapillary vasodilation, which, by increasing the number of open capillaries, reduces diffusion distance and restores some oxygenation to the deprived tissue.

LYMPH AND LYMPHATICS

Large molecules that reach the interstitial space are not effectively removed by back diffusion through the exchange vessels. If they were to accumulate in the interstitial space, they would exert sufficient effective interstitial oncotic pressure to upset transcapillary fluid exchange, and excessive fluid would accumulate in the interstitial space. Under normal conditions this accumulation does not occur.

Filtered fluid and other plasma constituents that reach the extravascular spaces are drained and returned to the circulatory system by way of the lymphatic network. In addition to this drainage function, the lymphatic system possesses concentrated areas of reticuloendothelial cells that remove bacteria and foreign material from the lymph circulation. This action is one of the body's major protective mechanisms against the invasion of harmful agents. The lymphatics also serve as a transport system for a number of materials, such as vitamin K and lipids absorbed from the intestine.

The lymphatic network originates in the tissue spaces as very thin, closed endothelial tubes (lymphatic capillaries), which have the same relation to the tissue spaces as have blood capillaries. However, lymphatic capillaries are far more permeable to large particles than are blood capillaries. Although the lymphatic capillaries consist of endothelial cells, their porosity is apparently so great as to provide little or no resistance to the passage of particles as large as plasma proteins into the lymphatic system. The lymphatic capillaries coalesce to form larger lymphatic vessels that possess valves to ensure a unidirectional flow of the lymph.

LYMPH FORMATION. *Lymph* is defined as the fluid that returns to the circulation from the tissue space by way of the lymphatics. Since lymph originates from plasma as the difference between filtration and absorption of fluid across the capillary wall, its composition is very similar to that of plasma, and all influences that modify transcapillary exchange also modify the formation and composition of lymph. Except for variation in protein concentrations and electrolytes associated with the Gibbs-Donnan equilibrium, lymph and plasma are almost identical. The average protein concentration of lymph is approximately 2 to 3 percent, compared with 6 percent in plasma. In addition, the concentrations of the different protein fractions in lymph differ from those in plasma, due to restricted diffusion and molecular sieving. Evidence of these restrictions is found in the ratio of albumin to globulin (A/G) in lymph and in plasma. The plasma A/G ratio is about 1.27, whereas the (A/G) ratio of lymph is 1.35.

In general, the protein concentration of lymph varies inversely with the rate of formation. With increased filtration from the capillaries, more fluid in relation to protein is cleared from the circulation; and although the quantity of protein filtered increases, the protein concentration in the interstitial compartment is reduced.

LYMPH FLOW. The rate of lymph flow in mammals is very low under normal circumstances, amounting to only about 10 percent of the filtered volume. Furthermore, it varies between tissues in accordance with local transcapillary dynamics. The flow of lymph in the thoracic duct is approximately 1.38 ml/kg per hour, which, in a 24-hour period, represents a volume of fluid equal to the plasma volume. In addition, 50 to 100 percent of the circulating plasma protein is returned to the circulatory system by the lymphatics during the same period. The preceding information demonstrates how this system is essential in maintaining the circulating blood volume.

Lymph flow is maintained by myogenic activity, as a consequence of contraction of neighboring musculature in which the lymphatic channels are compressed, as well as by the negative intrathoracic pressure. Although increased respiratory or muscular activity enhances lymph flow, the major limitation—and therefore the major determinant—of the flow is lymph formation. Consequently, any circumstance that increases the rate of filtration of fluid from the capillaries also increases the rate of lymph flow. Raising

capillary pressure by either arterial vasodilation or venous constriction enhances the rate of lymph flow.

Intravascular infusions of various solutions enhance the formation of lymph and lymph flow in two ways. First, the solutions increase the volume of fluid in the circulatory system and thereby raise arterial and venous blood pressure, so that capillary pressure is increased and filtration augmented. Moreover, the expanded volume of fluid in circulation dilutes the plasma proteins, effectively reducing the plasma colloidal osmotic pressure, thus enhancing filtration and reducing absorption.

Cardiac muscular contraction, peristaltic action of the intestinal smooth muscle, or voluntary muscular contraction massages the lymphatic channels, and the alternate compression and relaxation exerted by the contracting muscles propel lymph centrally, since the valves in the lymphatic channels prevent retrograde flow. Lymph flow is also augmented by increased tissue activity independent of external muscular contraction. The heightened tissue activity results in the accumulation of metabolites, presumably causing the arterioles and precapillary sphincters to dilate and the capillary endothelium to become somewhat more permeable, so that blood flow into the capillary network is enhanced at the same time that permeability increases. Eating results in lymph flows as high as 5.8 ml/kg per hour.

The activity of the microcirculation and the relationship between the systemic and lymphatic circulations are illustrated in Figure 12-5. At a cardiac output of 5.8 liters per minute, about 15 ml per minute is filtered across the capillary wall from the circulation, and 12 ml per minute is reabsorbed. The 3 ml remaining in the interstitial space is returned to the vascular system by way of the lymphatics. This lymphatic contribution does not seem important until one realizes that it amounts to about 4 liters of fluid per day.

Clearly, Figure 12-5 shows that diffusional exchange is the process whereby solute materials such as glucose are supplied to the tissue, since the amount of glucose that could be supplied through filtration is inadequate to satisfy the tissue utilization and is insignificant compared with the diffusional exchange. It is evident that one of the major functions of the lymphatic system is to return protein to the circulation. The lymphatics handle as much as 200 gm of protein per day, which is almost equal to the total intravascular mass of proteins.

EDEMA

Edema is the condition of excess accumulation of fluids in the tissue spaces. It is of functional significance in that the presence of excess fluid in the interstitial space modifies

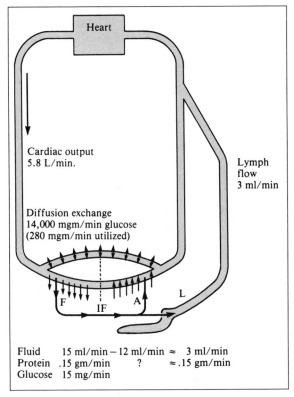

Heart

Cardiac output
5.8 L/min.

Lymph flow
3 ml/min

Diffusion exchange
14,000 mgm/min glucose
(280 mgm/min utilized)

F   IF   A   L

Fluid	15 ml/min − 12 ml/min	≈	3 ml/min
Protein	.15 gm/min	?	≈ .15 gm/min
Glucose	15 mg/min		

Fig. 12-5. The relationship between transcapillary exchange of fluid and solute and lymphatic function. (After E. M. Landis and J. R. Pappenheimer. Exchange of Substances Through the Capillary Walls. In W. F. Hamilton [ed.], *Handbook of Physiology.* Washington, D.C.: American Physiological Society, 1963. Section 2: Circulation. Vol. 2, p. 987.)

microvascular flow distribution and retards the exchange of nutrients and metabolites between cells and plasma. Edema results when net filtration of fluid from the circulation exceeds the lymphatic drainage. Edema formation requires either increases in venous pressure greater than 10 mm Hg or reductions in plasma oncotic pressure of at least 50 percent. Thus, there exists in the body a "margin of safety" against edema. This safety margin is manifested by the following: (1) increased P_i and reduced π_i caused by the filtered fluid; (2) increased lymphatic myogenic pumping due to the increased P_i, which minimizes the increase in interstitial volume and washes out interstitial protein; and (3) myogenic vasoconstriction of the arteriolar and precapillary sphincter smooth muscle, which reduces both capillary hydrostatic pressure and capillary surface area.

Although elevation in capillary pressure or reduction in

plasma oncotic pressure facilitates filtration and may eventually lead to edema, increased capillary permeability and lymphatic blockage cause the most rapid development of severe edema. With increased capillary permeability, plasma proteins permeate the capillary wall more easily and thereby reduce the plasma oncotic pressure, while at the same time they increase the tissue oncotic pressure. Therefore, the accumulation of fluid in the extravascular spaces is doubly augmented.

Capillary permeability increases under a variety of circumstances. Marked increases in capillary volume and pressure created by large intravascular infusions, large doses of vasodepressant drugs, advanced bacterial inflammation, and the production of toxic metabolic products increase the dimensions of the intercellular and intracellular openings. In addition, the toxic metabolites, agents such as histamine, and bacterial toxins appear to act directly on the capillary wall to increase its permeability. In each of these conditions, lymph flow is greatly enhanced. However, the capacity of lymphatic drainage is limited, and when capillary filtration exceeds lymphatic drainage, edema results.

The most exaggerated form of edema, *elephantiasis* of the limbs and scrotum, is caused by blockage of the lymphatics by filarial organisms. Under these circumstances, lymphatic drainage from the tissue spaces is retarded or interrupted. Nevertheless, filtration of fluid proceeds, and interstitial fluid accumulation becomes progressive as the interstitial space accommodates the increased volume through *stress relaxation.* Ultimately, the accumulation of extravascular fluid becomes so great that tissue pressure rises to a level at which it effectively counteracts the hydrostatic pressure in the capillary. To this extent the condition of edema may be said to be somewhat self-limiting. Unfortunately, because of the great distensibility of the extravascular space and because of the large capacity of body cavities in which the various viscera are contained, this point is not achieved until excessive quantities of fluid have been removed from circulation and circulating blood volume declines significantly. Moreover, on account of the marked diffusion gradient for protein between plasma and interstitial fluid, protein molecules continue to diffuse from the circulation into the interstitial spaces despite a reduction in gross filtration. Tissue oncotic pressure then rises and plasma oncotic pressure is reduced, with the result that edema formation may continue even after hydrostatic pressures have essentially been equilibrated. Ultimately, hydrostatic pressure and oncotic pressure equalize across the capillary wall, and no further net movement of fluid takes place. The point of equilibrium is seldom, if ever, achieved.

Edema also occurs in renal disease and is an outstanding

feature of the nephrotic syndrome. In conditions of nephrosis, excessive protein is lost from the circulation into the urine, thereby reducing the plasma protein concentration. The kidney can also contribute to the formation of edema by means of salt and fluid retention. With this retention, the volume of the circulatory and extravascular systems is increased and capillary pressure is elevated, a condition that facilitates the accumulation of fluid in the extravascular space. A more extensive discussion of fluid balance will be found in Chapter 23.

BIBLIOGRAPHY

Aukland, K., and Nicolaysen, G. Interstitial fluid volume: Local regulatory mechanisms. *Physiol. Rev.* 61:556–643, 1981.

Bennett, H. S., Luft, J. H., and Hampton, J. C. Morphological characteristics of vertebrate blood capillaries. *Am. J. Physiol.* 196:381–390, 1959.

Intaglietta, M., and Zweifach, B. W. Microcirculatory basis of fluid exchange. *Adv. Biol. Med. Phys.* 15:111–159, 1974.

Johnson, P. C. The Microcirculation and Local and Humoral Control of the Circulation. In A. C. Guyton and C. E. Jones (eds.), *Physiology, Series One: Cardiovascular Physiology.* London: Butterworth, 1974. Vol. 1.

Karnovsky, M. J. The ultrastructural basis of capillary permeability studied with peroxidase as a tracer. *J. Cell Biol.* 35:213–236, 1967.

Kedem, O., and Katchalsky, A. Thermodynamic analysis of biological membranes to non-electrolytes. *Biochim. Biophys. Acta* 27:229–246, 1958.

Landis, E. M., and Pappenheimer, J. R. Exchange of Substances Through the Capillary Walls. In W. F. Hamilton (ed.), *Handbook of Physiology.* Washington, D.C.: American Physiological Society, 1963. Section 2: Circulation. Vol. 2, pp. 961–1034.

Mayerson, H. S. The Physiologic Importance of Lymph. In W. F. Hamilton (ed.), *Handbook of Physiology.* Washington, D.C.: American Physiological Society, 1963. Section 2: Circulation. Vol. 2, pp. 1035–1073.

Pappenheimer, J. R. Passage of molecules through capillary walls. *Physiol. Rev.* 33:387–423, 1953.

Renkin, E. M. Transport of large molecules across capillary walls. *Physiologist* 7:13–28, 1964.

Ruszynák, I., Foldi, M., and Szabó, G. *Lymphatics and Lymph Circulation.* Elmsford, N.Y.: Pergamon, 1960.

Wiederhielm, C. A. Dynamics of capillary fluid exchange: A nonlinear computer simulation. *Microvasc. Res.* 18:48–82, 1979.

Yoffey, J. M., and Courtice, F. C. *Lymphatics, Lymph and Lymphoid Tissue.* Cambridge, Mass.: Harvard University Press, 1956.

Zweifach, B. W. Quantitative studies of microcirculatory structure and function: I. Analysis of pressure distribution in the terminal vascular bed in cat mesentery. *Circ. Res.* 34:843–857, 1974.

PUMPING ASPECTS OF CARDIAC ACTIVITY

13

Carl F. Rothe

The function of the circulatory system is to maintain an optimum environment for cellular function. The optimum environment requires control of concentrations of nutritive, hormonal, and waste materials, tensions of respiratory gases, and temperature. Because of continuous cellular activity, an optimum environment may be maintained only by an uninterrupted flow of blood to the tissues to renew nutrients and remove wastes. To maintain this continuous flow, then, is the role of the cardiovascular system, and it is the heart that pumps the blood. Failure of this pump is the prime cause of death among adults. The properties of heart muscle have been discussed in Chapter 3, and control of cardiac function will be covered in Chapter 15. This chapter deals with the specific anatomical and physiological properties of the heart that fit it for the role of a pump for the circulation of blood.

The arterial blood pressure of a normal person is held relatively constant. The rate of blood flow through individual tissues is controlled, at the tissue, by changes in the caliber of the resistance vessels (the small arteries and arterioles; see Chaps. 12 and 17). There is central autonomic nervous system modulation of both cardiac function and the peripheral tissue vasculature (see Chaps. 7 and 17). In consideration of normal and pathological function of the cardiovascular system, it is important to note (see Chap. 15) that the cardiac output—the flow of blood from the heart—is determined by the input pressure to the heart (the central venous pressure), cardiac contractility, and the arterial pressure load. Furthermore, although the cardiovascular system is a closed circuit, it is made up of distensible vessels that, in conjunction with the blood volume, influence the filling pressure to the heart. Finally, the cardiovascular system is dual, with a systemic (greater) circuit and a pulmonary (lesser) circuit. The flows through these are normally equal, but they have important differences in characteristics, as will be discussed.

There are four general causes of inadequate cardiac function even when the filling pressure is adequate and the outflow load is normal:

1. *Myocardial.* Pumping may be inadequate because the muscle does not contract vigorously enough. A common

cause of myocardial failure is myocardial ischemia—an inadequate blood supply to the heart muscle itself (see Chaps. 15 and 18).

2. *Anatomical.* Abnormalities of structure may be congenital (present at birth) or acquired from disease or accident. Valve damage is a typical example.

3. *Arrhythmia.* Impulse excitation may not be initiated or conducted adequately, or fibrillation occurs in which contractions are chaotic and asynchronized (see Chap. 14).

4. *Pericardial.* Fluids may accumulate between the heart and pericardium (tamponade), or the pericardium may become constricted and so impede adequate filling.

THE FETUS AND NEWBORN

Nutrient and waste exchange for the fetus occurs via the placenta, rather than the lungs and gastrointestinal tract. The "arterialized" umbilical venous blood (Fig. 13-1) is only about 80 percent saturated with oxygen. It goes to the liver, but most of it passes through the ductus venosus to the vena cava, where systemic venous blood is mixed with it. Blood supplying the body is only about 60 percent saturated. The umbilical flow accounts for about 55 percent of the total cardiac output of the fetus. About one-half of the total blood flow returning to the heart crosses from the right to the left atrium through the foramen ovale, bypassing the right ventricle. The pulmonary vascular resistance of the collapsed lung is high. The pulmonary arterial pressure is about 5 mm Hg greater than the aortic pressure, and so most of the right ventricular output passes through the ductus arteriosus into the aorta, bypassing the lungs. As a consequence, both ventricles share the load of the systemic circulation, while only about 10 percent of the total cardiac output goes through the lungs (Fig. 13-2).

Most of the changes from the fetal pattern of blood circulation (Fig. 13-2) to the adult pattern occur at birth when the placenta ceases functioning. The pattern for the neonate (newborn) is similar to that of the adult except for a patent

Fig. 13-1. Fetal circulation. (Modified from D. Longmore. *The Heart.* New York, McGraw-Hill, 1971. P. 73.)

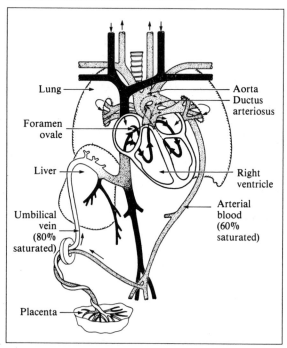

Fig. 13-2. Comparison of patterns of fetal, neonatal, and adult circulation.

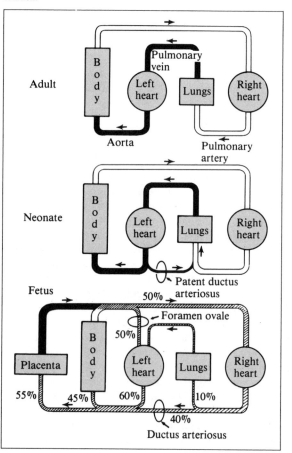

ductus arteriosus. The following complex switchovers occur at birth:

1. The umbilical blood flow is stopped, and the right atrial pressure decreases.
2. As the lungs inflate, the interstitial tissue, including the pulmonary microvasculature, is stretched and so the pulmonary vascular resistance is decreased, allowing blood to pass through the lungs more easily. The blood flow through the lungs increases (see Chap. 19). This increased flow into the left atrium causes its pressure to increase.
3. With the reversal of pressure gradient between the right and left atria, the flap valve at the foramen ovale closes, preventing the shunt flow from the right to the left heart.
4. With the decrease in pulmonary arterial pressure, the pressure gradient between the aorta and the pulmonary artery reverses, causing a reversal of the flow through the ductus arteriosus.
5. With respiration and functioning lungs, the oxygen tension of the arterial blood increases. The increased oxygen tension stimulates the smooth muscles of the ductus arteriosus to contract. However, the effect is slow, for flow through the ductus is not restricted until about 10 minutes after birth and normally requires about 2 days for complete occlusion. The smooth muscle of the ductus arteriosus is almost totally unresponsive to arterial oxygen tension throughout the first two-thirds of gestation, but thereafter it becomes progressively more sensitive.

At birth the right ventricle is as large as the left, as might be expected, since the right heart pumps about one-half of the total flow at aortic pressures. During the first year the right heart grows somewhat, but the left heart grows rapidly to more than double its weight.

FUNCTIONAL ANATOMY OF THE HEART

Anatomical drawings, such as those of Netter (1969), should be consulted for details of the orientation of the great vessels, chambers, and valves of the heart. The dense connective tissue between the atria above and the ventricles below is the fibrous skeleton of the heart, and on it are mounted the four heart valves (Fig. 13-3). Through this skeleton passes the bundle of His, the only structure maintaining the syncytial pathway between atria and ventricles to provide a conduction path for ventricular excitation.

ATRIAL ANATOMY. Above the fibrous skeleton sit the cup-shaped right and left atria. The thin-walled atria provide an easily expandable reservoir for the venous blood returning to the heart while the ventricles are emptying and the entrance valves are closed.

VENTRICULAR ANATOMY. The heavy-walled ventricles pump blood from the low-pressure venous system into the high-pressure arterial distribution system. The ventricles of the heart are a continuum of interdigitating muscle strands, rather than discrete bands or bundles of muscle. The myocardium contracts in the line of its fibers, which are generally oblique to the base-to-apex axis. The fiber orientation in the free wall tends to be parallel to the base-to-apex axis at both the endocardial (inside) and epicardial (outside) surfaces, but then smoothly but rapidly shifts toward a circumferential orientation in the middle of the wall (see Armour and Randall, 1970). The papillary muscles occupy only about 5 to 10 percent of the left ventricle during diastole, but by the end of systole they occupy nearly 30 percent of the remaining chamber volume, tending to obliterate the inflow tract and forming a smooth-walled infraaortic outflow tract.

THE VALVES OF THE HEART. Effective pumping action by the heart demands unidirectional flow from the venous side to the arterial side. This is the function of the four cardiac valves located in the fibrous skeleton of the heart at the entrances and exits of the two ventricular chambers. The *aortic* and *pulmonary* outlet valves, called semilunar valves, located at the exits of the left and right ventricles, are perhaps the simpler pair. They act to prevent the return of blood from the arterial system (Fig. 13-3). Both valves consist of three tough but flexible flaps or half-moon cusps attached symmetrically around the valve rings. A widening or outpouching of the aorta just above the valve ring leads to openings for the two coronary arteries that supply blood to the heart muscle (myocardium). The position of the three aortic valve cusps during their open phase is important, for if the valves were forced back to the aortic wall, blockage of the coronary orifice would occur, causing serious myocardial impairment.

The two atrioventricular (AV) valves (the *tricuspid* on the right side and the *mitral* on the left side) act to prevent the ventricles from ejecting blood back into the atria (Fig. 13-3). Both valves contain two large primary cusps. The tricuspid has an additional small cusp; hence the name. Because the pressure gradient available to fill the heart is low, the large size of the opening is important in not impeding filling. Furthermore, the total area of the cusps is much greater than

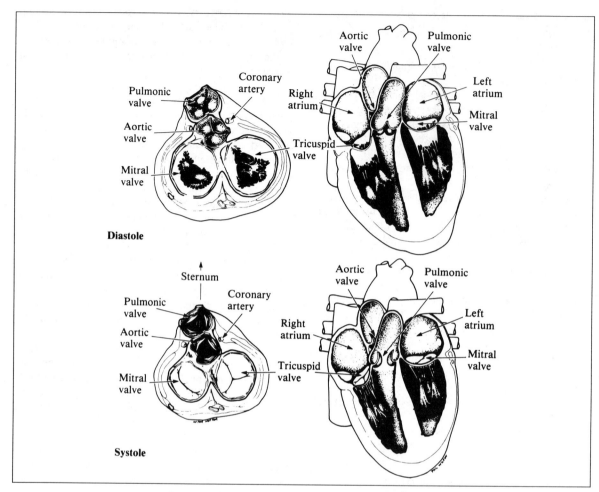

Fig. 13-3. Anatomy of the heart and its valves. (Base views redrawn after those of F. H. Netter. *Heart.* The CIBA Collection of Medical Illustrations. Summit, N.J.: CIBA, 1969, Vol. 5, Pp. 10–11.)

that of the orifice they guard. On account of this large size and the thin nature of the AV valves, added support is required during ventricular contraction. Thus, to the lower side of the cusps are attached the chordae tendineae leading to the papillary muscles on the ventricular walls. The papillary muscles function to prevent eversion of the flexible cusps into the atrium. They contract during ventricular ejection to support the valves. An inadequate papillary muscle tension during ventricular contraction allows bulging into the atria, called eversion.

The valves open passively in diastole whenever the atrial is greater than the ventricular diastolic pressure or, in systole, whenever ventricular pressure exceeds that in the artery. There are two mechanisms involved in valve closure.

The most obvious is a reversed pressure gradient causing a backflow of the blood, which closes the valve. The other mechanism is based on the Bernoulli principle (see Chap. 11). With the jet of high-velocity blood moving through the valve, the lateral pressure is reduced. In addition, eddies of blood swirl in behind (Fig. 13-4). As a result, the leaflets move toward the center of the flow stream. As filling slows, the momentum of the blood tends to continue the motion to close the valve. The timing of the cardiac cycle is such that normally the valves are about closed when the ventricle

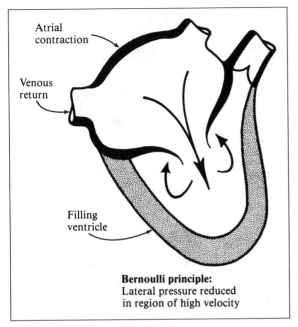

Fig. 13-4. Flowing blood into the heart aids closure of atrioventricular valves (Bernoulli principle).

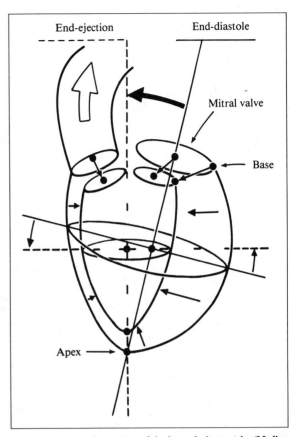

Fig. 13-5. Pattern of movement of the heart during systole. (Modified from J. E. Hinds, et al. Instantaneous changes in the left ventricular lengths occurring in dogs during the cardiac cycle. *Fed. Proc.* **28:**1351, 1969.)

contracts. This mechanism markedly reduces the stresses placed on the extremely thin—but tough—AV valves.

VENTRICULAR PUMPING ACTION. Motion of the heart during contraction is complex (Fig. 13-5). Not only does the heart become smaller and the apex move upward somewhat, but also the base moves toward the apex. In accelerating the mass of blood from the right heart into the arteries, there is an opposing force acting to move the heart in the opposite direction, just as a rocket moves in one direction as gas is ejected in the opposite direction. Because the myocardial fibers are shortening, the valve ring moves caudally, while the apex remains relatively stationary. As shown in Figure 13-5, the axis of the heart also shifts. These movements make observation of and surgery on the beating heart difficult. For the right heart, emptying is accomplished with a bellowslike action, in addition to simple reduction in diameter and length.

An outline of the sequence of events of the cardiac cycle is shown in Figure 13-6. Note the relation of the phases of systole (contraction) and diastole (filling) to valve action. After atrial contraction the ventricles contract, but the volume is constant (isovolumetric, or isovolumic) because

all valves are closed. Ejection follows. During the isovolumetric relaxation phase, the volume contained in the heart does not change, but the pressure rapidly decreases until it is less than that in the atria, allowing the AV valves to open.

When the right and left ventricles are compared as to function, the chief difference is that the left ventricle must pump a volume of blood through the high-resistance systemic (greater) circulation at relatively high pressures, while the right ventricle pumps this same volume through the low-resistance pulmonary (lesser) circulation at pressures that normally are only about one-sixth as high as for the left heart. This difference in function is reflected in the difference in structure, for the right heart has a thinner wall than the left. If the pressure load on the right ventricle is mark-

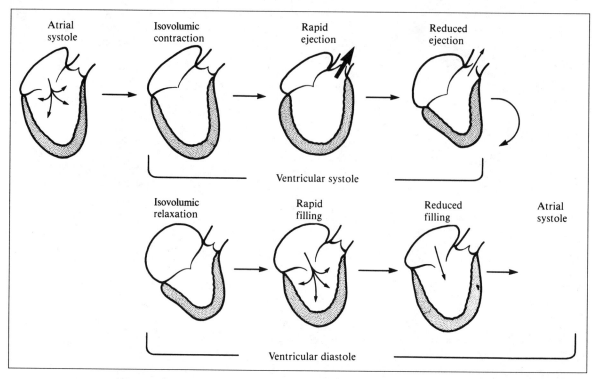

Fig. 13-6. Successive phases of the cardiac cycle. (Modified after C. J. Wiggers. *Physiology in Health and Disease.* Philadelphia: Lea & Febiger, 1949. P. 654.)

edly increased over a long period, the muscle mass of the right ventricle increases (it hypertrophies) and becomes similar in appearance and function to that of the left heart—a pressure pump.

EVENTS OF THE CARDIAC CYCLE

METHODOLOGY. The pumping action of the right and left ventricles, the presence of possible leaks in the walls between the atria or ventricles, and the adequacy of the blood supply to the heart itself (the coronary circulation) are studied by a technique in which a radiopaque (x-ray-absorbing) substance is injected into the bloodstream and x-ray motion pictures are taken. This technique is called cinefluorographic angiocardiography—cineangiography, for short (see, for example, Verel and Grainger, 1973, and Hurst, 1978). For studies of the dynamics of ventricular motion, the pictures are taken from two directions at the rate of at least 60 frames per second. The opaque material in the blood outlines the ventricular chambers and so can be observed in rapid sequence. Other studies have used radiographically opaque metal markers fastened onto the walls of the ventricles of experimental animals so that the motion

can be followed. In addition, cardiac function may be studied with echocardiography (see, for example, Fiegenbaum and Chang, 1972). This is a technique based on the transmission of low levels (milliwatts/cm^2) of ultrasound (frequency 1–10 megahertz) into the body and detection of the echos reflected from tissue and fluid interfaces. The signal is displayed on an oscilloscope screen. The junction between blood and the wall of the heart, valve leaflets, and the outline of the heart can thus be visualized. Ultrasonic techniques have also been developed to measure the velocity of blood at selected sites (see, for example, Cobbold, 1974, and Webster, 1978).

The complex geometry of the interior of the heart makes computation of stroke volume and residual volume from cineangiograms difficult. The simplest approximation is to consider the left ventricle a cylinder with a cone at the apex end. A better model is to consider the heart an ellipsoid. Computers, used to handle the more complex representations, have improved the accuracy of the determinations. However, the interior of the heart is so complicated and rough because of the papillary muscles that highly accurate estimates of instantaneous volume are not possible.

The events that are to be recorded occupy less than 1 second in the course of a normal heart beat; significant changes occur in milliseconds. Our current understanding suggests that the reasonable measurement of pressure, flow, and volume requires systems that have a uniform frequency response to at least 25 Hz. Types of transducer-recorder systems needed for these measurements are described by Geddes and Baker (1975), Cobbold (1974), and Webster (1978). Simultaneous recording of many variables gives insight into the chain of events concerned with the opening and closure of the valves and, in turn, the filling and emptying of the heart. The modern era in the study of circulatory dynamics in humans began about 1941, when Cournand and Ranges demonstrated the practicability and safety of catheterization of the right heart. Their studies were built on Forssmann's pioneer demonstration in 1929 of the feasibility of catheterizing the human heart; he used himself as the experimental subject. Catheterization of both the right and left heart are now standard diagnostic procedures. Hurst (1978) has provided useful presentations of both the normal and pathological patterns.

THE CARDIAC CYCLE. Understanding the normal cause-and-effect relationships between the variables of the heart is the basis for evaluating the significance of pathophysiological situations. To aid in discussing the cycle, it has been divided into the phases indicated in Figures 13-6 and 13-7.

Because a normal heart beats at about 80 beats per minute, a cycle requires about 750 msec and has been arbitrarily divided into fifteen 50-msec intervals. The average values for normal adults at rest are given at the bottom of Figure 13-7. Systole normally requires less than half the cycle time, but with a tachycardia of 180 beats per minute it occupies more than half the cycle and hence severely limits filling time (see Fig. 15-3). The description that follows is for the left ventricle. Right ventricular events are nearly synchronous and are very similar, except for the level of pressure generated (see Fig. 13-9).

1. Atrial Systole. Atrial systole is the beginning of the sequence of events. A pacemaker cell in the sinoauricular node depolarizes, and there is a spread of excitation across the atria, recognized in the electrocardiogram (ECG) as the P wave (Fig. 13-7, at the top). Atrial contraction follows, causing a development of pressure in the right atrium. This gives the a wave of the atrial pressure and central venous pressure. The increased pressure causes additional ventricular filling. A fourth heart sound is generated, related to the sudden atrial contraction. It is loud enough to be heard only when the left ventricle is distended. Finally, as a result of the filling due to contraction, the ventricular pressure is slightly increased to the value called the ventricular end-diastolic pressure.

EXCITATION OF THE VENTRICLES. The wave of excitation passes through the base of the heart via the AV node, the bundle of His, thence the Purkinje fibers, and finally the myocardium, leading to ventricular depolarization and the QRS complex of the electrocardiogram. As a consequence, ventricular contraction starts within about 10 msec. Then ventricular pressure starts to increase. With increased pressure, there is closure of the AV valves, and, as the valves are tensed, the first heart sound begins.

2. Isovolumic Contraction Phase. In response to the contraction of the muscle, the pressure in the ventricle increases very rapidly, but normally there is no change in intraventricular volume because both the valves to the ventricle are closed leak-tight. The rate of change of pressure (dP/dt) of about 1600 mm Hg per second is related to the intrinsic ability of the heart to contract and so provides an estimate of cardiac status compared to normal (see Chap. 15). (The value may be remembered by recalling that the pressure increases by about 80 mm Hg—from near zero to near the systolic level—within only one-twentieth of a second.) As the pressure in the ventricle exceeds that of the aorta, the outlet valve opens and flow starts, signaling the end of this phase. Note that the minimum systemic arterial pressure, called the diastolic blood pressure, occurs during early cardiac systole just as the valve opens.

3. Rapid Ejection Phase. The ventricular pressure now approaches its maximum and is greater than the aortic pressure. The magnitude of the difference in pressure is due primarily to the extra force required to accelerate the bolus of blood from the heart. If the outflow valve is stenotic (very small opening), the pressure gradient is increased because of viscous losses (resistance increase). The maximum outflow then follows while the contractile forces of the heart are still high. Blood spurts from the heart with a peak flow of 5 to 10 times the mean flow rate. Because of the high pressure in the ventricle during this phase, unless the papillary muscle tension can perfectly compensate for the pressure, there is a tendency for the valves to bulge into the atria early in the ejection phase. The bulge acts as a flow into the atria and thus increases the atrial pressure, giving the av wave. Soon after the increase in atrial pressure comes a decrease in pressure called by some the "x descent." Because of the inertia of the blood acting on the

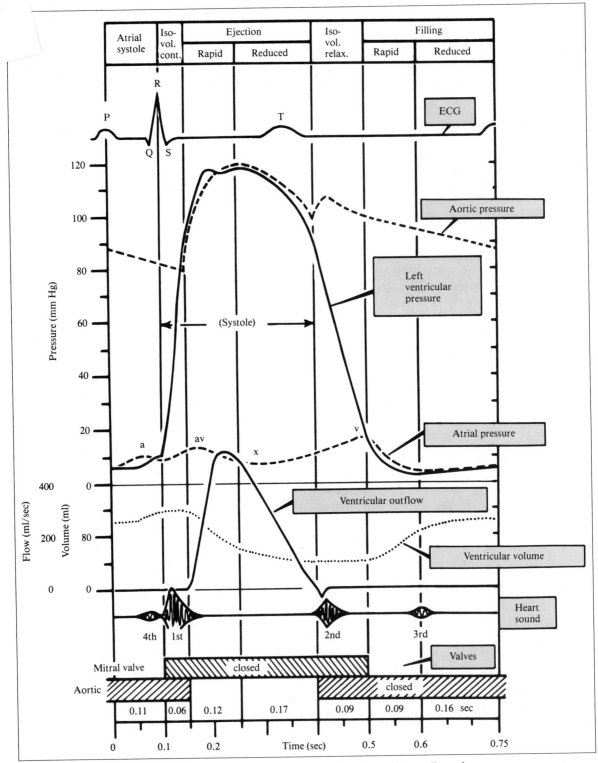

Fig. 13-7. Events of the cardiac cycle.

apex to hold it in position, the contracting heart exerts a pull on the valve rings and valves toward the apex. This force tends to move the valves out of the atrium, increase its volume, and so reduce the atrial pressure (Fig. 13-7). The reduced pressure also aids the return of blood to the atrium by providing a larger pressure gradient from the periphery to the heart. The ventricular volume decreases, slowly at first, because inertia prevents instantaneous attainment of peak velocities, but then increases progressively faster until the end of this phase.

SYSTOLIC ARTERIAL PRESSURE. The systolic arterial blood pressure (the maximum value) (see Chap. 16) may be used as the arbitrary separation point between the rapid and reduced ejection phases. The peak in arterial pressure follows the peak in flow from the heart because the distensible aorta accumulates some of the outflow, and thereby causes a dampening and lagging of the pressure pattern.

4. Reduced Ejection Phase. During the reduced ejection phase the contractile forces of the heart are decreasing and so is the intraventricular pressure. The aortic pressure is greater by a few millimeters of mercury than the ventricular pressure, but outflow continues because of the momentum of the rapidly moving bolus of blood. Repolarization of the myocardium occurs during the reduced ejection phase, giving the T wave of the ECG. Note that the correlation in time between the electrical and mechanical events is not exact, and that the membranes of cardiac muscle cells usually are repolarized before the contractile elements have returned fully to the relaxed state.

The stroke volume of the average adult is about 80 ml; the end-diastolic volume is about 120 ml. Under normal conditions a systolic reserve volume of about 40 ml remains in the ventricle at the end of systole. The *ejection fraction* is thus 2/3 [80/(40 + 80)]. The remaining volume provides a reserve that can be utilized if the vigor of contraction is increased or if the aortic pressure is decreased. The ejection fraction increases in exercise and with massive sympathetic activity to about 3/4 (see Chap. 29) and decreases to less than 1/2 in cardiac failure.

END OF SYSTOLE. The end of systole occurs at the beginning of the second heart sound (Fig. 13-7). The contractile forces of the muscle are now decreasing very rapidly. There is a reversal of arterial flow toward the valves. This closes the valves, and, because ventricular pressure is dropping so fast, the valves are tensed to such an extent that vibrations are set up, giving the second heart sound. Because the valves and aorta are elastic, they will recoil and cause a slight forward flow from the valve area and, as a consequence, a fluctuating pressure in the aorta. The notch in the aortic pressure is called the *incisura*. (The dicrotic notch, seen in a peripheral artery pressure wave, is caused by a different mechanism; see Chap. 16.)

5. Isovolumic Relaxation Phase. In the isovolumic relaxation phase, both valves are closed, and so, as the ventricular pressure decreases, there is no change in ventricular volume. The pressure decreases almost as rapidly as it had increased during the isovolumic contraction phase. Because blood has continued to flow into the atria from the venous bed, and the base of the heart is moving back toward its resting position, the atrial pressure increases to a maximum, called the *v* wave.

6. Rapid Filling Phase. The rapid filling phase is a period in which filling of the ventricle occurs quickly. The pressure in the ventricle is now less than in the atria, the AV valves open, inflow begins, and ventricular volume again starts to increase. Filling is rapid because the AV pressure gradient is relatively high. As the ventricle fills, the ventricular compliance acts to develop more and more back force to slow the filling, especially at large volumes at which the compliance is reduced (see Fig. 15-5). A third heart sound may sometimes be heard at the end of this phase and is attributable to a rapid decrease in the rate of filling, such as would occur if the heart tensed the pericardium.

7. Reduced Filling Phase. Demarcation between the rapid and reduced filling phases is arbitrary. Filling slows; indeed, if the diastolic period is abnormally long, a period of quiescence called *diastasis* ensues. During diastole the aortic outflow to the periphery continues because blood was stored in the elastic aorta, as one might store air in a balloon.

With the basic information available about this marvelously designed system, one should be able to follow the logical sequence of cause and effect, from excitation to contraction, to pressure development, and hence to outflow. In studying pathological conditions, it should be possible to reason back to the likely cause of the effect seen, such as a weakness of contraction, a valve problem, or an excitation conduction defect.

SYSTOLE/DIASTOLE. The term *diastole* comes from the Greek word meaning "separation" and thus is associated with filling of the ventricles as the walls separate. The end of diastole might be denoted as the time at closure of the AV valve when filling is stopped. The term *systole* comes from the Greek word meaning "a drawing together," or contraction, and thus is associated with the expulsion of blood by shortening of the fibers in the walls of the heart. In practice, however, the beginning of systole is usually said to coincide with the beginning of the first heart sound; the volume of the heart is maximum. (One might quibble by stating that systole starts with an electrical event associated with the Q wave of the ECG; or with the initiation of the muscular contractions; or with a development of tension preceding AV valve closure; or, even better, with the closure of the AV valve. However, all of these events are very difficult to measure and all occur within about 0.01 second.) The end of systole is even more confusing. By strict denotation it means the end of contraction. However, the end of contraction, or beginning of relaxation, is highly indeterminate. To be consistent with the definition of diastole—filling—and to leave no time undesignated, one might state that systole ends with the opening of the mitral valve, but clearly the ejection of blood has long since stopped. A functional definition of the end of systole is the beginning of the second heart sound; ventricular volume is at a minimum; outflow has stopped. Thus, ventricular systole is that interval between the beginning of the first heart sound and the beginning of the second, and diastole is the interval before the occurrence of the next first heart sound.

HEART AS AN IMPULSE FORCE GENERATOR. A muscle twitch tends to be sudden (see Chap. 3). The cardiac contraction lasts about one-third of a second, being intermittent between that of skeletal muscle (less than 0.1 second) and smooth muscle (more than 1 second). The cardiovascular system is so designed that normally blood flows relatively smoothly through peripheral capillaries of the body, even though it spurts from the heart. Not only does the suddenness of contraction place high stresses on the valves; it also provides the cardiologist with an index of the intrinsic contractile ability of the heart (see Chap. 15). It also requires mechanisms for storing the energy available for only a short time so that that energy can then be efficiently used throughout the remainder of the cycle to propel blood to the periphery (see Chap. 16).

During the early part of the rapid ejection phase, the intraventricular pressure is appreciably higher than that in the artery, because some of the pressure force is required to accelerate the bolus, in addition to that required to overcome the back pressure from the aorta and that dissipated in viscous friction (resistance). Under normal conditions this inertial component may account for only 2 to 10 mm Hg. Under conditions of vigorous contractions, as with heavy exercise or a high level of sympathetic nervous system activity, the acceleration is so great that the pressure difference between the ventricle and artery due to this factor is of

the order of 10 to 20 mm Hg. This phenomenon is a particularly important component of the right ventricular pressure pattern.

With vigorous contraction the maximum ventricular pressure occurs before the maximum flow, which in turn occurs before the maximum (systolic) arterial blood pressure. Under abnormal conditions, such as in heart failure, these peak values tend to be delayed and to occur at about the same time.

As a consequence of the momentum of the blood leaving the heart, some energy is available to continue flow into the periphery from the aorta during diastole, but most of this kinetic energy of motion is converted to potential energy during the ejection phases by the expansion of the elastic arteries (the Bernoulli principle; see Chap. 11). If the aorta is sclerotic (that is, hard and stiff), the pressure must be higher than normal to store a given bolus of blood (see Chap. 16). Because the heart ejects less blood if its pressure load is increased (see Chap. 15), the heart is less effective as a pump if the aorta is stiffer than normal.

Values of cardiovascular variables in the normal adult are given in Table 13-1.

THE HEART SOUNDS

AUSCULTATION OF THE HEART. To auscultate means "to listen to," and so the auscultation of the heart means to ascertain its condition by listening to its sounds. The mechanisms of hearing by the human ear (see Chap. 4) and the physics of sound and sound production should be reviewed as background for this section.

Vibrations from the heart are composed of unrelated very low frequencies (30–250 Hz), are of brief duration, and have amplitudes so low that the environment must be extremely quiet if one is to make a reasonable examination. Our ears are much less sensitive to sounds of these frequencies than to sounds in the speech range (500–2000 Hz; Fig. 13-8). By means of a phonocardiograph, a device consisting of a special microphone, an amplifier, and filters to attenuate extraneous frequencies, sounds of frequencies barely audible to the ear may be converted to a form that can be visualized (Fig. 13-9). Interpretation of the significance of heart sounds and murmurs is based on the temporal relationships between the heart sounds and the mechanical events of the cardiac cycle.

ORIGIN OF HEART SOUNDS. First Heart Sound. The first heart sound (S1), "lubb," is associated with the following dynamic phenomena: the beginning of cardiac muscle contraction; closing of the AV valves; rapid development of

Table 13-1. Typical resting cardiac variables in the adult

Variable	Right*	Left*
Heart rate	70 ± 3	
Atrial pressures (mm Hg)		
Mean	4.5 ± 3	8 ± 3
Ventricular pressures (mm Hg)		
End-diastolic	4.5 ± 4	9.5 ± 3
Peak systolic	26 ± 6	125 ± 15
Maximum rate of change (mm Hg/sec)	250 ± 100	1600 ± 500
Arterial pressures (mm Hg)		
Systolic	25 ± 7	125 ± 15
Mean	14 ± 4	95 ± 10
Diastolic	9 ± 4	80 ± 10
Cardiac output		
Sitting (liters/min)	5.5 ± 1.1	
Supine (liters/min)	6.8 ± 1.5	
ml/min per kg body weight	90 ± 20	
Cardiac index (liters/min/square meter body surface area)	3.5 ± 0.7	

*Mean ± standard deviation.

Primary source: P. L. Altman and D. S. Dittmer. *Biological Handbooks: Respiration and Circulation*. Bethesda, Md.: Federation of American Societies for Experimental Biology, 1971. Pp. 314–317, 319–320.

Fig. 13-8. Sounds of the heart and threshold for hearing. Note that only a small fraction of the cardiac sounds is audible.

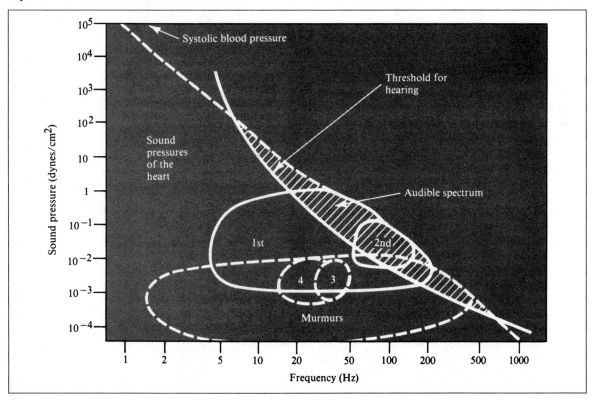

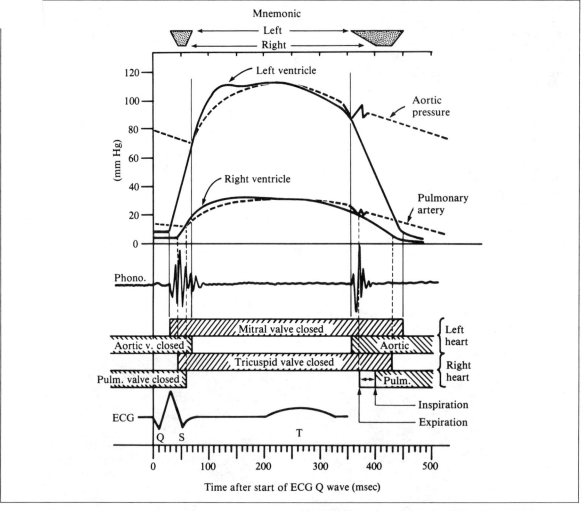

Fig. 13-9. Comparison of dynamics of the right and left heart. (Data in part from A. A. Luisada, et al. Changing views on the mechanism of the first and second heart sounds. *Am. Heart J.* 88:503–514, 1974; and J. W. Hurst [ed.]. *The Heart, Arteries and Veins* [3rd ed.]. New York: McGraw-Hill, 1974. Pp. 184, 185.)

pressure; opening of the semilunar valves; and outflow. Many components of the S1 have been proposed (Luisada, 1972). Although some vibrations can be recorded from contracting skeletal and cardiac muscle strips, their intensity is probably not sufficient to contribute to the sound heard on direct auscultation. Because the valve leaflets or cusps have such low mass, closure of valves is also probably silent. However, the sudden acceleration and deceleration of the cardiovascular structures (especially tensing of the valves and myocardium as the result of abrupt pressure change) cause vibrations in the cardiac structures that, when transmitted through the tissues of the body, are picked up from the precordium (chest surface near the heart) by the stethoscope (Luisada et al., 1974). The later part of the first heart

sound probably comes from the flow of blood past the valves and into the arteries.

The loudness of the first heart sound depends on (1) the rate of ventricular pressure rise—vigor of ventricular contraction; (2) the stiffness of the ventricle and valve structures; and (3) the positions of the mitral valve leaflets at the beginning of ventricular systole.

With sympathetic nervous system discharge and increased myocardial contractility, the rate of change in ventricular pressure is increased and a louder S1 may be heard.

Conversely, in myocardial failure, the sounds are reduced. With fibrosis, the AV valves become stiffer, leading to a sound that is louder and somewhat higher in frequency than normal. The sound originating from the right ventricle and tricuspid valve is much softer than that from the left ventricle because of the normally lower pressure gradients.

If the mitral valve is widely open at the time of ventricular systole, by the time the valve closes, a significant amount of blood will be moving toward the valve, causing some regurgitation and a rapid deceleration on closure. A relatively noisy sound will be produced as the valves are tensed by the then unusually high ventricular pressure. Under normal conditions the flow of blood past the AV valves tends to cause them to move together in accordance with the Bernoulli principle (see Chap. 11), so that they are nearly closed by the time ventricular systole occurs. However, if the P–R interval (i.e., the time between atrial and ventricular excitation) is shorter than normal (0.16 second), the valves will be open farther than normal at the onset of ventricular contraction, so that when they finally are closed, the sound is louder than normal. Maximum loudness occurs at a P–R interval of about 0.11 second. If the P–R interval is somewhat longer than the normal (e.g., 0.20 second), the first heart sound may be softer than normal because the valves are almost entirely closed when systole starts. However, if the P–R interval is very long (e.g., with a first-degree heart block), the S1 may again be louder than normal because the valves have time to close and then rebound open.

Second Heart Sound. The second heart sound (S2), "dup," is caused mainly by the tensing of the semilunar valves and the resulting vibrations of the valves, heart, and large arteries. It is normally of higher frequency and shorter duration than the S1; the following diastolic interval is usually longer than the preceding systolic interval. The sudden, very rapid ventricular relaxation at the end of systole causes a rapid deceleration of outflow. It appears that the second heart sound starts about 25 msec before the end of outflow and aortic valve closure. Blood in the roots of the aorta and pulmonary artery then rushes back toward the ventricular chambers, but this movement is abruptly arrested by closure of the semilunar valves. The momentum of the moving blood stretches the valve cusps, and the recoil causes oscillations to occur in both the arterial and ventricular cavities. Audible splitting of the S2 normally occurs during inspiration, as will be explained.

Third Heart Sound. The third sound (S3), if heard, occurs at the time of transition between rapid filling and reduced filling, when filling slows abruptly, giving a transient soft thud. The mechanism is not clearly understood. The sound is not normally heard in adults, but may be heard in many people under age 30; it is of low pitch and intensity. When the third heart sound is of pathological significance, it is associated with a dilated ventricle and high atrial pressure. Filling of the heart until the pericardium is tensed would provide a sudden deceleration that could cause audible vibrations.

Fourth Heart Sound. The fourth sound (S4) is almost always inaudible in normal adults and occurs at the time of the peak of atrial contraction. When heard, it is associated with a high atrial pressure, vigorous atrial contraction, and filling of the ventricle. Clinically it occurs when the ventricle is stiff, as in ventricular hypertrophy or ischemia. A fourth heart sound is not heard in mitral or tricuspid stenosis, since it originates in the ventricle and is associated with a distended ventricle.

SYNCHRONY OF CONTRACTION OF THE RIGHT AND LEFT HEART. Although the left ventricle is usually the first to start contracting and the last to start to fill, the timing differences between the two ventricles are of only a few hundredths of a second. Nonetheless, important clues to disease or malfunction are provided by the degree of asynchrony that may be detected when listening to the heart sounds with a stethoscope. The pattern of activity (Fig. 13-9) may be explained by the following sequence:

1. Excitation is initiated in the right atrium and is rapidly conducted to the ventricles.
2. Left ventricular contraction starts first, with closure of the mitral valve occurring within 50 msec of the start of the Q wave of the ECG. The right heart lags, because of the anatomy of the conduction system, by about 20 msec.
3. The pulmonary valve opens (100 msec after the Q wave), and pulmonary flow starts about 20 msec before aortic flow. Because pulmonary arterial diastolic pressure is so much lower than that in the aorta, it takes longer for the left ventricular pressure to develop to the level required to open the aortic valve (Fig. 13-9), even though the rate (dP/dt) is faster in the left ventricle than in the right. The isovolumetric period for the right heart is only about 15 msec compared with about 40 msec for the left.
4. Because of the higher pressure in the aorta, left ventricular outflow tends to end first. Thus, left ventricular flow *starts last* and *ends first*.
5. The mitral (left AV) valve opens after the tricuspid (right AV) valve—again, because more time is required for the pressure to drop from aortic pressure to left atrial pressure than for the right ventricular pressure to decrease.

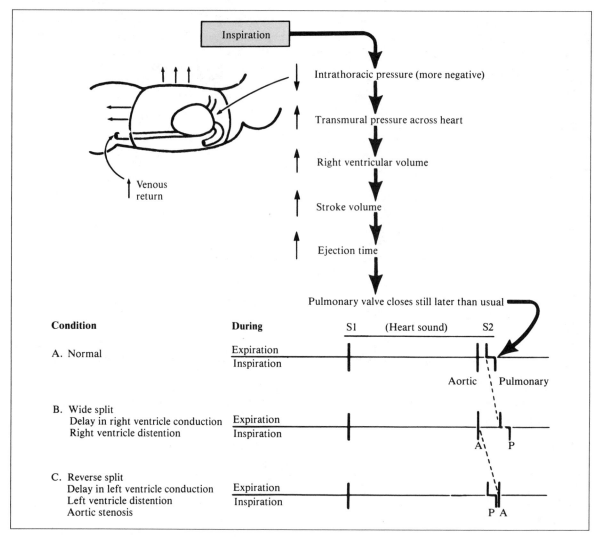

Fig. 13-10. Physiology of the split of the second heart sound.

Note the mnemonic at the top of Figure 13-9: The isovolumetric phases of the right heart, being shorter, occur *during* the corresponding phases of the left heart.

SPLIT SECOND HEART SOUND. In normal people the second heart sound is audibly split during inspiration because of the difference in time between the aortic and pulmonary valve closings (Fig. 13-10). Our ears so function that sudden sounds (e.g., a click) separated by more than 0.02 second can be discerned as separate, or split.

The mechanisms for the delay of the pulmonic valve closure during inspiration are outlined in Figure 13-10. An in-crease in stroke volume, or an increase in pressure (after-load), tends to prolong the ejection time of a ventricle. The more negative intrathoracic pressure on inspiration leads to increased right ventricular filling, prolonged ejection and right ventricular systole, and delayed valve closure and hence a delayed pulmonary second heart sound (P2). With right ventricular defects in conduction of excitation, an in-crease in right ventricular stroke volume from an atrial sep-tal defect, or right ventricular overload, the pulmonic valve closure is abnormally delayed, and the split in the second heart sound may be heard even during the expiratory pause—a *wide split*. On the other hand, blockage of con-

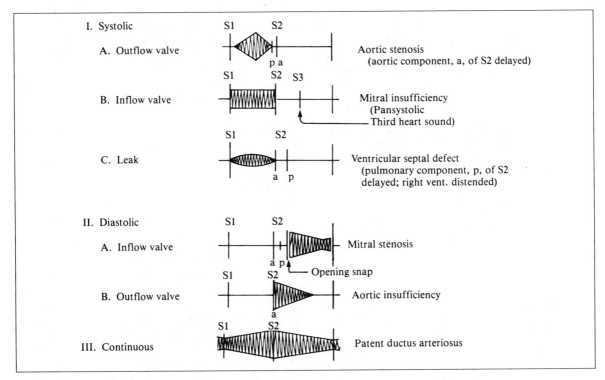

I. Systolic

    A. Outflow valve   Aortic stenosis (aortic component, a, of S2 delayed)

    B. Inflow valve   Mitral insufficiency (Pansystolic — Third heart sound)

    C. Leak   Ventricular septal defect (pulmonary component, p, of S2 delayed; right vent. distended)

II. Diastolic

    A. Inflow valve   Mitral stenosis — Opening snap

    B. Outflow valve   Aortic insufficiency

III. Continuous   Patent ductus arteriosus

Fig. 13-11. Patterns of heart murmurs.

duction along the left ventricular branch of the bundle of His, or left ventricular overload due to stenosis or severe hypertension may prolong the aortic component so that the split is *reversed* or *paradoxical,* that is, is heard during expiration but not during inspiration.

MURMURS. Murmurs are relatively prolonged audible vibrations that originate in either the heart or the great vessels as a result of high-velocity blood movement. Blood flow through vascular channels is normally laminar or streamline and silent. Turbulent, noisy flow develops under special circumstances. It may be *random,* with a "whooshing" sound of a wide, uniform-frequency spectrum; or *vortex shedding,* with a predominant frequency or pitch (Bruns, 1959; see also Chap. 11). The factors that determine whether flow is laminar or turbulent are (1) the size and smoothness of the channel, (2) the velocity of flow, (3) the fluid density, and (4) its viscosity. Since blood viscosity and density are reasonably constant in the large vessels, the factors usually of concern are the velocity of flow and degree of narrowing of flow channels (see Chap. 11).

It is probable that some turbulence is normally present (e.g., in the aorta or pulmonary artery just beyond the valve), but the sound produced is usually not intense enough to be heard with the stethoscope or is included with the first heart sound. Such minimal turbulence can be increased and result in an audible benign murmur when the velocity of blood flow is increased, as in exercise, fever, pregnancy, or hyperthyroidism.

When the diameter of a flow channel is reduced by disease or a local obstruction, a murmur may result at normal or even decreased flow rates. Likewise, a murmur is produced by regurgitation of blood through an incompetent valve. The significance of a murmur depends on (1) the location of its source; (2) when it is heard during the cycle—systolic or diastolic; (3) its quality and pitch; and (4) its intensity.

Systolic Murmurs. Systolic murmurs are those that originate with or after the first heart sound and end with or before the second heart sound.

In *aortic stenosis* (Fig. 13-11), the murmur is proportional to the rate of blood flow and so is loudest in midsystole, giving a diamond shape when visualized with a phonocardiograph. Blood is flowing in the normal direction, but the valve is narrowed to only a small orifice. When stenosis is

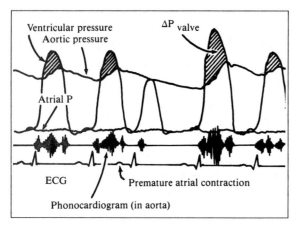

Fig. 13-12. Aortic stenosis. (Redrawn from H. L. Moscovitz, et al. *An Atlas of Hemodynamics of the Cardiovascular System.* **New York: Grune & Stratton, 1963. P. 79.)**

severe, the ejection of blood is delayed, and so the aortic (a) component of S2 is heard after the sound of the closure of the pulmonic (p) valve. On catheterization, a large difference in pressure ($\Delta P = 20$–100 mm Hg) can be measured across a stenotic aortic valve (Fig. 13-12).

In *mitral insufficiency,* the murmur is associated with regurgitation of blood through an insufficient valve (Fig. 13-11). Whether the valve is insufficient because of disease, dilated anulus, or ruptured chordae tendineae, the sound is similar because the mechanism is similar. When the pressure in the ventricle exceeds that in the corresponding atrium, the mitral and tricuspid valves close, producing the S1; but if closure is not complete, blood is forced back into the atrium under pressure, causing turbulence and a murmur. Usually, the murmur begins immediately after the S1 and extends throughout systole with a fairly constant intensity ending with the S2; it is called a *holosystolic* or *pansystolic* murmur, because it occurs throughout all of systole.

A *ventricular septal defect* is a hole in the wall between the right and left ventricles. A systolic murmur (see Fig. 13-11) will be heard during systole that will be like the murmur in mitral insufficiency, because of the similar mechanism. However, the position of maximum intensity on the precordium will be somewhat different. With the extra filling of the right ventricle—the leak plus the normal venous return—the right ventricular stroke volume will be greater than normal, and the pulmonic component of the S2 will be delayed, giving a split even during expiration. Valvular regurgitation and septal defects are diagnostically separated by injecting a radiopaque material into the left heart to visualize the leak,

called a *shunt.* Measuring the oxygen tension of the blood in the right atrium and right ventricle provides diagnostic information also, since blood leaking from the left side will have a high oxygen tension from passage through the lungs. Therefore, the blood in the right ventricle and pulmonary artery (distal to the shunt) will be better oxygenated than that in the right atrium (proximal to the shunt).

Diastolic Murmurs. In *mitral stenosis* (see Fig. 13-11), the murmur occurs during diastole and is proportional to the flow. It thus is loudest early in diastole immediately following the isovolumetric relaxation period when atrial pressure is high, and it also becomes loud during atrial systole ("presystolic"). The murmur is often heralded by a sound, the opening snap of the mitral valve. With mitral stenosis, the pulmonary venous pressure is abnormally high, as is the pulmonary wedge pressure. (This pressure is obtained by floating a small-diameter catheter through the systemic veins, right heart, and pulmonary artery until it is wedged into a small pulmonary artery. Because of the relatively few collateral vessels at this level and the low pressure gradient across the pulmonary microvasculature, the pressure beyond where the catheter is wedged in—the *wedge* pressure—has been found to be similar to that of the pulmonary veins and consequently the left atrium on the other side of the pulmonary capillaries. A systemic vein, and thence the pulmonary artery, is much easier and safer to catheterize than the left atrium.)

In *aortic insufficiency* (Fig. 13-11), the murmur is proportional to the flow, starting as soon as the valve partially closes and isovolumetric relaxation starts. It may last throughout diastole, but because the arterial blood can leak back into the heart as well as go through the normal pathways, the arterial diastolic pressure is low, and so the murmur usually dies out during diastole. The systolic pressure will be high, from the compensatory mechanisms acting to increase cardiac activity to eject a larger-than-normal stroke volume.

Continuous Murmurs. Murmurs may continue throughout both systole and diastole—for example, the murmur associated with a patent ductus arteriosus. Since the aortic pressure always exceeds the pulmonic, the flow is continuous, as is the resultant murmur. Furthermore, a valve may be so damaged as to be both stenotic and regurgitant.

Diagnosing possible pathological causes of murmurs requires consideration of numerous possibilities. There are four valves with two possible types of pathology: stenosis or insufficiency, or both. There may be septal defects at either the atrial or ventricular level, or both. There also may be a coarctation (narrowing) of the aorta, or patent ductus arteriosus, or an arteriovenous fistula leading to a murmur. The diagnostician must remember that the heart sounds are associated with sudden changes of motion and subsequent vibrations of elastic structures. Murmurs, on the other hand, are associated with high-velocity squirts of blood.

BIBLIOGRAPHY

Altman, P. L., and Dittmer, D. S. *Biological Handbooks: Respiration and Circulation*. Bethesda, Md.: Federation of American Societies for Experimental Biology, 1971.

Armour, J. A., and Randall, W. C. Structural basis for cardiac function. *Am. J. Physiol.* 218:1517–1523, 1970.

Bruns, D. L. A general theory of the causes of murmurs in the cardiovascular system. *Am. J. Med.* 27:360–374, 1959.

Cobbold, R. S. C. *Transducers for Biomedical Measurements*. New York: Wiley, 1974.

Constant, J. *Bedside Cardiology* (2nd ed.). Boston: Little, Brown, 1976.

Feigenbaum, H., and Chang, S. *Echocardiography* (3rd ed.). Philadelphia: Lea & Febiger, 1980.

Geddes, L. A., and Baker, L. E. *Principles of Applied Biomedical Instrumentation* (2nd ed.). New York: Wiley, 1975.

Hurst, J. W. (ed.). *The Heart, Arteries and Veins* (4th ed.). New York: McGraw-Hill, 1978.

Luisada, A. A. *The Sounds of the Normal Heart*. St. Louis: Green, 1972.

Luisada, A. A., MacCanon, D. M., Kumar, S., and Feigen, L. P. Changing views on the mechanism of the first and second heart sounds. *Am. Heart J.* 88:503–514, 1974.

Mirsky, I., Ghista, D. N., and Sandler, H. (eds.). *Cardiac Mechanics: Physiological, Clinical, and Mathematical Considerations*. New York: Wiley, 1974.

Netter, F. H. *Heart*. The CIBA Collection of Medical Illustrations. Summit, N. J.: CIBA, 1969. Vol. 5.

Parmley, W. W., and Talbot, L. Heart as a Pump. In R. M. Berne and N. Sperelakis (eds.), *Handbook of Physiology*. Bethesda, Md.: American Physiological Society, 1979. Section 2: The Cardiovascular System. Vol. 1, The Heart, pp. 429–460.

Shaver, J. A., and O'Toole, J. D. The second heart sound: Newer concepts. *Mod. Concepts Cardiovasc. Dis.* 46:7-16, 1977.

Verel, D., and Grainger, R. G. *Cardiac Catheterization and Angiocardiography* (2nd ed.). Edinburgh: Churchill Livingstone, 1973.

Webster, J. G. (ed.). *Medical Instrumentation. Application and Design*. Boston: Houghton Mifflin, 1978.

CARDIAC EXCITATION, CONDUCTION, AND THE ELECTROCARDIOGRAM

14

Kalman Greenspan

The excitation or stimulus for the heart arises from within the cardiac muscle itself. How and where this excitation occurs and how it is conducted throughout the entire myocardium are discussed in this chapter. The electrical events of the cardiac cycle as manifested in the electrocardiogram (ECG) and a simplified theory of electrocardiographic interpretation are presented.

RESTING AND ACTION POTENTIAL

The typical transmembrane potential of an inactive myocardial fiber (Fig. 14-1 A–C) varies between 80 and 100 mv, the interior of the cell being negative with respect to the cell's exterior. When the cell is excited, however, the potential difference is quickly lost and in fact becomes opposite in sign (+ 20 to + 40 mv), indicating that a reversal of polarity has occurred. This sudden initial upstroke (depolarization) is designated as phase 0 and includes the overshoot. Soon after the reversal of polarity the restorative processes start (repolarization), and the potential difference begins to de-

cline toward its resting value. This is characterized by fast repolarization (phase 1), which then slows down and results in a plateau (phase 2). Following the plateau there is once again a fairly rapid wave of repolarization (phase 3), until the resting transmembrane potential or diastolic period is attained (phase 4). The time course of the cardiac action potential differs also according to the fiber type; atrial cells (Fig. 14-1A) show a less prominent plateau. Action potentials obtained from the ventricular myocardium (Fig. 14-1B) and from the Purkinje network (Fig. 14-1C) located in and ramifying throughout these chambers show a more prolonged plateau. However, phase 0 is similar for atrial, ventricular, and Purkinje fibers, being rapid in upstroke.

Quite the contrary is seen in the configuration of the action potentials obtained for the fibers in the sinoatrial (SA) node (Fig. 14-1D) and the atrioventricular (AV) node (Fig. 14-2A). Phase 4 demonstrates a slow, continuous diastolic depolarization, so that during diastole the SA membrane potential progressively becomes less negative. The magnitude of the transmembrane "resting" potential is about

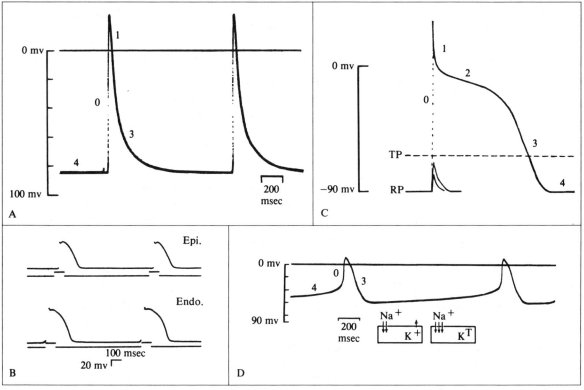

Fig. 14-1. Records of membrane potentials obtained from an atrial contractile fiber (A), ventricular epicardial (Epi) and endocardial (Endo) surfaces (B), Purkinje fiber (C), and sinoatrial pacemaker cell (D). Note the differences in rapid depolarization (phase 0), the plateau (phase 2), and repolarization (phase 3) between cardiac cells. Progressively increased cathodal stimuli are applied to a cardiac Purkinje fiber in C. With threshold stimulus the cell action potential was produced. In automatic fibers, progressive spontaneous self-depolarization occurs during diastole (D). Excitation occurs when the threshold level of depolarization is reached. Below, in diagrammatic form, are the possible ionic fluxes that could occur for the continuous slow diastolic depolarization. TP = threshold potential; RP = resting potential; K^T = reduction or cessation of K^+ permeability.

ticity* exhibit this characteristic sign of slow diastolic depolarization. The time course of these action potentials has great physiological significance in that cardiac rate and rhythmicity are dependent on the magnitude of the resting potential, the level of the threshold potential, and the slope or steepness of the diastolic depolarization. In turn, the slow diastolic depolarization is probably responsible for the property of automaticity, while the level of membrane potential will determine excitability and conduction.

PHYSIOLOGY OF PACEMAKER ACTION

As noted earlier, automaticity in the cardiac cell is demonstrated by a slow, continuing depolarization during the diastolic phase (phase 4) (no isoelectric period is observed during phase 4). It can be demonstrated that such automatic cells reside in the SA node (Fig. 14-1D), within the atrium,

−70 mv. Furthermore, phase 0 is not rapid; the overshoot (phase 1) is much reduced and in many instances is not present. Phases 1 and 2 blend with phase 3, and the entire process of repolarization is slow. Following the repolarization wave there is a progressive decline in the magnitude of the resting potential (an incline in the phase 4 slope), so that on reaching its threshold potential the cell becomes excited in a self-sustaining manner, resulting in full depolarization. All cells potentially capable of developing intrinsic automa-

*A myogenic property of certain cardiac cells defined as an inherent ability to develop spontaneous depolarization. Such cells are also called pacemaker cells.

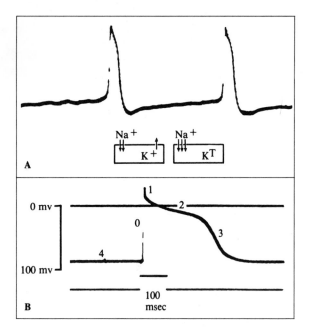

Fig. 14-2. A. Record of action potentials obtained from rabbit AV node (nodal = His region). Note the appearance of a spike similar to that in a His-type action potential, but the period of diastole (phase 4) shows an indication of continuous spontaneous depolarization. It may, however, represent a period of hyperpolarization that tends to level off. (See text for discussion.) Below the potentials are the possible ionic fluxes (decrease in K^+ conductance with an increasing Na^+ influx) that could account for the continuous depolarization. B. Action potential recorded from the bundle of His area. Note that there is no spontaneous self-depolarization; this activity requires a stimulus. The action potential is characterized by a rapid upstroke velocity, a sharp phase 1, and a prolonged plateau (phase 2).

and in the His-Purkinje system. True atrial and ventricular cells apparently do not meet the electrophysiological and pharmacological criteria for automaticity and hence are not considered as being composed of automatic cells.

The greatest rate of spontaneous depolarization is seen in the primary automatic cell of the SA node. The mechanism responsible for this slow diastolic depolarization has not been clearly defined, although there is some evidence of a concomitant progressive increase in membrane resistance during diastole that may be indicative of a decreased membrane permeability to K^+. Following depolarization of the primary automatic area in the SA node, the current so generated flows longitudinally and lowers the transmembrane potential of adjacent areas. When threshold level is attained, these areas in turn depolarize, and self-sustaining,

propagated action potentials spread over the entire myocardium. There is, however, a preferential sequence of activation (see Fig. 14-3). Activity is first initiated in the vicinity of the SA node. The velocity of conduction through pacemaker cells is difficult to measure, since it increases with distance from the primary pacemaker cell. In the rabbit heart the earliest activity is detected in the wall of the superior vena cava, a few millimeters away from the crista terminalis, and the activity propagates radially at the extremely low velocity of 0.05 meter per second. As the impulse approaches the crista terminalis, its conduction velocity markedly increases; in fact, activity from the crista terminalis spreads into the atrial roof and down to the coronary sinus at velocities ranging from 0.5 to 1.0 meter per second. From the atria, activity spreads through the AV node. There, after some delay, the sequence of activation is through the His bundle, bundle branches, peripheral Purkinje fibers, and then ventricular tissue proper.

Of great significance is the delay in the spread of excitation that occurs in the AV region. Microelectrode studies have aided immensely in the elucidation of the mechanism of nodal delay. From a functional point of view, the AV nodal region can be divided into AN (atrionodal), N (nodal), and NH (nodal-His) zones. Activity approaches the AN boundary region at right angles, and propagation becomes slower. Associated with the reduced conduction velocity is a decrease in the rise time of the action potential when the excitatory wave reaches the narrow, middle N zone; propagation is as slow as that seen in the SA pacemaker (0.05 m/sec). The upstroke of the action potential is slow, and the propagation time consumes about 30 msec of the total AV delay. After the N zone is passed and activity enters the NH zone (Fig. 14-2A), conduction velocity increases (1.0 to 1.5 m/sec), and the action potential starts to assume a configuration almost identical to that seen in the area of the bundle of His (Fig. 14-2B).

It has been postulated that conduction through the AN and NH regions is all-or-none, while the mechanism of propagation within the N layer node is one of decremental conduction. Decremental conduction means that, whatever the cause, there is a gradual decrease in the rising velocity and a gradual diminution in the amplitude of the action potential as the activity propagates over a given path. Associated with these phenomena is a decrease in conduction velocity. Accordingly, the N layer is the region where AV block is most likely to occur, since it is in this zone that conduction velocity is at its minimum. If decrement of conduction in the N zone is enhanced, as with acetylcholine or vagal stimulation, the action potential amplitude attained may not be sufficient to excite the next zone, and the propa-

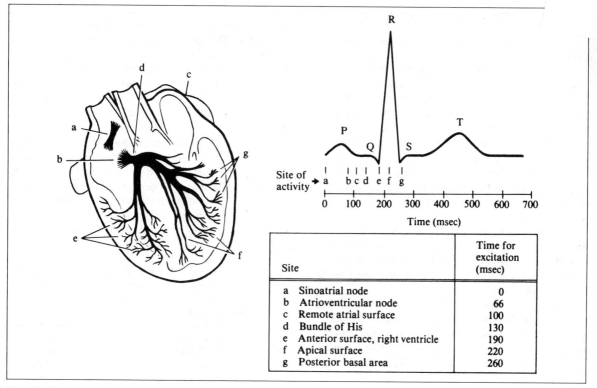

Fig. 14-3. Conduction pathways of the heart. The times, in milliseconds, following pacemaker stimulation required for excitation to occur are shown for various regions. An ECG is shown correlated with the conduction sequence.

Site	Time for excitation (msec)
a Sinoatrial node	0
b Atrioventricular node	66
c Remote atrial surface	100
d Bundle of His	130
e Anterior surface, right ventricle	190
f Apical surface	220
g Posterior basal area	260

gation wave will not pass to the bundle of His. The result is AV block.

The cells of the AV node require special consideration with regard to automaticity. At first glance they seem to show spontaneous depolarization (Fig. 14-2A), but the AV node does not contain automatic fibers. Following phase 3 there is a phase of hyperpolarization that rapidly returns to an isoelectric level and remains steady for the duration of phase 4. In addition, a foot appears just prior to the AV nodal action-potential upstroke. During rapid heart rates this foot of the nodal action potential merges with the early hyperpolarization, giving the appearance of diastolic depolarization (Fig. 14-2A). Thus, the so-called clinical nodal rhythms probably originate from the transitional fibers located between the NH area and the His area, or from the bundle of His itself. These areas possess automatic cells, and when one of the areas serves as the ventricular pacemaker, electrocardiographic interpretation would be a nodal rhythm, and should be referred to as a *junctional rhythm.*

Once excitation passes the N layer with a high enough amplitude, the wave will reach the bundle of His. As the activity enters the His area and the bundle branches, there is again an increase in conduction velocity (to 3–4 m/sec) attributable to the greater diameter of these fibers and their infrequent branching. Action potentials from these cells are characterized by a rapid upstroke, high amplitude, and long duration (Fig. 14-2B). Conduction then slows down (to 1 m/sec) as activity enters the ventricular muscle (Fig. 14-1B), presumably because of the smaller fiber diameter and frequent branching.

Generally, impulse transmission in cardiac tissue occurs at different velocities in the various parts of the heart. In any given fiber, however, changes in conduction velocity can be attributed to changes in the resting potential, threshold potential, rate of depolarization, action potential duration, degree of repolarization prior to the arrival of the next impulse, or changes in the core conductor properties of that tissue.

IONIC BASIS OF ELECTRICAL ACTIVITY

The resting potential of cardiac cells depends largely on the gradient of K^+ ions across the cell membrane. It is well established that K^+ concentration is 30 to 40 times greater inside cardiac cells than outside. The reverse is true for Na^+. This, coupled with the fact that the resting cell membrane is much more permeable to K^+ than Na^+, allows the cell to develop a potential difference across the cell membrane. This potential can be approximated using the Nernst relationship discussed in Chapter 1. In addition to the contribution of K^+, there is a small resting Na^+ flux that is sometimes called the background current. This current will modify the resting membrane potential, especially at high K^+ concentration.

The uneven distribution of Na^+ and K^+ ions is maintained by an energy-dependent pump that moves Na^+ out of the cell and K^+ into the cell. Up to this point the ionic basis of cardiac electrical activity is similar to that detailed for the squid axon in Chapter 1.

The action potentials in cardiac cells are complex, and vary among the tissues of the heart. To simplify this discussion, let us first divide cardiac action potentials into two major classes: (1) the fast-depolarizing action potentials (those described earlier for atrial muscle, the His-Purkinje system, and ventricular muscle); and (2) slowly-depolarizing action potentials (those described for the SA and AV nodes). The specific shape of the action potentials of a given cardiac cell is determined primarily by specific ionic fluxes through membrane channels that are gated.

The ionic changes responsible for the development of fast action potentials have been at least partially revealed by the use of various techniques to measure current flow across the cell membrane (Fig. 14-4). In the fast potentials the initial event occurring after a threshold stimulus is an explosive increase in Na^+ permeability. This permeability and a resultant rapid depolarization (phase 0) is believed to be mediated by the state of a population of fast Na^+ channels that can be activated in a fraction of a millisecond and are inactivated after several milliseconds. It appears that two gates, an "activation" gate and an "inactivation" gate, con-

Fig. 14-4. The fast action potential consists of a series of ionic currents occurring in a specific temporal course. The rapid stroke (phase 0) is primarily a Na^+ current. However, as the resting membrane potential moves positively (past about -60mv), the conditions are such that some Ca^{2+} current may flow. As the action potential reaches its peak, the fast sodium current (i_{Na^+}) proceeds toward inactivation. The inward chloride current (i_{Cl^-}), as well as the outward potassium current (i_{K^+}), then causes the initial early repolarization. The plateau (phase 2) is largely supported by the slow inward current (i_{si}), which is primarily a Ca^{2+} current.

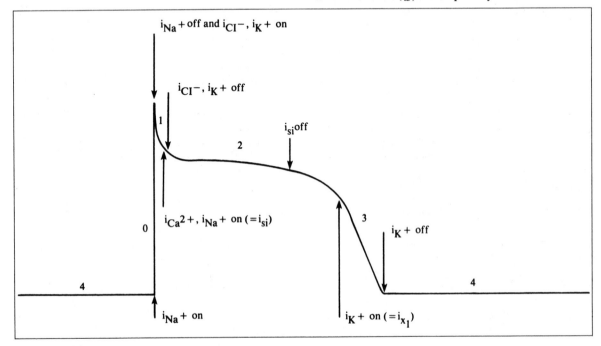

trol the fast Na^+ channel. The number of fast Na^+ channels that can be activated is voltage dependent, with a maximum activation occurring at a resting membrane potential greater than -90 mv. Hence, the rate of change of phase 0 (dV/dt) is dependent on the resting membrane potential.

After the phase 0 depolarization, cells with fast action potentials usually show a rapid but limited repolarization (phase 1). This voltage change appears to be due to a transient outward current that consists partially of Cl^- ions moving into the cell and partially of K^+ ions moving out of the cell.

At the end of phase 1 the action potential enters phase 2, or the plateau phase. At this point the membrane repolarizes very slowly, due to a "slow inward" current often called i_{si}. This current apparently results from an increase in permeability to both Na^+ and Ca^{2+}. The slow inward current uses a population of Na^+ and Ca^{2+} channels that are distinctly different from those associated with the fast Na^+ current. These channels can be identified because they are activated much more slowly and at less negative potentials. They are also inactivated at a slower rate. Minor alterations in the slow inward current have also been shown, caused by a time-independent K^+ current, the Na^+-K^+ exchange mechanism, and Cl^- flux.

The rapid repolarization of the cardiac cell s phase 3 is associated with the development of a current that is usually labeled as i_{x_1}. This current 1 due to an efflux of K^+ from the cardiac cell. Wi current i_{x_1} returns to zero, the cell has returned to its resting membrane potential.

Phase 4 ionic events are different for cells that show diastolic depolarization than for those that do not. The voltage clamp technique has demonstrated that in cells sharing spontaneous depolarization the ionic alteration during phase 4 is a slow inactivation of a separate K^+ channel, usually labeled i_{K_s}. This results in a decreased permeability to K^+ and a depolarization predicted by the Nernst relationship. Thus, pacemaker activity can be generated with a rate that will be proportional to i_{K_s} inactivation and phase 4 depolarization. In the primary pacemaker area, i_{Na^+} will be of minimal importance after threshold is reached. Cells showing a slower rate of diastolic depolarization will be captured and show an increasing contribution of i_{Na^+}. Cells that show no diastolic depolarization have no net current flow during phase 4.

Cells that show slowly-depolarizing action potentials are characterized by less negative resting membrane potentials and smaller action potentials (Fig. 14-5). The ionic events that underlie these potentials differ primarily by the virtual absence of the large Na^+ current that occurs in the initial phase of fast action potentials. This means that the depolarization seen in slow action potentials is largely dependent on

Fig. 14-5. The normal production of the "slow" action potentials is largely dependent on Ca^{2+} currents. Because of the low "takeoff" potentials (-60mv) found in these cells, the early i_{Na^+} is partially inactivated. This allows the i_{si} to play a dominant role in the development of depolarization (see text).

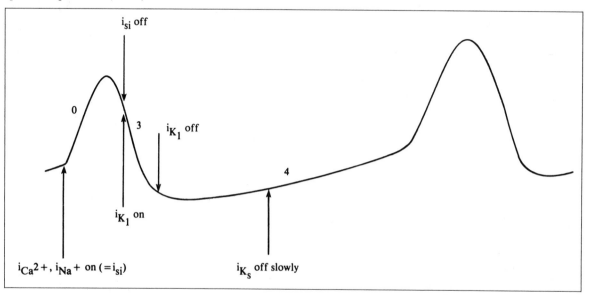

the slow inward current previously described. The slow action potentials are relatively insensitive to agents that affect the rapid Na$^+$ current (e.g., tetrodotoxin) and very sensitive to agents that affect the slow inward current (e.g., verapamil). Slow potentials are not only seen in the nodal areas, as previously mentioned, but also may play a role in conduction through damaged cardiac tissue. Since abnormal cardiac cells are often partially depolarized, the contribution of i_{Na^+} is drastically decreased, and the action potentials will be supported by i_{si} predominantly. This may explain the development of conduction-related arrhythmias.

EFFECT OF IONS ON THE RESTING POTENTIAL. A prominent effect of increasing the extracellular K$^+$ concentration is a reduction in the resting potential; the magnitude of the reduction bears an almost linear relation to the amount of K$^+$ added to the perfusion solution and can be approximately predicted by the Nernst equation (see Chap. 1). The depolarization effect of high K$^+$ concentration has been described for all fiber types of the heart in many species. In all these fiber types the initial effect of the reduction of the resting potential by increasing K$^+$ concentration is an enhancement of excitability. The reason is that the resting potential encroaches on the threshold potential level so that the stimulus requirement is lower. However, further increase in extracellular K$^+$ results in a loss of excitability, presumably through inactivation.

An interesting observation was made in relation to the K$^+$ and Ca^{2+} ratio. This ratio appears to be an important component of cardiac function; a similar interaction between K$^+$ and Ca^{2+} exists for the resting potential. Thus, the depolarizing effect of a high concentration of extracellular K$^+$ can be counteracted by simultaneously raising that of extracellular Ca^{2+} and augmented by lowering the Ca^{2+} concentration. With normal K$^+$ levels the alteration of Ca^{2+} to high or low levels does not affect the resting potential. This concept of a K$^+$ and Ca^{2+} antagonism in cardiac tissue is consistent with that observed in other viable cells.

EFFECTS OF IONS ON THE ACTION POTENTIAL OF NONAUTOMATIC CELLS

The change in the resting potential due to an alteration in extracellular K$^+$ concentration would be expected to affect the contour, amplitude, and duration of the action potential. Thus, following exposure to a high concentration of K$^+$ there is a reduction in maximum upstroke velocity (Fig. 14-6), amplitude, and action potential duration. These changes may result from the reduced resting potential, or perhaps they are due to a direct effect of the added K$^+$.

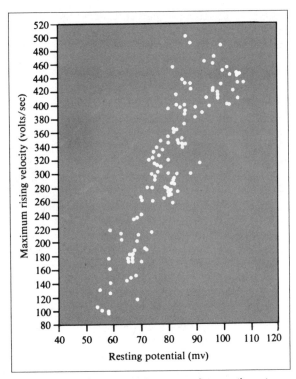

Fig. 14-6. Relationship between the transmembrane resting potential and the maximum rising velocity in canine Purkinje cells. As the potential is reduced, the maximum rate of rise decreases. The resting potential in the experiment was reduced by exposing the tissue to a high concentration of K$^+$.

The reduced upstroke velocity of phase 0 and the decreased amplitude are apparently indirect effects secondary to the reduced resting potential; the observation was made that if the resting potential is restored, as by anodal current, the effect of high-concentrate K$^+$ on phase 0 and upstroke amplitude is reversed. However, the enhanced repolarization induced by high K$^+$ concentration is not completely counteracted by anodal current. Thus, shortening of the action potential due to high K$^+$ concentration can occur independently of a reduction in resting potential, and the action of the cation is a direct effect, not a consequence of depolarization. Since during the normal period of repolarization there is an increase in the efflux of K$^+$, it would seem that the K$^+$ itself can act to develop a positive feedback, presumably by enhancing membrane permeability. Evidence has been advanced that an increase in extracellular K$^+$ concentration decreases membrane resistance. In fact, this effect is much more pronounced at a membrane potential of -40 mv, the potential level at which the plateau region for sheep Purkinje fiber occurs. Furthermore, both influx and efflux rates of K^+ increase with increasing K^+ concentrations. These results are highly suggestive of the possibility that K^+ increases membrane conductance or permeability for itself.

Since the phase of depolarization is determined by the movement of Na^+, the changes in action potential configuration can be predicted following changes in concentration of this ion in the extracellular compartments. Thus, a *reduction* in Na^+ concentration results in a decreased upstroke velocity and a decrease in the amplitude of the atrial action potential. Severe reduction of this ion (to 10 percent of normal) leads to a complete loss of excitability. The ventricular muscle behaves somewhat differently in a low-Na^+ medium in that phase 0 is not affected, although the duration is shortened on account of an increase in the steepness of phase 2. *Increasing* the Na^+ concentration has little direct effect on action potential configuration. An indirect effect may be observed as a result of changes in tonicity with the excess Na^+.

Calcium has little effect on the resting membrane potential and amplitude of the action potential of fast-depolarizing cells. However, the duration of the action potential is altered by the concentration of Ca^{2+} in the medium. In fact, the ventricular action potential following Ca^{2+} excess and subsequent intracellular influx becomes markedly prolonged. As might be expected from earlier discussions, atrial excitability (unlike ventricular excitability) is markedly decreased by Ca^{2+} depletion.

EFFECT OF TEMPERATURE ON THE ACTION POTENTIAL OF NONAUTOMATIC CELLS. The electrophysical properties of atrial and ventricular cells are strongly affected by changes in temperature. Temperature affects the action potential phases directly and indirectly through changes in heart rate. With hyperthermia, the rate is increased, and action potential duration is decreased. With cooling, depending on the severity, the heart rate is decreased, the action potential duration markedly prolonged, and the membrane potential reduced; these changes may ultimately lead to a complete loss of excitability. Temperature changes can produce these alterations directly and independently of heart rate.

FACTORS AFFECTING HEART RATE AND RHYTHM: AUTOMATIC CELLS

The heart beats in an extremely regular, continuous, and rhythmic fashion. It is a result of automaticity, a property present in those cells of the heart that exhibit the electro-

physiological pattern of slow diastolic depolarization (see Fig. 14-1D). This phenomenon confers on the cells the ability to depolarize spontaneously, independently of the extrinsic and intrinsic nerve supply. The cells of the heart that exhibit automaticity with the most rapid rate of depolarization and are responsible for depolarizing the rest of the heart are located in the SA node. Although myogenic in origin, the rate and rhythm may change spontaneously, or they may be influenced by temperature, inorganic ions, and neural and humoral activity. Any factor that changes the heart's rate and rhythm does so by affecting (1) the magnitude of the resting potential, (2) the threshold potential voltage that must be attained, and/or (3) the slope of diastolic depolarization. Study of Figure 14-7 reveals that any alteration either of these potentials or of the slope of depolarization alters the time required for reaching threshold and hence the rate of stimulus production.

EFFECT OF ACETYLCHOLINE AND THE VAGUS. Stimulation of the vagus or administration of acetylcholine (ACh) reduces the heart rate (bradycardia) and, with sufficient intensity or concentration, may lead to sinus arrest. The effects on the automatic cell of such maneuvers are as follows: First, repolarization is enhanced. Second, hyperpolarization takes place; in Figure 14-7 the resting potential becomes more negative. The magnitude of the hyperpolarization depends on the initial resting potential value; the smaller the resting potential, the greater the hyperpolarization produced by ACh. Third, there is a shift of the slope of diastolic depolarization; in Figure 14-7 the slope decreases. The net effect of cholinergic activity is an increase in the time required for the fiber to reach its threshold potential. The most prominent effect of cholinergic activity is on phase 4 depression; if severe enough, it can result in cardiac standstill. The mode of action of ACh appears to be via an increase in K^+ permeability, i_{K_S}, which in turn reduces the diastolic depolarization and thus leads to a reduction in heart rate (Fig. 14-8A).

CATECHOLAMINES AND SYMPATHETIC NERVES. The administration of epinephrine or norepinephrine, as well as stimulation of the cardioaccelerator nerve, increases the heart rate (tachycardia). The most striking effect is on the slope of phase 4 (Fig. 14-8A). On latent pacemakers, such as Purkinje fibers, sympathetic activity induces automaticity (Fig. 14-8B) and may lead to multifocal firing and ventricular fibrillation.

TEMPERATURE. Cooling of the SA node reduces spontaneous activity (depression of phase 4), and severe cooling may

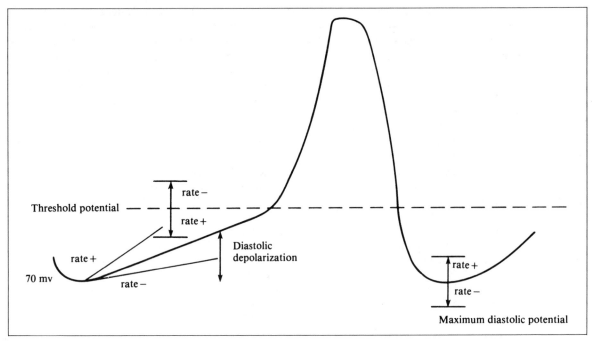

Threshold potential

rate −

rate +

Diastolic
depolarization

rate +

70 mv

rate −

rate +

rate −

Maximum diastolic potential

Fig. 14-7. The rate of spontaneous activation of a pacemaker cell is determined by three factors. As shown, heart rate alterations occur with the change in pacemaker rate due to shifts in (1) the slope of diastolic depolarization, (2) threshold potential (TP), and/or (3) maximum diastolic potential (MDP). Hence, affecting the levels of TP, MDP, and/or phase 4 diastolic slope will alter the time required to attain TP (rate + = rate increase or tachycardia; rate − = rate decrease or bradycardia).

result in complete cardiac standstill. Increasing temperatures increase the slope of diastolic depolarization in a manner similar to that seen with epinephrine (Fig. 14-8A). High temperature may also induce Purkinje fibers to become pacemaker cells, as is also seen with catecholamines (Fig. 14-8B).

IONS. Very few studies have been reported on the effect of K^+ on the electrophysiological properties of cells possessing automaticity. From the available information, it appears that a decrease in extracellular K^+ causes an increase in the slope of diastolic depolarization (probably due to a decrease in P_K^+), a decrease in the maximum diastolic potential, and a reduction in threshold. As shown in Figure 14-7, low K^+ results in a less negative maximum diastolic potential and a more negative threshold potential, and the slope of diastolic depolarization increases. All three of these phenomena accelerate activity in the normally automatic cells of the heart. Multifocal activity develops in SA node tissue perfused with a K^+-free solution. In normally quiescent but latent automatic fibers (Purkinje system), low K^+ concentration tends to initiate pacemaker activity and frequently leads to the production of extrasystoles. Eventually, however, such fibers exposed to an extracellular environment lacking in

K^+ will become arrested because of loss of resting potential.

On the other hand, an increase in extracellular K^+ has an opposite effect. The steepness of phase 4 is decreased, probably on account of an increase in membrane permeability to potassium. This would decrease the firing rate of the normal automatic cells. It has been stated, however, that the initial effect of increasing extracellular K^+ is an increased rate of firing from the SA node, which can occur, since high K^+ also decreases the magnitude of the maximum diastolic potential of the SA nodal cells. The firing rate increases or decreases following high K^+ exposure, depending on whether the cation effect is predominantly on the resting potential or on the diastolic depolarization.

Generally, K^+ affects all the cardiac cells in a similar fashion. However, the sensitivity of the cardiac cells to the cation is markedly different. Within the same heart it appears that the cells of the SA and AV nodes are more resis-

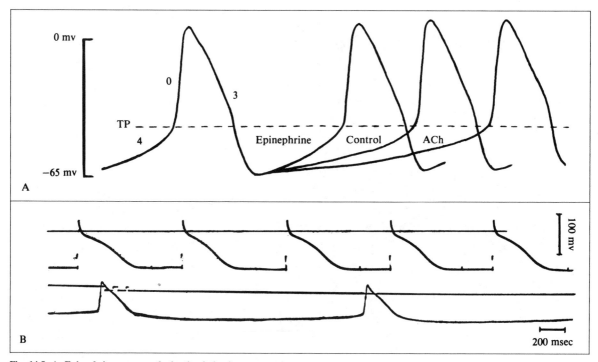

Fig. 14-8. A. Epinephrine or sympathetic stimulation increases heart rate by enhancing the slope of diastolic depolarization while parasympathomimetic activity or ACh depresses the steepness of phase 4 depolarization to result in the slowing of heart rate. B. Purkinje fiber action potentials (upper strip), which are converted to the pacemaker type (lower strip) with the influence of epinephrine.

tant to K^+ than are the contractile cells of the myocardium. In turn, the Purkinje fiber network is more sensitive to K^+ than are ventricular cells, while the bundle of His is highly resistant to increases of extracellular K^+. The sensitivity to K^+ also differs within the contractile tissue in that conduction fails earlier in the atrial fibers than in ventricular fibers. Of interest is the high resistance of the SA nodal cells to K^+. Increasing the perfusing solution with K^+ to five times greater than normal results in complete suppression of atrial activity, whereas the SA nodal cells show only a slight decrease in the magnitude of the resting potential and in the slope of diastolic depolarization. Only when the K^+ concentration is markedly increased (to 15 times normal) will the spontaneous activity of the pacemaker become arrested.

Other ions, such as Na^+ and Cl^-, have little effect on the automatic cells of the SA node unless the ionic changes are severe. However, Ca^{2+} certainly plays an important role in the inherent property of automaticity.

THE ELECTROCARDIOGRAM

In any fundamental consideration of the constitution of the ECG due to differences in the basic electrical properties of different cardiac tissue, the registration of the electrocardiographic complex consists of various wave forms. A representative complex is illustrated in Figure 14-9A. The P wave is constituted by the sum of atrial muscle depolarization. The QRS complex reflects ventricular muscle depolarization. The T wave represents ventricular muscle repolarization.

The relative magnitude of these electrocardiographic components reflects the relative mass of tissue represented. For example, the ventricular mass, which is responsible for the relatively large QRS complex, comprises by far the greatest bulk of the heart. The specialized conducting cells and pacemaker cells (SA node, specialized atrial fibers, AV junctional cells, and His-Purkinje cells) are not graphically registered in the conventional ECG because of relatively small total mass and their distance from the conventional recording electrodes. Atrial repolarization is infrequently observed, since it usually occurs simultaneously with ventricular depolarization, the wave form of the latter obscuring the atrial T wave.

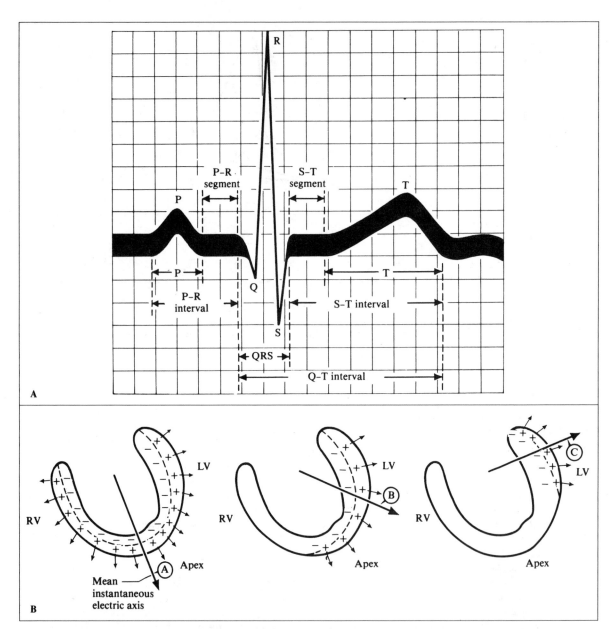

Fig. 14-9. (Upper panel) The normal ECG showing the time relationships. The distance between the vertical lines is 40 msec. The distance between the horizontal lines is 0.1 mv. (From G. E. Burch and T. Winsor. *A Primer of Electrocardiography* [6th ed.]. Philadelphia: Lea & Febiger, 1972. P. 320). (Lower panel) A. The mean instantaneous electrical axis at the beginning of ventricular depolarization. B. The mean instantaneous electrical axis at a later stage of ventricular depolarization. The vector is more to the left and with excitation proceeding toward the epicardium. C. The mean instantaneous axis late in ventricular depolarization. LV = left ventricle; RV = right ventricle.

Normally, the impulse originates in the SA node located in the right atrium at the opening of the great veins, and it traverses through the specialized paths to activate both atria. As a result, the P wave is inscribed on the ECG. Hence, the P wave represents the electrical record of atrial depolarization. Next, the impulse is delayed for some time in the AV junctional tissue and then is rapidly conducted through the bundle of His, the left and right bundle branches, and the Purkinje system. On the ECG this entire time is recorded merely as a straight line, the so-called isoelectric P–R segment. Finally, the myocardium is activated, the impulse going from endocardium to epicardium; on the ECG this is recorded as the QRS complex. The electrical status must then be reestablished, and this process of repolarization gives rise to the T wave on the ECG.

However, the surface ECG does not closely resemble the single-fiber action potential, the summated action potentials of which it is comprised. The source of such a marked disparity becomes evident when one considers that the surface ECG represents an algebraic sum of numerous action potentials, oriented in many directions and happening serially as well as simultaneously. If depolarization and repolarization occurred simultaneously in all fibers, the net electrical difference between two recording electrodes would be zero. But depolarization of ventricular fibers, initiated via the His-Purkinje conduction system, takes place initially in endocardial fibers and then proceeds radially to the epicardium (Fig. 14-9A–C). Thus, a QRS dipole vector is established by this temporal sequence of ventricular depolarization. As schematically illustrated in Figure 14-10, the addition of endocardial and epicardial action potentials yields an isoelectric S–T segment. It would seem reasonable, as well, to expect the T wave to be of equal amplitude and opposite direction to that of the QRS if repolarization occurred in the same sequence as depolarization. As shown in the figure, however, more rapid repolarization of the epicardium than of the endocardium yields an upright T wave in the presence of an upright QRS complex.

The most conventional method of recording the summation of action potentials involves the use of the three standard leads, as conceived by Einthoven in 1908. Viewing the heart as an electrical dipole situated in the center of an equilateral triangle (Fig. 14-11), one can perceive the derivation of the three standard leads. Lead I is recorded between the right and left arms, producing an upward deflection if the left arm is positive with respect to the right arm (i.e., when the projection of the dipole vector renders the left arm positive with respect to the right arm). Lead II is recorded between the right arm and left leg, yielding an upward deflection if projection of the cardiac vector on this lead

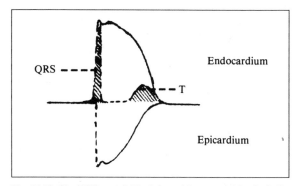

Fig. 14-10. The ECG record (shaded area) is generated by the bulk conduction of the sum of all the muscle action potentials of the heart. The typical record showing the QRS complex (ventricular muscle depolarization) and the T wave (ventricular muscle repolarization) occurs because there is a difference in the time course of the action potentials of the endocardium and epicardium. Such a lack of synchrony allows the current to flow, and hence a voltage can be measured at the body surface.

produces positivity in the left leg. Lead III is recorded between the left arm and left leg, yielding an upward deflection if the left leg is positive with respect to the left arm. Since Einthoven's triangle constitutes an electrical network, physical principles require that the sum of all electrical forces equal zero. Thus, Einthoven's law states that lead I voltage plus lead III voltage equals lead II voltage at any instant of the cardiac cycle. (Einthoven's modification of this basic principle involved a reversal of lead II polarity; thus, lead I + lead III = lead II.)

So the typical ECG of a cardiac cycle consists of three positive waves, the P, the R, and the T waves. Sometimes one can record a fourth wave, called the U wave, which is of unknown origin. It may represent afterpotentials, or it may represent repolarization of the Purkinje network. In any event, besides the positive waves there are two negative waves, the Q wave and the S wave. The Q and S waves with the R wave are collectively referred to as the QRS complex.

It is customary to quantitate these potentials generated from the heart, and we can carry out quantitation from the ECG because of the horizontal and vertical lines that are inscribed on the recording paper. The horizontal and vertical lines are 1 mm apart and represent time and amplitude respectively. Each millimeter of horizontal line is equal to 0.04 second, with every fifth horizontal line being wider and more heavily inscribed so that the distance between two adjacent heavy horizontal lines represents 0.2 second. The vertical lines are also wider after every fifth line, and the

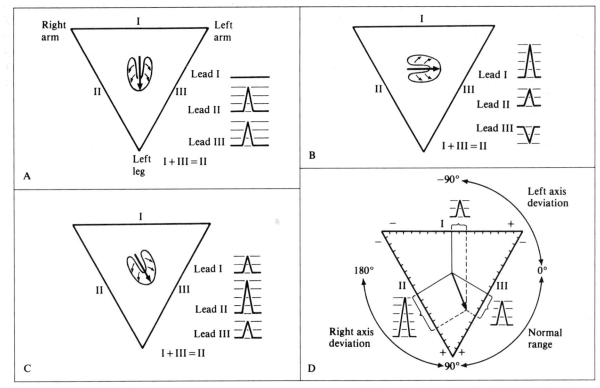

Fig. 14-11. Relationship of electrical axis of the heart to the recorded potentials. A. The vector of excitation, or axis, is directly vertical downward. The voltage changes seen at each lead are shown. B. The vector of excitation, or axis, is directly horizontal and to the left arm. C. The vector of excitation, or axis, is downward and toward the left arm. Lead II, which is most parallel to the axis, has the greatest deflection. D. To determine the mean electrical axis from records obtained for any two leads: Count off from the midpoint of each lead in the appropriate direction in the number of millimeters of deflection obtained. Drop perpendiculars from each point. The line drawn from the midpoint of the triangle to the point of intersection of the perpendiculars represents the axis. The arrowhead represents its direction, the length of the line represents its magnitude, and the degrees represent its orientation.

machine is so standardized that the addition of 1 mv from the circuit causes the pen to deflect 10 mm or 1 cm. We speak of these waves, intervals, and segments in terms of duration in seconds, and of amplitude in terms of millivolts or millimeters.

To analyze a single cardiac cycle: We first have the P wave, which represents atrial depolarization. The width of the P wave measured from the beginning to the end of the wave should not normally exceed 0.12 second. Its height does not usually exceed 2.5 mm. Note that, although the P wave represents atrial depolarization, we normally do not see an atrial repolarization or atrial T wave (Ta) because it is very small in voltage and is usually buried in the P-R segment and in the QRS complex. The P-R segment is measured from the end of the P wave to the beginning of the QRS complex, and its duration is about 0.08 second. The measurement of the beginning of the P wave to the beginning of the QRS complex is called the P-R interval or P-Q interval and represents the atrial depolarization time plus conduction through the atrial tissue, the AV node, and the specialized conduction system. Normally, this should not exceed 0.22 second.

The P-R interval is followed by an initial downward deflection called the Q wave, then an upward R wave followed by a downward S wave; normally, this QRS complex should not exceed 0.1 second. Since the complex represents ventricular depolarization or phase 0, a duration of this complex longer than 0.1 second is an indication of intraventricular conduction delay. That portion of the cycle between the end of the QRS and the beginning of the T wave is known as the S-T segment. It represents the depolarized state of the ventricle or phase 2. If one includes the T wave in the measurement, it is called the S–T interval, which represents the time from peak depolarization to complete ventricular repolarization. Measuring from the beginning of the Q wave to the end of the T wave, one obtains the Q–T interval, which normally lasts from 0.36 to an upper limit of 0.4 second and represents the entire time required for depolarization and repolarization of the ventricular muscle. This time varies with age, sex, and, obviously, heart rate. As a rough index the Q–T interval can be considered normal if its duration is less than half of an R–R interval. Finally, the T wave is equivalent to phase 3 of the action potential and represents ventricular repolarization. Its amplitude and duration depend on many factors. It is a very labile component and can be altered by emotional factors, smoking, drinking, and position of the heart; the significance of such changes is frequently difficult to evaluate.

We can best account for these phenomena in terms of the migration of a positive and a negative charge, collectively called a dipole. A dipole can be pictured as consisting of two points of opposite electrical charge separated by a very small distance. If the points are connected by a wire, then by convention it is understood that current will flow along the wire from the positive pole (anode) to the negative pole (cathode); actually, the electrons flow from the negative to the positive pole. We can then analyze the human ECG in terms of the body surface potentials, whose source is generated by a series of dipoles emanating from the heart and distributed throughout the body. So we consider the patient as a volume conductor and the electrical impulses originating in the heart as the source of potential difference. Adding and subtracting all of the many dipoles traveling in different directions within the heart at any given moment, we can represent, for practical purposes, the wave of depolarization as a single dipole. This can also be represented as a vector with direction, magnitude, and sense (that is, positivity or negativity). The vector travels in the direction of the depolarization, with the arrow indicating the positive field.

The term *cardiac vector* is used to designate all of the electrical events of the heart cycle. It has a known magnitude and direction. The *instantaneous vector* represents the net electrical forces being propagated through the heart at a given instant. A *mean vector* of any given portion of the heart cycle (such as the QRS interval) represents the mean direction and magnitude for that period in the heart cycle. The mathematical symbol of a vector is an arrow pointing in the direction of the net potential (that is, polarity), while the length of the arrow indicates the magnitude of the electrical force.

The summation of cardiac action potentials thus yields a net or resultant vector, with force and direction. Since the three standard leads reflect the projection of this cardiac vector on three separate axes, it is apparent that the cardiac vector can be reconstructed from the standard leads (as shown in Fig. 14-11). The orientation of the QRS axis, in the frontal plane, is usually inferior and somewhat to the left. Change in the mean QRS axis can be produced by mechanical alteration of the cardiac position, as well as by a wide variety of cardiac and pulmonary diseases that change the position of the heart. The electrical axis of the P and T waves may be calculated in a similar fashion.

LEAD SYSTEMS. Actually, the ECG recorded from the body surface is the result of the summation of all the conducted electrical impulses that are occurring in each cell of the heart. In clinical electrocardiology the electrical potential generated by the heart is recorded with bipolar leads or with unipolar leads.

With bipolar leads, both electrodes are in the electrical field generated by the heart, and the ECG records the difference between two points. The actual potential under either of the electrodes is not given. With unipolar leads, one of the electrodes is so constructed that it is electrically zero. This is called the *indifferent electrode*. The other connection of the unipolar lead sees the potential as absolute and is called the *exploring electrode.*

The important thing to remember in order to understand the clinical ECG is that it is the exploring electrode that determines the polarity of the wave. The ECG is so constructed that when the exploring electrode is in the positive field, the ECG records an upright deflection. If the exploring electrode is in the negative field, the ECG registers a downward deflection.

Bipolar Leads. Clinically three bipolar extremity leads are used. We think of the extremities merely as lead wires connected to the body, and the potentials are not altered if the electrodes are moved along the extremities. As previously stated, the conventional bipolar limb leads, designated as leads I, II, and III, record differences between right arm, left arm (LA), and left leg (L). In lead I the connection is between left arm and right arm and the LA lead is the exploring electrode. In lead II the connection is between the right arm and the left leg, and the L electrode is the exploring one. In lead III the connection is between left arm and left leg and again the L is the exploring electrode. The three limb leads are connected in what is known as Einthoven's triangle (Fig. 14-12A). If Einthoven's hypothesis is true, one should be able to hook up the three

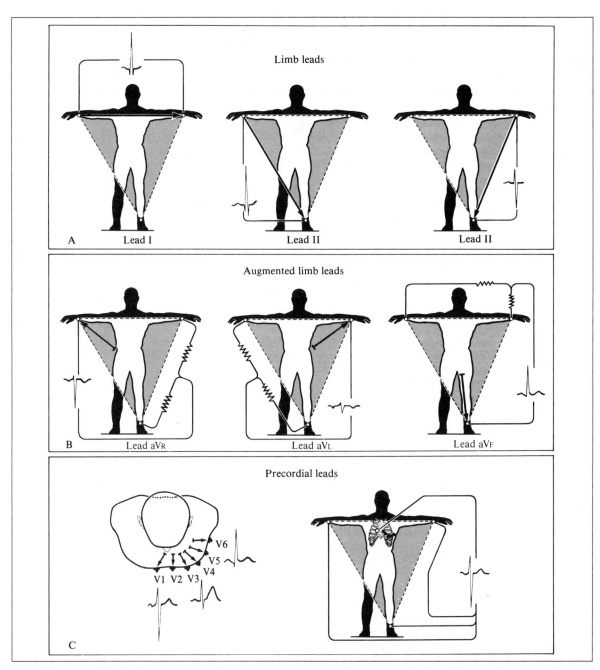

Fig. 14-12. A and B. Standard limb and augmented unipolar leads. Resistors are used to form an indifferent reference electrode in the unipolar leads. C. Unipolar precordial leads for V1–V6 resistors in the reference leads are omitted. The arrows indicate the vector required for an upward deflection of the recording machine. (From F. L. Abel. Functional Characteristics of the Heart. In E. E. Selkurt [ed.], *Basic Physiology for the Health Sciences*. Boston: Little, Brown, 1975. P. 362.)

extremity leads to an *indifferent electrode,* located so far away from the heart that its potential is considered to be zero. The indifferent electrode constitutes the central terminal of Wilson, while the exploring electrodes are the *unipolar extremity electrodes*—one attached to the right arm, called V_R; one to the left arm, V_L; and one to the left foot, V_F. The V indicates that the Wilson terminal is the reference electrode. In this system the potentials recorded are small, and it was therefore suggested that the potential at an extremity could be obtained simply by disconnecting one extremity electrode from the central terminal and recording the difference in potential between it and the remaining two extremity electrodes. Such a system gives potential amplitudes 1.5 times greater than a V lead amplitude (see below), and hence this system for recording is called the augmented unipolar system with the designations: aV_R (right arm), aV_L (left arm), and aV_F (left foot) (Fig. 14-12B).

Unipolar Lead System. The unipolar leads include six chest leads, called the V leads, recorded with the exploring electrode placed at the following positions: V1 is placed immediately to the right of the sternum at the fourth intercostal space. V2 is placed left of the sternum also at the fourth intercostal space. V4 is placed in the fifth intercostal space at the midclavicular line. V3 is placed between V2 and V4. V5 is placed over the fifth intercostal space in the anterior axillary line, and V6 is placed over the fifth intercostal space but at the midaxillary line.

In humans the right ventricle lies anteriorly, the left ventricle posteriorly. The septum is tilted slightly forward apically, and the base-to-apex axis of the ventricle is almost parallel to the diaphragm. The shape of the ventricular complex recorded at the body surface is determined by the pattern of ventricular activation, the particular ECG lead, and the position of the heart within the chest. Again, remember that when depolarization, represented as a dipole or vector, travels *toward* the *exploring* electrode, the latter is in the positive electrical field, and the ECG registers an upright or positive deflection. When the impulse travels *away* from the exploring electrode, the electrode finds itself in the negative field, and the ECG registers a downward or negative deflection. The initial phase of ventricular activity is usually directed from left to right in the septum, resulting from earlier and/or greater initial left-to-right activity. This activity produces a wave directed to the right, toward the head, and possibly slightly anteriorly. The wave will produce an initial negative deflection in leads I, II, and III, which accounts for the Q wave. At the same time, at the precordial leads, the leads on the right side of the chest, V1 and V2, face the positive side of the wave front and so record an upward deflection. The leads on the far left side of the chest, V5 and V6, record a negative deflection, since they see the negative side of the vector. The leads directly over the heart, V3 and V4, usually record biphasic complexes.

Soon after invasion of the septum begins, rapid conduction through the Purkinje system results in an irregular pattern of inside-out spread in the walls. Within the septum, activity is proceeding from left to right and from endocardium to epicardium. In all the limb leads, as well as in V1 and V6, the potential is near zero. In V2 and V5 the deflections are positive because activity proceeds toward the apex and free left wall. At about 40 msec after ventricular activation the overall activity is directed apically, posteriorly, and to the left. At this time all of the standard limb leads record positive deflections, and the leads on the left side of the chest "see" approaching activity, while the right side leads "see" negativity (Fig. 14-12C). The overlying lead, V3, is near zero. Finally, after depolarization of the apical regions is complete, the wave moves forward to the base of the left ventricle, particularly posteriorly, and from apex to base in the septum. This movement causes no potential in lead I and some negativity in II and III; a V_R is positive. V1 and V2 are returning to zero from negativity, and V4, V5, and V6, to zero from positivity.

To summarize, we can divide ventricular activation into two major phases. The first is initial depolarization of the septum from left to right, down and anteriorly. The second consists in depolarization of the ventricular walls from endocardium to epicardium. The depolarization of the free ventricular wall as registered on an ECG is predominantly that of the left ventricle. Early simultaneous depolarization of the right and left ventricles in opposite directions results in a net "zero" potential. The left ventricle, being about three times as thick as the right ventricle, continues to depolarize after the right ventricle is completely depolarized and generates a strong unopposed vector or electrical field directed to the left (Fig. 14-13).

In the foregoing discussion of the ECG we have considered all the lead systems. If one emphasizes only the three standard leads, such a limited frame of reference permits an evaluation of the cardiac vector only in relation to the frontal plane (i.e., laterally and superiorly-inferiorly). It should be noted that the mean cardiac vector courses anteriorly, as well as to the left and inferiorly, in the average person. These anteroposterior forces are projected on the horizontal and sagittal planes but not on the frontal plane. Conventional electrocardiography provides information regarding such an anteroposterior component by recording additional leads over the anterior thorax. Furthermore, besides re-

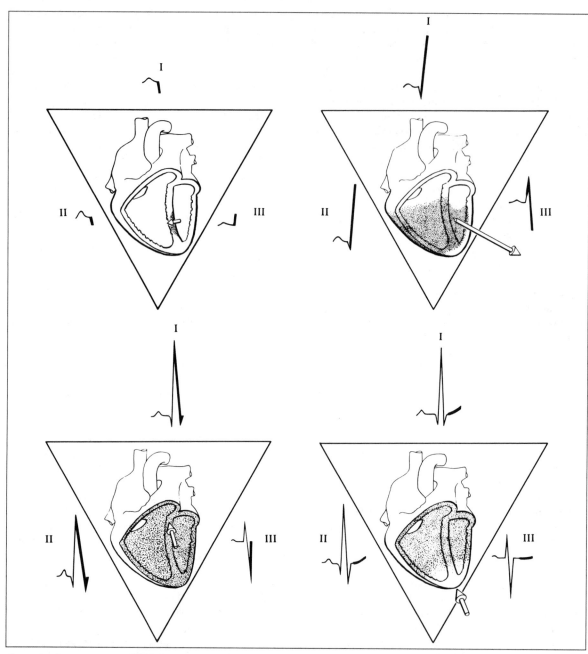

Fig. 14-13. Generation of the bipolar ECG signals by the underlying ventricular activity. Progressive depolarization to generate the QRS complex and early repolarization with the beginning of the T wave are shown (from left to right and from above downward). Stippled areas represent depolarization. The drawing on the lower right shows the beginning of repolarization of the apex. The resultant electrical vector (arrow) has components along the Einthoven triangle producing the signals shown in each limb lead. (From F. L. Abel. Functional Characteristics of the Heart. In E. E. Selkurt [ed.], *Basic Physiology for the Health Sciences.* Boston: Little, Brown, 1975. P. 363.)

cording an anteroposterior component, the precordial leads can disclose the electrical activity of localized areas of the myocardium, to which they are in closer proximity. This latter capacity is provided by the unipolar character of the precordial electrodes.

SPECIAL RECORDING TECHNIQUES. Under circumstances when the usual surface ECG does not provide sufficient information to aid in diagnosis, special recording systems are used. Thus, for a better resolution of left atrial activity an esophageal bipolar or unipolar lead is employed, while a bipolar or unipolar intraatrial lead is frequently used to define right atrial activity more clearly. In either case the utilization of such special lead systems occurs in association, or simultaneously, with the standard ECG.

The special lead systems are not replacement techniques. Furthermore, specific recording of electrical activity from the "specialized conducting system" may be obtained by the insertion, under appropriate anesthetization and visualization, of bipolar catheter electrodes. Usually, the catheter electrode technique is restricted in use to the assessment of the site(s) of abnormal electrical activity and to the determination of a specific site or sites of conduction block(s). In addition, special monitor chest leads may be *attached* to the patient when there is the need or desire for a continuous monitoring of cardiac activity. Frequently, as in the coronary care unit or hospital emergency room, these monitor leads are self-adhering, disposable disk electrodes, though extremity plates are sometimes used. The conventional placement of the disk electrodes is as follows:

1. In lead I the positive electrode is placed below the left clavicle with the negative electrode under the right clavicle;
2. In lead II the positive and negative electrodes are placed below the left pectoral muscle and the right clavicle respectively;
3. In lead III the positive and negative electrodes are placed beneath the left pectoral muscle and below the left clavicle, respectively. In all of these recordings, ground can be anywhere on the body surface.

Frequently, cardiac monitoring is done by connecting a positive electrode parasternally in the right fourth intercostal space, with a negative electrode placed near the left shoulder just under the outer third of the left clavicle: this is referred to as a monitor chest lead (MCL). It is important to note that in each of these telemetry techniques the *positive electrode* must be placed to the left and below the negative electrode placement. Otherwise, the deflections recorded on the paper strip will be reversed.

THE NORMAL ECG. The components of the surface ECG (see Figs. 14-9 and 14-14) represent circumscribed areas of depolarization and repolarization. Emanating from the area of the SA node, or so-called primary pacemaker, the wave of depolarization first traverses the atria, producing the P wave of the surface ECG, of 80- to 100-msec duration. The P wave vector is usually oriented toward the left arm and inferiorly, producing an upright deflection in the three standard leads (Fig. 14-14).

Following the P wave an isoelectric interval occurs, averaging 80 to 100 msec in duration and reflecting the delay in passage through the AV junctional area. The QRS complex reflects depolarization of the ventricles in a characteristic sequence. The first vector represents depolarization of the interventricular septal fibers, from left to right, often producing a small Q wave (i.e., an initial downward deflection of the QRS complex) in the three standard leads and V6, and a small R wave in lead V1. Subsequently, the principal vector becomes manifest, reflecting depolarization of the bulk of the left ventricle, leftward and inferiorly (along the positive axis of the limb leads). This is inscribed as a tall R wave (i.e., an upward deflection of the QRS complex) in leads I, II, III and the lateral precordial leads V4, V5, and V6. The precordial lead V3, anatomically positioned to the right of the bulk of ventricular tissue, records a deep S wave as the wave front is moving away from the electrode. (An S wave refers to a downward deflection following an upward deflection of the QRS complex.)

Terminally, the more basal region of the ventricles be-

Fig. 14-14. A normal human ECG showing a single complex in each of the 12 leads. Note the R wave amplitude progression in the precordial V1–V6 leads.

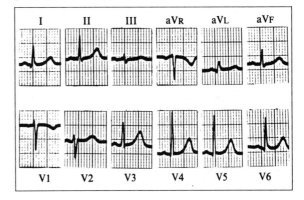

comes depolarized, producing a smaller S wave in leads II and III. The entire course of ventricular depolarization, thus completed, normally consumes 80 to 100 msec.

Following ventricular depolarization the algebraic sum of ventricular repolarization becomes manifest in the ECG. The S–T segment corresponds approximately to that portion of the single-fiber action potential referred to as the plateau or phase 2 (see Fig. 14-10). The T wave, in contrast, more closely corresponds to the period of rapid repolarization (phase 3) of the single-fiber action potential. As illustrated in Figure 14-1B, the epicardial fibers (the last to depolarize) exhibit more rapid repolarization than do the endocardial fibers. This paradoxical sequence results in the inscription of an upright T wave in the presence of a tall R wave, as previously noted.

Determination of Heart Rate. Normally, the paper speed is 25 mm per second. For rapid determination, one takes the number of 0.2-second periods between two R waves and divides it into 300, because there are three hundred 0.2-second intervals in 1 minute. For more accurate measurements one counts the number of heart cycles (R–R intervals) in a 3-second period and multiplies them by 20 to get the rate per minute.

Analyzing the Electrocardiogram. In attempting to analyze an ECG tracing, the following should be done:

1. First make sure that a 1-mv standardization input has produced a 1-cm deflection.
2. Be sure that each lead recorded has several beats.
3. Determine heart rate. If the heart is not in sinus rhythm, determine the atrial and ventricular rates separately.
4. Note whether the rate is regular or irregular.
5. Measure the P–R interval, Q–T interval, and QRS duration.
6. Note whether the relationships between P, QRS, and T waves are constant or variable.
7. Look at the complexes to see whether their shape and duration are normal.
8. Determine the electrical axis of the heart. This can be computed easily and offers much aid in the diagnosis of ECG abnormalities.

We compute the axis from two limb leads that have been recorded simultaneously. The algebraic sum of the positive and negative peaks in a lead is measured in millimeters and plotted along the proper side of the triangle. A perpendicular is then drawn at the termination of the line. This sequence is repeated for a second lead. The line forming the center of the triangle and the intersection of the two perpendiculars is the mean electrical axis.

In most normal persons, the axis falls between 0 and +90 degrees. If the axis is greater than +90, there is said to be right axis deviation. If the axis lies in the negative portion of circle, the person is said to have left axis deviation.

CHANGES IN ELECTROCARDIOGRAPHIC CONFIGURATION. Alterations in repolarization, producing electrocardiographic T wave changes, are the most frequently encountered of ECG abnormalities. While many of these changes are empirically regarded as abnormal, the mechanism by which they occur remains obscure. They are therefore usually referred to as nonspecific T wave changes. Susceptible to a variety of noxious factors, such as ischemia, drugs, or endocrine, metabolic, or electrolyte imbalance, repolarization may be diffusely or locally affected. More rapid repolarization of the endocardium, for example, may result in S–T segment depression and inversion of the T wave (by allowing epicardial repolarization to dominate the summated ECG complex), as can be inferred from Figure 14-10.

Damage to cardiac fibers may impede the course of normal repolarization and produce a so-called injury current, the area involved remaining electrically negative with respect to the unaffected region. This type of injury results in S–T segment shifts upward in the ECG leads overlying the damaged tissue (Fig. 14-15). As the damaged cells become electrically inactive, the S–T segment subsides. With electrical death of the myocardium, deep, wide Q waves develop. Such "electrically silent" areas are manifested by Q waves, because the dead myocardium acts as a "window"—the surface electrode records only the residual unopposed forces of depolarization, the way an intracavitary electrode would. Thus, the wave of depolarization travels away from the recording electrode. The changes demonstrated in Figure 14-15A are observed clinically in acute myocardial infarction, where the tissue is ischemic but still viable. The development of Q waves correlates with replacement of the dead tissue with fibrous tissue (Fig. 14-15B).

Conduction block within the Purkinje conducting system of the heart produces characteristic alterations of QRS configuration. Such disturbances in conduction may be due to either "functional" (e.g., rate dependent, prolonged refractory period) or lesions of the Purkinje system. This type of injury or lesion causes conduction of the impulse to be delayed as it courses over longer, more circuitous pathways and via the more slowly conducting ventricular muscle fibers. Lesions of one or the other of the main bundle branches, for example, yield characteristic electrocardiographic patterns, as illustrated in Figure 14-16. Right bun-

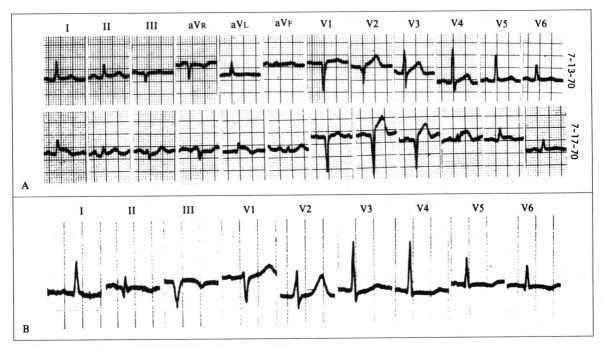

A
B

Fig. 14-15. A. The ECG recorded 7-13-70 demonstrates a normal S–T segment configuration. The ECG recorded 7-17-70 illustrates the progressive S–T segment elevation (leads I, aVL, V2, V3, V4, and V5) typical of acute myocardial infarction. **B.** Healed myocardial infarction is characterized by abnormally wide and deep Q waves (leads II and III).

Fig. 14-16. Right bundle branch block (RBBB) is characterized by right axis deviation, wide S waves in lead I and V6, and an RSR′ in V1. Left bundle branch block (LBBB) is typified by broad R waves in lead I and V6 and absence of the small R wave in V1.

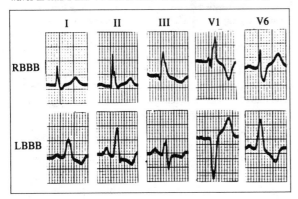

dle-branch block produces alteration of the QRS complex with widening due to late, unopposed depolarization of the right ventricle, represented by a wide terminal S wave in leads I and V6 and a secondary R wave (R′) in V1. The left bundle branch block pattern, on the other hand, occurs as the result of a disturbance in the initial course of ventricular depolarization. Normally, septal depolarization takes place from left to right on account of early branching of the left bundle. In left bundle branch block, however, the septum is depolarized in the opposite direction with loss of the so-called septal *q* wave often found in leads I, II, and V6. Late depolarization of the left ventricle then produces a widened, notched R wave in these leads.

CARDIAC DYSRHYTHMIAS. Automaticity The concept of dysrhythmia implies a disruption, transient or prolonged, in the normal sequence of an electrical activation of the heart. The propensity for creating such a disruption is inherent in cardiac tissue, as is evident from a consideration of the electrophysiological properties of various fiber types. As discussed previously, fibers possessing a capacity for automatic behavior are present in virtually every area of the heart. Specialized fibers are found in the atria; lower AV junctional fibers clearly manifest automaticity; and His-Purkinje fibers pervade the ventricular myocardium.

Normally, the more rapid rate of the SA pacemaker precludes the expression of activity by these more subordinate pacemakers, which have an inherently slower rate of spontaneous diastolic depolarization. Occasionally, however, a secondary pacemaker (ectopic* pacemaker) depolarizes prematurely and at a time when the surrounding tissue is not refractory, and thus may *initiate* depolarization. An impulse originating from specialized atrial tissue is referred to as an atrial premature excitation (APC). Its ventricular counterpart is known as ventricular premature excitation (VPC).†

The atrial ectopic beat (Fig. 14-17E) produces a P wave of atrial depolarization that may differ in amplitude and configuration from the sinus P; the QRS complex approaches the normal configuration (Fig. 14-17A), since the stimulus traverses the AV node and the normal ventricular conduction system. The premature complex, having depolarized the atrium and SA node region, however, is followed by a slight compensatory pause before reestablishment of SA excitability. Occasional spontaneous activity may originate from the AV junctional–His bundle area as well, producing similar ventricular excitation, although atrial depolarization here may be obscured by the QRS. Impulses originating from this area may activate the atrium in retrograde fashion, resulting in a change of the atrial electrical axis, or they may fail to excite the atrium (retrograde).

The NPC, originating from one or the other ventricle (Fig. 14-17F), elicits ventricular depolarization over abnormal pathways, producing a widened QRS complex of unusual configuration. This ectopic complex is not ordinarily associated with atrial depolarization; hence the SA rhythm is not disturbed. The ventricles are refractory to the next sinoatrial impulse, however, and a so-called *fully compensatory pause* results.

Although such occasional ectopic excitations may be observed in normal persons, under conditions that enhance the automaticity of subordinate atrial pacemaker fibers, ectopic activity may increase in rate sufficiently to usurp the function of the SA node. This circumstance may be evident either as frequent atrial premature impulses interrupting the sinus rhythm or as a persistent atrial rhythm, referred to as *atrial tachycardia*. A dysrhythmia of this type may vary in rate from 140 to 240 beats per minute (Fig. 14-17B).

Similarly, latent ventricular pacemaker activity may be enhanced to the point of producing ventricular tachycardia

---

*The term *ectopic* refers to an impulse or rhythm initiated by a cardiac pacemaker outside the SA node.

†The terms *APC* and *VPC* have, to date, been used to connote atrial or ventricular premature *contractions*. However, the term *contraction* so used is meaningless when applied to an ECG tracing, since the ECG is the recording of the *electrical* events. The latter may or may not lead to the actual mechanical event. Hence, *APC* and *VPC* as related to the ECG will be used here only for premature excitations or complexes.

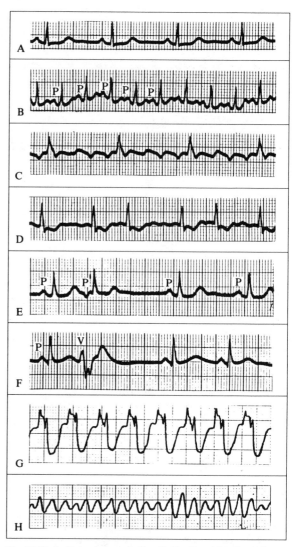

Fig. 14-17. A. Normal sinus rhythm. B. Rapid atrial tachycardia. C. Atrial flutter with a 4 : 1 ventricular response. The baseline demonstrates characteristic sawtooth configuration. D. Atrial fibrillation with an irregular ventricular response. The baseline shows no definitive P waves. E. ECG recording of an atrial premature beat (P'). F. Ventricular premature beat (V), followed by a compensatory pause and two normal sinus beats. G. Ventricular tachycardia characterized by wide aberrant QRS complexes and a rapid rate of 150/min. H. Ventricular fibrillation characterized by an undulating baseline without definable QRS complexes.

(Fig. 14-17G), by usurping SA dominance (i.e., beating more rapidly than the sinus node and precluding expression of sinus activity). The rate of such a pacemaker may vary from 100 to 150 beats per minute. This electrocardiographic rhythm is characterized by distorted and prolonged QRS (aberrant) waves, not preceded by atrial deflections. Relatively uncommon, ventricular tachycardia is observed primarily in the presence of severe cardiac disease or digitalis intoxication and is associated with deterioration of cardiac output.

Atrial Flutter and Fibrillation. More rapid ectopic pacemaker activity in the atrium may proceed to the point of atrial flutter (Fig. 14-17C). This is characterized by discrete depolarizations occurring at a rate of 200 to 350 per minute. Slow conduction through the AV node and a longer refractory period of this area do not permit a 1:1 ventricular response. Consequently, AV block of varying degree is produced, the ventricles often responding once to each two or four flutter waves. This type of dysrhythmia is most commonly observed in rheumatic and coronary heart disease.

Atrial fibrillation (Fig. 14-17D) has an even more rapid rate (300–450/min) and, in addition, uncoordinated atrial activation. Atrial contraction is totally ineffective, while the ventricular response to such an atrial rhythm is totally irregular (although the ventricular contractions per se are coordinated and effective). The mechanism of dysrhythmia is of obscure origin, possibly produced by rapidly firing single or multiple ectopic foci or by a continuous circus movement* initiated from one focus and perpetuated by a slow conduction velocity (permitting recovery of responsiveness in previously excited areas), a long conduction path, or a short refractory period.

Ventricular Flutter and Fibrillation. Ventricular flutter is a dysrhythmia considered by some cardiologists to be a separate entity, while others regard it as a phase preceding ventricular fibrillation. Ventricular rhythm during flutter is regular and is characterized by depolarizations that are oscillatory, at a rate of 175 to 250 per minute. The QRS complexes are wide and of large amplitude, and the S–T segment is indistinguishable from the T wave. The ventricular depolarization during flutter is probably from a single ectopic focus. It should be pointed out that ventricular flutter is seldom observed clinically.

Ventricular fibrillation is a dysrhythmia (Fig. 14-17H) characterized by rapid, uncoordinated ventricular activity

and a total irregularity of the QRS complexes. The condition is incompatible with life because of the ineffectiveness of ventricular contractions, for it is tantamount to ventricular arrest, with total loss of cardiac output. The mechanism of this type of dysrhythmia may be comparable to that underlying atrial fibrillation (although the latter is a *relatively* innocuous rhythm), associated with slowing of conduction velocity and the establishment of a circus movement.

Clinical and experimental experience with exogenous stimulation (i.e., an artificial pacemaker) has provided evidence that ventricular fibrillation may be elicited by a single stimulus at a time when the ventricle is extremely susceptible. This interval is referred to as the vulnerable period and is found toward the end of rapid repolarization (the latter portion of the T wave). These findings have stimulated the development of electrical pacing and defibrillating equipment synchronized to provide an appropriate stimulus well away from the vulnerable period.

CONDUCTION DISTURBANCES. Conduction disturbances may occur in all areas of the heart and may be related to slowing of conduction velocity itself or to an increase in conduction time secondary to a greater tissue mass. The latter results from depolarization traversing a greater mass at the same conduction velocity and may be noted in atrial or ventricular hypertrophy, producing a widening of the P wave or QRS complex respectively.

Conduction disturbances related to a decrese in conduction velocity are perhaps more commonly encountered in clinical practice and most frequently observed in regard to the transmission of depolarizing impulses from atria to ventricles via the AV junctional area. Fibers in the latter region exhibit a resting potential of low magnitude and consequently action potentials of low amplitude. Normally, these low-amplitude action potentials result in slowing of impulse propagation, producing the normal AV conduction delay. This type of conduction, in which action potential amplitude lessens and conduction velocity slows progressively because of successively less adequate action potential generation, is called *decremental conduction*. Thus, when physiological or pathological factors produce a further decrease in action potential amplitude, decremental conduction is enhanced, successive areas of fibers generating lesser and lesser action potentials, to the point of conduction failure. In the latter circumstance the fiber beyond the block remains excitable, but the stimulus is inadequate to effect a response. Illustrations of this phenomenon are found in Figure 14-18, in which AV conduction delay is seen to produce an increase in AV conduction time without other disturbances (first-degree AV block), intermittent block of atrial

*Unidirectional transmission of the impulse traversing in a circuitous manner (reentry) resulting in a continuous reexcitation.

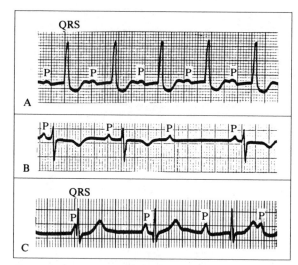

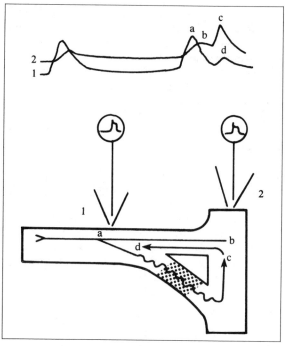

Fig. 14-18. A. First-degree block characterized by prolonged P–R intervals. There are no dropped beats. The maximal normal P–R interval is 0.22 second and in this case is 0.31 second. B. Second-degree AV block characterized by a dropped ventricular beat (after the third recorded P waves). After the dropped QRS complex, P–R is normal. C. Third-degree heart block or complete AV block. The P waves and QRS complexes are independent of each other. This is characterized by a simultaneous inscription of the P and QRS complexes of the first beat.

Fig. 14-19. Top panel shows the intracellular record from two electrodes placed in a Purkinje fiber shown in lower panel. The electrodes are positioned at locations 1 and 2. As an action potential invades site 1, it gives rise to action potential a (upper panel). This wave of depolarization passes from site 1 to site 2 fairly rapidly to give action potential b. A slower conduction pathway (shaded area) from site 1 to site 2 is also shown. As conduction occurs through this pathway, it reaches site 2 and gives rise to action potential c. Action potential c is then conducted back to site 1 to give action potential d. Hence, we have a reentry dysrhythmia resulting from drug-induced asynchronous conduction.

impulses (second-degree AV block), and complete cessation of atrioventricular conduction, with the atria and ventricles beating independently of one another (third-degree AV block).

Conduction disturbance is also the electrophysiological basis of reentrant dysrhythmias and can occur anywhere within the heart. In reentry the impulse is propagated at a given velocity that either speeds up or becomes reduced, so that the impulse enters other fibers with a conduction that is asynchronous. Hence, different areas of the heart will be activated at earlier or later times. When impulse conduction is sufficiently fragmented so that the tissue behind the impulse is repolarized, then that beat can turn around and "reenter." The end result is a production of echo or reverberated beats (Fig. 14-19). The phenomenon of reentry can therefore induce or help sustain a dysrhythmia initiated by automaticity.

BIBLIOGRAPHY

Burch, G. E., and Winsor, T. *A Primer of Electrocardiography* (6th ed.). Philadelphia: Lea & Febiger, 1981.

Fisch, C., Greenspan, K., and Knoebel, S. B. Electrophysiological Basis of Clinical Arrhythmias. In H. I. Russek (ed.), *Major Advances in Cardiovascular Therapy*. Paul D. White Symposium. Baltimore: Williams & Wilkins, 1978. Pp. 309–319.

Gadsby, D. C., and Witt, A. L. Electrophysiologic Characteristics of Cardiac Cells and the Genesis of Cardiac Arrhythmias. In R. D. Wilkerson (ed.), *Cardiac Pharmacology*. New York: Academic, 1981. Pp. 229–274.

Greenspan, K. Mechanisms of Ventricular Fibrillation. In L. S. Dreifus and W. Likoff (eds.), *Cardiac Arrhythmias: Pathophysiology, Pharmacology, and Treatment*. 25th Hahnemann Symposium. New York: Grune & Stratton, 1973. Pp. 195–202.

Greenspan, K., and Thies, W. H. Slow currents and the genesis of cardiac dysrhythmias. *Practical Cardiology* 4:48–65, 1978.

Hurst, J. W. *The Heart* (5th ed.). New York: McGraw-Hill, 1982.

Katz, L. N., and Pick, A. *Clinical Electrocardiography*. Philadelphia: Lea & Febiger, 1956.

CARDIODYNAMICS

15

Carl F. Rothe

The function of the heart is to propel enough blood into the aorta to maintain the blood pressure at a level sufficient to assure adequate flow of blood to all tissues. To understand these complex relationships, it is necessary to study the components of the system and the relationships between the following variables (Fig. 15-1):

1. The flow (F) of blood through the capillaries supplying the needs of cells depends on (1) the difference between the systemic arterial blood pressure and venous pressure (ΔP) and (2) the resistance (R) to the flow. This vascular resistance to flow depends on the internal diameter of the arterioles and precapillary sphincters (tissue vasomotor tone), as discussed in Chapters 11, 12, and 17.
2. The systemic arterial blood pressure (assuming that the central venous pressure is zero), is the product of (1) cardiac output and (2) the summed effects of all these peripheral resistance vessels, i.e., the total peripheral resistance. (These relationships are analogous to Ohm's law for electrical circuits.)

3. The cardiac output is the volume of blood pumped by the heart each minute and thus is the product of (1) the volume of each beat (the stroke volume,) and (2) the number of heartbeats per minute (the heart rate).
4. The stroke volume is the difference between (1) the volume of blood in the heart at the beginning of systole (the end-diastolic volume), and (2) the amount of blood that remains in the ventricles when the valves close at the end of systole (the end-systolic volume).

Four factors, then, are the primary determinants of systemic arterial blood pressure and cardiac output: (1) peripheral resistance to blood flow determined by the tissue vasomotor tone, (2) heart rate, (3) end-diastolic volume, and (4) end-systolic volume (Fig. 15-1). These factors are closely related. Some are intrinsic, being dependent on the physical characteristics of cardiac muscle and the vasculature, e.g., the Frank-Starling law of the heart. Other interrelationships are influenced by factors extrinsic to a tissue, e.g., neural and hormonal influences such as those from the presso-

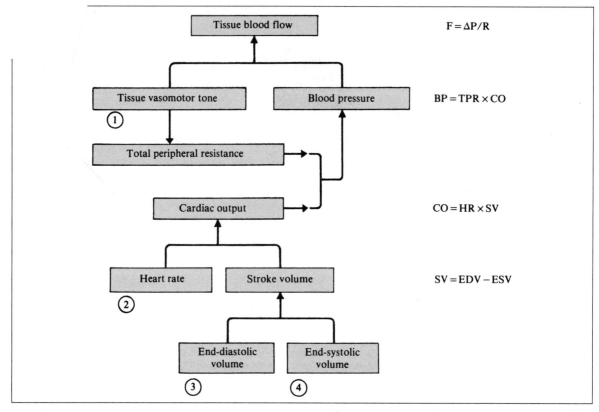

$$F = \Delta P / R$$

$$BP = TPR \times CO$$

$$CO = HR \times SV$$

$$SV = EDV - ESV$$

Fig. 15-1. Cardiovascular system dynamics. The relationships determining tissue blood flow, blood pressure, cardiac output, and stroke volume.

receptor reflex acting by way of the autonomic nervous system on all four variables (see Chap. 17). In the normal person the fine control of cardiovascular function is neural.

The three factors determining cardiac output (heart rate, end-diastolic volume, and end-systolic volume) will now be considered in detail. Factors influencing peripheral resistance are considered in Chapter 17.

HEART RATE

In the adult the normal heart rate at rest ranges between 40 and 100 beats per minute and averages about 70. In the newborn the heart rate is about 135 beats per minute and in the elderly, about 80 beats per minute. The heart of the trained athlete at rest often beats less than 50 times per minute. With complete block of ANS influences on the heart, the intrinsic heart rate is about 110 beats per minute. Vagal tone accounts for the normal heart rate of about 70 beats per minute at rest. The heart rate is widely variable,

for it is a quickly changeable factor in the maintenance of cardiovascular homeostasis. The heart rate is changed primarily by the autonomic nervous system (see Chaps. 7 and 17).

As shown in Figure 15-2, if the stroke volume is held constant, an increase in heart rate causes a directly proportionate increase in cardiac output (curve 1). However, since an increase in heart rate shortens the interval between beats and therefore tends to shorten the time for filling, the end-diastolic volume is decreased. (Consider the analogy of a bucket placed under a running faucet. If the time for filling is reduced by half, only half as much water is collected.) Unless the end-systolic volume decreases proportionately to the decrease in end-diastolic volume, there will be a consequent decrease in stroke volume. Indeed, at high heart rates, or in a denervated or isolated heart, the end-systolic volume does not decrease proportionately to the decrease in

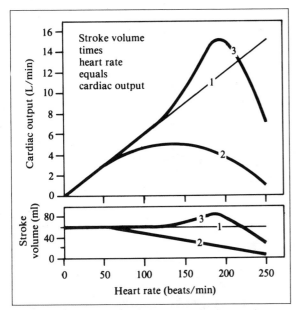

Fig. 15-2. The effect of heart rate on cardiac output. Since cardiac output is the product of stroke volume and heart rate, if the stroke volume remains constant, cardiac output will increase linearly with increases in heart rate (curve 1). If the stroke volume declines because of inadequate filling time or compensatory mechanisms, there is a peak output and then a decrease (curve 2), giving a relatively constant cardiac output as heart rate changes. As in the normal person, if the stroke volume is held constant or is increased by compensatory mechanisms such as increased contractility, the output increases to a higher peak before filling time becomes limiting (curve 3).

end-diastolic volume, and the result is a smaller stroke volume (Fig. 15-2, lower curve 2). Without autonomic nervous system (ANS) control, cardiac performance reaches a peak at moderate heart rates. Any further elevation of heart rate seriously limits filling time and markedly reduces cardiac output (Fig. 15-2, upper curve 2). Because ventricular filling occurs mostly in the first half-second and then tends to reach a plateau after 1 second, there is little decline in stroke volume until the heart rate exceeds about 60 beats per minute. Over the range of about 60 to 160 beats per minute, cardiac output tends to be independent of heart rate if ANS control is blocked, because the stroke volume decreases in proportion to the increase in heart rate above about 60 beats per minute. Furthermore, if the heart is artificially paced over this range and the autonomic reflexes are intact, cardiac output and blood pressure are little

changed because of compensatory mechanisms (baroreceptor reflexes, see Chap. 17). During heavy exercise, mechanisms mediated primarily by the sympathetic division of the ANS act to maintain the stroke volume at a constant level or even increase it (Fig. 15-2, curve 3) by (1) increasing the vigor of myocardial contraction to decrease the end-systolic volume and (2) by increasing left ventricular filling pressure. Thus, in the normal person, the output of the heart per minute tends to increase proportionally as the heart rate increases. However, an increase in heart rate does not *necessarily* mean that the cardiac output is increased, for the stroke volume may be concomitantly reduced.

As heart rate is increased by sympathetic division action, the duration of diastole is decreased proportionally more than that of systole (Fig. 15-3). At rest, the diastolic filling time accounts for about 50 percent of the cycle, but with a doubling of heart rate to 150 beats per minute, the diastolic time accounts for only about 30 percent of the cycle. More important, the actual time available is reduced from about 0.4 second to about 0.1 second, a reduction to 25 percent of the resting value! With even higher heart rates the impingement on filling time is even more serious, indicating the need for powerful mechanisms to maintain stroke volume.

The maximum effective heart rate in people is about 180 beats per minute. *Tachycardia,* as seen in atrial fibrillation, is also associated with a decrease in cardiac output, because in this case the time for filling is severely limited by the abnormally high heart rate. In *bradycardia,* following complete heart block, the stroke volume may be large, but the cardiac output is limited because of the very low heart rate.

Cooling of the pacemaker cells (e.g., during surgical hypothermia) causes a profound reduction in heart rate. This might be expected from the marked effect of temperature on enzymatic reactions. An increase of the body metabolism (e.g., during fever) acts to stimulate heart rate. A rise of body temperature of 1°C causes an increase in heart rate of about 12 to 20 beats per minute (7–11 beats/minute per 1°F). Generally the heart rate is closely associated with the overall metabolic rate. Any bodily response requiring an increased oxygen supply usually requires an increased cardiac output, which is accomplished as a rule by an increased heart rate. Examples include exercise and digestion. The act of standing upright also elicits an increased heart rate, as part of the homeostatic mechanism maintaining an adequate blood pressure (see Chap. 17).

The stroke volume, like the tidal volume in respiration, is not as large as the maximum volume of the heart. Consequently, there is usually both a diastolic volume reserve and a systolic volume reserve (Fig. 15-4). Discussion of the various factors influencing these volumes follows.

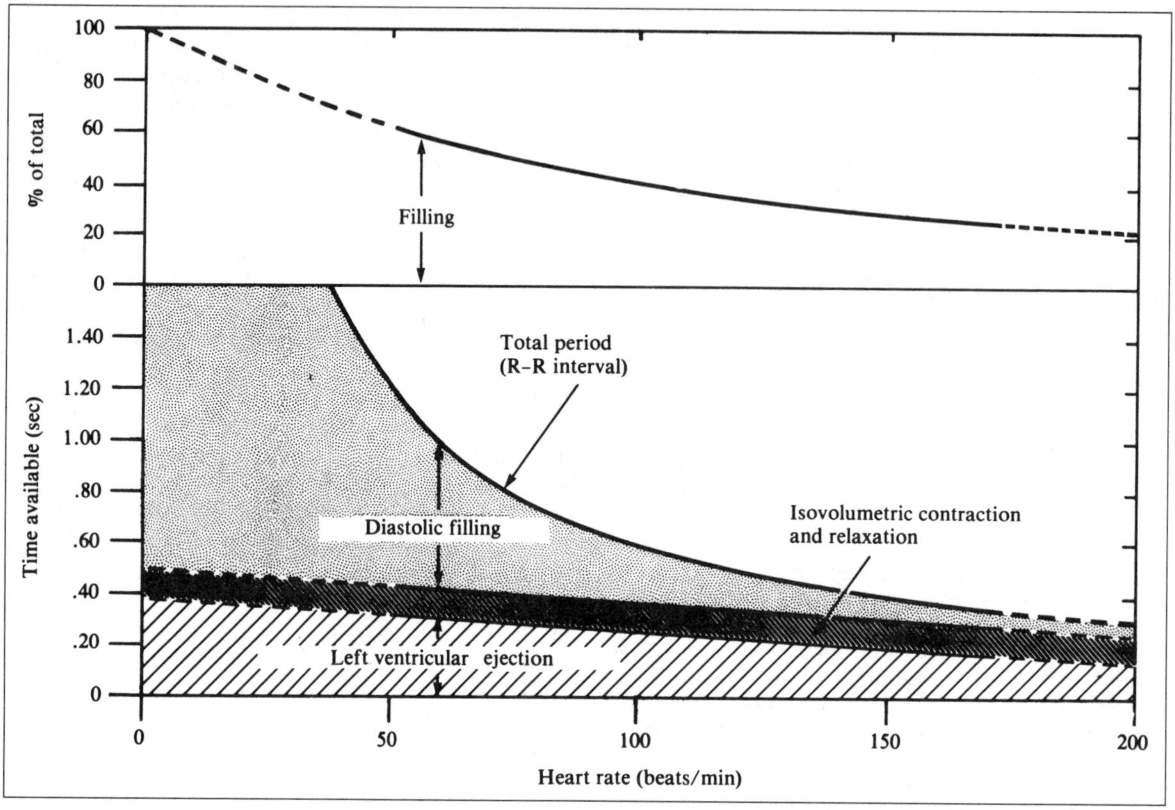

Fig. 15-3. Influence of heart rate on time available for left ventricular ejection and filling. Note the marked reduction in filling time at a high heart rate. (Values estimated from A. M. Weissler, W. S. Harris, and C. D. Schoenfeld. Systolic time intervals in heart failure in man. *Circulation* 37:152, 1968; and W. B. Jones and G. L. Foster. Determinants of duration of left ventricular ejection in normal young men. *J. Appl. Physiol.* 19:279, 1964.)

END-DIASTOLIC VOLUME

If the heart is to be an effective pump, the ventricles must fill before systole. The adequacy of filling is determined by the filling time, the effective filling pressure, the distensibility of the ventricles, and atrial contraction.

FILLING TIME. The *filling time* is dependent on the heart rate. As the heart rate increases, the diastolic period is reduced proportionally more than the systolic period, so that available filling time becomes an important factor at high heart rates. Under normal conditions, however, the filling time is adequate.

FILLING PRESSURE. The *effective filling pressure* is the pressure gradient between the inside of the ventricles and the pressure outside. It is this transmural pressure gradient across the ventricular walls, that causes the relaxed ventricles to distend during diastole (Fig. 15-5). It is closely dependent on the blood volume, capacitance vessel tone, and

gradient for venous return, since these influence the right atrial (central venous) pressure and pressure inside the right ventricle during diastole. The intrathoracic pressure outside the heart also is a factor influencing the transmural filling pressure. An increase in the central venous pressure or a more negative thoracic pressure acts to cause an increase in the transmural pressure and thus facilitates filling. Atrial contraction also contributes to ventricular filling. However, a stenotic atrioventricular (AV) valve adds resistance to flow and so impedes filling. If the blood volume is reduced (e.g., as a result of hemorrhage), or if there is loss of venomotor tone and hence dilation of the capacitance vessel (veins), the filling pressure drops and the cardiac output is seriously reduced.

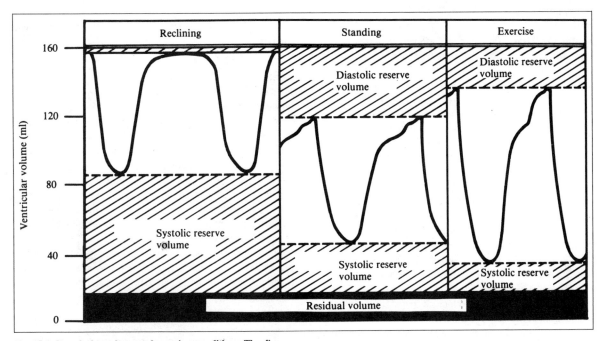

Fig. 15-4. Ventricular volume under various conditions. The *diastolic reserve volume* is the difference between the maximum volume and the end-diastolic volume. The *systolic reserve volume* is the difference between the end-systolic volume and the residual volume. The *residual volume* is what would remain in the heart if it ejected against a pressure of zero. (Modified from R. F. Rushmer. *Cardiovascular Dynamics* [4th ed.]. Philadelphia: Saunders, 1976. P. 273; and R. F. Rushmer. Work of the heart. *Mod. Concepts Cardiovasc. Dis.* 27:473, 1958.)

Fig. 15-5. Volume-pressure relationships of the heart depicting the Frank-Starling law. 1. Isovolumetric contraction phase. 2. Ejection. 3. Isovolumetric relaxation. 4. Filling. A normal cycle is indicated by a solid line. EDV = volume of the ventricle at the end of diastole; ESV = end-systolic volume.

CARDIAC DISTENSIBILITY. *Distensibility* is the property of a hollow organ that permits it to be distended or expanded (see Chap. 11). In Figure 15-5 the lower line is a typical distensibility, or pressure-volume, curve of a relaxed ventricle. A very small pressure is required to cause a marked increase in volume at low volumes. As the ventricle fills, its elastic fibers are stretched, and, because of the nonlinear stress-strain characteristics of these fibers, the ventricle becomes less distensible. Then a unit increase in pressure causes only a small change in volume. Myocardial distensibility does not appear to be significantly modified by sympathetic and parasympathetic division activity. However, it is decreased in hypertrophied or ischemic hearts—they are stiffer.

During the rapid filling phase, relaxation of the ventricle is so rapid that it appears to be active. An actual negative transmural pressure will be developed only if the previous

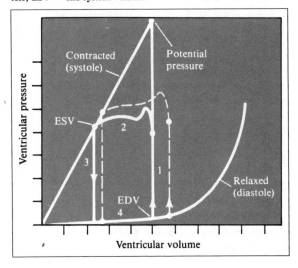

end-systolic volume was so small that it compresses elastic cardiac tissue.

ATRIAL CONTRACTION becomes especially important at high heart rates, when the diastolic time is limited. When the atria contract, blood is propelled in both directions: toward the veins and into the ventricles. However, the inertia of the blood tends to act as a valve. The cross-sectional area of the AV valve and ventricle is larger and the distance shorter than is the case with the veins leading to the atria; thus, the force required to accelerate a bolus of blood into the heart is less than that required to accelerate an equal volume to higher velocities back into the narrower veins. Furthermore, atrial contraction leads to reduced regurgitation by assisting AV valve closure (see Chaps. 11 and 13). During exercise the amount of blood pumped into the ventricles by the atria can account for as much as 40 percent of the stroke volume. But normally, atrial contraction is a minor factor in the filling of the ventricles. Lack of effective atrial contraction, such as occurs with atria fibrillation, does not jeopardize life but it does limit exertion.

Normally, homeostatic mechanisms compensate for moderate losses of blood volume that tend to decrease the end-diastolic volume. These mechanisms include an increased cardiac contractility and increased venomotor tone (see Chap. 17). A human can lose about 1 percent of the body weight of blood without any significant drop in blood pressure. Without the compensatory mechanisms, however, the effects of a decreased filling pressure become highly important.

PERICARDIUM. The *pericardium* provides a limit to the magnitude of the end-diastolic volume. Under normal conditions it is not essential in humans. The removal of an adherent, dense, scar-filled pericardium following pericarditis significantly improves the clinical condition of many patients recovering from this disease. The loss of the pericardium does not seriously limit their activity. The function of the pericardium is apparently to check acute overdistention during diastole and to provide a lubricated surface between the beating heart and the lungs. By limiting right ventricular filling, it protects against pulmonary congestion and left ventricular overload. This is important during various cardiac arrhythmias in which prolonged diastolic periods occur. It also aids in keeping the heart and great vessels aligned. Under chronic conditions of myocardial weakness or chronic overloading of the heart, the pericardium slowly distends as cardiac hypertrophy ensues.

If blood or fluids accumulate between the heart and pericardium, filling is impeded. This acute compression of the heart is called *cardiac tamponade*. It is a serious consequence of penetrating wounds of the heart, or of pericarditis, in which there is effusion of fluids from the inflamed membranes.

CARDIAC HYPERTROPHY. *Cardiac hypertrophy* occurs during a chronically increased pressure load on the heart, such as with aortic stenosis or severe hypertension. The increase in muscle fiber size (hypertrophy)—and, to some extent, an increase in number of fibers (hyperplasia) and capillaries—increases the ability of the heart to do work, but at some point deterioration of pumping ability may occur. A reduced blood capillary/muscle fiber ratio, increased oxygen diffusion distance, changes in morphology or in collagen concentration, or abnormalities in basic biochemical function of fibers are current theories explaining heart failure after longstanding cardiac hypertrophy.

The cause of hypertrophy is not clearly known, but it often follows a chronic increase in the size of the ventricles (dilation). The dilation would be expected to follow increases in filling pressure as a compensatory mechanism for increased pressure load (e.g., aortic stenosis), regurgitation, or myocardial weakness. Left ventricular failure leads to right ventricular hypertrophy because the pulmonary arterial pressure rises as a result of pulmonary congestion and so continuously overloads the right heart (see Chap. 18).

HETEROMETRIC AUTOREGULATION: THE FRANK-STARLING RELATIONSHIP. If cardiac muscle is stretched, it develops greater contractile tension on excitation. As Starling stated in 1914, "The law of the heart is therefore the same as that of skeletal muscle, namely, that the mechanical energy set free on passage from the resting to the contracted state depends . . . on the length of the muscle fibers." The vigor of contraction is a function of muscle fiber length. Stated another way, stroke volume tends to be directly proportional to diastolic filling; that is, the ventricle tends to eject whatever volume is put into it. Although many investigators, such as Frank in the 1880s, contributed to the study of this mechanism, Starling's formulation was such that it has become known as *Starling's law of the heart*. Sarnoff (1962) has coined the more descriptive term *heterometric autoregulation*. This implies that the performance of the heart is so regulated as to be an effective mechanism for the homeostatic control of circulatory function by intrinsic (auto) mechanisms that are not neural or hormonal in origin and are determined by changes (hetero) in myocardial fiber length (metric). (See also ventricular function curve and Fig. 15-9.)

There is no question concerning the validity of this relationship. However, it is but one factor among many controlling cardiac output. The relationship is described graphically in Figure 15-5. This diagram is similar to diagrams showing the tension developed by contracting skeletal muscle at various lengths. Systole starts at the end-diastolic volume (EDV) and first produces the isovolumetric change in ventricular pressure. If blood is prevented from leaving the ventricle, the ventricular pressure attains the *potential pressure*. However, as soon as the ventricular pressure exceeds the arterial pressure, blood ejection starts, the ventricular volume decreases, and so the pressure increase plateaus. As the velocity of myocardial fiber shortening increases, the force developed decreases. Systole ends at the end-systolic volume (ESV), after which isovolumetric relaxation occurs. The cycle is completed by the filling of the relaxed ventricle to the EDV. Increasing the effective filling pressure causes an increased EDV (Fig. 15-5, dashed line). The increase in EDV, by causing the myocardial fibers to contract with greater vigor, acts to increase cardiac work output. The ESV is about the same, and so there is a net increase in stroke volume at a constant pressure load.

As another example, if the total peripheral resistance increases so that the arterial pressure is increased (let us assume from 100 to 150 mm Hg), the systolic ejection is limited by the increased pressure head and the ESV is increased (say from 40 to 50 ml) because the usual amount of blood (80 ml/beat) was not ejected (now only 70 ml, a 10-ml decrease). The stroke volume and cardiac output would then be less, damming up the blood behind the ventricle. With a continuing right heart output the filling pressure increases over a period of several beats, causing the EDV to increase by about 10 ml, from 120 to 130 ml. The larger heart then contracts with more vigor and ejects the same volume per beat (130 − 50 = 80 ml) as before (120 − 40 = 80 ml). Note that the heart is beating with a larger ESV and EDV. The stroke volume may not be perfectly maintained, as in our example, because the venous inflow from the right heart is limited by the decreased distensibility of the heart at the larger volume. Thus, the EDV increases only partially from the 120-ml control EDV to the 130-ml volume, and so the expected 10-ml increase in EDV may not be fully achieved. An increased vigor of contraction via heterometric autoregulation increases stroke volume or pressure, or both, and therefore causes increased external work output by the heart (see Figs. 15-7 and 15-8). If a weak heart is overdistended, the peak pressure tends to plateau. Then, an increase in EDV is accompanied by a marked increase in ESV, and a smaller stroke volume is ejected. This sequence may happen in acute, uncompensated congestive heart failure. However, such a decrease in stroke

volume at very high EDVs is seldom seen in the normal heart, especially if the pericardium is intact.

From studies with isolated animal hearts, it had long been thought that cardiac output was controlled primarily by changes in stroke volume as determined by the EDV and the Frank-Starling relationship following increases in venous return. However, the importance of this phenomenon in intact, unanesthetized humans has been seriously questioned, since changes in right atrial pressures have been noted without concomitant changes in cardiac output. Evidence from human volunteers shows that if the sympathetic division is blocked by drugs, an increase in blood volume causing an increase in right atrial pressure is indeed followed by an increase in cardiac output. By using injections of radiopaque dyes in humans and animals, and special transducers to measure the dimensions of the ventricles in dogs, Rushmer and co-workers (1976) have found that the EDV is normally near maximum at rest in the reclining person or dog. The normal heart is large in the resting, reclining person; it apparently almost fills the pericardium (see Fig. 15-4). Under these conditions a marked increase in heart size in response to increased loads is not possible; hence, most of an increase in cardiac output must take place by means other than increased EDV and heterometric autoregulation. However, when a person stands, the EDV of the ventricles decreases, and so there is an end-diastolic reserve volume that can be utilized to increase stroke volume. Although the stroke volume increases during maximal exertion, Rushmer (1976) concluded that it remains "remarkably" constant over a wide range of voluntary exertion. Normally, the stroke volume increases only if the total oxygen consumption exceeds about eight times the resting rate. This neural control, which minimizes the importance of heterometric autoregulation (Starling's law), had not been seen to such a great degree in earlier animal experiments for the reason that most anesthetics depress the cardiovascular reflexes acting on the heart and cardiovascular system (see Chap. 17). Furthermore, the procedure of thoracotomy, almost always used in these studies, removed the negative intrathoracic pressure, with the result that the heart had a lower effective filling pressure. Thus, the heart was not adequately filled during diastole. The ESV of the heart is abnormally low under these conditions, and the heart rate is high and relatively constant. It is therefore not surprising that increases in filling pressures distended the heart, giving the marked increases in cardiac function explained by the Frank-Starling law.

Heterometric autoregulation is without question an important mechanism, in the normal person, for balancing the outputs of the right and left ventricles. For example, if the right ventricle pumps more blood than the left ventricle, the difference accumulates in the lungs, acts to increase pulmonary venous and left ventricular diastolic pressures, and so increases left ventricular filling. In accordance with Starling's law, the left ventricular output increases and so restores the balance. Heterometric autoregulation is deleterious during changes in body posture in humans, for unless other mechanisms act to maintain homeostasis, as the blood pools in the lower extremities there is a profound drop in cardiac output and hence in blood pressure (orthostatic hypotension). Likewise, during thoracic surgery, when the effective filling pressure to the heart is reduced, other mechanisms must provide compensation. The Frank-Starling relationship is of value as a defense against heart failure, for the increase in filling pressure (congestion) within certain limits acts to maintain the cardiac output.

END-SYSTOLIC VOLUME

End-systolic volume is that volume of blood remaining in the heart at the end of systole, at the time when the semilunar valves close. The primary factors determining the magnitude of the ESV are the pressure against which the ventricle is pumping and the strength of the myocardial fiber contraction.

PRESSURE LOAD. If the heart is to pump against a blood pressure higher than previously, the myocardial fiber tension must be greater—a result of the Laplace relationship (see Chap. 11). With an increase in tension the velocity of contraction is reduced. (See the discussion of the force-velocity relationship of muscle in Chap. 3.) As a consequence, less shortening will occur during the time of activation, and systole will terminate at a higher volume than before. The stroke volume will decrease unless the contractility of the myocardium is increased to maintain the velocity of contraction at the higher tension and pressures. Thus, *an increased arterial pressure (afterload) tends to reduce cardiac output.* Conversely, a reduction in systemic blood pressure results in a decrease in ESV and so in an increased stroke volume.

Another consequence of the Laplace relationship is that with increased ventricular volume, the myocardial tension required to develop the same pressure is increased, because the tension is proportional to the product of radius and pressure. This disadvantage of a large heart is offset by the fact that the large heart requires less shortening to eject a certain stroke volume than does the smaller heart. A change of volume (the amount ejected) of a sphere is related to its diameter cubed, and that of a cylinder or cone to its diame-

ter squared, but the circumference (degree of fiber shortening) is directly proportional to the diameter. This relationship holds true under most conditions, especially for the left ventricle, although the shapes of the two ventricles are so complex as to make a quantitative analysis uncertain.

There is a complex interaction between filling pressure, aortic (outflow) pressure, and cardiac performace, illustrated in Figure 15-6. Although the data shown are from dog experiments, the pattern is applicable to humans. Consider point N as the normal (Fig. 15-6). Increasing ventricular fiber length to point A increases left ventricular (cardiac) output, as predicted by the Frank-Starling relationship. (Ventricular fiber length, ventricular volume, end-diastolic pressure, and mean left atrial pressure are generally closely related.) At a constant filling pressure, an increase in mean arterial pressure (pressure load) results in a decrease in cardiac output (point C), whereas a decrease in arterial pressure results in an increased cardiac output (point D) with no change in cardiac contractility. Note that at high arterial pressures (200 mm Hg), increasing the filling pressure has relatively little influence on cardiac output, in sharp contrast to the marked influence of filling pressure on cardiac output at low (50 mm Hg) pressure loads. Note, too, that with low

Fig. 15-6. Effect of filling pressure (mean left atrial pressure) and output pressure (mean arterial pressure) on cardiac output (left ventricular output). (Modified from K. Sagawa. Analysis of the ventricular pumping capacity as a function of input and output pressure loads. Chapter 9 in E. B. Reeve and A. C. Guyton [eds.]. *Physical Bases of Circulatory Transport.* Philadelphia: Saunders, 1967. Pp. 141–147.)

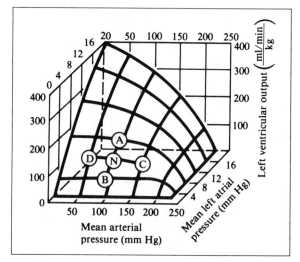

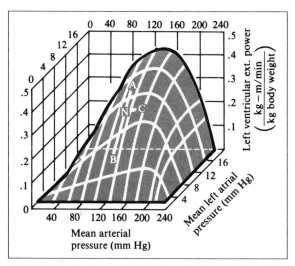

Fig. 15-7. Heart work per minute (left ventricular external work) as influenced by filling pressure and pressure load (mean arterial pressure). (Modified from K. Sagawa. Analysis of the ventricular pumping capacity as a function of input and output pressure loads. Chapter 9 in E. B. Reeve and A. C. Guyton [eds.]. *Physical Bases of Circulatory Transport.* Philadelphia: Saunders, 1967. Pp. 141–147.)

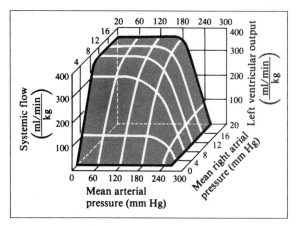

Fig. 15-8. Effect of mean *right* ventricular filling pressure and mean systemic arterial pressure on *left* ventricular pumping capacity with autonomic control blocked. (Modified from C. W. Herndon and K. Sagawa. *Am. J. Physiol.* 217:70, 1969.)

transmural filling pressures (e.g., 4 mm Hg) the cardiac output is low and relatively independent of pressure load.

The rate of doing work is a better index of cardiac function and is more closely related to cardiac reserves in disease states than is mere flow output. Figure 15-7 is a three-dimensional presentation of left ventricular power output (flow times pressure) as influenced by filling and output pressure—heart rate, and metabolic, neural, and hormonal influences constant. At a given arterial pressure, an increase in filling pressure leads to increased power output (the curve B-N-A), as well as flow output. If work per beat is used rather than work per minute (power), the relationship, along a line such as B-N-A, is called a *ventricular function curve* (see Myocardial Contractility).

HOMEOMETRIC AUTOREGULATION. In addition to the increased power output of the heart as a result of increased filling—heterometric autoregulation—the physiology of the heart is such that nearly the same stroke volume is ejected even if there is an increased aortic pressure; the work output is increased. This response requires a few beats to develop, but it is not dependent on changes in fiber length (i.e., homeometric) or on neural or hormonal factors (i.e., autoregulation). It is related in part to the increased coro-

nary (heart muscle) perfusion pressure. The relationship is shown in Figure 15-7 in which the arterial pressure is seen to increase from 100 mm Hg (point N) to 150 mm Hg (point C). At high mean arterial pressures (> 160 mm Hg) a reduction in external power output is seen with additional increases in pressure load. The increased output is not free, for when more work is done by increasing the arterial pressure load, the metabolic energy and hence oxygen requirements are increased. The systolic tension may be the principal determinant of not only homeometric autoregulation but also of oxygen consumption (see Cardiac Efficiency).

Finally, since the heart tends to eject the amount of blood returned, it is useful to consider the response of the *entire* cardiopulmonary bed to changes in filling pressure and systemic arterial pressure (Fig. 15-8). In contrast to the responses of the left ventricle alone (see Fig. 15-6), the entire system shows much less dependence on the systemic arterial pressure. Following an increase in systemic arterial pressure, left ventricular output is at first reduced. The right heart maintains its flow output, and so blood accumulates in the pulmonary bed and also increases the filling pressure of the left ventricle. This restores the left ventricular output. At equilibrium, left ventricular end-diastolic and end-systolic pressures and volumes are increased, as are the pulmonary vascular pressures. There is only a slight decrease in right ventricular output because of the small increase in pulmonary arterial pressure. However, at very high systemic arterial pressures the pulmonary vascular pressures increase so much that there is a precipitous decrease in the output of both ventricles. Note that at normal mean arterial pressures of 90 mm Hg and mean right atrial transmural filling pressures of 5 mm Hg, cardiac output will potentially change by about 50 ml/min per kg of body weight for a 1 mm Hg change in central venous (right atrial) pressure. This extreme sensitivity to filling is not very obvious because powerful reflex mechanisms act to maintain or change cardiac output without apparent changes in the central venous pressure (see Fig. 15-10).

MYOCARDIAL CONTRACTILITY

A description of cardiac performance is essential for diagnostic evaluation. If performance is inadequate, the physician must be able to distinguish between factors such as inadequate filling or valve malfunction and the intrinsic ability of the heart to contract. *Myocardial contractility* and *vigor of contraction* are vague terms, connoting the tendency to contract (fiber shortening) as well as the force developed by the muscle fibers to expel blood. Although *contractility* connotes the ability to contract, the meaning commonly is restricted to the *intrinsic ability of the heart to function other than that induced by changes in presystolic fiber length (preload) or arterial pressure (afterload)*. If myocardial contractility is increased, either the ESV will be decreased by the more vigorous contraction (thus increasing the stroke volume) or the same volume of blood will be ejected against a higher pressure by the more forceful beat. In either case, more work will be performed by the heart with each beat.*

One approach to evaluation of myocardial contractility is the determination of a *ventricular function curve* (Figs. 15-7 and 15-9). This is the relationship between ventricular stroke work (stroke volume times the change in pressure) and end-diastolic fiber length or pressure. The latter is often used because it is easier to measure (cardiac catheterization) and because the curvilinear relationship between the ventricular pressure just before systole and the fiber length is remarkably constant for a given heart. Even the mean atrial pressure has been used to derive satisfactory curves. In the usual method of determining the ventricular function curve, transfusion or hemorrhage is used to provide measurable changes in the end-diastolic pressure. In addition, cardiac output and heart rate are measured to provide stroke-volume data. The mean aortic pressure and cardiac output furnish sufficient information for evaluating cardiac

---

*The role of heart rate in cardiac power output is not even considered. Not only is it impossible, as yet, to measure contractility satisfactorily; there also is no clear and commonly accepted definition of the criteria the measurement must meet.

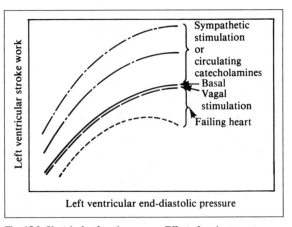

Fig. 15-9. Ventricular function curves. Effect of various agents on cardiac function curves relating stroke work (stroke volume × mean systolic blood pressure) and left ventricular end-diastolic pressure. Under normal conditions there is a small sympathetic division activity, giving a higher cardiac function curve than the basal level.

work, although the integral, during systole, of the instantaneous ventricular pressure times the instantaneous blood outflow provides a more accurate measure. Because such procedures as transfusion and hemorrhage take many seconds, neural and hormonal influences change during the time interval required; in the intact experimental subject, therefore, the curve so produced is not necessarily descriptive of the intrinsic contractility. Although a ventricular function curve provides the current standard of reference for myocardial contractility, it is limited, for the work output is a function of the pressure load and consequently the pressure load must be relatively normal throughout the study. Thus the complex surface shown in Figure 15-7 is a more realistic representation of contractility. Furthermore, it should be obvious that the determination of a ventricular function curve is not applicable under clinical conditions. Finally, since the curve is empirical, it cannot be expressed mathematically in a form in which each coefficient of a describing equation has biological meaning.

Four features of ventricular function curves should be considered: (1) Under a given set of conditions, increasing the fiber length leads to greater work output (heterometric autoregulation: the basal curve, Fig. 15-9). (2) There is a family of ventricular function curves. With sympathetic stimulation the heart functions on a curve different from the basal one, such that, at the same end-diastolic pressure, more work output (either increased pressure, flow, or some combination of both) is obtained. (3) Small changes in end-diastolic pressure of the *right* ventricle induce large changes in the output of the *left* heart (see Fig. 15-8). As Guyton et al. (1973) have emphasized, the right ventricular end-diastolic pressure is dependent on the factors influencing the flow of blood to the heart (see Chap. 16). Because the left ventricle can pump out only what is pumped to it by the right ventricle, and because of the sensitivity of the right ventricle to filling pressure, potent mechanisms must be available to maintain the optimal filling pressure to the heart. (4) It is important to realize that more work than before can usually be obtained from a weakened heart (i.e., lower function curve) if the filling pressure is increased at the same time. Thus, a single point on the curve provides only limited information about the condition of the heart. As the filling pressure is increased, a plateau in response is seen. In the failing heart, still further increases in end-diastolic pressure may result in less work than before, even with the same ventricular function curve (Fig. 15-9, bottom curve).

The peak rate of pressure development ($[dP/dt]max$) is a relatively easily measured and useful index of contractility. The maximum dP/dt usually occurs just as the aortic valve opens (see Fig. 13-7). Measuring rates of pressure rise in the ventricle is easier than developing a ventricular function curve or a force-velocity curve, or even measuring peak rates of flow under various pressure loads. The $[dP/dt]max$ increases in response to positive inotropes and is decreased by negative inotropes. It is markedly depressed in the failing heart. However, the maximum dP/dt changes somewhat with changes in preload (EDV) and afterload (arterial pressure). Numerous attempts have been made to compensate for these effects. To date there is no commonly accepted index of myocardial contractility based on the intraventricular pressure pattern.

The end-systolic ventricular pressure-volume relationship (see Fig. 15-5) tends to be a unique representation of the contractile state of the ventricle irrespective of the preload, afterload, or mode of contraction (Sagawa, 1981). The relationship tends to be linear, with a small positive intercept on the volume axis. The intercept increases in heart failure. The slope of the relationship is the end-systolic elastance (E_{max}), which is increased by positive inotropes and decreased by negative inotropes that reduce cardiac function. An increase in E_{max} will cause the ESV to be less than before and so lead to an increased stroke volume if the arterial pressure is not increased; or it will tend to maintain the ESV and thus stroke volume if the arterial pressure increases.

Yet another approach to estimating changes in contractility in experimental animals is to suture a strain-gauge arch across a slit in the myocardium to obtain an estimate of the tension developed by the myocardium at this site. (Ventricular filling and load must be the same during both the control and experimental periods.)

FACTORS INFLUENCING CONTRACTILITY. Myocardial contractility increases markedly in exercise; the heart beats more rapidly, and the blood pressure increases. In the normal person, ANS effects obscure the Frank-Starling relationship, so that the cardiac output is increased without a definite increase in EDV or right atrial pressure (see Fig. 15-10, equilibrium point C). The increase in contractility is a result of the *neural influences* on the heart and the release of circulating catecholamines.

Vagal stimulation decreases heart rate and therefore work done per minute. It reduces the strength of atrial contraction and decreases ventricular contractility somewhat (Fig. 15-9). The maximal depressing effect of the parasympathetic division on contractility is very much less than the augmenting effects of the sympathetic division, however.

The health and *metabolic conditions* of the myocardial cells are an obvious factor determining the strength of contraction. Hypoxia, hypercapnia, and acidosis are depressant. Ischemia from obstruction (occlusion) of a coronary artery leads to rapid deterioration of function because of an inadequate blood supply.

With a decrease in the time interval between heartbeats, between about 20 to 200 beats per minute, there is a progressive increase in vigor of contraction (the treppe or staircase phenomenon). Furthermore, although a premature depolarization results in a subsequent weakened contraction, the following beat is then more forceful than normal. This *postextrasystolic* potentiation is independent of ventricular filling. Both phenomena are probably related to the availability and/or transport of calcium between the cell membrane and the contractile machinery (Levy and Martin, 1974). The phenomena are intrinsic to the heart and so are considered by some researchers a form of homeometric autoregulation.

Inotropes are agents that cause a change in the contractility of the heart. Negative inotropes act on the heart to weaken the strength of contraction. Prime examples are the bacterial toxins released during many diseases, which reduce the ability of the heart to pump blood. Positive inotropes are agents such as digitalis and *circulating catecholamines* which act to strengthen the heartbeat. The infusion of norepinephrine to supplement the endogenously released catecholamines is of real, but limited, clinical value. In last-resort conditions the infusions of large amounts of norepinephrine may only produce more injury.

In summary, there are five classes of influence on the vigor of myocardial contraction:

1. *The metabolic condition* of the cells, which in turn is dependent on an adequate coronary blood flow, oxygen supply, nutrient supply, and lack of toxins.
2. *Heterometric autoregulation,* the Frank-Starling relationship, or the influence of the length of fibers at the beginning of contraction on cardiac performance.
3. *Central nervous system action,* including the effect of circulating catecholamines and the release of the autonomic mediators at the heart.
4. *Homeometric autoregulation,* the added power output of the heart in response to increases in pressure load without changes in end-diastolic fiber length or extracardiac influences.
5. *Interval between beats,* the increased vigor of contraction seen with moderate increases in heart rate.

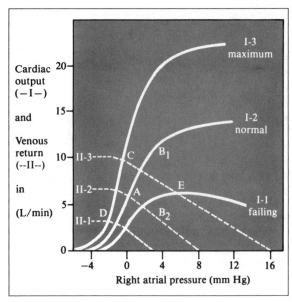

Fig. 15-10. Relationship between cardiac function and venous return. *Ventricular performance curves* (I), relating atrial pressure to cardiac output: 1. Failing heart weakened by negative inotropes or ischemia. 2. Normal relationship with moderate sympathetic influences. 3. Maximal sympathetic influence. *Venous return curves* (II): 1. After hemorrhage and mean circulatory pressure reduced to 3 mm Hg. 2. Normal with mean circulatory pressure of 8 mm Hg. 3. With maximum sympathetic influences or after a large transfusion of blood. *Operating points:* A. Normal. B₁ and B₂, outflow and inflow after a sudden increase in atrial pressure. C. At maximum sympathetic activity. D. Following hemorrhage. E. With failing heart showing increased central venous (right atrial) pressure.

RELATIONSHIP BETWEEN CARDIAC OUTPUT AND VENOUS RETURN

Right atrial pressure is an important variable influencing both cardiac output and venous return. Because the cardiac output must equal the venous return over long periods of time and yet may not be equal on a beat-to-beat basis, a study of the interrelationships helps one's understanding of cardiovascular function (see Guyton, et al., 1973, for details).

The curves labeled I-1 to I-3 in Figure 15-10 indicate the cardiac output (right ventricular outflow) under various conditions at given right atrial pressures. As filling pressure (right atrial pressure) is increased, cardiac output increases. The relationship is also shown in Figure 15-6, with the added variable of arterial pressure. As the level of sympathetic activity is increased (I-3 representing maximum activity versus I-2, a basal level), the cardiac output at a given atrial pressure (e.g., 2 mm Hg) is increased because the contractile ability of the myocardium is increased. (Note: The effective filling pressure is about 5 mm Hg greater than the atrial pressure because of the normal negative 5 mm Hg intrathoracic pressure.)

The curves in Figure 15-10 labeled II-1 to II-3 indicate the venous return (in liters/min) that occurs at various right atrial pressures (see Chap. 16). Such curves have been obtained experimentally in dogs, although the technique is difficult and rather traumatic. The point at zero venous return defines one end of the curve. It is measured by fibrillating the heart and rapidly pumping blood from the

arterial to the venous circuits until the pressures are equal. A steady value must be obtained within about 5 seconds before significant autonomic responses or fluid shifts have occurred. With the flow being zero, the pressure, called the *mean systemic filling pressure,* is the same throughout the entire systemic vasculature. Furthermore, this pressure is about the same as that in the peripheral venules and small veins (capacitance vessels) under normal conditions. Setting right atrial pressure at a value less than the mean systemic filling pressure, a pressure gradient is obtained, and the resulting flow is measured and plotted. (An external pump is used to maintain the cardiac output at the same rate as the venous return. Reflex compensation is blocked.) At atrial pressures much less than zero, the flow plateaus in value because the vessels entering the chest tend to collapse just outside the chest cavity. Providing a greater negativity does not increase flow, just as sucking harder on a collapsed soda straw does not yield more soda. The reciprocal of the slope of the venous return curve represents the venous resistance and is determined by the geometry of the venous vasculature. The zero-flow pressure intercept is determined by (1) the blood volume and (2) the pressure-volume relationship of the vasculature (see Chap. 11). Vascular capacitance may be reduced to increase the mean systemic filling pressure by increasing sympathetic activity that acts on the smooth muscle in the walls of the veins and venules—venomotor tone (see Chaps. 16 and 17).

The steady-state cardiac output and venous return must be equal. This condition will occur at the intersection, point A (Fig. 15-10), of the given cardiac output (curve I-2 in Fig. 15-10) and venous return (curve II-2). If the filling pressure is temporarily higher than the value at the intersection of the curves (e.g., 4 mm Hg), the cardiac output (point B_1) will be greater than the venous return (point B_2), because the right heart will be filled more, while the gradient for venous return is less. The volume of blood in the atrium will be rapidly decreased by the increased cardiac output and reduced venous return, and this will lead to a decrease in atrial pressure until inflow equals outflow. The converse is also true, for if cardiac output is less than venous return, blood accumulates, increasing the atrial pressure and cardiac output and decreasing the rate of venous return. At only one value of atrial pressure—in this case 0—will inflow equal outflow.

Even though cardiac output is greatly affected by a slight change in atrial pressure (e.g., curve I-3 between 0 and 1 mm Hg atrial pressure), a maximal sympathetic effect on both the heart (curve I-3) and peripheral venous bed (curve II-3) results in little or no change in atrial pressure, although the operating point is then at a much higher value, point C. Likewise, if only cardiac vigor is increased (curve I-3) without a change in the venous return relationship (curve II-2), there will be a small decrease in atrial pressure and only a slight increase in cardiac output.

Note that massive hemorrhage (curve II-1) greatly reduces the mean systemic filling pressure and so venous return. Even with massive sympathetic effects on the heart, the cardiac output is greatly reduced, point D. Because the venous return is reduced, the pressure drop from periphery to heart is also reduced ($\Delta P = R\dot{Q}$). The reduction in right atrial pressure with hemorrhage (from 0 to -2 mm Hg) is less than the reduction in mean systemic filling pressure (from 8 to 3 mm Hg).

With cardiac failure (curve I-1), an increased mean systemic filling pressure is required to maintain the cardiac output (point E). This increase is obtained by sympathetic discharge or, in the long term, by retention of water, which increases the blood volume.

The factors influencing cardiac output are summarized at the left of Figure 17-1.

CARDIAC OUTPUT AND ADAPTATION IN HUMANS

The cardiac output in normal people is about 5 liters per minute. Like many other physical functions of the body, it is associated with metabolism, which in turn is associated with the size of the animal. Therefore a size correction is of value for comparison of various people or animals. For animals of various species ranging in weight between dogs and horses, the cardiac output can be predicted as being about 100 ml/min per kg body weight. Body weight is one factor widely used to determine drug dosage, but the body surface area has been found to be an even better basis for comparison of, for instance, metabolic rate, blood flow, heart rate, size of organs, and respiratory rate. In people of widely different sizes an empirical equation to convert body weight (in kilograms) and height (in centimeters) to body surface area (in square meters) is

$$\text{Area} = 0.00718 \times \text{weight}^{0.425} \times \text{height}^{0.725}$$

In medical terms, an *index* often means the rate of a physiological function per minute per square meter of body surface area. Thus, the *cardiac index* is defined as the cardiac output per minute per square meter of body surface, and in humans it is about 3.5 liters, with values below 2 liters or above 5 liters being abnormal at rest. The determination of a person's cardiac output (see Chap. 11) provides an important diagnostic tool. A cardiac index below about 2 liters is a sign of some form of cardiovascular failure. Tachycardia from atrial fibrillation, or bradycardia from atrioventricular block, often results in seriously reduced cardiac output. The cardiac output is reduced in congestive heart failure and does not increase significantly following muscular effort. In patients in mild heart failure, the cardiac output may be within statistically normal limits, but for a given person it is lower than it would be if that person's heart were unimpaired.

Increases in cardiac output demonstrate compensatory mechanisms in the transport of oxygen to the tissues. In anemia, for instance, blood viscosity is decreased and the oxygen-carrying capacity is reduced. Thus, a marked increase in cardiac output might be expected because the blood flows more easily; but, more important, a greater blood flow is induced as a homeostatic adjustment to maintain the same oxygen transport, although the extraction of oxygen may also be increased. In pregnancy the blood vessels to the placenta and uterus act as an arteriovenous shunt to lower the total peripheral resistance. An adequate blood pressure is maintained by an increased cardiac output (see Chap. 17).

Hyperthyroidism (as in thyrotoxicosis), as well as fever or hyperthermia from high environmental temperature, causes an increase in metabolism that requires an increase in oxygen supply if homeostasis is to be maintained. Vessels dilate in response to hypoxia and so permit an increased flow. The cardiac output must increase if the blood pressure is to be maintained. On the other hand, at hypothermic temperatures of 15° to 20°C, the cardiac output is but 5 percent of normal. If the environmental temperature is high, so that the body temperature tends to rise, more blood is shunted to the skin to allow dissipation of more heat. This shunting of blood leads to an increase in the cardiac output.

The most pronounced increase in cardiac output occurs during severe exertion. The cardiac output of a normal person walking slowly is about 50 percent greater than basal. However, cardiac outputs of about 40 liters per minute have been measured in well-trained athletes during maximal exertion. Since this output is seven times normal, both heart rate and stroke volume were increased. (The stroke volume is maintained by the factors acting to increase the end-diastolic volume and the end-systolic volume is decreased by an increased cardiac contractility.)

Hypoxia, emotional excitement, insulin hypoglycemia, cutaneous pain—in fact, almost any mechanism acting to increase sympathetic division activity—increase cardiac output by increasing myocardial contractility and heart rate.

About the only normal cause of a decrease in cardiac output from basal flow is the act of changing from a recumbent to an upright position. On standing, the cardiac output decreases about 25 percent as a result of the redistribution of blood and the tendency toward blood pooling in the lower extremities. This decrease leads to orthostatic hypotension (low blood pressure on standing erect) unless compensatory mechanisms act to increase heart rate, myocardial contractility, peripheral resistance, and the filling pressure (venomotor tone).

CARDIAC POWER

If blood is to be pumped against a pressure head to the tissues, work must be done by the heart to drive a volume of fluid (V) against a pressure head (P). Thus, force (f) (in newtons) × distance (d) (in meters) = work (w) (in newton-meters) = pressure (P) (in newtons/m^2) × volume (V) (in m^3). Consequently, the effective work done per beat is the integral of instantaneous ventricular pressure times the instantaneous rate of outflow during systole. Adding these

over 1 minute gives the rate of doing work (power developed). The cardiac power output may be approximated as the product of mean arterial blood pressure times cardiac output.

It is informative to calculate the work done per minute by the left ventricle in a normal person. If 5.5 liters of blood is pumped per minute at a pressure of 95 mm Hg, the left ventricle will do 5.5×10^{-3} m^3/min × 12.66×10^3 newtons/m^2 or 69.6 newton-meters or joules of work per minute. (Recall that 1 mm Hg is 133.3 N/m^2 or pascal.) Since a newton-meter per second is a watt, the useful power output of the left heart is only about 1.2 watts, but nonetheless it is equivalent to lifting three large textbooks (7 kg) 1 meter per minute.

The right heart pumps the same volume as the left heart, but the mean pulmonary arterial pressure is about one-seventh of the mean aortic pressure, and thus the amount of work per minute done by the right heart is one-seventh that done by the left heart.

In addition to the potential energy, in the form of pressure-volume work, developed by the heart, metabolic energy is also converted to kinetic energy because of the velocity of movement of the blood (see Eq. 5 in Chap. 11). Usually, only about 2 percent of the useful work of the heart is in the form of kinetic energy. However, during exercise, when the velocity and rate of flow of blood are higher than normal, the kinetic energy may amount to 25 percent or more of the total useful work. Most of this kinetic energy is reconverted to useful potential energy (pressure times volume) by the distention of the elastic aorta. In effect, the elastic aorta catches the ejected blood and then throws it on into the arterial bed.

CARDIAC EFFICIENCY

During exercise, in order to propel more blood to the tissues against the same or higher pressure head, the heart must do more work. The extra work requires an increase in cardiac metabolism. The increase is produced either by a greater utilization of nutrients and oxygen or by an increase in the efficiency of the conversion of these materials to useful work. In humans, the biological oxidation of nutrients by 1 ml of oxygen gives about 20.2 joules (Nm) of energy. Thus, the heart would require about 3.5 ml of oxygen per minute to develop the 70 newton-meters per minute of power (1.2 watts) under basal conditions if it were 100 percent efficient in the conversion of oxygen and nutrients to useful work. However, the heart is only about 20 percent efficient. Thus, for 70 joules per minute of useful work, about 350 joules per minute of energy must be released. Therefore, the human adult left heart requires about 3.5/0.2 or 18 ml of oxygen per minute at rest and the right about 2.6 ml per minute. Because our basal requirement is about 250 ml of oxygen every minute, our hearts use about 8 percent of our total oxygen consumption.

Although the oxygen consumption is a good index of the metabolic release of energy under steady-state conditions, the oxygen consumption of the heart is not always closely related to the total useful work done in the intact animal because the efficiency of the heart changes. The oxygen consumption per minute by the heart is roughly proportional to the mean systolic arterial pressure times the duration of systole times the number of beats per minute. Cardiac oxygen consumption is more closely related to the intraventricular pressure developed than to the external work performed. The relationship may be explained if the following physiological characteristics of muscle are considered: Isometric contraction of muscle requires metabolic energy and oxygen, but results in no useful work, since no load is moved. Useful work is performed only if the heart ejects blood. However, when a muscle shortens under a load, only a relatively small amount of extra energy is required.

An increase in *blood pressure* has little effect on cardiac efficiency. The increased work, when the change is in pressure head or load, with a constant cardiac output and heart rate, requires a *proportionate increase in oxygen,* since pressure is a major factor in the determination of oxygen consumption. This is seen clinically in aortic stenosis. In this condition the resistance to flow of blood out of the heart is greatly increased because of the constricted aortic valve, but the cardiac output is held about normal by various compensatory mechanisms. When the pressure load is high, the oxygen consumption by the heart must be high. Overt myocardial hypoxia and heart failure are seen frequently in aortic stenosis. Hypertension also presents a serious load to the heart and rapidly leads to cardiac failure if the coronary blood flow is restricted by atherosclerosis.

If the *stroke volume* of the heart increases, with blood pressure and heart rate constant, the efficiency of the heart increases to as high as 40 percent, since the greater volume pumped requires but a *small increase in oxygen consumption.* Systole lasts slightly longer, but the effect is relatively small. In experiments with dogs the work output can be increased about 700 percent by changes in stroke volume with only about 53 percent increase in oxygen utilization. This result is in contrast to increased work by increased pressure, in which case the oxygen consumption parallels the increased work accomplished. Clinically, the heart does much work in mitral insufficiency (i.e., blood leaking back

to the atrium with each beat) or in intracardiac and extracardiac shunts, which involve a large bypass flow. However, early heart failure is surprisingly rare, considering the enormity of the defect in many cases. Aortic valvular insufficiency is an exception, since there is not only a massive back-and-forth movement of blood through the insufficient valve but also, at the region of the valve, a low blood pressure during most of diastole. Thus, the perfusion pressure of the coronary arteries is markedly reduced.

An increase in *heart rate,* blood pressure and cardiac output being constant, acts to decrease the efficiency. An increase in heart rate results in an increase in the rate of oxygen consumption without, necessarily, an increase in the amount of work per minute (pressure × flow) done. Since each beat requires a certain amount of energy whether or not useful work is performed, the energy expended increases and the efficiency drops. As a consequence, unless the heart of an athlete who must do much work adapts to provide a larger stroke volume at relatively low heart rates, the athlete will not excel. Likewise, the patient at the limit of cardiac reserve is not aided by excitement, which causes an increased heart rate and thereby a decrease in efficiency. Increases in heart rate can and do increase the cardiac output, but the price is a decrease in efficiency. Since the normal heart has adequate reserves of coronary blood flow, this is not serious.

In the failing heart, for reasons as yet unknown, the efficiency is decreased, and with exercise it appears to be decreased even more. Thus, adequacy of coronary flow is particularly important under these conditions. Unfortunately, partial or total occlusion of the coronary arteries is commonly the cause of the cardiac failure in the first place.

NUTRITION OF THE HEART

CORONARY BLOOD FLOW. The sustained ability of the heart to maintain a high blood flow to the peripheral tissues is limited primarily by the cardiac oxygen supply. Because the amount of oxygen extracted from the blood per minute is the product of blood flow and the difference between the arterial and venous oxygen contents per milliliter of blood, there are but three mechanisms by which the supply of oxygen to the myocardial cells may be increased.

1. *Increased oxygen extraction.* This is a minor source of additional oxygen for the myocardium, in contrast to peripheral tissue such as skeletal muscle. In skeletal muscle the arteriovenous difference in the oxygen content of the blood may be increased markedly by decreasing the oxygen content of the venous blood. The heart

normally extracts 70 to 90 percent of the oxygen from the arterial blood, leaving only 2 to 6 volumes per 100 volumes of blood in the coronary sinus venous blood—a low reserve. Arterial blood is normally 95 to 100 percent saturated.

2. *Increased myocardial efficiency.* The efficiency of the myocardial utilization of oxygen is determined by the type of work done by the heart and is generally independent of the oxygen needs of this organ.

3. *Increased coronary blood flow.* This is the only remaining possibility for an increased supply of oxygen to the heart. Since the heart can incur only a very limited oxygen debt, an increase in the coronary blood flow by dilation of the coronary arterioles is essential.

FACTORS REGULATING CORONARY BLOOD FLOW. In people at rest, the coronary blood flow is about 200 ml per minute (about 70 ml/100 gm of tissue). From data obtained from dogs, the flow rate can increase by at least nine times during severe stress. The myocardium accounts for about 4 percent of the cardiac output.

Metabolism. Myocardial hypoxia has a pronounced effect on coronary blood flow. Hypoxia can increase the coronary blood flow at least five times. The response to anoxia is rapid, since a 5-second occlusion of the coronary artery having little or no effect on contractility is followed by a definite increase in flow. This sensitivity and rapidity of response to tissue hypoxia is essential, since the increased oxygen required for increased cardiac work must be supplied promptly by increased coronary flow or there will be incipient cardiac failure. Indeed, the coronary blood flow to the heart is closely and directly correlated with the oxygen consumption by this organ. Although the steps are not completely established, they probably include tissue oxygen tension and adenosine released from the tissue that then acts to dilate the coronary arterioles. An increase in carbon dioxide tension or a decrease in pH has a similar, but lesser, dilating effect on the coronary blood vessels. Furthermore, the average coronary blood flow under constant metabolic conditions is relatively constant even if the aortic pressure is varied between 60 and 140 mm Hg, for the coronary bed shows strong autoregulation (see Chap. 17).

Mechanical Regulation of Coronary Flow. Since the coronary arteries are supplied from the aorta, and since flow is proportional to the perfusion pressure, the coronary flow is dependent on the mean systemic arterial blood pressure. The relationship is not simple, for when the heart contracts,

the intramural (intramuscular) pressures can exceed the arterial pressures and so close arteries and stop the blood flow. The normal contracting myocardium tends to impede flow. Although there is usually an inrush of blood very early in systole, during most of systole the high intramural pressure greatly reduces coronary arterial flow, so that only 10 to 40 percent of the total flow occurs during this period, even though the aortic pressure is highest then (Fig. 15-11). The venous discharge is hastened during systole because of the squeezing of the veins.

The anatomical arrangement of the arterial tree of the heart is such that if one of the coronary arteries is suddenly occluded, the pressure beyond the occlusion drops to about 30 mm Hg, and contractility of the affected myocardium decreases to useless levels. Collateral flow of about 10 percent of normal is present, but unlike skeletal muscle under similar circumstances, the myocardium will require several

Fig. 15-11. Effect of myocardial contraction on coronary venous outflow and arterial inflow at rest and during exercise. Note the hindrance to inflow during systole. Pressures given for timing reference. (Modified from D. E. Gregg, E. M. Khouri, and C. R. Rayford. Systemic and coronary energetics in the resting unanesthetized dog. *Circ. Res.* 16:105, 1965; and from R. F. Rushmer. *Cardiovascular Dynamics* [2nd ed.]. Philadelphia: Saunders, 1961. P. 217.)

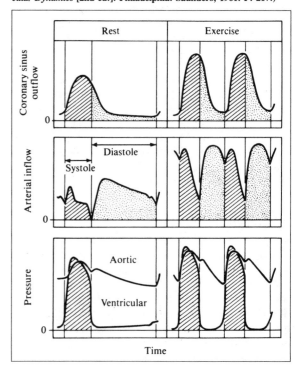

weeks for the blood flow to increase to adequate amounts. Thus, sudden coronary occlusion often results in a fatal myocardial infarction, whereas gradual occlusion may be followed by the development of an adequate collateral arterial blood supply.

Neural Control. The innervation of the coronary blood vessels is abundant, but the physiological function of the nerves is not clear, largely because of the difficulty in measuring blood flow under even approximately physiological conditions. Furthermore, it is hard to differentiate the direct neural effect on the resistance vessels and the concomitant effects of changed metabolism as a result of the neural action. Stimulation of the parasympathetic fibers acts to cause a reduction of the coronary blood flow, but in addition there is a decrease in heart rate and thus increased efficiency, decreased metabolism, and so a decreased oxygen consumption. One may therefore expect decreased coronary flow if the oxygen consumption is the primary regulator of this flow. An increase in the activity of the sympathetic nerves acting on the heart causes an increase in flow. Further, the heart rate increases, the contractility increases, more work is done, and thus the metabolism and oxygen consumption rise. Coronary vasodilation follows. Nonetheless, underlying these metabolic effects is sympathetic coronary alpha-receptor vasoconstriction. The dominant vagal tone in an unanesthetized person acts to keep the heart rate down, the efficiency up, and the coronary blood flow, in consequence, relatively low.

Circulating Epinephrine. Epinephrine causes an increase in coronary blood flow, probably on account of the increased metabolism, as occurs in skeletal muscle (see Chap. 17). Because epinephrine increases the vigor of contraction, heart rate, and metabolism, the resulting increased oxygen consumption probably elicits the arterial dilation. Angina in susceptible patients, following epinephrine injections or sympathetic nervous system action, has been attributed to the increased metabolism, with oxygen demands exceeding the supply. Since the increased blood flow produced by the vasodilating effects of the epinephrine is not adequate to provide for increased oxygen needs, hypoxia and ischemic pain result.

METABOLISM OF THE HEART. A continuous supply of nutrients and oxygen is required by the functioning heart. The primary nutrients during fasting are free fatty acids, but after eating or during exercise the primary nutrients are glucose and lactic acid. In vigorous exercise the venous oxygen saturation of active muscle is near zero, anaerobic metabolism is taking place, and lactic acid is released. Under these conditions, lactate is the primary substrate. It must be remembered, however, that utilization of lactate by the heart requires oxygen. When the heart itself is hypoxic, instead of utilizing lactic acid it produces it as a metabolite and develops a limited oxygen debt. In contrast to skeletal muscle, the mammalian heart can tolerate but little anaerobic metabolism. There seems to be no clinical situation in which the cardiac work capacity is limited by the lack of substrate for energy production.

ELECTROLYTES. In addition to intermediary metabolism, the heart depends on an optimal electrolyte environment (see Chaps. 3, 13, and 14). A plasma potassium concentration greater than about 8 mmol/L tends to lower the resting membrane potential and the heart rate and weakens the heart. Even higher concentrations will stop the heart in diastole. A deficit of calcium also causes a flaccid heartbeat. The effect of acids on the heart is complex. In dogs the lethal pH is about 6.0 and is independent of the anion, since both hydrochloric and lactic acids have similar effects. At a critically low pH there is rather sudden, sharp decrease in the contractility of the heart. Cardiac arrest in extreme diastole, similar to that caused by potassium, is seen as a terminal event.

BIBLIOGRAPHY

Berne, R. M., and Levy, M. N. *Cardiovascular Physiology* (4th ed.). St. Louis: Mosby, 1981.

Braunwald, E., and Ross, J., Jr. Control of Cardiac Performance. In R. M. Berne (ed.), *Handbook of Physiology*. Bethesda, Md.: American Physiological Society, 1979. Section 2: Circulation. Vol. 1, The Heart, pp. 533–580.

Braunwald, E., Ross, J., Jr., and Sonnenblick, E. H. *Mechanisms of Contraction of the Normal and Failing Heart* (2nd ed.). Boston: Little, Brown, 1976.

Guyton, A. C., Jones, C. E., and Coleman, T. G. *Circulatory Physiology: Cardiac Output and Its Regulation* (2nd ed.). Philadelphia: Saunders, 1973.

Levy, M. N., and Martin, P. J. Cardiac Excitation and Contraction. In A. C. Guyton and C. E. Jones (eds.), *Physiology, Series One. Cardiovascular Physiology*. London: Butterworth, 1974. Vol. 1, chap. 2.

Rushmer, R. F. *Cardiovascular Dynamics* (4th ed.). Philadelphia: Saunders, 1976.

Sagawa, K. The end-systolic pressure-volume relation of the ventricle: Definition, modifications and clinical use. *Circulation* 63:1223–1227, 1981.

THE SYSTEMIC CIRCULATION

16

Julius J. Friedman

GENERAL CHARACTERISTICS

The systemic circulation includes all the blood vessels that originate at the left ventricle and terminate at the right atrium. Those lying between the right ventricle and the left atrium constitute the pulmonary circulation (see Chap. 19). The systemic circulation is a closed system with the tissues arranged in parallel, except for the splanchnic circulation. In this system the total peripherovascular resistance (R_{total}) is determined by

$$1/R_{total} = 1/R_{muscle} + 1/R_{skin} + 1/R_{kidney} + \text{------}$$

Although the tissues are arranged in parallel, the resistance elements within each tissue are arranged in series, so that

$$R_{tissue} = R_{arteries} + R_{capillaries} + R_{veins}$$

Examination of some of the physical and functional properties of the systemic circulatory system can form a basis for better understanding of the physiological behavior of the system.

Figure 16-1 shows the aorta and large arteries to be heavy-walled vessels containing a relatively large proportion of elastin and collagen as well as smooth muscle.

The arteries have a relatively low compliance compared with the veins (Fig. 16-2), so that small changes in volume produce relatively large changes in pressure. These properties equip the arteries to serve as conduits that deliver blood, under high pressure, to the tissues. The terminal arteries and arterioles, which are especially well provided with smooth muscle, act as variable resistors to modulate tissue blood flow and blood flow distribution. The capillaries and venules, with walls essentially one endothelial cell thick, provide a porous barrier between the vascular and extravascular compartments across which exchange of water and solute molecules takes place. In addition to this function the venules act as a variable blood reservoir along with the small veins. The large veins and venae cavae, with larger diameters and thinner walls than their arterial

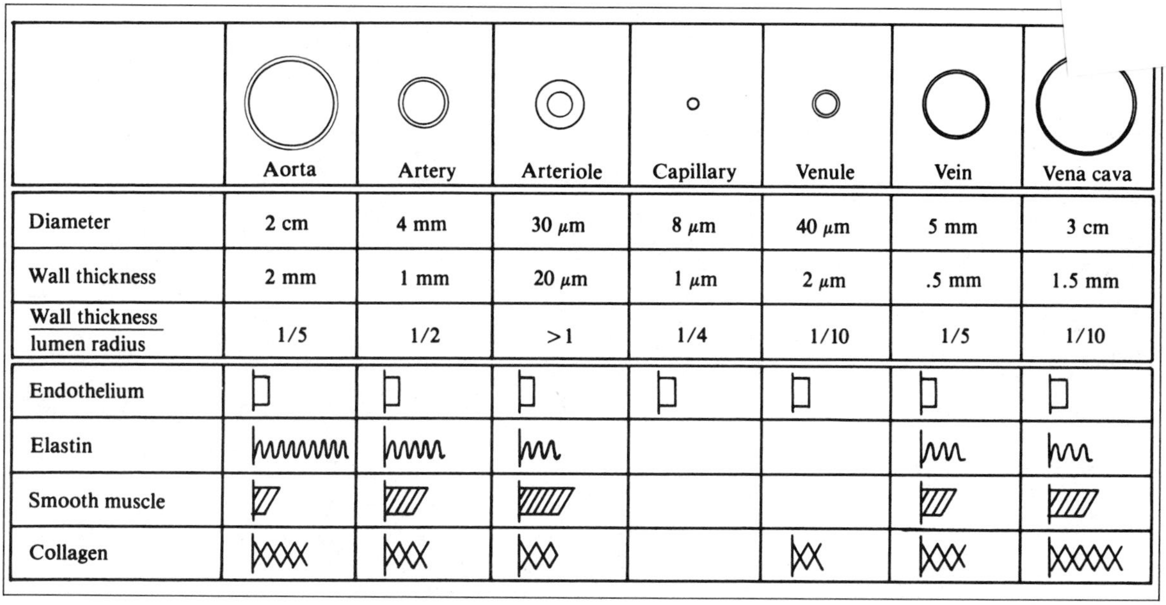

	Aorta	Artery	Arteriole	Capillary	Venule	Vein	Vena cava
Diameter	2 cm	4 mm	30 μm	8 μm	40 μm	5 mm	3 cm
Wall thickness	2 mm	1 mm	20 μm	1 μm	2 μm	.5 mm	1.5 mm
Wall thickness / lumen radius	1/5	1/2	>1	1/4	1/10	1/5	1/10
Endothelium	⊓	⊓	⊓	⊓	⊓	⊓	⊓
Elastin	⋀⋀⋀⋀⋀	⋀⋀⋀	⋀⋀			⋀	⋀⋀
Smooth muscle	▨	▨	▨			▨	▨
Collagen	⋈⋈⋈	⋈⋈	⋈⋈		⋈	⋈⋈	⋈⋈⋈

Fig. 16-1. Dimensions and structural constituents of the various vessels of the systemic circulation. (After A. C. Burton. Relation of structure to function of the tissues of the wall of blood vessels. *Physiol. Rev.* 34:619–642, 1954).

Fig. 16-2. The pressure-volume relationships of the arterial and venous systems.

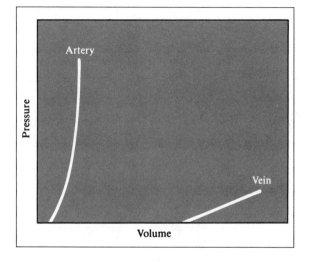

counterparts, are structured to conduct blood under low pressure from the tissues to the right atrium.

The relationships between blood pressure, cross-sectional area, blood flow velocity, and vascular capacity are presented in Figure 16-3. Mean blood pressure, about 90 mm Hg in the aorta, declines slowly through the arterial system. At the level of the terminal arteries and arterioles, the pressure drops sharply to an average of about 35 mm Hg at the capillary level and then falls slowly to about 10 mm Hg in the large veins. The longitudinal pressure profile is influenced by the level of vascular tone. As illustrated in Figure 16-4, vasoconstriction causes the profile to become more sigmoid, with a steep drop in pressure occurring at the arteriolar level; vasodilation, however, causes the profile to become less sigmoid, and the pressure drops somewhat more linearly.

The aorta of an 80-kg male is about 2 cm in diameter, with a cross-sectional area of about 3 cm^2. As it branches, the cross-sectional area increases slowly but progressively. Figure 16-3B shows that at the level of the arteriole the cross-sectional area increases dramatically to about 800 cm^2, and at the capillary-venular level reaches about 3500 cm^2. This large cross-sectional area then declines abruptly as the venous vessels coalesce to form the 3-cm vena cava containing a cross-sectional area of about 7 cm^2.

The volume of blood flowing through the various vascular segments is the cardiac output; the velocity of blood flow is

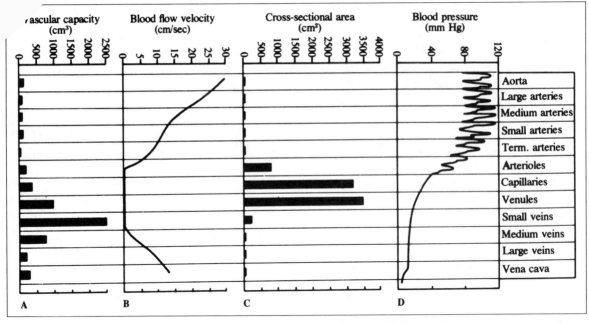

Fig. 16-3. Relationship between blood pressure, cross-sectional area, blood flow velocity, and blood volume in the various segments of the systemic vascular system.

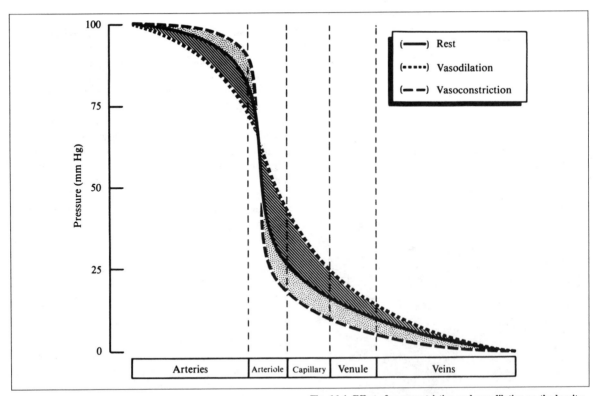

Fig. 16-4. Effect of vasoconstriction and vasodilation on the longitudinal blood pressure profile.

inversely related to the cross-sectional area. Figure 16-3C illustrates that the average velocity in the aorta is about 30 cm per second. The velocity falls progressively with branching until it reaches 0.026 cm per second at the capillaries. It then rises slowly and progressively through succeeding segments until it reaches about 14 cm per second in the vena cava. The high cross-sectional area of the capillaries and venules and the low velocity of blood flow through them are particularly well suited to the exchange function of these vessels.

The dimensions of the vascular segments provide them with the vascular capacities shown in Figure 16-3D. The arteries contain about 10 percent of the blood volume, the capillaries about 5 percent, the venules and small veins about 54 percent, and the large veins about 21 percent. The heart chambers hold the remaining 10 percent. Because the muscular venules and small veins possess large capacities and smooth muscle, they are effective reservoirs able to expand to accommodate large volumes of blood or to contract in order to transfer blood within them to the heart.

ARTERIAL BLOOD PRESSURES

Clinically, arterial blood pressures are routine measurements that reflect the status of the cardiovascular system.

The energy transferred to the arterial system by cardiac ejection generates a pressure pulse (Fig. 16-5). The rising limb of the pulse is the anacrotic limb; the declining limb is the catacrotic limb. The maximum pressure attained in the arterial system during cardiac ejection is the *systolic pressure* (SP). During the diastolic phase of the cardiac cycle, arterial pressure falls progressively. The minimal pressure achieved during diastole is the *diastolic pressure* (DP). The difference between systolic and diastolic pressures is the *pulse pressure* (PP). The *mean blood pressure* (MBP) is not the arithmetic average of systolic and diastolic pressures because of the unequal distribution of time for systole and diastole. A crude estimate of mean blood pressure is obtained by the following: MBP = (SP + 2 DP)/3. Ideally, mean blood pressure should be determined by integrating the pulse contour over the time interval of the cardiac cycle.

DETERMINANTS OF MEAN AND PULSE PRESSURES. Mean Arterial Pressure. The factors that determine the mean arterial blood pressure can be defined easily by an application of Ohm's law to the cardiovascular system (see Chap. 11), which provides the relationship

$$MBP = CO \times TPR$$

in which MBP is mean arterial blood pressure (mm Hg), and CO is cardiac output (liters/min), and TPR is total peripheral resistance (mm Hg per liter per min). Thus, mean blood

Fig. 16-5. Systolic, diastolic, and mean arterial blood pressures for an aortic and a femoral artery pressure pulse.

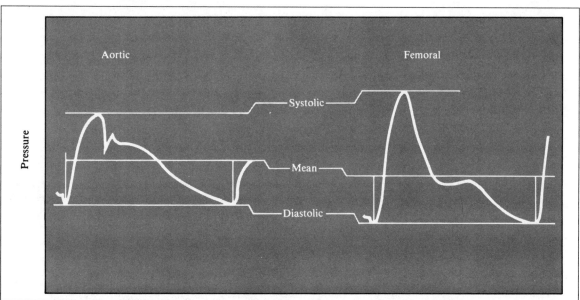

pressure can be modified by changes in either cardiac output, TPR, or both.

Pressure (P) in a vessel is a function of the volume in the vessel (V) relative to its ability to accommodate that volume, that is, its compliance (C). Thus, $P = V/C$. Since blood vessels contain some volume at zero pressure (Vo), the expression is more correctly represented as $P = (V - Vo)/C$.

Pulse Pressure. The change in arterial pressure that occurs during each cardiac ejection (dP/dt) depends on the change in arterial volume (dV) relative to the change in arterial compliance (dC), or $dP/dt = dV/dC$. The change in volume in turn is determined by the temporal relationship between inflow into ($\dot{Q}i$) and outflow from ($\dot{Q}o$) the arterial system, or

$$dV = \int_{t_1}^{t_2} (\dot{Q}i - \dot{Q}o)dt$$

Figure 16-6 illustrates the changes in arterial volume and pressure resulting from the changes in inflow and outflow during a cardiac cycle. The volume and pressure increase while inflow exceeds outflow, reach maxima when inflow and outflow are equal, and decline as outflow exceeds inflow.

Therefore, systolic pressure depends primarily on the stroke volume, peak cardiac ejection rate (cardiac contractility), residual arterial volume, and arterial compliance. Diastolic pressure depends on residual arterial volume, TPR, heart rate, and arterial elastic recoil.

Arterial Compliance. Arterial compliance or distensibility is the inverse of elasticity and is determined by the physical properties of the arterial wall. The elastin and smooth muscles in the vessel wall enable it to distend during systole and recoil during diastole. Thus, in accordance with the conservation of energy principle (see Bernoulli equation, Chap. 11), during systole the energy of cardiac ejection is distributed to the wall to cause distention, and to the blood to generate pressure and flow. Since only a portion of the energy introduced into the arterial system generates pressure, the level of systolic pressure achieved is lower than would occur in a rigid tube. During diastole, as the volume of blood in the arterial system declines and pressure falls, the rate of decline in pressure is reduced, and the diastolic pressure is supported somewhat by the energy of elastic recoil. As the arterial system becomes sclerosed, it becomes less compliant. In this condition, less energy is absorbed by the wall during systole, and systolic pressure rises to a greater extent. Less energy is restored to the blood by elastic recoil during diastole, and diastolic pressure falls, so that pulse pressure increases markedly.

The nature of arterial compliance can be better appreciated by examining the volume-pressure relationship (Fig. 16-7). The slope of these curves ($\Delta P/\Delta V$) represents the inverse of compliance (1/C). The figure shows that pressure increases with volume and that the extent of increase is directly related to age. This relationship reflects the decrease of compliance with increasing age, a consequence of atherosclerosis. It is also apparent in Figure 16-7 that the curves are not linear, indicating that arterial compliance is not constant.

Residual Arterial Volume. The volume-pressure relationships in Figure 16-7 illustrate the influence on arterial pressure of the residual arterial volume immediately prior to the next cardiac ejection. This volume, together with arterial compliance, determines the diastolic pressure; it also determines where on the volume-pressure curve, or at what initial compliance, the next pulse will begin.

The volume remaining in the arterial system at the end of diastole is determined by arterial runoff, which is tissue

Fig. 16-6. The effect of the relationship between arterial inflow and runoff on arterial volume and pressure.

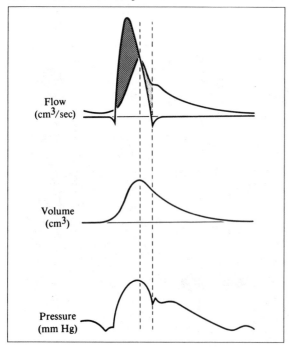

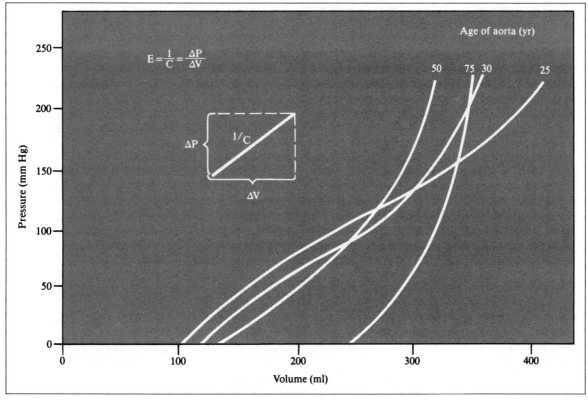

Fig. 16-7. The volume-pressure relationships of aortas obtained from subjects of different ages. (After P. Hallock and I. C. Benson. Studies on the elastic properties of human isolated aorta. *J. Clin. Invest.* 16:599, 1937.)

blood flow ($\dot{Q}$). This flow is determined by the perfusion pressure (PA − PV) and TPR : $\dot{Q}$ = (PA − PV)/TPR.

The effects of changes in stroke volume, TPR, and heart rate on the pulse pressures in a 25- and 50-year-old subject are illustrated in Figure 16-8. In the 25-year-old, with normal compliance, increasing stroke volume increases pulse pressure, whereas increasing either TPR or heart rate reduces the pulse pressure. The reduction in pulse pressure with increasing TPR is due to the effect of increased afterload on the heart. The reduction in pulse pressure with increasing heart rate is due primarily to reduced cardiac filling. In the 50-year-old subject, with reduced compliance, increasing stroke volume again increases pulse pressure but to a much greater extent. Whereas increasing TPR and heart rate reduces pulse pressure in the 25-year-old, the pulse pressure increases in the older subject. This modification is due to the steeper and curvilinear volume/

pressure curve of the 50-year-old, so that even small increases in aortic volume associated with increases in either TPR or heart rate shift the compliance to progressively lower values, thereby increasing pulse pressure progressively.

Pathological Pulses. Conditions that alter heart rate, stroke volume, TPR, and/or arterial compliance alter systolic and diastolic pressures. Frequently, these changes occur in combination as a result of the broad cardiovascular regulatory adjustment to stress. In such a situation the relationship of the characteristics of the pulse to the various determinants is not so clear-cut. However, a number of clinical disorders generate characteristic pulses, among them, aortic insufficiency, arteriosclerosis, aortic stenosis, and hypertension.

AORTIC INSUFFICIENCY. In aortic insufficiency the aortic valve does not close completely during diastole, and the blood regurgitates from the aorta into the left ventricle. The result is that the blood leaves the arterial system through two pathways, as peripheral arterial runoff and as backflow

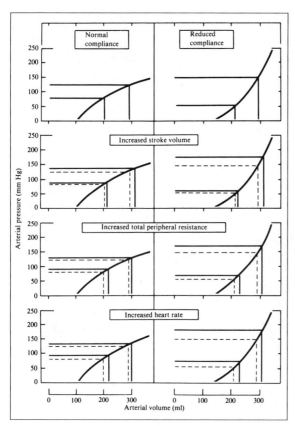

Fig. 16-8. The effect of increasing stroke volume, TPR, and heart rate in 25- and 50-year-old arterial systems.

characteristic slow, long anacrotic limb with a prominent anacrotic notch. Diastolic pressure is also reduced, but to a lesser extent than systolic pressure, so that pulse pressure is reduced.

HYPERTENSION. In hypertension, pulse pressure is markedly elevated along with mean blood pressure, because systolic pressure rises more than diastolic pressure. The rise in diastolic pressure is associated with an increase in TPR, and the greater rise in systolic pressure reflects the superimposition of a reduced arterial compliance. Hypertension can occur in a number of conditions. These are discussed in Chapter 18.

TRANSFORMATION OF PRESSURE PULSE DURING TRANSMISSION. Following cardiac ejection the arterial pressure pulse is transmitted throughout the arterial system at a velocity that varies with the stiffness of the arterial wall and the viscosity of the blood. The stiffer the wall and/ or the less viscous the blood, the more rapid the pulse transmission. Normally, the pulse velocity in the aorta is 3 to 5 meters per second, as compared with a mean blood velocity of 0.3 meter per second. The pulse velocity increases to 7 to 9 meters per second in the subclavian and femoral arteries and to 15 to 40 meters per second in the small arteries of the extremities. This increased velocity is due to the lower compliance and tapering of the smaller vessels. In hypertension or arteriosclerosis, pulse wave velocity may increase to three times normal.

As the arterial pressure pulse travels peripherally, it undergoes marked transformation due to pulse wave reflection and damping. The degree of reflection is directly related to the vascular tone. Thus, during maximal vasoconstriction, reflection is maximal, whereas during maximal vasodilation, reflection may be zero. The reflected wave travels up and down the arterial system until it is damped out. At the level of the femoral artery, where reflection and damping are prominent, the systolic pressure peaks more sharply and at a higher level than occurs in the base of the aorta (see Fig. 16-5). The incisura becomes damped and is replaced by the dicrotic notch; and the catacrotic limb of the pulse is amplified to produce the dicrotic wave.

SPHYGMOMANOMETRY. The use of direct pressure measurements has become fairly commonplace in the catheterization laboratory. However, routine clinical practice uses the simpler indirect method of sphygmomanometry. The basic idea of this method is illustrated in Figure 16-9. A cuff is placed over the brachial artery and inflated to a level above the expected systolic pressure, thereby occluding the underlying artery and interrupting pulse transmission. The pressure cuff is then slowly deflated at a rate of 2 to 3 mm

into the left ventricle, leading to a reduction in end-diastolic arterial volume and thus in diastolic pressure. The regurgitation also leads to an increased left ventricular end-diastolic volume, which produces an increased stroke volume and systolic pressure. In this manner, pulse pressure is markedly increased.

ARTERIOSCLEROSIS. In arteriosclerosis, sclerotic plaques are deposited on the arterial wall, thereby reducing compliance. As a result, damping is reduced and systolic pressure increases. Elastic recoil is also reduced and diastolic pressure falls. Thus, the pulse resembles that of aortic insufficiency. Differentiation between the two conditions is indicated by the diastolic murmur of aortic insufficiency.

AORTIC STENOSIS. In aortic stenosis the aortic valve becomes distorted and opens incompletely. The resulting increased resistance to cardiac ejection reduces stroke volume and systolic pressure. The aortic pulse displays a

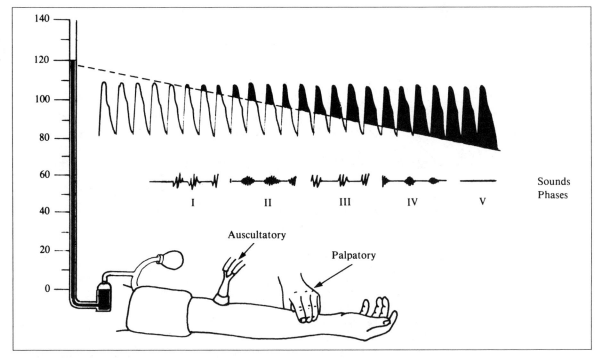

Fig. 16-9. Principles of sphygmomanometry illustrating the palpatory and auscultatory techniques for estimating human blood pressure.

Hg per beat while one either listens with a stethoscope over the brachial artery for a sound (auscultation) or palpates for a pulse at the radial artery (palpation). As the cuff pressure is lowered, the occluded brachial artery opens more and more with each pulse to allow blood to spurt through at high velocity. Turbulence is generated and the arterial wall oscillates, producing characteristic changes in sound (Korotkoff sounds). These sounds have been divided into several phases: In phase I, a sharp tapping sound is heard; in phase II, the sound assumes a blowing or swishing quality; in phase III, the sound becomes a soft thud; in phase IV, the sound suddenly becomes softer and develops a muffled quality; and then in phase V, the sound disappears.

Systolic pressure is assigned at the point at which the first sound is heard. Diastolic pressure is read at the point of silence. In the past it was read at the point of muffling, which, we know now, provides an overestimation. To avoid confusion, diastolic pressure is read at both points: (1) at muffling and (2) at silence. Thus, a normal auscultatory pressure reading would be 120/80/70.

On occasion, particularly in hypertension, a range of cuff pressure is encountered in which no obvious Korotkoff sound is emitted from the underlying artery. This absence of sound is referred to as the *auscultatory gap*. In such situations it is important either to inflate the cuff to a pressure high enough to exceed the range of the gap or to palpate the radial artery during inflation, since the pulse transmission continues within the pressure range of the gap.

FACTORS THAT INFLUENCE BLOOD PRESSURE. Population studies reveal that blood pressure varies with age, sex, weight, race, and socioeconomic status. In addition, as illustrated in Figure 16-10, arterial blood pressure varies considerably throughout the course of a day. This variability is the result of a person's physical, mental, and physiological activity, as well as his or her mental and emotional state. Consequently, it is important for a patient to be comfortable and at rest, and for repeated estimates of blood pressure to be run, before accepting a blood pressure reading as accurate.

Age. Figure 16-11 shows that normal systolic and diastolic pressures cover a wide range and vary with age. Normal mean systolic pressure increases from about 115 mm Hg at age 15 to about 140 mm Hg at 65 years of age, or about 0.5

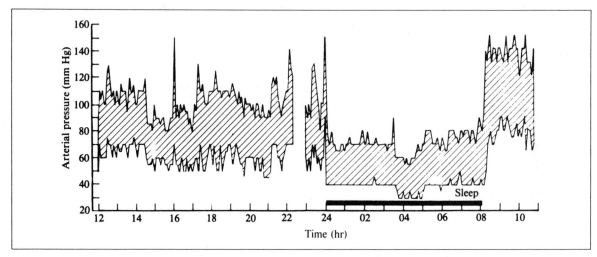

Fig. 16-10. A 24-hour record of arterial pressures plotted at 5-minute intervals. (From A. T. Bevan, A. J. Honour, and F. H. Stott. *Direct arterial pressure recording in unrestricted man. Clin. Sci.* 36:334, 1969.)

mm Hg per year. This increased systolic pressure probably reflects the progressive reduction in arterial distensibility that accompanies atherosclerotic vascular changes. Diastolic pressure increases from 70 mm Hg at age 15 to about 90 mm Hg at age 65, or about 0.4 mm Hg per year. The elevation in diastolic pressure probably reflects an increase in TPR, possibly due either to partial occlusive sclerotic changes in peripheral vessels or to sustained total body autoregulation (see Chap. 17) associated with elevated systolic pressure. Because of the wide range of normal pressures, estimates of arterial blood pressures, however well executed, should be viewed in this context, and diagnoses should be made only after repeated verification and then related to a person's medical history.

Sex. Systolic and diastolic pressures also vary with the sex of the subject. They are lower in women than in men under 40 to 50 years of age and higher in women over 50 years of age. This trend may be due to hormonal changes that take place at menopause.

Weight. Systolic and diastolic pressures are directly related to the weight of the subject. Although pressures in obese people are frequently overestimated because of improper cuff size, this relationship still holds when care is taken to avoid erroneous estimates.

Race and Socioeconomic Status. Systolic and diastolic pressures in blacks are consistently higher than those in whites at all ages and for both sexes. It has been suggested that genetic factors are responsible for this difference. There are, how-

ever, data that strongly implicate environmental factors, particularly the stress associated with the socioeconomic status of the individual.

Posture. When a person stands, gravity acts on venous return to reduce cardiac output and thoracic arterial pressure. Compensatory increases in heart rate and total peripheral resistance cause both systolic and diastolic pressures to rise.

Exercise. Systolic and diastolic pressures increase with exercise. Systolic pressure increases as a result of increased cardiac contractility. Diastolic pressure may decrease initially because of vasodilation of skeletal muscle vasculature. However, the increased heart rate limits runoff time and can cause diastolic pressure to rise.

VENOUS PRESSURE

The venous system is situated between the capillary-venular network and the right atrium. In this location, venous pressure can influence both capillary and cardiac function. Pressure in the venous system, as anywhere else, depends on the volume relative to the compliance. Because venous compliance is relatively high, venous pressure is quite low compared with arterial pressure. Since vascular pressures are determined with respect to a zero reference,

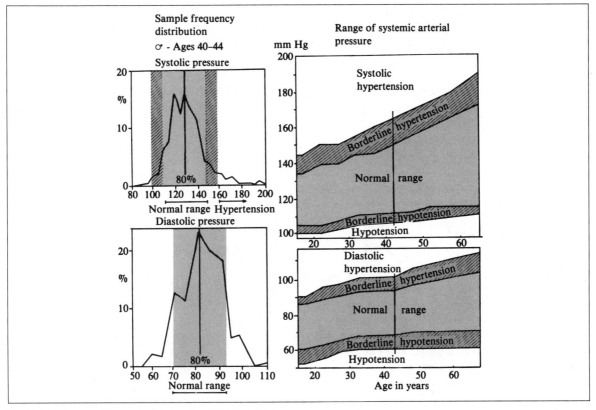

Fig. 16-11. The range of normal systemic arterial systolic and diastolic pressures and their variation with the age of the subject.
(From R. F. Rushmer. *Cardiovascular Dynamics* [4th ed.]. Philadelphia: Saunders, 1976. P. 194.)

measurements of low pressures are highly susceptible to the selection of the zero pressure reference point.

In the horizontal position the location of this point is not too critical, since the effect of gravity on the distribution of blood in the cardiovascular system is slight. However, in the standing position, gravity superimposes hydrostatic forces on blood pressures.

Guyton and colleagues (1973) suggested that the point of zero reference, the physiological reference point, is located at the base of the tricuspid valve. In a standing position the line passing horizontally through this point is called the *phlebostatic axis* and projects to the surface of the body at the point of intersection of the lateral border of the sternum and the fourth intercostal space. Gauer and Thron (1965), on the other hand, indicated that the location of the reference point—the "hydrostatic indifference point"—depends on the venous distensibility. In the horizontal position it is

located 5 cm below the diaphragm. In the standing position it moves closer to the lower extremities because of the high venous capacitance in the lower portions of the body.

Venous pressure can be measured directly by venipuncture with a needle connected to a suitable manometer. It can also be estimated indirectly by noting the level above the reference point at which engorged veins in a dependent hand collapse as the hand is slowly raised.

EFFECT OF POSTURE. In the standing position the force of gravity acts vertically on the vascular system. This influence modifies the distribution of pressure throughout the circulation, as illustrated in Figure 16-12. According to Pascal's law of hydrostatics, the pressure at the surface of an open vertical fluid column is equal to atmospheric pressure. As the point of measurement is moved below the surface, the hydrostatic pressure increases linearly with the height of the fluid column above the point of measurement. If the zero reference point were to be moved down into the fluid column, then all points above it would be negative in rela-

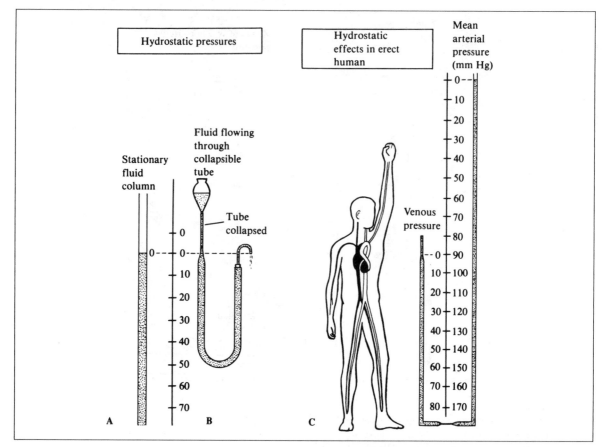

Fig. 16-12. The nature and significance of hydrostatic pressures. A. The pressure in a fluid column is dependent on its specific gravity and the vertical distance from the point of measurement to the meniscus. B. A collapsible tube is distended only so long as the transmural pressure is positive. The transmural pressure is zero in the partially collapsed portion of the tube. C. In the erect position, arterial and venous pressures at the ankle are increased by about 85 mm Hg. Above the heart, arterial pressure is reduced and effective venous pressure is zero. (After R. F. Rushmer. *Cardiovascular Dynamics* [4th ed.]. Philadelphia: Saunders, 1976. P. 221.)

tion to the reference point, precisely the circumstance in the body. The zero reference point is located either at the base of the tricuspid valve or below the diaphragm. Therefore, the hydrostatic pressures are superimposed on the vascular pressures to increase pressures below the zero reference point and reduce pressures above it.

The average distance from the level of the right atrium to the foot is approximately 130 cm, causing hydrostatic pressure at the foot to be 120 cm H_2O (90 mm Hg). In the reclining position the mean blood pressure in the dorsalis pedis artery is about 90 mm Hg. In the standing position this pressure becomes about 180 mm Hg (90 + 90). The pressure in the corresponding vein would be the sum of the reclining pressure of about 20 cm H_2O and the hydrostatic pressure of 120 cm H_2O, or 140 cm H_2O (100 mm Hg). This marked increase in pressure does not affect the perfusion pressure gradient, since arterial and venous pressures are affected equally. However, capillary pressure rises to over 100 mm Hg. Fortunately, the integrity of the capillaries is not endangered by the high pressure because, according to the Laplace relationship (see Chap. 11), their small radius allows for only a moderate wall tension to develop; moreover, tissue pressure becomes elevated too, although not so rapidly.

The pressure in the vessels above the heart is negative relative to the zero reference point. Thus, the pressure in a cerebral artery situated about 38 cm above the reference point would be about 62 mm Hg (100 − 38), and that in the accompanying cerebral vein would be as low as −28 mm Hg (10 − 38). The negative venous pressure results in a negative transmural pressure, and veins of the head and neck collapse. As blood volume and pressure build up behind the collapsed vessels, they open to allow blood to flow through and then collapse again as pressure falls. Venous flow from the head and neck, then, is intermittent in the erect position. Veins in the rigid cranium do not collapse because the hydrostatic influence is equal on all vessels balanced by equal extravascular influence.

VENOUS RETURN. Venous return represents the flow of blood from the periphery back to the right atrium. Since the cardiovascular system is a closed-tube network, and the system is, on the average, in a steady state, venous return and cardiac output are equal. In fact, in the short term, venous return substantially determines cardiac output in normal subjects.

Venous return is determined by the pressure gradient for venous flow and the resistance to flow imposed by the large venous system. The pressure gradient is represented by the difference between peripheral venous pressure and right atrial pressure. Peripheral venous pressure is the pressure generated by the flow of blood from the capillaries into the peripheral veins relative to the compliance of the veins. The compliance of the total vascular system, but primarily that of the venules and small veins, relative to the blood volume distending it, is revealed in the *mean systemic pressure* (mean circulatory pressure includes the pulmonary circulation). Mean systemic pressure represents the pressure generated by the total blood volume in the areflexic systemic vascular system when tissue blood flow is zero and the pressure throughout the system is equal. Mean systemic pressure varies directly with blood volume and inversely with the capacitance of the vascular system. When mean systemic pressure is high, peripheral venous pressure will be high, with the result that the pressure head for venous return is largely determined by mean systemic pressure.

The pressure in the right atrium is normally close to 0 mm Hg. Increasing right atrial pressure above zero, a condition occurring in instances of reduced cardiac function, results in reduction of the pressure gradient for venous return. This relationship is illustrated in a venous return curve (Fig. 16-13).

Mechanism for Maintaining Venous Return. Assumption of the erect position increases peripheral venous transmural pressure and thereby results in a shift of blood from the central circulation to the peripheral veins of the lower

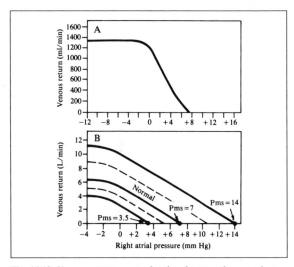

Fig. 16-13. Venous return curve, showing the normal curve when mean systemic pressure (Pms) is 7 mm Hg and showing the effect of altering the mean systemic pressure. (From A. C. Guyton. *Circulatory Physiology: Cardiac Output and Its Regulation.* Philadelphia: Saunders, 1963. P. 223.)

extremities. This shift causes venous return and cardiac output to decline, and, in the absence of compensation, syncope results. A number of mechanisms operate to maintain an adequate flow back to the heart.

THORACOABDOMINAL PUMP. Respiration is a factor that contributes to the maintenance of venous return. During inspiration, intrathoracic pressure decreases and abdominal pressure increases. The increased negative pressure in the thorax is transmitted to the right atrium, causing a fall in right atrial pressure. Increased intraabdominal pressure causes the peripheral venous pressure in this region to increase, augmenting the pressure gradient for venous return. During expiration the opposite effects occur.

When a person is placed on positive pressure respiration, intrathoracic pressure is elevated to positive values during inspiration, so that the thoracic pumping action is reversed and venous return is reduced.

SKELETAL MUSCLE PUMP. Another important means of maintaining venous return results from skeletal muscle contraction. The contraction of the skeletal muscles compresses the underlying or adjacent veins, forcing blood from them. The presence of venous valves oriented toward the heart ensures the flow of this displaced blood toward the heart, thereby increasing venous return. The contribution of the muscle pump to the maintenance of venous return and to the maintenance of a reduced peripheral venous pressure

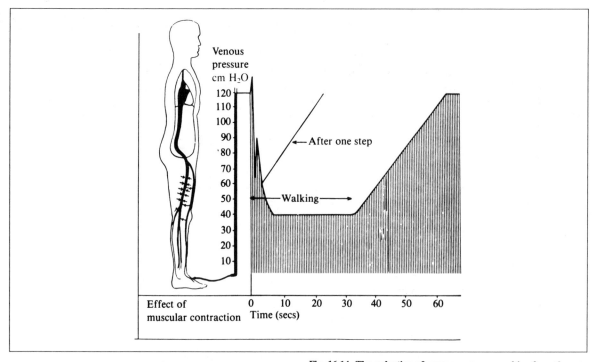

Fig. 16-14. The reduction of venous pressure resulting from the pumping action of muscles during walking. (From R. F. Rushmer. *Cardiovascular Dynamics* [4th ed.]. Philadelphia: Saunders, 1976. P. 225.)

during standing is seen in Figure 16-14. On quiet standing, venous pressure in the foot may rise as high as 125 cm H_2O. One single step can translocate enough blood toward the heart to reduce this pressure by 60 to 70 mm Hg. Repeated muscular contraction, such as occurs with walking, can discharge enough blood from the leg veins to maintain this pressure below 40 cm H_2O.

VENOMOTOR TONE. The venous blood reservoir is supplied with smooth muscles that receive sympathetic adrenergic innervation. In conditions in which the cardiac output is reduced, blood pressure will decline, generating a compensatory response through the baroreceptor system (see Chap. 17). One component of this compensatory response is an increase in venomotor tone that acts to reduce the capacitance of the venous reservoir, thereby displacing blood toward the right heart.

BIBLIOGRAPHY

Berne, R. M., and Levy, M. N. *Cardiovascular Physiology* (5th ed.). St. Louis: Mosby, 1983.

Folkow, B., Mellander, S., and Sweden, G. Veins and venous tone. *Am. Heart J.* 68:397–408, 1964.

Folkow, B., and Neil, E. *Circulation*. New York: Oxford University Press, 1971.

Gauer, O. H., and Thron, H. Postural Changes in the Circulation. In W. F. Hamilton and P. Dow (eds.), *Handbook of Physiology*. Washington: American Physiological Society, 1965. Section 2: Circulation. Vol. 3.

Guyton, A. C. *Textbook of Medical Physiology* (6th ed.). Philadelphia: Saunders, 1981.

Guyton, A. C., Jones, C. E., and Coleman, T. E. *Circulatory Physiology: Cardiac Output and Its Regulation* (2nd ed.). Philadelphia: Saunders, 1973.

Kannel, W. B. Role of blood pressure in cardiovascular morbidity and mortality. *Chest* 65:5–24, 1974.

Rushmer, R. F. *Cardiovascular Dynamics* (4th ed.). Philadelphia: Saunders, 1976.

CONTROL OF THE CARDIOVASCULAR SYSTEM

Carl F. Rothe and Julius J. Friedman

If tissue is to receive the oxygen and nutrients required for metabolism and to have waste products removed, the flow of blood through the tissue must be adequate. This flow through each organ is regulated in part by local and in part by central mechanisms that change the vascular resistance to blood flow in accordance with tissue requirements. For the system to be effective the systemic arterial pressure should be held reasonably constant by homeostatic mechanisms. With a constant blood pressure the flow of blood through an organ or tissue is inversely related to its vascular resistance (see Chaps. 11 and 12 and Fig. 17-1). To maintain constant systemic blood pressure as the flow through a tissue increases, either the cardiac output must be increased or the flow through some other region of the body must be reduced. Effective treatment of hypertension and many other cardiovascular diseases requires a knowledge of the mechanisms controlling blood pressure.

Figure 17-1 indicates the sequence of cause and effect of the cardiovascular factors determining tissue perfusion. (For simplification, no attempt was made to indicate the many control loops.) Tissue perfusion may be inadequate because of either an inadequate perfusion pressure gradient or excessive resistance due to vasoconstriction. The arterial blood pressure may be inadequate because the cardiac output is low or there is a generalized reduction in total peripheral resistance due to vasodilation. The cardiac output in turn is dependent on stroke volume and heart rate. Stroke volume is dependent on many factors (see Chap. 15), including the effective filling pressure of the right heart—its transmural pressure. Filling pressure is also dependent on many factors (see Chaps. 15 and 16), including the relationship between vascular blood volume and vascular capacitance. Control of blood volume is complex (see Chaps. 16, 23, and 30). Although the cardiovascular system will function without autonomic system activity, optimal operation, especially during stress, is possible only by the modulation and neurogenic stimulation of the heart and vascular smooth muscle provided by the autonomic nervous system as part of a control system. The effectors for the control of the cardiovascular system are in large measure dependent

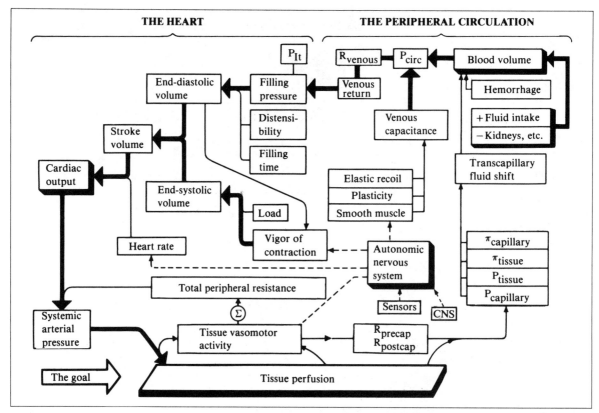

THE HEART   **THE PERIPHERAL CIRCULATION**

Fig. 17-1. Factors influencing cardiovascular function. The sensors influencing the autonomic nervous system include the arterial pressoreceptors, the chemoreceptors, and the atrial pressoreceptors. The ratio of vascular resistance (R) between the precapillary (Precap) and postcapillary (Postcap) segments of the tissue, in conjunction with the rate of blood flow and venous pressure, determines the capillary pressure. The hydrostatic (P) and oncotic (π) pressures influence the rate of transcapillary fluid shifts. The passive elastic recoil of the veins, and so the tendency to redistribute blood from the periphery to the heart, is dependent on the venous transmural presssure, which in turn is dependent on flow and venous resistance. P_{It} = the intrathoracic pressure; R_{venous} = resistance to venous return; P_{circ} = mean circulatory pressure (pressure in small veins). Venous return is in L/min.

upon the vigor of contraction of the heart muscle and the level of smooth muscle activity of the arteries and veins.

The organization of our understanding into "simplified" mathematical models has become exceedingly complicated because of the wide variety and closeness of interrelationships of the system (see, for example, Guyton et al., 1972). Such models, however, provide clues to relationships that had not been noted before, suggest experiments to test hypotheses, and are instrumental in helping us understand disease processes. Working with one section of the cardiovascular system at a time in detail or studying the integrated effects of many such modules together leads to progress.

The homeostatic control of blood pressure involves the integrated activity of the cardiovascular system acting over the short term (minutes), in conjunction with the systems concerned with body fluid and electrolyte balance acting over the long term (days). The nature of body fluid and electrolyte balance and its influence on blood pressure are discussed in Chapters 18, 22, and 23.

HOMEOSTATIC CONTROL OF BLOOD PRESSURE

The control of cardiovascular activity to maintain constancy of blood pressure is a classic example of a homeostatic, negative feedback system (see Chap. 7). The *effector mechanism* is dual in that blood pressure is increased by an increase in either cardiac output or peripheral resistance. The blood pressure is sensed by special organs called

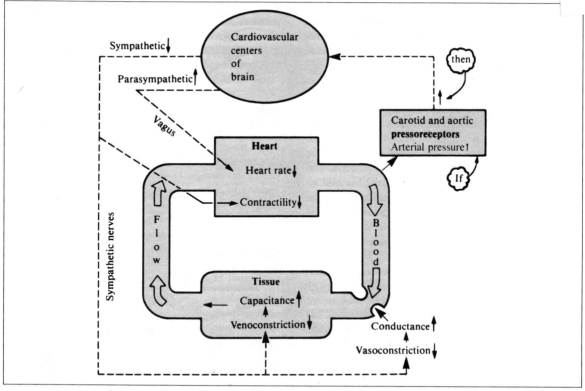

Fig. 17-2. Control of systemic arterial blood pressure by the arterial pressoreceptor reflex. Starting the loop at the pressoreceptors, *if* there is an increase in arterial pressure, *then* the firing rate of the pressoreceptors increases, with the consequences indicated.

pressoreceptors (baroreceptors), located primarily in the carotid sinuses and the arch of the aorta. Impulses from these receptors are transmitted via afferent sensory nerves to the *cardiovascular centers*, located in the brainstem. After complex interactions, signals, acting to correct deviations in blood pressure, pass by way of the efferent nerves (the autonomic nervous system) to the peripheral vasculature and to the heart.

The *arterial pressoreceptor reflex* controlling the systemic arterial blood pressure is shown schematically in Figure 17-2. An increase in blood pressure stretches the pressoreceptors in the walls of certain arteries, stimulating them to increase the frequency of discharge along afferent nerves. This acts by way of the cardiovascular centers to *inhibit* sympathetic division action on the heart and peripheral vasculature and to *increase* parasympathetic division action on the heart by way of the vagus nerves. Cardiac contractility and heart rate decrease, and the degree of

constrictor tone on both the resistance and capacitance vessels is reduced. Thus, the tetralogy of reduced cardiac vigor, bradycardia, vasodilation, and venodilation tends to follow an increase in blood pressure. The combination of these responses results in a decrease in blood pressure. The hypertension is buffered; hence the name *buffer reflex*. This is a negative feedback response, since the reduction in blood pressure is elicited by an increase above normal.

The arterial pressoreceptor reflex is also important in counteracting a decline in blood pressure. When pressure declines, fewer afferent impulses from the pressoreceptors go to the cardiovascular centers. Vagus tone is decreased, and there is less inhibition of the sympathetic division. As a result, cardiac action, vasoconstriction, and venoconstriction increase. Other names for the arterial pressoreceptor reflex include *carotid sinus reflex, baroreceptor reflex,* and *depressor reflex*.

The proportion of change of the four effector mechanisms (heart rate, contractility, vasoconstriction, venoconstriction) is not the same under various stresses (see, for example, Downing, 1979; Korner, 1974 and 1979; and Rothe, 1983). With arterial pressure deviations from normal, heart rate and muscle vascular resistance changes predominate. In the conscious dog, carotid sinus nerve stimulation causes bradycardia and vasodilation, especially in the limbs, but little change in kidney or skin vascular resistance or in cardiac output. Because cardiac output is determined by many factors (see Fig. 17-1 and Chap. 15), a negligible change in cardiac output does not necessarily mean no change in cardiac vigor of contraction. Indeed, cardiac contractility and venoconstriction also decrease with carotid sinus stimulation (increased arterial blood pressure), but the degree in the normal person is not clear. Reflex vasoconstriction, of whatever cause, is not uniform throughout the body. For example, blood flow through the intestine may be reduced with little change in other organs. Under some stressful conditions, splanchnic blood flow may be reduced by vasoconstriction with but little change in renal blood flow. With further stress, renal perfusion may be greatly decreased.

CARDIOVASCULAR CONTROL CENTERS

The neural centers for the control of the cardiovascular system are complex and not completely defined. They are *not* discrete structures but tend to be scattered and intermixed within the neural tissue. The concept of "centers," however, aids in the understanding of the regulatory process. These groups of neurons integrate neural impulses both from peripheral sensors and from the higher brain centers.

MEDULLARY CONTROL. Neurons in the medulla oblongata are responsible for the integration of afferent impulses and the origination of efferent impulses for the homeostatic control of blood pressure. *Vasoconstrictor centers* in the medulla oblongata are responsible for the neurogenic component of basal vasoconstrictor and venoconstrictor tone. *Cardiostimulator centers* increase cardiac activity. Electrical stimulation of these *pressor areas* located in the lateral reticular formation causes an increase in blood pressure by increasing vasoconstriction, heart rate, and probably cardiac vigor. The vasoconstrictor neurons can discharge without afferent stimuli, for vasomotor discharge does not cease and the blood pressure may not decrease even if incoming impulses from peripheral pressoreceptors are stopped. There is a continuous "tonic" activity of neuronal discharge *modified* by impulses from the depressor area, chemoreceptors, and higher centers. The medullary respiratory center neurons are intermingled with those of the pressor area. There are not only similarities in the cardiovascular and respiratory response to varying conditions but also interactions between the systems, both under normal conditions

and following various stresses. Arterial pressoreceptor discharge inhibits not only cardiac and vascular activity but also respiration. The respiratory system may influence the cardiovascular system, as seen by fluctuations in arterial blood pressure and a complex pattern of heart rate changes (*sinus arrhythmia*) related to respiratory activity.

Cardioinhibitor and *vasodepressor* areas are located in the caudal and medial reticular formation of the medulla in association with the dorsal nucleus of the vagus. Stimulation of these areas results in a reduction in arterial blood pressure. The response is produced by inhibition of the constrictor tone—not activation of vasodilator fibers—and by vagal slowing of the heart. Like the vasoconstrictor centers, the cardioinhibitor centers are tonically active under basal conditions; cutting the vagus nerves releases the inhibitory tone, and the heart rate increases.

HYPOTHALAMIC CONTROL. The hypothalamus, being a generalized center of control of the autonomic nervous system, modifies the activity of the bulbar region. There is probably no tonic activity coming from this region, because section at the level of the pons does not necessarily reduce blood pressure, and the baroreceptor reflex is little influenced by decerebration. However, the suprabulbar centers are of great importance in reflex cardiovascular control. The cardiovascular adjustments to emotions such as rage are mediated here. Redistribution of blood flow and characteristic patterns of cardiovascular response to exercise are integrated at this level. For example, stimulation of certain highly discrete areas results in a cardiovascular response closely similar to that seen during exercise, including changes in heart rate, blood pressure, cardiac action, and peripheral vascular tone. It has been possible to stimulate sympathetic fibers at this level that influence a particular organ with minimal effects in other organs; the neurogenic outflow, though generally increased, is nonuniformly distributed to all organs, and further, the receptivity of the organs differs.

The hypothalamus is involved in the distribution of blood flow for the control of body temperature. Lesions in this region impair the ability of an animal to protect itself from extreme environmental temperature, because sweating and changes in tone of the resistance and capacitance vessels of the skin no longer occur in response to the appropriate conditions. Temperature control is primarily by way of the sympathetic division, since the parasympathetic division has little direct effect except for that on heart rate.

CEREBRAL CORTICAL CONTROL. Impulses affecting the cardiovascular system originate in the forebrain and may have an important bearing on psychosomatic disorders. The sympathetic vasodilator outflow to skeletal muscle apparently originates in the cerebral cortex and passes through the hypothalamus and medulla, where the efferent discharge pattern may be modified. Electrical stimulation of the cortex may evoke both autonomic and somatic effects.

SPINAL CONTROL. Control centers in the spinal cord are apparently of minor importance for the maintenance of an adequate circulation under normal conditions. In cases of circulatory depression the cord is of greater importance. Transection of the thoracic spinal cord causes an immediate and profound fall in blood pressure. However, after a time a certain degree of control of the blood pressure may be regained. Once the blood pressure in a spinal animal has returned to near-normal levels, total destruction of the cord results in a permanent reduction in pressure. Thus, there are neurons in the cord responding to pressoreceptor impulses or reduced blood flow and hypoxia that initiate impulses along vasoconstrictor fibers. Stimulation of cutaneous receptors by pain or cold induces segmentally arranged vasoconstriction of the intestinal vessels of spinal animals, a phenomenon indicative of the action of spinal centers on the cardiovascular system.

CARDIOVASCULAR SENSORS AND REFLEX CONTROL OF THE CARDIOVASCULAR SYSTEM

Effective cardiovascular control depends on information provided by sense organs that transmit information to the control centers in the brain. Our understanding of the reflex control of the cardiovascular system is far from complete, however.

Stimulation of the various cardiovascular and pulmonary *mechanoreceptors* (stretch, tension, pressure) leads to *reflex inhibition* of the activity of circulatory and respiratory systems. This is a first approximation, but exceptions to this general rule are few.

ARTERIAL PRESSURE RECEPTORS. The sensors for the control of the systemic blood pressure by the arterial pressoreceptor reflex are sensitive nerve endings that respond to stretching of the walls of arteries as the transmural pressure is increased. The pressoreceptors are located not only in the carotid sinuses and arch of the aorta but also along the common carotid arteries. The rate of firing of the pressoreceptors at various pressures has been measured (e.g., Fig. 17-3). The carotid artery receptors are effectively quiescent below pressures of about 60 mm Hg, although some may be firing at pressures as low as 30 mm Hg. As the transmural

pressure is increased beyond 60 mm Hg, the frequency increases progressively. The *change* in impulse frequency per millimeter mercury pressure change is maximum at about normal blood pressure—the receptors are most sensitive and induce the highest control gain at this pressure. A plateau is reached at about 160 mm Hg. Additional increases in pressure do not appreciably increase the rate of receptor discharge. Consequently, when the blood pressures decrease below about 60 mm Hg or increase above about 160 mm Hg, little further compensatory response is elicited by this reflex system. The threshold for the aortic pressoreceptors is higher than that for the carotid receptors, suggesting that their function is primarily to limit excessively high cardiac and arterial pressures rather than to elicit reflexes to restore a low blood pressure.

The frequency of pressoreceptor discharge is also determined by the *rate* of stretch of the receptors as well as by the average magnitude. Thus, there are bursts of activity with each pulse (Fig. 17-3A). In this manner the rate of change of pressure (and so the *pulse pressure*) determines, in part, the level of sympathetic division influence on the heart and peripheral vasculature.

The arterial pressoreceptors are not required to obtain a near-normal blood pressure and heart rate (Cowley et al., 1973). Although they fire during each heartbeat at normal blood pressures and show little adaptation over a period of a few hours, after complete carotid sinus and aortic arch denervation, the blood pressure returns to within about 10 mm Hg of normal in a few days. In contrast to the normal animal, the blood pressure becomes highly variable with activity, but the heart rate is relatively constant.

The role of the pressoreceptors in hypertension is not yet clear. The frequency of impulses is lower than normal in some forms of this disease. However, current evidence suggests that the adaptation of the pressoreceptors over a period of days and weeks is such that the response curve is simply shifted to a higher pressure (to the right, Fig. 17-3B). In most forms of hypertension the pressoreceptors neither attenuate the disease nor cause it (see Chap. 18).

CHEMORECEPTORS. The peripheral chemoreceptors are specialized nerve endings in the carotid and aortic bodies found near the carotid sinus and arch of the aorta. These nerve endings are sensitive to a decrease in oxygen tension, an increase in carbon dioxide tension, or an increase in the acidity of the blood. Poisons that inhibit oxidative pathways, such as cyanide or fluoride, stimulate the chemoreceptors. Chemoreceptor stimulation causes an increase in pulmonary ventilation (in this respect the carotid body is much more important than the aortic body in humans) and

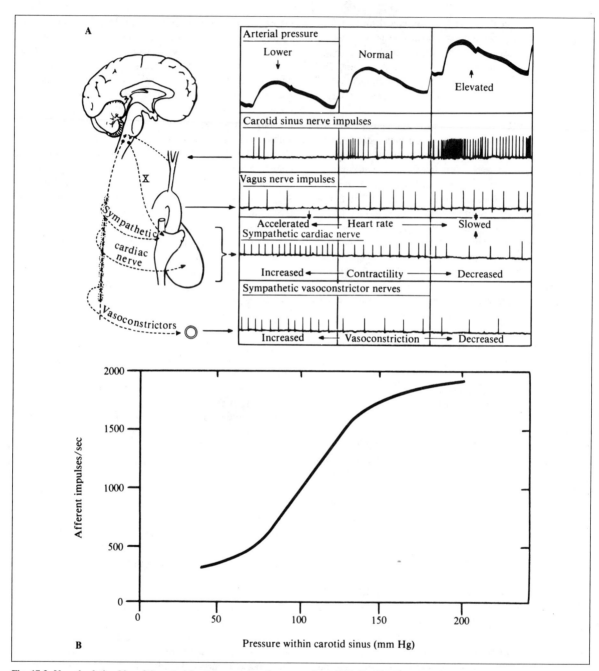

Fig. 17-3. Neural relationships of the arterial pressoreceptor reflex. A. Effect of arterial pressure pattern on carotid sinus stretch receptor discharge and the impulse rate of efferent nerves. (From R. F. Rushmer. *Cardiovascular Dynamics* [4th ed.]. Philadelphia: Saunders, 1976. P. 188.) B. Rate of discharge of afferent nerves at various static sinus pressures. (After W. Kalkoff. Pressorezeptorische Aktionspotentiale und Blutdruckregulation. *Verh. Dtsch. Ges. Kresilaufforsch.* 23:397, 1957.)

an increase in blood pressure by peripheral vasoconstriction. Hypoxia generally has little effect on heart rate, in part because peripheral hypoxia causes primarily a slowing of the heart, while the concomitant increase in respiratory activity secondarily tends to speed up the heart reflexly in proportion to the change in respiratory minute volume.

The carotid body is one of the most vascular tissues of the body. The tissues weigh only about 2 mg in humans, but the blood flow through them is about 20 ml/min per gram of tissue—4 times that through the thyroid, 30 times that through the brain, and about 200 times that through the body as a whole. In addition, the chemoreceptor cells consume oxygen at the rate of about 90 ml/min per kg of tissue, about three times the rate of the brain. The carotid chemoreceptors respond to decreases in oxygen tension and somewhat to decreases in oxygen content. The high blood flow through this tissue is apparently necessary to supply the metabolic needs of the chemoreceptor cells, for a decrease in blood flow through the carotid and especially the aortic bodies, caused by decreased arterial blood pressure, results in chemoreceptor stimulation. Simultaneous arterial hypoxemia and hypotension are synergistic and thus produce a vigorous response. Although the peripheral arterial chemoreceptors show little activity at normal arterial oxygen and carbon dioxide tensions, they are sensitive to changes from normal. A few minutes after circulatory arrest, however, aortic chemoreceptor activity rapidly declines (Paintal, 1973). If the arterial oxygen availability is very low, their response fails.

The *diving response* is a complex reflex induced by submersion of the nostrils or even by the expectation of diving under water. Extreme bradycardia and peripheral vasoconstriction follow in animals such as the seal, whale, or duck. The cardiac output is only a small fraction of normal. Oxygen from myoglobin stores and then anaerobic metabolism are utilized for skeletal muscle metabolism. In humans the response seems to occur in the infant more readily than in an older person.

CAROTID SINUS REFLEX. Occlusion of the carotid arteries on the cardiac side of the pressoreceptors and chemoreceptors produces a *carotid sinus reflex*. The increase in sympathetic division activity follows from both a reduction of stimulation of the arterial pressoreceptors, because the blood pressure within the sinuses is reduced, and a stimulation of chemoreceptors, because the supply of oxygen to these tissues is decreased. These two receptor systems thus help assure an adequate flow of well-oxygenated blood to the central nervous system. Both sets of receptors are important for the homeostatic compensation following hemor-

rhage. Because of different degrees of cerebrovascular anastomoses and the vertebral arterial supply, the pressure in the sinuses is not reduced to zero following bilateral carotid occlusions.

CEREBRAL ISCHEMIA. If the flow of blood through the brain is severely restricted, there is a delayed but profound sympathetic division response. This ischemic response acts at blood pressures below those that have no *more* effect by way of the arterial pressoreceptor reflex. The brain becomes hypoxic and hypercapnic below blood pressures of about 50 mm Hg. At these perfusion pressures the chemoreceptors along the carotid arteries are of course also stimulated by the stagnant anoxia. In response, there is a marked elevation of systemic blood pressure to as high as 350 mm Hg. The resulting degree of sympathetic vasoconstrictor discharge gives renal vasoconstriction of such a magnitude that urine flow stops. The mechanism of action probably results from an increase in carbon dioxide tension and a decrease in oxygen tension in the medullary region of the brain.

The *Cushing reflex* is a corollary of the cerebral ischemic response. If the cerebrospinal fluid pressure is greater than the systemic arterial blood pressure, the flow of blood to the brain is stopped, because the fluid pressure acting in the rigid skull occludes the arteries. Under these conditions the central ischemic response produces an increase in blood pressure. This tends to restore cerebral blood flow. A very high blood pressure following a head injury or cerebral vascular accident indicates cerebral hemorrhage and the operation of this mechanism.

With severely depressed cardiovascular function or serious metabolic acidosis, rhythmic variations in arterial blood pressure may be seen. These fluctuations, called *Mayer waves*, have a period of 15 to 60 seconds and are independent of respiration. As the blood pressure declines, central and/or peripheral chemoreceptors excite reflex vasomotor activity, which acts to increase the blood pressure. Chemoreceptor activity then declines, followed by a decrease in blood pressure. The response is oscillatory because of lags in the system (see Chap. 21).

PULMONARY AND CARDIAC RECEPTORS. *Right atrial* and *central vein receptors* respond to distention. If vagal tone is high, giving a relatively slow heart rate, distention of these areas leads to an increased heart rate (*Bainbridge reflex*)— an exception to the generalization that mechanoreceptor stimulation inhibits the cardiovascular system.

Left atrial volume receptors respond to increased transmural pressure, such as occurs from an increased left atrial volume or a more negative intrathoracic pressure. Impulses transmitted to the brain may reduce the secretion of antidiuretic hormone (ADH) (vasopressin) and so increase the loss of water from the body. Atrial distension may also induce capacitance vessel dilation. Reflex hypotension and bradycardia are also sometimes seen following left atrial distension. In response to hemorrhage and a reduction of left atrial pressure, ADH is released from the hypothalamus to induce water retention. The amounts of ADH secreted into the bloodstream are such that there may be a direct pressor effect of importance in the normal control of arterial blood pressure.

Ventricular mechanoreceptors respond to ventricular wall tension and distension. There are also receptors (possibly the same nerve endings) that respond to ischemia and drugs such as veratridine or nicotine. Ischemia may sensitize the receptors to distension. The afferent sensory fibers traverse both the vagal and sympathetic tracts. In contrast to the atrial receptors, the ventricular receptors are apparently active only under extreme conditions. The classic response to stimulation is cardiovascular inhibition (bradycardia and hypotension—the Bezold-Jarisch reflex) that tends to protect the ventricle from overload. However, cardiac reflexes mediated by afferent fibers coursing along sympathetic nerves, or by afferent fibers providing spinal input from the cardiac region to the brain, are mainly excitatory or pressor, in contrast to the depressor affect of most cardiopulmonary receptors (see, for example, Brown, 1979).

Pulmonary arterial pressoreceptors affect the cardiovascular system in a manner similar to that of the systemic arterial pressoreceptors, but to a smaller degree; for example, an increased pulmonary arterial pressure induces bradycardia, hypotension, and hypopnea. Pulmonary artery distension under some conditions causes systemic, but not renal, vasoconstriction.

Pulmonary congestion, pulmonary edema, pulmonary emboli, and strong irritants stimulate type J (nociceptive) pulmonary alveolar receptors (see Paintal, 1973), giving rise to a sensation of dyspnea, an inhibition of somatic muscle activity, hypotension, and bradycardia. These receptors may be important during maximal exercise or in congestive heart failure.

Sinus arrhythmia is the association of heart rate changes with respiration and is seen in normal, healthy, relaxed subjects with a high vagal tone and slow heart rate. With inspiration there is a transient increase in heart rate (tachycardia) followed by a bradycardia (slowing). Several mechanisms may be involved, as follows:

1. *An initial increase in heart rate.*
   a. The respiratory center neurons may also act on the cardioexcitatory neurons.
   b. Lung stretch receptor afferent nerve activity tends to suppress parasympathetic slowing of the heart, as well as inhibit generalized sympathetic vasoconstrictor tone.
   c. Inspiration lowers the intrathoracic pressure to increase right atrial filling and so may elicit the Bainbridge reflex.
   d. With a decreased intrathoracic pressure, the left heart pumps from a lower pressure, and so the systemic arterial pressure transmitted to the carotid pressoreceptors is decreased. This would elicit an arterial baroreceptor reflex increase in heart rate.
2. *Subsequent decrease in heart rate.*
   a. The increased right heart filling—and so output—eventually leads to an increased cardiac output and so to increased arterial blood pressure. This stimulates arterial baroreceptor activity to cause bradycardia.
   b. The nerve activity that acts to inhibit inspiratory effort may also stimulate the cardioinhibitory (vagal) neurons to slow the heart.

Because these mechanisms require a relatively fixed time, as the respiration rate increases, the mechanisms overlap. Moreover, expiration tends to elicit a similar heart rate response—transient tachycardia, then bradycardia.

Almost all *sensory nerves* have some connection with the cardiovascular reflex system. Some of the reflexes involve centers in the spinal cord only, since vasoconstrictor and vasodilator reflexes can be elicited in animals with sectioned spinal cords. Generally, a painful stimulus is followed by a rise in blood pressure. The conscious realization of pain, causing anxiety and stimulation of the adrenal medullae, also adds to the response. On the other hand, severe cutaneous pain, painful stimulation of the gastrointestinal or genital tracts, stretching of hollow organs, or the stimulation of deep visceral pain receptors may elicit the opposite response: a fall in blood pressure. However, urinary bladder distension in humans with sectioned spinal cords can cause an increase in systolic arterial pressure to over 300 mm Hg. Emotional stress may be followed by fainting; called *vasovagal syncope*. The heart rate slows dramatically and there appears to be vasodilation of the resistance vessels. Cerebral hypoxia occurs as the blood pressure declines, leading to a loss of consciousness.

RESPONSE TIME. By recording both receptor activity (e.g., nerve impulse from pressoreceptors) and effector activity (e.g., blood flow through organs such as the kidney or skeletal muscle, heart rate, venoconstriction) we now have a better, but far from complete, understanding of the time course of the various components. Within 0.2 second after pressoreceptor distension, the activity can be recorded in the efferent fibers of both the sympathetic and parasympathetic neurons. Whereas heart rate then changes within less than a second (one heartbeat), little change is seen in the smooth muscle activity of the resistance vessels for about 3 seconds and in the capacity vessels for about 7 seconds. The complete, steady-state response to a step-change in receptor activity requires at least 20 seconds and often 2 minutes or more. The arterial chemoreceptors also respond rapidly (within less than 1 second) to changes in arterial blood composition, but the final response is delayed, depending on the effector system involved. The description of the time course of these reflex responses is complicated by the fact that they are nonlinear (the magnitude of response to an input of a given magnitude is different, depending on the initial input level) and they are not symmetrical (the response to an input may vary in time or magnitude, depending on whether the receptor input was increased or decreased).

NEURAL CONTROL OF CARDIAC OUTPUT

The primary determinants of cardiac output at a given pressure load are the heart rate, filling pressure, distensibility, and contractility (see Chap. 15 and Fig. 17-1). *Neural impulses* from the cardioregulatory centers in the lower brain act to modify these variables.

The *heart rate* is determined primarily by the balance between the inhibitory effects on the pacemaker of acetylcholine (ACh) released by the vagus nerves of the parasympathetic division, and the excitatory effects of norepinephrine released by the sympathetic nerve endings (see Chaps. 7 and 14). The sympathetic and parasympathetic divisions tend to be antagonistic, since an increase in heart rate, caused by an increase in sympathetic division activity, can be reduced to normal by adequate stimulation of the vagus nerves. In fact, massive vagal stimulation stops the heart for many seconds. In the normal resting person, a continuous vagal tone reduces the heart rate to about 70 beats per minute. The sympathetic outflow is low. The intrinsic heart rate of the adult with all autonomic input blocked is about 110 beats per minute. Blocking of parasympathetic activity with atropine or sectioning of the vagi results in a marked increase in heart rate. During exercise or anesthesia, on the other hand, the sympathetic accelerator fibers exert a stimulating influence on the heart. Sympathectomy by sev-

ering the sympathetic outflow in the thorax from T2 to T5 (see Fig. 7-1) in humans is usually followed by a reduced cardiac response to exertion. In dogs with denervated hearts, the cardiac response to moderate exercise is slowed but little reduced—in part because of circulating catecholamines. The heart rate increases if the blood volume is either increased or decreased. Thus, with volume expansion, cardiac output is increased in conscious dogs with but little change in stroke volume. However, following hemorrhage the stroke volume decreases so much that even with a reflex increase in heart rate the cardiac output decreases.

Myocardial contractility is increased by an increase in the rate of discharge of the sympathetic division nerves going to the heart. An increase in myocardial contractility results in an increase in cardiac vigor: The pressure developed by the ventricles increases; the size of the heart, particularly at the end of systole, tends to decrease; and the rates of change of pressure and size increase. The parasympathetic division has little effect on ventricular contractility, in contrast to its marked effect on heart rate. The atria are richly innervated by the parasympathetic nerve endings, and there appears to be a decrease in atrial contractility as a result of vagal discharge.

Cardiac distensibility is apparently little affected by the autonomic nervous system, although some experiments indicate that sympathetic division activity may increase ventricular distensibility somewhat and so permit easier filling. The rate of relaxation, like the rate of contraction, is increased by positive inotropes.

The *filling pressure* of the right heart is determined primarily by the peripheral factors affecting the return of venous blood and the degree of negative intrathoracic pressure (see Chap. 15 and 16). The degree of constriction of the venous capacitance vessels is important. The filling of the left ventricle is determined primarily by the action of the right heart and the left atrium.

ADRENAL MEDULLARY INFLUENCE. Catecholamines, released from the adrenal medullae, provide a slowly acting but effective adjunct to the autonomic innervation of the heart. Cardiac output is thereby augmented by means of an increased myocardial vigor and heart rate.

PERIPHERAL CIRCULATION

The peripheral circulation is that segment of the circulatory system that is concerned with the transport of blood to the tissues, blood flow distribution within the tissues, exchange between blood and tissue, and storage of blood.

The maintenance of normal tissue activity as well as specialized tissue function requires an adequate supply of oxygen and other nutrients. Figure 17-4 illustrates the blood flow and the oxygen availability, consumption, and reserve for the major circulations of the body. The oxygen made available to the tissues each minute is determined by the tissue blood flow times the arterial blood oxygen content (ml O_2/100 ml blood) and is represented by the areas of the rectangles. The amount of oxygen consumed by each tissue is estimated as the tissue blood flow times the arteriovenous (A-V) oxygen difference and is represented by the numbers in the cross-hatched segments. The remaining area of the rectangle, tissue blood flow times venous oxygen content, represents an oxygen reserve for the tissue.

The level of tissue metabolic activity substantially determines the amount of oxygen consumed. The increased oxygen consumption associated with increased metabolic activity is reflected in either a higher blood flow, or lower venous oxygen content, or both.

Except for the heart and brain, tissues generally possess a large *effective* reserve of oxygen in the venous blood, which enables them to endure modest reductions in blood flow without suffering a severe oxygen deficit. In case of reduced flow, more oxygen is extracted from the blood, venous oxygen content decreases, and the A-V oxygen difference increases. Neither blood flow distribution nor oxygen consumption is determined by organ size. For example, the kidneys, which represent 0.5 percent of body weight, receive 20 percent of the cardiac output, whereas *resting* muscle, which accounts for approximately 40 percent of body weight, receives only 20 percent of the cardiac output. Tissue blood flow is determined by overall tissue function, which includes normal tissue maintenance as well as specialized tissue function. Because of their specialized functions of waste clearance and thermoregulation, kidney and skin receive blood flow in excess of their oxygen requirement. The coronary and cerebral circulations, on the other hand, receive relatively low blood flows compared with their high metabolic rates and oxygen consumption. This low blood flow is reflected in a relatively high A-V oxygen difference.

The parallel architectural arrangements of the tissue circulations provide a means by which blood flow to a tissue can have variations independent of changes in blood pressure. Figure 17-5 illustrates the nature of the redistribution of the cardiac output to skeletal muscle and the extent of change in the oxygen consumption of skeletal muscle during graded exercise in an average subject and in an athlete. With the heightened work of exercise, the cardiac output and total body oxygen consumption increase, with a progressively greater percentage of the cardiac output being distributed to the exercising muscle and a greater percentage of total oxygen consumption accounted for by the exercising muscle. The figure also illustrates the marked rise in capacity and utilization of oxygen that results from training.

Figure 17-6 shows that in order to provide the exercising skeletal and cardiac muscles with a progressively increasing share of the cardiac output, blood flow is diverted from the liver, intestine, kidneys, and eventually the skin. At moderate levels of exercise, skin blood flow increases as a result of higher body temperature. With maximal exercise, the demand by muscle exceeds the need for temperature regulation, and skin blood flow declines. The cerebral circulation remains relatively constant at all levels of exercise (autoregulation).

The flow of blood through tissues is determined by the perfusion pressure and by the resistance to the flow of blood (see Chap. 11). Since the perfusion pressure, blood viscosity, and length of tissue resistance vessels are relatively constant, the radius of the resistance vessels is the primary determinant of local blood flow. The radius of the resistance vessel is controlled in part by *central* influences manifested by neural and by humoral mechanisms, and in part by *local* factors, such as oxygen tension, metabolites, intrinsic reflexes, and autoregulation. Central and local mechanisms continuously interact to modify the distribution of blood flow.

CENTRAL CONTROL OF THE VASCULATURE

Central control of the peripheral circulation is effected through the reflex activation of the sympathetic division of the autonomic nervous system and through the action of circulating humoral agents.

VASCULAR RESISTANCE. The primary mechanism of central control of tissue blood flow is provided by the discharge of *sympathetic vasoconstrictor fibers* that cause smooth muscle contraction resulting in decreased vessel lumen radius and increased resistance to flow. The order of sensitivity of the vascular elements to sympathetic adrenergic influence is: precapillary sphincter > terminal arteriole > small artery > small vein.

Fig. 17-4. Blood flow and oxygen consumption of 70-kg man. The areas of the rectangles represent the amount of oxygen available to the tissues. The cross-hatched segments identify the amount utilized by the tissues. The numbers within the cross-hatched areas are the milliliters of oxygen consumed per minute. The remaining areas of the rectangles are the oxygen reserves of the tissues. (After R. F. Rushmer, *Cardiovascular Dynamics* [4th ed.]. Philadelphia: Saunders, 1976. P. 151.)

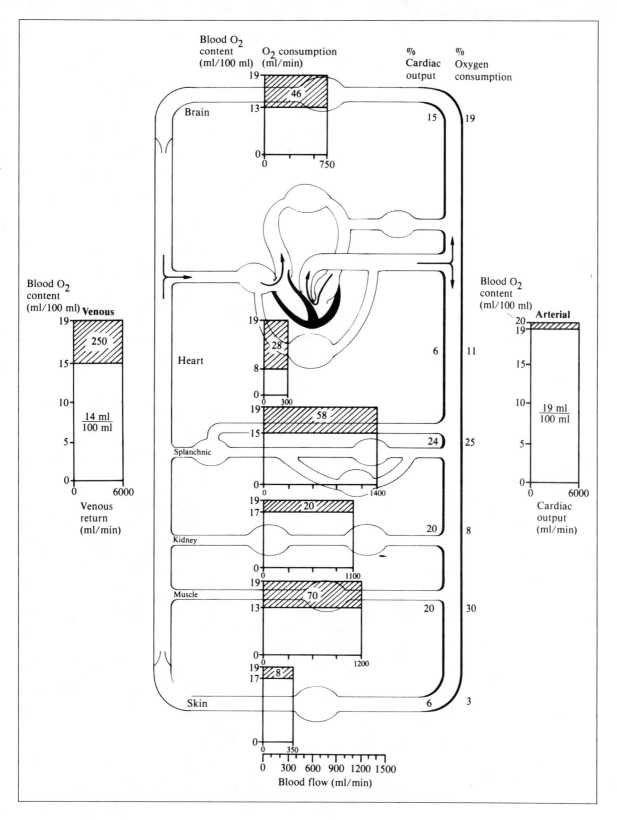

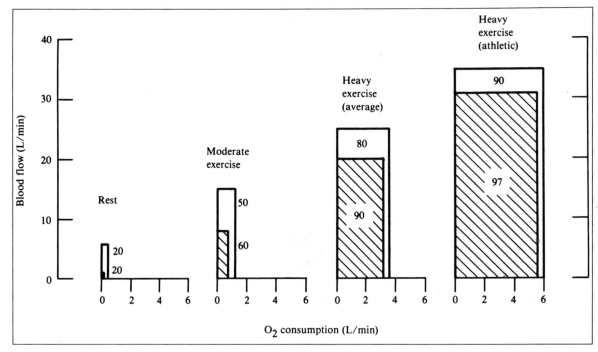

Fig. 17-5. The changes in muscle blood flow and oxygen consumption during graded exercise. It appears that the athlete in heavy exercise is a heart-lung-muscle organism. The upper numbers represent the percentage of cardiac output; the lower numbers represent the percentage of total body oxgyen consumption.

Figure 17-7, curve 1, illustrates the effect of sympathetic adrenergic stimulation on skeletal muscle flow at constant perfusion pressures. The resting frequency of nerve fiber discharge to skeletal muscle is about 0.5 to 1.0 impulses per second. This discharge provides a basal level of vasoconstrictor tone from which adjustment of vascular resistance can occur for the maintenance of circulatory homeostasis. Maximal resistance responses develop at a frequency of about 10 impulses per second. Inhibition of this basal vasoconstrictor tone results in vasodilation.

The relative responsiveness of the vasculature of different tissues varies, due in part to differences in the amount of vascular smooth muscle, the innervation ratio, the ratio of the vessel wall thickness to the vessel lumen radius (W/r), and the level of resting vascular resistance relative to the reactive range (from maximum vasoconstriction to maximum vasodilation). Figure 17-8 shows that at higher W/r the same degree of shortening of the smooth muscle will produce greater lumen reduction, resulting in a greater increased resistance to blood flow. Figure 17-9 shows that at the normal perfusion pressure, resting tissue blood flows, and thus the initial vascular resistances, of the various tissues are markedly different. It also reveals that the tissues have different reactive ranges. Thus, liver and muscle have

high initial resistances relative to their reactive ranges and can therefore elicit only a relatively slight additional vasoconstriction; kidney, which has a low initial resistance relative to its reactive range, can elicit profound vasoconstriction. The intestinal circulation responsiveness is between these two. Skin, because of its specialized function in temperature regulation, can elicit both marked vasoconstriction and vasodilation despite its high initial resistance. The coronary and cerebral circulations possess little capacity for vasoconstriction.

Variations in vascular tone modify the relationship between perfusion pressure and tissue blood flow. Figure 17-10, curve D, shows this relationship in resting muscle. At a pressure of 100 mm Hg, resting muscle blood flow is about 3 ml/min per 100 gm, signifying that a high degree of vascular tone is present in this tissue. Maximal sympathetic vasoconstriction (curve E) reduces blood flow at all perfusion pressures. At a pressure of 100 mm Hg the flow is about 0.3 ml/min per 100 gm of tissue. Maximal vasoconstrictor inhibition (curve B) shifts the relationship to the left, and, at

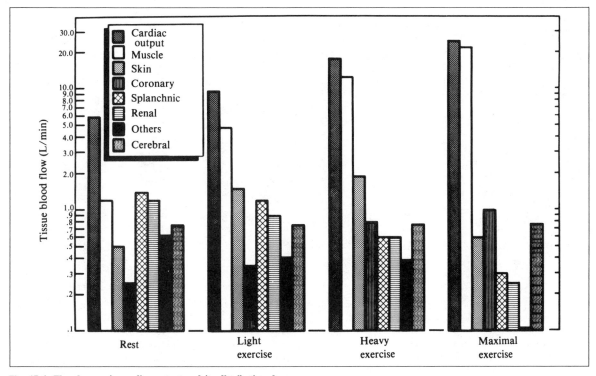

Fig. 17-6. The changes in cardiac output and its distribution during graded exercise.

the same perfusion pressure of 100 mm Hg, flow is increased about sixfold compared with the normal resting condition. Normally, the tissue vasculature functions well within this range of vascular tone.

Sympathetic vasoconstriction is not the only determinant of the basal vasculature tone, since an intrinsic level of vascular smooth muscle tone is present after denervation. Blockade of the vascular innervation permits a twofold to fivefold increase in flow through skeletal muscle (Fig. 17-10, curve B), while skeletal muscular contractions may elicit a further increase to over 20 times the normal flow (Fig. 17-10, curve A).

Although the sympathetic vasoconstrictor outflow is somewhat diffuse and has a general effect on the overall peripheral resistance, important changes in the distribution of blood flow may be brought about by these fibers. For example, uncomfortable environmental temperatures, hypoxia, digestion, or muscular exertion results in *discrete* cardiovascular adjustments in the distribution of the cardiac output. Sympathetic vasoconstriction during stress, such as severe hemorrhage and hypotension, can markedly reduce

the flow of blood through the skin, skeletal muscle, renal, and splanchnic vascular beds. This blood is redistributed to other more immediately vital tissues. The characteristic cold skin, muscular weakness, and anuria so often seen following hemorrhage are due in part to this redistribution.

VASCULAR CAPACITANCE. Reflex changes in vascular capacitance (see Chap. 11) are part of the high-pressure baroreceptor control of arterial blood pressure. A decrease in carotid sinus pressure of 25 mm Hg causes a decrease in systemic capacitance of about 3.5 ml per kilogram body weight. Most of the change is in unstressed venous volume, although sympathetic stimulation causes some reduction in venous compliance. A reflex-induced (i.e., *active*) reduction of systemic vascular capacitance tends to redistribute blood from the periphery to the heart to increase right ventricular filling and thus restore or even increase cardiac output. The maximum range of reflex change in vascular capacity, at a constant mean circulatory pressure of about 10 mm Hg, is about 15 ml per kilogram body weight (Fig. 17-11). In the 30 seconds needed to change blood volume and measure the mean circulatory filling pressure, reflex changes in the venous smooth muscle occurred. At low blood volumes, reflex

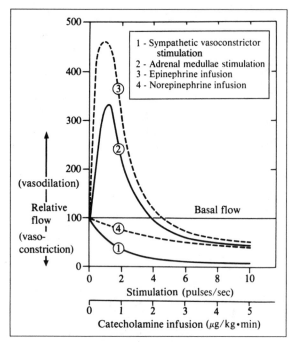

Fig. 17-7. The effect of neurogenic and humoral adrenergic influences on skeletal muscle blood flow. (Data from O. Celander. The range of control exercised by the sympathico-adrenal system. *Acta Physiol. Scand.* 32 [Suppl. 116]:48, 1954.)

venoconstriction approached that of maximal venoconstriction. The slope of the 30-second reflex curve thus does not represent the true vascular compliance under control conditions.

Blood also may be redistributed to the thorax by passive elastic recoil of the capacitance vessels if the distending pressure within them is decreased. This occurs when flow is reduced during vasoconstriction (increased vascular resistance). Because there is a small, but finite, venous resistance to blood flow, a reduced flow causes less pressure drop across the venous resistances between the capacitance vessels and right atrium than at normal flow. Therefore, the distending pressure is reduced. Passive elastic recoil is particularly important as a compensatory mechanism during hemorrhage. With reflexes intact, about 35 ml of blood per kilogram of body weight must be removed before the mean circulatory filling pressure and cardiac output are reduced to near zero; about 10 ml per kilogram comes from maximal active reflex venoconstriction from the basal volume and about 25 ml per kilogram, from passive recoil of the veins. Because blood pools in our legs when we stand, most of our reflex venoconstriction reserve is then utilized.

The capacitance vessels are sparsely innervated by sympathetic adrenergic fibers that discharge at a frequency of less than 1 Hz to provide a basal level of venomotor tone. The veins seem to be insensitive to tissue metabolites.

Veins of the skin are under sympathetic control related to temperature regulation. They are little influenced by the baroreceptors or chemoreceptors. The veins of the splanchnic bed, however, are greatly influenced by baroreceptor reflexes. Stimulating the low-pressure volume receptors and the peripheral chemoreceptors influences the capacitance vessels, but the effect seems to be minor.

ADRENOMEDULLARY INFLUENCE. Direct neural control of the vascular smooth muscle may be supplemented by circulating catecholamines such as epinephrine and norepinephrine released into the bloodstream through stimulation of the adrenal medulla. The effects on the vasculature are minor in comparison with direct neural control. Stimulation of the splanchnic innervation to the adrenal medulla over a range of frequencies results in a biphasic change in muscle vascular resistance (see Fig. 17-7, curve 2). At frequencies below 3 impulses per second, vasodilation occurs, whereas frequencies from 3 to 10 impulses per second produce progressive vasoconstriction. The basis for this variable effect lies in the fact that adrenomedullary stimulation liberates primarily epinephrine with some norepinephrine. Norepinephrine produces vasoconstriction by alpha-receptor stimulation at all doses (see Fig. 17-7, curve 4). In skeletal muscle, epinephrine in low concentrations (below 2 μg/min-kg causes dilation by stimulation of beta receptors (see Fig. 17-7, curve 3, and Fig. 7-4). At doses greater than 2 μg/min-kg, alpha-receptor stimulation predominates, resulting in vasoconstriction.

The precapillary sphincter is considerably more sensitive to circulating catecholamines than are the arterioles, which in turn are more sensitive than venules (Korner, 1974).

The dog with a denervated heart is capable of near-maximal exercise. However, if circulating catecholamines are also blocked (by propranolol), running performance is drastically reduced. With intact innervation of the heart, the circulating catecholamines add little to exercise performance. Thus, either an intact sympathetic innervation to the heart or an intact sympathoadrenal system permits near-maximal exercise performance.

SYMPATHETIC VASODILATOR FIBERS. Skeletal muscle blood flow increases with the thought of exercise prior to the onset of muscle contraction. This behavior has been attributed to sympathetic vasodilator fibers that originate in the motor cortex and synapse in the hypothalamus and col-

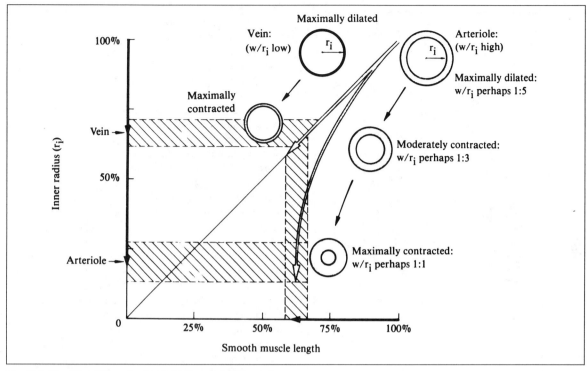

Fig. 17-8. The effect of the ratio of wall thickness (w) to the vessel lumen radius (r) on the change in vessel radius produced by smooth muscle contraction. (After B. Folkow and E. Neil. *Circulation*. London: Oxford University Press, 1971. P. 48.)

licular region. Impulses pass through the ventrolateral part of the medulla oblongata to the lateral spinal horns, with final distribution from the sympathetic ganglia to the blood vessels. These fibers also innervate the external genitalia. They have no direct effect on the control of blood pressure associated with pressoreceptor activity. The enhanced blood flow to skeletal muscle appears to bypass the capillary exchange network, as evidenced by a reduced oxygen consumption. The functional significance of this system is uncertain, although it could establish a blood flow reserve that would be immediately available on initiation of the metabolic demand of muscle contraction. Figure 17-10, curve C, shows that maximal sympathetic cholinergic vasodilation increases muscle blood flow up to five times the resting level.

PARASYMPATHETIC VASODILATOR ACTION. The erectile tissues of the external genitalia are innervated by cholinergic vasodilator fibers of both the sympathetic and parasympathetic divsions.

Activation of various exocrine glands (e.g., salivary, sweat) by parasympathetic fiber discharge leads to increased blood flow. However, this increase does not represent direct vascular influence. Instead, increased glandular activity triggers the release of a proteolytic enzyme that acts on a plasma globulin to form a potent vasodilator, *bradykinin*. This response, following parasympathetic stimulation, may therefore be referred to as secretomotor vasodilation.

No known parasympathetic vasoconstrictor fibers participate in the regulation of blood pressure. Although the vagal slowing of the heart is followed by a vasoconstriction of the coronary vesels, this process is probably an indirect consequence of the decreased metabolism.

DORSAL ROOT VASODILATION. In addition to the autonomic efferent fibers, stimulation of the skin (or the peripheral end of a sectioned dorsal root fiber) often produces a dilation of the adjacent superficial blood vessels. Probably, this dilation results from impulses arising from receptor sites (pain) and passing up through sensory afferent fibers until a branch in the neuron is reached. Impulses then go back over the branch to a vasomotor ending. For this reason they are called *antidromic vasodilator* impulses. The

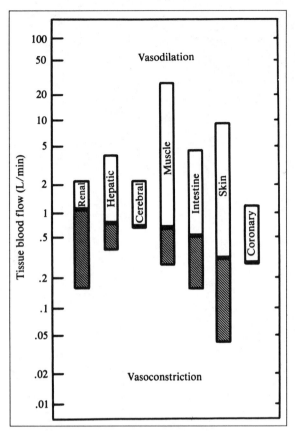

Fig. 17-9. The resting blood flows and the reactive ranges (maximum vasoconstriction to maximum vasodilation) of the major tissue circulations.

afferent and efferent parts of this *axon reflex* are formed by the branching of a single nerve fiber. Effective stimulation is probably a result of the release of histamine from tissue trauma. Only in areas rich in pain fibers, such as mucous membrane and skin, does vasodilation occur from this mechanism. The axon reflex may be of functional importance in the development of the *flare* following mechanical trauma to the skin and possibly in the development of inflammation about an infected area. The increase in blood flow would presumably aid the healing process.

LOCAL CONTROL OF TISSUE BLOOD FLOW

PASSIVE RESPONSE TO CHANGES IN TRANSMURAL PRESSURE. In order to appreciate the effectiveness of the local regulation of tissue circulation, one must understand the passive behavior of a vascular bed based on the distensible properties of the walls and the transmural pressure. One can gain some insight into the nature of a vascular bed by examining its *pressure-flow relationship*. For a rigid tube it is essentially linear. That is to say, each elevation in pressure will produce a proportionate elevation in flow. However, blood vessels are distensible tubes, and elevations of pressure in passive vascular beds produce greater-than-proportionate increases in flow (see Fig. 11-4), depending on the existing vascular tone (see Fig. 17-10). Ultimately, as the collagen fibers become tensed at high intramural pressures, the blood vessel will behave like a rigid tube.

In a tissue vascular system the nonlinear pressure-flow relationship (see Fig. 17-10) is also due to recruitment of additional vascular channels as perfusion pressure is elevated.

ACTIVE RESPONSE AND AUTOREGULATION OF BLOOD FLOW. In certain vascular beds, including the cerebral, coronary, muscle, intestinal, renal, and hepatic arterial circulations, tissue blood flow is maintained relatively constant over a range of perfusion pressures by means of active local adjustment of vascular tone. This phenomenon is called *autoregulation*, an intrinsic control mechanism that operates independent of neurogenic or systemic humoral factors, although it can be modified by them. It represents the local control of blood flow related to the functional needs of the tissue; in Figure 17-12, curve 1, it is seen as a deviation from the passive pressure-flow pattern.

Autoregulation is characterized by the phenomenon illustrated in Figure 17-12, curves 1 and D. Elevation of perfusion pressure from the control level results in a passive initial increase of blood flow, followed shortly by a reduction of flow to about the previous control level. Since flow is thus maintained relatively constant in the presence of an increased perfusion pressure, vascular resistance necessarily increases. Conversely, reduction of perfusion pressure from the control level produces an initial pressure fall in tissue blood flow that is followed within about 30 seconds by an upward adjustment of flow to the control level. In this case a maintained flow during reduced perfusion pressure reflects a reduction in vascular resistance. Therefore, autoregulation represents the adjustment of vascular resistance to keep blood flow constant over a range of perfusion pressures. The mechanism of autoregulation is complex, based on the response of the vascular smooth muscle to changes in transmural pressure, intrinsic metabolic factors, or other local determinants of vascular reactivity. A number of theories have been proposed to account for the phenomenon of autoregulation.

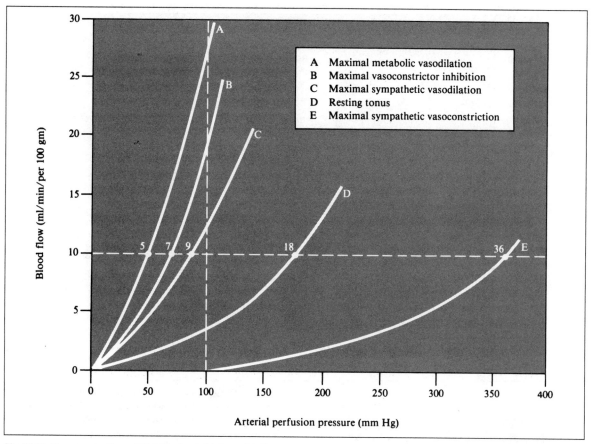

Fig. 17-10. The pressure-flow relationship for skeletal muscle. The numbers indicate the peripheral resistance at a flow of 10 ml/min per 100 gm tissue. (From E. M. Renkin and S. Rosell. Effects of different types of vasodilator mechanisms on vascular tonus and on transcapillary exchange of diffusible material in skeletal muscle. *Acta Physiol. Scand.* 54:244, 1962.)

Metabolic Theory. Numerous lines of evidence implicate metabolic factors in the local control of tissue blood flow. Metabolic activity increases in exercising muscle and contributes to the increase in muscle blood flow (exercise hyperemia). Similarly, restoring tissue blood flow after a brief stoppage results in a marked transient increase in blood flow (reactive hyperemia). The extent of hyperemia is directly related to the level of oxygen consumption and inversely related to the oxygen reserve of the particular tissue.

Numerous metabolic factors have been implicated in hyperemia, such as potassium ion, pH, plasma osmolality,

PO_2, PCO_2, and adenosine. However, no single factor has been able to reproduce the increases in blood flow seen in hyperemia. Combinations of factors, though, acting at different times and to different degrees, have been able to approximate the changes in blood flow. In exercise, for example, K^+ and pH initially increase; then, shortly thereafter, plasma osmolality increases. However, the effectiveness of $K+$ and hyperosmolality dissipate with time. Finally, a reduction in PO_2 leads to an increase in adenosine and a decrease in pH, which appear to maintain the hyperemia.

Myogenic Theory. Elevating perfusion pressure produces an increase in transmural pressure and thus in the wall tension of small arteries and arterioles. Tension on the vascular wall provides the stimulus for vascular smooth muscle contraction, which reduces the diameter, increases the resistance, and impedes the flow. The reverse effect occurs

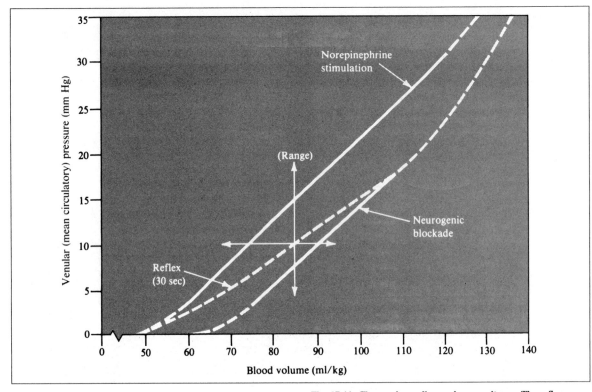

Fig. 17-11. Changes in cardiovascular capacitance. The reflex curve is the mean circulatory pressure found 30 seconds after a rapid hemorrhage or after transfusion to the blood volume indicated. Reflexes were blocked with hexamethonium. Infusion of norepinephrine at 1.5 μg/min-kg caused nearly maximal venoconstriction. (Modified from J. A. Drees and C. F. Rothe. Reflex venoconstriction and capacity vessel pressure-volume relationships in dogs. *Circ. Res.* 34:365, 1974.)

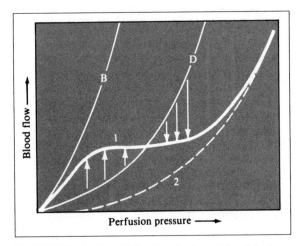

Fig. 17-12. Pressure-flow relationships of autoregulating skeletal muscle. B = passive, distensible vasculature at very low constant tone (compare with Fig. 17-10, curve B); D = passive, distensible vasculature at constant resting tone (compare with Fig. 17-10, curve D); 2 = passive, distensible vasculature with increased constant vasculature tone; 1 = autoregulating vascular bed.

when perfusion pressure is reduced. Thus, adjusting vascular resistance in the same direction as the change in perfusion pressure tends to maintain flow at a constant level. Organs in which a myogenic type of autoregulation occurs include the kidney, intestine, and the arterial supply of the liver.

Other Theories. The suggestion has been made that, in addition to increasing blood flow, elevating perfusion pressure increases capillary hydrostatic pressure and enhances filtration. The resulting increase in *tissue hydrostatic pressure*, by reducing the transmural pressure, should partially collapse the microcirculation to increase vascular resistance and thereby restore blood flow to control. That this sequence is seen in the encapsulated kidney only at excessive tissue pressures, and in the brain with its rigid cranium only when cerebrospinal pressure exceeds arterial pressure, indicates that it is not a physiological phenomenon.

It has been proposed that autoregulation could represent the participation of *neurogenic reflexes*. However, autoregulation has been demonstrated in isolated denervated preparations under the influence of local anesthesia. Consequently, this theory does not appear to be feasible.

Summary of Mechanisms of Autoregulation. Probably no single factor is responsible for the autoregulation observed in various tissues. It is more than likely that the interaction of multiple factors is responsible for autoregulation in any one kind of tissue, and each factor may have a greater effect in one tissue than in another.

INTERACTION BETWEEN CENTRAL AND LOCAL INFLUENCES. Vascular smooth muscle elements are continuously under central and local influences that may act synergistically or antagonistically to modify vascular tone. At the level of the arteriole the smooth muscle distribution favors central neurogenic activity, whereas more distally, at the level of the precapillary sphincters, local influences predominate. The smooth muscle of postcapillary resistance vessels seems to be under central neurogenic control.

Thus, a framework exists that can adjust peripheral vascular tone to provide for both systemic and local cardiovascular homeostasis. This feature of the control system may be reflected in the effect of sympathetic stimulation on the series resistance vessels in muscle. Sympathetic stimulation produces small-artery, arteriolar, and venous vasoconstriction to increase peripheral resistance and venous return, thereby contributing to blood pressure regulation. The reduced tissue blood flow resulting from the vasoconstriction leads to a reduced transmural pressure as well as an accumulation of metabolites at the precapillary sphincter, both of which produce precapillary sphincter dilation. This dilation provides a more uniform distribution of local blood flow and permits a more complete exchange between the vascular and extravascular compartments, despite the reduced total blood flow. Ultimately, the level of vascular tone and the sensitivity and reactivity of the vascular elements depend on the interaction of the numerous vasoactive influences (neurogenic, metabolic, and myogenic). At any moment the state of vascular tone represents the net effect of vasoconstrictor and vasodilator factors acting on the vascular smooth muscle. Thus, vasoconstriction can result from either increased vasoconstrictor influence or reduced vasodilator influence.

EXTRANEURAL CONTROL OF THE CARDIOVASCULAR SYSTEM

The physical characteristics of the vasculature and heart also act to control cardiac output. Right ventricular output—and so overall cardiac output—is highly dependent on filling and so on the central venous pressure (see Chap. 15). The vasculature is normally distended, and any situation that reduces the distending pressure leads to movement of blood from the periphery toward the heart. Thus, a transient decrease in cardiac output—and thus flow through the peripheral tissue—will lead to a smaller pressure drop across the venous resistances, less distending pressure, a passive recoil of the capacitance vessels, and therefore a redistribution of blood from the periphery to the thorax. This redistribution acts to increase central venous pressure and subsequently cardiac filling. As a consequence the cardiac output is restored.

An increase in blood volume tends to increase cardiac output, arterial blood pressure, and renal blood flow and thus excretion of water in the kidney. This water loss subsequently acts to reduce blood volume and so cardiac output. Thus renal function is a part of cardiovascular homeostasis.

A reduction in arterial blood pressure or constriction of the resistance vessels of a tissue bed results in a decrease in flow. Not only is the distending pressure of the capacitance vessels decreased, but the capillary pressure also decreases. This leads to fluid resorption from the interstitial space (see Chap. 12) and thereby provides a powerful compensatory mechanism acting to restore the cardiac output after hemorrhage or other forms of blood volume deficit.

Finally, a reduction in peripheral resistance, such as occurs during vigorous exercise, acts to increase venous return, increase central venous pressure and so cardiac output. As cardiac contractility increases, the central venous pressure tends to be reduced (see Chap. 15, Relationships between cardiac output and venous return). The increased flow tends to increase capacitance vessel pressure and cause some blood accumulation, but the volume change also reduces the resistance to venous return by way of the Poiseuille relationship (see Chap. 11). Because of the action of the muscle pump (see Chap. 16), the relatively low compliance of skeletal muscle veins and the reduced venous resistance, blood accumulation in active muscle during exercise is minimized.

CARDIOVASCULAR SYSTEM RESERVE

The cardiovascular reflexes provide the widely varying flow of blood needed to meet the metabolic demands of the various tissues of the body, but blood flow to the tissue is limited because of the limitations of the cardiovascular system. A person cannot run at maximum speed (high muscle metabolism) after a full meal (splanchnic bed dilation) on a hot day (skin dilation). Heart disease requires the utilization of cardiovascular reserves to provide even the basal blood flow (see Chap. 18).

Control of the cardiovascular system utilizes four primary cardiovascular reserves in the maintenance of circulatory homeostasis.

VENOUS OXYGEN RESERVE. During normal resting conditions, about 4 ml of oxygen is extracted from the arterial blood per 100 ml of blood flow through the tissue (Fig. 17-4). Thus, a reserve of about 16 volumes of oxygen per 100 volumes of blood (16 volumes percent) is in venous blood. During severe exertion, when the oxygen utilization may be over 10 times the basal level, the extraction of oxygen from the blood is increased to over 12 volumes percent.

A person with heart failure has a decreased cardiac output, but adequate amounts of oxygen can be transported to tissues because most cells can extract a larger-than-normal fraction of the oxygen from the blood by increasing the oxygen A-V difference. Unfortunately, following cardiac failure, the reduced flow to the peripheral tissue is not uniformly distributed. The blood flow through heart muscle and the brain tends to be maintained because there is little neurally induced vasoconstriction in response to the fall in blood pressure following the decrease in cardiac action. Peripheral vasoconstriction causes a disproportionate reduction flow of blood through the kidneys. As renal blood flow is reduced, the oxygen consumption is reduced and function impaired. Thus, although an increase in A-V oxygen difference provides a source of oxygen to tissues, its benefits are limited in the patient with cardiac failure because of the encroachment on renal function. In the normal person, however, this reserve provides for a twofold to threefold increase in oxygen utilization by active muscle (see Fig. 17-5).

HEART RATE RESERVE. During moderate exercise the heart rate is increased by the cardiovascular control system, and the stroke volume is held relatively constant. This response provides for an increase in cardiac output. However, the heart rate increase limits the filling time; hence, if the stroke volume is to be maintained, the venous filling pressure or the effective distensibility of the heart must be increased by homeostatic mechanisms, or the end-systolic volume must be decreased parallel to a decrease in end-diastolic volume.

An increase in heart rate does not provide a particularly useful reserve for the cardiac patient, because a rapidly beating heart is less efficient in the use of oxygen than a slowly beating one (see Chap. 15). If the heart failure is primarily a result of inadequate coronary blood flow in the first place, this loss of efficiency becomes quite significant. Furthermore, an increase in heart rate tends to limit the coronary blood flow (see Chap. 15). Thus, an increase in heart rate is not an efficient means of increasing cardiac output by the failing heart, although it is a highly important mechanism for the normal person, providing a twofold to threefold increase in cardiac output.

SYSTOLIC VOLUME RESERVE. In the normal person the stroke volume can be increased from about 70 ml to over 100 ml per stroke. Athletes with cardiac outputs of over 30 liters per minute and heart rates of about 200 beats per minute must have stroke volumes of about 150 ml. This increase in stroke volume during *maximal* exertion by well-trained people is provided by a decrease in end-systolic volume and an increase in end-diastolic volume (see Fig. 15-4). Normally, only about 60 percent of the end-diastolic volume is ejected (the ejection fraction). If the systemic arterial pressure does not decrease, the end-systolic volume reserves can be utilized only if the myocardial contractility increases so that the added volume can be ejected. The loss of myocardial contractility in cardiac failure is such that sympathetic activity from the control system is relatively ineffective in utilizing this reserve.

DIASTOLIC VOLUME RESERVE. As the central venous pressure increases in heart failure, the heart becomes distended. The increase in diastolic volume, produced by stretching the myocardial fibers, tends to cause the release of more energy in accordance with Starling's law of the heart, so that more blood tends to be pumped. As fluids accumulate, the failing heart becomes larger and more distended. Because even the normal resting, reclining person has a relatively large heart, and therefore a small diastolic volume reserve, there is little reserve for heart failure.

BIBLIOGRAPHY

Abboud, F. M., Heistad, D. D., Mark, A. L., and Schmid, P. G. Reflex control of the peripheral circulation. *Prog. Cardiovasc. Dis.* 18:371–403, 1976.

Berne, R. M., and Levy, M. N. *Cardiovascular Physiology* (5th ed.). St. Louis: C. V. Mosby, 1983.

Brown, A. M. Cardiac Reflexes. In R. M. Berne and N. Sperelakis (eds.), *Handbook of Physiology*. Bethesda, Md.: The American Physiological Society, 1979. Section 2: The Cardiovascular System. Vol. 1, *The Heart*, pp. 677–689.

Celander, O. The range of control exercised by the sympathico-adrenal system. *Acta Physiol. Scand.* 32 (Suppl. 116):1–132, 1954.

Cowley, A. W., Jr., Liard, J. F., and Guyton, A. C. Role of the baroreceptor reflex in daily control of arterial blood pressure and other variables in dogs. *Circ. Res.* 32:564–576, 1973.

Downing, S. E. Baroreceptor Regulation of the Heart. In R. M. Berne and N. Sperelakis (eds.), *Handbook of Physiology*. Bethesda, Md.: American Physiological Society, 1979. Section 2: The Cardiovascular System. Vol. 1, *The Heart,* pp. 621–652.

Drees, J. A., and Rothe, C. F. Reflex venoconstriction and capacity vessel pressure-volume relationships in dogs. *Circ. Res.* 34:360–373, 1974.

Fishman, A. P., and Richards, D. W. (eds.). *Circulation of the Blood: Men and Ideas*. New York: Oxford University Press, 1964. Chap. 7.

Folkow, B., and Neil, E. *Circulation*. London: Oxford University Press. 1971.

Guyton, A. C., Coleman, T. G., and Granger, H. J. Circulation: Overall regulation. *Annu. Rev. Physiol.* 34:13–46, 1972.

Johnson, P. C. Principles of Circulatory Control. In P. C. Johnson (ed.), *Peripheral Circulation*. New York: Wiley, 1978.

Korner, P. I. Control of Blood Flow to Special Vascular Areas: Brain, Kidney, Muscle, Skin, Liver and Intestine. In A. C. Guyton and C. E. Jones (eds.). *International Review of Physiology. Cardiovascular Physiology*. London: Butterworth, 1974. Vol. 1, Chap. 4, pp. 123–162.

Korner, P. I. Central Nervous Control of Autonomic Cardiovascular Function. In R. M. Berne and N. Sperelakis (eds.), *Handbook of Physiology*. Bethesda, Md.: American Physiological Society, 1979. Section 2: The Cardiovascular System. Vol. 1, *The Heart*, pp. 691–739.

Little, R. C. *Physiology of the Heart and Circulation* (2nd ed.). Chicago: Year Book, 1981.

Mellander, S., and Johanson, B. Control of resistance, exchange and capacitance functions in the peripheral circulation. *Pharmacol. Rev.* 20:117–196, 1968.

Paintal, A. S. Vagal sensory receptors and their reflex effects. *Physiol. Rev.* 53:159–227, 1973.

Rothe, C. F. Reflex control of the veins and the capacitance vessels. *Physiol. Rev.* 63:1281–1342, 1983.

Rushmer, R. F. *Cardiovascular Dynamics* (4th ed.). Philadelphia: Saunders, 1976.

Shepherd, J. T., and Vanhoutte, P. M. *The Human Cardiovascular System*. New York: Raven, 1979.

Smith, J. J., and Kampine, J. P. *Circulatory Physiology—the Essentials*. Baltimore: Williams & Wilkins, 1980.

18

HYPERTENSION

Julius J. Friedman

Arterial hypertension is a condition of sustained elevated systemic arterial blood pressure. No clear-cut line of demarcation exists between normal and hypertensive blood pressures, since blood pressure is influenced by many factors. The minimum level of systemic arterial pressure considered to be hypertensive has been arbitrarily set at 140/90 mm Hg. While hypertension usually involves elevations in mean and pulse pressures, the rise in diastolic pressure has been regarded as the critical criterion, suggesting that hypertension involves an increased peripheral resistance. In recent years, greater attention has been directed at the level of systolic pressure, because it is thought to be responsible for the development of such secondary cardiovascular diseases as arteriosclerosis, congestive heart disease, atherothrombic brain infarction, and nephrosclerosis.

Hypertension is a highly prevalent condition. In an extensive study conducted in Framingham, Massachusetts, over an 18-year period involving over 5200 men and women, black and white, between the ages of 30 and 62 years, 18 percent of the men and 16 percent of the women had blood pressures greater than 160/95, and 41 percent of the men and 48 percent of the women had pressures in excess of 140/90. The frequency of hypertension was greater in the older people who were studied.

Hypertension is a serious cardiovascular disease. It is responsible for approximately 10 to 15 percent of the deaths in people over 50 years of age. Associated with the elevated blood pressure is an increased morbidity as well as mortality. Insurance studies reveal that at age 45, a systolic pressure above 150 mm Hg will reduce the life expectancy of men by 11.5 years and of women by 8.5 years.

The incidence of hypertension is higher in blacks than in whites at every age, and blacks suffer a greater mortality from this disease. Whereas 16 white men in 100,000 die of hypertension annually, 66 black men in 100,000 are so affected. Thus, the mortality for blacks is four times higher than that for whites. The same ratio applies to black women. This higher prevalence of hypertension in blacks has been attributed to genetic factors; however, socioeconomic status must be taken into consideration as well.

The incidence of cardiovascular disease is greater in hypertensives than in normotensives. Hypertensives suffer from seven times more stroke, four times more congestive heart failure, three times more coronary heart disease, and twice as much occlusive peripherovascular disease.

In approximately 20 percent of the cases of hypertension, the elevated blood pressure is a clinical sign of a specific disease (e.g., renal disease) and not a disease entity in itself. Such conditions of elevated blood pressures are referred to as *secondary hypertension*. In the remaining 80 percent of the cases, the development of hypertension cannot be attributed to any known origin and probably represents a specific disease state. This form of hypertension is called *primary* or *essential hypertension*.

CLASSIFICATION OF HYPERTENSIVE VASCULAR DISEASE

To gain insight into the nature of essential hypertension, the various diseases that cause an elevation in blood pressure have been studied both clinically and experimentally. Those studied include cardiovascular, neurogenic, endocrine, and renal disorders.

CARDIOVASCULAR HYPERTENSION. Systemic arterial blood pressure is determined by cardiac output, total peripheral resistance, and arterial compliance. Consequently, a number of changes in the cardiovascular system can take place as a result of elevated arterial blood pressure. However, elevations of arterial pressure that result from increases in cardiac output alone, or from increases in peripheral resistance not including the renal vasculature, are found to be relatively short term. The elevated pressure increases the urine volume load and produces natriuresis, which in time reduces the extracellular fluid volume and thus blood pressure.

Increases in cardiac output can produce sustained elevated arterial pressure by means of autoregulatory increases in peripheral resistance, including the afferent arterioles of the kidney. Guyton (1974) found that total peripheral resistance was elevated within 10 minutes of an increase in cardiac output. In 30 to 60 minutes, resistance had increased to three times the increase in cardiac output. Over a period of days and weeks a 1 to 5 percent increase in cardiac output

resulted in as much as a 50 percent increase in peripheral resistance.

NEUROGENIC HYPERTENSION. The centers that regulate and integrate cardiac and vasomotor activity are situated in the brain. Consequently, impairment of cerebral function in such a manner that cardiac and vasomotor activity is enhanced produces an elevation of blood pressure. Cerebral function may be impaired by cerebral ischemia due to either cerebral arterial occlusion or severely increased intracranial pressure, or by the development of a tumor in various regions of the brain. In each case the resulting hypertension may be relieved by appropriate treatment of the organic disturbance.

Another form of neurogenic hypertension may be initiated by carotid sinus baroreceptor denervation. With such denervation the gain of the baroreceptor regulatory loop is reduced, leading to increases in cardiac activity and vasomotor tone. A form of carotid sinus denervation can occur by means of baroreceptor adaptation to a maintained elevation of arterial pressure. In a matter of days following arterial pressure elevation, the baroreceptor discharge frequency approaches the control rate despite the maintained elevation of arterial pressure, and the pressure then becomes regulated at this new level. Restoration of the arterial pressure to the original level will result in a readaptation. The adaptation differs from the "resetting" of the baroreceptors that apparently accompanies the change in physical properties of the baroreceptor vessels following sustained hypertension.

The sympathetic nervous system has been implicated in the development of hypertension. Extensive sympathectomy or ganglionic-blocking agents produce a reduction in blood pressure. Whether this sympathetic contribution is a primary one, reflecting altered function, or a secondary one, representing some regulatory activity, is not clear.

ENDOCRINE HYPERTENSION. The adrenal medulla contains chromaffin cells that secrete the catecholamines epinephrine and norepinephrine. These cells may develop tumors (pheochromocytomas) that secrete excessive quantities of catecholamines into the circulation, either continuously or periodically; the blood pressure then is elevated by increased cardiac output and peripheral resistance. Surgical removal of the tumorous tissues relieves the condition and usually restores blood pressure to normal levels.

The adrenal cortex has also been implicated; however, the specific role of this tissue and its secretions in the pathogenesis of hypertension remains uncertain. Many cases of hypertension can be relieved by restriction of the sodium content of the diet; a high intake of sodium tends to aggravate the condition. Aldosterone, a substance secreted by the adrenal cortex, stimulates the kidney to retain sodium ion, causing water retention. In conditions of adrenal hyperplasia with primary aldosteronism, hypernatremia with volume expansion develops, and the blood pressure becomes elevated. This condition is reversible, and removal of the adrenal gland or administration of spironolactone, an aldosterone antagonist, usually restores the blood pressure and electrolyte pattern to normal levels. This form of hypertension is described as volume-dependent, or low-renin, hypertension and is characterized by hypernatremia and hypokalemia associated with volume expansion, and by minimal sympathetic activity due to reflex inactivation. Volume-dependent hypertension usually responds well to diuretic therapy.

In almost all forms of chronic experimental hypertension, aortic smooth muscle contains an elevated concentration of sodium and potassium ions and an increased volume of water, which may contribute to increased blood pressure in a number of ways. Ion and water accumulation in the walls of resistance vessels could cause swelling of the walls and encroachment on the lumen, thereby increasing peripheral resistance and blood pressure. The increase in the ratio of vessel wall thickness to lumen diameter results in greater-than-normal changes in vascular resistance in response to normal degrees of smooth muscle contraction. In addition, alterations of the ionic composition of vascular smooth muscle could, by disturbing the membrane potential, affect the sensitivity of resistance vessels to circulating vasoconstrictor substances and autonomic nervous influences. Also, ionic changes in the muscular elements of the vascular wall may alter the state of actomyosin to produce an increased contractility. Thus, by producing changes in the physical state of resistance vessels and by increasing excitability and contractility of vascular smooth muscle, alterations in electrolyte balance could contribute significantly to the development of hypertension.

RENAL HYPERTENSION. Permanent hypertension may be induced by a variety of manipulations that effectively reduce either renal blood flow, functional renal tissue, or tubular sodium delivery. Reduction of renal blood flow by arterial occlusion or by renal compression consistently produces hypertension. The relationship of the kidney to hypertension is independent of nervous activity because complete renal denervation does not prevent the development of hypertension. Therefore, renal hypertension is attributed to a humoral mechanism. As early as 1898 it was recognized that an extract of kidney cortex was capable of producing an

increase in blood pressure when injected into a normal subject. The agent responsible for this effect is *renin*.

When renal ischemia exists, the juxtamedullary cells adjacent to the afferent arterioles secrete increased amounts of renin. However, renin itself has no vasoactive properties. It is a proteolytic enzyme that acts on a plasma protein, alpha-2 globulin, to produce a polypeptide, angiotensin I, which is also vasoinactive. A converting enzyme, present in plasma, then converts angiotensin I to the active form, angiotensin II, (formerly known as *angiotonin* or *hypertensin*). Angiotensin II increases total peripheral resistance through widespread arteriolar vasoconstriction. This agent also produces venoconstriction and reduced vascular capacity, thereby increasing venous return, cardiac output, and blood pressure.

However, in most forms of hypertension, the angiotensin II concentration in plasma is not very great. Unless angiotensin II acts in some indirect manner to increase either the sensitivity or the reactivity of the vascular smooth muscle, or both, its level in the plasma is insufficient to maintain an elevated pressure. Angiotensin II can indirectly influence vascular reactivity as well as the volume of extracellular fluid through its influence on the zona glomerulosa of the adrenal cortex (by stimulating the zona glomerulosa to secrete aldosterone). Aldosterone acting on the kidney causes sodium retention, which increases extracellular fluid volume. This expanded volume, together with the lessened vascular capacity produced by angiotensin II venoconstriction, results in blood pressure augmentation. In view of the multiple influence of angiotensin II, it is not surprising to find that peptide-blocking agents that prevent the conversion of angiotensin I to angiotensin II lower blood pressure in renal hypertension. This form of hypertension is referred to as "high-renin" hypertension and is characterized by reduced extracellular volume due to pressure natriuresis. The condition is characterized by malignant hypertension in which the elevated angiotensin II levels are accompanied by secondary aldosteronism. In high-renin hypertension, adrenalectomy, aldosterone antagonists, or diuretics are of questionable value.

SELF-PERPETUATING NATURE OF HYPERTENSION. Hypertension is characterized by an increase in pulse pressure as well as mean blood pressure. Normally, one would expect an increased peripheral resistance to cause pulse pressure to decline (see Chap. 16). The elevated pulse pressure in hypertension is attributed to a reduction of large-artery distensibility, which reflects the progressive arteriosclerotic process in hypertension, as well as the nonlinear pressure-volume relationship of the arterial system. The sclerotic process is not restricted to the large arteries. When it involves the small arteries and arterioles, these vessels undergo medial hypertrophy that increases the mass of the vascular smooth muscle, causing encroachment on the lumen. Thus, their lumina are permanently reduced and peripheral resistance remains elevated even after extensive treatment, including radical sympathectomy. A similar process is proposed as the basis for the "resetting" of the baroreceptors. In this manner the condition of hypertension enhances and perpetuates its own development—an example of positive feedback.

ESSENTIAL HYPERTENSION. *Essential hypertension* refers to the sustained elevation in systemic arterial blood pressure for which there is no discernible origin. The hypertension may be "benign," in that it develops slowly and progressively over many years, or "malignant," in which case it develops rapidly in a brief period of time. Attempts to identify the cause of essential hypertension have been fruitless. After exclusion of hypertension due to renal disease, adrenal dysfunction, or cardiovascular and neurogenic alterations, no consistent explanation is obvious at this time.

Subjects with essential hypertension exhibit an elevated peripheral resistance that is rather uniformly distributed throughout the body and is not a result of changes in central vasomotor activity. Investigators have considered the participation of altered pressure regulation effected by a "resetting" of the baroreceptor mechanism. While the regulatory mechanism of the carotid sinus is active in both normotensive and hypertensive patients, the pressoreceptors of hypertensives fire intermittently, and at lower frequencies, at pressures that elicit continuous, higher-frequency firing in normotensives. Thus, the carotid sinus pressor mechanism may, by becoming adapted (set) to a new, elevated pressure level, perpetuate a higher pressure.

An alternative interpretation of altered pressoreceptor activity concerns the organic alteration of the carotid sinus wall. Sclerosis of the carotid sinus wall reduces its distensibility and limits the deformation of the receptors within the sinus wall. Consequently, larger pressure changes are required to activate the pressoreceptor mechanism to the same extent, and over a wide range the pressure would vary as though the sinus were denervated. Whether this alteration precedes essential hypertension or is caused by hypertension is not clear.

There is evidence from studies of sympathetic blockade that the sympathetic nervous system contributes to the hypertensive state through an increased reactivity of resistance vessels. Patients with essential hypertension exhibit more marked pressor responses to sympathetic influences,

both neurogenic and humoral, than do normal subjects. Whether these reactive characteristics are due to altered neurotransmitter function, modification of cyclic nucleotide metabolism, or structural alterations is uncertain. However, the tendency toward hyperactivity is apparently genetically determined, because hypertensive patients almost always have a family history of the disease.

An increasing body of evidence suggests that the interrelationship between the kidney and the adrenal cortex is significant in severe hypertension. Renal damage releases renin, which results in the formation of angiotensin, and this agent stimulates the secretion of aldosterone by the adrenal zona glomerulosa (see Chap. 34). It may be that angiotensin, acting on vessels made hyperreactive by electrolyte alterations, is a significant feature of essential hypertension.

Guyton (1974) proposed the hypothesis that long-term regulation of arterial pressure is dependent on kidney function, which regulates the extracellular fluid volume and subsequent pressure by means of the relationship between blood volume and vascular capacity. Guyton suggested that the baroreceptor system is a rapidly acting short-term regulator of pressure through adjustments of cardiac output and total peripheral resistance.

Hypertension is caused by increased afferent arteriolar resistance that reduces renal blood flow and urine output. The resulting increase in extracellular fluid volume increases blood volume, venous return, cardiac output, and arterial blood pressure. It is stipulated that this feedback loop is designed to restore the urine volume load by increasing blood pressure and renal blood flow; however, when the autoregulation involves the afferent arteriole, a vicious positive feedback cycle is generated and sustained hypertension ensues.

In summary, it is unlikely that any one of the factors that have been cited is the sole causative agent. In all probability, essential hypertension involves the participation of a number of factors to varying degrees in different cases. This multiple-factor concept is particularly attractive in view of the conflicting experimental evidence and unpredictable pharmacological responses of the disease. Furthermore, there are probably many cases of hypertension classified as "essential" that may be due to other causes. They may represent diseases not yet identified, an imbalance in the relationship of endogenous vasoactive substances, a modification of the participation of regulatory factors in the maintenance of normal blood pressure, or a vasoactive substance that has not yet been characterized. Figure 18-1 relates the various participating systems to one another.

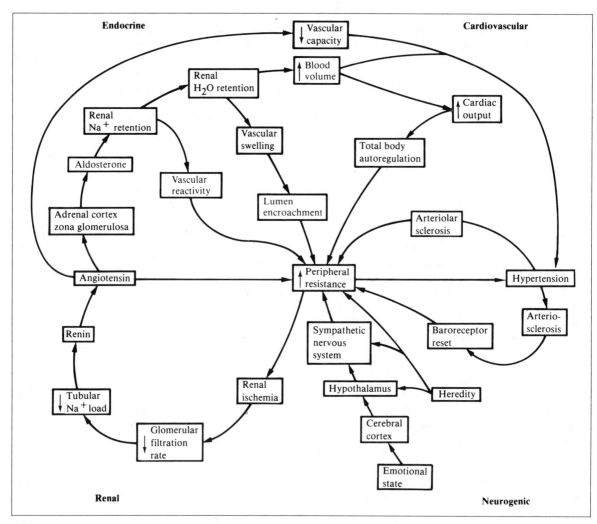

Fig. 18-1. Possible participation of various factors in the development of hypertension.

CONGESTIVE HEART FAILURE

Ewald E. Selkurt

Heart failure may be simply defined as a state in which the heart fails to maintain an adequate circulation for the needs and demands of the body despite what appears to be satisfactory filling pressure. When this failure is accompanied by an abnormal increase in blood volume and interstitial fluid, the condition is known as *congestive* heart failure (CHF). In Chapter 17, four types of cardiovascular reserves were discussed, which should be reviewed in terms of their role in the failing heart. These were (1) the venous blood oxygen reserve, (2) the heart rate reserve, (3) the systolic volume reserve, and (4) the diastolic volume reserve. As was explained, each has limitations as a reserve for the heart in failure. Perhaps the most valuable is the diastolic volume reserve, which acts by way of increased filling pressure. In the early stages of heart failure these reserves may be adequate; accordingly, persons at rest or engaged in activity requiring minimal exertion manifest minimal symptomatology. However, increasing venous pressure distends the ventricle beyond the point of critical diastolic stretch, so that ultimately the contractile force declines and the output falls. The diminished cardiac reserve shows up during increased exertion and is evidenced by breathlessness, palpitation (perceptible heartbeat), and fatigue with degrees of exercise that were formerly tolerated quite easily. As the condition progresses, these symptoms appear even at rest and are involved in several of the sequelae that result from heart failure and become a part of the overall clinical syndrome.

CAUSES OF HEART FAILURE AND ITS PHYSIOLOGICAL CONSEQUENCES

Circulatory failure occurs for a number of reasons that can be only briefly outlined here. They may be grouped into several categories: First, failure may occur because of *interference with systemic venous return* due to (1) factors usually remote from the heart, as in hemorrhage and shock or the type of peripheral vascular collapse noted in syncope or (2) factors in the heart region which impair venous return, such as pericardial tamponade, constrictive pericarditis, and tricuspid stenosis. Venous congestion is usually a feature of the latter group. Second, failure may result from *interference with the pumping or filling of the ventricle* due to anatomical abnormalities, either inherited or acquired through disease such as rheumatic fever. This type of failure is seen in semilunar valvular stenosis or insufficiency, mitral stenosis or insufficiency, and tricuspid insufficiency, and is also associated with systemic arterial hypertension and pulmonary arterial hypertension. Third, *primary diseases of the myocardium* lead to failure. This happens with myocardial infarcts created by coronary occlusion. Last, *chronic stimulus to high output* of one or both ventricles may ultimately cause failure. This happens with drastic reduction in peripheral vascular resistance, as in beriberi, systemic arteriovenous (A-V) fistula, Paget's disease (in which increased blood flow through the affected bone acts as an A-V fistula), anemia, and some types of congenital cardiovascular malformations—for example, those that have large left-to-right shunts. Despite the high output in the last group of examples, the blood supply to the various tissues and organs is inadequate, particularly in terms of the metabolic needs. On this basis, hyperthyroidism may lead to high-output failure. In all categories, circumstances occur that lead ultimately to failure of cardiac reserves.

PATHOLOGICAL CHANGES. Sudden occlusion of a branch of coronary artery (done experimentally in dogs) reduces arterial inflow to that region of the heart; after 10 to 15 seconds, the cells become cyanotic, abnormal electrocardiographic changes appear, and the cells cease contracting. Metabolism shifts from the aerobic to the anaerobic form.

Characteristic structural changes are seen in severely ischemic myocardial cells after 30 to 40 minutes of coronary occlusion. These changes include virtual depletion of glycogen stores, relaxation of myofibrils, and nuclear and mitochondrial changes (swelling). On reperfusion by release of the occlusion, damaged cells swell enormously as the first phase of development of the irreversible phase of ischemic injury. Thus, the beginning of cell destruction is heralded by the inability of the cell to maintain its normal volume. Impairment of the pump(s) (e.g., sodium) which maintain normal electrolyte balance of the cells is implied.

Ischemic Injury. The sequence of events that occurs with coronary artery occlusion is outlined below. The extent of the reduction in flow resulting from occlusion would obviously determine whether an immediate fatal outcome would result or recovery were possible.

1. Decrease in arterial flow
2. Hypoxia → anaerobic metabolism
3. Decrease in cellular energy level

cessation of
specialized functions

IRREVERSIBILITY

Cell death

The pathological changes characteristic of chronic heart failure include hydropic degeneration of the myocardial fibrils followed by necrosis and fibrosis, particularly when associated with degenerative infections and metabolic heart disease. Ultimately, however, the defect must be sought in the ability of the myocardium to utilize energy and to perform work.

METABOLIC CHANGES. From a biochemical point of view, heart muscle cell failure could result from abnormalities in *energy production,* in *conservation,* or in *utilization* of the energy. Figure 18-2 summarizes the steps involved in the transfer of foodstuffs to cardiac work.

Meerson (1962, 1969), studying the responses of experimental heart failure by production of aortic stenosis, divided the course of failure into three convenient stages (Table 18-1): First, a transient breakdown stage; a second long-term stage of relatively constant hyperfunction; and a third long-term stage of exhaustion and fibrosis. The first

stage is one in which the myocardium compensates for the increased load; the second is one in which increased myocardial function allows a return toward normal circulation; and the third stage reflects irreversible damage, so that function declines again and clinical symptoms return, leading to death.

The primary change in abnormalities of the myocardium is an abnormality of the contractile proteins of the failing myocardium. This involves a fall in myosin adenosine triphosphatase (ATPase) activity. This in turn reduces myocardial contractility by lessening shortening velocity, especially at light loads (thus reducing the heart's rate of ejection (Katz, 1977).

Hypertrophy of the pressure-overloaded heart (stage 2) is associated with an increased myofibrillar mass. This adds another factor to the precarious balance between energy

Fig. 18-2. *Production.* Metabolism of precursors to acetyl coenzyme A (Ac-CoA), which condenses with oxalacetate (OAA) to form citrate. This goes through the steps of the Krebs cycle to yield eight hydrogen atoms or electrons ($H = H^+ + e$). These represent the energy content of the acetyl, which contributes four hydrogen atoms plus four from the water added to the cycle intermediates in effecting the oxidation of the fragment. This step yields "energy-rich" electrons for the transport of oxygen. *Conservation.* This is associated with the electron flow along the hydrogen transport chain, which is depicted as coupling the phosphorylation of adenosine diphosphate (ADP) to the oxidation of the cytochrome enzymes. *Utilization.* The high-energy phosphate bond of adenosine triphosphate (ATP) is channeled into the contractile mechanism and results in the performance of mechanical work. (From R. E. Olson and D. A. Piatnek. Conservation of energy in cardiac muscle. *Ann. N.Y. Acad. Sci.* **72:**467, 1959.)

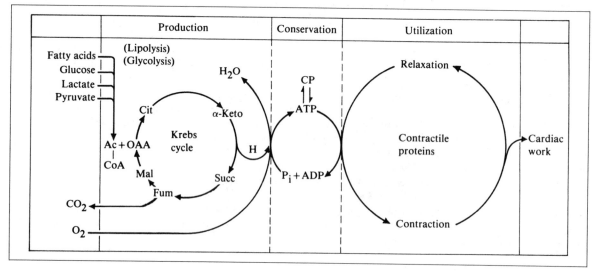

Table 18-1. Meerson's three stages of cardiac response to hemodynamic overloading of the left ventricle

First short-term stages of damage
 Clinical: Left heart failure, pulmonary congestion.
 Pathological: Dilatation of the left ventricle.
 Histological: Swelling and separation of the myofibrils.
 Biochemical: Glycogen and ATP levels decreased, phosphocreatine level markedly decreased, lactate production slightly increased. Protein synthesis, RNA, and mitochondrial mass (especially the inner membranes) increased.

Second long-term stage of relatively constant hyperfunction
 Clinical: Relief of symptoms.
 Pathological: Hypertrophy.
 Histological: Increased size of cardiac fibers, minimal fibrosis.
 Biochemical: Glycogen, ATP, phosphocreatine levels normal. Lactate production increased. Protein synthesis, RNA level normal, DNA level decreased. Myofibrillar mass increased relative to that of the mitochondrial mass.

Third long-term stage of exhaustion and fibrosis
 Clinical: Reappearance of heart failure.
 Pathological: Fibrous replacement of muscular tissue.
 Histological: Disproportionate appearance of connective tissue, fatty dystrophy; muscle cell nuclei become pyknotic.
 Biochemical: As in second stage, except decline in protein synthesis and marked decline in DNA levels.

Source: A. M. Katz. *Physiology of the Heart.* Copyright © 1977 by Raven Press, Publishers/New York. P. 412. (See Meerson, 1962, 1969.)

production and energy utilization. The enlarged myocardial mass tends to deplete further energy reserves. However, the synthesis of myosin molecules with low intrinsic ATPase activity has a certain value in that it increases the amount of mechanical energy derived from each mole of ATP utilized—at the expense, however, of a decrease in the overall rate of ejection. Although activity is slowed, the energy demands for contraction are more easily met.

It is well known that a high K^+ concentration depresses the activity of cardiac muscle in vitro. Increased ionic strength reduces the interaction between actin and myosin when cardiac muscle is subjected to various inotropic interventions. On the other hand, although there is a positive correlation between K^+ loss and enhanced myocardial contractility, the interrelationship is quite indirect. Normally, the balance between extracellular sodium and intracellular potassium shows little, if any, change. It seems acceptable, in conclusion, that there is little direct effect of potassium in the intact heart, and it probably does not contribute to a fall in myocardial contractility in the chronically overloaded heart.

Calcium is important in the contraction and relaxation phases of cardiac action (see Chap. 14). Its indispensable role in maintaining the force of cardiac contraction (actomyosin system) is well known. A possible role of calcium in heart failure has been explored. Heart failure can be experimentally produced in laboratory animals by perfusion of the heart with cobalt or nickel ions, which mimics the metabolic and mechanical effects of calcium deficiency, presumably by interfering with the role of Ca^{2+} in contraction. Such a failure state can be corrected by addition of calcium to the cardiac perfusion. The rate of uptake of calcium by the sarcoplasmic vesicles of humans with CHF remains to be determined, however. Studies in this area may become available as heart transplants become more common.

The catecholamine content of the heart is depleted in both experimental heart failure and CHF in humans. Norepinephrine is synthesized by tyrosine hydroxylase and other enzymes in the heart from tyrosine via dopa and dopamine. Epinephrine cannot be synthesized, however, and that found in the heart is taken up from the circulation. Free catecholamines, derived either from a rapidly turning-over pool of storage particles (by nerve stimulation or anoxia) or from the circulation, become active in the contraction process at receptor sites on the heart muscle, to play an important role in oxidative metabolism, glycolysis, and contractility of the heart and to produce the inotropic effect. Nevertheless, the absence of a simple relationship between catecholamine depletion and impaired mechanical function of the heart suggests that depletion is not a basic and primary event in heart failure. This does argue strongly, however, against the use of antiadrenergic drugs in patients with established heart failure.

In certain other pathological conditions the possibility remains that the defect resides in energy production or conservation. For example, in hemorrhagic shock and myocardial infarction, general and local anoxia leads to defective energy production. In anemia, myocardial failure of the high-output type could result because insufficient oxygen is drawn in for substrate metabolism. Likewise, in beriberi heart disease, also associated with high-output failure, a deficiency in thiamine pyrophosphate (cocarboxylase) produces a breakdown of certain decarboxylation reactions that results in interference with normal myocardial energy liberation and conservation.

HEMODYNAMIC ALTERATIONS. It has been customary in the past to consider the hemodynamic alterations in terms of two general aspects of heart failure classified as *forward failure* and *backward failure.* Forward failure implies alterations and sequelae that result from reduction in cardiac output and peripheral blood flow. The concept of backward failure is concerned with loss of contractile power and distension of the ventricle, with ultimate inability to maintain cardiac output. Venous pressure rises, and with it come serious developments in terms of capillary fluid filtration and the formation of edema fluid (see Chap. 12). On a functional basis it is unrealistic to categorize the consequences of heart failure into these two classifications, and the designations should be dropped.

Cardiac failure can occur on both sides of the heart, but it usually happens on the left. If failure occurs on one side initially, it may in time affect the other side. For example, with left-sided failure, left atrial pressure will increase, with back pressure in the pulmonary system, causing the pulmonary artery pressure to increase two or more times. This loads the right ventricle, eventually leading to failure of the right side of the heart. Primary damage on the right side reduces the overall cardiac output, resulting in diminished coronary blood flow, to the ultimate disadvantage of the myocardium.

In chronic left-sided failure without concomitant failure of the right side, blood continues to be pumped into the pulmonary artery, and because it is not adequately pumped out by the left ventricle, pulmonary filling pressure rises and systemic filling pressure falls. As the blood volume in the lung increases, pulmonary artery pressure rises, and the pulmonary capillary pressure exceeds the plasma colloid osmotic pressure, which is 25 to 30 mm Hg, so that fluid filters out of the capillaries into the interstitial spaces and alveoli. Thus, pulmonary vascular congestion and edema are the consequences. However, if the pulmonary capillary pressure stays below 25 mm Hg, the lungs will remain "dry" (Fig. 18-3A).

An additional facet of left heart failure concerns the reduction of renal function (e.g., decreased glomerular filtration rate), with salt and water retention. In the chronic phase, this adds to edema fluid accumulation in the systemic circulation.

In unilateral right heart failure, blood is pumped normally by the left ventricle, but is not pumped adequately from the systemic circulation into the lungs. As a consequence, blood accumulates in the right side, causing systemic congestion and edema formation (Fig. 18-3B).

PULMONARY CONGESTION. If pulmonary venous pressure rises beyond about 30 mm Hg, edema fluid begins to form and accumulate in the lung tissue and in the alveoli, impairing oxygen exchange. Engorgement of the lungs with blood and the accumulation of edema fluid seriously damage the mechanism of ventilation and increase the work of breath-

ing, leading to undue breathlessness. Fluid accumulation leads to sounds heard by auscultation that are called *rales*. The lung tissue may become tough and lose its resilience because of connective tissue proliferation in the parenchyma. Alveolar membranes become thickened and edematous, further increasing the distance between the alveolar air and blood. *Vital capacity* (see Chap. 19) is reduced not only by the extra blood and interstitial fluid in the lungs but also by the usual cardiac hypertrophy that occurs. Finally, there may be accumulation of excessive amounts of mucus and secretions in the small bronchioles and airways, further impeding air exchange. Thus, *dyspnea* is commonly seen. This condition is defined as rapid, shallow breathing of which the patient is conscious, as distinguished from *tachypnea* (rapid breathing) or *hyperpnea* (deep breathing). Dyspnea is not due solely to decreased oxygen and increased carbon dioxide in the blood; it is caused in large part by stimulation of vagus nerve stretch receptors in the lung.

Further manifestations of abnormal respiration are noted, perhaps of more serious consequence. These include *orthopnea, paroxysmal nocturnal dyspnea,* and *Cheyne-Stokes respiration.* Orthopnea is difficulty in breathing in the recumbent position as contrasted with the erect. It probably results from the shifting of blood from the dependent parts of the body to the vessels of the lung, exacerbating the subjective sensations and increasing the respiratory effort, in all likelihood because of enhanced reflex activity from the lung receptors. Paroxysmal nocturnal dyspnea is similar but occurs at night. It is also associated with the redistribution of blood that occurs during recumbency, plus depression of the central nervous system centers during sleep. The subject awakens with a feeling of suffocation, sits up gasping for air, and exhibits excessive pallor of the skin and profuse sweating. The skin may appear cyanotic because of the reduced hemoglobin of the circulation. Coughing and wheezing accompany the attack. The entire episode may last 10 to 20 minutes.

Basically, Cheyne-Stokes breathing presumably results from the depression of the respiratory center that is a consequence of the impaired blood flow following decreased cardiac output. Local vascular disease may contribute to this condition in the elderly. The depressed respiratory center is presumed to be insensitive to normal carbon dioxide tension but may still respond to raised carbon dioxide tension of the blood, and reflexly to anoxia (see Chap. 21). With normal PCO_2 and PO_2, breathing stops. During apnea the arterial PO_2 falls and the PCO_2 rises, directly and indirectly stimulating the respiratory center to renew the breathing. However, the hyperpnea leads to a blowoff of carbon diox-

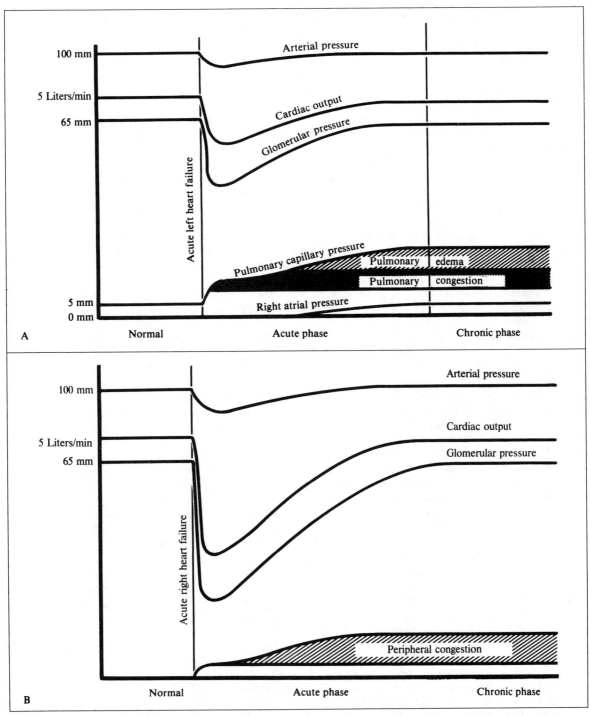

Fig. 18-3. A. The overall effects, acute and chronic, of left-sided heart failure. B. The overall effects, acute and chronic, of right-sided heart failure. (A and B from A. C. Guyton. *Textbook of Medical Physiology* [6th ed.]. Philadelphia: Saunders, 1981.)

ide and improvement of the oxygen tension, with the result that apnea occurs again. This sequence is repeated cyclically. A more detailed discussion of abnormal breathing will be found in Chapter 21.

SYSTEMIC CONGESTION. When failure extends to the right side of the heart, other sequelae become manifest. These may be the indirect consequence of left heart failure and the resulting pulmonary congestion and pulmonary artery hypertension, which lead to ultimate impairment of right ventricular function. They may be more directly the consequence of primary lung disease with pulmonary hypertension or pulmonic valvular stenosis. Pulmonary embolism may lead to the symptomatology, as may certain types of atrial septal defects.

The immediate clinical manifestation of this change is elevation of systemic venous pressure (values of 25–40 cm H_2O). Venous engorgement is evidenced by the concomitant distension of the jugular veins. These are distended even in the erect position, which is not the case in the normal person (see Chap. 16). The elevation of venous pressure results not only from the inability of the right heart to pump adequately but also, in chronic heart failure, from the increase in blood volume—an increase both of the erythrocytes (hypoxemic stimulation of bone marrow) and of plasma volume (fluid retention). Unquestionable evidence has been obtained that this is the case by using tagged albumin (^{131}I) and tagged erythrocytes (^{57}Fe and ^{59}Fe) simultaneously. In addition, several investigators believe that there is enhanced venoconstriction, a reflex response to reduced cardiac output. Ganglionic-blocking agents (Dibenamine, Arfonad) that decrease sympathetic nerve activity cause significant reductions in venous pressure in the CHF patient, while only negligible changes in venous pressure occur in the normal subject given the drugs. This finding has been taken as presumptive evidence of increased venomotor activity in the heart patient.

A consequence of the elevated venous pressure is that the liver is engorged and becomes tender and painful in response to pressure. Ultimately, as a combined effect of venous congestion and reduced inflow, this leads to *cirrhosis* (fibrotic degeneration) of the liver, evidenced by increased bilirubin in the blood and jaundice. More specifically, decreased indocyanine green and Bromsulphalein (BSP) clearance result, indicating reduced hepatic blood flow. Splenomegaly is also seen. Peripheral venous congestion (i.e., in the skin) leads to the appearance of cyanosis; this occurs partly as a result of dilation of venules but also because of the probably impaired oxygenation of the arterial blood and greater oxygen extraction by the tissues. Thus, reduced he-

moglobin in cutaneous vessels exceeds the 5 gm per 100 ml of blood necessary for the appearance of visible cyanosis.

MECHANISM OF PERIPHERAL EDEMA. A number of factors are set into play in CHF that favor the accumulation of fluid in the interstitial spaces of various parts of the body, leading to edema. This is first manifested by an increase in the subject's weight—a gain of 10 to 20 pounds—before any objective signs are noted. Then, the ankles and dependent parts begin to swell, and pitting edema is observed. Pitting edema is detected by a firm finger pressure applied to the edematous region. Fluid is expressed, leaving a depression that slowly refills. In the bedridden patient, sacral edema is prominent. Accumulation of the edema fluid in the lung bed (*hydrothorax*) has already been discussed. Fluid accumulation also occurs in the abdomen. This is called *ascites*. Peripheral edema fluid is low in protein (about 0.5 gm/100 ml), but ascitic fluid may approach the protein content of plasma (5–6 gm/100 ml).

The production and accumulation of edema fluid in CHF constitute a complicated process and are not simply the result of elevation in venous pressure leading to increased capillary filtration pressure. Since intake of fluid is apparently normal, accumulation of fluid must be caused by abnormal retention; hence, the kidney plays a paramount role. An accumulation not only of water but also of salt, predominantly sodium chloride, to maintain the isotonicity of the fluid, is involved in this mechanism. Thus, although the first step in the formation of edema fluid is increased capillary filtration, a mechanism for overall accumulation of water and contained salts must be provided.

The mechanism for renal retention of salt and water has not been completely settled. Until the student becomes more familiar with renal physiology (Chap. 22), only brief details can be given. It has been noted that glomerular filtration is usually reduced in CHF, and some authorities feel that this action may be important for the retention of salt and water. Simply stated, when the load of sodium salts ($NaCl$ and $NaHCO_3$) and water offered to the renal tubular cells is reduced, more complete resorption of the smaller load occurs. Although such a mechanism can be experimentally demonstrated, it perhaps oversimplifies the situation in CHF.

It has been demonstrated that, underlying the overall reduction in renal blood flow in heart failure, redistribution of regional renal blood flow occurs in which outer renal cortical flow is substantially reduced, whereas medullary flow may be little, if at all, modified. The ^{133}Xe washout method was employed for this analysis (Kilcoyne et al., 1973). (See Chap. 22 for additional information on regional renal blood flow.) The shift of blood flow from the outer cortex to the juxtamedullary nephrons may invoke function of proportionally greater numbers of nephrons having higher filtration-resorptive capabilities for sodium than do cortical nephrons, so that, per volume of renal blood flow through this zone, sodium is more avidly retained.

The reduced cortical blood flow due to impaired cardiac output could result in increased renin production by the juxtaglomerular apparatus and, in turn, increased angiotensin production. The outer cortex of the kidney has a particularly high renin content and hence might be particularly susceptible to the effects of reduced blood flow. It has been speculated that recirculated angiotensin II, which is a potent renal vasoconstrictor, might promote further reduction in cortical blood flow.

But explanations for sodium retention based purely on renal hemodynamic mechanisms are probably an oversimplification, because humoral factors, such as hormones secreted by the adrenal cortex, especially aldosterone, are involved. It is well known that angiotensin stimulates aldosterone release by the adrenal cortical glomerulosa cells (see Chap. 34). It has been clearly demonstrated by Leutscher's group (Camargo et al., 1965; Cheville et al., 1966) that there is enhanced production of aldosterone by the adrenal cortex in heart failure patients with edema. In advanced CHF, aldosterone secretion may be three to four times normal. Plasma aldosterone is elevated also because the liver's capability to metabolize the hormone is impaired by hypoxia, a result of the impaired hepatic blood flow in heart failure (reduced arterial inflow and increased hepatic venous pressure, reflecting elevated right atrial pressure). The kidneys, too, play an important role in the metabolism and excretion of aldosterone and its conjugates, so that when kidney blood flow and filtration rate are reduced, this avenue of metabolism and excretion is damaged and retention of aldosterone is favored. The increased levels of plasma aldosterone promote enhanced resorption of sodium by the distal convoluted tubules of the nephron (see Chaps. 22 and 34).

However, continued injection of aldosterone in the normal experimental subject, although it may cause transient sodium retention, is followed by a later readjustment of renal mechanisms, so that sodium and attendant water is again lost to restore homeostasis. Thus, hyperaldosteronism itself cannot be the sole mechanism involved but is accepted as an important contributing factor. Possible mechanisms for triggering increased release are discussed in Chapter 34. Davis (1965) has suggested that an additional factor (or factors) of a humoral or hemodynamic nature, distinct from any identified at present, is involved in CHF and potentiates the aldosterone influence. The possibility that a sodium-"losing" hormone is involved in the overall homeostasis has been entertained by some investigators. Reduction in this hormone would favor salt retention.

In conclusion, primary mechanisms involved in CHF have been presented, and the sequelae of the primary derangements have been discussed. Although physicians are concerned with improving the well-being of their patients by allaying the sequelae—e.g., removing edema fluid by administering diuretics to institute renal loss of salt and water—they continually strive to correct, insofar as possible, the basic derangements. An illustration is the use of drugs such as digitalis to improve myocardial pumping action.

PHYSIOLOGY OF SHOCK

Ewald E. Selkurt

Shock is defined as an abnormal physiological condition resulting from inadequate propulsion of blood into the aorta and therefore inadequate flow of blood perfusing the capillaries of various tissues and organs. It is typified by deterioration of cellular functions, particularly of vulnerable organs such as the kidney, liver, and heart. It is further characterized by a progressive failure of the circulation eventuating in an *irreversible state.*

CLASSIFICATION OF TYPES OF SHOCK; CAUSES AND MANIFESTATIONS

Circulatory shock describes a syndrome that is characterized by protracted prostration and hypotension, pallor, coldness and moistness of the skin, collapse of superficial veins, reduced sensibilities, and suppression of urine formation. It is important to recognize that all these symptoms may be present without development of the potentially irreversible physiological condition to which the term should be limited. In consequence, some authorities have recommended two subdivisions: primary shock and secondary shock. Primary shock is essentially a bout of acute hypotension. It tends to be transient, and there is usually a natural tendency to recovery. Secondary shock is typified by deterioration of cellular function and progressive failure until an irreversible state is reached, culminating in death. In arriving at a diagnosis, one must therefore relate the existing clinical signs and symptoms to the antecedent events, gauging the probabilities as to whether they indicate development of true shock or are merely a temporary circulatory derangement. Although better clinical criteria could be desired, progressive reduction of arterial pressure and pulse pressure measured at frequent intervals constitutes the best available evidence of the existence of a physiological state of shock. Collapse of the veins is an important sign.

Investigators in the field, both in the clinic (patient) and laboratory (experimental animal) now recognize various entities in shock, which are different enough in the causal (physiological manifestations) phases, treatment, and recovery to warrant subclassifications. These are: (1) hypovolemic shock, (2) ischemic shock, (3) cardiogenic shock, (4) traumatic shock, and (5) endotoxic and septic shock. Although hypovolemic shock (body tissue and organ an-

oxia) is probably the least complex, in terms of causation and treatment, it cannot be overemphasized that hypovolemia and inadequate circulation are basic to all forms of shock. This and other forms of shock reveal alterations in cellular metabolism and function as the result of enhanced release of humoral agents. Some of these agents are directly toxic or work indirectly, often in conjunction with other such agents, leading to complete metabolic breakdown and death.

SYMPTOMATOLOGY. Clinical symptoms are referable to the skin, the mucous membranes and the countenance, the neuromuscular system, the circulatory and respiratory systems, and metabolism. The skin is pale and cold because of peripheral vasoconstriction. This is accompanied by sweating because of massive stimulation of the sympathetic division of the autonomic nervous system. The mucous membranes especially tend to be bluish as a result of the enhanced oxygen extraction from the blood (stagnant anoxia). The countenance appears haggard and drawn, in part because of the removal of interstitial fluid from the skin due to hypotension, with resultant reduction in capillary hydrostatic pressure, which normally opposes the inward-drawing force of the plasma proteins. The reflexes tend to be depressed and responses to noxious stimuli reduced. The blood pressure is decreased, and the pulse is rapid and thready as a result of reduced cardiac output. The respiration is rapid and shallow, on account of both reflex and chemical stimulation. Body temperature may be decreased, partly because of decreased general metabolism resulting from a serious reduction in oxygen supply. All these conditions are symptomatic of the loss of circulating blood volume.

In the case of septic or toxic shock (gram-negative bacteremia; e.g., that arising from infection by *Brucella melitensis, Escherichia coli*), additional striking symptoms are noted (Shubin, et al., 1977). These typically begin with a shaking chill followed by fever 2 to 24 hours after mechanical manipulation (of the splanchnic bed) or other trauma (e.g., ischemia) during surgery. Impaired cerebral function may result. Vomiting and diarrhea are frequent findings. The blood picture shows leukopenia and thrombocytopenia. Serum transaminase levels increase, reflecting the extent of cellular damage. Abnormalities of the S–T segment and T wave reflect a reduction in coronary artery perfusion.

When first seen, a patient with gram-negative bacteremia may have fever and warm extremities (pyrogenic phase). Cardiac output is increased, as is pulse pressure. This is followed in a later phase by reduced cardiac output and arteriolar vasoconstriction. The altered mechanism of the portal circulation (hepatic vasoconstriction) favors pooling of blood in the intestinal venous capacitance bed, reducing effective circulatory blood volume. Tissue oxygenation is curtailed, and blood lactate levels rise (Shubin et al., 1977), contributing to the acidosis noted in shock.

PREDISPOSING CAUSES. In most cases there is a loss of blood or plasma fluid with a consequent prolonged period of hypotension. Some of the more important causes are (1) hemorrhage; (2) extensive edema or serous effusion due to mechanical, chemical, or thermal capillary damage; (3) seepage of wounds or burn surfaces; (4) abstraction of fluids through excessive sweating, vomiting, diarrhea, or drainage of secretions after surgical operations; (5) disturbances of electrolyte and water balance, causing withdrawal of water from the bloodstream; and (6) unusual engorgement of the capillaries with blood, reducing effective blood volume. The last is frequently the causative factor in the shock that occurs during overwhelming infections, such as those from gram-negative bacteria.

STAGES OF SHOCK

INITIAL OR HYPOTENSIVE STAGE. Primary loss of blood or fluid leads to a disparity between the circulating blood volume and circulatory capacity. This results in reduction of venous return and in an acute decrease in cardiac output, with reduced arterial and pulse pressures and ultimate impairment of circulation and lowered oxygenation of the tissues of the body. Compensatory mechanisms are instituted rapidly, as follows: Hypotension acts via the carotid sinus and aortic arch baroreceptors to cause reflex vasoconstriction and is supplemented by the action of catecholamines released in increased amounts that augment peripheral arteriolar and venous tone. The same reflexes augment the cardiac vigor and rate. The fall in blood pressure also reflexly causes an increase in the depth and rate of respiration, which favors venous return and ultimately cardiac filling—the so-called abdominothoracic pump mechanism discussed in Chapter 16. Increased activity of venopressor mechanisms may contribute to venous return. Further repletion of blood volume results from an increase in plasma volume through the reduction of capillary pressure, which leads to resorption of tissue fluid and ultimately plasma replacement. Thus, hemodilution is often noted during this stage.

Increased production and release of aldosterone and antidiuretic hormone (ADH) accompany the bleeding (Fig. 18-4). These can also be viewed as initial compensatory mecha-

nisms, in the light of favoring salt and water retention by kidney, which is further favored by the reduced glomerular filtration resulting from the hemorrhagic hypotension. The increase in renal sympathetic nerve activity (upper panel, Fig. 18-4) facilitates afferent arteriolar constriction, reducing glomerular filtration even more (see Chap. 22). Blood catecholamine concentration may increase more than threefold (to about +320 percent) (Chien, 1967). The ADH increase results from reduced stimulus of the left atrial volume receptors; reduced firing of these receptors (second panel, Fig. 18-4) releases from inhibition the neurohypophyseal secretion of ADH (see Chap. 30).

Chien and Usami (1974) have confirmed the ADH release in response to hemorrhage and provided convincing evidence that the afferent fibers are in the vagus and carotid sinus nerves. However, extravagocarotid factors may play a role in causing ADH release after hemorrhagic hypotension.

Increasing evidence of involvement of the prostaglandin (PG) system is accumulating (Lefer, 1970). Prostaglandins are primarily found in the kidney (mostly in the medulla). The important ones are: PGE_2, PGI_2 (prostacyclin) both dilators; and vasoconstrictors, such as $PGF_{2\alpha}$ and thromboxane (TxA_2). Although the prostaglandins are usually fairly completely metabolized in the lung, more are produced in shock and more pass through the pulmonary vascular bed to act systemically (Selkurt, 1979). Increased renin production, with an ultimate increase in the potent vasoconstrictor angiotensin II, adds another significant factor to the vasoconstrictor side of the balance sheet.

If vasodilator substances predominate, it would seem to outbalance the constriction in hypovolemia, and thus the net effect would favor decompensation, as seen in the late phase of shock (i.e., irreversible phase).

PROGRESSIVE OR IMPENDING STAGE. The arterial pressure, previously stabilized by compensatory mechanisms, may decrease slowly and progressively. There is no further vasoconstriction; in fact, total peripheral resistance (TPR) may begin to lessen. The low arterial pressure, together with vasoconstriction, exerts additional deleterious effects on the various organs and tissues because of reduction in capillary flow. The generalized tissue hypoxia leads to impairment of enzyme systems and to anaerobic metabolism, with an accumulation of lactates and pyruvates and with failure of resynthesis of organic phosphates. Reduction of alkali reserve occurs, and there is migration of potassium from tissue cells. The vital organs are variously affected. This stage has been called *oligemic shock.*

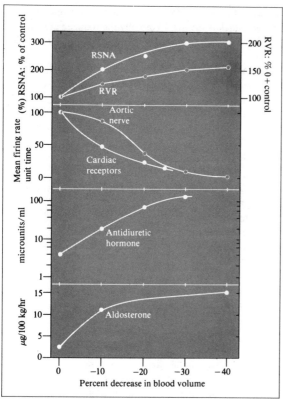

Fig. 18-4. Data showing (from above downward) the measured responses to a graded hemorrhage in terms of the increase of autonomic (sympathetic) drive as the measured change in renal sympathetic nerve activity (RSNA) and in renovascular resistance (RVR). With loss of the first 10% of the blood volume, the sympathetic drive from baroreceptors in the low-pressure system increases sharply as the firing rate of the cardiac (atrial) receptors falls to one-half (second panel). As the loss exceeds 10%, there is an increasing influence from the arterial system, as exemplified by the aortic nerve activity. Receptor activity is expressed in percentage of resting value; receptor drive is an inverse function of mean firing rate.

Meanwhile, the plasma aldosterone and antidiuretic hormone (ADH) concentrations increase, effecting water and salt retention. Volume-regulating hormones are expressed as micrograms of aldosterone released into the plasma, and as microunits per milliliter of ADH. The increase in ADH is presumed to result from cardiac receptors in the left atrium. When stimulated by volume increase, afferent vagal fibers that ascend to the neurohypophysis inhibit ADH release. When atrial volume is decreased, as with hemorrhage, the reverse trend occurs; i.e., ADH output increases. The increase in aldosterone release results, in general terms, from decreased blood flow and triggering of the juxtaglomerular apparatus renin-angiotensin mechanism (see Chap. 34). (For data source, see O. H. Gauer, J. P. Henry, and C. Behn. Regulation of extracellular fluid volume. *Annu. Rev. Physiol.* 32:562, 1970. Aldosterone was measured in anesthetized dogs, and ADH in unanesthetized dogs—note the logarithmic scale. Atrial [cardiac receptor] firing, RSNA, and RVR activity were measured in anesthetized dogs, aortic nerve activity in the anesthetized cat. RSNA, RVR, and aortic nerve activity data were supplied by W. V. Judy.)

ORGANIC ALTERATION IN THE PROGRESSIVE STAGE. Kidney. Ischemia and the resultant stagnant anoxia result in failure of glomerular filtration, and the reduced blood flow causes eventual tubular damage, evidenced by impaired extraction of para-aminohippuric acid. These renal alterations lead to anuria and retention of metabolic products (urea, creatinine, uric acid). Furthermore, they contribute to an upset in the acid-base balance in the direction of metabolic acidosis because of failure of the normal mechanisms for hydrogen ion secretion by the kidney, and the accumulation in the blood of phosphates and lactates. When crushing injuries to the limb accompany shock, there is more profound renal injury because the release of myohemoglobin and products of muscle autolysis contribute to renal failure; renal tubules, particularly the distal and collecting, are blocked by debris and hemoglobin deposits. The term *lower nephron nephrosis* has been applied to this situation. A serious consequence of shock is that delayed renal failure may ultimately lead to death in uremia even though the patient may have been treated with transfusion, with adequate recovery of blood pressure. In any event, even after treatment, the kidney may, for a prolonged period, show residual damage characterized by loss of concentrating power, evidenced in excretion of dilute urine of low specific gravity. The loss of concentrating ability is largely due to the loss of the gradient of osmolality in the kidney (see Chap. 22), due to "washing out" of the high papillary concentration needed for urinary concentration by the ADH mechanism. The washout is a result of marked reduction or cessation of glomerular filtration, with persistence of blood flow through the medulla (Selkurt, 1969).

The frequent finding of dilute urine of low specific gravity and positive free water clearance ($+CH_2O$, Chap. 22) by the shock kidney invokes the possibility that other factors are involved beyond a simple washout of the gradient. Since a possible deficiency of output of ADH could not be a factor (Selkurt, 1974), an apparent loss of sensitivity of the epithelium of the collecting duct to ADH action is indicated. Strong evidence is accumulating that another category of humoral agents, the prostaglandins PGE_2 and PGI_2 are involved in the apparent ADH antagonism, as well as in other manifestations of impaired tubular function in shock, e.g., impaired electrolyte resorption (Na^+, Ca^{2+}, and K^+). It is important to emphasize that these are not direct effects of hypoxia in the tubular epithelium, but rather the consequence of the influence of hypotension and altered renal medullary blood flow on the synthesis of prostaglandins by cells in the medulla that produce them from fatty acid precursors (see Chaps. 22 and 30). In addition, the prostaglandins may directly influence renal tubular electrolyte resorptive processes, acting primarily on the thick ascending limb of Henle and distal tubule.

The prostaglandins of the renal medulla could play a significant role in altered renal function because of their strategic location. Lipid-containing granules in the cytoplasm of the interstitial cells of the renal medulla and papilla give evidence of the sites of synthesis. These cells lie horizontally in the interstitial space between the vertically oriented loops of Henle and vasa recta (see Chap. 22). The granules contain lipids that presumably serve as precursors for synthesis of the prostaglandins. When released, these substances quite possibly diffuse into the neighboring blood vessels to be carried up to the juxtamedullary zone by ascending vasa recta to relax the vascular smooth muscle and sphincters, and thus increase medullary blood flow. This action could favor the washout of the gradient. Ultimately, some prostaglandins leave via the renal vein. Entering the loop of Henle, they could influence tubular cell activity of the ascending limb of Henle and finally the distal convolution and collecting duct, with regard to opposing ADH action. Some finally appear in the urine.

Interestingly, infusion of PGE_1 or PGE_2 directly into the renal artery elicits many of the electrolyte transport and urinary volume changes commonly observed in shock, such as increases in fractional osmolar and electrolyte clearance (Na^+, Ca^{2+}, K^+). If the kidney is initially concentrating the urine (see Chap. 22), this process may be reversed to production of dilute (hypotonic) urine of relatively greater volume.

The mechanism that triggers synthesis and/or release is still under investigation. One effective means of activating increased appearance in renal vein blood is brief renal ischemia. Renal vasoconstriction by nerve stimulation also enhances release. Strong evidence likewise exists for humoral mediators that promote synthesis and release of prostaglandins, i.e., catecholamines and angiotensin II. Release of both is stimulated by hemorrhage, as has been discussed earlier (Fig. 18-4).

Liver. The circulatory anoxia impairs the normal mechanisms of metabolic turnover of such substances as pyruvates, lactates, and amino acids released from traumatized tissue, by disorganizing the hepatic cell enzyme systems. The reason for the accumulation of the amino acids in the blood is impairment of the liver's normal ability to form urea. There may be early glycogenolysis and glycogen depletion and hyperglycemia, but later impairment of gluconeogenesis and ultimate hypoglycemia. In connection with the disorganized enzyme systems, it has been observed that the adenosine triphosphate (ATP) and adenosine diphosphate (ADP) concentrations are reduced in the shock liver.

Cation transport of the liver is impaired in hemorrhagic shock in rats (Sayeed et al., 1974). Thus, sodium-potassium transport is deranged, and sodium accumulation within the liver cells is accompanied by water entry, since cell volume regulatory capability is lost. As a result the cell swells in an early phase of cell destruction. Reduced concentrations of ATP and ADP account for the membrane electrolyte pump failure, in part. Early in shock (one-half hour of hypotension) the changes could be reversed by reinfusion of shed blood and Ringer's lactate, but not in intermediate (1 hour) or late (2 hours) shock.

Heart. Eventually, the myocardium may become depressed because of reduced oxygen supply and possibly accumulation of toxic materials. A reduced coronary flow because of hypotension occurs despite compensatory dilation of the vascular supply of the myocardium.

Two types of lesions are characteristic of heart damage in shock, subendocardial hemorrhage and necrosis and "zonal lesions" (Hackel et al., 1974). The latter are described as a zone of hypercontraction at the end of the myocyte adjacent to an intercalated disk, accompanied by shortening of the sarcomeres, fragmentation of the Z band, distortion of the myofilaments, and mitochondrial displacement. Such lesions are also seen in certain muscular dystrophies and excessive catecholamine stimulation.

In endotoxic shock, myocardial depression may also be an entity involved. Experimental evidence has varied, however, some finding no myocardial dysfunction (McCaig et al., 1979) or none until about 4 to 5 hours after endotoxic shock (Hinshaw et al., 1974).

Figure 18-5 summarizes the current understanding of myocardial dysfunction in endotoxic shock (Hinshaw, 1979). Coronary hypoperfusion and depressed responses to beta-adrenergic stimuli are primary factors in causing cardiac dysfunction. Intracardiac disturbances perform a major role, but their causes are obscure. Development of edema of the myocardium occurs in the contractile elements and mitochondria. Possibly, ionic imbalances (K^+ versus Ca^{2+}) occur. Depressed diastolic filling and contractility reflect the loss of cardiac power and efficiency of contraction.

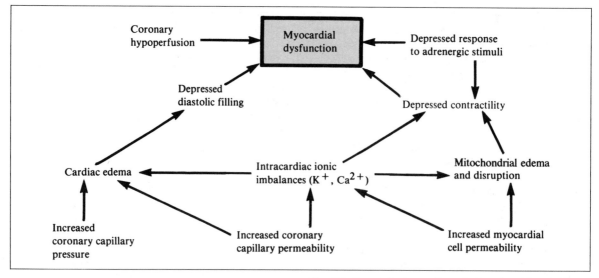

Fig. 18-5. Suggested adverse effects of endotoxin on the myocardium. (After L. B. Hinshaw. Myocardial function in endotoxin shock. *Cir. Shock* [Suppl. I]:50, 1979.)

Brain. The blood supply to the brain appears to be reasonably well maintained because of compensatory vasodilation and at the sacrifice of flow to the other organs. For example, the blood flow to the skin, kidneys, and splanchnic bed is diverted by more intense vasoconstriction to the brain, pulmonary, and coronary circulations. Brain samples indicate that the metabolic condition appears to be good in terms of ADP, glycogen, and phosphocreatine contents. However, occasionally, death by respiratory failure occurs during this stage, suggesting greater susceptibility of the respiratory centers.

Green and Rapella (1968) reported that cerebral blood flow in the dog was well maintained during severe hemorrhagic hypotension, and in the subsequent normovolemic phase, suggesting maintained autoregulatory control. Other workers, studying cerebral blood flow in the rat, dog, and monkey, do not entirely agree. Although cerebral blood flow is well maintained during small to moderate hemorrhage, it decreases significantly during severe hemorrhage. However, the long-term effect indicated that some vasoconstrictor substance(s) had been released that affected cerebral blood flow. Possibilities include thromboxane and the vasoconstrictor prostaglandin ($PGF_{2\alpha}$). Also, serotonin and angiotensin II are probably involved in the alteration in circulation. The vasoconstrictor changes produced by the latter agents favor formation of thrombi and plugging of the vessels, further impairing the circulation.

In the dog the cerebral involvement in gram-negative endotoxic shock is shown by an early decrease in cerebral blood flow to 63 percent of control (Parker and Emerson, 1977). With prolonged hypotension (4 hours) it decreases to 39 percent of control. The early decrease is due to fall in perfusion pressure. Later, however, a progressive increase in cerebrovascular resistance occurs, accompanying the phase of arterial hypotension. Acidosis and hypoxia thus may contribute to the late phase of cerebral ischemia.

IRREVERSIBLE STAGE. There is further decrease in cardiac output. Blood pressure may be very low, and pulse pressure may be barely detectable. Respiration becomes depressed. Cardiac deceleration may follow because of default of reflexes or weakening of the myocardium. In some instances the TPR begins to decrease, indicating relaxation of compensatory vasoconstriction. This may be the result of failure of the vasomotor centers due to prolonged hypoxia, but more probably it is caused by the production and release of vasoactive substances in the blood that directly or indirectly depress blood pressure or heart action, or both. It is of extreme significance that large infusions of blood and blood substitutes given at this time may prove to be of temporary benefit only. With transfusion, blood pressure and other hemodynamic alterations may be reasonably well restored toward normal for a time, but the restorations are likely not to be maintained. The state is called *normovolemic shock* and may progress into irreversible failure and death.

A major manifestation of the irreversible stage is the observation in experimental animals that the capillaries, particularly in the splanchnic bed, dilate by relaxation of the precapillary sphincters. Thus, there is an eminent possibility that stagnant blood will be impounded in these areas The capacious pulmonary vascular bed does not appear to be an important site of pooling, however (Abel et al., 1967). Furthermore, release of proteases, histamine, and other substances during the hypotensive phases seems to favor increased capillary permeability, which may allow fluid to escape into the extravascular spaces. There is evidence that fluid moves from the interstitial compartment into cells. This movement may be responsible for the hemoconcentration that is seen in the terminal phases of shock. In the dog especially, the intestinal bed shows marked congestion of the capillaries and venules, with extreme extravasation of fluid and even blood into the intestinal lumen; hence, significant amounts of transfused blood and fluid may ultimately be lost into the intestinal lumen and be discharged from the body as bloody diarrhea. However, this mechanism is not important in humans and the monkey.

Experimental investigation indicates that myocardial depression contributes to and even accelerates the fatal end: Cardiac output decreases despite terminal elevation of ventricular filling pressure, suggesting weakening of the myocardium; the response to equivalent states of diastolic distension is lessened when an infusion is given; and the S-T segments of the electrocardiogram are altered.

EXPERIMENTAL PRODUCTION OF SHOCK

The experimental approaches to the study of the basic mechanisms of shock have been many and varied, and space does not permit a detailed description and evaluation of the numerous techniques. Generally, these techniques are designed to simulate the development of shock as it is caused in humans. One of the most commonly employed is hemorrhage, illustrated in Figure 18-6.

The dominant effect of hemorrhage (curve 1) is the reduction in cardiac output that results (curve 3) because of reduced filling pressure (curve 4). Mean blood pressure declines (curve 2), with a decrease in pulse pressure, reflecting reduced stroke volume. Reflex stimulation of respiration results (curve 5), supplemented by chemical stimulation from developing acidosis later in the oligemic phase. Circulatory compensatory mechanisms include an increase in heart rate, particularly if the control rate is slow (curve 6), and enhanced vasoconstriction, so that TPR increases markedly at first (curve 11). The increase in TPR is both neurogenic and humoral. Regional and organ changes in

vascular resistance vary. The resistance actually *decreases* in the coronary circulation (curve 7) and also in the cerebral circulation (not shown); it shows only minor increase in the splanchnic circulation (curve 8) but increases significantly in the kidney (curve 9) and muscles (curve 10; the dashed line shows the relationship with cold block of the nerve). Later, in oligemic shock, TPR declines somewhat. An apparent reduction of plasma protein concentration (not shown) suggests that a supplementary compensatory mechanism is the influx of interstitial fluid into the capillaries.

Following transfusion, cardiac output and arterial pressure are temporarily restored, then begin a progressive downward trend. Cardiac output declines because of decrease in filling pressure. Although heart rate is restored almost to the control rate, there is a terminal slowing.

The respiratory rate slows somewhat at first, but increases later in normovolemic shock, reflecting in part the underlying disturbance in acid-base balance. Susceptible animals may die abruptly during this phase (Fig. 18-6) because of respiratory failure. Respiratory changes are, however, variable; a slowing late in normovolemic shock is often observed.

Following transfusion, TPR returns toward control but then begins a secondary rise. Coronary resistance returns to the control level but may show a decrease terminally. Renal resistance is temporarily restored toward the control level, then progressively increases. Splanchnic resistance shows a phase *below* control that has not been satisfactorily explained, then gradually rises back to the control value. Muscle resistance remains high, although it manifests a downward trend later in normovolemic shock.

Several conclusions can be drawn. Failure of the circulation is not the result of failure of vascular reflex compensatory mechanisms, in view of the well-maintained TPR. Myocardial failure cannot be implicated as an important factor early in normovolemic shock. When extra fluid is given to restore filling pressure, cardiac output is restored. Thus, continued loss of effective circulating blood volume is the reason for the gradual decline in blood pressure. Late in normovolemic shock, despite an attempt to maintain coronary circulation by compensatory dilation during hypotension, flow is impaired because of the reduced cardiac output and blood pressure. Alterations in myocardial metabolism have been demonstrated as a consequence of the impaired flow. Moreover, toxic products of a vasodepressor nature developed in shock probably have an influence on the depression of the myocardium.

Other methods of producing shock include crushing and contusion of muscles, fractures of bones, burns and scalds, intestinal obstruction, and prolonged intestinal exposure

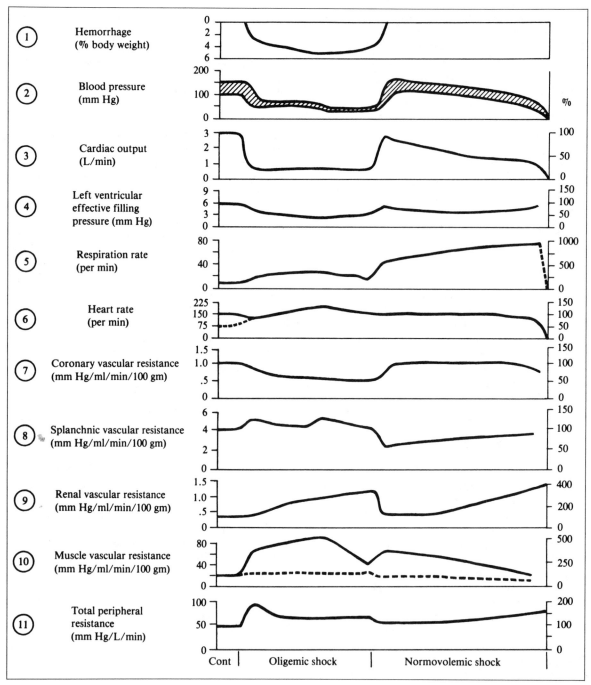

1. Hemorrhage (% body weight)

2. Blood pressure (mm Hg)

3. Cardiac output (L/min)

4. Left ventricular effective filling pressure (mm Hg)

5. Respiration rate (per min)

6. Heart rate (per min)

7. Coronary vascular resistance (mm Hg/ml/min/100 gm)

8. Splanchnic vascular resistance (mm Hg/ml/min/100 gm)

9. Renal vascular resistance (mm Hg/ml/min/100 gm)

10. Muscle vascular resistance (mm Hg/ml/min/100 gm)

11. Total peripheral resistance (mm Hg/L/min)

Cont | Oligemic shock | Normovolemic shock |

Fig. 18-6. Cardiodynamic trends in experimental hemorrhagic shock in dogs. Values to the left are absolute units. Values to the right represent percentage change from control (Cont) (set at 100%).

and manipulation. Less directly, prolonged ischemia of the limbs, induced by tourniquets and followed by release, or ischemia and release of the intestinal circulation can be used for this purpose. Anoxia of the splanchnic bed, experimentally induced by ischemia, has received considerable attention lately as a method of producing shock, for several reasons. In addition to the products of ischemia, the intestinal bed is the site of toxic amines (e.g., tyramine), which might be released into the general circulation with deleterious effects on the cardiovascular system. Ischemia of the intestine created by clamping of the superior mesenteric artery for an hour or two, then release, leads to irreversible shock with much of the symptomatology characteristic of this condition. The organism becomes particularly susceptible because of concomitant impairment of normal liver function, which therefore cannot detoxify the products that may reach the portal circulation. Experimental evidence exists that vasodepressor substances do pass into the systemic circulation as a consequence of intestinal hypoxia, not only with ischemic shock but also during hemorrhagic shock.

THEORIES OF THE MECHANISM OF SHOCK

Several theories of the causes of irreversibility of shock have been based on findings in experimental animals—the dog, rat, and monkey. A key mechanism is fluid loss from the effective circulation. This can follow directly, as by hemorrhage, or indirectly by a loss of fluid resembling plasma, as occurs in burns. There may be local fluid loss at the site of trauma, or it may be caused by prolonged vomiting and diarrhea. The loss of circulating volume triggers a series of mechanisms (described previously) that lead to the secondary sequelae of reduced cardiac output, reflex vasoconstriction, and overall impairment of flow in the various tissues and organs, with consequent secondary effects due to deranged metabolism.

NEUROGENIC FACTORS. The role of the nervous system has received much attention. The effect of afferent impulses from the sites of trauma, burns, etc., on the vasomotor and respiratory systems has been explored, together with the possibility that failure of the ensuing reflex vasoconstriction accounts for the late irreversibility of shock. Current evidence indicates that TPR is well maintained throughout the course of oligemic shock (reduced blood volume) and through most of normovolemic shock (following transfusion). Sympathetic nerve impulse discharge remains high, in the main, and catecholamine output is never decreased. Thus, the progressive fall in blood pressure after transfusion

must result in large measure from other factors. Among these, fluid loss has to be given serious consideration.

CHANGES IN BODY FLUIDS. The problem of fluid loss can be considered from the broad point of view of the application of known techniques for the measurement of vascular volume, interstitial volume, and intracellular water during shock, to see whether significant alterations in these volumes indeed occur. Closely related is the consideration of specific organs that might be involved in such disruption by virtue of being particularly susceptible, such as the intestine, or perhaps important because of their mass, such as skeletal muscle.

Many studies have dealt with the changes in plasma volume and red cell mass in the posttransfusion phase of hemorrhagic shock. Most of them have been done on the dog, but more and more data are accumulating for humans and other primates. Very typically in the dog, the plasma volume and red cell volume return to normal immediately after transfusion, but both plasma volume and red cell volume later decrease progressively.

During oligemic hypotension, blood is diluted by fluid influx. This is restored on transfusion, but late in the posttransfusion phase, protein concentration and hematocrit values appear to increase, presumably because fluid is lost again. Although plasma loss and hemoconcentration are generally found during the late oligemic period and after retransfusion in the dog, it is important to note that hemoconcentration is not found in humans, even after severe hemorrhage. If anything, the findings indicate that in humans subjected to hemorrhagic shock there is hemodilution rather than extravasation of plasma fluid. Admittedly, this is often complicated by the therapeutic regimen. In a critical experiment done in another species, the sheep, Gillett and Halmagyi (1966), by the use of ^{51}Cr-labeled cells and ^{131}I-labeled plasma for determination of total blood volume, found it virtually unchanged in irreversible shock. Such findings emphasize that fluid loss alone, when it occurs, need not be the sole factor responsible for circulatory failure in shock.

Other relevant questions pertain to the changes in extracellular and intracellular fluid volumes. Extracellular water, as measured by distribution of inulin or ^{22}Na or ^{35}S sulfate measurements, interestingly shows a decrease in the late hypotensive and late posttransfusion phases in the splenectomized dogs. The typical finding was a small reduction in extracellular water in the posttransfusion phase. The magnitude of reduction, regardless of the label used, was quite modest, ranging from about 5 to 8 percent. Occasionally, larger changes were observed. For example, in several of the dogs examined by Shizgal et al. (1967), changes were greater than 10 percent and averaged about 20 percent. The question was: What happened to the extracellular fluid water under the circumstances of "irreversible shock"? The possibility that the loss of extracellular fluid water could be accounted for by movement into the cells was suggested by the work of Slonim and Stahl (1968) and Matthews and Douglas (1969). The notion of increased intracellular water volume has led to the speculation that there may have been a failure of the sodium pump as a result of the hypoxia during the oligemic stress, and that now water moved into the cells because of the failure of the pump to move sodium out.

MYOCARDIAL FAILURE. The decline of systemic arterial pressure on account of inadequate cardiac output in the phase of deterioration in shock raises the question: Why does the cardiac output decline after the reinfusion of the shed blood? Rothe (1970) sought an answer by examining ventricular stroke work in dogs during hemorrhage and the posttransfusion phase. In some animals, massive overtransfusion (twice their original blood volume) was resorted to, supplying an adequate volume and ensuring a good filling pressure. In the animals with more severe stress (hypotension prolonged until a 20 to 40 percent uptake of shed blood from the arterial reservoir was indicated), definite evidence of depression of cardiac function was observed in the left ventricle. This persisted even with infusion of 160 ml per kilogram of blood, twice the original bleeding volume.

HUMORAL FACTORS. The humoral theories of the mechanism of shock are concerned with the formation of substances influencing the cardiovascular system or release of such substances from the site of injury. Some vasoactive substances may be looked on as participating in the earlier compensatory mechanisms, for they are vasopressor in action. The increased outpouring of the catecholamines (epinephrine and norepinephrine) has already been cited. Evidence exists that renin or a reninlike substance appears in greater amounts following hemorrhage. Renin release could lead to formation of the pressor substance angiotensin. Increased discharge of vasopressin has been reported. Serotonin (5-hydroxytryptamine), vasoconstrictor in action, may be involved.

Other substances may be toxic to the cardiovascular system and contribute to later circulatory collapse. Some examples are potassium ion, which has a depressing effect on the myocardium, and histamine, which is deleterious to the peripheral circulation. Other products of tissue damage that are vasodepressor in action have been studied. These include the adenosine compounds and vasoactive polypeptides such as bradykinin. Lefer (1970) proposed a myocardial depressant factor (MDF), probably a polypeptide or glycopeptide of low molecular weight (about 1000). Recent investigation suggests that the myocardial depressant factor may not itself act directly on the myocardium, but rather in conjunction with some plasma factor. The pancreas is an important source of MDF.

The vasoactive polypeptides (kinins) are formed by the proteolytic action of proteases on certain plasma proteins precursors, splitting off the vasoactive peptides (e.g., bradykinin, MDF). The toxic, shock-producing principle, anaphylotoxin, is also probably produced by a proteolytic step. It differs from other kinins in that it needs the intermediation of histamine release.

The combined action of proteases and vasodilator kinins may explain the increase in capillary permeability in shock, for the proteases weaken the endothelium by their protein-digesting capabilities, while the kinin dilator action further promotes capillary filtration.

Prostaglandins. Mention has been made of the influences that favor increased synthesis and release of these substances. Among the prominent prostaglandins released by the kidney are the important dilators, PGI_2 and PGE_2. Evidence of increased output of PGE in the hypotensive phases has been adduced (Johnston and Selkurt, 1976). Arterial concentration rises also, suggesting that the ability of the lung to metabolize prostaglandins is impaired in hemorrhagic hypotension. Markedly reduced pulmonary blood flow would be expected to take out of action the cells that would normally metabolize prostaglandins. Possible shunts open that bypass the pulmonary sites of metabolism. Inasmuch as renovascular resistance progressively increases during hypotension, the obvious conclusion is that the accumulation of pressor substances (cathecholamines, angiotensin, and vasopressin) overrides any vasodepressor action of the prostaglandins during this phase. On transfusion, arterial PGE concentration is transiently reduced, but later rises again during normovolemic shock because of increased synthesis and/or reduced metabolism by the lungs. The sharp rise seen late in the posttransfusion phase may be contributory to reduced TPR and the downward trend in mean arterial blood pressure noted at this time, heralding the approaching death of the experimental animal. When the potent prostaglandin synthetase inhibitor indomethacin is administered to the animal, in this phase of shock, concomitant with the reduction in release of PGE, renal blood flow decreases and renovascular resistance increases (Selkurt, 1974).

The question may be raised whether an increase in circulating PGE is, in fact, detrimental to the animal by virtue of its potent dilator action. This idea would be in accord with the notion that use of dilator drugs (see Treatment of Shock) can be beneficial in promoting better organ and tissue perfusion (if blood volume and cardiac output are adequate). It has been demonstrated that PGE_1 improves the coronary flow and myocardial contractility of dogs in hypovolemic shock and hence improves cardiac output (Priano et al., 1974). However, PGE must be infused into the left ventricle to escape metabolism in the lung for this beneficial effect to be discerned.

TREATMENT OF SHOCK. Comments on the treatment of shock are pertinent here because of the physiological implications. The obvious immediate treatment is transfusion of blood or a suitable substitute, depending on the nature of fluid loss. The value of this procedure depends on the lapse of time since the fluid or blood loss occurred, the quantity transfused, and the character of the transfusate.

To avert irreversible shock, the transfusion must be given as soon as possible after the initial blood loss and the development of hypotension. With regard to the quantity infused, the general principle is to give blood until the blood pressure (arterial and venous) and other essential cardiovascular signs (e.g., heart rate) have returned to normal. One of the most reliable of indices that can be rather simply employed is monitoring of the central venous pressure; a catheter in the superior or inferior vena cava is most useful for this purpose.

It was the practice in the treatment of battle casualties during the Korean and Viet Nam wars to transfuse quantities of blood that greatly exceeded the apparent immediate or obvious loss, and the results were highly satisfactory. With a weakened myocardium, however, there is some danger of pulmonary congestion if pulmonary venous pressure is raised rapidly by excessive transfusion. It is well to remember that transfusion of 1 liter of blood does not mean that the circulating blood volume is necessarily increased by 1 liter. In the development of irreversible shock, dilated capillary channels might accumulate stagnant blood drawn out of the effective circulation, or rapidly leak out fluid.

The transfusate should conform to a number of requirements. It should be nontoxic and well retained by the vascular system. If a foreign substance, it should not be deposited in the tissues and impair their function. It should not hemolyze or agglutinate cells. It should, of course, be isotonic and of proper pH. It should contain colloidal material to provide colloid osmotic pressure similar to that of the plasma protein. Hemorrhage is best treated with transfusion of fresh, properly matched blood. If there has been loss of fluid and electrolytes by vomiting and diarrhea, saline infusion (Ringer's) is the fluid of choice; with burns, plasma, dextran, or polyvinyl pyrollidone. The latter two are synthetic polysaccharide plasma expanders that have proved safer than those formerly used, e.g., gelatin, gum acacia, and pectin. If anemia supervenes as a sequela to severe burns, blood transfusion is necessary.

The use of vasopressor agents has often been considered in the treatment of shock with hypotension. Norepinephrine (levarterenol [Levophed]) and metaraminol (Aramine) have been utilized. Use of these drugs in the treatment of hypovolemic (oligemic) shock is of course not justified, in view of the high level of circulating catecholamines. However, they may be helpful allies in the treatment of shock caused by myocardial infarction or bacteremic or hypersensitivity reactions, and they usually have a favorable effect against neurogenic shock.

Under the thesis that tissue and organ blood perfusion is impaired, it appears to be therapeutically advantageous to use vasodilator drugs, coupled with adequate restoration of blood volume, in the treatment of shock. Phenoxybenzamine (Dibenzyline) is the agent for which the largest clinical experience is available. Chlorpromazine, although a somewhat different pharmacological entity, produces equivalent and apparently equally effective peripheral vasodilation.

A limitation on the effectiveness of vasodilator drugs would appear to be the effective circulating blood volume. If the volume is severely depleted, blood perfusion of vital organs (e.g., the brain) might conceivably be further impaired by opening other peripheral vascular beds. Therefore, it cannot be overemphasized that for the most effective use of dilator drugs they must be coupled with adequate blood volume restoration.

TREATMENT OF ENDOTOXIC SHOCK (Special Features). In recognition that fluid infusion is of primary importance in endotoxic shock, other aspects of therapy can also be considered. Treatment with an antibacterial agent should not be delayed until the results of blood, urine, sputum, and wound exudate cultures are available, but should be begun immediately. Gentamicin sulfate has been recommended (Shubin et al., 1977).

Corticosteroid analogues in large doses (50 times that required for adrenal replacement) tend to prevent nonspecific cellular injury. They also increase cardiac output and decrease peripheral resistance (advantageous if blood volume is normalized). Methylprednisolone sodium succinate also has been recommended (Shubin et al., 1977).

Among other vasoactive drugs to be considered are dopamine hydrochloride and beta-adrenergic isoproterenol, a potent myocardial stimulant. Isoproterenol must be used judiciously, however, because it may cause tachycardia and cardiac arrhythmias.

BIBLIOGRAPHY

HYPERTENSION

Guyton, A. C. Hypertension: A disease of abnormal control. *Chest* 65:328–338, 1974.

Kannel, W. B. Role of blood pressure in cardiovascular morbidity and mortality. *Prog. Cardiovasc. Dis.* 17:5–24, 1974.

Laragh, J. H. (ed.). Symposium on hypertension: Mechanisms and management. *Am. J. Med.* 55:261–414, 1973.

Page, I. H. Arterial hypertension in retrospect. *Circ. Res.* 34:133–142, 1974.

Page, I. H., and McCubbin, J. W. The Physiology of Arterial Hypertension. In W. F. Hamilton and P. Dow (eds.), *Handbook of Physiology*. Washington, D.C. American Physiological Society, 1965. Section 2: Circulation. Vol. 3, Chap. 61, pp. 2163–2208.

CONGESTIVE HEART FAILURE

Abel, F. L., and McCutcheon, E. P. *Cardiovascular Function. Evaluation of Ventricular Performance* (Chap. 8). Boston: Little, Brown, 1979, pp. 197–216.

Blumgart, H. L. (ed.). Symposium on congestive heart failure. *Circulation* 21:95–128, 218–255, 424–443, 1960.

Braunwald, E. (ed.). Symposium on myocardial metabolism. *Circ. Res.* 35 (Suppl. 3): 1–215, 1974.

Camargo, C. A., Dowdy, A. J., Hancock, E. W., and Leutscher, J. A. Decreased plasma clearance and hepatic extraction of aldosterone in patients with heart failure. *J. Clin. Invest.* 44:356–365, 1965.

Cheville, R. A., Leutscher, J. A., Hancock, E. W., Dowdy, A. J., and Nokes, G. W. Distribution, conjugation and excretion of labeled aldosterone in congestive heart failure and in controls with normal circulation. *J. Clin. Invest.* 45:1302–1316, 1966.

Chidsey, C. A., Kaiser, G. A., Sonnenblick, E. H., Spann, J. F., Jr., and Braunwald, E. Cardiac norepinephrine stores in experimental heart failure in dogs. *J. Clin. Invest.* 43:2386–2393, 1964.

Chidsey, C. A., Weinbach, E. C., Pool, P. E., and Morrow, A. G. Biochemical studies of energy production in the failing human heart. *J. Clin. Invest.* 45:40–50, 1966.

Davis, J. O. The Physiology of Congestive Heart Failure. Chapter 59 in W. F. Hamilton and P. Dow (eds.), *Handbook of Physiology*. Washington, D.C.: American Physiological Society, 1965. Section 2: Circulation. Vol. 3, Chap. 59, Pp. 2071–2122.

Honig, C. R. *Modern Cardiovascular Function*. Boston: Little, Brown, 1981, pp. 173–179.

Katz, A. M. *Physiology of the Heart*. New York: Raven, 1977. Chap. 21: Heart Failure; Chap. 22: The Ischemic Heart. Pp. 397–418; pp. 419–433.

Kilcoyne, M. M., Schmidt, D. H., and Cannon, P. J. Intrarenal blood flow in congestive heart failure. *Circulation* 47:786–797, 1973.

Lindenmayer, G. E., Sordahl, L. A., and Schwartz, A. Reevaluation of oxidative phosphorylation in cardiac mitochondria from normal animals and animals in heart failure. *Circ. Res.* 23:439–450, 1968.

Meerson, F. Z. Compensatory hyperfunction of the heart. *Circ. Res.* 11:250–258, 1962.

Meerson, F. Z. The myocardium in hyperfunction, hypertrophy and heart failure (Monograph 26, American Heart Association). *Circ. Res.* 25 (Suppl. 2): 1–163, 1969.

Opie, L. H. Metabolism of the heart in health and disease: Part I. *Am. Heart J.* 76:685–698, 1968; Part II. 77:100–122, 1969; Part III. 77:383–410, 1969.

Pool, P. E., Chandler, B. M., Spann, J. F., Jr., Sonnenblick, E. H., and Braunwald, E. Mechanochemistry of cardiac muscle: IV. Utilization of high-energy phosphates in experimental heart failure in cats. *Circ. Res.* 24:313–320, 1969.

Stoner, C. D., Ressallat, M. M., and Sirak, H. D. Oxidative phosphorylation in mitochondria isolated from chronically stressed dog hearts. *Circ. Res.* 23:87–97, 1968.

Wood, P. H. *Diseases of the Heart and Circulation* (3rd ed.). Philadelphia: Lippincott, 1968.

PHYSIOLOGY OF SHOCK

Abel, F. L., Waldhausen, J. A., Daly, W. J., and Pearce, W. L. Pulmonary blood volume in hemorrhagic shock in the dog and primate. *Am. J. Physiol.* 213:1072–1078, 1967.

Chien, S. Role of sympathetic nervous system in shock. *Physiol. Rev.* 47:214–288, 1967.

Chien, S., and Usami, S. Rate and mechanism of release of antidiuretic hormone after hemorrhage. *Circ. Shock* 1:71–80, 1974.

Emerson, T. E., and Parker, J. L. Cerebral hemodynamics during endotoxin shock in the dog. *Circ. Shock* 3:21–38, 1976.

Gillett, D. J., and Halmagyi, D. F. J. Blood volume in reversible and irreversible posthemorrhagic shock in sheep. *J. Surg. Res.* 6:259–261, 1966.

Green, H. D., and Rapela, C. E. Cerebral vascular responses to localized and systemic hypotension induced by hemorrhage and shock. D. Shepro and G. F. Fulton (eds.), in *Microcirculation as Related to Shock*. New York: Academic, 1968. Pp. 93–119.

Hackel, D. B., Ratliff, N. B., and Mikat, E. The heart in shock. *Circ. Res.* 35:805–811, 1974.

Hinshaw, L. B. Myocardial function in endotoxin shock. *Circ. Shock* (Suppl. 1):43–51, 1979.

Hinshaw, L. B., Archer, L. T., Black, M. R., Elkins, R. C., Brown, P. P., and Greenfield, L. J. Myocardial function in shock. *Am. J. Physiol.* 226:357–366, 1974.

Johnston, P. A., and Selkurt, E. E. Effect of hemorrhagic shock on renal release of prostaglandin E. *Am. J. Physiol.* 230:831–838, 1976.

Lefer, A. M. Role of myocardial depressant factor in the pathogenesis of hemorrhagic shock. *Fed. Proc.* 29:1836–1847, 1970.

Lefer, A. M. Role of the prostaglandin-thromboxane system in vascular homeostasis during shock. *Circ. Res.* 6:297–303, 1979.

Lefer, A. M., Saba, T. M., and Mela, L. M. (eds.) *Advances in Shock Research.* New York: A. R. Liss, Inc., 1979. Vol. 1.

Matthews, R. E., and Douglas, G. J. Sulphur-35 measurement of functional and total extracellular fluid in dogs in hemorrhagic shock. *Surg. Forum* 20:3–5, 1969.

McCaig, D. J., Kane, K. A., Bailey, G., Millington, P. F., and Paratt, J. R. Myocardial function in feline endotoxin shock: Correlation between myocardial contractility, electrophysiology, and ultrastructure. *Cir. Shock* 6:201–211, 1979.

Parker, J. L., and Emerson, T. E., Jr. Cerebral hemodynamics, vascular reactivity, and metabolism during canine endotoxin shock. *Circ. Res.* 4:41–53, 1977.

Priano, L. L., Miller, T. H., and Traber, D. L. Use of prostaglandin E_1 in treatment of experimental hypovolemic shock. *Circ. Shock* 1:221–230, 1974.

Rothe, C. F. Heart failure and fluid loss in hemorrhagic shock. *Fed. Proc.* 29:1854–1860, 1970.

Sayeed, M. M., Wurth, M. A., Chaudry, I. H., and Baue, A. F. Cation transport in the liver in hemorrhagic shock. *Circ. Shock* 1:195–207, 1974.

Selkurt, E. E. Primate kidney function in hemorrhagic shock. *Am. J. Physiol.* 217:955–961, 1969.

Selkurt, E. E. Status of investigative aspects of circulatory shock. *Fed. Proc.* 29:1832–1835, 1970.

Selkurt, E. E. Current status of renal circulation and related nephron function in hemorrhage and experimental shock: I. Vascular mechanisms; II. Neurohumoral and tubular mechanisms. *Circ. Shock* 1:3–15, 89–97, 1974.

Selkurt, E. E. Role of the Kidney and Lung in the Handling of Prostaglandin E in Hemorrhagic Shock. In W. Schumer, J. J. Spitzer, and B. E. Marshall (eds.), *Advances in Shock Research.* New York: A. R. Liss, 1979. Vol. 2, Pp. 159–177.

Shizgal, H. M., Lopez, G. A., and Gutelius, J. R. Extracellular fluid volume changes following hemorrhagic shock. *Surg. Forum* 18:35–36, 1967.

Shubin, H., Weil, M. H., and Carlson, R. W. Bacterial shock. *Am. Heart J.* 94:112–114, 1977.

Slonim, N., and Stahl, W. M., Jr. Sodium and water content of connective versus cellular tissue following hemorrhage. *Surg. Forum* 19:53–54, 1968.

Wiggers, C. J. *Physiology of Shock.* New York: Commonwealth Fund, 1950.

VENTILATORY AND GAS EXCHANGE FUNCTION OF THE LUNG

19

Rodney A. Rhoades

Cellular organisms can readily obtain O_2 and excrete CO_2 simply by diffusion between the cell and its surrounding environment. Humans and other higher organisms need increasingly complex specialized cardiovascular and respiratory systems to supply the cells with adequate nutrients and O_2 and to remove CO_2 and other waste products. Previous chapters have shown that a major function of the cardiovascular system is to provide an adequate amount of blood at any given moment to all of the tissues of the body. In this chapter it will be demonstrated that one of the primary functions of the lung is to *arterialize mixed venous blood* coming from the right side of the heart.

The process of arterializing (i.e., taking up O_2 and excreting CO_2) mixed venous blood in the lung is accomplished by *gas exchange*, which includes a number of different steps. The first is *ventilation*, a cyclic process of inspirations and expirations in which gas is delivered to and extracted from the alveoli. The second is *diffusion*, a process of moving gases across the alveolar-capillary membrane. The third is *pulmonary blood flow*, which involves carrying gases out of

the lung. The fourth is matching of ventilation and blood flow in various regions of the lung to provide efficient gas exchange. The term *ventilation perfusion ratio* is often used to assess regional matching of ventilation and blood flow. The fifth step is *gas transport*, the carrying of O_2 and CO_2 by the blood.

The steps involved in gas exchange are shown in Figure 19-1. This chapter will cover these five steps, while Chapter 20 will deal with the mechanical aspects of respiration, i.e., the forces involved in supporting the movement of the lungs and chest wall. Chapter 21 will deal with the control of breathing, namely, those mechanisms that regulate gas exchange function of the lung.

LUNG STRUCTURE-FUNCTION RELATIONSHIPS

The primary function of the lung is for gas exchange, i.e., to allow O_2 to move from the lung into the venous blood and CO_2 to move from the venous blood to the lung. The lung has secondary functions as well. It plays an important role

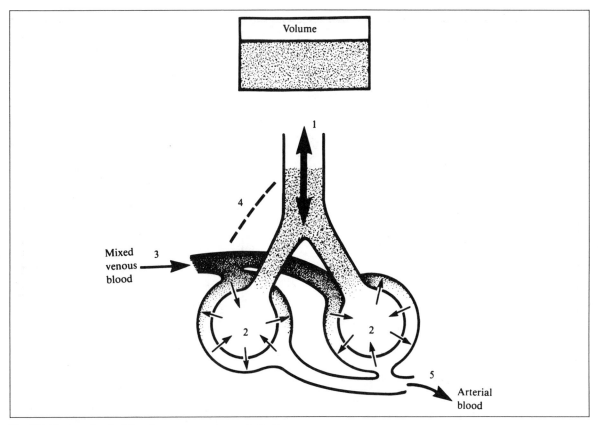

Fig. 19-1. Processes involved in pulmonary gas exchange. 1. Ventilation. 2. Diffusion across alveoli. 3. Pulmonary blood flow. 4. Matching of ventilation and blood flow. 5. Gas transport. (Adapted from J. H. Comroe. *Physiology of Respiration* [2nd ed.]. Chicago: Year Book, 1977. P. 1.)

in regulating pH and in filtering debris and air emboli in the circulation, and it serves as a blood reservoir. In addition, the lung metabolizes many vasoactive hormones, including prostaglandins, histamine, serotonin, and angiotensin. The lung also has a formidable defense mechanism to protect itself as well as the body against hostile environments. However, since the cardinal function of the lung is gas exchange, the architectural design is reflected in this function.

The right and left lung are housed in the chest, or *thoracic cavity,* and are separated by the *mediastinum.* The lung is a highly organized structure consisting basically of three components: *airways, blood vessels,* and the *elastic connective component.* The elastic connective component consists primarily of elastin and collagen that is synthesized by pul-

monary fibroblast. The conducting airways consist of a series of branching tubes that penetrate the lung.

The idealized architecture of the human airways is seen in Figure 19-2. Airway generation can be divided into two zones, the *conducting zone* and *respiratory zone*. From Figure 19-2, one can see that in the conducting zone the branching tubes become more numerous, shorter, and narrower. The first four generations are subjected to the full effect of thoracic pressure and contain a considerable amount of cartilage to prevent airway collapse. In the trachea and main bronchi the cartilage consists of U-shaped rings. In the lobar and segmental bronchi the cartilaginous rings give way to small plates of cartilage. In the bronchioles, cartilage disappears altogether, and the airways are embedded in the parenchyma. The bronchioles are suspended somewhat like guy wires on a bell tent, and it is the elasticity of lungs that keeps these airways open. Thus, the diameter of the airways from the bronchioles on down are influenced by lung volume, while airways above the bron-

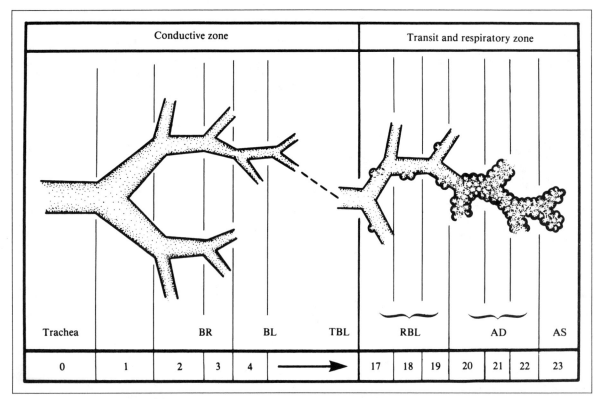

Conductive zone						Transit and respiratory zone						
Trachea			BR	BL	TBL	RBL			AD			AS
0	1	2	3	4	→	17	18	19	20	21	22	23

Fig. 19-2. Idealized architecture of the human airways. The numbers are "generations" of proliferations. BR = bronchi; RBL = respiratory bronchioles; BL = bronchioles; AD = alveolar ducts; TBL = terminal bronchioles; AS = alveolar sacs. (Adapted from E. R. Weibel. *Morphometrics of the Human Lung*. Berlin: Springer-Verlag, 1963. P. 111.)

chioles are influenced primarily by thoracic pressure. The major function of the conducting zone is to warm and humidify inspired air, distribute air to all regions of the lung, and provide the first line of defense against dust particles and irritant gases. The conducting zone has its own separate circulation, the *bronchial circulation,* which comes off the descending aorta and drains into the pulmonary veins.

The last seven generations constitute the respiratory zone and is the site of gas exchange. The branching ends into blind alveolar sacs. The adult lung contains about 300 million alveoli but the number ranges from 200 to 600 million, correlating with the height of the subject. The number of alveoli increase from birth until adolescence. Alveolar diameter is not uniform throughout the lung. In an upright person, alveoli are larger in the upper part and are smaller at the base of the lung. Because of the architectural design, a tremendous amount of surface area is housed in a relatively small volume in the chest. If the surface area of alveolar membrane were stretched out as a continuous sheath, it would have a surface area about the size of a tennis court

and represents one of the largest biological membranes in the body. The alveolar capillary membrane is exceedingly thin (approximately 0.5 μm), and because of its contact with the external environment it is very susceptible to injury from irritant pollutants. The alveolar- capillary membrane is referred to as the *blood-gas barrier* and is the site of gas exchange. Air is brought to one side of the barrier and blood to the other side. Because of its large surface area, the lung is very efficient in gas exchange.

Like the conducting zone, the respiratory zone also has its own separate and distinct circulation, the *pulmonary circulation.* Blood flow is very fast in pulmonary circulation, with red cells staying in the capillaries less than 1 second. Nevertheless, blood cells are capable of traversing two to three alveoli before they are collected in the veins.

VENTILATION

Ventilation deals with the aspect of gas exchange that involves the delivery of air to, and the extraction of air from, the alveoli. Before examining gas movement in the lung, we first need to be aware of the variety of symbols that respiratory physiologists use when quantitative aspects of lung functions are to be manipulated algebraically. Some of the basic symbols used in pulmonary physiology are shown in Table 19-1, broken down into primary and secondary symbols with examples shown. These symbols will be used in the chapters on respiration, and using Table 19-1 as a reference will avoid confusion.

SPIROMETRY AND LUNG VOLUMES. First, we need to consider some of the basic lung volumes before examining gas movement in detail. These volumes include both static (anatomical) and dynamic (functional) volumes. Measurement of these volumes, especially the dynamic volumes, is frequently used to assess the ventilatory function of the lung. Figure 19-3 is a schematic rendering of a spirometer, a simple gas volume recorder. The spirometer that is frequently used in the laboratory consists of a double-walled cylinder into which a bell is immersed in water to form a seal. The bell is attached by a pulley to a pen that writes on a rotating drum. As the subject inhales and exhales through the mouthpiece, the bell moves up and the pen down. To determine lung volumes, the subject is seated and breathes quietly from a spirometer. Volume is plotted on the ordinate or y axis and time on the abscissa or x axis. With inspiration the pen shows an upward deflection on the *spirograph* record and with expiration, a downward deflection.

Tidal volume (V_T) is the volume of air inhaled or exhaled with each breath. The volume of air remaining in the lungs at the end of a normal expiration is the *functional residual capacity* (FRC). This volume is maintained by the opposing inward recoil force of the lung and outward recoil force of the chest wall while the respiratory muscles are at rest. The significance of this will be seen in Chapter 20.

A point of terminology is important at this juncture. When referring to the subdivisions of lung volume the terms *volume* and *capacity* are both used. *Capacity* is used when a volume can be broken down into two or more smaller volumes; for example, *expiratory reserve volume* (ERV) and *residual volume* (RV) = FRC. *Inspiratory capacity* (IC) is the maximal volume of air that can be inhaled from FRC and is made up of two subdivisions: V_T and *inspiratory reserve volume* (IRV). The IRV is the maximum amount of air that can be inspired from the end-inspiratory position; ERV is the maximal amount of air that can be expired from

Table 19-1. Symbols used in pulmonary physiology

GASES

Primary symbols		Examples	
P	= gas pressure	$P_{A}O_2$	= alveolar O_2 pressure
F	= fractional concentration in dry gas phase	PcO_2	= mean capillary O_2 pressure
f	= respiratory frequency (breaths/unit time)	V_A	= volume of alveolar gas
V	= gas volume	$\dot{V}O_2$	= O_2 consumption per minute
$\dot{V}$	= gas volume/unit time	F_IO_2	= fractional concentration of O_2 in inspired gas
D	= diffusing capacity		
R	= respiratory exchange ratio	DO_2	= diffusing capacity for O_2 (ml O_2/mm Hg per minute)
STPD	= 0°C, 760 mm Hg, dry	R	= $\dot{V}CO_2/\dot{V}O_2$

Secondary symbols	Examples	
E = expired gas	$\dot{V}_A$	= alveolar ventilation per minute
I = inspired gas		
A = alveolar gas	V_T	= tidal volume
T = tidal gas	V_D	= volume of dead space gas
B = barometric		
D = dead space gas	P_B	= barometric pressure
	V_ECO_2	= volume of expired gas
	F_ICO_2	= fractional concentration of CO_2 in inspired gas

BLOOD

Primary symbols		Examples	
$\dot{Q}$	= volume of blood	Qc	= volume of blood in pulmonary capillaries
$\dot{Q}$	= volume flow of blood/unit time	$\dot{Q}c$	= blood flow through pulmonary capillaries per minute
S	= % saturation of Hb with O_2 or CO	CaO_2	= ml O_2 in 100 ml arterial blood
C	= concentration of gas in blood phase	$S\bar{v}O_2$	= saturation of Hb with O_2 in mixed venous blood

Secondary symbols	Examples	
a = arterial blood	$P\bar{v}O_2$	= partial pressure of O_2 in mixed venous blood
v = venous blood		
c = capillary blood	PcCO	= partial pressure of CO in pulmonary capillary blood
	$PaCO_2$	= partial pressure of CO_2 arterial blood

Overbar (-) indicates a mean value; dot (`) indicates a time derivative.
Source: American College of Chest Physicians: American Thoracic Society. Pulmonary terms and symbols. A report of the ACCP-ATS Joint Committee on Pulmonary Nomenclature. *Chest* 67:583–593, 1975.

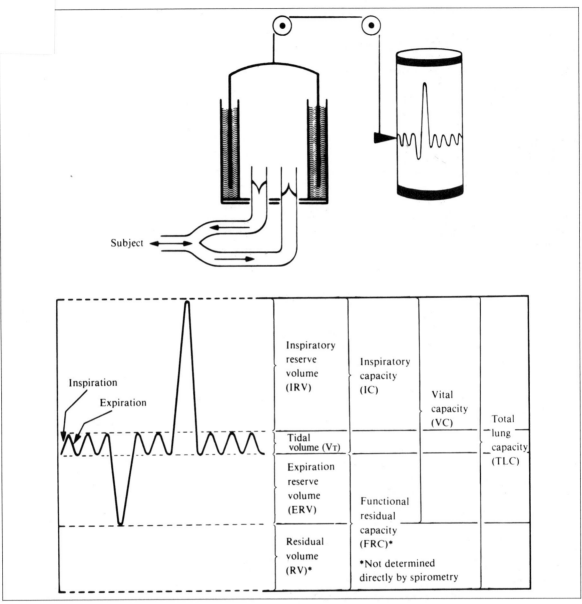

Fig. 19-3. Spirometer and lung volumes. Upper panel shows a cross section of a spirometer. Lower panel shows lung volumes and subdivisions.

Table 19-2. Definitions of standard lung volumes and capacities

Term	Definition
V_T (tidal volume)	Volume of air inhaled or exhaled with each breath
VC (vital capacity)	Volume of air that can be expired after maximal inspiration
IC (inspiratory capacity)	Maximal volume of air that can be inspired from resting expiratory level
FRC (functional residual capacity)	Volume of air in lungs at resting end-expiratory level
RV (residual volume)	Volume of air in lungs at end of maximal expiration
IRV (inspiratory reserve volume)	Volume of air that can be inspired from end-tidal inspiration
ERV (expiratory reserve volume)	Maximal volume of air that can be expired from resting expiratory level
TLC (total lung capacity)	Volume of air in lungs at end of maximal inspiration

the end-expiratory position. *Total lung capacity* (TLC) is the volume of air contained in the lung following a maximal inspiration, and *vital capacity* (VC) is the maximal amount of air that can be expired following a maximal inspiration. These lung volumes are defined in Table 19-2.

When expiration is performed as rapidly as possible, this volume is called *forced vital capacity* (FVC), one of the most commonly used lung volume measurements to assess ventilatory function. In the measurement of FVC, the subject inspires maximally to TLC, then exhales into a spirometer as forcefully, rapidly, and completely as possible (Fig. 19-4). The volume forcibly exhaled in 1 second is termed the *1-second forced expiratory volume* (FEV_1). This volume has the least variability of the measurements obtained from the forced expiratory maneuver and is considered to be one of the most reliable measurements of lung function. Another useful measurement is the FEV_1 expressed as a percentage of the FVC (FEV_1/FVC%). Normally, FEV_1 is about 80 percent of the FVC. In lung disease, two general patterns can be distinguished: In restrictive patterns, the lung is stiffer (such as seen in pulmonary fibrosis), which results in a decrease in both FEV_1 and FVC. However, the ratio FEV_1/FVC% is normal or increased. In obstructive patterns, the airways are partially obstructed (e.g., as in bronchial asthma), and FEV_1 is reduced proportionately more than FVC, giving a low FEV_1/FVC (Fig. 19-4).

An additional measurement commonly used to assess

lung ventilatory function is the mean rate of airflow over the middle half of the FVC (between 25 and 75 percent), as seen in Fig. 19-4. In the past this value has been termed maximal midexpiratory flow rate but now is referred to as *forced midexpiratory flow* (FEF 25–75%). The FEF 25–75% is obtained by identifying the 25 and 75 percent volume points and determining the time it takes to exhale the volume between these points. The volume obtained from the y axis divided by the time from the x axis gives the flow in liters per second. Another useful way of looking at forced expiration is with a *flow-volume curve,* as seen in Figure 19-5. At large volumes (close to TLC) maximal expiratory flow is effort dependent and consequently varies with the degree of force exerted by the subject. In contrast, maximal effort is not required to achieve maximal flow rates at intermediate and at low lung volumes. The FEV 25–75%, which does not take into account the first portion of expiration, provides a good measure of airway resistance. With airway obstructions (as seen with asthma and bronchitis), the RV and total lung volume are increased. Despite larger lung volumes, expiratory flow is markedly reduced. In restrictive disorders, all lung volumes are reduced, but airflow is usually normal or slightly reduced. These changes are summarized in Figure 19-6. Some of the ramifications of these changes will be discussed in Chapter 20.

MEASUREMENT OF FUNCTIONAL RESIDUAL CAPACITY. Because the lung cannot be emptied completely by maximal expiration, there are two lung volumes that cannot be measured directly by spirometry. These include FRC and RV. In practice, RV is obtained by subtractng ERV from FRC. For the measurement of FRC, three techniques are used: *closed-circuit helium dilution, open-circuit nitrogen washout,* and *the body plethysmograph.* The *closed-circuit helium dilution* method (Fig. 19-7) involves the dilution of helium, an inert, insoluble gas. The subject is connected to a spirometer filled with 10 percent helium in air. Starting precisely at the end-expiratory position, the subject begins to breathe from the spirometer. The CO_2 is reabsorbed by soda lime, and the O_2 consumed is replaced by adding O_2 to the spirometer. After several breaths the helium concentrations in the lungs and spirometer become the same. Since negligible amounts of helium have not been lost from the lungs, the volume of helium in the system does not change. Consequently, the initial concentration of helium ($He_{initial}$) multiplied by the volume of gas in the spirometer at the beginning of the test ($V_{spirometer}$) equals the final concentration of helium (He_{final}) multiplied by the volume of gas in the spirometer at the end of the test plus the volume of air in the lung (FRC). The equation can be written as follows:

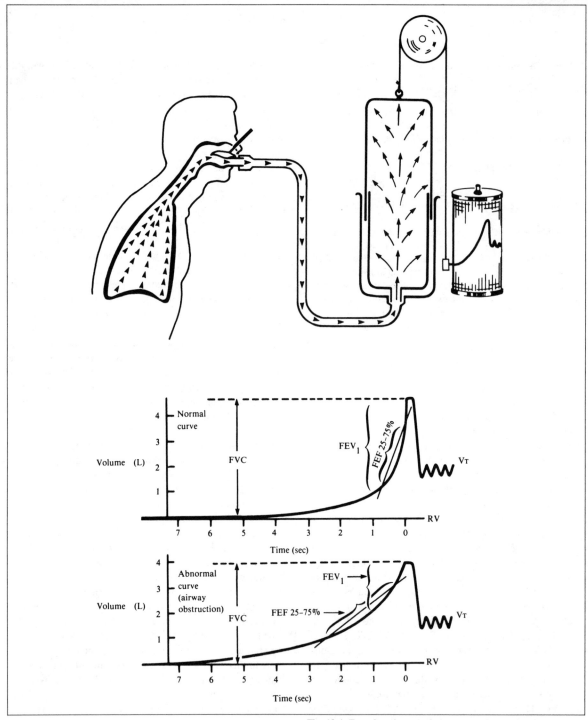

Fig. 19-4. Forced expiratory vital capacity. Subject inspires maximally to total lung capacity and then exhales into a spirometer as forcefully and as completely as possible. V_T = tidal volume; FVC = forced vital capacity; FEV_1 = forced expiratory volume in 1 second; FEF = forced midexpiratory flow (25–75%).

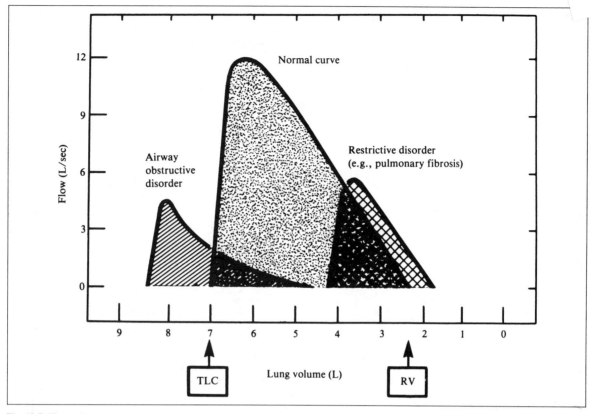

Fig. 19-5. Flow-volume curve obtained by measuring flow rate versus lung volume during forced expiration from maximal inspiration. In both obstruction and restriction the *maximal flow rate* is decreased. Note that in obstruction, despite larger lung volumes (TLC), the maximal flow rate is decreased. In contrast, with restriction, all lung volumes are reduced.

$$He_{initial} \times V_{spirometer} = He_{final} \times (V_{spirometer} + FRC)$$

Solving for FRC, the equation is written as

$$FRC = \frac{V_{spirometer} \times (He_{initial} - He_{final})}{He_{final}}$$

In the *open-circuit nitrogen washout* technique (Fig. 19-8), nitrogen is completely displaced from the lungs during a period of 100 percent O_2 breathing. The subject breathes 100 percent O_2 for about 7 minutes, and the expired gas is collected in a large spirometer. The nitrogen concentration in the spirometer is determined, and the volume of exhaled nitrogen is calculated. In the example shown in Figure 19-8, at the end of the test, the spirometer contains 40,000 ml of gas and the N_2 concentration was 5 percent. Therefore, the spirometer contains $0.05 \times 40,000 = 2000$ ml of nitrogen, all which came from the lungs. Since 2000 ml of nitrogen in the lung existed as 80 percent nitrogen, the unknown lung volume (FRC) was $2000 \times 100/80 = 2500$ ml.

These two techniques, helium dilution and nitrogen washout, have a common limitation: Equilibration may not occur in subjects in whom regions of the lung are poorly ventilated (e.g., plugged airways). These areas of the lung using helium dilution or nitrogen washout would give falsely low values for FRC.

An entirely different approach to measuring FRC that overcomes this problem of trapped gas in the lung is the use of the *body plethysmograph*, commonly called the body box. This is an airtight box in which a subject is seated (Fig. 19-9). The pressure inside the box can be measured very accurately. The subject breathes in and out against a closed mouthpiece at a particular volume. During inspiration against a closed mouthpiece, the pressure inside the lung decreases and the volume increases. Because the body box

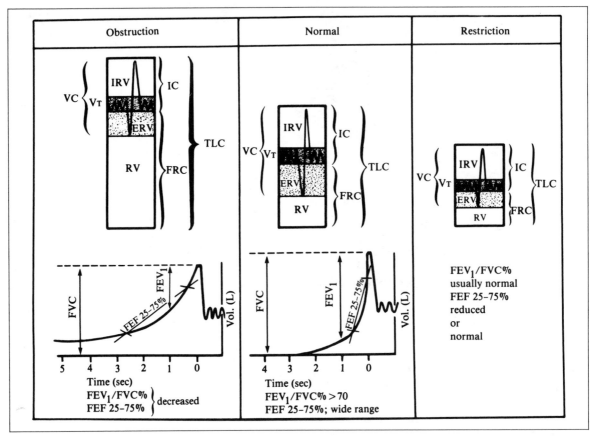

Fig. 19-6. Pulmonary function tests showing changes in lung volumes and forced vital capacities with obstructive and restrictive lung diseases. Note that $FEV_1/FVC\%$ is normal with restriction, since both FEV and FVC are decreased. Note also that $FEF_{25-75\%}$ may be reduced or normal with restrictive disorders.

is airtight, the resulting increase in lung volume is reflected by an increase in pressure within the plethysmograph. An expiratory effort against a closed mouthpiece produces the opposite effects. Thus, pressure change in the box is employed to measure a volume within the lung (FRC). In Figure 19-9 the lung volume is determined by applying Boyle's law, which states that for a gas at a constant temperature, the product of pressure and volume is constant.

ALVEOLAR VENTILATION AND MINUTE VOLUME. So far, we have been concerned only with volume or spaces. However, gas exchange is a dynamic process, and we must therefore consider volume inspired per unit time. If the respiratory system is moving 500 ml into the lung with each breath (V_T), and if respiratory frequency (f) is 15 breaths per minute, then the total volume inspired that enters the lung each minute is $500 \times 15 = 7500$ ml/per minute. This volume of air expired per minute is known as *minute volume* ($\dot{V}$).

$$\dot{V} = V_T \cdot f, \text{ or } \dot{V} = V_E/time$$

It is important to realize that not all of the air taken into the lungs reaches the alveoli, i.e., the site of gas exchange. Thus, during ventilation the V_T is distributed between the conducting airways and alveoli. For each 500 ml inhaled, approximately 150 ml remains in the conducting airways and is not involved in gas exchange. This volume is known as *dead-space volume* (V_D). Thus, the volume of "fresh

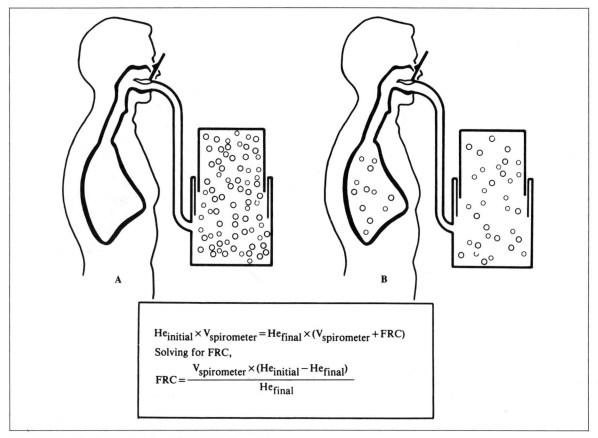

$$\text{He}_{initial} \times V_{spirometer} = \text{He}_{final} \times (V_{spirometer} + \text{FRC})$$

Solving for FRC,

$$\text{FRC} = \frac{V_{spirometer} \times (\text{He}_{initial} - \text{He}_{final})}{\text{He}_{final}}$$

Fig. 19-7. Functional residual volume measured by helium dilution. A. The spirometer is filled with mixture of He and O_2. B. The subject breathes from the spirometer until the helium concentration equilibrates between lung and spirometer.

air'' entering the alveoli is $(500 - 150) \times 15 = 5250$ ml per minute and is termed *alveolar ventilation* ($\dot{V}_A$).

$$\dot{V}_A = (V_T - V_D) \cdot f$$

Alveolar ventilation is of fundamental importance because it represents the amount of fresh air that participates in gas exchange.

Since V_T is distributed between both the conducting airways and alveoli, it can be expressed as $V_T = V_D + V_A$ (Fig. 19-10). (Note: V_A is the volume of alveolar gas in tidal volume and not the total volume of alveolar gas in the lung.)

Multiplying each term by respiratory frequency (f) gives

$$V_T \cdot f = V_D \cdot f + V_A \cdot f$$

If we substitute

$$\dot{V}_E = \dot{V}_D + \dot{V}_A$$

where $\dot{V}$ is volume of gas expired per minute, and rearrange, then

$$\dot{V}_A = \dot{V}_E - \dot{V}_D$$

It is important to note that $\dot{V}_A$ represents not the volume of air entering the alveoli per minute but the volume of *fresh* air entering them. Deep breathing causes a greater fraction of the V_T to enter the alveoli, whereas shallow breathing

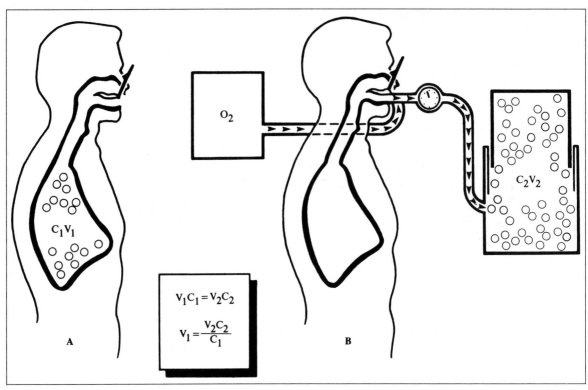

Fig. 19-8. Lung volume determined by nitrogen washout method. A. Breathing air prior to the test. B. After breathing 100% O_2 for 7 minutes. V_1 = volume of the lungs (to be determined); C_1 = concentration of nitrogen in the lungs at the beginning of the test (assumed to be 80%, as in room air); V_2 = volume of gas in the spirometer at the end of the test (measured); C_2 = concentration of nitrogen in the spirometer at the end of the test (measured). (Adapted from C. A. Guenter, M. H. Welch, and J. C Hogg. *Clinical Aspects of Respiratory Physiology.* Philadelphia: Lippincott, 1978. P. 104.)

causes a smaller fraction of the V_T to enter the alveoli. Alveolar ventilation is difficult to determine because the anatomical dead space is not easily measured. Often, dead space is approximated for a seated subject by assuming that dead space (ml) is equal to the subject's weight in pounds (e.g., a subject who weighs 170 lb would have a dead-space volume of 170 ml). The value for dead space assumed in this manner is fairly reliable.

Alveolar ventilation can also be measured in a normal subject from CO_2 concentration in the expired gas (Figure 19-11). Carbon dioxide is formed in the tissues and is carried by the venous blood to pulmonary capillaries, where it enters the alveoli and is expired. Because no gas exchange occurs in the conducting airways, and since the inspired air contains essentially no CO_2, then we can assume that all the CO_2 in the expired air originates from the alveoli. Therefore,

$$\dot{V}_{CO_2} = \dot{V}_A \cdot F_{A}CO_2$$

where $\dot{V}_{CO_2}$ is the volume of CO_2 expired per minute, and

$F_{A}CO_2$ is the fractional concentration of CO_2 in alveolar gas. Rearranging,

$$\dot{V}_A = \frac{\dot{V}_{CO_2}}{F_{A}CO_2}$$

From this relationship, alveolar ventilation can be obtained by dividing CO_2 output by the fractional concentrations of CO_2 in the alveolar gas. The alveolar CO_2 concentration can be obtained by sampling the gas at the mouth. The last portion of the V_T during expiration, called *end-tidal volume*, contains CO_2 from the alveolar region.

It is important to note that the partial pressure of CO_2 is

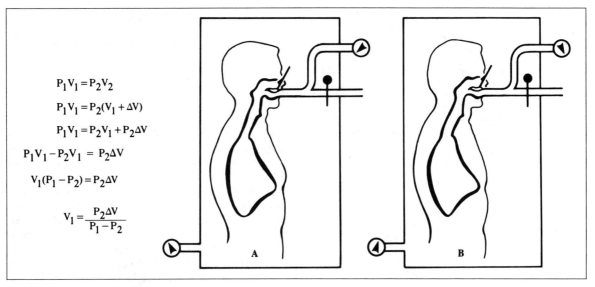

$$P_1V_1 = P_2V_2$$

$$P_1V_1 = P_2(V_1 + \Delta V)$$

$$P_1V_1 = P_2V_1 + P_2\Delta V$$

$$P_1V_1 - P_2V_1 = P_2\Delta V$$

$$V_1(P_1 - P_2) = P_2\Delta V$$

$$V_1 = \frac{P_2\Delta V}{P_1 - P_2}$$

Fig. 19-9. Measurement of FRC with a body plethysmograph. A. Resting expiration. B. Inspiratory effort. P_1 = alveolar pressure during resting expiration (measured at the mouth); V_2 = volume of lungs during inspiratory effort (unknown); $\Delta V = V_1 - V_2$ (measured as box pressure change). (Adapted from C. A. Guenter, M. H. Welch, and J. C. Hogg. *Clinical Aspects of Respiratory Physiology*. Philadelphia: Lippincott, 1978. P. 105.)

denoted as PCO_2 and is proportional to its fractional concentration in the alveoli (F_ACO_2), or

$$P_ACO_2 = F_ACO_2 \cdot K$$

where K is a constant. Substituting,

$$\dot{V}_A = \frac{\dot{V}_{CO_2} \cdot K}{P_ACO_2}$$

Since P_ACO_2 is in equilibrium with the partial pressure of CO_2 in the blood ($PaCO_2$), arterial blood can be used to calculate $\dot{V}_A$, as follows:

$$\dot{V}_A = \frac{\dot{V}_{CO_2}}{PaCO_2}$$

It is important to recognize the inverse relationship between $\dot{V}_A$ and $PaCO_2$ (Fig. 19-12). If alveolar ventilation is halved, $PaCO_2$ will double (assuming a steady state and that CO_2 production remains unchanged). A decrease in alveolar ventilation is termed *hypoventilation*. A clinical example of hypoventilation is seen in a comatose patient following a drug overdose. Conversely, increased ventilation is referred to as

Fig. 19-10. Distribution of tidal volume (VT) between conducting airways and alveoli.

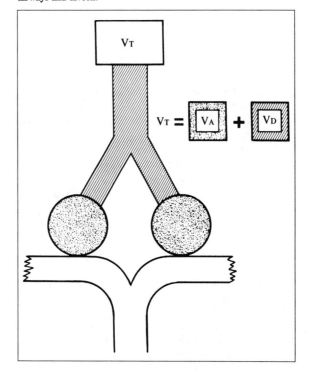

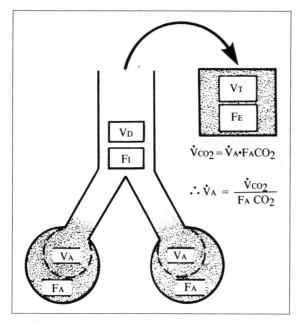

Fig. 19-11. Tidal volume (VT) is a mixture of gas from the anatomical dead space (VD) and a contribution from the alveolar gas (VA). The concentrations of CO₂ are shown by stippling. F = fractional concentration; I = inspired; E = expired.

hyperventilation. The two examples shown in Figure 19-12 illustrate the importance of $\dot{V}_A$ as an important determinant in regulating $PaCO_2$.

If alveolar ventilation increases, alveolar PO_2 will also increase. Doubling alveolar ventilation, however, will not lead to a doubling of alveolar PO_2. The quantitative relationship between alveolar ventilation and alveolar PO_2 is more complex than that for alveolar PCO_2 for two reasons: First, the inspired PO_2 (PIO_2) is not zero. Second, the respiratory exchange ratio R ($\dot{V}_{CO_2}/\dot{V}_{O_2}$) is usually less than 1, which means that more oxygen is removed from the alevolar gas phase per unit time than CO_2 is produced. The alveolar PO_2 can be calculated by using the *alveolar gas equation,* as follows:

$$PaO_2 = PIO_2 - PaCO_2 \left[FIO_2 + \frac{1 - FIO_2}{R} \right]$$

where FIO_2 is the fraction of oxygen, PIO_2 is the inspired PO_2, and R is the *respiratory exchange ratio.* (The derivation of this formula is outside the scope of this chapter.) In a normal resting individual breathing air at sea level whose $PaCO_2$ is 40 mm Hg and whose R is 0.82 (normal), the calculation would be

$$PaO_2 = 149 \text{ mm Hg} - 40 \text{ mm Hg} \left[0.21 + \frac{1 - 0.21}{0.82} \right]$$

$$PaO_2 = 149 \text{ mm Hg} - 40 \text{ mm Hg} (1.2) = 102 \text{ mm Hg}$$

Thus, if one knows $PaCO_2$ and multiplies by the correction factor (1.2), one can obtain an average alveolar PO_2 from a rather complex-looking equation.

Dead Space. From Figure 19-10, we have seen that dead space is the volume of gas in the conducting airways. Dead space is not just limited to airways. Any time that inspired air does not participate in gas exchange also constitutes dead space. For example, if inspired air is distributed to alveoli that have no blood flow, this constitutes dead space (Fig. 19-13). Thus, dead space may be either anatomical or alveolar in nature. The sum of the two types of dead space is called *physiological dead space* (physiological dead space = anatomical + alveolar). Alveolar dead space is not confined to alveoli in which no blood flow occurs. Gas exchange units that have reduced blood flow utilize less inspired air than usual for normal gas exchange (Fig. 19-13). Thus, any portion of alveolar air that is in excess of that needed to maintain normal gas exchange constitutes alveolar dead space.

The measurement of dead-space volume is difficult and requires an indirect approach. Dead-space volume can be measured either by a single-breath method (Fowler method) or by the Bohr equation. The Bohr equation measures physiological dead space and is based on the idea that all of the expired CO_2 originates from the alveolar region and none from the conducting airway. Therefore, the volume of CO_2 in the tidal volume can be expressed as

$$VT \cdot FeCO_2 = VA \cdot FaCO_2$$

where VA represents only that portion of alveolar air that makes up the tidal volume. As stated earlier, VT is distributed between alveolar and conducting airways.

$$VT = VA + VD$$

Rearranging,

$$VA = VT - VD$$

Substitute

$$VT \cdot FeCO_2 = (VT - VD) \cdot FaCO_2$$

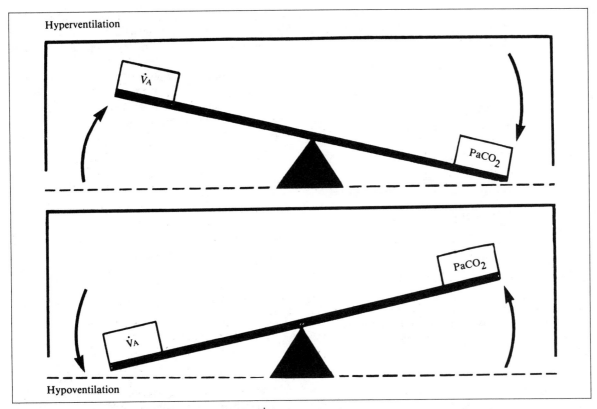

Fig. 19-12. Inverse relationship between alveolar ventilation ($\dot{V}A$) and arterial $PaCO_2$.

Therefore,

$$\frac{V_D}{V_T} = \frac{F_ACO_2 - F_ECO_2}{F_ACO_2} \quad \text{(Bohr equation)}$$

Since $FCO_2 = PCO_2 \cdot K$, and $PACO_2$ and $PaCO_2$ are in equilibrium, the Bohr equation can be rewritten as

$$\frac{V_D}{V_T} = \frac{PaCO_2 - P_ECO_2}{PaCO_2}$$

The normal ratio of dead space to tidal volume is in the range of 0.20 to 0.35 during breathing at rest. The ratio tends to increase with age but decreases with exercise.

Anatomical dead space is measured by using Fowler's single-breath method (Fig. 19-14). The subject is connected to a device in which the percentage of nitrogen can be measured at the mouth as the subject breathes. When a subject is breathing air, the percentage of nitrogen between inspirations and expirations changes very little because nitrogen is not utilized by the body. To perform the measurement, the subject, who previously has been breathing air, takes a single inspiration of O_2 and then exhales. During inspiration the nitrogen concentration drops to zero (Fig. 19-14). At the end of inspiration the dead space is filled with O_2. At the beginning of expiration the first gas to come from the mouth is 100 percent O_2 which has entered and left the dead space without any exchange with alveolar gas; therefore, the nitrogen meter continues to record 0 percent nitrogen (Fig. 19-14). At mid expiration the nitrogen concentration increases as the dead space is washed out by alveolar gas. Finally, an almost uniform gas concentration is reached and represents pure alveolar gas. During the single-breath test the dead space is found by plotting nitrogen concentration against expired volume and drawing a vertical line such that area A equals area B (Fig. 19-14B). This vertical line represents the "theoretical front" between conducting airways and alveolar gas. Thus, the volume up to the vertical line is the volume of gas in the conducting airways (anatomical dead space). Anatomical dead space may be decreased with bronchial obstructions (asthma, bronchiolitis).

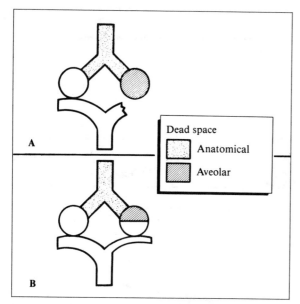

Fig. 19-13. Alveolar dead space. A. No blood flow to an alveolar region. B. Reduced blood flow to an alveolar region. Note that physiological dead space = anatomical dead space + alveolar dead space.

Dead space, anatomical dead space, alveolar dead space, and *physiological dead space* are terms that often lead to confusion. Basically, the terms denote the volume of inspired gas that does not participate in arterializing the venous blood. In one case there is a fraction of our V_T that does not reach the alveoli (anatomical dead space). In another, because a fraction of the V_T reached the alveoli with no blood flow (alveolar dead space), and in a third because too much of the V_T reached the alveoli in proportion to capillary blood flow (again, alveolar dead space). Physiological dead space represents the sum of anatomical and alveolar dead space. In normal subjects, physiological dead space is approximately the same as anatomical dead space. It is only in lung disease that physiological dead space increases because of the inability of the lung to match blood flow and ventilation in various lung regions. This relationship between alveolar ventilation and pulmonary blood flow in the lung will be discussed in more detail later in this chapter.

GAS DIFFUSION AND TRANSPORT

In the previous section, we examined the ventilatory aspects of lung volumes and gas movements from the atmosphere, conducting airways, and into alveoli. We now come

to the process of transferring gas across the blood-gas barrier and the transport of these gases by the blood. Before turning to gas transfer and transport, we need to review briefly some of the basic physical gas laws pertaining to these two processes in respiration. The first is *partial pressure. Dalton's law* states that each gas in a mixture exerts a pressure according to its own concentration, and is independent of other gases present. The pressure of each gas is *partial pressure* or sometimes is referred to as *gas tension.* The total pressure, or barometric pressure, is the sum of the individual partial pressures of the gases present and can be written as

$$P_B = P_x + P_y + P_z$$

where P_B is barometric (total) pressure and $P_x + P_y + P_z$ are partial pressures of individual gases. The partial pressure of any gas can be determined by measuring barometric pressure and fractional concentration of the gases. For example the partial pressure of oxygen (PO_2) is determined as follows:

$$PO_2 = P_B \cdot FO_2$$

where PO_2 is the partial pressure of O_2 in millimeters of mercury, P_B is barometric pressure in millimeters of mercury, and FO_2 is the fractional concentration of O_2. The partial pressure of O_2 at sea level is 159.6 mm Hg = 760 mm Hg × 0.21.

An exception to Dalton's law is the partial pressure of water vapor, known as *water vapor pressure.* Water vapor pressure at saturation is influenced exclusively by temperature rather than by total pressure. At normal body temperature (37°C), water vapor pressure is 47 mm Hg. As long as body temperature remains constant, the vapor pressure also remains constant, even when the partial pressures of other gases are expanded or compressed by changes in barometric pressure. Thus, allowances for water vapor pressure must be made when the partial pressures of gases are to be determined inside the airways and lung. Dalton's law may be modified as follows for gases inside the body:

$$P_x = (P_B - PH_2O) \cdot F_x$$

where P_x is the partial pressure of gas x, PH_2O is water vapor pressure, and F_x is the fractional concentration of gas x. The composition including partial pressures of the respired gas is seen in Table 19-3. Note the influence of water vapor pressure on inspired PO_2 (tracheal air).

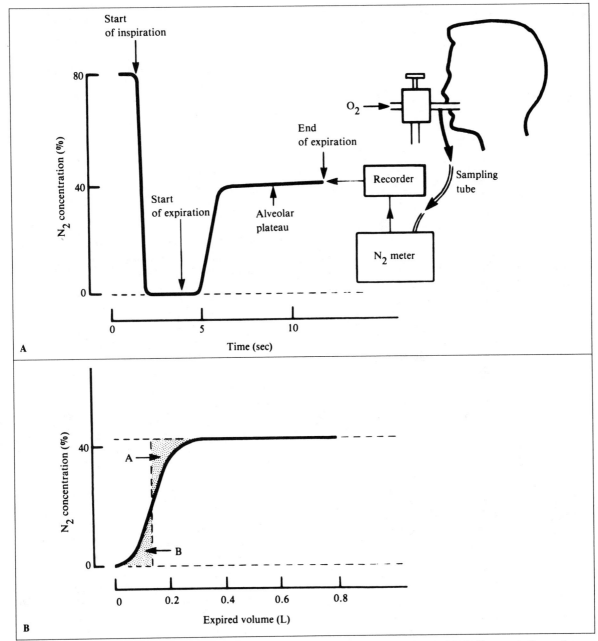

Fig. 19-14. Fowler's method for determining anatomical dead space. A. A subject takes a single breath of 100% O_2, holds his or her breath for a second, and then expires. During inspiration the N_2 meter records zero. At the beginning of expiration, pure O_2 is expired (%N_2 zero), which comes from the dead space. During the next phase of expiration, the N_2 concentration rises, which represents a washout of the remainder of dead-space air and alveolar air. In the latter part of expiration the N_2 concentration reaches a plateau and represents pure alveolar air. B. N_2 concentration is plotted against expired volume, and the dead space is the volume of air up to the vertical broken line, which makes the two shaded areas equal. (Adapted from J. B. West. *Respiratory Physiology* (2nd ed.). Baltimore: Williams & Wilkins, 1979. P. 18.)

Table 19-3. Composition of Respired Gases

Gas	Dry Air		Moist Tracheal Air		Alveolar Air	
	Partial Pressure (mm Hg)	Percent of Total	Partial Pressure (mm Hg)	Percent of Total	Partial Pressure (mm Hg)	Percent of Total
Nitrogen	600.7	79.02	563.6	74.16	568.0	74.74
Oxygen	159.1	20.95	149.2	19.63	105.0	13.82
Carbon dioxide	0.2	.03	0.2	.03	40.0	5.26
Water vapor	0.0	0.00	47.0	6.18	47.0	6.18
Total	760.0	100.00	760.0	100.00	760.0	100.00

GAS DIFFUSION. Because of their random motion, gas molecules will distribute themselves randomly throughout a given space until the partial pressure is the same everywhere. The process by which gas moves from a region of high partial pressure to one of low pressure is known as *diffusion*. *Graham's law* states that the rate of diffusion and the speed of the gas molecules are inversely proportional to the square root of its density. Thus, a gas with a lower molecular weight will diffuse faster than a gas with a higher molecular weight. *Henry's law* states that the concentration of gas that dissolves in a given solvent is directly proportional to the partial pressure of the gas, temperature remaining constant. This relationship can be written as

$$C = K \cdot P$$

where C is the concentration of gas in a liquid (ml/100 ml of blood), K is the solubility constant whose value depends on the particular gas-liquid system and temperature, and P is the partial pressure of the gas. Gases diffuse in solution just as they do in the gas phase. That is, gas moves from a region of higher partial pressure to one of low partial pressure. Also, the rate of diffusion is inversely related to the square root of the molecular weight.

Alveolar Diffusion. Gases cross the blood-gas barrier by simple diffusion, described by *Fick's law,* which states that the volume of gas per unit time diffusing across a tissue barrier is directly proportional to the surface area, a diffusion constant, and the partial pressure differences across the tissue and is inversely proportional to tissue thickness (Fig. 19-15):

$$\dot{V} \text{ gas} = \frac{As \cdot D \cdot (P_1 - P_2)}{T}$$

where As is surface area, D is the diffusion constant, $P_1 - P_2$ is partial pressure difference, and T is thickness or diffusion distance. The diffusion constant (D) is often called *diffusivity* of the gas and is proportional to solubility and inversely related to the square root of the molecular weight of the gas.

$$D = \frac{\text{solubility}}{\sqrt{mw}}$$

The diffusivity is of functional importance. First, let us examine the effect of the molecular weight of a gas. From the preceding equation it is intuitively obvious that heavier gas molecules will move more slowly and consequently will diffuse more slowly than gases of lighter molecular weight. For example, CO_2 diffuses 85 percent as fast as O_2 in the gas phase under the same pressure gradient ($P_1 - P_2$) and diffusion path (thickness). This is calculated as follows:

$$\frac{\text{Rate for } CO_2}{\text{Rate for } O_2} = \frac{\sqrt{mw\ O_2}}{\sqrt{mw\ CO_2}} = \frac{\sqrt{32}}{\sqrt{44}} = 0.85$$

The rate in which gases diffuse through a liquid phase depends not only on the molecular weight of the gas but also on its solubility in liquid. In the gas phase the concentration of a gas is directly proportional to the partial pressure alone. However, in the liquid phase the concentration of the same gas is directly proportional to the partial pressure times its solubility. Therefore, a more soluble gas will have a greater concentration difference than one that is less soluble and will tend to diffuse more easily. If we again compare O_2 and CO_2, CO_2 is about 23 times more soluble than O_2 in plasma. The diffusibility (D) for these two gases can be compared as follows:

$$\frac{DCO_2}{DO_2} = \frac{\sqrt{32}}{\sqrt{44}} \times 23 = \frac{20}{1}$$

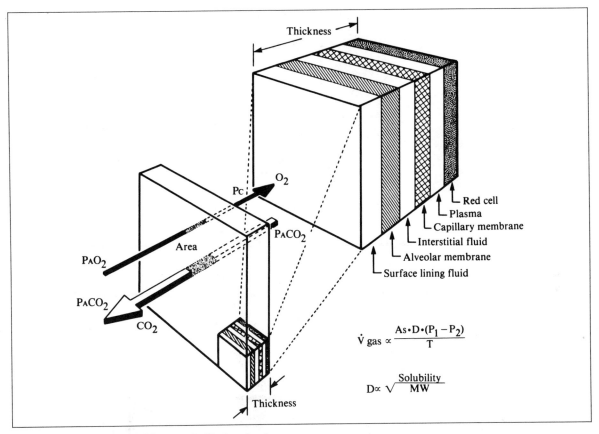

Fig. 19-15. Diffusion of O_2 and CO_2 across the blood-gas barrier. The amount of gas transferred is proportional to surface area (As), a diffusion constant (D), partial pressure gradient ($P_1 - P_2$), and inversely proportional to thickness (T) or diffusion distance. The constant, D, is proportional to the gas solubility and inversely related to the square root of its molecular weight.

Thus, CO_2 diffuses 20 times as easily as O_2 in a liquid phase, whereas it diffuses only 0.85 times as easily as O_2 in the gas phase.

The effect of diffusion of O_2 and CO_2 between alveolar gas and blood on the changes in PO_2 and PCO_2 in the capillary blood can be seen in Figure 19-16. At rest, the time that blood spends in the pulmonary capillary in which it is exposed to alveolar gas is about 0.75 second. The alveolar PO_2 and PCO_2 are 100 and 40 mm Hg respectively. The blood entering the pulmonary capillary (mixed venous blood) has a PO_2 of 40 mm Hg and a PCO_2 of 46 mm Hg. As mixed venous blood enters the capillaries, there is a large O_2 diffusion gradient (partial pressure difference between alveolus and blood) as compared with that of CO_2. Diffusion of the two gases proceeds along the concentration gradient until their respective partial pressures reach equilibrium (partial pressure in the capillary equals partial pressure in the alveolar gas). The equilibrium is achieved in about 0.3 second for O_2 and sooner for CO_2. The reason CO_2 can equilibrate sooner, even though it has a lesser partial pressure gradient ($P_1 - P_2$) is its higher diffusibility, which is about 20 times that of O_2.

DIFFUSION AND PERFUSION LIMITATIONS. Although gases cross the blood-gas barrier by simple diffusion, the rate at which gas transfer occurs across alveoli is also affected by pulmonary blood flow. Let us consider how blood flow in the pulmonary capillaries affects the transfer of respiratory gases, O_2 and CO_2, across the alveolar-capillary membrane. First, let us examine a gas (nitrous oxide) that diffuses across the blood-gas barrier and dissolves in the plasma but does not readily combine with hemoglobin in the red cell.

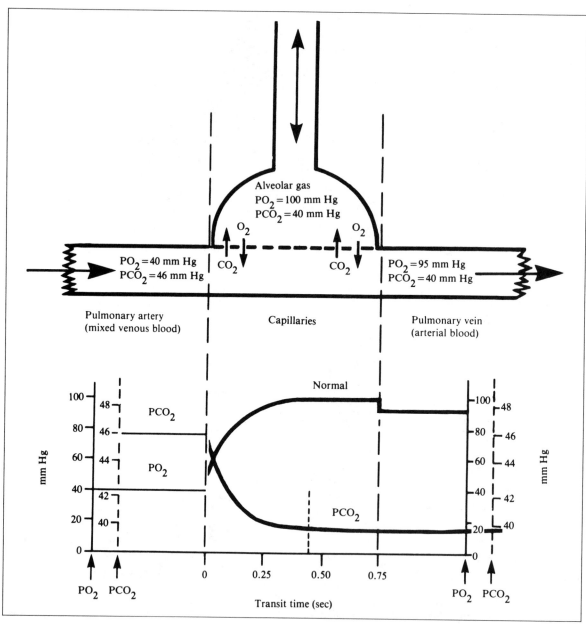

Fig. 19-16. Transfer of O_2 and CO_2 between the alveolus and capillary.

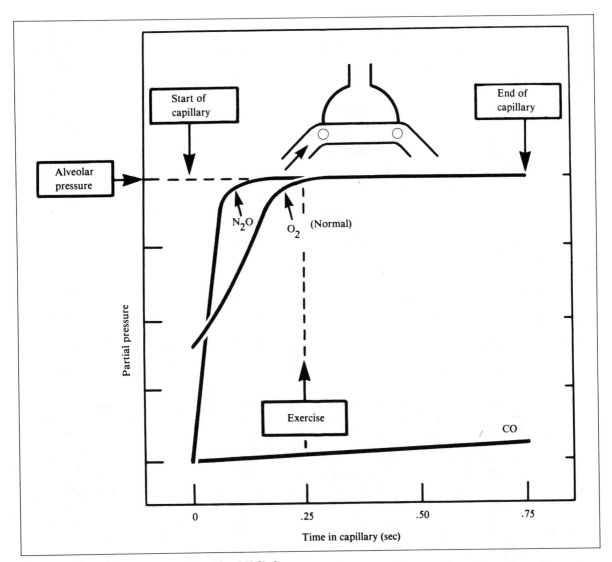

Fig. 19-17. Transfer of O_2 and two test gases (CO and N_2O) along an ideal pulmonary capillary. For N_2O the partial pressure in the blood rapidly reaches equilibrium with that of the alveolar gas very early in the capillary, so that the transfer of this gas is limited by blood flow (perfusion). In contrast, the partial pressure of CO in the blood is virtually unchanged, so that the transfer, in this case, is limited by diffusion. Oxygen transfer is perfusion limited but can be diffusion limited with thickening of the blood-gas barrier. (Adapted from J. B. West. *Respiratory Physiology*. Baltimore: Williams & Wilkins, 1979. P. 24.)

Figure 19-17 again shows the time course required for the plasma and red cells to move through the capillary. The time required for the blood to traverse the capillary is sometimes referred to as *transit time*. It is important to realize that transit time changes with cardiac output; for example, in exercise, in which cardiac output increases blood flow in the capillaries but the transit time decreases because less time is required for the blood to pass through. If a person breathes a low concentration of nitrous oxide, the nitrous oxide diffuses across the lung by Fick's law. As blood enters the capillary, the nitrous oxide diffuses across and is dissolved in the plasma, and the partial pressure in the plasma rises very rapidly and virtually reaches the partial pressure of nitrous oxide on the alveolar side by the time the plasma is only one-tenth of a way along the capillary. At this point the pressure gradient ($P_1 - P_2$) is zero, and no additional nitrous oxide is transferred unless blood flow is increased. Thus, the amount of nitrous oxide gas taken up by the plasma is almost entirely dependent on blood flow and not on the diffusion properties of the blood-gas barrier. The transfer of nitrous oxide is therefore limited by blood flow (*perfusion limited*).

Now let us consider another foreign gas such as carbon monoxide (CO), which also readily diffuses across the blood-gas barrier. Carbon monoxide has a very strong affinity for hemoglobin in the red cell. As the red cell enters the pulmonary capillary, CO moves very rapidly across the blood-gas barrier into the red cell. As it does so, CO is readily bound to hemoglobin. Consequently, CO can be taken up within the cell with little increase in plasma PCO. Thus, as the blood moves along the capillary, the PCO in the plasma changes very little. Consequently, the partial pressure gradient across the blood-gas barrier for CO never reaches equilibrium. The amount of CO transferred to the blood is therefore limited by the diffusion properties of the blood-gas barrier and not by blood flow.

Now let us consider O_2. In Figure 19-17 the time course lies between that of carbon monoxide and nitrous oxide. Oxygen also combines with hemoglobin, but not as readily as carbon monoxide, because it does not have as strong a binding affinity. When venous blood enters the pulmonary capillary, the PO_2 is already about four-tenths that of alveolar pressure because all of the oxygen was not utilized by the tissues. As venous blood traverses the pulmonary capillary, the rise in PO_2 is much greater than in PCO because of differences in the binding affinity. Under resting conditions the capillary PO_2 reaches alveolar PO_2 when the blood is about one-third of the way along the capillary. At this point there is no net transfer of oxygen. Under these conditions, oxygen transfer is like nitrous oxide transfer and limited by blood flow in the capillary (*perfusion limited*). Thus, the major way to increase the transfer of oxygen to the blood is to increase capillary blood flow, such as occurs with exercise, when cardiac output is elevated. Note that with increased cardiac output the pulmonary transit time is decreased.

Normally, the transit time at rest is about 0.75 second. With heavy exercise the time may be reduced to 0.33 second. With the decrease in transit time there is less time available to oxygenate the blood. The important thing to note from Figure 19-17 is that when a normal subject exer-

cises and breathes air there is still adequate time to oxygenate the blood. Consequently, end-capillary PO_2 rarely falls with exercise. In abnormal situations in which there is a thickening of the blood-gas barrier so that O_2 diffusion is impaired, end-capillary PO_2 may not reach equilibrium with alveolar PO_2. In this case there is a measurable difference between alveolar gas and end-capillary PO_2.

Another way of emphasizing the transfer properties of the lung is to lower O_2 partial pressure gradient. This can be accomplished by a person's breathing either a low concentration of O_2 or breathing air at altitude. In Figure 19-18 the normal O_2 diffusion gradient is 15 mm Hg (100–85 mm Hg mean pulmonary capillary). With hypoxia, PaO_2 is reduced from 100 to 50 mm Hg. Hypoxia will also reduce venous PO_2 (PvO_2) as well, so that the partial pressure of venous blood entering the pulmonary capillary is around 20 mm Hg. The important thing to note in hypoxia is that the O_2 diffusion gradient across the blood gas-barrier (P_1–P_2) is reduced. In Figure 19-18B the partial pressure difference for O_2 is reduced from 15 to 11 mm Hg; O_2 will be transferred across the blood-gas barrier at a slower rate, and consequently the rise in capillary PO_2 is less. Another factor that causes the rise in capillary PO_2 to be less in hypoxia is that O_2 does not bind to hemoglobin in linear fashion. It binds at a faster rate at lower O_2 tension, which causes the rise in capillary PO_2 to be slower. For both these reasons, end-capillary PO_2 may fail to reach equilibrium with alveolar PO_2. Thus, exercising at high altitude is one way in which diffusion impairment of O_2 occurs in a normal subject (i.e., transfer of O_2 is limited by diffusion rather than by blood flow). A patient with an already thickened blood-gas barrier, when breathing a low O_2 concentration, will show evidence of diffusion impairment, which may not be manifested when breathing air at sea level. On the other hand, CO_2 essentially always achieves equilibrium between alveolar gas and blood because of its greater diffusivity.

LUNG DIFFUSION CAPACITY (TRANSFER CAPACITY). In measuring the diffusion characteristics of the lung, it is clearly evident that the test gas to be used must be limited by diffusion and not by perfusion. Carbon monoxide is one obvious choice of gas to use. Besides selecting a test gas that is limited by diffusion, there are three additional variables we need to know before diffusion characteristics of the lung can be measured. These include lung surface area (As), blood-gas thickness (T), and diffusion coefficient for the lung (D). With a highly complex organ like the lung, readily measuring As, T, and D directly in an individual is impossible. To circumvent this problem, Fick's law can be rewritten as seen in Figure 19-19, where DL combines these

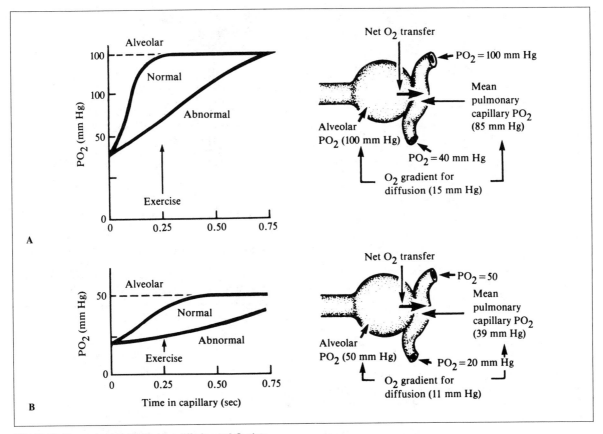

Fig. 19-18. Oxygen tension gradient for diffusion and O_2 time course in the capillary when diffusion is normal and abnormal. **A.** A normal O_2 gradient for diffusion when alveolar PO_2 is normal. **B.** Reduced alveolar PO_2 (hypoxia). With hypoxia, that O_2 gradient is reduced and causes reduced oxygenation. Note that in both cases exercise reduces the transit time.

three terms, constituting what is termed *diffusion capacity* (DL).

To determine the diffusing capacity for O_2, then, one must measure VO_2, alveolar oxygen partial pressure (PAO_2), and capillary O_2 partial pressure (PcO_2). The difficulty in measuring DLO_2 is that Pc oxygen tension changes all along the capillary. A far easier approach, which will provide the same information, is to use CO to determine the diffusing capacity of the lung. Carbon monoxide offers two distinct advantages in that (1) there is essentially no CO in the mixed venous blood, and (2) the affinity of CO for hemoglobin is 210 times greater than that of oxygen, which causes the partial pressure of CO to stay essentially zero in the pulmonary capillaries. Thus, one needs only to measure the up-

take of carbon monoxide and $PACO$. The equation then becomes

$$DL = \frac{\dot{V}CO}{PACO}$$

The diffusion capacity of oxygen and carbon monoxide can be compared by incorporating their solubility and molecular weight:

$$\frac{DLO_2}{DLCO} = \frac{\sqrt{mwCO}}{\sqrt{mwO_2}} \times \frac{\text{solubility } O_2}{\text{solubility CO}} = 1.25$$

Thus, $DLO_2 = 1.25\ DLCO$.

Two of the most common techniques for making this measurement include the *single-breath technique* and *steady-state method*. In the single-breath test the patient inspires a single dilute mixture of CO and then holds his or her breath for approximately 10 seconds. By determining the percentage of CO in the alveolar gas at the beginning

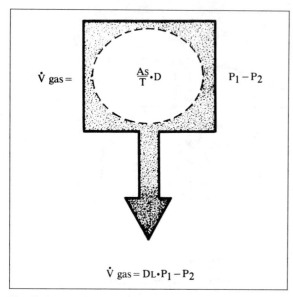

$$\dot{V}\,gas = \frac{A_S}{T} \cdot D \quad P_1 - P_2$$

$$\dot{V}\,gas = D_L \cdot P_1 - P_2$$

Fig. 19-19. Rearrangement of Fick's law.

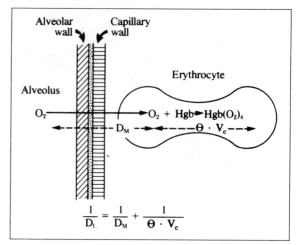

$$\frac{1}{D_L} = \frac{1}{D_M} + \frac{1}{\Theta \cdot V_c}$$

Fig. 19-20. The diffusing capacity of the lung (DL) consists of two components: (1) that due to the diffusion process itself and (2) the time required for O_2 to react with the Hb. The rate of reaction of O_2 with Hb is described by Θ, which gives the rate (ΘV_c) in ml of O_2/min/mm Hg/100 ml of blood. D_M = diffusion through the membrane(s); V_c = volume of capillary blood.

and the end of 10 seconds and by measuring lung volume, one can calculate $\dot{V}_{CO}$. The single-breath test is very reliable and is often used in pulmonary testing to measure diffusion capacity. In the second technique (steady state), a person breathes a low concentration of CO (0.1 percent) for short periods (several minutes) until a steady state is reached, at which time expired gas is collected for CO uptake measurements. P_ACO is determined simultaneously. End-tidal samples gives satisfactory P_ACO values for a healthy person but not for patients with nonuniform gas distribution. In these patients the Bohr equation can be used to estimate P_ACO. The steady-state method is commonly used for exercise studies. Normal resting values for D_LCO in healthy persons are 25 ml/min/mm Hg and can increase with exercise. D_LCO is decreased by conditions that thicken the blood-gas barrier, such as edema.

Since diffusion is gas flow per unit pressure difference, D_L can be viewed as analogous to conductance. The reciprocal, $(1/D_L)$ is analogous to resistance. The sites of resistance for diffusion of gases are at the membrane and in the red blood cell (the time required for O_2 or CO to bind chemically with hemoglobin). Thus, the transfer of O_2 can be regarded as occurring in two stages: (1) diffusion across the blood-gas barrier and (2) reaction with hemoglobin (Fig. 19-20). Each stage can be regarded as contributing to the resistance of oxygen movement. To obtain the total resistance, we can add the resistance offered by the membrane, including

plasma as well as the cytosol in the red cell, and the reaction to hemoglobin.

$$\frac{1}{D_L} = \frac{1}{D_M} + \frac{1}{\Theta \cdot V_c}$$

The resistance offered by membranes and red cell components are approximately equal. Thus, when capillary blood volume falls, total resistance increases, with a concomitant decrease in diffusion capacity. A decrease in D_LCO may therefore not necessarily indicate a defect in the alveolar-capillary diffusion of gases, but in many instances may be due to a low capillary blood volume. A low D_L indicates primarily that gas and blood do not meet efficiently.

GAS TRANSPORT

OXYGEN TRANSPORT SYSTEM. Oxygen is essential to the mechanisms by which the cell converts metabolic substrates to energy. Oxygen serves as a "sink" by which electrons flow along the respiratory chain in the mitochondria. The primary function of the O_2 transport system is *oxygenation*, which is the process in which molecular O_2 is passively extracted from the lung and delivered to the tissues. There are two distinct steps in oxygenation: (1) O_2 transport, the process by which oxygen is carried by the blood to the tissue, and (2) O_2 utilization by the cell.

Oxygen transport is the means by which blood carries O_2 from gas exchange units (alveoli) to metabolically active tissues. Oxygen is transported in two forms, dissolved and in chemical combination with hemoglobin. Because of its low solubility, O_2 that is physically dissolved is present in very low concentration. Though the quantity of O_2 is very small, it is extremely important because it establishes the partial pressure of the gas in the blood (PO_2). Arterial O_2 tension is an important indicator of the efficiency of oxygenation. The amount of O_2 that is physically dissolved is proportional to oxygen tension (partial pressure) by the following relationship: [Dissolved O_2] = $\alpha \cdot (PO_2)$, where α is the solubility coefficient of oxygen in plasma (0.03 ml/liter/mm Hg). For example, 3 ml of O_2 is dissolved in each liter of blood with a normal arterial O_2 tension of 100 mm Hg; 3 ml of O_2 per liter is totally inadequate for human metabolic needs.

Hemoglobin and Oxygen Saturation. Most of the O_2 transported is carried by hemoglobin (Hb) rather than in the dissolved form. The amount of O_2 carried by hemoglobin in the red cell is approximately 65 times as great as the amount dissolved. Thus, in a resting subject, about 95 percent of the total O_2 transported is carried by the Hb. During exercise, about 99 percent of the O_2 delivered to the tissue is carried by Hb.

Oxygen binds chemically to Hb to form oxyhemoglobin (Hb + O_2 $\rightleftharpoons$ HbO_2); 1 gm of Hb will bind with 1.39 ml of O_2. The reaction is reversible and is a function of partial pressure. The maximal amount of O_2 that can be carried by the Hb is called O_2 capacity. The carrying capacity for a normal adult is about 20.8 ml of O_2 per 100 ml of blood (1.39 ml/gm Hb × 15 gm Hb/100 ml blood). Oxygen saturation is given by

$$O_2 \text{ saturation} = \frac{O_2 \text{ combined with HB}}{O_2 \text{ capacity of Hb}} \times 100$$

Oxygen saturation is not a measure of concentration but a ratio of concentrations, i.e., the ratio of the quantity *actually bound* to the quantity that can be *potentially bound*. The potential bound is equivalent to the oxygen capacity. Oxygen saturation is the second index of oxygen transport. Arterial blood is about 97 percent saturated with O_2 (20.1 ml O_2/20.8 ml O_2/100 ml of blood). The third index to O_2 transport is O_2 content and is equal to the actual amount bound to Hb plus the amount dissolved.

There are three closely related indices to O_2 transport: arterial oxygen tension, O_2 saturation, and O_2 content. The relationship between PO_2, oxygen saturation, and oxygen content can best be seen by the O_2-Hb dissociation curve (Fig. 19-21). The sigmoid shape of the curve is due to the fact that the four binding sites on the hemoglobin molecule

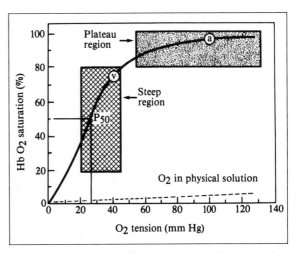

Fig. 19-21. Oxygen-hemoglobin dissociation curve. The "driving" pressure for loading O_2 onto Hb is arterial O_2 tension, which is represented by the amount of O_2 physically dissolved in solution: a = arterial; v = venous; P_{50} = O_2 tension required to saturate 50% of the hemoglobin.

interact with one another. That is to say, when the first site is bound to a molecule of O_2, the binding of the next site is facilitated; the binding of O_2 to the next site facilitates the third; and so on. As a result of this, the relationship between PO_2 and O_2 saturation is sigmoid over a PO_2 of 0 to 100 mm Hg. Thus, the affinity to bind with O_2 increases progressively as more O_2 is added on.

The shape of the O_2-Hb dissociation curve has several physiological advantages. The plateau region of the curve can be called the "association part" and occurs in the lung. At the flat portion of the curve, the important advantage is that saturation and O_2 content remain fairly constant despite wide fluctuations in alveolar PO_2 (± 20 mm Hg). For example, if PaO_2 is only 80 mm, Hb would still be over 95 percent saturated; if PaO_2 were to rise to 120 mm Hg, Hb would be only slightly more saturated (98 percent). For this reason, O_2 content cannot be raised appreciably by increased ventilation (hyperventilation). The steep phase, or what might be called the "dissociation phase," of the curve allows large quantities of O_2 to be released (dissociated) at the lower capillary PO_2 that prevails at the tissue. In brief, the blood can saturate Hb under high partial pressures in the lung, yet can give up (dissociate) the greatest amount of O_2 with small changes in PO_2 at the tissue level.

The relationship between PO_2, O_2 saturation, and O_2 content is important to grasp. For example, an anemic patient who has an Hb concentration of 7 gm per 100 ml of blood can have a normal arterial PO_2 of 95 mm Hg and a normal O_2 saturation. For an anemic person to have a normal O_2 saturation (98 percent) means that O_2 content is reduced (10 ml/100 ml).

The affinity in which Hb binds to O_2 is affected by several factors. Changing the binding affinity shifts the O_2-Hb dissociation curve to the right or left of normal. The PO_2 at which 50 percent of the Hb is saturated is used as an indicator to determine the relative position of the curve. The symbol P_{50} denotes the partial pressure required to saturate 50 percent of the Hb. A normal P_{50} ranges from 26 to 28 mm Hg. A high P_{50} reflects a low affinity of Hb for O_2, whereas a low P_{50} signifies an increase in Hb for oxygen (Fig. 19-22). It is important to realize that a shift in the P_{50} in either direction has only a small effect on the "loading" of oxygen in the normal lung because such loading occurs at the plateau phase. However, that is not the case for the steep phase.

Three main factors that affect the binding affinity of Hb are temperature, CO_2 tension, and pH. A fall in pH, a rise in PCO_2, and a rise in temperature all shift the curve to the right. Most of the CO_2 effect, which is known as the *Bohr effect*, can be attributed to pH. An increase of $[H^+]$ concentration lowers the affinity of Hb to bind with O_2. A shift of the O_2-Hb dissociation curve to the right is physiologically advantageous at the tissue level (steep phase) because the affinity is lowered (increased P_{50}), which enhances the unloading of O_2 for a given PO_2 in the tissue capillary. A simple way of remembering these shifts is that an exercising muscle is hot, acid, and has a high PCO_2, all of which favor unloading more O_2 to the metabolically active muscle cells.

A shift of the O_2-Hb curve to the left increases the affinity of Hb for O_2, but the ability to *release* O_2 is lowered. Conditions that cause the curve to shift to the left are shown in Figure 19-22 (increased pH, decreased PCO_2, and a decrease in body temperature). Organic phosphate, particularly 2,3-diphosphoglycerate (2,3-DPG), is another factor that profoundly affects the O_2-Hb affinity. In the red cell, 2,3-DPG is much higher than in other cells because the erythrocyte utilizes a shunt that converts much of the formed 1,3-DPG to 2,3-DPG. An increase in 2,3-DPG facilitates unloading of O_2 from the red cell at the tissue level (shifts the curve to the right). An increase in red cell 2,3-DPG occurs with hypoxia (e.g., ascent in altitude, anemia, chronic lung disease), as well as with exercise.

Carbon monoxide is an odorless, colorless, and poisonous gas that interferes with O_2 transport by competing at the same binding sites as O_2 to form carboxyhemoglobin (HbCO). The reaction (Hb + CO $\rightleftharpoons$ HbCO) is reversible and is a function of CO tension. This means that breathing higher concentrations of CO will shift the reaction to the

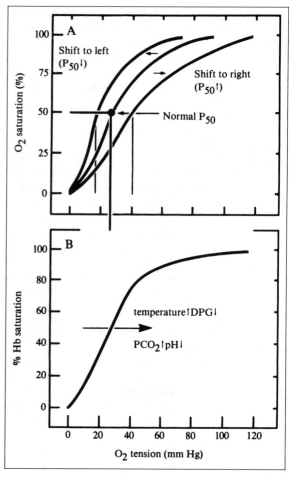

Fig. 19-22. Basic shifts of the O_2-Hb dissociation curve from changes in the affinity of Hb for O_2. A: the effect on P_{50}; B: the curve shifting to the right, with an increase in $PCO_{2'}$ temperature, 2,3-diphosphoglycerate (DPG_1), and a decrease in pH. Note that a shift to the right (P_{50}) is physiologically "advantageous" in that oxygen is more readily released from Hb in the tissues.

right, and breathing fresh air will shift the reaction to the left and release CO from the Hb. The dissociation curve for CO has a shape similar to that of O_2-Hb curve. The major difference is that CO has a binding affinity about 210 times to that of O_2. This means that CO will bind with the same amount of Hb as oxygen, but with a partial pressure 210 times lower than O_2. For example, breathing normal air (21 percent O_2) contaminated with 0.1 percent CO would cause *half* the Hb to be saturated with CO and *half* with O_2. Thus, with the increased affinity of Hb for CO a person can inadvertently

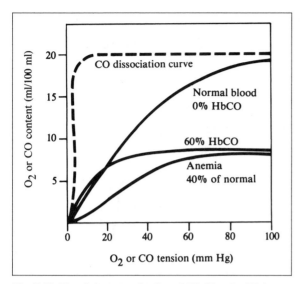

Fig. 19-23. Dissociation curves for O_2 and CO. Note that Hb is completely saturated with CO at low partial pressure (dotted line). Note also that CO displaces the O_2-Hb dissociation curve to the left, as compared to the O_2-Hb dissociation curve for anemic blood.

be exposed to a small amount of CO in the air and have considerable amount of CO bound to Hb. A good example of this is a person driving a car with closed windows and with a leaky exhaust system. The important thing to realize is that with CO exposure, arterial PO_2 and Hb concentrations are normal, but O_2 content is grossly reduced.

Figure 19-23 compares the dissociation curves for O_2 and CO. Note that the presence of CO shifts the O_2-Hb curve to the left, making it more difficult to unload O_2 at the tissue levels. Because CO is an odorless, colorless, and a nonirritating gas, exposure to trace amounts is extremely dangerous. Since arterial PO_2 is normal, there is no feedback mechanism to increase ventilation, and since HbCO is cherry red, there is no evidence of cyanosis. Thus, a person can be exposed to a lethal concentration of CO until enough COHb is bound to cause *anoxia* (lack of O_2). Since the brain is the organ first affected by the lack of O_2, CO can cause a person to collapse without ever being aware of the danger. The best way to treat CO poisoning is to give 100 percent O_2 or 95 percent O_2 with 5 percent CO_2. Since O_2 and CO compete for the binding site on the Hb molecule, breathing a high O_2 concentration favors the formation of HbO_2. The addition of 5 percent CO_2 to the inspired gas stimulates ventilation and increases the release of CO from Hb. Remember that CO is not stored in the body but can associate to or dissociate from Hb as a function of PCO.

Oxygen Utilization. The amount of O_2 available to the tissue per minute is determined by O_2 *delivery*, which is a product of the O_2 content of the blood and blood flow.

$$O_2 \text{ delivery} = CaO_2 \times \dot{Q}$$

where CaO_2 is arterial O_2 content and $\dot{Q}$ is flow. Since O_2 content is fairly constant in a normal person, the major determinant to O_2 delivery is a change in blood flow. Thus, when a tissue needs more O_2, O_2 delivery is augmented by increasing flow, primarily by elevating cardiac output. Two other resources are available to increase the cardiovascular reserve further to enhance O_2 delivery. These include regional perfusion and capillary perfusion. If a tissue becomes hypoxic, blood flow may shut down in other areas on a regional basis, thereby increasing flow to critical organs. Usually, hypoxia and local tissue acidity can cause local vasodilation, leading to a preferential increase in perfusion. Another reserve of the cardiovascular system is the ability of tissue to vary its capillary density, which can profoundly change tissue perfusion.

The amount of O_2 consumed by a tissue is given by

$$O_2 \text{ consumed} = \dot{Q} \times (CaO_2 - CvO_2)$$

where CvO_2 is the O_2 content of venous blood draining the tissue. The ratio of the amount consumed to the amount available (O_2 delivery) is termed *oxygen utilization* and is given by

$$O_2 \text{ utilized} = \frac{O_2 \text{ consumed}}{O_2 \text{ delivered}} = \frac{\dot{Q}(CaO_2 - CvO_2)}{\dot{Q}CaO_2}$$

$$= \frac{CaO_2 - CvO_2}{CaO_2}$$

The utilization of O_2 varies from organ to organ. It is 10 percent in the kidney and 60 percent in the heart, and over 90 percent of the available O_2 is utilized in exercising muscle.

CARBON DIOXIDE TRANSPORT SYSTEM. Carbon dioxide is carried in the blood in three forms: (1) physically dissolved, (2) in the bicarbonate ion, and (3) in carbamino proteins (mainly Hb). Approximately 10 percent is carried in the dissolved form, 60 percent as bicarbonate ion, and 30 percent as carbamino protein. Carbon dioxide, like O_2, dissolves in solution. According to Henry's law, CO_2 is about 20 times more soluble in plasma. Figure 19-24 illustrates the overall process of CO_2 transport. A gradient exists between the cell and the plasma. The driving pressure of this gradient

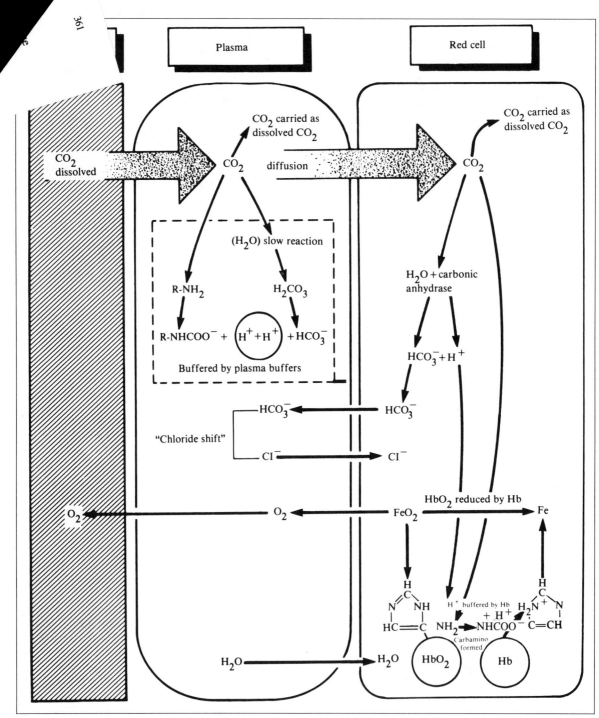

Fig. 19-24. Mechanisms of CO_2 transport. The amount of CO_2 physically dissolved in the plasma is small but extremely important, because this determines the driving pressure of CO_2 transport.

is measured as CO_2 tension (PCO$_2$). *Bicarbonate ion* is formed in the blood by the following reaction:

$$CO_2 + H_2O \xrightleftharpoons[\text{anhydrase}]{\text{carbonic}} H_2CO_3 \rightleftharpoons H^+ + HCO_3^-$$

This reaction occurs very slowly in the plasma, but it can be accelerated about 250-fold in the red cell by a zinc-containing enzymatic catalyst, *carbonic anhydrase*. Carbonic anhydrase is present in red cells but not in the plasma. The enzyme is also found in the renal tubular cell, gastrointestinal mucosa, and muscle. However, it is found in highest concentration in the red cell.

Carbonic acid readily ionizes within the red cell. HCO_3^- diffuses out of the red cell, but the H^+ ion cannot readily move out because of the relative impermeability of the membrane to H^+. To maintain electrical neutrality, Cl^- diffuses into the red cell from the plasma (Fig. 19-24). The chloride movement is known as the *chloride shift*. Since the chloride ion is osmotically more active than HCO_3^-, the red cell tends to swell on the venous side, and when the red cell passes through the pulmonary capillaries, they return to normal. Note in Figure 19-24 that most H^+ formed from the ionization of H_2CO_3 is buffered by Hb: $H + HbO_2 \rightleftharpoons H \cdot Hb + O_2$. This reaction is favored to the right at the tissue level because deoxygenated Hb is a better proton acceptor (less acid) than the oxygenated form. Remember that as H^+ binds to Hb, it reduces O_2 binding and shifts the O_2-Hb dissociation level to the right (Bohr effect). Thus, the unloading of O_2 from the Hb in the tissues favors the carriage of CO_2, while in the pulmonary capillaries the oxygenation of Hb favors the unloading of CO_2. The buffering of H^+ by Hb to increase the ability to carry O_2 is referred to as the *Haldane effect*. The third form in which CO_2 is carried is in the form of *carbamino proteins*. These are formed by dissolved CO_2 reacting with free amine groups (NH_2) on protein to form carbamino groups by the following reaction:

$$CO_2 \text{ (dissolved)} + NH_2 \rightleftharpoons \text{protein NH} \cdot COOH$$
$$\rightleftharpoons \text{protein } NH_2 \cdot COO + H^+$$

The chief protein for this reaction is the Hb molecule in the red cell to form carbaminohemoglobin. Deoxygenated Hb can bind much more CO_2 than oxygenated Hb. Again, the deoxygenating Hb in the tissue capillaries favors the loading of CO_2, while the oxygenation in the pulmonary capillaries enhances the unloading of CO_2 (Haldane effect).

It is important to remember that although major reactions occur in the red blood cell regarding CO_2 transport, the bulk of the CO_2 is actually carried in the plasma in the form of bicarbonate ion.

Carbon Dioxide Dissociation Curve. A CO_2 dissociation curve can be constructed in a fashion similar to that of an O_2-Hb dissociation curve (see Fig. 19-21). The relationship between PCO$_2$ and CO_2 content (total) of the blood is shown in Figure 19-25. Again, note the Haldane effect. If one compares the CO_2 dissociation with the O_2 dissociation curve (Fig. 19-26), major differences can be observed. First, there is much more CO_2 than O_2 in every liter of blood. Second, the CO_2 dissociation curve is much more linear and steeper. The increased linearity and steepness of the CO_2 dissociation have very important physiological implications in gas exchange—a topic discussed later. As a final comment, CO_2 transport by the blood plays a major role in acid-base regulation. This important subject will be discussed in detail in Chapter 24.

SHUNTS AND VENTILATION/PERFUSION RELATIONSHIPS

SHUNTS. In the last few sections we have emphasized the equilibration for both PCO$_2$ and PO$_2$ that occurs between the alveoli and the blood in the pulmonary capillaries. However, the blood that leaves the lung (via the pulmonary veins) and is returned to the left side of the heart has a markedly lower PO$_2$ than alveolar gas. Usually, arterial PO$_2$ (PaO$_2$) is around 95 mm Hg, while end-tidal gas or alveolar gas PO$_2$ (PAO$_2$) is approximately 100 mm Hg (Fig. 19-27). This alveolar-arterial difference is referred to as the *alveolar-arterial O_2 gradient*, or $(A - a)O_2$ gradient. Arterial PO$_2$ has a normal range between 95 to 85 mm Hg. The $P(A - a)$ O_2 gradient increases with age, and a PaO$_2$ of 85 mm Hg and a $(A - a)O_2$ gradient of 15 mm Hg can be normal in an elderly person. Thus, a normal $(A - a)O_2$ gradient can range from 5 to 15 mm Hg. The $(A - a)O_2$ gradient arises because even in normal persons (1) a small percentage of venous blood bypasses the alveolar-capillary gas exchange units, thereby forming a *right-to-left shunt,* and (2) ventilation and blood flow among alveoli are not evenly matched. The shunt and mismatching of ventilation and flow lead to *venous admixture* (admixture of venous with pulmonary end-capillary blood) and are the two main factors that cause PaO$_2$ to be less than PAO$_2$, thereby forming a $P(A-a)O_2$ gradient even in normal, healthy persons.

Shunt is defined as any mechanism in which venous blood (low PO$_2$) from the pulmonary artery bypasses ventilated areas of the lung and mixes with arterial blood (pulmonary veins). In normal, healthy persons, about 2 percent of the cardiac output is involved in this normal shunting of "right-sided blood" to the "left side" without being arterialized by the lungs. One cause for the shunt is attributable to the bronchial circulation, which supplies the conducting air-

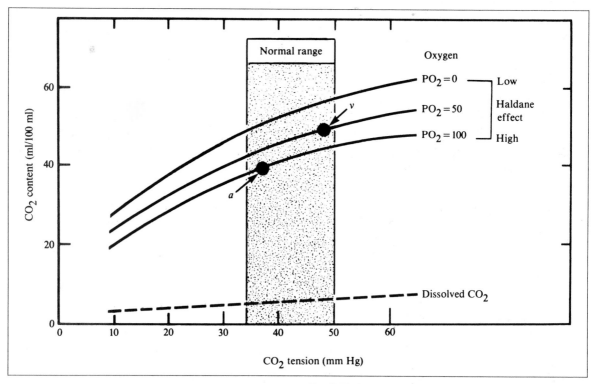

Fig. 19-25. Carbon dioxide dissociation curves at different O_2 tensions. Note that oxygenated blood carries less CO_2 for the same PCO_2 (Haldane effect).

ways and returns blood to the left heart via the pulmonary veins. Respiratory infection can often increase bronchial circulation, resulting in a concomitant increase in shunting. The second cause for shunting is that a small amount of blood drains directly from the myocardium of the left ventricle via the thebesian veins. This blood, along with blood from the bronchial circulation, contributes to the lowering of PaO_2.

When a shunt is caused by the addition of mixed venous blood from the right ventricle to the left ventricle without coming in contact with the ventilated region, it is possible to calculate the amount of shunt flow (Fig. 19-28). By using the Fick principle, one can derive an equation to calculate shunt flow ($\dot{Q}s$) to total flow ($\dot{Q}t$):

$$\frac{\dot{Q}s}{\dot{Q}t} = \frac{C\bar{c}O_2 - CaO_2}{C\bar{c}O_2 - C\bar{v}O_2}$$

where a, $\bar{v}$, and $\bar{c}$ refer, respectively, to systemic arterial, mixed venous, and mean end-capillary O_2 content per milliliter of blood. The O_2 content of end-capillary blood can be calculated from alveolar PO_2 and the O_2 dissociation time.

It is important to note that cardiac output need not be

known to calculate the fraction of cardiac output that is shunted past alveoli. An important feature of a shunt is that giving a person a 100 percent oxygen to breathe does not dramatically elevate arterial PO_2. This is because the shunted blood, which bypasses alveoli, is not exposed to the higher alveolar PO_2. Some increase in arterial PO_2 occurs because of the O_2 added to capillary blood from ventilated alveoli. It is important to note that this modest elevation in arterial PO_2 is due primarily to the increased O_2 in the dissolved form and not to that bound to the red cell because the blood that leaves the ventilated alveoli is already 98 percent saturated. Thus, administering 100 percent O_2 is a very sensitive way of detecting a shunt, because when the PO_2 is high, a small amount of mixed venous blood (low O_2 content) has a dramatic effect on arterial PO_2 due to the flat slope of the O_2 dissociation curve (Fig. 19-29).

Another important feature of a shunt is that arterial PCO_2 is seldom elevated, even though the shunted blood has a high PCO_2. The reason for this is that PCO_2 is a very powerful stimulus to respiration and results in increased ventila-

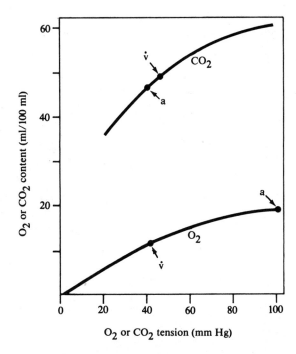

Fig. 19-26. Oxygen and CO_2 dissociation curves. Note that the O_2 curve has both a plateau and a steep region between the normal arterial (a) and mixed venous (v̄) points, while the CO_2 curve is essentially rectilinear between these two points. Note also that CO_2 content is much greater than O_2 content and that the CO_2 dissociation curve is much steeper.

tion. This lowers arterial PCO_2 by blowing off excess CO_2 from unshunted ventilated regions until arterial PCO_2 returns to normal.

VENTILATION/PERFUSION RATIO. The second cause of venous admixture that leads to $P(A-a)O_2$ gradient is due to improper matching of ventilation and blood flow within the lung. Each lung receives about 2 liters per minute, giving a total alveolar ventilation ($\dot{V}A$) of 4 liters per minute. Each lung also receives about 2.5 liters per minute of blood flow, giving a total blood flow ($\dot{Q}$) of 5 liters per minute. Thus, the overall ventilation/perfusion ($\dot{V}A/\dot{Q}$) ratio is about 0.8. Because of gravity, the lung in a normal upright person is not perfused or ventilated evenly. At the top part of the lung (apex) both blood flow (perfusion) and ventilation are much less than that of the base (Figure 19-30). Thus, at normal lung volumes, both blood flow and ventilation increase markedly from the apex to the base. Another important

point to make from Figure 19-30 is that the regional differences for ventilation are less marked than those for blood flow. For these reasons, alveolar ventilation and blood flow are poorly matched in the top and bottom regions of the lung. This means there are regions in the lung that have ventilation/perfusion ratios higher (apex) and lower (base) than 0.8 (Fig. 19-30). This points out that *total ventilation* and *total blood flow* are relatively unimportant in gas exchange. The crucial factor is *matching* of ventilation and blood flow in an individual gas exchange unit.

The effect of ventilation/perfusion ratios on alveolar gas composition is shown in Figure 19-31. In a lung unit with a normal ventilation/perfusion ratio (Fig. 19-31B), inspired air has a PCO_2 of 0 mm Hg and a PO_2 of 150 mm Hg; mixed venous blood entering these units has a PCO_2 of 45 mm Hg and PO_2 of 40 mm Hg. In alveolar regions in which the $\dot{V}A/\dot{Q}$ is 0.8, the PAO_2 is 100 mm Hg, and the $PACO_2$ is 40 mm Hg. Now suppose that ventilation of a unit is reduced, but blood flow is unchanged. When this occurs, PAO_2 in the unit will fall, and $PACO_2$ will rise, as seen in Figure 19-31A. In this situation, blood flow is in excess of ventilation, which results in "wasted" blood and constitutes a shuntlike effect.

Thus, venous admixture can occur by two ways: One is by an anatomical shunt (e.g., bronchial circulation), and the other is by a low ventilation/perfusion ratio, which causes a shuntlike effect. The sum of these two (anatomical plus shuntlike) equals a *physiological shunt* and is analogous to physiological dead space. Thus, the shunt equation can be rewritten to represent physiological shunt:

$$\frac{\dot{Q}sp}{\dot{Q}t} = \frac{C\bar{c}O_2 - CaO_2}{C\bar{c}O_2 - C\bar{v}O_2}$$

where $\dot{Q}sp$ is physiological shunt. A special case of $\dot{Q}sp$, called *anatomical shunt* ($\dot{Q}san$) is defined by the above equation when an individual breathes 100% O_2 for 20 to 30 minutes. Alveoli with very low $\dot{V}A/\dot{Q}$ ratios will have enough alveolar PO_2 to completely saturate hemoglobin during the 100 percent O_2 breathing period. These low $\dot{V}A/\dot{Q}$ alveolar units would no longer contribute to the calculated $\dot{Q}sp/\dot{Q}t$ and the new calculated shunt would include only the anatomical shunt. The similarities between shunts and dead space are shown in Table 19-4.

It is important to realize that a low ventilation/perfusion ratio (Fig. 19-31A) can occur in two ways. One is by gravitational effects (primarily in the base of the lung). The second is by *regional hypoventilation* (alveolar ventilation below normal in a region of gas-exchange units). Regional hypoventilation can occur as a result of a partially plugged or restricted airway. If the airway is totally blocked (e.g., by

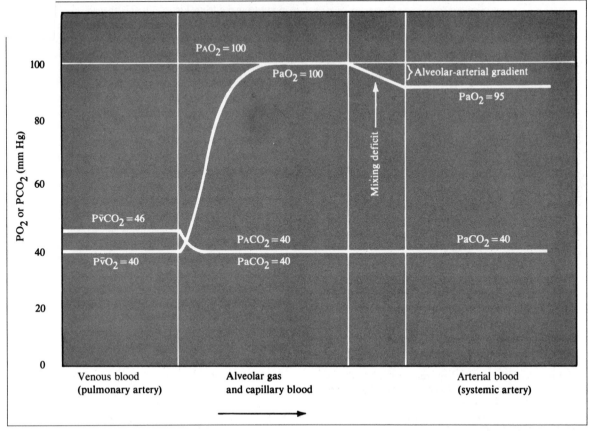

Fig. 19-27. Oxygen and CO_2 tension in mixed venous, pulmonary capillary, and arterial blood. Note the alveolar-arterial oxygen gradient. (Adapted from F. F. Kao. *Respiratory Physiology*. New York: Elsevier, 1972. P. 94.)

aspiration of a seed), the ventilation/perfusion ratio becomes zero, and that region acts like an anatomical shunt.

Now let us consider the opposite condition, when there is an obstruction of blood flow with no alteration in ventilation (Fig. 19-31C). In this situation, PAO_2 rises and $PACO_2$ falls. When the ventilation/perfusion ratio reaches infinity, alveolar gas composition will equal that of inspired gas. Regions where ventilation/perfusion ratios are greater than 0.8 tend to have "wasted" alveolar ventilation, and this leads to an increase in alveolar dead space.

Figure 19-32 presents a model that shows how variations in $\dot{V}A/\dot{Q}$ affect gas exchange. Both ventilation and blood flow are gravity dependent and consequently decrease down the lung. The gradient for blood flow is steeper than that for ventilation, so that the ventilation/perfusion ratio increases up the lung. As a result, $\dot{V}A/\dot{Q}$ ratios are higher at the apex and decreases down the lung, and the differences in gas exchange stem from variations in these ratios. Oxygen tension changes much more than CO_2 between the top

and bottom of the lung. Also, it is important to note the large difference in pH occurs down the lung from the variation of CO_2. The decreased O_2 uptake seen in the apex is due primarily to very low blood flow in that region, while differences in CO_2 excretion between the apex and base are more closely related to ventilation. For this reason the differences between CO_2 at the apex and base of the lung are much less than with O_2, since CO_2 exchange is more related to ventilation. With an increase in cardiac output, such as occurs in exercise, the distribution of flow becomes more uniform, and the $\dot{V}A/\dot{Q}$ decreases in the apex and consequently assumes a more important share of gas exchange.

In addition, there is evidence that differences in $\dot{V}A/\dot{Q}$ ratios tend to affect the localization of some types of diseases. One good example is tuberculosis, which tends to

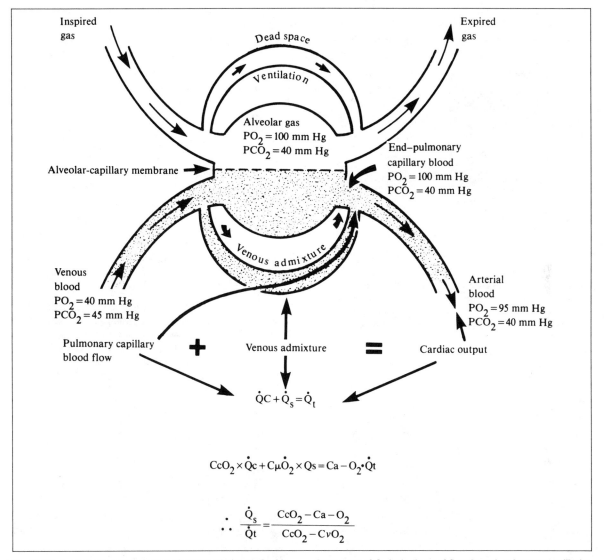

$$CcO_2 \times \dot{Q}c + C\mu\dot{O}_2 \times Qs = Ca - O_2 \cdot \dot{Q}t$$

$$\therefore \frac{\dot{Q}_s}{\dot{Q}t} = \frac{CcO_2 - Ca - O_2}{CcO_2 - CvO_2}$$

Fig. 19-28. A schematic representation of venous admixture. In this functional relationship of gas exchange in the lungs, equilibrium is obtained between alveolar gas and pulmonary end-capillary blood. Note that alveolar gas is mixed with dead-space gas to give expired gas, while pulmonary end-capillary blood is mixed with shunted venous blood to give arterial blood. The shunt equation is similar to the Bohr equation and is based on the assumption that the total amount of O_2 in 1 minute of arterial blood flow equals the sum of the amount of O_2 in 1 minute of flow through pulmonary capillaries and the amount of O_2 in 1 minute of flow through the shunt. The amount of O_2 in 1 minute of flow equals the product of blood flow and O_2 content in the blood. $\dot{Q}t$ = cardiac output; Qc = capillary flow; CaO_2 = arterial oxygen content; $C\bar{c}O_2$ = mean O_2 content in the pulmonary capillary; $C\bar{v}O_2$ = mixed venous O_2 content. **(Adapted from J. F. Nunn. *Applied Respiratory Physiology* (2nd ed.). Boston: Butterworth, 1977. P. 247.)**

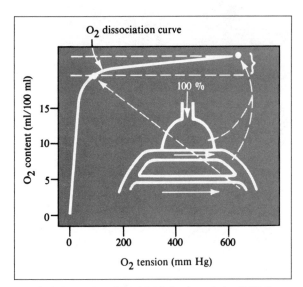

Fig. 19-29. Depression of arterial PO2 by shunt during 100% O2 breathing. Because the O2 dissociation curve is so flat when the PO2 is very high, the addition of a small amount of shunted blood with its low O2 concentration greatly reduces the PO2 of arterial blood. (Adapted from J. B. West. *Respiratory Physiology* **(2nd ed.). Baltimore: Williams & Wilkins, 1979. P. 57.)**

localize in the apical region of the lung because it provides a more favorable environment for this organism (higher oxygen, less acidic).

One final comment is that anesthesia has an effect on $(A - a)O_2$ gradient. Routine uncomplicated anesthesia in adults usually results in an increased $P(A - a)O_2$. The primary dysfunction is not due to a reduction of diffusion capacity as once thought, but to a reduction in cardiac output that leads to increased venous admixture, i.e., shunting, to areas with low $\dot{V}A/\dot{Q}$ ratios. Thus, the reason for the increased $P(A - a)O_2$ gradient is that a decreased cardiac output, with other things being equal, causes increased shunting and the development of a low $\dot{V}A/\dot{Q}$ ratio, which leads to a concomitant decrease in arterial PO_2.

HYPOXEMIA AND HYPERCAPNIA

One of the primary consequences of abnormal gas exchange is that it alters arterial PO_2 and PCO_2 tension. If the impair-

Fig. 19-30. Distribution of blood flow (perfusion) and ventilation in an upright lung. Both ventilation and perfusion are gravity dependent and decrease from the bottom to the top of the lung. Note that the curve for regional blood flow is steeper than ventilation, so that the ventilation/perfusion ratio increases up the lung. (Adapted from J. B. West. *Respiratory Physiology* **(2nd ed.). Baltimore: Williams & Wilkins, 1979. P. 61.)**

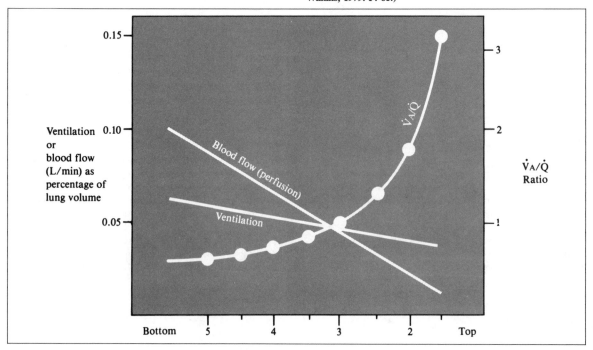

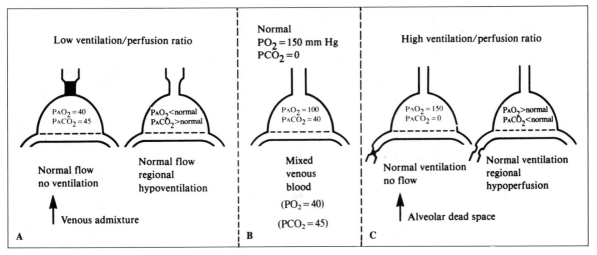

Fig. 19-31. Conditions that alter ventilation/perfusion ratios. Note the effect of altered ventilation/perfusion ratios on alveolar gas composition.

Table 19-4. Comparison of shunt and dead space

Shunt	Dead Space
Anatomical ($\dot{Q}$san)	Anatomical (VDan)
+	+
"Shunt like effect" ($\dot{Q}$srel) ($\dot{V}A/\dot{Q}$ variation)	Alveolar (VDA) ($\dot{V}A/\dot{Q}$ variation)
Physiological ($\dot{Q}$sp) (calculated total venous admixture)	Physiological (VD)

ment of gas exchange is severe enough, it will lead to a low arterial PO_2 (*hypoxemia*) and/or to increased arterial PCO_2 (*hypercapnia*). The normal range for arterial PO_2 is 85 to 95 mm Hg and for PCO_2, 35 to 45 mm Hg. Hypoxemia occurs when arterial PO_2 is *below* 85 mm Hg, and hypercapnia is present when arterial PCO_2 is *greater* than 45 mm Hg.

The mechanisms that produce hypoxemia can be listed in order of importance, as follows

1. Ventilation/perfusion mismatch
2. Anatomic shunt
3. Generalized hypoventilation
4. Diffusion block

DIFFUSION BLOCK. For all practical purposes, diffusion block at the alveolar-capillary membrane occurs only under unusual circumstances, such as seen with specific environmental insults (e.g., asbestosis).

GENERALIZED HYPOVENTILATION. *Generalized hypoventilation* is defined as a decrease in *total alveolar ventilation* as contrasted with regional hypoventilation. This decrease may be due to a decrease in total ventilation (such as in a comatose condition following a drug overdose). A decrease in alveolar ventilation means that alveolar ventilation is too low to maintain a *normal arterial PCO_2*. Thus, generalized hypoventilation is better defined as alveolar ventilation insufficient to maintain a normal arterial PCO_2. When arterial PCO_2 rises, arterial PO_2 must fall. Thus, a person with generalized hypoventilation has a high $PaCO_2$ and a low PaO_2. Since arterial PCO_2 is high, regional hypoventilation is complicated by a low arterial pH. An important feature of generalized hypoventilation is that a normal $P(A-a)O_2$ gradient is present. Therefore, if a patient has a high $PaCO_2$, low pH, and low PaO_2 but a normal $P(A-a)O_2$, then hypoxemia is due entirely to generalized hypoventilation. It should be pointed out that the condition of a patient suffering from generalized hypoventilation (acidosis, hypoxemia, and hypercapnia) can be reversed solely by mechanical ventilation.

SHUNTING. A second most important cause of hypoxemia is shunting of blood through nonventilated regions of the lung,

$\dot{V}_A$ $\dot{Q}$ (L/min)		$\dot{V}_A/\dot{Q}$	PO_2 PCO_2 (mm Hg)		O_2 content (ml/100 ml) CO_2		pH
0.25	0.07	3.6	130	28	20	42	7.50
0.8	1.3	0.62	88	42	19	49	7.36

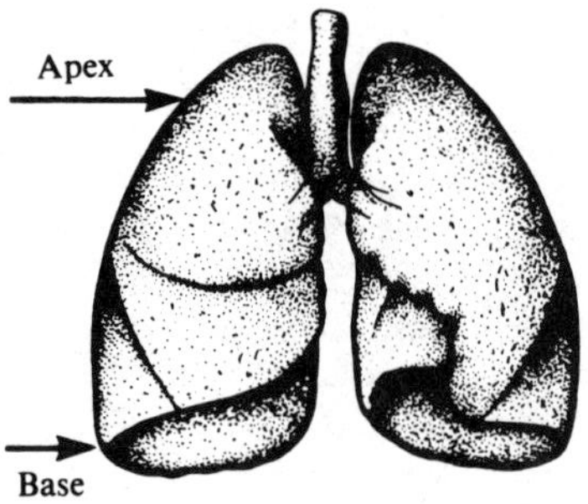

Fig. 19-32. Regional differences in gas exchange down the lung. Only the values for the apex and base are shown.

such as in congenital heart disease with a septal defect or in pulmonary edema. As was mentioned earlier, an important feature of an anatomical shunt is that if a patient is given 100 percent O_2 to breathe, the arterial PO_2 does not rise significantly, because the shunted blood does not pass through ventilated alveoli and is never exposed to the high alveolar PO_2. Thus, the venous admixture continues to depress the arterial PO_2.

VENTILATION/PERFUSION MISMATCH. The most important cause of hypoxemia is ventilation/perfusion mismatch (in approximately 90 percent of cases). The reason that ventilation/perfusion mismatch has such a profound effect on arterial PO_2 and PCO_2 has been alluded to before, but will be discussed here in more detail. The first reason why uneven ventilation and blood flow lead to difficulty in oxygenating arterial blood is seen in Figure 19-32. Note that PO_2 at the apex is about 40 mm Hg higher than at the base. However, a major share of the cardiac output goes to the base of the lung, where the alveolar PO_2 is low. For the same reason, arterial PCO_2 tends to be high because a major share of the blood leaving the lung comes from the base, where ventilation is the lowest. The second reason that ventilation/perfusion inequality has such a profound influence is the effect of venous admixture on lowering the arterial PO_2. This can be illustrated by Figure 19-33, which depicts O_2 content from three groups of alveoli with low, normal, and high ventilation/perfusion ratios. Because of the sigmoid shape of the O_2-Hb dissociation curve, blood perfusing high $\dot{V}_A/\dot{Q}$ alveoli will add relatively little O_2 to the blood, while blood

Fig. 19-33. Depression of arterial by ventilation/perfusion mismatch. Regions with high ventilation/perfusion ratios add very little O_2 to the blood, compared with the decrement caused by regions with low ventilation/perfusion ratios.

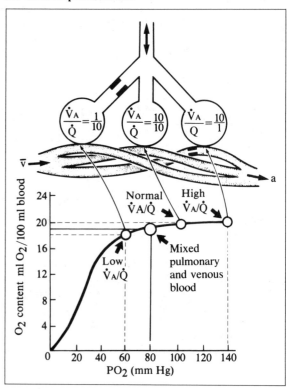

perfusing the low $\dot{V}A/\dot{Q}$ alveoli will have significantly less O_2 content than normal. The blood equilibrated with a $P_{A}O_2$ of 60 mm Hg (low $\dot{V}A\dot{Q}$) will have an O_2 content of 18.1 ml per 100 ml of blood, while that coming into equilibrium with a $P_{A}O_2$ of 140 mm Hg (high $\dot{V}A/\dot{Q}$) will have an O_2 content of 19.9 ml per 100 ml. These two blood flows will mix in the pulmonary veins.

If we assume there were equal flows of blood perfusing these two units, then the O_2 content of the mixed pulmonary venous blood would be halfway between the contents of the two streams, or 19.0 ml per 100 ml. This would give an arterial PO_2 of about 79 mm Hg, which is below an alveolar PO_2 of 100 mm Hg. In the preceding example, however, we have made the assumption that equal blood flow was perfusing these two groups of alveoli, which is not the case. As has been mentioned, proportionately more blood flow comes from alveoli with low ventilation/perfusion ratios than from high $\dot{V}A/\dot{Q}$ ratios. Thus, the actual O_2 content would be weighted much more heavily toward low $\dot{V}A/\dot{Q}$. Even though gas and blood reach equilibrium with one another in the pulmonary capillary, arterial PO_2 would be even farther away from alveolar PO_2.

Oxygen breathing is a very effective means of raising arterial PO_2 in a person whose hypoxemia is caused by ventilation/perfusion inequality, and it is often used to distinguish between shunt and ventilation/perfusion inequality as the cause of hypoxemia. For example, if in a patient given 100 percent O_2 the arterial PO_2 remains below 150 mm Hg, suspect an increase in anatomical shunt. If, on the other hand, arterial PO_2 rises above 150 mm Hg, ventilation/perfusion mismatch is the *only* important cause of the hypoxemia.

BIBLIOGRAPHY

Comroe, J. H., Jr., Forster, R. E., Dubois, A. B., Briscoe, W. A., and Carlson, E. *The Lung: Clinical Physiology and Pulmonary Function Tests* (2nd ed.). Chicago: Year Book, 1962.

Guenter, C. A., Welch, M. H., and Hogg, J. C. *Clinical Aspects of Respiratory Physiology*. Philadelphia: Lippincott, 1978.

Nun, J. F. *Applied Respiratory Physiology* (2nd ed.). London-Boston: Butterworth, 1977.

Saunders, K. B. *Clinical Physiology of the Lung*. Philadelphia: Lippincott, 1977.

Weibel, E. R. Morphological bases of alveolar-capillary gas exchange. *Physiol. Rev.* 53:419–482, 1973.

West, J. B. *Respiratory Physiology* (2nd ed.). Baltimore: Williams & Wilkins, 1979.

20

Rodney A. Rhoades

Maintenance of ventilation (i.e., the process of delivery of gas into and out of the alveoli) requires rhythmic inflation and deflation of the lung. To accomplish gas flow within the lung, the respiratory muscles must do work to stretch the elastic components of the lungs and chest as well as to overcome resistance to airflow. In this chapter the following components of lung mechanics will be presented: muscles of ventilation, elastic and surface properties, interdependence, airway resistance, uneven ventilation, and the work of breathing.

MUSCLES OF VENTILATION

In normal quiet breathing the inspiratory muscles increase the lung volume above its "equilibrium volume" (functional residual capacity [FRC]) and then relax, allowing the elastic recoil of the lung and chest to expire passively. Thus, in normal quiet breathing, inspiration is active and expiration passive. By this statement we mean that inspiratory muscles contract to elevate the respiratory system above its equilibrium position (FRC) and then relax and allow the elastic recoil of the system to return back to equilibrium position passively (expiration). Three sets of muscles are involved in inspiration and include the diaphragm, external intercostals, and accessory muscles (Fig. 20-1).

The main inspiratory muscle of inspiration is the *diaphragm*. The diaphragm is a dome-shaped muscle between the thoracic and abdominal cavity, and its contraction enlarges the thoracic cavity. This accomplished in two ways. First, when the diaphragm (which is attached to lower ribs, sternum, and vertebral column) contracts, the abdominal contents are pushed downward. Second, the lever action of the diaphragm on the abdominal contents moves the rib cage upward and outward (Fig. 20-1).

The effectiveness of the diaphragm for changing the thoracic cavity is related to the strength of its contraction and to its dome shaped configuration when relaxed. With a normal tidal volume the dome diaphragm moves about 1 to 2 cm, but, with forced inspiration, a total excursion of 10 to 12 cm can occur. The effectiveness of the diaphragm in enlarg-

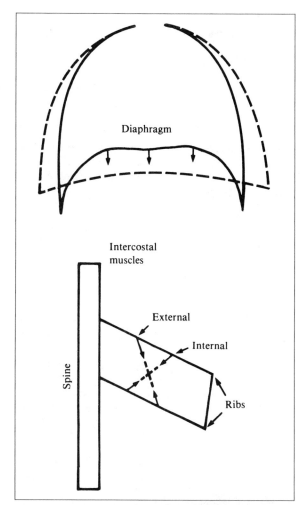

Fig. 20-1. The dome-shaped diaphragm contracts on inspiration, and the abdominal contents are pushed downward and forward. At the same time, because of the diaphragm's architectural connection to the rib cage, as the diaphragm contracts, the rib cage is lifted upward. On forced inspiration the external intercostal muscles contract, and the rib cage is lifted upward and forward, thereby greatly increasing thoracic volume. The internal intercostals have the opposite effect on thoracic volume. (Adapted from J. B. West. *Respiratory Physiology* [2nd ed.]. Baltimore: Williams & Wilkins, 1979. P. 87.)

ing the thoracic cavity can be impeded by obesity, pregnancy in the latter part of gestation, and tight clothing around the abdominal wall. Damage to the phrenic nerve (the diaphragm is innervated by two phrenic nerves, one to each lateral half) can also lead to paralysis of the diaphragm. When one of the phrenic nerves is damaged, the innervated portion of the diaphragm moves up rather than down during inspiration.

The *external intercostal* muscles are the next most important muscles of inspiration. They are innervated by intercostal nerves. Their contraction raises the anterior end of the rib cage, causing the rib cage to move upward and outward (Fig. 20-1). The last group of muscles involved in inspiration are referred to as *accessory muscles*, which are active only when breathing is greatly increased, as in muscular exercise. These muscles include the scalenes in the neck, which elevate the upper rib cage, the sternocleidomastoids, and the trapezius.

Although expiration is passive during normal tidal volumes, on exercise or on forced expiration it can become rather active. The principal expiratory muscles include the muscles of the abdominal wall (namely, the rectus abdominus, external oblique muscles, and transversus abdominus). Contraction of the abdominal wall pushes the diaphragm upward into the chest, thereby reducing thoracic volume. These accessory muscles are important and necessary for such functions as coughing, straining, vomiting, and defecating.

PRESSURES OF THE VENTILATORY APPARATUS

In the previous chapters dealing with the cardiovascular system, the units of pressure were usually expressed in millimeters of mercury. In Chapter 19, dealing with gas tension, the units were also expressed as millimeters of mercury. However, pulmonary physiologists frequenty use centimeters of water (cm H_2O) as a convenient unit in lung mechanics because the pressures in and around the lung are so low. Another concept dealing with pressures in the lung that needs a brief review is that pressures are often measured *relative to atmospheric pressure*. For example, an alveolar pressure of 5 cm H_2O means that the pressure in the alveoli is 5 cm H_2O *greater* than atmospheric pressure. Remember that *absolute pressure* would be atmospheric pressure plus 5 cm H_2O. Conversely, the pressure between the lung and chest wall (*intrapleural pressure*) is approximately -5 cm H_2O, the negative sign indicates that the intrapleural pressure is 5 cm *less* than atmospheric pressure. If no other information is given, you should always assume

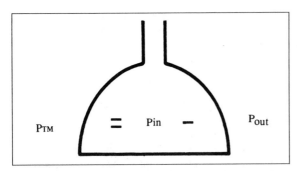

Fig. 20-2. Transmural pressure (Ptm) in the lung is defined as the pressure inside (P_{in}) the lung minus the pressure outside it (P_{out}). Ptm = P_{in} − P_{out}.

that the pressure measured is relative to atmospheric pressure.

Not only do pulmonary physiologists make measurements of pressures relative to atmospheric pressure, but they also must make pressure across walls or across the lung (*transmural pressures*). *Transmural pressure* (Ptm) is defined as P_{in}-P_{out} (Fig. 20-2). The three major transmural pressures to consider in the ventilatory apparatus are seen in Figure 20-3. The first is *transpulmonary pressure* (P_{tp}), which is the pressure difference across the lung and is

measured by subtracting intrapleural pressure (Ppl) from alveolar pressure (PA).

$$P_L = P_A − Ppl$$

When the measurement is made under static conditions (no airflow), then transpulmonary pressure provides a quantitative measure of the elastic recoil properties of the lung. The *elastic recoil pressure* is frequently used to assess the elastic recoil of the lung. Note that when no airflow occurs, elastic recoil pressure equals transpulmonary pressure and is positive, which tends to make the lung smaller. Elastic recoil pressure for the lung is

$$Pel = P_A − Ppl$$

A second transmural pressure important in the pulmonary system is *trans–chest wall pressure* (P_w) and is defined as

$$P_w = Ppl − P_B$$

Trans–chest-wall pressure provides a quantitative measure of the elastic recoil properties of the chest wall under re-

Fig. 20-3. The three basic transmural pressures to consider in the pulmonary system. (1) Transpulmonary pressure ($P_L = P_A − Ppl$); (2) transairway pressure (Pta = $P_B − Ppl$); and (3) trans–chestwall pressure (Pw = Ppl − P_B). PA = alveolar pressure.

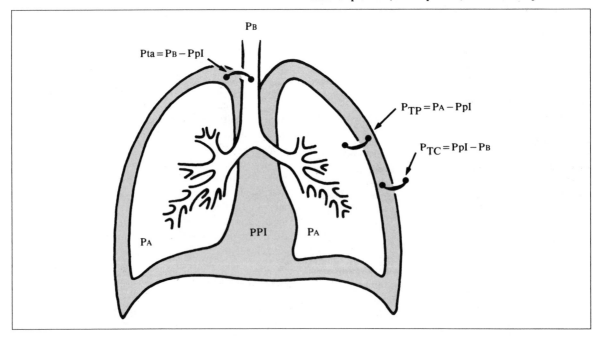

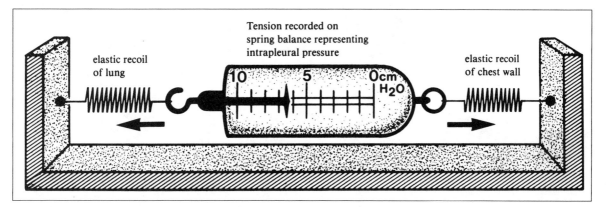

Fig. 20-4. Intrapleural pressure can be seen as analogous to tension from two springs pulling with equal force but in opposite directions.

laxed conditions with no airflow. The third transmural pressure is transairway pressure (PTA) and is defined as

$$P_{TA} = P_B - P_{pl}$$

The transairway pressure is very important in keeping the airways open during expiration and will be discussed at the end of this chapter.

In addition to transmural pressures, another important pressure to consider in lung mechanics that is *intrapleural pressure* (Fig. 20-3), the pressure in the fluid-filled space between the lung and chest wall. The inward elastic recoil of the lung is balanced by outward elastic recoil of the chest wall to produce an intrapleural pressure of about -5 cm H_2O at FRC. The equilibrium at FRC is analogous to two springs joined together but pulling with equal force but in opposite directions, which creates a negative pressure (subatmospheric) in the pleural cavity. At equilibrium the tension of each spring is the same but in opposite directions. Thus, FRC is sometimes referred to as the resting or equilibrium volume because the elastic force of lung and chest are equal (Fig. 20-4). Intrapleural pressure is usually determined by measuring *esophageal pressure,* since the esophagus passes through the pleural space. Because intrapleural pressure is subatmospheric, one might expect that gas or fluid would tend to accumulate in this potential space. Fluid accumulation does not occur because the difference in pressure between the pleural capillaries and pleural "space" is less than the oncotic pressure of the plasma proteins. Gas is absent because the sum of the partial pressures of gas in the venous blood is less than atmospheric pressure. Therefore, there is a pressure gradient to absorb gases and to prevent fluid accumulation in the normal lung.

PRESSURE CHANGES DURING BREATHING

The pulmonary system consists of lungs surrounded by a chest wall. Included as part of the chest wall is not only the rib cage, but also the diaphragm and abdominal wall. The lung fills the thoracic cavity, and is in contact with the chest wall. As a result the lung and chest wall tend to act in unison. At rest (end of a normal expiration) the respiratory muscles are relaxed, and the lung and chest recoil are equal but in opposite directions (Fig. 20-5A). As seen earlier, these opposing forces create an intrapleurance pressure of approximately -5 cm H_2O in the potential space between the lung and chest. The pressure difference between the opening of the tracheobronchial tube (mouth) and alveoli is zero, and consequently no airflow occurs. During inspiration, (Fig. 20-5B) the inspiratory muscles contract and the chest expands, causing intrapleural pressure to become more negative. This, in turn, increases transpulmonary pressure and causes the lung to expand. As the lung expands, alveolar pressure becomes subatmospheric with respect to pressure at the opening of airways. The pressure gradient between the mouth and alveoli causes air to flow down the tracheobronchial tree. Airflow continues until alveolar pressure again reaches atmospheric pressure, so that a pressure gradient no longer exists between airway and alveoli. At the end of inspiration the intrapleural pressure is more negative, and consequently the volume of air in the lung is greater than at the beginning of inspiration. Since alveolar pressure has returned to atmospheric pressure, the difference between alveolar pressure and intrapleural pressure (transpulmonary pressure) increases. It is important to remember that the pressure responsible for inflating the lungs is an increase in transpulmonary pressure. During expiration (Fig. 20-5C) the inspiratory muscles relax, the elas-

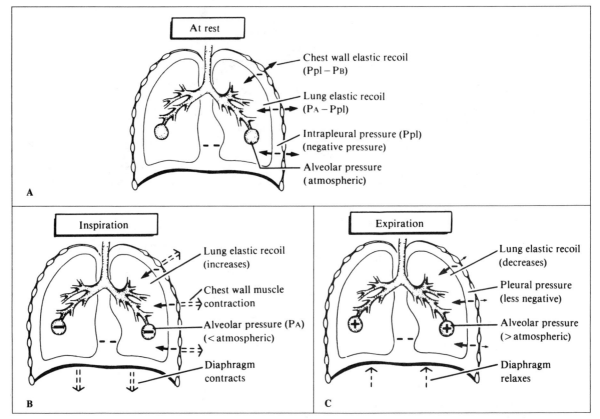

Fig. 20-5. Pressures and forces during a breathing cycle. A. Forces at rest. B. During inspiration. C. During expiration.

tic recoil of lungs causes the lung volume to decrease and consequently increases alveolar pressure above the pressure at the airway opening (mouth). As a result of the pressure gradient between alveoli and airway opening, air flows out of the lung.

The relationship between airflow, alveolar pressure and pleural pressure is presented in Figure 20-6. As seen in the figure, before inspiration occurs, alveolar pressure equals airway opening pressure (atmospheric), and consequently there is no airflow. During inspiration, pleural pressure becomes more negative, and alveolar pressure drops below atmospheric pressure and then returns to zero (atmospheric) at the end of inspiration. Note that peak flow occurs during midinspiration and midexpiration. Also, it is noteworthy that if airway resistance remains constant throughout the breathing cycle, the profile of flow and alveolar pressure curves would be identical. Figure 20-6 shows lung volume increasing on inspiration and decreasing on expiration. Although not shown in Figure 20-6, transpul-

monary pressure increases as lung volume becomes larger. Remember, too, that not only is transpulmonary pressure the pressure that actually inflates the lung, but it also provides an estimate of the elastic recoil force of the lung when airway resistance is zero.

ELASTIC AND SURFACE PROPERTIES OF THE LUNG

Elastic properties of a structure are evaluated by how distensible and how stiff the material is. The term *elasticity* means the ability of a stretched material to returned to its unstretched position. The stiffer the material, the greater the ability to return to its unstretched position and hence greater the elasticity. The opposite is true for distensibility. A material that is easily stretched is very distensible but not very elastic. A measure of distensibility of a material is *compliance,* which is defined as the volume change per unit pressure change. The reciprocal of compliance is *elastance.*

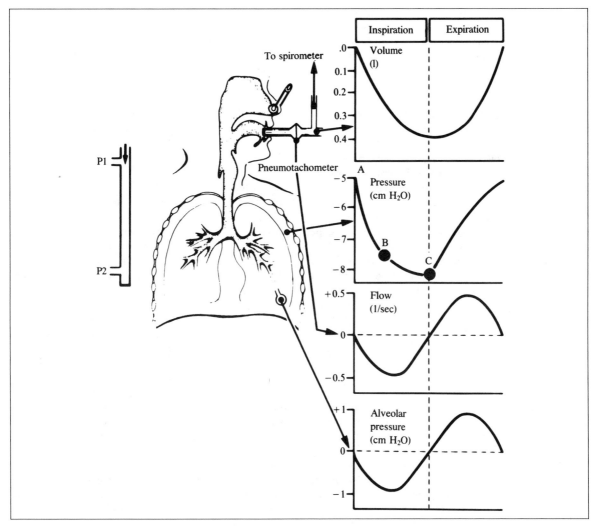

Fig. 20-6. Pressure and flow relationships during breathing.
(Adapted from J. B. West, *Respiratory Physiology* [2nd ed.]. Baltimore: Williams & Wilkins, 1979. P. 103.)

Thus, compliance provides an estimate of distensibility, and elastance provides a measure of stiffness. For example, a spring that is very compliant is easily stretched but not very elastic (i.e., the ability to return to its unstretched portion). Conversely, a spring that has a high elastance (low compliance) is very difficult to stretch but is very elastic, because the stiff spring literally "snaps" back to its unstretched position. Unfortunately, *compliance* is used far more commonly than *elastance* in pulmonary physiology and often leads to confusion when used for the elastic prop-

erties of the lung. The important thing to remember is that compliance is equivalent to distensibility and that the reciprocal of compliance is elastance and is equivalent to stiffness. Thus, a stiff lung has low compliance, and is difficult to inflate. By the same token a stiff lung readily recoils back to its unstretched position and thus has a high degree of elasticity.

PRESSURE-VOLUME CURVES OF THE LUNG

The elastic properties of the lungs are determined by measuring changes in lung volume relative to changes in transpulmonary pressure (Fig. 20-7). Note that the elastic properties are determined when no airflow occurs. Volume changes can be measured quite simply with a spirometer. Changes in transpulmonary pressure at each series of lung volume with no airflow can be obtained by subtracting alveolar pressure from pleural pressure. Under conditions of no airflow, alveolar pressure equals the pressure at the airway opening (mouth pressure). Pleural pressure is measured indirectly by the pressure in the esophagus by means of a balloon catheter. In practice, a pressure-volume curve is produced by first inspiring maximally to total lung capacity and then expiring slowly while airflow is periodically interrupted and measurements of lung volume and transpulmonary pressure recorded (Fig. 20-7). The volume change per unit pressure change ($\Delta V/\Delta P$) is called *static compliance* because no airflow is occurring. Thus, the slope of the pressure-volume curve at any given point is lung compliance at that point. Since the pressure-volume curve of the lung is nonlinear, lung compliance is not the same at all lung volumes, but is lower at high lung volumes and higher at low lung volumes. In the midrange of the pressure-volume curve lung compliance is about 0.2 liters/cm H_2O. This value approximates the volume change of tidal volume at resting FRC.

Compliance clearly depends on lung size. For example, a mouse lung will have a smaller volume change per unit change in transpulmonary pressure than will a whale lung. Similarly, a person who has had one lung surgically removed will also have reduced compliance. To overcome the difficulty of comparing the elastic properties of lungs of different sizes, lung compliance is normalized by dividing compliance by FRC, to give *specific compliance*. The specific compliance of an adult lung is about 0.08 liter/cm H_2O (0.2 liter/cm H_2O/2.5 liters) and about 0.06/cm H_2O (0.005 liter/cm H_2O/0.08 liter) for a newborn. Normalizing lung compliance allows comparisons between lung sizes, and when specific compliance is used, a similar value of about 0.08/cm H_2O is found for a wide spectrum of mammalian lungs, ranging from mouse to elephant.

The compliance of the lung and chest wall provides important information about the functional state of the pulmonary system. Lung injury can alter the functional state, and changes in compliance can be used to assess the degree of lung injury. For example, fibrosis caused by sarcoidosis or chemical injury makes the lung stiffer (less compliant). Similarly, pulmonary congestion (vascular engorgement) and collapsed alveoli make the lung less compliant. Because persons with decreased lung compliance must generate a greater transpulmonary pressure to receive the same

volume of air, they must do more work to inspire than those with normal lung compliance.

Emphysema destroys alveolar septal tissue and causes a loss of lung elasticity. Thus, emphysema increases lung compliance. This means that a patient with emphysema can bring air into the lung very easy, but has great difficulty in exhausting air from the lung because of the loss of elastic recoil. Note that a patient with emphysema who has lost elastic recoil has an increased resting volume (FRC).

ELASTIC PROPERTIES OF THE CHEST WALL

Just as the lung has elastic properties, so does the chest wall. The chest wall is pulled in by the inward elastic recoil of the lung, while lung expansion is aided by the outward elastic recoil of the chest wall. The importance of the chest wall's elastic recoil becomes more apparent when the chest wall becomes punctured, and air is introduced into the pleural space (*pneumothorax*), as shown in Figure 20-8. When air is introduced between the chest and lung, the intrapleural pressure becomes atmospheric, and the lungs collapse. Collapsed lungs no longer "pull" the chest wall inward, and the chest springs out to approximately 70 percent of the total lung capacity.

The relationship between the elastic properties of the lung and chest wall can be visualized in what is known as a *relaxation pressure-volume curve* (Fig. 20-9). Note that the pressure in Figure 20-9 is airway pressure and not transpulmonary pressure. The relaxation pressure (airway pressure) is obtained when the subject is completely relaxed, (i.e., with the glottis open and respiratory muscles relaxed and no attempt to expand or contract the lung and chest). This can be accomplished by having the subject inhale or exhale to a given volume as measured by a spirometer, closing the valve, and then measuring the airway pressure with the subject relaxed.

The solid line in Figure 20-9 represents the relaxation pressure-volume curve for the lung plus the chest wall. Note that at FRC the relaxation pressure (airway pressure) is zero. This occurs because the recoil force of the lung and of the chest are equal but opposite in direction. At volumes above FRC the relaxation pressure is positive, which means that the inward elastic recoil of lung is greater than the outward recoil of the chest. At volumes below FRC, airway pressure is negative, which means the chest "wants" to spring out.

The elastic recoil forces of the lung and chest can be separated by measuring that of lung alone and subtracting the lung from the total (chest plus lung) to obtain the contri-

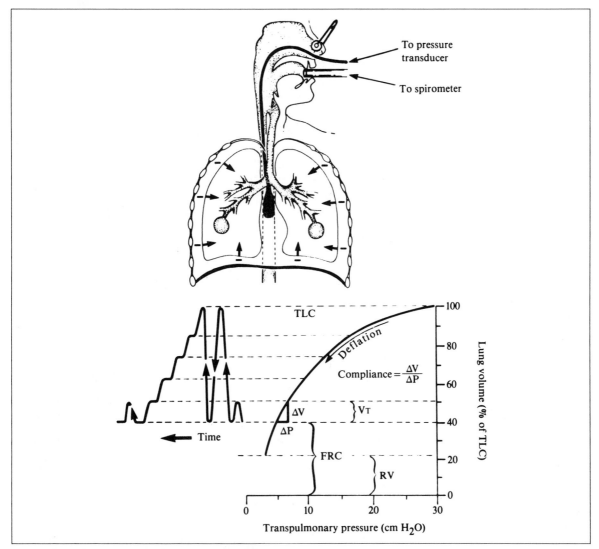

Fig. 20-7. Measurement of a pressure-volume curve. Flow during expiration is periodically interrupted, and lung volume and transpulmonary pressure are measured. Intrapleural pressure is determined from an esophageal balloon connected to a pressure transducer. Transpulmonary pressure is determined from the difference between alveolar pressure and pleural pressure. Note that when there is no airflow, alveolar pressure equals airway opening pressure.

bution from the chest wall. This pattern is shown in Figure 20-9 by the two dashed lines. The relaxation pressure-volume for the lung alone is obtained by inflating the lung by positive pressure. Notice that at FRC a positive pressure of 5 cm H_2O develops as the lung attempts to collapse.

Next, examine the dashed line for the chest alone in Figure 20-9. Note that the relaxation pressure is zero at about 70 percent of TLC. This represents the equilibrium or resting position of chest wall—it does not "want" to spring either inward or outward. At this point, the elastic recoil

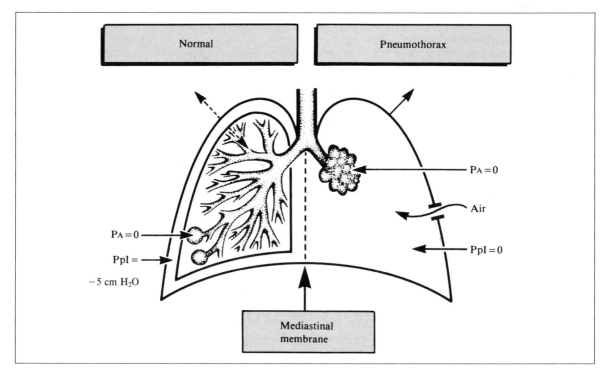

Fig. 20-8. Pneumothorax. The tendency of the normal lung to recoil inward is balanced by the tendency of the chest wall to recoil outward; as a result, intrapleural pressure is subatmospheric (-5 cm H_2O). When there is a hole in the chest wall, or nitrogen is introduced surgically into the pleural space, intrapleural pressure becomes atmospheric (zero), the lung collapses, and the chest wall springs outward.

pressure of the chest wall is zero. If the lung volume is increased above 70 percent of TLC and further expands the thorax, the chest wall will have a tendency to recoil inward, resisting further expansion and favoring a return to the equilibrium position.

The slope of all the curves in Figure 20-9 represents compliance. The compliance of the total respiratory system (lung plus chest) is one-half the compliance of either the lung or chest wall, which means that the total system is twice as stiff as either of its components. Therefore, expanding the total system requires a greater pressure change than expanding either the lung or the chest alone. Total compliance can be obtained by adding the compliance of the lung and chest in series, reciprocally.

$$\frac{1}{\text{Compliance (total)}} = \frac{1}{\text{compliance (lung)}} + \frac{1}{\text{compliance (chest)}}$$

$$0.1 \text{ liter/cm } H_2O = \frac{1}{0.2} + \frac{1}{0.2}$$

It is important to note that lung compliance obtained from a relaxation-pressure-volume curve is a special case to show the interaction between the elastic properties of the lung, chest wall, and total system and should not be confused with lung compliance obtained normally with transpulmonary pressure.

FACTORS DETERMINING THE ELASTIC BEHAVIOR OF THE CHEST WALL AND LUNGS

The elastic behavior of the chest wall depends on the rigidity of the thoracic cage and on its shape. Thus, chest wall compliance can be changed by arthritic conditions, obesity, and muscle relaxants. Remember that one component of chest wall elasticity also depends on the diaphragm and abdominal structure. Thus, abdominal disorders or skeletal muscle disorders will also affect chest wall compliance.

Lung elasticity depends on two components: (1) the elastic elements of the lung itself and (2) the surface tension at the air-liquid interface at the alveolar surface.

The inherent elastic components of the lung arises from

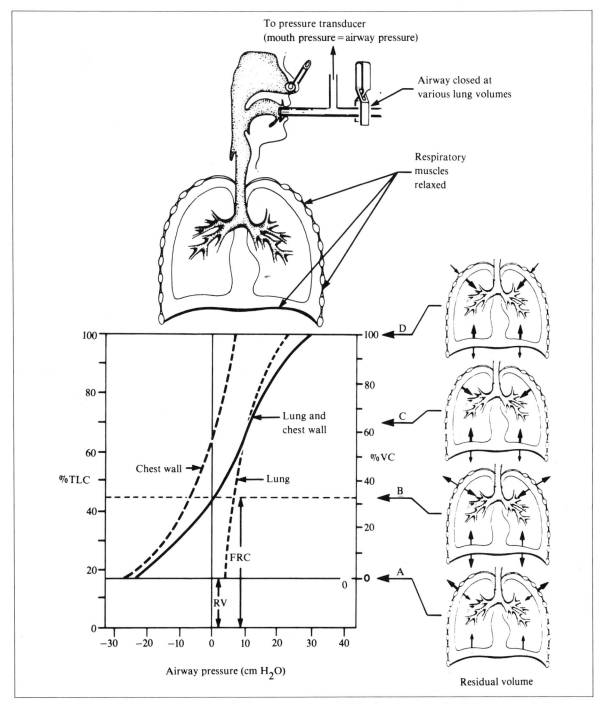

Fig. 20-9. Relaxation pressure–volume curve. Side panel shows the airway pressure of various lung volumes with the airway closed and no airflow. At residual volume (A), elastic recoil of the chest wall directed outward is much larger than elastic recoil of the lung directed inward. At FRC (B) the elastic recoil of the chest wall and of the lung equal each other, and airway pressure is zero. At approxi- mately 70% of total lung capacity (TLC) (C) the recoil of the chest wall is zero (equilibrium position of chest wall). At TLC (D) the elastic recoil of the chest wall and lung is directed inward to prevent over expansion. The elastic recoil of the lung and the chest (solid line) is the algebraic sum of the two components.

the architectural arrangements of elastin and collagen fibers around the alveoli, bronchioles, and pulmonary capillaries. Expanding the lung seems to occur through an unfolding and rearrangements of individual fibers and is somewhat analogous to stretching a nylon stocking. That is, the architectural arrangements allow for the stocking to be stretched even though the elongation of individual fibers is minimal.

SURFACE FORCES. The second factor responsible for the elastic properties of the lung is surface tension at the air-liquid interface on the inner surface of the alveolar lining. Surface tension creates a force (dynes) sometimes referred to as surface force. Surface force markedly affect lung compliance. The elastic component and surface component of the lung can be seen by examining a pressure-volume curve from an excised lung inflated first with air and then with saline (Fig. 20-10). In Figure 20-10, volume is plotted as function of transpulmonary pressure in an excised lung, with a stepwise inflation followed by a stepwise deflation, first with air and then with saline.

Two important conclusions can be drawn by comparing these two pressure-volume curves. First, the slope of the saline curve is much steeper than the air curve (deflation limb). Thus, when surface tension is minimized, the lung is far more compliant (more distensible). To say it another way: When the air-liquid interface is eliminated, the lung is less stiff, and most of the elastic recoil pressure is elimi-

nated as well. The second conclusion is that most of the elastic recoil force is due primarily to surface tension. This can be easily realized by examining the area under each curve. Since the area under each curve represents a work component, force (change in pressure) × distance (change in volume), one can readily separate the elastic force and the surface forces. The area under the saline curve represents force due to the elastic properties of the lung alone, and the area under the air curve represents total recoil force (sum of elastic and surface). By subtracting the area under the saline curve from the total area, one can readily see that approximately two-thirds of the elastic recoil force in the lung is due to surface forces. Thus, it is important to remember that lung elastic recoil is due to both the elastic elements and surface forces of the lung.

Surface tension has some important ramifications in lung function. First, let us briefly consider some of the basic properties of surface tension and then examine the relationship between surface properties and lung function. Surface tension is a force (dynes) acting across an imaginary line 1 cm long at an air-liquid interface. At the air-liquid interface there is an imbalance of intermolecular force (more liquid molecules pulling down than gas molecules pulling up), causing the surface to shrink to the smallest possible area. The resulting force at the surface is referred to as *surface tension*. Thus, surface tension is really a manifestation of attracting forces between gas-liquid molecules. Since surface tension is the force (dynes) acting across an imaginary line 1 cm long at the surface, the unit of force per unit length is dynes/cm.

In a sphere, surface tension has another important property in that it develops pressure. The relationship between surface tension and pressure inside a sphere can be seen in an alveolus (Fig. 20-11A). The pressure developed in a sphere like an alveolus is given by Laplace's law, which states that the pressure is equal to twice the surface tension (T) divided by the radius (R).

$$P = 2T/R$$

Since alveoli are connected in parallel and vary in diameter, Laplace's law assumes some functional importance in the lung. This is seen in Figure 20-11A, in which alveoli are connected in parallel and have the same surface tension but vary in diameter. As a result an unstable condition occurs, because pressure in the smaller alveolus is much greater than the larger alveolus. This causes alveoli to be unstable, especially at low lung volumes, in which the smaller alveoli collapse, a phenomenon known as *atelectasis,* and the larger alveoli become overdistended. Thus, the question

Fig. 20-10. Pressure–volume curve of an excised lung that was first inflated/deflated with air and then saline. Surface tension is minimized with saline, resulting in increased compliance.

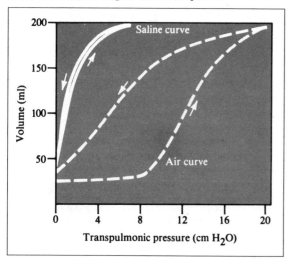

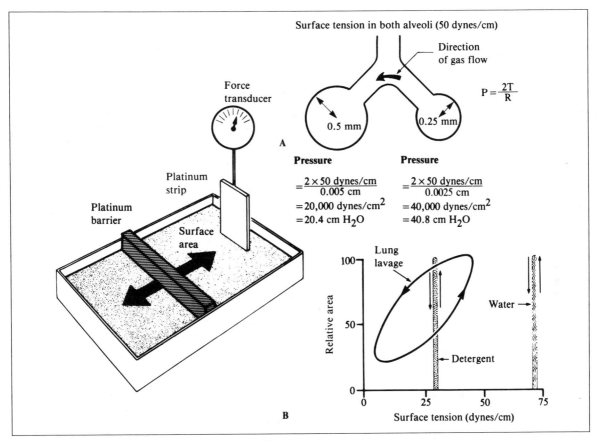

Fig. 20-11. Surface tension and alveolar pressure. A. The pressure relationship of two interconnected alveoli that vary in size but have a constant surface tension. B. Plots of surface tension–area relationships. Note that the area change corresponds to the change in alveolar surface area during inflation-deflation. Lung washings (lavage) show not only that surface tension changes as a nonlinear function of area, but also shows that very low surface tension is exhibited at low lung volumes.

arises how the lung keeps alveoli open at all, and how alveoli of different sizes coexist when interconnected.

Another complicating factor is that at FRC the pressure required to keep the lung inflated is about 5 cm H_2O. If we assume a surface tension of 50 dynes/cm (a measured value for most biological fluids) and a population of alveoli whose radii are about 5×10^{-3} cm (mean diameter for alveoli at FRC), then transpulmonary pressure, seen from Figure 20-11, would be $2 \times 50/0.005 = 20,000$ dynes/cm². If we convert this value to cm H_2O (1 cm H_2O = 980 dynes/cm²), then PTP would equal approximately 20 cm H_2O, which is

about four times higher than the normal value for a human lung at FRC.

This raises some questions: Why is the lung not unstable at low lung volumes, and why is pressure 5 instead of 20 cm H_2O as we previously calculated, assuming surface tension remained constant at 50 dynes/cm? The answer to these questions lies in the fact that the surface tension is not constant at 50 dynes/cm as seen in other biological fluids. The alveolar surface lining is coated with a special surface-reducing agent that lowers surface tension. The surface reducing agent, termed *surfactant,* not only lowers surface tension but also changes surface tension with a change in alveolar surface area (as lung diameter changes). The functional importance of surfactant is demonstrated using a surface tension balance (Fig. 20-11B). Surfactant is harvested from a lung washing (saline lavage) and layered in a Teflon trough. Surface tension is measured by dipping a platinum strip, connected to a force transducer into the liquid trough.

Table 20-1. Composition of pulmonary surface-active material

Chemical components	(%)	Phospholipid component	(%)	Fatty acid component of phosphatidylcholine*	(%)
Lipid	88	Phosphatidylcholine	83	Palmitic [16:0]	74
Protein	10	Phosphatidylethanolamine	3	Palmitoleic [16:1]	9
Hexose	<2.0	Sphingomyelin	0.4	Myristic [14:0]	6
Hexosamine	<0.5	Phosphatidylserine	0	Stearic [18:0]	4
Nucleic acid	<0.5	Phosphatidylinositol	3	Oleic [18:1]	4
		Phosphatidylglycerol	9	Other	3
		Unidentified phospholipid	<2		
				Total Saturation	84

*The first number of the fatty acid designation is chain length; the second is the number of double bonds.

Surface film is repeatedly compressed and expanded, simulating lung expansion-compression.

Figure 20-11B compares the surface tension properties of the three different substances. When distilled water is placed in the trough, surface tension is 72 dynes/cm and *independent* of surface area. When a detergent is added, surface tension is drastically reduced, but it is still independent of area. However, when fresh lung lavage is placed into the trough, surface tension is not only reduced but also changes in a nonlinear fashion with surface area. Thus, surfactant has detergentlike properties but also has another separate and distinct property that changes surface tension with area. The latter is especially important, because as the alveoli decrease in diameter (decrease surface area), surfactant decreases surface tension faster than alveolar diameter changes. Therefore, it is only possible for alveoli of different diameters, connected in parallel, to coexist and have stability at low lung volumes when surface tension changes in the same direction as the radius but faster than the radius changes.

The composition of surfactant is given in Table 20-1. Basically, the material is a lipoprotein with traces of hexose and hexosamine. The principal agent responsible for the surface-reducing properties, which exhibits the same properties as the alveolar lining layer when spread on an aqueous surface of a surface tension balance, is a *phosphatidylcholine (lecithin)*. Another phospholipid that exhibits similar properties but is found in much lower concentrations is phosphatidylglycerol.

Lung phosphatidylcholine has two features unique to lung surfactant. One is its high concentration, and the other is its fatty acid composition. Phosphatidylcholine is not unique to the lung but is found predominantly in membranes of most organs, including the membranes of the lung. In the lung,

phosphatidylcholine is present in much higher concentration than elsewhere. The fatty acid composition of membranous phosphatidylcholine consists predominantly of a saturated fatty acid (no double bonds) in the C-1 position and unsaturated fatty acid in the C-2 position. This makes for "a loose and floppy" arrangement to enhance exchange of material across the cell membrane. However, the phosphatidylcholine that is synthesized and secreted onto the surface lining of the lung, is *saturated* in both the C-1 and C-2 position. The predominant fatty acid is palmitic acic (Table 20-1). Thus, surfactant is often called *dipalmitoyl phosphatidylcholine* (DPPC).

The nonpolar fatty acids (palmitate) are hydrophobic (water repelling), are very rigid, and allow for close packing of the molecule on the gas-liquid interface. The polar end of the molecule consists of the glycerol backbone and a choline group. The molecule orients itself perpendicularly to surface of the gas-liquid interface, with the polar end immersed in liquid phase and the nonpolar portion sticking up into the gas phase (Fig. 20-12). A monolayer of DPPC has considerable mechanical stability and works by generating a film pressure that opposes surface tension at the gas-liquid interface. When the molecules are compressed (lung deflation), film pressure rises, resulting in a concomitant decrease in surface tension. At low lung volumes, when DPPC is compressed together tightly, it is squeezed out of the surface and forms miscelles. On expansion (inflation) a new film is spread onto the alveolar surface lining. This phenomenon is one reason why the expansion limb of the surface tension area curve is not the same as that of the compression limb (Fig. 20-11). During quiet and shallow breathing, in which the surface area remains fairly constant, the spreading of surfactant is impaired. Often, a deep sigh or yawn is generated to overcome this and results in the lung's

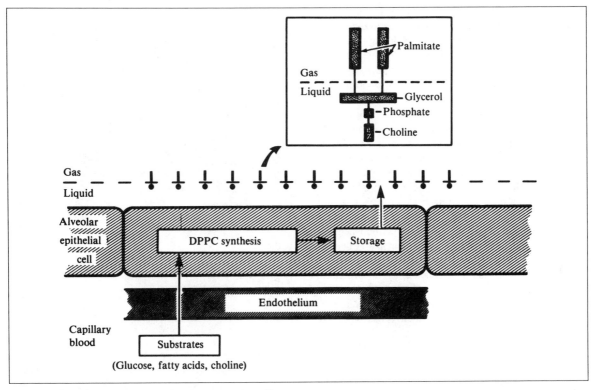

Fig. 20-12. Biosynthesis and secretion of surfactant monolayer film onto the alveolar surface lining.

inflating to a larger volume, causing new surfactant molecules to spread onto the gas-liquid interface.

Following anesthesia, patients are often encouraged to breathe deeply to enhance the spreading of surfactant. Patients who have undergone abdominal or thoracic surgery find it painfully difficult to breathe deeply and sometimes have the complication of lung-unit collapse, with resultant infection occurring.

Surfactant is synthesized in the alveolar type II cell. The alveolar epithelium consists basically of two cell types; the alveolar type I and alveolar type II cell (alveolar type II pneumocyte). The ratio of these cells in the epithelium lining is on the order of 1/1. However, the type I cell occupies approximately two-thirds of the surface area. Type II cells seem to form aggregates around the alveolar septa. They are rich in mitochondria compared with type I cells and are metabolically quite active. Electron-dense *lamellar inclusion bodies* are a distinguishing feature of the type II cell. These lamellar inclusion bodies are rich in DPPC and are thought to be the storage site of surfactant.

The overall process of surfactant production is shown in Figure 20-12. Substrates are taken up from the circulating blood (namely, glucose, palmitate, and choline) and synthesized by the alveolar type II cell. Surfactant is first stored in the lamellar inclusion bodies and then subsequently discharged onto the alveolar surface. The turnover of surfactant is high because of the continual renewal of DPPC molecules added to the alveolar surface during each expansion of the lung. The high rate of replacement of DPPC probably accounts for the active lipid synthesis that occurs in an organ whose primary function is gas exchange.

The appearance of surfactant is very important in the perinatal period. Maturation of the fetal lung is a critical event in the survival of the neonate. Preparation of the lung for the gas-exchange process at birth requires structurally intact and open alveoli as well as adequate amounts of surfactant to reduce surface tension and impart stability to the alveoli. Failure of proper lung maturation during the perinatal period is a major cause of death in the neonate. In many cases the lung is structurally intact but is still functionally immature. That is, the lung has opened and intact alveoli but inadequate amounts of surfactant to reduce surface tension and stabilize surface forces during breathing.

Since the lung is one of the last organs to develop, the synthesis of surfactant appears rather late in gestation. In the human, surfactant appears about the thirty-fourth week (term is 40 weeks). Regardless of the total duration of gestation in any mammalian species, the process of lung maturation seems to be "triggered" about the time gestation is 80 to 90 percent complete. Furthermore, there is a very close correlation between the biochemical and physiological events of fetal lung maturation (Fig. 20-13). With the appearance of surfactant, as seen by increased production of DPPC, there is a concomitant decrease in surface tension and corresponding rise in lung stability. Furthermore, the close correlation of these events with the period when gestation is about 85 to 90 percent complete indicates that the fetal lung is endowed with a special regulatory mechanism to control the timing and appearance of surfactant. Occasionally, infants are born prematurely, or have hormonal disturbances (e.g., diabetic pregnancies) that interfere with the control and timing of lung maturation. As a result these infants have immature lungs at birth, which often leads to a condition termed *respiratory distress syndrome* (RDS). This syndrome affects 20,000 to 45,000 newborns a year in the United States.

The immature lung is characterized by a deficiency of surfactant and high surface tension. The sequence of events that leads to RDS as a result of inadequate surfactant is shown in Figure 20-14. In the figure, one can readily see that an abnormal surface lining causes grave problems for the newborn. The high surface tensions make the lungs stiffer (lowers compliance), which make breathing very difficult and labored. An inordinate amount of energy is spent on just inflating the lungs, and on expiration these stiff lungs deflate to very low volumes (abnormally low FRC). Moreover, surface tension exhibits a small change with a change in surface area, which causes the alveoli to become very unstable and collapse (atelectasis) at low lung volumes. Finally, high surface tension causes pulmonary edema. These two events alone (atelectasis and edema) rapidly lead to impaired gas exchange and cause infants to become hypoxemic, hypercapnic, acidotic, and exhausted.

The hypoxemic, hypercapnic, and acidotic state leads to further complications by causing capillary endothelial damage. The endothelial damage is reflected morphologically as a pink, glistening, fibrinous alveolar membrane (leading to RDS's alternative name, hyaline membrane disease). The endothelial damage further impairs gas exchange. Unless special precautions are taken (the infant is given positive

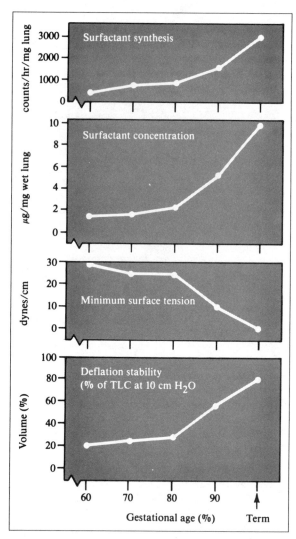

Fig. 20-13. Correlation between biochemical and physiological events during fetal lung maturation. (Adapted from P. M. Farrell and M. Hamosh. *Clinics in Perinatology: The Respiratory System.* Philadelphia: Saunders, 1978. P. 199.)

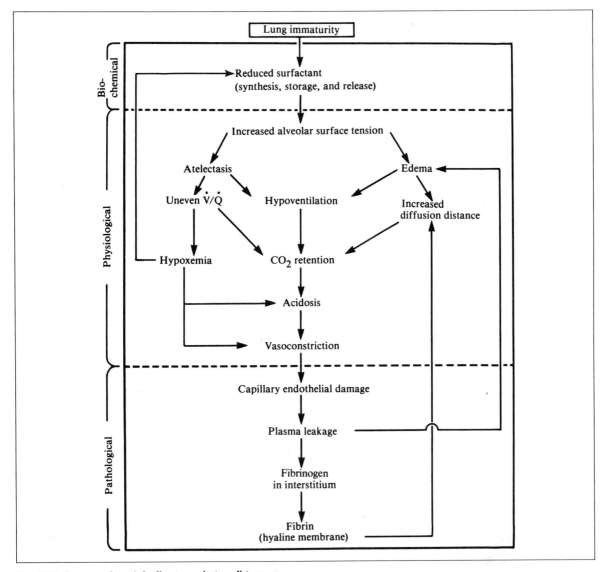

Fig. 20-14. Sequence of events leading to respiratory distress syndrome (hyaline membrane disease).

pressure ventilation), the odds of survival are 50/50. The outcome is usually dependent on a race between the worsening of gas exchange on the one hand and maturation of the surfactant system on the other.

In summary: Two decades ago, very little was known about surfactant and its functional importance. The physiological advantages of surfactant include (1) lowering surface tension in the alveoli, making the lung more compliant and thereby reducing the work required to inflate the lung; (2) promoting alveolar stability at low lung volumes; and (3) keeping the lung dry (preventing edema). The inward-contracting force that tends to collapse alveoli also tends to "pull" fluid from the capillaries. Pulmonary surfactant reduces this tendency by lowering surface forces. Some pulmonary physiologists think that this may be one of the major roles of surfactant. Inadequate surfactant probably has its most dramatic effect during the perinatal period.

INTERDEPENDENCE

Another mechanism that has been shown to play a role in promoting alveolar stability is *interdependence,* or mutual support of adjacent lung units. This is illustrated in Figure 20-15, in which alveoli (except those next to the pleural surface) are all interconnected with surrounding alveoli and thus support each other. Studies have shown that this type of structural arrangement, with many connecting links, pre-

Fig. 20-15. Schema of alveoli depicting how each unit is supported by adjacent units. Termed *interdependence*, this support plays an important role in maintaining alveolar stability.

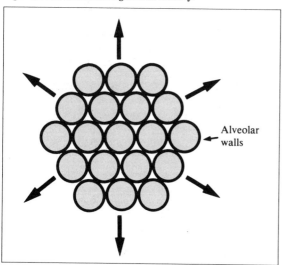

vents the collapse of adjacent alveoli. For example, if a lung unit tended to collapse, large expanding forces would be developed by surrounding units. Thus, interdependence can play a role in preventing atelectasis as well as opening up lungs that have collapsed. However, interdependence seems to be more important in the adult than in the neonate, because the neonate has fewer interconnecting links.

AIRWAY RESISTANCE

Ventilation (airflow) in the lung occurs as a result of a pressure gradient between the airway opening (mouth) and the alveoli. The pressure difference between these two ends of the airway depends on flow and resistance. The relationship between pressure, flow, and resistance is

$$Pressure = flow\ (\dot{V}) \times resistance\ (R)$$

Rearranging, it becomes

$$R = \frac{P\ (cm\ H_2O)}{\dot{V}\ L/sec}$$

Thus, the units for resistance is cm H_2O/L per sec. Resistance only assumes significance when there is airflow.

Basically, there are two major types of airflow in the lung (Fig. 20-16). *Laminar flow,* which occurs at low flow rates and is characterized by streamline flow with a flow profile that is parabolic; i.e., flow in the center of the airway is faster than flow adjacent to the walls. With laminar flow the pressure difference is directly proportional to flow times resistance.

$$P = \dot{V} \times R$$

Resistance in laminar flow is directly proportional to tube length (l) and fluid viscosity (n) and inversely proportional to the *fourth power* of the radius (r) of the airway.

$$R = \frac{8\ nl}{\pi r^4}$$

Note that if the radius of an airway is halved, the resistance increases by a factor of 16. To say this another way, if airway radius is halved, the pressure gradient required to maintain the same ventilation (or airflow) must be increased 16-fold. Note also that resistance is proportional to length. Thus, doubling length will only double resistance. Conversely, if driving pressure is constant, flow will fall one-half when length is doubled.

The second type of airflow (*turbulent flow*) occurs at high

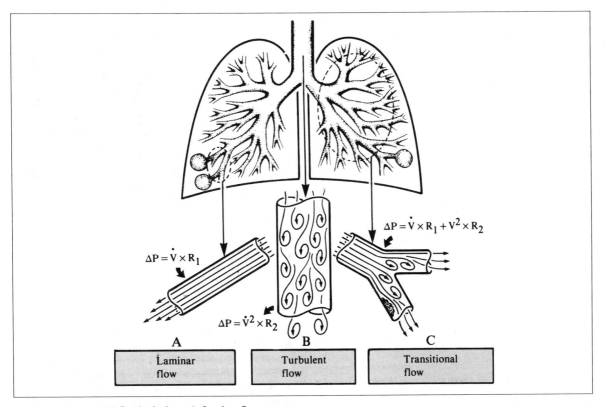

Fig. 20-16. Patterns of airflow in the lung. A. Laminar flow occurs at low flow rates and appears mainly in small peripheral airways. Driving pressure for laminar flow is proportional to the viscosity of the gas. B. Turbulent flow occurs in large airways and trachea, where flow rates are high. Driving pressure is proportional to square of flow and is also dependent on the density of the gas. C. Transitional flow occurs in larger airways, particularly at sites of branching and narrowing. Driving pressure is proportional to both gas viscosity and density.

flow rates and is characterized by air molecules moving laterally and colliding with one another. Whether flow will be laminar or turbulent is determined to a large extent by *Reynold's number (Re)*. Reynold's number is a dimensionless number and depends on flow velocity (Ve), tube diameter (D), fluid density (p) and fluid viscosity (n) according to the equation

$$Re = \frac{p \cdot Ve \cdot D}{n}$$

In turbulent flow the pressure-flow relationship changes. Airflow is no longer directly proportional to the driving pressure, as seen with laminar flow. Rather, the driving pressure required to produce a given flow is proportional to the square of flow.

$$\Delta P = V^2 \cdot R_1$$

In a straight, smooth, and rigid tube, turbulence occurs when Reynold's number exceeds 2000. In the lung, turbulence will tend to occur with high velocity in a large-diameter tube. Thus, turbulence regularly occurs in a large airway, namely, the trachea. Also, the driving pressure is more influenced by gas density than by gas viscosity. Thus, breathing helium tends to reduce turbulence because of its low density.

A mixture of these two types of flow (*transitional flow*) occurs in larger airways, particularly at branches and at sites of narrowing. With transitional flow the parabolic profile of laminar flow usually becomes blunted, and the streamline flow is interrupted and minor eddy formation may be developed. The driving pattern for transitional flow is dependent on both flow rate and its square.

$$\Delta P = \dot{V} \cdot R_1 + \dot{V}^2 \cdot R_2$$

Also, the driving pressure required to produce a given flow rate is dependent on both the viscosity and the density of the gas.

As seen in Figure 20-16, laminar flow occurs only in very small airways, where flow through any given airway is slow. Flow in the remainder of the tracheobronchial tube is mainly transitional.

MEASUREMENT OF RESISTANCE. Total airway resistance is defined as the ratio of driving pressure to an airflow. For total airway resistance (Raw), the driving pressure would be the pressure difference between the mouth (P_{mouth}) and the alveoli (P_{alv}), according to the equation

$$Raw = \frac{P_{mouth} - P_{alv}}{\dot{V}}$$

Resistance is expressed as cm H_2O/L per sec. Mouth pressure is easily obtained, but alveolar pressure can be obtained only indirectly by body plethysmography. The patient, seated in the plethysmograph pants at a frequency of two to three breaths per second, and airflow is measured with a pneumotachometer. On inspiration the lung expands and box pressure rises. During expiration the lung contracts and box pressure decreases. Alveolar pressure can be calculated from the difference in box pressure.

SITE OF RESISTANCE. A major site of resistance to airflow occurs in the upper respiratory tract. Resistance to airflow through the nasal passage is so high that breathing through the nose accounts for about 50 percent of total resistance. The major sites of airway resistance in the tracheobronchial tube are in the medium-sized bronchi (lobar and segmental) and on down to about the seventh generation (Fig. 20-17). Based on Poiseuille's equation, with the radius raised to the fourth power, one would expect the major site of resistance to be located in the narrow airways, namely, the bronchioles. For a long time, pulmonary physiologists thought this to be true until recent measurements showed that it was not. The largest increase in resistance occurs in the upper airways, down to approximately the seventh generation.

Between 10 and 20 percent of the total airway resistance can be attributed to the bronchioles (airways less than 2 mm in diameter). The reason for this apparent paradox is that as the airways penetrate the periphery of the lung, they be-

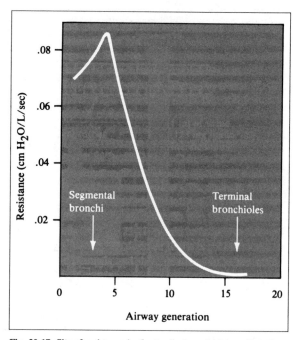

Fig. 20-17. Site of resistance in the tracheobronchial tree. Note that the major site of resistance is in the intermediate-sized airways (lobar segmental bronchi down to about the seventh generation) and that very little resistance occurs in the smaller peripheral airways.

come narrower, as well as more numerous. However, the bronchioles branch far more rapidly than their diameters decrease. Thus, the resistance of each individual bronchiole is relatively high, but their branching in parallel results in a large total cross-sectional area, causing the *combined resistance* to be low. The fact that small airways make up such a low percentage of total airway resistance causes a serious problem, because a considerable degree of airway disease can be present in the bronchioles before any unusual airway resistance can be detected.

FACTORS AFFECTING AIRWAY RESISTANCE. Lung Volume. The airways, like lung tissue, exhibit elastic properties and are capable of being compressed or distended. Since airways (the bronchi and smaller airways) are embedded in the lung parenchyma, they are interconnected by "guy wires" to surrounding tissue. Thus, a major factor that affects airway diameter, especially bronchioles on down, is lung volume. As the lung enlarges, airway diameter increases and airway resistance (Raw) falls. Conversely, at low lung volumes, airways are compressed, the diameter decreases,

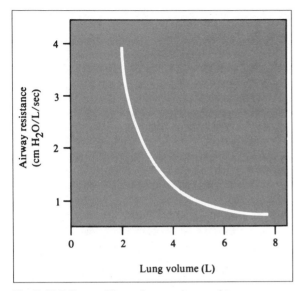

Fig. 20-18. Influence of lung volume on airway resistance.

and airway resistance rises. Figure 20-18 shows the affect of lung volume on airway resistance. Note that the relationship between lung volume and airway resistance is nonlinear. At lower lung volumes, Raw rises rapidly.

Elastic Recoil. Airway resistance is also influenced by lung elastic recoil. Elastic recoil has its effect primarily through changes in intrapleural pressure. Airway diameter varies with transmural pressure applied to the airway, (i.e., the difference between the pressure inside the airway and that ouside it). The pressure outside the airway is intrapleural pressure, even in airways that are embedded in the lung parenchyma. (see Fig. 20-3). If elastic recoil is reduced, then at any given lung volume the transmural airway pressure is reduced, causing compression of the airways and leading to greater airway resistance. Thus, many patients with emphysema (destruction of alveolar walls and loss of elastic recoil) tend to breathe at higher lung volumes. The greatest problem for an emphysema patient is not getting air into the lung but getting air out. Breathing at higher lung volumes makes the lung stiffer (increases elastic recoil), which enables the person to exhaust the air from the lung, but also reduces airway resistance. Patients with diseased airways (e.g., bronchial asthma) also tend to breathe at higher lung volumes because of "guy wires" that tend to distend airways, and the higher elastic recoil increases transmural airway pressure, which prevents airway compression during expiration.

Muscle Tone. Another factor that affects airway resistance is bronchial smooth muscle tone. A change in smooth muscle tone will change the diameter of airways. The smooth muscles are under autonomic control. Parasympathetic stimulation causes bronchial constriction, while sympathetic stimulation causes dilation. Such drugs as isoproterenol and epinephrine, which stimulate beta-adrenergic receptors in the bronchioles, cause dilation. These drugs are often used to treat asthmatic attacks by alleviating bronchial constriction.

Gas Density and Viscosity. Gas density and viscosity are additional factors that affect airway resistance. This is seen most dramatically in deep-sea diving, in which the diver must breathe air whose density is enormously increased. As a result of increased resistance, a large pressure gradient is required just to move a volume of air equivalent to tidal volume. A helium-oxygen mixture is often breathed during diving as substitute for air because it is less dense and consequently makes breathing easier. The fact that density has a greater effect on airway resistance than viscosity again indicates that flow in the medium-sized airways (the main site of resistance) is not purely laminar but transitional as well.

FLOW-VOLUME RELATIONSHIP

An assessment of flow and resistance properties of the airways can be examined further from a *flow-volume curve,* which shows the relationship between airflow and lung volume during a maximal expiration (Fig. 20-19). The flow-volume curve was discussed briefly in Chapter 19 (see Fig. 19-5). The curve is generated by having a subject inspire maximally to TLC and then exhale to residual volume as forcibly, rapidly, and completely as possible. As seen from Figure 20-19, during forced expiration, flow rises very rapidly to a maximal value, and, as lung volume decreases, it decreases linearly over most of expiration. A family of flow-volume curves can be generated by repeating full expiratory efforts over the entire vital capacity. If a person exhales from TLC, but this time makes much less effort overall, a curve like curve B in Figure 20-19 would be obtained. The maximal flow rate would be somewhat reduced, but over much of expiration the flow-volume curve would be almost superimposed on the curve obtained from forced expiration (curve A). Finally, the same subject breathes slowly from TLC and then, halfway through expiration, forcibly expires as fast as possible (curve C). The latter part of the flow-volume curve is superimposed on the first curve when the

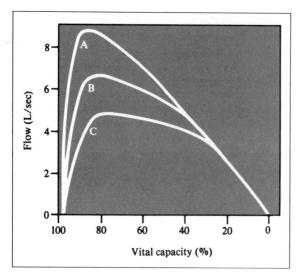

Fig. 20-19. Expiratory flow–volume curves. A. Forced expiration from total lung capacity. B. Expiration is initially slow and then forced. C. Expiration is slow until about halfway through expiration, when expiration is forcible. Note that a high lung volume, airflow during expiration increases progressively with effort, while at intermediate and low lung volumes, airflow is independent of effort.

subject exhales forcibly, rapidly, and as completely as possible.

These curves show that at large volumes close to TLC, airflow increases progressively with increasing effort. However, at intermediate and low lung volumes, expiratory flow reaches maximal levels with only moderate efforts and thereafter increases no further despite increasing efforts. Thus, airflow is influenced by both lung volume and effort. At lung volumes greater than 75 percent of vital capacity, airflow increases with an increase in intrapleural pressure (more positive), and thus is effort dependent. At volumes below 75 percent of vital capacity, flow levels off as intrapleural pressure exceeds atmospheric pressure. This means that over most of expiration, flow is virtually independent of effort, whether great or small. Since airflow remains fairly constant despite an increase driving pressure, this would indicate that resistance to flow increases proportional to the increase in driving pressure. The increase in resistance occurs because the large airways are compressed, which effectively limits flow rate.

The mechanism for airway compression is seen in Figure 20-20. In preinspiration, intrapleural pressure is -5 cm H_2O, and airway pressure is zero (no airflow). The transmural airway pressure (Paw $-$ Ppl), the pressure that holds

the airways open, is 5. At the start of inspiration, intrapleural pressure decreases to -7 cm H_2O, and alveolar pressure falls to -2 cm H_2O. Since lung volume is negligible at this point, the difference between alveolar pressure and intrapleural pressure is still 5 cm H_2O. However, there is a pressure drop from mouth to alveoli because of resistance to airflow, so the transmural airway pressure will vary. If, at a particular point, the pressure inside the airway is -1 cm H_2O, the pressure holding the airway open would be 6 cm H_2O $[(-1-(-7) = 6)]$. At the end of inspiration, intrapleural pressure decreases further (-10 cm H_2O), and airway pressure is again zero because of no airflow. At the end there is a pressure of 10 cm H_2O holding the airways open.

On forced expiration a striking change occurs. Intrapleural pressure rises above atmospheric pressure and can increase up to $+30$ cm H_2O. The pressure difference between alveolar space and intrapleural pressure is still 8 cm H_2O because, again, at the beginning of expiration the change in lung volume is negligible. Note that there is a pressure drop along the airway because of airway resistance. Thus, as one moves up the airway toward the mouth, the pressure gradient across the airways tends to collapse the airway. For example, at a point inside the airway where the pressure is 19 cm H_2O, the transmural pressure would be -11 cm H_2O, which would tend to close the airway. Thus, during forced expiration, intrapleural pressure rises above atmospheric pressure and increases alveolar pressure. Airway pressure falls progressively (due to resistance to airflow) from the alveolar region toward the airway opening (mouth).

At a point along the airway the progressive fall in airway pressure equals the intrapleural pressure. This is termed the *equal pressure point* (Fig. 20-21). At the equal pressure point the pressure surrounding the airways and the pressure inside the airways are equal (Note: Transmural airway pressure would be zero.) Downstream (toward the airway opening from the equal pressure point), the transmural airway pressure falls below intrapleural pressure and becomes negative. As a result, airways downstream are subjected to dynamic compression.

The equal pressure point divides the airway into two segments, an upstream segment (from alveolar space to equal pressure point) and a downstream segment (equal pressure point to airway opening). Once maximal expiratory flow is achieved, the position of the equal pressure point becomes fixed. A further increase in intrapleural pressure with forced expiration will simply produce more compression on the downstream segment but will have little effect on airflow in the upstream segment.

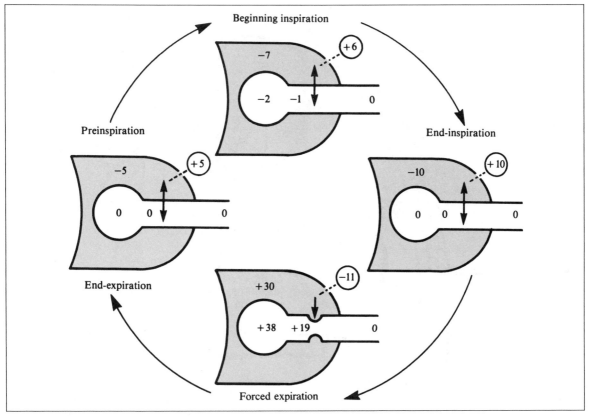

Beginning inspiration

Preinspiration

End-inspiration

End-expiration

Forced expiration

Fig. 20-20. Transmural airway pressure during forced expiration. Note that transmural airway pressure holds the airways open except during forced expiration, when airways are compressed.

Two fundamental conclusions can follow from Fig. 20-21. First, no matter how forceful the expiratory effort, flow cannot be increased, because as intrapleural pressure rises, alveolar pressure rises as well, with the driving pressure (alveolar–intrapleural pressure) remaining constant. This explains why 75 percent of the expired vital capacity is not dependent on effort. Second, maximal flow rates will be determined mainly by the elastic recoil of the lung, because it is this force that generates the alveolar–intrapleural pressure difference. As lung volume decreases, so does the elastic recoil force. The decrease in elastic recoil force is one of the reasons that maximal flow falls so rapidly at low lung volumes.

The importance of elastic recoil can best be seen when a normal lung is compared with an emphysematous lung, which has an abnormally high compliance (Fig. 20-22). In both instances, with forced expiration, intrapleural pressure

rises to $+30$ cm H_2O. The normal stiff lung (Fig. 20-22), with its elastic recoil, has added 10 cm H_2O to produce an alveolar pressure of $+40$ cm H_2O (the transpulmonary pressure at this volume is 10 cm H_2O). As a result of airway flow resistance there is a progressive fall in airway pressure. Because of the elastic recoil, the normal lung has "added" pressure to the airway that keeps its pressure above intrapleural pressure at all points in the airway, and no collapse occurs. However, in the emphysematous lung, intrapleural pressure rises to $+30$ cm H_2O with expiratory effort, but the lung has lost its elastic recoil (abnormally high compliance) and adds only 5 cm H_2O, which results in a pressure of $+35$ cm H_2O in the alveolar region. With expiratory effort, flow precedes along the progressive pressure gradient. However, the pressure inside the airway falls below intrapleural pressure well before the airway leaves the chest cavity, resulting in collapsed airways. As a result, flow stops in these collapsed airways, and airway pressure rises to equal alveolar pressure, causing the airways to open

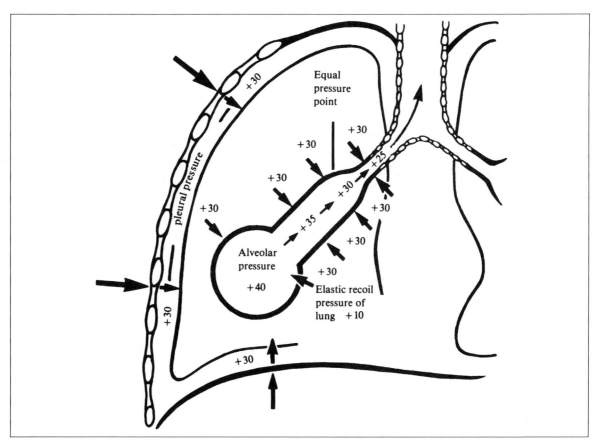

Fig. 20-21. Equal pressure point during forced expiration. Contraction of the expiratory muscles during forced expiration increases intrapleural pressure above atmospheric pressure (+30 cm H_2O). Alveolar pressure (sum of intrapleural pressure and elastic recoil pressure) is higher still. Note that airway pressure falls progressively from the alveolus to the mouth to overcome airway resistance. At the equal pressure point of the airway, pressure inside the airway equals the pressure outside (intrapleural pressure). Beyond the equal pressure point, as pressure inside the airway falls below intrapleural pressure, the airways become compressed.

again. This process continually repeats itself in the emphysematous lung, leading to a condition called a "wheeze."

UNEVEN VENTILATION

In an upright person, intrapleural pressure is more negative at the apex of the lung than at the base (Fig. 20-23). These differences between intrapleural pressure at the apex and the base of the lung are due to the effects of gravity. The gravitational effects occur because the lung is approximately 80 percent water, and gravity exerts a "downward pull" on it, which causes intrapleural pressure to become negative at the apex and less negative near the base.

Because of these gravitational effects on intrapleural pressure, the transpulmonary pressure is greater at the apex of the lung than at the base. At FRC the higher transpulmonary pressure at the apex causes the alveoli to be more expanded than at the base, causing regional differences in

compliance. At any given volume from FRC and above, the apex of the lung is stiffer than the base. It is clear from Figure 20-23 that the base of the lung has both a larger change in volume for the same pressure change and a smaller resting volume that at the apex. Consequently, greater ventilation occurs at the base. Thus, as one takes a breath, a greater portion of the tidal volume will go to the base of the lung. Although the base of the lung is relatively poorly expanded compared with the apex, it is better ventilated.

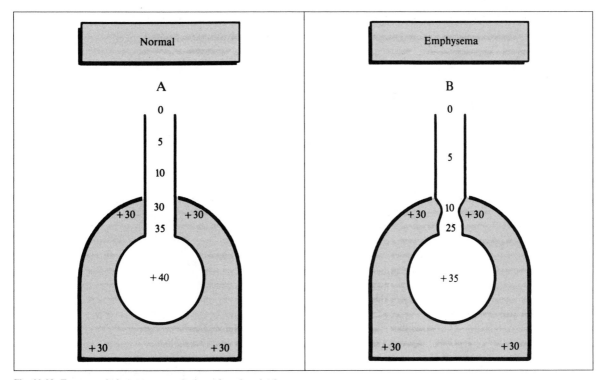

Fig. 20-22. Transmural airway pressure during a forced expiration in a normal lung and an emphysematous lung. Airway pressure in the normal lung during expiration is greater than intrapleural pressure, and airways therefore remain open. In the emphysematous lung the elastic recoil pressure is low and adds very little pressure to the alveolar region. As a result, pressure inside the airways become less than intrapleural pressure, and they collapse. Pressures are expressed in cm H₂O.

At low lung volumes a change occurs in the distribution of ventilation (Fig. 20-24). At lung volumes approaching residual volume, the intrapleural pressure at the base of the lung actually exceeds the pressure inside the airway, leading to airway closure in the periphery. At residual volume, then, the base of the lung is not being expanded but compressed, and ventilation in the base is impossible until intrapleural pressure falls below atmospheric pressure. By contrast, the apex of the lung is in a more favorable portion of the compliance curve and ventilates well. The first portion of the breath taken in from residual volume would enter alveoli in the upper region (apex). Thus, at low lung volumes the distribution of ventilation is inverted; i.e., the upper regions are better ventilated than the low regions.

AIRWAY CLOSURE

Not all of the expired air is squeezed out of the compressed region in the lung near the base of the lung (Fig. 20-24). Small airways, particularly the respiratory bronchioles, close first and trap gas in the distal alveoli. In normal subjects this airway closure occurs only at low lung volumes. However, in older subjects, airway closure occurs at somewhat higher volumes. For example, at age 65, airway closure may be present at FRC. Airway closure occurs at higher lung volumes with increasing age because of the loss of elastic properties, which causes intrapleural pressure to be less negative.

Airway closure can be assessed by the *single-breath nitrogen washout test* illustrated in Figure 20-25. The concentration of nitrogen at the mouth is measured and plotted against expired lung volume following a single full inspiration of 100 percent oxygen from residual volume to TLC. The first portion of inspiration, which consists of dead-space gas rich in nitrogen, goes to the upper regions of the lungs, while the remainder of inspiration (containing only O₂) goes preferentially to the basal region of the lung. As a

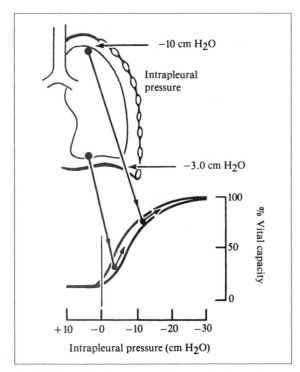

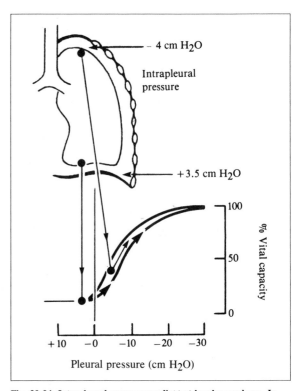

Fig. 20-23. Intrapleural pressure gradient at resting lung volume (FRC). Because of the gravitational pull, intrapleural pressure in the upright lung is more negative at the apex of the lung. As a result, the lower part of the lung is relatively more compressed at rest, but expands more on inspiration than the apex. (Adapted from J. B. West. *Respiratory Physiology* [2nd ed.]. Baltimore: Williams & Wilkins, 1979. P. 96.)

Fig. 20-24. Intrapleural pressure gradient at low lung volume. Intrapleural pressure at the apex is less negative and is actually positive at the base. As a result, airways close in this region, and no air enters. (Adapted from J. B. West. *Respiratory Physiology* [2nd ed.]. Baltimore: William & Wilkins, 1979. P. 97.)

result, the concentration of O_2 is greater in the alveolar region near the base of the lung than in that in the apex. During subsequent expiration the first portion of expired air consists of dead-space air and contains no nitrogen (phase I). Then, as alveolar gas containing a mixture of nitrogen and oxygen begins to wash out, the concentration of nitrogen in the expired air rises steeply (phase II). The nitrogen concentration then reaches a plateau (phase III). At low lung volumes, when airways at the base close, only regions in the apex continue to empty. Since the alveoli at the top part of the lung have a higher nitrogen concentration than at the base, there will be an abrupt increase in nitrogen concentration at the time airways close in the base (phase IV).

The volume at which airways close at the base, causing an abrupt increase in the slope of the nitrogen-volume curve, is referred to as *closing volume*. In young adults the closing volume is about 10 percent of vital capacity. With

age, closing volume is about 40 percent of vital capacity (equivalent to FRC). There is evidence that diseases of the small airway promote premature airway closure and result in a higher closing volume. For example, cigarette smokers show evidence of increased closing volume when the results of other tests of lung function, particularly airway resistance, are still normal. Although the closing-volume test can be applied in the early diagnosis of small-airway disease, it is important to remember that the test is nonspecific; that is, loss of lung elasticity alone will also increase the closing volume.

Remember from Figure 19-14 that the single-breath nitrogen washout test can be used to determine anatomical dead space. In addition, the single-breath test can be used to assess the degree of uneven ventilation. In a normal person the plateau phase (phase III) is fairly flat because the nitrogen coming from the alveolar region is uniformly diluted by

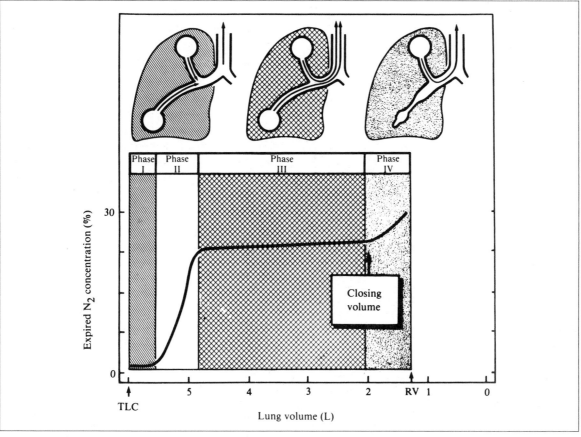

Fig. 20-25. Measurement of closing volume using the single-breath nitrogen washout test. A single breath of 100% O_2 is inhaled from residual volume to total lung capacity and then slowly expired. Expired N_2 concentration at the mouth is plotted against lung volume. **Phase I.** The first portion of the breath exhaled comes from dead space and contains only O_2. **Phase II.** Contains mixed dead space and alveolar gas. **Phase III.** Contains alveolar gas from both apex and base. **Phase IV.** Preferential emptying of alveolar gas from the apex after airway closure at the base. The upper lung has higher N_2 concentration because the apex received the first part of the tidal volume, which was dead air (rich in N_2).

the 100 percent oxygen taken in. When the distribution of ventilation is nonuniform, gas coming from different alveolar regions will have different nitrogen concentrations. This will produce a rise in nitrogen concentration during phase III. The reason for the rise is that poorly ventilated regions receive relatively little O_2 concentration and therefore have a relatively high nitrogen concentration. These same poorly ventilated regions also empty last (poor elastic recoil) and thereby cause a rise in nitrogen concentration in phase III.

WORK OF BREATHING

To sustain ventilation, energy is expended in moving the lung and chest. The concept of work is useful in expressing the energy expended in breathing. The work involved in breathing can be measured as force × distance. Force can be measured in terms of change in pressure, and distance can be measured in terms of change in volume. Thus, the

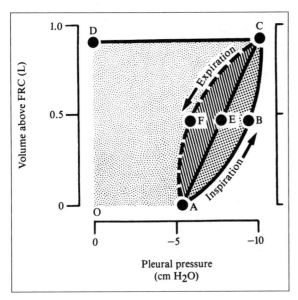

Fig. 20-26. Pressure–volume curve of the lung illustrating the work of breathing. The trapezoid (area OAECD) represents work done to overcome elastic forces during inspiration, and the dotted area (ABCEA) represents work done to overcome viscous forces (airway resistance and tissue resistance).

pressure-volume curve can be used to estimate the work required for breathing.

The work of breathing is shown conceptually in Figure 20-26, in which intrapleural pressure is plotted on the x axis and volume at any given pressure, on the y axis. Point A on the graph represents intrapleural pressure at the end of expiration, when airflow is zero and volume is at FRC. Point C represents intrapleural pressure at the end of inspiration, when flow is again zero, and the volume is 1 liter above FRC. Line AC would therefore be the compliance of the lung. (The line is assumed to be straight because lung compliance over this volume range is linear.) The total work of the lung (pressure × volume) is given by the area OABCD. Of this, the work done to overcome just the elastic force is given by the trapezoid OAECD. The dotted area ABCEA represents the work needed to overcome airway resistance plus tissue resistance. The higher the airway resistance or the higher the flow rate, the more negative the intrapleural pressure excursion between A and C becomes and the larger the area. The cross hatched area (AECFA) is work

required to overcome airway resistance (plus tissue resistance) during expiration. This area usually falls within the trapezoid area (OAECD) and represents the portion of energy stored in the lungs at the end of inspiration. The differences between areas AECFA and OAECD represents the elastic energy that is not recovered during expiration and is dissipated as heat.

Breathing is most economical when the balance of elastic force and resistive forces yields the lowest work. It is important to remember that the faster the breathing rate or the higher the flow rates, the larger is the work area (ABCEA) needed to overcome airway and tissue resistance. In contrast, the larger the tidal volume, the larger is the work area (OAECD) needed to overcome elastic force. Thus, in patients with stiff lungs (low compliance), tidal volumes tend to be small and rapid, while patients with severe airway obstruction tend to breathe more slowly in order to reduce the work. Despite this tendency, patients with obstructive disease still expend a considerable portion of their basal energy for respiratory function.

Efficiency of breathing can be calculated by the work performed and the oxygen used during inspiration and expiration as follows:

$$\text{Efficiency} = \frac{\text{work}}{\text{oxygen used}}$$
$$\%$$

The efficiency of breathing is about 5 to 10 percent, which means that most of the energy for respiration is dissipated in the form of heat.

BIBLIOGRAPHY

Altose, M. D. The physiological basis of pulmonary function testing. *Ciba Clin. Symp.* 31:1–39, 1979.

Comroe, J. H., Jr., Forster, R. E., Dubois, A. B., Briscoe, W. A., and Carlson, E. *The Lung: Clinical Physiology and Pulmonary Function Tests* (2nd ed.). Chicago: Year Book, 1962.

Nunn, J. F. *Applied Respiratory Physiology* (2nd ed.). Boston: Butterworth, 1977.

Otis, A. B. The Work of Breathing. In W. O. Fenn and H. Rohn (eds.), *Handbook of Physiology*. Washington, D.C.: American Physiological Society, 1964. Section 3: Respiration. Vol. 1.

Radford, E. B. *Tissue Elasticity*. Washington, D.C.: American Physiological Society, 1957.

West, J. B. *Respiratory Physiology* (2nd ed.). Baltimore: Williams & Wilkins, 1979.

CONTROL OF BREATHING

21

Rodney A. Rhoades

The volume of air that is moved during ventilation is regulated by a complex system. The control of ventilation is essentially automatic, incapable of appreciable voluntary control, and is housed in the brainstem. Breathing works completely without our conscious effort during sleep, under anesthesia, or while awake. Our basic rhythm (approximately 15 breaths/min) can be modified by a variety of stimuli, ranging from a change in our metabolic activity to a change in our external environment. Although the mechanism that regulates ventilation is not fully understood, ventilation is nevertheless exquisitely matched to meet our metabolic needs. A prime example of this is seen with exercise, in which ventilation is so finely "tuned" that arterial PO_2 (PaO_2), PCO_2, and pH remain fairly constant until exhaustive work rates are achieved. The ventilatory control system is also responsible for hyperventilation of high altitudes.

The control system regulating ventilation consists of several linked components of chemoreceptors and mechanoreceptors, a respiratory control center, and an ef-

fector system made up of respiratory muscles (Fig. 21-1). This chapter will concentrate on the control of basic respiratory rhythm, special pulmonary receptors, and the regulation of alveolar ventilation during normal and altered physiological states.

BRAINSTEM REGULATORY CENTERS

The basic or inherent rhythm of breathing is determined by special clusters of neurons (centers) housed in the brainstem. The centers that are absolutely essential for maintaining the basic respiratory rhythm are located in the medulla. A section through the caudal portion of the medulla leads to apnea (cessation of breathing). There are two additional centers, located in the pons, that modify the basic rhythm primarily by changing the intrinsic discharge pattern in the medullary region (Fig. 21-2).

MEDULLARY CENTER. The basic rhythm is controlled by the medullary center. Within this center there are two sepa-

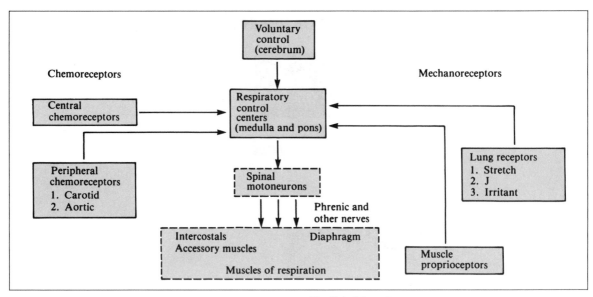

Fig. 21-1. Schema for the control of breathing.

rate and distinct groups of neurons, referred to as *expiratory* and *inspiratory neurons,* which are responsible for the cyclic pattern of inspiration. The inspiratory neuron has an intrinsic rhythm that sets the basic breathing rate. When the inspiratory neuron fires, inspiration is initiated by way of phrenic and intercostal nerve discharge. At the same time the inspiratory neurons inhibit the expiratory neurons. Expiratory neuron activity then builds up and inhibits the firing of inspiratory neurons, thereby leading to expiration. During expiration the inspiratory neurons become dormant, and then they suddenly and automatically "turn on" again after a few seconds to initiate inspiration.

If all incoming nerve fiber connection to these centers are cut or blocked, the inspiratory neurons still emit a repetitive burst of activity that causes rhythmic inspiratory patterns. Thus, there seems to be an inherent intrinsic excitability of the inspiratory neurons that sets the pace for the basic rhythm.

APNEUSTIC CENTER. The *apneustic center* is located in the pons region of the brain, and its function is not as well defined. This center provides the signal that normally terminates inspiration (inspiratory cutoff switch). Stimulation of this center results in *apneusis,* in which there are large inspiratory gasps or inspiratory "spasms."

PNEUMOTAXIC CENTER. The *pneumotaxic center* is also located in the pons and, like the apneustic center, acts

to modulate the medullary center. Stimulation of the pneumotaxic center leads to an accelerated respiratory rate, and its function seems to modulate the output from the inspiratory and expiratory neurons.

PULMONARY RECEPTORS

Neural information from the pulmonary receptors is transmitted to the brainstem to modify ventilation. The pulmonary receptors are divided into three major types: pulmonary stretch receptors, pulmonary irritant receptors, and juxtapulmonary capillary receptors, or J receptors. The pulmonary stretch receptors lie in the walls of smooth airways and are activated by stretch. Their afferent pathways travel via the vagus, and when these receptors are stretched by lung inflation, they cause a slowing of respiratory frequency. The slowing of frequency is accomplished by prolonged expiration, which is referred to as the *Hering-Breuer reflex.* This reflex also causes a concomitant increase in heart rate, bronchodilation, and vasoconstriction. Recent studies have shown that the Hering-Breuer reflex appears to be more important in animals than in humans. The reflex does, however, manifest itself in the human neonate.

The pulmonary irritant receptors are found in airway epithelium and activated by both particulate matter (dust) as well as chemical irritants (noxious gases) such as sulfur dioxide. The receptors, when activated, elicit a cough reflex and bronchoconstriction. Recent studies have shown that

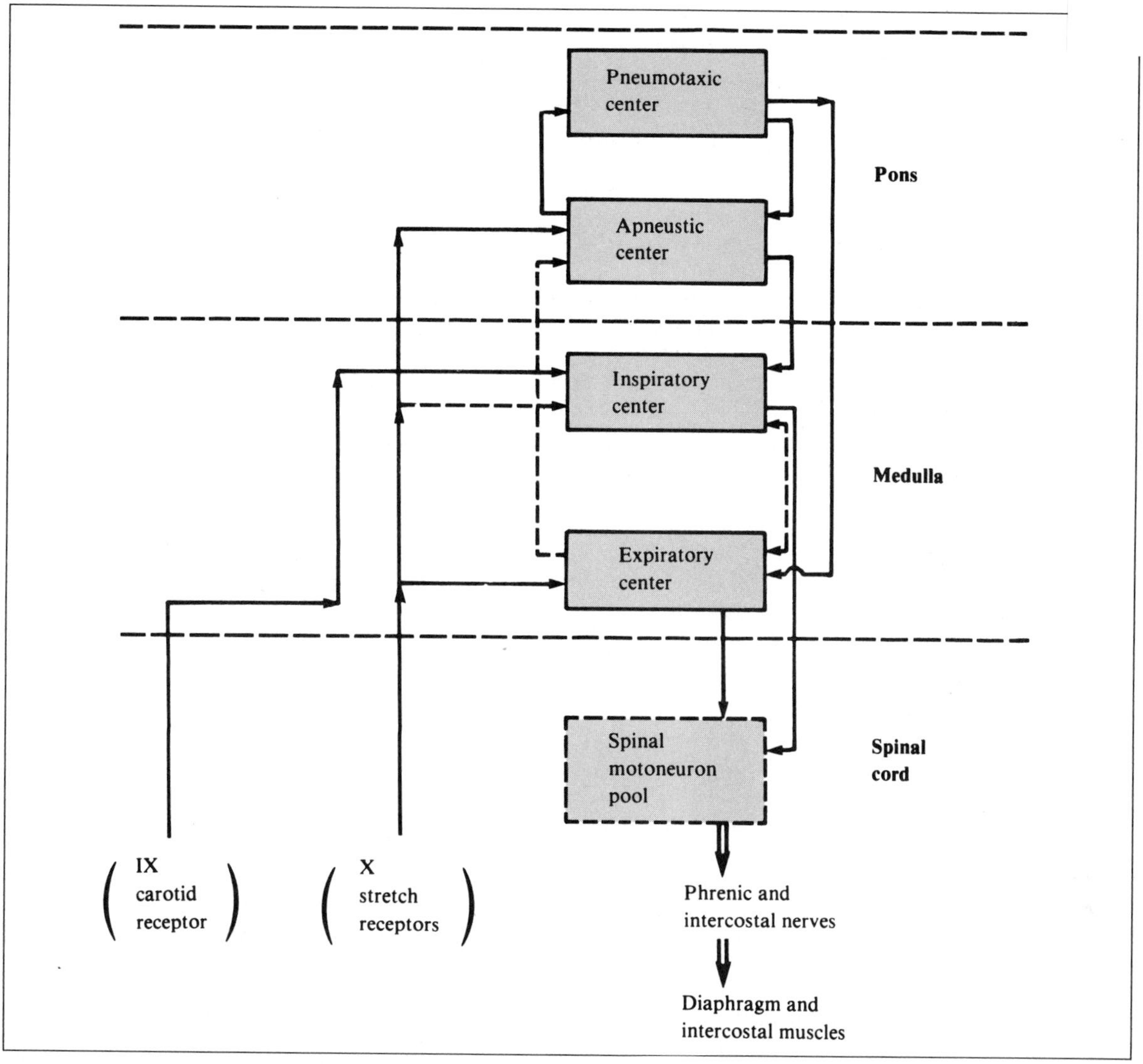

Fig. 21-2. Central respiratory centers in the brainstem. IX = input from the glossopharyngeal nerve; X = input from the vagus nerve; solid arrows = stimulatory influences; dashed arrows = inhibitory influence; double arrow = efferent motor pathways to muscles of breathing.

these receptors are also excited by histamine, which suggests that they may be involved in the reflex bronchoconstriction that is triggered by histamine release during an asthmatic attack.

The J receptors are located in the walls of the pulmonary capillaries and have been shown to be stimulated by interstitial distortion, such as edema, pulmonary congestion, and air emboli, and by low lung volumes. The reflex generated from the J receptors lead to hypotension, bradycardia, and dyspnea (labored breathing).

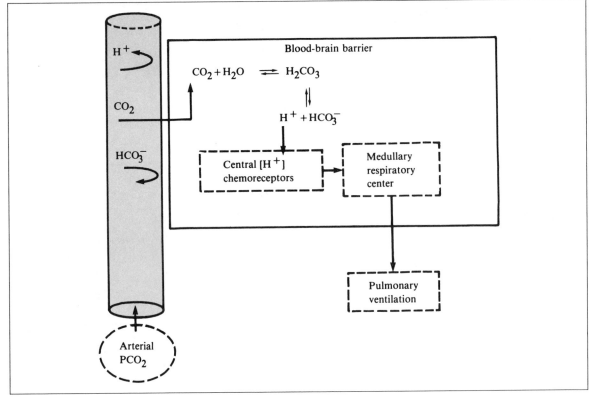

Fig. 21-3. Medullary chemical control of ventilation.

CHEMICAL CONTROL

In addition to neural control, pulmonary ventilation is controlled chemically, namely, by CO_2, H^+, and O_2 levels in the arterial blood. Changes in these three substances in the blood are sensed by chemoreceptors that are located both centrally and peripherally (see Fig. 21-1). The receptors located centrally are referred to as *medullary chemoreceptors* and are found adjacent to the expiratory and inspiratory neurons on the ventrolateral surface of the medulla. These chemosensitive areas in the medullary region have been located physiologically but not anatomically; that is, discrete areas are sensitive to changes in $[H^+]$ and PCO_2 that evoke a powerful stimulation of ventilation, but no discrete neurons have been identified. The medullary chemosensitive areas are stimulated by changes in $PaCO_2$ but not by changes in O_2 or arterial pH. The reason for this is that the blood brain-barrier is permeable to CO_2 but not to H^+ (Fig. 21-3). The CO_2 diffuses across the blood-brain barrier and

forms carbonic acid, which readily disassociates to form H^+ and HCO_3^-. Since the cerebrospinal fluid (CSF) is an ultrafiltrate, its protein concentration is extremely low, and consequently the buffering capacity of the CSF is extremely poor. Thus, any change in $PaCO_2$ has a profound effect on $CSF[H^+]$, and it is the $CSF[H^+]$ that actually stimulates the chemosensitive areas. Consequently, some pulmonary physiologists think the medullary chemoreceptors should be called $[H^+]$ receptors.

It is important to note that arterial HCO_3^- will rise with elevated CO_2. The blood-brain barrier is impermeable to bicarbonate (Fig. 21-3). If, however, $PaCO_2$ is maintained at high levels, the CO_2 will diffuse across the blood-brain barrier and will elevate the $CSF[HCO_3^-]$. The rise in bicarbonate in the CSF will tend to return pH to normal in a day or so, thereby minimizing ventilatory stimulation from arterial $PaCO_2$ even though the CO_2 levels in the blood remain high.

Hypoxia does not stimulate the medullary chemoreceptors, but it has have an indirect effect. At low PO_2, cerebral

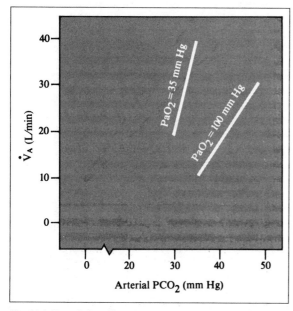

Fig. 21-4. Potentiating effect of hypoxia on ventilatory response to CO_2.

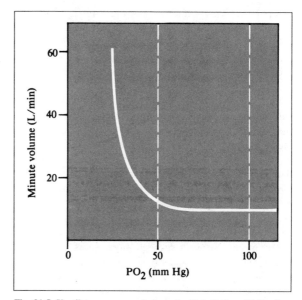

Fig. 21-5. Ventilatory response to hypoxia. Note that ventilation is essentially constant at values above an arterial PO_2 of 50 mm Hg.

tissue becomes anaerobic, which leads to an accumulation of lactic acid in the CSF, resulting in an indirect stimulation of ventilation by way of an increase in $[H^+]$. Hypoxia acts indirectly in another manner by having a potentiating effect on arterial CO_2. This is seen in Figure 21-4, where a lower arterial oxygen tension results in a stronger ventilatory response for the same arterial Co_2 tension.

PERIPHERAL CHEMORECEPTORS

The peripheral chemoreceptors are located in two regions, in the arch of the aorta and near the bifurcation of the common carotid artery, hence their name, *aortic* and *carotid chemoreceptors*. These receptors are fast responding and are stimulated by arterial CO_2, O_2, and pH. Of the two chemoreceptors, the carotid bodies are almost exclusively responsible for changes in ventilation. The carotid bodies transmit impulses to the medullary center via the carotid sinus nerve and glossopharyngeal nerve, causing a change in both tidal volume and frequency. The carotid body is stimulated by low PaO_2 (hypoxemia), low pH (acidosis), and elevated $PaCO_2$ (hypercapnia). The carotid body is not as sensitive to O_2 as it is to $[H^+]$ and CO_2. Arterial PO_2 must be decreased to about 50 mm Hg before respiration is dramatically increased (Fig. 21-5).

As seen in Figure 21-5, ventilation is essentially constant at values above an arterial tension of 50 mm Hg. Thus, it appears that the carotid chemoreceptors are designed to guard against hypoxia in the tissues rather than regulating respiration. The carotid bodies, unlike O_2, respond somewhat linearly to arterial CO_2 and pH over a physiological range (CO_2, 30–60 mm Hg; pH, 7.25–7.45). Of the three substances, CO_2 provides the most powerful stimulation to respiration. A rise of only 2 to 4 mm Hg is capable of doubling resting minute ventilation. The reason that CO_2 is such a potent stimulant to respiration is that, unlike arterial O_2 and $[H^+]$, it acts on both the medullary and peripheral chemoreceptors to stimulate ventilation in the same direction. Of the two receptors, the medullary chemoreceptor is the most sensitive to CO_2. About 80 percent of the increase in ventilation resulting from elevated CO_2 comes from the medullary receptors, and about 20 percent comes from the carotid bodies.

Low CO_2 has an effect opposite to that of high CO_2 in that it depresses respiration. The depressant effect of low CO_2 can be so strong that it can override a hypoxic and/or acidotic drive to stimulate ventilation. Its depressant effect on respiration can potentially lead to a serious problem in swimmers who hyperventilate before diving under water. Hyperventilation blows off large amounts of CO_2, thereby lowering $PaCO_2$ without any dramatic increase in O_2 content or PO_2. Hyperventilation prior to swimming under water allows the person to stay under much longer because the low arterial CO_2 depresses the ventilatory drive in the medullary center. However, the muscles are simultaneously utilizing oxygen at an accelerated rate and thus lowering arterial PO_2.

This situation is serious for two reasons. First, only the carotid bodies and not medullary receptors are sensitive to hypoxia. Second, the arterial PO_2 must fall to around 50 mm Hg before any stimulation to respiration occurs. When hypoxic stimulation does occur at this low PO_2, it is overridden by the depressant effect of the low CO_2 on the medullary center. Consequently, the brain can become hypoxic before CO_2 can be built up again to provide a strong stimulus for breathing, and the swimmer looses consciousness under water.

The mechanism by which chemoreceptors respond to changes in O_2, CO_2, and pH is not exactly known. The ultrastructure of the carotid body consists of several types of so called glomus cells (sympathetic and parasympathetic nerve cells) that are richly innervated by afferent neurons. It is not clear whether it is the glomus cells or the nerve-fiber endings themselves that act as chemosensitive receptors. Recent studies have shown, however, that the glomus cells contain large amounts of catecholamines, especially dopamine. One current theory is that the nerve-fiber endings act as the chemoreceptors and that the glomus cells, rich in dopamine, are inhibitory in nature and act to modulate impulse generation by the afferent endings.

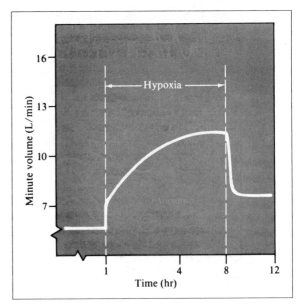

Fig. 21-6. Time course of minute ventilation before, during, and after chronic hypoxic exposure (8 hr) in a normal subject.

RESPONSE TO HYPOXIA

The ventilatory response to hypoxia is mediated directly through the peripheral chemoreceptors, namely, the carotid body. The stimulus for the hypoxic drive for hyperventilation is *arterial PO_2* rather than O_2 content. Thus, ventilation would not be stimulated in patients with anemia, because their arterial PO_2 is normal, and the carotid chemoreceptors are not activated.

Hypoxic hyperventilation appears to occur in two stages. In the first stage there is an immediate increase in ventilation, which lowers PCO_2 and elevates pH. As a result, this initial increase in ventilation is rather small. For example, if a person breathes 12 percent oxygen at sea level, which is equivalent to breathing air at 14,000 feet, this will lead to an immediate rise in ventilation, but the increase will be about 15 percent or less of the subject's total ventilation. Ventilation increases so little because the hypoxic hyperventilation lowers arterial PCO_2 and depresses the medullary

chemoreceptors. As a result the peripheral and the medullary chemoreceptors work against each other. The decrease in arterial pH is also antagonistic to the hypoxic drive.

The second phase of hypoxic hyperventilation is referred to as *ventilatory acclimatization*. This occurs with chronic hypoxic (exposure for more than 3 hours). In the second phase, with prolonged hypoxia, ventilation continues to rise in the face of a continued decrease in PCO_2; with both the ventilatory increase and the decrease in alveolar PCO_2 tending to plateau at around 8 to 10 hours (Figure 21-6). Another feature of ventilatory acclimatization seen is that after removal from the hypoxic environment, neither alveolar PCO_2 nor ventilation returns to normal levels (Fig. 21-6). In many instances a few days at sea level are required before ventilatory acclimatization to chronic hypoxia disappears.

The importance of ventilatory acclimatization is that it minimizes a decrease in arterial PO_2 and prevents O_2 desaturation. The mechanism that allows ventilatory acclimatization to occur with chronic hypoxia is not completely understood. The initial rapid hyperventilation that occurs with the onset of hypoxia is mediated through the carotid body. However, the mechanism responsible for the second phase, i.e., the continual hyperventilation that occurs over a day or so in the presence of hypocapnia, is not well defined. Recent studies have shown that even though the blood and CSF are alkaline, the medullary chemoreceptors may be sufficiently acid to drive respiration because of increased lactic acid production from hypoxic brain tissue.

RESPONSE TO EXERCISE

With increased muscular activity, O_2 consumption and CO_2 production increase. The control of ventilation is adjusted to meet the increased metabolic demand. One of the striking features of exercise hyperventilation is that it occurs in the presence of a normal arterial PCO_2, PO_2, and pH. The details of the control of breathing and other ventilatory responses to exercise will be presented in Chapter 29.

BIBLIOGRAPHY

Comroe, J. H., Jr., Forster, R. E., Dubois, A. B., Briscoe, W. A., and Carlson, E. *The Lung: Clinical Physiology and Pulmonary Function Tests* (2nd ed.). Chicago: Year Book, 1962.

Cohen, M. I. Neurogenesis of respiratory rhythm in the mammal. *Physiol. Rev.* 59:1105–1173, 1979.

Mitchell, R. A. Control of Respiration. In E. D. Frohlich (ed.), *Pathophysiology* (2nd ed.). Philadelphia: Lippincott, 1976. Pp. 131–147.

West, J. B. *Respiratory Physiology* (2nd ed.). Baltimore: Williams & Wilkins, 1979.

RENAL FUNCTION

22

Ewald E. Selkurt

The kidney plays a dominant role in maintaining the constancy of the internal environment. It does so in the first instance by the regulation of the water content of the body with the aid of the antidiuretic hormone (ADH) released from the hypothalamo-neurohypophyseal system. This regulation is closely integrated with the maintenance of proper salt balance in the body, i.e., the proper proportion of sodium, potassium, calcium, chloride, phosphate, bicarbonate, and sulfate. In the control of electrolyte balance, other endocrine regulations are important (see Chap. 34).

A major factor in proper electrolyte balance is the renal adjustment of the acid-base balance. Since the problem is primarily one of ridding the body of the acids created by metabolism, the kidney favors excretion of acid radicals through an ion exchange mechanism whereby hydrogen ion is secreted in exchange for sodium, making the urine acidic. Second, the kidney forms ammonia, which is secreted by the tubular cells, replacing valuable cation. Thus, the renal mechanism for acid-base regulation is basically a cation

conservation mechanism. It will be described in greater detail in Chapter 24.

The kidney has the further responsibility of the excretion of waste, such as urea, uric acid, creatinine, and creatine. It must perform this function and others while conserving valuable foodstuffs, such as glucose and amino acids. These must be reabsorbed selectively while the undesirable waste products are eliminated. Certain additional functions include detoxification, e.g., the conjugation of glycine and benzoic acid to form the more innocuous hippuric acid, which is rapidly secreted by the tubular cells into the urine. Other metabolic functions, such as amino acid oxidation and deamination, also occur in the tubular cells.

In summary: The problem of excretion by the kidney involves filtration at the glomeruli of all but cellular constituents and the plasma proteins; further, it is the function of the tubules to reabsorb selectively the necessary valuable substances and to reject waste products and undesirable excesses of anything taken into the body. In addition, certain substances may be added to the urine by tubular secretion.

FUNCTIONAL ANATOMY OF THE KIDNEY

The functional unit of the kidney is the nephron, and there are approximately 1¼ million nephrons per kidney in the human. The nephron is diagrammed in Figure 22-1 with the associated blood supply. It consists of the Malpighian corpuscle and the attached tubular system. The Malpighian corpuscle includes Bowman's capsule, which consists of an inner *visceral* layer and an outer *parietal* layer. The filtering area of the capsule has been estimated to be about 0.8 mm^2, which would represent 2 m^2 of filtering surface for both kidneys. The visceral layer of Bowman's capsule is in intimate contact with the capillary loops important in filtration; the unit is called the *glomerulus*. Bowman's capsule continues into the *proximal convoluted tubule*. The proximal segment has the widest diameter of any portion of the nephron. It is made of truncated pyramidal cells characterized by a *brush border* on the inner or luminal aspect. At their basal margins they have striations perpendicular to the basement membrane. These striations are related to the distribution of the mitochondria in the cell. The proximal convoluted tubule turns toward the medulla to become the *loop of Henle*. The descending limb is at first of a thickness comparable to that of the proximal convoluted tubule but is then replaced by highly attenuated cells of the thin segment, which is characterized by a flattened epithelium. This makes a hairpin turn in the medulla, then turns up toward the cortex into the ascending thick limb of the loop. The thin segment is found only in mammals and in a small percentage of the nephrons of birds. The average length varies considerably in different mammals and, indeed, varies considerably within the human kidney, depending on the location of the nephron (Fig. 22-1).

The ascending limb of the loop of Henle has cells that are at first cuboidal but become more columnar as they approach the cortex. This becomes the *distal convoluted tubule,* the cells of which lack brush borders but exhibit basal striations. This segment joins the treelike system of *collecting tubules* and *ducts.*

RENAL CIRCULATION

ANATOMICAL ASPECTS. The renal artery divides into *interlobar arteries,* which subdivide into primary, secondary, and tertiary *arcuate arteries,* from which spring *interlobular arteries.* The *afferent arterioles* arise from the interlobular arteries; in the dog, the afferent arterioles usually supply one glomerulus but, rarely, may branch to supply two to four glomeruli, with a total of 200,000 per kidney. This number compares with estimates ranging from 600,000 to 1,700,000 in each human kidney.

The glomerular capillaries converge in the efferent arterioles and then in a second capillary bed around the convoluted tubules, which drains into the *interlobular veins.* These drain into the *arcuate veins* and *interlobar veins.*

Although Figure 22-1 suggests a one-for-one relationship of the efferent arterioles to the peritubular capillary network, this is an oversimplification. While such a relationship may prevail in the superficial cortex, studies by Beeuwkes (1980) in the dog kidney show that it does not apply to the mid-cortex and deep cortex. In the bulk of the cortex, proximal tubules were found to be dissociated from the efferent vessels of the same glomerulus. Thus, the blood supply to a given nephron may come from a common capillary bed derived from free anastomoses of the efferent arteriolar branches of other glomeruli. This finding has implications for the analysis of single-nephron function and its relationship to its peritubular environment.

Arterioarterial and *venovenous* anastomoses have been found in the kidneys of humans and dog. Although arteriovenous anastomoses have been said to exist, their occurrence is infrequent.

JUXTAGLOMERULAR APPARATUS. In the normal human kidney the afferent arteriole, as it approaches the glomerulus, shows a significant increase in the number of cells in the media, thickening the arteriolar wall to form an asymmetrical cap (*polkissen*) on one side of the glomerulus. The polkissen is composed of small, spindle-shaped, afibrillar cells and juxtaglomerular (JG) cells. They may have a particular role in regulation of renal blood flow (RBF). These act as baroreceptors. In rats they show changes in granularity with variations in systemic blood pressure. Some investigators feel that renin, which plays a part in production of angiotensin, the pressor principle in renal hypertension, is formed in these cells.

A portion of the distal tubule that lies opposite the JG cells shows a distinct modification characterized by condensation of nuclei on one side of the tubule in an epithelial plaque, the *macula densa*, which together with the JG cells, is called the *juxtaglomerular apparatus* (JGA) or *complex*. It is of interest to note that the macula densa is contiguous with both afferent and efferent arterioles.

It has been suggested that the JG cells and macula densa form a regulatory system responding to changes in the composition of the distal tubular fluid, particularly the sodium

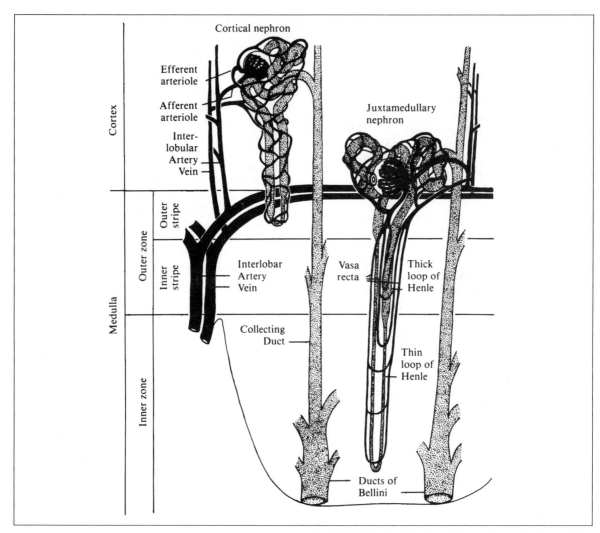

Fig. 22-1. Cortical and juxtamedullary nephrons of the human kidney, with representative blood supply. (From R. F. Pitts. *Physiology of the Kidney and Body Fluids* [3d ed.]. P. 8. Copyright © 1974 by Year Book Medical Publishers, Inc., Chicago. Used by permission.)

concentration. This in turn regulates renin production (Nash et al., 1968).

In 1968, Thurau proposed a hypothesis that a local control system regulating the caliber of the afferent arteriole involved angiotensin II production within the JG cells in response to sodium chloride concentration in the distal tubular fluid. The notion was subsequently revived (Thurau and Mason, 1974) with the presentation of some evidence that renin, converting enzyme, and substrate angiotensinogen are found in the JGA.

Another control system involves the *mesangial cells* of the JGA. Heretofore, they have been considered to be the

structural supporting framework of the glomerulus. Recent evidence (Caldicott et al., 1981) suggests a possible role in regulating glomerular function via the influence of the renin-angiotensin system. Receptors in the mesangial cells react to angiotensin II by reducing glomerular size and hence filtration rate.

BLOOD SUPPLY TO THE MEDULLARY ZONE. The glomeruli and associated tubules that lie deep in the cortex adjacent to the medulla (*juxtamedullary glomeruli*) have distinctive modifications of their circulation compared to the cortical glomeruli and nephrons. They are typified by long loops of Henle dipping into the medullary zone and papillary portions of the kidney, and have associated modified vascular structures. The efferent arterioles from these glomeruli not only break up into the typical capillary supply to the convoluted portions but also form long hairpin loops of thin-walled blood vessels, the *vasa recta*, which accompany the loops of Henle. Although thin walled, they have a diameter several times greater than that of the typical peritubular capillary. In humans the juxtamedullary glomeruli constitute about 20 percent of the total number of glomeruli in each kidney.

MEASUREMENT OF RENAL BLOOD FLOW. Both direct and indirect methods have been used to measure RBF. Direct methods employ flowmeters of various types, which record either arterial inflow or venous outflow. The indirect methods are an application of the *renal clearance principle*. This is based on the clearance ratio $C = U\dot{V}/P$, in which U is the urinary concentration in milligrams per milliliter, $\dot{V}$ is the minute urine volume, and P is the plasma concentration (usually systemic vein) in milligrams per milliliter. The derivation and meaning of the plasma clearance ratio is explained in greater detail under Glomerular Filtration Rate.

Fick Principle Method. The requisite for the clearance of a substance to measure plasma flow is that it be entirely removed (or nearly so) from the plasma in one transit through the kidney. Clearance is verified by examination of the concentration in arterial plasma (A_c) and in the renal vein (V_c), and by application of the *extraction ratio*

$$E = \frac{A_c - V_c}{A_c}$$

Hence, if V_c equals 0, E will equal 1.0. E is less than unity to the extent that the material is not removed by urinary excretion. It is clear that if the renal clearance $U\dot{V}/P$ is divided by the extraction ratio

$$\frac{U\dot{V}/P}{(A_c - V_c)/A_c} \text{ or } \frac{C}{E}$$

the resultant quotient will yield the total plasma flow. The latter is an expression of the Fick principle. For most substances, systemic vein concentration (P) and arterial concentration (A_c) are identical, so that these cancel out to yield the familiar Fick equation,

$$RPF = \frac{U\dot{V}}{A_c - V_c}$$

where RPF is renal plasma flow. To obtain total RBF, the hematocrit measurement of the blood is introduced as follows:

$$RBF = \frac{C}{E} \times \frac{1}{1 - \text{hemat.}}$$

Several substances are so efficiently removed by combined processes of glomerular filtration and active tubular (proximal) transport (secretion) at low plasma concentrations that the renal vein concentration is very low (i.e., extraction is nearly complete, and E close to unity). These include iodopyracet (Diodrast) (D) – E_D is 0.74 – and para-aminohippurate (PAH) – E_{PAH} is 0.80 to 0.85 in the dog and 0.90 in humans. Then, C_D or C_{PAH} is nearly equivalent to plasma flow. The fact that extraction is not complete is interpreted as indicating that a small fraction of blood does not perfuse the excretory tissue: capsule, inert supportive tissue, and medullary tissue (loops of Henle, collecting ducts), calycine mucosa, and pelvis. On this basis, Smith (1951) referred to the measured clearance as the "effective" plasma or blood flow.

The Fick principle can be employed with any substance cleared by the kidney that shows a measurable arteriovenous (A-V) difference. Obviously, the smaller the A-V difference, the more prone to error the calculation will be. Thus, phenol red, urea, mannitol, and inulin have been employed, but they have considerably smaller A-V differences than do Diodrast and PAH.

Other Indirect Methods. An application involving uptake and then washout of the radioactive gases krypton (^{85}Kr) and xenon (^{133}Xe) injected into the renal artery has been used in dogs and humans. The washout curve can be broken down into different components, signifying differential blood flow rates through different compartments. Following this procedure, computations of cortical blood flow and medullary blood flow have been made (Thorburn et al., 1963; Ladefoged, 1966). Representative values for the human kidney are as follows: cortex, 3.6 to 6.3 ml/min/gm of cortical tissue; outer medulla, 1.25 ml/min/gm of tissue.

The distribution of 18- to 36-μm radioactive microspheres (^{85}Sr, ^{145}Ce, ^{169}Yb) in the renal substance after intraarterial injection yields information on regional cortical flow; four zones have been discerned from outer to inner cortex, giving information on juxtamedullary blood flow. The microspheres do not pass the glomeruli and hence are excluded from the medulla (McNay and Abe, 1970). In dogs at normal blood pressure, the zonal flows, outer cortex to inner cortex, averaged 4.18, 6.80, 3.07, and 1.68 ml per gram per minute, respectively.

RENAL BLOOD FLOW VALUES. Total RBF averages about 350 ml per minute in the dog (average weight, 15 kg) and about 1200 ml per minute in humans, or about one-fifth of the cardiac output. Renal blood flow averages about 3.5 to 4.0 ml/min/gm of kidney weight.

OXYGEN UTILIZATION. The renal venous blood contains considerably more oxygen than does that draining other tissues. The resulting small A-V oxygen difference (1.7 volumes/100 ml in humans and about 3.0 volumes/100 ml in the dog and cat) remains constant over a wide range of renal blood flows, although it may increase at very low flow rates. Thus, oxygen consumption (normally 0.06–0.10 ml/gm/min in the dog) is related to flow, so that when flow is reduced, as in shock, the organ ordinarily does not increase its extraction but apparently suffers curtailment of oxidative metabolism.

Warburg kidney tissue slice studies have been done in the guinea pig, dog, and cat. The average Qo_2 values are as follows: in cortex, 21.3 mm^3/mg dry tissue/hr; outer medulla, 15.1; and inner medulla, 6.2. These findings lead to the conclusion that the structures of the inner medulla (loops of Henle and collecting ducts) do not have important energy-requiring functions or, alternatively, operate in part anaerobically. This zone is important in the countercurrent urinary concentration process. The blood flow to the medulla has been calculated to be small, perhaps less than 10 percent of the total—0.7 to 1.0 ml/min/gm in the outer medulla and only about 0.2 ml/min/gm in the inner medulla compared with 4.0 in the cortex (Fig. 22-2). The cortex, on the other hand, contains the segments of the nephron involved in active sodium transfer, and it is the sodium-reabsorptive mechanism that largely accounts for the renal oxygen utilization.

The important relationship of renal oxygen utilization to sodium transport is depicted in Figure 22-3. In Figure 22-3A, utilization in ml O_2/min/100 gm of kidney weight is related to an experimentally varied range of RBF and related glomerular filtration rate (GFR). Note the inflection and linear increase at about 225 ml/min/100 gm of kidney weight, where GFR presumably begins, with attendant filtration and reabsorption of sodium salts by the nephrons. The increase is linear with the progressive increase in RBF and related GFR. To the left of 225 ml per minute, baseline consumption of oxygen by renal tissue metabolism is implied. In

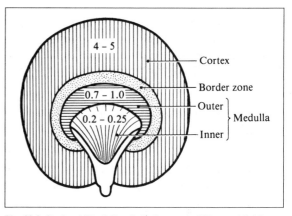

Fig. 22-2. Regional blood flow (ml/min per gm kidney weight) in the kidney. (From K. Thurau. In J. R. Brobeck [ed.], *Best and Taylor's The Physiological Basis of Medical Practice* [10th ed.]. P. 225. Copyright © 1979, The Williams & Wilkins Co., Baltimore.)

Figure 22-3B, oxygen utilization is expressed in mEq/min/100 gm of kidney weight on the ordinate, as related to absolute sodium reabsorption in mEq/min/100 gm of kidney weight.

AUTONOMY OF THE RENAL CIRCULATION. Rein, in 1931, in his study of regional blood flow in dogs during changes in systemic arterial pressure, was struck by the relative constancy of RBF as compared with flow in other tissues. This observation has been confirmed many times since. An example of the constancy of RBF with variations in arterial perfusion pressure appears in Figure 22-4. The relative constancy of flow as revealed by this figure (see bottom panel) is an example of *autoregulation* of the circulation. The mechanism is intrinsic, for it is manifested by the denervated kidney. The constancy of flow is maintained through a range of about 80 to 200 mm Hg of arterial perfusion pressure. Note the relative constancy of the dependent functions (GFR, intrarenal pressures). The autoregulating mechanism seems to be dominated by changes in afferent arteriolar resistance (AR, top panel).

Several hypotheses have been advanced to explain renal autonomy. Of those considered in Chapter 17, the myogenic has been most popular, although the tissue pressure and metabolic theories have received some support. (For a detailed account, see Selkurt, 1963.) Evidence for the myogenic type of response appears to be offered in Figure 22-4, but participation of metabolic factors cannot be ruled out in this type of experiment.

When arterial pressure is suddenly increased, RBF in-

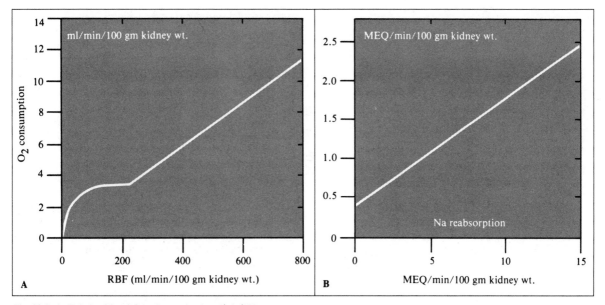

Fig. 22-3. A. Relationship of O_2 consumption in ml/min/100 gm kidney weight to renal blood flow. (From K. Kramer and P. Deetjen. Beziehungen des O_2-Verbrauchs der Niere zu Durchblutung und Glomerulus Filtrat bei Änderung des arteriellen Druckes. *Pflügers Arch. Physiol.* 270:782–790, 1960.) B. Relationship of O_2 consumption in mEq/min/100 gm kidney weight to sodium production. (From P. Deetjen and K. Kramer. Die Abhängigkeit des O_2-Verbrauchs der Niere von der Na-Rückresorption. *Pflügers Arch. Physiol.* 273:636–650, 1961.)

creases, but in 30 to 60 seconds internal adjustments occur, probably in the afferent arterioles, which bring the flow down to the original level despite maintenance of elevated pressure (Fig. 22-4). When pressure is suddenly dropped, an upward adjustment can be observed.

The dynamic reactivity implied in these fairly rapid adjustments corresponds to the type of reactivity anticipated from smooth muscle of the vasculature; hence the term *myogenic*. Isolated arterial segments have been shown to manifest similar responses. Moreover, that a vital phenomenon is involved is supported by the action of a number of agents known to impair smooth muscle activity that eliminate autoregulation: papaverine, potassium cyanide, procaine, ethyl alcohol, and the anesthetic chloral hydrate. Cooling removes autoregulation, and it is depressed by hemorrhage and anoxia.

The metabolic theory of autoregulation (see Chap. 17), as it may apply to the kidney, must take into account two important humoral components: the renin-angiotensin system (see Chap. 34) and the prostaglandins. Renin, synthe-

sized in the JGA of the cortical afferent arterioles, converts the angiotensinogen of the plasma to angiotensin I; this is converted by angiotensin-converting enzyme to the vasopressor angiotensin II. An intrarenal regulatory role has been proposed for the renin-angiotensin system (Thurau et al., 1968). This system may play a role in renal autoregulation of blood flow (see pp. 409, 419). Of importance is newer evidence indicating that prostaglandins can activate the renin-angiotensin system, supporting the feedback concept (Schnermann and Briggs, 1981).

The prostaglandins, presumably synthesized in specialized interstitial cells of the renal medulla from fatty acid precursors (e.g., arachidonic acid) include the potent vasodilators PGE_2, $PG\dot{I}_2$, and PGD_2. Reduced RBF, resulting from neurogenic or humoral vasoconstriction, as with hemorrhage (see Chap. 18), or renal ischemia enhances synthesis and release of prostaglandins. The strategic location of the cells of origin with respect to the medullary blood supply (vasa recta) supplies a possible mechanism for intrarenal distribution and circulatory regulation. The importance of these substances for regulating juxtamedullary and medullary blood flow was demonstrated by Kirschenbaum et al. (1974). Presumably, the control points are the arterioles and vasa recta sphincters of the juxtamedullary circulation. In keeping with the autoregulatory role, it is conceivable that initially enhanced RBF, as with a sudden increase in arterial perfusion pressure, would "wash out"

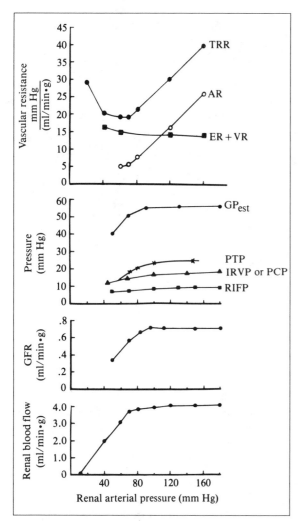

Fig. 22-4. Representative relationships between renal arterial pressure and various indices of function in the dog. All of these exhibit autoregulatory behavior, suggesting that the requisite changes in renovascular resistance occur at preglomerular sites. TRR = total resistance; AR = afferent arteriolar resistance; ER = efferent arteriolar resistance; VR = venous resistance; GP$_{est}$ = estimated glomerular pressure; PTP = proximal tubular pressure; IRVP = intrarenal venous pressure; PCP = peritubular capillary pressure; RIFP = renal interstitial fluid pressure; GFR = glomerular filtration rate. (From L. G. Navar, D. J. Marsh, R. C. Blantz, J. Hall, D. W. Ploth, and A. Nasjletti. Intrinsic control of renal hemodynamics. *Fed. Proc.* 41:3023, 1982.)

these dilator substances, eventuating in relative vasoconstriction. Recent work has demonstrated that certain prostaglandins (e.g., PGI$_2$) are also synthesized in the cortex, possibly by the endothelium of cortical blood vessels. The collecting tubular cells in the cortex and medulla are also capable of prostaglandin synthesis. Thus, any consideration of a role for prostaglandins in the autoregulation of RBF or in other types of RBF regulation must take these facts into account. It should be added that enhanced production and release of prostaglandins by the kidney does not usually have a systemic influence, since the lungs normally largely metabolize the prostaglandins that arrive in the mixed venous blood (especially PGI$_2$, PGE$_2$, and PGF$_{2\alpha}$, a vasoconstrictor).

In summary: It seems likely that any single theory thus far advanced would oversimplify the complex adjustments that the kidney undergoes in its efforts to maintain the constancy of its blood flow. The possibility has been considered that several of the mechanisms that have been discussed operate together in a composite mechanism.

FACTORS THAT MODIFY THE RENAL CIRCULATION. Despite the inherent autonomy of the renal circulation, a number of factors, neurogenic and humoral, can decrease or increase kidney blood flow.

Innervation and Neurogenic Regulation. It is generally agreed that the major nerve supply to the kidney has its origin largely from the levels thoracic (T) 12 through lumbar (L) 2 of the sympathetic nervous system in humans, and in the dog, from T4 to L11, but most abundantly from T10 to T12. In relation to its size, the kidney receives a more profuse and widespread nerve supply than almost any other viscus. The thoracolumbar sympathetic supply to the kidney provides it with a rich source of vasoconstrictor fibers, but presumptive evidence of dilator fibers is at hand, as will be discussed. The vagus nerve sends fibers to the kidney, but there is no evidence for vasodilator fibers in this circuit. Hence, the extrinsic vasomotor status of the kidney is maintained largely by variations in vasoconstrictor tone.

Evidence favors the notion that extrinsic regulation is low or absent in the basal state, to be invoked only in emergency states of heightened sympathetic nervous system activity. The principal evidence is that when one kidney is denervated, or denervated and transplanted, function is equal in the experimental and control kidneys in dogs and humans. This includes concordance of GFR (inulin or creatinine clearance), plasma flow (Diodrast or PAH clearance), diuretic activity, and electrolyte excretion. These findings are in accord with the concept of autonomy of the renal circulation.

Fluorescence histochemistry has revealed that catecholaminergic nerves extend to the afferent and efferent glomerular arterioles (Barajas, 1981). Innervation of the tubules, apparently arising from the arteriolar nerve bundles, has been observed in the monkey. In the rat, however, such innervation is confined to tubular areas in the juxtamedullary region. Histofluorescence techniques have revealed dopamine-containing nerves in the glomerular vascular pole of the dog kidney. By contrast, the nerves associated with the arcuate arteries contain mostly norepinephrine.

The cholinergic fibers, on the other hand, appear to innervate predominantly efferent arterioles, most prominently those of the juxtamedullary nephrons, and the sphincters of the vasa recta (Moffat, 1967; McKenna and Angelakos, 1968). A role in the regulation of the medullary circulation is possible for these vasodilator fibers. The cholinergic fibers may function to maintain blood flow during reduction of arterial pressure and may participate in alleged renal vasodilator reflexes. In concert with adrenergic constrictor fibers to the afferent arterioles, dilator fibers to efferent arterioles may provide a more versatile and effective control of glomerular filtration pressure, especially in juxtamedullary glomeruli.

Humoral and Pharmacological Factors. The catecholamines L-epinephrine (Arterenol) and Norepinephrine (Levarterenol) are active vasoconstrictors of the renal vasculature. Norepinephrine infused intraarterially causes rather selective vasoconstriction in the cortex, implying that all the adrenergic terminals are here (Hollenberg et al., 1968). Angiotensin is another important vasoconstrictor. Several well-known and physiologically important humoral agents that cause renal vasodilation are acetylcholine (ACh), serotonin (in low dosage), and prostaglandins (found in high concentration in the renal medulla, and also in the cortex). Two constrictor prostaglandins produced by the kidney are $PGF_{2\alpha}$ and thromboxane (TxA_2). When ACh is infused intraarterially, RBF increases in both cortex and medulla, but mostly in the medulla, implying that the greater proportions of the ACh-sensitive receptors are here.

Another field receiving attention currently is the kallikrein-kinin system, formed and stored intrarenally as is renin (Baer and McGiff, 1980). When released, these agents act on plasma globulins to liberate the decapeptides angiotensin I and lysyl-bradykinin. An enzyme converts the decapeptides to the more active substances, the octapeptide angiotensin II and the nonapeptide bradykinin. Although the kinins (dilators) and angiotensins (constric-

tors) have opposing effects on the renal circulation, they share an important property, namely, the ability to promote prostaglandin synthesis and to alter the metabolism of prostaglandins. In addition, release of renin and kallikrein is partially controlled by a prostaglandin mechanism. The interactions of the renin-angiotensin and kallikrein-kinin systems with prostaglandins can result in major modifications of the effects of these hormonal systems on renal hemodynamics and excretion of salt and water, e.g., attenuation of the vasoconstrictor-antidiuretic action of angiotensin and enhancement of the vasodilator-diuretic action of kinins.

RESPONSE OF RENAL BLOOD FLOW TO PHYSIOLOGICAL STRESS. Exercise. Renal blood flow is decreased in proportion to the severity of exercise. Both the amount of work involved and the duration of the exercise influence the results. As an example, subjects who had run the 440-yard dash at full speed had reductions of 18 to 54 percent below control, which remained for 10 to 40 minutes after exercise. It has been estimated that a saving of 0.5 to 1.0 liter of blood is made available to active tissue by renal vasoconstriction.

In highly conditioned animals (Alaska sled dogs) undergoing heavy exercise, RBF did not decrease (Van Citters and Franklin, 1969). Ultrasonic flowmeters, chronically implanted, were monitored by telemetry. In contrast to exercising humans, in the sled dogs the compensatory redistribution of blood flow was not a significant reserve mechanism during exercise. However, it is well to point out that arterial blood pressure rose during exercise; in the face of a constant RBF, increased vascular resistance in this organ is implied. It could be extrinsic or autoregulatory in nature.

Posture and Orthostatic Hypotension. In normal young males, C_{PAH} in the sitting position is 91 percent of that in the supine; in the erect position, it is 85 percent. Motionless standing, or tilting of the subject, leads to progressive venous stagnation, reduced cardiac output, and neurogenic vasoconstriction, until the cerebral circulation becomes inadequate, when syncope occurs. Kidney blood flow is reduced to about one-half during the compensatory phase.

Renal Hypoxia and Ischemia. Hypoxia due to low oxygen intake or simulated altitude induces a slight increase in RBF at first, but with more severe hypoxia this decreases. Apparently, mild hypoxic states, unaccompanied by significant reflex vascular alterations, manifest a slight renal hyperemia, but more severe hypoxic states or asphyxia triggers vasoconstrictor reflexes in which the kidneys participate.

Acute Renal Ischemia. Very brief periods of ischemia (1–5 min) result in a flow overshoot (*reactive hyperemia*) after release. Longer periods of ischemia in dogs (20 min) produce slight, inconsistent results on renal blood flow. Ischemia lasting 30 minutes to 2 hours results in marked, persistent reduction in RBF. Since there is tubular damage, one must employ caution in interpreting clearance methods (e.g., C_{PAH}), although employment of the Fick principle (C_{PAH}/E_{PAH}) serves to correct for the effects of tubular damage in the blood flow estimates. Reduction in RBF after 2 hours of complete renal ischemia has been found to persist for 24 hours to 2 weeks after release.

Hypercapnia and Acidosis. Hypercapnia and acidosis cause renal vasoconstriction that is centrally mediated (enhanced activity of the vasoconstrictor centers).

Hemorrhagic Hypotensive Shock. Acute hemorrhage provokes responses in the renal circulation that are typical of general compensatory mechanisms brought into play: reflex vasoconstriction, and shunting of blood to other tissues to compensate for blood loss. If blood loss is great enough, there is a shutdown of renal excretory function, which if prolonged might have serious consequences to the organism. Moreover, a prolonged period of anoxic hypotension impairs the function of the tubular epithelium, adding to the problem of shock the probability of renal failure and uremia (see also Chap. 18).

In *tourniquet* and *traumatic shock,* as well as in *hemorrhagic shock,* blood flow is markedly reduced as a result of enhanced neurogenic vasoconstrictor activity and of an increase in renal vasoconstrictor substances such as epinephrine. Evidently, under these circumstances the renal circulation is subordinate to the welfare of the body as a whole.

Blood Flow in Renal Disease. Relevant information on RBF in acute and chronic nephritis, nephrosis, and hypertensive kidneys has accrued from studies employing C_D and C_{PAH}. Varying degrees of tubular damage, reflected in a decrease in *tubular excretory maxima* of Diodrast and PAH (Tm_D and Tm_{PAH}) and extraction (E_D and E_{PAH}), have complicated interpretation. When allowance is made for the influence of impaired tubular function, it is apparent that RBF may actually be elevated in early nephrosis and the acute, inflammatory phase of nephritis. Later, as the disease becomes chronic, blood flow is undeniably reduced to varying degrees and may be very low in chronic nephritis accompanying the resultant disorganization of the kidney vascular pattern. The accompanying renal hypertension seen in nephritis is a manifestation of reduced renal blood flow.

RENAL LYMPHATIC SYSTEM. The lymphatic circulation is prominent in the cortex, where the lymphatic capillaries begin blindly. Lymphatics within the cortex follow the course of the interlobular vessels and drain in a centripetal fashion toward the arcuate vessels. Some evidence has been presented that lymphatics occur in the medulla of the human kidney, but their presence or absence in other species, including the dog, is controversial. The medullary lymphatics, when they occur, drain the vasa recta system, join the cortical branches at the arcuate level, then pass out with the interlobar vessels toward the renal pelvis. After converging at the hilus of the kidney, the lymphatic vessels course as perivascular channels to the cisterna chyli and the thoracic duct. Lymph flow can be increased by renal venous pressure increment and ureteral blockade.

THE GLOMERULUS AND ITS FUNCTION

The glomerular capillaries are not simple loops but form a freely branching anastomotic network (Fig. 22-5). More specifically, larger *through channels* exist with an associated capillary network of smaller anastomotic channels. This arrangement may afford a structural basis for the skimming of plasma relatively freed of cells into the network of small capillaries, where the actual filtration proceeds, whereas the greater mass of blood cells directly and rapidly flows through the glomerular lobules to the efferent arterioles as an axial stream. This arrangement facilitates filtration processes by slowing the flow and reducing turbulence and by possibly exposing more of the plasma that is to be filtered to the effective filtration surface.

The details of the filtering structures have been clarified by the use of the electron microscope (Fig. 22-6), which has revealed that the capillary endothelium, called the *lamina attenuata* or *lamina fenestrae* (0.05 μm thick), has pores 400 to 900 Å in size. The pores are too large to restrain the plasma constituents.* Instead, they expose the ultrafil-

---

*It is now recognized (Brenner and Beeuwkes, 1978) that glomerular filtration of charged particles will be modified by negatively charged sialoproteins of the endothelium and basement membrane. These will repel negatively charged (anionic) particles and impede their filtration relative to neutral substances. On the other hand, filtration of cationic substances is slightly greater. In this manner the endothelial pores might play a role in modifying GFR. Recall that circulating albumin is negatively charged.

The presence of albumin in the urine is called *albuminuria.* There is evidence that in nephritis the negative charges in the glomerular wall are dissipated, and that albuminuria can occur for this reason without an increase in the size of the pores in the membrane.

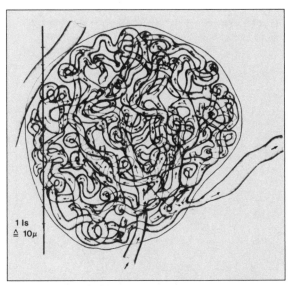

Fig. 22-5. Modern concept of glomerular function; one space on the scale at left equals approximately 10 μ. Glomerular details of a rat kidney are demonstrated by photographs taken from a television screen, showing flow direction in individual loops as determined by off-line video analysis. Note frequent bidirectional flow. Attention is also called to the afferent arteriole (lower pole) and efferent arteriole (to right). Note significant constriction of efferent arteriole as it leaves Bowman's capsule. This would be an aid in maintaining a high glomerular filtration pressure. Also note slight constriction of afferent arteriole at site of entry into the glomerulus. (From M. Steinhausen et al. *Kidney Int.* 23:802, 1983).

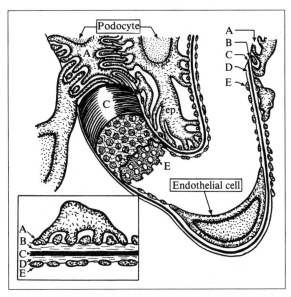

Fig. 22-6. Glomerular capillary and associated portions of Bowman's capsule. Epithelial cells (ep.), the *podocytes* of the visceral layer of the capsule, form layer A. Layer C is the lamina densa and with layers B and D (cement layers) forms the basement membrane. Layer E is the capillary endothelium (lamina fenestrae). (From D. C. Pease. Fine structure of the kidney seen by electron microscopy. *J. Histochem. Cytochem.* 3:297, 1955. Copyright © 1955, The Williams & Wilkins Co., Baltimore.)

tration membrane, the *glomerular basement membrane* (0.1 μm thick), to the free flow of plasma by removing the endothelial cytoplasmic barrier. Although this membrane exhibits differences in stratification, it is probably a homogeneous layer and appears to be the limiting membrane for restraint of plasma proteins. The visceral layer of Bowman's capsule is organized into extremely specialized cells, the *podocytes* (foot cells), which possess thousands of foot processes (*pedicels*) resting on the basement membrane. The spacing between the pedicels may be narrow enough (100 Å) to be a limiting dimension in restriction of plasma proteins. It has been designated by some authorities a *slit pore*.

DYNAMICS OF GLOMERULAR FILTRATION. The forces involved in glomerular filtration are the same as those that were discussed in Chapter 12 under the consideration of transcapillary molecular exchange. Glomerular filtration rate is proportional to what is designated as *net filtration pressure*. This is expressed as

$$GFR = K\,(P'' - P' - \pi'' + \pi')$$
$$60 \quad 10 \quad 30 \quad 0^* \;\;(mm\;Hg)$$

where K is a proportionality coefficient embodying the volume of fluid filtered per minute per mm Hg pressure. Net filtration pressure (P_f) equals 20 mm Hg in equation 1. The double primes refer to the blood side of the capillary wall, and the single primes refer to the glomerular capsule fluid side. P denotes hydraulic pressure (P″, glomerular capillary pressure; P′, the capsular pressure); π″ denotes colloid osmotic pressure of plasma proteins. π′ is the colloid osmotic pressure in the capsular fluid, usually zero, or nearly so. The forces shown in millimeters of mercury are values for the human kidney estimated from direct and indirect measurements made in other animals.

It has been calculated that an average functional pore diameter of 75 to 100 Å exists in the glomerular membrane, a pore size that would permit only minute quantities of

*Assumes no filtration of plasma protein.

Substance	Mol. wt	Dimensions in angstrom units		Filtrate/Filtrand
	Grams	Radius from diffusion coefficient	Dimensions from x-ray diffraction	
Water	18	1.0		1.0
Urea	60	1.6		1.0
Glucose	180	3.6		1.0
Sucrose	342	4.4		1.0
Inulin	5,500	14.8		0.98
Myoglobin	17,000	19.5	← 54 → ‡ 8	0.75
Egg albumin	43,500	28.5	← 88 → 22	0.22
Hemoglobin	68,000	32.5	← 54 → 32	0.03
Serum albumin	69,000	35.5	← 150 → 36	<0.01

Fig. 22-7. Glomerular permeability to molecules of different dimensions. (From R. F. Pitts. *Physiology of the Kidney and Body Fluids* [3d ed.]. P. 52. Copyright © 1974 by Year Book Medical Publishers, Inc., Chicago. Used by permission.)

albumin to be filtered. The approximation of the "effective" pore diameter was arrived at by analyzing the glomerular sieving of molecules of various sizes, as illustrated in Figure 22-7. Note that serum albumin is at the critical level of filtration, with an average diameter of 70 to 75 Å based on its diffusion coefficient, and 150 by 36 Å based on x-ray diffraction. The membrane resistivity to movement of fluid could thus be described by Poiseuille's equation and characterized as bulk filtration, with fluid and solutes driven by hydrostatic pressure.

Consideration of the filtering membranes suggests that the large pores ("fenestrae") (up to 900 Å in diameter) of the capillary endothelium perform no filtering action for diffusible substances but do, in fact, freely expose the plasma to the basement membrane, which can be considered as playing a possible role in filtration. (Recall the role of "charged" proteins, page 414). The basement membrane has been viewed as a hydrated gel, the supporting structures being protein micelles manifesting a thixotropic behavior, providing a possible sievelike mechanism. Since the membrane is relatively homogeneous and pore free as viewed by electron microscopy, the concept of bulk filtration through "pores" would of necessity imply a "virtual" pore—i.e., some physical structure operating in a sieving action comparable to that of a pore.

Another likely site of filtration restraint has been emphasized, namely, the slit pores between the pedicels of the podocytes. Figure 22-8A shows the electron microscopic detail of the human kidney, revealing slit-pore membranes between the pedicels (foot processes). The slit pore has a highly ordered structure, as shown in a tangential section (Fig. 22-8B), with an irregular central filament with crossbars to the adjacent foot processes. These irregularly spaced crossbars enclose individual rectangular, electron-lucent spaces with a calculated dimension of 50 × 120 Å. The alternation of crossbars gives the entire slit a zipperlike appearance. It was calculated that these pores occupy 2.7 percent of the glomerular capillary surface (Schneeberger et al., 1975). Probably, the estimate of the slit-pore diaphragm of the human kidney would normally be at the limit of passage of serum albumin, accepting the prolate ellipsoid size of about 150×36 Å, which would necessitate an "end-on" passage, to leak out. Even the size estimated from physiological experiments indicating a hydrodynamic sphere with a 36-Å radius, would require for passage a hit comparable to a rimless shot in basketball.

The forces involved in glomerular filtration are indicated in Figure 22-9. The figure shows data obtained by direct micropuncture of the glomerular capillaries and pressure measurement with a micropipette transducer in a unique strain of Wistar rats with accessible surface glomeruli. Note the progressive rise of the calculated plasma oncotic pressure along the length of the glomerular capillary, a consequence of ultrafiltration of the plasma, leaving the albumin behind in increasing concentration. Observe that filtration equilibrium is usually attained before the blood leaves the glomerular capillaries (Brenner et al., 1971; Deen et al., 1974). Comparable measurements have been made in the anesthetized squirrel monkey (Maddox et al., 1974). The pressure gradient was found to be very similar to that of the rat, when normalized to the same aortic blood pressure. In absolute terms, with a mean aortic pressure of 115 mm Hg, pressure fell across the preglomerular resistance vessels to 48.5 mm Hg in the glomerular capillaries, then to about 8 mm Hg in the renal vein across the peritubular capillaries. Net filtration pressure (P_f) at the beginning of the glomerular capillaries averaged 12.4 mm Hg.

In the Munich-Wistar rat, glomerular capillary pressure measurements have now been made under a variety of conditions (Blantz, 1980). Glomerular capillary hydrostatic pressure (P_G) ranged from 43 to 50 mm Hg at mean arterial pressures of 100 to 125 mm Hg. In other strains of rats, P_G values ranged from 55 to 60 mm Hg. Tubular pressures under these conditions were 10 to 16 mm Hg, which requires a pressure gradient of approximately 35 mm Hg from glomerular capillary to Bowman's space.

Data from the unanesthetized dog are now available (Marchand, 1981). Micropuncture of the glomerular capillaries yielded an average value of 56 ± 3 mm Hg (SEM), with an average systemic arterial pressure of 127 mm Hg. Net filtration pressure in the dog averages 18.0 mm Hg (Osswald, 1979). Filtration equilibrium is not attained in this species. In this respect the human kidney resembles that of the dog.

In summary: It can be stated that the filtration mechanism of the kidney is almost ideal, since it provides a very permeable membrane to filtration and also supplies relatively large forces for ultrafiltration that are capable of a high degree of regulation by extrinsic and intrinsic (autoregulating) mechanisms.

Micropuncture studies of the capsular space have revealed that the qualifications for ultrafiltration are met; the membranes are largely impermeable to protein, but water and solutes pass freely, and the latter are found in approximately equal concentration on both sides of the glomerular membrane, allowing for slight differences in electrolyte concentration due to the Gibbs-Donnan effect (see Chap. 1). Thus the classic investigations of Richards in amphibia (1935) and of Walker et al. (1941) in mammals (see Pitts, 1974) have shown that glucose, inorganic phosphate, creatinine, urea, uric acid, sodium and chloride, total osmotic pressure, and pH are approximately the same in the capsular fluid as in the plasma water. When the substance inulin was given, (its clearance is taken to be a measure of GFR), it was found in equal concentration on both sides of the membrane. Harris et al. (1974) successfully punctured superficial glomeruli of rats (Munich-Wistar strain) and confirmed that the glomerular fluid/plasma ratio of inulin was 1.00.

REGULATION OF GLOMERULAR FILTRATION. Glomerular filtration pressure regulation is probably dominated by the adrenergic constrictor activity of the afferent arterioles. Evidence is accumulating that adrenergic control of the efferent arterioles is also involved (Barajas, 1981). In renal clearance studies, insight into flow regulation is derived from the use of the ratio of GFR to the concurrent total plasma flow. The ratio, called the *filtration fraction* (FF), is as follows: FF = C_{IN}/RPF. The clearance of PAH is often employed in the designation of FF. Thus, FF = C_{IN}/C_{PAH}. The FF averages about 0.35 in the rat, 0.30 in the dog, and 0.20 in humans. For example, if afferent arteriolar vascular tone increased more than that of the efferent arterioles, FF would decrease. The converse would be true; i.e., FF increases when the afferent arterioles dilate proportionately more than the efferent arterioles. The student should be able to reason out other permutations.

Blantz (1980) found that under a wide variety of physiological conditions, afferent and efferent vascular resistances tended to change in parallel and not independently. The conditions imposed included hydropenia, volume-expanded states, infusion of angiotensin II, increased ureteral pressure, and so on. However, the specific mechanism(s) that links changes in afferent to efferent arteriolar resistance have not been completely elucidated. This type of regulation could, however, imply that adrenergic influences operate in concordance on both sets of arterioles to establish and maintain the final FF.

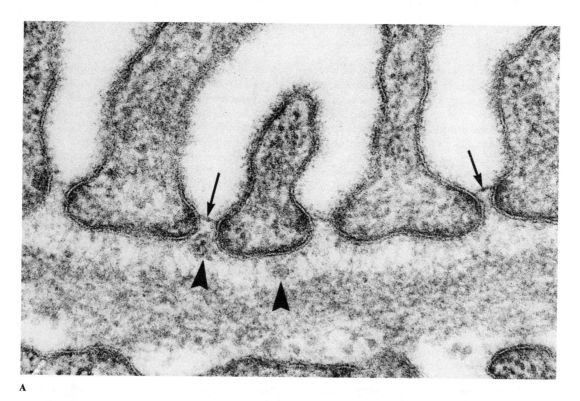

A

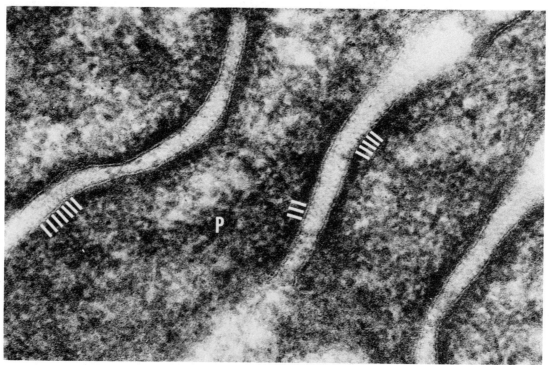

B

Recently, experimental approaches have been devised that utilize isolated glomeruli (in vitro), examined by measuring changes in the volume of the isolated glomerulus in response to variations in colloid osmotic pressure of the bathing medium. Another approach is direct perfusion of the glomerulus, through a range of pressures utilizing microcatheters inserted into the afferent arteriole and outflow measured from the efferent arterioles. The advantages of these approaches are that the glomerulus is free of systemic influences (hemodynamic, humoral, neurogenic, and possibly species variation). Regional glomerular variation (cortex versus medulla) can also be assessed (Osgood et al., 1982).

FEEDBACK CONTROL OF GLOMERULAR FILTRATION RATE. The burgeoning of micropuncture techniques in recent years has prompted investigation of feedback control systems at the single-nephron level, e.g., single-nephron GFR. The approach commonly used is to perfuse single-nephron segments with appropriate solutions to see if changes occur in GFR (Fig. 22-10). The critical control point studied has been the interplay of single-nephron GFR and the JG region—more specifically, the macula densa. This is viewed as the sensor, with feedback control of the contractile elements of the arterioles with which the macula densa is in intimate relation.

The basic observation is that increases in the flow rate of fluid through the loop of Henle and the macula densa zone are associated with decreases in the rate at which fluid is filtered by the glomerulus in the same nephron. Aspects of the negative feedback response that have been characterized quantitatively include the magnitude of the changes in glomerular function, the rates of tubule fluid flow that elicit the changes, and the sensitivity of the response in different physiological circumstances. Some information about the sensing step in the feedback pathway in rats has come from microperfusion experiments employing changes in the composition of fluids used to perfuse the loop of Henle or changes in the electrical driving forces across the wall of the distal tubule.

Taken together, the results of experiments using these different approaches indicate that changes in loop of Henle flow rate may be sensed by a process that requires transport of some constituent of distal tubule fluid, possibly the chloride ion, from the distal tubule into some compartment within the JGA (Wright, 1981a). Another view is that distal tubule fluid osmolality or distal tubule solute delivery may participate in the initiation of feedback signals (Navar et al., 1981). In any case, there is now evidence that the renal vasoconstriction and reduction in GFR can be mediated by the local action of angiotensin II (Tucker and Blantz, 1980).

MEASUREMENT OF GLOMERULAR FILTRATION RATE. Direct methods are not applicable to the measurement of filtration rate in the mammalian kidney, and indirect methods have been devised. These involve an understanding of the *renal plasma clearance concept*. The rate at which a substance (X) is excreted in the urine is the product of its urinary conentration (U_x, mg/ml) and the volume of urine per minute ($\dot{V}$). The rate of excretion ($U_X \cdot \dot{V}$) per minute depends on the concentration of X in the plasma (P_X, mg/ml). It is therefore reasonable to relate $U_X\dot{V}$ to P_X, and this relation is called the *clearance*, or

$$\frac{U_X \cdot \dot{V}}{P_X}$$

or, more generally, $U\dot{V}/P$. The calculated value has the dimensions of volume related to time and is in reality the *smallest volume of plasma from which the kidneys can obtain the amount of X excreted per minute*. It must be understood that the kidneys do not usually clear the plasma completely of a substance in transit through them, but clear a larger volume incompletely. The clearance therefore is not a real but a *virtual* volume. When substances are being cleared simultaneously, each has its own clearance rate, depending on the amount reabsorbed from the glomerular filtrate or added to it by tubular secretion. The former has the lower clearance, the latter, the higher. Those cleared only by glomerular filtration are intermediate, and their clearance, in effect, measures the GFR in milliliters per minute.

Fig. 22-8. A. Human glomerular capillary wall from a kidney fixed by perfusion with tannic acid-glutaraldehyde (X168,000). The slit-pore membrane (arrows) consists of a single electron-dense line in the middle of which is a small black dot measuring 105 Å in diameter. Irregular collections of electron-dense granules present in the lamina rara externa (arrowheads) may represent serum proteins that were incompletely removed by the preliminary washout procedure. The middle layer of the glomerular basement membrane (lamina densa) shows a dense meshwork of irregular fibrils ranging from 35–100 Å in thickness. The lower layer is the lamina rara interna. B. Tangential view of slit-pore membranes present between epithelial foot processes (P) (X 175,000). Areas in which the cross-bars are clearly delineated are indicated by a series of short white lines. In these areas the central filament is also more clearly visible. Photographs supplied through the courtesy of Dr. Eveline E. Schneeberger. (From E. E. Schneeberger et al. The isoporous substructure of the human glomerular slit diaphragm. *Kidney Int.* 8:49,50, 1975. Reprinted by permission.)

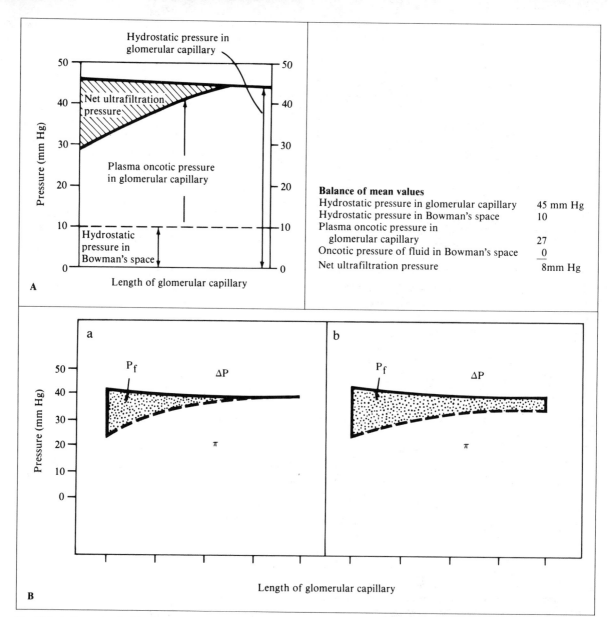

Balance of mean values

Hydrostatic pressure in glomerular capillary	45 mm Hg
Hydrostatic pressure in Bowman's space	10
Plasma oncotic pressure in glomerular capillary	27
Oncotic pressure of fluid in Bowman's space	0
Net ultrafiltration pressure	8mm Hg

Fig. 22-9. A. Forces involved in glomerular ultrafiltration in rats. As shown, ultrafiltration pressure declines in glomerular capillaries, mainly because plasma oncotic pressure rises. (In contrast, in extrarenal capillaries the decline in ultrafiltration pressure is due mainly to a decrease in intracapillary hydrostatic pressure.) The pattern for the rise in plasma oncotic pressure as function of capillary length is not known precisely; hence, the mean values for plasma oncotic and net ultrafiltration pressure listed in the table are approximations. It is not yet known at what point in the capillary the sum of hydrostatic pressure in Bowman's space and of plasma oncotic pressure exactly balances the hydrostatic pressure in the glomerular capillary. If, as shown here, balance occurs before the end of the capillary is reached, ultrafiltration would not take place over the entire length of the glomerular capillary. (From H. Valtin. *Re-* *nal Function: Mechanisms Preserving Fluid and Solute Balance in Health* (2nd ed.). Boston: Little, Brown, 1983. P. 44.) B. Pressure gradients are shown along the length of the glomerular capillaries at different blood flow rates: in *a,* at low flow rates; in *b,* at high flow rates. Note the failure to attain equilibrium. The figures represent pressure values measured in the rat and monkey. Measurements in the dog kidney, and indirect estimates in humans, resemble *b* even at normal blood flow rates; i.e., equilibrium is not attained at normal or resting values of renal blood flow. (Recall that the normal values for P and P_f are higher in these species.) ΔP = hydrostatic pressure gradient across the glomerular capillary wall; π = colloid osmotic pressure of plasma; P_f = net filtration pressure (stippled area).

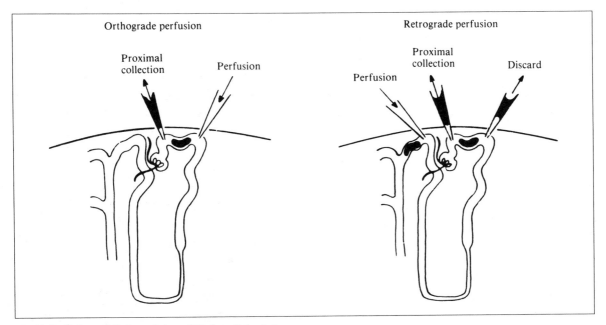

Fig. 22-10. (Left panel) Orthograde loop of Henle perfusion technique showing microperfusion pipette in the late proximal tubule and collection pipette in the early proximal tubule. (Right panel) Retrograde loop of Henle perfusion technique showing a microperfusion pipette in the early distal tubule, and collection pipettes in the late proximal tubule for collection of perfusate and in the early proximal tubule for timed collection of glomerular filtrate. (J. Briggs. The macula densa sensing mechanism for tubuloglomerular feedback. *Fed. Proc.* 40:100, 1981.)

The best-known substance that can be infused into the blood to provide a clearance equal to glomerular filtration rate is *inulin,* a polymer of fructose containing 32 hexose molecules (mw 5200). Strong evidence exists that it (1) is neither reabsorbed nor secreted by the tubular cells (Fig. 22-11), (2) is nonmetabolized in the body, (3) has no physiological influences, and (4) is freely filterable at the glomerular membranes. Several lines of evidence indicate that the clearance of inulin is at the level of *glomerular filtration,* i.e., is an index of the total volume of fluid filtered by the glomerular membranes per unit of time. When simultaneous clearances of inulin and other sugars such as glucose, xylose, and fructose, are obtained, the clearances of the sugars are less than the clearance of inulin. The differences in clearance between these and inulin depend on their particular reabsorptive mechanisms. However, when the transfer mechanism for sugar is inhibited by the glucoside *phlorizin,* the reabsorption of sugars is blocked, and their clearances become identical with that of inulin; i.e., under these cir-

cumstances, the sugars are cleared only by the physical process of filtration. Evidence that inulin is not secreted by tubular cells is somewhat more indirect. One factor against tubular secretion is that inulin is not found in the urine of the aglomerular kidneys of certain marine forms even when tremendously high plasma concentrations are attained by intravenous injection of inulin.

The clearance rate of inulin (C_{IN}) in humans is 120 to 130 ml per minute. This is taken to be the GFR (also C_F [clearance of filtrate]), or the amount of plasma that is filtered through the glomeruli per minute.* Substances other than inulin have been used to provide an estimate of GFR, the most common of which is creatinine. It has been shown that exogenous (true) creatinine clearance is approximately equal to that of inulin in the dog, rabbit, sheep, seal, frog, and turtle and thus provides a measure of GFR in these

*Glomerular filtration rate is expressed in terms of milliliters of plasma filtered through glomerular capillaries per minute. This system introduces an ambiguity in terminology, for plasma is 93 to 94 percent water and some 6 to 7 percent protein, and only the aqueous phase is filtered. All of the true crystalloids of plasma are dissolved in the aqueous phase; yet their concentrations are commonly expressed in milligrams per 100 ml (or mEq/liter) of plasma, not of plasma water. Thus, all of the crystalloids in 125 ml of plasma are filtered each minute when the GFR is 125 ml per minute. However, only 0.93 × 125 ml = 116 ml of water is filtered. The student should keep in mind that GFR is defined as *volume of plasma filtered per minute,* not as volume of plasma water, although obviously only the water and its dissolved crystalloids are actually filtered.

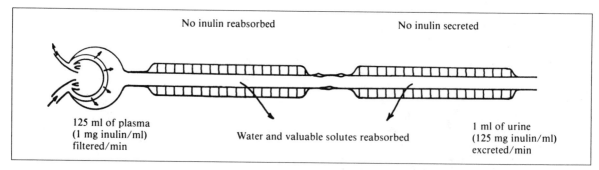

No inulin reabsorbed   No inulin secreted

125 ml of plasma   1 ml of urine
(1 mg inulin/ml)   (125 mg inulin/ml)
filtered/min   Water and valuable solutes reabsorbed   excreted/min

Fig. 22-11. Diagram of nephron illustrating requisites of a substance such as inulin for measurement of GFR. (From R. F. Pitts. *Physiology of the Kidney and Body Fluids* (3d ed.). P. 62. Copyright © 1974 by Year Book Medical Publishers, Inc., Chicago. Used by permission.)

animals. However, in addition to being filtered through the glomeruli, creatinine is secreted in part by the tubular cells in humans and other primates; hence, exogenous creatinine clearance cannot be used as an estimate of filtration rate.

A creatininelike chromogen is found in human plasma that gives the same chemical reaction for colorimetric analysis (Jaffe's reaction) as does true creatinine, but whose calculated clearance approximates that of inulin. The "noncreatinine chromogen" is apparently not excreted into the urine, so that the plasma level of measured creatinine (true plus false) yields the lower calculated clearance. The so-called endogenous creatinine clearance is widely used in patients as an estimate of GFR. An application appears in Figure 22-12, in which blood urea in developing uremia is related to the endogenous creatinine clearance, a measurement of GFR. Note that GFR must be reduced below 60 ml per minute during renal failure before blood urea rises above normal limits (20–50 mg/100 ml) (BUN, about 10–25 mg/100 ml). With a low protein intake, the increase occurs only below 30 ml per minute.

Other substances that have a clearance close to that of inulin in humans, dogs, and other animals are mannitol and thiosulfate.

A knowledge of the GFR permits quantitation of the amount of any substance freely filtered (C_{IN}, ml/min × P_X, mg/ml). When dealing with a substance that is reabsorbed, one may subtract 1 minute's excretion, $U_X \dot{V}$, from the filtered load to find the amount reabsorbed by the tubules in milligrams per minute.

$$T_R = C_{IN} \cdot P_X - U_X \cdot \dot{V}$$

Furthermore, under the same principle, the knowledge of GFR permits the calculation of tubular secretory activity, as follows:

$$T_S = U_X \cdot \dot{V} - C_{IN} \cdot P_X$$

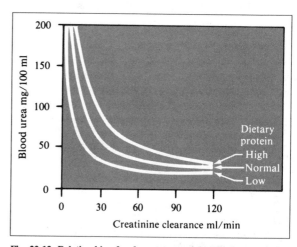

Fig. 22-12. Relationship of endogenous creatinine clearance in humans to blood urea. (From H. DeWardener. *The Kidney: An Outline of Normal and Abnormal Structure and Function* (3d ed.). Boston: Little, Brown, 1967. P. 28.)

A number of the substances studied for tubular secretion are bound to plasma protein in varying degrees and cannot be freely filtered. The filtered moiety therefore requires correction for the binding. This introduces the factor f, which represents the fraction of the substance not bound by the plasma proteins and therefore free to be filtered. Factor f also depends on the compound (substrate) concentration and plasma protein (albumin) concentration. Factor f may vary with the species (e.g., it is 0.92 for PAH in the dog and 0.83 in humans). Also, it differs for various substances. Thus, the f value is 0.40 for phenol red and 0.72 for Diodrast. The estimates are made by dialysis experiments. The completed equation is

$$T_S = U_X \cdot \dot{V} - C_{IN} \cdot P_X \cdot f$$

It should be added that plasma binding does not affect the fraction of clearance that involves tubular secretion. Such mechanisms operate by the action of transfer systems that rapidly remove the substances from the plasma water, to establish a gradient whereby the plasma binding is readily broken down, so that the substances continue to be rapidly removed by tubular secretion. However, the diffusion process across the glomerular membranes does not affect the protein binding, and it is the moiety that is free in plasma water that is filtered.

In summary: Clearances less than that of inulin represent

mechanisms involving filtration and reabsorption; clearances higher than that of inulin represent mechanisms involving glomerular filtration plus some degree of tubular secretion with allowance for possible plasma binding.

TUBULAR REABSORPTION OF ORGANIC SUBSTANCES

GLUCOSE. Utilizing micropuncture methods in experimental animals. Walker and associates, as early as 1937, identified the proximal tubules as the site of glucose reabsorption. More recent experiments (Rohde and Deetjen, 1967) utilizing sensitive enzymatic methods for determination of glucose have confirmed that the major portion of reabsorption normally goes on in the initial portion of the proximal tubule (Fig. 22-13). Recent studies have shown that reabsorptive capacity is highest in the early, and lowest in the late, proximal tubule. With increasing delivery of glucose to the tubule, successive segments are engaged. If plasma glucose levels then rise high enough to saturate the total reabsorptive capacity, correspondingly greater amounts of glucose are excreted.

Fig. 22-13. The decrement in glucose concentration along the proximal tubule, as observed in micropuncture in rats. The range of normal plasma glucose is indicated by the vertical bar. From H. Rohde and P. Deetjen. Die Glucoseresorption in der Rattenniere. *Pflugers Arch. Physiol.* 302:223, 1967.

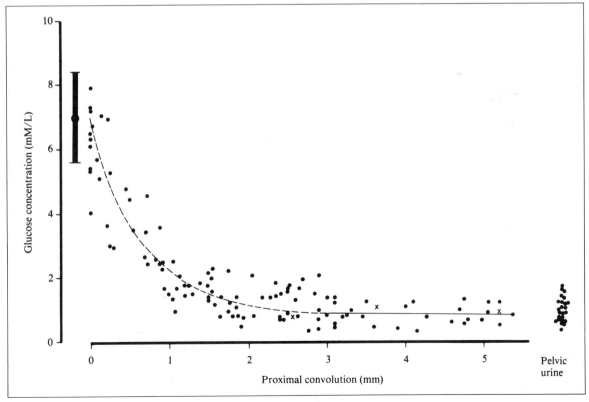

The distal tubules are normally not committed to glucose handling, although the possibility exists that glucose may be utilized for metabolic processes in the renal medulla.

When plasma glucose is elevated in humans to about 180 to 220 mg per 100 ml, the so-called *threshold*, glucose appears in the urine. As the concentration is further raised by continued infusion of glucose, the reabsorptive mechanism becomes progressively saturated until the rate of reabsorption becomes constant and maximal. This indication of saturation of the transport system is referred to as the *tubular maximum* or Tm—in this instance Tm_G. In the human, Tm_G has a value of 340 mg per minute; it is about 200 mg per minute in the dog.

Figure 22-14 illustrates the principle of saturation of the reabsorptive system by a progressive elevation of plasma glucose concentration experimentally. It will be noted that $U_G\dot{V}/P$ or C_G increases progressively. This approaches asymptotically the clearance of inulin as the plasma levels increase further beyond the point of saturation of the tubular mechanism. To put it another way, the clearance becomes more and more preponderantly the consequence of glomerular filtration, and the reabsorbed moiety becomes a fractionally smaller part of the total excretory component.

A theory of the kinetics of the glucose transport system is diagrammed in Figure 22-15. A carrier substance, X, is present in the luminal membrane of proximal tubular cells in a fixed and limited amount. The carrier combines reversibly with D-glucose, G, from the tubular fluid, to form the complex G · X within the membrane. The complex moves to the cytoplasmic side of the membrane, where G is split off. From the cytoplasm, G moves out of the basal end of the cell by another step, apparently not rate limiting, possibly by facilitated diffusion. Carrier X needs to be regenerated via an intermediate step, Y, which handles metabolically abnormal types of sugar (e.g., L-glucose) that are transported to and secreted into the lumen. Then, Y is restored to X, involving an enzymatic step and introduction of energy, via adenosine triphosphate (ATP). A sodium dependency, as well as phlorizin sensitivity, has been demonstrated at the carrier site, as in the intestine.

It now seems clear that phlorizin competitively inhibits glucose reabsorption by binding with great affinity at the membrane site that binds or orients glucose (Diedrich, 1963). Phlorizin structurally is glucose with the C_1 hydroxyl replaced with phloretin; hence, competitive binding is not surprising. Other hexoses (galactose, fructose, and slightly, xylose) compete with glucose for the transport site. The disaccharide sucrose shows very little reabsorption in the in vivo system.

Recent work has identified *co-transport* of glucose with

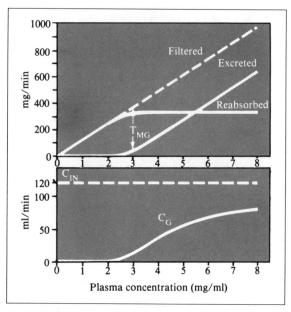

Fig. 22-14. Glucose tubular reabsorption and clearance of glucose as related to increase in plasma concentration. C_{IN} = filtration rate; C_G = glucose clearance; T_{MG} = tubular maximal rate of transport of glucose.

sodium ions as the process that not only provides the mechanism for translocating glucose into the cells of the proximal tubule but also accounts for the driving force that moves glucose from the low concentration in the lumen to the higher one in the cells (Ullrich, 1976). Part of the sodium crossing the luminal membrane of the cells of the proximal tubule does so by combining with a carrier molecule that binds both sodium ion and glucose. Since there is a gradient favorable to the movement of sodium across this membrane, the electrochemical potential difference for sodium provides the driving force for the movement of glucose into the cell against its gradient of chemical potential. Although the transport of glucose shows all the characteristics of active transport previously described, the current view is that it is ultimately the active transport of sodium by the peritubular membrane, the sodium pump, that is responsible for the reabsorption of glucose (Fig. 22-16). This is demonstrated by the consequences of inhibition of this primary active transport process. If sodium extrusion across the peritubular membrane is blocked, the sodium concentration in the epithelial cells rises, and the favorable gradient of electrochemical potential for sodium across the luminal membrane is lost. With this attenuation of the gradient for sodium, net transport of glucose ceases. The exit of glucose

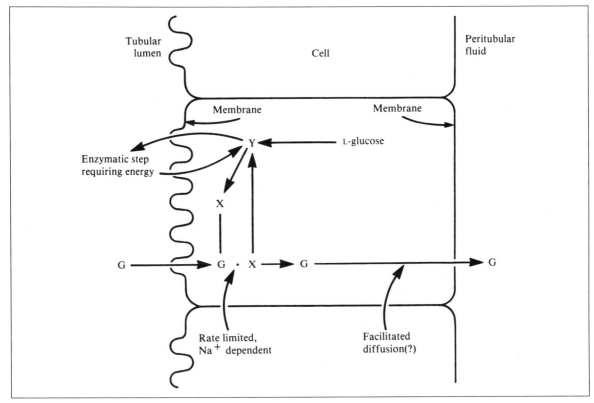

Fig. 22-15. Model of glucose transport illustrating countertransport of L-glucose.

from cell to interstitial fluid takes place along a favorable concentration gradient for glucose and hence is passive in nature. A similar process involves the movement into the cells of sodium with amino acids and of sodium with phosphate.

AMINO ACIDS. Amino acid clearance is low in humans (1–8 ml/min). Much of the earlier characterization of the mechanism has been done in the dog. More recently, rat kidney studies have supplied additional information, utilizing isolated nephrons, kidney slices, and micropuncture techniques. Several amino acids (glycine, arginine, lysine, proline, and hydroxyproline) demonstrate relatively poor reabsorption with a small Tm; others, like histidine, methionine, leucine, isoleucine, tryptophan, valine, threonine, and phenylalanine, are so effectively reabsorbed that saturation is not achieved by plasma concentrations that do not cause severe nausea or other physiological disturbances.

There are probably no less than three renal tubular mechanisms for reabsorption of amino acids (Pitts, 1974). One transports lysine, arginine, ornithine, cystine, and possibly histidine; a second handles aspartic and glutamic acid; a third has been shown to involve proline, hydroxyproline, and glycine. Co-transport with sodium is also involved in the reabsorption of amino acids. The driving force is the extrusion of Na^+ in exchange for K^+ into peritubular and lateral intercellular spaces.

The amino acid transport mechanisms display saturability, substrate specificity, and, in vitro, Michaelis-Menten kinetics. An interesting series of "inborn errors of membrane transport" is manifested by some of the amino acids—cystinuria, hyperprolinemia, and hydroxyprolinemia, among others. Hydroxyprolinemia is associated with mental deficiency.

ASCORBIC ACID. Ascorbic acid has a Tm of 2.0 mg per minute in humans (about 1.5 mg/100 ml of filtrate). In the dog, which is able to synthesize ascorbic acid, reabsorption is 0.5 mg per 100 ml of filtrate. The Tm represents the net activity of a three-component system (filtration, proximal

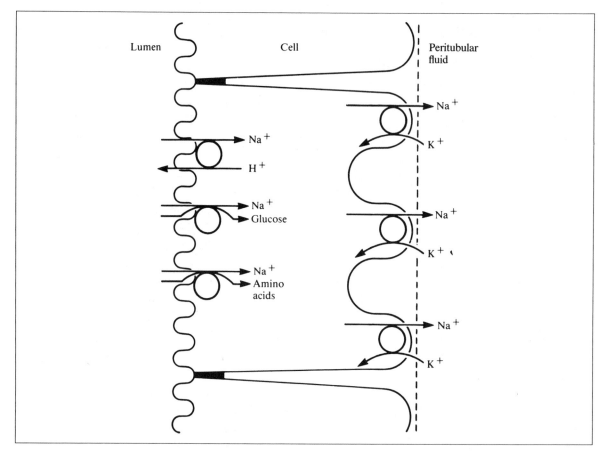

Fig. 22-16. Model of proximal tubular cell illustrating co-transport of glucose and amino acids with sodium.

tubular reabsorption, and distal tubular secretion). Distal secretion is promoted by pretreatment with adrenal steroid (deoxycorticosterone acetate), which stimulates distal sodium-potassium exchange. Additional stimulation of distal secretion results from increasing the filtered load of sodium, by alkalinization of the urine with $NaHCO_3$, and administration of acetazolamide. Evidently, there is a linkage between the ascorbic acid secretory site and the distal sodium-potassium exchange mechanism. This may be related to the vitamin's acidic properites.

UREA. Urea, a major product of protein metabolism, is filtered and reabsorbed to varying degrees (40–70 percent) throughout the proximal nephron and again in the medullary collecting tubule and duct (Figs. 22-17 and 22-28). The thick ascending limb, distal convolution, and early collecting tubule are urea impermeable. Urea's reabsorption is inversely related to the rate of urine production, because reabsorption is largely a process of back diffusion in humans, dogs, rabbits, and chickens. In such a reabsorptive mechanism the return of urea from the tubular urine, where it is in relatively high concentration on account of water uptake by the nephron, proceeds by the process of diffusion across the tubular epithelium into the peritubular capillaries, where it is in relatively low concentration. With rapid urine flow, the time for diffusion is limited. However, active reabsorptive mechanisms for urea (involving a transfer system and expenditure of energy) operate in the kidneys of elasmobranchii. Protein-depleted rats on a low-protein, high-salt diet show urea/inulin clearance ratios as low as 0.01, with papillary tissue urea concentration about three times that of the final urine, suggesting an adaptive development of an active reabsorption mechanism for urea, necessary to maintain the high papillary solute concentration required for

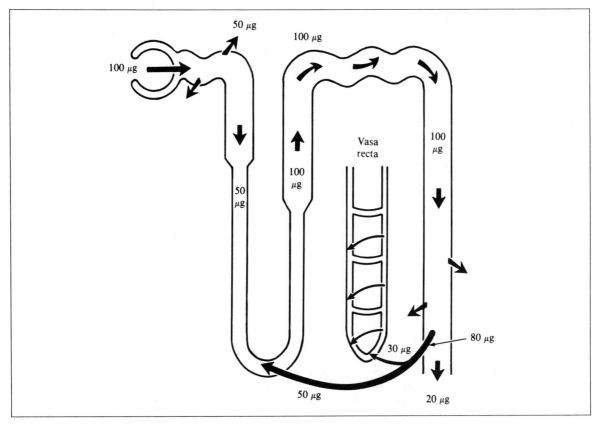

Fig. 22-17. Tubular handling of urea. The model shows a representative amount of 100 µg entering the glomerulus. (After K. J. Ullrich, K. Kramer, and J. W. Boylan. Current knowledge of the counter-current system in the mammalian kidney. *Prog. Cardiovasc. Dis.* 3:395–431, 1961. By permission.)

Fig. 22-18. Relationship of urea clearance (C_U) to urine flow. C_{IN} = inulin clearance.

the countercurrent mechanism (Truniger and Schmidt-Nielsen, 1964). On the other hand, in the amphibian (anuran) kidney the tubules *secrete* urea by an active process.

The past importance of urea clearance stems from the fact that it has been used as a clinical indication of glomerular filtration, which it reflects reasonably well at high rates of urine flow, when its reabsorption is minimal. The relationship of the clearance of urea to the rate of urine flow is shown in Figure 22-18. Urea plays an important role in the development and maintenance of the gradient of osmolality, cortex to papilla, needed for urinary concentration. This aspect will be considered in greater detail in a later section.

URIC ACID. In mammals, uric acid appears as a consequence of metabolism of purine bases. In most mammals it

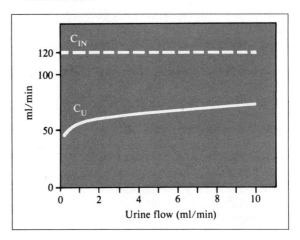

is oxidized to allantoic acid, but not in primates or the Dalmatian coach dog. It was originally assumed that a Tm characterized its reabsorption in humans. But under conditions of injection of a uricosuric drug (sulfinpyrazone). which blocks tubular reabsorption of uric acid, combined with vigorous mannitol diuresis, the ratio of excreted urate over filtered urate exceeded unity (up to 1.23) in humans, demonstrating tubular secretion. Thus, a three-component system of filtration, reabsorption, and secretion was suggested, so that the amount excreted was normally the *net effect* of these operations.

The three-component mechanism for renal urate handling was generally accepted until new data from humans and chimpanzees began to appear in the early 1970s, indicating that postsecretory reabsorption also played a role in renal urate handling, thus establishing a fourth component. Although definitive data in humans still are not available, the four-component system for renal urate handling is most consistent with the available data. This system may be summarized as involving the following:

1. Filtration of plasma urate at a rate approaching that of inulin.
2. Virtually complete reabsorption of filtered urate.
3. Subsequent secretion of urate at a rate approximating that of filtration by a system susceptible to modulation by various hormones, metabolites, and drugs.
4. Reabsorption of a major percentage of secreted urate by a high-capacity, variable-affinity transport mechanism, where reabsorptive transport may be substantially influenced by drugs and changes in extracellular fluid volume.

Thus, in health, the renal contribution to urate homeostasis is mediated primarily by the balance between the rate of urate secretion and reabsorption. It is likely that the three components of this system involving reabsorptive and secretory transport may be predominantly associated with the three anatomically distinct segments of the proximal tubule. Net transtubular flux of urate in all three segments is most likely the product of both active and passive bidirectional transport. Reabsorptive transport predominates within segments one and three, whereas secretory transport represents the major flux in segment two (Rieselbach, 1982).

It should be added that there are other uricosuric agents besides sulfinpyrazone, all of which have been used in the treatment of gout, a disease characterized by deposition of uric acid crystals in the joints. These include cinchophen, salicylate, acetylsalicylic acid, the mercurial Salyrgan, carinamide, and Benemid (probenecid).

CREATINE. Creatine is a product of muscle metabolism that disappears from the urine of humans after adolescence. It is filtered and reabsorbed in concentrations below 0.5 mg per 100 ml. At higher concentrations, reabsorption is incomplete and excretion is enhanced. No Tm has been demonstrated. At higher plasma levels, the creatine/C_{IN} ratio becomes constant at 0.8.

OTHER SUBSTANCES. Other substances involved in tubular reabsorption have been demonstrated in dogs and humans. These include acetoacetic acid, β-hydroxybutyric acid, lactic acid, and guanidoacetic acid. Maximal rates of transport have been demonstrated for β-hydroxybutyric and lactic acids. α-Ketoglutarate, utilized metabolically by the renal tubular cells, is both reabsorbed (Tm limited) and secreted (Balagura and Pitts, 1964). (See also relation to PAH secretory mechanism, under Tubular Secretion.)

Tubular reabsorption of various substances is limited by a maximal rate, and this varies markedly from one substance to another. In some instances (the sugars and certain amino acids) a transport mechanism may be shared in common and involve the reabsorption of several related substances, but there appear to be many transport systems that operate wholly independently of one another. Besides glucose and the amino acids, these include lactic acid, uric acid, acetoacetic acid, and the vitamin mechanisms.

Evidence for another co-transport mechanism involving sodium and lactate has been obtained by Murer et al. (1980) in studies with isolated proximal tubule cell vesicles. They found that concentrative uptake of L-lactate occurred only in the presence of inwardly directed sodium concentration differences.

There is considerable evidence that most of these substances are reabsorbed either in their entirety or in part in the proximal convoluted tubule. Information on localization has been obtained by micropuncture of the tubules and by the stop-flow technique. The technique is as follows: The ureter of the experimental animal is clamped during vigorous diuresis, bringing the movement of urine in the nephrons to a halt. Stoppage of the tubular movement of urine for a brief interval (about 8 min) permits the tubular reabsorptive and secretory processes to proceed to greater completion. Fractional collection of the urine that issues on release of the ureter permits allocation of observed changes in tubular urine concentration to various segments of the nephron. This is aided by the injection of a marker substance (such as inulin or creatinine) during the period of ureteral blockade to tag the entry of new glomerular fluid or, in effect, to mark the end of the stagnant urine column at the approximate level of the glomerulus. The knowledge of enzymatic and cellular mechanisms involved with these transport systems is rather limited at the present time. Further investigative effort is necessary to clarify the multiplicity of mechanisms involved.

TUBULAR SECRETION

Evidence has accumulated from methods such as the stop-flow technique that the proximal convoluted tubule is the site of active secretion of some physiologically occurring substances as well as certain foreign substances injected into the circulation.

TUBULAR SECRETION OF FOREIGN SUBSTANCES. Para-aminohippurate. At low concentrations, PAH is almost completely cleared from the plasma by a combination of glomerular filtration and efficient tubular secretion. Hence, C_{PAH} measures approximately 90 percent of the total plasma flow through the kidneys. In humans, the clearance is 600 to 700 ml per minute and corrected for hematocrit gives the *effective RBF*.

When plasma levels are elevated to the range of 30 to 50 mg per 100 ml, the secretory mechanism becomes saturated, and the Tm can be discerned. Tm for PAH in humans is about 77 mg/min/1.73 m^2 of surface area. In the dog it is about 33 mg/min/1.73 m^2 of surface area. The relationship of the mechanism of urinary excretion as it relates to plasma concentration appears in Figure 22-19. As the tubular secretory mechanism becomes saturated and is exceeded by progressive increments in plasma PAH, C_{PAH} is observed to decrease. The reason is that with saturation of the tubular transfer system, the total $U\dot{V}$ and the calculated $U\dot{V}/P$ become progressively more a function of glomerular filtration; hence $U\dot{V}/P$ approaches that function asymptotically. The portion of $U\dot{V}$ contributed by tubular secretion becomes a constant but fractionally smaller part of the total $U\dot{V}$.

The mechanism of transfer of PAH and other substances secreted by the tubular cells has been approached by studies involving isolated nephrons and kidney slices. This approach offers the opportunity to study the effect of substances that accelerate or depress accumulation of PAH in the slices as observed in the Warburg apparatus. It has been shown that the accumulation of PAH in tissue slices is dependent on aerobic metabolism by the fact that it is suppressed by lack of oxygen, by chilling, or by any metabolic poisons that inhibit respiration. Uptake is also suppressed by 2,4-dinitrophenol and by related compounds, which do not inhibit the uptake of oxygen but uncouple phosphorylation. Accumulation was increased greatly by the addition of

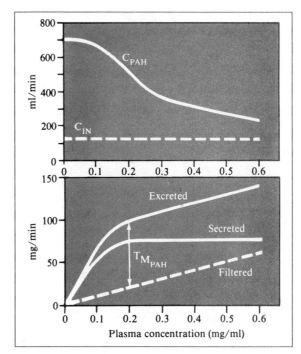

Fig. 22-19. The relationship of para-aminohippurate clearance (C_{PAH}) and tubular secretion to an increase in plasma concentration. C_{IN} = inulin clearance.

acetate to the medium, and to a lesser extent by the addition of lactate and pyruvate. Acetate and lactate increase Tm_{PAH} markedly when administered intravenously.

The site of secretion of the organic acid PAH is the proximal convolution, and its active transport depends on generation of high-energy phosphate compounds (e.g., ATP) by oxidative metabolism. It is not understood how the energy is coupled to the transport process. Uptake of PAH from the blood side of the tubules probably involves an active pump at the peritubular cell membrane. Foulkes (1963) concluded, from a kinetic analysis in the rabbit kidney, that intracellular PAH is in a free, not bound, form. Transfer from cell to lumen probably involves a facilitated diffusion mechanism, with no energy-requiring step at the luminal membrane. In the mammal the large volume of glomerular filtrate renders unnecessary an active concentration step at the luminal membrane.

The purpose of the PAH (organic acid) transport system has been a subject of speculation. (1) It may be to rid the body of foreign organic acids and end products of metabolism, such as glucuronides, sulfate esters, and urates. Interestingly, the liver, ciliary body of the eye, and choroid

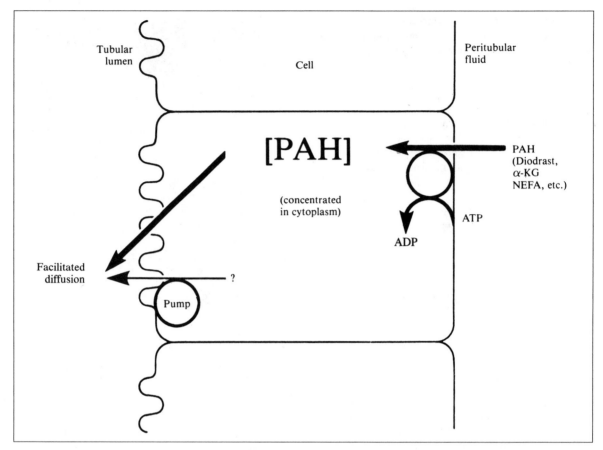

Fig. 22-20. Model of PAH transport system in the proximal tubular cell. α-KG = α-ketoglutarate; NEFA = nonesterified fatty acids.

plexus of the brain actively transport these compounds in an outward direction. (2) This transport system may serve to transport metabolic substrates, such as α-ketoglutarate and nonesterified fatty acids to the site of utilization in the proximal tubular cells (Fig. 22-20).

Other Substances. Additional foreign substances secreted by the tubules are *Diodrast* (iodopyracet), *phenol red,* and *penicillin.* Each has its characteristic Tm; e.g., Diodrast Tm averages 50 mg per minute in humans, and phenol red Tm is 36 mg per minute. The transfer systems for the different substances so far described appear to be mutually interrelated, because competitive depression of Tm is noted when they are simultaneously cleared at high plasma levels. Such blocking agents as carinamide and probenecid inhibit the secretion of these substances.

A group of *organic bases* is actively secreted by the proximal tubule: tetraethylammonium (TEA), tetramethylam-

monium, tetrabutylammonium, mepiperphenidol (Darstine), tolazoline (Priscoline), and hexamethonium. The Tm's are small (e.g., Tm_{TEA} is 1.0 to 1.4 mg/min/m^2 of surface area in dogs). The mechanism is distinct from that of organic acids such as PAH.

TUBULAR SECRETION OF PHYSIOLOGICAL SUBSTANCES. Creatinine. Creatinine is derived from creatine in muscles. Exogenous creatinine is cleared by glomerular filtration plus tubular secretion in humans and other primates. The Tm is small—about 16 mg per minute. It is also secreted by tubules of certain fish, alligators, chickens, goats, rats, cats, and guinea pigs. Animals from which it is cleared only by glomerular filtration have been mentioned previously.

It has been disclosed that creatinine transport is shared by both the organic acid and organic base transport mechanisms in the monkey and guinea pig (Selkurt et al., 1968; Arendshorst and Selkurt, 1970). This finding may be ascribable to the amphoteric nature of creatinine.

Naturally Occurring Organic Bases. Naturally occurring organic bases include N-methylnicotinamide, guanidine, piperidine, thiamine, histamine, and choline. N-methylnicotinamide, a metabolic derivative of nicotinic acid, has a clearance up to three times GFR, suggesting tubular secretion.

EXCRETION OF ELECTROLYTES

Virtually all segments of the nephron are concerned with the electrolyte economy of the body, to help preserve the proper balance of cations and anions. Reabsorptive mechanisms involve active pumps and energy expenditure.

ELECTRICAL POTENTIALS. A knowledge of electrical potentials across the tubular membrane is necessary in order to decide which ions are actively transported. The origin of such potentials was discussed in Chapter 1 and is based on application of the Nernst equation. The student will find that a brief review of transmembrane potentials is helpful in the understanding of the nature of tubular ion transport. Thus, strong evidence for active transfer of Na^+ is that it moves out of the proximal tubule against a gradient of electrochemical potential.

CATIONS. Sodium. About 99 percent of filtered Na^+ is reabsorbed by active transport mechanisms (Table 22-1). Estimates vary from 65 to 80 percent reabsorption in the proximal tubules and the remainder in the ascending portion of the loop of Henle, the distal convoluted tubules, and the collecting duct. Reabsorption in the proximal tubule is isosmotic. Reabsorption of NaCl and $NaHCO_3$ here is primary, by an active process; water follows passively and is not influenced by ADH. Another site of NaCl reabsorption is in the ascending thick portion of the loop of Henle—significant because this is relatively water-impermeable,

and, as a consequence, the urine becomes hypotonic to plasma at this site.

Evidence of the active nature of sodium reabsorption has been cited in terms of its being pumped against an electrochemical gradient. Other evidence is that the TF/P (tubular fluid/plasma concentration) of Na^+ has been observed to go below 1.0 in the face of brisk osmotic diuresis.

No Tm for sodium has been discerned. In humans, about 0.2 mEq per minute is excreted. Increased loading of the kidney tubules by Na^+ is followed by increased total reabsorption, but with decreased efficiency, since the total load is not quite so effectively reabsorbed. As a result, urinary excretion of Na^+ increases. The hormone aldosterone, secreted by the zona glomerulosa of the adrenal cortex, is necessary for efficient reabsorption. The distal convoluted tubule and collecting tubule are the most important site of aldosterone activity, although some evidence suggests a possible proximal tubular action also.

SODIUM TRANSPORT MECHANISMS. Figure 22-21 represents a model of a proximal tubular cell. The K^+ content of the cell is high (about 140 mEq/liter), while Na^+ and Cl^- content is low (10 mEq/liter). The cell is electrically negative with respect to the lumen and peritubular fluid. The Na^+ can diffuse along an electrochemical gradient from lumen to cell; its concentration within the cell is kept low by a pump that extrudes it into the peritubular fluid. Diffusion of K^+ and Cl^- is the major determinant of the potential across the peritubular membrane. The peritubular membrane pump functions also to return K^+ that has diffused out.

The potential across the luminal cell membrane is produced by a current × resistance drop as current flows from the lumen into the cell. The electromotive force is created by the diffusion potential of Na^+ and the sodium pump at the peritubular membrane. The circuit is closed by the existence of intercellular channels in parallel with the cellular pathways. The active reabsorption of Na^+ maintains a small electrical potential difference across the proximal tubular epithelium. The lumen early in the nephron is 3 to 4 mv negative with respect to the surrounding interstitial fluid

---

Table 22-1. Reabsorption of sodium and related anions

Plasma conc. (mEq/L)		Donnan factor	GFR	Filtered (mEq/min)	Excreted (mEq/min)	Reabsorbed (mEq/min)	Percent reabsorption
Na^+	140	0.95	125	16.6	0.135	16.465	99.3
Cl^-	103	1.05	125	13.5	0.135	13.37	99.1
HCO_3^-	27	1.05	125	3.55	<0.002	3.54	99.9

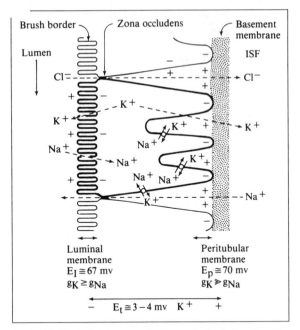

Fig. 22-21. Electrolyte transport by the proximal tubular cell. ISF = interstitial fluid; E = electrical potential difference; l = luminal membrane; p = peritubular and lateral membranes; E_t = transepithelial potential. (From L. P. Sullivan, *Physiology of the Kidney*. Philadelphia: Lea & Febiger, 1974. P. 60.)

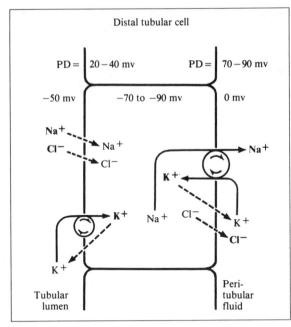

Fig. 22-22. Model of distal tubular cell transport mechanisms for Na^+, K^+, and Cl^-. Note that K^+ movement in the luminal membrane has the potential of being bidirectional. PD = potential difference. (From R. F. Pitts. *Physiology of the Kidney and Body Fluids* (3rd ed.). Copyright © 1974 by Year Book Medical Publishers, Inc., Chicago. Used by permission.)

(Fig. 22-21). This electrical gradient and a small chemical gradient cause passive Cl^- diffusion out across the tubular wall, which has a high conductance to chloride. The diffusion proceeds importantly through the zona occludens. HCO_3^- is removed by a process involving H^+ secretion, which will be discussed more fully in a later section. Present thinking includes the possibility of a passive back flux of Na^+ into the lumen because of a high conductance of the tubular wall to Na^+ (Fig. 22-21). This may occur through the zonula occludens. Nevertheless, the *net* flux of Na^+ out of the tubules is large, as has been stated.

Seely and Chirito (1975) demonstrated that in the rat there is a progressive increase in potential difference from the initial negative values to positive values ($+2$ to $+4$ mv) in later segments of the proximal tubule. This correlates with a rise in the tubule fluid to plasma (TF/P) chloride ratio to 1.3. The late luminal positivity appears to be a consequence of the development of a chloride diffusion potential. This results from the preferential reabsorption of HCO_3^- with Na^+, along with osmotically obligated water.

The potential across the peritubular membrane of the

distal tubule is greater than that of the proximal convoluted tubule, as is the transtubular potential (-50 mv) (Fig. 22-22). Na^+ still diffuses into the cell from the lumen, but along a less steep gradient, since the luminal fluid content of sodium has been considerably reduced to this point (to about 10 percent of the filtered sodium at the beginning of the distal tubule, and to about 2 percent at the end). Finally, the collecting duct reabsorbs almost all the remaining sodium.

The loop of Henle is also concerned with sodium and chloride transport. This topic will be discussed under Excretion of Water: Urinary Concentration and Dilution Mechanisms.

ENERGY SUPPLY OF ACTIVE SODIUM TRANSPORT. Extrusion of sodium from the cell against an electrochemical gradient requires expenditure of metabolic energy. The energy for this process is thought to be supplied by the high-energy phosphate bonds in ATP, which is a product of the cell's metabolism. It will be recalled (Chap. 1) that an enzyme located in the cell membrane can release this energy by catalyzing the splitting of ATP to form adenosine diphosphate and inorganic phosphate.

Potassium. Potassium clearance is usually well below that of inulin, suggesting fairly complete reabsorption. Under certain circumstances, however, surrounding the giving of large amounts of potassium salts of foreign anions, more potassium is excreted than is filtered. Thus, it appears that a three-component system operates with proximal tubular reabsorption and distal secretion. This conclusion is supported by micropuncture analysis. K$^+$ is actively reabsorbed by a pump in the tubular membrane of the proximal tubular cell. Figure 22-22 shows that the theoretical K$^+$ pump can operate either way in the distal tubule, reabsorbing K$^+$ under conditions of low intake or secreting K$^+$ during potassium loading.

The micropuncture studies of Giebisch et al. (1964) show this process more clearly (Fig. 22-23). Progressive reabsorption of filtered K$^+$ is indicated in the proximal tubule. In the distal tubule the K$^+$ concentration ratio drops to low levels with a low-K diet (Fig. 22-23A), suggesting continued reabsorption through the collecting duct. In Figure 22-23B (normal diet), the rising concentration ratio along the distal segment demonstrates a secretory process. This is magnified under the influence of a high-K diet (Fig. 22-23C), when the K/IN[TF/P] ratio exceeds unity, indicating net secretion.

The distribution of potassium ions is determined largely by the transtubular potential gradient. In potassium deficiency the transtubular potential at the secretory site is low, and little potassium is excreted. With a high-K diet, the converse is true.

K$^+$ secretion is also dependent on the availability of sodium. Although K$^+$ secretion was originally thought to be effected by a carrier system in exchange for reabsorbed sodium, it has been more recently viewed as an effect of luminal Na$^+$ concentration on the electrical potential, which in turn influences K$^+$ secretion. Thus, a low sodium concentration is associated with a low potential gradient and a lower rate of K$^+$ secretion, and vice versa (Wright, 1974).

In summary: K$^+$ urinary excretion is modified by at least four systemic variables. In addition to variations in plasma K$^+$ and availability of sodium, an increase in plasma aldosterone increases K$^+$ excretion, and vice versa. Changes in acid-base balance influence K$^+$ excretion (see Chap. 24). Thus, acute metabolic acidosis decreases and acute metabolic alkalosis increases the rate of renal K$^+$ excretion (Wright, 1981).

Calcium. Excretion of calcium is complicated by the fact that a significant part of its plasma content is combined with

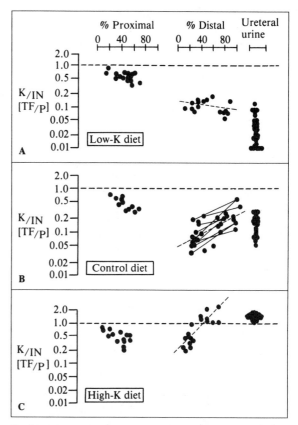

Fig. 22-23. Potassium excretion along the nephron of rats on a control diet (B), on a low-K diet (A), and on a high-K, low-Na diet (C). Potassium/inulin [TF/P] concentration ratios given as a function of tubular length. The K/IN [TF/P] ratio corrects for the influence of tubular water reabsorption. (From G. Malnic, R. M. Klose, and G. Giebisch. Micropuncture study of renal potassium excretion in the rat. *Am. J. Physiol.* 206:682, 1964.)

plasma proteins. However, urinary excretion is nominally low (about 8.5 μEq/min), suggesting efficient tubular reabsorption of the moiety that is filtered. Parathyroid hormone, secreted by the parathyroid gland, promotes tubular reabsorption of calcium (see Chap. 32).

Magnesium. Magnesium not bound by plasma proteins is filtered and reabsorbed. Excretion of about 5 to 6 μEq per minute occurs in humans.

As revealed by micropuncture studies, magnesium is not as effectively reabsorbed in the proximal convoluted tubule as are Na$^+$ and Ca^{2+}, with allowance being made for some plasma binding of Mg^{2+}. A summary of studies made in the dog (Quamme et al., 1978) indicated that 29 percent of the filtered Mg^{++} is reabsorbed in the proximal convolution, and 63 percent is reabsorbed in the loop of Henle. Thus, it would appear that the loop of Henle plays a major role in Mg^{2+} reabsorption in the dog.

The dog has a Tm of 150 μg/min/kg of body weight at a load/Tm ratio of about 2, with manifestation of considerable splay (Massry and Coburn, 1973).

Other Electrolytes. In the category of other electrolytes, NH_3 and H^+ can be considered. Ammonium is synthesized in the tubular epithelium from glutamine and other amino acids and is secreted into the tubular urine. It exchanges in this process with Na^+ as a mechanism in acid-base regulation and Na^+ conservation. Finally, the tubule is able to generate and secrete H^+ as a means of acidifying the urine by the aid of the enzyme carbonic anhydrase. (For further details, see Chap. 24.)

ANIONS. Chloride. Because chloride is the chief indifferent anion that accompanies Na^+ through the kidney tubular epithelium, this renal mechanism is conditioned by sodium reabsorption. Thus, the Cl^- is pulled by its electrical charge as a result of the active transfer of Na^+. However, there is evidence (see Chap. 23) that Cl^- may behave independently of Na^+.

It is now accepted that the thick ascending limb of Henle pumps Cl^- actively (Kokko, 1974; Burg and Stoner, 1974). Khuri et al. (1974) found intracellular concentrations of Cl^- equal to 42.3 ± 3.1 mEq per liter in late distal tubule cells of the rat—much higher than a passive equilibrium distribution would predict. Hence, they also postulated an active transport of Cl^- across the luminal cell border, since transport must proceed against both a chemical and an electrical gradient. An electrically neutral NaCl pump involving a double carrier combining with Na^+ and Cl^- simultaneously was considered as a possible alternative to the anion pump.

Bicarbonate. Data on bicarbonate reabsorption are expressed in milliequivalents per 100 ml of glomerular filtrate with an apparent physiological relationship between HCO_3^- reabsorption and filtration rate. As plasma HCO_3^- increases, the quantity filtered increases in direct proportion. Essentially all of the filtered HCO_3^- is reabsorbed until the load exceeds 2.8 mEq per 100 ml of filtrate, when frank excretion of HCO_3^- begins. The quantity reabsorbed remains constant at this value despite further increases in HCO_3^- plasma concentration, the excess being excreted. Thus, there is a grossly maximal reabsorptive capacity that is reached at a filtered load of about 2.8 mEq/100 ml glomeru-

lar filtrate/min. The critical value of plasma HCO_3^- required to cause frank excretion is about 27 mEq per liter in humans. The role of bicarbonate reabsorption in acid-base regulation will be considered in Chapter 24.

Phosphate. In the plasma, about 80 percent of phosphate occurs as $HPO_4^=$ with about 20 percent as $H_2PO_4^-$. Both are usually combined with Na^+. The ultimate ratio of $H_2PO_4^-/HPO_4^=$ in the urine is determined by the pH. In reabsorption of phosphate a Tm is manifested that has a magnitude of about 0.13mM per minute. Reabsorption is in the proximal convoluted tubule. Excretion in humans is from 7 to 20μM per minute. Excretion is modified by PTH action (see Chap. 32), which tends to block reabsorption.

Sulfate. Sulfate is actively reabsorbed in both dogs and humans and shows a well-defined, although small, Tm. It is 0.05mM per minute in the dog and 0.110mM per minute in humans.

EXCRETION OF WATER: URINARY CONCENTRATION AND DILUTION MECHANISMS

The mechanisms that concern tubular reabsorption of water are intimately related to the handling of osmotic constituents, primarily NaCl and urea. Another factor is the action of the ADH that is elaborated by the supraoptic and paraventricular nuclei of the hypothalamus. Finally, the composite mechanisms are integrated in the light of a *countercurrent multiplier system*, operating particularly in the nephrons that project long loops of Henle and vasa recta into the papillary zones of the renal medulla. The anatomical arrangement is especially exemplified by the kidneys of the golden hamster and kangaroo rat but is also applicable to some of the nephrons of the white rat, the dog, and the human kidney.

The basis for this theory is the finding that the osmotic constituents of the kidney are arranged so that they are isosmotic with plasma in the cortex, with the concentration rising three to four times and even more in certain species, at the tip of the papilla. Using a cryoscopic method that depends on a polarizing microscope to detect tiny crystals of ice formed during the cooling of sections of tissue of 30-μm thickness, Wirz et al. (1951) found that points of equal osmotic pressure form shells concentric to the tip of the papilla and parallel to the interzonal boundary. Uniform distribution of osmotic constituents among the loops of Henle, vasa recta, and collecting tubules has been proved by micropuncture in several animal species. Thus, as the osmotic concentration progressively increases from the cortex to the papillary tip, at any given level it is approximately equal in the loop of Henle, the vasa recta, and the collecting tubule.

PRINCIPLE OF THE COUNTERCURRENT SYSTEM. The arrangement of the loop of Henle system and vasa recta, which causes the currents of the fluids in the two limbs of this hairpin to flow in opposite directions, has provided a foundation for the theory of the *hairpin countercurrent system*. Basically, this operates because of the principle of countercurrent exchange, which has several applications in thermodynamics. Heating engineers have used the principle of countercurrent exchange in furnace exhaust and air-intake ducts. If the ducts lie side by side, warm air moving from the inside to the outside passes along the duct that brings the cold air from the outside. The heat is conducted across to the incoming cold air, raising its temperature, while the air being exhausted to the outside is cooled. Thus, overall heat conservation results (Fig. 22-24A). The principle is utilized for conservation of heat in the living organism by means of the countercurrent arrangements of the blood vessels in the limbs (see Chap. 28 for further discussion). When the two systems are connected at one end, an opportunity for a *countercurrent multiplier system* is created. This is illustrated by the use of a physical model in Figure 22-24B. Suppose that two parallel circuits are separated by a semipermeable membrane. A solution is driven through the circuit by the hydrostatic pressure (P) of an elevated reservoir. Because the membrane is permeable to water, but not to solute, water is driven across by the hydrostatic force, indicated by the vertical arrows. As fluid leaves the upper tube, the solute (C_1) becomes progressively concentrated toward the end. On making the turn, the highly concentrated solution (C_2) becomes rediluted as it picks up the water flowing through the membrane.

The principle applies to the kidney, but the forces are different, involving changes in osmotic pressure based on the movement of ions and solutes, rather than hydrostatic pressure forces. The analysis of a model corresponding to the loop of Henle and the collecting duct in Figure 22-25 serves to clarify one possible mechanism. In Fig. 22-25A, if it is considered that the membrane is unidirectionally permeable (i.e., that NaCl can move from ascending limb across to the descending limb), the opportunity for countercurrent multiplication exists. In the classic view, the Na^+ from the ascending limb added to the isotonic fluid of the descending limb increases the tonicity of the fluid approaching the hairpin turn. This process continues as the hypertonic fluid rounds the turn and is multiplied over and over again, and the Na^+ is, in effect, trapped in relatively high

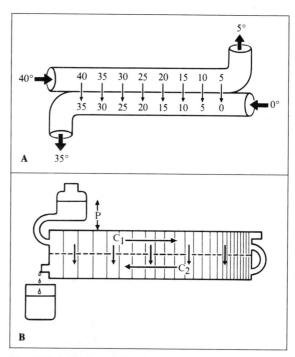

Fig. 22-24. A. Diagram of the principle of countercurrent exchange. B. Model illustrating the principle of the countercurrent multiplier. (B from B. Hargitay and W. Kuhn. Model illustrating the countercurrent multiplier. *Z. Elektrochem.* 55:539, 1951.)

concentration at the tip. Thus, the fluid would be isotonic entering the decending limb, hypertonic at the bend, and hypotonic issuing at the upper end of the ascending limb.

Final Concentration of Urine. The mechanism for final concentration of the urine becomes more apparent if a tube representing the collecting tubule of the nephron is added to this sytem. If the upper loop to the right, analogous to the distal convoluted tubule, and the descending tube, analogous to the collecting tubule and duct, are made water permeable, then urinary concentration becomes possible. It will be seen later that this occurs because of the action of ADH.

Removal of water to the interstitium of the kidney (from the upper right-hand loop analogous to the distal convoluted tubule) establishes isotonicity in this region, but now water is drawn from the collecting tubule and duct to the zone of hyperosmolality established by the countercurrent multiplier system, so that the fluid issuing from the descending right-hand limb, analogous to the collecting duct, becomes hypertonic.

Although this model illustrates the basic principles in-

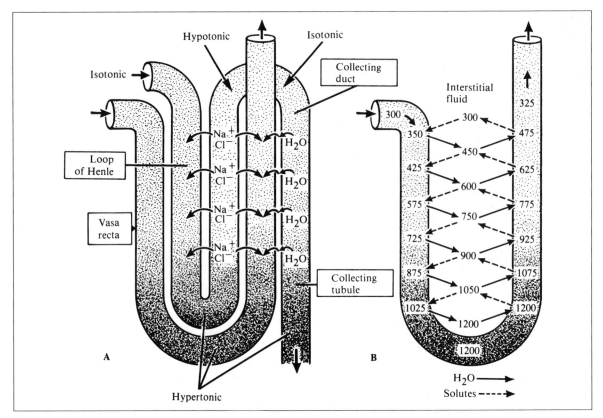

Fig. 22-25. A. Relationship of vasa recta to the loop of Henle and the collecting duct, to illustrate role in removal of sodium and water. B. Operation of the countercurrent vascular loops in preserving the osmotic gradient in the renal medulla.

volved in the countercurrent multiplier as it is utilized in the concentration of the urine, some immediate technical difficulties become apparent from the use of the simplified model. One is that the volume of the ascending limb of the loop of Henle and the distal convoluted tubular system would be increased by the return of water from the collecting duct. An extremely important consideration is that interstitial fluid is interposed between the loops of the tubule, and that the vasa recta lie in juxtaposition to the loop of Henle system and the surrounding interstitial fluid. Another important facet is the finding that water leaves the descending limb of the loop of Henle, as indicated by micropuncture studies showing that the inulin U/P ratio increases to the tip of the papilla in the hamster kidney, and probably also in the rat and rabbit, proving that water has left the descending limb of the loop. This would contribute to the operation of the countercurrent multiplier system by producing an increment of osmotic pressure in the descending limb.

Last, consideration should be given to the distribution of osmolality in the vasa recta system. Since their course par-

allels the loop of Henle system, and they are freely permeable to water and solutes, the vasa recta should manifest the same stratification of osmotic pressure. This then becomes the final important link in the mechanism, for it permits the water taken out of the collecting tubule and duct to be picked up by the plasma moving through the vasa recta system and carried away into the venous channels of the kidney. The force is the colloid osmotic pressure of the plasma protein. The active reabsorption of sodium by the collecting tubular cells may contribute to hyperosmolality of the vasa recta system.

The vasa recta operate also as *countercurrent diffusion exchangers* in terms of preserving the osmotic gradient in the renal medulla (Fig. 22-25B). Blood enters the loop with a concentration of 300 mOsm per liter. As it dips down, water diffuses out, and osmotically active particles (NaCl, urea, ammonia) diffuse in from the medullary interstitium, increasing the concentration in the descending portion of the loop as this process is repeated. Then, as blood traverses the loop and ascends, osmotically active particles diffuse out, and water enters, thus reducing the gradient. The net effect is to trap solutes in the tip of the vascular loop, in equilibrium with the loop of Henle, in a concentration about four times that of the entering blood. Because of the short-circuiting effect of water across the tops of the vascular loops, red cells and plasma proteins are also concentrated, the latter aiding in picking up water reabsorbed from the collecting duct, facilitated by ADH action.

PRINCIPLES OF OPERATION IN THE NEPHRON SYSTEM. The situation as it exists in the nephron system is illustrated in Figure 22-26, based on tubular puncture techniques. Osmolality values have been arbitrarily assigned as they apply to the human kidney and become the basis for a more extensive recapitulation of the mechanism. The greatest proportion of the filtered sodium and attendant anions is reabsorbed by an active process in the proximal convoluted tubule (Fig. 22-26,A). Water follows passively, not requiring the mediation of ADH, so that the proximal tubular fluid remains essentially isosmotic. Micropuncture studies in the hamster and rat have revealed that 67 percent of the filtrate is absorbed in the proximal convolution (Lassiter et al., 1961). The descending limb of the thin segment of the loop of Henle does not transport salt actively.

Another site of active NaCl reabsorption is the ascending limb of the loop of Henle, especially the thick portion, which is relatively water impermeable. According to one view, chloride, by an active mechanism, and Na$^+$, as a result of the electrochemical gradient established, are transported into the interstitium of the outer medulla until a gradient of perhaps 200 mOsm per kilogram of water has been established between the fluid of the ascending limb and the interstitium. This single effect is multiplied as the fluid in the thin descending limb comes into osmotic equilibrium with the interstitial fluid by the diffusion of water out of and possibly the diffusion of some NaCl into the descending loop, thus raising the osmolality of the fluid making the hairpin turn to be presented to the ascending limb. Progressive concentration of NaCl and other solutes in the tubular fluid of this segment of the nephron could result largely, if not entirely, from water diffusion into the hypertonic interstitium. In any event, an increasing osmotic gradient is established in the direction of the tip of the papilla (Fig. 22-26,B), and yet at no level is there a large osmotic difference between the luminal and interstitial fluid. An additional 10 to 20 percent of the filtered water is reabsorbed in the pars recta and descending limb of the loop of Henle (passive

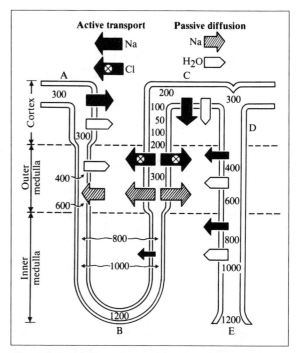

Fig. 22-26. Diagram of nephron showing sites of sodium, chloride, and water reabsorption. Figures represent the distribution of osmolality values. It seems likely that equilibrium (return of osmolality to 300) is probably not achieved in the distal convoluted tubule of the primate kidney. Although osmolality values are shown in the simplified diagram to be approximately equal in the thin descending and thin ascending limbs of corresponding levels of the loop of Henle, the ascending limb values actually average approximately 100 mOsm lower than the descending, as reported by Jamison et al., 1967. Although Na$^+$ transport is shown as being reabsorbed by passive diffusion in the thick ascending limb of Henle, recent findings suggest that Na$^+$ transport is also active here (see Eveloff et al., 1981, below [p. 440]).

efflux), and 10 percent in the distal convolution (Fig. 22-26,C), aided by ADH, leaving about 5 to 10 percent of the original volume to enter the collecting tubules. The model in Figure 22-27 illustrates the principle.

In contrast to the ascending limb, the epithelium of the collecting tubules (see Fig. 22-26,D) in the presence of ADH is also believed to be water permeable, resulting in diffusion of water out of the collecting tubules and ducts into the hyperosmotic medullary interstitium. The volume of fluid remaining in the collecting ducts becomes correspondingly concentrated (Fig. 22-26,E). The degree to which distal tubular fluid is restored to isotonicity (300 mOsm/liter) appears to vary with the species. This is achieved in the rat and

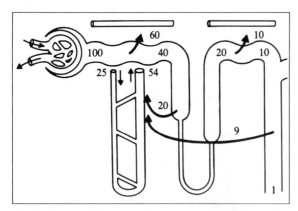

Fig. 22-27. The fluid balance of the concentrating kidney, assuming a GFR of 100 ml/min and a total medullary flow of 25 ml/min. Net water flow is indicated by the heavy arrows. The tubes at the top represent cortical peritubular capillaries. (From P. Deetjen, W. Boylan, and K. Kramer. *Physiology of the Kidney and Water Balance.* New York: Springer-Verlag, 1975.)

hamster kidney before the distal convolution is left, but not in the dog and monkey kidney, where restoration to isotonicity is delayed until the collecting tubule (see Fig. 22-26,D).

For the urine to be significantly concentrated, the flow through the loops of Henle must exceed considerably the flow through the collecting ducts. This change is accomplished by the action of ADH, which facilitates diffusion of water out of the distal convolution (see Fig. 22-26,C) into the interstitium of the cortex, thereby reducing the volume and increasing the osmolality to the isosmotic level before the fluid is presented to the collecting tubules. The rapid flow of blood through the cortex facilitates this removal.

Although the sodium salts are important, urea probably also functions importantly in this system, from present evidence. It has been proposed that the urea diffuses out of the medullary collecting tubules and the collecting ducts, to be concentrated in the loop of Henle by countercurrent exchange, adding to the gradients established in the loop. Ammonia produced in the proximal tubule also becomes concentrated in the papilla by operation of the medullary countercurrent exchange system.

To work most effectively, the medullary blood flow should be low, and experimental evidence at the present time indicates that it is. Probably the osmotic equilibration of the plasma in the vasa recta with the medullary interstitial fluid is due not only to the diffusion of solute into the descending and out of the ascending limbs but also largely to the diffusion of water in the opposite direction. This short-

circuiting across the tops of the vascular loops may be in part the cause of the apparently rich content of cells and plasma protein that has been found in the vasa recta at the tip of the papilla.

The importance of the vasa recta is that the blood entering the medulla picks up not only water that diffuses out from the thin descending limb of the loop of Henle but also water from the collecting ducts. The water movement to the vasa recta system is the result of the favorable gradient established by the electrochemical potential created by the colloid osmotic pressure of the plasma proteins. If indeed, as seems to be the case, the plasma proteins are concentrated in this zone, the greater effectiveness of this mechanism becomes apparent.

The next question to consider is the mechanism in the diluting kidney. This is the kidney that operates in the absence of ADH, as occurs following water ingestion. In rats in which the final urine was quite dilute, Wirz (1956) found that the fluid remained dilute throughout the distal convolution and the collecting ducts. Thus, an important role of ADH is that it renders the epithelium of the collecting tubules and ducts and perhaps the distal convolution freely permeable to water, so that the osmotic gradients established can operate in the presence of the hormone and are reduced or eliminated in the absence of the hormone. Studies with tritiated water (HTO) show that ADH acts by increasing the permeability of the luminal surface of the cellular membranes of the distal convolution and/or collecting ducts to water diffusion by increasing the number of so-called pores within the luminal portion of the membrane of the epithelial cells.

Antidiuretic hormone may have an additional role in regulating medullary blood flow. It is well known that ADH is a vasoconstrictor. Reduction in medullary blood flow should favor a further increase in osmolality of the papillary tip and thus favor further concentration of the urine. On the contrary, increased blood flow through the medullary circuit would tend to "wash out" the hypertonic zone, thus blunting the capacity for water uptake from the collecting tubule as outlined above.

PASSIVE EQUILIBRATION MODEL OF COUNTERCURRENT SYSTEM. The classic description of the countercurrent mechanism has been given above in some detail. The theory was developed largely in the absence of direct measurements of electrolyte and osmotic constituents in the loop of Henle, particularly in the outer medulla, the site of the medullary thick ascending limb (MTAL) of Henle. A successful technique of perfusing the loop of Henle in vitro (Kokko, 1974; Burg and Stoner, 1974) has permitted mea-

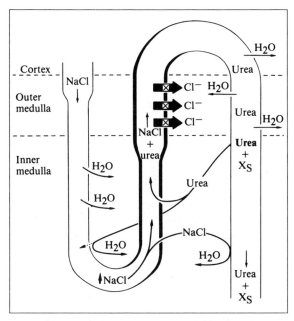

Fig. 22-28. Passive equilibration model of the countercurrent system. For explanation, see text. x_s = solutes in excess of urea. (From J. P. Kokko et al. Experimental and theoretical evidence for the passive equilibration model of countercurrent multiplication system. *Proc. 5th Int. Congr. Nephrol.* 2:93, 1974.)

surement of transtubular potentials, net fluid flux, and net movement of solute (e.g., urea, and Na^+ and Cl^-). The investigating groups used the isolated loop of Henle of the New Zealand rabbit. An important unexpected finding was that Cl^- was actively pumped in the MTAL, accompanied passively by Na^+; as expected, a lumen positive transtubular potential was found here (+ 6.7 mv—Kokko, 1974). The correlated demonstration that the thin descending limb (DL) of Henle had a high degree of osmotic water permeability, but a low permeability to NaCl and urea, led to a revision of the classic concept, referred to by Kokko as the "passive equilibration model of the counter-current multiplication system." The model achieves the increasing gradient of osmolality and osmotic equilibration principally by water extracted from the thin DL without requiring net entry of solute (NaCl, urea). The important interstitial osmotic ingredient is urea, diffusing in from the distal collecting tubule and duct (Figs. 22-17 and 22-28). Nevertheless, NaCl attains a high concentration at the bend of the thin segment because of outward diffusion of water (Fig. 22-28). In the water-impermeable thin ascending limb, NaCl now diffuses outward, down its gradient, to increase the interstitial solute

concentration of the inner medulla, along with the urea diffusing into the interstitium from the collecting duct. The active pumping of Cl^-, in the water-impermeable MTAL, accompanied by Na^+, further contributes to the osmotic gradient for water abstraction at this level, but leaves behind a hypotonic luminal fluid.

Urea entering the ascending limb of Henle is concentrated in its passage through the urea-impermeable distal convoluted tubule and early collecting tubule by water abstraction (favored by ADH action), so that as the tubular fluid enters the inner medulla, a more favorable gradient for outward diffusion of urea is created (Fig. 22-28). The final concentration or dilution of the urine depends, as before, on the relative concentration of ADH resulting from the existing state of fluid balance.

In summary: The main features of the "passive equilibration model" that distinguish it from the classic model are as follows: (1) Solute entry (chiefly NaCl) into the thin descending limb is not required (urea enters slightly at the tip); (2) water abstraction occurs in the relatively solute-impermeable thin descending limb by water diffusing down its concentration gradient; (3) NaCl diffuses passively out of the thin ascending limb, rather than being actively pumped; (4) the energy-requiring single effect is localized in the MTAL, and is initiated there by the active pumping of Cl^-, with Na^+ following passively; (5) greater emphasis is placed on the urea "recycling" mechanism of the distal nephron and, hence the important contribution of urea to the mechanism of development of the osmotic gradient.

SODIUM-CHLORIDE CO-TRANSPORT IN THE THICK ASCENDING LIMB OF HENLE'S LOOP

The Kokko model (Fig. 22-28) emphasized the dominant role of active chloride transport, with Na^+ presumably following passively. However, the most recent view is that both Na^+ and Cl^- are actively transported, the initial reabsorptive step being a sodium-chloride co-transport pump across the luminal border of the MTAL (medullary thick ascending limb) cell (Fig. 22-29). Strong support for this concept is provided by the work of Eveloff et al. (1981) using isolated rabbit cell culture techniques and specific sodium-chloride reabsorption blocking agents (furosemide, ouabain). The rate of oxygen consumption was the criterion used.

The following model was postulated: Na^+ enters the MTAL cell together with Cl^- via a co-transport system that can be inhibited by furosemide. The Na^+ is pumped out of the cell by Na^+-K ATPase, which maintains the sodium gradient. This step can be blocked by ouabain. The driving

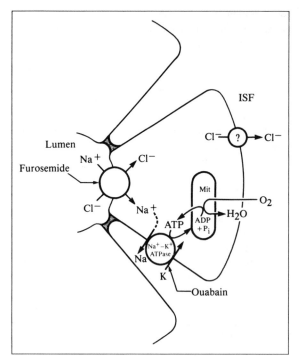

Fig. 22-29. Model for sodium-chloride co-transport in the MTAL in the rabbit kidney medulla. The sodium-chloride cotransport would occur across the luminal membrane of the cell energized by the favorable electrochemical potential difference for Na⁺ into the cell. Na⁺ would be excluded from the cell via the Na⁺-K⁺ ATPase located in the basal-lateral membranes. Cl⁻ exit from the cell may occur via Na⁺ independent mechanism or following a favorable electrochemical difference. Mit = mitochondrion; Pᵢ = inorganic phosphate; ISF = interstitial fluid. (From J. Eveloff, E. Bayerdorffer, P. Silva, and R. Kinne. Sodium chloride transport in the thick ascending limb of Henle's loop. *Pflugers Arch.* 389:268, 1981.)

force for the co-transport system would be a favorable "downhill" electrochemical potential difference for Na⁺ into the cell across the luminal membrane, which concomitantly drives the "uphill" accumulation of Cl⁻ into the cell. The sodium gradient into the cell would be maintained by the Na⁺-K⁺ ATPase. Chloride transport in the MTAL is thus not a primary active transport system, but a secondary active system that derives its energy from the Na⁺ gradient into the cell.

Although conceptually satisfying as an energy-conserving biological system, the Kokko model presents difficulties. It is questioned whether or not this system makes a quantitatively significant contribution. For example, as pointed out by Marsh and Azen (1975), although its mechanism might

serve to maintain the gradient of osmolality, it is hard to see how it can develop the gradient fully, or restore it after washout of the medulla.

Recent thinking (Eveloff et al., 1981) has reemphasized the importance of an active sodium pump in the MTAL and coupled transport of chloride, also viewed as active (Fig. 22-29). Conclusions were derived from evidence supplied by the sodium transport–blocking agents furosemide and ouabain. Assuming luminal localization of the sodium-chloride carrier (which is supported by furosemide's being effective only when applied to the luminal surface of the tubule) and the basal-lateral localization of the Na-K ATPase, sodium and chloride would enter the cell at the luminal membrane on a co-transport system sensitive to furosemide.

Recent studies by Greger and Schlatter (*Pflugers Arch. Physiol.* 396:325, 1983) in the cortical thick ascending limb suggest that the luminal membrane NaCl co-transporter may be involved with the transport of K⁺. Thus, the luminal pump actually transports 1 Na⁺, 1 K⁺, and 2 Cl⁻ into the cell; K⁺ recycles back into the lumen. Chloride leaves the cell across the basolateral cell membrane in part via an electroneutral KCl co-transporter and in part by a Cl⁻ conductive pathway.

This model of sodium-chloride co-transport in the luminal membranes of the MTAL agrees with the data obtained in the intact perfused MTAL tubule; i.e., chloride transport is inhibited by ouabain and furosemide and was sensitive to the sodium concentration of the incubation medium.

CONCEPT OF OSMOLAR CLEARANCE AND FREE WATER CLEARANCE. The highest concentration of urine achieved by the human kidney is about 1200 to 1400 mOsm per liter, compared with other mammalian kidneys that can concentrate to higher maxima (e.g., dog, 2500; white rat, 3000; and kangaroo rat, 5000). The lowest concentration observed is about 40 mOsm per liter. In terms of the more commonly employed clinical measurement, the *specific gravity*, this concentration involves a range between about 1.035 and 1.002 for human urine as compared with 1.010 for plasma. Understanding of the mechanism of change in urinary concentration requires the introduction of several concepts that involve the expressions for *free water clearance* and *osmolar clearance*. The volume of water required to contain all the urine solutes in an isosmotic solution with contemporaneous plasma is designated as the osmolar clearance (C_{Osm}). It is given by the equation

$$C_{Osm} = \frac{U_{Osm} \dot{V}}{P_{Osm}}$$

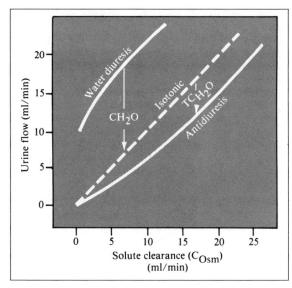

Fig. 22-30. Relationship between urine flow and solute clearance in water diuresis and formation of concentrated urine.

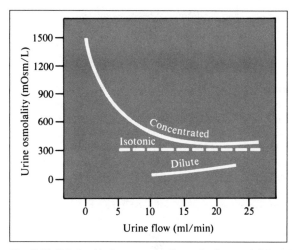

Fig. 22-31. Relationship between urine flow and urine osmolality when urine flow is modified by changing solute excretion. See text for explanation.

Free water clearance (C_{H_2O}) is the difference between an osmolar clearance and the rate of urine flow, namely,

$$C_{H_2O} = \dot{V} - C_{Osm}$$

When C_{H_2O} is positive, as in water diuresis, the urine is more dilute than the plasma; when C_{H_2O} is negative, as in dehydration, the urine is more concentrated than the plasma. Negative C_{H_2O} is more conveniently expressed by the designation $T^c_{H_2O}$.

To eliminate negative terms, the equation for $T^c_{H_2O}$ can be written

$$T^c_{H_2O} = C_{Osm} - \dot{V}$$

The mechanism of formation of osmotically concentrated urine in hydropenic subjects has been studied by giving additional ADH by vasopressin infusion. The subjects were then given an infusion of hypertonic mannitol solution to produce a progressively increasing osmotic diuresis. Figure 22-30 illustrates the result of plotting of osmolar clearance against urine flow. The dashed line in the figure represents the isosmotic parameter; i.e., the line that would be generated if the osmolar clearance equaled the urine volume at all values, in which case the urine would always be isosmotic with the plasma. The actual line is shifted to the right in the concentrating kidney, indicating that the urine is

hypertonic. The vertical distance between these lines, labeled $T^c_{H_2O}$, represents the amount of pure water that, if added to the concentrated urine, would restore it to isosmolarity, or conversely, that which must have been removed from the isosmotic tubular fluid to result in the concentrated urine. In this case, the value for $T^c_{H_2O}$ is 5 ml per minute at the higher rates of flow and clearance. The removal of the same volume of pure water from varying volumes of originally isosmotic fluid would lead to a progressive fall in urinary concentration as solute load and flow increase during osmotic diuresis. Thus, removal of 5 ml of pure water from 10 ml of isosmotic tubular fluid would give a final U/P ratio equal to 2. However, removal of 5 ml from 30 ml of tubular fluid would concentrate the urine to only 1.2, because the amount of solute concentrated in 5 ml of tubular urine would now be dissolved in 25 ml (hence, U is one-fifth of the original concentration instead of twice, and U/P is 1.2 instead of 2.0). Any substance that will institute an osmotic diuresis will produce the same effect. Increasing osmotic diuretic activity would thus result in increased amounts of isosmotic fluid passing from the proximal system to the loop of Henle and distal system. With removal of a fixed amount of pure water, the U/P ratio would decline (see Fig. 22-31).

At very low solute loads and low urine flows, the value for $T^c_{H_2O}$ decreases, the limiting factor being an osmolar ceiling above which the urine cannot be concentrated (at about 1400 mOsm/liter), representing an osmotic U/P ratio of approximately 4.7 (Fig. 22-31).

When it is dilute, the urine flow exceeds the osmolar

clearance, and the difference becomes the *free water clearance*, C_{H_2O}, which represents the *volume of pure water that originally contained the salt actively reabsorbed in the ascending limb of Henle, the distal convoluted tubule, and the collecting duct.* The capacity of these segments to reabsorb sodium salts therefore sets the limit on the amount of water that can be freed and excreted *above that containing the remaining urinary solutes in isotonic solution.*

The role of ADH is primarily to convert the dilute urine of the distal system to isotonicity and then to hypertonicity. The relative quantitative importance of the dilution and concentration processes is illustrated by considering the excretion of 2 liters of isotonic urine in 24 hours. If the subject were to produce maximally diluted urine (absence of ADH), he or she would excrete 20 liters of urine per 24 hours. Therefore, if ADH were to effect a rise in urinary concentration only to isotonicity, 18 liters of water would be conserved. If the solute were to be excreted in maximally concentrated urine (maximal ADH effect), the urine volume would be reduced to 500 ml, with a further saving of only 1.5 liters. Viewed in this light, it is clear that water conserved by not putting out dilute urine is quantitatively much more important than that conserved in forming concentrated urine.

Discussion of the renal role in body water balance and the control systems related thereto will be deferred until the latter part of Chapter 23, after the facts concerning body water and electrolyte distribution have been considered.

BIBLIOGRAPHY

Arendshorst, W. J., and Selkurt, E. E. Renal tubular mechanisms for creatinine secretion in the guinea pig. *Am. J. Physiol.* 218:1661–1670, 1970.

Baer, P., and McGiff, J. C. Hormonal systems and renal hemodynamics. *Ann. Rev. Physiol.* 42:589–601, 1980.

Balagura, S., and Pitts, R. F. Renal handling of α-ketoglutarate by the dog. *Am. J. Physiol.* 207:483–494, 1964.

Barajas, J. The juxtaglomerular apparatus: Anatomical consideration in feedback control of glomerular filtration rate. *Fed. Proc.* 40:78–86, 1981.

Beeuwkes, R., III. The vascular organization of the kidney. *Annu. Rev. Physiol.* 42:531–542, 1980.

Blantz, R. C. Segmental renal vascular resistance: Single nephron. *Annu. Rev. Physiol.* 42:573–588, 1980.

Brenner, B. M., and Beeuwkes, R. The renal circulation. *Hosp. Pract.* 13:35, 1978.

Brenner, B. M., Troy, J. L., and Daugharty, T. M. The dynamics of glomerular ultra-filtration in the rat. *J. Clin. Invest.* 50:1776–1780, 1971.

Briggs, J. The macula densa sensing mechanism for tubuloglomerular feedback. *Fed. Proc.* 40:99–103, 1981.

Burg, M., and Stoner, L. Sodium transport in the distal nephron. *Fed. Proc.* 33:31–36, 1974.

Caldicott, W. J. H., Taub, K. J., Margulies, S. S., and Hollenberg, N. K. Angiotensin receptors in glomeruli differ from those in renal arterioles. *Kidney Int.* 19:687–693, 1981.

Deen, W. M., Robertson, C. R., and Brenner, B. M. Glomerular ultrafiltration. *Fed. Proc.* 32:14–20, 1974.

Diedrich, D. F. The comparative effects of some phlorizin analogs on the renal reabsorption of glucose. *Biochim. Biophys. Acta* 71:688–700, 1963.

Eveloff, J., Bayendorffer, E., Silva, P., and Kinne, R. Sodium chloride transport in the thick, ascending limb of Henle's loop. *Pflugers Arch.* 389:263–270, 1981.

Foulkes, E. C. Kinetics of p-aminohippurate secretion in the rabbit. *Am. J. Physiol.* 205:1019–1024, 1963.

Gottschalk, C. W., and Mylle, M. Micropuncture study of the mammalian urinary concentrating mechanism: Evidence for the countercurrent hypothesis. *Am. J. Physiol.* 196:927–936, 1959.

Harris, C. A., Baer, P. G., Chirito, E., and Dirks, J. H. Composition of mammalian glomerular filtrate. *Am. J. Physiol.* 227:972–976, 1974.

Hollenberg, N. K., Epstein, M., Rosen, S. M., Basch, R. I., Oken, D. E., and Merrill, J. P. Acute oliguric renal failure in man: Evidence for preferential renal cortical ischemia. *Medicine* (Baltimore) 47:455–474, 1968.

Jamison, R. L., Bennett, C. M., and Berliner, R. W. Countercurrent multiplication by the thin loops of Henle. *Am. J. Physiol.* 212:357–366, 1967.

Khuri, R. N., Agulian, S. K., Bogharian, K. Electrochemical potentials of chloride in distal renal tubule of the rat. *Am. J. Physiol.* 227:1352–1355, 1974.

Kirschenbaum, M. A., White, N., Stein, J. H., and Ferris, T. F. Redistribution of renal cortical blood flow during inhibition of prostaglandin synthesis. *Am. J. Physiol.* 227:801–805, 1974.

Kokko, J. P. Membrane characteristics governing salt and water transport in the loop of Henle. *Fed. Proc.* 33:25–30, 1974.

Ladefoged, J. Renal cortical blood flow and split function test in patients with hypertension and renal artery stenosis. *Acta Med. Scand.* 179:641–651, 1966.

Lang, F., Greger, R., and Deetjen, P. Handling of uric acid by the rat kidney. *Pflugers Arch.* 338:295–302, 1973.

Lassiter, W. E., Gottschalk, C. W., and Mylle, M. Micropuncture study of net transtubular movement of water and urea in nondiuretic mammalian kidney. *Am. J. Physiol.* 200:1139–1147, 1961.

Maddox, D. A., Deen, W. M., and Brenner, B. M. Dynamics of glomerular ultrafiltration: VI. Studies in the primate. *Kidney Int.* 5:271–278, 1974.

Marchand, G. R. Direct measurement of glomerular capillary pressure in dogs. *Proc. Soc. Exp. Biol. Med.* 167:428–433, 1981.

Marsh, D. J., and Azen, S. P. Mechanism of NaCl reabsorption by hamster thin ascending limb of Henle's loop. *Am. J. Physiol.* 228:71–79, 1975.

Massry, S. G., and Coburn, S. W. The hormonal and nonhormonal control of renal excretion of calcium and magnesium. *Nephron* 10:66–112, 1973.

McKenna, O., and Angelakos, E. T. Acetylcholinesterase-containing nerve fibers in the canine kidney. *Circ. Res.* 23:645–651, 1968.

McKenna, O., and Angelakos, E. T. Adrenergic innervation of the canine kidney. *Circ. Res.* 22:345–354, 1968.

McNay, J. L., and Abe, Y. Pressure-dependent heterogeneity of renal cortical blood flow in dogs. *Circ. Res.* 27:571–587, 1970.

Moffat, D. B. The fine structure of the blood vessels of the renal medulla with particular reference to the control of the medullary circulation. *J. Ultrastruct. Res.* 9:532–545, 1967.

Morel, F., Mylle, M., and Gottschalk, C. W. Tracer microinjection studies of effect of ADH on renal tubular diffusion of water. *Am. J. Physiol.* 209:179–187, 1965.

Murer, H., Barac-Nieto, M., and Kinne, R. Renal proximal tubular transport of lactate. *Renal Physiol.* 2:145, 1980.

Nash, F. D., Rostorfer, H. H., Bailie, M. D., Wathen, R. L., and Schneider, E. G. Renin release: Relation to renal sodium load and dissociation from hemodynamic changes. *Circ. Res.* 22:473–487, 1968.

Navar, L. G., Bell, R. D., and Ploth, D. W. Role of feedback mechanism in renal autoregulation and sensing step in feedback pathway. *Fed. Proc.* 40:93–108, 1981.

Osgood, R. W., Reineck, H. J., and Stein, J. H. Methodologic considerations in the study of glomerular ultrafiltration. *Am. J. Physiol.* (Renal, Fluid and Electrolyte Physiol.) 2:F1-F7, 1982.

Osswald, H., Marchand, G. R., Haas, J. A., and Knox, F. G. Glomerular dynamics in dogs at reduced arterial pressure. *Am. J. Physiol.* 5:F25–F29, 1979.

Pennell, J. P., Lacy, F. B., and Jamison, R. L. An in vivo study of the concentrating process in the descending limb of Henle's loop. *Kidney Int.* 5:337–347, 1974.

Pitts, R. F. *Physiology of the Kidney and Body Fluids* (3rd ed.). Chicago: Year Book, 1974.

Quamme, G. A., Wong, N. L. M., Dirks, J. H., Roinel, N., De Rouffignac, C., and Morel, F. Magnesium handling in the dog kidney: A micropuncture study. *Pflugers Arch. Physiol.* 377:95–99, 1978.

Rohde, R. and Deetjen, P. Mikropunktionsuntersuchungen zum Glucose-Transport im proximalen Tubulus der Ratteniere bei freiem Fluss. *Pflugers Arch. Ges. Physiol.* 302:219, 1967.

Richards, A. N. and Walker, A. M. Urine formation in the amphibian kidney. *Am. J. M. Sc.* 190:727, 1935.

Rieselbach, R. E. The role of the kidney in urate metabolism. *Newsletter,* Council on the Kidney in Cardiovascular Disease, American Heart Association. 4:2–8, 1982.

Schneeberger, E. E., Levey, R. H., McCluskey, R. T., and Karnovsky, M. J. The isoporous substructure of the human glomerular slit diaphragm. *Kidney Int.* 8:48–52, 1975.

Schnermann, J., and Briggs, J. P. Participation of renal cortical prostaglandins in the regulation of glomerular filtration rate. *Kidney Int.* 19:802–815, 1981.

Seely, J. F., and Chirito, E. Studies of the electrical potential difference in rat proximal tubule. *Am. J. Physiol.* 229:72–80, 1975.

Selkurt, E. E. Renal Circulation. In W. F. Hamilton and P. E. Dow (eds.), *Handbook of Physiology.* Washington, D.C.: American Physiological Society, 1963. Section 2: Circulation. Vol. 2, chap. 43, pp. 1485–1489.

Selkurt, E. E., Wathen, R. L., and Santos-Martinez, J. Creatinine excretion in the squirrel monkey. *Am. J. Physiol.* 214:1363–1369, 1968.

Smith, H. W. *The Kidney: Structure and Function in Health and Disease.* New York: Oxford University Press, 1951.

Sullivan, P. L. *Physiology of the Kidney.* Philadelphia: Lea & Febiger, 1974.

Thorburn, G. D., Kopald, H. H., Herd, J. A., Hollenberg, M., O'Morchoe, C. C. C., and Barger, A. C. Intrarenal distribution of nutrient blood flow determined with krypton[85] in the unanesthetized dog. *Circ. Res.* 13:290–307, 1963.

Thurau, K., and Mason, J. The Intrarenal Function of the Juxtaglomerular Apparatus. In K. Thurau (ed.), *Physiology. Series One. Kidney and Urinary Tract Physiology.* Baltimore: University Park Press, 1974. Vol. 6, chap. 11, pp. 357–389.

Thurau, K., Valtin, H., and Schnermann, J. Kidney. *Annu. Rev. Physiol.* 30:441–524, 1968.

Truniger, B., and Schmidt-Nielsen, B. Intrarenal distribution of urea and related compounds: Effects of nitrogen intake. *Am. J. Physiol.* 207:971–978, 1964.

Tucker, B. J., and Blantz, R. C. Studies on the mechanism of reduction in glomerular filtration rate after benzolamide. *Pflugers Arch. Physiol.* 308:211–216, 1980.

Ullrich, K. J. Renal mechanisms of organic solute transport. *Kidney Int.* 9:134–148, 1976.

Valtin, H. *Renal Function: Mechanisms Preserving Fluid and Solute Balance in Health* (2nd ed.). Boston: Little, Brown, 1983.

Van Citters, R. L., and Franklin, D. L. Cardiovascular performance of Alaska sled dogs during exercise. *Circ. Res.* 24:33–42, 1969.

Walker, A. M., Bott, P. A., Oliver, J., and MacDowell, C. The collection and analysis of fluid from single nephrons of the mammalian kidney. *Am. J. Physiol.* 134:580–595, 1941.

Wirz, H. Der osmotische Druck in den corticalen Tubuli der Rattenniere. *Helv. Physiol. Pharmacol. Acta* 14:353–362, 1956.

Wirz, H., Hargitay, B., and Kuhn, W. Lokalisation des Konzentrierungsprozesses in der Niere durch direkte Kryoskopie. *Helv. Physiol. Pharmacol. Acta* 9:196–700, 1951.

Wright, F. S. Potassium Transport by the Renal Tubule. In K. Thurau (ed.). *Physiology. Series One. Kidney and Urinary Tract Physiology.* Baltimore: University Park Press, 1974. Vol. 6, chap. 3, pp. 79–105.

Wright, F. S. Characteristics of feedback control of glomerular filtration rate. *Fed. Proc.* 40:87–92, 1981.

Wright, F. S. Potassium transport by successive segments of the mammalian nephron. *Fed. Proc.* 40:2398–2402, 1981.

BODY WATER AND ELECTROLYTE COMPOSITION AND THEIR REGULATION. MICTURITION

23

Ewald E. Selkurt

FLUID VOLUMES. The principle of measurement of fluid volumes was introduced in Chapter 10, where the technique was explained in terms of its application to the determination of plasma volume. The same principle is employed for the measurement of total body fluid and extracellular volumes, based on the use of a substance that distributes itself throughout the compartment to be measured. The equation is modified from that used for plasma volume, $V = A/C$, as follows:

$$V = \frac{A - E}{C}$$

where V is the volume in liters; A is the amount of the substance administered; and E is the amount excreted in the urine at the time that C, the concentration per liter, is determined. The introduction of E, a subtraction for the amount excreted in the urine, is necessary because substances com-

monly used for these calculations are removed from the body by the kidney.

TOTAL BODY WATER. Total body water (water in the *cellular* plus *extracellular* compartments) is measured as the volume of distribution in the body of an appropriate indicator after a single intravenous injection. The requirement is that the substance must distribute itself uniformly in all the fluid spaces. Apparently the most reliable and the most commonly used indicators today are antipyrine and the heavy isotopes deuterium oxide and tritium oxide. Confirmatory evidence has been supplied by desiccation of human cadaver material to constant weight at a temperature of approximately 105°C. On the basis of determinations with the indicators, total body water ranges from 500 to 600 ml per kilogram of total body weight (50–60 percent) for the average adult human subject. Desiccation studies reveal a figure of 640 ml per kilogram. The total body water volume for a 70-kg subject is about 42 liters. Some authorities feel it is more desirable to express the total body water in terms of

Table 23-1. Body water distribution in a healthy young adult male*

Compartment	Percent of body weight	Percent of total body water	Liters
Plasma	4.5	7.5	3
Interstitial-lymph	12	20	8.5
Dense connective tissue and cartilage	4.5	7.5	3
Inaccessible bone water	4.5	7.5	3
Transcellular	1.5	2.5	1
Total extracellular	27	45	19
Functional extracellular	21	35	14.5
Total body water	60	100	42
Total intracellular	33	55	23

*All figures rounded to nearest 0.5.
Source: I. S. Edelman and J. Leibman. Anatomy of body water and electrolytes. *Am. J. Med.* 27:260, 1959.

the lean body mass, since adipose tissue is relatively deficient in water. On this basis, the generally accepted figure is 70 percent of the lean body mass.

The *extracellular* fluid (ECF) compartment can be directly determined by the use of a substance that distributes itself throughout this particular compartment. Although there is some controversy as to the exact space such a substance measures, at the present time mannitol appears to be best for measuring this space, and the distribution of sucrose, thiosulfate, and inulin supplies confirmatory evidence. One of the technical problems involved is whether or not the fluid of the connective tissue and bone should be considered part of the ECF. The tracer substances, mannitol, inulin, sucrose, and others, do not penetrate connective tissue fluid and bone water readily, nor is it likely that they enter the cerebrospinal fluid (CSF) freely; hence, the space measured may actually be smaller than the true physiological extracellular space. The average figure of 180 ml per kilogram includes the plasma compartment, the interstitial-lymph space, and transcellular fluids. If dense connective tissue and cartilage, and inaccessible bone water, are added, the total becomes 270 ml per kilogram (Table 23-1). *Transcellular* water is that concerned with the transport activity of specialized cells (salivary, pancreas, liver, mucous membrane of respiratory and gastrointestinal tract, CSF, and ocular fluid). The *intracellular* water can be estimated as the difference between total body water and the total extracellular water. This would yield a volume of 23 liters.

As indicated in Chapter 10, the plasma volume can be measured by the distribution of suitable indicators such as

T-1824 or iodinated albumin, giving a plasma volume of about 4.5 percent of body weight, or 3.2 liters for the average-sized adult. The approximate breakdown of the various volumes of total body water appears in Figure 23-1.

COMPOSITION OF THE EXTRACELLULAR FLUID. The principal constituents of ECF are shown in Table 23-2. Note that the concentrations of the interstitial fluid are corrected for the Gibbs-Donnan effect (see Chap. 1). Thus, the essential differences between the serum water and interstitial fluid concentrations are the small inequalities resulting from the Gibbs-Donnan equilibrium effect. The exception is the much lower amount of protein found in the interstitial fluid. The important cation is Na^+, and the important anions are Cl^- and HCO_3^-. The last column supplies, for comparative purposes, the approximate values found in the intracellular fluid (ICF).

Specialized Extracellular Fluids: Aqueous Humor and Cerebrospinal Fluid. The aqueous humor of the eyeball and the CSF are examples of specialized interstitial fluids that cannot be described as simple filtrates from the plasma. Not only is each fluid different in composition from plasma dialysate, but also, in some respects, they differ from each other. In general, CSF has higher Cl^- concentrations than plasma dialysate, while the concentrations of K^+ and Ca^{2+} are significantly lower in this fluid. Na^+ is slightly lower in CSF than plasma ($R_{CSF} = 0.98$), but is higher than in the dialysate (0.98 compared with 0.945). The aqueous humor and CSF have similarities (lower urea and glucose concentrations than plasma) but differ from each other in concentrations of

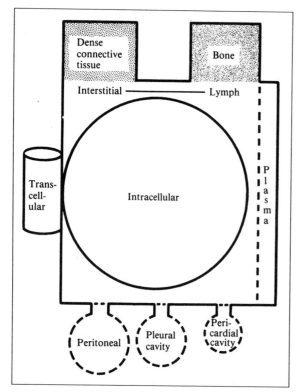

Fig. 23-1. Compartments of total body water. About two-thirds of body water is intracellular water; the remainder is the water of ECF. The dashed line separating plasma from interstitial fluid represents the capillary membranes. The three cavities drawn with circular dashed lines represent potential extracellular spaces that, however, are normally negligible in volume. (From R. H. Maffly. The Body Fluids: Volume, Composition and Physical Chemistry. In B. M. Brenner and F. C. Rector [eds.], *The Kidney*. Philadelphia: Saunders, 1976. P. 78.)

Cl^- and K^+. The ionic distribution between plasma and CSF in humans is presented in Table 23-3.

Because the solutes are not distributed as one might expect in an ultrafiltrate, the suggestion has been made that these fluids are secretions. Reasons have been advanced for not viewing the differential concentrations as the result of simple filtration through a highly selective membrane (Davson, 1967). Thus, the osmotic pressure required to maintain the difference in concentration of potassium between plasma and CSF can be calculated to be 54 mm Hg. The pressure required to maintain the difference in concentration of protein is 30 mm Hg. Consequently, the total pressure available must be 84 mm Hg, a figure well beyond capillary pressure. In other words, the supply of energy

necessary to maintain this difference would be inadequate if the hydrostatic pressure in the capillaries were the sole force involved. Moreover, the concentration of Na^+ and Cl^- is higher in the CSF than in dialysate, as are HCO_3^- and ascorbic acid in the aqueous humor. These facts appear to deny simple filtration. The conclusion that active transport is involved appears warranted.

FORMATION AND DRAINAGE: AQUEOUS HUMOR. The ciliary body of the eye is considered the source of the aqueous humor. The highly vascular ciliary processes have been specifically assigned the role of production of fluid. The principal drainage route is the *canal of Schlemm*, which connects via very fine ducts (*collectors*) with the intrascleral venous plexus.

CEREBROSPINAL FLUID. The CSF is elaborated by the choroid plexuses that project with many folds and villi into the roofs of the third and fourth ventricles, and into the sides of the lateral ventricles. The fluid passes from the lateral ventricles to the third ventricle through the foramen of Monro, then to the fourth ventricle via the aqueduct of Sylvius. It leaves the fourth ventricle through the foramina of Magendie and Luschka, to reach the subarachnoid spaces, here expanding to form the *cisterna magna*. From this region it passes into the dural sinus. Differing from the eye, wherein the canal of Schlemm enters directly via fine ducts into the venous system, the CSF is separated from the blood by the mesothelial lining covering the arachnoid villi.

Transport Systems Operating in Formation of Cerebrospinal Fluid. The major interest in the choroid plexus had centered around the role of this structure in the elaboration of CSF. This crucial function of the plexus clearly has been supported by the demonstrated architecture of its tissue. The most significant finding, however, has been the proved ability of plexus tissue to transport bidirectionally a number of biologically important substances. The current knowledge of some of the substances that are transported across the choroidal ependyma is summarized in Figure 23-2.

In the case of the ions, it is believed that actively pumped sodium is the main transported ion responsible for the diffusional transport of water from blood to CSF across the lamina epithelialis. Magnesium and calcium appear also to be actively transported, as does the potassium ion. The potassium ion transport system is unique in that it saturates at one-half of the blood concentration level. The anion transport systems have been studied in depth and appear to operate in the direction of CSF to blood. This fact explains the efficiency of the CSF as a "sink" for adjacent brain tissue, since a concentration differential gradient is main-

Table 23-2. Composition of the body fluids

Substance	Serum (mEq/L)	Serum water (mEq/L)*	Interstitial fluid (mEq/L)†	Intracellular fluid (mEq/L)
Na^+	138	148	141	10
K^+	4	4.3	4.1	150
Ca^{2+}	4	4.3	4.1	—
Mg^{2+}	3	3.2	3	40
Cl^-	102	109	115	15
HCO_3^-	26	28	29	10
PO_4^{3-}	2	2.1	2	100
SO_4^{2-}	1	1.1	1.1	20
Organic acids	3	3.2	3.4	—
Protein	15	16	1	60

*Correction to concentration per liter of serum water is based on the value 93 percent water.
†Values derived by use of Gibbs-Donnan factor of 0.95 for cations and 1.05 for anions in serum water.

Table 23-3. Distribution of various ions between lumbar cerebrospinal fluid and plasma in human subjects

Substance	Concentration (mEq/kg H_2O)		
	Plasma	CSF	R*$_{CSF}$
Na	150.00	147.00	0.98
K	4.63	2.86	0.615
Mg	1.61	2.23	1.39
Ca	4.70	2.28	0.49
Cl	99.00	113.00	1.14
HCO_3	26.80	23.3	0.87
Br	2.45	0.90	0.37
Inorganic P (mg/100 ml)	4.70	3.40	0.725
Osmolality	289.00	289.00	1.00
pH	7.397	7.307	—
PCO_2	41.10	50.50	—

*R represents the following ratio: concentration in CSF–water/concentration in plasma-water.
Source: H. Davson. *Physiology of The Ocular and Cerebrospinal Fluid*. Boston: Little, Brown, 1976. P. 229.

tained between these two compartments by the active transport of these anions out of CSF. Although the evidence seems reasonably complete for the bidirectional transport of glucose by facilitated diffusion, the situation for amino acids is unclear at present. The lower part of Figure 23-2 shows a number of other transport systems that are of some pharmacological interest.

The histological similarity of the ependymal cell to the proximal tubular cell of the kidney is worthy of note (Fig. 23-3). Thus, the choroid epithelial cells have a "brush border" and basal infoldings. Cytoplasmic organelles, similar to those in the kidney, may suggest a role in water transport. From a functional aspect, transepithelial potentials are of the same order of magnitude as is seen in the proximal tubular cells.

At this time it can be stated that from both an anatomical and physiological standpoint the choroid plexus indeed functions as a miniature kidney. The secretory product has been called the "neural urine" by some. Clearly, the CSF is continuous with, and similar in composition to, the interstitial fluid of the brain. Therefore, the blood-CSF barrier (represented by the choroid plexus) and the blood-brain barrier both influence the neural environment.

COMPOSITION OF INTRACELLULAR FLUID. The predominant cations of the intracellular fluid (ICF) are K^+ and Mg^{2+}. The predominant anions are organic PO_4^{3-}, SO_4^{2-}, and proteins (see Table 23-2). It should be stated that these are only approximations of the composition of cell fluid and are based on the findings in muscle. Inadequate data are available to form conclusions concerning the complex forms in which the materials exist, the valence of the organic anions, the degree of dissociation of the compounds, and the extent to which the cations are bound, and so on. In any event, the osmolar balance between the ICF and ECF is operationally equivalent. A bar graph summary of the distribution of cations and anions for the fluids of the body is presented in Figure 23-4.

FLUID EXCHANGE

EXCHANGE BETWEEN EXTRACELLULAR AND INTRACELLULAR COMPARTMENTS. Most cell membranes are appar-

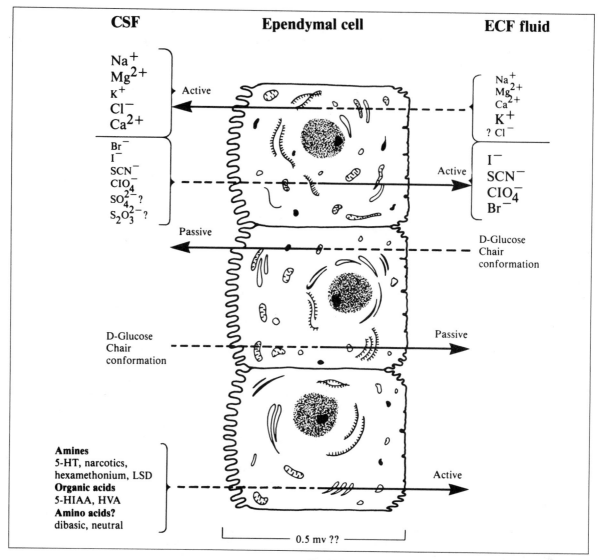

Fig. 23-2. Summary of transport systems operating across the choroid plexus. Active and passive represent active transport and facilitated diffusionary processes respectively. The size of ion symbols represents usual relative concentration levels in the CSF as compared with those in blood or choroidal ECF. 5-HT = 5-hydroxytryptamine; 5-HIAA = hydroxyindolacetic acid; HVA = homovanillic acid; LSD = lysergic acid diethylamide. (From M. Pollay. Transport Mechanisms in the Choroid Plexus. Fed. Proc. 33:2068, 1974.)

ently completely permeable to water, and the total exchange is enormous. The *net exchange* of water is governed by the osmotic pressure changes in the two compartments. When osmotic pressure changes, water moves across the cellular membrane to maintain an isosmotic state. The situation is diagrammed in Figure 23-5. Figure 23-5A shows the normal situation, i.e., concentrations of 300 mOsm per liter both within the cell (C) and in the extracellular compartment (E). Figure 23-5B shows the result of adding 300 mOsm to compartment E. Water moves out of the cell

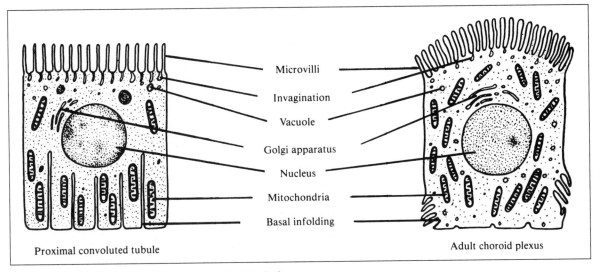

Fig. 23-3. Structural similarities of the ependymal epithelial cell of the choroid plexus and a typical proximal convoluted tubular cell.

(space C) to dilute the concentration of the materials in E, so the final concentration in each compartment will be somewhat less than 600 mOsm per liter. The net effect will be cellular dehydration. Figure 23-5C illustrates what will happen if the osmolarity of the extracellular compartment is reduced to 150 mOsm per liter. The cellular fluid is hyperosmolar, and water moves in. Both compartments will become isosmotic, and a value greater than 150 and less than 300 mOsm per liter will be attained, because of the resulting movement of water from compartment E into C. The cell swells.

EXCHANGE BETWEEN VASCULAR AND INTERSTITIAL COMPARTMENTS. When fluid moves between the vascular and interstitial compartments, the total exchange of water and salts in solution is enormous, because the capillaries are highly permeable to water and contained solutes, and the area of exchange is great. The total surface area of the capillary bed (muscle) in humans has been estimated at 6000 square meters. The limiting factor to diffusion is, in fact, the rate of circulation of the blood through the tissue. However, the net exchange is small because of the normal maintenance of volume of the interstitial and plasma compartments within relatively fixed limits. The protein concentration of the plasma becomes very important in the net exchange of fluid. It will be recalled that the osmotic pressure exerted by the proteins opposes the hydrostatic filtering forces that exist in the capillaries, so that, according

to the Starling hypothesis (see Chap. 12), the filtration of water is opposed by the retaining action of the plasma protein. Stated simply, the fluid that has been filtered outward at the arteriolar end of the capillaries is returned by the effective gradient of plasma protein at the venular end. The importance of the role of the lymphatics in the overall net water exchange between the capillaries and the interstitial space should be emphasized. The lymphatic system is important in draining off excesses of interstitial fluid (Fig. 23-6).

ELECTROLYTE EXCHANGE

EXCHANGE BETWEEN EXTRACELLULAR AND INTRACELLULAR COMPARTMENTS. As was indicated earlier in this book (see Chaps. 1–3, 14, and 22), the exchange of electrolytes between the ICF and ECF compartments is a complicated process. It has been discussed from the standpoint of the function of nerve and muscle cells. The difference in ionic concentration between these two compartments, which reaches tremendous proportions in some instances, must be viewed in terms of mechanisms and operations that require the expenditure of energy. In brief summary: Some of the ions under consideration move against the concentration gradient in a transport system that has the following characteristics: (1) a carrier compound that forms a reasonably stable complex with an ion and (2) a chemical system that can react with this ion carrier compound so as to release the ion from its association with the carrier. Some ions may be held in high concentration in the cell, at least in part

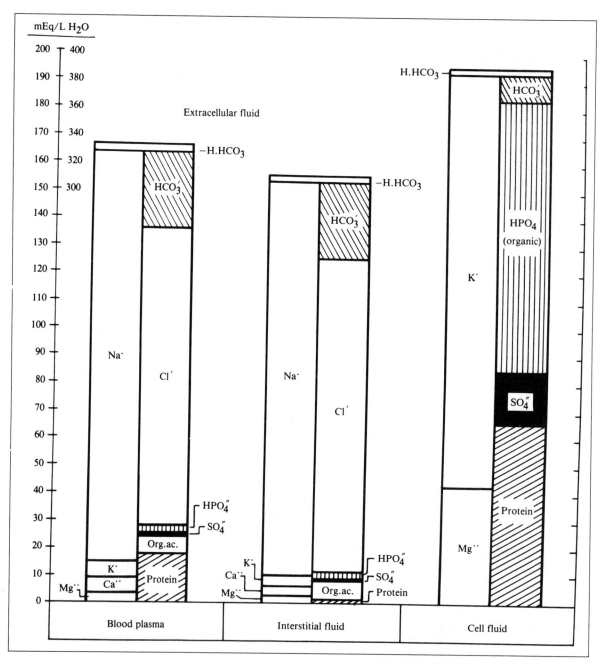

Fig. 23-4. Ion distribution in the fluid compartments of the body. (From J. L. Gamble. *Chemical Anatomy, Physiology and Pathology of Extracellular Fluid* [6th ed.]. Cambridge, Mass.: Harvard University Press. P. 5. Chart II. Copyright © 1942, 1947, 1954 by the President and Fellows of Harvard College.)

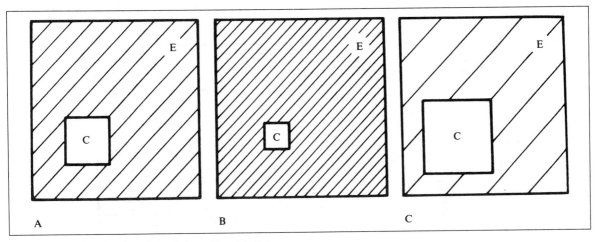

Fig. 23-5. Regulation of water exchange between the extracellular and intracellular compartments. The representative volume of ECF (E) is taken as 1 L. C = intracellular volume. A. Normal (300 mOsm/L in both E and C). B. Effect of addition of 300 mOsm to E (initially 300 mOsm/L). C. The osmolarity of E is reduced to 150 mOsm/L; C is 300 mOsm/L. The result is net water gain by the cell as fluid moves into the relatively hyperosmolar zone (C).

in an intimate and stable relationship with other large complex molecules—for example, the relationship between potassium and myosin in muscle cells.

The thorough treatment of this topic in the chapters cited obviates the need for further discussion here. The reader is referred to these chapters for detailed presentation of the facts.

EXCHANGE BETWEEN VASCULAR AND INTERSTITIAL COMPARTMENTS. The movement of electrolytes and other solutes between the vascular compartment and the interstitial compartment is limited only by the movement of the water in which they are in solution. Superimposed on this is the slight influence exerted by the Gibbs-Donnan effect.

WATER BALANCE

Water balance is achieved between the forces responsible for *intake* of water, as manifested by thirst, and the metabolic water and contained water of ingested food-stuffs and *water loss* through various organs of the body, namely, the skin, the lungs, the gastrointestinal tract, and the kidneys.

WATER INTAKE. Thirst is a subjective sensory impression that results in ingestion of water. Physiological factors that seem to be involved are primarily concerned with increased osmolarity of the ECF brought about either by salt excess or by water deficit. A decrease in circulating vascular volume may be involved. In addition, influences that are related to habits of salt and water intake have an effect. Basic to stimulation is the sensation of dryness of the mucous membranes of the mouth and pharynx. The central nervous system areas involved are not known precisely, although it is believed that they lie in or near the ventromedial nuclei of the hypothalamus. The basic physiological stimuli for thirst are summarized in Figure 23-7.

WATER LOSS. Average daily losses are as follows: insensible loss (vaporization through the lungs and skin), 800 to 1200 ml; urine, 1500 ml; stool, 100 to 200 ml. In addition, there may be variable water loss through sensible perspiration (sweat), depending on environmental conditions and degree of physical activity.

Insensible and Sensible Perspiration. Insensible water loss is modified by the effective osmotic pressure of the body fluids. Thus, there are data suggesting that the rate of insensible loss is inversely related to the concentration of sodium in the ECF. It follows that the rate of water loss from the skin and lungs by this mechanism may be related to vapor pressure of the body fluids. Insensible perspiration goes on at a reasonably constant rate when body temperatures remain constant. About half is lost by the skin, half by the lungs.

Sensible perspiration may vary from 0 to 2 liters per hour and more. The output of sensible perspiration is greatly increased by elevated temperatures and with exercise (see Chap. 28). Centers regulating sweating are located in the

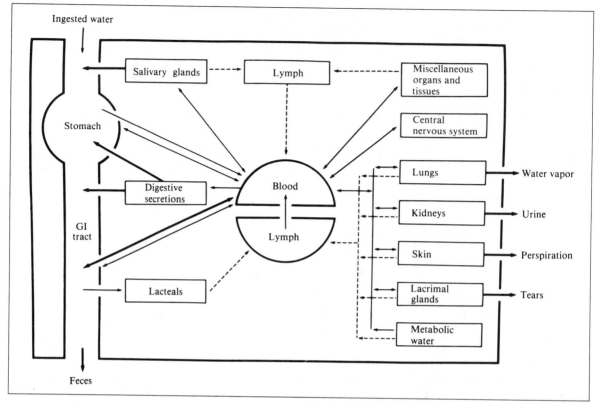

Fig. 23-6. The various aspects of water balance in the body. The heavy arrows denote the principal channels for water loss. Dashed lines signify lymphatic drainage from the interstitial fluid compartments of various organs and tissues. The solid lines signify exchange of water among blood, tissues, and organs. The metabolic water produced daily by certain intracellular chemical reactions represents only a very small fraction of the total body water. Note also that, from a topologic standpoint, fluid entering the gut now is outside of the body. (From D. Jensen. *Principles of Physiology.* New York: Appleton-Century-Crofts, 1976. P. 797.)

anterolateral portions of the hypothalamus. These respond to elevation in blood temperature or reflex afferent nerve stimulation (e.g., gustatory nerve stimulation). Sensible perspiration is a hypotonic fluid. The important solutes in sweat are sodium (48 mOsm/liter) and chloride (40 mOsm/liter). The concentrations of other solutes (urea, potassium, and ammonia) are less than 9 mOsm/liter.

Gastrointestinal Tract Contribution. Gamble (1958) estimated that the volume of digestive secretions provided by an average adult in 1 day is about 8 liters, to which gastric and intestinal mucosal secretions contribute about 5.5 liters. The electrolytic composition of these secretions differs considerably in the various regions of the gastrointestinal tract (see Chap. 26). However, the average solute concentration in the various secretions is reasonably close to that of the ECF except for saliva, which is distinctly hypotonic. Although the amount of fluid lost from the gastrointestinal tract is normally small, persistent and excessive vomiting, diarrhea, or drainage from an intestinal fistula very quickly causes a serious reduction in volume of the ECF compartment. In addition, a serious disturbance in electrolyte balance is produced. This results from both dehydration and differential electrolyte loss. For example, because potassium is found in these fluids in somewhat higher concentration than in ECF, a potassium deficit would develop. Since disproportionate concentrations of HCO_3^-, Cl^-, and H^+ occur, it is to be expected that disturbances in acid-base balance result from excessive vomiting or loss of intestinal secretions.

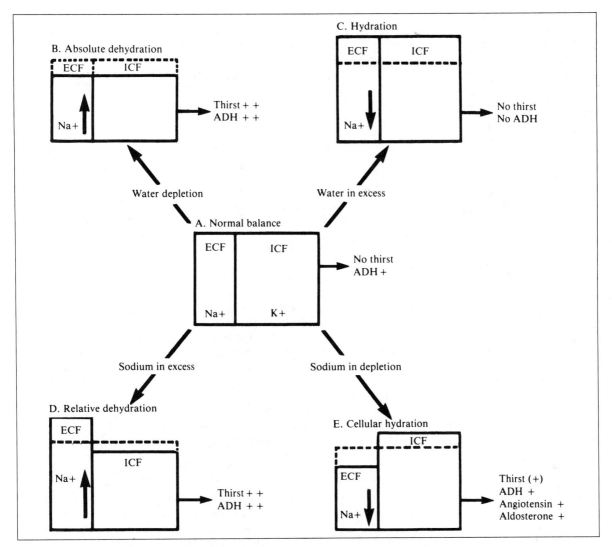

Fig. 23-7. Diagram summarizing stimuli and inhibitors of thirst. Changes in body fluid distribution during negative and positive water and salt balances are shown. Compensatory factors aimed at restoring fluid balance are indicated on the right of the diagrams. ECF = extracellular fluid compartment; ICF = intracellular fluid compartment; ADH = antidiuretic hormone. (From B. Anderson. Thirst—and brain control of water balance. *Am. Sci.* 59:410, 1971.)

SUMMARY OF FACTORS THAT REGULATE URINE VOLUME

The kidney serves as the buffering organ for adjustment of the fluid balance in the sense that it is able to retain water under circumstances of dehydration. Alternatively, it can lose large excesses of water when it is necessary to maintain isotonicity and normal volumes in the fluid compartments of the body. Extremes of volumes have been registered between 500 ml for 24 hours to 20 liters or more, the latter in the absence of regulation by antidiuretic hormone (ADH) in diabetes insipidus.

The output of urine is the net result of kidney operation on the original 170 liters of fluid filtered per 24-hour period by the glomeruli. The factors that govern the release of ADH are discussed in greater detail in Chapter 30. In brief, the important controls involve the osmolarity of the ECF and the volume of the body fluids. Increases in osmolar concentration of the plasma ECF supplying the zones in the hypothalamus where the osmolar receptors are located cause increased output of ADH and ultimate water conservation by the kidney, according to the findings of Verney (1947). Excessive intake of water causes a reduction in the osmolarity of the ECF, inhibits the output of ADH by the hypothalamoneurohypophyseal system, and leads to water diuresis. Second, changes in the volume of the ECF are perceived by volume receptors and baroreceptors, which cause the requisite correction in total fluid volume, by intermediation of ADH and the kidney.

A related important limiting factor on the renal regulation of water content of the body is the solute load that must be excreted. The solute load is composed of metabolites, such as urea, creatinine, uric acid, and excess electrolytes. The maximal osmolar concentration of human urine, 1400 mOsm per liter, obtains only when the solute loads for excretion and obligatory urine volume are both low, the latter being about 0.5 ml per minute. With increasing solute loads, in spite of presumed maximal ADH activity, urine volume increases, and urine osmolarity actually declines progressively. The reason is that the ADH mechanism has attained, in effect, a "ceiling" of operation, so that water is progressively lost with increasing solute loss (see Chap. 22).

MODIFICATION OF GLOMERULAR FILTRATION RATE. It is well known that changes in glomerular filtration rate (GFR) may produce parallel changes in urine volume. Changes in GFR can be brought about by changes in systemic arterial blood pressure, which in turn produce corresponding changes in the glomerular filtration pressure. An obvious example is the oliguria or anuria following hemorrhage. In this circumstance, arterial pressure and, in turn, glomerular capillary pressure are significantly reduced. Besides the hydrostatic filtering pressure, another important factor is the colloid osmotic pressure that opposes filtration. Hence, if plasma protein concentration is altered, changes in GFR and subsequent changes in urine volume take place. For example, rapid infusion of isotonic salt solution decreases the plasma protein concentration by dilution and increases the effective filtration pressure.

Certain drugs affect glomerular pressure by acting on the afferent or efferent arterioles. As an example, the xanthines (caffeine, theobromine, and theophylline) dilate the afferent arterioles and increase GFR. They also secondarily impair tubular reabsorption of Na^+.

The changes in filtration rate involve changes in solute load to the tubules, which must be considered in explaining the final effect on urine volume. Thus, the extent of solute excretion governs the amount of water that will follow because of osmotic obligation.

INTRARENAL CONTROL OF URINE VOLUME BY PHYSICAL FACTORS. It will be recalled from Chapter 22 that the bulk reabsorption of water filtered by the glomeruli occurs by passive (osmotic) accompaniment of active sodium reabsorption in the proximal convoluted tubules. A review of the structure and arrangement of the proximal tubular epithelial cells (see Fig. 22-21) would be helpful at this time. Attention is particularly directed to the basal infoldings and intercellular channels and their relationship to the sodium and potassium pumps.

Involvement of physical forces in the movement of salt and water from the tubular epithelium through the interstitium and into the capillaries is illustrated in Figure 23-8. This brings into the picture the Starling forces: the balance of hydrostatic and oncotic forces between the interstitium and the capillary lumen. It is clear from application of facts considered in Chapter 12 that an increase in hydrostatic pressure in the capillaries would decrease isotonic absorption and tend to impede the osmotic water flow. Ultimately, uptake of salt and water would be impaired. With reduction of hydrostatic pressure, the reciprocal effect could be anticipated.

Changes in oncotic pressure would similarly influence the mechanism (i.e., dilution of peritubular capillary plasma proteins would tend to impair the osmotic water flow and favor urinary loss; conversely, an increase in plasma oncotic pressure would favor uptake and eventuate in decreased urinary excretion of salt and water.

Numerous examples are at hand in which experimental manipulation of physical forces has resulted in changes in electrolyte and water excretion (Earley and Schrier, 1973). Critical experiments would involve alteration of so-called physical factors while GFR and related filtration of water and sodium salts remained constant. These manipulations include changes in renal arterial pressure, assuming that some part of the change in arterial pressure is transmitted to the peritubular capillary bed. It has been demonstrated that an increase in renal arterial pressure (with constant GFR) increases sodium and water excretion. Agents (e.g., acetylcholine, prostaglandins) that produce renal vasodilation also

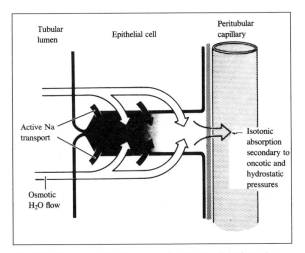

Fig. 23-8. Representation of movement of tubular fluid from the proximal tubular lumen to the peritubular capillary, constructed to conform to the anatomical characteristic of the proximal tubule and peritubular capillary circulation. Sodium is actively transported into the extracellular spaces between adjacent epithelial cells, creating a hypertonic pool in this extracellular, intratubular area. Water moves osmotically into the intercellular channel, creating elevated hydrostatic pressure within the intercellular space. The channel is tightly closed at the luminal end of the cell, and the epithelial basement membrane and capillary endothelium have a high hydraulic conductivity. The increased hydrostatic pressure within the channel, together with the oncotic pressure of the peritubular plasma, would drive the column of reabsorbate into the capillary circulation. The continual osmotic flow of water into the channel results in the delivery of virtually isotonic reabsorbate to the capillary circulation. (From L. E. Earley and R. W. Schrier. Intrarenal Control of Sodium Excretion by Hemodynamics and Physical Factors. In J. Orloff and R. W. Berliner [eds.], *Handbook of Physiology.* Washington, D.C.: American Physiological Society, 1973. Section 8: Renal Physiology. Chap. 22, P. 735.)

promote salt and water loss, particularly in conjunction with increased renal arterial pressure. A decrease in renal arterial pressure with a constant GFR (due to autoregulation) causes reduced urinary output of sodium and water.

Saline infusion would operate in part through this mechanism to cause diuresis; dilution of plasma proteins would amplify the effect. Intravenous infusion of hyperoncotic albumin, with an increase in peritubular capillary plasma oncotic pressure, causes a net decrease in urine volume and solute excretion, including sodium (Elpers and Selkurt, 1963).

FACTORS AFFECTING TUBULAR REABSORPTION OF WATER. The role of ADH in the regulation of tubular water handling has been discussed. Control mechanisms are detailed in Chapter 30. Mechanisms that inhibit ADH production and release include reduction of plasma osmotic pressure—for example, by the oral ingestion of large quantities of water or infusion of hypotonic NaCl or isotonic glucose solutions. (When glucose is metabolized, the water of the solution is freed.) Figure 23-9 shows typical water diuresis resulting from ingestion of 1 liter of pure water. Typically, the specific gravity of the urine rapidly changes from about 1.030 to 1.002 at the peak of diuresis.

The effect of ingestion of 1 liter of isotonic saline can also be considered. The modest diuresis following saline ingestion has been explained by a small increment in GFR. Specific gravity of the urine shows little if any change, for body fluid osmolarity is not altered, and ADH output should not change. In well-hydrated subjects, saline ingestion produces a greater diuresis, with a decrease in urine specific gravity. It is probable that ADH output is decreased in this circumstance by activation of volume receptors located in the left atrium, or by altering of the carotid sinus baroreceptor influence on ADH release (see Chap. 30).

Among the factors stimulating ADH secretion and thereby inducing antidiuresis are water deprivation and dehydration, which increase the plasma osmotic pressure, in turn acting on the osmolar receptors in the hypothalamus. Also, as a result of reducing plasma volume and, in consequence, stimulation of left atrial volume receptors and carotid sinus baroreceptor activity, ADH output is further enhanced. Pain and emotional stress likewise cause an increase in ADH output. Various agents, such as nicotine, acetylcholine, morphine, and the anesthetics barbiturates, ether, and chloroform, all act on the hypothalamoneurohypophyseal system to increase the output of ADH and lead to antidiuresis.

Another means of tubular regulation of urine volume concerns the role of solutes delivered to the tubular cells, as discussed previously. *Osmotic diuresis* results if the filtered load of a given solute exceeds the reabsorptive capacity for it of the cells at the site of reabsorption in the nephron. The unreabsorbed solutes osmotically retain some of the water filtered at the glomeruli and carry it on through the rest of the system. Common osmotic diuretics include certain organic substances such as mannitol, urea, sucrose, and glucose. Hypertonic NaCl and Na_2SO_4 are others. The acid-forming salts ammonium chloride, ammonium nitrate, and calcium chloride have been used as diuretic agents, alone or in conjunction with others (e.g., mercurials). The am-

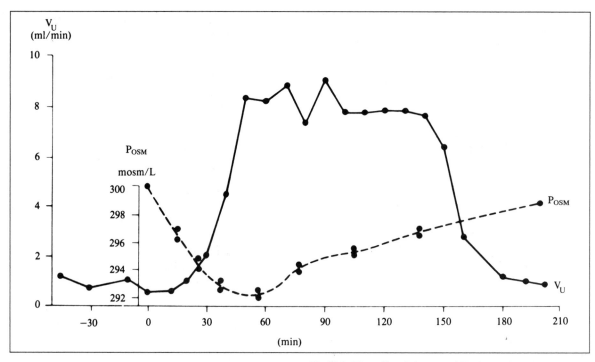

Fig. 23-9. Water diuresis in a man following the ingestion of 1 L of pure water at time 0. The rise in urine volume (V_u) follows a decrease in plasma osmolality. (From P. Deetjen, J. W. Boylan, and K. Kramer. *Physiology of the Kidney and of Water Balance.* New York: Springer-Verlag, 1975. Pp. 83, 123.)

monium salts act in part by increasing urea formation, which acts as an osmotic diuretic. For example, NH_4Cl is converted to urea and hydrochloric acid, and the latter is buffered as follows:

$$HCl + NaHCO_3 \rightarrow NaCl + H_2CO_3$$

H_2CO_3 dissociates to carbon dioxide and water. The resulting acidosis is compensated in part by the release of carbon dioxide by the lungs. There is, however, an accelerated excretion into the urine of the residual Cl^- accompanied by Na^+, and the loss of NaCl is accompanied by an equivalent loss of water. Oral ingestion of $CaCl_2$ acts by giving up Cl^- to be lost with Na^+. The poorly absorbed Ca^{2+} tends to stay in the intestine combined with other anions.

OTHER DIURETIC AGENTS. Carbonic Anhydrase Inhibitors. Carbonic anydrase inhibitors are heterocyclic sulfanilamide derivatives of which acetazolamide (Diamox) is the one commonly employed. These compounds are effective inhibitors of carbonic anhydrase in vitro, and the mechanism involved may be related to the mechanism for producing diuresis. It is probable that the drugs reduce the availability of hydrogen ions to both proximal and distal tubular ion exchange mechanisms. The agents promote the renal loss of NaCl and $NaHCO_3$. More indirectly, increased loss of Na^+ favors enhanced K^+ secretion by supplying greater amounts of the ion needed for the exchange mechanism to operate in the distal nephron. Loss of all these ions promotes loss of water by their osmotic diuretic action.

Benzothiadiazine Diuretics. Sulfonamyl benzothiadiazines such as chlorothiazide (Diuril) have a weak carbonic by anhydrase-inhibiting action and act chiefly by blocking tubular reabsorption of sodium and chloride ions; potassium loss is also favored. Since they do not interfere with the production of maximally concentrated urine during hydropenia, it has been conjectured that they exert their action predominantly in the distal convoluted tubule.

Aldosterone Antagonists. The antialdosterone substances are spirolactones, the most effective of which is spironolactone (Aldactone). They are structurally similar to aldosterone and act by competitive inhibition, thus blocking aldosterone action and favoring Na^+ loss.

Two Potent Loop of Henle Electrolyte Transport Inhibitors. Ethacrynic acid (Edecrin) is a potent diuretic that acts to inhibit electrolyte transport in the proximal tubule and throughout the loop of Henle. In hydropenic subjects the ability to concentrate the urine is lost, and in hydrated patients the ability to dilute the urine is impaired. Thus, the loop loses its ability to dilute or concentrate, and hence its ability to maintain the gradient of osmolarity in the medulla. Increased Cl^-, K^+, and H^+ excretion also results, because increased amounts of NaCl reach the distal tubule, allowing for greater ion exchange there. Chemically, ethacrynic acid is an unsaturated ketone of aryloxyacetic acid.

Furosemide (Lasix) is a sulfonamide derivative that also acts to block electrolyte transport in the ascending limb of the loop of Henle, as well as proximal sites. Since it has been established that Cl^- is the ion actively pumped in the ascending limb of the loop, then ethacrynic acid and furosemide presumably act here by blocking Cl^- transport, and sodium is lost with chloride.

ELECTROLYTE BALANCE

SODIUM INTAKE. The average adult diet contains 10 to 12 gm of NaCl per day, either in the food (higher in meat diets than in vegetable diets) or added to it. Examples of excessive ingestion of salt or salt craving have been cited in the literature. This is usually associated with deficiency of adrenocortical secretion, as in Addison's disease. As an extreme example, Strauss (1957) cites the case of a 34-year-old patient with Addison's disease who put an approximately one-eighth-inch layer of salt on his steak, used nearly one-half glass of salt for his tomato juice, put salt on oranges and grapefruit, and even made lemonade with salt. In fact, salt craving is often an early manifestation of Addison's disease that is of considerable diagnostic significance.

However, excesses of sodium *(hypernatremia)* are less common clinically than hyponatremia. Whereas diseases that cause dehydration are usually associated with sodium losses in excess of water losses, excess sodium is almost always associated with an excess accumulation of body water, except in hypernatremia due to cerebral lesions. In cases of intracranial damage, sodium concentration is sacrificed to maintain water volume, and hypernatremia without edema may result.

There will be increased levels of blood sodium in conditions associated with increased adrenocortical activity.

High production of aldosterone by the adrenal cortex will favor extraction of excess amounts of sodium from tubular urine by the renal tubular cells (see Chap. 22). For the same reason, patients who are on steroid therapy for prolonged periods may have abnormal sodium and water retention (see Chap. 34). Although cortisone is a glucocorticoid (not mineralocorticoid), glucocorticoids can cause some sodium retention. This is less a problem with some of the newer steroid preparations than was the case a few years ago, but patients on steroid therapy should nonetheless be observed for signs of sodium and water retention and other untoward reactions.

Aldosterone is normally inactivated by the liver. In some types of liver disease, aldosterone may not be properly degraded, so that there may be abnormal retention of sodium and water (see Chaps. 18 and 34).

After treatment of diabetic coma, when cellular sodium is being returned to the extracellular compartment, there may be a temporary rise in the blood sodium level.

SODIUM LOSS. Total body sodium is 58 mEq per kilogram of body weight, 25 mEq per kilogram as bone sodium, of which about half is exchangeable (Edelman and Leibman, 1959). The intake of Na^+ is normally quite variable, as is loss by sensible perspiration, so that it falls on the kidney to regulate the homeostasis of its content in the body. Excesses of Na^+ are excreted by the kidney. Conversely, if the dietary intake is reduced, the urine will become virtually sodium free in several days in an attempt to maintain sodium balance.

Excessive loss of Na^+ from the extracellular compartment (due to renal reabsorptive impairment in renal disease), gastrointestinal losses (e.g., diarrhea), or excessive sweat loss, leads to low blood Na^+ levels (hyponatremia). The accompanying loss of fluid volume contributes to the observed physiological changes. The symptoms of hyponatremia are weakness, lassitude, apathy, and headache. The decrease in vascular volume leads to faintness, and in extreme cases to orthostatic hypotension, tachycardia, and shock.

Gastrointestinal symptoms associated with sodium loss include anorexia, nausea, and vomiting. Thirst is usually absent. There is a loss of turgor and elasticity of the skin. Normally, when the skin is pinched and picked up, it returns to its previous shape immediately when it is released. When there is a sodium loss and attendant water loss, the fold in the skin may remain for 30 seconds or longer. In severe sodium loss there is confusion, stupor, and eventual coma.

POTASSIUM INTAKE. The major cation for maintaining proper pH of the ICF is potassium. All cellular activities involving electrical phenomena, such as skeletal and cardiac muscle contraction and nerve impulse conduction, are dependent on the gradients of K^+ and Na^+ across cell membranes. In considering the shifts of K^+ between the extracellular and intracellular compartments, it is well to keep in mind the quantitative disproportion that exists. Thus, a total of 350 mEq in the ECF compartment contrasts with 3500 mEq in the ICF (Edelman and Leibman, 1959). Therefore, small losses out of the body cells or small uptakes can cause relatively significant changes in the concentration of potassium in the ECF. Tissue repair and growth require K^+; conversely, protein breakdown releases K^+ from the cells into the extracellular compartment, from which it is lost in the urine. The transfer of glucose from the extracellular to the intracellular phase requires K^+.

Since there is a narrow range of normal blood levels of K^+ (3.5–5.0 mEq/liter), relatively small changes in blood K^+ can cause serious problems. Thus, in *hyperkalemia* (blood K^+ levels of 8 mEq/liter), significant changes occur in the configuration of the electrocardiogram (loss of P wave, changes in the QRS and T waves). Above 11 mEq per liter, ventricular fibrillation is imminent. In addition, neuromuscular symptoms, numbness, tingling, and eventually flaccid muscle paralysis are seen in hyperkalemia.

Increased blood K^+ results from kidney disease, in which excretion of the cation is diminished. In intestinal obstruction, blood K^+ is elevated because 10 percent of the dietary intake of K^+ normally is eliminated in the stool. In Addison's disease, deficiency of aldosterone (see Chap. 34), impairs the kidney's ability to excrete K^+, and hyperkalemia is seen; increased loss from tissue cells also occurs.

POTASSIUM LOSS. The urine always contains potassium, and if none is administered or ingested in the diet, a deficit of this cation will develop. Ingestion of about 3 gm of KCl a day appears to be adequate under normal circumstances. Normally, about 10 percent of K^+ loss occurs in the stool, the remainder in the kidney. Renal regulation involves the adrenocortical hormones. Potassium loss may be excessive in certain types of chronic renal diseases associated with polyuria and in the diuretic phase of recovery from acute lower nephron diseases. In a condition called renal tubular acidosis, abnormal amounts of HCO_3^- are excreted in the urine along with accompanying fixed cations, including K^+. In acidosis, K^+ is replaced in the cells by Na^+ and H^+. In severe metabolic acidosis, as in diabetic acidosis, K^+ leaves the cells to enter the ECF compartment and is excreted in the urine. Contributing to cellular loss of potassium is cellu-

lar protein catabolism associated with increased gluconeogenesis.

Decreased blood K^+ results from prolonged high gastrointestinal suction. Diuretics may lead to excessive renal excretion of K^+. Increased renal loss also accompanies the excessive administration of bicarbonate. As stated, in alkalosis, K^+ enters the cells, lowering the blood concentration.

Symptoms of hypokalemia tend to be nonspecific. They include malaise, apathy, muscular weakness, and intestinal distention leading to paralytic ileus (a result of decreased smooth muscle tone). There is usually postural hypotension, and diastolic blood pressure is low. The terminal event is heart block; the heart stops in *systole*.

CHLORIDE INTAKE. Chloride intake is governed by the intake of NaCl, as discussed under Sodium Intake. Although chloride movement across cell membranes is ordinarily viewed as a passive mechanism secondary to that of sodium, some important exceptions occur. The active transport of chloride by the gastric mucosa is an outstanding example. The pumping of Cl^- by the ascending limb of Henle is another important example (see Chap. 22). Keynes (1963) disclosed that the axoplasm of the giant squid axon actively takes up labeled chloride. The uphill inward transport was inhibited by dinitrophenol, and the lack of effect of ouabain indicated that the influx was not linked to transport of cations by the sodium pump. Evidence for an active chloride pump in amphibian cornea was supplied by Zadunaisky (1965), who showed that the short-circuiting current could account for the active transport of Cl^- from the aqueous to the tear side of the cornea. This mechanism appears to have a function in maintaining corneal transparency (by keeping the intracellular ion concentration low and thus preventing swelling).

The most common cause of excess blood chloride (hyperchloremia) is dehydration. Increased amounts of Cl^- are reabsorbed by the kidney with the intensive water reabsorption triggered by dehydration. This favors hyperchloremic acidosis.

CHLORIDE LOSS. In general, Cl^- behaves similarly to Na^+ in regard to renal handling and sweat loss. However, there are exceptions. During acid-base alterations, proportions of Cl^- and HCO_3^- associated with Na^+ may be noted in the plasma and in the urine, differing from the normal relationship. It has also been observed that mercurial and thiazide diuretics cause greater loss of Cl^- than of Na^+ in the urine.

Abnormal loss of Cl^- because of excessive vomiting or gastric suction may result in alkalosis. When chloride is lost as the hydrochloric acid of the gastric juice, the stomach is obliged to replace the lost HCl; it does so by removing Cl^- from circulating NaCl and combining it with H^+ (see Chap. 24). The remaining Na^+ and OH^- in the bloodstream combine with circulating CO_2 to increase blood levels of sodium bicarbonate ($NaHCO_3$), which causes the alkalosis.

Mercurial and other diuretics (ethacrynic acid, furosemide) interfere with enzyme systems needed for Cl^- reabsorption by the kidney. The excreted Cl^- takes Na^+ and water with it; with excessive loss, salt depletion and hypochloremic alkalosis ensue. The type of alkalosis results because with relatively more circulating Na^+ than Cl^-, Na^+ tends to accumulate as bicarbonate.

CALCIUM AND MAGNESIUM BALANCE. Because of their low concentration, the role of calcium and magnesium ions is small in the control of volume and osmolarity of the body fluids, and in acid-base equilibrium. Calcium, a major constituent of bone, is important primarily because of its influence on cell membrane permeability, neuromuscular excitability, transmission of nerve impulses, blood coagulation, and activation of certain enzyme systems. It is both protein bound and ionized in the plasma; the extent of ionization is determined by the acid-base equilibrium, being increased in acidosis and decreased in alkalosis. The action of parathyroid hormone in renal regulation of Ca^{2+} balance is also related to PO_4^{3-} handling. The hormone favors increased tubular Ca^{2+} reabsorption, while apparently depressing tubular reabsorption of PO_4^{3-}, with increased urinary loss. This leads to a secondary rise in serum Ca^{2+}. Continued mobilization from bone leads to further hypercalcemia and ultimate renal loss by overloading the reabsorptive mechanism (see Chap. 32).

Like calcium, magnesium is in either a bound or an ionized phase in the plasma. Its physiological action is noted mostly in connection with the functioning of the neuromuscular and cardiovascular systems. Thus, magnesium excess results in depression of the central nervous system, with loss of tendon reflexes, drowsiness, and finally coma. An excess may cause bradycardia and depression of conduction through the conducting tissue and myocardium.

DISTURBANCES OF FLUID AND ELECTROLYTE BALANCE

DEHYDRATION. When water loss exceeds intake, the interstitial fluid yields water as long as possible, but desiccation of cells eventually occurs (*dehydration*). Water deficits are incurred in normal persons under two common conditions: excessive loss of sweat and prolonged deprivation of water.

During preliminary stages of negative water balance, the skin and muscle give up fluid first, so that vital organs are protected. Following depletion of interstitial fluid, the plasma water is increasingly removed. The blood becomes more concentrated (*anhydremia*). With prolonged loss, or deprivation, or both, intracellular water is also lost.

Characteristic symptoms result from dehydration. Water deficit gives rise to a shrunken appearance of the face and body. The skin loses its elasticity and becomes hard and leathery. There is rapid loss of body weight. When the deficiency reaches such a degree that the water is no longer sufficient for removal of the heat of metabolism, high fevers may occur. As the condition worsens, circulatory failure develops. Anuria results, and acid products are retained, leading to acidosis. Cerebral disturbances, excitement, delirium, and coma terminate in death.

Clinical dehydration may be a consequence of (1) failure of absorption from the alimentary tract (as in pyloric stenosis or high intestinal obstruction); (2) excessive loss from copious sweating, prolonged vomiting, diarrhea, and excessive diuresis; or (3) drainage from wounds or burns.

Water loss under these conditions also involves loss of electrolytes, predominantly NaCl. If fluid balance is therefore restored with only pure water, a hypotonic state results in the extracellular compartment. As a consequence, water moves in abnormal amounts into the cells, now relatively hyperosmotic, producing symptoms referable to *water intoxication*, among them the well-known heat cramps. Obviously, in treatment, electrolytes (chiefly NaCl) must be given as well as water.

SURVIVAL WITH WATER SHORTAGE. Humans are more susceptible to the effects of water deprivation than most land animals. A value of 20 percent of normal body weight represents the maximal weight loss (as water) that can be tolerated with survival. This compares with 40 percent in dogs and cats and 50 percent in birds and reptiles. Under the most favorable conditions (shade and moderate environmental conditions), survival without water intake extends from 11 to 20 days.

Extreme situations are met in survival at sea in a lifeboat or raft and in the desert. The drinking of seawater is not recommended because the salt concentration may be as high as 3.5 percent, and the human kidney cannot elaborate urine with a salt content greater than 2 percent. Hence, body water is lost in association with the excess salt excreted, and dehydration actually becomes more severe. It is of interest to note that the salinity of the sea or ocean has a bearing on the effect of drinking saltwater. Certain bodies of water (e.g., the Black Sea and Caspian Sea) have only about one-half the salt content of the Atlantic Ocean; the Baltic

Sea is almost isotonic and so could be tolerated as a fluid replacement.

As was discussed in Chapter 22, an important function of the renal medulla is concerned with its capability for concentration of the urine, as established by its gradient of osmolality, from the cortex to the papillary tip. Aided by the osmotic gradient and ADH activity, maximal conservation of tubular urine is achieved. In humans, the highest U/P_{Osm} ratio attained is about 4.0. This is in marked contrast to the gradients achieved in animals adapted for desert existence (kangaroo rat, jerboa, desert rat), in which ratios as high as 15 to 20 are found, establishing maximal water salvage by the kidney.

DESALINATION OF SEAWATER. A chemical procedure (preferred over a distillation apparatus), the so-called Permutit method, has been the favored method at seawater desalinization. Tablets of silver zeolite are placed in a plastic vessel containing seawater; silver chloride and sodium, and magnesium zeolites, are precipitated out, and water suitable for drinking is filtered off.

Manned flight in outer space presents a special problem in water supply and conservation. Because of payload limitations, transportation of water would be restricted. The only practical expedient is to develop techniques for recovery of water from the kidney, lungs, and skin. For example, urinary water could be reclaimed by freeze-drying followed by filtration through activated charcoal.

EXCESS WATER LOADS. A condition of cellular overhydration may result also from attempts to produce diuresis by forcing hypotonic fluids, particularly in the presence of renal impairment. On the other hand, prolonged and excessive diuresis may result in excessive loss of extracellular solutes, and water will migrate from a zone of relative hypotonicity into the cells. The symptoms of water intoxication that follow are particularly referable to the central nervous system: salivation, nausea and vomiting; restlessness, asthenia, muscle tremors, ataxia, and tonic and clonic convulsions that may eventuate in stupor and death.

EDEMA

The fundamental bases for formation of edema fluid have been discussed in Chapter 12 in connection with capillary fluid dynamics. Since edema formation represents an abnormal accumulation of ECF, particularly that of the interstitial compartment, added facets of the problem of edema are presented here.

CARDIAC EDEMA. Cardiac edema was discussed in detail in Chapter 18.

CIRRHOTIC EDEMA. The edema of cirrhosis is localized in the abdominal cavity (ascites). Degeneration of the parenchymal cells of the liver leads to impairment of circulation through the liver, interfering particularly with the low-pressure portal vein system. The resulting portal hypertension favors the exudation and accumulation of fluid in the abdomen, both from the capillaries within the liver and from the capillaries in the intestine. However, portal hypertension is not the only factor in this type of edema. Impaired liver function results in hypoalbuminemia, contributing to the effective outward gradient for filtration of fluid. The increased titer of ADH in the urine of cirrhotic patients is evidence of greater antidiuretic activity, presumably due to failure of the liver to inactivate ADH. There is also evidence that aldosterone activity is enhanced under these circumstances, acting to favor salt and water retention. Other circumstances characterized by obstructive conditions in the liver give a similar picture (e.g., sclerosis of the hepatic vessels or compression of the portal vein by tumors, aneurysms, and the acute icteric phase of infectious hepatitis).

RENAL EDEMA. The basic causes of renal edema are (1) a decrease in oncotic pressure of the plasma because of loss of albumin via the kidney; (2) a coexisting increase in systemic capillary permeability with loss of protein to the interstitial spaces; (3) coexisting congestive heart failure of the hypertensive type when associated with chronic nephritis; and (4) Na^+ retention. In chronic nephrosis, in the nephrotic stage of glomerulonephritis, and in the amyloid kidney, the kidneys excrete 10 to 20 gm of protein a day into the urine because of glomerular damage. Plasma albumin decreases from a normal of about 5 percent to about 2 percent, resulting in a marked reduction of oncotic pressure. The protein content of the edema fluid is low (about 0.1 percent). With acute glomerulonephritis there is widespread capillary injury in all parts of the body; hence, the protein content of the edema fluid may be high (over 1 percent). The edema is a combination of decreased oncotic pressure of the plasma and increased oncotic pressure of the interstitial fluid. Hypertensive heart disease accompanies chronic diffuse glomerulonephritis, pyelonephritis, and polycystic renal disease. With this, the ultimate cardiac decompensation contributes factors that are characteristic of congestive heart failure, including elevated venous pressure and sodium retention by mechanisms previously described.

NUTRITIONAL EDEMA. Nutritional edema can be caused by faulty metabolism or nutritional deficiency. Although vitamin lack—e.g., lack of vitamin B in beriberi and of vitamin C in scurvy—is contributory, perhaps most important is protein lack in the diet. Inadequate protein intake leads to hypoproteinemia and a reduction in the oncotic pressure of the plasma proteins, favoring outward movement of fluid from the capillaries into the interstitial spaces. The decrease in tissue pressure that results from the wasting of tissue may favor this outward movement of fluid, and a contributing factor may be the changes in capillary permeability caused by the attendant vitamin deficiency. Most authorities feel that the general reduction in cardiac activity and blood flow impairs kidney function and leads to secondary Na^+ retention as a final contributory factor.

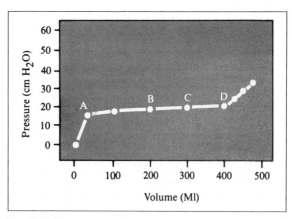

Fig. 23-10. The pressure-volume relationship of the human bladder during filling. The steeper rise in pressure above 400–450 ml indicates filling beyond physiological capacity. A = first feeling of fullness; B = first desire to void; C = sensation of discomfort; D = urgency, associated with bladder contraction.

FUNCTION OF URETER AND URINARY BLADDER; MICTURITION

FUNCTION OF URETERS. Urine collected in the kidney pelvis passes through the ureters to the bladder, not only by virtue of forces of gravity in the erect position but also by contraction of the muscle layers of the ureter. These muscular contractions are necessary to develop sufficient pressure to overcome the gradually increasing tension in the bladder as urine accumulates. The peristaltic waves observed probably originate in the muscles but are modified by the action of the splanchnic nerves via the renal plexus to the upper portions of the ureters, and via the hypogastric plexus to the lower portions. These are excitatory fibers; inhibitory fibers have been described in the inferior mesenteric plexus. Afferent sensory fibers are present, as evidenced by the excruciating pain felt on passage of calculi through the ureter. The ureters enter the base of the bladder obliquely, thus forming a valvular flap that passively prevents reflux of urine (Waldeyer's sheath).

BLADDER FILLING AND CONTINENCE. The bladder consists of a sphere and a cylinder (the neck extending into the urethra) that have a special intrinsic properties. The smooth muscle of the bladder can be considered as a sheet of muscle in the form of a reticulum or webwork that extends without interruption down into the urethra as the muscular wall. There is a particularly heavy concentration of elastic connective tissue fibers in the wall of the urethra. The smooth muscle and elastic tissue exert continuous tension in an autonomous manner, with a negligible expenditure of energy (see Chap. 3).

When the bladder fundus is being distended with fluid,

the smooth muscle fibers of the bladder wall are at first stretched, and caused in turn to contract and increase their tension. Thus, measurement of intravesical (bladder) pressure will demonstrate an initial increase with initial filling of the bladder with fluid; but then, as filling continues, the intravesical pressure will remain approximately constant until bladder capacity is reached (Fig. 23-10). At capacity (250–450 ml) the intravesical pressure rises more steeply. The ability of the bladder fundus to maintain a relatively low intravesical pressure is a type of accommodation, which allows easy filling from the ureters.

The urine is kept from leaking out of the bladder in part by resistance offered principally by the upper 3 cm of the bladder neck and urethra (Fig. 23-11A). The bladder neck can be viewed as an "internal sphincter," although authorities disagree as to whether or not a true sphincter exists here. The urethra aids the sphincteric action by the continuous intrinsic autonomous tension exerted by the smooth muscle and connective tissue in its wall (the "external sphincter"). These tissues keep the uretha compressed, so that its lumen is sufficiently collapsed to prevent urine from flowing out of the bladder fundus under low or moderate pressures. To oppose higher intravesical pressures, the urogenital diaphragm and *levator ani* aid in the act of urinary bladder continence. These muscles increase the efficiency of the urinary sphincter by compressing the urethra circumferentially and elongating it by pulling it cephalad. The striated muscle can act on a voluntary basis,

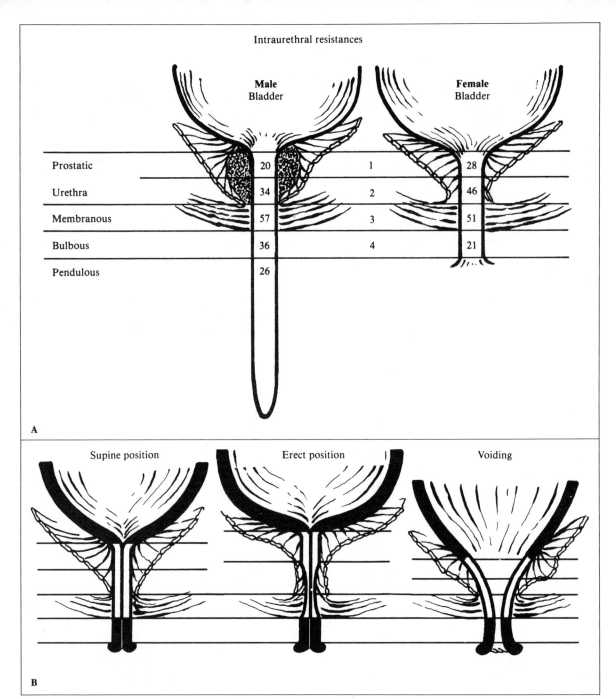

Fig. 23-11. A. Resistance to pressure of a given zone (in cm H$_2$O) in the male and female urethra. Note that the middle 2 cm of the female urethra corresponds to the distal part of the prostatic urethra and membranous urethra of the male. This zone is contiguous with the periurethral striated muscle. B. Effect of erect versus supine posture on the length of the urethra, and effect of voiding when the periurethral striated muscle is relaxed. In addition, the urethrovesical junction is pulled open into a funnel-shaped structure by active contraction of the vesiculourethral smooth muscle sheet. The net result is a functional shortening and widening of the posterior urethra. (A and B modified from J. Lapides, Ch. 33. In M. F. Campbell and J. H. Harrison [eds.], *Urology*. Philadelphia: Saunders, 1970. Vol. 2, pp. 1348–1349.)

or contraction can take place reflexly as part of postural reflexes. Thus, one can willfully compress the urethra and prevent urinary incontinence when the bladder is full and the patient has an urgent desire to urinate. Or the urinary sphincter can be compressed and elongated reflexly, as in assuming the erect posture, coughing, or straining (Fig. 23-11B).

It is important to note that the urinary sphincter is the intact urethra, and that the striated muscles surrounding the urethra are of secondary importance in that they can increase the efficiency of the urinary sphincter but cannot substitute for it (Fig. 23-11B). The striated muscles serve also to interrupt urination rapidly (1–2 sec), when there is an urgent need, by compressing and elongating the urinary sphincter until the vesical smooth muscle stops contracting (10–20 sec).

MICTURITION. As the volume increases further, the tension starts to rise, in part because tonic contractions of the bladder musculature begin. When bladder volume reaches a capacity of about 150 to 250 ml, stretch and tension (proprioceptive) receptors are stimulated to send out the afferent impulses responsible for the sensation of distension and the desire to urinate. They also cause the act of micturition when spinal reflexes are released from cerebral control. Voluntary control can be exerted until the intravesical pressure increases to about 100 cm H_2O, at which point involuntary micturition begins.

As with the ureter, the extrinsic nerves of the bladder are important in regulating its normal evacuation. They modify changes of tonus, enhance or depress the vigor of rhythmic contractions, and alter the tonus of the external and internal sphincters. This innervation is by way of the parasympathetic *pelvic* nerves, sympathetic *hypogastric* nerves, and somatic *pudendal* (pudic) nerves (Fig. 23-12). The pelvic nerves function to maintain tonus of the bladder, and the pudic nerves mediate impulses that contract the striated musculature of the external sphincter. The role of the sympathetic innervation is not so well understood. Afferent fibers run chiefly in the pelvic nerves, although they are also found in the hypogastric and pudendal nerves. They enter the spinal cord at the sacral level, then go on up to the hypothalamus and ultimately to the cortex, where voluntary control resides.

Micturition is normally a voluntary act. Corticospinal impulses sent to the lumbosacral region of the cord cause intensive contraction of the bladder, opening of the sphincters, and relaxation of the perineum and thus lead to voiding (see Fig. 23-11B).

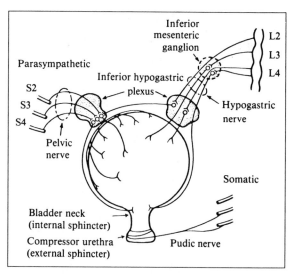

Fig. 23-12. Innervation of the bladder and urethral sphincters. (From K. Koizumi, and C. M. Brooks. The Autonomic System and Its Role in Controlling Body Functions. In V. B. Mountcastle [ed.], *Medical Physiology* [14th ed.]. St. Louis: Mosby, 1980. P. 912.)

The urinary sphincter decreases its resistance during urination by relaxation of the periurethral striated muscle either reflexly or voluntarily. This results in some decrease in the length of the urethra and tension of the urethral wall against its lumen. A further decrease in urethral length and an increase in the caliber of the urethral lumen are effected by active contractions of the smooth muscle of the bladder and urethra. When the bladder begins to contract down on a bolus of urine, the urethrovesical junction and proximal portion of the urinary sphincter are pulled open by active contraction of the muscle sheet that is continuous from the bladder into the posterior urethra (see Fig. 23-11B). The urinary sphincter does not open by passive relaxation but by active contraction of the vesicourethral muscle fibers.

Measurement of intravesical pressure during normal micturition indicates that pressures varying between 25 and 50 cm H_2O are attained at the height of urination. Yet, when the bladder is at rest and storing urine, it may require intravesical pressures greater than 150 to 250 cm H_2O to overcome the resistance of the posterior urethra or urinary sphincter so that urinary flow occurs. During urination it is preferable to have low intravesical pressures obtaining, for high pressures, as seen in obstructive uropathy, predispose to dilatation of the urinary tract, infection, and renal deterioration.

In the infant the normal urinary sphincter can maintain continence between voidings, but it cannot prevent the reflex voiding contractions. Thus, the normal baby will not dribble urine continously but will wet at intervals during forceful bladder contractions. Voluntary control of urination is gained as the child becomes toilet trained and as the corticoregulatory tract begins to function.

BIBLIOGRAPHY

Burke, S. R. *The Composition and Function of Body Fluids*. St. Louis: Mosby, 1980.

Christensen, H. N. *Body Fluids and Their Neutrality*. New York: Oxford University Press, 1963.

Davis, J. O., Ayers, C. R., and Carpenter, C. C. J. Renal origin of an aldosterone stimulating hormone in dogs with thoracic caval constriction and in sodium depleted dogs. *J. Clin. Invest.* 40:1466–1474, 1961.

Davson, H. *Physiology of the Cerebrospinal Fluid*. London: Churchill, 1967.

Davson, H. *Physiology of the Ocular and Cerebrospinal Fluids*. Boston: Little, Brown, 1956.

Deetjen, P., Boylan, J. W., and Kramer, K. *Physiology of the Kidney and of Water Balance*. New York: Springer-Verlag, 1975. P. 123.

Diamond, J. M. Tight and leaky junctions of epithelia: A perspective on kisses in the dark. *Fed. Proc.* 33:2220–2223, 1974.

Earley, L. E., and Schrier, R. W. Intrarenal Control of Sodium Excretion by Hemodynamic and Physical Factors. In J. Orloff and R. W. Berliner (eds.), *Handbook of Physiology*. Washington, D.C.: American Physiological Society, 1973. Section 8: Renal Physiology.

Edelman, I. S., and Leibman, J. Anatomy of body water and electrolytes. *Am. J. Med.* 27:256–277, 1959.

Elpers, M. J., and Selkurt, E. E. Effects of albumin infusion on renal function in the dog. *Am. J. Physiol.* 205:153–161, 1963.

Fishman, A. P. (ed.). *Symposium on Salt and Water Metabolism*. New York: New York Heart Association, 1959.

Gamble, J. L. *Chemical Anatomy, Physiology and Pathology of Extracellular Fluid* (7th ed.). Cambridge, Mass.: Harvard University Press, 1958.

Hays, R. M., and Levine, S. D. Pathophysiology of Water Metabolism. In B. M. Brenner and F. C. Rector (eds.), *The Kidney*. Philadelphia: Saunders, 1981. Vol. 1, p. 772.

Keynes, R. D. Chloride in the squid giant axon. *J. Physiol.* (Lond.) 69:690–705, 1963.

Maffly, R. H. The Body Fluids: Volume, Composition, and Physical Chemistry. In B. M. Brenner and F. C. Rector (eds.), *The Kidney*. Philadelphia: Saunders, 1981. Vol. 1, p. 78.

Maxwell, M. H., and Kleeman, C. R. *Clinical Disorders of Fluid and Electrolyte Metabolism*. New York: McGraw-Hill, 1962.

Meyers, F. H., Jawetz, E., and Goldfien, A. *Review of Medical Pharmacology* (3rd ed.). Los Altos, Calif.: Lange, 1972.

Muntwyler, E. *Water and Electrolyte Metabolism and Acid-Base Balance*. St. Louis: Mosby, 1958.

Pollay, M. Transport mechanisms in the choroid plexus. *Fed. Proc.* 33:2064–2068, 1974.

Smith, K. *Fluids and Electrolytes: A Conceptual Approach*. New York: Churchill Livingstone, 1980.

Strauss, M. B. *Body Water in Man*. Boston: Little, Brown, 1957.

Verney, E. B. Antidiuretic hormone and the factors which determine its release. *Proc. R. Soc. Lond. [Biol.]* 135:25–106, 1947.

Zadunaisky, J. A. Chloride active transport in the isolated frog cornea. Abstracts of the 9th Annual Meeting of the Biophysics Society, San Francisco, February, 1965. P. 9.

RESPIRATORY AND RENAL REGULATION OF ACID-BASE BALANCE

24

Ewald E. Selkurt

The chemical constituents of the extracellular fluid in which a cell normally lives are regulated within narrow limits to provide an optimal chemical environment for cellular function. One of the most precisely regulated constituents of body fluids is the hydrogen ion (H^+), and it is the regulation of this chemical species to which the term *acid-base regulation* actually applies.

Deviation from the normal range of hydrogen ion concentration markedly affects many metabolic reactions, accelerating some and depressing others, primarily because of the effect of hydrogen ions on enzyme activity. Therefore, the concentration of hydrogen ions in body fluids must be regulated to an optimal value if homeostasis is to be ensured.

SOURCES OF HYDROGEN IONS

The chief acid product of metabolism is CO_2 (H_2CO_3), of which an adult produces about 288 liters per day or 26 Eq, which is equal to 2.6 liters of concentrated HCl. Adults subsisting on a mixed diet produce in addition to CO_2 a substantial quantity of nonvolatile (fixed) acids. For example, 100 gm of protein produces approximately 60 mEq of sulfate by the oxidation of proteins and a quantity of phosphate, requiring 50 mEq of base for its neutralization at pH 7.4. An additional 50 mEq of base is required to neutralize the phosphate from 100 gm of fat (phospholipids). The acids produced are mainly sulfuric and phosphoric but include some hydrochloric, lactic, uric, and β-hydroxybutyric acid. Approximately one-half of these metabolically produced acids are neutralized by base in the diet, but the remainder must in some fashion be neutralized by the buffer systems of the body.

BUFFER SYSTEMS OF THE BODY

Several buffer systems prevent the organism from being overwhelmed by its own acid products. The one that reacts most readily is the blood buffer system. This depends only on chemical processes that occur in the blood to attenuate the change in hydrogen ion concentration. Ultimately, the buffering capacity of the tissues is brought into play also, since blood pH influences interstitial and intracellular pH as well.

For long-term stability of pH it is necessary to have an effective means of excreting the acids formed in the metabolic processes. This stability consists of the appropriate responses of the respiratory control center and renal tubules to changes in the pH and CO_2 of the arterial blood. Moreover, these acids must be excreted as acids or as ammonium salts by the kidney, since, if they were excreted as neutral salts, the buffers of the body would be rapidly depleted. This possibility becomes increasingly apparent when it is recognized that the total cation available in the blood as buffer salts does not exceed 150 mEq. In the body as a whole, including the protein salts and phosphates of the tissues, the total cation available is 1000 mEq.

The importance of the buffering capacity of the tissues cannot be overemphasized. If acid is produced metabolically, or infused intravenously experimentally, the immediate neutralization is by the blood buffers. Thus, the assumption is made that the acid is largely neutralized by these buffers (in plasma and erythrocytes). However, several groups of investigators have shown that only 15 to 20 percent is neutralized by blood buffers, that a somewhat larger proportion is neutralized by interstitial fluid, and that a major fraction is neutralized by buffers in tissue cells. Of this fraction, 15 percent of the extracellular hydrogen ions were buffered by exchange with cellular potassium and 36 percent by exchange with cellular sodium (Pitts, 1974). It is probable that some fraction of the sodium ions exchanged for H^+ may actually have come from the apatite crystals of the bone.

This type of buffering, which may be called *biological buffering*, consists in ionic shifts that protect extracellular pH. When acid or base is added to the extracellular space, approximately half of the added ions eventually diffuse into cells where they are probably buffered in the chemical sense. These ions or others that affect acid-base equilibrium (e.g., H^+, OH^-, HCO_3^-) are exchanged across the cell membrane for intracellular ions or are accompanied into cells by ions of opposite charge. For example, if an acid (H^+) is added to the body, some of it will be buffered chemically within the extracellular fluid. Some of the hydrogen ions will also diffuse across cell membranes (Fig. 24-1). Since H^+ is a positively charged ion, this diffusion requires either that a negatively charged ion such as chloride accompany it, or that other positively charged ions cross the cell membrane in the opposite direction. Both processes occur, although the movement of cations out of the cell is quantita-

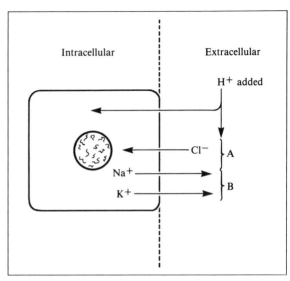

Fig. 24-1. Biological (physiological) buffering: diffusion into cells. Buffering by diffusion into cells accompanied by ions of opposite charge (A) and buffering of extracellular pH with exchange across the cell membrane with ions of like charge (B).

tively much more important. In any case, extracellular pH is defended by removal of some of the added acid from this compartment. Movements occurring in the opposite direction with alkalosis fall also into this type of buffering.

BLOOD BUFFERS. The Brønsted-Lowry concept of acids and bases finds wide acceptance in the consideration of acid-base balance (Muntwyler, 1968). According to this view, an acid is a proton (H^+) donor, and a base is a proton acceptor. On this basis, the reaction indicating the acidity of an acid, A, takes the form

$$A \rightleftharpoons B + H^+$$

where B is a base because it can accept a proton to form the acid, A, the proton donor. Some examples are given in Table 24-1. Important buffer acids are shown in Table 24-2.

Buffers are substances that tend to stabilize the pH of a solution. A buffer is a mixture of either a weak acid and its conjugate base or a weak base and its conjugate acid. A buffer is effective only when there are appreciable quantities of the alkali salt and either the acid or the base from which it derives. It is most effective when the acid and the base are present in equal quantities. Buffers are usually thought of in terms of buffer pairs. Acid-base buffer pairs of most importance in the body consist of weak acids associ-

Table 24-1. Proton donors and acceptors

Acid		Proton	Conjugate base
HCl	$\rightleftharpoons$	H^+	Cl^-
H_2SO_4	$\rightleftharpoons$	H^+	HSO_4^-
NH_4^+	$\rightleftharpoons$	H^+	NH_3

Table 24-2. Most important buffer acids at physiological pH

Proton donor		Proton acceptor		Proton
H_2CO_3	$\rightleftharpoons$	HCO_3^-	+	H^+
$H_2PO_4^-$	$\rightleftharpoons$	$HPO_4^=$	+	H^+
H protein	$\rightleftharpoons$	Proteinate$^-$	+	H^+
$HHbO_2$	$\rightleftharpoons$	HbO_2^-	+	H^+
HHb	$\rightleftharpoons$	Hb^-	+	H^+

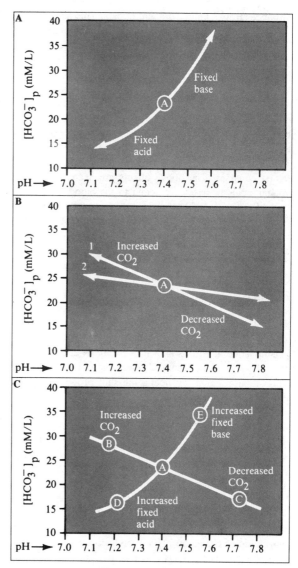

Fig. 24-2. A. Effect of fixed acid and fixed base on pH and HCO_3^- content of separated plasma. **B.** Effect of changes in CO_2 on pH and HCO_3^-. Line 1 is true plasma (plasma equilibrated in the presence of red cells and measured anaerobically). Line 2 is separated plasma (removed from red cells). **C.** Combination of the effects shown in A and B on true plasma. Note the increased slope of the buffer line with true plasma. (Reprinted from *The ABC of Acid-Base Chemistry* [6th ed.], by H. W. Davenport, by permission of The University of Chicago Press. Copyright 1947, 1949, 1950, 1958, 1969 and 1974, by The University of Chicago.)

ated with a salt of their conjugate base. Examples of common buffer pairs in the blood are $NaHCO_3/H_2CO_3$ and Na_2HPO_4/NaH_2PO_4. The proteins of the blood, hemoglobin, oxyhemoglobin, albumin, and globulin, are also extremely important buffers. Because of the amphoteric nature of proteins, each species represents a buffer pair. The major protein buffer in the blood is hemoglobin because of its high concentration. It is also considerably more important than the bicarbonate and phosphate buffers. The bicarbonate, however, has proved to be an accurate and easily determined index of acid-base balance.

MECHANISM OF BUFFER ACTION. The mechanism of buffer action may be understood from consideration of the bicarbonate system outlined in equation 1.

$$CO_2 + H_2O \rightleftharpoons H_2CO_3 \rightleftharpoons H^+ + HCO_3^- \tag{1}$$

If hydrogen ions are added to the system, some will combine with bicarbonate ions and drive the reaction to the left. Thus, not all the hydrogen ions added will stay in solution in the ionic form. Conversely, if base is added, some of the H^+ will be removed, shifting the reaction to the right and causing the H_2CO_3 to dissociate into H^+ and HCO_3^-. Thus, there is an inverse relationship between H^+ and HCO_3^- concentration when acid or base is added. This relation may be plotted as shown in Figure 24-2. The pH-bicarbonate diagram used here and in other portions of this chapter was introduced by Davenport (1974). It has proved extremely useful in describing the underlying causes of acid-base disturbances from a limited amount of data. This figure represents the buffer properties of plasma, of which bicarbonate is the principal buffer. The initial point A represents the normal condition of the plasma in arterial blood, namely, a pH of 7.4 and a HCO_3^- concentration of 24 mmol per liter. When an acid is added to plasma, the pH decreases, and, as a result of the reactions previously described, the HCO_3^- concentration falls also. When base is added to plasma, both the pH and the HCO_3^- concentrations increase. The effects of adding acid and base to blood or plasma are as shown in Figure 24-2A only when the CO_2 tension is maintained constant. The above effects are seen when nonvolatile acids or bases are added to the blood. These are called *fixed* acids and bases, since they are not removed by respiratory activity. It should be emphasized that the terms *volatile* and *nonvolatile* and *fixed* are useful physiological terms but have no strict meaning in a chemical sense.

The pH of the blood can also be influenced greatly by the CO_2 level. However, the effects of CO_2 on the pH and HCO_3^- concentration are quite different from those of the fixed acids and bases, as may be seen by referring to equation 1. If CO_2 is added or removed, the concentration of H^+ and HCO_3^- changes in the same direction. The behavior of plasma alone in vitro is shown in Figure 24-2B. Thus, the effect of CO_2 can be readily distinguished from that of fixed acid or base simply by noting the shift produced on the graph.

BUFFERING EFFECT OF WHOLE BLOOD. The changes produced in *separated plasma* studied in vitro will differ in some important respects from those seen in plasma that is part of whole blood (*true plasma*) (Fig. 24-2B). When fixed acid or base is added to plasma from whole blood, all the buffers participate in the process of attenuating the pH shift. As a result, the directional changes in the pH-bicarbonate diagram will be the same as in separated plasma, but the magnitude of change in pH will be decreased, per unit of acid or base added. Thus, when a *fixed acid* is added to whole blood, the buffering power of the red blood cell, which is due to the hemoglobin, also tends to attenuate the change in pH. If the amount of acid added were plotted against the pH of the solution, the slope of the buffer line would be steeper for whole blood as compared with separated plasma.

When CO_2 is added to whole blood, there is also a lesser effect on pH than would be the case with separated plasma. In contrast to the effects of fixed acid, the slope of the buffer line (Fig. 24-2C) changes because H^+ and HCO_3^- are formed in equal quantities but some of the H^+ is neutralized by other buffer systems while all the HCO_3^- remains. As a result, in whole blood more bicarbonate ions are formed per unit change in pH than in separated plasma. It should be borne in mind that the concentration of HCO_3^- being considered is that of the plasma portion of the blood only.

The concentration of H_2CO_3 and dissolved CO_2 in the plasma is much smaller than that of $NaHCO_3$. The ratio between these is approximately 1/20. Ordinarily, this would mean that such a buffer would be rather ineffective, particularly when base is added to the blood. However, since CO_2 is continuously manufactured and to some degree stored in the tissues, an additional supply is readily available. This makes it unnecessary for a large quantity of CO_2 to be present to permit the buffer system to operate effectively.

The curves of Figure 24-2C represent the response of true plasma to fixed acid or base and changes in CO_2 level. The four variations from the normal buffer point represent the four types of acid-base disturbance seen clinically: *respiratory acidosis* (CO_2 excess), *respiratory alkalosis* (CO_2 deficit), *metabolic acidosis* (fixed acid excess), and *metabolic alkalosis* (fixed base excess). By plotting the pH and HCO_3^- levels of a sample of arterial blood on the graph, it is possible to determine whether the subject is normal or in one of these states of acid-base disturbance.

To carry these considerations further, it is necessary to analyze their quantitative aspects. The pH-bicarbonate diagram is actually a graphic representation of the Henderson-Hasselbalch equation:

$$pH = pK + \log \frac{[HCO_3^-]}{[H_2CO_3]} \qquad (2)$$

The pK of the bicarbonate system is known, and the pH and HCO_3^- concentration can be measured, but the H_2CO_3 concentration, which is extremely low, cannot be directly measured. Thus, the utility of equation 2 as it now stands is limited. However, reference to equation 1 reveals that H_2CO_3 concentration is directly dependent on the concentration of dissolved CO_2, which is in turn dependent on its partial pressure and solubility coefficient of CO_2 in plasma. When these factors are substituted into the equation, the following is obtained:

$$pH = pK + \log \frac{[HCO_3^-]}{a\ PCO_2} \qquad (3)$$

where the solubility coefficient, a, $= 0.0301$ and pK $= 6.10$. Substituting, equation 3 becomes:

$$pH = 6.10 + \log \frac{[HCO_3^-]}{0.0301\ PCO_2} \qquad (4)$$

Equation 4 now relates three measurable quantities; if any two are known, the third may be calculated. Nomograms are available to facilitate such calculations.

ABNORMAL ACID-BASE BALANCE STATES. The following sections illustrate how the buffer systems operate in abnormal states (acidosis, alkalosis). Compensatory mechanisms come into play to attenuate or restore the disturbance in the acid-base balance depending on the nature and degree.

Respiratory Acidosis. Since the directional changes on the graph of any form of acid-base disturbance are known, it is possible to determine the type and degree of any abnormality that may exist. Such a representation is shown in Figure 24-3. Point A represents the normal pH and HCO_3^- values. With a disturbance in acid-base balance these values will be shifted according to the type of disturbance. For example, if CO_2 elimination by the lungs is inadequate (e.g., obstructive lung disease, damage to muscles of respiration), the PCO_2 of the alveolar and arterial blood will be increased. This situation is called *respiratory acidosis* and is indicated by point B on the diagram. If the deviation from normality is caused only by respiratory acidosis, the point will lie somewhere on the normal buffer line. The change in pH and HCO_3^- produced represents simply the interaction of the increased CO_2 and the blood buffers. As compensatory mechanisms come into play (in this instance the kid-

ney), point B will move away from the normal buffer line in a manner that will be described later. The situation illustrated is *uncompensated respiratory acidosis*.

Respiratory Alkalosis. There are also clinical disturbances in which chronic hyperventilation occurs. Examples are hysteria and salicylate poisoning in children. With such hyperventilation, *respiratory alkalosis* is produced, as represented by point C in Figure 24-3. Since it lies on the normal buffer line, it is *uncompensated respiratory alkalosis*. Because compensation by the kidney is slow, this uncompensated state will be occasionally observed in the clinic.

Metabolic Acidosis. Metabolic acidosis is produced by conditions such as ketosis induced by diabetes mellitus or excessive loss of alkaline fluids from the lower digestive tract, as occurs in diarrhea. When such a condition occurs, the pH of the blood decreases and the HCO_3^- concentration decreases also (point D, Fig. 24-3). It can therefore be differentiated from respiratory acidosis. If the PCO_2 is not changed as a result of the acidosis, point D will lie on the same PCO_2 isobar as the normal point, and the term *uncompensated metabolic acidosis* is used. However, reduction of pH has a stimulatory effect on respiration that will reduce the PCO_2 levels. The effect of this respiratory compensation is to increase pH and reduce HCO_3^- levels, as will be seen later.

Metabolic Alkalosis. Metabolic alkalosis is produced by conditions such as persistent vomiting, which causes loss of a considerable amount of acid. In metabolic alkalosis (point E, Fig. 24-3) the pH and HCO_3^- increase together, enabling it to be distinguished from respiratory alkalosis. If there is no respiratory response to the change in pH, the condition is *uncompensated metabolic alkalosis*.

In some instances there may be a mixture of acid-base abnormalities that can also be detected by use of the pH-bicarbonate diagram. For example, point F in Figure 24-3 represents a combination of metabolic and respiratory acidosis, and point G is metabolic and respiratory alkalosis.

Summary of Changes in Plasma Bicarbonate–Carbonic Acid Buffer System in Disturbances of Neutrality. Figure 24-4 attempts to summarize the primary and secondary changes in the HCO_3^-/H_2CO_3 buffer system with various disturbances in acid-base balance.

Fig. 24-3. Effect of acid-base disturbances on the pH and HCO_3^- level of the plasma, plotted on a pH-bicarbonate diagram. Note that each type of disturbance occupies a unique place on the diagram. (Reprinted from *The ABC of Acid-Base Chemistry* [6th ed.], by H. W. Davenport, by permission of The University of Chicago Press. Copyright 1947, 1949, 1950, 1958, 1969 and 1974, by The University of Chicago.)

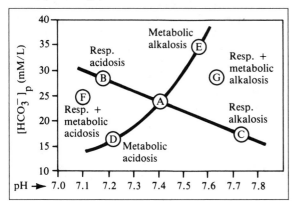

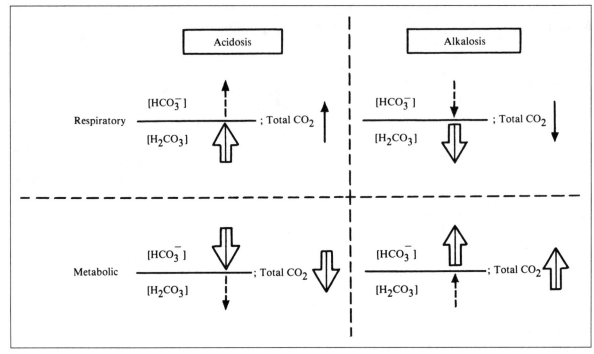

Fig. 24-4. Primary and secondary changes in the bicarbonate-carbonic acid buffer system descriptive of the various types of disturbances in acid-base balance. Note that total CO_2 does not always fall in acidosis or rise in alkalosis. (From H. N. Christensen. *Body Fluids and Their Neutrality*. New York: Oxford University Press, 1963. P. 122.)

1. In *respiratory acidosis,* the retention of CO_2 and rise in H_2CO_3 are the primary cause. The kidney is the important compensatory organ, secreting H^+ and conserving bicarbonate.
2. In *respiratory alkalosis* the net loss of carbonic acid by overbreathing is primary, and any tendency to compensate would decrease the bicarbonate ion concentration secondarily. The decrease is achieved by the kidney.
3. In *metabolic acidosis* the entering hydrogen ion reacts with the bicarbonate ion to produce a primary decrease in its concentration (shown by the large arrow). Respiratory activity causes a smaller, secondary decrease in the carbonic acid level (shown by the dotted arrow). Because the numerator is decreased more than the denominator, the pH must be lowered. In this instance the total CO_2 must be distinctly decreased. The total CO_2 content is a commonly employed laboratory determination, in which the serum is acidified and the CO_2 all brought into the gaseous form and measured. The compensatory mechanisms are discussed in greater detail later.
4. Likewise, in *metabolic alkalosis,* the bicarbonate ion is primarily increased, the carbonic acid being secondarily

elevated and to a lesser degree. The total effect on the CO_2 content is an elevation. Renal compensation involves loss of $HCO_3{}^-$.

Note that the overall direction in the change of the CO_2 content is downward in both metabolic acidosis and respiratory alkalosis and upward in metabolic alkalosis and respiratory acidosis. The reliability of any association between low CO_2 content and respiratory acidosis is vitiated. Reliance on the serum CO_2 content (or capacity) alone as an assay of acid-base balance obviously may lead to mistaken conclusions regarding acid-base balance disturbances of both the respiratory and the metabolic types.

RESPIRATORY AND RENAL COMPENSATION IN pH-BICARBONATE DIAGRAM. It has already been indicated that the respiratory system acts as a buffer system by responding to changes in blood pH in such a manner as to maintain the pH very nearly constant. This is accomplished only by altering the CO_2 level of the blood. Thus, any change in pH and bicarbonate induced by the respiratory system occurs along a buffer line, whether a respiratory disturbance of acid-base balance or a respiratory compensation for a metabolic disturbance is being considered.

The kidney can influence acid-base balance by conserving hydrogen ions or removing them from the bloodstream. The renal mechanisms that perform this operation will be considered later. The kidney acts to alter the relative amounts of fixed acid or base in the blood. Any changes of this nature in the blood follow the same path as those produced by metabolic acids or bases—that is, up or down a PCO_2 isobar.

These compensatory mechanisms may now be considered in terms of their effect on the pH-bicarbonate diagram. The pathways followed by the compensatory mechanisms are shown in Figure 24-5A. When *respiratory acidosis* occurs, for example (point B), renal compensation follows, in the form of an abnormally large excretion of acid by the kidney. With this, HCO_3^- tubular reabsorption is increased. Plasma Cl^- is lost to the extent that HCO_3^- concentration is raised. This acid loss causes movement of the point up the PCO_2 isobar toward a normal pH. At point B_1 the pH has been only partially restored to its normal value; at point B_2 it has been fully restored—hence the designations *partially compensated* and *fully compensated respiratory acidosis*. The kidney is able under some circumstances to bring the pH back to 7.4 as indicated by the fully compensated state. This achievement shows that the kidney has a highly effective mechanism for hydrogen ion disposal. The response is, however, quite slow. Several days may be required for it to compensate for a chronic condition.

In *respiratory alkalosis* the kidney also acts to compensate for the abnormality, this time by conserving hydrogen ions produced in metabolic processes and by increased excretion of HCO_3^-. With increased HCO_3^- loss, Cl^- is conserved, raising plasma Cl^- concentration to replace lost HCO_3^-. Point C in Figure 24-5A then moves to positions C_1 and C_2 as the degree of compensation increases. Again, it is possible that the compensation may be complete and a normal pH will be restored.

When *metabolic acidosis* occurs (point D, Fig. 24-5A), respiratory activity is immediately increased. As a result there is a very rapid partial compensation (point D_1). However, respiratory compensation is never complete enough to bring the pH back to the normal value. Apparently, an appreciable pH difference is required to stimulate respiration. It is evident that if pH were returned precisely to its original value, the stimulus for increased respiration would cease and the respiratory compensation would disappear. The

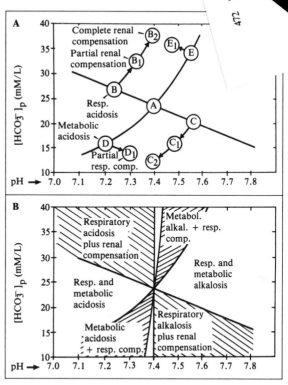

Fig. 24-5. A. Effects of compensation on the pH and bicarbonate content. B. Zones of acid-base disturbance and compensation. (Reprinted from *The ABC of Acid-Base Chemistry* [6th ed.], by H. W. Davenport, by permission of The University of Chicago Press. Copyright 1947, 1949, 1950, 1958, 1969 and 1974, by The University of Chicago.)

compensatory patterns of the kidney and the respiratory system therefore differ in the respect that respiratory compensation is never complete. The kidney may later again restore pH to normal.

A similar pattern is seen in *metabolic alkalosis* (points E, E_1, Fig. 24-5A). Under the influence of the high pH of the arterial blood, respiratory activity diminishes, allowing CO_2 to accumulate. The compensation therefore occurs along a buffer line that is parallel to but displaced from the normal buffer line. The kidney is again needed for full compensation.

In Figure 24-5B are shown the areas of renal and respiratory compensation. Any set of values of pH and bicarbonate for a given sample of arterial blood must fall within one of these areas or those of combined metabolic and respiratory acidosis or alkalosis. Thus, the patient's acid-base status may be determined by plotting the data on the graph and noting the area in which the point falls.

RESPIRATORY REGULATION OF BODY FLUID pH. All of the body's chemical buffers are tied into respiration through the bicarbonate buffer system, as evidenced in the following sequence of reversible reactions:

$$H^+ + HCO_3^- \rightleftarrows H_2CO_3 \rightleftarrows H_2O + CO_2$$

The conversion of carbonic acid to water and CO_2, and vice versa, is greatly accelerated by the presence of the enzyme *carbonic anhydrase* (CA) within the erythrocyte. An increase in the concentrations of any of the reactants or products drives the reactions in the direction that tends to reduce the increase until equilibrium occurs. For example, an increase in the concentration of hydrogen ions drives the reactions to the right, with the subsequent formation of carbonic acid and a reduction of hydrogen ion concentration toward normal. Conversely, a decrease in the concentration of any of the reactants or products drives the reactions in the direction that tends to increase the concentration toward normal. A decrease in the hydrogen ion concentration, for example, drives the reactions to the left. The subsequent dissociation of carbonic acid elevates hydrogen ion concentration toward normal. Through the elimination of CO_2 by the respiratory system, we may effect changes in body fluid pH: An increase in blood PCO_2 tends to elevate body fluid hydrogen ion concentration (decreases pH), and a decrease in blood PCO_2 tends to lower body fluid hydrogen ion concentration (increases pH).

Alveolar Ventilation and Body Fluid pH. Increases of body fluid hydrogen ion concentration, specifically in the blood and cerebrospinal fluid, result in a reflex acceleration of respiratory rate and an increase in tidal volume (see Chap. 21). The net effect is to increase the rate of CO_2 elimination from the body, thereby driving the carbonic acid reactions to the right, with the subsequent return of hydrogen ion concentration toward normal. Severe decreases in body fluid pH may increase alveolar ventilation to about five times its normal value. Conversely, increases in body fluid pH tend to depress respiratory activity, allowing CO_2 and hence hydrogen ions to increase, but the respiratory response is not as marked as it is to a decrease in pH. Severe increases in body fluid pH may decrease alveolar ventilation by 50 percent of normal.

It may be correctly inferred from the foregoing that not only does the pH of body fluids affect respiratory activity but also that alterations (voluntary or otherwise) in alveolar

ventilation have a pronounced effect on the pH of body fluids. Voluntary apnea (holding one's breath), for example, results in a decrease of pH, while voluntary hyperventilation elevates body fluid pH. Doubling the normal resting rate of alveolar ventilation elevates extracellular pH by about 0.23, whereas reducing the rate to one-fourth normal reduces extracellular pH by 0.4. As long as the rate at which CO_2 is being eliminated from the body keeps up with the rate of CO_2 production, there will be no net change in the carbonic acid/bicarbonate ratio of body fluids, and hence under normal circumstances a stable pH is maintained.

Because of the respiratory system's ability to regulate body fluid pH through the elimination of CO_2, it is sometimes referred to as a "physiological buffer system" (Fig. 24-6). As is the case with the body's chemical buffers, the respiratory system cannot eliminate actual hydrogen ions from the body, nor can it eliminate bicarbonate ions. The excretion of CO_2 in expired air results in the elimination of potential, not actual, hydrogen ions. The kidneys, however, are able to excrete hydrogen ions as well as effectively regulate the concentration of bicarbonate in body fluids. Al-

Fig. 24-6. Stepwise illustration of the pulmonary response to acid invasion (not carbonic acid). Enough acid is administered at the bottom arrow to cause conversion of almost half the HCO_3^- to H_2CO_3. (At the same time, other buffer anions also accept H^+.) An explosive increase in respiration results, so that the newly formed H_2CO_3 is swept out as CO_2, and the hypothetical pH of 6.1 (second stage) is never reached. The third stage shows the transient picture when the PCO_2 has returned to a normal value. (From H. N. Christensen. *Body Fluids and Their Neutrality.* New York: Oxford University Press, 1963. P. 118.)

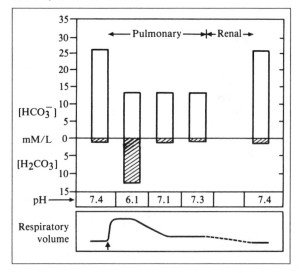

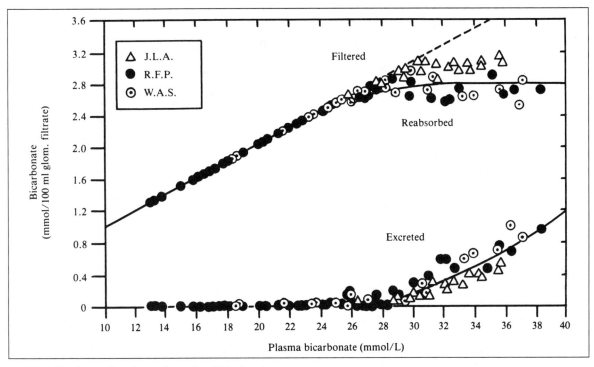

Fig. 24-7. Filtration, reabsorption, and excretion of bicarbonate as functions of plasma concentration in normal man. (From R. F. Pitts, J. L. Ayer, and W. A. Schiess. The renal regulation of acid-base balance in man. III. The reabsorption and excretion of bicarbonate. *J. Clin. Invest.* 28:35–49, 1949.)

though the renal response to changes in body fluid pH is not as quick as the response of either the chemical buffer systems or the respiratory system, the kidneys play an extremely important role in maintaining acid-base balance because of their ability to conserve or eliminate various acids and bases selectively, thus maintaining effective buffer ratios in body fluids.

RENAL ROLE IN ACID-BASE BALANCE. In a quantitative sense the major role of the kidneys in the maintenance of acid-base balance is conservation of circulating stores of bicarbonate. In humans, some 5100 mEq of bicarbonate ions is reabsorbed per day from the glomerular filtrate. In exchange, an approximately equivalent number of hydrogen ions are secreted. This is a key mechanism, along with the operation of the lung, in determining the proper ratio of carbonic acid and the bicarbonate ion and ultimately of all the body buffers. In addition, the kidney generates titratable

buffer acid and produces ammonia, these two mechanisms aiding in *base conservation.*

RENAL HANDLING OF BICARBONATE. The characteristics of the renal reabsorption and excretion of bicarbonate in humans are revealed in Figure 24-7. If the plasma bicarbonate is reduced below about 26 mEq per liter, virtually all of the filtered bicarbonate is reabsorbed and none is excreted; if bicarbonate is infused and plasma concentration is elevated above 28 mEq per liter, a limited but constant amount (2.8 mEq/100 ml filtrate) is reabsorbed. All bicarbonate filtered in excess of this is excreted in the urine. Thus, the plasma value of 26 to 28 mEq per liter in man suggests a so-called renal bicarbonate threshold. The term *threshold* is used advisedly, for the statement that bicarbonate is practically absent from the urine at plasma concentrations below 26 mEq per liter is true only so long as the respiratory center is free to regulate pulmonary ventilation in response to the prevailing level of plasma CO_2. During hyperventilation, for instance, the urine may become alkaline and contain considerable amounts of bicarbonate, although the plasma concentration may be well below the so-called normal threshold.

The reason is that when CO_2 is blown off by the lungs,

the cations, such as sodium, are balanced by anionic groups of plasma proteins, which are nonfilterable at the glomerulus. But Na^+ is filtered and combines with HCO_3^- in the tubular urine, and the formed $NaHCO_3$ makes the urine alkaline even though the plasma bicarbonate is relatively low. The bicarbonate is synthesized by the epithelial cells of the tubules, aided by CA (see below) and combines with reabsorbed Na^+ and passes into the interstitial fluid (Fig. 24-8).

PROXIMAL TUBULAR MECHANISM. The proximal tubular mechanism is specialized to reabsorb the bulk of the filtered bicarbonate from the tubular urine against a relatively low gradient. The basic elements of the mechanism are illustrated in Figure 24-8. Sodium and bicarbonate ions enter the proximal tubule in the glomerular filtrate. Sodium ions diffuse into the tubular cells down a concentration and electrical gradient and are actively extruded into the peritubular fluid by a pump, that maintains intracellular sodium concentration at a low value. Hydrogen ions move from the interior of the cell to the tubular lumen in exchange for sodium ions against an electrical gradient, necessitating active transport. Whether the passive influx of sodium ions and active efflux of hydrogen ions are carrier linked is not known. However, they are linked operationally. Hydrogen ions associate with bicarbonate ions to form carbonic acid, which decomposes into CO_2 and water. The CO_2 diffuses into the cell, where it undergoes hydration to form carbonic acid; this reaction is catalyzed by CA. Subsequent dissociation provides the hydrogen ions, which are exchanged for sodium ions across the luminal membrane, and the bicarbonate ions, which diffuse down a potential gradient into the peritubular fluid. According to this concept, bicarbonate ions are reabsorbed indirectly by conversion to CO_2 within the tubular lumen. The key element of the reabsorptive process is the exchange, across the luminal membrane, of intracellular hydrogen ions for sodium ions in the tubular fluid. Some 90 percent of the filtered bicarbonate is reabsorbed in the proximal tubule.

A critical role is played by the enzyme CA, a zinc-containing metalloprotein of molecular weight 30,000. It occurs in high concentration in renal tubules, particularly in the proximal convolutions, where it is found in the brush border as well as in the cytoplasm. The evidence for CA action comes from the fact that sulfonamide derivatives, which have the ability to inhibit CA, block H^+ secretion and produce an excess of sodium in the urine (see Chap. 23). Carbonic anhydrase is not present on the luminal surface of the distal tubules, and none is present in the collecting duct.

The presence of CA on the brush border of proximal

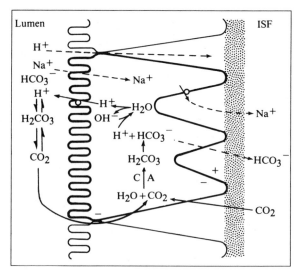

Fig. 24-8. Ion exchange mechanism in the proximal tubular cell, emphasizing the role of carbonic anhydrase (CA). ISF = interstitial fluid.

tubules facilitates the bulk reabsorption of bicarbonate, for it reduces by a factor of 10 the gradient against which hydrogen ions must be secreted and correspondingly reduces the energy cost of bicarbonate conservation. It is maintained by some authorities that hydrogen ions are actively secreted by a redox (reduction-oxidation) ion pump. The mechanism is similar to that involved in the secretion of gastric acid.

DISTAL TUBULE AND COLLECTING DUCT MECHANISMS. The mechanisms for reabsorption of bicarbonate in distal tubules and collecting ducts are qualitatively the same as that in proximal tubules. They differ quantitatively in the following respects: (1) Together they account for only 10 percent of total bicarbonate reabsorption, and (2) they are capable of salvaging essentially all of the bicarbonate remaining in the tubular fluid when plasma concentration is normal or low. To accomplish the latter end, the collecting ducts, specifically, must be able to secrete hydrogen ions against much steeper gradients than either the proximal or the distal tubules. Also, sodium ions may have to be pumped from lumen into cells to achieve their nearly complete removal in states of sodium depletion. Accordingly, a more tightly coupled hydrogen-sodium exchange pump may be present in collecting ducts.

FACTORS AFFECTING BICARBONATE TRANSPORT. At least four factors influence bicarbonate transport: (1) changes in

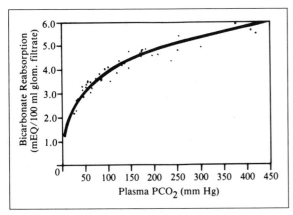

Fig. 24-9. Relationship between PCO_2 and bicarbonate reabsorption. (From F. C. Rector, D. W. Seldin, A. D. Roberts, and J. S. Smith. The role of plasma CO_2 tension and carbonic anhydrase activity in the reabsorption of bicarbonate. *J. Clin. Invest.* 39:170, 1960.)

the CO_2 tension of the arterial blood, (2) variations in the body stores of potassium, (3) variations in the plasma level of chloride, and (4) variations in the secretion of adrenocortical hormones.

Effect of Changes in the PCO_2 of Arterial Blood. Figure 24-9 shows the effects of acute variations in arterial PCO_2 on bicarbonate reabsorption in the dog. When animals are hyperventilated, so that PCO_2 values are reduced to less than the normal 40 mm Hg, the reabsorption of bicarbonate is reduced. When animals breathe CO_2-air mixtures, so that the PCO_2 is increased to values above the normal 40 mm Hg, the reabsorption of bicarbonate is increased. At very high levels of PCO_2, bicarbonate reabsorption more than doubles. Chronic exposure of animals to high partial pressures of CO_2 still further enhances the tubular reabsorption of bicarbonate, the major response occurring within 48 hours.

Variations in Body Stores of Potassium. Figure 24-10 illustrates the effects, in the dog, of the intravenous infusion of potassium salts on the rate of tubular reabsorption of bicarbonate. When the plasma concentration of potassium falls below the normal value of 4 mEq per liter, bicarbonate reabsorption is elevated above the normal value of 2.4 to 2.6 mEq per 100 ml of filtrate. When, in consequence of the infusion of potassium salts, the plasma potassium concentration exceeds 4 mEq per liter, bicarbonate reabsorption is reduced below the normal value. The mecha-

nism of the influence on bicarbonate reabsorption concerns the intracellular concentration of potassium. When infused, it enters the cells in exchange for H^+, which leaves the cells and is buffered by the plasma HCO_3^-, whose concentration then falls. Presumably, the renal tubular cells respond in a similar manner, accumulating K^+ from the peritubular fluid and extruding H^+ into it. As a result of reduced intracellular H^+ following potassium infusion, fewer hydrogen ions are secreted into the lumen, and less bicarbonate is reabsorbed. The converse would operate with reduced potassium in the extracellular fluid. It is no longer held that the relationship between hydrogen ion and potassium ion secretion represents competition for a common secretory mechanism.

Hydrogen ion secretion and thus bicarbonate reabsorption are increased by hypokalemia and decreased by hyperkalemia along the entire length of the nephron, on account of the changes in cellular hydrogen ion concentration previously described. Because filtered potassium is almost entirely reabsorbed in the proximal tubule, no competition between potassium ions and hydrogen ions for secretory transport could occur at this site. The potassium that is excreted is secreted in the distal part of the nephron. The secretory mechanism is passive, and the distribution of potassium ions is determined largely by the transtubular potential gradient (see Chap. 22).

Relationship to Plasma Chloride Concentration. When the body is depleted of chloride and hypochloremia develops, plasma bicarbonate increases and bicarbonate reabsorption increases, sustaining a higher plasma concentration of HCO_3^-. Conversely, when body chloride stores are expanded and hyperchloremia develops, plasma bicarbonate decreases. Bicarbonate reabsorption is reduced, and plasma concentration is maintained at a subnormal level. When the chloride deficit is replaced, or when the chloride excess is excreted, the plasma level of bicarbonate rapidly returns to normal. The mechanism is not well understood but probably relates to the volume of the extracellular compartment. If extracellular volume is expanded by infusion of isotonic solutions, reabsorption of bicarbonate decreases. If extracellular volume is contracted by dialysis, reabsorption of bicarbonate increases.

Influence of Adrenocortical Hormones. In Cushing's syndrome and in hyperaldosteronism, the plasma HCO_3^- is moderately elevated. In Addison's disease the plasma concentration of bicarbonate is moderately reduced. The mechanism of action in response to variation in output of the adrenocortical hormones is not known at this time but could

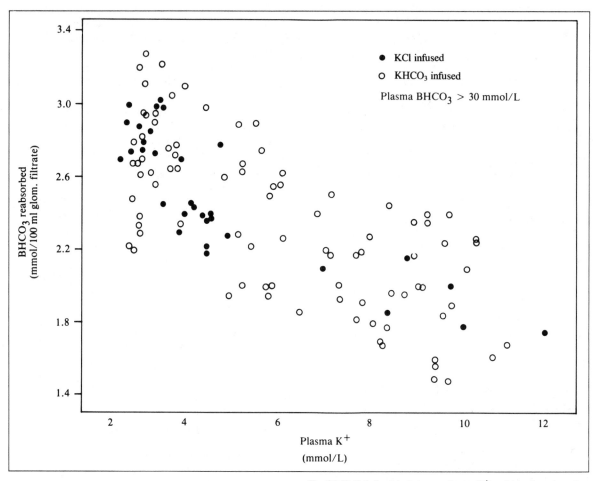

Fig. 24-10. Relationship between plasma K^+ and bicarbonate reabsorption. (From G. R. Fuller, M. B. MacLeod, and R. F. Pitts. Influence of administration of potassium salts on renal tubular reabsorption of bicarbonate. *Am. J. Physiol.* 182:115, 1955.)

be secondary to the hormonal influence on renal potassium handling (see Chap. 34).

RENAL MECHANISMS FOR BASE CONSERVATION

Two mechanisms exist for base conservation: (1) *acidification of the urine* and (2) *ammonia synthesis*. In acidification of the urine the neutral salts of weak buffer acids are converted to acid salts or even free acids. For example, NaH_2PO_4, uric acid, and so on can be excreted as titratable acid of the urine, by reducing the glomerular filtrate from a pH of 7.4 to urine having a pH as low as 4.5. The general equation for this process is

$$Na_2HPO_4 + HHCO_3 \rightarrow \underset{\text{(excr.)}}{NaH_2PO_4} + \underset{\text{(reabs.)}}{NaHCO_3}$$

The other main mechanism is ammonia synthesis. Where there are strong acids (H_2SO_4, HCl), they are finally excreted as ammonium salts.

$$Na_2SO_4 + 2H^+ + 2NH_3 \rightarrow \underset{\text{(excr.)}}{(NH_4)_2SO_4} + \underset{\text{(reabs.)}}{2Na^+}$$

Replacement of sodium by hydrogen ions means that the acid radicals can ultimately be excreted as free acids. There is a limit to the amount of acid that can be excreted in this form, however, because the kidneys cannot excrete urine with a pH less than about 4.5. Excretion of strong acids by this means is accompanied by their full equivalent of base. (Recall that a base is a proton acceptor.) Weaker acids can be excreted in the free form to an extent determined by their pK values. Thus, on application of the Henderson-Hasselbalch equation,

$$pH = pK + \log \frac{\text{concentration of base}}{\text{concentration of acid}}$$

Recall also that pK is the pH at which there are equal quantities of acid and the accompanying base. If the pK is the same as the pH of the urine, one-half of the acid will be present in the free form. In the case of a weak acid whose pK is 1 pH unit above the pH of the urine, nine-tenths of the acid will be free and one-tenth combined with base. However, with a strong acid whose pK is 1 pH unit below the pH of the urine, only one-tenth of the total amount of acid can be excreted in the free form, so that nine-tenths of its equivalent of base will have to accompany it. Stronger acids with pK values of more than 1 unit below the minimal urinary pH must always be excreted with more than nine-tenths of the equivalent base.

Those weaker acids that can be excreted to a considerable extent in the free form in the urine within the physiological range of pH are known collectively as the *buffer acids* of the urine. The most important is NaH_2PO_4, but amino acids and creatinine (pK 4.97) are also in this category. The pK for the dissociation of the second hydrogen ion of phosphoric acid is 6.8; therefore, each equivalent of phosphate in a urine of pH 4.8 requires one equivalent of base to cover it, whereas in plasma at pH 7.4 it requires nearly two equivalents. Hence, by excreting urine of maximal acidity the kidneys can salvage one equivalent of fixed base for each equivalent of acid phosphate in the urine, as well as additional amounts corresponding to the other buffer acids that are present. The process of acidification of the kidney is reversed when the urine is titrated with alkali back to the pH of the plasma, 7.4, to determine the "titratable acidity." This reflects the amount of base the kidneys have conserved by acidifying the urine. Such base conservation may range from about 30 mEq per day to as high as 150 mEq during diabetic acidosis, when the urine contains large quantities of β-hydroxybutyric acid (pK 4.7), and acetoacetic acid.

MECHANISM OF ACIDIFICATION OF THE URINE. Pitts and Alexander (1945) examined older theories of phosphate and bicarbonate reabsorption with simultaneous measurement of glomerular filtration rate and found that both fell far short of accounting for the possible titratable acidity of the urine, in each case accounting for only 20 percent or less, the limitation being the amount of acid in the glomerular filtrate

(Pitts, 1974). To explain the much greater availability of the acid, it was postulated that the tubular cells are able to exchange ions between the tubular urine and the bloodstream. Secretion of H^+ as a basis for an ion exchange mechanism was postulated, and this is now accepted.

Ion Exchange Mechanism. According to the ion exchange mechanism, H^+ is secreted by the tubular cells into the lumen in exchange mainly for Na^+ from the glomerular filtrate. The source of H^+ is the CO_2 of the metabolic activity of the tubular cells or that brought in by the blood. The mechanism was discussed in relation to the bicarbonate reabsorptive mechanism.

As shown in Figure 24-11, the filtrate is moderately acidified in the kidney of the rat as it flows along the proximal tubule, but the major acidification occurs beyond the distal tubule, i.e., in the collecting duct. This is where the major fraction of the titratable acid represented by buffers of low pK, such as β-hydroxybutyrate (pK_A, 4.7), is formed. In this portion of the nephron a relatively small quantity of H^+ is pumped against a high gradient, whereas in the proximal tubule a large quantity of H^+ is exchanged against a low gradient. Titratable acid with a high pK, as represented by phosphate, is probably formed chiefly in the proximal tubule, but also in the distal tubule. An example of the ion exchange theory is diagrammed in Figure 24-12. It can be discerned that the $NaHCO_3$ formed in this manner actually represents "new" bicarbonate, as contrasted to "reclaimed HCO_3," as discussed earlier for the bicarbonate reabsorptive mechanism in the proximal tubule.

The overall rate of secretion of H^+ is determined by the following factors: (1) the degree of acidosis, (2) the quantity of buffer acid excreted, and (3) the strength of the buffer; i.e., how strongly it resists giving up Na^+ for H^+ (e.g., phosphate with a pK of 6.8 permits greater secretion of H^+ than creatinine with a pK of 4.97). The significance of the last two factors is illustrated in Figure 24-13.

Ammonia Production. It has been stated that the ability to excrete strong acid in free titratable form is limited because the urine pH cannot go below about 4.5. Another mechanism for conservation of base is created by secretion of NH_3 into the urine, which binds hydrogen ion and forms the ammonium ion NH_4^+. The ability of the kidney to form ammonia was discovered by Nash and Benedict, who found more NH_3 in the venous blood than in the arterial blood entering the kidney.

According to Van Slyke et al. (1943), the most important mechanism for formation of ammonia is its formation from

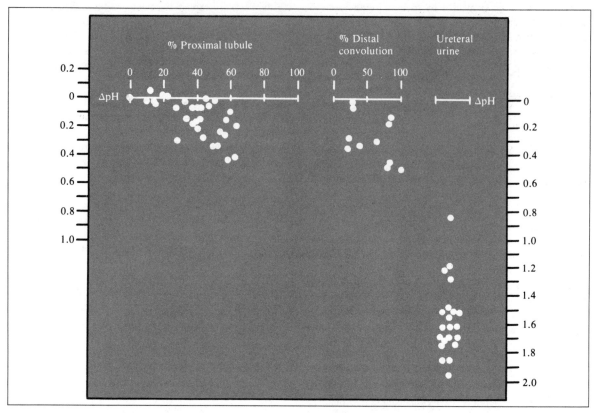

Fig. 24-11. Change in pH of tubular fluid along the nephron of the rat. Measurements were made under equilibrium conditions. (From C. W. Gottschalk, W. E. Lassiter, and M. Mylle. Localization of urine acidification in the mammalian kidney. *Am. J. Physiol.* **198:** 582, 1960.)

plasma glutamine with the aid of glutaminase.

$$\text{Glutamine} \xrightarrow{\text{(glutaminase)}} \text{glutamic acid} + NH_3$$

Thus, glutamine appears to be the immediate precursor of ammonia normally produced by the kidney. Other amino acids (L-asparagine, DL-alanine, L-histidine, L-aspartic acid, glycine, L-leucine, L-methionine, and L-cysteine), when infused into acidotic dogs, also increase the secretion of ammonia. These acids are transformed to α-ketoglutarate by transaminases to form glutamate, and this in turn is converted to additional glutamine by the glutamine synthetase system.

α-ketoglutarate + α-amino acid → glutamate + α-keto acid
glutamate + ATP + NH_3 → glutamine + ADP + PO_4

Additional ammonia can theoretically come from glutamate directly (by undergoing deamination within the tubular cells containing a specific glutamic dehydrogenase), or from the glutamine formed, as above.

In summary: Glutamine and amino acids contribute amino and amide groups to a nitrogen pool from which ammonia is formed and diffuses into the urine, where it is trapped as ammonium ion (Fig. 24-14). The mechanism of action is as follows: Each molecule of NH_3 (a proton acceptor) secreted binds one hydrogen ion and permits one sodium ion to be reabsorbed. Thus, it is to be expected that the pH of the urine will determine the rate of diffusion of NH_3 into the tubular lumen. The mechanism is understood better by the following series of equations:

$$NH_3 + H^+ \rightleftharpoons NH_4^+ \tag{1}$$

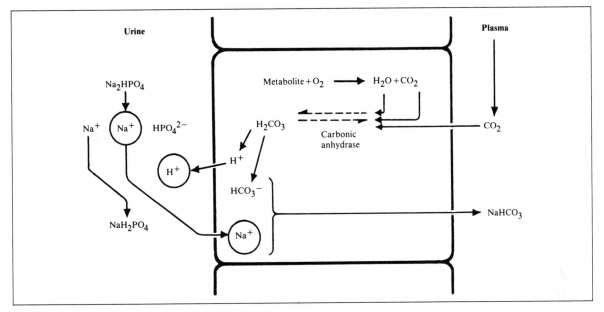

Fig. 24-12. The mechanism of formation of titratable acid and conservation of base.

The ionization constant, K_A, is

$$K_A = \frac{[H^+][NH_3]}{[NH_4^+]} \qquad (2)$$

$$pH = pK_A + \log \frac{[NH_3]}{[NH_4^+]} \qquad (pK_A = 8.9) \qquad (3)$$

It is accepted on good evidence that the cells generally are permeable to the molecular species NH_3 but not to the ionic species NH_4^+, perhaps because NH_3 is lipid soluble, whereas NH_4^+ is water soluble. Since NH_3 is presumably in the same concentration on both sides of the cell membrane because of its easy diffusibility, the numerator of the last term of the equation can be considered to be a constant; or, to put it simply, the log of NH_4^+ will vary inversely with the pH. Then, as the urine becomes increasingly acid, the mass law operates to increase the fraction of the total ammonia produced in the distal segment that is captured in the urine as nondiffusible NH_4^+ and excreted as such. The capture of NH_4^+ in the urine serves to neutralize free acid and thus permits a larger quantity of acid radical to be excreted as NH_4^+ salt at a given urine hydrogen ion concentration than would otherwise be possible. Stated another way, the excretion of NH_4^+ salt reduces the quantity of Na^+ that would otherwise accompany the acid radical into the urine. The

tubular excretion of NH_4^+ thus serves to conserve an equivalent quantity of Na^+ for the body.

Site of Production of Ammonia. Micropuncture data in the rat reveal that the proximal tubule is an important potential source of final urine ammonia, and that increased proximal ammonia addition occurs in response to chronic ammonium chloride acidosis (which leads to metabolic acidosis and increased renal glutaminase activity). Distal tubular ammonia addition is also important in the final ion exchange mechanism (Fig. 24-15). The collecting duct was found to be a significant source of final urine ammonia in the hamster kidney but not in the rat kidney.

SUMMARY OF BASE CONSERVATION MECHANISMS. The micropuncture techniques of Gottschalk et al. (1960) revealed that in the nondiuretic rat the pH of the proximal tubular urine decreased below that of the plasma (see Fig. 24-11). This result is in keeping with the finding, also by micropuncture techniques, that considerable reabsorption of bicarbonate goes on in the proximal convoluted tubules. The drop in pH in the proximal system is not as great as in the distal nephron (less than one-half a pH unit) and is explained in terms of the bulk exchange that proceeds here. In terms of bicarbonate reabsorption in the proximal tubule, the amount of H^+ exchanged here must be considerably greater (3500 mEq/day) than that secreted in the distal con-

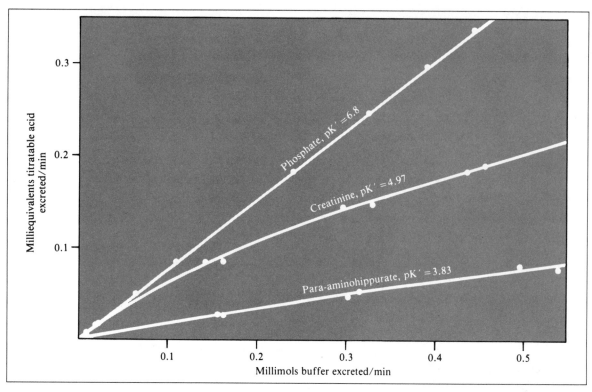

Fig. 24-13. Relationship between buffers of different pK' values and excretion of titratable acid. (From W. A. Schiess, J. L. Ayer, W. D. Lotspeich, and R. F. Pitts. The renal regulation of acid-base balance in man. II. Factors affecting excretion of titratable acid by the normal human subject. *J. Clin. Invest.* 27:59, 1948.)

voluted tubule and collecting duct (100 mEq/day) but proceeds on a one-for-one basis ($Na^+ \rightleftharpoons H^+$), so pH does not change significantly in the proximal segment. The reason is that the H^+ that reacts with filtered HCO_3^- is recycled back into the cell and therefore does not ordinarily contribute to the net elimination of acid.

Only that small fraction of the total secreted H^+ that is bound to a urinary buffer, either as titratable acid or as NH_4^+, serves to rid the body of metabolically produced acids. The factors influencing the distribution of H^+ between filtered HCO_3^- and non-HCO_3^- buffers determine to what extent the potentially available H^+ will be utilized to conserve filtered HCO_3^- and to what extent, to generate new HCO_3^-. If, for example, metabolic acid is administered, $NaHCO_3$ will be decomposed, resulting in a fall in the HCO_3^- in glomerular filtrate. Consequently, less H^+ will be utilized in the reabsorption of filtered HCO_3^- and more available for excretion as titratable acid and urinary ammonia. If there is a plentiful supply of filtered buffers (e.g., phosphate, creatinine), titratable acid will be produced. If, on the other hand, there is a paucity of filtered buffers, the removal of H^+ from the renal cell will be restricted, and an

intracellular acidosis will supervene. The intracellular acidosis will stimulate the formation of new buffer in the form of ammonia, which is then added to the tubular fluid, thus permitting H^+ secretion to return toward its normal rate.

The particular importance of the distal nephron mechanism stems from the fact that the amount of H^+ that can be pumped into the urine is greater than the excretion of base. The base conservation thus facilitated is abetted by capitalizing on the NH_3 production mechanism. However, because ammonia is synthesized in the proximal convolution, and pH under certain conditions decreases along this segment, some overlap of function between proximal and distal convolutions can be anticipated.

SUMMARY OF THE REGULATORY MECHANISM. Aside from the buffering effects of body cells when hydrogen ion is

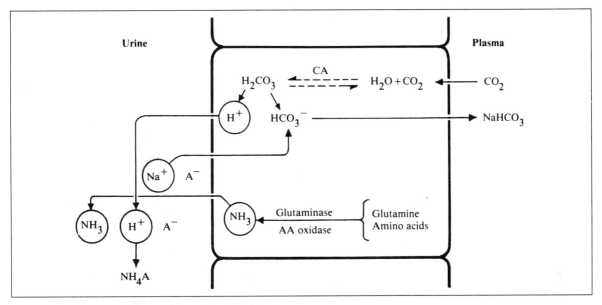

Fig. 24-14. Mechanism of production of ammonia by the kidney and its role in base conservation.

Fig. 24-15. Sites of secretion of ammonia in untreated rats as compared to rats made acidotic by NH_4Cl ingestion. (From S. Glabman, R. M. Klose, and G. Giebisch. Micropuncture study of ammonia excretion in the rat. *Am. J. Physiol.* 205:131, 1963.)

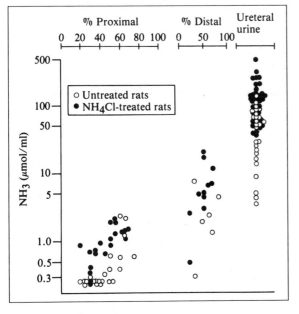

ingested, there will be buffering by the major plasma buffer system, $H^+ + HCO_3^- \leftrightarrow H_2CO_3 \leftrightarrow H_2O + CO_2$ (Fig. 24-16). If one adds hydrogen ion to this system, this equation shifts to the right, resulting in a lowering of bicarbonate and increased production of CO_2. The CO_2 is then carried to the lungs and expired. This action serves to minimize the change in the ratio HCO_3^-/PCO_2, or pH, and is referred to as *compensation*. Despite respiratory compensation, the organism still has excess hydrogen ion; and because the equation has moved to the right, there has been a lowering of serum bicarbonate. The kidney normally, then, has the dual roles of eliminating the excess hydrogen ion and regenerating bicarbonate.

Although the regeneration of bicarbonate is a continuous operation in the nephron, for convenience it can be summarized as a two-step phenomenon.

1. The first step may be regarded as almost complete reabsorption of the filtered bicarbonate taking place in the proximal tubule and *preventing* further reduction in serum bicarbonate. The reabsorption of bicarbonate is not a direct process. The filtered HCO_3^- is converted to water and CO_2 by the H^+ that has been secreted into the lumen. The CO_2 diffuses back into the cell, shifting the reaction toward the formation of H_2CO_3 and HCO_3^-. This intracellularly generated HCO_3^- is delivered to the extracellular fluid with sodium that was reabsorbed in exchange for the secreted hydrogen ion. The result of this phase is the prevention of

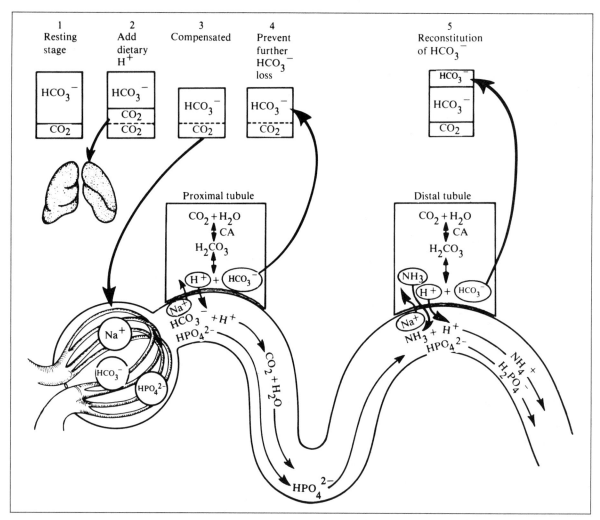

Fig. 24-16. Summary of the role of the kidney in regulation of acid-base balance. This is exemplified by the renal handling of the daily dietary acid (hydrogen ion) load of about 50 mEq. (From S. Papper. *Clinical Nephrology.* Boston: Little, Brown, 1971. P. 68.)

further lowering of serum bicarbonate and the loss of some hydrogen ion.

2. Although most of the filtered HCO_3^- is converted to water and CO_2, H^+ secretion continues, primarily in the distal nephron. Here the H^+ is buffered by filtered buffers such as HPO_4^{2-} (titratable acid) to form $H_2PO_4^-$, or by ammonia (NH_3) and excreted as ammonium ion (NH_4^+). Ammonia is secreted by the distal tubular cells in response to acidosis. These two essential buffer mechanisms allow for elimination of hydrogen ion without lowering the urinary pH to levels that may damage the tissues. The hydrogen that is secreted is derived from cell H_2CO_3, leaving bicarbonate behind in equimolar amounts. This bicarbonate, along with the sodium reabsorbed in exchange for secreted hydrogen, moves into the extracellular fluid and reconstitutes the serum bicarbonate level. Again, it is as if more bicarbonate is reabsorbed to regenerate the depressed bicarbonate. In renal tubular cells the hydration of CO_2 and consequent dissociation into H^+ and HCO_3^- is catalyzed by CA.

Table 24-3. Examples of acid-base disturbances

Arterial pH	Primary disturbance	Primary acid-base defect	Example of blood acid-base values	Compensatory mechanisms
Normal	None	None	pH 7.35 – 7.45 PCO_2 35 – 48 mm Hg $[HCO_3^-]$ 23 – 28 mEq/L	
Acidemia (low pH)	Respiratory acidosis (chronic obstructive lung disease)	Retention of CO_2	pH 7.30 PCO_2 80 mm Hg $[HCO_3^-]$ 39 mEq/L	Increased respiration; renal retention of bicarbonate, increased excretion of titratable acid and ammonium ions
	Metabolic acidosis (uncontrolled diabetes mellitus)	Excess keto acid production	pH 7.22 PCO_2 30 mm Hg $[HCO_3^-]$ 12 mEq/L	Increased respiration; renal retention of bicarbonate, increased excretion of titratable acid and ammonium ions
Alkalemia (high pH)	Respiratory alkalosis (voluntary hyperventilation)	Excess CO_2 loss	pH 7.62 PCO_2 19 mm Hg $[HCO_3^-]$ 19.5 mEq/L	Decreased respiration; increased renal excretion of bicarbonate
	Metabolic alkalosis (vomiting)	Loss of acid gastric juice	pH 7.57 PCO_2 45 mm Hg $[HCO_3^-]$ 42 mEq/L	Decreased respiration; increased renal excretion of bicarbonate

Source: From R. G. Pflanzer. Acid-Base Regulation. In E. E. Selkurt (ed.), *Basic Physiology for the Health Sciences*. Boston: Little, Brown, 1981. P. 463.

At the end of this process, the daily dietary acid load is excreted, buffered mostly with ammonia and phosphate, and the serum bicarbonate is reconstituted.

CLINICAL ABNORMALITIES OF ACID-BASE HOMEOSTASIS. Since the hydrogen ion concentration of the blood ultimately affects the hydrogen ion concentration of all body fluids, and because blood is readily accessible for chemical analysis, disorders of acid-base balance will be described in terms of deviations from the normal pH value of blood.

Acidemia is defined as an acidic condition of the blood signified by a pH value less than 7.35. The physiological processes that are causing acidemia define the term *acidosis* (literally, "a condition of becoming acidic"). *Alkalemia* is defined as an alkaline condition of the blood signified by a pH value greater than 7.45. The physiological processes that are causing the alkalemia define the term *alkalosis* (literally, "a condition of becoming alkalotic").

Examples of acid-base disturbances are given in Table 24-3.

Clinical Evaluation. Clinical evaluation of the acid-base status of a patient involves the determination of arterial blood pH, PCO_2, and $[HCO_3^-]$. All three variables are mutually re-

lated by means of the Henderson-Hasselbalch equation. The application has been discussed earlier.

Clinically, six possible combinations of alkalosis and acidosis may be observed in patients, depending on etiology and compensatory processes:

1. Respiratory acidosis + metabolic acidosis, arterial pH below 7.4.
2. Respiratory alkalosis + metabolic alkalosis, arterial pH above 7.4.
3. Primary respiratory acidosis + secondary metabolic alkalosis, arterial pH below 7.4.
4. Primary respiratory alkalosis + secondary metabolic acidosis, arterial pH above 7.4.
5. Primary metabolic acidosis + secondary respiratory alkalosis, arterial pH below 7.4.
6. Primary metabolic alkalosis + secondary respiratory acidosis, arterial pH above 7.4.

Physiological compensation for major disturbances of acid-base equilibrium is rarely complete; therefore, the observed values of arterial PCO_2 and $[HCO_3^-]$ indicate which component (respiratory or nonrespiratory) is causing acidosis and/or alkalosis, and the value of arterial pH indicates which process is primary and which is secondary or compensatory.

Effects of Acidosis and Alkalosis. When the pH of body fluids falls below normal values, all of the body's systems become affected to varying degrees. Most noticeable is the depressant effect of acidosis on the activity of the central nervous system. When arterial pH falls to near or below 7.0, neuromuscular coordination becomes erratic (e.g., the person may stagger as though drunk); the condition is compounded by disorientation, and eventually coma ensues, followed by death. A classic example is that of the patient with uncontrolled, severe diabetes mellitus who, because of excessive metabolism of lipids with resultant elevation of blood keto acids (metabolic acidosis), experiences a fall in arterial pH (diabetic acidemia), loss of neuromuscular coordination and normal orientation (often mistaken for drunkenness, since it is usually combined with excretion of acetone by the lungs), and finally coma.

The principal effects of severe alkalosis are hyperexcitability of both the central and peripheral nervous systems. An increase in the excitability of peripheral nerves is manifested by tetany of skeletal muscle. Central nervous system involvement is reflected in extreme nervousness, overreaction to normal stimuli, and, in some instances (e.g., epilepsy), convulsions.

BIBLIOGRAPHY

Berliner, R. W. Ion Exchange Mechanisms in the Nephron. In A. P. Fishman (ed.), *Symposium on Salt and Water Metabolism.* New York: New York Heart Association, 1959. P. 892.

Davenport, H. W. *The ABC of Acid-Base Chemistry* (6th ed.). Chicago: University of Chicago Press, 1974.

Dick, D. A. T. *Cell Water.* Washington, D.C.: Butterworth, 1966.

Gamble, J. L. *Chemical Anatomy, Physiology and Pathology of Extracellular Fluid.* Cambridge, Mass.: Harvard University Press, 1954.

Gamble, J. L., Jr. *Acid-Base Balance: A Direct Approach.* Baltimore: Johns Hopkins University Press, 1982.

Gottschalk, C. W., Lassiter, W. E., and Mylle, M. Localization of urine acidification in the mammalian kidney. *Am. J. Physiol.* 198:581–585, 1960.

Hayes, C. P., Jr., Mayson, J. S., Owen, E. E., and Robinson, R. R. A micropuncture evaluation of renal ammonia excretion in the rat. *Am. J. Physiol.* 207:77–83, 1964.

Masoro, E. J., and Siegel, P. D. *Acid-Base Regulation: Its Physiology, Pathophysiology and the Interpretation of Blood Gas Analysis.* New York: Churchill Livingstone, 1977.

Maxwell, M. H., and Kleeman, C. R. *Clinical Disorders of Fluid and Electrolyte Metabolism.* New York: McGraw-Hill, 1962.

Muntwyler, E. *Water and Electrolyte Metabolism and Acid-Base Balance.* St. Louis: Mosby, 1968.

Pitts, R. F. *Physiology of the Kidney and Body Fluids* (3rd ed.). Chicago: Year Book, 1974.

Pitts, R. F., and Alexander, R. S. The nature of the renal tubular mechanism for acidifying the urine. *Am. J. Physiol.* 144:239–254, 1945.

Smith, H. W. *Principles of Renal Physiology.* New York: Oxford University Press, 1956.

Valtin, H. *Renal Dysfunction: Mechanisms Involved in Fluid and Solute Imbalance.* Boston: Little, Brown, 1979.

Van Slyke, D. D., Phillips, R. A., Hamilton, P. B., Archibald, R. M., Futcher, P. H., and Hiller, A. Glutamine as source material of urinary ammonia. *J. Biol. Chem.* 150:481–482, 1943.

Welt, L. G. *Clinical Disorders of Hydration and Acid-Base Equilibrium* (2nd ed.). Boston: Little, Brown, 1959.

GASTROINTESTINAL SYSTEM

25

CIRCULATION

Ewald E. Selkurt

The act of chewing and swallowing is carried out by the voluntary and involuntary muscles of the upper gastrointestinal (GI) tract. The blood flow through these tissues is closely related to the metabolic requirements of the muscle, as discussed in Chapter 29. When the food reaches the stomach, the digestive process begins; this ultimately involves a number of organs—stomach, small and large intestine, liver, and pancreas—which together constitute what is known as the splanchnic region. While having no digestive function, the spleen is also included in this domain.

GENERAL CONSIDERATIONS. The preceding functions require a highly organized vascular bed. After passing through the GI tract circulation the vessels funnel into the liver, which is coupled in series with the GI tract via the portal vein (Fig. 25-1). The combined vascular beds of the liver, the stomach, and the intestines are called the *splanchnic circulation*. In a 70-kg man the splanchnic bed totals 3.5 to 4.0 kg. Of this, the liver weighs 1.5 kg and the GI tract almost 2 kg. The spleen weighs 200 gm and the pancreas, 60 to 80 gm. The GI tract is about half smooth muscle and half glandular and mucosal tissue.

At rest, the splanchnic vascular bed receives 20 to 25 percent of the cardiac output, 1250 to 1500 ml per minute in a 70-kg man, or 700 ml/min/m² of surface area.

HEPATIC CIRCULATION

The liver is the main "chemical factory" of the body, and its oxygen usage is nearly 20 percent of that of the whole human body at rest. Its oxygen supply, which is largely from the portal vein, is reinforced by the hepatic artery,

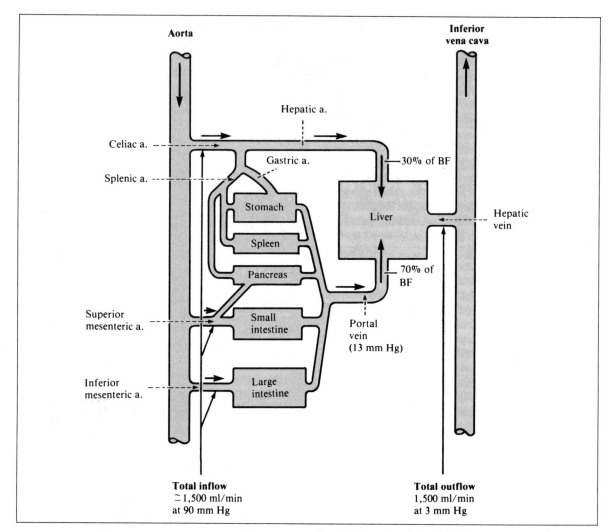

Fig. 25-1. Gastrointestinal and hepatic circulation. Note that the liver receives 30 percent of its blood supply from the hepatic artery and 70 percent from the portal vein.

which provides 30 percent of the blood flow (and 40–50 percent of the oxygen used). Portal venous and hepatic arterial blood mix in the hepatic sinusoids, drain into the hepatic veins, then enter the inferior vena cava. The vascular bed of the liver is thus coupled both in series (portal vein system) and in parallel (hepatic artery) to the GI tract.

ANATOMICAL ASPECTS. The reader should review the anatomical features of the liver before continuing with this section. Certain details of the vasculature are presented here that have a direct bearing on the physiological function of the liver.

Hepatic Artery. The role of the hepatic artery is to supply oxygenated blood to the parenchymal cells. This is not an end artery but has many collateral connections. Anastomotic connections with the portal vein occur at several levels, in the interlobular spaces or to terminal branches of the portal vein just before these branches enter the sinusoids. Terminally, intralobular hepatic arterioles and arterial capillaries empty into the sinusoids lined by the parenchymal cells. These branches supply the interior sinusoids in the central part of the lobule as well as peripheral sinusoids. Other branches supply the capsule, interlobular septa, and tissue of the bile ducts.

Portal Vein. The blood supplied through the portal vein comes from the GI tract, pancreas, and spleen. Although functioning primarily to carry digested foodstuffs from the intestine to the liver, the portal vein also supplies oxygen to the liver. When the portal flow is shunted away from the liver to the inferior vena cava (*Eck's fistula*), deranged hepatic function develops because of the impaired circulation. Central portions of the lobule atrophy and become filled with fat.

Sinusoids. Conducting portal vein branches give rise to small distributing veins. These, as well as smaller axial portal vein branches, give off inlet venules that enter the lobules through holes in the limiting lamina of the hepatic parenchyma. Beyond this, the inlet venules branch into sinusoidal arborizations. The terminal ends of the portal vein branches split directly into sinusoids, the "exchange" vessels of the liver. The sinusoids are slightly ampullate in shape just before they join the draining veins, giving the impression of a sphincter. Also, they appear narrowed at the point of inlet from the venules.

The presumption that sphincters are located at the entry and exit of the sinusoids, permitting regulation of flow and storage of blood, has given rise to the concept of *intermittence of flow*, which has been observed in the frog and in the rat. Spontaneous local variations in blood flow of the human liver have been observed with a Hensel calorimetry needle. This temperature-recording device, inserted into the liver substance, shows temperature changes that reflect local alterations in blood flow.

Veins. The sinusoids empty into the central vein of the lobule, which connects with the sublobular veins. These in turn converge into the hepatic veins, which discharge into the inferior vena cava. Strong muscular sphincters have been found in the hepatic vein of dogs, but not in humans. They may aid in regulation of hepatic venous outflow.

SPLANCHNIC CIRCULATION: METHODS FOR MEASUREMENT OF THE BLOOD FLOW OF THE LIVER AND SPLANCHNIC BED

Measurement methods are direct (venous outflow, electromagnetic flowmeter, ultrasonic flowmeter, rotameter, bristle flowmeter) or indirect, utilizing the Fick principle (see Chap. 11). An application of the latter method suitable for human subjects utilizes the removal of dyes from the plasma, cleared only by the specific organ being studied—in this case, the liver. Referring to Figure 25-1, we see that dye extraction by the liver will measure the blood flow per minute of the entire splanchnic bed. Hence, splanchnic blood flow (SBF) and hepatic blood flow (HBF) are essentially equal.

Dyes used for this purpose are sodium sulfobromophthalein (Bromosulphalein) (BSP) and indocyanine green (Cardio-green). A catheter directed into the hepatic vein is essential for measuring the arteriovenous difference. (Systemic venous blood concentration suffices for the arterial concentration of the dye, since these concentrations are essentially in equilibrium.) The rate of dye removal into the bile is determined indirectly (Bradley, 1963) by provisional infusion of dye at varying rates until a constancy of plasma concentration is attained (i.e., dye is being infused at a rate equal to removal by the liver). (Slight changes in the slope of the dye concentration can be corrected to give the true rate of removal, from a knowledge of the blood volume.) The method assumes removal of the dye only by the liver; extrahepatic removal would lead to obvious overcalculation of splanchnic blood flow. The calculation of blood flow is then made as follows:

$$\text{SBF (or HBF)} = \frac{I}{C_A - C_{HV}}$$

$$= \frac{R}{C_A - C_{HV}} \times \left(\frac{1}{1 - \text{hemat}}\right)$$

C_A represents the systemic dye concentration (mg/ml), C_{HV} represents the hepatic vein concentration, and $C_A - C_{HV}$ represents extraction by the liver. When arterial concentration (C_A) becomes constant, the dye infusion rate, I (mg/min), equals removal rate (R). Since the method analyzes for plasma clearance, total blood flow requires adding the red cell volume. (For determination of the hematocrit [hemat], see Chap. 10.)

In actuality, the method measures "effective" blood flow, in the sense used for the indirect measurement of renal blood flow (see Chap. 22), since it measures blood flow to the parenchymal tissue, which does the secreting of the dye.

Other indirect methods for whole organ flow and the organs' constituent tissues have employed the inert gas "washout" principle (see Chap. 22). The labeled gases commonly used are ^{133}Xe and ^{85}Kr, which have proved to be particularly useful for determining the regional flows in such organs as the kidney and GI tract. ^{42}K uptake and aminopyrine clearance have also been useful in the assessment of gastric blood flow.

Table 25-1 gives a quantitative summary of basic circula-

Table 25-1. Weight, share of cardiac output, and blood flows of the gastrointestinal organs in an "average" 15-kg dog

Organ	Weight (gm)	Fraction of cardiac output (%)	Blood flow (ml/min)	Perfusion rate (ml/min/100 gm)
Stomach	100	1.9	50	50
Intestine	270	6.5	180	70
Colon	50	1.6	40	80
Pancreas	30	0.7	18	60
Gallbladder	2	0.04	1	40

Source: After J. P. Delaney and J. Custer. Gastrointestinal blood flow in the dog. *Circ. Res.* 17:400, 1965. By permission of the American Heart Association, Inc.

tory variables for three important splanchnic vascular beds. The data were derived from experiments done in dogs and cats, but values in humans are of comparable magnitude (Lundgren, 1978).

SPLANCHNIC BLOOD VOLUME. Estimates employing tagged red cells (^{32}P) and labeled albumin (^{131}I) techniques have been variable, but splanchnic blood volume appears to be approximately 25 percent of the total blood volume (see Chap. 23). About one-half is in the GI tract, and the remainder is equally divided between the liver and spleen.

This relatively high volume, plus rich sympathetic nerve innervation (see Chap. 7), provides an important reservoir depot for "autotransfusion" into the central circulation during hemorrhage. In particular, the spleen serves as a major blood reservoir, perhaps less so in humans than other species.

SPLANCHNIC HEMODYNAMIC CHARACTERISTICS. The small intestinal tract is characterized particularly by the large villous surface specialized for absorption and fluid exchange. The total blood flow has been reported in a range of 20 to 80 ml/min/100 gm, depending on the species of experimental animal. It can increase to 250 to 300 ml/min/100 gm with maximal dilation (Folkow and Neill, 1971). The vascular bed of the small intestine demonstrates a significant degree of autoregulation (Johnson, 1964). Thus, in a range of arterial pressure from 100 to 35 or 40 mm Hg, the blood flow and capillary pressure change but little.

In meeting the diverse needs of the gastrointestinal tract, a basic constraint is placed on the system, since it is necessary to maintain a reasonably constant capillary pressure.

The same is true to some degree of all vascular beds, but the filtration coefficient of intestinal and liver capillaries is high; thus, small changes in capillary pressure can produce rapid filtration and edema formation. Effective regulation of the peripheral circulation requires mechanisms that provide adequate blood flow to meet the needs while maintaining a nearly constant capillary pressure. Clearly, autoregulation of blood flow in the intestine helps in maintaining the constancy of the capillary pressure.

Local regulation of blood flow (autoregulation) is manifested in the liver in somewhat the same manner as has been described for the small intestine. However, because of its dual blood supply, the liver possesses the unique feature of reciprocity between portal and hepatic arterial inflow. If portal inflow is reduced, hepatic artery inflow increases, so that a more nearly constant blood supply to the liver tissues is maintained. Apparently, a reduction of portal inflow reduces pressure in the terminal portion of the hepatic arteriolar bed, causing the arterioles to dilate. A myogenic mechanism is probably responsible (Hanson and Johnson, 1966). While this phenomenon is usually termed reciprocity of flow, it is not truly reciprocal, since a reduction of hepatic arterial inflow does not alter portal flow, which is determined by the resistance vessels of the organs of the digestive tract.

The wide range of flow possible in the small intestine is illustrated in Figure 25-2. In the mucosa, there is an especially rich vascularization of the secretory crypts. This zone (secretory crypts) can increase its flow tremendously to values rivaling those in the flow in actively secreting salivary glands (700 ml/min/100 gm). In the cat the absorptive villous portion of the mucosa can attain flows up to 150 to 200 ml/min/100 gm (Folkow and Neil, 1971).

BLOOD FLOW DURING DIGESTION. In trained, unanesthetized dogs, in which mesenteric blood flow was being measured with chronically implanted ultrasonic or electromagnetic flowmeters, presentation and ingestion of food caused an average increase of 62 percent in cardiac output and 33 percent in arterial pressure ("anticipation-ingestion phase"). Mesenteric flow began to increase 5 minutes after eating and reached a peak of 132 percent above the control level in 30 to 90 minutes (average 55) during the "digestion" phase. It gradually returned to the control value in 2 to 6 hours postprandially. In the resting dog a slight compensatory vasoconstriction occurred in the inactive muscle bed (Vatner et al., 1970). Coronary blood flow increased transiently early in the digestion phase (about 5 min following eating).

Cholinergic nerves, local mucosal nerves, active reab-

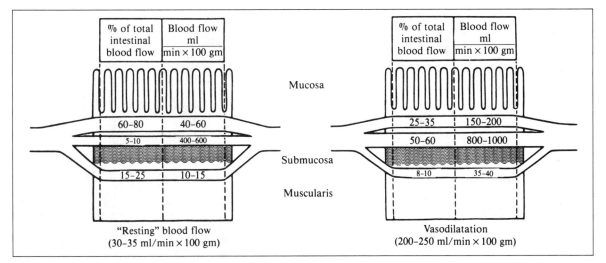

Fig. 25-2. Blood flow distribution in the small intestine of the cat at "rest" and intense vasodilatation produced by isopropyl norepinephrine. The three vascular pathways depict in essence the villous mucosal circulation, the submucosal circulation, especially that of the secretory crypts, and the muscularis circulation. Note the huge flows in the submucosa, presumably reflecting the rich vascularization of the secretory crypts. Regional flows were estimated by the washout of radioactive krypton (^{85}Kr). (From O. Lundgren. Studies in blood flow distribution and countercurrent exchange in the small intestine. *Acta Physiol. Scand.* [Suppl.] 303:3–42, 1967.)

sorptive processes, and local or circulating humoral substances (gastrin, secretin, cholecystokinin) have been suggested as possible mediators of postprandial intestinal hyperemia. Of interest is the finding (Chou et al., 1976), that the hyperemia is confined to the area of the intestine whose mucosa was exposed to food with relatively low fat and protein concentration. Higher concentrations of fat and protein, however, cause the effect to extend to other sections of the small intestine not exposed to food.

COUNTERCURRENT MECHANISM OF THE VILLUS. The vessels of the villus are arranged in the form of complex hairpin loops (Fig. 25-4). With this pattern as a foundation, Lundgren (1967, 1978) presented evidence for a countercurrent exchange mechanism in the villi.

The intestinal *countercurrent exchanger* mechanism can be approached either from the luminal side of the small intestine (Fig. 25-4A) or from the tissue side (Fig. 25-4B). Sodium, for example, is concentrated at the tip of the villus and falls progressively toward its base (Haljamäe et al., 1973). Estimated values range from about 700 to 900 mEq

per liter at the tip. One view is that actively absorbed sodium is short-circuited from the epithelial capillary network to the central artery of the villus, resulting in recirculation and concentration of the sodium toward the villus tip. This process would call for "fenestration" or interendothelial pores of the central arterial vessel. The effect would be multiplied along the villus length, resulting in increasing sodium concentration in the capillaries of the villus and a high concentration at the tip, as in the renal papillae. Also, an increase in sodium concentration and osmolarity in the capillaries of the villus could cause transfer of water from the central artery to the capillary network, eventuating finally in a similar sodium multiplication.

Such a "trapping" phenomenon could lead to a "damping" of entrance into the bloodstream (Fig. 25-4, *impedance of absorption*) of other rapidly absorbed materials, particularly small lipid-soluble particles. When such substances are carried with the descending bloodstream of the hairpin loop, they pass very close to the ascending bloodstream and will pass across to it, down their concentration gradient, by cross-diffusion. Consequently, high concentrations will be reached in the upper and outer parts of the villi in the steady-state situation, and they will actively leave the villi by venous drainage only slowly. This mechanism would hinder development of too excessive peak concentrations in the systemic blood.

Conversely, rapidly diffusing molecules in the arterial bloodstream, such as oxygen, will tend to leave the arterial ascending limbs of the villi and by cross-diffusion become "short-circuited" over to the venous descending limbs (Fig.

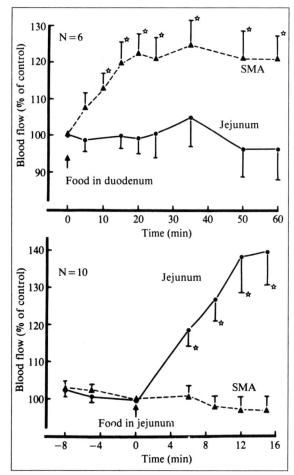

Fig. 25-3. Blood flows (mean ± SE) in the superior mesenteric artery (SMA) and an isolated in situ jejunal segment following infusion of digested food into the duodenum of dogs at 4 ml/min or following placement of 10 ml of digested food into a jejunal segment. Values are percent of control. Mean control flow in the SMA was 14.5 ± 0.4 ml/min/kg body weight; in the jejunal segment it was 0.60 ± 0.05 ml/min/gm tissue weight. Asterisks indicate values significantly different from control value ($P < 0.05$). (From C. C. Chou, C. P. Hsieh, Y. M. Yu, P. Kvietys, R. Pitman, and J. M. Dabney. Localization of mesenteric hyperemia during digestion in dogs. *Am. J. Physiol.* 230:587, 1976.)

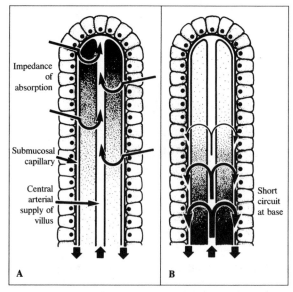

Fig. 25-4. The functional implications of the intestinal countercurrent exchanger schematically illustrated. The intervascular distance is greatly exaggerated for the sake of clarity. The central lacteal of the villus is not shown. Functionally, events in A and B proceed simultaneously. The shading of the interstitium in A is intended to depict the countercurrent exchange for substances such as sodium. The shading in B is to show the countercurrent exchange of oxygen at the base of the villus. (It is understood that concentration in the capillary is approximately equal to that in the interstitium at any level—not shown.) (From O. Lundgren. Studies in blood flow distribution and countercurrent exchange in the small intestine. *Acta Physiol. Scand.* [Suppl.] 303:16, 30, 1967.)

25-4). Thus, the oxygen tension at the tip of the villi would be lower than that at their base. The consequence of the countercurrent system may be one reason for the rapid turnover of epithelial cells at the tip of the villi, where oxygen delivery is probably poor. It is known that there is a steady and considerable production of new epithelial cells from basal parts of the villi, which are gradually displaced toward the tip. There may be other important functional consequences of the hairpin arrangement of the blood flow in the villi that are still unknown. In any case, when the villous bloodstream becomes rapid, as in very intense dilation, less time is available for cross-diffusion, and countercurrent "trapping" may then be less efficient (Folkow and Neil, 1971).

REGULATION OF SPLANCHNIC BLOOD FLOW. Autonomic Nervous System: Sympathetic Division. Adrenergic fibers of the sympathetic nervous system (SNS) to the intestine, as studied by the fluorescence technique, show a greater density on the arterial than the venous side. On sympathetic nerve stimulation, blood flow to the small intestine is markedly reduced. However, after 2 to 4 minutes, with continued stimulation, blood flow increases and reaches a "steady state" not far below the initial control value. This is called "autoregulatory escape" from vasoconstrictor fiber influence. It is not due to fatigue of the neurogenic elements involved in constriction, because the veins remain constricted during this phase. This constriction of capacitance vessels is a compensatory aid in blood loss, since they contain about 40 percent of regional blood content of the system which is expelled during vasoconstriction. In humans, about 200 ml of blood can be mobilized by venous vasoconstriction.

The sympathetic nerves also control the number of perfused capillaries, which is reduced during the steady-state phase of vasoconstriction. This implies a "precapillary-sphincter" type of control, although such sphincters cannot easily be demonstrated anatomically. However, implicit in this type of control is that there is control of flow redistribution. Thus, some capillaries are overperfused, but flow ceases in others; hence, a "physiological shunting" to intestinal tissues more vitally in need of oxygen can result. The villous blood flow is therefore well maintained at the prestimulatory control level despite overall reduction in intestinal blood flow. The crypt part of the mucosa participates in the generalized vasoconstriction, but the inference is that, because of maintained villous circulation, the overall rate of absorption is not critically affected by the sympathetic system (Lundgren, 1978).

The nervous control of intramural blood flow in the stomach and colon has been less well studied, although both beds manifest "autoregulatory escape." A more generalized vasoconstriction has been noted. In the stomach this may lead to impairment of gastric secretion.

Sympathetic stimulation supplying the liver results in vasoconstriction. Dilation of the liver vessels cannot be detected by vagal stimulation. It is therefore doubtful that specific vasodilator fibers run to the liver.

Autonomic Nervous System: Parasympathetic Division. Parasympathetic fibers (vagus, pelvic nerve) innervate all organs in the abdomen and are presumed to be vasodilator in nature. Such cholinergic fibers would be expected to manifest a reciprocal control with the SNS. This has been shown not to be the case in the small intestine, when there is no apparent parasympathetic vascular control, e.g., during vagal or pelvic nerve stimulation. However, proper stimulation

(voltage and frequency) of the parasympathetic supply to the stomach can, in fact, cause an increase in gastric blood flow. This increase occurs mainly in the mucosa. Atropine reduces or abolishes this change. The augmented flow may involve increased kinin release at the stimulation site.

The upper colon has some vagal innervation and pelvic nerve innervation of the lower half. Pelvic nerve stimulation yields here a response pattern characterized by hyperemia, increased net secretion, and increased motility.

Hormonal Control. The hormone *gastrin's* main effect on the stomach is augmented secretion and increase in blood flow. Other so-called vasodilator metabolites have been considered as possible agents to account for these effects. Histamine is a likely candidate and the prostaglandins and kinins need to be investigated.

In the small intestine, secretin and cholecystokinin induce hyperemia. They may act directly on vascular smooth muscle or via production of other dilator metabolites previously mentioned.

In the liver, epinephrine in low doses causes increased blood flow. However, since this agent stimulates hepatic metabolism, dilator products of metabolism may be involved. Norepinephrine, whose metabolic influences on the liver are very weak, causes pure vasoconstriction. Histamine is a potent dilator of the liver circulation. Interestingly, insulin injection results in hepatic hyperemia, probably on a metabolic basis; possibly, increased epinephrine release may be the consequence of insulin-induced hypoglycemia.

SALIVARY GLANDS. Reflecting a capability for high rates of formation of saliva is the range of possible blood flows in the salivary glands, e.g., from a basal rate of 10 to 60 ml/min/100 gm to 700 ml/min/100 gm at a pressure head of 100 mm Hg. The parasympathetic nerves (chorda tympani branch of cranial nerve VII) stimulate the glands and increase blood flow (Fig. 25-5), with an accompanying increase in capillary permeability. A hormonal factor may also enter into the mechanism, for it has been shown that a proteolytic enzyme, *kallikrein,* splits off a polypeptide from a globulin, *kininogen,* to form *lysin-bradykinin,* from which the final active form, *bradykinin,* is derived (Fig. 25-5). The SNS influence can exert a powerful vasoconstrictor action, with a resultant sparse but mucus-rich secretion.

SPLEEN. The spleen has an important blood reservoir function whose content may be released into the systemic circu-

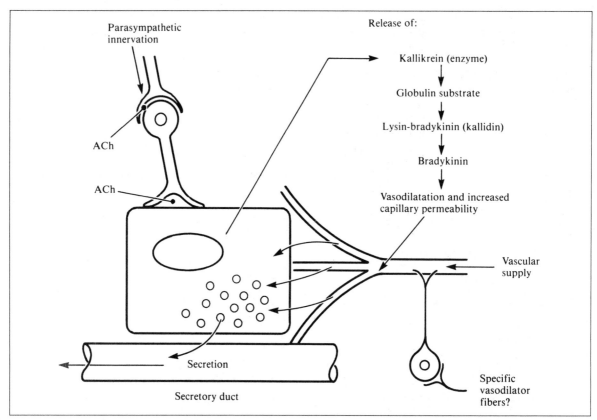

Parasympathetic
innervation

Release of:

Kallikrein (enzyme)

Globulin substrate

Lysin-bradykinin (kallidin)

Bradykinin

Vasodilatation and increased
capillary permeability

ACh

ACh

Vascular
supply

Secretion

Secretory duct

Specific
vasodilator
fibers?

Fig. 25-5. The vasodilatation is due to the presence of kallikrein in the saliva, which acts as an enzyme to split off from a globulin the powerful vasodilator lysin-bradykinin (kallidin). This, in turn, can be rapidly transformed into bradykinin. ACh = acetylcholine. (From B. Folkow and E. Neill. *Circulation*. London: Oxford University Press, 1971.)

lation by various stimuli, e.g., exercise, hemorrhage, severe hypoxia, some anesthetics (such as chloroform), and other drugs and agents. Because of its high hematocrit (80 percent higher than that of arterial blood), it represents a small emergency store of oxygen, as well as providing replenishment of blood lost by hemorrhage.

The blood is stored in the sinusoids (venous sinuses) of the red pulp of the spleen. Sphincters at the sinus inlets and outlets provide a control mechanism for storage or discharge into the splenic vein and ultimately into the portal vein.

The control is by way of the SNS outflow in the splanchnic nerve branches of the SNS and is activated reflexly. Norepinephrine is the final mediator for arteriolar vasoconstriction, plus contraction of the splenic capsule. The efferent sphincters become relaxed, aiding in the release of stored cells.

LYMPHATIC CIRCULATION OF THE GASTROINTESTINAL SYSTEM

ANATOMICAL ASPECTS. The stomach has an abundant supply of lymphatic vessels, which originate as blind projections among the tubular glands and muscularis mucosae to form a plexus here. A second plexus is formed in the submucosa. These plexuses return lymph to the *cisterna chyli* and *thoracic duct*.

The lacteals of the villi of the small intestine also represent a site of blind beginnings of a plexus that lies on the inner surface of the muscularis mucosa. Tributaries arise in thin-walled lymphatic capillaries surrounding the solitary and aggregated follicles of Peyer's patches.

The lymphatic vessels of the colon and rectum form plexuses in the mucosa, leaving eventually through the muscle layers.

In the liver the ultimate functional unit, the lobule, is not well supplied with lymphatic capillaries. However, lymphatics are found in the periphery of the lobule that carry the highly proteinized liver lymph to collecting trunks, which join the thoracic and right lymphatic ducts.

Lymphatics are seen only in the capsule and thickest trabeculae of the spleen.

FUNCTIONAL ASPECTS OF LYMPH FLOW. The thoracic duct is the important recipient of the splanchnic lymphatic vessel outflow. In humans the thoracic duct flow is in the range of 1.0 to 1.6 ml/kg body weight/hr (Yoffey and Courtice, 1970). It can increase to 5.8 ml with eating and drinking. In 24 hours a turnover of fluid roughly equivalent to the plasma volume is noted, with a relatively high turnover of total circulating proteins.

Lymphatic flow in the small intestine, as anticipated, varies with the rate of digestive processes. The lacteals are particularly important because of the products of fat digestion that enter them (triglycerides and some cholesterol and phospholipids).

In the large intestine, lymphatics are mainly involved in extravascular circulation of proteins and migration of lymphocytes. Clinically, their regional drainage is important in the spread of tumors.

The liver's contribution of lymph to the thoracic duct flow is variable, depending on the species and dietary regimen. In experimental animals, a high degree of permeability of the hepatic capillaries to plasma protein makes the liver a significant contributor to overall lymph protein turnover.

MOVEMENTS OF THE DIGESTIVE TRACT

Leon K. Knoebel

The GI system (Fig. 25-6) is a collection of heterogeneous organs specialized to handle ingested food. A major component of the system is the digestive tract, which is essentially a tube about 5 meters in length and of variable cross-sectional area, running from the mouth to the anus. The digestive tract includes the mouth, pharynx, esophagus, stomach, and small and large intestines. Other components of the GI system, all of which lie outside the digestive tract, are the salivary glands, pancreas, and biliary system (liver, gallbladder, and bile ducts).

The major function of the GI system is to provide nutrients for body cells. These nutrients are derived from the food we eat. Food enters the body from the external environment through the oral cavity and is propelled by muscular contractions through the different regions of the digestive tract. A number of digestive juices are secreted at various points along the route, and the enzymes contained in the secretions catalyze the conversion of complex molecules to simple molecules. These products of digestion and other substances are absorbed into the blood and lymph, which carry them to the cells of the body. Unab-

Fig. 25-6. Organs of the gastrointestinal system.

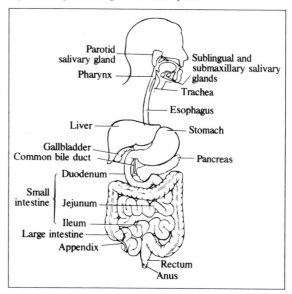

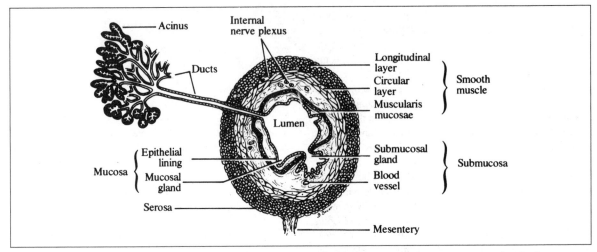

Fig. 25-7. Structure of the gastrointestinal system showing a cross section of the digestive tract and an exocrine gland external to the tract.

sorbed food residues and a number of waste products are moved to the end of the tract and eliminated from the body.

It is thus apparent that four major processes occur in the GI system: muscular movements, secretion, digestion, and absorption. In this chapter the motor functions of the digestive tract are discussed, leaving the secretory, digestive, and absorptive functions of the GI system for discussion in Chapter 26. However, before one proceeds to a more detailed review of the roles played by the various GI processes involved in GI function, some general features of the system as a whole will be considered.

GENERAL STRUCTURE OF THE GASTROINTESTINAL SYSTEM

With some local variations, the general structural characteristics of the regions of the digestive tract that extend from the esophagus to the anus are similar. These structural features are represented diagrammatically in Figure 25-7. The structures from outside inward are: the serosa, the outer membrane of the tract; the longitudinal muscle layer; the circular muscle layer; the submucosa; and the mucosa. An additional small muscle layer, the muscularis mucosa, lies between the submucosa and mucosa.

The longitudinal muscle layer, which runs lengthwise along the tract, the circular muscle layer, which encircles the tract, and the muscularis mucosa are composed of smooth muscle fibers. These smooth muscle layers perform most of the motor functions of the digestive tract. The submucosa, which is located between the muscularis mucosa and the circular muscle layer, consists mostly of connective

tissue, some secretory gland cells, and blood and lymph vessels.

The essential features of the mucosa are exocrine glands and an epithelial lining. The various exocrine glands of the digestive tract, namely, those of the stomach and small and large intestine, contain a number of types of secretory cells, each of which elaborates its own specific secretion. The epithelium of the small intestine plays the major role in the transport of nutrients from the lumen into the blood and lymph vessels present in the mucosa and submucosa. In addition, the epithelial cells of the small intestine contain enzymes that are important in the completion of the digestion of certain constituents of food.

The salivary glands and parts of the pancreas also contain one or more types of exocrine secretory cells, the acini, which are arranged in clusters (Fig. 25-7). The acini lead into a system of small ducts that converge into larger ducts. Finally saliva is conducted into the oral cavity and pancreatic juice into the upper small intestine through one or more main ducts. A system of ducts also delivers the exocrine secretion (bile) of the parenchymal cells of the liver into the upper small intestine.

CONTROL OF GASTROINTESTINAL FUNCTION

With only a few exceptions, the processes of digestion and absorption are not regulated. On the other hand, the motility of, and secretion by, the organs of the GI system are regulated by neural and hormonal mechanisms.

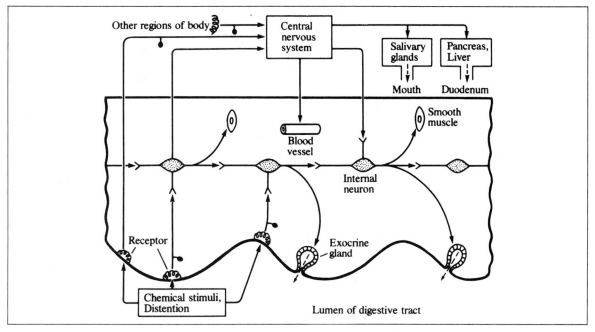

Fig. 25-8. Neural pathways for the control of motor and secretory activity of the gastrointestinal system.

NEURAL CONTROL. The neural pathways for the control of motor and secretory activity of the GI system are summarized in Figure 25-8. There are two major interconnecting neuronal networks or plexuses in the wall of the digestive tract from esophagus to anus that are important in the regulation of smooth muscle contraction and secretion by exocrine glands (see Fig. 25-7). The *myenteric plexus of Auerbach* lies between the longitudinal and circular smooth muscle layers and plays a role in the regulation of smooth muscle contractile activity. *Meissner's plexus* lies in the submucosa and is important in the control of secretion. Many of the nerve cells of these plexuses lie entirely in the wall of the digestive tract, and for this reason the tract has an intrinsic nervous system of its own. These intrinsic nerves include sensory neurons, interneurons, motor neurons, and secretory neurons. Consequently, the digestive tract possesses all the elements required for short-reflex regulation of its activity. These short reflexes endow the digestive tract with a degree of autonomy in the sense that muscle contraction and exocrine gland secretion can be controlled to an extent in the absence of regulatory influences that originate outside the wall of the digestive tract.

The sensory neurons of the intrinsic plexuses have receptors that are excited either by various intraluminal chemical stimuli or by a change in degree of wall distention. Impulses generated in these sensory neurons travel over a variable number of interneurons and finally reach the effector muscle and secretory cells through motoneurons and secretory neurons. The consequence of this is a modification of motor and secretory activity, not only at the point stimulated, but also some distance above and below this region, as a result of conduction of impulses through the numerous synaptic junctions of the nerve networks. Modification of activity can be in the direction of either excitation or inhibition, the actual effect depending on the mediator that is released at the effector cell by motoneurons and secretory neurons. If the mediator is acetylcholine, as is usually the case, the result is excitation of secretion by exocrine glands and contraction of visceral smooth muscle.

The extrinsic innervation of most of the GI system is supplied by the two divisions of the autonomic nervous system. Both parasympathetic and sympathetic nerves carry afferent fibers to the central nervous system from sensory receptors of the GI system. In addition, the autonomic nerves regulate the GI system's activity by providing the efferent limbs of a variety of long reflex arcs that can be initiated by stimulation of receptors in the GI system or any part of the body. Thus, a major function of the extrinsic nerves is to correlate activities, not only between different regions of the GI system, but also between the GI system and other parts of the body.

Since both parasympathetic and sympathetic efferent fibers synapse with cell bodies of the intrinsic nerve plexuses, the regulatory influences of the activities of these nerves are often manifested through these nerve networks. In other words, a function of these extrinsic nerves is to modify activity that can be initiated and maintained in the wall of the digestive tract itself, and this modification can be in the direction of augmentation or suppression of activity, depending on which extrinsic nerves are involved. The vagus nerve (parasympathetic preganglionic), besides carrying many afferent fibers from the GI system, is the major motor and secretory nerve of this organ system. Most vagal efferent fibers are cholinergic and excitatory. However, some are inhibitory, but in this instance there is uncertainty regarding the nature of the mediator(s). The splanchnic nerves (sympathetic postganglionic), in addition to synapsing with cell bodies of the intrinsic nerve plexuses, often directly innervate effector cells, such as vascular smooth muscle and some gland cells. The norepinephrine liberated by these adrenergic fibers depresses excitability of the nerve cells of the intrinsic plexuses and thereby inhibits short reflexes mediated through the plexuses and long reflexes over the vagus nerve. In contrast, activation of adrenergic nerves causes contraction of vascular smooth muscle (vasoconstriction) in the gut.

HORMONAL CONTROL. Certain regions of the digestive tract contain endocrine gland cells dispersed as single cells among the epithelial cells of the mucosa. Hormone-containing granules are concentrated at the base of these endocrine cells in close proximity to the blood capillaries. Various stimuli cause these hormones to be released into the portal blood from whence they are transported to the heart and finally into the systemic circulation, which carries them back to the GI system. Here, they exert their specific excitatory or inhibitory motor and secretory effects, usually on a region of the GI system other than from which they were secreted.

Three hormones of gastrointestinal origin are important in the regulation of the activities of this organ system: *gastrin,* whose major source is the gastric pyloric antrum, and *secretin* and *cholecystokinin,* which are secreted by endocrine cells in the mucosa of the upper small intestine. All of these hormones belong to the class called polypeptide hormones (see Chap. 30), chains of amino acids of varying length. Gastrin is a chain of 17 amino acids, and since the last 4 amino acids of this chain display all the physiological properties of the whole molecule, this terminal tetrapeptide

is the active fragment of natural gastrin. Cholecystokinin consists of 33 amino acids with the entire biological activity residing in the terminal octapeptide moiety. There are 27 amino acids in secretin, all of which are required for biological activity.

Secretion of all three GI hormones occurs in response to either intraluminal chemical or mechanical stimuli acting directly on the various endocrine cells without the intervention of nerves. Release of gastrin is mediated, in addition, through neural pathways similar to those involved in stimulation of smooth muscle and secretory cells; i.e., gastrin is released by intraluminal stimuli acting through short reflexes in the intrinsic nerve plexuses and by long reflexes over the vagus nerve.

The importance of the GI hormones in the control of GI motility and secretion will be discussed here and in the next chapter. It should be pointed out that at present there is a list of candidate GI hormones—substances extracted from the mucosal lining of the gut that produce some responses in the GI system when injected into the body but have not yet met all the criteria required to achieve full status as hormones.

MOTOR FUNCTIONS OF THE DIGESTIVE TRACT

Basically two types of muscular activity are involved in the digestive and absorptive functions of the digestive tract: propulsive movements and mixing movements. In addition, changes occur in the state of the tonus of the musculature of all parts of the digestive tract. In the case of propulsion, the rate of movement of contents through the various regions of the digestive tract depends on the functions served by the different organs of the tract. For example, the passageway from mouth through pharynx and esophagus is simply a conduit for conveying food to the stomach, and transit through these regions is rapid. On the other hand, transit from the stomach through the small and large intestines is slow, which is consistent with the time required for the completion of the digestive and absorptive processes that occur in these organs. The mixing movements of the stomach and small intestine promote digestion by mixing the digestive juices with the food that enters from above. In addition, mixing movements facilitate absorption from the small intestine and proximal large intestine by bringing fresh portions of the intestinal contents into contact with the absorbing surfaces.

The muscular movements of the digestive tract are concerned mostly with the activity of the smooth muscle, which extends from the distal esophagus through most of the large intestine. It should be noted, however, that skeletal muscle activity is of primary importance at either end of the tract: i.e., the mouth through the proximal esophagus at the upper end and the external anal sphincter at the lower end.

CHEWING

When solid food enters the mouth, chewing occurs. This process is important from a number of standpoints. As food is moved about in the mouth, the taste buds are stimulated, and the odors that are released stimulate the olfactory epithelia. These events are significant because much of the satisfaction of eating is derived from these stimuli. Reflex secretion of saliva also occurs during the chewing of food. The food is mixed with saliva, which softens and lubricates the food mass and thereby facilitates swallowing. In addition, chewing reduces the food to a particle size convenient for swallowing.

A crushing force of 100 to 160 pounds can be exerted by the molars and 30 to 80 pounds by the incisors of humans. Since a force of 115 to 173 pounds is sufficient to crack hazelnuts, the maximum biting force that can be generated by the muscles of mastication is far greater than that required for the chewing of ordinary food. The force of biting is not the major factor in determining the efficiency of the chewing process. The occlusive contact area between the molars and premolars is much more important in this respect.

Although the act of chewing is under voluntary control, it is also partly reflex in nature. That reflexes can be involved is shown by the fact that an animal decerebrated above the mesencephalon will chew reflexly when food is put in the mouth. The process of mastication is carried out by the combined action of the muscles of the jaws, lips, cheeks, and tongue. The movements of these skeletal muscles are coordinated by impulses over cranial nerves V, VII, and IX–XII. Once chewing is accomplished to the chewer's satisfaction, the food mass or bolus is ready for swallowing.

SWALLOWING

Swallowing is the process by which material is transported from the mouth through the pharynx and esophagus into the stomach. The act of swallowing has been divided into three stages on the basis of the regions through which a bolus passes on its way to the stomach: the mouth, the pharynx, and the esophagus. The forces involved in moving a bolus of food through these parts of the digestive tract are the pressure gradients generated by the smoothly coordinated sequence of muscular contractions that occur in response to the stereotyped all-or-none swallowing reflex.

SWALLOWING MOVEMENTS. Many of the events that occur during swallowing can be visualized by means of x-ray motion pictures of a human subject swallowing a radiopaque suspension of barium sulfate. Furthermore, the pressures at various points along the route, both at rest and during a swallow, can be measured by pressure-sensing devices.

Oropharyngeal Stages. The coordinated contractions of a variety of skeletal muscles participate in moving a bolus of food from the mouth through the pharynx into the esophagus (Fig. 25-9). This series of events occurs in about 1 second.

During the *first stage* the bolus is passed from the mouth through the isthmus of the fauces into the pharynx. The food mass, either liquid or solid, is rolled toward the back of the tongue, and the front of the tongue is pushed up against the hard palate. Respiration is inhibited briefly. At the same time, the mylohyoid muscles contract rapidly and force the bolus into the pharynx.

In the *second stage,* the bolus is passed through the pharynx into the esophagus, taking about 0.2 second. X-ray motion pictures show that a number of events occur simultaneously. The continued contraction of the mylohyoid muscles and the position of the tongue prevent the re-

Fig. 25-9. Passage of a bolus from the mouth through the pharynx and upper esophagus during a swallow.

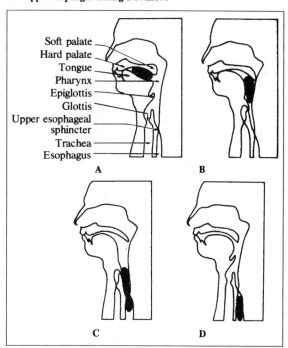

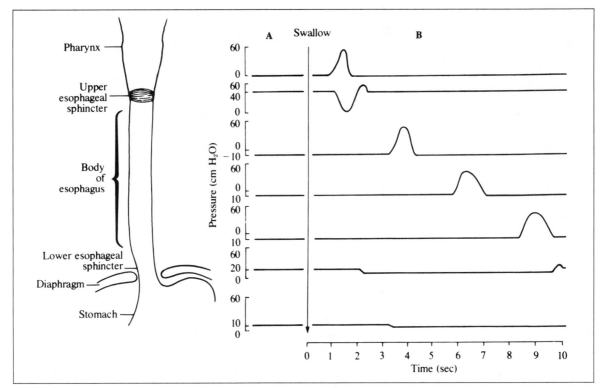

Fig. 25-10. Intraluminal pressures in the pharynx, upper esophageal sphincter, three regions of the body of the esophagus, lower esophageal sphincter, and stomach. A. At rest. B. Following a swallow.

entrance of food into the oral cavity. The soft palate is elevated and shuts off the posterior nares. Food is prevented from entering the larynx by the elevation of the larynx and by the approximation of the vocal cords, both of which serve to close the glottis. The epiglottis may or may not be pressed down over the laryngeal orifice, but even if it is, it probably acts only as an auxiliary mechanism to keep food from entering the respiratory passages, since the epiglottis can be removed without resultant abnormalities in swallowing. As these openings close, the pharyngeal constrictors contract and force the bolus into the esophagus. Respiration resumes.

Esophageal Stage. The esophagus, which is located mostly in the thorax, is a muscular tube that conducts material from the pharynx to the stomach during the *third stage* of swallowing. The upper third of the human esophagus consists of skeletal muscle, and the lower third consists of smooth muscle. The middle third is a mixture of the two muscle types.

Pressures in the lumen of the pharynx, esophagus, and stomach, both at rest and following a swallow, are shown in Figure 25-10, which should be referred to in the ensuing discussion. When there is no swallowing activity, pressure in the mouth and pharynx is atmospheric, whereas that in the lumen of the body of the esophagus is the same as intrathoracic pressure, which is subatmospheric (see Chap. 19). Consequently, it might be anticipated that, during the normal course of respiration, air would flow from the mouth and pharynx into the esophagus in response to the pressure gradient that exists between these two regions. However, when the esophagus is at rest, pressure in the upper 1 to 3 cm of esophagus is about 40 cm H_2O. This relatively high pressure indicates the existence of a mechanism that produces closure of this region between swallows, thereby impeding the flow of air into the esophagus. This is accomplished by the *pharyngoesophageal,* or *upper esophageal, sphincter,* which is closed by the tonic contraction of a surrounding band of skeletal muscle, the cricopharyngeal muscle. Following a swallow, pressure drops to atmospheric when the sphincter relaxes, and the bolus is forced into the esophagus by the pressure generated in the pharynx by the various muscular activities that occur during the oropharyngeal stage of swallowing. Pressure in the upper esophageal sphincter then rises above resting level as the result of contraction of the skeletal muscle of this region, and this prevents reflux of food from the esophagus to the pharynx. Pressure then gradually subsides to the resting level as this muscle relaxes.

If the bolus is liquid, it is shot through the esophagus by the initial force of swallowing and travels by gravity to the stomach in about 1 second. If semisolid, the bolus is propelled down the esophagus by *peristalsis,* a type of muscular contraction that is the major propulsive movement of the digestive tract. The main feature of esophageal peristalsis is a contraction of the circular muscle of the upper esophagus that passes as a wave over the entire esophagus to the stomach at the rate of 2 to 4 cm per second. The wave of contraction, which actually begins in the pharynx, is thought by some researchers to be preceded by a wave of inhibition. However, since the esophagus is normally relaxed at rest, it is difficult to detect any further esophageal relaxation. On the other hand, since the resting tone of the upper sphincter is high, this structure does relax as the wave of inhibition passes over it. The wave of contraction that follows closes the upper sphincter and forces the bolus ahead of it toward the stomach, the transit time generally being 7 to 9 seconds.

Although there is no well-differentiated muscular structure in the region where the esophagus joins the stomach, a zone of high pressure about 3 to 6 cm in length extends up from the intraabdominal esophagus through the diaphragmatic hiatus to the intrathoracic esophagus. This is the *gastroesophageal,* or *lower esophageal, sphincter.* This region is tonically contracted between swallows and exerts a pressure of about 10 to 20 cm H_2O (Fig. 25-10). Since pressure in the sphincter is somewhat higher than intragastric pressure, which is also above atmospheric pressure, the possibility of reflux of gastric contents into the esophagus is minimized. This is important, because when acid contents of the stomach enter the esophagus, an unpleasant sensation, *heartburn,* may be perceived. The extent to which the sphincter is contracted—and thus the magnitude of the pressure developed here—depends at least in part on the magnitude of intragastric pressure, because an increase in intragastric pressure activates a vagal reflex that increases sphincter tone and pressure. Consequently, this neural mechanism helps prevent reflux when intragastric pressure is high, as, for example, when the stomach is full. There is also a mechanical effect that helps to prevent reflux. Any circumstance that causes compression of the abdominal contents will produce an increase in intraabdominal pressure and thus in intragastric pressure. Abdominal compression occurs, for example, when one bends or coughs, or when the diaphragm descends during a deep inspiration. Reflux does not ordinarily occur in these circumstances because the increase in intraabdominal pressure is transmitted not only to the stomach, but also to the subdiaphragmatic esophagus. The result is that pressure in the subdiaphragmatic sphincter remains higher than that in the stomach, because pressure in the subdiaphragmatic sphincter was higher to begin with; i.e., the same pressure barrier as before is maintained.

Figure 25-10 shows that, almost immediately following the initiation of a swallow, pressure at the gastroesophageal junction drops and remains low during the time a peristaltic wave is traversing the lower esophagus. Presumably, the wave of inhibition causes relaxation of the gastroesophageal sphincter, so that when a solid bolus is propelled down the esophagus by the wave of contraction, it can easily enter the stomach. Occasionally, a liquid bolus can be seen to accumulate momentarily above the sphincter, because a liquid bolus travels so rapidly that it sometimes arrives at the sphincter before relaxation is sufficient to allow entrance of the bolus. Once pressure in the lower esophagus falls to resting level, the pressure in the gastroesophageal junction rises and remains elevated for about 10 seconds before falling to resting level once again.

THE SWALLOWING REFLEX. The coordination of swallowing depends on neural mechanisms. The first stage may be initiated voluntarily, but is usually a reflex action. The remainder of the swallowing response and all other movements of the GI tract with the exception of defecation are independent of the will.

A swallowing center located in the medulla is activated by stimulation of receptors in the mouth, pharynx, and esophagus (Fig. 25-11). This center sends impulses over a number of efferent nerves in the proper sequence to the numerous skeletal and smooth muscles involved in the swallowing response, and the complete act of swallowing occurs. The glossopharyngeal and hypoglossal nerves are concerned mainly with the oropharyngeal stages, whereas the vagus is the nerve that is important in coordinating upper esophageal sphincter activity, the orderly progress of the esophageal peristaltic wave, and lower esophageal sphincter activity. A biphasic complex of inhibition and then excitation is probably mediated by the vagus after a swallow. This results in the passage of a wave of relaxation followed by a wave of contraction over the upper sphincter, esophagus, and gastroesophageal sphincter.

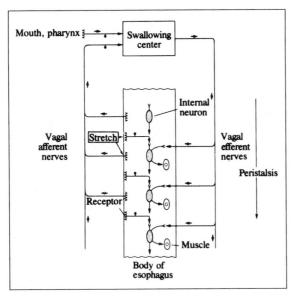

Fig. 25-11. Long and short reflex regulation of esophageal peristalsis.

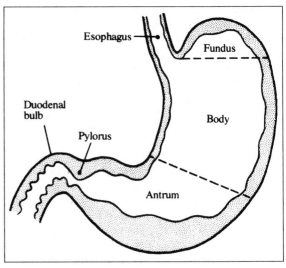

Fig. 25-12. Physiological anatomy of the stomach.

Vagal preganglionic fibers synapse with the cholinergic fibers of the myenteric plexus of Auerbach in the esophageal wall. It is through this nerve network that the vagus nerve manifests its motor effects. Actually, the myenteric plexus itself is capable of coordinating peristaltic activity, as shown by the fact that peristalsis can occur in a vagotomized esophagus, although it is weaker than normal. Thus, it is likely that both long and short reflexes play a role in regulating esophageal peristalsis (Fig. 25-11).

Primary peristalsis is that which follows the oropharyngeal stages of swallowing. Peristalsis can also be elicited by local stimulation (stretch) of the esophageal wall and without being preceded by the oropharyngeal stages. This *secondary peristalsis* is mediated by long reflexes over the vagus and by short reflexes in the myenteric plexus. Secondary peristalsis is important when food remains in the esophagus following the passage of a primary peristaltic wave. Secondary peristalsis facilitates the removal of such residues from the esophagus.

ABNORMALITIES OF SWALLOWING. A condition known as *dysphagia* arises when there are abnormalities of any of the three stages of swallowing. Abnormalities may result from a variety of disorders associated with the nerves and muscles involved in these processes. There are abnormal motility patterns of the esophagus and gastroesophageal junction in

achalasia. The esophageal musculature may show uncoordinated and spastic contractions, and the gastroesophageal sphincter often fails to relax following a swallow. Food has difficulty entering the stomach and tends to accumulate above the sphincter, causing dilation of the esophagus. This abnormality is apparently due to a lack of or degeneration of the nerve cells making up the plexus of Auerbach.

GASTRIC MOTILITY

Included in the motor functions of the stomach (Fig. 25-12) are the storage of ingested food for variable periods of time and the discharge of gastric contents into the small intestine at a rate that is optimal for intestinal digestion and absorption. In addition, the mixing movements of this organ aid in the conversion of large, solid food particles to a finely divided, liquid form prior to evacuation.

THE EMPTY STOMACH. During periods of fasting the volume of contents present in the human stomach is only about 50 ml or less. In this circumstance there is not much tension in the gastric wall, and the intragastric pressure is low (5–10 cm H_2O). Gastric motor activity is minimal early in fasting, but as a fast is prolonged, gastric contractions become progressively more vigorous. When the stomach is devoid of food, the sensation of hunger is often perceived. As a part of the overall sensation of hunger, some people feel a sense of emptiness accompanied by occasional sharp pangs referred to the abdominal region. Although the matter

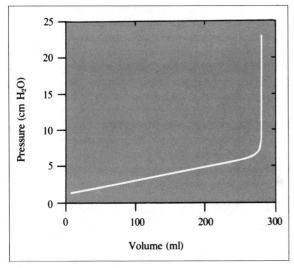

Fig. 25-13. Intragastric pressure during the addition of fluid to the stomach of a rabbit.

has been widely debated over the years, the so-called hunger pangs may be directly associated with the more vigorous contractions of the empty stomach. Thus, these gastric contractions may contribute to the sensation of hunger. The control of food intake as related to the sensations of hunger and satiety is discussed in Chapter 8.

GASTRIC FILLING. One function of the stomach is the storage of variable volumes of contents, and it is subserved by the smooth muscle of the fundus and body of the stomach. These regions possess the ability to adapt to the volume of contents they contain in that relatively large volumes can be introduced into them with little increase in gastric wall tension and intragastric pressure (Fig. 25-13). A neural mechanism is at least partly responsible for this volume adaptation. With each swallow there is a slight relaxation of the reservoir region of the stomach (see Fig. 25-10). This *receptive relaxation* is the result of a long reflex that has its afferent and efferent pathways in the vagus nerve. A property of smooth muscle that also contributes to this volume adaptation is that of plasticity (see Chap. 3); i.e, smooth muscle fibers, when stretched, either rearrange internally or slide past one another in such a manner that increases in wall tension and intraluminal pressure are minimized.

MECHANICS OF GASTRIC EMPTYING. Peristaltic contractions of the stomach are largely responsible for mixing ingested food with gastric secretions and for generating the forces required for gastric emptying. When food enters the stomach, peristaltic waves begin to travel over the body, pyloric antrum, pylorus, and duodenal bulb at a frequency of about three per minute. The waves of contraction that move over the relatively thin muscle of the body at a rate of 1 cm per second are weak. However, when these waves reach the pyloric antrum, where the muscle is much thicker (see Fig. 25-12), the contractions become much more vigorous and increase in speed to 3 to 4 cm per second, so that the terminal antrum contracts as a unit—the so-called antral systole. At this time, pressure in the antrum rises momentarily above that in the duodenal bulb by 20 to 30 cm H_2O, and the contents begin to move through the pylorus. When the contraction reaches the pylorus, this structure contracts, and after this, the duodenal bulb contracts. The contents continue to escape from the stomach as the gradually contracting pylorus closes, until the increased resistance offered by this musculature prevents further evacuation. Only a few milliliters of gastric contents enter the duodenum with each wave. Material trapped in the antrum is squirted back forcibly into the body of the stomach by the systolic contraction of the antrum, an action that mixes food particles with the gastric secretions. The pylorus remains contracted somewhat longer than does the duodenal bulb, preventing to some extent regurgitation of duodenal contents into the stomach. The contraction of the duodenal bulb helps to propel the contents down the intestine. All these structures finally relax again, and the cycle of emptying is repeated with the coming of the next peristaltic wave. This activity continues for variable periods of time until the stomach is finally emptied of its contents.

There is a zone of elevated pressure at the pylorus, the thick band of circular smooth muscle and connective tissue that separates the stomach from the duodenum. This suggests the existence of a sphincter in this region. Actually, the degree of constriction of the pylorus, by altering the resistance to flow of material from stomach to duodenum, does influence the rate at which the stomach empties. In general, there is a reciprocal relationship between the vigor of antral peristalsis and the tone of the pylorus: Factors that increase antral peristalsis usually decrease pyloric tone, and vice versa.

In summary: The driving force behind the emptying process is the pressure differential between the antrum and duodenum. If the pressure is higher in the antrum, and high enough to overcome resistance offered at the pylorus, the contents will leave the stomach. Since the muscular activity of the pyloric antrum determines to a large extent the magnitude of the pressure differential, and since the pyloric muscle tone determines the resistance to flow from stomach to duodenum, gastric emptying is regulated largely by mechanisms that control the muscular activities of these regions of the stomach.

CONTROL OF GASTRIC MOTILITY. The basic mechanisms controlling the movements of the stomach, small intestine, and large intestine share certain features. These include an autonomous, myogenic control system whose activity is modified by impulses over both extrinsic autonomic nerves and neurons located entirely in the wall of the digestive tract as a part of the intrinsic nerve plexuses of this organ system.

Neural Control. The extrinsic innervation of the stomach is by the vagus (parasympathetic) and splanchnic (sympathetic) nerves. These autonomic nerves, in addition to providing afferent fibers from the stomach, serve as the efferent limbs of a variety of reflex arcs that can be initiated from many visceral and somatic receptors throughout the body and can cause either excitation or inhibition of gastric movements. Thus, the major function of the extrinsic nerves of the stomach is to correlate activities between the stomach and other parts of the GI tract as well as other regions of the body.

In general, stimulation of the parasympathetic (cholinergic) nerves usually produces increased muscular activity, whereas stimulation of sympathetic (adrenergic) nerves most often results in inhibition. These generalities can be applied not only to the stomach but also to most of the smooth muscle of the digestive tract. As noted later, the vagus nerves play the dominant role in the neural regulation of gastric motility. Although the sympathetic nerves do participate in the reflex regulation of gastric motor activity, they play only a minor part in the regulation that occurs during the normal course of digestion and absorption.

When the extrinsic nerves of the human stomach are cut, gastric peristalsis is at first very much reduced in strength or even absent, gastric emptying is delayed, and food is retained in the stomach for relatively long periods of time. This is the result of the loss of the excitatory influences that are normally mediated by the vagus nerves. However, after a period of recovery, motility and rate of gastric emptying tend to return to normal. In this instance, gastric peristalsis is initiated and maintained entirely by impulses originating in the intrinsic nerve plexuses in the gastric wall. These nerve plexuses contain all the components necessary for short-reflex regulation of gastric motor activity and, in fact, are required for normal gastric peristalsis, because this pattern of activity occurs only when these nerve nets are functional. Since efferent fibers of extrinsic nerves to the stomach synapse on cell bodies in the intrinsic plexuses, the motor regulatory influences of the vagus and splanchnic nerves are frequently manifested through these nerve nets.

Electrical Activity of Gastric Muscle. Two major types of potential changes can be recorded from gastric smooth muscle (see Chap. 3). One wave shape consists of single or multiple short-lived oscillations that represent typical action potentials. The muscle action potential is the electrical event that is associated with contraction; i.e., contraction occurs only when preceded by action potentials.

The second potential change has a smaller amplitude (10–15 mv) and is of much longer duration (1–4 sec). This electrical event represents a relatively small and slow depolarization followed by repolarization of the muscle cell membrane. This so-called *slow wave* appears under a recording electrode about three times every minute. The reason it does so is because it originates continually at this frequency in the longitudinal muscle of the upper stomach, spreads by conduction to the circular muscle, and is propagated downward through the gastric muscle as a ring around the stomach. Because this slow wave shows this continuous cyclic activity, it is referred to as the *basic electrical rhythm* of the stomach.

The slow wave differs from action potentials in that contraction may or may not occur in response to a slow wave. However, there is a relationship between the basic electrical rhythm and action potentials and thus contraction, in that when action potentials occur, they do so during the slow wave. This would be expected because muscle excitability would be higher at this time, i.e., closer to threshold. It is at this point where any prevailing excitatory influences, such as release of acetylcholine by nerve endings or muscle stretch, would most likely cause threshold to be reached, muscle action potentials to be fired, and contraction to occur. Since the slow wave propagates as a circumferential ring and encompasses a certain population of cells at any one moment, the muscle cells in that ring have an increased probability of discharge at about the same time. Thus, a function of the basic electrical rhythm is to synchronize spiking activity in a specific region of the stomach, the result being the coordinated and efficient mechanical effort of peristalsis.

Excitatory and inhibitory stimuli, by virtue of their ability to enhance or inhibit spiking activity, are of primary importance in determining the strength with which the gastric muscle responds to the exciting signals represented by the basic electrical rhythm. For example, the vagal stimulation induced by feeding increases the strength of gastric contrac-

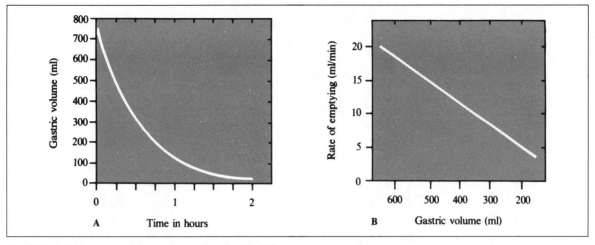

Fig. 25-14. A. Basic pattern of the gastric emptying of a 750-ml liquid meal. B. Rate of emptying of the same 750-ml liquid meal.

tions, so that they become typical strong peristaltic contractions. If available stimuli are insufficient to elicit action potentials, the slow wave continues its cyclic activity, but is not accompanied by contractions.

REGULATION OF GASTRIC EMPTYING. Both the force with which the stomach contracts to expel its contents into the small intestine and the tone of the pylorus are under the combined influence of a variety of excitatory and inhibitory mechanisms. The mechanisms involved in regulating the vigor of antral peristalsis are shown in Figure 25-15, which should be referred to in the discussion that follows.

A factor that influences gastric emptying is the consistency of gastric contents. For example, large chunks of meat take a much longer time to leave the stomach than does meat that is finely ground. Liquids are generally evacuated more rapidly than solids. These observations indicate that the contents of the stomach must be in a finely divided and liquid form prior to evacuation.

There is a basic pattern of gastric emptying of a meal (Fig. 25-14A) in that the rate of evacuation decreases progressively as the stomach empties (Fig. 25-14B). Furthermore, the greater the size of a meal, the greater is the initial rate of emptying. These facts show that the rate of gastric emptying varies directly with the volume contained in the stomach at any given time. This effect is related to the degree of distention of the gastric wall. Stimulation of mechanoreceptors in the wall of the stomach by distention results in augmentation of antral peristalsis in proportion to the degree of dis-

tention and also in relaxation of the pylorus. This occurs through activation of long reflexes over the vagus and short reflexes through the intrinsic nerve plexuses. These neural influences, along with any contractile activity that is elicited directly in response to stretch of gastric muscle, are major mechanisms involved in providing the excitation required to empty the stomach of its contents (Fig. 25-15). In addition, peristalsis is augmented by the hormone gastrin, which is released from the pyloric antrum by the intraluminal stimuli that are present during digestion of a meal and also by the vagal stimulation that occurs at this time.

Many of the mechanisms that participate in the regulation of gastric emptying are initiated in the duodenum. Gastric contents must come into contact with the duodenal mucosa in order to maintain the normally slow gastric evacuation. It will be obvious from the discussion to follow that practically any stimulation of the duodenum tends to check gastric emptying. These inhibitory mechanisms are utilized to prevent the absorptive powers of the intestinal mucosa from being overwhelmed by a flood of ingested materials and to prevent undue chemical, mechanical, and osmotic irritation of the duodenum.

Both the chemical and the physical properties of the contents that enter the duodenum have a profound influence on gastric emptying. With regard to the major foods, the presence in the duodenum of the digestion products of carbohydrate, protein, and especially fat impedes gastric emptying. Solutions of pH 3.5 or less greatly retard gastric evacuation. The osmotic pressure of the gastric contents is also an important factor in gastric emptying. The introduction of hypotonic and, especially, hypertonic solutions into the duo-

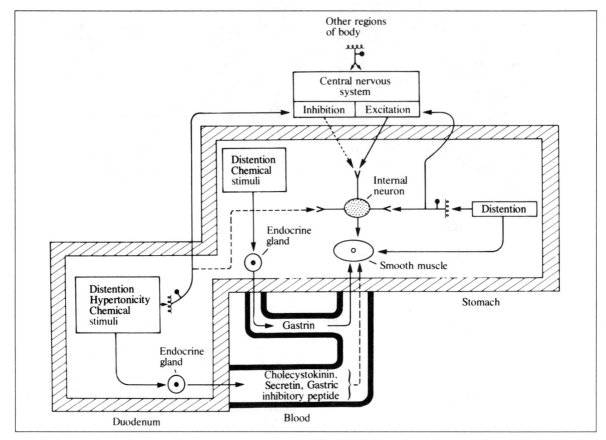

Fig. 25-15. Pathways involved in the regulation of the gastric evacuation of a meal. The solid arrows indicate excitation; the dashed arrows indicate inhibition.

denum causes inhibition of gastric motility, suggesting that the osmotic environment most favorable to the duodenum is one close to isotonicity. Carbohydrate and protein probably exert at least part of their inhibitory influences by the formation of osmotically active particles produced by the digestion of these foods. Distention of the duodenum by increasing intraluminal pressure 10 to 15 mm Hg inhibits gastric emptying.

There is a variety of both duodenal endocrine cells and receptors, such as osmoreceptors, mechanoreceptors, and hydrogen ion and other chemoreceptors, that respond to intraluminal stimuli to produce hormonal and reflex inhibition of gastric motor activity (Fig. 25-15) and enhancement of pyloric tone. Hormonal suppression of gastric motor activity and contraction of the pylorus result when fat and protein digestion products release cholecystokinin and acid releases secretin from the duodenal mucosa into the circula-

tion. Fat and protein digestion products also release *gastric inhibitory peptide,* which inhibits gastric motility. It is a candidate hormone, i.e., a humoral agent that has not yet met all criteria required to achieve full status as a hormone. The inhibition of gastric emptying by the duodenal stimuli previously listed also depends to varying degrees on neural mechanisms collectively called the *enterogastric inhibitory reflex.* Included are long reflexes over the vagus and short reflexes through the intrinsic nerve plexuses (see Fig. 25-15).

Although the regulation of gastric emptying is controlled from the duodenum to a large extent, gastric motility may be stimulated or inhibited reflexly from any sensory region of the body (Fig. 25-15). Gastric emptying is delayed when the ileum is full (ileogastric reflex) and when the anus is mechanically distended (anogastric reflex). Stimulation of visceral and somatic pain receptors results in inhibition of gastric movements. Various emotional states, such as anger, fear, anxiety, and resentment, produce changes in gastric motor activity, but the direction of the changes is not always predictable.

VOMITING. The act of vomiting accomplishes the purpose of rapidly emptying the stomach of its contents. Vomiting is generally preceded by profuse secretion of saliva, sweating, rapid heart rate, and a feeling of nausea. A forced inspiration is made, and the glottis and nasal passages are closed by the contraction of the appropriate muscles. The body of the stomach, the gastroesophageal sphincter, and the esophagus relax, and the pyloric antrum contracts. The diaphragm descends on the stomach, and there is a forcible contraction of the abdominal muscles. The pressure generated propels the gastric contents into the esophagus through the pharyngoesophageal sphincter and into the mouth. The major force for vomiting is supplied by the contraction of the skeletal muscle of the diaphragm and abdomen, rather than by contraction of the gastric musculature. If an animal is given curare, an agent that paralyzes striated muscle, vomiting cannot occur.

Vomiting is an extremely complex reflex act and is coordinated by a center located in the medulla. Afferent impulses arrive at this center from many regions of the body. The most potent stimuli arise from the sensory nerve endings of the fauces and the pharynx. Other prominent receptor areas include almost any part of the digestive tract, other abdominal viscera, and the labyrinths during motion sickness. Loss of acid gastric contents by prolonged vomiting can result in profound disturbances in fluid and acid-base balance (see Chap. 24).

MOTILITY OF THE SMALL INTESTINE

Ingested food, which is liquefied and partially digested in the stomach, passes into the small intestine, where the major part of digestion and absorption occurs. Excretory products and food residues are moved into the colon. Different types of movements that accomplish these functions can be observed and recorded in the small intestine.

MOVEMENTS OF THE SMALL INTESTINE. The sequence of events in the type of movement most frequently seen in the small intestine and known as *segmenting contractions* is illustrated diagrammatically in Figure 25-16. Although segmenting contractions probably help to move the contents

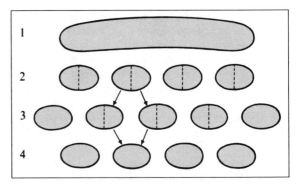

Fig. 25-16. Segmentation movements of the small intestine. An animal is fed a radiopaque meal, and the intestine is observed by means of x-rays. 1. A string of a barium-impregnated meal can be seen when the gut is quiet. 2. The circular muscle of the intestine then contracts at a number of places, dividing it into a series of segments. 3. Each segment can be seen to divide, with adjoining halves coming together to form new segments. 4. The original pattern is established once again.

down the small intestine, this type of activity serves primarily to mix the contents with the digestive juices and to facilitate absorption by bringing fresh portions of contents into contact with the absorbing epithelium. The process can be repeated about 11 times a minute in the upper intestine of humans and continues for variable periods of time in a section of intestine.

Another type of movement observed in the small intestine is peristalsis, whose function is to propel contents along the intestine. Under normal circumstances the wave moves slowly at a rate of 1 to 2 cm per second and travels only 4 to 5 cm. Thus, this type of motor activity provides the slow transit necessary to permit completion of the absorptive processes. Occasionally peristalsis occurs as a swift movement that sweeps along the entire intestine (peristaltic rush). Such rapidly propagated contractions occur when the excitability of the intestine is high, as, for example, when the intestine is irritated by toxic agents. Diarrhea and malabsorption may result from the rapid transport of contents through the small intestine.

CONTROL OF INTESTINAL MOTILITY. As in the stomach, there is a close association between the electrical and mechanical activities of the small intestine in that contractions follow action potentials, and action potentials occur during intestinal slow waves. Slow waves originate in the longitudinal muscle in a pacemaker region close to the entrance of the bile duct, pass to the circular muscle by muscle connections between the two layers, and move down the intestine as a circumferential ring. If a stimulus is present in a certain region of the gut and is sufficient to bring excitability to threshold during a slow wave as it traverses the region, spikes are fired, and contraction results.

The velocity of, and direction taken by, peristalsis is the same as that of the slow wave. Consequently, the slow wave determines both the rate at which a wave of contraction moves down the intestine and its polarity. Slow-wave frequency decreases progressively from 11 to 12 per minute in the duodenum to 7 to 9 per minute in the terminal ileum, suggesting the possibility that different regions of the intestine have their own pacemakers. On those occasions when every slow wave in a particular region of intestine is accompanied by action potentials, the maximal rate of segmenting contractions is attained, and a very rhythmic contraction pattern is observed. Thus, it is apparent that the frequency of the segmenting contractions is also determined by the intestine's basic electrical rhythm.

Although the intestinal basic electrical rhythm determines the frequency of segmenting contractions and the velocity and direction of peristalsis, other factors determine whether or not action potentials, and thus contraction, will occur. Local mechanical and chemical stimulation by the intestinal contents is probably largely responsible for the initiation and continuance of movements of the small intestine. Factors that increase intestinal motility are increased intraluminal pressure; hypotonic, hypertonic, and acid solutions; and products of digestion. These stimuli act mainly by activation of local cholinergic reflexes through the intrinsic nerve plexuses of the intestinal wall; intestinal contractions become very weak if these plexuses are blocked by drugs or have degenerated.

The small intestine retains relatively normal movements following extrinsic denervation. However, intestinal motility is influenced by impulses over extrinsic nerves. Stimulation of the vagus generally causes increased intestinal motility, whereas sympathetic stimulation usually produces inhibition. The importance of the extrinsic nerves in modifying intestinal movements is shown by the fact that intestinal motility can be altered reflexly by stimulation of many sensory areas. Distention of any part of the intestine inhibits the whole intestine. This effect, known as the *intestinointestinal inhibitory reflex*, does not occur after section of the splanchnic nerves. Trauma to organs outside the digestive tract, such as irritation of the peritoneum or urinary tract, causes intestinal inhibition. The atonic or flaccid gut, a condition known as *paralytic ileus*, which often follows surgery in these areas, is a result of reflex inhibition. Feed-

ing increases motility. This is the result of a reflex initiated when food enters either the stomach or the intestine and is termed a *gastrointestinal,* or *intestinointestinal, excitatory reflex.*

Evidence is accumulating for the involvement of GI hormones in the regulation of intestinal motility. In general, gastrin and cholecystokinin stimulate intestinal movements, whereas secretin is inhibitory. Serotonin, whose concentration is relatively high in the intestinal mucosa, and certain of the prostaglandins stimulate intestinal contractions. However, the role played in the control of intestinal movements by these endogenously released chemical agents is not yet clear.

ILEOCECAL SPHINCTER. The terminal ileum receives unabsorbed food residues about 3.5 hours after the beginning of gastric evacuation. Contents arrive at a region immediately proximal to the cecum, where the last 2 to 3 cm of the muscular coat is thicker than the rest of the ileum. This is the *ileocecal sphincter,* which is normally closed. A zone of high pressure about 20 cm H_2O above atmospheric pressure and approximately 4 cm in length is present at the ileocecal junction. Distention of the lower ileum produces a drop in pressure in the ileocecal sphincter, whereas distention of the cecum leads to increased pressure in the sphincter. The implication is that, when contents are present in and distend the lower ileum, the sphincter reflexly relaxes to allow them to be driven into the colon by the propulsive movements of the distal small intestine, and then reflexly contracts, preventing regurgitation into the ileum. The musculature of the ileum is generally not very active, but activity is increased when food enters the stomach (*gastroileal reflex*).

THE GALLBLADDER AND BILE DUCTS

Bile is manufactured and continuously secreted by the liver. This secretion passes from the liver by way of the hepatic duct into the common bile duct. When the upper intestine is devoid of food, the bile is diverted by way of the cystic duct into the gallbladder, where it is stored for variable periods of time. In addition to acting as a storage organ, the gallbladder provides a safety factor for regulating pressures in the biliary system. However, this organ can be removed without interfering with normal digestion and absorption.

FILLING AND EVACUATION OF THE GALLBLADDER. Gallbladder function can be evaluated by watching this organ fill or evacuate its contents. A radiopaque substance, such as tetraiodophenolphthalein, which is secreted by the liver into the bile, is administered. The outline of the gallbladder and bile ducts can then be visualized by means of x-rays. The gallbladder can be seen to fill during periods of fasting. Since relatively little bile enters the duodenum under this circumstance, there must be a mechanism available to prevent the passage of bile into the small intestine and at the same time divert it into the gallbladder. It is widely accepted that there is a sphincter (*sphincter of Oddi*) in the region where the common bile duct enters the lumen of the small intestine. If a catheter is inserted into the common bile duct here, the gallbladder will not fill because the sphincter is unable to exert its restraining influence. The resistance offered to the flow of bile by the sphincter of Oddi has been investigated by measuring the pressure required to force bile through the region. Such pressure is a measure of the sphincteric resistance, and in fasting, unanesthetized dogs it can be as high as 30 cm H_2O. It is probably this resistance that prevents bile from entering the intestine during fasting.

When food is taken, the gallbladder can be seen to empty over a period of time by means of a series of rather sluggish contractions. The intrabladder pressure rises to about 20 to 30 cm H_2O, and under most circumstances the pressure exerted by the contracting gallbladder is enough to overcome the sphincteric resistance. Further, it is believed that the gallbladder and sphincter act as a functional unit; i.e., when the gallbladder contracts, the sphincter relaxes. The entrance of bile into the gut is also influenced by the state of contraction of the duodenal musculature. During active contraction of the duodenum, the flow of bile is decreased or completely blocked by compression of the duct, but during the relaxation phase, resistance to flow is reduced. Therefore, as the result of duodenal movements, bile may be observed to enter the intestine in squirts.

Contraction of the gallbladder occurs reflexly when a meal is eaten. The efferent limb of the reflexes involved is in the vagus nerve. However, evacuation of the gallbladder is primarily under the control of the hormone cholecystokinin, which stimulates contraction of the gallbladder and relaxation of the sphincter. A variety of substances in intestinal contents release cholecystokinin from the duodenal mucosa, the most notable being fat and its digestion products. Proteins are effective, but carbohydrates are not. Agents that promote emptying of the gallbladder are called *cholecystagogues*.

MOTILITY OF THE COLON; DEFECATION

A firm fecal mass is formed in the colon by the absorption of water from the contents that enter from above. This rela-

tively dehydrated material is stored in the large intestine for variable periods of time and is ultimately evacuated to the outside of the body by defecation.

COLONIC MOVEMENTS. The large intestine in humans is inactive for a large proportion of the time. However, when material is present in the proximal colon, segmenting contractions (haustral churnings) occur at a much lower frequency than in the small intestine and produce a limited back-and-forth movement of contents. This is the major type of movement of the large intestine, and by exposing the contents to the mucosa it promotes the absorption of water, thereby facilitating the formation of a firm fecal mass in the proximal colon.

At infrequent intervals of three to four times a day in humans, a strong contraction of the proximal colon drives contents into the distal colon, where material accumulates distal to the pelvirectal flexure. These propulsive *mass movements* may represent a powerful segmenting contraction in association with extensive distal relaxation. That they are rather strong contractions is shown by the fact that the pressure in a segment undergoing such a contraction may reach a peak of 100 cm H_2O. Peristalsis also aids in moving contents in a downward direction.

It generally takes about 18 hours for contents to reach the distal colon after leaving the small intestine, and they are stored in the distal colon for varying lengths of time until defecation occurs. This may be 24 hours or longer following the ingestion of food. Although the rectum is normally empty, contents are occasionally shifted into the rectum after one of the mass contractions, and the resultant distention of the region elicits the desire to defecate. The act of defecation is partly voluntary and partly involuntary. The involuntary movements are concerned with smooth muscle; i.e., the distal colon contracts and the internal anal sphincter relaxes. Relaxation of the external anal sphincter, which consists of striated muscle, is voluntary. Other voluntary movements that can supply one-half of the force involved in evacuation of the rectum are a contraction of the abdominal muscles and forcible expiration with the glottis closed (straining movements).

REGULATION OF COLONIC MOTILITY. The proximal colon possesses a large degree of autonomy and functions in a relatively normal manner in the absence of its extrinsic motor innervation, which is derived from the vagus nerve. Movements here are probably largely initiated by distention of the colonic walls, which stimulates contractile activity by triggering short reflexes through the intrinsic nerve plexuses. However, colonic muscular activity also can be modified by impulses over extrinsic nerves. For example, when food enters the stomach or duodenum, a mass contraction occurs in the proximal colon. These *gastrocolic* and *duodenocolic reflexes* are usually most evident after the first meal of the day and are often followed by the desire to defecate. Part of this effect may be due to the release of gastrin, which has an excitatory effect on the colonic musculature.

The distal colon is somewhat more dependent on its extrinsic nerve supply, and movements in this region, including the act of defecation, disappear after transection of these nerves. However, weak movements do return after a time, and there is a semblance of the act of defecation. Defecation as it normally occurs is under voluntary control. There are subsidiary centers, since a certain region of the medulla can be stimulated to cause defecation. If the spinal cord is transected in the thoracic region, defecation can still occur without voluntary control after spinal shock has passed off. On the other hand, if the sacral cord is destroyed, defecation becomes very imperfect. This portion of the cord presumably serves as a local reflex center for the act of defecation. Distention of the rectum causes impulses to be sent over afferent fibers in the pelvic nerve to the sacral cord, which relays impulses over parasympathetic nerve fibers to the distal colon and anal sphincters. If defecation is inhibited by higher centers, the rectum relaxes, the stimulus of distention disappears, and defecation is postponed. The efferent fibers to the distal colon and internal anal sphincter are in the pelvic nerve, while those to the external anal sphincter are in the pudendal nerve. During periods of nonactivity, the internal and external anal sphincters are maintained tonically contracted by impulses over the lumbar sympathetics and pudendal nerve respectively.

BIBLIOGRAPHY

CIRCULATION

Bradley, S. E. The Hepatic Circulation. In W. F. Hamilton and P. E. Dow (eds.), *Handbook of Physiology*. Washington, D.C.: American Physiological Society, 1963. Section 2: Circulation. Vol. 2, chap. 41, pp. 1387–1438.

Chou, C. C., Hseih, C. P., Yu, Y. M., Kvietys, P., Pittman, R., and Dabney, J. M. Localization of mesenteric hyperemia during digestion in dogs. *Am. J. Physiol.* 230:583–589, 1976.

Davenport, H. W. *Physiology of the Digestive Tract* (5th ed.) Chicago: Year Book, 1982.

Delaney, J. P., and Custer, J. Gastrointestinal blood flow in the dog. *Circ. Res.* 17:394–402, 1965.

Fara, J. W., Rubinstein, E. H., and Sonnenschein, R. R. Intestinal hormones in mesenteric vasodilation after intraduodenal agents. *Am. J. Physiol.* 223:1058–1067, 1972.

Folkow, B. and Neil, E. *Circulation*. New York: Oxford University Press, 1971.

Grim, E. The Flow of Blood in the Mesenteric Vessels. In W. F. Hamilton and P. E. Dow (eds.), *Handbook of Physiology*. Washington, D.C.: American Physiological Society, 1963. Section 2: Circulation. Vol. 2, chap. 42, pp. 1439–1456.

Haljamae, H., Jodal, M., and Lundgren, O. Countercurrent multiplication of sodium in intestinal villi during absorption of sodium chloride. *Acta Physiol. Scand.* 89:580–593, 1973.

Hanson, K. M., and Johnson, P. C. Local control of hepatic arterial and portal venous flow in the dog. *Am. J. Physiol.* 211:712–719, 1966.

Hilton, S. M., and Lewis, G. P. The relationship between glandular activity, bradykinin formation and functional vasodilation of the sub-mandibular salivary gland. *J. Physiol.* (Lond.) 134:471–483, 1956.

Jacobson, E. D. (ed.). Symposium on the gastrointestinal circulation. *Gastroenterology* 52:332–471, 1967.

Jacobson, E. D., Swan, K. G., and Grossman, M. I. Blood flow and secretion in the stomach. *Gastroenterology* 52:414–420. 1967.

Johnson, P. C. (ed.). Autoregulation of blood flow. *Circ. Res.* 15 (Suppl. 1):1–291, 1964.

Johnson, P. C., and Hanson, K. M. Capillary filtration in the small intestine of the dog. *Circ. Res.* 19:766–772, 1966.

Lundgren, O. Studies in blood flow distribution and countercurrent exchange in the small intestine. *Acta Physiol. Scand.* (Suppl.) 303:3–42, 1967.

Lundgren, O. The Alimentary Canal. In P. C. Johnson (ed.), *Peripheral Circulation*. New York: Wiley, 1978. Pp. 255–283.

Selkurt, E. E. Gastrointestinal Circulation: Part D: Physiology. In D. I. Abramson (ed.), *Blood Vessels and Lymphatics*. New York: Academic, 1962. Chap. 11, pp. 333–340.

Swan, K. G., and Jacobson, E. D. Gastric blood flow and secretion in conscious dogs. *Am. J. Physiol.* 212:891–896, 1967.

Vatner, S. F., Franklin, D., and Van Citters, R. L. Mesenteric vasoactivity associated with eating and digestion in the conscious dog. *Am. J. Physiol.* 219:170–174, 1970.

Yoffey, J. M., and Courtice, C. F. *Lymphatics, Lymph and the Lymphomyeloid Complex*. New York: Academic, 1970.

MOVEMENTS OF THE DIGESTIVE TRACT

Atanassova, E., and Papasova, M. Gastrointestinal motility. *Int. Rev. Physiol.* 12:35–69, 1977.

Bortoff, A. Myogenic control of intestinal motility. *Physiol. Rev.* 56:418–434, 1976.

Code, C. F. (ed.). *Handbook of Physiology*. Washington, D.C.: American Physiological Society, 1968. Section 6: Alimentary Canal. Vol. 4.

Cohen, S. Gastrointestinal Motility. In R. K. Crane, E. D. Jacobson, and L. L. Shanbour (eds.), *International Review of Physiology: Gastrointestinal Physiology III*. Baltimore: University Park Press, 1979. Vol. 19, pp. 107–149.

Davenport, H. W. *A Digest of Digestion* (2nd ed.). Chicago: Year Book, 1978.

Duthrie, H. S. (ed.) *Gastrointestinal Motility in Health and Disease*. Baltimore: University Park Press, 1978.

Grossman, M. I. Neural and hormonal regulation of gastrointestinal function: An overview. *Annu. Rev. Physiol.* 41:27–33, 1979.

Johnson, L. R. (ed.) *Gastrointestinal Physiology*. St. Louis: Mosby, 1977.

Phillips, S. F., and Devroede, G. J. Functions of the Large Intestine. In R. K. Crane, E. D. Jacobson, and L. L. Shanbour (eds.), *International Review of Physiology: Gastrointestinal Physiology III*. Baltimore: University Park Press, 1979. Vol. 19, pp. 263–290.

Weisbrodt, N. W. Gastrointestinal Motility. In A. C. Guyton, E. D. Jacobson, and L. L. Shanbour (eds.), *International Review of Physiology: Gastrointestinal Physiology I*. Baltimore: University Park Press, 1974. Vol. 4, pp. 105–138.

Weisbrodt, N. W. Neuromuscular organization of esophageal and pharyngeal motility. *Arch. Intern. Med.* 136:524–531, 1976.

SECRETION AND ACTION OF DIGESTIVE JUICES; ABSORPTION

26

Leon K. Knoebel

Substances contained in ingested food that are important in the nutrition of the body include carbohydrates, proteins, lipids, vitamins, inorganic salts, and water. Many of the organic constituents of the diet are structurally complex and are not readily absorbed from the digestive tract in their natural states. However, the gastrointestinal (GI) system has a variety of exocrine glands that secrete digestive juices into the lumen of the digestive tract, and the enzymes contained in these secretions, along with those in the wall of the small intestine, convert complex organic molecules to smaller molecules. These digestion products are transferred from the intestinal lumen across the wall of the small intestine into the blood and the lymph, which in turn distribute them to the cells of the body. This chapter discusses secretion by the exocrine glands of the GI system, describes the actions of the digestive enzymes, and considers the manner in which the end products of digestion and other substances are absorbed into the circulation.

GASTROINTESTINAL SECRETION

The GI system contains a variety of exocrine secretory cells. The secretion that enters the gland lumen from a secretory cell is the *primary secretion* of that cell, and its composition is characteristic of the cell. A few primary secretions consist entirely of water and inorganic ions, but most contain one or more organic constituents in addition. The most notable of the organic components are the digestive enzymes of the salivary glands, stomach, and pancreas; the bile salts, bile pigments, and cholesterol secreted by the liver; and the mucin produced by all regions of the digestive tract.

Plasma is the ultimate source of the digestive secretions in the sense that it supplies the constituents necessary for their elaboration. The function of the secretory cells is to modify plasma in some way to produce a secretion. In the case of the organic constituents, the secretory cells synthesize these from raw materials provided by the plasma, and in some instances the finished products are stored in the cells until an appropriate stimulus causes them to be secreted. Water and some of the inorganic ions of the various secretions are the result of direct movement from plasma through and out of the secretory cells at the time they are stimulated to secrete. On the other hand, certain ions that are present in some secretions are produced by the metabolic activities of the secretory cells, but even in such instances, plasma supplies the reactants that participate in the formation of these ions.

Because most exocrine glands contain more than one type of secretory cell, the composition of the total glandular secretion as it enters the lumen of the digestive tract depends in part on the relative activities of the various types of secretory cells. Furthermore, the composition of the mixture of primary secretions may be modified during passage through the ducts as a result of the movement of water and inorganic ions to and from the blood across the duct epithelium. This is the case particularly for the salivary glands, pancreas, and liver, each of which has a relatively long duct system connecting the glandular secretory cells with the lumen of the tract.

The transfer of water and solutes from plasma across the secretory cell membranes into the duct lumen is accomplished by means of a variety of active and passive transport processes (see Chap. 1). Energy is required for the process of secretion. This is attested to by the fact that during active secretion both the oxygen consumption of the digestive glands and their utilization of energy-supplying substrates are increased above resting levels. The energy provided by the increased metabolic activities of the secretory cells is used to support the various activities of the transport systems, which are the primary events associated with secretion.

Although the mechanisms involved in the formation of a secretion are not well understood, it is known that the various types of secretory epithelia are stimulated to secrete by either nervous or humoral mechanisms, and in most instances by both (see Chap. 25). The exact manner in which nervous and humoral factors act on secretory cells to cause secretion is unknown.

SALIVARY SECRETION

There are three pairs of human salivary glands, the *parotid, sublingual,* and *submaxillary* glands. These glands are composed of secretory cells arranged in acini whose secretions empty into a system of small ducts and then into larger excretory ducts that conduct saliva into the oral cavity. There are two types of cells in the acini of the sublingual and submaxillary glands: (1) serous cells, which secrete water, inorganic salts, and the digestive enzyme *salivary amylase,* which initiates the digestion of high-molecular-weight carbohydrates, and (2) mucous cells, which, in addition to water and inorganic salts, secrete a number of glycoproteins collectively called *mucin.* Mucin in water produces a solution of high viscosity known as *mucus,* which gives a thick, viscous quality to certain types of saliva. Since the parotid gland contains only serous cells, its secretion is very watery and of relatively low viscosity.

Humans secrete between 1.0 and 1.5 liters of saliva every day. The rate of secretion is very low during sleep, increases during the awake state in the absence of apparent stimuli to a level of about 0.5 ml per minute, and under maximal stimulation may attain a level of 7 ml per minute. By virtue of its moistening and lubricating properties, saliva facilitates chewing, swallowing, and speech. Dehydration results in decreased secretion of saliva, and this contributes to the sensation of thirst. Saliva has a cleansing action that is important in the prevention of dental caries; in the absence of salivary secretion the incidence of tooth decay is higher than normal. Saliva accomplishes this function by diluting noxious substances and by flushing food particles and bacteria from the mouth. In addition, thiocyanate ions and certain proteolytic enzymes in saliva are bacteriostatic. An important ion of saliva is bicarbonate, whose salivary concentrations are relatively high. This ion neutralizes acids produced by oral bacteria, thereby preventing these acids from dissolving the enamel of teeth.

COMPOSITION AND FORMATION OF SALIVA. Both the volume and the composition of saliva secreted in response to different stimuli are quite variable and depend on the strength and nature of the stimulation. In general, saliva contains the usual electrolytes of the body fluids, the principal ions being sodium, potassium, chloride, and bicarbonate. At low rates of secretion the concentrations of all ions except potassium are low relative to plasma (Fig. 26-1), and saliva is decidedly hypotonic. As the secretory rate increases, the concentrations of sodium, chloride, and bicarbonate rise, and potassium concentration declines a little. At maximal secretory rates, potassium and bicarbonate concentrations are higher than those of plasma, whereas the concentrations of sodium and chloride are less than the plasma levels of these ions. The overall result of these changes is that, as secretory rate increases, osmolarity increases and at maximal rates of secretion approaches that of

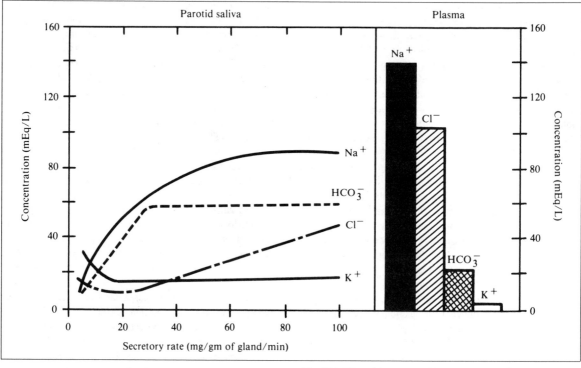

Fig. 26-1. Electrolyte composition of human parotid saliva as a function of secretory rate. (From F. Bro-Rasmussen, S. Killmann, and J. H. Thaysen. The composition of pancreatic juice as compared to sweat, parotid saliva, and tears. *Acta Physiol. Scand.* 37:97, 1956.)

plasma. The pH of saliva rises from 6.2 to 7.4 with increasing rates of secretion as a result of the concomitant increase in bicarbonate concentration.

A number of theories have been offered in explanation of the changes in osmotic pressure and electrolyte composition of saliva that occur with variations in flow rate. One proposes that the primary secretion of the acinar cells is approximately isotonic and has an electrolyte compositon similar to that of plasma. It is further proposed that, during passage through the salivary ducts, the composition of the primary secretion is modified by the transport of ions into and from the duct lumen. Sodium is actively transported from the duct lumen to the blood and is accompanied passively by chloride; however, potassium is actively secreted into duct lumen, but at a rate slower than that of sodium. At low rates of secretion these transport systems have more time to manifest their modifying influences on the composition of the duct fluid; i.e., flow through the ducts is slow. In this circumstance, sodium and chloride concentrations and osmolarity would be at their lowest values, whereas potassium concentrations would be high. Saliva is decidely hypotonic at low rates of secretion, which is explained by (1) the greater amount of sodium leaving the luminal fluid than is

replaced by potassium and (2) the relative impermeability of the duct epithelium to water. As secretory rate increases, there is less time for the sodium and potassium transport systems to act on the duct fluid, and sodium and chloride concentrations and osmolarity increase progressively to their highest values at the maximal secretory rate, and potassium concentration falls. The fact that, over a wide range of secretory rates, bicarbonate concentration is higher than that in plasma suggests that there is an active transport system for this ion that is directed from blood to duct lumen.

The content of organic matter in the saliva, the most notable constituents being mucus and amylase, is variable and seems to depend on the type of stimulus applied. For example, dogs with permanent fistulas of the submaxillary gland have been stimulatd to secrete reflexly by introducing either meat powder or dilute hydrochloric acid into the mouth. It was found that even though the volumes of secretion, electrolyte concentrations, and blood flow through the gland were the same for both stimuli, the concentration of organic matter in the saliva was considerably greater for meat powder as compared with acid.

It is interesting that a response is often well adapted to the function it must perform. To cite one example: If a dog is given fresh meat, a viscous saliva containing much mucus is secreted that lubricates the bolus, thereby facilitating its passage into the stomach. If the same dog is given dry meat powder, a large volume of a watery secretion is produced and washes the powder out of the mouth.

REGULATION OF SALIVARY SECRETION. The secretion of saliva is controlled exclusively by nerve impulses. Saliva is produced in response to impulses acting on salivary centers in the medulla oblongata. These impulses originate mostly from stimulation of sensory nerve endings in the mucous membranes of the mouth, but they can be initiated from many parts of the body, i.e., from the eyes, the nose, and other regions of the gastrointestinal tract as part of the vomiting reflex. The salivary centers, in turn, send impulses over the parasympathetic and sympathetic nerve supplies of the parotid, submaxillary, and sublingual glands, causing them to secrete their specific juices.

In decerebrate animals, stimulation of taste buds by introducing substances into the mouth results in the secretion of saliva. Higher nerve centers are not required for this secretion of saliva, and such reflexes have been called *unconditioned reflexes*. There are also salivary reflexes that require previous experience and involve higher nerve centers. These *conditioned reflexes* were well demonstrated in Pavlov's experiments on salivary secretion in the dog. For example, when a bell was rung every time food was brought to a dog, before long the dog salivated to the bell alone. This type of reflex requires that the cerebral cortex be intact, and it comes into play, for example, when one sees or thinks of appetizing food.

Contrary to their usual antagonistic actions on effector organs, both parasympathetic and sympathetic stimulation are excitatory to salivary secretion. However, the parasympathetic nerves really play the dominant role in normal reflex secretion, because the volume of saliva secreted in response to parasympathetic activity is much greater than that resulting from sympathetic stimulation; the mouth is dry when parasympathetic nerves are nonfunctional.

SALIVARY DIGESTION. The one digestive enzyme present in saliva is salivary amylase, which in the resting gland is stored in zymogen granules in the serous acinar cells. The action of the enzyme is to degrade complex polysac-

charides, such as starch and glycogen, through several intermediate stages (dextrins) to maltose. Small quantitites of glucose are also formed. Since salivary amylase does not function in an acid medium such as is present in the stomach, and since food remains in the mouth for such a short time, it might be thought that the action of this enzyme is extremely limited. However, the bolus of food does not disintegrate immediately on entry into the stomach, and digestion can continue inside the bolus for fairly long periods of time.

GASTRIC SECRETION

The combined activities of the various secretory cells of the human gastric mucosa result in the secretion of 2 to 3 liters of gastric juice every day. Gastric juice is most commonly regarded as the juice secreted by the exocrine glands in the mucosa of the body and fundus of the stomach. Three types of secretory cells are present in these glands (Fig. 26-2). The *parietal (oxyntic) cell* secretes a nearly isosmotic solution of hydrochloric acid (HCl) plus *intrinsic factor,* whereas the *chief cell* is the source of the protein-digesting enzyme *pepsin.* Cells located in the neck of these glands and also in the glands in the cardiac and pyloric regions of the stomach secrete an alkaline fluid that contains a soluble mucus. In addition to the glandular cells, the surface epithelial cells of the stomach secrete an alkaline juice containing an insoluble mucus.

It is impossible to make a single statement concerning the composition of gastric juice, since so many types of secretory epithelia are available for its production. The composition of gastric juice is determined by the relative activities of the different types of secretory cells. These in turn depend on the mechanism, or combination of mechanisms, stimulating the various cells to secrete their specific juices.

SECRETIONS OF THE PARIETAL CELL: HYDROCHLORIC ACID. The parietal cell is unique in that it secretes an almost isosmotic solution of strongly dissociated HCl. Although HCl is not essential for body function, it aids in the gastric breakdown of connective tissue and muscle fibers, activates pepsin, provides an optimal low pH for peptic digestion, and kills bacteria that enter the digestive tract with food. Because of its corrosive actions, HCl is a predominant factor in the formation of ulcers in various regions of the upper digestive tract.

Mechanism of Hydrochloric Acid Secretion. The mechanism of HCl secretion is an extremely efficient active transport process, because the hydrogen ion concentration of parietal cell

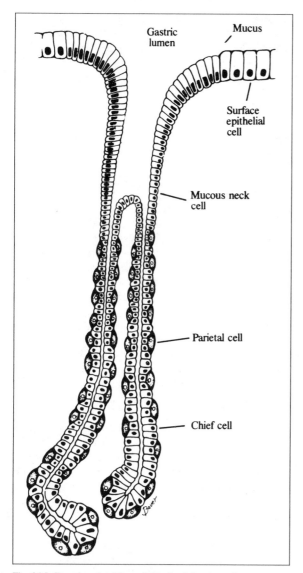

Fig. 26-2. Exocrine gland from the body of the stomach.

tempts to explain the mechanism of HCl secretion. One is illustrated in Figure 26-3. Hydrogen ions are thought to originate from certain metabolic processes of the parietal cell, and these, in conjunction with the energy released by the hydrolysis of ATP, are actively secreted by a pump into the lumen of the stomach. Whenever a hydrogen ion is secreted, an alkaline ion remains behind in the parietal cell, and this base, which can be represented as a hydroxyl ion, must be buffered in order to maintain cellular pH constant. This is accomplished in the following manner: Carbon dioxide, which is either removed from plasma or formed by the metabolic processes of the parietal cell, reacts with water to form carbonic acid under the influence of the enzyme carbonic anhydrase, whose concentration is relatively high in the parietal cell; the hydroxyl ions are buffered by reacting with carbonic acid, with the resultant formation of water and bicarbonate ions. Chloride ions are removed from plasma and secreted along with hydrogen ions to form HCl. Chloride ions do not just move passively along with hydrogen ions, because the surface of the mucosa is negative relative to the serosa; i.e., chloride ions move across the gastric mucosa against a potential difference. In addition, since chloride ions move from a concentration in plasma of about 108 mMol per liter to a concentration of about 170 mMol per liter in gastric juice, they are transported against a chemical as well as an electrical gradient. The transport of this ion is therefore an active, energy-requiring process, probably involving a pump distinct from, but coupled with, that utilized in the transport of hydrogen ions.

To maintain electrical neutrality, for all chloride ions that move from plasma into gastric juice, an equivalent number of bicarbonate ions move from the parietal cell into the plasma. As a result of this movement of bicarbonate ions, the blood leaving the stomach becomes more alkaline during the secretion of gastric acid. However, the acid-base balance of the body is unchanged because, as will be seen later in this chapter, the pancreas removes bicarbonate from the plasma and secretes into the duodenal lumen an amount of this ion about equal to that put into the plasma by the stomach.

Control of Hydrochloric Acid Secretion. The rate of parietal cell secretion depends on the balance between the excitatory and inhibitory mechanisms that influence the parietal cell at any given time. Both neural and hormonal mechanisms are involved in regulating parietal cell secretion.

Two substances in the body act on the parietal cell to stimulate its secretion. One is acetylcholine (ACh), which is liberated by intrinsic neurons that end on the parietal cell, and the other gastrin, which manifests its excitatory in-

secretion (about 150 mEq/liter) is about 2 million times that of the plasma from which it is derived; i.e., transport is uphill against a concentration gradient. A transport process such as this requires the expenditure of energy, and this is provided in the form of high-energy phosphate bonds of adenosine triphosphate (ATP), which are derived from oxidative metabolism and aerobic glycolysis of the parietal cell.

Many theories have been proposed over the years in at-

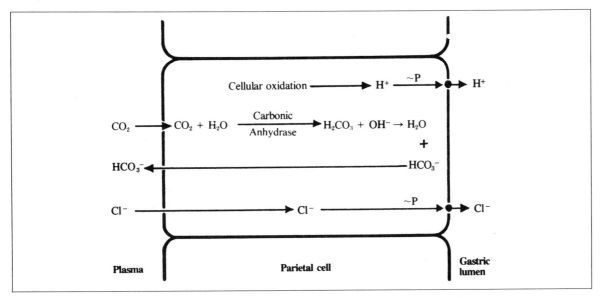

Fig. 26-3. A possible mechanism for the secretion of gastric HCL.

fluence following its release into the blood from endocrine cells (G cells) in the mucosa of the pyloric antrum. Acetylcholine and gastrin act synergistically; i.e., the secretory response to both together is greater than the sum of the two acting alone. Excitation of acid secretion can be divided into the *cephalic, gastric,* and *intestinal phases,* based on the region in which a stimulus acts to cause secretion (Fig. 26-4).

Excitation of receptors in the head, namely, those associated with the taste, smell, and sight of food, activate the reflex secretion of acid that characterizes the cephalic phase of gastric secretion. Since the vagus nerve provides the efferent limb of the long reflex arcs involved, this nerve plays the central role in mediating the cephalic phase of gastric secretion. Vagal impulses, acting through the intrinsic nerve plexuses, cause the release of ACh, not only at the parietal cell but also at the G cell, which responds by secreting gastrin (Fig. 26-4).

The stimuli provided by certain foods are not always sufficient to produce the cephalic secretion, because this response is usually weak or fails if food is not taken with appetite. The fact that food must be agreeable to elicit a strong cephalic response is supported by the observation that there is little gastric secretion during certain types of emotional upset, presumably because no food is palatable under these circumstances. On the other hand, certain emotional stresses can result in excessive secretion, presumably

as the result of hyperactivity of the cephalic phase. It would appear that impulses from higher centers in the brain can modify the rate of secretion of gastric acid by influencing the vagal release of ACh and gastrin. The higher centers include the cortex and the hypothalamus, the latter being, among other things, a center for coordinating impulses associated with emotional expression.

The gastric phase begins when food enters the stomach and continues as long as food remains there. The major stimuli that trigger this phase of gastric secretion are the contact of certain chemical agents with, and distention of, different regions of the stomach (Fig. 26-4). For example, substances that either are present in food or are a product of the digestion of food directly stimulate the G cell to release gastrin. The most potent of these stimulating agents are protein digestion products supplied by gastric digestion. Gastrin release also occurs in response to short reflexes in the antral wall that are activated when this region of the stomach is distended. In this instance, mechanoreceptors in the antral wall are stimulated, and impulses are transmitted over intrinsic nerve fibers causing the release of ACh at the G cell and excitation of gastrin secretion. Part of the gastric juice secreted during the gastric phase is the result of stimulation of the body and fundus of the stomach. Chemical agents, such as protein digestion products, act directly on the parietal cell to stimulate its secretion. Distention of the body and fundus of the stomach activates short reflexes

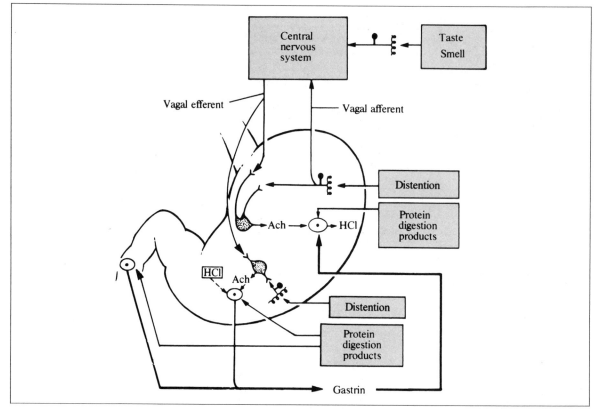

Fig. 26-4. Pathways involved in the regulation of gastric secretion of HCl. The solid arrows represent excitation: the dashed arrow represents inhibition.

through the intrinsic nerve plexuses and long reflexes over the vagi, both of which result in the release of ACh at the parietal cell.

Stimulation of acid secretion at a low rate is continued during the intestinal phase. Since there are some G cells in the mucosa of the duodenum, this response is at least partly due to the release of gastrin by intraduodenal stimuli, such as amino acids and peptides. Other unidentified humoral agents may also contribute to this phase.

A powerful stimulus to parietal cell secretion besides ACh and gastrin is histamine, a substance that is present in relatively large quantities in the gastric mucosa. The role of histamine in the control of gastric acid secretion has not been clarified. It has been suggested that histamine is the final link in the secretion of HCl; i.e., all neural and humoral stimuli act by causing the release of histamine, and that this is the substance that stimulates the parietal cell to secrete. However, it is possible that histamine liberated in the region of the parietal cell sensitizes this cell to ACh and gastrin,

thereby potentiating the effects of these neural and hormonal stimuli.

It is apparent that a host of mechanisms are available for the stimulation of gastric acid secretion. However, there are also mechanisms that inhibit the secretion of HCl. For example, acid solutions in contact with the mucosa of the pyloric antrum inhibit HCl secretion by the body of the stomach. The extent of this inhibition is proportional to the hydrogen ion concentration of the antral contents, with complete inhibition occurring at pH 2. Thus, HCl acts as an inhibitor of its own secretion, the function of such an autoregulatory mechanism being to prevent excessive secretion of acid. Although it has been suggested that acid inhibition is the result of the release of an inhibitory hormone from the antrum, the bulk of the evidence is in favor of the idea that acid acts by suppressing the release of gastrin (Fig. 26-4); i.e., the G cell is sensitive to hydrogen ions.

Since the pH of the contents of the antrum is low at different times during a day, this suppression mechanism plays an important role in regulating gastric acid secretion. For example, during interdigestive periods, the pH of the small volume of contents that is present in the stomach at this time is low, and the release of gastrin is suppressed. When a meal is eaten, the acid that is present is buffered by the constituents of food, antral pH rises, the inhibition is removed, and gastrin is released in response to the usual stimuli. Gastric juice is secreted at a high rate, which continues until the buffering power of whatever food remains in the stomach is exhausted. At this time, pH decreases, and the resultant acidification of the antrum brings the gastric phase to an end.

A variety of substances are known to inhibit gastric secretion when they contact the duodenal mucosa. Examples are fat digestion products, acid, and hypertonic solutions. The mechanisms involved are the same neural and hormonal pathways that are operative under the same circumstances in the inhibition of gastric motility (see Fig. 25-15); i.e., both gastric secretion and motility are inhibited by activation of enterogastric reflexes and release of secretin, cholecystokinin, and gastric inhibitory peptide from the duodenal mucosa. Although the effect of the intravenous administration of cholecystokinin is to stimulate acid secretion, the actual result of the liberation of this hormone from the intestinal mucosa in response to a meal is one of inhibition. The reason for this is that there is competitive inhibition of gastrin by cholecystokinin in the process of gastric acid secretion. Since both gastrin and cholecystokinin possess the same terminal tetrapeptide, the two molecules compete for the same acid-stimulating receptor sites on the parietal cell. However, cholecystokinin is a weak stimulant compared with gastrin. Consequently, when both hormones are released in response to a meal, the overall gastric secretory response is less than that of gastrin alone, because gastrin cannot manifest its full excitatory effects in the presence of cholecystokinin; i.e., cholecystokinin denies receptor sites to gastrin.

SECRETIONS OF THE PARIETAL CELL: INTRINSIC FACTOR. An important substance secreted by the human parietal cell is intrinsic factor, the only gastric secretory component that is essential for life. This mucoprotein is necessary for the absorption of vitamin B_{12}, which is required for the formation of normal red blood cells. Absorption of vitamin B_{12} by a special transport process in the lower small intestine requires that this vitamin be combined with intrinsic factor. When intrinsic factor is not secreted by the stomach, vita-

min B_{12} absorption is defective, and pernicious anemia results.

SECRETION OF PEPSIN. Pepsin is secreted in the form of an inactive molecule, *pepsinogen,* that is synthesized by and stored as zymogen granules in the chief cells. Pepsinogen is also secreted by the various mucous cells of the gastric mucosa and by Brunner's glands of the upper small intestine. Once secreted, a small fragment is split from inactive pepsinogen to form active pepsin. This conversion occurs in the presence of either HCl or small amounts of pepsin. Thus, it is likely that in the gastric lumen, HCl initiates the breakdown of pepsinogen, and the small amounts of pepsin so liberated then carry the process to completion.

Pepsin is an endopeptidase and splits linkages in the interior of both proteins and polypeptides. Since this occurs most effectively at pH 2 to 3, the acid parietal cell secretion provides a favorable environment for optimal activity of this enzyme. Pepsin is inactivated when the pH is greater than 5. As the result of the peptic digestion of protein, mostly peptides of varying chain length along with a few amino acids are found in the stomach following ingestion of protein.

Since the most effective stimulus for the secretion of pepsinogen is ACh, the long and short reflexes that occur during the cephalic and gastric phases and result in the release of ACh in the vicinity of the pepsinogen-secreting cells are of major importance in determining pepsin levels in the gastric contents. Stimulation of pepsinogen secretion by gastrin, although effective, is weak compared with stimulation by ACh. However, there is good correlation between acid and pepsin outputs in the human in that most stimuli that increase the secretion of acid also increase secretion of pepsinogen. An exception is the hormone secretin that inhibits secretion of acid but stimulates pepsinogen secretion.

SECRETION OF MUCUS. All regions of the stomach contain cells that secrete an alkaline fluid containing mucus. A layer of mucus, 1.0 to 1.5 mm thick, coats the gastric wall and serves as a protective barrier against various forms of damage to the gastric mucosa. It provides protection against mechanical injury by serving as a lubricant, and against chemical injury by virtue of its neutralizing properties. Mucus holds the alkaline fluid within its gel-like structure, and when acid diffuses into the gel, it can be neutralized before coming into direct contact with the epithelium. When pepsin diffuses into the mucus barrier, this enzyme is inactivated in the medium of high pH, so that the chance of attack on the protein structure of the underlying epithelium is minimized. Prime stimuli for the secretion of mucus are chemical, mechanical, and thermal irritation of the gastric mucosa. Both vagal and sympathetic stimulation also excite mucus secretion.

ULCERS. Under certain circumstances the corrosive actions of HCl produce erosions of the gastric mucosa (gastric ulcer) and, more frequently, of the mucosa of the upper small intestine (duodenal ulcer). These erosions can extend into, or even all the way through, the underlying tissues, and damage to blood vessels in the wall can result in bleeding. However, even though HCl is more or less continually present in the stomach, mechanisms are usually available to provide protection sufficient to prevent cellular destruction of this magnitude.

The most important mechanism protecting the stomach from chemical damage by acid is what has been called *the gastric mucosal barrier*. This barrier is created by the permeability characteristics of the gastric mucosal cell membranes such that the entrance of ions to the interior of these cells, even an ion as small as the hydrogen ion of HCl, is very restricted; i.e., HCl does not ordinarily enter these cells in quantities sufficient to destroy them. Another protective mechanism is dilution and neutralization of the highly acid secretion of the parietal cell. Food, particularly protein, and, to a lesser extent, the alkaline secretions of the salivary glands and gastric nonparietal cells participate in the dilution and neutralizing mechanism. In addition, as will be seen later in this chapter, the pancreas secretes an alkaline juice into the duodenum that very efficiently neutralizes acid that enters the upper small intestine from the stomach, thereby protecting the lining of the intestine from damage. Finally, an important factor in the prevention of injury to the gastric mucosa is the rapidity (1–3 days) with which the cells lining the stomach are regenerated by cell division.

Ulcer may result when excessive secretion of acid overwhelms the protective mechanisms. For example, the persistence of certain emotions, particularly those of a stressful nature, is associated with ulcer formation. In this instance it is suspected that hypersecretion of gastric juice results from a high level of vagal stimulation through higher centers in the brain that coordinate impulses associated with emotion. Ulcers may also be formed in the presence of a normal gastric secretory pattern when membrane permeability is increased by bacterial infection, mechanical trauma, or interference with the blood supply to localized regions of the gastric mucosa. The increase in membrane permeability allows HCl to enter and destroy cells.

Ulcer therapy is based on physiological considerations. Attempts are usually made either to reduce the secretion of HCl or to minimize the potentially destructive actions of this secretory component. One approach has been to perform gastric vagotomy in order to eliminate the excitatory influences on secretion that are normally mediated by the vagi. Limiting features of this procedure are the decreased gastric motility and retention of food in the stomach resulting from loss of vagal excitation of gastric motor activity. Either the pyloric antrum can be removed to eliminate gastrin as an excitant of gastric secretion, or portions of the body of the stomach can be excised to reduce the mass of acid-secreting cells. Less traumatic, nonsurgical procedures include rest and an environment free from worry (reduced emotional stress), elimination of certain acid-stimulating agents from the diet, and the ingestion of various commercial alkalis or buffers (neutralization of gastric acid).

PANCREATIC SECRETION

In addition to containing islets of alpha and beta endocrine cells that secrete the hormones glucagon and insulin into the circulation (see Chap. 31), the pancreas contains exocrine secretory cells arranged in acini that are connected by small intercalated ducts to larger excretory ducts. The larger ducts converge into one or two main ducts that deliver the exocrine secretion of the pancreas to the duodenum.

The pancreas secretes digestive enzymes that make an extremely important contribution to the digestion of food. Since the pancreatic enzymes that are secreted into the lumen of the small intestine require a pH close to neutrality for best activity, the acidity of the contents entering the duodenum from the stomach must be neutralized. Pancreatic juice contains a relatively high concentration of sodium bicarbonate, and this alkaline salt is largely responsible for the neutralization of gastric acid.

Pancreatic juice consists of two distinct components. One of these is an aqueous juice that makes up the largest part of the volume secreted by the pancreas and has a high concentration of sodium bicarbonate, but with very little enzyme activity. The cells responsible for the secretion of the 1000 to 1500 ml of aqueous juice produced every day are those lining the intercalated ducts. The second component is a juice of very small volume that contains the digestive enzymes of the pancreas. This enzyme juice is elaborated by the acinar cells, which synthesize the digestive enzymes of the pancreas and store them as zymogen granules prior to secretion.

SECRETION OF WATER AND BICARBONATE. The composition of the aqueous juice has been well established (Fig. 26-5). This secretion has the same osmotic activity as plasma

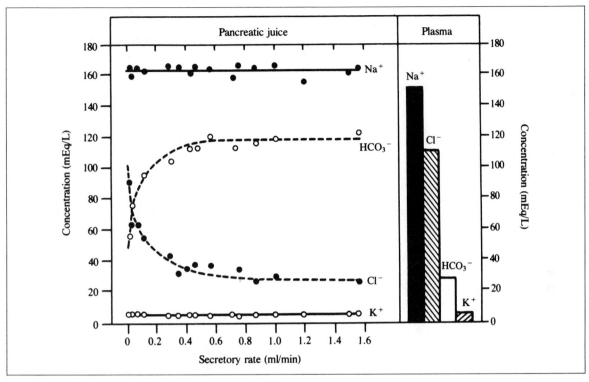

Fig. 26-5. Relation between rate of secretion and concentration of sodium, potassium, chloride, and bicarbonate in the pancreatic juice of the dog (after secretin injection). (From F. Bro-Rasmussen, S. Killmann, and J. H. Thaysen. The composition of pancreatic juice as compared to sweat, parotid saliva, and tears. *Acta Physiol. Scand.* 37:97, 1956.)

and, because of its high bicarbonate content, is alkaline, with a pH ranging from 7.6 to 8.2. The principal cations are sodium and potassium, which are present in the same concentrations as in plasma at all rates of secretion. The outstanding feature of the aqueous juice is that its bicarbonate concentration is high relative to plasma. The concentration of this anion increases with increased rates of secretion and ranges from about 60 mEq per liter at low secretory rates to 140 mEq per liter at high rates. The other anion is chloride, whose concentration is low relative to plasma. However, the sum of the chloride and bicarbonate concentrations at any rate of secretion is the same as the sum of these concentrations in plasma; i.e., as bicarbonate concentration rises with increased secretory rate, chloride concentration falls, so that the osmolarity remains unchanged and the same as that of plasma.

A number of theories have been suggested regarding the mechanism of aqueous juice secretion. One suggests that the primary cellular secretion is an isosmotic solution consisting mostly of sodium bicarbonate along with a little potassium bicarbonate, and that the bicarbonate ion is actively transported into the duct lumen by the cells of the intercalated ducts (Fig. 26-6). That the transport of bicarbonate is active is indicated by the movement of this ion from a concentration in plasma of 27 mEq per liter to concentrations as high as 140 mEq per liter in pancreatic juice; i.e, transport is against a concentration gradient. The bicarbonate of the aqueous juice is formed in the duct cells by the hydration of carbon dioxide, which either enters the cells from plasma or is generated by the metabolic processes of the cells. This reaction is catalyzed by the enzyme carbonic anhydrase, whose concentration is relatively high in the duct cells. The hydrogen ion that results from the reaction

$$CO_2 + H_2O \xrightarrow[\text{anhydrase}]{\text{carbonic}} H_2CO_3 \longrightarrow HCO_3^- + H^+$$

is transported to plasma, probably in exchange for sodium.

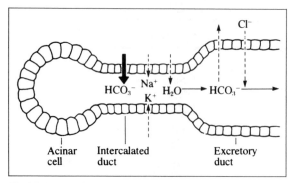

Fig. 26-6. Mechanism of secretion of the aqueous component of pancreatic juice. The large solid arrow represents active transport; the dashed arrows represent passive transport.

That a sodium pump is involved in this exchange is suggested by the fact that agents that inhibit epithelial transport of sodium also interfere wih pancreatic secretion of bicarbonate. As bicarbonate is pumped into the duct lumen, in order to maintain electrical neutrality, sodium and potassium follow bicarbonate passively in the same proportion in which these cations are present in plasma, and water moves passively with the electrolytes in order to satisfy osmotic dictates.

The variations in bicarbonate and chloride concentrations that occur with alterations in rate of secretion can be explained as follows: As the bicarbonate solution moves through the duct system (Fig. 26-6), the secreted bicarbonate exchanges passively with plasma chloride, the extent to which this exchange occurs being determined by the rate of secretion. The greatest exchange would take place at lower flow rates, because there would be more time available for the exchange to occur; in this instance the concentration of bicarbonate in pancreatic juice as it entered the intestine would be relatively low, with that of chloride relatively high.

SECRETION OF ENZYMES. Pancreatic juice is the most versatile and active of the digestive secretions. Its enzymes are capable of almost completing the digestion of all foods in the absence of other digestive secretions.

Trypsin, a proteolytic enzyme, is secreted as an inactive precursor, *trypsinogen.* Trypsinogen is initially converted to active trypsin by *enterokinase,* an enzyme present in the mucosal cells of the small intestine. The trypsin so formed in turn activates trypsinogen and the other proteolytic enzymes of pancreatic juice. One of these is *chymotrypsin,* which is secreted as inactive *chymotrypsinogen.* Both tryp-

sin and chymotrypsin split certain linkages in the interior of proteins and peptide chains. Pancreatic juice also contains *carboxypeptidase,* which is secreted as inactive *procarboxypeptidase.* This enzyme liberates amino acids with free carboxyl groups from the end of peptide chains. The combined actions of pepsin and the various protein-digesting enzymes of pancreatic juice are to degrade food proteins to a mixture of small peptides and amino acids.

A trypsin inhibitor present in the pancreas, combines with trypsin in the ratio of one molecule of inhibitor to one molecule of trypsin. Since the end product of the combination is enzymatically inactive, the inhibitor prevents the pancreas from digesting itself during those times when small amounts of trypsin are activated in it. However, if large amounts of trypsin are activated, the inhibitor is overcome, and the pancreas may be damaged or destroyed completely.

The pancreas supplies the enzyme *pancreatic lipase,* which is primarily responsible for digestion of fats. The action of this enzyme is to split the ester linkages in triglycerides, to produce mostly free fatty acids and 2-monoglycerides.

The pancreas also secretes an *amylase,* whose action is similar to that of salivary amylase. Starch and glycogen are degraded to maltose along with small amounts of glucose. Some pancreatic amylase escapes into the blood during pancreatic secretion, because stimulation of the pancreas results in a rise in plasma amylase. Plasma amylase levels have been used to evaluate the functional status of the pancreas. For example, trauma to this organ elevates plasma amylase levels, presumably by increasing the escape of pancreatic amylase into the blood. During autolysis of the pancreas by tryptic digestion, both plasma and urinary amylase show marked rises.

Additional enzymes secreted by the pancreas include ribonuclease, deoxyribonuclease, elastase, cholesterol esterase, and phospholipase.

CONTROL OF SECRETION. The secretion of pancreatic juice is regulated by both neural and hormonal mechanisms. These mechanisms are summarized in Figure 26-7.

Reflex Control. Pancreatic secretion is a part of the cephalic phase of digestion, which is demonstrated by the fact that the smell, sight, and chewing of food induce a secretion of small volume but rich in enzymes. The efferent limb of the reflex arcs involved in this cephalic response is the vagus nerve. There is also neural control of pancreatic secretion during the gastric phase of digestion. When the body of the stomach is distended by food, the pancreas responds by increasing its output of enzymes. The reflex involved is the

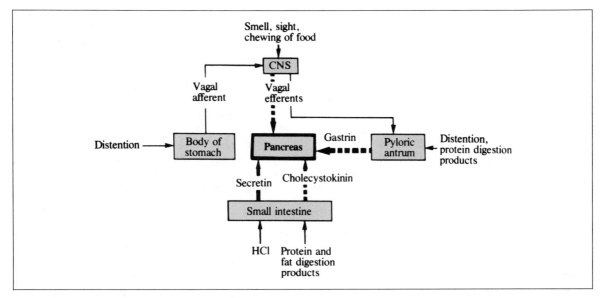

Fig. 26-7. Neural and hormonal mechanisms for the secretion of pancreatic juice. The heavy solid arrow represents aqueous juice; the heavy dashed arrows represent enzyme juice.

gastropancreatic reflex, whose afferent and efferent pathways are both in the vagus nerve.

Hormonal Control. Hormones play the central role in the control of secretion by the pancreas. One of these, secretin, was the first of the many hormones of the body to be discovered. This polypeptide hormone is released into the blood from the duodenal mucosa during the intestinal phase of digestion and is carried to the pancreas, where it stimulates the secretion of a large volume of bicarbonate-rich juice; i.e., secretin stimulates secretion of the aqueous component of pancreatic juice. Although peptides, amino acids, and fatty acids cause the release of secretin, the most potent stimulus for secretin release is HCl, the threshold pH being 4.5. Thus, the main function of the secretin mechanism is in the neutralization of acid gastric contents by stimulating the release of alkali from the pancreas.

In addition to the neural controls for the regulation of the secretion of pancreatic enzymes and the intestinal hormonal control for the secretion of water and electrolytes, there is also an intestinal hormone for the secretion of enzymes. This polypeptide hormone, which was formerly called *pancreozymin,* but is now called *cholecystokinin,* is released from the intestinal mucosa when it is contacted by the peptides, amino acids, and fatty acids that are present in the intestinal contents as the result of the digestion of protein and fat.

There is a strong pancreatic enzyme secretory response to the release of gastrin from the pyloric antrum. It is not surprising that gastrin stimulates pancreatic enzyme secretion, because it possesses the same terminal tetrapeptide as cholecystokinin. Thus, secretion of pancreatic enzymes is stimulated by the gastrin released as the result of impulses over the vagus nerves during the cephalic phase of digestion and by direct stimulation of the antrum during the gastric phase.

PATHOPHYSIOLOGY OF THE EXOCRINE PANCREAS. The digestion of food is incomplete when secretion of pancreatic enzymes into the small intestine is defective. Certain types of pancreatic disease impair the ability of the pancreas to synthesize and secrete enzymes in quantities sufficient to ensure adequate digestion, and thus absorption, of certain constituents of the diet. For example, under certain circumstances, activated pancreatic enzymes liberated from the acinar cells into the surrounding tissue may attack the pancreas itself, producing the condition *pancreatitis.* This process of self-digestion results in a reduced output of enzymes. Lack of enzymes may also be due to obstructed flow of pancreatic juice into the duodenum. Under these circumstances there is an increased amount of carbohydrate and protein in the feces. However, the dominant effect of pancreatic enzyme deficiency is *steatorrhea,* the appearance of excessive amounts of undigested fat in the feces, i.e., up to 60 to 70 percent of ingested fat. The loss of these nutrients can be partly corrected by oral administration of pancreatic enzymes.

BILE SECRETION

Bile is secreted continuously by the parenchymal cells of the liver. It first enters the bile canaliculi and flows from these minute channels through a converging system of small ducts into the hepatic duct and then into the common bile duct, which enters the duodenum at or near the site where the pancreatic duct joins the organ. During periods when there is no food in the upper digestive tract, most of the bile secreted by the liver is diverted into the gallbladder, where it is stored and concentrated by the absorption of fluid. Following a meal the amount of bile entering the duodenum is increased as the result of contraction of the gallbladder and enhanced secretion by the liver. Bile is the vehicle for the excretion of certain end products of hemoglobin metabolism, the bile pigments, and provides the bile salts, which play an important role in the intestinal absorption of fat.

COMPOSITION OF BILE. The amount of bile secreted every day by the human liver ranges from 500 to 1000 ml. The composition of liver bile is given in Table 26-1. The bile salts are the major secretory components of bile, and the bile pigments are the chief excretory components. Other organic constituents present in lesser amounts are cholesterol, lecithin, fatty acids, and mucin. Inorganic ions of bile include sodium as the major cation and chloride and bicarbonate as the predominate anions.

The amount of bile secreted by the liver every day is far greater than the capacity of the gallbladder, which is about 50 ml. However, this inequality of volumes is compensated for by the considerable absorptive capacity of the epithelium of the gallbladder. During the time bile is stored in the gallbladder, the epithelial cells lining the inner surface of this organ actively transport sodium from lumen to blood. The active transport of sodium is accompanied by the passive movement of chloride and bicarbonate, and the osmotic gradient created by the transfer of inorganic ions causes water to follow these solutes as they leave the gallbladder lumen. The result of this removal of isosmotic fluid from bile is that the organic constituents, which are not absorbed, are concentrated fivefold to tenfold in the gallbladder (Table 26-1).

Both liver bile and gallbladder bile have the same osmotic pressure as plasma. This is the case even though the sum of the concentrations of all of the solutes of bile is considerably greater than that of plasma. For example, the concentration of sodium in bile is much higher than plasma sodium concentration. This apparent discrepancy is reconciled by the fact that certain constituents of bile, namely the bile salts, lecithin, and cholesterol, form macromolecular aggregates (*micelles*) that have low osmotic activity compared with the monomolecular dispersion of these molecules. The bile salts are weak organic acids and at the pH of bile are present in the micelles as anions. Since each bile salt molecule in a micelle requires a cation (sodium) to balance its negative charge, the ability of cations sequestered in micelles to develop osmotic pressure is minimized.

BILE PIGMENTS. The major pigment of bile is *bilirubin,* a water-insoluble end product of the degradation of the porphyrin moiety of hemoglobin by the reticuloendothelial system. The 0.5 to 1.0 gm of bilirubin produced in a human every day is extracted by the parenchymal cells of the liver from the blood, where it is maintained in solution by attachment to albumin. This pigment is actively secreted into the bile following its conversion by the liver cells to the water-soluble bilirubin diglucuronide. Intestinal bacteria convert bilirubin into other pigments, and small quantities of these are absorbed from the intestine and excreted by the kidneys. Most of the bile pigments are excreted in the feces, to which they impart a characteristic brown color.

BILE SALTS. Bile salts are related structurally to cholesterol, from which they are synthesized by the parenchymal cells of the liver. The liver synthesizes the primary bile acids, cholic and chenodeoxycholic acid, whereas the secondary bile acids, deoxycholic and lithocholic acid, are products of intestinal bacterial metabolism of primary bile acids. The bile acids are conjugated with the amino acids glycine and taurine to form the more water-soluble bile salts, such as glycocholic and taurocholic acid.

In addition to synthesizing bile salts, the liver cells possess an active transport system for the secretion of these

Table 26-1. Composition of human bile

Constituent	Liver bile %	Gallbladder bile %
Water	97.48	83.98
Mucin and pigments	0.53	4.44
Bile salts	0.93	8.70
Fatty acids	0.12	0.85
Cholesterol	0.06	0.87
Lecithin	0.02	0.14
Mineral salts	0.83	1.02

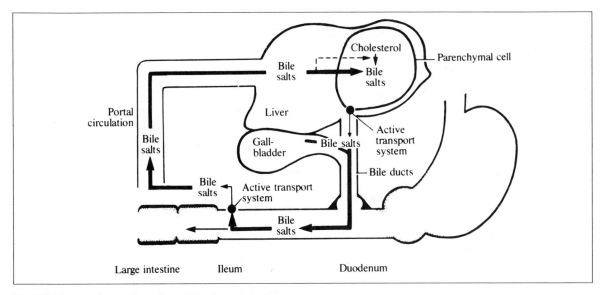

Fig. 26-8. The enterohepatic circulation of bile salts as shown by the heavy solid arrows. The dashed arrow represents inhibition of bile salt synthesis.

molecules into the bile canaliculi. This transport system enables these cells to secrete bile salts into bile in concentrations many times greater than their concentration in the plasma from which they are derived. There is an upper limit to transport capacity at high plasma bile salt concentrations; i.e., the transport system becomes saturated.

Following their secretion and subsequent appearance in the small intestine, the bile salts play an essential role in the absorption of fat. The function of the bile salts in this absorptive process will be discussed later in the chapter. Following their participation in the absorption of fat in the proximal small intestine, the bile salts are absorbed from the terminal ileum by an active process specialized for the transport of these substances. They are then returned by the portal vein to the liver, where they are removed from the blood by the liver cells and resecreted in the bile. The recycling of bile salts between liver and small intestine is called the *enterohepatic circulation of bile salts* (Fig. 26-8). This circulation is interrupted during fasting, when the bile salts are stored in the gallbladder.

Since about 4 to 8 gm of bile salts is secreted in response to a meal, and since the total body content of bile salts averages only 3.6 gm, the body pool of bile salts must circulate about once or twice during the digestion of a meal. A total about 12 to 30 gm of bile salts is secreted every day in response to all food eaten; hence, each bile salt molecule

circulates from four to nine times, the actual extent of the recirculation depending on the type and amount of food ingested. About 5 percent of the bile salts escapes into the large intestine with each circulation (Fig. 26-8), and the result is a loss of 0.5 to 1.0 gm of bile salts in the feces every day. However, since the liver cells are capable of synthesizing bile salts at rates up to 3 to 5 gm per day, synthesis of new bile salts by the liver easily makes up this loss. Although synthesis of bile salts is a continuous process, the rate at which these molecules are synthesized depends on the amount of bile salts returned to the liver in the enterohepatic circulation. When the portal blood concentration of bile salts is low, the rate of synthesis is high, and vice versa. This control of synthesis is through inhibition by circulating bile salts of 7-alpha-hydroxylase, the enzyme that catalyzes the rate-limiting step in the series of reactions leading to bile salt synthesis (Fig. 26-8).

CONTROL OF SECRETION. Any substance that stimulates an increased secretion of bile is a *choleretic,* and the process by which secretion occurs is known as *choleresis.* Humoral agents are the major choleretics involved in the regulation of bile secretion.

The bile salts, which act directly on the secretory cells of the liver to stimulate secretion of bile in proportion to their concentration in the portal venous blood, are potent chemical choleretics and are the major stimuli for the secretion of bile. Certain studies have quantitated the influence of the bile salts on the rate of flow of water and electrolytes into

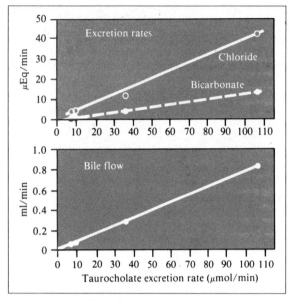

Fig. 26-9. Relationships between secretion rate of bile salt and output of chloride, bicarbonate, and water (bile flow) in the dog. Sodium taurocholate was infused intravenously at rates of 8, 40, 118, and 8 μmol/min (in that order) to obtain the four points. Output of water was directly proportional to the taurocholate secretion rate, as was the output of chloride and bicarbonate ions. (From H. O. Wheeler. Water and Electrolytes in Bile. In C. F. Code [ed.], *Handbook of Physiology*. Washington, D.C.: American Physiological Society, 1968. Section 6: Alimentary Canal. Vol. 5, p. 2417.)

the bile. Figure 26-9 shows that bile flow increases in direct proportion to the rate of taurocholate secretion, and there is a similar relationship for the output of chloride and bicarbonate. These data suggest that both the flow of bile (water output) and electrolyte output are dependent on the rate at which bile salts are secreted. The explanation for this is that the active transport of bile salts provides the primary osmotic force for the secretion of bile. When bile salts are actively transported into the bile canaliculi, a certain amount of water follows by virtue of osmotic drag. In addition, since there is a diffusion gradient for inorganic ions, these ions move into the canaliculi accompanied by more water, the final result being the formation of a secretion isosmotic to plasma.

During periods between meals, the bile salts are sequestered in the gallbladder, the portal venous concentration of these substances is low, and stimulation of bile flow by bile salts is minimal. When food is being digested, the bile salts are circulated between the small intestine and liver, the

portal venous concentration of bile salts is raised, and the extent to which these substances are actively transported is increased, with the result that liver bile flow is stimulated.

Any stimulus that is known to release secretin and to cause the pancreas to secrete also stimulates the secretion of bile. The response of the liver to secretin consists of an increased output of water and electrolytes, particularly sodium bicarbonate, without any increase in the output of organic constituents. This is analogous to secretin stimulation of the pancreas, which has little effect on enzyme output but a decided effect of the water-electrolyte component. As is also the case for the pancreas, bicarbonate concentration rises and chloride concentration falls as the secretory rate is increased in response to secretin. Another analogy between these two organs is that the site of action of secretin on the biliary system, like the pancreas, is the epithelial cells lining the ducts. The increase in bile flow produced in response to secretin is small compared with that produced by the bile salts. However, the secretin-stimulated alkaline fluid is important in maintaining the bile salts in water-soluble form, because these molecules precipitate out of solution when the pH falls below 4.

The major pathways involved in the excitation of secretion of bile by the liver, as well as those for contraction of the gallbladder, are summarized in Figure 26-10. In addition, impulses over the vagi produce an increased liver bile flow. This may be due in part to a direct stimulatory effect on the secretory cells of the liver. However, since gastrin elicits a small secretory response in the liver, vagal stimulation probably manifests some of its excitatory effect through the release of gastrin.

ABNORMALITIES OF THE BILIARY SYSTEM. *Cholelithiasis* (the presence of gallstones in the gallbladder or bile ducts) is one of the most common diseases of the gastrointestinal system. Cholesterol, a normal constituent of bile, is the major component of most gallstones. Cholesterol itself is virtually insoluble in water. However, two other constituents of bile, the bile salts and lecithin, together possess the capacity to dissolve cholesterol by forming water-soluble micellar aggregates that contain all three of these molecules (*mixed micelles*). Whenever cholesterol precipitates out of solution, small crystals of cholesterol are formed, which can gradually grow into large gallstones. This can happen when the capacity of the bile salts and lecithin to dissolve cholesterol is exceeded: when, for example, the liver secretes higher-than-normal quantities of cholesterol, or the amount of bile salts secreted is less than normal.

When the bile ducts are obstructed either by a gallstone

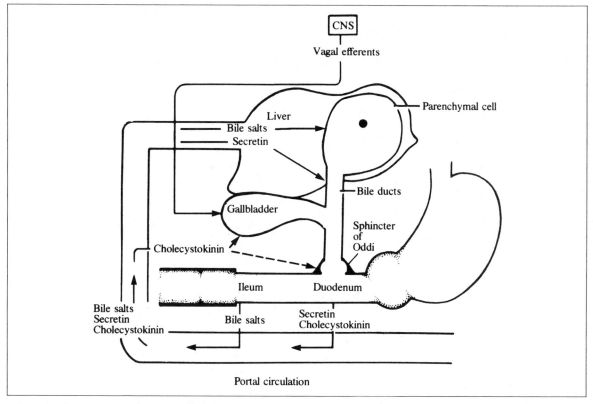

Fig. 26-10. Pathways in the regulation of secretion of bile by the liver and expulsion of bile into the duodenum. The solid arrows represent excitation; the dashed arrow represents inhibition.

or a tumor, flow of bile into the small intestine is retarded or stopped. Since secretion by the liver continues, pressure in the biliary system rises to higher-than-normal levels, and severe pain may be experienced as the result of stimulation of receptors in the wall of the biliary tree. High pressures also result in the leakage of bile from the biliary system back into the circulation, and the bile pigments, rather than being excreted, accumulate in the blood and tissues, giving the skin a yellow color (jaundice). Another consequence of obstructed flow of bile into the intestine is excessive secretion of fat in the feces, because the quantity of intestinal intraluminar bile salts is not sufficient to ensure normal absorption of fat.

SECRETION OF INTESTINAL JUICE

The secretion of intestinal juice is the result of the secretory activities of a variety of cell types whose combined activities produce 1 to 2 liters of juice per day. Two types of intestinal glands contribute to this secretion: Brunner's glands, which are present in the duodenum, and the crypts of Lieberkühn, which are scattered through the entire length of the small intestine.

Brunner's Glands. The major contribution to the secretion of the upper duodenum is probably made by Brunner's glands, which lie in the submucosa and contain only mucous cells. The volume of juice secreted here is low during fasting and increases somewhat following feeding. However, it is known that the upper duodenum has a greater resistance to ulceration as compared with the remainder of the small intestine, probably because the mucus secreted by this region is thick and forms a protective coating over the underlying epithelium.

Crypts of Lieberkühn. Four types of cells are present in the tubular glands of the mucosa that constitute the crypts of Lieberkühn. These are (1) undifferentiated cells that migrate up to the villi to become columnar epithelial absorptive cells; (2) mucus-secreting goblet cells; (3) argentaffin cells, which in addition to synthesizing serotonin probably have an endocrine function; and (4) Paneth cells, whose function is unknown.

Experiments dealing with the secretion of intestinal juice are complicated by the fact that absorption proceeds at the same time as secretion. Consequently, both the volume and the composition of the juice recovered from the lumen of different regions of the small intestine are the net result of both processes. This fluid is isosmotic to plasma and contains the usual electrolytes of plasma and some organic matter consisting of mucus, enzymes, and cellular debris. The concentrations of sodium, potassium, and calcium are relatively constant throughout the small intestine and similar to those in plasma. Although the total anion concentration is also much the same as in plasma, bicarbonate concentration is lower and chloride concentration is higher in jejunal fluid than in ileal fluid.

The small quantity of enzymes that has been reported to be present in intestinal juice, rather than being a component of an exocrine secretion, is probably derived from epithelial cells that are shed from the intestinal mucosa and disintegrate in the intestinal lumen. This idea is supported by the observation that the amount of enzymes present in juice collected from the intestine is usually related to the content of cellular debris. The process of extrusion of the epithelium is a continuous one, resulting in complete renewal of the intestinal epithelium every 1 to 2 days. In any event, only two of the enzymes found in intestinal juice are thought to play a significant role in intraluminal digestive processes. These are enterokinase, which has been mentioned in connection with the activation of trypsin, and intestinal amylase, which is similar in action to salivary and pancreatic amylases.

A proposed function of intestinal secretion is that of providing fluid in large enough quantities to allow efficient intestinal digestion and absorption of food; i.e., the water of intestinal juice serves as a solvent and as a medium of suspension and transport for the solids that are either dissolved or suspended in the contents of the small intestine.

CONTROL OF SECRETION. Vagal stimulation increases the rate of secretion of mucus throughout the length of the small intestine. However, local mechanical and chemical stimulation of any region of the small intestine is the major stimulus for intestinal secretion. This excitation of secretion is probably mediated largely through short reflexes in the intestinal wall. Intestinal secretion is also stimulated by secretin and cholecystokinin and is inhibited by catecholamines.

COLONIC SECRETION

The colonic mucosa has numerous tubular glands containing goblet cells that secrete mucus. The fluid secreted by these glands is very viscous and alkaline (pH 8.0–8.4). Stimulation of the parasympathetic fibers of the pelvic nerves, chemical irritation, and mechanical irritation increase the volume of colonic secretion. The function of the secretion is to lubricate and facilitate the passage of feces and to protect the mucosa of the colon from mechanical and chemical trauma. The alkaline secretion serves to neutralize irritating acids formed by bacterial action, a process that occurs to a considerable extent in the colon. When there is excessive secretion of mucus by the colon, this substance can constitute the major portion of a bowel movement. This condition is called *mucous colitis* and is thought to be caused by excessive parasympathetic activity, most often attributable to emotional tension.

INTESTINAL ABSORPTION

Intestinal absorption is the movement of water and dissolved materials, such as various digestion products, vitamins, and inorganic salts, from the lumen of the small intestine through the barrier imposed by the semipermeable intestinal membrane and into the blood and lymph. The structure of the intestinal absorptive surface is shown in Figure 26-11. A major feature of this surface is the *villus*, which is a small, fingerlike projection lined with epithelial cells. The luminal aspect of each epithelial cell is covered with much smaller projections, the *microvilli*. Collectively, the microvilli of the epithelial cells lining the small intestine are known as the *brush border membrane*. This luminal membrane contains both a variety of digestive enzymes and a number of transport systems that are specialized for the absorption of specific components of the diet. Anatomically, the small intestine is well suited to the task of absorption, because the surface area available for absorption is greatly increased, not only by the presence of the villi but also by the microvilli (Fig. 26-12).

Certain chemical reactions occurring in the epithelial cells supply energy for the normal functioning of those intestinal transport systems that are capable of moving certain digestion products and inorganic salts from the intestinal lumen to the blood against electrochemical gradients (active transport). Such transfer is impeded when the oxidative metabolism of the mucosal cells is inhibited. Another driving force involved in the transport of molecules across the intestinal epithelium is the difference in concentration of a substance

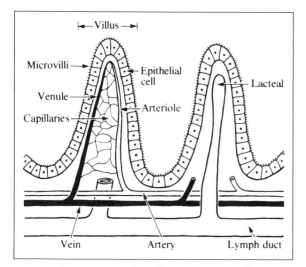

Fig. 26-11. Structure of the absorptive surface of the small intestine.

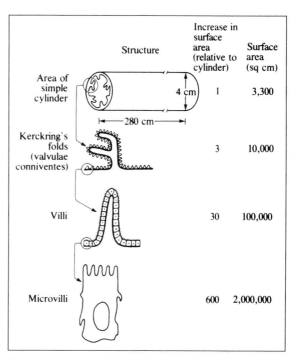

Fig. 26-12. Three mechanisms for increasing the surface area of the small intestine. (From T. H. Wilson. *Intestinal Absorption*. Philadelphia: Saunders, 1962. P. 2.)

on the two sides of the epithelium. In this instance, the rate of transfer is dependent not only on the magnitude of the diffusion gradient but also on the molecular size and the lipid solubility of the substance being transported. Chapter 1 supplies a more detailed description of the mechanisms involved in the movement of substances across biological membranes.

Practically all substances capable of being absorbed disappear from the lumen of the small intestine by the time mid-jejunum is reached. The ileum does not participate in absorption to a large extent because the more proximal regions of the small intestine are so efficient in this regard that usually only small quantities of absorbable material enter the ileum. However, the distal small intestine has the capacity to absorb, and it does so in situations in which food escapes absorption upstream; i.e., the ileum is a reserve absorptive area. Actually, about 50 percent of the small intestine can be removed without interfering with absorption. It should be recalled, on the other hand, that vitamin B_{12} and the bile salts are absorbed specifically in the terminal ileum, and removal of this region of intestine will result in defective absorption of these substances.

The contents recovered from the terminal ileum contain no digestible carbohydrate, very little lipid, and only 15 to 17 percent nitrogen-containing substances. Most of this material can be accounted for as bacteria, desquamated epithelial cells, the remains of various digestive secretions, and undigested and unabsorbed residues of food, such as cellu-

lose and connective tissue. Although the intestinal epithelium does act as a barrier to many materials, it also permits—and in many instances facilitates—the absorption of a large variety of substances. The following discussion will deal with the absorption of carbohydrates, proteins, lipids, water, and electrolytes.

ABSORPTION OF CARBOHYDRATES. In terms of calories, carbohydrate makes up approximately half of the American diet. The major digestible carbohydrate of food is the polysaccharide plant starch. This large molecule is composed of straight and branched chains of glucose, and it is reduced by the actions of salivary and pancreatic amylase mostly to the much smaller disaccharide molecule maltose. Humans do not secrete an enzyme capable of digesting cellulose, a plant polysaccharide that also consists of glucose molecules, but joined by linkages different from those of starch. Consequently, this complex carbohydrate is excreted in the feces. Other carbohydrates that are present in the diet in smaller, but significant, quantities are the disaccharides sucrose or table sugar (glucose-fructose) and lactose or milk sugar (glucose-galactose).

It is widely accepted that carbohydrates are absorbed in the form of monosaccharides. This idea is supported by the observation that, following ingestion of a wide variety of carbohydrates, only monosaccharides appear in the portal blood, which is the route of absorption for carbohydrates. The reason polysaccharides and disaccharides are not absorbed as such is that the intestinal epithelium is impermeable to carbohydrates of high molecular weight, and furthermore the absorptive surface is lacking in special transport systems for these molecules. In any event, the major form in which carbohydrate is presented to the intestinal epithelium for absorption is as disaccharides, which either are present as such in ingested food or result from the amylase

digestion of starch. However, the activities of the enzymes that hydrolyze disaccharides are very low in intestinal contents. On the other hand, it has been shown that the brush border of the mucosal cells contains all the disaccharidases that are present in these cells. They include maltase, lactase, and sucrase. Prevailing opinion is that disaccharides, such as maltose, lactose, and sucrose, are hydrolyzed to their constituent monosaccharides (glucose, galactose, and fructose) on the surface of the brush border of the epithelial cells during the absorptive process (Fig. 26-13A).

Evidence accumulated over the years demonstrates that, among the naturally occurring monosaccharides glucose

Fig. 26-13. A. Luminal and microvillus membrane digestion of carbohydrates. B. Carrier-mediated coupled transport of glucose and sodium across the intestinal epithelial cell.

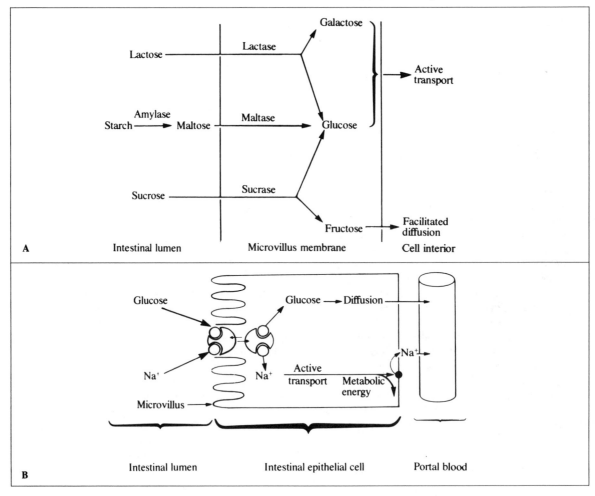

and galactose are absorbed by an active, energy-requiring process. For example, both sugars are absorbed against a concentration gradient, and their transport is slowed by anaerobic conditions and by metabolic inhibitors. Present thinking is that a membrane carrier located on the luminal border of the epithelial cells transports glucose and galactose across this membrane unchanged. In this regard, an intestinal transport maximum has been demonstrated for both sugars; i.e., a transport saturation phenomenon such as this is typical of carrier-mediated transport. It is also of interest that actively absorbed sugars have certain structural features in common, and it might be imagined that a particular structure is required to "fit" into a binding site on the carrier. The evidence further indicates that the same carrier is involved in the transport of both glucose and galactose, because the presence of one of these sugars inhibits the transport of the other sugar—there is competition between the sugars for binding sites on the carrier.

A recent proposal is that there is a coupling of the sodium and sugar transport systems of the intestine (see Cotransport, Chap. 1). It has been shown in studies in vitro that active transport of glucose ceases when the medium is devoid of sodium, and that as the sodium content is increased, absorption of glucose increases. Furthermore, cardiac glycosides that are known to inhibit sodium transport also inhibit glucose absorption. An assumption is that there is a mobile carrier in the brush border of the epithelial cell, and that this carrier has a binding site that has substrate specificity for actively absorbed sugars.

In addition to a binding site for sugar, it is theorized that there is a specific binding site on the same carrier for sodium (Fig. 26-13B), and that the carrier affinity for sugar is greatest when sodium is bound to the carrier. Moreover, when the carrier is loaded with both sugar and sodium, it travels across the membrane, the driving force being the concentration difference that exists for sodium from the luminal to the interior surface of the brush border membrane. Both sugar and sodium are released to the cell interior, and the sugar diffuses from the cell interior through the basolateral membrane into the blood. However, to ensure continued translocation of the sodium-loaded and sugar-loaded carrier across the brush border membrane, and to maintain the release of sugar to the cell interior, it is essential that a low intracellular sodium concentration be preserved. This function is fulfilled by an energy-dependent pump that moves sodium from the inside to the exterior of the cell through the basolateral membrane. Since the sodium pump requires energy derived from metabolism for normal function, it is here that the energy requirement for active sugar transport is manifested. When sodium-pump activity is abolished by

anaerobic conditions or metabolic inhibitors, sugar does not accumulate in the cell, because the increased cellular sodium content reduces the sodium gradient across the brush border membrane, thereby decreasing the driving force for movement of the carrier.

Since fructose is not transported against a concentration gradient, this monosaccharide is absorbed passively. Transport of fructose into the cell is by facilitated diffusion (see Chap. 1). In addition, there is an enzyme system in the intestinal epithelial cell that transforms some fructose to glucose. The partial conversion of fructose to glucose enhances entry of fructose into the cell by maintaining a relatively high diffusion gradient between the intestinal lumen and the cell interior.

ABSORPTION OF PROTEINS. Protein that is available for absorption from the small intestine is derived not only from food but also from the desquamated cells and the many enzymes of the various secretions that enter the lumen of the digestive tract. The protein of endogenous origin constitutes a sizable fraction of the total protein presented for digestion and absorption.

The intestine of many newborn mammals can absorb by pinocytosis considerable quantitites of intact protein molecules. Although the capacity to absorb proteins is lost shortly after birth, this process is important in a number of species because absorption of antibodies in the colostrum confers passive immunity against infection. From a nutritional standpoint, absorption of native proteins in the adult is insignificant. However, minute amounts of proteins are occasionally absorbed, as shown by the fact that some adults have allergic reactions to various food proteins. When a protein is absorbed unchanged, it sensitizes the individual to future doses of the same protein.

The combined action of the proteolytic enzymes that are secreted into the lumen of the digestive tract is to reduce large protein molecules to small peptides and amino acids. There are enzymes in the brush borders of the mucosal cells that hydrolyze peptides. Thus, different peptides are split to their constituent amino acids at the surface of the microvilli, and the amino acids so formed are then transported into the portal blood. Small peptides also cross the brush border membrane to the cell interior, where they are hydrolyzed prior to passage into the circulation.

In addition to specific transport systems for the transfer of small peptides across the brush border membrane, the intestinal epithelial cell is equipped with special mechanisms that aid in the transport of amino acids across this membrane. Proof for the active transport of amino acids has been provided by the observations that they are transported across the intestinal epithelium against a concentration difference, and that this movement is prohibited by anaerobic conditions and metabolic inhibitors.

There is more than one transport system for amino acids. For example, neutral amino acids compete with one another for absorption, but their absorption is not inhibited by basic amino acids. Furthermore, basic amino acids compete with one another for absorption, and although a few neutral amino acids do inhibit the absorption of some basic amino acids, these results have suggested that there is one transport system for neutral amino acids and another having a primary affinity for basic amino acids. Other transport systems exist for other groups of amino acids, such as the dicarboxylic amino acids and the imino acids.

It appears that the mechanism of transport of amino acids is similar to that of the actively transported sugars. As is the case for these sugars, brush border membrane carriers are involved in the transfer of amino acids across the intestinal epithelial cell, and these carriers are sodium dependent. It is of interest that glucose and galactose compete with certain amino acids for absorption. The implication is that the carrier mechanisms for sugars and some amino acids are shared.

ABSORPTION OF LIPIDS. The lipids of the diet are mostly in the form of triglycerides but also include phospholipids, cholesterol, and vitamins A, D, E, and K. Studies in vitro have shown that pancreatic lipase, which acts only at the lipid-water interface, preferentially catalyzes the hydrolysis of the 1- and 3-ester bonds of triglycerides, with the result that there is the consecutive formation first of 1,2-diglycerides and 2,3-diglycerides and then of 2-monoglyceride with the liberation of two fatty acids (Fig. 26-14). Since lipase attacks the 2-ester linkage with considerable difficulty, there is relatively little complete hydrolysis

of triglycerides to glycerol and fatty acids. Consequently, the major digestion products of triglycerides are monoglycerides and free fatty acids. Considerable evidence has accumulated in support of the idea that fat traverses the luminal membrane as such.

Long-chain (16- and 18-carbon) fatty acids are the most common type of dietary fatty acid, and these, as well as the triglycerides, diglycerides, and monoglycerides that contain them, are water insoluble. Without the availability of some special mechanism, triglycerides and their digestion products would exist as large oil droplets in the aqueous medium that constitutes intestinal contents, and optimal absorption of fat could not occur. It is known that lipid-soluble substances, by virtue of their solubility in the lipid portion of cell membranes, are able to diffuse through these membranes. However, these substances must be present in some form that assures ready entrance into the cell; i.e., to gain access to the surface of a cell, lipids must be in a water-soluble and freely diffusible form. In this regard, studies that have characterized the physical and chemical states of the intraluminal lipids recovered following the feeding of triglycerides have shown that the bile salts play the major role in preparing lipids for absorption by the small intestine.

If intestinal contents are centrifuged at high speed, an oily top phase and a clear bottom phase are obtained. The oily phase is derived from droplets of lipid that are present in intestinal contents, and this phase contains practically all the intraluminal triglycerides and diglycerides along with some monoglycerides and free fatty acids. Although some of these droplets are quite large, others are part of an extremely stable oil-in-water emulsion consisting of particles having an average diameter of only 5000 Å. This fine emulsion is produced by the combined action of a number of emulsifying agents in the lumen of the small intestine. These include monoglycerides and free fatty acids that result from

Fig. 26-14. Digestion of triglycerides by pancreatic lipase.

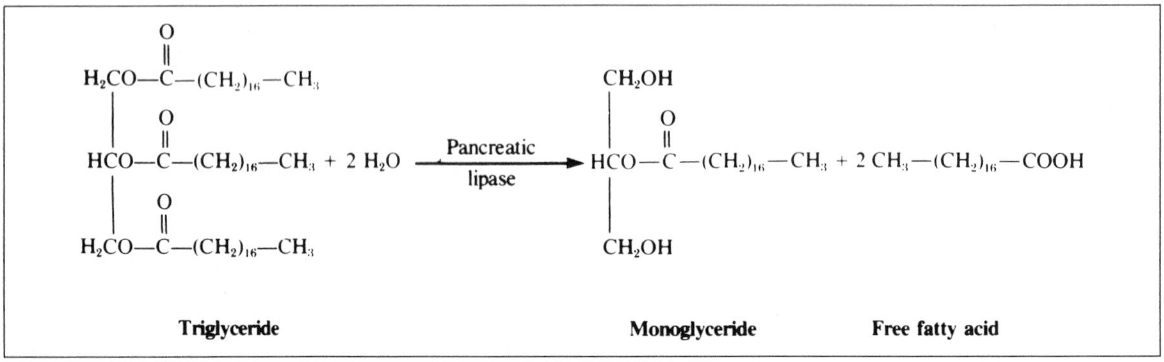

Triglyceride   Monoglyceride   Free fatty acid

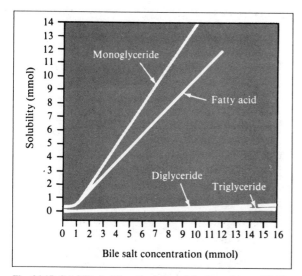

Fig. 26-15. Solubility in bile salt solution of the four classes of lipids in the lumen of the small intestine during the digestion and absorption of triglycerides. The values were obtained with triolein, diglyceride, monoolein, and oleic acid under conditions similar to those present in the jejunal lumen during digestion (pH, 6.3; 37°C; Na$^+$, 0.15 mol). (From A. F. Hofmann. Function of Bile in the Alimentary Canal. In C. F. Code [ed.], *Handbook of Physiology*. Washington, D.C.: American Physiological Society, 1968. Section 6: Alimentary Canal. Vol. 5, p. 2516.)

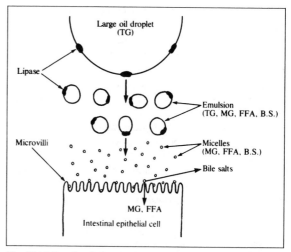

Fig. 26-16. Intraluminal and cell penetration phases of triglyceride absorption. TG = triglycerides; DG = diglycerides; MG = monoglycerides; FFA = free fatty acids; BS = bile salts.

the lipase hydrolysis of triglycerides, the bile salts, phospholipids and their digestion products, and proteins. The formation of such an emulsion enhances the action of pancreatic lipase by exposing a greater lipid surface area on which this enzyme can act and is the major factor in allowing rapid completion of the digestive process.

The clear phase contains mostly monoglycerides and free fatty acids, and these lipids, along with the bile salts, are present as water-soluble aggregates of micellar dimensions. At low concentrations the bile salts form molecular solutions, but as their concentration is increased, a point, the critical micellar concentration, is reached at which the bile salts aggregate to form micelles. The arrangement of bile salt molecules in a micelle is such that the hydrophilic groups are oriented outward into the water phase, and the hydrophobic groups are oriented inward and grouped together to form a nonpolar core. It is within this nonpolar core that monoglycerides and free fatty acids (and cholesterol and fat-soluble vitamins) are effectively dispersed and solubilized (Fig. 26-15). The product of this solubilization process is a mixed micelle whose average diameter is about 40 Å. Since the solubility of triglycerides and diglycerides in bile salt

micelles is very low (Fig. 26-15), these lipids are present almost entirely in the oil phase of intestinal contents. Also, since there is a limit to which available bile salt micelles can solubilize monoglycerides and free fatty acids, some of these lipids exist in the oil phase with the triglycerides and diglycerides.

The phyical and chemical states of intestinal intraluminal lipids are represented diagrammatically in Figure 26-16. The situation visualized is that, when triglycerides in the form of large droplets enter the duodenum from the stomach, pancreatic lipolysis is initiated, free fatty acids and monoglycerides are formed, and the presence of these digestion products, along with the bile salts and other emulsifying agents, results in the formation of a finely dispersed emulsion. However, since free fatty acids, monoglycerides, and bile salts have the capacity to form mixed micelles, a micellar phase also ensues. Micellar solubilization of monoglycerides and free fatty acids (and cholesterol and fat-soluble vitamins) facilitates entry of these lipids into the absorptive cells by providing an aqueous concentration of intraluminal lipid, thereby creating a gradient for diffusion through the lipid portion of the cell membrane. The monoglyceride and free fatty acid molecules in the micellar aggregates are in physicochemical equilibrium and rapidly exchange with their corresponding monomers that are dispersed in minute quantities (monoglyceride, 10^{-6} moles; free fatty acid, 10^{-5} moles) in the surrounding water molecules. Present thinking is that the monomers of monoglyceride and free fatty acid in this molecular solution represent the form in which these lipids penetrate the intestinal brush border. There is an equilibrium between the emulsion and the micellar and monomeric phases in that, as the monomeric and micellar phases are depleted by absorption, they are continually replenished because lipolysis of triglycerides and diglycerides of the emulsified fat continues with the production of more monoglycerides and free fatty acids. The process continues until triglyceride digestion is complete.

It is of interest that the bile salts, which are required for micelle formation, are not absorbed to any great extent until they reach the terminal ileum. Thus, they apparently perform their fat-dispersing function quite efficiently throughout the entire length of the small intestine in that a minimal amount of bile salts is necessary to prepare a relatively large quantity of fat for absorption.

The lymph provides the major route of absorption for fatty acids having 14 or more carbon atoms. Because the triglycerides of the diet are composed chiefly of 16- and 18-carbon fatty acids, most of the absorbed fat is transported into the lymph. Analysis of lymphatic fat of dietary origin shows it to be almost completely in the form of triglycerides. Consequently, once monoglycerides and free fatty acids enter the intestinal epithelial cells, they are resynthesized to triglycerides by intracellular enzyme systems.

Two pathways have been determined for the intracellular synthesis of triglycerides from absorbed digestion products. One involves the conversion of fatty acid to triglyceride and the other, monoglyceride to triglyceride—in accord with the concept that triglycerides are absorbed as free fatty acids and monoglycerides. It should be noted that the resynthesis of monoglycerides and free fatty acids to triglycerides maintains a concentration gradient for the continued diffusion of these lipids from the lumen of the intestine. The newly synthesized triglycerides are aggregated into droplets, which become progressively larger during passage through the cell. These fat droplets are stabilized by enclosure in a layer of phospholipid and protein. The final product is known as a *chylomicron*. Chylomicra vary in size from 500 to 10,000 Å in diameter. They enter the central lacteal of the villus and are carried by the lymphatic system to the general circulation. The cellular phase of fat absorption is illustrated in Figure 26-17.

Triglycerides containing fatty acids with chain lengths less than 12 carbon atoms are both hydrolyzed and absorbed more rapidly than the longer-chain triglycerides, and incorporation into bile salt micelles is not required for absorption. Thus, there is less interference in the absorption of the shorter-chain triglycerides under conditions of decreased pancreatic lipase and bile salt levels in the intestinal lumen, and it has been considered helpful to feed medium-

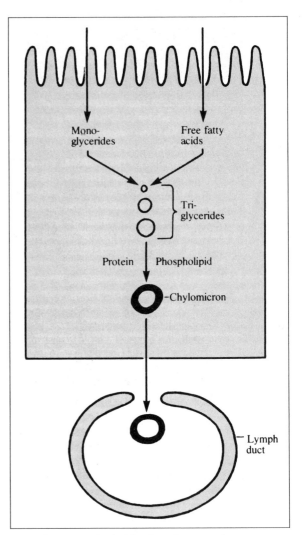

Fig. 26-17. Cellular phase of fat absorption.

chain triglycerides (8 to 10 carbon atoms in length) to patients with either pancreatic insufficiency or bile salt deficiency. Fatty acids with less than 12 carbon atoms are absorbed into the portal blood, without being synthesized to triglycerides. This situation may be due to low activity of the esterifying enzymes of the intestinal mucosa toward these fatty acids, which in turn may be related to the fact that the shorter-chain fatty acids are water soluble.

ABSORPTION OF WATER AND ELECTROLYTES. Water and inorganic ions, which are present in the lumen of the digestive tract as the result of ingestion and secretion of these substances, are absorbed primarily from the small intestine and to a lesser extent from the colon. Approximately 8 liters of water enters the small intestine every day. About 1 liter is ingested, and the remaining 7 liters is derived from the various secretions of the GI system. Obviously, any malfunction of the water and electrolyte absorptive mechanisms of the intestine can quickly result in serious depletion of body water and salt.

Small Intestine. It is currently accepted that the absorption of water, a process that occurs along the entire length of the small intestine, is by the physical process of osmosis. In support of this idea is the finding that when solutions of different osmolarities are introduced into the lumen of the small intestine, the osmotic activity of these solutions is adjusted to that of blood, in that water is rapidly absorbed from hyposmotic solutions and hyperosomotic solutions are diluted by entrance of water into the gut lumen; i.e., water movement occurs passively in response to the solute concentration gradient that exists across the intestinal epithelium.

The contents that enter the duodenum from the stomach can be either hyperosmotic or hyposmotic. However, the isosmotic condition is reached rather quickly in the upper small intestine. For example, if the duodenal contents are hyperosmotic, water moves from blood to duodenal lumen, and solute moves from duodenal lumen to blood in response to the water and solute concentration gradients that exist across the intestinal wall. Once the contents of the upper small intestine become isosmotic with blood, they tend to remain so throughout the remainder of the intestine. This is the result of isosmotic absorption of water and solute in the more distal regions of the gut, with water movement being secondary to the transport of solute. In other words, as various solutes are transported from lumen to blood, the intestinal contents tend to become hyposmotic. This creates a diffusion gradient for water, and water moves from lumen to blood in response to the osmotic gradient produced by the transport of solute. The result is that, as contents move down the small intestine, volume decreases progressively, but the isosmotic condition is maintained.

Water absorption is dependent on the transport of a variety of solutes. The solute of greatest importance in accounting for water absorption is sodium, the most abundant solute of the intestinal contents. Sodium transport is active and polarized. Sodium ion diffuses down an electrochemical gradient from the lumen, across the brush border membrane into the intestinal epithelial cell, and also moves across this membrane via the sugar and amino acid carriers discussed previously. Sodium then is extruded against an elec-

trochemical gradient into the interstitial space by a sodium pump located in the basolateral membrane of the cell. Probably, the electrical potential created by the movement of sodium across the intestinal epithelial cell (the serosal surface positive to the mucosal surface—see the discussion of sodium transport across frog skin in Chap. 1) is responsible for the simultaneous movement of anions.

It should be emphasized that movement of water and inorganic ions is not a one-way process in that movement occurs only from lumen to blood. Not only are there large fluxes of water and ions from lumen to blood, but there are also large fluxes simultaneously in the opposite direction, from blood to lumen. However, because the active transport of solute is inwardly directed, the flux from lumen to blood is normally somewhat greater than that from blood to lumen, so that the net result of these two large fluxes is absorption. If flux from lumen to blood is reduced for some reason, the result will be the net movement of fluid into the gut lumen. If this condition persists, and if this fluid is lost either by vomiting or by excretion from the colon, vascular collapse and death can occur.

That the intestine possesses special mechanisms for the transport of certain divalent cations is demonstrated by the observations that both calcium and iron are transferred from mucosa to serosa against a concentration difference, and that such movement is abolished by procedures that interfere with oxidative metabolism. The absorption of calcium requires that this element be in a water-soluble and ionized form in the lumen of the intestine. Carbonate and phosphate ions, especially in alkaline solution, tend to form insoluble salts with calcium and inhibit the absorption of this ion. Insoluble calcium soaps are formed in the presence of large quantities of fatty acids and are excreted in the feces. Both parathyroid hormone and vitamin D are necessary for optimal transport of calcium. A deficiency of dietary vitamin D leads to impaired calcium absorption, and the unabsorbed calcium forms insoluble salts with phosphorus. Since the salts are not readily absorbed, the absorption of phosphorus is also decreased under these conditions. Phosphorus can be rapidly absorbed under normal circumstances, as shown by the fact that radioactive inorganic phosphorus appears in the circulation within 5 minutes after introduction into the small intestine.

With very few exceptions, the extent to which different substances are absorbed from the digestive tract is not regulated, because few mechanisms are available for this purpose. However, the intestinal absorption of iron represents the absorption of a substance that is regulated in that this cation is absorbed only when the body becomes deficient in iron. Although the average daily intake of iron is about 20 mg, the amount absorbed is only a small fraction (5 percent) of that ingested and is just sufficient to replace the daily loss of iron. In the male the loss occurs entirely through desquamation of the epithelial cells of the small intestine. The amount of iron absorbed by women is about double that of men because of the additional iron lost during menstrual bleeding. Ferrous iron of the mucosal cells is in equilibrium with circulating iron and with ferritin, a protein-iron complex in the mucosal cells. When the iron of the blood is decreased below normal values, this element is liberated from the mucosal stores of ferritin-iron, and more iron is absorbed from the gut to replenish these stores.

Large Intestine. The absorption of ingested carbohydrates, proteins, lipids, and other nutrients is complete by the time material is discharged from the ileum into the proximal colon. This material consists of cellular debris, connective tissue, cellulose, and significant quantities of water and inorganic salts. The large intestine absorbs sodium, chloride, and water from, and adds potassium and bicarbonate to, its contents. In addition, this organ contains a variety of microorganisms, some of which synthesize certain vitamins. Although these nutrients are absorbed from the large intestine, they provide only a small part of the daily vitamin requirement.

Sodium and water are absorbed from the large intestine by processes similar to those of the small intestine; i.e., sodium is actively transported into the blood, and water follows passively in response to the osmotic gradient created by the removal of sodium from the intraluminal fluid. Water is absorbed to the extent that only about 100 ml is lost in the feces every day. Although the colon does absorb water, it is not the water-conserving organ of the digestive tract. About 8000 ml of water is absorbed every day by the small intestine and only about 300 to 400 ml by the colon. It is frequently a misfunction of the small intestine that causes diarrhea. In this instance, large quantities of fluid enter the colon from the small intestine and overwhelm the colonic absorptive mechanism, which is not nearly as efficient as that of the small intestine. The colon has the capacity to absorb 2500 ml of water daily.

BIBLIOGRAPHY

Code, C. F. (ed.) *Handbook of Physiology*. Washington, D.C.: American Physiological Society, 1967. Section 6: Alimentary Canal, Vols. 2 and 3.

Davenport, H. W. *Physiology of the Digestive Tract* (4th ed.). Chicago: Year Book, 1977.

Davenport, H. W. *A Digest of Digestion* (2nd ed.). Chicago: Year Book, 1978.

Gerolami, A., and Sarles, J. C. Biliary secretion and motility. *Int. Rev. Physiol.* 12:223, 1977.

Hofmann, A. F. Functions of Bile in the Alimentary Canal. In C. F. Code (ed.), *Handbook of Physiology*. Washington, D.C.: American Physiological Society, 1968. Section 6: Alimentary Canal. Vol. 5, pp. 2507–2533.

Jerzy Glass, G. B. *Gastrointestinal Hormones*. New York: Raven, 1980.

Johnson, L. R. (ed.) *Gastrointestinal Physiology*. St. Louis: Mosby, 1977.

Johnson, L. R. (ed.) *Physiology of the Gastrointestinal Tract*. New York: Raven, 1980.

Preshaw, R. M. Pancreatic Exocrine Secretion. In E. D. Jacobson and L. L. Shanbour (eds.), *MTP International Review of Science: Physiology*. Baltimore: University Park Press, 1974. Vol. 4, pp. 265.

Sachs, G., Spenney, J. G., and Rehm, W. S. Gastric secretion. *Int. Rev. Physiol.* 12:127, 1977.

Schneyer, L. H. Salivary Secretion. In E. D. Jacobson and L. L. Shanbour (eds.), *MTP International Review of Science: Physiology*. Baltimore: University Park Press, 1974. Vol. 4, pp. 183.

Schultz, S. G., Frizzell, R. A., and Nellans, H. N. Ion transport by mammalian small intestine. *Annu. Rev. Physiol.*, 36:51, 1974.

Sernka, T. J., and Jacobson, E. D. *Gastrointestinal Physiology - the Essentials*. Baltimore: Williams & Wilkins, 1979.

Silk, D. B. A., and Dawson, A. M. Intestinal absorption of carbohydrate and protein in man. *Int. Rev. Physiol.* 19:151–204, 1979.

Simmonds, W. J. Absorption of Lipids. In E. D. Jacobson and L. L. Shanbour (eds.), *Physiology. Series One: Gastrointestinal Physiology*. Baltimore: University Park Press, 1974. Vol. 4.

Soll, A., and Walsh, J. N. Regulation of gastric acid secretion. *Annu. Rev. Physiol.*, 41:35, 1979.

Wilson, T. H. *Intestinal Absorption*. Philadelphia: Saunders, 1962.

27

Leon K. Knoebel

Metabolism is the sum of all transformations of both matter and energy that occur in biological systems. Metabolism endows all cells with the properties of excitability, growth, and reproduction and, for various types of cells, makes possible such specialized processes as contraction, conduction, secretion, and absorption. Thus, metabolism is the basis for all physiological phenomena that one can observe or measure. Transformations of matter concern the chemical reactions that occur in the body, and these fall mainly in the province of the biochemist. However, transformations of matter are accompanied by transformations of energy. The following discussion will be confined to this facet of metabolism, energy metabolism.

TRANSFORMATIONS OF ENERGY

A diagrammatic scheme of biological transformations of energy is given in Figure 27-1, and it is suggested that the reader refer to this illustration during the subsequent discussion. Five major forms of energy are encountered in the living organism: chemical, mechanical, osmotic, electrical, and thermal energies. The cells of the body are able to use energy from one source only: the chemical energy that is liberated by chemical reactions. Chemical energy of the body can be transformed into mechanical, osmotic, electrical, and thermal energies, but these transformations are irreversible.

CHEMICAL ENERGY. The chemical energy that is necessary to sustain the processes of life is provided by the breakdown of carbohydrates, proteins, and lipids contained in the cells of the body. These large organic molecules are oxidized through a series of enzymatically catalyzed reactions to much smaller molecules, principally carbon dioxide and water, and as the reactions proceed, chemical energy is released for the performance of work by the cells. Reactions that involve the degradation of complex molecules to simple molecules with the liberation of chemical energy are designated *catabolic reactions* (catabolism).

The immediate source of chemical energy for the cells is

535

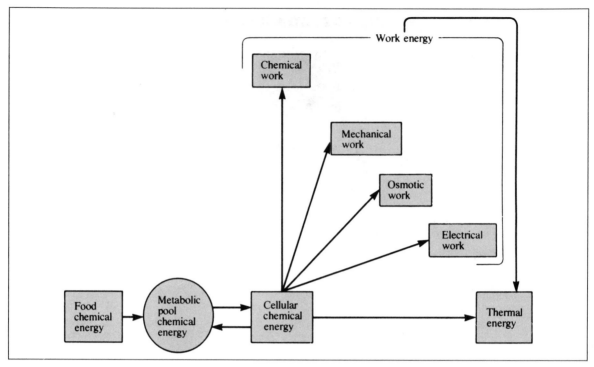

Fig. 27-1. Biological transformations of energy.

the *metabolic pool,* which can be thought of as existing throughout the extracellular and intracellular fluids of the body. A variety of carbohydrates, proteins, lipids, and the short-chain fragments of catabolism of these molecules are present in the pool and all can be readily used by cells as a source of energy. Furthermore, there is a free interchange (*dynamic state*) between the cells of the body and the metabolic pool of energy-containing substances in that the chemical energy of certain cells can be mobilized through the metabolic pool and utilized as a source of energy by other cells of the body.

As the chemical energy of the body is constantly being depleted, it must be replenished in order to maintain the energy status of the body at normal levels. The chemical energy of the body is replaced by the chemical energy of ingested food. Since the absorption products of ingested food are identical to some of the substances in the metabolic pool, the body can make free use of these products for replenishing cellular energy reserves.

It should be pointed out that the chemical energy released by the catabolism of carbohydrates, proteins, and lipids is not directly used by cells to perform various types of work.

Instead, it is first incorporated into high-energy phosphate compounds, the most notable of which is adenosine triphosphate (ATP). It is the hydrolysis of ATP that serves as the immediate source of chemical energy for the performance of work.

WORK ENERGY. Chemical energy is utilized for the purpose of doing work. Energy supplied in the form of work energy enables various types of cells to maintain the processes of life. In the category of work energy are four forms of energy mentioned previously: mechanical, osmotic, electrical, and chemical. The conversion of chemical energy to work in the form of mechanical energy is exemplified by the muscle that shortens and lifts a load. The transformation of chemical energy to work in the form of osmotic energy occurs when a cell maintains a concentration gradient of a substance across its membrane, as, for example, the intestinal epithelial cell that transports glucose from the intestinal lumen where the concentration of glucose is low to the cell interior where glucose concentration is relatively high. Both osmotic and electrical work are done at the expense of chemical energy in the maintenance of the unequal distribution of inorganic ions that exists across a nerve or muscle cell membrane. In this instance, not only are individual ionic concentration gradients maintained but, since there is also a separation of charges across the membrane (membrane potential), electrical work is done. Chemical energy can also be used to do chemical work by supplying the energy necessary for synthetic reactions. When complex molecules are synthesized from simple molecules, energy is stored and chemical work is done. *Anabolism* is the term used to designate this type of metabolic activity. Anabolic reactions are always accompanied by catabolic reactions because, in order to synthesize and store energy in a large molecule, chemical energy must be supplied by the breakdown of other molecules; i.e., chemical work is done at the expense of chemical energy.

THERMAL ENERGY. The efficiency of the body in converting chemical energy to work energy is in the neighborhood of only 20 percent. A large part of the chemical energy expended must therefore be converted to a form of energy other than work, and it appears as thermal energy or heat. In addition to the thermal energy that is directly liberated by chemical reactions, work energy is also ultimately converted to heat. For example, the heart does mechanical work in pumping blood. However, the work energy of the heart is converted to heat in overcoming friction as blood passes through the circulatory system. The chemical work that is done when a protein in synthesized is lost as heat when that protein is eventually catabolized. Other examples can be cited for other functions. Thus, the chemical energy of the body ultimately appears as heat, generated either directly from chemical reactions or indirectly from the dissipation of work energy.

Thermal energy cannot be used for doing work, because cells have no mechanism available for this purpose. Heat derived from metabolic processes may be used to maintain the body temperature at a level that is optimal for the enzymatically regulated reactions occurring in the body. Yet, much of the time, more heat is generated than can be used for this purpose, and the body has the problem of getting rid of excess heat (see Chap. 28).

ENERGY BALANCE

According to the first law of thermodynamics, energy can be neither created nor destroyed. This law can be applied to living systems in that studies can be made of energy balance as well as of energy transformations; i.e., the total amount of energy taken in by the body must be accounted for by the energy put out by the body. The relationship between the factors involved in the balance between input and output of energy is given in the following equation:

Energy input = energy output

Chemical energy of food = heat energy + work energy
$$\pm \text{ stored chemical energy}$$

All energy of the body is ultimately derived from one source, the chemical energy of food, whose intake is regulated primarily by hypothalamic centers in the brain (see Chap. 8). The major energy output of the body is in the form of heat, whose rate of production varies from one time to another depending on prevailing circumstances. For example, the heat production of an exercising person is greater than that of the same person at rest. The eating of food is followed by an increase in the amount of heat produced relative to that generated during fasting. Heat production is also influenced by the environmental temperature, circulating levels of a variety of hormones, age, sex, and body size. The role of each of these and additional factors in determining the rate at which heat is produced will be considered in more detail in this and subsequent chapters.

In addition to releasing energy as heat, the body puts energy out in the form of work. When the amount of energy ingested as food is sufficient to balance the amount of energy put out in the form of heat plus work, the chemical energy of the body remains constant. However, this situation is the exception rather than the rule. For example, the chemical energy stores of the body of the growing child increase markedly as body mass is increased over a period of years. The increase in chemical energy stores that is associated with growth is mainly in the form of protein. When an adult becomes obese, the chemical energy stores of the body are increased as the result of an excess deposition of body fat. On the other hand, the chemical energy of the body decreases under conditions of nutritional insufficiency. During early starvation, endogenous chemical energy is lost primarily as the result of the catabolism of fat, and if starvation is extended, protein is also used as an energy source. Even over the course of a day there are fluctuations in the chemical energy stores of the body of a normal person, and these are reflected by small changes in body weight.

It is obvious from these considerations that another factor, stored chemical energy, must be included in the energy balance equation in order to provide for changes in the chemical energy of the body. If intake of food energy is greater than the energy put out as heat and work, the body stores of energy increase, and storage energy is positive in order to balance the equation. If more energy is released in the form of heat and work than is ingested, the body stores of energy are depleted, and storage energy is negative in the equation.

To study the energy balance of a person under the conditions normally encountered in life, three of the variables present in the energy balance equation must be measured so that the fourth can be determined by difference. Although all four variables can be measured, it is not always convenient to measure them. This is especially the case in the clinic, where the facilities are usually inadequate and time precludes such an approach. However, a measurement of energy balance can be simplified by eliminating some of the variables. For example, the energy balance of a subject in the postabsorptive state can be determined, thereby excluding chemical energy derived from food. Moreover, when voluntary movement is restricted, energy in the form of work can be disregarded. With these two variables eliminated, the equation representing energy balance is simplified:

$$- \text{ Stored chemical energy} = \text{heat energy}$$

This equation simply states that a person in the resting, fasting condition uses a certain amount of stored chemical energy (as shown by the minus sign) with the resultant production of a certain amount of heat. Under these conditions the chemical energy that is utilized from the metabolic pool is completely converted to heat. Furthermore, since the subject is at rest and postabsorptive (*in the basal state*), the chemical energy is used solely for the purpose of maintaining the vital activities of the body, i.e., for the maintenance of such functions as heart action and respiration.

Under the circumstances that have been described, only two variables are concerned in the energy balance of the body. If one can be determined, the other is known, because they are equal. The rate at which chemical energy is expended by the body is known as *metabolic rate*. Since the rate at which heat is produced is equal to the rate of expenditure of chemical energy, the former is a direct measure of metabolic rate, and metabolic rate is usually expressed in terms of rate of heat production. Metabolic rate can be assessed by determining directly the amount of heat produced per unit time by an individual. However, the more frequently used approach is to calculate the rate of heat production from oxygen consumption, i.e., an indirect determination of heat production. A few of the methods used for these determinations will be considered in the following sections.

ENERGY UNITS

The unit of energy that has been most commonly used in the study of energy metabolism and will be employed in the following discussion is the kilocalorie (kcal). In terms of heat, 1 kcal (1000 cal) is the amount of energy required to raise the temperature of 1 kg of water 1°C. It has recently been suggested that the joule be used as the measure of energy, and for purposes of conversion, 1 kcal is equal to 4187 joules. The rate of energy conversion (power) has been expressed as kilocalories per hour, but it can also be stated in terms of watts (joules/sec) by multiplying kilocalories per hour by 1.16.

DIRECT CALORIMETRY

The heat production of humans and experimental animals can be determined by *direct calorimetry*. The subject is placed in an insulated chamber (calorimeter) through which cool water is circulated. The rate at which the water flows through the chamber is adjusted so that the temperature of the calorimeter is kept constant. In addition, a meter measures the volume of water that flows through the chamber in a given time, and the temperature of the water entering and leaving the chamber is determined. Thus, a knowledge of the volume of water passing through the chamber and the rise in the temperature of the water enables one to calculate in kilocalories the amount of heat transmitted from the subject to the water.

Although this measurement accounts under most circumstances for most of the heat produced by the subject, another avenue of heat loss from the body is represented by the water that leaves the body from the skin and the respiratory membranes in the form of water vapor. For every gram of water vaporized, 0.58 kcal is lost from the body as heat (heat of vaporization). Water vapor produced by the subject is collected in a suitable chemical absorbent and determined by weight. The amount of heat dissipated by evaporation is calculated by multipying the weight of water vapor by the heat of vaporization. The rate at which heat is lost by evaporation plus the amount of heat absorbed per unit time by the circulating water represents the rate at which the subject produces heat. This method for the assessment of energy exchange is tedious and difficult to perform; although it was used extensively in the past, simpler techniques have now replaced it.

INDIRECT CALORIMETRY

The chemical energy utilized from the body stores is measured by *indirect calorimetry*. By measuring oxygen consumption, carbon dioxide production, and urinary nitrogen excretion, and by knowing certain experimentally predetermined factors, one can determine accurately the type and amount of substances utilized in the metabolic pool. Furthermore, as a measure of metabolic rate, the total amount of heat produced per unit time by the oxidation of these substances can be calculated; i.e., an indirect measurement of heat production is made.

CALORIC VALUE OF FOODS. Both carbohydrates and fats are completely oxidized to carbon dioxide and water in the body, and the same is true when these substances are combusted in vitro. Since the initial reactants and the final products are the same in both instances, the amounts of energy released as heat in vivo and in vitro are identical. Consequently, it is a relatively simple matter to determine the caloric value of carbohydrates and fats in the body by measuring the heat of combustion of these materials in a bomb calorimeter. A bomb calorimeter is a metal chamber in which is placed a weighed amount of substance. Contact is made between the substance and an iron wire, the chamber is sealed, and oxygen is introduced into the chamber under high pressure. The calorimeter is immersed in a water bath of known volume and temperature, and when an electric current is sent through the wire, the material is ignited and complete combustion occurs. Depending on the substance combusted, a certain amount of heat is liberated. The heat of combustion of the substance (kcal/gm) can be calculated from the volume of water in the bath and the rise in temperature of the water.

The heat of combustion of various carbohydrates differs. That of glucose is 3.7 kcal per gram, whereas that of starch is 4.2 kcal per gram. However, an average value for carbohydrate is generally taken as 4.1 kcal per gram. The energy content of fat, which is more than double that of carbohydrate, varies, depending on the constituent fatty acids, but an average value of 9.3 kcal per gram is most often used.

Protein differs from carbohydrate and fat in that it is not completely oxidized in the body. Certain end products of protein metabolism appear in the excreta, for the most part in the urine as urea. Urea contains chemical energy, and if it were ignited in a bomb calorimeter, heat would be liberated. Thus the amount of heat that is produced by the oxidation in vivo of 1 gm of protein is less than that produced when the same amount of protein is combusted in a bomb calorimeter

(5.6 kcal/gm). By taking into account the energy lost in the excreta, it has been calculated that the caloric value of protein in vivo is 4.3 kcal per gram.

ENERGY EQUIVALENT OF OXYGEN. To calculate heat production from oxygen consumption, it is necessary to relate the number of kilocalories released by the combustion of carbohydrate, fat, and protein to the volume of oxygen consumed during the oxidation of each of these substances in the body. More specifically, it is necessary to know how many kilocalories of heat are produced when 1 liter of oxygen is used to oxidize each type of substance. This is known as the *energy equivalent of oxygen*, which is expressed as kilocalories per liter of oxygen. Carbohydrate and fat react with definite amounts of oxygen to produce certain quantities of carbon dioxide and water. Using glucose (molecular weight 180) as representative of the oxidation of carbohydrate, one can write the following equation:

$$C_6H_{12}O_6 + 6\ O_2 \rightarrow 6\ CO_2 + 6\ H_2O$$

In this instance, 180 gm of glucose reacts with 6 moles of oxygen or 134.4 liters of oxygen (6 moles oxygen $\times$ 22.4 liters oxygen/moles oxygen). The same volume (134.4 liters) of carbon dioxide is formed. One gram of glucose reacts with 0.75 liter of oxygen (134.4 liters oxygen $\div$ 180 gm glucose) to produce 0.75 liter of carbon dioxide. Since 3.7 kcal is liberated when 1 gm of glucose is oxidized, it follows that the energy equivalent of oxygen is 5 kcal per liter of oxygen (3.7 kcal/gm $\div$ 0.75 liter oxygen/gm).

A similar calculation can be made for a typical fat. The following equation represents the oxidation of a triglyceride (molecular weight 860), whose fatty acids are oleic, palmitic, and stearic acids:

$$C_{55}H_{104}O_6 + 78\ O_2 \rightarrow 55\ CO_2 + 52\ H_2O$$

In this case, 860 gm of fat combines with 1747 liters of oxygen to form 1232 liters of carbon dioxide. It can be calculated that 1 gm of fat combines with 2.03 liters of oxygen to form 1.43 liters of carbon dioxide. Since 9.3 kcal is liberated by the oxidation of 1 gm of fat, the energy equivalent of oxygen is 4.7 kcal per liter of oxygen (9.3 kcal/gm $\div$ 2.03 liters oxygen/gm).

Determining the energy equivalent of oxygen is more difficult for proteins, since they are structurally complex and are incompletely oxidized in the body. However, by using the empirical formula for a typical protein and taking into account the amount of urea that would be formed by

Table 27-1. Metbolic values for carbohydrates, fats, and proteins

Unit of measurement	Carbohydrates	Fats	Proteins
Kilocalories per gram	4.1	9.3	4.3
Liters of CO_2 per gram	0.75	1.43	0.78
Liters of O_2 per gram	0.75	2.03	0.97
Respiratory quotient	1.00	0.70	0.80
Kilocalories per liter of O_2	5.0	4.7	4.5

the oxidation of this protein, one can calculate the volumes of oxygen consumed and carbon dioxide produced during the oxidation of 1 gm of protein. These values are 0.97 liter of oxygen per gram and 0.78 liter of carbon dioxide per gram. Since 4.3 kcal is released when 1 gm of protein is oxidized, 1 liter of oxygen is equivalent to 4.5 kcal (4.3 kcal/gm ÷ 0.97 liter oxygen/gm). These data are summarized in Table 27-1.

RESPIRATORY QUOTIENT. It is obvious that there are variations in the volumes of oxygen consumed and carbon dioxide produced by the individual oxidations of 1 gm of carbohydrate, fat, or protein. Furthermore, the ratio of the volume of carbon dioxide produced to the volume of oxygen consumed during the oxidation of each type of food also varies (Table 27-1). This ratio (vol. CO_2/vol. O_2) is called the *respiratory quotient* (RQ). A calculation of the RQ from the rate of oxygen consumption and carbon dioxide production of a subject furnishes qualitative information regarding the substances utilized in the metabolic pool. If only carbohydrate is oxidized, an RQ of 1.00 is obtained, whereas the specific oxidation of fat results in an RQ of 0.70. Values that fall between these extremes represent the oxidation of various mixtures of carbohydrates, fats, and proteins. This will be discussed in more detail in the following section. The RQ of a subject on an ordinary mixed diet is about 0.85, and that of a fasting subject is about 0.82.

The RQ is based solely on the rate of exchange of the respiratory gases, since it is calculated from measurements of oxygen consumption and carbon dioxide production. Consequently, this ratio can be affected by factors other than oxidative metabolism. For example, much carbon dioxide is excreted during hyperventilation. This tends to increase the RQ, and values as high as 1.5 to 1.7 are obtained during severe exercise (see Chap. 29). On the other hand, carbon dioxide is retained during hypoventilation, and the RQ may fall below 0.70. An RQ obtained under these circumstances simply reflects the state of the respira-

tion and supplies no information about the composition of the metabolic mixture being oxidized. Large amounts of carbon dioxide are also excreted in conditions of metabolic acidosis, and the RQ is high, whereas the opposite effect is obtained during metabolic alkalosis (see Chap. 24).

It is apparent that an RQ determined under these circumstances is not a reliable measure of oxidative metabolism. A better term to apply to this ratio is *respiratory exchange ratio* (see Chap. 20). This has the advantage that the ratio need not always be considered to represent an index of oxidative metabolism; i.e., it is a ratio based simply on the rate of exchange of the respiratory gases. However, if the measurements on which the ratio is based are made under well-controlled conditions, the respiratory exchange ratio can be used to evaluate oxidative processes.

CALCULATION OF METABOLIC RATE. It is possible to determine quantitatively the type and amount of substances oxidized in the metabolic pool when oxygen consumption, carbon dioxide production, and urinary nitrogen excretion are measured and when the data given in Table 27-1 are available. Furthermore, the energy liberated as heat by the individual oxidations of carbohydrates, fats, and proteins can be calculated, and from these figures the total amount of heat produced per unit time can be used as a measure of metabolic rate. The principles of indirect calorimetry can be exemplified by considering a resting human subject in the postabsorptive state on whom the following measurements have been made: (1) urinary nitrogen excretion, 0.5 gm per hour; (2) oxygen consumption, 16.0 liters per hour; and (3) carbon dioxide production, 13.5 liters per hour.

Practically all nitrogen that is derived from the oxidation of protein is excreted in the urine. Since the amount of nitrogen contained in a typical protein molecule represents, on the average, 16 percent of the total weight of a protein molecule, the weight of protein oxidized per unit time is determined by multiplying the urinary nitrogen excretion by a factor of 6.25.

0.5 gm N/hr × 6.25 gm protein/gm N = 3.1 gm protein/hr

Since the oxidation of 1 gm of protein produced 4.3 kcal,

3.1 gm protein × 4.3 kcal/gm protein = 13.4 kcal/hr

The quantity of heat produced by the oxidation of carbohydrate and fat is determined from the oxygen consumption and carbon dioxide production. The total volumes of oxygen consumed and carbon dioxide produced during the oxidation of all three foods are measured. However, it is necessary to determine the amounts of these gases that are involved only in the metabolism of carbohydrates and fats. Thus one must know the volumes of oxygen and carbon dioxide involved in the oxidation of protein, and they are calculated in the following manner:

3.1 gm protein/hr $\times$ 0.97 L O_2/gm protein = 3.0 L O_2/hr

3.1 gm protein/hr $\times$ 0.78 L CO_2/gm protein = 2.4 L CO_2/hr

The total respiratory exchange is then corrected for the amounts of oxygen consumed and carbon dioxide produced by the oxidation of protein.

16.0 L O_2/hr − 3.0 L O_2/hr = 13.0 L O_2/hr

13.5 L CO_2/hr − 2.4 L CO_2/hr = 11.1 L CO_2/hr

The ratio of the volume of carbon dioxide produced to the volume of oxygen consumed during the oxidation of mixtures of carbohydrates and fats is called the *nonprotein RQ*. This is calculated as follows:

11.1 L CO_2/hr ÷ 13.0 L O_2/hr = 0.85

Recall that when pure carbohydrate is oxidized in the body, the RQ is 1.00, and when only fat is oxidized, the RQ is 0.70. The nonprotein RQ determined in this experiment shows that a mixture of carbohydrate and fat is oxidized. The energy equivalent of oxygen for carbohydrate is 5.0 kal per liter of oxygen and for fat is 4.7 kcal per liter of oxygen. Every mixture of carbohydrate and fat between the extremes of pure carbohydrate and pure fat has a specific nonprotein RQ and energy equivalent of oxygen. These relationships have been determined and are given in Table 27-2. At the nonprotein RQ (0.85) determined for the subject under consideration, the consumption of 1 liter of oxygen results in the production of 4.862 kcal. Consequently, the heat produced by the oxidation of this particular mixture of carbohydrate and fat is

13.0 L O_2/hr $\times$ 4.862 kcal/L O_2 = 63.2 kcal/hr

The amount of heat liberated by the oxidation either of carbohydrate or of fat can also be calculated from data supplied in Table 27-2. At a nonprotein RQ of 0.85, carbohydrate supplies 51 percent and fat 49 percent of the heat produced during the oxidation of this mixture of carbohydrate and fat.

0.51 $\times$ 63.2 kcal/hr = 32.2 kcal/hr (from carbohydrates)

0.49 $\times$ 63.2 kcal/hr = 31.0 kcal/hr (from fats)

Table 27-2. Energy equivalent of a liter of oxygen at various nonprotein respiratory quotients

Nonprotein respiratory quotient	Kilocalories/L oxygen	Kilocalories derived from	
		Carbohydrate (%)	Fat (%)
0.70	4.686	0	100.0
0.71	4.690	1.10	98.9
0.72	4.702	4.76	95.2
0.73	4.714	8.40	91.6
0.74	4.727	12.0	88.0
0.75	4.739	15.6	84.4
0.76	4.751	19.2	80.8
0.77	4.764	22.8	77.2
0.78	4.776	26.3	73.7
0.79	4.788	29.9	70.1
0.80	4.801	33.4	66.6
0.81	4.813	36.9	63.1
0.82	4.825	40.3	59.7
0.83	4.838	43.8	56.2
0.84	4.850	47.2	52.8
0.85	4.862	50.7	49.3
0.86	4.875	54.1	45.9
0.87	4.887	57.5	42.5
0.88	4.899	60.8	39.2
0.89	4.911	64.2	35.8
0.90	4.924	67.5	32.5
0.91	4.936	70.8	29.2
0.92	4.948	74.1	25.9
0.93	4.961	77.4	22.6
0.94	4.973	80.7	19.3
0.95	4.985	84.0	16.0
0.96	4.998	87.2	12.8
0.97	5.010	90.4	9.58
0.98	5.022	93.6	6.37
0.99	5.035	96.8	3.18
1.00	5.047	100.0	0

Source: After G. Lusk, Animal calorimetry. Analysis of oxidation of mixtures of carbohydrate and fat. *J. Biol. Chem.* 59:42, 1924.

The amounts of carbohydrates and fats that are oxidized can also be calculated from these data.

32.2 kcal/hr ÷ 4.1 kcal/gm carbohydrate = 7.9 gm carbohydrate/hr

31.0 kcal/hr ÷ 9.3 kcal/gm fat = 3.3 gm fat/hr

It is apparent that less than half as much fat as carbohydrate need be oxidized to supply approximately the same amount of heat.

The total heat production is the heat liberated by the oxidation of proteins plus that liberated by the oxidation of carbohydrates and fats.

13.4 kcal/hr + 63.2 kcal/hr = 76.6 kcal/hr

It can be further calculated that, of the total heat production, 42 percent is derived from carbohydrates, 41 percent from fats, and 17 percent from proteins. Evidently all three major foods are used to supply energy for the maintenance of the vital activities of the body. The total heat production is a measure of metabolic rate; i.e., it is equivalent to the rate at which the chemical energy of the body stores is metabolized.

DETERMINATION OF METABOLIC RATE FROM OXYGEN CONSUMPTION. The method of indirect calorimetry described in the preceding sections requires a moderate amount of equipment and considerable time. Since the determination of metabolic rate is a clinical tool of some importance, it has been necessary to devise simpler methods. One means of assessing metabolic rate in the clinic is based on a single measurement, that of oxygen consumption. It was found, using the more complicated techniques of indirect calorimetry, that under standard conditions of resting and fasting the average RQ of a large group of normal subjects was 0.82. On the basis of these determinations, the assumption is now made that the RQ of a subject in the basal state is 0.82. At a nonprotein RQ of 0.82, the energy equivalent of oxygen is 4.825 kcal. Heat production is determined by measuring oxygen consumption over a short interval of time, usually 6 minutes, converting this value to an hourly basis, and multiplying it by 4.825 kcal per liter of oxygen. This approach considerably simplifies the determination of metabolic rate.

One type of apparatus that is commonly used in the clinic to measure oxygen consumption is shown in Figure 27-2. This apparatus consists of a spirometer bell that is arranged to write on a moving kymograph by means of a pulley system. The bell is filled with oxygen, and during a determination a motor-blower draws oxygen through the inspiratory tube and returns the expired gas mixture through the expiratory tube. The expiratory tube is connected to a canister of soda lime in order to remove all carbon dioxide present in the expired air. As the subject consumes oxygen and as carbon dioxide is removed from the system, individual respiratory excursions and the drop of the bell are recorded. The difference between the height of the bell at the beginning and the end of the experiment is a measure of the volume of oxygen consumed during this time. To determine the distance the bell drops, a line is drawn that best follows either the peaks or the troughs of respiration as recorded, and the vertical distance between the top and the bottom of this sloping line is measured. Since the spirometer is calibrated in terms of volume per unit distance fall of the bell, the total volume of oxygen consumed during the experiment is calculated by multiplying this conversion factor by the distance the bell falls. The oxygen consumption is corrected for water vapor tension, reduced to standard conditions of temperature and pressure, and converted to liters of oxygen consumed per hour. Some subjects find it hard to breathe normally through a mouthpiece and valves, and one serious criticism of this method is that the accuracy of the determination depends on the ability of a subject to breathe regularly. If the rate and/or the amplitude of respiration are irregular, it is difficult to decide precisely where to draw the sloping line.

Obviously, a determination of metabolic rate by this method introduces certain errors into the calculations. Since a nonprotein RQ is used to determine the energy equivalent of oxygen, protein metabolism is neglected. However, the RQ (0.80) and the energy equivalent of oxygen (4.5 kcal/liter of oxygen) for the oxidation of protein are only slightly less than the average values used. Furthermore, protein accounts for a relatively small part of the total energy exchange. Consequently, the error introduced by neglecting protein oxidation is small and has been calculated to be only about 1 percent. In addition, the assumption is made that the RQ of a subject is normal; but if the RQ is not 0.82, it is not correct to use 4.825 as the energy equivalent of oxygen. Still this use introduces only a small error into the calculations, because the energy equivalent of oxygen varies so slightly between the extreme respiratory quotients of 1.00 and 0.70. By using an RQ of 0.82, which falls midway between the extremes, there can be at most an error of only 3.5 percent.

The errors inherent in the short method are not significant enough to outweigh the advantages gained. Moreover, since a large number of studies have been carried out using this method, good standards are available for comparison. One drawback to the method is that it does not provide informa-

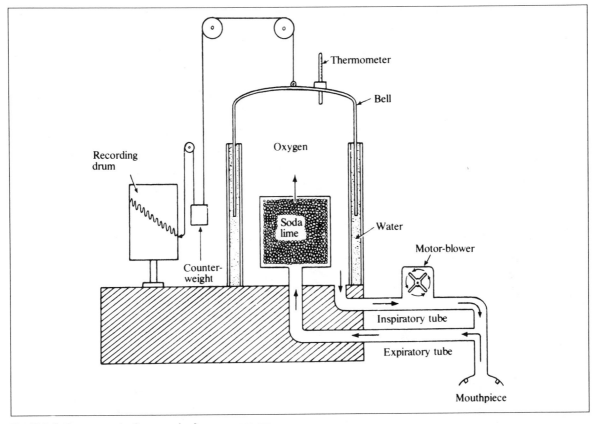

Fig. 27-2. Sanborn apparatus for measuring human oxygen consumption.

tion regarding the type and amount of substances oxidized in the body stores. The only information that can be obtained by measuring oxygen consumption is the total heat production of a subject. However, this value can be extremely useful as an aid to diagnosing certain disorders.

BASAL METABOLIC RATE

The methods and calculations of indirect calorimetry described have dealt with the energy exchange of a fasting and resting subject. The term applied to the exchanges of energy that occur under these conditions is *basal metabolism* or *basal metabolic rate* (BMR). Clinically, BMR is determined 12 to 14 hours after the last meal, usually in the morning after at least 8 hours of sleep. There should be no voluntary muscular movement during the test and no muscular exertion within half an hour to an hour prior to the test. In addition, there should be no stress due to extremes of environmental temperature, and the subject should be both mentally and physically at rest. Many factors that produce variable influences on metabolic rate are eliminated by carrying out the determination under these circumstances. The heat production so determined is a measure of the energy exchange required to maintain the vital activities of the body. The basal metabolic state does not represent the minimal functional activity of the body, since energy exchange is about 10 percent lower during sleep. This drop has been attributed to the more complete muscular relaxation achieved during sleep.

Once a subject's basal heat production has been determined in terms of kilocalories per hour, it is necessary to ascertain whether or not this value is normal. Consequently, normal values must be available for comparison. If a comparison is to be valid, the heat production of a subject should be compared with the average heat production of a large population of normal subjects having physical and biological characteristics similar to those of the subject.

Among the more important factors to consider when making the comparison are the body size, sex, and age of a subject.

BODY SIZE. The total energy exchange of a large animal is greater than that of a small animal. For example, the heat production of an elephant is many times that a mouse. Likewise, a 21-year-old man weighing 140 kg has a greater total energy exchange than a man of the same age weighing 70 kg. Consequently, when comparisons of metabolic rate are made between individuals of the same or different species, it is essential that there be some basis for comparison which takes body size into consideration. Various criteria of body size that have been considered include body weight, body surface area, and lean body mass.

Metabolic rate is not directly related to body weight, since the caloric output per kilogram of body weight of a small animal is greater than that of a large animal. However, energy exchange is thought to be proportional to certain power functions of body weight (kcal/$kg^{0.67}$/hr or kcal/$kg^{0.73}$/hr). When heat production is expressed in this manner, large and small animals have approximately the same metabolic rates.

The trend commonly followed in the clinic today is to express metabolic rate in terms of body surface area (kcal/m^2/hr). This concept is based on the idea that heat is lost at the surface of the body and that an amount of heat equivalent to what is lost must be produced in order to maintain a constant body temperature; i.e., an animal presumably produces heat in proportion to its surface area. Some relationships between heat production and body surface area are given in Table 27-3. Individuals of the same species (e.g., dogs) show considerable variation in body weight, but metabolic rate, when expressed in terms of surface area, is approximately the same for all. A like relationship is evident when different species are compared. Although a hog weighs about 7000 times more than a mouse, the caloric outputs are similar when based on body surface area. The same is true for other species, such as humans, dogs, and guinea pigs. Surface area obviously cannot be measured every time metabolic rate is determined. However, the surface areas of people of many different sizes and shapes have been measured by determining the total area of pieces of paper required to cover each body surface. It has been possible from these measurements to derive an equation that can be used to calculate body surface area from the weight and height of a subject, both of which are easily measured variables. Furthermore, surface area can be con-

Table 27-3. Relationship between metabolic rate and body surface area

Dogs		Various animals		
Body weight (kg)	kcal/m^2/day	Animal	Body weight (kg)	kcal/m^2/day
31.20	1036	Hog	128.00	1074
24.00	1112	Man	64.00	1042
19.80	1207	Dog	15.00	1039
18.20	1097	Guinea pig	0.50	1246
9.61	1183	Mouse	0.018	1185
6.50	1153			
3.19	1212			

Source: Data from M. Rubner. Ueber den Einfluss der Körpergrösse auf Staff und Kreftwechsel. *Z. Biol.* 19:535, 1883; Die Quelle der Fhierischen Wärme. *Z. Biol.* 30:73, 1894.

veniently determined from nomograms that have been constructed from this equation.

There has been some opposition to the use of surface area for relating energy exchanges of individuals of different sizes, since heat loss is affected by factors other than total surface area. Posture and the insulating effects of hair and clothing influence heat loss to varying degrees by modifying the actual surface exposed to the environment. It has been suggested that a better standard might be the weight of the active tissue of the body, i.e., of those tissues that are actively engaged in metabolic processes and are primarily responsible for the production of heat. Since adipose tissue is relatively inactive with respect to energy exchange, lean body mass might serve as a better reference for comparing metabolic rates of individuals of different sizes. Methods for estimating the weight of the body tissues exclusive of adipose tissue have been proposed.

SEX. Since the BMR of women is 6 to 10 percent lower than that of men of the same size and age, it is necessary to have standard values for both males and females. The BMR of the female increases markedly during pregnancy as the result of the additional metabolic activity of the fetus.

AGE. It is well known that BMR (kcal/m^2/hr) is at its peak during the early years and gradually declines throughout the remainder of life (Table 27-4). On the basis of size, the growing animal has a higher metabolic rate than an adult, because the rate of turnover of the body tissues is greater in

Table 27-4. The Mayo Foundation normal standards of basal metabolic rate (kcal/m²/hr)*

Males		Females	
Age	BMR	Age	BMR
6	53.0	6	50.6
7	52.5	6½	50.2
8	51.8	7	49.1
8½	51.2	7½	47.8
9	50.5	8	47.0
9½	49.4	8½	46.5
10	48.5	9–10	45.9
10½	47.7	11	45.3
11	47.2	11½	44.8
12	46.7	12	44.3
13–15	46.3	12½	43.6
16	45.7	13	42.9
16½	45.3	13½	42.1
17	44.8	14	41.5
17½	44.0	14½	40.7
18	43.3	15	40.1
18½	42.7	15½	39.4
19	42.3	16	38.9
19½	42.0	16½	38.3
20–21	41.4	17	37.8
22–23	40.8	17½	37.4
24–27	40.2	18–19	36.7
28–29	39.8	20–24	36.2
30–34	39.3	25–44	35.7
35–39	38.7	45–49	34.9
40–44	38.0	50–54	34.0
45–49	37.4	55–59	33.2
50–54	36.7	60–64	32.6
55–59	36.1	65–69	32.3
60–64	35.5		
65–69	34.8		

*The normal limits are usually taken as ± 10 percent, and divergence beyond these limits indicates an abnormal BMR for a subject of this age and sex.
Source: From W. M. Boothbay, J. Berkson, and H. L. Dunn. Studies of the energy of metabolism of normal individuals: A standard of basal metabolism, with a nomogram for clinical application. *Am. J. Physiol.* 116:471, 1936.

a growing animal, and this is reflected by an increased production of heat. The size of the body increases during growth, and much energy is stored by the synthesis of carbohydrates, fats, and proteins. Since relatively more synthetic activity occurs in a growing animal than in an adult, relatively more energy is liberated as heat by the catabolic reactions that necessarily accompany the synthetic reactions. As a result, standards for energy exchange must be based on age as well as on sex.

STANDARD VALUES. Standard normal values based on age, sex, and body surface area have been established for basal metabolic rate. One such set of standard values is given in Table 27-4. When a value in kcal/m²/hr is obtained for a subject, it is then compared with the normal value obtained from subjects of the same age and sex. The usual clinical method of reporting BMR is in terms of the percentage deviation from the standard value. For example, if the BMR of a male subject, age 28, is determined to be 38.3 kcal/m²/hr and is compared with the standard value of 39.8 kcal/m²/hr,

$$\% \text{ deviation} = \frac{38.3 - 39.8}{39.8} \times 100 = -3.7\%$$

FACTORS THAT AFFECT METABOLIC RATE

EFFECT OF FOOD. When a resting subject ingests food, heat production increases over the basal level. The increased production of heat that results from eating is called the *specific dynamic action* of foods. Specific dynamic action varies with the type of food ingested. If protein is eaten, the heat production over a period of hours rises above the basal level by 25 to 30 percent of the energy equivalent of the protein ingested. Heat production also increases when either carbohydrates or fats are ingested, but to a lesser extent. For example, suppose that the basal heat production of a subject is 75 kcal per hour. The total heat production of this subject over a period of 4 hours should be 300 kcal. If the subject is fed an amount of protein equivalent to 300 kcal, enough energy should be available in the food to balance that lost as heat over the 4-hour period. However, a measurement of metabolic rate over the 4 hours following the ingestion of the protein would show that the heat production of the subject is 380 kcal or, on the average, 95 kcal per hour. It can be calculated that 80 more kilocalories were produced during the 4-hour interval than can be accounted for on the basis of the basal heat production. In terms of the energy equivalent of the protein fed (80 kcal ÷ 300 kcal ×

100), 27 percent of the energy equivalent of the ingested protein appears as extra heat.

These results show that there cannot be kilocalorie-for-kilocalorie substitution of food energy for energy stored in the body reserves. In the example cited, proteins spare some of the body reserves of energy, but obviously some of the reserve energy is needed to utilize the ingested proteins. Consequently, in order to spare the body stores of energy more completely, the total caloric intake of protein and of other foods should be somewhat greater than the basal energy exchange.

It is not certain why heat production increases following the ingestion of food. The consensus is that the excess heat is the result of intermediary metabolism of absorbed foods. If amino acids are injected intravenously into dogs, heat production increases. However, if the animal's liver is first removed, heat production increases only slightly. The results suggest that the liver is the major site of the extra heat production, and the importance of the liver with respect to the processes of intermediary metabolism of amino acids and other foods is well established.

OTHER FACTORS. A number of factors in addition to those that have been described influence metabolic rate, and the relationships between some of them and metabolic rate are discussed in detail in other chapters. However, they are briefly summarized here.

Exercise or skeletal muscle activity causes by far the most marked effect on metabolic rate (see Chap. 29). Table 27-5 shows the energy expenditure for various activities by a normal 20-year-old male. Metabolic rates up to 350 kcal/m^2/hr can be maintained for long periods by a person who is in good physical condition. Higher levels of activity are possible for only relatively short durations.

A variety of hormones influence energy metabolism. Of considerable importance are the catecholamines epinephrine and norepinephrine, which are released by stimulation of the sympathetic nervous system, and thyroxine, a hormone of the thyroid gland. Maximal release of catecholamines can increase metabolic rate 30 to 80 percent by virtue of the stimulatory effects of these substances on the catabolism of carbohydrates and fats. Thyroxine exerts a considerable influence on the rates at which cellular oxidations occur, and excesses or deficits of circulating thyroid hormone modify metabolic rate (see Chap. 33). Basal metabolic rate can be 25 to 40 percent lower than normal in hypothyroidism and 40 to 80 percent higher in hyperthyroidism.

Table 27-5. Energy expenditure for various activities

Activity	kcal/m^2/hr
Rest	
Sleeping	35
Lying awake	40
Sitting upright	50
Light activity	
Writing, clerical work	60
Standing	85
Moderate activity	
Washing, dressing	100
Walking (3 mph)	140
Housework	140
Heavy activity	
Bicycling	250
Swimming	350
Lumbering	350
Skiing	500
Running	600
Shivering	to 250

Source: From A. C. Brown. Energy Metabolism. In T. C. Ruch and H. D. Patton (eds.), *Physiology and Biophysics* (20th ed.). Philadelphia: Saunders, 1973. Vol. 3, Chap. 4.

Temperature, both of the environment and of the body, modifies metabolic rate (see Chap. 28). One of the responses of the body to a cold environment is shivering, which consists of a series of rapid contractions of skeletal muscle. When the contractions occur at a maximal frequency, a metabolic rate as high as 250 kcal/m^2/hr can be attained (Table 27-5). The heat generated by shivering aids in the maintenance of body temperature at a normal level. Metabolic rate increases with a rise in body temperature and decreases when the temperature of the body falls. For example, the metabolic rate of a person at rest with a 40°C fever would be increased by 33 percent as compared with a normal body temperature of 37°C. The rise is due to a direct effect of temperature on the rates at which the chemical reactions of the body proceed.

The nature of the diet seems to have little influence on metabolic rate, but prolonged malnutrition may cause a decrease of 20 to 30 percent, presumably because of a deficiency of cellular nutrients. Intense mental work does not modify metabolic rate. However, the increased muscular tone and increased sympathetic activity accompanying stressful emotional states can produce increases of 5 to 20 percent.

BIBLIOGRAPHY

Brody, S. *Bioenergetics and Growth*. New York: Reinhold, 1945.

Carlson, L. D., and Hsieh, A. C. L. *Control of Energy Exchange*. New York: Macmillan, 1970.

Clark, W. M. *Topics in Physical Chemistry* (2nd ed.). Baltimore: Williams & Wilkins, 1952.

DuBois, E. F. *Basal Metabolism in Health and Disease* (3rd ed.). Philadelphia: Lea & Febiger, 1936.

DuBois, E. F. Energy metabolism. *Annu. Rev. Physiol.* 16:125–134, 1954.

Garry, R. C. (ed.). Energy expenditure in man (symposium). *Proc. Nutr. Soc.* 15:72–99, 1956.

Kleiber, M. Body size and metabolic rate. *Physiol. Rev.* 27:511–541, 1947.

Newsholme, E. A., and Start, C. *Regulation of Metabolism*. New York: Wiley, 1973.

Richardson, H. B. The respiratory quotient. *Physiol. Rev.* 9:61–125, 1929.

Swift, R. W., and French, C. E. *Energy Metabolism and Nutrition*. New Brunswick, N.J.: Scarecrow, 1954.

Tepperman, J. Energy Balance. Chapter is in *Metabolic and Endocrine Physiology*. (4th ed.) Chicago: Year Book, 1980. Pp. 267–293.

Wilhelmj, C. M. The specific dynamic action of food. *Physiol. Rev.* 15:202–220, 1935.

BODY TEMPERATURE REGULATION

28

William V. Judy

Humans and other *homeothermic* animals are capable of maintaining a nearly constant internal thermal environment even when exposed to a wide range of climatic conditions. Just as the various physiological systems work together to regulate intracellular and extracellular fluid volume, electrolyte concentration, and pH levels, they also strive to maintain a near constant internal temperature to ensure optimal cellular function.

Thermal regulation is necessary in warm-blooded animals because cellular biochemical and enzymatic reaction rates are temperature dependent. At the optimal body temperature, many cellular enzymatic reaction rates occur between 20 to 50 kcal per mole. If, however, the cellular temperature were to drop by 10°C, these reaction rates would decrease approximately 2.5-fold; that is, metabolic activity in humans changes about 25 percent for every degree centigrade change in body temperature. Since cooling body tissues reduces cellular metabolic activity and minimizes tissue hypoxia, induced hypothermia has become an acceptable practice in some types of surgery requiring blood flow oc-

clusion. Body temperature can be lowered several degrees for short intervals (several hours) without serious cellular damage, but it cannot be increased greatly, because temperature in humans is regulated at a point near the maximal tolerable level. Since temperature increases biological processes, their influence on cellular activity can lead to deterioration or the burning out of cells. Thus, to ensure the optimal cellular thermal environment, there must be some physiological regulatory mechanisms affecting cellular heat production, heat conductance from the cell, and heat loss or gain by the body surfaces. If heat loss mechanisms become inoperative, body temperature rises, stimulating enzymatic reaction rates and further heat production, and a vicious cycle is created, leading quickly to cellular destruction. If humans had no means of conserving body heat, their temperatures would parallel atmospheric temperature, as occurs in cold-blooded animals (poikilotherms); and when body temperature fell, so would metabolic activity and functional capabilities.

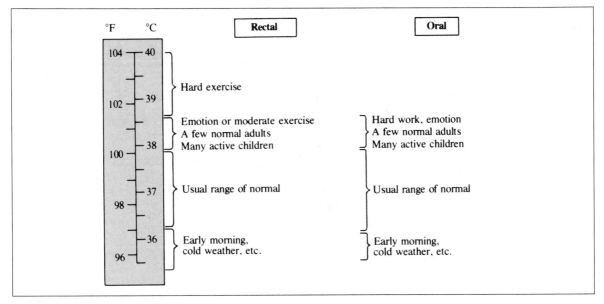

Fig. 28-1. An estimate of the ranges in body temperatures found in normal persons. (From E. F. Dubois, *Fever and the Regulation of Body Temperature*, 1948. Courtesy of Charles C. Thomas, Publisher, Springfield, Illinois.)

BODY TEMPERATURE

ORAL AND RECTAL TEMPERATURE. The measured and compared indices of body temperature are oral, rectal, and, in some cases, axillary temperatures. Oral and axillary temperatures are about 0.65°C (1.0°F) lower than rectal temperature and are subject to even greater variations on account of their close proximity to surrounding air (ambient) temperatures. It is difficult to specify one temperature as normal, since body temperature varies considerably between individuals (Fig. 28-1). In one study, rectal temperatures in a group of healthy subjects varied from 34.2 to 37.6°C (93.7 to 99.7°F) with a mean of 36.9°C (98.4°F). Rectal temperature, approximately 37°C, refers to deep central body areas such as the brain, heart, lungs, and abdominal organs. This level is considered the normal temperature for most mature individuals, but as shown in Figure 28-1, "normal" has a wide range. The vital organs produce about 72 percent of total body heat during rest but only 25 percent during exercise (Fig. 28-2). It is interesting that the vital organs and glands constitute only 8 to 10 percent of total body weight yet produce about three-fourths of the total resting body heat. At all times there must be a thermal gradient between deep body or major heat production areas and the skin in order that heat can be easily transported or conducted from internal to external areas.

SKIN TEMPERATURE. Skin temperatures, unlike deep body temperature, show considerable variations between areas (Table 28-1) and with changes in atmospheric temperatures. Mean skin temperature for the average person at a comfortable room temperature (24 to 25°C) is about 33°C. Surfaces covering the areas of high resting heat production have the highest skin temperatures (34.6°C). Surfaces covering the large muscle masses of the arms and legs have a mean temperature of 30.8°C, and areas that cover very little muscle (hands and feet) have the lowest mean resting skin temperature: 28.6°C.

The skin, unlike internal core tissues, can function well at temperature extremes. For example, skin temperature may fluctuate ± 10 to 12°C around its normal mean without damage; however, prolonged skin temperatures as low as 18°C will lead to tissue anoxia, pain, and tissue damage. Sustained high temperatures of 45°C will lead to tissue burns, accumulation of subcutaneous fluid, and pain.

Heat loss from the various body surfaces is also nonuniform; in general, areas with the highest resting temperature have the lowest heat loss per unit surface area. For example, the extremities, with their lower skin temperatures, have greater heat losses than the warmer trunk and head surfaces. The extremities act as major heat dissipation areas from which heat loss can be increased to maintain

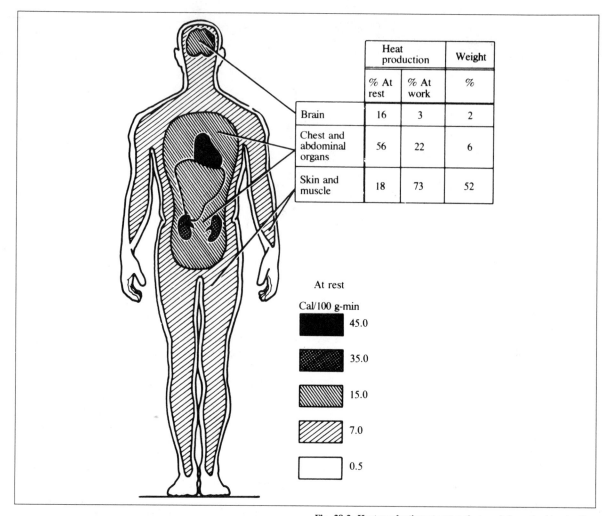

	Heat production		Weight
	% At rest	% At work	%
Brain	16	3	2
Chest and abdominal organs	56	22	6
Skin and muscle	18	73	52

At rest

Cal/100 g-min

██	45.0
▓▓	35.0
▨▨	15.0
▨▨	7.0
☐	0.5

Fig. 28-2. Heat production at rest and at work in various parts of the body expressed in cal/100 gm tissue/min. (Adapted from J. Aschoff and R. Wever. Kern und Schale im Warmehaushalt des Menschen. *Naturwissenschaften* 45:478, 1958.)

thermal comfort by either cutaneous vasodilation or removal of clothing. Heat loss may be decreased from these areas by vasoconstriction and the addition of clothing.

MEAN BODY TEMPERATURE. The temperature differences (gradients) between the major internal organ areas, skeletal muscle masses, and the skin have led to the concept that *mean body temperature* ($\bar{T}_b$) is equal to the fractional sums of core temperature, represented by rectal temperature ($\bar{T}_r$), and shell temperature, represented by mean skin temperature ($\bar{T}_s$). At a comfortable room temperature (24–25°C), two-thirds of the body mass is considered to be at core temperature and one-third at shell temperature (Fig. 28-3);

therefore, $\bar{T}_b$ may be expressed by the equation

$$\bar{T}_b = 0.67 \, \bar{T}_r + 0.33 \, \bar{T}_s$$

The relative importance of this relationship is that it allows estimation of body heat lost to or gained from the environment and the rate and amount of heat storage that normally take place in peripheral tissues, or shell. The relationship is, however, subject to error, since the proportion of body mass at core and shell temperatures changes as a function of

Table 28-1. Regional skin temperatures and heat loss for thermal comfort at rest

Region	Area (m²)	Ideal temp. (°C)	Heat loss at ideal temp. (kcal/m²/hr)
Head	0.20	34.6	20.0
Chest	0.17	34.6	48.3
Abdomen	0.12	34.6	37.5
Back	0.23	34.6	53.9
Buttocks	0.18	34.6	46.2
Thighs	0.33	33.0	36.0
Calves	0.20	30.8	73.0
Feet	0.12	28.6	83.3
Arms	0.10	33.0	84.0
Forearms	0.08	30.8	107.5
Hands	0.07	28.6	228.6
Total body	1.80	33.0 (mean)	59.4

Source: The Development of Water Conditioned Suits. Royal Aircraft Establishment Technical Note No. Mech. Engin. 400. London: Ministry of Aviation. Unpublished data originally supplied by D. McK. Kerslake (used by permission).

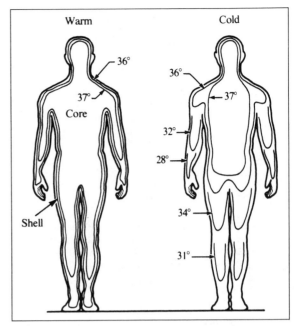

Fig. 28-3. Isotherms (surfaces connecting points of equal temperature) in the body. Isotherms in a warm environment (left) and in a cold environment (right) in degrees centigrade. The innermost isotherm may be considered the boundary of the body core; the core includes most of the body in hot environments. The outer isotherm is representative of skin or shell temperature. When heat must be conserved, the core contracts to the proportions indicated on the right. In severe cold exposure the combined effect of vasoconstriction and countercurrent heat exchange results in the pattern of isotherms shown in the limbs. (After J. Aschoff and R. Wever. Kern und Schale im Warmehaushalt des Menschen. *Naturwissenschaften* **45**:481, 1958.)

environmental temperatures. At high ambient temperatures the proportion of body mass at core temperature increases, as do mean skin temperature and body heat loss. The result will be a small increase in $\overline{T}_b$ and the peripheral heat stores due to rapid heat transfer from the core to the shell through peripheral vasodilation.

At low ambient temperatures the proportion of body mass at core temperature is reduced, as are mean skin temperature and heat loss. Mean body temperature will decrease, although $\overline{T}_r$ may increase slightly because of peripheral vasoconstriction resulting from direct thermal stimulation or increased sympathetic nerve activity. Thus, the thermal gradients or the isotherms shown in Figure 28-3 are changed by peripheral vasomotor activity, which controls the rate of heat transfer away from vital internal organs during exposure to normal, hypothermic, or hyperthermic conditions.

FACTORS INFLUENCING BODY TEMPERATURE. As Figure 28-1 shows, many factors influence body temperature. Most mature individuals show a *diurnal rhythm* (circadian rhythm)—i.e., a 24-hour cycle during which deep body temperature fluctuates ± 0.5°C around the person's normal mean temperature (Fig. 28-4). It is usually lowest in the morning during sleep, slightly higher in the early waking hours, and highest in early afternoon or midafternoon dur-

ing peak activity. The diurnal temperature pattern is probably of a controlled nature, and the depression during sleep appears to be effected through physiological cooling mechanisms and not to be dependent on metabolic depression. These cyclic patterns persist regardless of activity or disease states. They are not well established in small children, and they may be shifted in time or reversed in adults by changing day-night, work-rest schedules or by fast international time zone changes. Therefore, the temperature measured will be influenced by the time of day at which it is taken.

Other factors influence body temperature. Rectal, oral, and skin temperatures all increase during exercise, prolonged exposure to high ambient temperatures, emotional stress (pleasure and displeasure), febrile disease states (fe-

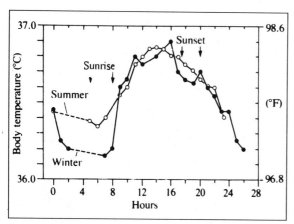

Fig. 28-4. Circadian rhythm in body temperature. Body temperature is lowest in early morning, gradually increasing during the day, and reaches a maximum in midafternoon, during peak activity. (From T. Sasaki. Effect of rapid transposition around the earth on diurnal variation in body temperature. *Proc. Soc. Exp. Biol. Med.* 115:1130, 1964.)

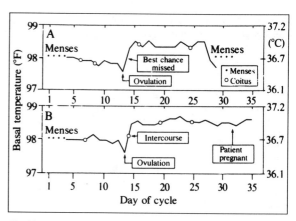

Fig. 28-5. Body temperature, recorded daily before arising, of a woman during two menstrual cycles. The second cycle is followed by pregnancy. (From Selle, W. A., *Body Temperature: Its Changes with Environment, Disease and Therapy*, 1952. P. 10. Courtesy of Charles C. Thomas, Publisher, Springfield, Illinois.)

ver), and in nonfebrile disease states (hyperthyroidism). Similarly, body and skin temperatures fall during prolonged exposure to severe cold (e.g., frostbite), during prolonged inactivity (e.g., in sleep), and in metabolic disorders such as hypothyroidism (myxedema). Patients with peripheral circulatory obstruction will have marked reduction of muscle and skin temperature in the afflicted area, as well as pain.

Age influences body temperature. Children tend to have higher rectal and oral temperatures (37.5–38.0°C) than adults, and their temperatures are more variable, since they do not have well-established diurnal patterns. Newborn and premature babies are prisoners of their thermal environments because they do not have well-developed thermoregulatory systems. If they are not protected, their temperatures will fluctuate with ambient temperature; therefore, they must be kept in controlled temperature and humidity areas to maintain an optimal thermal environment.

Most women show body temperature changes related to the menstrual cycle (Fig. 28-5). A few days before menstruation, core temperature drops 0.6°C, and the new temperature is maintained until just before ovulation, when it drops another 0.2°C. After ovulation, temperature rises to a normal level, where it remains until about the twenty-eighth day, when the onset of menstruation again takes place. If pregnancy occurs the normal premenstrual temperature drop is inhibited by the continual presence of the hormone progesterone. Temperature also has an influence on concep-

tion, because high body temperature reduces the viability of sperm; for this reason the male testes, unlike the female ovaries, are located outside the abdominal cavity.

PRINCIPLES OF PHYSIOLOGICAL HEAT TRANSFER

Heat is constantly being produced internally, transported to the body surfaces, and exchanged with the environment. Production is through basal metabolic activity, specific dynamic action of food, and muscle contraction (see Chap. 27). Of the total metabolic cost in the basal or active state, 80 percent is heat and only 20 percent is utilized for contractile energy. Basal heat production is about 72 kcal per hour for a 70-kg man and may increase by as much as 20 times during exercise (1440 kcal/hr). Resting humans can increase their heat production approximately fivefold through shivering when exposed to cold; however, they cannot willingly decrease or turn off basal metabolic heat production when exposed to hyperthermic conditions. Thus, humans have thermoregulatory capabilities that enable them to generate, distribute, and dissipate heat.

To characterize temperature regulation quantitatively, a measure of the change in body heat stores must be utilized. This measurement is the heat capacity, or specific heat, defined as the ratio of heat supplied (or removed) to the corresponding temperature rise (or decrease) or

$$\text{Specific heat} = \frac{\Delta \text{kcal/kg}}{\Delta \text{T}}$$

When 1 kcal is added to 1 kg of water, the temperature is raised 1°C. Thus, the specific heat of water is 1, and the kilocalorie as an energy unit is defined. The average specific heat of the body is about 0.83 because of the high water content with lesser proportions of protein and lipids of lower specific heat. Thus, mean body temperature should change 1°C for every 0.83 kcal of heat added or subtracted per kilogram body weight from the body heat store. The change in heat content rather than the overall total content is the important factor to the physiologist, and it can be estimated by

$$S \text{ (in kcal/hr)} = \frac{0.83 \, (\bar{T}_{b_1} - \bar{T}_{b_2}) \times \text{body weight}}{\text{time (hr)}}$$

where $\bar{T}_{b_1}$ and $\bar{T}_{b_2}$ are the mean body temperatures at the beginning and the end of the time period.

Body heat stores (body fluids, fat, and muscle) are basically good thermal insulators and poor conductors, therefore providing protective insulation when exposed to cold stress. If, however, humans had no effective or rapid means of transferring heat around these insulative stores or to them from internal production sites, they would be detrimental in both normal and hyperthermic environments. To have thermal regulation, there must be a balance between heat production or gain and heat loss. If, for example, all heat loss channels were somehow completely stopped, the basal heat production of a 70-kg man would raise body temperature to the upper (maximal) thermal limit of 40 to 42°C in 3 to 5 hours. This increase would require the addition of about 0.83 kcal per kilogram to heat stores each hour. Similarly, a 70-kg man with a metabolic activity three times basal would have a heat gain of approximately 2.4 kcal/kg/hr, which would raise body temperature about 3°C per hour. Considering the stimulating influence of temperature on cellular metabolism, this simple mathematical manipulation becomes even more astounding and shows the importance of internal heat transfer and external heat exchange in the maintenance of thermal equilibrium.

INTERNAL HEAT TRANSFER. Internally produced heat must be readily removed from the cell to prevent excessive heat accumulation. Heat is transferred from the site by *thermal conduction* or *conductance* between adjacent cells and fluids down a thermal gradient, and by *forced convection*, in which heat is conducted across the vessel wall to the vascular fluid and swept away to cooler areas by the circulating blood. Thermal conductance between the core and the shell depends on the steepness of the thermal gradient; however, since the core is a large area of constant temperature, con-

ductance is very slow on the inside and much greater at the core-shell junction, where circulatory adjustments occur readily. If there were no peripheral or skin blood flow, the amount of heat conductance from the internal areas to the periphery would be about 5 to 10 kcal/°C gradient/hr. Assuming a thermal gradient of 4°C ($\bar{T}_r = 37$°C and $\bar{T}_s = 33$°C), 20 to 40 kcal per hour would be transported by tissue conductance alone. This value would indicate that anywhere between 38 and 72 kcal per hour would be added to the body heat stores for a 70-kg man in the basal condition (heat production = 72 kcal/hr) and 176 to 196 kcal per hour for a normally active person. Thus, body tissues are poor thermal conductors and good insulators. For example, the thermal conductance of a 1-cm-thick piece of perfused beefsteak is about equal to that of a 1-cm-thick piece of cork (conductivity = 18 kcal/hr/°C gradient). Internal heat transport by forced convection requires that a thermal gradient exist between circulating blood and the tissues themselves. This is accomplished by having the cooler venous blood return from the periphery or shell area in vessels adjacent to warm arterial blood from the core areas. The circulatory system, then, provides a major heat transport system that maintains a proper cellular thermal environment, as well as proper fluid, electrolyte, nutrient, and oxygen environments.

EXTERNAL HEAT TRANSFER TO THE ENVIRONMENT. The major mechanisms of heat exchange between the skin and the environment are *radiation, conduction, convection,* and *evaporation.* Since these mechanisms are technically difficult to separate, they are classified together as *nonevaporative* heat exchange mechanisms. When skin temperatures are greater than the environmental temperature or that of nearby objects, these pathways are characterized as *heat loss* channels. When skin temperatures are lower than the temperature of the surrounding air or nearby objects, the nonevaporative heat exchange mechanisms become *heat gain* channels through which body temperatures can actually be increased. The relative importance of evaporative and nonevaporative heat exchange mechanisms as means of maintaining deep body temperature varies with the ambient temperature (Table 28-2).

When a nude male is exposed to various air temperatures (calorimeter temperature, Table 28-2) between 20 and 40°C, the evaporative heat loss, mean skin temperature, and thermal conductance (heat flow between deep body areas and skin surface) increase greatly, whereas rectal temperature increases only slightly (1°C), and nonevaporative heat loss decreases to zero. These data show that deep body temperature is maintained fairly constant at the various ambient

Table 28-2. Steady-state evaporative and nonevaporative heat loss, rectal and mean skin temperature, and thermal conductance at various ambient temperatures

Thermal regulatory variables	Calorimeter temperature (°C)				
	20	25	30	35	40
Evaporative (% of total)	17	30	50	93	100
Nonevaporative (% of total)	83	70	50	7	0
Rectal temp. (°C)	36	36.3	36.5	36.8	37
Mean skin temp. (°C)	27.8	30.0	33.5	35.3	35.7
Thermal conductance (cal/m^2/hr/°C)	11	16	18	35	39

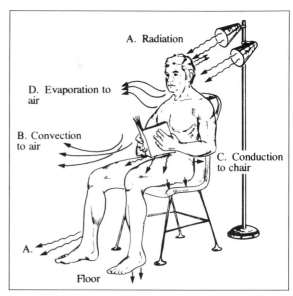

Fig. 28-6. Mode of heat transfer with the environment. A. *Radiation*—heat transfer by electromagnetic waves from warm to cooler objects. B. *Convection*—heat transfer by molecules of air or fluid moving between areas of unequal temperature; convective heat exchange is facilitated by air or fluid movements (forced convection). C. *Conduction*—heat exchange between objects in contact; this includes transfer to air. D. *Evaporation*—heat loss by water molecules diffusing from the body surface and by sweating.

temperatures by changes in the amount of heat transferred from deep body areas to the skin and exchanged with the environment by evaporative and/or nonevaporative mechanisms.

Radiation. Radiation is heat transfer by electromagnetic waves between objects that are not in contact (Fig. 28-6A). All dense objects radiate heat; the greater the temperature difference between two objects, the larger will be the amount of heat radiated from the warmer to the cooler object. At low ambient temperatures (20°C, 68°F), radiative heat loss in humans may account for 70 percent of total heat production. At high temperatures (35°C, 95°F), the body may actually gain heat through radiation. The warmth that is felt from a heat lamp or the sun's rays is an example of radiant heat. Radiant heat loss is greatly influenced by the total exposed surface area. Normally, not all body surface areas are effective in radiative heat exchange with the environment. Surfaces under the arms, between the fingers, and between the legs radiate heat to the opposing skin surface, and usually there is no net heat loss by the exchange. However, a man who extends his arms and spreads his legs and fingers may increase his effective radiating area from 75 to 85 percent. The spread-eagle stance is commonly observed in many mammals and birds on hot summer days or nights as they increase their effective radiant surface and heat loss. Conversely, humans and animals exposed to cooler environments will curl up to reduce radiative heat loss.

Convection. Convection is heat transfer by movement of molecules of a gas or liquid between two locations of differ-

ent temperatures (Fig. 28-6B). When air temperature is less than that of the skin, heat is transferred from the skin by conduction or radiation to the gas or fluid medium on the skin surface. This medium gains heat, becomes less dense, rises, and is replaced by cooler air in a continuous process. Convective heat loss is aided by air of fluid currents (forced convection). At normal room temperatures and air movements, convective heat loss accounts for only a very small fraction of nonevaporative heat loss; however, forced convection may become quite effective.

Conduction. Conduction is heat exchange in the form of kinetic energy between atoms or molecules of objects in contact (Fig. 28-6C). It includes heat transfer to air or fluid molecules as well as to dense objects with which one may be in contact. Actually, little heat is exchanged with the environment by this route unless one is literally in *cold* water. Again, as with convection and radiation, the amount of conductive heat exchange is proportional to the temperature difference between objects.

Evaporation. Evaporation is a constantly operating heat loss mechanism. Small amounts of water are continually being diffused from the skin surface, respiratory passages, and mucous membranes of the mouth. These exchanges are called *insensible* perspiration and account for 20 to 25 percent of basal heat production loss. Profuse sweating becomes the most effective heat loss channel at high ambient temperatures and during heavy work. Under such conditions, the amount of heat lost by evaporation is inversely related to the humidity of the surrounding air, and for sweating to be effective as a cooling mechanism, sweat must be rapidly evaporated.

PARTITIONAL HEAT EXCHANGE

The relationships between body heat production and heat lost to or gained from the environment at any specific environmental temperature are conveniently demonstrated by balance statements. When mean body temperature and the quantity of heat in body heat stores remain constant, humans are considered to be at a thermal steady state or equilibrium. Under such conditions, heat sources must equal heat losses.

Sources = losses
Metabolism + heat stores = evaporative + nonevaporative losses

If heat sources become greater than heat losses, mean body temperature rises; if losses exceed sources, temperature falls. The relationships between heat sources and losses are different at various ambient temperatures, and the primary mode of thermal regulation changes from one temperature range to another. The modes of temperature regulation are *metabolic, vasomotor,* and *sweating.* Metabolic mechanisms are most effective in cold environments, vasomotor mechanisms in comfortable environments, and sweating in hot environments.

In cool environments the thermally unprotected (nude) person is subject to body cooling through both evaporative and nonevaporative mechanisms. Evaporative heat loss under these conditions is insensible loss only; it is low, is fairly constant, and cannot be greatly decreased to prevent loss of stored heat. Nonevaporative heat loss is the major heat loss channel in a cool environment because mean skin temperature exceeds ambient temperature, and the amount of heat loss is proportional to the degree of cold stress (Table 28-2). Thermal steady states may be established by increasing metabolic heat production or decreasing nonevaporative heat loss. Naturally, humans protect themselves by putting on clothing to reduce nonevaporative heat loss, but the physiological response of the unprotected individual is to increase metabolic heat production and thus replace the heat given up by body heat stores. In addition to basal heat production, cellular metabolic activity may be increased by hormonal (thyroxine, catecholamines), mechanical (muscle activity or shivering), or thermal (direct temperature effects on cell) means.

In warm or hot environments, the primary regulatory mechanism used to maintain thermal equilibrium is evaporative heat loss, or *sweating.* When the ambient temperature exceeds mean skin temperature, nonevaporative heat loss mechanisms become heat sources. The problem, therefore, is to prevent body temperature from rising because of heat accumulation. At high temperature, metabolism is increased slightly as a result of direct thermal effects on cellular activity; however, sweating is initiated to maintain thermal equilibrium by effective cooling of heat stores. The sweating rate is directly related to the degree of heat stress, and its effectiveness in cooling the body is inversely related to the moisture content of the air (humidity).

At comfortable temperatures, thermal equilibrium is achieved primarily through *vasomotor regulation.* Here, nonevaporative heat loss is altered by changing the skin to the ambient temperature thermal gradient. This is accomplished by sympathetic nervous system control of skin and extremity blood flow. Thus, thermal equilibrium is maintained without sweating or increasing metabolism.

The partitional heat exchange diagram in Figure 28-7 summarizes the avenues of heat loss or gain. The values shown are examples that will change when exposure time, relative humidity, posture, clothing, or air velocity are varied. The figure indicates the steady-state relationships between heat sources and heat losses (sinks) at various ambient temperatures. It should be noted that balance statements are equal at all points; that is, the algebraic sums of the *sources* above the zero line and *sinks* (losses) below the zero line at any specific temperature are equal.

HEAT TRANSFER WITHIN THE BODY

INTERNAL HEAT DISTRIBUTION. As stated previously, heat is transferred from internal production sites to body surfaces by combined conduction and circulatory-assisted convection (forced convection). Conductive heat transfer between deep body cells is slow and small because the temperature difference (thermal gradient) between adjacent cells is small and body tissues are relatively poor thermal conductors. Circulatory-assisted convection is rapid and is the major means of internal heat transfer. It involves a form of bulk movement of heat by fluids flowing between body areas of unlike temperatures. Heat is conducted to the capil-

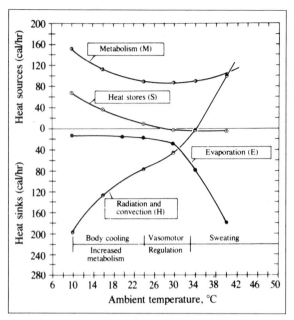

Fig. 28-7. Partitional heat exchange of a human subject at different ambient temperatures. All heat sources are above the zero line; all heat sinks are below the line. Note the three modes of thermoregulation (metabolism, vasomotor regulation, and sweating) and their functional temperature ranges. (From R. W. Bullard, Temperature Regulation. In E. E. Selkurt [ed.], _Physiology_ (4th ed.). Boston: Little, Brown, 1976. P. 687.)

lary wall, where convective exchange occurs between the vessel wall and the blood by the action of fluid flow. The circulatory system transports the heat-carrying blood to cooler tissue areas, where the exchange process is reversed and heat is gained by these tissues from the blood.

EFFECTOR SYSTEMS.

Circulation. The circulatory system is essential to body temperature regulation as it is to all body systems. During resting or stress conditions, it distributes heat throughout the body, controls the proportions of body mass at core and shell temperatures, and sets up a countercurrent exchange mechanism in the extremities that reduces the amount of heat available for exchange with the environment. The centrally located vital organ systems (see Fig. 28-2) have not only the highest resting metabolic heat production but also the greatest resting blood flow per unit of body weight. This flow acts as an effective cooling mechanism by moving large quantities of centrally produced heat to cooler, less metabolically active peripheral tissues. Thus, an optimal thermal

environment is maintained for all cells by keeping most of the body mass at a relatively constant (37°C) temperature.

Effective Body Insulation. The vasomotor responses of systemic arterioles perfusing cutaneous capillaries control the body proportions at core and shell temperatures (see Fig. 28-3). The systemic capillaries do not extend into the superficial epidermis; therefore, heat transferred through this outermost layer is by tissue conduction only. At comfortable temperatures, vasomotor activity establishes a thermal steady state without sweating or increasing metabolism. At the upper limits of thermal comfort, cutaneous arterioles dilate, increasing capillary blood flow (100-fold in some areas), and expose large quantities of warm blood to the superficial capillary and the epidermis and increases heat loss as much as 20-fold. When the cutaneous arterioles are constricted, capillary blood flow and the number of perfused vessels are greatly reduced, conduction distance is increased, and heat loss is reduced. By these vasoconstrictor and dilator actions, the circulatory system regulates the amount and rate of heat transferred to the body surface, and the heat loss from the skin, by changing the size of the body shell that separates or insulates deep body core areas from the environment.

Countercurrent Heat Exchange. The route taken by venous blood in the extremities on its return to the heart influences the amount of heat lost from these areas. Venous flow may be shifted from superficial vessels to deep venous plexuses (venae comitantes) that are adjacent to the arterial inflow vessels. The shift sets up a conductive heat exchange between warm arteries and cool veins known as the _countercurrent heat exchange,_ which is illustrated for the human arm in Figure 28-8. In essence, cool venous blood is warmed as it returns to the heart, and arterial blood is cooled as it proceeds down the extremities. For example, in a cool environment, warm arterial blood (37°C) is pumped down the brachial artery, but before it reaches the radial and ulnar arteries, it is reduced to 33°C; and when it reaches the superficial capillaries, its temperature is 24°C. The temperature decrease is due to conductive heat transfer to adjacent venous vessels as blood returns countercurrently to the heart. Thus, less heat is transported to open cutaneous vessels; the capillary surface thermal gradient is reduced, as is heat exchange with the environment; and the venous blood transports heat back to deep body areas as a means of heat conservation.

At the upper end of vasomotor control, peripheral vasodilation shunts large quantities of blood to the periphery. Cooler venous blood returning to the heart does not gain as

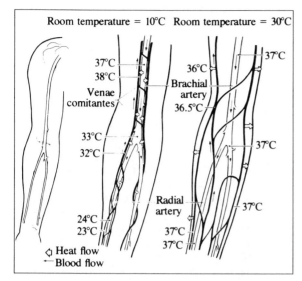

Fig. 28-8. The countercurrent heat exchange in the human arm. Venous flow returning to the heart is shifted from superficial vessels at a room temperature of 30°C to deep vessels at a room temperature of 10°C. (From R. W. Bullard. Temperature Regulation. In E. E. Selkurt [ed.], *Physiology* (4th ed.). Boston: Little, Brown, 1976. P. 689.)

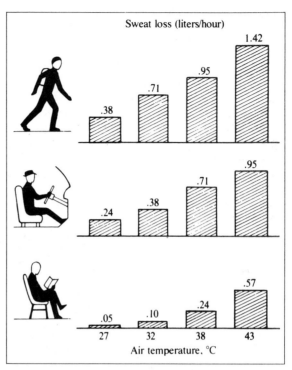

Fig. 28-9. Sweating rates for a clothed man in various activities and at various ambient temperatures. (From E. F. Adolph et al., *Physiology of Man in the Desert*. New York: Interscience, 1947. P. 4.)

much heat because the return flow is directed away from the deep veins. Therefore, vascular control mechanisms initiate blood flow shunting that effectively channels blood through vascular structures in the extremities as required to facilitate heat loss or conservation.

Sweating. When body heat increases because of either muscle activity or decreased nonevaporative heat loss, sweating is initiated as a cooling mechanism. Sweat is a true secretion of mostly water and sodium chloride, with an osmolar concentration well below that of plasma (see Chap. 23). Release of sweat from the subcutaneous glands (eccrine glands) is controlled by the sympathetic postganglionic cholinergic division of the autonomic nervous system (see Chap. 7) and occurs simultaneously in most skin areas. The pumping action of contractile elements in the gland duct forces sweat to the body surface, and this action increases as both skin and rectal temperatures rise.

In the strictest sense, sweating is a type of forced convection, because body heat is conducted to the glandular fluids and transported to the surface by ductile pumping. Pumping rates of 1 to 2 contractions per minute occur at low sweat rates and 15 to 20 per minute during profuse sweating. This becomes a very effective cooling mechanism when one considers that humans have an estimated 2.5 million sweat glands spread under the skin. As a single organ system, the sweat glands have a combined ductile-surface exposure area of 90 cm² (1 square yard), and about 40 ml of fluid is required to fill the sweat gland ductile system. Sweating rates for various work levels at different ambient temperatures are shown in Figure 28-9. A man working in severe heat and high humidity can produce 4 liters of sweat per hour for short periods. However, his ability to sweat and maintain high outputs is developed by continuous heat exposure. Persons who are not accustomed to high heat loads cannot sustain high sweat rates. The sweating mechanism is initiated at lower ambient temperatures in adults (32–34°C) than in full-term infants (35–37°C), and prematures do not show well-developed sweat patterns but increase their respiratory rates as a means of losing heat.

The effectiveness of sweating as a cooling mechanism is reduced when the air water content (humidity) is high. Sweat must be evaporated easily to be effective in cooling the body surface and enhancing heat loss. Therefore, evaporation is quicker in hot dry climates than in hot humid climates, because the driving force for evaporation is the difference between the vapor tension of water on the skin surface and that in air.

Shivering. Shivering is a compensatory heat production mechanism that is initiated when peripheral vasoconstriction is inadequate to prevent heat loss. When exposed to sustained cold stress, humans increase their metabolic heat production by voluntary muscle contraction or involuntary shivering. In voluntary muscle contraction, approximately 20 percent of the chemical energy released in the contractile process is converted to work, whereas in shivering almost all of the energy is converted to heat because the muscles do no work. A seminude human will double his or her heat production during a 60-minute exposure to 5°C (41°F) temperature by shivering; however, the effectiveness of shivering in maintaining body heat is very small. At maximal shivering heat production (eightfold), thermal equilibrium in the thermally unprotected human (nude) cannot be attained during prolonged severe cold stress. Newborns and prematures do not demonstrate shivering thermogenesis when exposed to cold, but they become restless and irritable and seem to have an increased sensitivity to cold. Their metabolic heat production (oxygen consumption) may be increased by as much as 100 percent over control values, which is a much greater value than adults can achieve.

Shivering consists of synchronous contraction and relaxation of small antagonistic muscle groups. The nervous pathways involved are somatic efferent neurons and their proprioceptive and tension feedback system. The amount of heat produced by shivering is influenced by the posterior hypothalamus, and the intensity depends on interactions between cortical, anterior hypothalamic, and cerebellar areas. The rhythmicity is influenced by the cerebellum and the frequency, by interaction in the spinal cord.

THE THERMOREGULATORY SYSTEM

The thermoregulatory system, like other physiological control systems, has at least three major components: sensory receptors, central integrator or controller, and effector organ systems (Fig. 28-10). The sensory thermoreceptors supply skin and deep body temperature information to the central integrator, which compares this information to a standard reference or *set-point* value. On the basis of the difference between the thermoreceptor inputs and the set-point input, the central integrator supplies output information to effector systems controlling heat production or loss, thus regulating body temperature around the set-point value. For example, if the peripheral or central thermoreceptor's input information indicates a temperature less than that of the set point, effector heat conservation or production mechanisms are activated. If input information signifies that body temperature is greater than the set-point value, heat loss mechanisms are activated. The information that the receptors transmit to the central nervous system is not temperature per se, but afferent nervous activity (discharge frequency) proportional to the steady state, and to changing temperatures.

Fig. 28-10. Basic components of the thermoregulatory system, showing receptors (skin and central receptors), hypothalamic controller (or integrator) with central set point, and effector systems regulating heat production, conservation, and loss.

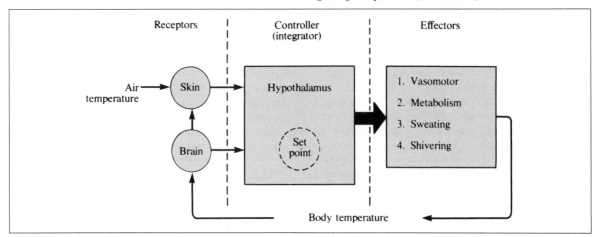

THERMORECEPTORS. Thermoreceptors that monitor skin or shell temperatures are located just beneath the skin. They are classified as cold and warm receptors and are not uniformly distributed over the body. For example, a larger number of cold receptors is found below the skin of the face and hands than below that of the chest and legs. Warmth receptors have been histologically defined as *Ruffini's end-organs,* cold receptors as *Krause end-bulbs.* However, current information indicates that the *naked nerve endings* are the major peripheral thermoreceptors. Additional thermal receptors are present in the tongue, the respiratory tract, and deep body areas such as the viscera and spinal cord. Both groups of cutaneous receptors are rapidly adapting, in that their discharge rates are proportional to the rate of temperature change as well as to the steady-state temperature. Figure 28-11 shows the average static discharge rate of cutaneous thermal receptors (warm and cold fiber populations) as a function of skin temperature. These data were taken from the noses of cats at different fixed levels of temperature. Cold fibers are identified by increasing activity as the tissue is cooled, and warm fibers are those whose activity level increases in response to warming. Afferent signals from the cutaneous receptors enter the spinal cord at all levels and ascend to the thalamic portion of the brain through the lateral spinothalamic tract, along with pain fibers.

Central thermoreceptors are found in the hypothalamus, and they are sensitive to local "core" temperature (Fig. 28-12). Cold receptors are located in the preoptic area of the anterior hypothalamus (heat loss center), and heating or stimulation of that area initiates *antirise* responses (e.g., vasodilation, sweating, panting). Warm receptors are found in the posterior hypothalamus (heat production), and local cooling or other stimulation of the area initiates *antidrop* responses (vasoconstriction, epinephrine release, and shivering). It is obvious that the thermoregulatory system utilizes both central and peripheral temperature information to determine the function of effector systems; however, the relative importance of the receptors in the overall temperature regulation picture is unclear. When the skin is warm, approximately 1°C of local hypothalamic cooling is required to evoke a strong cold-protective response. Since this level of cooling is far greater than that which would occur normally, the peripheral receptors must play a major role in calling forth protective mechanisms. Similarly, local skin temperature changes will greatly alter cutaneous blood flow and sweating rates without measurable change in hypothalamic temperature.

HYPOTHALAMIC INTEGRATION AND TEMPERATURE CONTROL

Experimental evidence points to the *hypothalamus* and its *preoptic areas* as the center of body temperature regulation. These areas, through their various interactions, function as a thermostat operating dually to prevent both excessive body heating and cooling; that is, the mechanisms controlling heat production or conservation and heat loss are regulated by hypothalamic influence over effector systems. The nervous pathways that transmit hypothalamic directions to the various effectors are the somatic and autonomic nervous systems. The sympathetic portion of the autonomic nervous system plays a very large part, since it controls vasomotor responses. The somatic nervous system controls voluntary function, such as the gross muscle movements required to add or remove clothing and the minute muscle movements associated with muscle tone and shivering.

The overall integrated control of various effector systems is shown in Figure 28-13. This is an expansion of Figure 28-10 and shows the peripheral thermoreceptors and their afferent input pathways on the left, the central receptor areas and hypothalamic integrator center in the middle, and the effector systems and their appropriate efferent output pathways on the right. The effector systems are designated according to their *antirise* or *antidrop* effects on body temperature. Antidrop outflow for skin vasoconstriction and piloerection ("gooseflesh") is mediated through sympathetic centers (posterior hypothalamus). The antirise efferent outflow responsible for skin vasodilation acts by an

Fig. 28-11. Static discharge rate of cutaneous cold and warm fiber populations in the nose of the cat as a function of skin temperature. (From H. Hensel and R. D. Wurster. Static behavior of cold receptors in the trigeminal area. *Pflugers Arch.* **313:154, 1969.)**

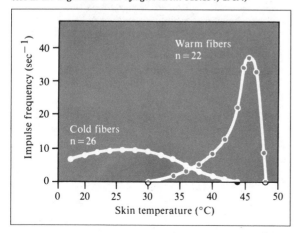

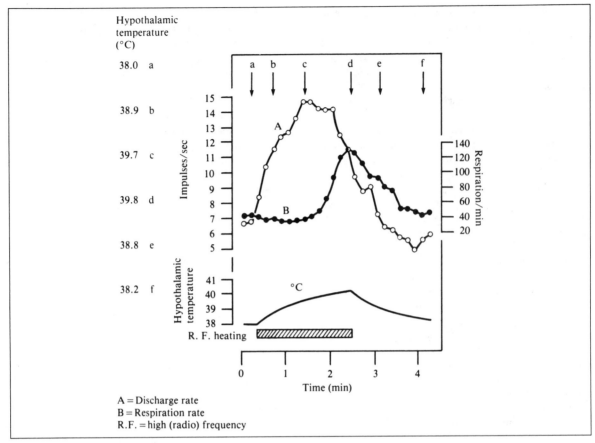

Fig. 28-12. Electrical activity of thermal-sensitive units in the preoptic area of the anterior hypothalamus, and respiratory rate in response to local heating and cooling of these central thermal-sensitive areas in anesthetized dogs. A = discharge rate; B = respiratory rate; R.F. = radio frequency. (From T. Nakayama, J. S. Eisenman, and J. D. Hardy. Single unit activity of anterior hypothalamus during local heating. *Science* 134:560, 1961. Copyright © 1961 by the American Association for the Advancement of Science.)

inhibition of sympathetic constrictor activity, possibly through inflow, and is also mediated by the sympathetic system; however, this is a sympathetic cholinergic rather than adrenergic response. Animals (e.g., dogs, cats, rabbits) that do not have sweat glands lose large quantities of heat through the respiratory passages by panting, fluid evaporation from the tongue, and salivation. These mechanisms are parasympathetic in origin and the only clearly defined parasympathetic contribution to the thermoregulatory process.

The shivering efferent pathways are not completely defined at present. Continuous somatic efferent activity from the central nervous system to the lower motoneurons initiates shivering. Proprioceptive impulses from the muscles that feed back on the spinal cord inhibit the motor input activity, thereby turning shivering off. This mechanism accounts for the shivering rhythm, which may occur at rates of 10 to 20 per second.

The physiological responses to cold and heat also involve the systems that stimulate appetite and thirst, and these drives are initiated through higher-center activity. Appetite is related to the antidrop metabolism through the *specific dynamic action* of food. Thirst is related to the antirise sweating and internal heat distribution by bulk fluid flow. Fluid consumption must be stimulated to maintain body fluid levels in the event of high sweat activity. The loss of body fluids (dehydration) causes body temperature to rise, and in severe prolonged dehydration, fever is prevalent.

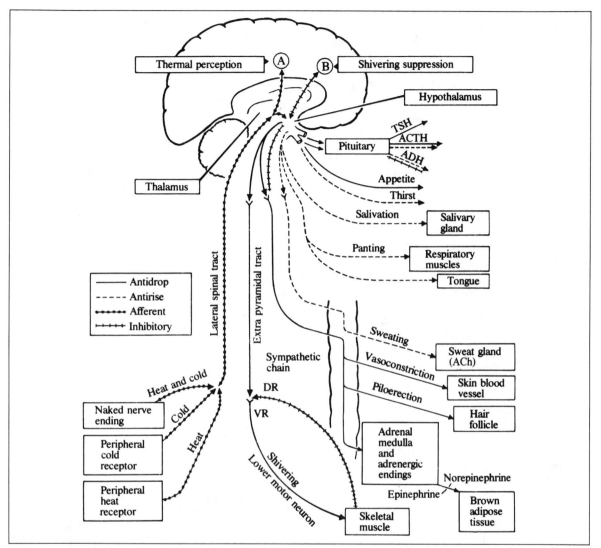

Fig. 28-13. Integration of temperature regulation mechanisms. A. The peripheral thermoreceptors and their afferent pathways. B. The central thermoreceptors and integrator system in the brain, and the effector systems with the afferent inputs. DR and VR represent dorsal and ventral root of the spinal cord for shivering feedback control. TSH = thyroid-stimulating hormone; ACTH = adrenocorticotropic hormone; ADH = antidiuretic hormone; ACh = acetylcholine. (From A. C. Burton. Integration of temperature regulation mechanisms. *J. Appl. Physiol.* 6:65, 1953.)

The hypothalamus has direct and indirect influence on the neuroendocrine system. Posterior hypothalamic activity controls the level of autonomic sympathetic activity (cholinergic) to the adrenal medulla and thus the release of epinephrine. Similarly, posterior hypothalamic activity increases sympathetic postganglionic activity and the release of the neurotransmitter norepinephrine. Catecholamines have antidrop vasoconstrictor and metabolic influences. The calorigenic effects of epinephrine and norepinephrine have been shown to be major stimulants of *nonshivering thermogenesis* in cold-adapted small animals, but the evidence for such action in humans is incomplete. Many of the releasing factors that control the synthesis or release of anterior and posterior pituitary hormones are stimulated through hypothalamic activity during various environmental stress conditions. Thyroid-stimulating hormone and adrenocorticotropic hormone (ACTH), released from the anterior pituitary, facilitate antidrop metabolism by stimulating thyroxine and glucocorticoid release. Thyroxine's caloric activity is synergistic with epinephrine in cold exposure or adaptation. Antidiuretic hormone from the posterior pituitary and ACTH from the anterior have antirise influences through their control of fluid and electrolyte reabsorption by the kidney.

In summary: Physiological responses to cold or hot stress conditions involve many neuroendocrine hormonal changes. In fact, almost all tissues and organ systems are influenced directly by the environment or indirectly by the responses of other systems to environmental changes.

ACCLIMATION TO ENVIRONMENTAL CONDITIONS

Physiological adjustments occurring during continual exposure to hot or cold environments that enhance performance or survival are collectively termed *acclimation*. The two extreme environmental thermal stresses (hot and cold) induce specific physiological and morphological changes in animals; however, the degree of change among animals is not uniform. For example, acclimation to cold is more pronounced in smaller mammals than in larger ones. The ability of humans to survive under extreme environmental temperatures has been aided by the development of artificial thermoregulatory aids (clothes, shelter, air conditioning, living habits). Nevertheless, considerable physiological changes do occur under some circumstances.

ACCLIMATION TO HEAT. When humans are abruptly exposed to a high heat stress (40°C), their skin temperature, rectal temperature, sweating rate, and heart rate increase. If they try to work, their tolerance and performance are limited. However, after 6 to 9 days of working at high temperatures, adaptive changes occur: Heart rate, rectal temperature, and skin temperature decrease, while sweat rate increases (Fig. 28-14). This acclimation is achieved by physiological adjustments in the sweating mechanism. A person who is acclimated to working in hot environments has a lower sweating threshold, sweats at a faster rate with increased evaporation, and, because of lowered skin temperature, has increased heat conduction from deep body areas to the skin. As acclimation changes take place, instead of heat oppression a feeling of comfort and well-being develops.

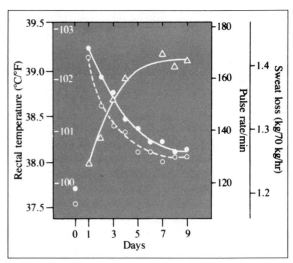

Fig. 28-14. Typical average rectal temperatures (●), pulse rates (○), and sweat rates (△) of a group of men during the development of acclimation to heat. On day 0, the men worked for 100 minutes at an energy expenditure of 300 kcal/hr in a cool climate. The exposure was repeated on days 1 to 9, but in a hot climate (48.9°C). (From L. D. Carlson and A. C. L. Hsieh. *Control of Energy Exchange.* London: Macmillan, 1970. P. 98.)

The increases in sweating and skin blood flow are the true adaptive responses to heat stress, whereas the decreases in heart rate and rectal temperature occur as a result of the human's ability to cool the body and to tolerate the work stress. Humans cannot, by acclimation, decrease their sweating rate and water consumption.

COLD ACCLIMATION. Acclimation to cold stress by animals is anatomical and physiological. New fur coats are grown, as well as fat layers for insulative protection. Hormonal changes occur that enhance metabolic activity without shivering. This *nonshivering thermogenesis* is moderated through an increase in norepinephrine release from sympathetic postganglionic fibers. Thyroxine levels in blood also rise, and thyroxine seems to have a synergistic action with norepinephrine.

Naturally, humans do not grow a new fur coat (although they may buy one), and there is evidence that their insulative fat layer decreases; however, they have a higher utilization of thyroxine and increased urine levels of norepinephrine during prolonged cold exposure, indicating increased production. Aborigines (Australia) sleep almost nude, exposing themselves to near freezing temperatures, show no metabolic increase during sleep, but allow the body to cool at night, thus conserving energy. Body heat is replenished in the early morning by exercising while running naked (streaking) across the Australian bush country. The ama (Japanese and Korean shell divers) show considerable seasonal temperature regulatory variations. They have lower skin blood flow and heat conductance to the skin during winter, and they show what appears to be nonshivering thermogenesis.

FEVER

Fever is an elevation of normal body temperature that is *not* related to work, exposure to hyperthermic conditions, or breakdown of the thermoregulatory system. It involves the mechanisms that establish the central reference or set point and those that regulate temperature around that level.

Common causes of fever are infection, primary neurological disorders, and dehydration. Fever associated with infection may be prolonged or intermittent. The chemical agents that give rise to fever are classified as pyrogens: bacterial pyrogens (endotoxins from gram-negative bacilli) or endogenous pyrogens (leukocyte extracts). The exact mechanism of action of these agents is not known, but indications are that they affect the firing rates of preoptic neurons, which set the temperature reference point at a new

Fig. 28-15. Time course of typical febrile episode. The actual body temperature lags behind the rapid shifts in set points (A). Note that regulation is maintained during the fever but is less precise, so that temperature fluctuations are generally greater than normal (B). (From G. Brengelmann. Temperature Regulation. In T. C. Ruch and H. D. Patton [eds.], *Physiology and Biophysics* [20th ed.]. Philadelphia: Saunders, 1966. Vol. 3, Chap. 5, P. 132.

high level. Figure 28-15 depicts the events and time course of a typical febrile episode. The presence of the febrile agent in the blood shifts the central set point to a higher level. Hypothalamic integrator centers compare peripheral or central thermoreceptor temperature information with the new set-point temperature and register that the body temperature is too low. At the onset of fever, the person responds physiologically as if he or she were in the cold and commonly has a chill and gooseflesh. Peripheral vasoconstriction occurs, nonevaporative heat loss is reduced, and temperature starts to rise. Feeling cool, the person may protect himself or herself by turning up the heat, putting on more clothes, and so on. Some time later (10–60 min), shivering may occur, with a resultant heat production and temperature rise until it reaches the new set point. Regulation at the new level will persist until the so-called crisis occurs. When the febrile agent is no longer effective or present, the set point is shifted to the lower normal value, and the physiological responses are similar to those that take place during heat exposure; i.e., both evaporative and nonevaporative heat loss; body temperature returns to the normal level; the febrile episode is over.

Clinically, the presence of fever is a reliable and common sign of disease, but over prolonged periods it is considered a disadvantage because of adverse effects on the nervous system and other organ systems. Historically, the crisis was considered a major recovery sign, but today, the febrile state is reduced by antipyretic drugs long before the terminal or crisis stage occurs. Before the advent of modern drugs, fever therapy was used for treating various diseases. Pa-

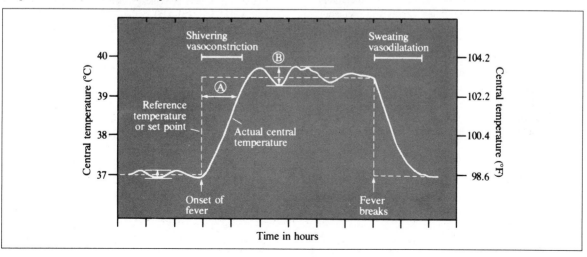

tients with neurosyphilis were treated by raising body temperature to 40 or 41°C. They were inoculated with tertian malarial parasites or typhoid to raise deep body temperatures. Needless to say, such treatment was not without its own adverse effects. Thanks to the simplest of antipyretic drugs (aspirin) many of the long and sometimes traumatic ordeals experienced by patient and physician have been alleviated.

DISEASES THAT AFFECT TEMPERATURE REGULATION

Pathophysiological conditions that prevent heat loss or heat production and lead to thermal discomfort influence temperature regulation. Morphological and physiological disorders that prevent the transfer of heat from active cells to the skin, and those that reduce the heat exchange with the environment, reduce the human's ability to tolerate heat stress. Similarly, disorders that lower metabolic activity and increase heat exchange with the environment will lower human tolerance to cold stress. The categories of heat illnesses generally recognized are (1) skin disorders (including sunburn and prickly heat); (2) heat syncope (fainting) resulting from the combined effects of extensive peripheral vasodilation, orthostatic hypotension, and cerebral ischemia; (3) heat stroke, primarily a result of inability to sweat, causing brain temperatures to rise to critical levels (41°C) and bringing on unconsciousness, coma, or convulsions—and death if body temperature is not promptly lowered; and (4) heat exhaustion and heat cramps, primarily a malfunction related to water and electrolyte metabolism. Sedative and tranquilizing drugs suppress or interfere with temperature regulation in that they cause vasodilation, decreased blood pressure, and lower heart rates. Such effects will predispose a person to heat syncope and lower his or her tolerance to cold stress.

Idiopathic malignant hyperthermia during anesthesia, although uncommon, is a striking condition that carries a high mortality. Patients may react to an anesthetic agent by increasing metabolism, thereby raising body temperature to critical levels. Temperature rises very rapidly (6–8°C/hr), and rigorous cooling methods must be used to reduce body heat.

Cold illnesses or disorders influencing a human's ability to tolerate cold stress generally involve peripheral circulatory or cardiac defects. Hypersensitivity of the arteries in the fingers to cold (Raynaud's disease) reduces hand and finger blood flow to such critical levels that pain, local asphyxia, and gangrene will occur if the stress is continued. Inflammation of arteries and veins in the extremities (Buerger's disease) will result in permanent luminal obstruction, reducing blood flow and heat transfer to, and heat loss from, the afflicted segment. Cold exposure in some persons induces *hemagglutination,* i.e., clumping of red blood cells in small superficial vessels, which reduces cutaneous blood flow and heat transfer. The incidence of hemagglutination is higher in subjects who have had *frostbite* and are thus hypersensitive to cold. This increased sensitivity to cold is probably due to anatomical rather than physiological changes, since naked nerve endings and thermoreceptors are closer to the skin surface than Krause end-bulbs.

BIBLIOGRAPHY

Adolph, E. F., et al. *Physiology of Man in the Desert.* New York: Interscience, 1947.

Bullard, R. W. Temperature Regulation. In E. E. Selkurt (ed.), *Physiology* (4th ed.). Boston: Little, Brown, 1976.

Burton, A., and Edholm, O. G. *Man in a Cold Environment.* New York: Hefner, 1969.

Carlson, L. D., and Hsieh, A.C. L., *Control of Energy Exchange.* London: Macmillan, 1970.

Dill, D. B. (ed.). *Handbook of Physiology.* Washington, D.C.: American Physiological Society, 1964. Section 4: Adaptation to the Environment.

DuBois, E. F. *Fever and the Regulation of Body Temperature.* Springfield, Ill.: Thomas, 1948.

Greenfield, A. D. M. The Circulation Through the Skin. In W. F. Hamilton and P. E. Dow (eds.), *Handbook of Physiology.* Washington: American Physiological Society, 1963. Section 2: Circulation. Vol. 2, pp. 1325–1353.

Hammel, H. T. Regulation of internal body temperature. *Annu. Rev. Physiol.* 30:641–710, 1968.

Hardy, J. D., Gagge, A. P., and Stolwijk, J. A. J. *Physiological and Behavior Temperature Regulation.* Springfield, Ill.: Thomas, 1970.

Kuno, Y. *Human Perspiration,* Springfield, Ill.: Thomas, 1956.

Selle, W. A. *Body Temperature: Its Changes with Environment, Disease, and Therapy.* Springfield, Ill.: Thomas, 1952.

Strom, G. Central Nervous Regulation of Body Temperature. In J. Field (ed.), *Handbook of Physiology.* Washington, D.C.: American Physiological Society, 1960. Section 1: Neurophysiology. Vol. 2, pp. 1173–1196.

PHYSIOLOGY OF EXERCISE

29

William V. Judy

Skeletal muscle contraction is the primary physiological event in exercise, with all other body systems playing a supportive role. The complex events associated with the excitation-contraction coupling mechanism of skeletal muscle were described in Chapter 3, and those concerning the conversion of food substances into chemical energy required by muscle cells to produce mechanical energy and heat were described in Chapter 27.

Scores of enzymatic reactions are required in proper sequences before muscle contraction can occur. Activity changes and integration of all body systems occur simultaneously to support active muscle, thereby allowing the continuation of repetitive contraction-relaxation for extended periods. Exercise places demands on all body systems, beginning with conscious or subconscious thought processes, and involves impulses from the central nervous system (CNS) that initiate coordinated muscle, cardiovascular, pulmonary, and other system activity. Greater activity of circulatory and respiratory function is necessary during exercise to provide oxygen and nutrients to active muscle cells and to remove CO_2, other metabolites, and heat. A firm understanding of exercise physiology permits the physician, nurse, and life scientist to evaluate the maximal physical exertion levels of different age groups, and with such knowledge to designate the type and amount of activity patients with cardiac, pulmonary, and other diseases should be allowed during convalescence and rehabilitation.

METABOLIC ASPECTS OF EXERCISE

ENERGY SOURCE DURING EXERCISE. The primary energy source for skeletal muscle contraction is high-energy phosphate units. These are produced from both aerobic and anaerobic metabolism (Fig. 29-1). The aerobic (oxidative) metabolism of glucose or glycogen, lipids, and proteins in the Krebs citric acid cycle produces 38 high-energy units, CO_2, and water when the electron transport is complete. Anaerobically, these units are produced from glycolysis, which consists in the phosphorylation and breakdown of glycogen to pyruvate or lactate. A net yield of only 2 high-

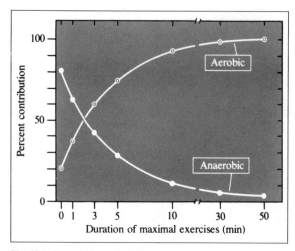

Fig. 29-1. The relative contribution of aerobic and anaerobic energy processes during various periods of maximal exercise. Note that the energy for long-term exercise is provided aerobically, whereas that required for maximal short-term exercise is provided anaerobically.

energy units results from a breakdown of each 6-carbon unit compared with the 38 produced aerobically. These units are stored in skeletal muscle as molecules of *adenosine triphosphate* (ATP) and *creatine phosphate* (CP). The actual energy required for the sliding of the thin actin filament over the thick stationary myosin filament is provided by the hydrolysis of ATP. When hydrolysis occurs, mechanical energy of contraction, thermal energy (heat), *adenosine diphosphate* (ADP), and inorganic phosphate are produced.

Approximately 20 percent of the total energy released is the mechanical energy of contraction; the remaining 80 percent is thermal energy. The high thermal energy release during exercise places additional demands on the thermoregulatory system, specifically, the cardiovascular system, which transports heat to the body surface, and the sweat glands, which play a major role in the heat transfer from the body surface to the environment (see Chap. 28).

Continuation of the contractile process requires that the tissue stores of ATP be maintained. These are quickly regenerated from the second high-energy source stored in muscle, CP. This high-energy unit in the presence of ADP yields new ATP units and creatine. As exercise becomes more severe or extended, aerobic metabolism gives way to anaerobic conditions because of the lack of available O_2 at the cellular level. In this event, CP is produced anaerobically from glucose and glycogen in addition to pyruvic and lactic acid. Through this pathway, tissue ATP stores are replenished anaerobically from the available CP. The CP

content of resting human quadriceps muscle is about 8 mg per 100 gm of tissue (dry weight), and the ATP content is about 2 mg per 100 gm of tissue. During light to moderate exercise the CP content falls to 2 to 4 mg per 100 gm of tissue, and during heavy exercise it falls to near zero while the ATP level remains fairly constant. Thus, CP breakdown must quickly serve to replenish tissue ATP stores.

Normally, aerobic ATP production is the major source of mechanical energy during light exercise; however, in short-term maximal exertion or in events in which exhaustion occurs, anaerobic ATP production is the major energy source (Fig. 29-2). In long-term or extended work or exercise such as a cross-country run, aerobic energy production must dominate. In such cases the person, through training or experience, learns to pace himself or herself so that the cardiopulmonary delivery of O_2 to the active skeletal muscle cells meets the energy demand. A well-conditioned athlete should have considerable metabolic reserve and therefore be able to perform maximally for a short interval through anaerobic energy production. The amount of reserve should determine the duration of the final maximal exertion or "kick" associated with most distance running, provided that sufficient energy (substrate sources) is still available at the cellular level.

ENERGY REQUIREMENTS DURING EXERCISE. The amount of energy required to do a particular task is proportional to the intensity and duration of the event (Fig. 29-3). Work or exercise classified as *light* to *moderate,* which includes most everyday occupations except manual labor, requires an energy expenditure of three times the basal (resting) requirements, i.e., about 4 kcal per minute or 1900 kcal in an 8-hour working day. *Hard work,* which includes manual labor (e.g., heavy industry, farming, mining), requires up to 9 kcal per minute or 4300 kcal in an 8-hour workday. Assuming that 500 kcal is expended during sleep and 1400 kcal in the nonworking hours, the total daily energy requirements are about 3800 and 6200 kcal for the moderate and heavy work levels respectively. Persons accustomed to hard work can maintain the required energy output for extended periods. Greater activity levels or *maximal* working ability can be sustained for only short intervals before exhaustion occurs. Exhaustion may result from the lack of available energy reserves, insufficient delivery of O_2 to the cellular level, high thermal loads, or accumulation of anaerobic metabolites such as lactic acid in active muscles. Metabolites diffuse from the cellular to extracellular compartments, disturbing the normal homeostatic equilibrium.

The energy (kilocalories) required to do a specific task can be easily equated to the work produced or done in

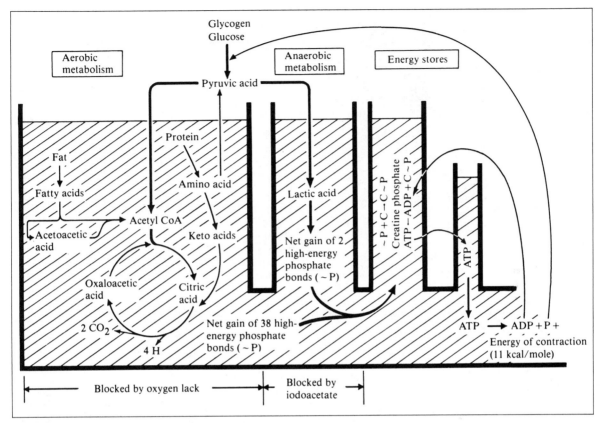

Fig. 29-2. Energy sources for muscle contraction. The adenosine triphosphate (ATP) and creatine phosphate (CP) may be considered storage depots for high-energy phosphate bonds. The anaerobic glycolysis and aerobic metabolism function to restore these depots. ADP = adenosine diphosphate. (From R. W. Bullard. Physiology of Exercise. In E. E. Selkurt [ed.], *Physiology* (4th ed.). Boston: Little, Brown, 1976. P. 702.)

moving a known mass a specific distance. For example, if you are a 50-kg person who would like to climb a hill 100 meters high to enjoy the view, you would do approximately 5000 kg-meters of work: 5000 kg-meters = 50 kg × 100 meters. Since to do 427 kg-meters of work requires about 1 kcal of energy, you need 11.7 kcal to climb the hill.

$$\frac{5000 \text{ kg-ms}}{427 \text{ kg-ms/kcal}} = 11.7 \text{ kcal}$$

The same task would require about 16.4 kcal for a 70-kg person. Assuming that all this energy was produced aerobically, about 2.5 gm of glucose and 2.3 liters of O_2 would be necessary for the 50-kg person and approximately 3.5 gm of glucose and 3.5 liters of O_2 for the 70-kg person (1 gm glucose = 4.7 kcal; 1 liter O_2 = 5.0 kcal). However, since the hydrolysis of ATP to contractile energy is only one-fifth (20 percent) of the total energy production, it will take about five times the original amount of glucose and oxygen to produce the 11.7 and 16.4 kcal of energy, or 58.5 and 72.0 kcal total energy costs respectively.

The 20 percent mechanical energy produced from the hydrolysis of ATP shows that human *mechanical* or *gross* efficiency (ME) is about 20 percent, where efficiency is calculated from the output/input energy ratio. Therefore, the percent ME is calculated as

$$\text{Percent ME} = \frac{\text{External work (kcal)}}{\text{Total energy required (kcal)}} \times 100$$

Exercise or work efficiency varies with the speed of muscle contraction, the load against which the muscle contracts, the fatigue level, training, and the metabolic level being used by the body to perform the work. Calculated

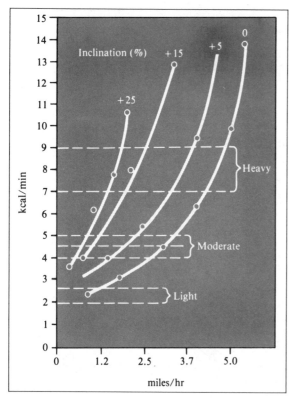

Fig. 29-3. Steady-state energy cost for various work loads, and the effect of speed and incline on energy cost of walking on a motor-driven treadmill. (Data from D. B. Dill. The economy of muscular exercise. *Physiol. Rev.* 16:269, 1936 and from R. Margaria. Sulla Fisiologia, e specialmente sul consumo energetico della Marcia e della corsa a varie velocita ed inclinazioni del terreno. *Att des Lancei* 7:299, 1938.)

efficiency can be improved to a certain extent by training or practice. Trained athletes develop coordinated muscle activity or rhythms that allow them to obtain the maximal functional output from muscle contraction. On the other hand, untrained persons will waste considerable energy through unnecessary muscle movements even though they may be in good physical condition. Studies have shown that the trained swimmer may have an efficiency at least six times that of an untrained swimmer. Mechanical efficiency is also related to the rate of work or velocity of muscle contraction. Very slow and extremely fast movements are inefficient. For example, climbing a stairs at 50 steps per minute is more efficient for a young person than doing it at a slower or faster rate. The voluntary control of static muscle

contraction or "holding back" generates a good deal of thermal energy but not useful work and therefore reduces efficiency. At faster rates of movement, efficiency is reduced because greater forces are required for acceleration and deceleration of the limbs and because the person is fighting, more and more, the intrinsic or viscous resistance of the contractile and supportive tissue elements.

The product of force and velocity is the rate of work or power output; it can be derived from the force-velocity curve for the human forearm as shown in Figure 29-4. The maximal rate of power production that a muscle can sustain is shown with peak power output corresponding to loads of about 30 percent of isometric maximum. Similarly, when a person bicycles at a maximal work rate, the leg extensors produce tensions of up to about 30 percent of isometric maximum. The same results have been found for the isolated frog skeletal muscle, and there is a definite inverse relationship between the energy cost of exercise and mechanical efficiency.

Exertion, as in most sports, requires even greater energy output. Superior athletes can increase their energy output 13 to 14 times their basal metabolic rate. Basal metabolic rate for males is approximately 72 kcal per hour (38 kcal/m²/hr) and slightly lower for females (see Chap. 27). The total energy output for male athletes may exceed 1000 kcal per hour (72 kcal/hr × 14 = 1008 kcal/hr). Similarly, well-conditioned female athletes (swimmers, for example) can achieve energy outputs of 780 to 870 kcal per hour. Less demanding sports, such as archery, require outputs of only 105 kcal per hour. Obviously, the total energy required to climb the hill previously mentioned (58.5 kcal) now seems minor. However, if you assume that you climbed the 100 meters (328 feet) in 8 minutes, it would require about 440 kcal to continue to climb for an hour. These values are based on continuous exercise at fairly constant rates.

To be sure, most contact sports do not require sustained higher energy outputs, nor do most track and field events. The actual number of calories used in the performance of a particular event is not very large when compared with a total daily requirement. With the exception of long-distance running (marathon race or cross-country run), the total caloric output for most sustained (4–10 min) athletic events requires only 400 to 800 kcal (200–400 kcal/m²). Therefore caloric requirements are not themselves a limiting factor in exercise tolerance. However, a high caloric intake is necessary for training, since daily training events may last 2 to 4 hours. It takes only about 800 to 1000 kcal for a person to walk at a rate of 2 miles per hour for 6 hours (see Fig. 29-3; zero incline). Assuming that this is the only caloric utilization in excess of daily requirements, caloric intake should

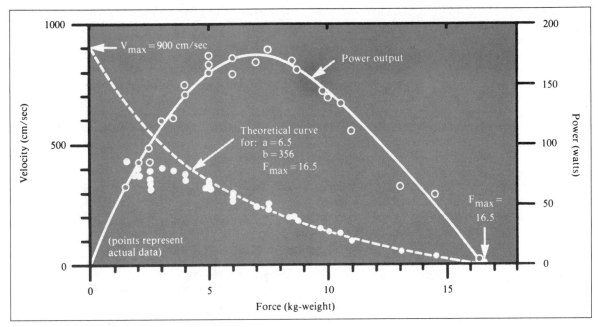

Fig. 29-4. Force-velocity and power-output relationships of the human forearm muscle. (From R. Meiss. Unpublished data.)

Table 29-1. Oxygen consumption (L/min)

Individual	Basal	Exercise Light	Exercise Heavy
Male athlete	0.25	0.8–1.2	3.5–5.0
Male normal	0.25	0.7–1.0	2.0–3.0
Woman athlete	0.23	0.65–1.0	2.5–3.2
Woman normal	0.23	0.50–.90	1.8–2.2

include this amount plus what is needed to repair or return body tissues to some equilibrium condition.

OXYGEN REQUIREMENTS DURING EXERCISE. Representative O_2 consumptions are shown in Table 29-1. These are shown for basal conditions, light exercise, and heavy exercise. Note that the basal O_2 consumption (250 ml/min) is approximately the same for all persons; for females it is slightly lower. An untrained person can increase O_2 consumption 3-fold (750 ml per minute) during light exercise and 8- to 12-fold (2 to 3 liters per minute) during heavy exercise. Well-trained athletes may increase their O_2 consumption 16 to 20 times (4 to 5 liters/min), whereas superior individuals or champions may increase theirs 25 times (6.2 liters/min). Normally, O_2 consumption capacity is expressed in units of body weight (ml O_2/kg/min), and average values of 44 to 51 ml are observed for untrained males and 35 to 43 ml for untrained females. Values between 50 and 60 ml are obtained for average males in good condition. Males normally have aerobic O_2 consumption capacities during exercise about 20 to 25 percent greater than females. The highest measured O_2 consumption on record is 85 ml for a champion cross-country male skier and 65 ml for the champion female skier. Naturally, age is a very important factor in one's exercise tolerance. Oxygen consumption values given thus far have been for 20- to 29-year-olds. With age, maximal O_2 consumption decreases for both males and females; for example, the range is 21 to 42 and 25 to 44 ml for 50 to 60-year-old females and males respectively.

Similarly, training increases the exercise efficiency and stabilizes the amount of O_2 required to do a specific task. For example, after 2.5 to 3.0 months of training to walk on a treadmill at a rate of 8 miles per hour (zero incline) for 10 to 15 minutes, the mean O_2 consumption for nine men decreased from 47 liters/kg/min to less than 44. The same subjects on an initial exposure to running 7 mph (9 percent grade) had a mean O_2 consumption of about 53 liters/kg/min, and they were exhausted after 5 minutes of effort. However, after 2 months of training they could run at the same rate and level for 15 minutes without exhaustion because their aerobic metabolic limit was increased to above 56 liters/kg/min. The physiological factors that may limit human O_2 consumption are (1) the rate of cardiovascular transport to the active tissue, (2) the O_2 utilization by the active cells, and (3) O_2 diffusion capacity in the lungs.

In addition to physical fitness and age, disease and induced inactivity greatly influence the maximal O_2 consumption capacity of humans. Maximal O_2 consumption is decreased during long-term inactivity. Young and old patients confined to bed rest or inactivity lose their physical stamina and become easily fatigued with minimal exertion. How quickly they recover depends on the type and severity of their illness, their age, and their willingness to work hard and to endure the extreme exhaustion that accompanies minimal activity after prolonged bed rest. For example, healthy young men confined to bed rest for 20 days showed a 29 percent decrease in maximal O_2 consumption capacity and a 23 percent decrease in the volume of air they could move out of the lungs per minute (maximal expiratory volume). Although these men were not ill, they recovered slowly (10–20 days). After 60 days of physical training they were able to increase their O_2 consumption capacities and maximal expiratory volumes 20 and 10 percent respectively above their pre-bed rest control levels. Naturally, patients recovering from long-term illness or even minor surgery would not be able to achieve the same degree of recovery, nor would they be able to do so in such a short time. Recovery and the establishment of a high maximal O_2 consumption capacity is a slow process, and the upper limit of O_2 consumption is *one* of the best indices of physical fitness. But having a high maximal O_2 consumption does not necessarily mean that a person will have a high performance or be a superstar.

There is no good correlation between maximal O_2 consumption and performance. Many can, through physical fitness, obtain a high O_2 consumption but have mediocre to poor performances. Other factors, less objectively measured, are also important: motivation, ability, efficiency, and training. During maximal exertion, the main thing as far as O_2 intake is concerned, is its availability at the cellular level. If the amount of O_2 consumed equals that required for all metabolic activity, then aerobic ATP production is furnishing the necessary energy, and a *pay-as-you-go* condition exists. Such a condition occurs only during light to moderate work. As stated previously, when O_2 delivery to the active cells falls behind the demand, anaerobic ATP production is the major contributor of energy. This can be considered a *buy-now, pay-later condition,* because the energy stores used must be paid back after the end of exercise. The magnitude of the energy stores

used anaerobically that must be paid back is equivalent to the O_2 *debt.*

Oxygen Debt During Exercise. The O_2 debt incurred during moderate to heavy exercise is shown in Figure 29-5. During the initial minutes, O_2 consumption increases abruptly and plateaus at some steady maximal intake level. After exercise has stopped, consumption declines slowly for several minutes to an hour. The O_2 debt is defined as the O_2 consumed after exercise above the pre-exercise control or basal consumption. In Figure 29-5, area A represents the basal consumption utilizing aerobic energy production; area B represents the incurred debt where the rate of O_2 uptake attains a value appropriate for the level of O_2 expenditure, the so-called steady state. The repaid debt is represented by area D, where the extra O_2 consumed is to repay that incurred in B. Characteristically, the debt is paid rapidly at first and then at a slower rate.

The increased O_2 debt, as measured by the rate of O_2 consumption after stopping an exercise event, probably has three components: (1) a fast component that is paid back in the initial 2 minutes, probably by means of a small net decrease in O_2 stored in venous blood and muscle hemoglobin (100 ml O_2 in moderate exercise; 250 ml O_2 in exhausting work); (2) the rapid resynthesis of high-energy phosphate bonds of ATP and CP; and (3) the slow removal of lactic acid formed from pyruvate in the aerobic breakdown of glycogen. Although the O_2 debt is generally attributed to the cost of oxidization and reconversion of lactic acid to replenish the tissue stores of high-energy phosphates, the correlation between O_2 debt and blood lactate content is low until the O_2 consumption reaches 2.5 to 3.0 liters per minute (Fig. 29-6). Note that at low levels of work ($\dot{V}_{O_2}$ less than 2.5 liters/min) an O_2 debt occurs without a significant increase in blood lactate. This is known as the *alatic O_2 debt,* which is generally attributed to a restoration of the depleted body O_2 stores. With more severe exercise, the O_2 debt and blood lactate levels increase proportionally, and there is a good correlation between alactic O_2 debt and the amount of O_2 required to regenerate the CP store to its original level.

The buildup of an O_2 debt to a critical level appears to be an important but poorly understood factor limiting the duration of heavy exercise and representing one form of exertion. A person's tolerance to O_2 lack and ability to reach a high debt before exercise is stopped is another important criterion of physical fitness or stamina. Untrained persons can tolerate a maximum of only 10 liters of O_2 debt before stopping on account of fatigue and, in some cases, nausea and vomiting. The debt ceiling for superior athletes may be

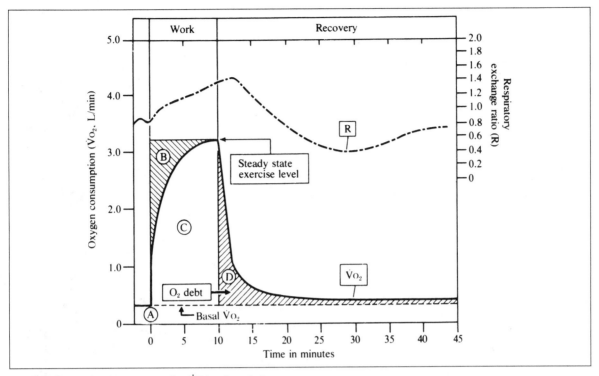

Fig. 29-5. Changes in oxygen consumption ($\dot{V}O_2$) and respiratory exchange ratio (R) before, during, and after heavy exercise. Area A represents basal or resting O_2 consumption; area B, the incurred O_2 debt; area C, the period in which the O_2 uptake attains a value appropriate to the O_2 expenditure; and area D, the repaired O_2 debt. (From R. W. Bullard. Physiology of Exercise. In E. E. Selkurt [ed.], *Physiology* [4th ed.]. Boston: Little, Brown, 1976. P. 702.)

as high as 17 to 18 liters. The determinant of the maximal obtainable debt, whether muscle glycogen or CP depletion or other factors, has not been fully evaluated. It is obvious that if the exercise is of long duration (distance running), the person cannot allow the continuous accumulation of the debt but must work aerobically or at a rate low enough for cardiopulmonary delivery of O_2 to active cells to parallel the O_2 requirement or *cost* (see Fig. 29-2). The O_2 cost is that consumed above the control or the basal state during exercise plus the O_2 debt. If the total O_2 consumption exceeds that required to perform a particular task in a specific time interval, that task should be easily performed. If the cost is greater than the total consumption, the participant should not be able to finish it within the specific time (e.g., running a 4-minute mile).

FUEL OF EXERCISE AND RESPIRATORY EXCHANGE RATIO.
Both carbohydrates and lipids are utilized as fuel during exercise in proportions varying with work intensity, duration, and availability of stored fuels. At low rates of activity, lipid supplies a large portion of the required energy. In heavy exercise (i.e., that consuming above 70 percent of maximal O_2 consumption), the fuel consists of carbohydrates derived primarily from glycogen stored in skeletal muscle. It appears that the body can substitute carbohydrate for lipid, but lipid cannot be substituted for carbohydrate, particularly at high rates of work. During severe exercise, muscle fatigue is probably due to exhaustion of muscle glycogen stores; however, individuals can continue to exercise at a lower rate by utilizing available lipids. In short-term maximal exertion, such as running the 100-yard dash in less than 10 seconds, the delivery of O_2 by the cardiopulmonary system to the activity cell is the major limiting factor. In this case a tremendous O_2 debt (5–6 liters) is incurred in a short-term effort. The metabolic substrates for aerobic energy production are plentiful. However, their utilization is limited by the available O_2. On the other hand, the availability of tissue oxidative metabolic substrates is probably the

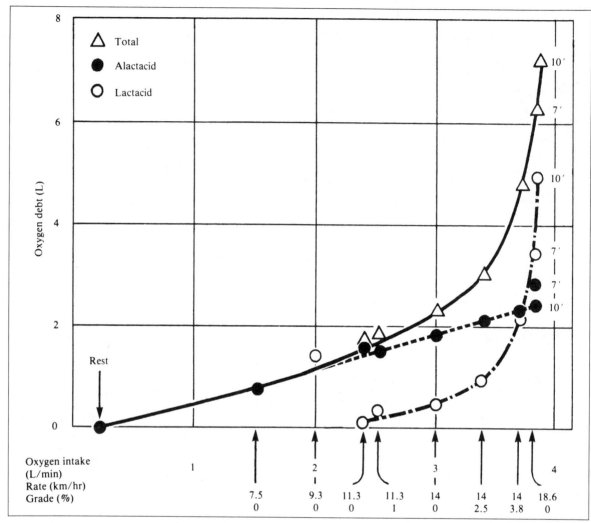

Fig. 29-6. Amounts of alactacid and lactacid O_2 debt as a function of O_2 consumption (intake) and work rate on a treadmill. (From R. Margaria, H. I. Edwards, and D. B. Dill. Possible method of contracting and paying the oxygen debt and role of lactic acid in muscular contraction. *Am. J. Physiol.* **106:**708, 1933.)

major limiting factor in long-term athletic events such as the cross-country and marathon run. In these events, athletes through training build up their maximal O_2 consumption capacity and learn to pace themselves so that they work primarily by oxidative metabolism. Fatigue and exhaustion result from depletion of readily available metabolic substrate stores of carbohydrates or glucose. It is possible to build up or overfill these metabolic substrates in the tissue stores by dietary selectivity 10 to 20 hours before the exercise event. For example, many present-day Olympic-caliber distance runners voluntarily increase their carbohydrate consumption by eating spaghetti, crackers, or pizza after their last training event.

Recall that the fuel oxidized during basal steady-state conditions can be determined by the respiratory quotient (volume of expired CO_2/O_2 consumed, Chap. 27). During the non-steady-state condition of exercise this ratio is called the *respiratory exchange ratio* (R) and does not indicate the fuels being oxidized. During exercise, R indicates immediate alterations of respiratory gas exchange. Just prior to the onset of exercise, some people hyperventilate, anticipating the respiratory demands to follow. This action results in a pumping out of CO_2 stores and may temporarily increase R (see Fig. 29-5). Further increase in R may be seen during exercise as lactic acid begins to build up in active cells and diffuses into the extracellular spaces. Lactic acid acting as a fixed acid forms carbonic acid, which in turn breaks down into water and Co_2, as follows:

Muscle activity

$$lactic\ acid\ +\ Na_2CO_3 \rightleftharpoons Na\ lactates\ +\ H_2CO_3$$
production

$$H_2CO_3 \rightleftharpoons H_2O\ +\ CO_2\ expired$$

The CO_2 is then expired (compensation for metabolic acidosis), thereby raising R. The H^+ content of body fluids increases during exercise to a point where pH may drop 0.6 to 0.7 unit before compensation occurs. Tolerance to low pH seems to be an acquired characteristic developed during training for distance running. Trained athletes can tolerate plasma pH levels of 6.7 to 6.8, normally considered incompatible with cellular function. The R may keep increasing for a short time following exertion, while O_2 consumption drops rapidly, but lactic acid is still diffusing from cells and forming CO_2. As recovery progresses and lactic acid is removed from body fluids, these reactions are reversed, thereby retraining CO_2 in bicarbonate formation and replacing lactate. With retention of CO_2, R values become extremely low. Complete recovery is indicated by a return to normal levels.

BODY TEMPERATURE IN EXERCISE. Body temperature increases during prolonged exercise in response to a 10- to 20-fold increase in metabolic activity. Since 80 percent of the total energy expenditure is thermal energy or heat, this energy must be dissipated from the body surface or added to the body heat stores (see Chap. 28). Body temperatures up to 39 to 40°C have been observed frequently during long-term exertion, even in cool environments. Deep body temperature (rectal or core temperature) of champion athletes

may reach the fever range (40–41°C, 105–106°F) after prolonged exertion. Temperature increases during exercise are independent of environmental temperature except at extreme ranges; however, they are dependent on the metabolic or work load. Normally, rectal temperature levels off at a few degrees above normal regardless of the external environment. The temperature rise is not due to a deficiency in the thermoregulatory mechanism except possibly in very hot and humid conditions in which the efficiency of losing heat by sweating is reduced (see Chap. 28).

Whether increased body temperature enhances physical performance is not known. Athletes normally warm up before events to increase muscle blood flow, relax the muscles, and prepare them (they hope) for the forthcoming stress by thermal stimulation of cellular metabolism. In addition, heat buildup in active muscle promotes the unloading of O_2 in the tissue by shifting the O_2-hemoglobin dissociation curve to the right (see Chap. 19). Similarly, increased temperature enhances the diffusion rates of respiratory gases, decreases the viscosity of blood, and relaxes vascular smooth muscle; both of the latter will increase muscle blood flow by reducing flow resistance.

An advantage of high body temperature during exercise is that it increases the thermal gradient between the body surface and the environment. This rise reduces the strain on the thermoregulatory system by enhancing convective and radiative heat loss and lowering the heat-removal demands placed on the sweating mechanisms. If temperature did not rise during exercise, skin blood flow would have to be greatly increased to carry the large amounts of internally produced heat to the body surface. The consequent reduction in muscle blood flow might deprive active cells of required nutrients and O_2.

Heat is not always beneficial to athletes; it may be their worst enemy. To compete successfully in hot environments, athletes must be acclimated to working in heat; therefore, they must have a well-developed sweating mechanism. Recall that the threshold and the rate of sweating are physiological mechanisms that improve with working or training in heat (see Chap. 28). Athletes who train or compete in cold climates have less well-developed or responsive sweating mechanisms. Many athletic events have been decided by the differences in heat tolerance of the participants. For example, note the differences between the Minnesota Vikings (trained in the north) and the Miami Dolphins (trained in the south) during a football game in Miami in mid-January with an environmental temperature of 80°F. The northern team, having less conditioned sweat mechanisms, may suffer from overheating, since their sweating mechanisms are slow or poorly tuned. On the other hand, the Dolphins, accustomed to working in heat, will have a definite environmental advantage.

The combination of high environmental temperature and greatly elevated heat production rapidly brings on the deterioration of performance. Excessive sweating, however, promotes dehydration. This, accompanied by the rapid reduction of intravascular volume of 10 to 15 percent due to increased filtration into the interstitial compartment, could drastically influence tissue perfusion if not corrected by the use of the old-fashioned water bucket or modern Gatorade. Exercise in cold environments (below 20°C) may result in an increase in plasma volume after an extended athletic event. The lack of sweat loss plus metabolic water production during exercise initially produces excessive interstitial fluid, which rapidly shifts to the vascular compartment after exercise.

RESPIRATORY ASPECTS OF EXERCISE

MINUTE VENTILATION AND RESPIRATORY RATE. The increased cellular activity during exercise (i.e., increased O_2 consumption and CO_2 production) places extra demands on the respiratory system. Changes occur in the mechanics of breathing, gas exchange, gas transport, and control of respiration. Pulmonary ventilation (minute volume), which is normally about 5 to 6 liters per minute, may exceed 150 liters per minute in severe short-term exercise. The change is attributed to a fourfold increase in respiratory rate (15 to 50 breaths/min) and a sixfold increase in tidal volume (0.5–3.0 liters/breath). Such changes require greater energy output from the respiratory muscles, increasing the work of breathing 100-fold (Fig. 29-7). During exercise, the increase in minute volume is proportional to the metabolic demand (O_2 consumption and CO_2 production). Figure 29-8 shows the ventilatory response of normal untrained adults and well-conditioned athletes. Note that athletes have a more efficient ventilation in that they are able to consume more O_2 and remove more CO_2 per unit volume of expired air each minute.

Similarly, athletes have smaller increases in ventilation rate during exercise than do normal untrained adults (Fig. 29-9). The rate of breathing increases linearly with pulmonary ventilation; however, athletes have to take fewer breaths to attain the same minute ventilation. It may be concluded that respiratory rate is the major contributor to increased minute volume for the normal untrained person during exercise, whereas tidal volume is the major contributor for the athlete. This difference may be accounted for by two factors: The athlete may have (1) larger upper

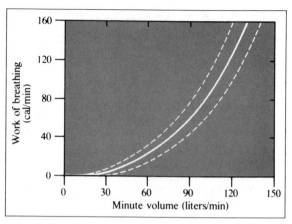

Fig. 29-7. The work of breathing as a function of pulmonary ventilation (minute volume). Dashed lines represent estimated range. Under resting conditions, with a minute volume of 4 to 5 liters/min, the energy cost for the active portion of breathing is less than 1 percent of resting metabolic activity. During exercise this value increases to only 3 to 5 percent of the maximal aerobic power a person is able to develop. (Redrawn from R. Margaria, G. Milic-Emili, J. M. Petit, and G. Cavagna. Mechanical work of breathing during muscular exercise. *J. Appl. Physiol.* 15:356, 1960.)

respiratory airways and therefore lower airflow resistance and greater lung volume change per breath; and (2) greater chest wall and lung compliance and therefore larger lung volume change per unit change in the intrapulmonary pressure (see Chap. 20).

LUNG VOLUME CHANGES. Lung volume changes during exercise are shown in Figure 29-10; note that lung gas volume decreases slightly because of increased pulmonary blood flow and volume. The major changes occur in tidal volume and inspiratory reserve volume. Tidal volume increases at the expense of inspiratory reserve volume, which is correspondingly reduced. The expiratory reserve volume and the residual volume change very little during exercise, provided there is no major change in body position. Changes in body position from standing to supine decrease intrathoracic volume and thereby reduce total lung volume through the influence of reduced gravitational forces on the lung and abdominal viscera. In prolonged and severe exercise, a slight increase in reserve volume may occur, and this may be due in part to more blood in the lung. Pulmonary blood flow increases during exercise, and blood volume distribution changes occur, facilitating gas exchange. These aspects will be presented later in the discussion of exercise and ventilation-perfusion.

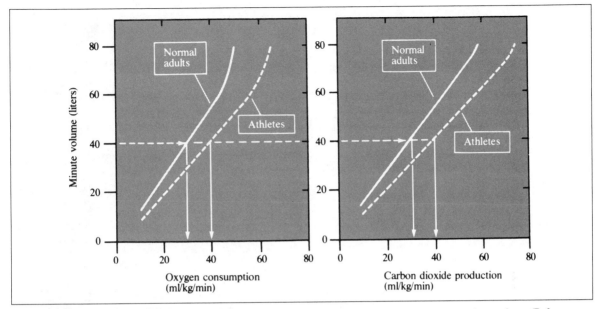

Fig. 29-8. Pulmonary ventilation (minute volume) as a function of O_2 consumption and CO_2 production during exercise. Note that at a ventilation of 40 L/min the athlete consumes approximately 10 ml O_2/kg/min and produces about 10 ml CO_2/kg/min (25 percent) more than an untrained adult at the same minute volume. (Redrawn from R. Margaria and P. Cerretelli. The Respiratory System and Exercise. In H. B. Falls [ed.], *Exercise Physiology.* New York: Academic, 1968. P. 44.)

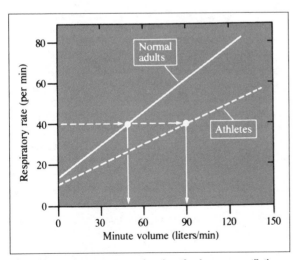

Fig. 29-9. Respiratory rate as a function of pulmonary ventilation (minute volume) during exercise. Note that at a rate of 40 breaths/min, athletes have a much greater ventilation (90–100 percent) than do untrained adults. (Redrawn from I. Brambilla, P. Cerretelli, and G. Brandi. *Bull. Soc. Ital. Biol. Spec.* 34:1820, 1958.)

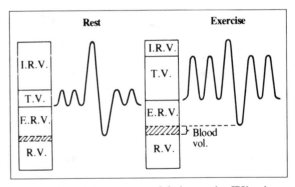

Fig. 29-10. Lung volumes at rest and during exercise. IRV = inspiratory reserve volume; TV = tidal volume; ERV = expiratory reserve volume; RV = residual volume; and shaded area = blood volume. Note that the greatest change occurs in TV, which increases at the expense of IRV.

FORCES INVOLVED IN THE WORK OF BREATHING. In the moving of gas in and out of the lung, work is done to overcome the resistance to air displacement from one position in the lung to another and to change the position, size, and shape of the lung, rib cage, and abdomen. The driving forces required to do this work are classified as *elastic* (static) and *frictional* (dynamic) (see Chap. 20). The static forces are those required to overcome (1) the elasticity of the chest wall and lung, (2) the surface tension of the alveoli, and (3) the gravitational forces on different respiratory structures. The dynamic forces are those required to overcome (1) the viscous resistance of the tissue and (2) the resistance to laminar and turbulent airflow in the pulmonary airways. In health persons the main opposing forces involved in exercise hyperventilation are the dynamic forces required to overcome airflow resistance, and most of the increased work is spent in overcoming this resistance. Recall that the forces required to expand the chest and lung are stored and utilized completely in making passive expiration. During exercise, additional muscle groups are recruited during inspiration and account, in part, for increased work. In normal breathing, the diaphragm, external intercostal muscles, and scaleni muscles (in some individuals) provide the necessary active inspiratory forces, whereas expiration is passive because of the recoiled nature of the lung and thorax. At ventilatory volumes of 50 liters per minute or greater, the sternocleidomastoid, trapezius, and pectoralis muscles of the chest support inspiration; during expiration (at volumes > 30–40 liters/min), muscles of the anteroabdominal wall and internal intercostals provide an active expiratory force. Figure 29-11 shows the increased work involved in exercise hyperventilation.

Although there is a very large increase in the work of breathing during exercise, the overall energy required for ventilation is a small fraction of the total energy expenditure. For example, at a ventilation rate of 130 liters per minute, about 0.6 to 0.8 kcal per minute is used for moving air in and out of the lungs. Assuming that this is associated with a maximal aerobic energy production of 13.3 kcal per minute (800 kcal/hr), the work of breathing at this level is only 3 to 5 percent of the total energy expenditure. The percentage can definitely increase in more severe exercise, or when there is some airway obstruction, or when negative pressure breathing occurs. Respiratory rate also affects the work of breathing in that work is elevated at very low and very high rates and attains a minimal value at an intermediate optimal rate. The body tends spontaneously to regulate ventilation at an optimal frequency, which is increased proportionally with ventilation and O_2 consumption (Fig. 29-11).

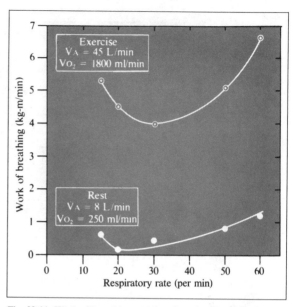

Fig. 29-11. Work of breathing as a function of respiratory rate and exercise. The minimal point for each condition represents the normal spontaneous rate. Forced hypoventilation or hyperventilation at any exercise level will increase the work of breathing. To convert kg-m/min to kcal/min, multiply by 2.34×10^{-3}. $\dot{V}_A$ = alveolar minute ventilation; $\dot{V}O_2$ = O_2 consumption. (Redrawn from G. Milic-Emili and J. M. Petit. *Arch. Sci. Biol.* [Bologna] 43:326, 1959.)

During exercise, airflow resistance in and out of the lung is reduced by switching from nose to mouth breathing. The size and irregular surfaces of the nasal passages create rather highly resistive airflow passages. In addition, as the velocity of airflow in these passages increases, turbulence is induced, which also increases airflow resistance. To compensate, a person who reaches a minute volume of 30 to 40 liters per minute switches from nose to mouth breathing to lower the work of breathing.

PULMONARY GAS DIFFUSION AND VENTILATION-PERFUSION DURING EXERCISE. The diffusion capacity of O_2 and CO_2 across the alveolar-capillary boundaries increases during exercise. Oxygen diffusion capacity rises from an averge pressure gradient of about 20 to 25 ml per millimeter of mercury at rest to about 80 during exercise. This change is related to the greater number of open capillaries in the lung, increased gas diffusion gradients, and large increases in pulmonary capillary blood volume. Recall that only a fraction of the pulmonary capillaries are open during rest—those in the lower middle or base of the lung (see Chap. 20). Pulmo-

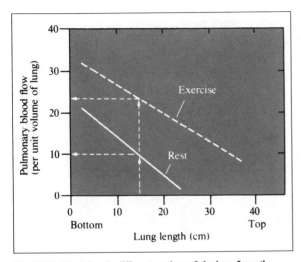

Fig. 29-12. Blood flow in different sections of the lung from the apex (top) to the base (bottom) at rest and during projected exercise. Note that at a 15-cm distance from the bottom, more than twice the resting amount of blood is in the lung during exercise. (Redrawn from J. B. West and C. T. Dollery. Distribution of blood flow and ventilation-perfusion ratio in the lung, measured with radioactive CO_2. *J. Appl. Physiol.* 15:408, 1960.)

nary blood flow distribution during rest and exercise is shown in Figure 29-12. Note that during exercise the apex of the lung has twice the resting blood flow; therefore, there is a larger lung area for effective gas exchange. During exercise the ventilation/perfusion ratio ($\dot{V}/\dot{Q}$) triples, as does the diffusion capacity, and ventilation and cardiac output increase 15 and 5 times respectively. Although ventilation increases, its distribution in the lung is proportionally equal to that during rest; therefore, with the increased flow and blood volume distribution to the apex a greater gas-blood exchange area occurs, facilitating diffusion (see Chap. 19).

CONTROL OF PULMONARY VENTILATION DURING EXERCISE. No single factor has been designated as the primary controlling influence of respiration during exercise. Recall from Chapter 21 that normal respiratory control revolves around the interaction between CNS inspiratory and expiratory centers, pulmonary stretch receptors, and peripheral and central chemoreceptors sensitive to blood gas tension and H^+ concentrations. In exercise, the control mechanisms involved in resting conditions appear to be inadequate, in fact, neither increased arterial CO_2 nor decreased O_2 leads to the high ventilation values observed during exercise. The possibility of other controlling stimuli of a different nature originating from active muscle or from higher

CNS areas has been proposed. According to the several theories advanced, the stimuli fall into two main categories, *humoral* stimuli and *neural* stimuli. Figure 29-13 shows the time course of ventilation changes before, during, and after exercise, divided into fast (F,F′) and slow (S,S′) components. The fast components occur at the onset (F) and at the end of exercise (F′), and the slow components occur after the onset (S) and after stopping (S′). The fast components should realistically be related to nervous stimuli, since they occur faster than metabolic changes. The slow components are related to humoral stimuli, which have to be transported to the CNS respiratory centers by the cardiovascular system. The exact nature of both humoral and neural control of ventilation in exercise has not been determined; however, they do not seem to work independently but, occurring together, have a cumulative influence on ventilation.

HUMORAL STIMULI. Increased Arterial Carbon Dioxide Concentration. Carbon dioxide production is greatly increased in muscle exercise and could lead to an increase in arterial CO_2 concentration of sufficient magnitude to stimulate CNS respiratory centers. Actual measurements of arterial CO_2 concentration indicate that changes in exercise are slight. Small increases often occur during light or moderate exercise, and no change or even a slight decrease may occur during heavy exercise. If CO_2 alone were the primary stimulus, larger increases in blood than those that have been measured would have to take place. For example, inhalation of an 8 percent CO_2, a 20 percent O_2, and a 72 percent nitrogen mixture markedly increases blood CO_2 content and raises ventilation to 100 liters per minute. The striking point is that trained athletes may attain ventilations of 120 liters per minute during exercise without any detectable change in arterial CO_2 content.

Increased Hydrogen Ion Concentration. The acidification of blood (increased H^+ concentration) due to lactic acid production by active muscle could stimulate respiratory centers and account for exercise hyperventilation. However, only during very strenuous exercise are significant amounts of lactic acid produced, and it has been shown that distance runners (marathon runners) attain only slight increases in blood lactic acid yet have tremendously high ventilation rates. After exercise, lactic acid levels may increase slightly while ventilation rate rapidly falls. It is, therefore, unlikely that blood H^+ concentration alone provides the major stimulus for exercise hyperventilation.

Decreased Arterial Oxygen Concentration. Arterial hypoxemia (decreased PO_2) cannot account for more than a few mm Hg

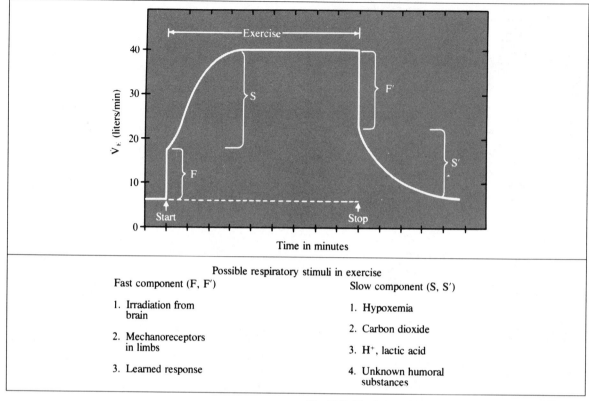

Possible respiratory stimuli in exercise

Fast component (F, F')

1. Irradiation from brain

2. Mechanoreceptors in limbs

3. Learned response

Slow component (S, S')

1. Hypoxemia

2. Carbon dioxide

3. H⁺, lactic acid

4. Unknown humoral substances

Fig. 29-13. Ventilatory responses in exercise. (Redrawn from P. Dejours. The Regulation of Breathing During Musuclar Exercise in Man, a Neuro-humoral Theory. In J. C. Cunningham and B. B. Lloyd [eds.], *The Regulation of Human Respiration.* Oxford, England: Blackwell, 1963. P. 545.)

changes in arterial oxygen tension (PaO_2) during exercise since arterial blood O_2 saturation is not appreciably affected in strenuous exercise. Recall that anoxia affects respiration by eliciting chemoreceptor activity from the aortic and carotid bodies (see Chap. 21). The threshold of the receptors is at an arterial blood O_2 tension of approximately 60 mm Hg. Since such PO_2 levels are not reached during strenuous exercise, this mechanism alone cannot provide the necessary stimulation for the hyperventilation.

Circulating Catecholamines. Because the epinephrine and norepinephrine concentrations in blood and urine increase with the intensity of work, and because injected catecholamines produce marked increases in ventilation in resting subjects, these substances may play some role in exercise hyperpnea. The actual mechanism of the response is at present poorly understood, but it may involve changes in the threshold or sensitivity of chemoreceptors or respiratory centers. It must be concluded that humoral stimuli alone cannot explain the

hyperpnea of exercise, but these factors, combined with neurogenic stimuli, may have a major part in respiratory control during exercise.

NERVOUS STIMULI. Irradiation of Impulse from Higher Central Nervous System Centers. It has been suggested that the sudden increase in ventilation just before and during the onset of exercise is due to impulses arising from the motor cortex. These impulses, which pass through the reticular formation, may irradiate impulses to the respiratory center and evoke a ventilatory increase. There is no doubt that some ventilatory responses are of a volitional nature and learned by experience. They are obvious when one watches a short-distance runner or swimmer hyperventilate just prior to the start of an athletic event. Tests with hypnotized subjects who exercised without full awareness of work intensity have resulted in a lower ventilatory response than would be predicted by metabolic rates.

Peripheral Receptors. Muscle contraction and movement of the extremities may stimulate receptors in joints (proprioceptors) and muscle spindles, which in turn send afferent signals to the respiratory centers. It has been observed that passive limb movement evokes a ventilatory response roughly proportional to the number of joints involved in the movement. Such a mechanism may be related to the overall rhythmic pattern or smoothness a superior runner or swimmer adheres to during a race.

Lung Stretch Receptor Activity. Lung stretch receptors, when stimulated by filling of the lung, send afferent information to the respiratory center, which inhibits inspiration (inhibitoinspiratory reflex) and indirectly activates expiration (see Chap. 21). The excitatoinspiratory reflex may also be activated. The same receptors may play a role in the hyperpnea of exercise by augmenting respiratory rate. However, since voluntary ventilation is not self-perpetuating, this mechanism is not considered a prime stimulus of exercise hyperpnea.

Sensitivity of Respiratory Centers. An increase in respiratory center sensitivity to the amount of CO_2 in the arterial blood could account for exercise hyperpnea, but this would have to be a drastic sensitivity change, since arterial CO_2 content in heavy exercise does not seem to decrease much. It has, however, been observed that if one tries to hold his breath during moderate work, his breaking point occurs at a lower alveolar CO_2 than when at rest.

Body Temperature. Hyperthermia increases ventilation without exercising as a means of losing body heat (see Chap. 28). In exercise the ventilatory increase is greater than that observed during hyperthermia and occurs before rectal temperature rises. Since temperature rise increases cellular activity, the hyperthermia of exercise may increase the activity of joint and muscle receptors as well as respiratory center cells.

CARDIOVASCULAR RESPONSE DURING EXERCISE

MICROCIRCULATION. Before and during exercise, microcirculatory (capillary) changes occur that facilitate the delivery of substances to and from the active skeletal muscle cell. It is estimated that only 20 to 50 percent of the capillaries in resting skeletal muscle are open at any one time. During exercise, closed capillaries open as blood flow

increases, giving a total estimated capillary surface of 300 to 600 m^2, which considerably shortens the diffusion distance from the capillary to the cell energy unit (mitochondrion). Since the increase in the number of capillary openings and in muscle blood flow is about the same (20 percent of resting value), the velocity of blood flow through the capillaries should not change. Thus transcapillary exchange should be further enhanced by maintenance of the precapillary to postcapillary transit time at approximately 1 second.

Another factor influencing the delivery of substances to and from the active cell is capillary permeability, i.e., the size and number of membrane pores per unit of capillary surface area. Recall that capillary permeability is not ordinarily affected by the buildup of local vasodilator substances in active muscle (see Chap. 12), but the amount of fluid filtered across the capillary membrane for each 1 mm Hg of transcapillary pressure is determined by the total surface area of the open capillaries.

Capillary flow increases during exercise as systolic and mean arterial pressures rise, therefore increasing blood flow through active muscle. Venous pressure is little affected, and, as a consequence, filtration forces exceed absorption forces (see Chap. 12). This imbalance causes fluid to leave the vascular compartment, upsetting the dynamic equilibrium between outflow and inflow water balance. Mean capillary pressure may rise as much as 10 mm Hg. Assuming a maximal filtration coefficient of 0.04 ml/mm Hg/min/100 gm of tissue, circulating blood volume would then be reduced approximately 20 percent (1 liter) in only 10 minutes if the involved muscle mass weighed 10 kg. Indeed, intravascular to extravascular fluid shifts occur, as shown by the hemoconcentration (Hct = 56 to 60 percent) and plasma volume reductions of 10 to 15 percent. These observations are more pronounced in long-term exercise events, such as distance running, and are more obvious when performed in cold environments. Naturally, compensatory responses must be activated or circulatory insufficiency would eventually occur.

As fluid enters the interstitial (nonvascular) spaces, the hydrostatic pressure in this compartment rises, and as hemoconcentration occurs in the vascular compartment, plasma colloid osmotic pressure rises. These changes tend to increase reabsorption (see Chap. 12) and should therefore help to maintain effective circulating blood volume; however, complete compensation may not occur because of interstitial protein accumulation and dehydration of 3 to 5 percent of body weight by excessive sweat loss. In addition, fluid return to the systemic circulation is enhanced by the lymphatic circulation, which is aided by the pumping action of contracting and relaxing muscles.

The reduction of circulating blood volume in exercise is mainly compensated for by two mechanisms, one fast and the other slow. The fast mechanism involves the adjustment of the size and volume of the capacitance vessels (venules and veins); therefore, a reduction in their diameters will greatly increase effective circulating blood volume. These vessels are controlled by changes in vasomotor tone through sympathetic vasoconstrictor activity (see Chap. 7). Constriction of these vessels during exercise and the force exerted on them by contracting muscle set the volume of the venous compartment and hence venous return and cardiac filling. The slow mechanism (fluid reabsorption) is too slow to aid in early exercise. It obviously has a considerable delay, because hemoconcentration of 20 percent may occur early in exercise (10–20 min).

CARDIAC OUTPUT. Cardiac output rises during exercise roughly in proportion to the increase in O_2 consumption (Fig. 29-14); however, it does not change to the same extent as pulmonary ventilation and metabolic rate. For example, during moderate to heavy exercise, ventilation may change from 8 to 100 liters per minute (12-fold), O_2 consumption from 0.38 to 3.8 liters per minute (10-fold), and cardiac output from 5 to 30 liters per minute (6-fold). The deviation from linearity at the higher O_2 consumption in Figure 29-14 is probably related to thermal regulation at high work levels. As body temperature rises during exercise, skin blood flows must increase so that the high internal heat loads may be transferred to the skin and thus be dissipated (see Chap. 28). Cardiac output at rest and during exercise is greatly influenced by body position. Resting output may be 1 to 2 liters per minute greater in the supine position because of the lack of gravitational pooling effects of dependent areas. During heavy exercise, cardiac output changes are less in the supine position, but they are nevertheless greater than in the upright position.

In the transition from rest to exercise, cardiac output increases rapidly at first, then plateaus (Fig. 29-15). The rate of initial increase and the peak level are set by the severity of the stress. After cessation of exercise, output decreases gradually toward resting levels, as does O_2 consumption. The recovery similarities for cardiac output and O_2 consumption are undoubtedly related to the incurred O_2 debt. Heart rate and stroke volume influence cardiac output greatly, and the increased output during stress is the product of these two variables.

Stroke Volume. Stroke volume, like cardiac output, depends on cardiac filling; therefore, it is strongly influenced by body position. It is higher in the resting supine position than in standing but changes less during supine exercise. Normal values range from 70 to 150 ml per beat in the resting supine male, depending on body size and whether or not the man is an athlete. Athletes are at the upper range of resting stroke volumes, and females usually have 25 to 30 percent less than males at any position during rest or stress. During exercise in the standing position, stroke volume increases rapidly as venous return is increased by venoconstriction,

Fig. 29-14. Cardiac output at rest (minimal point) and during exercise (increasing O_2 consumption) for males and females. (Redrawn from P. O. Åstrand, T. E. Cuddy, B. Saltin, and J. Stenberg. Cardiac output during submaximal and maximal work. *J. Appl. Physiol.* 19:271, 1964.)

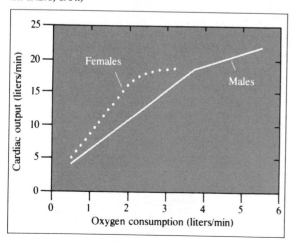

Fig. 29-15. Time course of cardiac output changes during and following light and heavy exercise, measured by impedance cardiograph. (From W. V. Judy and F. D. Nash. Unpublished data.)

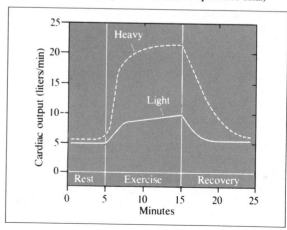

muscle pumping, and thoracoabdominal pumping mechanisms. It reaches a maximal level (180–190 ml/beat) in a short interval (5–10 min), then plateaus, as does cardiac output. In exercise events lasting several hours, stroke volume may decrease 10 to 20 percent below the maximal value; however, cardiac output is maintained fairly constant on account of increases in heart rate. During supine exercise, stroke volume increases only slightly and is maintained at heart rates of 200 per minute. Evidently, the time available for cardiac filling is sufficient at such high rates when the venous return is aided by muscle pumping, thoracoabdominal pumping, venoconstriction, and reduced gravitational forces on the cardiovascular system.

Heart Rate. There is little difference between resting supine and standing heart rates or their changes during light or heavy exercise in the respective positions. Athletes have lower resting rates than nonathletes, who do well to increase theirs twofold to threefold before complete exhaustion. The rate may increase in normal persons from 75 to 100 beats per minute during light exercise, to 130 in moderate exercise, and to nearly 180 in heavy exercise. Rates of 180 per minute are considered the upper limit for nonathletes, but 200 per minute have often been observed in athletes and children. During light exercise the initial rate increase may be exaggerated, and it is subsequently reduced to a lower steady-state level. In heavy exercise there is a tendency for the rate to increase progressively until adequate cardiac output is achieved. After cessation of exercise, the rate slowly returns to or below control levels, the rate of return again being proportional to the severity of the exercise and the person's physical condition.

BLOOD PRESSURE IN EXERCISE. Pressures in the systemic arterial system increase during heavy exercise, although total peripheral resistance decreases fourfold to fivefold (pressure = cardiac output × total peripheral resistance). Systolic pressure increases more than diastolic pressure and changes at a rate of about 8 mm Hg per 0.5 liter of O_2 uptake (Fig. 29-16). It seldom rises above 180 mm Hg in normotensives but has been observed to rise above 200 mm Hg (50 percent greater than at rest) in some athletes. Diastolic pressure does not change much; it may even decrease slightly in light exercise. Since systolic pressure increases more than diastolic pressure, pulse pressure must increase proportionally and mean pressure somewhat less (Fig. 29-16). Venous pressure increases only slightly at the onset but drops to or below resting values during exercise. Therefore, central venous and right atrial pressures are not increased, even though venous return and cardiac filling are augmented by

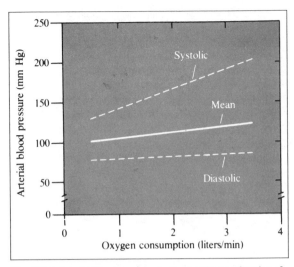

Fig. 29-16. Arterial blood pressure during exercise as a function of O_2 consumption. Note the widening of pulse pressure, basically related to a systolic pressure rise and a slight increase in diastolic pressure that reflects the decrease in total peripheral resistance.

the mechanisms previously described. The initial venous pressure rise at the onset of exercise may be due to sympathetic vasoconstrictor activity on the capacitance vessels (veins), causing them to narrow and thus reducing their total volume and increasing blood availability to the heart, or facilitated venous return due to skeletal muscle and abdominothoracic pumping before the maximal opening of the muscle capillaries.

Pressures in the pulmonary circulation also increase slightly during exercise. Although pulmonary pressures are low compared with those in the systemic arteries (see Chap. 13), small increases may greatly influence pulmonary blood flow distribution by forcing open some of the normally closed capillaries in the middle and upper areas of the lung. Similarly, opening of the capillaries accounts for the marked increase in total lung capillary blood volume during exercise. Volume changes from 60 to 96 ml have been estimated for nonathletes and from 70 to 200 ml for athletes, from rest to exercise respectively. Training or physical conditioning seems to have little influence on various vascular pressures during resting or active conditions.

REGIONAL BLOOD FLOW DISTRIBUTION. The changes in blood pressure, cardiac output, and cellular oxygen demand during exercise necessitate rather dramatic alterations in blood flow distribution in the body. Although cardiac output increases fivefold to sixfold during maximal exercise, the

Table 29-2. Blood flow distribution during rest and exercise for a well-conditioned athlete

Area	Rest		Exercise			
			Light		Heavy	
	ml/min	%	ml/min	%	ml/min	%
Splanchnic	1,400	24.0	1,100	12.0	300	1.0
Renal	1,100	19.0	1,100	12.0	900	4.0
Brain	750	13.0	750	8.0	750	3.0
Coronary	250	4.0	350	4.0	1,000	4.0
Skeletal muscle	1,200	21.0	4,500	46.0	22,000	85.5
Skin	500	9.0	1,500	14.0	600	2.0
Others	600	10.0	400	4.0	100	0.5
Cardiac output	5,800	100.0	9,700	100.0	25,650	100.0

bulk of this volume flows through active skeletal muscle, including the respiratory muscles. Blood flows and the percentage of cardiac output for various organs and tissues during rest, light exercise, and heavy exercise are estimated in Table 29-2. Note that brain and kidney flow is maintained during exercise, while splanchnic flow is decreased, and skeletal muscle, heart, and skin flows are increased. Obviously, some large vascular beds, as in the viscera, decrease their blood flow so that flow through more vital or active areas may be maintained or increased. Renal flow initially decreases but returns to control during exercise as blood pressure rises, since renal function is damaged by extended periods of low perfusion.

Coronary flow, which is about 250 ml per minute (60–70 ml/min/100 gm tissue) at rest, may increase four- to fivefold or more during exercise. This rise shows the demand of active cardiac cells for O_2 that must be provided to the heart, since myocardial resting extraction of O_2 from capillary blood is 75 to 85 percent, and heart muscle *cannot* utilize anaerobic glycogenolysis as a major source of energy for the contractile process.

Skin blood flow decreases during the initial moments of exercise, then increases as body temperature rises. In resting adults, total skin flow is about 400 to 500 ml per minute, but when the vessels are fully dilated, as in severe hyperthermia or exercise, it may increase to 3 liters per minute. The vessels of the skin are so constructed and controlled that they play a major role in body temperature regulation (see Chap. 28). Blood flow through vascular beds during rest and exercise is influenced by the mechanisms that change the diameters of the resistance vessels (arterioles) and the tightness of the precapillary sphincter. Remember

that several mechanisms may be involved: (1) autoregulation, (2) sympathetic tone, (3) circulating hormones, and (4) local metabolites (see Chap. 12). The complex interaction between them makes it difficult to analyze the significance of any one during exercise; however, the dominating factor seems to be sympathetic nervous tone, with some small hormonal and metabolite influence. During rest, there is a high sympathetic tone in muscle arterioles, as signified by the large number of poorly perfused capillaries. Similarly, sympathetic influence on the arterioles of more functional systems such as the heart, kidney, and intestines is small. Therefore, these vessels are open, less resistive, and more easily perfused by the arterial driving pressure.

Circulatory patterns start to change before the onset of exercise in that muscle blood flow increases, whereas that to skin, kidney, and intestine decreases. These changes are due to intense sympathetic activity initiated in higher CNS areas that causes constriction of skin, kidney, and intestinal arterioles and dilation of those in muscles (see Chap. 7). Since the constricted vessels become the most resistive— and the dilated vessels become the least resistive— channels to blood flow, pressure created by the pumping action of the heart will force more blood through the less resistive muscle vasculature. This neurogenic mechanism is therefore able to shift or shunt blood flow to more demanding or active areas very quickly. The thought of exercise activates cortical motor areas, which in turn activate hypothalamic and medullary sympathetic constrictor and dilator areas. It has been shown that stimulation of these areas in anesthetized animals increases heart rate and myocardial contractility, constricts resistance vessels in less active organs, dilates vessels in active muscle and heart, and mobilizes extra energy through influences on the liver to release glucose and on metabolism. Inhibition of sympathetic vasoconstrictor activity to resting muscle also enhances blood flow; it is thus conceivable that sympathetic constrictor activity to muscle during exercise is inhibited, thereby allowing flow to increase.

Hormonal control of circulation or blood flow distribution is mainly concerned with epinephrine released from the adrenal gland as a part of the massive sympathetic discharge occurring just before and during exercise. The action of this substance mimics that of sympathetic constrictor influence on less vital systems and dilation of muscle vasculature at low blood levels (see Chap. 7). The most effective single vasodilator mechanism at the tissue level is that of local metabolites. No one single agent or factor has been confirmed as the primary vasodilator during exercise. Of the many metabolic products, CO_2, lactic acid, histamine, and others have been investigated. Histamine is the most potent vasodilator, but its concentration in muscle decreases during exercise. Several substances may work together to produce local vasodilator control. The local aspects of muscle vasodilation and blood flow distribution during exercise must be emphasized because there is no evidence that metabolites released from contracting muscle circulate to other constricted muscles and organs, causing them to undergo vasodilation.

An additional factor influencing muscle blood flow during exercise is the force of muscle contraction against the open blood vessels. Figure 29-17 shows the effects of intermittent contraction on leg blood flow. During rhythmic exercises such as running or walking, rather sharp decreases in flow occur during contraction as the external pressure around the vessels forces blood out of them. During relaxation, equally sharp increases in flow occur, since the vascular bed is fully dilated and arterial pressure elevated. Note that the minimal flow during contraction exceeded the controls, and that after exercise ceased, flow was 8- to 10-fold greater than control flow. During recovery, flow gradually decreased over 2 or more minutes. This sequence shows how dilated the muscle bed was, the importance of muscle contraction on squeezing blood through the active muscle, and the role of muscle contraction in maintaining venous return.

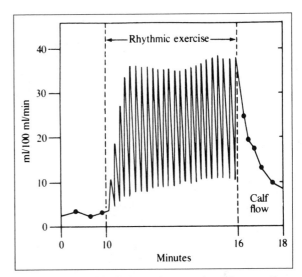

Fig. 29-17. Effect of rhythmic muscle contraction on human calf muscle blood flow. (Redrawn from H. Barcroft and H. J. C. Swan. *Sympathetic Control of Human Blood Vessels.* London: Arnold, 1953. Chapter 5, p. 48.)

EXERCISE IN HEALTH AND DISEASE

It is not necessary to have superior muscle strength or physical stamina to be healthy, nor is it true that exercise and physical fitness will protect one from diseases. Most diseases have no human boundaries. Exercise or physical activity is, however, used in medical science as a diagnostic tool and—perhaps more important—as a therapeutic aid during convalescence and rehabilitation.

Humans adapt to their normal daily environment and activity by building up muscle strength and exercise stamina according to their daily physical demands. Naturally, then, a wide range of physical fitness can be found among healthy people. However, even the most physically fit, such as athletes and astronauts, when placed in a less stressful environment will adapt to the required stress. Bed rest or space flight removes many of the normal daily stresses and leads to physical deconditioning. In addition to the previously stated reduction in maximal O_2 consumption capacity that occurs during bed rest, marked changes occur in skeletal and smooth muscle, bone mineral content, body fluids, and fluid distribution. As little as 2 days of bed rest or space flight results in an initial diuresis due to vascular fluid displacement to the thoracic cavity; left atrial volume increases, leading to inhibition of antidiuretic hormone release from the posterior pituitary and a resulting reduction in kidney fluid reabsorption. If the condition is prolonged, there is loss of muscle tone and physical stamina, and eventually muscle atrophy occurs. In both bed rest and space flight the lack of gravitational stress on bone induces demineralization, so that calcium is lost. The heart and blood vessels lose their reflex responsiveness, and *cardiovascular deconditioning* is the outcome. Inactivity may lead to constipation, to thrombosis due to collapsed venous segments, and eventually to psychiatric and neuropsychiatric disorders.

These responses are consistent findings for all age groups. The recovery or readaptation to a more stressful environment may be slow and traumatic. For this reason, during the last 30 years, exercise has become a major therapeutic aid in health and disease. Classically, many physicians advocated complete bed rest during convalescence from almost all afflictions. For countless patients, such as healthy mothers, this was an unnecessary and sometimes harmful indulgence. For others, such as those stricken by rheumatic fever, heart attacks, and partial paralysis, it meant a life of inactivity.

Today, the physiological and psychological advantages of even the most modest exercise (toe wiggling, coughing) are well documented in the practice of medicine. The medical scientist is aware that the inactivity of bed rest will prolong the recommended therapy and may eventually cause fatal complications. For example, the commonly occurring pulmonary congestion and edema observed in bedridden patients can lead to pneumonia and death. There is risk of pulmonary thromboembolism. The simple exercises of changing body position and coughing can alleviate this problem. Today, new mothers are up and moving around 6 to 8 hours after delivery. Patients who have undergone appendectomies and other surgery are allowed to sit up as early as 4 hours after surgery. In case the patient must remain in bed, exercise is still encouraged. Toe wiggling, knee bending, arm stretching, and muscle contraction are encouraged on hourly schedules as preventive as well as therapeutic measures, because the sudden exertion of standing or walking after extended inactivity may lead to muscle soreness and fatigue. In many instances this discomfort may reduce the patient's willingness or desire to exercise in order to regain his or her strength.

Exercise in medicine leading to physical conditioning may take two forms. The first is kinesiological reconditioning, i.e, exercise of a specific muscle group or joint by active or passive movement, massage, or heat in an attempt to restore the function of that unit. The second is physiological reconditioning, in which the entire body is conditioned by graded physical exercise. Here, calisthenics, modified sports, or occupational pursuits are used to restore the physical endurance of the whole body. The growing list of medical specialties, occupational and rehabilitative medicine, physical therapy, occupational therapy, and the appearance of activity rooms and exercise facilities in both old and new medical institutions show the marked awareness of exercise as a major therapeutic tool in health and disease.

RELATION OF METABOLIC, RESPIRATORY, AND CIRCULATORY CHANGES DURING EXERCISE

Exercise, or the ability to perform a specific muscular task, is not simply the end product of the excitation-contraction coupling of the actin and myosin filaments of skeletal muscle. As stated previously, it involves scores of enzymatic reactions occurring in the proper sequence, the CNS coordination of all major systems, and local control of microvascular and cellular activity. Therefore, exercise can be described as the sum total of all body systems working together to support active muscle. Psychologically, it starts with the conscious or subconscious thought of having to do a physical task and ends with the final satisfaction or relief that the task is over.

The physiological manifestations of exercise discussed in this chapter were primarily concerned with metabolic, respiratory, and circulatory adjustments. Briefly, multiple systems interactions occur; some organs and systems are deactivated or shutdown, while others are stimulated to increased activity to meet the requirements for continuous muscle contraction. The metabolic increases are supported by similar changes in the systems responsible for O_2 and metabolite transport to and from the active cell. The changes in metabolic activity are closely paralleled by changes in pulmonary ventilation (see Fig. 29-8). However, cardiac output does not change in a parallel fashion (see Fig. 29-14) but seems to lag behind the metabolic demand. Some investigators feel that the slightly slower response of the cardiovascular system is the weak link in meeting the metabolic demands of active muscle. This, however, is compensated for by several factors, which together provide a more efficient O_2 delivery to tissues. The ventilatory increase is supported by an increased O_2 diffusion capacity in the lung, and a similar event occurs at the active cellular level.

In the lung, increased gas and blood volume raises the diffusion capacity by increasing the total capillary-alveolar surface area. Therefore, more red blood cells become O_2 enriched, and the total O_2 content of blood is greatly increased. At the level of the muscle cells, O_2 delivery is enhanced by the increased metabolism. This lowers the tissue O_2 content, thereby raising the diffusion gradient, which favors quicker and more extensive gas delivery to the cellular mitochondria. In addition, the accumulation of CO_2, lactic acid, heat, and the pH decrease in and around the cell promote the unloading of O_2 by shifting the hemoglobin dissociation curve to the right (see Chap. 19). Another major compensatory factor—and perhaps the most important—is the enhanced delivery of blood to the active cells by an increase in the number of open capillaries. Arteriolar dilation, along with an increase in total capillary blood volume, notably increases the surface area across which gas can diffuse. At the same time, the distances across which O_2 must travel from the capillary to the cell—and CO_2 from the cell to the capillary—is sharply reduced by the increase in open capillaries.

Metabolism, respiration, and circulation tell only part of the story of exercise physiology. It should be obvious that nervous, hormonal, and enzymatic changes are also involved, and that their interaction with metabolism, respiration, and circulation is most important to the total physiological system before, during, and after exercise. For a more thorough and advanced review of the physiology of exercise, the reader is encouraged to consult the bibliography that follows.

BIBLIOGRAPHY

Asmussen, E., and Nielsen, M. Pulmonary ventilation and effects of oxygen breathing in heavy exercise. *Acta Physiol. Scand.* 43:365–378, 1958.

Åstrand, P. O., and Rodahl, K. *Textbook of Work Physiology.* New York: McGraw-Hill, 1977.

Barcroft, H. Circulation in Skeletal Muscle. In W. F. Hamilton (ed.), *Handbook of Physiology.* Washington, D.C.: American Physiological Society, 1963. Section 2: Circulation. Vol. 2, pp. 1353–1368.

Beregard, B. S., and Shepherd, J. T. Regulation of circulation during exercise in man. *Physiol. Rev.* 47:178–209, 1967.

Chapman, C. B. (ed.) Physiology of muscle exercise. *Circ. Res.* 20 (Suppl I): 1967.

Dejours, P. Control of Respiration in Muscular Exercise. In W. O. Fenn and H. Rann (eds.), *Handbook of Physiology.* Washington, D.C.: American Physiological Society, 1964. Section 3: Respiration. Vol. 1, pp. 631–648.

Ekelund, L. G. Exercise. *Annu. Rev. Physiol.* 31:85–116, 1969.

Falls, H. B. (ed.). *Exercise Physiology.* New York: Academic, 1968.

Harris, P. Lactic acid and the phlogiston debt. *Cardiovasc. Res.* 3:381–390, 1969.

Johnson, W. R., and Buskirk, E. R. *Structural and Physiological Aspects of Exercise and Sport.* Princeton, N.J.: Princeton Book Co., 1980.

Kao, F. F. An Experimental Study of the Pathways Involved in Exercise Hyperpnea Employing Cross Circulation Techniques. In J. C. Cunningham and B. B. Lloyd (eds.), *The Regulation of Human Respiration.* Oxford, Engl.: Blackwell, 1963.

Wilmore, J. H. Acute and Chronic Physiological Response to Exercise. In E. A. Amsterdam, J. H. Wilmore, and A. N. DeMaria, *Exercise in Cardiovascular Health and Disease.* New York: York Medical Books, 1977. Pp. 53–69.

THE HYPOPHYSIS: NEUROENDOCRINE AND ENDOCRINE MECHANISMS

30

Ward W. Moore

The degree of complexity of functions observed in humans has been attained in part by the evolution of the two primary integrating systems, the nervous system and the endocrine system. Each system communicates information between cells of the body, and each participates in the coordination and regulation of all the body systems. The endocrine system functions to maintain the volume and concentration of substances in the internal environment (the body fluids) at relatively constant levels in the face of wide changes in the body's activity and fluctuations in the external environment.

The endocrine system controls and integrates many body functions, including organic metabolism and energy balance, electrolyte metabolism and fluid balance, growth, and reproduction. It transmits information by means of chemical messengers, the *hormones*. Chemically, the hormones may be divided into three groups: steroids, polypeptides, and amino acid derivatives. The fundamental microscopic anatomy of the glandular elements of the endocrine glands is similar to that of the exocrine glands, with two excep-

tions: The endocrine gland does not possess a duct system, and each glandular cell has a surface that abuts a capillary where diffusion can occur. The morphology of the system is such that the secretions can be released directly into the extracellular fluid and circulation. Thus, the hormones are dispersed without direction in the blood and are able to reach all tissues and cells. However, they act only on genetically conditioned and differentiated cells of the body, the *target cells*.

The target cells for each hormone contain specialized molecules termed *receptors*. The receptors are protein in nature and contain a site or sites to which the particular hormones can bind and subsequently mediate the specific cellular actions. The hormone receptors serve two primary roles: (1) They permit cells to accept or not accept particular molecules and thus form a mechanism that permits the target cell to distinguish a particular signal from the myriad of hormones and molecules that impinge on it; and (2) they relay the signal to the target cell in such a manner that chemical changes are generated in them and produce appropriate and regulated metabolic responses. The known receptors are located at one of three locations in the target cell: on the plasma membrane, in the cytosol, or in the cell nucleus.

Emphasis will be placed in this and succeeding chapters on the mechanisms that regulate the rate of secretion of hormones and the role of hormones in homeostasis. The hormone-secreting elements to be considered are the secretory or glandular cells of the hypothalamus, the hypophysis, or pituitary gland, the pancreas, the parathyroid glands and derivatives of the ultimobranchial body, the thyroid gland, the adrenal cortex, the gonads, and the placenta.

METHODS OF STUDY

The evidence required to show that an organ functions as an endocrine gland involves the demonstration of specific effects in the absence of the organ and specific physiological restorative responses following exogenous administration of the hormone by transplantation of the glandular tissue or by injection of suitable extracts of the gland. A further step needed to verify the endocrine activity of a gland calls for the administration of extracts of the gland to inact animals with the reproduction of exaggerated effects of the hormone. Other requirements are concerned with evaluation of the chemical and physical characteristics of the hormone, descriptions of its effects, the factors that regulate its rate of secretion, its molecular structure and synthesis, its receptors, and its molecular mode of action. The molecular structures of many of the hormones are known, and most of the hormones, including the polypeptides and proteins, have been synthesized. Our knowledge of the molecular actions of the hormones is rapidly expanding.

FUNCTIONS OF HORMONES

The effects of hormones fall into three general groups. First, they influence reactions that aid in the maintenance of a constant internal environment. Thus, they regulate the rates at which carbohydrates, fats, proteins, electrolytes, and water are deposited in or removed from the tissues of the body. For example, insulin participates in regulating the chemical and/or physical factors that ensure an adequate supply of glucose to most extrahepatic tissues by increasing the permeability of the cell membrane to glucose. Second, the hormones have a morphogenic action; this includes effects on growth, differentiation, development, maturation, trophic actions, and aging processes. Examples of the morphogenetic actions are the effects of the ovarian or testicular hormones (under the influence of the pituitary gland) during the growth and development of the accessory sex organs and secondary sex characteristics at puberty. Finally, the hormones regulate autonomic activity, as well as certain central nervous system (CNS) activities and behavioral patterns. For instance, maternal behavioral patterns are often linked to the presence of various hormones. Also, the sensitivity of effector cells to the catecholamines epinephrine and norepinephrine is altered by the circulating levels of some hormones.

It must be emphasized that the action of all hormones is not to initiate chemical reactions, but to alter the rates of preexisting reactions without contributing significant amounts of either matter or energy to the process. The hormones can function at peak efficiency only when the temperatures and the concentrations of substrate, cofactors, and hydrogen ion are at optimal levels. It follows that, in experimental tests utilizing hormones the subjects should be in nutritional, temperature, and fluid balance and as free as possible from psychological stimuli if representative results are to be expected.

The mechanisms by which hormones exert their control are varied. Information is available concerning such mechanisms in most of the hormones. In general, some hormones may be said to alter the rate of protein (enzyme) synthesis or activity and/or membrane permeability, and some hormones may act directly on membranes and alter their permeability characteristics. In each case the hormones appear to change the activity of rate-limiting steps on a metabolic pathway.

The cellular response to all hormones depends on the presence of receptors in the target cells. For these receptors to receive a hormonal signal they must have a relatively high affinity for the hormone. They also must have the property of specificity for the hormone if they are able to distinguish its signal from noise. The sensitivity of the receptors in the target cells to hormones may also be affected by the following: the concentration of the receptors; the endocrine, metabolic, and genetic states of the cell; the stage of the cell cycle; and the differentiation of the cell. At present three types of hormone receptors are known: (1) receptors for the polypeptide hormones, located on the external surface of the plasma membrane of the target cells; (2) receptors for the steroids, located in the cytosol; and (3) receptors for the thyroid hormones, located in the target cell nucleus.

The receptors for the catecholamines (epinephrine in particular) and the polypeptides are located on the external surface of the cell membrane. The work of Sutherland and his co-workers led to the finding that many hormones bind to a receptor on the cell membrane, which results in the activation of an enzyme, adenylate cyclase. After activation by the hormone-receptor complex, the adenylate cyclase increases the rate of conversion of adenosine triphosphate (ATP) to cyclic adenosine 3',5'-monophosphate (cyclic AMP). The cyclic AMP then acts as a "second messenger" and triggers a sequence of events in the cell that leads to the overall response of the cell to the hormone. Any action of cyclic AMP is halted by its breakdown into an inactive noncyclic AMP by the enzyme phosphodiesterase. In many tissues the effects of cyclic AMP have been shown to be exerted by activation of protein kinases. They catalyze the transfer of a phosphate group from ATP to certain enzymes and lead to activation (or inactivation) of the enzyme. This may lead to an elaborate cascade of events in which enzymes are converted in sequence from inactive to active forms. The cascade of events triggered by cyclic AMP in most instances amplifies the effects of a hormone many fold. Thus, one molecule of epinephrine could theoretically generate the production of several million glucose molecules by its effect on the glycolytic pathway in the liver. In addition to the induction of enzymes, the various hormones, through their second messenger, adenylate cyclase, induce the release of hormones stored in the gland and changes in the permeability of cell membranes to amino acids, ions, or water, or modify the rate of enzyme-controlled reaction.

Hormones may also act through other types of second messengers such as cyclic 3',5'-guanosine monophosphate (cyclic GMP) or by the induction of changes in the concentration of free calcium in the cell. The latter may be accomplished by increasing the net flux of calcium into the cell, by freeing bound calcium in the cytosol, or by favoring the movement of calcium into organelles such as the mitochondria or endoplasmic reticulum.

The steroid hormones, those from the adrenal cortex (cortisol and aldosterone) and the gonads (androgens, estrogens, and progestins), modify cell metabolism by inducing changes in the rate of enzyme formation of their target cells. Hormones of this type, by virtue of their lipid solubility, may penetrate the cell membrane by diffusion or by specific uptake mechanisms. On entering the target cell they are rapidly bound by their receptor, and the binding results in a change in the hormone-receptor complex. This interaction is termed *activation* or *transformation* and permits the complex to bind to nuclear chromatin. Control is exerted by regulating the transcription of deoxyribonucleic acid (DNA) and thus regulating the formation of messenger ribonucleic acid (mRNA). Interaction between the hormone-receptor complex and the nuclear chromatin regulates the level of a specific messenger RNA (mRNA). This leads to an increase in the rates of synthesis of proteins translated off those mRNAs, and the changes in protein synthesis then mediate the response to the hormone. For example cortisol, the primary glucocorticoid produced by the adrenal cortex, acts on the hepatocytes of the liver to increase the concentration of several amino transferases. The increase in the enzymes results in increased glucose production from noncarbohydrate sources.

As noted, the receptors that bind the thyroid hormones are located in the cell nucleus, specifically in the chromatin. The hormone of the thyroid gland that appears to be bound to the nuclear receptor is triiodothyronine (T_3), not tetraiodothyronine (T_4), or thryoxine, the predominant secretory product of the thyroid gland: T_4 penetrates the cell and is converted to T_3 under the influence of a deiodinase, and the T_3 binds to the receptor to form a T_3-receptor complex that binds to DNA. The complex then acts in a manner similar to that of the steroid hormone–receptor complex, resulting in the formation of a specific mRNA and a change in the rate of protein (enzyme) synthesis.

Recently, it has been demonstrated that some polypeptides hormones may gain access to cells by a concentrative mechanism that employs a set of specialized organelles. First, the hormones bind to receptors on the cell surface, and the hormone-receptor complexes form clusters in the membrane and enter the target cells by a process termed *receptor-mediative endocytosis*. One hormone that enters cells via this pathway is insulin.

PRINCIPLES OF HORMONE ASSAY

The endocrine glands produce and store only minute amounts of the hormones they secrete, and the plasma contains amounts in even smaller concentration. The physiological activity of the hormones was, in many cases, the only guide to their presence, and these effects formed the basis for bioassay procedures. In such cases the functional effect of a hormone may be used as the basis for the quantitive estimation of the amount of hormone present.

The level of activity of many of the endocrine glands can be assessed by observing the morphology of the glandular cells themselves. Such criteria as the size and shape of cells, as well as the quantity, quality, and characteristics of intracellular granules, supply indices of the level of activity. Similarly, the morphology of cells and functional activity of tissues on which the hormones act are used to assess the activity of an endocrine gland.

The hormonal content of each of the endocrine glands can be estimated by means of biological or physiological techniques, or both. In addition, the level at which a gland is functioning can be assessed by analyzing body fluids for the secretion of the gland involved. However, as noted, only very small amounts of hormone are present in the plasma, and small amounts can be recovered in the urine. The extent to which the urinary concentration of hormones and their metabolites reflects the amount in the plasma is unknown in many cases, and it has become much too convenient to assume that such concentrations truly reflect the plasma levels. It is the body fluids, exclusive of the urine, from which the hormones must eventually exert their action.

Radioimmunological methods are currently applied in the quantative assay of the protein and polypeptide hormones in the plasma. The radioimmunoassay (RIA) is sensitive, precise, specific, and accurate. Radioimmunoassays can be applied to very small quantities of plasma and depend on highly specific reactions between a protein or polypeptide hormone, the antigen, and its specific antibody. The assay is based on competition between unlabeled hormone and radioactively labeled hormone for specific sites on the antibody. Initially, radioactively labeled ("hot") hormone is complexed with antibody, and then known amounts of unlabeled ("cold") hormone are added and serve as standards. Some of the "cold" hormone displaces some of the "hot" hormone from the antibody. The free or unbound hormone is then separated from the complexed or antibody-bound hormone, and the amount of radioactivity in both samples is determined. The lower the concentration of "cold" hormone, the more labeled or "hot" hormone will remain bound to the antibody. A standard curve can be prepared by using different concentrations of unlabeled hormone and comparing them with an unknown or plasma samples. Assays of this type are now available for the determination of plasma levels of insulin, glucagon, calcitonin, parathyroid hormone, growth hormone, adrenocorticotropic hormone (ACTH), thyroid-stimulating hormone (TSH), luteinizing hormone, follicle-stimulating hormone, prolactin, melanocyte-stimulating hormones (melanotropins), and the hormones of the hypothalamus.

The principles of the radioimmunoassay may also be applied in the determination of the plasma levels of the nonprotein hormones, such as the steroid hormones and the hormones that are derivatives of amino acids. The method is based on the presence of specific binding proteins (globulins) for the steroids and the thyroid hormones in the plasma. Normally, a large fraction of these hormones is specifically bound to the carrier globulins, and the carrier globulins are substituted for the antibodies in this type of assay procedure. The competitive-binding assays thus provide sensitive assay procedures for the thyroid hormones and the gonadal and adrenal steroids. Current techniques for the plasma determination of the catecholamine epinephrine include a combination of high-pressure liquid chromatography and electrochemical detection.

REGULATION OF ENDOCRINE ACTIVITY

A variety of stimuli are capable of altering the rate of secretion of hormones from endocrine glands. Changes in both the internal and external environment precipitate factors that lead to changes in the rate of hormone secretion. Generally speaking, hormones are not secreted at constant rates—most are secreted in short bursts. Consequently, the plasma concentrations of a hormone may fluctuate rapidly over a brief period of time. In the presence of appropriate stimuli the bursts may appear more frequently; in the absence of such stimuli the bursts appear less frequently. The rate of secretion of some hormones (e.g., cortisol) vary over a 24-hour period. Their circadian pattern of secretion is related to the subject's sleep-wake cycle. Others (e.g., as the ovarian hormones) are secreted in relationship to a 28-day (lunar) pattern.

The hormones, with the exception of the polypeptide hormones, circulate in the plasma bound to specific plasma proteins. The amount of hormone that is free is very small and is in equilibrium with the bound fraction. It is the free hormone that exerts its effects on the target cells. Thus, altering the concentration of the specific plasma protein may effect the amount of free hormone in the plasma.

Either directly or indirectly, a normally functioning system of endocrine glands is dependent on a normal CNS and its many processes. The release of the secretions of two of the endocrine glands, both arising from embryonic neural tissue, is regulated directly by their innervation; i.e., their efferent nerves are secretomotor. These two structures are the adrenal medulla and the neurohypophysis, and both undergo atrophy following denervation. The feedback relations of these glands with the CNS is depicted in Figure 30-1A.

The anterior pituitary (AP), which has been considered the master gland, is in fact the target of various stimuli, both neural and humoral, that stimulate or inhibit its activity. It is certain that the primary control of the endocrine system resides within the CNS, but this regulation is not expressed via secretomotor fibers to the anterior pituitary. Rather, it is expressed through the secretion of hormones released by nerves of the hypothalamus into the plasma, and they either stimulate or inhibit the release of the respective tropic hormones from the AP (see Fig. 30-1B). It has been shown that the sympathetic mediators epinephrine and norepinephrine increase the responsiveness of the thyroid gland to a con-

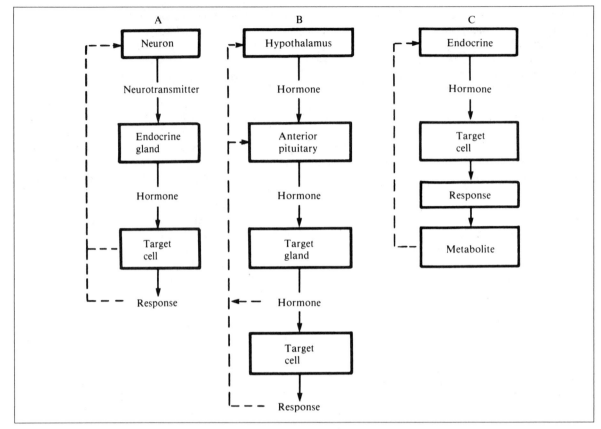

Fig. 30-1. Types of negative feedback systems involved in regulating the rate of hormone synthesis and secretion. See the text for specific examples.

stant dose of thyroid-stimulating hormone. Thus, release of the mediators at the vasomotor endings may act at the level of the glandular cells to increase their sensitivity to the tropic hormones.

The endocrine glands should not be considered individually but viewed as an integrated system, because a given hormone seldom acts independently and generally affects other endocrine glands and alters their rate of secretion. In this action the hormone may assist or oppose the effect of another hormone. Such interrelationships are illustrated by the following observations: (1) The administration of thyroxine inhibits secretion of TSH by the AP gland and at the same time stimulates the consumption of oxygen by liver cells. (2) The two pancreatic hormones, insulin and glucagon, induce opposite effects on the concentration of glucose in the plasma; the former lowers it and the latter elevates it. In addition, the functional state of an endocrine gland may be altered by a hormone secreted by a gland other than the AP; e.g., the administration of thyroid hormone may in-

crease the functional activity of the adrenal cortex. Such interrelationships between hormones of various glands can account for many of the paradoxical effects observed in experimental and clinical studies.

The AP, under the influence of hypothalamic hormones, secretes several hormones whose effects may vary considerably. Several endocrine glands are directly affected by the AP hormones: the gonads, thyroid, and adrenal cortex, which are referred to as the target glands. They can secrete small amounts of their hormones in the absence of the pituitary, but normal tropic hormone secretion by the AP is necessary for normal target gland function. An increase in the rate of secretion of a tropic hormone results in an increase in the rate of secretion from the respective target gland. The secretions of the target gland in turn tend to produce countereffects that oppose the secretion of its particular tropic hormone by the AP and thus inhibit the initial source of stimulation. For example, when there is an increase in the rate of secretion of TSH by the AP, the rate of secretion of thyroxine from the target gland, the thyroid gland, increases. The increasing secretion of thyroxine then inhibits the secretion of TSH. The end result is a balance of forces, and the levels of thyroid function and of heat production by the cells of the body are maintained within relatively narrow limits. This serves to illustrate the concept of the negative feedback system illustrated in Figure 30-1B. In addition to the tropic hormones, which exert their primary effects on specific glands, the AP also secretes a growth hormone that acts through intermediates, the somatomedins, on many cells of the body. The hormones of the thyroid gland act on all cells of the body.

The rate of secretion of some hormones is relatively independent of the pituitary, being regulated by the concentration of nonhormonal metabolites in the plasma. An example of this type of regulation concerns the secretion of parathyroid hormone. A low serum calcium concentration stimulates the secretion of parathyroid hormone, and, conversely, hypercalcemia induces a decrease in its rate of secretion. In each case the calcium acts directly on the gland to alter its secretion rate. The secretion rates of at least three other hormones calcitonin, insulin, and glucagon, are controlled primarily be the concentrations of nonhormonal metabolites. Such feedback relationships are illustrated in Figure 30-1C.

In addition to the effects of the tropic hormones of the AP and their reciprocal relationships between the target glands and the pituitary, the hormones of the target glands may affect one another. For instance, for normal ovarian function the level of thyroid activity must be optimal.

HYPOPHYSIS (PITUITARY GLAND)

The hypophysis of the adult human weighs about 0.5 gm and is a compound gland of ectodermal origin, arising from two different sources. One part, the *neurohypophysis,* arises from the ventral floor of the diencephalon and remains connected to the hypothalamus throughout life by means of its stalk. The glandular portion of the hypophysis, the *adenohypophysis,* stems from oral ectoderm as Rathke's pouch, which is an outgrowth from the roof of the mouth. This outgrowth meets the embryonic neural portion of the gland and then loses its connection with the oral epithelium.

The adenohypophysis is described as consisting of three parts: the *pars distalis* (often referred to as the anterior pituitary or AP), the *pars intermedia,* and the *pars tuberalis*. The neurohypophysis is also divided into three parts: The *infundibulum* (median eminence), *infundibular stem* (neural

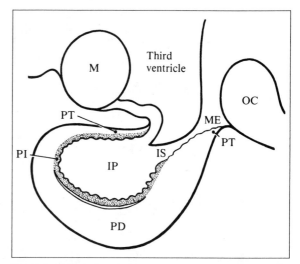

Fig. 30-2. Midsagittal section through the hypophysis. The parts of the adenohypophysis or anterior pituitary are the pars distalis (PD), the pars intermedia (PI), and the pars tuberalis (PT). The neurohypophysis or posterior pituitary contains the median eminence (ME) of the tuber cinereum, the infundibular stem (IS), and the infundibular process (IP) or neural lobe. The mamillary body (M) and the optic chiasm (OC) are also depicted.

stalk), and *infundibular process* (neural or posterior lobe). The hypophyseal stalk includes the neural stalk plus the sheath portions of the adenohypophysis, the pars tuberalis (Fig. 30-2).

The neurohypophysis receives its blood supply from three sources. (1) The infundibulum is connected to the circle of Willis via the *superior hypophyseal arteries;* (2) the *middle hypophyseal arteries* supply both the infundibular stem and process; and (3) the *inferior hypophyseal arteries* link the intracavernous carotid arteries to the infundibular process (Fig. 30-3). All three portions of the neurohypophysis are linked together by a continuous capillary bed. Two capillary beds arise in the infundibulum and are directed toward the hypothalamus. One, an *external plexus,* is superficial; another, an *internal plexus,* is deep. Both plexuses are continuous with capillary beds of the hypothalamus. The capillaries of the two plexuses are also linked to the AP by the *long portal vessels.* The neurohypophysis is also linked to the AP by the *short portal vessels,* vessels extending between the infundibular process and the AP. Efferent vascular pathways from the neurohypophysis link it (1) via capillaries to the hypothalamus, (2) to the AP via the portal vessels, and (3) to the cavernous sinus via the neurohypophyseal veins. The AP receives no direct arterial

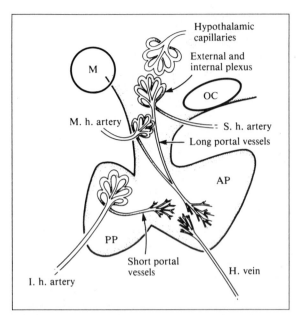

Fig. 30-3. The blood supply of the anterior and posterior pituitary (AP, PP). Note that the AP does not receive a direct arterial blood supply. M = mamillary body; OC = optic chiasm; S. h. = superior hypophyseal; M. h. = middle hypophyseal; and I. h. = inferior hypophyseal. See the text for further explanation.

supply, and its entire afferent vascular connections are via the two types of portal vessels (Figure 30-3).

Physiological studies have demonstrated high concentrations of AP hormones in the long portal vessels, which can be prevented by removal of the infundibular process. The vascular connections described permit both hypothalamic and AP secretions to be conveyed to the capillary beds of the neurohypophysis, and the hormones can thus leave the neurohypophysis via several routes: an efferent route directed to the AP, another directed to the systemic circulation via the cavernous sinus, and other efferent routes directed toward the brain.

The hypophysis secretes at least thirteen hormones, each being polypeptide or protein in nature. Some of the pituitary hormones, the tropic hormones, regulate the functional capacities of other endocrine glands: ACTH, TSH, follicle-stimulating hormone (FSH), luteinizing hormone (LH), or interstitial cell–stimulating hormone, and prolactin. Growth (somatotropic) hormone affects nearly all cell types, whereas the actions of vasopressin or antidiuretic hormone (ADH), oxytocin, and two melanocyte-stimulating hormones are restricted to particular cell types. Other hormones secreted by the AP are β-lipotropin, γ-lipotropin,

and β-endorphin. The precise functions of these three hormones have not been elucidated, and they are being extensively investigated.

The well-protected hypophysis is one of the most inaccessible organs of the body, being located in the sella turcica, a depression in the sphenoid bone. The gland is encapsulated by the dura mater, but the hypophyseal stalk penetrates the dura through the diaphragma sellae. The neurohypophysis is characterized by its rich innervation of hypothalamic origin, whereas the adenohypophysis is characterized by rich vascularization.

NEUROHYPOPHYSIS

Strictly speaking, the neurohypophysis is not an endocrine organ because it lacks glandular elements and serves merely as a storage place for certain secretions of the hypothalamus. However, it plays an essential role in the release of the stored hormones.

Seen microscopically, the neurohypophysis is composed of four basic structures: unmyelinated nerve fibers, glial cells, and capillaries surrounded by argyrophilic connective tissue. The nerve fibers have their origin in the cells of the supraoptic and paraventricular nuclei.

It is generally accepted that the hormones of the neurohypophysis, vasopressin (ADH) and oxytocin, are synthesized with their binding proteins, the neurophysins, in the hypothalamic nuclei. They are packaged into neurosecretory vesicles and move by axonal transport from the neural cell bodies down the infundibular stalk to nerve terminals in the infundibular process or the neural lobe of the pituitary. The neurosecretory vesicles are stored in the neural lobe and are released into the circulation following appropriate physiological stimulation. Neurophysin I, or estrogen-stimulated-neurophysin, is considered to be the carrier protein for oxytocin, and neurophysin II, or nicotine-stimulated neurophysin, is the carrier for ADH. The neurophysins are polypeptides with molecular weights of about 10,000 and they are also released into the circulation following appropriate stimulation. The concept of one neuron, one hormone, one neurophysin appears to be a reasonable hypothesis with respect to the so-called neurohypophysial hormones. It appears proper to refer to this system as the hypothalamoneurohypophysial system (HNS) and regard it as an organ of internal secretion.

Four different activities have been attributed to the hormones of the HNS. First, in 1895, Oliver and Schafer observed that the intravenous administration of extracts of the whole pituitary resulted in a marked rise in systemic arterial blood pressure (pressor effect). The effect was purely one of vasoconstriction, and Howell showed later that this effect was obtainable only if neural lobe extracts were administered. Second, Henry Dale demonstrated that extracts that caused contraction of the uterus (oxytocic effect) could be prepared from this organ. Third, it was demonstrated that neural lobe extracts were helpful in ameliorating the symptoms of diabetes insipidus (antidiuretic effect). Finally, extracts of the HNS have been shown to induce contractions of the myoepithelial cells of the alveoli of the lactating mammary gland and cause evacuation of milk from the alveoli (milk letdown effect).

The brilliant research of Du Vigneaud and his associates resulted in the isolation of two distinct polypeptides from the neural lobe. Later work led to the synthesis of the two compounds. Each has a molecular weight of about 1000 and contains eight amino acids, six of which are common to both hormones. One compound, vasopressin, possesses a high degree of antidiuretic activity and no oxytocic activity. The other, oxytocin, shows little antidiuretic activity and high oxytocic activity.

Fig. 30-4. **Factors that regulate the secreton of ADH by the hypothalamoneurohypophyseal system (HNS). + = stimulation and − = inhibition.**

The most obvious consequence of neurohypophysectomy is the induction of polyuria. This is characterized by the excretion of a high-volume, low-concentration urine. As a result of the large volume of water lost, there is an intense thirst and a high water intake (polydipsia). Because ADH regulates the excretion of free water from the kidney, it plays an important role in the maintenance of water balance. The site of action of ADH on the nephron is discussed in Chapter 22.

The rate of secretion and release of ADH is under nervous control and is subject to changes in the effective osmotic concentration of the extracellular fluid, the effective volume of the plasma, exteroceptive stimuli, and psychological stimuli (Fig. 30-4).

Various drugs, such as nicotine, acetylcholine, morphine, barbiturates, ether, and chloroform, act on the HNS to increase ADH secretion. On the other hand, alcohol inhibits it.

Verney has shown that the injection of hypertonic solutions of NaCl, glucose, or sucrose into the carotid artery results in a diminution of water diuresis. Injection of urea (which is ineffective osmotically at the cell) is not an effec-

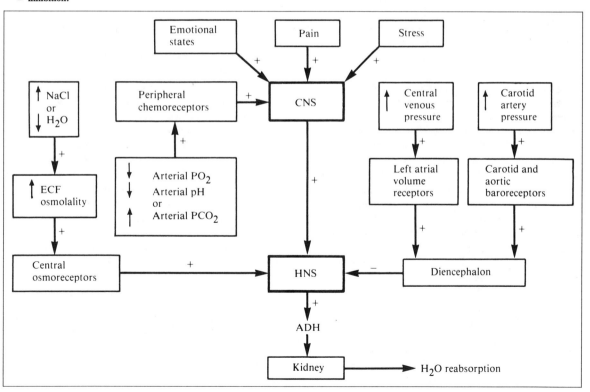

tive inhibitor of water diuresis. Subsequent work has shown that a sustained (about 30 min) increase of only 2 percent in the effective osmotic pressure causes an increase in ADH release and a decrease in free water clearance. Similarly, the ingestion of a large volume of water lowers the osmotic concentration, and the secretion of ADH is depressed. It has been suggested that certain receptors are sensitive to differences in the osmotic concentrations between the extracellular and intracellular fluid. These receptors, the *osmoreceptors,* are located in the anterior hypothalamus, near or within the supraoptic nuclei, and therefore transmit their stimuli to the neural lobe over this tract. Stimulation of the area around the supraoptic nucleus causes inhibition of water diuresis. Similarly, it has been shown that the injection of hypertonic saline solution increases the electrical activity of single neurons in and adjacent to the supraoptic nuclei. Normally, the plasma ADH concentration is very low or undetectable by radioimmunoassay when the plasma osmolality is reduced to 280 mOsm per kilogram (mOsm/kg) or less, and the urine is maximally diluted (50–60 mOsm/kg). When the plasma osmolality rises above 280 mOsm/kg, the plasma ADH concentration increases in direct proportion to the plasma osmolality. At a plasma osmolality of about 295 mOsm per kilogram the ADH concentration reaches a level ($\sim$ 5 pg/ml) at which the kidney is stimulated to produce a maximally concentrated urine (1200–1250 mOsm/kg). The conscious desire to drink is also stimulated when the plasma osmolality reaches 295 mOsm per kilogram.

Superimposed on the control mechanism whereby changes in the effective osmotic pressure of extracellular fluid affect ADH secretion by the HNS are alterations induced by various reflexes. These alterations may be due to excitation of the HNS by emotional stress or pain.

Studies pioneered by Henry and Gauer and their coworkers have firmly established that *volume receptors* and *pressure receptors* play a prominent role in the control of vasopressin secretion during isotonic contraction or expansion of the effective blood volume. The best available evidence shows that the monitoring system for such reflexes has an intravascular origin.

It has been shown that distention of the left atrium or the pulmonary vein within the pericardium is invariably followed by a diuresis. The diuresis occurs in spite of the fact that the manipulations markedly diminish the cardiac output. Distention of the pulmonary artery, the right atrium, or the pulmonary vein outside the pericardium does not lead to either a diuresis or an antidiuresis. Procedures such as negative pressure breathing, the infusion of iso-oncotic or hyperoncotic albumin solutions, isotonic saline infusion, or

changing from the sitting to the supine position, each of which results in an increased left atrial volume, lead to a decrease in plasma ADH and an increase in urine volume. These procedures are effective only when the vagus is functional, and associated with each is increased neuronal activity of afferent fibers running in the vagal trunk. That volume and not pressure is the stimulus is shown by the fact that bursts of activity in the nerve are minimal during the "a" wave of the atrial pressure cycle, which is a time when the pressure is high, and the firing rate is greatest during the "v" wave of the atrial pressure cycle, when atrial volume is maximal. Conversely, such situations as venous occlusion of the legs, hemorrhage, orthostasis, positive pressure breathing, and changing from the supine to the sitting position, all of which lead to a decreased left atrial volume, result in increased plasma ADH and reduced urine flow.

The evidence that has been presented suggests that activation of the left atrial stretch receptors inhibits the release of ADH from the HNS (Fig. 30-4). Share and his coworkers have shown that the circulating level of ADH increases following isotonic contraction of the blood volume, following bilateral vagotomy, or following cartoid occlusion. However, the blood ADH concentration decreases during left atrial distention. They have also shown that carotid sinus baroreceptor activity and carotid body chemoreceptor activity influence ADH secretion. Activity of the baroreceptors appears to be inversely related to ADH secretion, whereas chemoreceptor activity (induced by hypoxia, hypercapnia, or a decrease in pH) appears to be directly related to ADH secretion. The ADH concentration in the blood of the human subject in the standing position is elevated when compared to that in the subject in the recumbent position. Similarly, the person acutely exposed to a cold environment shows a decrease in blood ADH, whereas an increase follows acute exposure to a hot environment. Therefore, the rate of receptor discharge from the left atrial receptor is a function of the central blood volume. When the central blood volume is high, ADH secretion is reduced; when it is low, ADH secretion is increased. Marked increases in plasma ADH may occur in the presence of high plasma osmolar concentrations if the mean left atrial pressure, or volume, or both is elevated. The volume receptor and osmoreceptor regulatory mechanisms provide, in concert, a very precise control over ADH secretion and protect the volume and concentration of the ECF within narrow limits.

The role of ADH in the regulation of renal tubular water handling and homeostasis of the extracellular fluid has been discussed in Chaps. 22 and 23. Recent evidence suggests that the action of ADH on the distal renal tubules is mediated through cyclic AMP. However, the nature of the change in permeability to water remains obscure. In humans, the distal renal tubules and collecting ducts are extremely sensitive to changes in the plasma ADH concentration. The concentrating power of the kidney may be increased from a urine/plasma ratio of 1 up to a maximum of 4 by increasing the plasma ADH from 1 to 5 picogram per milliliter.

It appears that a type of negative feedback mechanism provides reciprocity between the kidney, gastrointestinal tract, lungs, skin, and sweat glands and the HNS and thus ensures a relatively constant extracellular fluid with respect to its effective volume and osmolarity. This feedback mechanism is directly dependent on its connections with the CNS.

The regulation of the secretion of oxytocin, its role in sperm transport, and its effects on uterine motility and milk ejection are discussed in Chapter 35.

ADENOHYPOPHYSIS

The glandular portion of the pituitary is made up of irregular masses and columns of epithelial cells that are supported by a delicate framework of connective tissue. The groups and columns of cells are separated by sinusoids. Two main types of cells are evident. One type, the chromophobes, does not show any conspicuous stainable cytoplasmic granules. The other type, the chromophils, contain cytoplasmic granules that take up stains readily. The latter are considered to be secretory in nature. Many cytologists believe them to be daughter cells of the chromophobes, but it is also possible that the chromophobes represent resting cells. The chromophils may be further divided, according to the stainability of their granules, into *acidophils* (alpha cells) and *basophils* (beta cells). Recently, highly specific immunocytochemical methods, combined with light and electron microscopic techniques, have shown that there are five secretory cell types in the AP.

HORMONES OF THE ADENOHYPOPHYSIS. The 11 hormones of the AP may be divided into three groups, based on their structural features. One group, the simple peptides, includes the corticotropins, the melanotropins, the lipotropins, and the endorphins. These are derived from a common precursor molecule, prepro-opiocortin, synthesized by a specific type of basophil, termed the *corticotrope.*

A second group, the simple proteins, includes somatotropic hormone (STH) (somatotropin) and prolactin (PRL). They have considerable structural homology and some common biological properties. Both are monomers, both

contain two S-S bonds in the same locations, and their amino acid sequences indicate that they have many identical amino acids. Both hormones are secreted by acidophilic cells: STH by somatotropes, and PRL by lactotropes. A hormone secreted by the placenta, human placenta lactogen, has a chemical structure similar to both STH and PRL, and the three hormones have overlapping biological effects.

The third group, the glycoproteins, include TSH, LH, and FSH. The three hormones are dimers, and their alpha subunits are nearly identical. The primary structures of the beta subunits of the three hormones are different and provide hormonal specificity to the entire glycoprotein molecule. For instance, if the beta subunit of TSH is combined with the alpha subunit of LH, it generates thyrotropic activity. The LH molecule contains 211 AA (96 in the alpha and 115 in the beta), FSH has 210 residues (92 alpha and 118 beta), and TSH has 215 (96 alpha and 119 beta). Another placental hormone, human chorionic gonadotropin, is also a glycoprotein, has alpha and beta subunits similar to those of LH and shares many biological actions with LH. A basophilic cell, a *thyrotrope,* secretes TSH, and LH and FSH are secreted by a single basophilic cell type, called the *gonadotrope.*

The hypothalamic hormones regulate the secretory activity of the anterior pituitary gland. Several hormonal peptides are synthesized by hypothalamic neurons and are released into both the internal and external plexuses in the infundibulum. From there, they move with the plasma into the long portal vessels and are transported to the AP, where they bind to specific receptors on their target cells (see Fig. 30-3). The presence of six such hypothalamic hormones have been established, four of which have been characterized and synthesized. One, thyrotropin-releasing hormone, is a tripeptide and stimulates the secretion of both TSH and PRL. A second, gonadotropin-releasing hormone is a decapeptide and it stimulates the secretion of both LH and FSH. A third hormone, somatostatin, is a tetradecapeptide and inhibits the release of STH, TSH, and PRL. The catecholamine dopamine is the fourth substance secreted by the hypothalamus, and it inhibits PRL secretion. The two remaining hormones, corticotropin-releasing factor and growth hormone–releasing hormone, have not been fully characterized, but are presumed to be polypeptides. Present evidence suggests that each of the six hormones acts on its target cells via the adenylate-cyclic AMP system. Thyrotropin-releasing hormone, gonadotropin-releasing hormone, growth hormone–releasing hormone, and corticotropin-releasing hormone stimulate adenohypophysial cyclic AMP accumulation in their target cells and parallel changes in specific hormone release, whereas opposite effects have been found with somatostatin and dopamine. Somatostatin is also secreted by cells in the endocrine pancreas and various portions of the gastrointestinal tract, and it may perform a key role in the regulation of organic metabolism and energy balance. The functions and the regulation of the secretions of each of the tropic hormones of the AP will be considered later in greater depth in the appropriate sections.

EFFECTS OF HYPOPHYSECTOMY. Several notable morphological and functional alterations result from total hypophysectomy in the young animal. These are as follows:

1. Failure of the gonads to mature, with resultant infantile sexual development and sterility because of lack of LH and FSH.
2. Atrophy of the thyroid gland and the characteristics of thyroid insufficiency because of lack of TSH.
3. Atrophy of the adrenal cortex and signs of hypoadrenalism without salt loss because of ACTH deficiency.
4. Cessation of growth, failure to attain an adult stature, a decided tendency toward hypoglycemia, hypersensitivity to insulin, and a loss of body nitrogen accompanied by diminished fat catabolism because of lack of STH.

However, if optimal environmental conditions are maintained, removal of the pituitary is not incompatible with life.

Following hypophysectomy or during the development of hypopituitarism, the earliest and most frequent deficiencies observed are failure of growth and gonadotropic hormone secretion and resultant gonadal failure. Evidence of thyroid insufficiency next appears and is followed by signs of adrenocortical insufficiency. This sequence of malfunction generally prevails regardless of the species. In several species it has been shown that ablation of increasing amounts of AP tissue results in gonadal, thyroid, and adrenocortical failure in that order. The removal of up to 80 percent of the AP is compatible with normal gonadal, thyroidal, and adrenal cortical function. Thus, it appears that a large safety factor is present, and the same is true of all the other endocrine glands.

If properly treated, the patient whose hypophysis has been removed retains a normal appearance, is able to gain weight and attain a positive nitrogen balance, can repair bone, and can be maintained in good health indefinitely. It follows that the administration of hormones of the AP to intact subjects results in marked development of the gonads and accessory sex glands, adrenal and thyroidal hypertrophy with signs of hyperadrenalism and hyperthyroidism, an increase in growth rate with the laying down of nitrogen,

an increase in the length of the long bones, and a tendency toward diabetes mellitus.

FUNCTIONAL ROLE OF THE ADENOHYPOPHYSIS. The adenohypophysis occupies a crucial position in the body economy because it directly influences the output of hormones from the adrenal cortex, thyroid, and gonads through the elaboration of its tropic hormones. It acts directly on all body structures through STH, and it also indirectly affects the output of the hormones from the pancreas and parathyroids as a result of its action, both direct and indirect, on various metabolic pathways.

The effects of an excess or a deficiency of the tropic hormones of the AP are thus mediated through the various target glands. The tropic hormones act only on preexisting chemical reactions and do not initiate them. However, the tropic hormones are essential for complete morphological and physiological development of their target glands. In general, following the administration of hormones, functional changes in target glands are induced prior to anatomical changes.

According to the negative feedback concept of control systems, the tropic hormones are released in amounts varying with a person's functional state and not at a constant rate. In all probability, at least two types of negative feedback systems operate between the tropic hormone secretion of the AP and secretion of hormones by the target gland. Initially, it was believed that the hormones of the target glands acted directly on the AP to inhibit the tropic hormone output. In this instance, when the hormone secretion rate of the target gland was high, tropic hormone secretion would be inhibited, and when the concentration of target gland hormone was low, inhibition would be removed and tropic hormone output would be increased–and so on. However, it has been shown that various target gland hormones may also act on specific hypothalamic neurons, alter hypothalamic hormone secretion, and indirectly influence the synthesis and secretion of a tropic hormone. Changes in both the internal and external environment can affect AP hormone secretion by regulation of the hormone output via the hypothalamus and the hypophyseal portal vessels.

GROWTH (SOMATOTROPIC) HORMONE. Growth hormone, as the name implies, plays a central role in the growth of the organism. Human growth hormone is a single-chain polypeptide with a molecular weight of 21500 and containing 191 amino acid residues. The hormone's complete sequence of amino acids has been determined. It is derived from a larger precursor polypeptide, pro-growth hormone or "big" growth hormone. Growth hormone exerts many biological effects, but it is only recently that its effects could be studied in humans, because the growth hormones of subprimate species were ineffective in humans. Early studies of the role of growth hormone in human physiology were limited to clinical observations in acromegaly, gigantism, and dwarfism. The hormone affects several variables of protein, fat, and carbohydrate metabolism, and in contrast to the other adenohypophysial hormones, it requires a target gland, the liver, and it also acts on certain target cells directly. The plasma level of growth hormone in the adult in a basal state, as determined by radioimmunoassay, is about 2 to 3 mg per milliliter. It has a plasma half-life of approximately 20 minutes.

Growth hormone plays a prominent role in protein metabolism and the regulation of growth. It is a protein anabolic hormone and it induces a positive nitrogen balance. Many of its effects are mediated by a group of small polypeptides called somatomedins. The somatomedins are synthesized by the liver, and their rates of secretion are regulated primarily by growth hormone. Growth hormone excess is associated with increases in plasma somatomedins. The stimulatory effect of growth hormone on the heptic production of somatomedins may be potentiated by insulin, and by the growth hormone–like peptides, PRL and human placental lactogen, whereas these effects may be antagonized by adrenal steroids, such as cortisol, by estrogens, and by systemic illnesses. Good nutrition also appears essential for optimal production of the somatomedins by the liver. Six somatomedins have been characterized, and they are transported in the plasma bound to carrier proteins, which have relatively long half-lives (2–3 hr).

The effects of the somatomedins are generally anabolic in nature. They bind to receptor sites on the membrane of their target cells, but their mechanism of action is not known. Their tissue effects include the following: (1) cartilage-stimulating activity (amino acid transport; synthesis of DNA, RNA, protein, and collagen; and formation of proteoglycans [chondroitin sulfate]); (2) mitogenic activity in cartilage, fibroblasts, and liver; and (3) insulinlike activity in muscle and fat cells. The insulinlike effects of the somatomedins on muscle and fat cells, i.e., glucose transport, glycogen formation, and glucose incorporation into lipid, are only about a hundredth as potent as those of insulin.

Probably the most striking gross effect of STH is that on the skeleton. Hypersecretion of STH or its exogenous administration can lead to gigantism if instituted prior to closure of the epiphysial plates of the long bones. However,

the presence of high levels of STH in the adult (after closure of the epiphysial plates) results in acromegaly. This condition is characterized by enlargement of the skeleton, especially of the skull and hands and feet, skin, subcutaneous tissue, and viscera. Acromegaly is generally the result of an acidophilic adenoma of the AP. The disease is accompanied by sellar enlargement, gonadal atrophy, changes in the visual fields, and diabetes mellitus.

Although the effects of growth hormone are mediated by the somatomedins with respect to cartilage and bone, growth hormone has several direct effects on other target cells. It stimulates the formation of somatomedins in the liver, promotes incorporation of amino acids into other proteins by the liver and other tissues, and leads to an overall increase in new protein formation. It also leads to a reduction in the rate of lipid synthesis, promotes the mobilization of fatty acids from adipose tissue, and increases fatty acid oxidation. The first indication that growth hormone played a role in carbohydrate metabolism was the observation that hypophysectomy, when performed on an animal that has undergone pancreatectomy, led to the amelioration of the symptoms of diabetes mellitus. Growth hormone is diabetogenic because it increases hepatic glucose production. In general, it directly antagonizes the effects of insulin on carbohydrate metabolism in muscle and fat cells. Chronic growth hormone administration or secretion may lead to the development of permanent diabetes mellitus.

The hypothalamus plays an important role in the regulation of STH secretion. Neurons of the ventromedial hypothalamus release two hormones into the hypophyseal portal circulation that affect STH secretion: (1) growth hormone–releasing hormone, which stimulates growth hormone secretion by somatotropes; and (2) a growth hormone-inhibiting polypeptide, somatostatin, which inhibits growth hormone secretion.

It is presumed that all factors that influence the rate of growth hormone secretion by the somatotropes exert their effects through changing growth hormone–releasing hormone and somatostatin secretion rates. Glucose (hyperglycemia) fatty acids, and beta-adrenergic agents act on the overall system to inhibit STH secretion. Conditions in which actual decreases occur in the substrate for energy production, such as low blood glucose (hypoglycemia) and fasting, stimulate STH secretion. It is also stimulated by arginine, alpha-adrenergic agents, and vasopressin. Various areas of the CNS may act on the hypothalamic elements to stimulate or inhibit growth hormone secretion; several types of stressful stimuli and factors such as exercise and sleep stimulate growth hormone secretion. However, during

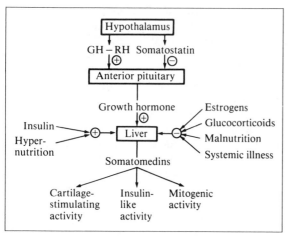

Fig. 30-5. Factors that regulate the secretion and functions of growth hormone and the somatomedins. Plus signs indicate stimulation; minus signs indicate inhibition.

rapid-eye-movement sleep, STH secretion is inhibited. It may also inhibit its own secretion by acting directly on the hypothalamus, presumably by stimulating somatostatin secretion and depressing growth hormone–releasing hormone secretion. This type of control is referred to as the "short" feedback mechanism.

Figure 30-5 illustrates the primary factors involved in the regulation of the secretion and function of STH and the somatomedins.

OTHER ORGANS WITH ENDOCRINE FUNCTIONS

Several hormones are secreted by glandular cells of the gastrointestinal tract. In fact, the first hormone described was secretin, described by Bayliss and Starling in 1902. The function of these hormones has been discussed in Chapter 26. The kidney secretes two substances that may be called hormones. One, renin, will be discussed in Chapter 35 in conjunction with the factors that regulate aldosterone secretion by the adrenal cortex. Another, erythropoietin, is stimulated by hypoxia, and it in turn stimulates hemoglobin synthesis and the production of red blood cells by the bone marrow. The pineal gland secretes a substance, melatonin, that may affect the state of gonadal maturation. Tumors of the pineal in young boys have been demonstrated to induce precocious puberty. The thymus secretes several hormones that regulate the development and function of immuno-competent T lymphocytes and thus play an important role in the maintenance of immune function and homeostasis.

These substances include thymosin, thymopoietin, thymus humoral factor, and serum thymic factor.

The prostaglandins are ubiquitous and are composed of a series of closely related unsaturated fatty acids that have 20 carbon atoms, contain a cyclopentane ring, and affect many systems, tissues, and cells. Some of their effects include inhibition of platelet aggregation, vasodilation of vascular beds, induction of luteolysis in the ovary, induction of uterine contractions, relaxation of bronchial smooth muscle, and induction of fever at the hypothalamic level. Their diverse effects and precise roles in homeostasis remain to be determined.

The endorphins and the enkephalins, in addition to being synthesized as part of the prepro-opiocortin molecule by the AP, are also secreted by the *arcuate nucleus* of the hypothalamus. The large precursor molecule is also secreted by the arcuate nucleus. Various lines of evidence strongly support the concept that the endorphins, such as β-endorphin, modulate the sensory input for pain and have opiatelike effects that act via opiate receptors in the CNS. In addition, such peptides have been linked with appetite in that they induce an increase in food intake. They are also involved with thermoregulation because they induce hypothermia, and they influence reproduction because they influence testosterone feedback inhibition of LH secretion in the male and inhibit gonadotropin secretion and lead to menstrual dysfunction in the female. They affect responses to "flight or fight" situations because they act centrally to increase sympathetic outflow to the adrenal medulla and increase plasma epinephrine and norepinephrine levels. Their distribution throughout the brain is extensive, and they influence many neuroendocrine and neural events. They apparently have significant roles in both central and peripheral neurotransmission. The precise functional role(s) of these peptides is currently being extensively investigated.

BIBLIOGRAPHY

Baxter, J. D., and Funder, J. W. Hormone receptors. *N. Engl. J. Med.* 301:1149–1161, 1979.

Bergland, R. M., and Page, R. B. Pituitary-brain vascular relations: A new paradigm. *Science* 204:18–24, 1979.

Bloom, F. E. Neuropeptides. *Sci. Am.* 245:148–168, 1981.

Gauer, O. H., Henry, J. P. and Behn, C. The regulation of extracellular fluid volume. *Annu. Rev. Physiol.* 32:547–584, 1970.

Guillemin, R., and Burgus, R. Hormones of the hypothalamus. *Sci. Am.* 227:24–33, 1972.

Hays, R. M. Antidiuretic hormone. *N. Engl. J. Med.* 295:659–665, 1976.

Hayward, J. N. Functional and morphological aspects of hypothalamic neurons. *Physiol. Rev.* 57:574–658, 1977.

King, A. C., and Cuatrecasas, P. Peptide hormone–induced receptor mobility, aggregation and internalization. *N. Engl. J. Med.* 305:77–88, 1981.

Kreiger, D. T., and Martin, J. B. Brain peptides. *N. Engl. J. Med.* 304:876–885, 944–951, 1981.

Martin, J. B. Neural regulation of growth hormone secretion. *N. Engl. J. Med.* 288:1384–1393, 1973.

Moore, W. W. Antidiuretic levels in normal subjects. *Fed. Proc.* 30:1387–1394, 1971.

Pastin, I. Cyclic AMP. *Sci. Am.* 227:97–105, 1972.

Phillips, L. S., and Vassilopoulou-Sellin, R. Somatomedins. *N. Engl. J. Med.* 302:371–380, 438–445, 1980.

Segar, W. E., and Moore, W. W. Regulation of antidiuretic hormone release in man. *J. Clin. Invest.* 47:2143–2151, 1968.

Seif, S. M., and Robinson, A. G. Localization and release of neurophysins. *Annu. Rev. Physiol.* 40:345–376, 1978.

Tepperman, J. *Metabolic and Endocrine Physiology.* Chicago: Year Book, 1980.

Weitzman, E. D. Circadian rhythms and episodic hormone secretion in man. *Annu. Rev. Med.* 27:225–243, 1976.

Yalow, R. S. Radioimmunoassay: A probe for the fine structure of biologic systems. *Science* 200:1236–1245, 1978.

ENDOCRINE FUNCTIONS OF THE PANCREAS

31

Ward W. Moore

The relationship between the pancreas and altered carbohydrate metabolism was first shown by the classic experimental work of Von Mehring and Minkowski in 1889. They observed that removal of the pancreas resulted in the production of all the features of diabetes mellitus. This disease state has been known for centuries, and it is the most frequent and probably the most studied metabolic disorder.

The endocrine component of the pancreas, the islets of Langerhans, secretes three hormones, insulin, glucagon, and somatostatin, and each functions in the regulation of carbohydrate metabolism. The most prominent pancreatic hormone, insulin, was named and its major biological properties were elucidated before Banting and Best obtained the first stable insulin preparation in 1922. The net effects of insulin lead to the storage of potential fuel, because it promotes the synthesis of glycogen, triglycerides, and protein and favors their storage. In 1926, Murlin discovered that some of the extracts of pancreatic tissue also contained a substance that induces an increase in the plasma concentration of glucose and proposed the name glucagon for this factor. The net effects of glucagon lead to the mobilization of the primary energy sources, glucose and fatty acids, from their storage sites in the liver and adipose tissue. Somatostatin, described in the late 1960s, in addition to its effects on the anterior pituitary, appears to have direct inhibitory effects on both insulin and glucagon secretion and on the physiology of the gastrointestinal system.

ANATOMY

The pancreas is derived from gut endoderm and remains connected to the small intestine by means of the duct of Wirsung and the duct of Santorini. The glandular tissue appears in the third month of intrauterine life and develops as side buds from the ducts and duct branches. The acinar tissue remains connected to the duct system and becomes the exocrine portion of the pancreas. Small islets of cells proliferate from the ducts, differentiate, lose their connection with the duct system, and become the endocrine portion of the pancreas.

Langerhans first described small nests or islets of highly vascularized cells that are independent of the duct system of the pancreas. These islets maintain their functional integrity following ligation of the pancreatic ducts, whereas the acinar tissue becomes atrophic. The total volume of the islet tissue makes up about 1 to 3 percent of the entire pancreas. Three major cell types have been shown to be present in the islets of Langerhans, and they have been differentiated on the basis of their solubility and affinity for certain stains. They have been termed alpha, or A, cells, beta, or B, cells, and delta, or D, cells. The A cells, the source of glucagon, form an outer layer of the cortex of each islet and represent about 25 percent of the total islet-cell population. The B cells occupy the central portion of the islet, constitute about 65 percent of its mass, and secrete insulin. The D cells are interspersed between the A and B cells, form about 10 percent of the islet cells, and secrete somatostatin.

CHEMISTRY OF INSULIN AND GLUCAGON

In 1954, Sanger and associates established the number and sequence of amino acids (AA) for insulin and showed that it consists of two dissimilar polypeptide chains. One chain, chain A, contains 21 amino acids and a disulfide (-S-S-) bridge, and another, chain B, contains 30 amino acids. The chains are linked by two -S-S- bridges. Subsequently, it has been shown that insulin arises from two precursors, preproinsulin and proinsulin. Preproinsulin is synthesized in the endoplasmic reticulum of the B cell as a single polypeptide chain of over 100 residues. Proinsulin is formed after removal of a 16 amino acid fragment, the signal peptide, from the C-terminal of the polypeptide. The amino acids are arranged in a folded manner, so that there are three disulfide bridges within the molecule, two between the A and B chains and one within the A chain.

Proinsulin is a polypeptide chain containing a sequence of about 30 amino acid residues that is not present in insulin. In this peptide chain the connecting peptide (C-peptide) joins the carboxyl end of the B chain and the amino terminal of the A chain of the insulin molecule. The proinsulin is transferred to the Golgi complex, packaged in granules, and converted to insulin by proteolytic cleavage and removal of the C-peptide. The secretion of insulin is accomplished by the release of the granules by emiocytosis. At the time of secretion, all three peptides, proinsulin, insulin, and C-peptide, are released into the plasma. Insulin is the only one with significant biological activity, although C-peptide is released in an equimolar amount. The conversion of preproinsulin to proinsulin and then to insulin is depicted in Figure 31-1. The biological activity of insulin appears to be a function of the entire molecule rather than of specific groupings within the molecule, because potency is lost following slight alterations in the molecular structure. All the insulins cross species barriers with respect to their physiological activities. Insulin is readily inactivated either by acid hydrolysis or by proteolytic enzymes and hence is rendered ineffective in the gastrointestinal tract and must be administered parenterally.

Insulin is degraded rapidly in the body; less than 0.1 percent may be recovered in the urine following the administra-

Fig. 31-1. Conversion of preproinsulin to insulin. A = A chain of insulin molecule; B = B chain; C = connecting peptide, or C = peptide; D = the signal peptide; -S-S- = disulfide bridge.

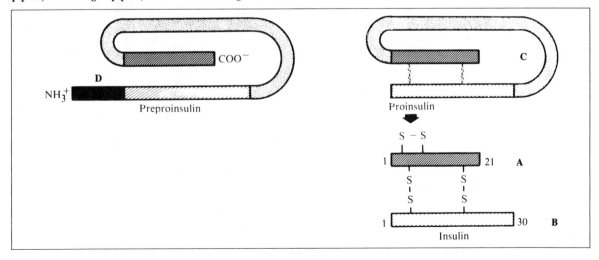

tion of large doses. The factor responsible for the hepatic inactivation and degradation of insulin is probably a series of proteolytic enzymes that has been termed the *insulinase system*. The hormone also becomes firmly bound to its receptors in the target tissues, such as liver, muscle, adipose tissue, and mammary gland.

The biosynthesis of glucagon, like that of insulin, involves two precursors, preproglucagon and proglucagon. Proglucagon has a molecular weight at least twice that of glucagon and is immunoreactive with glucagon, but lacks the biological activity of glucagon. It is synthesized in the endoplasmic reticulum, transported to the Golgi complex, and packaged into A cell granules. Its secretion is similar to that of insulin. Glucagon is composed of one long chain consisting of 29 amino acid residues, has a molecular weight of 3485, and has been synthesized in vitro. Like insulin, it is cleared by proteolytic enzymes, and the integrity of most of the molecule appears to be required for activity, because none of its degradation products have hyperglycemic activity.

Normal fasting plasma levels in humans have been determined by radioimmunoassay for both insulin and glucagon. The fasting plasma level of insulin is about 8 μU per milliliter, with a range of 5 to 20 μU per milliliter, and that of glucagon is approximately 75 pg per milliliter, with a range of 35 to 200 pg per milliliter. Insulin has been shown to have a half-life of about 30 minutes, whereas that of glucagon is about 10 minutes. C-peptide is also normally secreted into the plasma, and its fasting level approximates 1 pmole per milliliter. Very small amounts of both proinsulin and proglucagon may be found in the plasma.

METABOLIC INTERRELATIONSHIPS

The pancreas, through its hormones, plays a prominent role in the regulation of metabolism. Disturbed utilization and regulation of carbohydrate metabolism result from a relative or absolute deficiency in insulin. Similarly, an insulin deficiency causes marked defects in the synthesis, storage, and utilization of fats and proteins. Any essential fault in carbohydrate metabolism necessarily involves the metabolism of protein and fat as well, because the metabolic pathways through which the organism derives energy from food sources are known not to be separate and distinct, but intermingled. Carbohydrate is the active fuel of the body and is ordinarily the primary source of energy for the cell, but fatty acids can be utilized by several types of cells. Carbohydrates also contribute part of their substance to fatty acid and amino acid formation. Consequently, the proper utilization and formation of protein and fat are dependent on car-

bohydrate metabolism. The digestion and absorption of each of these foodstuffs has been discussed in Chapter 26.

CARBOHYDRATE METABOLISM. Following absorption of the nutrient sugars from the gastrointestinal tract, they are transported directly to the liver. This organ converts the sugars to a phosphorylated hexose and acts as a central clearing house for their disposition. The hexose phosphate that results from various transformations may be converted to glycogen (glycogenesis) and stored in the liver or it may be broken down (glycolysis) to intermediates that may be used either as sources of energy or in synthetic processes, or glucose may be released into the general circulation.

Most extrahepatic tissues (e.g., brain, muscle, adipose tissue) utilize glucose for their supply of energy. For such tissues to utilize glucose, it must permeate the cell membrane and be trapped by the cell. Immediately after entering a cell, glucose is phosphorylated to glucose 6-phosphate by hexokinase or glucokinase and thus trapped, because the reaction is irreversible, and the cell membrane is impermeable to sugar-phosphate esters. After glucose is trapped in the hepatic cell as glucose 6-phosphate, it may follow one of four primary avenues (Fig. 31-2): (1) It may be converted under the influence of glycogen synthase I to glycogen and stored; (2) it may be broken down via the hexose monophosphate shunt; (3) it may undergo anaerobic glycolysis to form lactic acid via the Embden-Meyerhof pathway; and (4) it may, under aerobic conditions, form pyruvate, which may be either decarboxylated and react with coenzyme A (Co A) to form acetyl Co A or be carboxylated to form malate. Therefore, pyruvate may serve as precursor to both ingredients necessary for the tricarboxylic acid (TCA), or Krebs, cycle. The TCA cycle is a major source of carbon dioxide and contributes to the production of high-energy phospate bonds via oxidative phosphorylation. Glucose 6-phosphate may also be dephosphorylated by glucose 6-phosphatase and released from the liver cell into the extracellular fluid (ECF) as free glucose.

The energy for biological work is made available from the degradation and transformation of various substances. Compounds such as adenosine triphosphate (ATP) that contain high-energy bonds are produced in the process of metabolism of energy-yielding substances (exergonic processes). The cleavage or hydrolysis of the high-energy bonds is then coupled with energy-requiring processes (endergonic) for biological work: synthetic and secretory processes, muscle contraction, and nerve conduction, and so on. Compounds such as ATP thus act as the currency that maintains the metabolic economy of the cells and of the

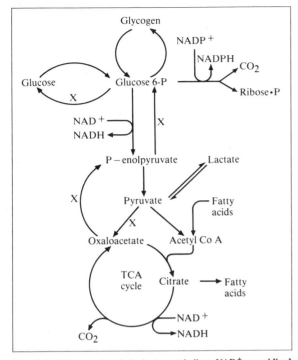

Fig. 31-2. Pathways of carbohydrate metabolism. NAD^+ = oxidized form of nicotinamide adenine dinucleotide; NADH = reduced form of nicotinamide adenine dinucleotide; $NADP^+$ = oxidized form of nicotinamide adenine dinucleotide phosphate; NADPH = reduced form of nicotinamide adenine dinucleotide phosphate; TCA cycle = tricarboxylic acid cycle; x = gluconeogenesis steps in liver.

organism. High-energy phosphate bonds may be generated in glycolysis, in the hexose monophosphate shunt, in the formation of acetyl Co A, and in the TCA cycle. Various monographs and biochemistry textbooks contain detailed maps of enzymatic function and biological oxidations.

Acetyl Co A plays an important role in intermediary metabolism because it is the compound that is common to carbohydrate, protein, and fat metabolism. Furthermore, through this compound the metabolic products of each metabolic pathway enter the common pathways that lead to oxidative processes furnishing the organism with energy. When acetyl Co A is oxidized to carbon dioxide and water, phosphorylation is coupled with oxidation, and 12 moles of ATP are formed per revolution of the TCA cycle (per mole of acetate utilized); 38 moles of ATP are formed by complete oxidation of 1 mole of glucose by way of the glycolytic and TCA pathways. This amounts to an energy storage of a considerable magnitude; 1 mole of ATP stores about 11,000 cals of free energy.

FAT METABOLISM. A small amount of fat is synthesized in the liver, but most of it is synthesized in adipose tissue. Because 1 gram of fat yields about 9 kcal as compared with 4 kcal per gram of protein or carbohydrate, the fat depots of the body constitute a concentrated store of energy. The adipose depots and their functional cells, the adipocytes, are metabolically extremely active. They furnish about 40 percent of the energy content in a normal person and represent the major source of stored energy.

The fatty acids synthesized in adipocytes derive their carbon from acetyl Co A and malonyl Co A. Lipogenesis, the synthesis of long-chain fatty acids, results from carboxylation of acetyl Co A with the formation of malonyl Co A. Repeated condensations of malonyl Co A plus decarboxylation and subsequent reduction in the presence of nicotinamide adenine dinucleotide phosphate (NADPH) lead to the formation of fatty acids. The free fatty acids can then combine with glycerol and form triglycerides. The degradation of fatty acids (lipolysis) proceeds stepwise by the removal of two carbons at a time, which results in the formation of acetyl Co A. Thus, fatty acids are broken down to acetyl Co A units and are eventually channeled into the TCA cycle. It is quite evident that any excess of caloric intake, as either carbohydrate, fat, or protein, can be channeled via acetyl Co A to result in increased lipogenesis, and that the bulk of storage fat depots must increase if the rate of lipogenesis exceeds that of lipolysis.

Acetyl Co A may also be converted into ketones instead of entering the TCA cycle. The circulating ketone bodies, acetoacetic acid, β-hydroxybutyric acid, and acetone, are formed in the liver from the condensation of two acetyl Co A molecules, and their production rate increases as the supply of fatty acids to the liver increases. Oxidation of the ketone bodies occurs in extrahepatic tissue, and their circulating levels in the plasma increase as their production exceeds their utilization (Fig. 31-3).

PROTEIN METABOLISM. The amino acids absorbed from the gastrointestinal tract either are used by the metabolic mill for the synthesis of protein or contribute their carbon chains for the synthesis of fatty acids, for glucose (gluconeogenesis), and for glycogen formation or as an energy source. Eight amino acids must be supplied in the human diet (none of these can be synthesized): isoleucine, leucine, lysine, methionine, phenylalanine, threonine, tryptophan, and valine. Other amino acids, such as alanine,

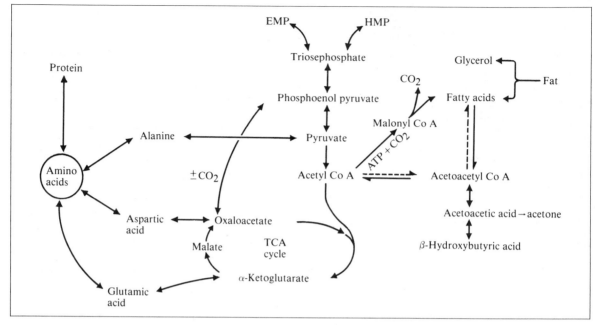

Fig. 31-3. Relationships of certain phases of lipid and protein with carbohydrate metabolism. ATP = adenosinetriphosphate; TCA cycle = tricarboxylic acid cycle; EMP = Emden-Meyerhof pathway; HMP = hexose monophosphate shunt.

glycine, glutamic acid, and aspartic acid, may be synthesized from specific intermediates of the glycolytic pathway or TCA cycle.

All or a portion of the carbon chain of an amino acid may be converted to glycogen or to ketone bodies. Amino acids that yield pyruvate are channeled into carbohydrate metabolism and classified as *glycogenic,* whereas those that form acetoacetyl Co A and contribute primarily to the reactions of fat metabolism are *ketogenic.* The great majority of amino acids are glycogenic.

The need for utilization of dietary protein during growth is great in humans, and there is a continuing and constant need for protein in order to maintain and repair cells throughout a person's entire life span. As is indicated in Figures 31-2 and 31-3, liver and muscle glycogen stores may be replaced indirectly with glucose derived from protein, and fat may be synthesized and stored in adipose tissue as a result of ingesting protein calories in excess of needs.

ROLE OF INSULIN IN METABOLISM. Numerous effects on the body have been attributed to the administration of insulin. However, many of these alterations in body functions are probably subsidiary to the effect of insulin on glucose transfer from the extracellular fluid (ECF) into certain cell types. Insulin facilitates the transfer of glucose across the

cell membrane and thus lowers the concentration of glucose in the plasma (hypoglycemia). Tissues such as skeletal and cardiac muscle, fibroblasts, the mammary gland, the anterior pituitary, and adipose tissue require insulin for the transfer of glucose across their cell membranes. The permeability of nervous tissue, erythrocytes, intestinal mucosal cells, renal tubules, and hepatic cells to glucose is independent of the action of insulin. Insulin also increases the rate of transfer of amino acids and fatty acids through the cell membranes, particularly in cardiac and skeletal muscle.

Insulin can be considered to be the dominant hormone in the "fed state" and is a potent anabolic agent. It promotes the synthesis of glycogen in both liver and muscle; it promotes the incorporation of amino acids into proteins in muscle, adipose tissue, and liver; it suppresses lipolysis and promotes lipogenesis in adipose tissue; and it stimulates ribonucleic acid (RNA) and deoxyribonucleic acid (DNA) synthesis.

Thus, overall, insulin promotes the storage of energy. It is also necessary for normal growth. It is the only hormone that produces hypoglycemia; all others induce hyperglycemia, an increase in the plasma glucose concentration.

EFFECTS OF PANCREATECTOMY. The many and varied effects of insulin on metabolism are best illustrated by the derangements in metabolism that occur in its absence in diabetes mellitus. Insulin deficiency brings about a decrease in the rate of translocation of glucose from the ECF into the insulin-sensitive cells, a decrease in glucose utilization by cells, and a decrease in glycogen formation. These effects, coupled with increased glycogenolysis and gluconeogenesis, lead to an increased blood glucose level. The oxidation of glucose via the hexose monophosphate shunt is decreased in relation to the glucose oxidized over the Embden-Meyerhof and TCA cycles in hepatic and adipose tissue. However, to gain entry into the cells where it can be utilized, the concentration of glucose in the ECF must be significantly elevated. When the degree of hyperglycemia reaches a critical level, the amount of glucose filtered at the renal glomerulus exceeds the tubular maximum (Tm) for glucose. The excessive filtered load of glucose acts as an osmotic diuretic, glucose spills over into the urine, and the urine volume increases markedly (polyuria). Large amounts of Na^+ and Cl^- are lost as a consequence of the polyuria. The ECF volume is reduced, and dehydration, thirst, and increased fluid intake (polydipsia) occur because of the large amounts of water and salt that are excreted in the urine. The decrease in carbohydrate utilization brings about an essential carbohydrate starvation, and, by an unknown mechanism, centers within the hypothalamus are affected so as to augment the food drive (see Chap. 8), and increased food intake (polyphagia) results in uncontrolled diabetes mellitus.

The rate of lipogenesis decreases to about 5 percent of normal. This decline in fatty acid synthesis may be caused by a deficiency in nicotinamide adenine dinucleotide phosphate (NADPH). However, the decreased lipogenesis may be related to a deficiency in end products of carbohydrate metabolism that are necessary for the formation of fatty acids. The rate of catabolism of fats is increased greatly in insulin deficiency, and as the fat stores are mobilized, the fat content of the plasma increases. With decreased carbohydrate utilization, most of the energy source of the diabetes must be derived from the fat of adipose tissue. In this state, lipolysis is increased to such an extent that acetyl Co A is produced in excess of the amount that can be handled by the TCA cycle, and acetoacetyl Co A tends to accumulate in the liver. The excess is channeled at an increasing rate to *ketone body* production. When the rate of production of ketone bodies exceeds the capacity of extrahepatic cells to oxidize them, *ketonemia* and *ketonuria* result.

The ketone bodies are acids and are buffered by the bicarbonate buffer system, and they markedly decrease the blood pH and consequently lead to an acidosis. Ketones are excreted mainly as sodium and potassium salts, so that the loss of base contributes further to acidosis. Compensatory reactions to the metabolic acidosis include hyperventilation and a resultant increase in loss of carbon dioxide via the lungs. This is triggered by stimulation of the respiratory centers by the lowered pH of the blood. Thus, the HCO_2^-/CO_2 ratio returns toward normal, and the pH is elevated. The respiratory changes may compensate for some of the base lost in mild diabetic ketoacidosis, and, along with renal compensation (increase in loss of fixed anions by excreting them with H^+ or NH_4^+), complete compensation may be attained and a normal pH restored. However, in severe insulin insufficiency, *uncompensated metabolic acidosis* occurs. The presence of an excess of ketone bodies depresses the oxygen consumption of the central nervous system and results in depression of its activity. If the depression becomes marked, mental disorientation and coma follow. Lack of insulin also effects protein metabolism. There is an increase in the rate of protein catabolism, and this, coupled with impaired protein synthesis, leads to negative nitrogen balance, impaired growth, delayed healing, and increased susceptibility to infections. The main source of the glucose (other than exogenous sources) excreted in the patient with uncontrolled diabetes is probably protein. For protein to be converted to glucose, it must be hydrolyzed to amino acids, which are deaminated and converted to keto acids. The latter may be converted to pyruvate or to ingredients of the TCA cycle. The glycolytic pathway must then be traversed in a reverse direction to form glucose. Because of the increase in protein catabolism and gluconeogenesis, increases occur in plasma urea, urinary nitrogen, and glucose excretion. These contribute further to the osmotic diuresis and the tendency to dehydration.

The effects observed in insulin deficiency appear to result from a diminution in the utilization of glucose for oxidative purposes, lipogenesis, and glycogenesis. The concentration of glucose in the ECF becomes elevated because the rate of transport of the hexose across the cell membrane of most tissues (muscle, adipose tissue) is reduced. A loss of large amounts of available glucose follows because the renal Tm for glucose is exceeded. The relative carbohydrate deficit precipitates lipolysis and proteolysis, which lead to metabolic events whereby the carbon chains of fatty acids and amino acids are utilized as sources of energy. Parts of the substances also may be lost in the urine as ketones, glucose, and nitrogen. The body economy of the uncontrolled diabetic patient resembles that produced by fasting, famine, or chronic starvation, and even though the diabetic patient may avoid acidotic crises, he or she may succumb eventually to depletion of the body tissues.

EFFECTS OF INSULIN. Insulin is bound at the cell membrane at specific receptor sites of its target cells. Insulin receptors are those molecules in the plasma membrane of cells that possess a high degree of affinity and specificity for binding insulin. The manner by which insulin increases the translocation of glucose and certain amino acids from the ECF into sensitive cells appears not to involve changes in enzymatic activity and remains to be elucidated. Current evidence indicates that insulin gains access to the cell interior by receptor-mediated endocytosis. Two membrane-bound enzyme systems have been implicated in the transmission of the insulin signal: the adenylate cyclase–cyclic AMP system and an Mg^{2+}-activated $(Na^+ + K^+)$ system. Insulin has been shown to suppress adenyl cyclase activity in many cells, promote changes in intracellular enzyme activities that depress lipolytic activity, and stimulate glycogenic activity. Insulin stimulates membrane adenosine triphosphatase activity and accentuates the accumulation of both K^+ and Mg^{2+} in the intracellular compartment of many cell types. It has been proposed that as Mg^{2+} accumulates at intracellular loci, it acts as a second messenger to influence intracellular enzyme function.

Insulin increases the rate of glucose uptake by muscle cells and adipose tissue and thus induces a hypoglycemia and makes more glucose available for utilization. Insulin decreases glucose 6-phosphatase activity and promotes the synthesis of glucokinase in the liver, thereby permitting more glucose 6-phosphate to enter the glycogenic and glycolytic pathways and allowing less free glucose transfer from the liver to the ECF. The rate of transformation of glycogen synthase from the D to L form, which accelerates the rate of glycogen synthesis, is also stimulated by insulin. In addition to increasing the synthesis of glucokinase, insulin increases the rate of synthesis of phosphofructokinase and pyruvate kinase. It increases hexose phosphate shunt activity, and more glucose is oxidized via this pathway. Because of the initial actions, the glucose concentration of the ECF is depressed, the amount of glucose filtered at the renal glomerulus becomes less than the Tm for glucose, glucosuria ceases, and the diuretic effect of glucose is lost. Thus, less glucose, water, and salt are lost from the ECF.

The effect of insulin on glucose assimilation in muscle is much more rapid than are its effects on the two hepatic enzymes glucokinase and glucose 6-phosphatase. However, the slower response of the two enzymes to insulin may serve as an important buffer against abrupt changes in blood glucose concentration. Severe hypoglycemia might be expected to be a consequence of an elevated insulin secretion that would follow the ingestion of a high-carbohydrate meal in the normal person if the activity of the two enzymes paralleled glucose assimilation by extrahepatic tissue.

The administration of insulin to the diabetic patient returns the rates of lipogenesis and lipolysis to normal levels. The rate of fatty acid synthesis is increased, because the greater assimilation of glucose increases the availability of the intermediates of carbohydrate metabolism that are utilized in fat synthesis. The rate of lipolysis is decreased, thus reducing ketone body production to a level at which extrahepatic tissue oxidation of these substances is equal to or less than their production. Therefore, ketonemia and the resultant ketonuria are inhibited. As the circulating ketone bodies return to normal, the alkaline reserve returns to normal, less base is lost via the urine, and electrolyte metabolism returns to normal.

The rates of protein synthesis and gluconeogenesis are returned to normal levels following administration of insulin to the diabetic patient. The stimulating effect of insulin on protein synthesis is dependent on adequate glucose metabolism, which supplies the intermediates and cofactors necessary for the process. Insulin also stimulates the transport of amino acids into cells and thus makes them available to enter into synthetic reactions. Decreased gluconeogenesis results in a decrease in urinary nitrogen excretion, and the contribution of protein to the formation of blood glucose declines.

All the metabolic alterations of the diabetic patient are readily reversed by insulin. These effects of insulin increase the utilization of glucose by peripheral tissue, promote the formation of hepatic and muscle glycogen, stimulate the synthesis of fat and protein, and decrease lipolysis and gluconeogenesis. The administration of excessive amounts of insulin or hypersecretion of insulin causes hypoglycemia. Because the central nervous system is largely dependent on blood glucose for its nutrition, it reacts as if it were deprived of oxygen (see Chap. 6).

REGULATION OF INSULIN SECRETION. The development of an accurate radioimmunoassay for plasma insulin has provided a major advance in the understanding of insulin secretion and diabetes mellitus. The major stimulus for insulin secretion by the B cells of the islets of Langerhans is glucose. Figure 31-4 depicts the effects of a carbohydrate meal on both the plasma insulin and glucagon concentrations in a normal subject. The amount of insulin secreted from isolated islet cell preparations in vitro can be increased or decreased by perfusing the preparations with fluid containing high or low glucose concentrations respectively. Several other sugars, as well as amino acids, fatty acids, beta keto

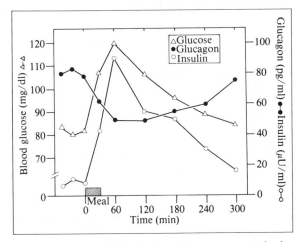

Fig. 31-4. Changes in plasma insulin and glucagon concentration in response to a carbohydrate meal.

acids, glucagon, and agents that generate islet cell cyclic adenosine monophosphate (AMP) also stimulate insulin secretion. Several drugs, such as alloxan and streptozocin, may specifically destroy the B cells of the islets and lead to diabetes mellitus.

Neurogenic mechanisms may also play a role in insulin secretion, because stimulation of vagal fibers to the pancreas results in insulin secretion. Similarly, acetylcholine and several cholinergic drugs promote insulin secretion. The alpha-adrenergic agents inhibit insulin secretion, and the beta-adrenergic agents stimulate it.

The intestinal hormones gastrin, secretin, and pancreozymin-cholecystokinin (PZ-CCK) each stimulate insulin secretion directly. Somatostatin has a profound inhibitory influence on both insulin and glucagon secretion, and it blocks the stimulatory influence of many agents on both the A cells and B cells. Its effects on its neighboring islets cells may be a direct one, and it has been hypothesized that there is a paracrine cell-to-cell interaction within each islet. Nevertheless, the plasma concentration of somatostatin is of great enough magnitude to exert an effect on the B cells.

The insulin level in the plasma may also play a role in the regulation of its own secretion, because if the B cells are perfused with insulin prior to or during glucose infusion, degranulation of the B cells is less than that observed following the infusion of glucose alone, in spite of the presence of marked hyperglycemia in both cases.

The endocrine portion of the pancreas is independent of direct control by the pituitary gland. However, the endo-

crine functions of the pancreas can be indirectly altered by several hormones. The anterior pituitary gland has indirect influences on insulin secretion. Carbohydrate metabolism is affected by growth hormone, adrenocorticotropic hormone, and thyroid-stimulating hormone. The effects of the latter two hormones on carbohydrate metabolism are indirect because their influence is mediated via the hormones of their respective target organs.

Houssay demonstrated that hypophysectomy reduces the effects of pancreatectomy in the dog. Furthermore, animals from which the hypophysis has been removed are hypoglycemic and extremely sensitive to the action of insulin. Growth hormone causes a decided aggravation of diabetes in the diabetic patient who has undergone hypophysectomy. In acromegaly and gigantism there is a strong tendency toward hyperglycemia and the diabetic state. The administration of growth hormone over prolonged periods produces degeneration of the B cells and results in the production of diabetes. The effect of growth hormone on insulin secretion is indirect and stems from the ability of the hormone to elevate blood glucose. Growth hormone decreases the peripheral utilization of glucose, increases the release of glucose from the liver, increases the release of a pancreatic hyperglycemic factor, and affects insulin binding to its receptors, thus leading to the diabetic state.

Adrenalectomy ameliorates the symptoms of diabetes, whereas the administration of cortisol-like compounds increases the severity of the symptoms of the disease. The diabetogenic effect of these hormones results from their causing a marked increase in gluconeogenesis, lipolysis, and ketogenesis and a decrease in the utilization of glucose. Therefore, their administration promotes the secretion of insulin because they produce hyperglycemia.

Diabetes may be produced by the administration of thyroid hormones in the animal whose pancreas has been partially removed. It results from B cell degeneration and subsequent decline in insulin secretion. Diabetes increases in severity in hyperthyroid states and decreases in hypothyroid states. The effects of thyroid hormones on carbohydrate metabolism include an acceleration in the rate of gluconeogenesis and increased glucose oxidation in tissue, both of which contribute to a decrease in the rate of insulin secretion.

Epinephrine is a hyperglycemic agent by virtue of its ability to increase the rate of hepatic and muscle glycogenolysis. The lactic acid produced as a result of muscle glycogenolysis is liberated into the circulation and used in the hepatic synthesis of glucose. Hypoglycemia is a potent stimulus to the sympathetic nervous system and increases catecholamine release. This might provide an important buffer mechanism against acute episodes of hypoglycemia. Epinephrine also increases the release of free fatty acids from adipose tissue. This ability is shared with thyroxine and cortisol, but insulin decreases the release of free fatty acids from adipose tissue. Many agents influence the release or action of insulin, or both. Some are capable of producing temporary or permanent diabetes by virtue of their ability to antagonize the action of insulin or to destroy the B cells, the source of insulin.

ROLE OF GLUCAGON IN METABOLISM. Glucagon, in contrast to insulin, is a catabolic agent and can be considered a dominant hormone in the fasted state. It acts to deplete stored glucose and fatty acids in liver and adipose tissue respectively and increases the supply of these energy sources for oxidation.

Glucagon has a rapid hyperglycemic effect. It is a powerful hepatic glycogenolytic agent and has a potent effect on gluconeogenesis. It also has a lipolytic action in that it stimulates lipase activity in both liver and adipose tissue. Thus, it stimulates free fatty acid release from the adipocyte and increases fatty acid uptake and oxidation in liver and muscle. Glucagon is a potent protein catabolic agent, and it increases the rate of uptake and deamination of amino acids in the liver. However, it decreases the rate of uptake of amino acids by muscle cells. The net outcome of the lipolytic and proteolytic effects of glucagon is to supply more fatty acids and amino acids for the gluconeogenic pathway. Glucagon, along with other hormones such as cortisol, appears to be an important hormone in stressful states (e.g., severe injury, famine, and "flight or fight" situations) in that it helps to maintain a ready supply of glucose, fatty acids, and even ketone bodies for energy. Changes in glucagon and insulin secretion rates appear effectively to combine and supply a steady flow of fuel to the appropriate tissues.

Glucagon stimulates insulin secretion by the B cells and somatostatin by the D cells. It increases the rate of secretion of epinephrine from chromaffin tissue in the adrenal medulla. It also has two important effects on the heart. They are direct and not mediated via catecholamine release at postganglionic sympathetic nerve endings. Glucagon acts directly on the sinoatrial node and causes an increase in heart rate. In addition to this chronotropic effect, glucagon also exerts a positive inotropic effect on the myocardium and papillary muscle. All of the actions of glucagon are mediated by cyclic AMP.

Control of Glucagon Secretion. The plasma glucose concentration is the primary regulator of glucagon secretion. A fall in

the concentration stimulates the secretion of glucagon, and a rise inhibits secretion. The response of the plasma glucagon concentration to plasma glucose levels in normal subjects is shown in Figure 31-4, which also depicts the relationship between circulating insulin and glucagon levels. There is a reciprocal relationship between the rates of secretion of glucagon and insulin, and both are regulated by the concentration of glucose perfusing the endocrine pancreas.

Glucagon secretion is also stimulated by several amino acids, the gastrointestinal hormones pancreozymin-cholecystokin and gastrin, beta-adrenergic agents, cortisol, and by exercise. Somatostatin is a potent inhibitor of glucagon secretion by the A cell. Insulin, secretin, alpha-adrenergic agents, free fatty acids, and ketones all inhibit glucagon secretion.

SOMATOSTATIN. Somatostatin, in addition to being secreted by hypothalamic neurons, is also secreted by the D cells of the islets of Langerhans. It has the identical tetradecapeptide structure of the somatostatin secreted by the hypothalamus. In addition to its role in the regulation of secretion of several hormones of the anterior pituitary, it exerts several actions on the endocrine pancreas, gastrointestinal tract, and exocrine pancreas. It appears to exert its action by preventing the generation of cyclic AMP in its target cells.

Somatostatin has been shown to reduce both basal insulin and glucagon levels in humans, and it also inhibits the normal insulin and glucagon response to a variety of stimuli. The effects on plasma insulin and glucagon stem from direct actions on the B cells and A cells respectively. Somatostatin decreases gastric acid and pepsin secretion and inhibits the secretion of pancreozymin-cholecystokinin, secretin (and thus decreases pancreatic bicarbonate and volume secretion), and gastric inhibitory peptide. It has been theorized that somatostatin is a paracrine hormone in the alimentary tract and coordinates endocrine and exocrine secretion.

It has been postulated that somatostatin, by virtue of its inhibitory effects on both insulin and glucagon secretion, has an important role in the regulation of metabolism, and that this role is made possible by paracrine cell-to-cell interactions with the pancreatic islets.

BIBLIOGRAPHY

Bray, G. A., and Campfield, L. A. Metabolic factors in control of energy stores. *Metabolism* 24:99–117, 1975.

Cahill, G. F., and McDevitt, H. O. Insulin-dependent diabetes mellitus. *N. Engl. J. Med.* 304:1454–1464, 1981.

Daughaday, W. H., Herington, A. C., and Phillips, L. S. The regulation of growth by endocrines. *Annu. Rev. Physiol.* 37:211–244, 1975.

Felig, P., and Wahren, J. Fuel homeostasis in exercise. *N. Engl. J. Med.* 293:1078–1084, 1975.

Hedeskou, C. J. Mechanism of glucose-induced insulin secretion. *Physiol. Rev.* 60:442–509, 1980.

Krahl, M. E. Endocrine function of the pancreas. *Annu. Rev. Physiol.* 36:331–360, 1974.

Lehninger, A. L. *Biochemistry*. New York: Worth, 1975.

Stryer, L. *Biochemistry* (2nd ed.). San Francisco: Freeman, 1981.

Unger, R. H., Dobbs, R. E., and Orci, L. Insulin, glucagon, and somatostatin secretion in the regulation of metabolism. *Annu. Rev. Physiol.* 40:307–343, 1978.

Unger, R. H., and Orci, L. Glucagon and the A cell. *N. Engl. J. Med.* 304:1518–1524 and 1575–1580, 1981.

HORMONAL CONTROL OF CALCIUM METABOLISM

32

Ward W. Moore

Several hormones are involved in the regulation of calcium and phosphorus metabolism. Two of these hormones, parathyroid hormone (PTH) and calcitonin (CT), have effects on the calcium activity of the extracellular fluid (ECF) and on bone. The parathyroid glands secrete PTH in response to a low serum calcium activity and induce a hypercalcemic action. Cells that are derivatives of the ultimobranchial bodies and are located in the thyroid gland secrete CT in response to hypercalcemia and induce a mild hypocalcemia. A third hormone, 1,25-dihydroxycholecalciferol [1,25(OH)$_2$D$_3$], is derived from vitamin D in the skin, liver, and kidneys. Like PTH, it induces a hypercalcemia. The three hormones act together on target tissues, which include bone, intestine, and kidney, to maintain a constant level of calcium and phosphate in the ECF and regulate bone metabolism. Other hormones, such as the estrogens, androgens, growth hormones, and cortisol, also effect calcium, phosphorus, and bone metabolism.

ANATOMY AND BIOCHEMISTRY

The parathyroid glands are small, paired bodies found in the region of the thyroid gland. In total, they weigh 20 to 40 mg. There are usually four glands, which develop from the third and fourth pairs of branchial pouches. The blood supply to the glands is through the inferior thyroid arteries, although the glands may receive some blood from the superior thyroid arteries. Each gland is surrounded by a connective tissue capsule, and septa divide the glands into lobules. Seen microscopically, the glands resemble hyperplastic thyroid tissue. Two types of epithelial cells are present: the chief·cells, the usual source of the hormone of the gland, and the oxyphil cells, which have the potential of secreting PTH and generally do not appear until after puberty. The functional significance of the oxyphil cells is obscure.

The ultimobranchial glands are derived from the terminal branchial pouch and in nonmammalian vertebrates appear as separate bodies. However, in mammals the cells become embedded in the thyroid, and they are referred to as the parafollicular or "light" or "C" cells. The thyroid "C" cells are the major source of CT in humans.

The hormone of the parathyroid glands, PTH, is a straight-chain polypeptide containing 84 amino acids. The biologically active portion of the molecule has a sequence of 34 amino acids, and this portion also has full immunological activity. Synthesis of PTH is via two successive cleavages of amino terminal sequences from a larger polypeptide of 115 amino acid residues, preproparathyroid hormone (pre-pro-PTH), in the chief cells. The first occurs in the endoplasmic reticulum, and removal of 25 amino acids forms pro-PTH (90 residues). Cleavage of pro-PTH prior to release in the circulation forms PTH. It is known that the hormone acts on three target tissues; namely, bone, kidney, and gastrointestinal tract. Calcitonin contains 32 amino acids residues and has a disulfide bridge between amino acids 1 and 7. The entire molecule is required for complete biological activity. Calcitonin has been shown to exert a slight effect on the kidneys and on bone. Both PTH and CT have been synthesized, and radioimmunoassays are available for both. Normal subjects with plasma calcium concentrations between 9 and 10 mg per 100 ml have a PTH plasma concentration between 200 and 800 pg per milliliter and a CT concentration of 50 to 150 pg per milliliter. Both hormones have relatively short half-lives (about 10 min).

The precursors of the most biologically active form of vitamin D, $1,25(OH)_2D_3$, are dietary in nature and include 7-dehydrocholesterol and cholecalciferol. Humans depend heavily on the ultraviolet rays of the sun to convert 7-dehydrocholesterol to vitamin D_3 in the skin. Vitamin D_3 is transported to the liver; bound to a specific binding globulin; and hydroxylated at carbon 25 to form 25-hydroxycholecalciferol $[25(OH)D_3]$. The $25(OH)D_3$ is released by the liver; bound to a plasma globulin, transported to the kidneys, and, under the influence of the enzyme $25(OH)D_3$-1-alpha-hydroxylase is hydroxylated at the 1 position to form $1,25(OH)_2D_3$. This enzyme is stimulated by PTH and is also facilitated by hypocalcemia and hypophosphatemia. Figure 32-1 illustrates the pathway of $1,25(OH)D_3$ biosynthesis. Estrogens, prolactin, and cortisol also stimulate renal $25(OH)D_3$-1-alpha-hydroxylase activity. The $1,25(OH)_2D_3$ exerts its effects on the small intestine and on bone.

FUNCTIONS OF CALCIUM AND PHOSPHORUS

Calcium is an indispensable mineral constituent of all body tissues and participates in several physiological processes, among which are coagulation of the blood, maintenance of cardiac rhythmicity, membrane permeability, neuromuscular excitability, secretion, and formation of bone and teeth. Primary attention here is focused on the last function. The other roles of calcium are discussed in preceding chapters.

The normal adult ingests about 1000 mg of calcium per day, 600 mg of which is absorbed in the upper gastrointestinal tract. Its absorption is facilitated by hydrochloric acid, $1,25(OH)_2D_3$, PTH, and protein. Excessive fats, excessive quantities of phosphate and oxalate, alkali, and chelating agents interfere with the intestinal absorption of calcium. Approximately 500 mg of calcium is secreted into the gastrointestinal tract, and about 900 mg per day is lost in the feces. A small amount of calcium, 100 mg per day, is excreted by the kidneys.

Following absorption, the greater part (about 99 percent of a total of 1000 gm) of the total body calcium is stored in the skeleton, providing an enormous reservoir of calcium that turns over very slowly. A small fraction of the total body calcium is present in other tissue: 10 gm in cells other than bone and about 1 gm in the ECF. The total plasma calcium (9–11 mg/100 ml, or 5 mEq/liter) is made up of two major fractions. The first is the protein-bound or nondiffusible fraction, which contributes about 40 percent of the total plasma calcium concentration. This fraction is bound mainly to plasma albumin, and the relationship between the calcium ion, Ca^{2+}, and the concentration of albumin in the plasma is represented by the following simple mass action expression:

$$K = (Ca^{2+})(\text{albumin})/(\text{calcium albuminate})$$

where K is the constant. The second, the nonprotein-bound or diffusible fraction, is ultrafiltrable and constitutes approximately 60 percent of the total plasma calcium. Roughly 95 percent of the diffusible fraction is in the form of Ca^{2+}, and the remaining 5 percent is in the form of complexed calcium. Consequently, the concentration of Ca^{2+} is about 50 percent (5 mg/100 ml) of the total plasma concentration. It is this concentration of Ca^{2+} that is exquisitely controlled by hormonal mechanisms.

Phosphorus plays an important role in biological systems because it is involved in the transfer of energy in the intermediary metabolism of foodstuffs, aids in maintenance of the pH of body fluids, and is an important constituent of bone. Phosphorus absorption from the gastrointestinal tract is facilitated by acids and an excess of fat and is decreased by high calcium and other cations and by alkaline salts. The average adult diet contains approximately 1000 mg of phosphorus per day, and about 70 percent of it is absorbed by the gut. The main avenue of phosphorus excretion is the

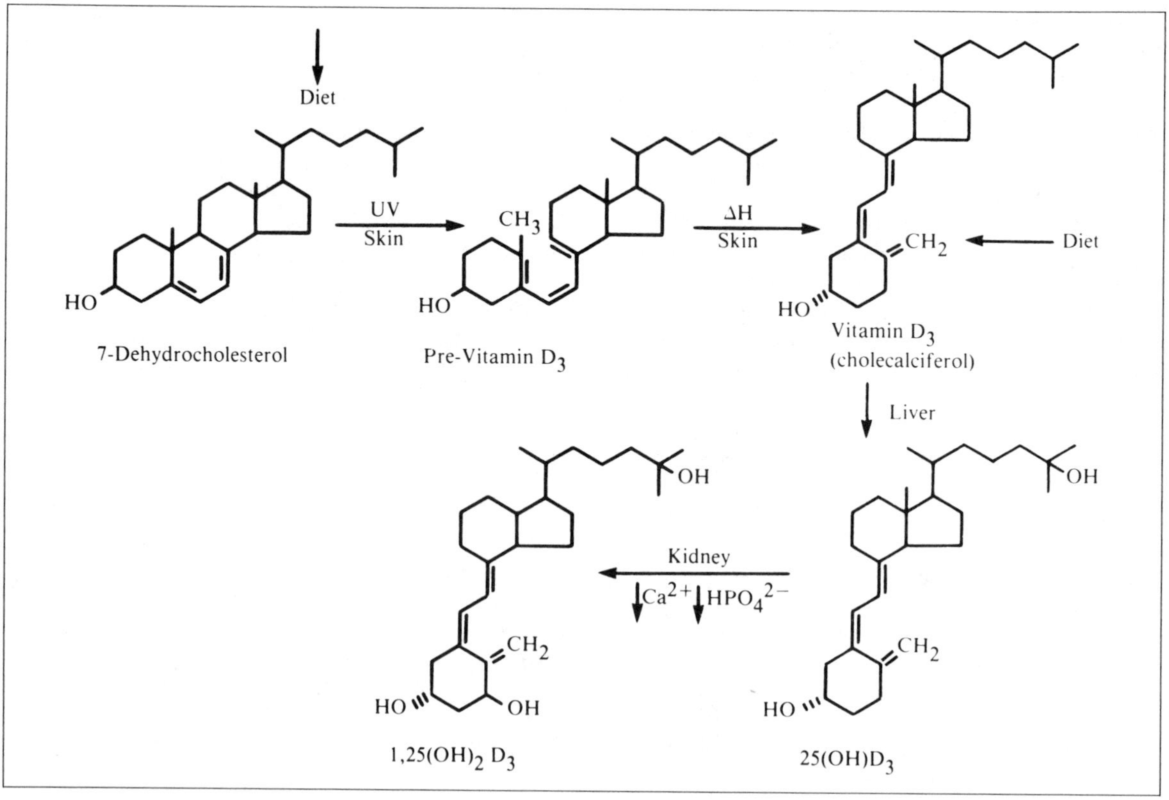

Fig. 32-1. Biosynthetic pathway in the metabolism of the primary active form of vitamin D, 1,25(OH)$_2$D$_3$. UV = ultraviolet. See the text for further explanation.

kidney, and some appears in the feces. A person in phosphorus balance loses about 350 mg in the stool each day, and 650 mg is excreted in the urine. About 80 percent of the total body phosphorus is in the skeleton, about 11 percent is in muscle, and the rest is present in the body fluids or distributed in other tissues as organic compounds. The total phosphorus present in the serum is about 12 mg/per 100 ml. This may be divided into three portions: lipid phosphorus (8 mg/100 ml), ester phosphorus (1 mg/100 ml), and inorganic phosphorus (3 mg/ml or 2 mEq/liter).

BONE

The manner in which hormones regulate the calcium and phosphorus levels of the ECF and participate in the control of skeletal growth requires a brief discussion of the development and maintenance of bone. Bone serves two primary functions. One is purely a mechanical function, in that it serves as an essential support structure and as a protective element for various organs and tissues of the body. To fulfill this function it must be strong, light, mobile, and capable of

orderly growth, response to stress, and repair. Second, bone serves a metabolic function in that it provides a large and ready supply of calcium, phosphorus, magnesium, sodium, and carbonate ions to the ECF and other tissue.

Bone is well vascularized and is metabolically active. It is constantly being remodeled throughout life by the processes of formation and resorption. During periods of bone growth the bone mass increases because bone formation exceeds bone resorption. Studies using the isotopes of calcium and phosphorus have shown that the skeletal calcium of a normal child is completely turned over approximately once each year, while calcium turnover occurs in an adult about once every 5 years. Bone, in addition to its ground substances, is composed of an organic matrix of connective tissue, collagen, onto which is deposited a complex salt of calcium and phosphate, hydroxyapatite [3Ca$_3$(PO$_4$)$_2$·Ca(OH)$_2$], and calcium carbonate.

Skeletal growth is initiated by the formation of a cartilage template that is subsequently mineralized and replaced by bone. Three cell types in bone are involved in bone formation or accretion and bone resorption: osteoblasts, osteoclasts, and osteocytes. The osteoblasts have an important function in bone formation and are located on the surface of bone that is being formed. Protein (collagen) synthesis is one of the major functions of this cell type. Acid, alkaline, and neutral phosphatases are all associated with the surface of the osteoblasts, which probably function in the nucleation and calcification of the organic matrix. The role of the osteoblasts appears to be essential in the formation of calcifiable fibers in a suitable ground substance and the alignment of the fibers to form the bone template. The events that initiate calcification of collagen are poorly understood. The mechanisms of nucleation probably involve an enzymatic phosphorylation of collagen. When sufficient concentrations of Ca^{2+} and HPO_4^{2-} are present in the ECF, each interacts with specific sites on the collagen fibers and forms hydroxyapatite crystals. These crystals grow and, during the process of growing, constitute the exchangeable mineral of bone, and an equilibrium is established between the Ca^{2+} and HPO_4^{2-} of the ECF and bone. However, as crystal formation increases, more water is excluded, and because the crystals do not release their ions readily, part of the bone structure becomes a relatively inert, solid, nondiffusible mass. About 99 percent of the mineral of bone is in this state, the so-called nonexchangeable mineral of bone. All bone would achieve this state if it were not constantly being remodeled.

The osteoclasts perform a role in bone resorption, and this includes the dissolving of bone mineral and the destruction of bone collagen. It is believed that bone resorption results from increased acid phosphatase activity, acid production, and proteolytic activity, which lead to the release of Ca^{2+} and HPO_4^{2-} into the ECF and the breakdown of collagen into its constituent amino acids. Hydroxyproline is one of the amino acids unique to collagen, and its rate of excretion in the urine is a reliable index of the rate of collagen breakdown. It has been estimated that about 500 mg of calcium from old bone is resorbed under the influence of the osteoclasts each day. Naturally, this old bone must be replaced by an equivalent amount of new bone formation, under osteoblastic activity, if equilibrium is to be attained. Conceptually, one may regard the ions of the ECF as being in simple equilibrium with those of exchangeable bone. Thus, the exchangeable bone acts as a buffer in controlling the concentration of Ca^{2+} and HPO_4^{2-} in the internal environment.

The osteocyte is a bone cell that is surrounded by calcified matrix and is required for the homeostatic regulation of bone metabolism. The osteocytes are capable of promoting considerable bone resorption as well as indirectly participating in bone formation and are probabaly very important in the rapid mobilization of calcium.

EFFECTS OF PARATHYROID HORMONE

Parathyroid hormone has three primary sites of action: the skeleton, kidney, and gastrointestinal tract. Its actions on each of its target tissues lead to an increase in the concentration of Ca^{2+} in the ECF, and thus it protects the organism from hypocalcemia. In addition, PTH permits the extensive remodeling of bone while maintaining a normal level of calcium ion in the plasma. It is a very potent hypercalcemic agent. The effects of PTH are mediated by stimulation of adenylate cyclase and accumulation of cyclic adenosine monophosphate in its target tissue.

The administration of PTH or its excessive secretion in parathyroid hyperplasia causes demineralization of bone and a consequent softening of the skeletal system. The effect of PTH is primarily on the osteoclastic resorption of bone. It also inhibits osteoblastic collagen synthesis, and this is due largely to a decrease in collagen synthesis, not to an increase in the breakdown of collagen.

Prolonged exposure of bone to high levels of PTH increases osteoblastic activity and local bone growth. However, this local anabolic effect of PTH is probably not direct and probably reflects a coupling of increased bone formation to the increased bone resorption. The net effects of the hormone are the removal of calcium from bone, its transfer to the ECF, the induction of hypercalcemia, a decrease in collagen formation, and increased hydroxyproline in the ECF. The elevated plasma Ca^{2+} level then usually leads to increases in urinary calcium, phosphate, and volume, a tendency toward increased coagulability of blood, and a decrease in the excitability of nerve and muscle cells.

That a primary action of the hormone is in bone is shown by the following observations: First, transplantation of parathyroid tissue to membranous bone of the skull leads to bone resorption at the site of contact with the transplant, while bone deposition occurs on the opposite surface. Second, the addition of parathyroid tissue or PTH to bone grown in tissue culture leads to increased osteoclast activity and bone resorption. Third, a prompt fall in the plasma calcium level occurs following parathyroidectomy in the anephric animal. These observations indicate that PTH, through its effects on bone cells, increases the rate of bone resorption. Parathyroid hormone induces a variety of changes in osteoclasts and osteocytes: increases in numbers, enhanced lysosomal activity, enhanced collagenase activity, and enhanced organic acid formation. These activities lead to the destruction of bone matrix. The concentration of hydroxyproline in plasma and urine is a reliable indicator of the rate of destruction of the collagen portion of the matrix. The greater the concentrations, the greater the rate of breakdown of collagen.

Parathyroid hormone has three actions on the kidney: It induces phosphaturia, decreases calcium excretion, and enhances $25(OH)D_3$-1-alpha-hydroxylase activity, thus promoting $1,25(OH)_2D_3$ synthesis. The hormone acts on the renal proximal tubule to reduce the reabsorption of phosphate. This effect of PTH involves a parallel inhibition of sodium and bicarbonate reabsorption in the proximal tubule. The decrease in the clearance of calcium induced by PTH probably results from the action of the hormone on the distal nephron. Many factors other than PTH and CT can affect the renal handling of both HPO_4^{2-} and Ca^{2+}. They include dietary intake, plasma concentrations, filtered loads, the renal handling of sodium, adrenal steroids, and growth hormone among others.

The outstanding metabolic changes that follow parathyroidectomy are hypocalcemia, hyperphosphatemia, and hypocalciuria. As a result of the fall in plasma Ca^{2+}, symptoms attributed to increased neural and muscular excitability are evident. However, the initial change observed following parathyroidectomy is an increase in renal calcium excretion; the subsequent decrease occurs only after a significant fall in plasma calcium.

Parathyroid hormone acts to increase the Ca^{2+} concentration in ECF. Four of its actions are direct: (1) enhanced movement of skeletal Ca^{2+} to plasma; (2) enhanced reabsorption of Ca^{2+} from renal tubular fluid; (3) enhanced activity of renal $25(OH)D_3$-1-alpha-hydroxylase; and (4) decreased reabsorption of inorganic phosphate from renal tubular fluid. Another effect, enhanced intestinal Ca^{2+} absorption, is indirect and mediated by vitamin D_3. Figure 32-2 illustrates the effects of PTH on its target organs.

EFFECTS OF CALCITONIN

Calcitonin, in contrast to PTH, tends to protect the organism from hypercalcemia, and it reduces the concentration of both calcium and phosphate in the plasma. Calcitonin acts on two primary target tissues, bone and kidney, and induces its effects by increasing the removal of Ca^{2+} and HPO_4^{2-} from the ECF, or by decreasing the rate of entry of the ions into the ECF, or both. Its effects are mediated by cyclic adenosine monophosphate.

Calcitonin inhibits the osteoclastic resorption of bone. As a result of its effects on the osteoclast it reduces the net rate of movement of calcium from bone to ECF and induces hypocalcemia. In addition, the rate of collagen destruction is decreased, and the plasma concentration and urinary excretion of hydroxyproline are depressed. In vitro studies have shown that this is a direct effect on bone. As previously noted, PTH added to bone in tissue culture leads to a decrease in osteoblast activity and an increase in osteoclast activity, hydroxyproline release, and demineralization.

Fig. 32-2. Regulation of PTH secretion and the effects of PTH and $1,25(OH)_2D_3$ on their target organs and the level of Ca^{2+} in the ECF. Both mineralization and demineralization of bone is promoted by $1,25(OH)_2D_3$, which may therefore add Ca^{2+} to, or remove it from, the ECF Ca^{2+} pool.

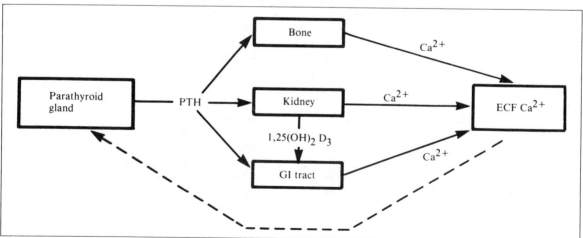

However, CT added to PTH-stimulated bone in vitro has no effect on osteoblast activity but reduces osteoclast activity and causes a decrease in the release of calcium. In vivo studies on the rate of removal of calcium from PTH-sensitive pools, the urinary excretion of hydroxyproline, and the urinary excretion of ^{85}Sr and ^{45}Ca following CT administration indicate that the primary effect of CT, if any, is to prevent bone resorption. Some evidence indicates that CT may decrease the rate of bone formation or accretion.

Calcitonin reduces the renal reabsorption of both calcium and phosphate, and the increase in the renal clearances of these ions leads to hypocalcemia and hypophosphatemia. Micropuncture studies indicate that CT inhibits the resorption of phosphate in the proximal tubule and allows increased loss of phosphate in the urine. With the use of isolated kidney tubules, it has been shown that CT inhibits the extrusion of calcium from the cell. In contrast, PTH increases the influx of calcium into renal cells. Thus, the net effect of the two hormones, PTH and CT, on the renal handling of calcium and phosphate is the same. Calcitonin also acts to reduce the reabsorption of sodium, magnesium, and potassium by cells of the renal proximal tubules, and it also induces a saline diuresis and losses of ECF volume and body weight. These losses of sodium and volume result in increase in renin and aldosterone secretion by the kidney and adrenal cortex respectively.

The precise role of CT in human physiology remains to be determined. It has been termed by many to be a hormone searching for a function.

EFFECTS OF VITAMIN D

Vitamin D, like PTH, is a hormone that provides protection against hypocalcemia. It acts at two primary sites, bone and intestine, and is necessary, along with PTH, for normal skeletal mineralization. In all likelihood, vitamin D—or its active form, $1,25(OH)_2D_3$—acts on receptors in the cytosol of its target cells, and its actions are mediated in a manner similar to that of other steroid hormones.

Vitamin D_3 stimulates the mobilization of bone mineral to increase the supply of Ca^{2+} and HPO_4^{2-} to the plasma. It stimulates osteoclastic bone resorption and inhibits osteoblastic bone formation. However, in severe vitamin D deficiency, both bone matrix synthesis and bone mineralization are depressed. It appears that vitamin D also leads to mineral deposition within the skeleton, and it is the lack of this effect that represents the most overt manifestation of vitamin D deficiency in humans. Both mineralization and demineralization are promoted by vitamin D.

Vitamin D, through its most biologically active metabolite $1,25(OH)_2D_3$ stimulates the movement of Ca^{2+} against an electrochemical gradient in the gut and promotes its absorption. Phosphate absorption also accompanies the Ca^{2+} translocation, but $1,25(OH)_2D_3$ also directly stimulates a HPO_4^{2-} transport system that is independent of the calcium transport system. Thus, the net results of this hormone on the gut is to elevate both the plasma Ca^{2+} and HPO_4^{2-} to permit normal skeletal formation as well as making these ions available for their other normal physiological activities.

REGULATION OF SECRETION OF PARATHYROID HORMONE, CALCITONIN, AND VITAMIN D

The rates of secretion of PTH and CT are independent of the anterior pituitary and the central nervous system. Their secretion rates are controlled primarily by the Ca^{2+} concentration in the plasma. Studies involving perfusion of isolated glands have revealed that the rate of secretion of either hormone can be altered by increasing or decreasing the Ca^{2+} concentration of the perfusate. The secretion rates of PTH and CT are reciprocally related.

There is a simple inverse linear relationship between plasma Ca^{2+} and plasma PTH. Studies conducted over a range of plasma Ca^{2+} of 4 to 12 mg/per 100 ml (2–6 mEq/liter L) showed that Ca^{2+} regulates PTH secretion through a proportional control mechanism. When the concentration of Ca^{2+} rises, the parathyroid glands are inhibited and secrete less hormone, and when the Ca^{2+} falls, the glands are stimulated to secrete more hormone. Similarly, high-calcium diets lead to parathyroid hypoplasia, and low-calcium diets result in hypertrophy and hyperplasia of parathyroid glands. Thus, there is a negative feedback system between the parathyroid glands and circulating Ca^{2+} level (Fig. 32-2). Following a depression of Ca^{2+} levels, PTH is secreted, and it acts on the renal cells to increase 25(OH)-1-alpha-hydroxylase activity, and the resultant $1,25(OH)_2D_3$ promotes the intestinal absorption of Ca^{2+}, acts on the kidney tubules to increase reabsorption of Ca^{2+} and loss of HPO_4^{2-} from the glomerular filtrate, and acts on bone cells to increase osteoclastic resorption of bone, leading to a release of Ca^{2+} from bone. The hormones also increases the rate of phosphate released by bone, which is offset by increased urinary excretion of phosphate. The final result is an increased plasma Ca^{2+} and decreased HPO_4^{2-}. When the plasma Ca^{2+} concentration reaches a critical level, the parathyroid glands are inhibited, PTH secretion decreases, and the changes are reversed. Adenylate cyclase and the accumulation of cyclic adenosine triphosphate in the chief cells of the parathyroid gland appear to be the common denominator and the mechanism through which all agents, including hypocalcemia, stimulate PTH secretion. Other agents, such as epinephrine, may stimulate PTH secretion. Studies indicate that substances such as $1,25(OH)_2D_3$ or its metabolites feedback on the parathyroids and inhibit PTH secretion. Various other hormones indirectly stimulate PTH secretion. These include CT, cortisol, and growth hormones.

Parathyroid hormone is responsible for the hour-to-hour regulation of Ca^{2+} concentration in plasma. Studies utilizing isotopic calcium have shown that an amount of calcium equivalent to the total blood calcium is replaced every minute in young animals. Two mechanisms are responsible for this rapid turnover. The faster mechanism is ion exchange between blood and the exchangeable mineral of bone and is independent of hormonal action. As a result of this exchange the serum calcium concentration rarely falls below 6 mg per 100 ml, even after parathyroidectomy. The second mechanism is primarily a function of PTH and involves a calcium-mobilizing effect of the hormone on bone. This effect maintains the serum calcium level at about 10 mg per 100 ml. In addition, an increased release of HPO_4^{2-} into plasma occurs. If the effect of PTH on bone were the only means of regulating the Ca^{2+} level of the serum, the feedback system would lead to wide oscillations in the Ca^{2+} level of the serum, for this effect is relatively slow in onset. The indirect effect [via $1,25(OH)_2D_3$] of PTH on gastrointestinal tract absorption of calcium is more rapid than that on bone and tends to produce the oscillations in the plasma Ca^{2+} level. However, the kidney is capable of responding rapidly to PTH and tends to maintain the level of serum Ca^{2+} within narrow limits. It also facilitates the excretion of HPO_4^{2-} by inhibiting its tubular resorption of HPO_4^{2-} and offsets its rise, produced by resorption of bone.

A similar relationship exists between plasma Ca^{2+}, CT secretion, and plasma CT, except that the linear relationship is direct. It holds for Ca^{2+} levels between 8 and 20 mg per 100 ml (4–10 mEq/liter). Thus, an increase in plasma calcium above 8 mg per 100 ml increases CT secretion, and the increase is in direct proportion to the plasma Ca^{2+}. A negative feedback system also operates between CT secretion and the circulating Ca^{2+}. Following elevation of Ca^{2+} levels, CT is secreted and it acts to remove Ca^{2+} from the ECF or prevent the addition of Ca^{2+} to the ECF via its effects on bone and kidney. Calcitonin secretion is stimulated following a meal, an effect probably mediated by the gastrointestinal hormones gastrin, pancreozymin-cholecystokinin, and enteroglucagon. This effect may protect the skeleton from excessive bone resorption during periods of dietary sufficiency.

The level of $1,25(OH)_2D_3$ in the plasma is affected by several factors. One is the presence of its precursors in the diet and their absorption from the gastrointestintal tract. Any situation that leads to malabsorption of its two precursor steroids will lead to a deficiency of $1,25(OH)_2D_3$. Similarly, in the absence of ultraviolet (UV) rays from the sun, the skin cannot complete the photometabolic conversion of 7-dehydrocholesterol to pre–vitamin D_3, and a deficiency of $1,25(OH)_2D_3$ will occur. Other factors that are necessary for the complete conversion of vitamin D_3 to the metabolically active hormone $1,25(OH)_2D_3$, include hypocalcemia, hypophosphatemia, PTH, prolactin, and the estrogens.

The three hormones PTH, CT, and $1,25(OH)_2D_3$ exert precise homeostatic control over the plasma Ca^{2+}. Parathyroid hormone and $1,25(OH)_2D_3$ prevent or protect the organism from hypocalcemia, while CT prevents or protects against hypercalcemia. Parathyroid hormone and CT participate in the processes of bone remodeling, with PTH promoting the osteoclastic resorption of bone and CT promoting osteoblastic formation. The paramount function of $1,25(OH)_2D_3$ is to permit normal skeletal mineralization. Its action in causing the dissolution of bone serves to provide Ca^{2+} and HPO_4^{2-} for the accretion of mineral in new bone.

EFFECTS OF OTHER HORMONES

Several hormones other than PTH, CT, and $1,25(OH)_2D_3$ influence bone formation and Ca^{2+} and HPO_4^{2-} metabolism. A major effect of growth hormone, acting via the somatomedins, promotes cartilage cell proliferation and synthesis of bone matrix, stimulates osteoblastic function, and leads to an increase in linear bone growth while the epiphyses are open. It increases bone formation after the epiphyses close. Growth hormone also decreases the renal loss of HPO_4^{2-} but increases Ca^{2+} loss. Because of the effect of growth hormone on the renal handling of Ca^{2+}, there is a tendency for it to cause hypocalcemia. This leads to an increase in PTH secretion and thus to an increase in the rate at which bone is remodeled.

Insulin has effects on bone that are similar to those of growth hormone. These effects may be partly through its stimulatory effect on hepatic somatomedin production. However, insulin receptors are present in cartilage, and, in vitro, insulin has been shown to stimulate both cartilage and bone growth directly.

The thyroid hormones are necessary for normal bone differentiation and development. They increase the osteoclastic resorption of bone directly and enhance the rates of turnover and remodeling of bone. In hyperthyroid states in children, linear bone growth is increased, while in hypothyroid states, bone growth and maturation are subnormal.

At puberty the estrogens and androgens promote bone formation, mineralization, and acceleration of linear growth and epiphyseal closure. They are necessary for a normal nitrogen, Ca^{2+}, and HPO_4^{2-} content of bone in the adult; in their absence a low bone mass is present. In women at menopause, a time during which estrogen secretion is low or even absent, there is a loss of bone mass, which results from an imbalance between bone formation and bone resorption. The imbalance results primarily from increased resorption rather than a decrease in the formation of bone.

Cortisol, the major hormone of the adrenal cortex, has a marked effect on bone and inhibits its growth. This results from a decrease in the formation of osteoblasts and the effect of cortisol on protein catabolism. Cortisol stimulates the rate of protein breakdown, decreases the rate of collagen synthesis, and leads to a decrease in the total mass of calcified bone. It also decreases the rate of absorption of Ca^{2+} from the gut and increases the rate of urinary Ca^{2+} excretion. These changes lead to hypocalcemia and promote PTH secretion.

In summary: Negative feedback mechanisms for the regulation of plasma Ca^{2+} and HPO_4^{2-} concentration involve the parathyroids, cells derived from the ultimobranchial body, skin, bone, gastrointestinal tract, and kidneys. The renal regulator is rapid in response, but has limited capacity. The gastrointestinal regulator is slower to respond than is the kidney and also has limited capacity. On the other hand, the bone regulator is slow to respond to PTH and $1,25(OH)_2D_3$, but responds rapidly to CT. Bone has an unlimited capacity for responding to the three hormones. Other hormones also play significant roles in bone metabolism and calcium and phosphorus homeostasis.

BIBLIOGRAPHY

Arnaud, C. D. Calcium homeostasis: regulatory elements and their integration. *Fed. Proc.* 37:2557–2560, 1978.

Austin, L. A., and Heath, H., III. Calcitonin: Physiology and pathophysiology. *N. Engl. J. Med.* 304:269–278, 1981.

Canelis, E. The hormonal regulation of bone formation. *Endocr. Rev.* 4:62–77, 1983.

DeLuca, H. F. Recent advances in the metabolism of vitamin D. *Annu. Rev. Physiol.* 43:199–209, 1981.

Fraser, D. R. Regulation of vitamin D metabolism. *Physiol. Rev.* 60:551–613, 1980.

Habener, J. F. Regulation of parathyroid hormone secretion and biosynthesis. *Annu. Rev. Physiol.* 43:211–223, 1981.

Habener, J. F., and Potts, J. T., Jr. Biosynthesis of parathyroid hormone. *N. Engl. J. Med.* 299:580–585, 635–644, 1978.

Haussler, M. R., and McCain, T. A. Basic and clinical concepts related to vitamin D metabolism and action. *N. Engl. J. Med.* 297:974–983, 1041–1050, 1977.

Holick, M. F., and Clark, M. B. The photobiogenesis and metabolism of vitamin D. *Fed. Proc.* 37:2567–2574, 1978.

Martin, K. J., Hruska, K. A., Freitag, J. J., Klahr, S., and Slatopolsky, E. The peripheral metabolism of parathyroid hormone. *N. Engl. J. Med.* 301:1092–1098, 1979.

Raisz, L. G., and Kream, B. E. The regulation of bone formation. *N. Engl. J. Med.* 309:29–35 and 83–89, 1983.

THYROID PHYSIOLOGY

33

Ward W. Moore

FUNCTIONS OF THE THYROID GLAND

The thyroid gland plays an important part in a homeostatic control system that aids in maintaining an optimal level of oxidative metabolism and heat production in the organism. Other parts of this regulatory mechanism include the central nervous system, anterior pituitary, general circulation and certain plasma proteins, and the metabolic machinery of all the cells of the body. The thyroid gland also has a vital role in the growth, differentiation, and development of the individual. In its role, the thyroid gland excels in three characteristic functions: (1) trapping of iodide, (2) synthesis of organic iodine and the formation of iodotyrosines and iodothyronines, and (3) storage and secretion of the iodothyronines. The growth and overall function of the thyroid gland are regulated by the adenohypophyseal hormone *thyroid-stimulating hormone* (TSH), and the secretion rate of TSH is regulated by the hypothalamic hormone *thyrotropin-releasing hormone* (TRH). Two hormones, *3,5,3'5'-tetraiodothyronine* (thyroxine, T_4) and *3,5,3'-triiodo-thyronine* (T_3) are the primary secretory products of the thyroid gland. They produce all of the hormonal effects attributed to the thyroid gland.

ANATOMY

The human thyroid first appears in the 3-week-old embryo as a ventral diverticulum from the midventral floor of the pharynx between the first and second pharyngeal pouches. By the fifth week a small, hollow sac has developed that remains joined to the pharynx by the thyroglossal duct. The latter atrophies during the sixth week, and with this change the thyroid loses its central cavity and assumes a bilobed form. Discontinuous cavities become manifest within the gland by the eighth week. These represent the beginnings of the follicles of the adult gland. The follicles soon become filled with colloid substance.

The human thyroid is made up of two lobes linked together by a thin band of tissue, the isthmus. Normally, the lobes are closely attached to the lateral aspect of the trachea, and the isthmus fits over the anterior surface of the second and third cartilaginous rings. The adult gland weighs about 20 to 30 gm and possesses an enormous potential for growth in goiters weighing up to several hundred grams. The thyroid gland receives its blood supply from paired superior and inferior thyroid arteries arising from the external carotids and subclavians respectively . The rate of blood flow through the gland is very high (5–7 ml/gm/min). The gland is innervated by laryngeal and pharyngeal branches of the vagus and by cervical sympathetic ganglia. These fibers are vasomotor, and the transplanted or denervated gland can function normally.

The characteristic and dominant structural feature of the adult thyroid gland is the follicle or acinus, which is lined by epithelium and filled with colloid. The colloid is a viscid, homogeneous substance and clear in appearance in the fresh state. The follicular epithelium consists of two cell types, the follicular epithelium and the C cells. The latter were discusseed in the preceding chapter and are the source of calcitonin. The follicular or lining epithelium rests on a basement membrane that is surrounded by a rich capillary network. The normal epithelium varies in appearance with the state of activity of the gland; generally speaking, the height of the cells is low when the gland is resting and high when it is active. The ultrastructure of the follicular cells is similar to that observed in other secretory or absorptive cells. The basal portion of the cells shows an extensive endoplasmic reticulum that contains wide, irregular tubules, and numerous microvilli are present on the apical portion of the cell and extend into the colloid. The active cell is characterized by an enlarged Golgi apparatus, pseudopodia formation at the apical surface, the appearance of colloid droplets in the apical portion of the cell, and increased numbers of lysosomes. In addition, the active gland typically displays an increase in follicular cell height, a decrease in the amount of colloid, and increased vascularity. The cells lining the follicle have the unique capacity both to release their secretions into the follicular lumen and to permit the passage of the active principles of the cells from colloid to bloodstream.

SYNTHESIS AND STORAGE OF THYROGLOBULIN

It has been shown that the biosynthetic mechanisms involved in the formation of hormones are autonomous. Nevertheless, the rate of formation of the hormones is regulated by extrathyroid factors. The most important factors are TSH, which is secreted by the anterior pituitary, and the availability of iodide. However, intrathyroid mechanisms also regulate hormonogenesis. Several key steps are involved in the synthesis, storage, and secretion of thyroid hormones (Fig. 33-1).

Iodine is an essential element in the synthesis of the thyroid hormones. The primary source of iodine is dietary, and the average daily intake of an adult approximates 500 μg per day. A minimal intake of 100 to 150 μg per day is required for normal thyroid function. Most of the iodine lost from the body is in the form of iodide (I^-), but a small portion is lost as T_4 and T_3 and their metabolites in the urine, feces, and sweat. A person in iodine balance excretes about 475 μg of iodide in the urine per day, and about 25 μg is lost in the stool. Iodide is filtered at the glomerulus, and the filtered iodide is largely reabsorbed by the tubules. Renal iodide clearance is normally about 35 ml per minute. The thyroid hormones and their metabolites are present in bile as glucuronides, and as they pass down the small intestine, part is reabsorbed, with the remainder appearing in the feces. Most of the fecal iodine is in the form of organic iodine. Roughly 25 per cent of the ingested iodine enters the thyroid per day; of this, about 80 μg is secreted as T_3 and T_4, and about 40 μg is returned to the extracellular fluid each day. The plasma concentration of iodide is about 1.0 μg/100 ml in a normal person with a normal iodine intake.

The thyroid hormones are derived from their prohormone thyroglobulin (TG). Human TG is an iodine-containing glycoprotein with a molecular weight of 670,000. It contains about 350 residues of carbohydrate (galactose, mannose, N-acetylglucosamine, sialic acid, and fucose) per mole, and these contribute about 9.7 percent of the total weight of the compound. In addition to containing about 5800 residues of amino acids per mole, TG contains five types of iodoamino acids: 3 monoiodotyrosine (MIT), 3,5-diiodotyrosine (DIT), 3,5,3'-triiodothyronine (T_3), 3,3',5'-triiodothyronine (reverse T_3 or RT_3), and 3,5,3',5'-tetraiodothyronine (thyroxine or T_4). The structures of these compounds are shown in Figure 33-2.

The synthesis of TG occurs sequentially in three independent stages. The amino acids enter the basal portion of the thyroid cell and are assembled into polypeptide chains in the rough endoplasmic reticulum. As the protein migrates from the basal to the apical portion of the cell, various carbohydrate moieties are added in the endoplasmic reticulum and Golgi apparatus. Iodination of tyrosyl groups and the formation of the iodothyronines T_3 and T_4 occur within the matrix of the thyroglobulin molecule near the apical portion of the cell after the incorporation of the amino acids and carbohydrates into the glycoprotein has been completed. The total iodine content of thyroglobulin in general reflects the physiological state of the gland and varies from about 0.2 to 1.1 percent. The iodine content in goiter is

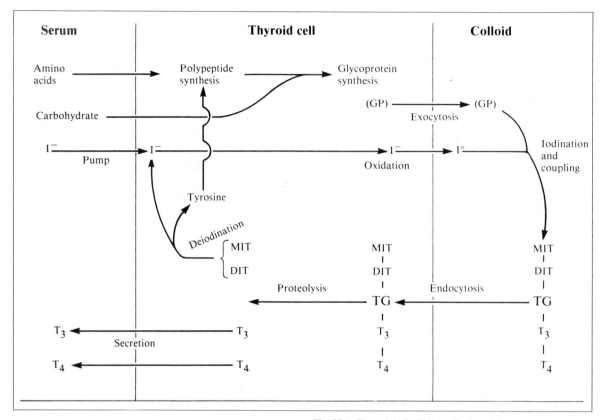

Fig. 33-1. Thyroglobulin (TG) synthesis and storage, and secretion of the thyroid hormones, triiodothyronine (T_3) and thyroxine (T_4). The iodination of the tyrosyl residues of this molecule occurs at the apical surface of the thyroid cell. GP = glycoprotein; MIT = monoiodotyrosine; DIT = diiodotyrosine.

generally less than 0.1 percent. The protein synthesis necessary for TG synthesis follows the same patterns found in other tissues. Similarly, all the intermediary metabolic processes carried out in the gland are not unique to the gland. The uniqueness of the thyroid gland lies in the fact that in order for it to synthesize an adequate amount of thyroid hormone, it must concentrate iodide (I^-) from the plasma and incorporate it into the existing TG molecule.

Normally, a concentration gradient for iodide (I^-) of about 25/1 is maintained between the thyroid and the serum. This is referred to as the thyroid/serum (T/S) ratio. Iodide is actively transported into the thyroid cell against an electrochemical gradient at the basal membrane, and a potential difference of -40 to -50 mv is maintained across the basal cell membrane. The transport mechanism is a carrier-mediated, energy-requiring process that is related to the function of the Na^+-K^+–dependent adenosine triphosphatase (ATPase) system. The iodide-concentrating mechanism is referred to as the *iodide pump*.

That the iodide pump is of physiological importance is best illustrated following the administration of thiocyanate, perchlorate, and other, similar anions. These anions compete with iodide at the site of the concentrating mechanism and thus depress the uptake of iodide. It should be pointed out that other tissues, particularly the salivary glands, gastric mucosa, mammary gland, and placenta, also process iodide-concentrating mechanisms. Inhibition of the iodide pump in these organs has been observed following treatment with anions of the perchlorate group.

The best-known regulator of the iodide pump is TSH, and it decreases (makes less negative) the membrane potential across the basal cell membrane of thyroid cells, whereas hypophysectomy increases it. Removal of the anterior pituitary depresses the T/S ratio to 3 to 7, and, under certain circumstances, the T/S ratio may reach 250 to 300. The thyroid is able to concentrate iodide in the absence of TSH

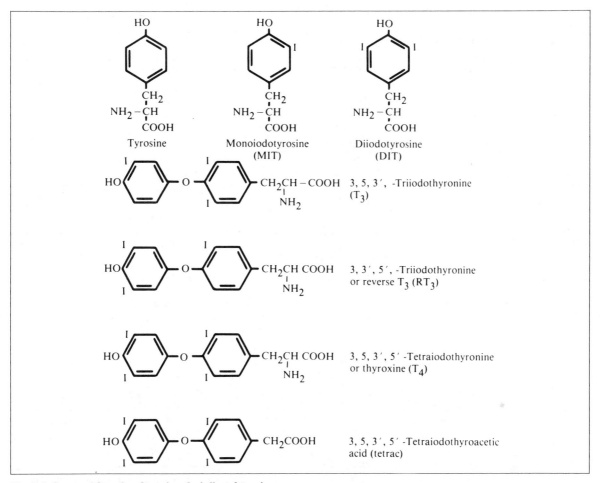

Fig. 33-2. Structural formulas of tyrosine, the iodinated tyrosines, the iodinated thyronines that are present in the thyroglobulin molecule. A metabolite of thyroxine (T_4), tetraiodothyroacetic acid (TETRAC), is also shown.

and therefore retains some degree of autonomy. Thyroid-stimulating hormone is the most important factor in regulating the iodide pump, and its action, as well as that of the long-acting thyroid stimulator (LATS), is mediated by cyclic adenosine monophosphate (cyclic AMP). Large doses of iodide saturate the pump and reduce the T/S ratio. The iodide-concentrating mechanism of the salivary glands, gastric muscosa, mammary gland, and placenta is not influenced by the removal of the anterior pituitary, or the addition of TSH, or by dietary iodide.

The kidneys, as well as the thyroid gland, remove iodide from the blood. The thyroid gland of a normal person ac-

cumulates about one-third of a tracer dose of ^{131}I within 24 hours, and most of the remaining iodide is lost in the urine. In the absence of the thyroid, less than 2 percent of a dose of ^{131}I is retained by the body, and nearly all of the remainder is lost via the urine.

Over 90 percent of the thyroid iodine is organically bound, and therefore it must appear as free iodine (I_2), the reactive form. Iodide is converted to iodine ($2 I^- \rightarrow I_2$), which can enter into organic combination. Iodine formation from iodide occurs through the action of a peroxidase in the microvilli of the apical portion of the cell. Agents such as propylthiouracil depress the amount of iodine available for iodination of tyrosyl radicals. Following the use of extremely high doses of propylthiouracil, only traces of iodotyrosines may still be formed, but significant amounts of iodotyrosines and iodothyronines are still formed after small doses.

Thyroglobulin acts as a substrate for a series of reactions leading to the formation of thyroid hormones. The steps involved are the formation of MIT and DIT, followed by the oxidative coupling of MIT and DIT, with a loss of alanine within the TG molecule. Following the formation of elemental iodine, the oxidized form is bound to tyrosyl groups that are in peptide linkage within the TG molecule. The result is the formation of the iodotyrosines MIT and DIT within the molecule. The formation of T_3, RT_3, and T_4 is thought to be catalyzed by the action of a peroxidase or similar enzyme on the iodotyrosines in peptide chains. It is probable that the phenolic group of DIT moieties of two adjacent TG molecules become attached and form a molecule of T_4 on one TG molecule and an alanine side chain on the other; T_3 and RT_3 may be formed in a similar manner, involving MIT and DIT, or by removal of iodine at either the 5' or 5 position. In the euthyroid person, each TG molecule contains about 11 molecules of MIT and DIT and 2 molecules of T_3 and T_4.

In general, it can be stated that TSH increases the rate of organification of iodine and iodothyronine formation. However, the mechanism or mechanisms by which TSH accelerates organification of iodide are now known. Administration of TSH decreases the MIT/DIT and T_3/T_4 ratios and increases the absolute amount of T_3 and T_4 in TG. It follows that TSH might play a role in stimulating the oxidative coupling of MIT with DIT and DIT with DIT in the thyroglobulin molecule.

There is an intrathyroid regulation of the synthesis of iodotryosines and iodothyronines. This is seen most clearly in a person who ingests an iodine-deficient diet. In such a situation the MIT/DIT and T_3/T_4 ratios within the TG molecule are markedly increased, but the concentration of each compound is decreased. If the elevated T_3/T_4 ratio in the gland should be maintained in the serum, an important physiological adjustment in cases of iodine deficiency would be demonstrated. However, it has not been determined whether or not the more potent mixture of T_3 and T_4 is carried over to the serum under these circumstances.

In summary: thyroglobulin is synthesized in the endoplasmic reticulum and Golgi apparatus. It is transported to the apical cell surface enclosed in vesicles that are formed in the Golgi apparatus, and the vesicles are then moved into the follicular lumen by exocytosis and incorporated into the colloid. The completion of the synthetic process, the formation of the iodoamino acids, occurs in the microvilli at the apical portion of the follicular cell. The hormones are then stored as TG in the colloid.

PROTEOLYSIS OF THYROGLOBULIN AND SECRETION OF THE HORMONES

The manner in which the thyroid hormones are stored is unique among the endocrine glands. T_4, T_3, and RT_3 are contained, via peptide linkage, within the large TG molecule. This is stored in an extracellular site, the follicular colloid, and the rate of its formation and removal is in part dependent on TSH. Following TSH administration, droplets or particles of colloid appear in the apex of the follicular cells as a result of an endocytotic process. These vesicles migrate toward the base of the cell and fuse with lysosomes that have migrated from the base of the cell. The lysosomes contain enzymes that can hydrolyze TG. T_4 and T_3 release occurs by TG lysis within the fused droplet and lysosome or "derived lysosome." MIT and DIT are also released into the thyroid cell by the hydrolysis of TG. The ingestion of colloid by endocytosis occurs at the periphery of the lumen, and the most recently iodinated TG is immediately adjacent to the apical cell border. Consequently, a higher proportion of recently iodinated thyroglobulin than of older thyroglobulin should be ingested. Thus, a last-formed, first-secreted or "last come, first served" phenomenon exists.

The iodotyrosines MIT and DIT are not normally secreted into the plasma because they are deiodinated rapidly by dehalogenases in the thyroid cells. Thus, the iodide may be reclaimed immediately by the gland. However, the iodothyronines, T_3 and T_4 are resistant to the dehalogenases and diffuse from the gland as a result of gradients that may be as high as 100/1 between the gland and the blood. The diffusion gradient of thyroxine is also aided in consequence of competitive binding favoring certain plasma proteins over the thyroid proteins. Thyroid-stimulating hormone leads to a rapid increase in thyroid cyclic AMP. The effects of this hormone on thyroid hormone synthesis and secretion are probably mediated via the "second messenger." Thyroid-stimulating hormone accelerates the rate of proteolysis of TG and the rate of release of T_3, RT_3, and T_4, and following diffusion into the bloodstream, each is bound by specific plasma proteins.

Several observations indicate that an intrathyroid factor participates in the regulation of the release of organic iodine from the gland. When patients are placed on chronic propylthiouracil ingestion, the daily output of hormonal iodine falls in proportion to the store of hormonal iodine in the gland. However, an increase in circulating TSH (which must occur in this instance) should cause an increase in proteolysis of TG, with a resultant increase in the release of thyroid hormones from the gland. Under this circumstance, one would expect an increase in thyroid hormone output instead of a decline, and it must follow that some mechanism within the thyroid induces conservation of thyroid iodine. Apparently, the three intrathyroid factors that regulate thyroid function lead to physiological adjustments that aid in the maintenance of an adequate level of thyroid iodine stores. At the same time they lead to the secretion of a mixture of T_3 and T_4 of increased potency when challenges that tend to decrease hormonogenesis are put to the thyroid-pituitary-hypothalamus axis.

It is obvious that there are many points in the synthesis and release processes at which failure could cause decreases in thyroid activity. Thus, the exogenous supply of iodine may be inadequate, or a deficiency of one of the enzymes needed for trapping, oxidation, coupling, or hydrolysis may result in reduced hormone production. A deficiency of TSH, because of its effects at several stages of hormonogenesis, would cause a deficiency of thyroid hormone.

TRANSPORT OF THYROID HORMONES IN THE CIRCULATION

The thyroid gland of a normal adult secretes approximately 90 μg of T_4 per day, 5 μg of T_3, and less than 3μg of RT_3. Thus, T_4 accounts for more than 90 percent of the total thyroid hormone output. However, a large portion of T_4 is converted to T_3 and to RT_3 in peripheral tissues, chiefly the liver and kidney. The total production rate of T_3 is roughly 30 μg per day; 20 percent comes directly from thyroid secretion and 80 percent from the peripheral conversion of T_4. Recent studies in humans have shown that RT_3 is also a major product of T_4 metabolism, and about 40 μg is produced peripherally from T_4 each day. Several factors, such as caloric restriction, hepatic disease, systemic illness, and advancing age, depress the peripheral conversion of T_4 to T_3. An additional amount of T_4, about 15 μg per day, is turned over with some being glucuronated and excreted in the feces, and the remainder is excreted as its acetic acid derivative tetraiodothyroacetic acid, or TETRAC (Fig. 33-2). The half-life of each of the thyroid hormones is long; 7 days for T_4 and 1.5 days for T_3.

In plasma, T_4 and T_3 are primarily bound to thyroxine-binding globulin (TBG), and alpha globulin of hepatic origin. Thyroxine-binding prealbumin and albumin also bind T_3 and T_4, but they play lesser roles in the plasma transport of the thyroid hormones than does TBG. T_4 binds much more avidly to TBG and thyroxine-binding prealbumin than does T_3, and this may partly account for the fact that the T_4/T_3 ratio in plasma is about 50/1, even though the total produc-

tion rate of T_4 is only three to four times that of T_3. The relationship between the thyroid hormones and their binding proteins in plasma may be expressed by the following equation:

$$\text{Free } T_4 + \text{free TBG} \leftrightarrow T_4\text{—TBG complex}$$

At equilibrium the equation is shifted markedly to the right, for the "free" (unbound) fraction of T_4 is only 0.03 percent of the total, and the "free" fraction of T_3 is 0.3 percent of the total.

Several factors increase the hepatic production of TBG and thus affect the rate at which T_3 and T_4 may be taken up by their target cells. These include estrogen therapy or pregnancy, hypothyroidism, and various genetic or familial conditions. On the other hand, TBG synthesis is depressed by androgens and cortisol and in liver disease, acromegaly, and severe illness and stress. Several drugs, such as the salicylates, penicillin, heparin, and some of the oral hypoglycemic agents, decrease the binding of T_3 and T_4 to TBG and thyroxine-binding prealbumin. Obviously, either elevated or depressed TBG synthesis will markedly affect the availability of T_3 and T_4 for uptake by their target cells and thus effect the euthyroid state. The normal concentration of total T_4 in human plasma averages about 8.5 μg/per 100 ml (with a normal range of 5–13 μg/100 ml). The comparable values for T_3 are 120 ng per 100 ml (normal range of 90–200 ng/100 ml). The values for "free T_4" and "free T_3" are much less, and the normal values approximate 2.0 and 0.4 ng per 100 ml respectively. Under ordinary circumstances, only small amounts of thyroglobulin (5–10 ng/100 ml) and iodinated tyrosines (0.1–0.2 μg/100 ml) appear in the plasma.

EFFECTS OF THE THYROID HORMONES

The thyroid gland is not indispensable to life, but the presence of adequate amounts of its hormones is necessary for normal heat production, oxygen consumption, growth, differentiation, and development and for a person's general well-being.

The effects of T_3 and T_4 have been studied by observing structural, functional, and biochemical changes that accompany spontaneous and experimental hyperthyroidism and hypothyroidism. These hormones control many metabolic processes; influence oxygen consumption and heat production; participate in the metabolism of carbohydrates, fats, proteins, and nucleic acids; and influence the metabolic activities of other hormones. In general, excesses of T_3 and T_4 promote accelerated rates of turnover of substances, whereas deficiencies promote reduced turnover rates. Extrathyroid conversion of T_4 to T_3 occurs and appears to be necessary for most of the metabolic activity of the hormones in the target cells. Thus, T_3 is probably the primary active thyroid hormone. It has about four to five times the biological potency of T_4, and its onset of action is more rapid than that of T_4.

The receptors of the thyroid hormones are in the cell nucleus of the target cells. T_3 and T_4 readily penetrate their target cells, and intracellular dehalogenases convert T_4 to T_3. The T_3 then binds to receptors in the nuclear chromatin, and its binding to the receptors induces changes in specific messenger ribonucleic acids. These changes then lead to alterations in the rates of synthesis of the particular proteins that mediate or reflect the thyroid hormone response. The thyroid hormones may also have actions at extranuclear sites, specifically at the level of the mitochondria. It is these organelles at which the thyroid hormones exert their effects on oxygen consumption and oxidative phosphorylation.

The most prominent effect of the thyroid hormones is their effect on respiratory exchange. Hyperthyroidism is characterized by increased respiratory exchange and heat production; these effects are reversed in hypothyroidism.

The increased oxygen consumption and heat production (*calorigenic effect*) that follow the administration of thyroid hormones are marked. The calorigenic effect can also be demonstrated in certain excised tissues obtained from animals with experimentally induced hyperthyroidism. However, when thyroid hormones are added to excised tissues taken from a normal animal, little or no increase in oxygen consumption results. The oxygen consumption of the anterior pituitary, in conrast to other tissues, is not affected by thyroid insufficiency or excess. Similar observations have been made concerning the brain and spleen. The observation that heat production is less in the absence of a hypophysis than in the absence of a thyroid indicates that the thyroid hormones are not the only hormones involved in the regulation of heat production. Rather, it is regulated by a balance of hormonal factors, including growth hormone, insulin, and the adrenocortical and medullary hormones, all of which are calorigenic.

Normally, 100 to 150 μg of thyroxine daily is required to maintain an adult human in the euthyroid state. Disturbances of the heat-regulating mechanism are readily observed in patients who have undergone thyroidectomy. The normal person responds to a cool environment by increasing the output of thyroid hormones, stimulating an increase in the basal metabolic rate (BMR). It has been shown that the thyroxine requirement is inversely related to the environmental temperature. However, the twofold to threefold increase in heat production over the resting level that follows sudden exposure to a cool environmental temperature is not due to increased thyroid activity, for shivering and epinephrine release account for most of the increase in heat production. It should be borne in mind that many mechanisms other than hormonal mechanisms are involved in thermoregulation, and earlier chapters (Chaps. 27, 28) on energy metabolism and temperature regulation should be consulted for a discussion of factors that contribute to thermoregulation.

The effects observed after the administration of thyroid hormones or after thyroidectomy have a considerable latent period. When one considers heat production, for instance, after thyroidectomy, one notes that the BMR falls steadily until it reaches a minimum of 35 to 50 percent of normal in about 40 days. On the other hand, a change in the BMR is not detectable until approximately 36 hours have passed after a large dose of T_4 has been administered to a normal person, and a maximal effect is not observed until 3 to 5 days later.

Many effects observed in hyperthyroid and hypothyroid states are in all probability the results of both direct calorigenic effects and reflex responses to them. Most of these effects change in proportion to the BMR and therefore are assumed to stem from the calorigenic effect of the hormones. Changes observed in hyperthyroid states are a decrease in heat tolerance, nervousness, restlessness, insomnia, weight loss; an increase in fasting nitrogen excretion; increases in the rates at which carbohydrate and fatty acids are absorbed from the gastrointestinal tract; and a diminished serum cholesterol that results from increased cholesterol catabolism. Cardiac output and plasma volume increase, and pulmonary ventilation rises. Increases in glomerular filtration rate, tubular maxima for para-aminohippurate and glucose, and urine volume occur. Thus, all systems are accelerated, and many adaptations take place, with the increase in thyroid hormone output leading to an increase in oxygen consumption, and the organism provides adequate means of satisfying the increased demand for oxygen.

On the other hand, in hypothyroid states, varying degrees of anemia and decreased bone marrow activity commonly occur. The hypothyroid person does not tolerate the cold, is quiet, sleeps more than normal, and exhibits muscular fatigue and weakness. Weight gain usually occurs and is generally in proportion to the decline in BMR. Sensitivity to external stimuli decreases, and tendon reflex time increases. Certain other changes, opposite to those noted in hyperthyroidism, are observed in all systems (e.g., slowed heart rate, decreased cardiac output, decreased ventilation), and thus all systems are slowed. In severe hypothyroidism there is an accumulation of mucopolysaccharide-rich fluid, particularly in the dermis, leading to a generalized nonpitting thickening of the skin, myxedema. In juvenile hypothyroidism, sexual development and maturation are retarded, and the onset of puberty is delayed. The thyroid hormones are essential for normal reproduction and lactation.

The thyroid hormones also participate in growth, differentiation, and development of cells, organs, and systems; when the thyroid is absent or their production is subnormal, several abnormalities in structure and function are noted. The thyroid hormones are not required for early fetal growth, and the fetus does not synthesize either TSH or the iodothyronines until the tenth to eleventh week of gestation. Maternal TSH, T_4, T_3, and RT_3 does not cross the placenta, and there is no evidence that maternal thyrotropin-releasing hormone normally affects fetal pituitary-thyroid function. Thus, differentiation and development during this period are independent of the thyroid hormones and are controlled solely by genetic mechanisms. The lack of T_4 does not interfere with growth prior to birth, because birth weights and lengths of athyroid infants at full term are within normal limits. However, the differentiation and maturation of the fetal central nervous system and skeleton are dependent on T_4 prior to birth. Infants with hypothyroidism at birth (cretins) have delayed development of ossification centers and often show epiphyseal dysgenesis. Their neural development and mental functions are also impaired. The earlier the onset and the more marked the degree of fetal hypothyroidism, the more marked are the alterations in skeletal and nervous development at birth.

A euthyroid state in postnatal life, in contrast to that in prenatal life, is necessary for normal growth. Postnatal hypothyroidism leads to marked growth retardation. Skeletal growth is subnormal, and the height and bone age of the cretin lags behind that of the normal child. The lagging development of the skeletal system can usually be overcome with replacement therapy. However, if treatment is delayed, a dwarfed stature persists even though accelerated growth rates are attained. An excess of T_4 in children enhances the rate of linear growth and maturation, and thus the height and bone age of hyperthyroid children exceeds that of normal children of the same age. Large excesses of T_4 do, however, impair the ultimate linear growth of bone.

The mechanism of growth stimulation by T_4 has not been elucidated. The observation that T_4 has a mitogenic action suggests a vital role for this hormone in cell division and differentiation. It should be pointed out that the effects of T_4 on growth are optimal only in the presence of growth hormone.

Probably no tissue suffers more than does nervous tissue from lack of thyroid hormone during fetal and early post-

natal life. This deficiency results in mental deficiency of unusual severity. In adult hypothyroidism the observed mental deficiencies are manifested as reduced mental alertness. Studies concerned with thyroid hormone and central nervous system structure and function have shown that a deficiency of thyroid hormone results in a delayed appearance and decreased content of myelin in fiber tracts, a decrease in the size and number of cortical neurons, and decreased cerebral vascularity. These changes may be reversed if thyroid hormone is given very early in postnatal life; if not, irreversible damage results. It has also been observed that brain succinic dehydrogenase decreases in concentration following thyroidectomy of rats at birth. Such activity can be restored to normal with thyroid hormone if treatment is instituted within 10 days. All these observations might provide an explanation for the irreversible retardation of mental development in cretins. On the other hand, the presence of thyroid hormone in excess does not cause acceleration of growth and development and may even retard them.

REGULATION OF THYROID FUNCTION

Many factors influence the rate of thyroid hormone secretion, but the common denominator in all states of thyroid activity is TSH (Fig. 33-3). It is the only hormone that normally stimulates thyroid growth and hormonogenesis directly. Its effects on the thyroid gland are mediated by cyclic AMP. Every type of abnormal growth of the thyroid and/or change in hormone production and secretion by the gland must be accompanied by a change, either relative or absolute, in TSH secretion by the anterior pituitary. The secretion of TSH by the anterior pituitary in turn is regulated directly by at least two hypothalamic hormones, TRH and somatostatin, and both directly and indirectly, by T_3 and T_4.

Thyroid-stimulating hormone is secreted by the basophilic thyrotropes of the anterior pituitary and is a glycoprotein. It is composed of two subunits, an alpha unit and a beta unit. The alpha unit is very similar in structure to the alpha units of follicle-stimulating hormone, luteinizing hormone, and human chorionic gonadotropin. Its biological specificity is conferred by its beta unit. The plasma concentration in the euthyroid person averages about 4 µU per milliliter (normal range = 1–8µU/ml).

The actions of TSH are many, and, as stated previously, it affects every stage of thyroid hormonogenesis, storage, and release. It also decreases the volume of colloid and causes hypertrophy and hyperplasia of the follicular cells of the thyroid and therefore is goitrogenic. Any maneuver that

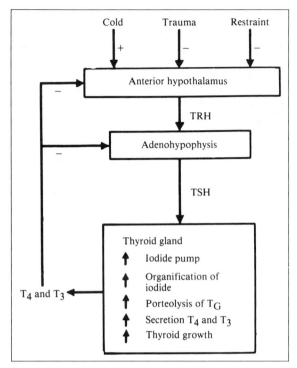

Fig. 33-3. Factors that influence the secretion of thyrotropin-releasing hormone (TRH), thyroid-stimulating hormone (TSH), and the thyroid hormones. TG = thyroglobulin; + = stimulation; − = inhibition.

leads to a decrease in the plasma level of "free T_3" and "free T_4" results in an increase in TSH secretion. In primary hypothyroidism the plasma TSH may be as high as 400 μU per milliliter, with the severity of the defect of T_3 and T_4 secretion being correlated with the plasma TSH concentration.

Conversely, any procedure that elevates the "free T_3" and "free T_4" concentrations in the plasma induces a decrease in TSH secretion. In normal persons a balance is established by a negative feedback system that maintains an optimal level of thyroid function. An exception to this generalization occurs in Graves' disease, a form of hyperthyroidism. In this disease a substance called long-acting thyroid stimulator is found in the plasma. This is an immunoglobulin, an IgG, that enhances thyroid iodide uptake, increases thyroid hormone synthesis and secretion, and is goitrogenic. Along with TSH it also induces hypertrophy of retroorbital tissue, which leads to the condition known as exophthalmos. Long-acting thyroid stimulator is distinctly different from TSH and is not dependent on hypothalamic

or anterior pituitary function, but, like TSH, its actions are mediated by cyclic AMP. Goiter or thyroid enlargement will result from any condition leading to a marked and sustained imbalance between TSH secretion and T_3 and T_4 secretion if the balance favors TSH secretion.

The secretion rate of thyroid hormones in "normal" persons on an iodine-deficient diet is generally normal and is maintained at normal levels only by an increase in TSH output. It is obvious that iodine plays a very important role in the synthesis of thyroxine, since, in addition to being part of the T_3 and T_4 molecules, it decreases the output of thyroid hormone either by inhibiting TSH secretion or by decreasing TSH activity. Therefore, at the outset of iodide deprivation the output of T_4 is decreased, so a degree of inhibition of release of TSH is removed. The output of TSH then rises, increasing the avidity of the thyroid for iodine and making more iodide available for the synthesis of thyroid hormone. Thus, a marked, sustained iodine deficiency indirectly maintains an increased TSH secretion, which results in hypertrophy and hyperplasia of thyroid cells. The increase in the rate of iodide uptake from the low iodine levels can then maintain a level of thyroid hormone production within normal limits, as judged by the observations that all criteria that assess the effects of thyroid hormones appear to be satisfied. Nevertheless, simple goiter is produced. However, if the increase in TSH output cannot restore normal thyroid hormonogenesis over a prolonged period, the gland will eventually become exhausted and atrophy. The exhaustion is reversible if the supply of iodide is increased.

The ingestion of large amounts of iodide can also strongly influence thyroid hormonogenesis. It decreases the T/S iodide ratio through an intrathyroid mechanism and by inhibiting TSH action. It decreases the rate of organification of iodide in the thyroid and may lead to the development of hypothyroidism and goiter. It has no effect on the rate of release of hormone in the normal person, but decreases the rate of hormone release in the hyperthyroid person or a normal person treated with TSH. The manner in which high levels of circulating iodide influence thyroid physiology is not known.

The secretion of TSH is stimulated by TRH, which is produced by neurons of the anterior hypothalamus. Its secretion is inhibited by somatostatin and the iodothyronines. Several types of experimental procedures have shown that the hypothalamus is intimately concerned with TSH secretion. First, transplantation of the anterior pituitary to extrasellar sites leads to a reduction in TSH secretion, and thyroid function is decreased, approaching, but not reaching, that observed in the animal that has had a hypophysectomy. Second, hypophyseal stalk section results in a decreased rate of TSH secretion. Third, lesions of the anterior hypothalamus, particularly in the supraoptic region, depress thyroid function—a reflection of decreased TSH secretion. Finally, electrical stimulation of the anterior hypothalamus or rostral portion of the infundibulum results in an increase in TSH secretion by the thyrotropes. Both hypophyseal stalk section and lesions of the anterior hypothalamus lead to a reduction in the total TSH content of the anterior pituitary and a decrease in plasma TSH concentration. However, the pituitary TSH concentration and the appearance of the thyrotropes remain normal. Alterations that follow each procedure are the formation of TG and secretion of T_4 and T_3 at levels comparable to those observed following hypophysectomy. Another feature observed is the failure to respond to stimuli that lead to either an increase or decrease in TSH secretion in the intact animal.

Thyrotropin-releasing hormone is a tripeptide amide that is synthesized and released by neurons of the anterior hypothalamus into the hypophyseal portal vessels, and by activating adenylate cyclase and increasing cyclic AMP in the thyrotropes, it exerts a tonic stimulatory effect on TSH secretion. Thyrotropin-releasing hormone also stimulates prolactin secretion by the lactotropes of the anterior pituitary. The infusion of minute amounts of T_3 or T_4 into the anterior pituitary results in an inhibition of TSH release by the thyrotrope, and it has been demonstrated that this inhibitory action results from blocking the pituitary response to TRH. Experiments also indicate that T_3 and T_4 may inhibit TRH synthesis and secretion by acting directly on the neurons in the anterior hypothalamus.

Whatever the case, the secretion of TSH is related to the secretion of T_3 and T_4 through a negative feedback system. Present evidence indicates that the principal feedback mechanism on pituitary TSH secretion is a direct one exerted by the thyroid hormones. However, for full stimulatory or inhibitory adaptations by external stimuli, the functional integrity of the hypothalamohypophysial system must be maintained.

Other neural factors can influence the secretion of TRH from the hypothalamus and TSH from the anterior pituitary. The area of the anterior hypothalamus whose neurons release TRH is adjacent to areas containing temperature-sensitive receptors involved in integrating mechanisms for the regulation of the body temperature (see Chap. 28). Cooling of the temperature-sensitive receptors, in addition to activating heat-production and heat-conserving mechanisms, stimulate TSH secretion and thyroid function. Conversely, fever induced by the usual means tends to inhibit

TSH secretion and depress thyroid function. Exposure to cold in adults leads to a modest increase in TSH secretion, but in the newborn, increases in TSH and thyroid hormone secretion are marked. Thus, peripheral cold receptor activity stimulation can influence TRH and TSH secretion rates.

Experimentation has shown that stressful stimuli powerfully inhibit the uptake of ^{131}I and the release of thyroid hormones. The alterations in thyroid function occur prior to the withdrawal of TSH and result from the increased circulating corticosteroids. These hormones enhance urinary iodide loss and slow thyroid circulation, thus contributing to the reduced thyroid iodide uptake. Chronic stress produces depressed thyroid function, and this is apparently mediated by the action of cortisol in depressing the release of TRH. Emotional stress has long been thought to precipitate a marked increase in the rate of TSH release. However, current studies indicate that if emotional stress has any effect on TSH secretion and thyroid function, it is one of inhibition (see Fig. 33-3).

BIBLIOGRAPHY

Bjorkman, U., Ekholm, R., Elmqvist, L. G., Ericson, L. E., Melander, A., and Smeds, S. Induced unidirectional transport of protein into the thyroid follicular lumen. *Endocrinology* 95:1506–1517, 1974.

Cavalieri, R. R., and Rapoport, B. Impaired peripheral conversion of thyroxine to triiodothyronine. *Annu. Rev. Med.* 28:57–65, 1977.

Ingbar, S. H., and Braverman, L. E. Active form of thyroid hormone. *Annu. Rev. Med.* 26:443–449, 1975.

Jackson, I. M. D. Thyrotropin-releasing hormone. *N. Engl. J. Med.* 306:145–155, 1982.

Larsen, P. R. Thyroid-pituitary interaction. *N. Engl. J. Med.* 306:23–32, 1982.

O'Connor, J. F., Wu, G. Y., Gallagher, T. F., and Hellman, L. The 24-hour plasma thyroxine profile in normal man. *J. Clin. Endocrinol. Metab.* 39:765–771, 1974.

Roti, E., Grudl, A., and Beaverman, L. E. The placental transport, synthesis, and metabolism of hormones and drugs which affect thyroid function. *Endocr. Rev.* 4:131–149, 1983.

Sterling, K. Thyroid hormone action at the cell level. *N. Engl. J. Med.* 300:117–123, 173–177, 1979.

Sterling, K., and Lazarus, J. H. The thyroid and its control. *Annu. Rev. Physiol.* 39:349–371, 1977.

Utiger, R. D. Serum triiodothyronine in man. *Annu. Rev. Med.* 25:289–302, 1974.

Van Herle, A. J., Vassart, G., and Dumont, J. E. Control of thyroglobulin synthesis and secretion. *N. Engl. J. Med.* 301:239–249, 307–314, 1979.

FUNCTIONS OF THE ADRENAL GLANDS

34

Ward W. Moore

The adrenal glands consist of two distinct organs, the adrenal cortex and the adrenal medulla or the sympathoadrenal system. The adrenal cortex is essential to life, and its normal functional integrity is dependent on adrenocorticotropic hormone (ACTH), which is secreted by the anterior pituitary gland, corticotropin releasing factor (CRF), which is secreted by the hypothalamus, and the renin-angiotensin system. The cells of the adrenal cortex synthesize and secrete into the plasma several steroid compounds. Many are metabolically inactive, but some are extremely active biologically and may selectively possess lipolytic, gluconeogenic, protein catabolic, electrolytic, androgenic, and estrogenic properties. Generally, those adrenal steroids that have a pronounced effect on intermediary metabolism are referred to as *glucocorticoids,* while those whose main effect is on salt and water metabolism are called *mineralocorticoids*. The functional integrity of the adrenal cortex must be normal if a person is to withstand the stresses of life, such as injury, surgery, disease, or any challenge that causes an imbalance in homeostatic mechanisms. The adrenal medulla secretes two hormones, epinephrine and norepinephrine. Both are catecholamines, and they have effects that trigger metabolic and hemodynamic changes ranging from hyperglycemia to hypertension.

ANATOMY

In embryological development the cortical tissue is the first to appear. It becomes evident at about the fifth week of intrauterine life in the human and arises from mesoderm in the urogenital zone. During the seventh week the mass of presumptive cortical cells is invaded by cells migrating from neural crest material. These cells are surrounded by the mesodermal cells and form the adrenal medulla.

Normally, the paired adrenal glands of the normal adult weigh about 10 gm. Each adrenal receives its blood supply from several small arterial branches that arise from the aorta and the renal and phrenic arteries and has a total blood flow of about 0.5 ml/gm/min. The venous drainage of each gland is through a single central vein; the left adrenal vein empties into the renal vein, and the right adrenal vein empties directly into the vena cava.

The adrenal cortex consists of epithelioid cells arranged in continuous cords or sheets separated by capillaries. Structurally, the cortex may be divided into three zones. The *zona glomerulosa*, the outer zone next to the fibrous capsule, has cells arranged in irregular masses. The middle zone, the *zona fasciculata*, has cords or sheets of cells that are straight and radially disposed. The inner cortical zone, the *zona reticularis*, has cords that form an irregular network. The ultrastructure of the human adrenocortical cells reveals an abundance of mitochondria, and smooth endoplasmic reticulum is plentiful. These structures are prominent in all steroid-synthesizing cells. All the cortical cells show evidence of accumulation, storage, and secretion of lipid. The chief lipid constituent is cholesterol, the precursor of the adrenocortical hormones.

The fetal adrenal gland may be larger than the normal adult gland, and, in relation to fetal size, it is immense. The fetal adrenal cortex is composed primarily of large, ovoid cells arranged adjacent to the medulla. These cells are referred to as the *fetal zone*, and they produce a steroid, dehydroepiandrosterone sulfate, that serves as a precursor for the placental production of estrogens. The importance of this unit, the fetal-placental unit, is discussed in Chapter 35.

The cells of the adrenal medulla are grouped in clumps and irregular cords around the capillaries. The cells are large and ovoid, contain granules that stain with chromium salts, and are referred to as *chromaffin cells,* or pheochromocytes. These cells are the source of the medullary hormones epinephrine and norepinephrine and are modified postganglionic sympathetic nerve cells. The adrenal medulla is regulated directly through its sympathetic nerve supply.

ADRENAL MEDULLA

The adrenal medulla is composed of chromaffin cells, or pheochromocytes that secrete epinephrine and norepinephrine in an approximate ratio of 4/1 (i.e., 4 moles of epinephrine for each mole of norepinephrine). The glands do not fully differentiate or mature until about the third year of life. Most of the blood supply to the paired glands arrives by way of a portal system from the adrenal cortex, and thus the total blood supply has very high concentrations of the steroids secreted by the adrenal cortex. Many lines of evidence support the concept that normal adrenal medullary function is dependent on normal adrenocortical function. The two glands work closely together to protect individuals during stressful situations.

The hormones of the adrenal medulla are derived from the amino acid tyrosine. The first synthetic step is the rate-limiting step and involves the conversion of tyrosine to 3, 4-dihydroxyphenylalanine (dopa). Dopa is then decarboxylated and converted to dopamine. The intermediate, dopamine, has been localized in nerve terminals in several areas of the central nervous system and functions as a neurotransmitter in those areas. It also inhibits the secretion of prolactin by the anterior pituitary. Dopamine then undergoes beta-hydroxylation to form norepinephrine. Epinephrine is then derived from norepinephrine by the addition of a methyl group under the influence of phenylethanolamine-*N*-methyltransferase. The catecholamines are metabolized primarily by two enzymes, monoamine oxidase and catechol-*O*-methyltransferase, and the primary excretory product of the adrenal medullary hormones is vanillylmandelic acid. The biosynthetic and metabolic pathways are illustrated in Figure 34-1.

Epinephrine and norepinephrine exert their effects by binding to receptor sites on the cell membrane of their target cells. The receptors have been classified into two major categories on the basis of their potencies when compared with that of pharmacological agonists and antagonists. These have been termed alpha and beta receptors and were discussed in Chapter 7.

The sympathetic nervous system and the adrenal medulla play major roles in the response of the body to many stressful and emergency situations. The effects produced by the activation of the sympathoadrenal system enable the individual to make adjustments that have been called responses for "flight or fight." These adjustments include increases in heart rate and cardiac output and alterations in vascular resistance, so that less of the cardiac output is channeled to the skin, gastrointestinal tract, and kidneys, and a greater portion is channeled to the skeletal muscle and brain. Additional changes induced are decreases in gastrointestinal tract motility, alterations in the respiratory tract that decrease the resistance to airflow, and alterations in the central nervous system that drive ventilation and maintain an awake and alert state. In the "flight or fight" situation, as a result of increases in sympathoadrenal activity, all body systems respond and meet the demands placed on them by the emergency. The precise mechanisms by which these demands are met have been discussed in previous chapters.

The rapid and profound rise in sympathoadrenal activity that occurs in emergency situations, in addition to the general systemic effects, produces changes in metabolic pathways that provide a rapid and available source of fuel to all tissues and increases the overall rate of metabolism. These metabolic effects of the catecholamines are mediated by

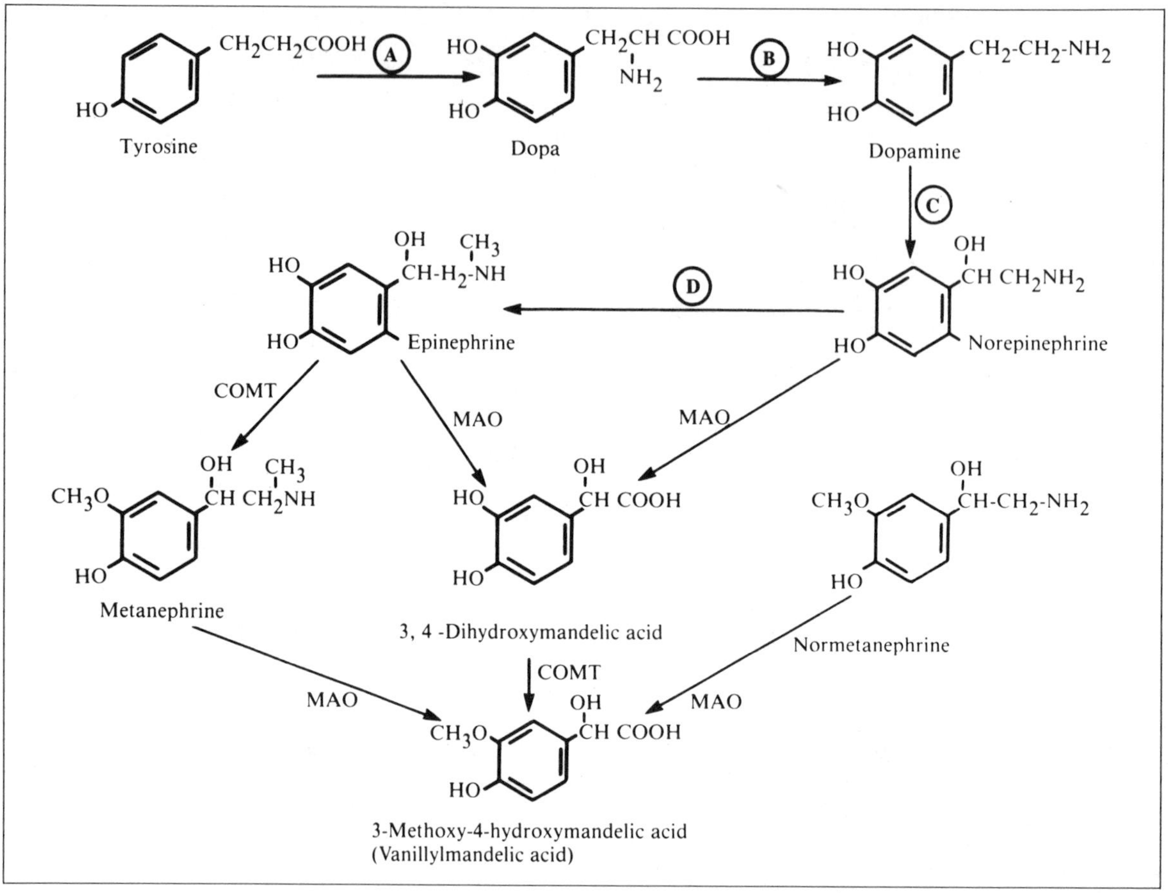

Fig. 34-1. Biosynthetic and metabolic pathways of the adrenal medullary catecholamines. The enzymes involved are tyrosine hydroxylase (A); aromatic L-amino acid decarboxylase (B); dopamine β-hydroxylase (C); and phenylethanolamine-N-methyltransferase (D). COMT = catechol-O-methyltransferase; MAO = monoamine oxidase.

cyclic adenosine monophosphate (cyclic AMP). The catecholamines, epinephrine in particular, are potent hyperglycemic agents. Their effect is accomplished by accelerating the delivery of glucose from the liver to the circulation through both glycogenolysis and gluconeogenesis and by reducing the clearance of glucose from the circulation. They also promote glycogenolysis in skeletal muscle, but the muscle cell is unable to form glucose and free it to the circulation. Instead, the glucose 6-phosphate produced as a result of the increase in glycogenolysis proceeds down an anaerobic pathway to form lactic acid. The lactic acid can diffuse out of the cell into the plasma and is transported to the liver, which then may convert lactate to glucose via the reverse glycolytic pathway. As a consequence of these events following the increase in epinephrine secretion, there are abrupt increases in the plasma glucose and lactic acid levels and decreases in the amounts of glycogen stored in the liver and muscle. As the liver metabolizes lactic acid,

the plasma lactate falls, the liver glycogen content rises, and the plasma glucose concentration slowly returns to normal as the glucose is taken up by extrahepatic tissue.

The catecholamines are also very potent lipolytic agents. Their effect is primarily at the level of the adipocyte, where they activate lipolytic enzymes that accelerate the breakdown of triglyceride to free fatty acids. Epinephrine is also ketogenic, partly as a result of the increased delivery of free fatty acids to the liver and partly through a hormone-mediated effect on the liver.

Activation of the sympathoadrenal system is triggered in many ways besides by stress, and some of these include hypoglycemia, exercise, postural hypotension, diabetic ketoacidosis, surgery, and myocardial infarction. The plasma concentration of epinephrine in a resting supine subject is approximately 40 pg per milliliter, and a slight increase (50–100 pg/ml) in the concentration induces changes in heart rate, systolic blood pressure, and plasma free fatty acids, and slightly higher elevations lead to rises in the plasma glucose, lactate, and ketone bodies. Moderate exercise leads to plasma epinephrine concentrations greater than 100 pg per milliliter and heavy exercise produces levels in excess of 400 pg per milliliter. The increasing levels of epinephrine in the plasma thus permit increases in oxygen, blood, and metabolic substrate delivered to tissue in a time of need.

BIOCHEMISTRY OF ADRENAL STEROIDS

Many steroids have been isolated from the adrenal cortex. All are derived from cholesterol, and the major ones of

Fig. 34-2. Structure of the basic steroid nucleus showing the ring designation and system used for numbering the carbon atoms. Also shown are the three major steroids secreted by the adrenal cortex.

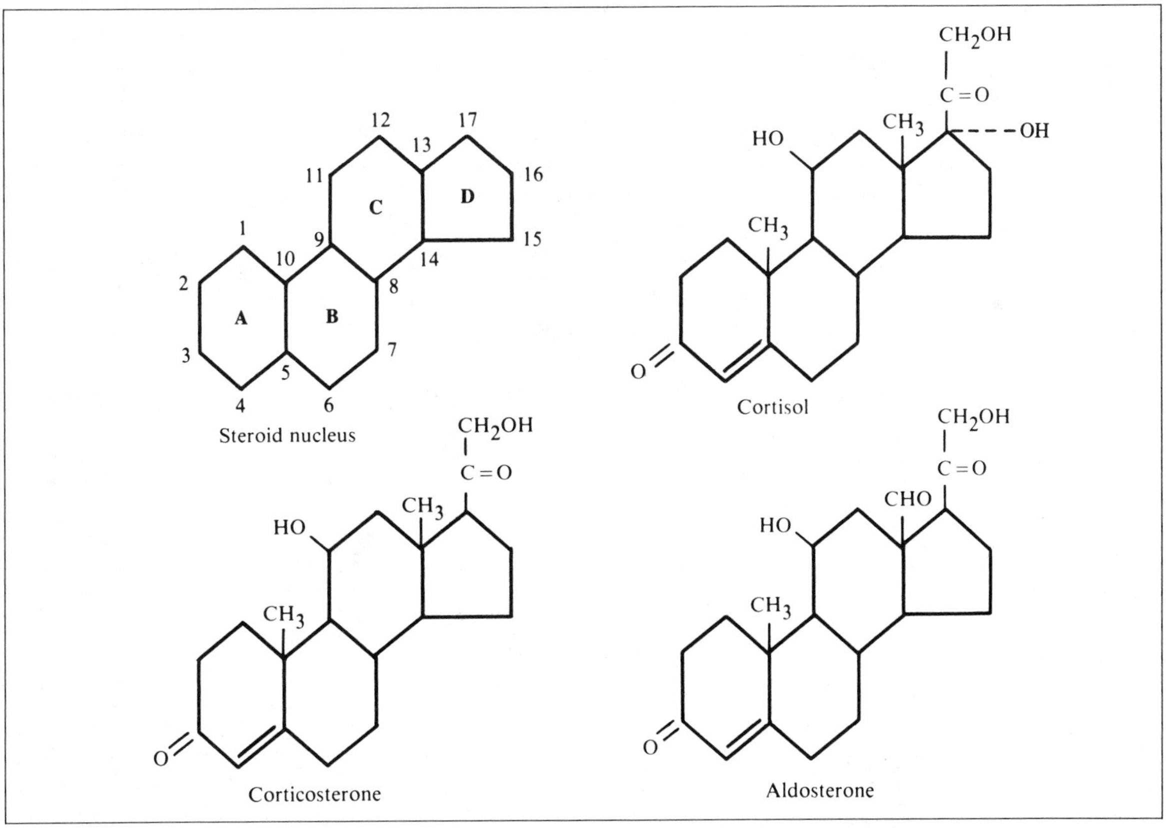

biological importance are depicted in Figure 34-2, along with the structural formula of the steroid nucleus, cyclopentanoperhydrophenanthrene.

The reaction sequence of cortical steroid synthesis (Fig. 34-3) from cholesterol is initiated when cholesterol is cleaved to yield a 21-carbon compound with a ketone group at the 20 position to form Δ^5-pregnenolone. The latter, under the influence of 3β-hydroxydehydrogenase, is converted into progesterone. Progesterone occupies a key position in adrenocortical hormone biosynthesis, and in the human the greater part of the progesterone is subjected to three successive hydroxylating steps catalyzed by the enzymes 17α-hydroxylase, 21-hydroxylase, and 11β-hydroxylase to form 17α-hydroxyprogesterone, 17α-hydroxydeoxycorticosterone, and cortisol (hydrocortisone) respectively. The last, cortisol, is the predominant circulating adrenal steroid in the human. Each of the three cortical zones possesses the 11- and 21-hydroxylases, but the 17α-hydroxylase is present only in the zona fasciculata and zona reticularis. The amount of progesterone that escapes hy-

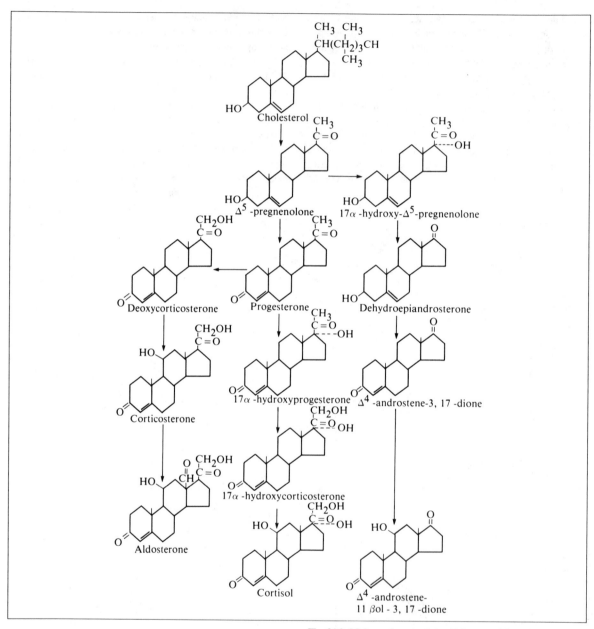

Fig. 34-3. Major pathways of steroid hormone biosynthesis in the human adrenal cortex.

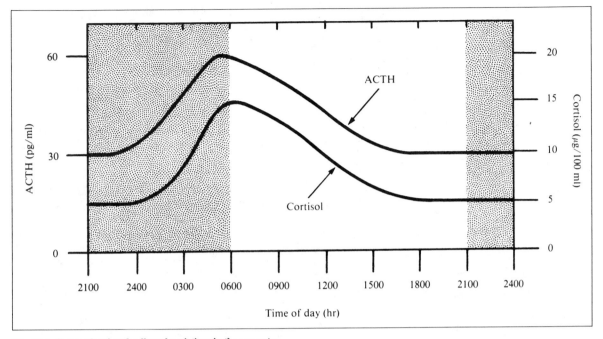

Fig. 34-4. Curves showing the diurnal variations in the concentrations of ACTH and cortisol in the plasma.

droxylation at the 17α position undergoes hydroxylation at the 21 position to form deoxycorticosterone. This is hydroxylated at the 11 carbon to form corticosterone.

The rates of secretion of both cortisol and corticosterone are not constant throughout the day. They are secreted in response to ACTH in a diurnal or circadian manner. The concentration of cortisol in the plasma is much higher in the morning (0600 hrs) than in the evening (1800 hrs). The diurnal curve of ACTH and cortisol is shown in Figure 34-4. Corticosterone is a direct precursor of aldosterone. It occurs as the result of a substitution of an aldehyde group at the 18 carbon of corticosterone under the influence of "18-aldolase," an enzyme located only in the zona glomerulosa. The concentration of aldosterone in human peripheral venous plasma is about 0.1 μg per 100 ml.

The secretion of androgenic substances by the adrenal cortex can be of extreme importance in the developing fetus. The effects of androgens on sex differentiation of the fetus are described in Chapter 35.

In the adrenal cortex, Δ⁵-pregnenolone plays a pivotal role in the formation of the 19-carbon, androgenic steroids. Δ⁵-pregnenolone is hydroxylated at the 17α position to form 17α-hydroxy-Δ⁵-pregnenolone, and as a result of the cleav-

age of the side chain of 17α-hydroxy-Δ⁵-pregnenolone, the 19-carbon-17-ketosteroid dehydroepiandrosterone is formed. The enzyme 3β-hydroxydehydrogenase catalyzes the formation of Δ⁴-androstenedione from dehydroepiandrosterone. Δ⁴-androstenendione may be formed in the adrenal cortex directly from 17α-hydroxyprogesterone. The adrenal androgens are biologically very weak, but may be converted to the very potent androgenic steroid testosterone in many tissues of the body.

The estrogenic steroids secreted by the adrenal cortex generally follow the synthetic pathway that leads from Δ⁵-pregnenolone to progesterone, 17α-hydroxyprogesterone, and Δ⁴-androstenedione. Δ⁴-Androstenedione is transformed into estrone following aromatization of ring A and demethylization of carbon 10 leaving an 18-carbon compound. Estradiol can be formed from estrone by transforming the 17-keto group to a 17-hydroxy group. The adrenal cortex normally secretes minimal amounts of estrogenic steroids.

The steroids secreted by the adrenal cortex are excreted mainly as inactive forms, although a small amount of free active steroid may appear in the urine. Cortisol, for instance, is released by the adrenal cortex and is transported in the blood partly bound to an alpha globulin, transcortin, and partly in the free state. Only the free cortisol is meta-

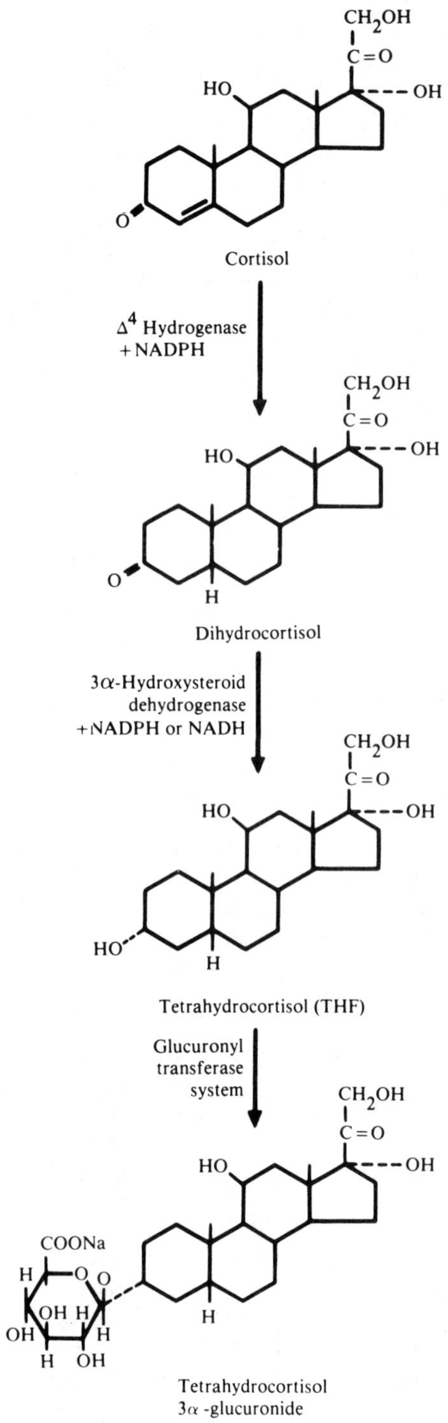

Cortisol

Δ^4 Hydrogenase
+ NADPH

Dihydrocortisol

3α-Hydroxysteroid
dehydrogenase
+ NADPH or NADH

Tetrahydrocortisol (THF)

Glucuronyl
transferase
system

Tetrahydrocortisol
3α-glucuronide

bolically active. The blood level of cortisol is constantly reduced as a result of degradation to an active form in the liver. In this organ and others, the adrenal hormones undergo further changes to form the inactive tetrahydro forms (Fig. 34-5). The active hormones are rendered inactive by saturation of ring A and substitution of hydroxyl groups at carbons 3, or 20, or both. The metabolites are rendered water soluble by conjugation with glucuronic acid in the liver prior to excretion by the kidney. The adrenal cortex of the normal human adult secretes approximately 15 to 30 mg of cortisol per day, 2 to 5 mg of corticosterone, and 0.05 to 0.20 mg of aldosterone.

The major urinary metabolites of the adrenal androgens are dehydroepiandrosterone, the androsterones, and the etiocholanolones, and they are excreted as 17-ketosteroids that have been rendered inactive by saturation of ring A and hydroxyl substitution at carbon 3. These are excreted as sulfates at the rate of about 15 mg per day in the adult male and 10 mg per day in the adult female. The adrenal cortices normally start to contribute to the urinary 17-ketosteroids at an early age in both sexes. The rate of excretion increases with age from about 1 to 2 mg per day at the age of 2 until it reaches the adult level at about 23 to 25 years of age. Part of the adult level is of testicular or ovarian origin, depending on the sex.

FUNCTIONS OF THE ADRENOCORTICAL STEROIDS

Cortisol and aldosterone are the physiologically significant steroids normally secreted by the human adrenal cortex. Cortisol, a glucocorticoid, exerts its effects on the metabolism of protein, carbohydrate, and fat and on inflammatory and immunological processes. Aldosterone, a mineralocorticoid, is concerned primarily with electrolyte metabolism. The effects of the androgenic and estrogenic steroids will be discussed in Chapter 35.

The effects of the adrenal steroids are best illustrated by the changes that take place after removal of the adrenals in animals. Following adrenalectomy, animals invariably die if no supportive means are instituted. The observation that such animals can be maintained indefinitely by the adrenal steroids is taken as proof that the cortical portion of the gland, not the medullary portion, is essential for life. Symp-

Fig. 34-5. Major pathway for the inactivation and excretion of cortisol. NADPH = the reduced form of nicotinamide adenine dinucleotide phosphate; NADH = the reduced form of nicotinamide adenine dinucleotide.

toms that appear in these animals include loss of appetite, asthenia, gastrointestinal disturbances, hemoconcentration, hypotension, renal failure, and reduced body temperature.

GLUCOCORTICOID EFFECTS. Severe defects in protein, carbohydrate, and fat metabolism are observed in patients with adrenal insufficiency. In the person lacking cortisol there is a tendency for the liver to lose glycogen. The person's ability to mobilize the precursors of glycogen from the peripheral tissues is reduced, the rate at which glucose is utilized by the peripheral tissues is elevated, and the deposition of body fat is reduced. Therefore, the patient whose adrenals have been removed or who has hypoadrenalism is prone to the development of hypoglycemia and is extremely sensitive to the action of insulin. Such a patient also exhibits an elevated plasma nonprotein nitrogen, which results not from increased protein breakdown but from renal decompensation that develops as a result of circulatory failure.

Cortisol, on the other hand, is diabetogenic. It can reverse all the changes in carbohydrate and fat metabolism observed in adrenal insufficiency. Following the administration of cortisol to a fasting subject, a marked increase in liver glycogen occurs within hours. This is accompanied by an increase in the plasma sugar and a marked increase in the excretion of nonprotein nitrogen. The increase in total body carbohydrate in the "steroid diabetic" must therefore be the result of decreased glucose oxidation and/or an accelerated liver glycogen formation from tissue protein (gluconeogenesis). Various studies have shown that cortisol markedly stimulates protein breakdown in peripheral tissue and the uptake of amino acids by the liver. It has been observed that cortisol has a protein catabolic effect with respect to muscle and lymphatic tissue, and it leads to an overall negative nitrogen balance. However, cortisol exerts a strong protein anabolic effect on hepatic tissue. This differential effect on the liver may explain the diabetogenic effect. Myocytolytic and lymphocytolytic effects are induced by cortisol. In addition, it induces an increase in enzyme (protein) synthesis by hepatic cells. One can produce hyperglycemia and glucosuria in intact animals fed a high-carbohydrate diet by the prolonged administration of glucocorticoids. Conversely, the metabolic symptoms of diabetes mellitus may be alleviated by adrenalectomy.

Glucocorticoids also mobilize fat from its storage place in adipose tissue and increase the free fatty acid levels in plasma. Together with greater mobilization of fat there is a redistribution of fat depots that produces a "moon face" (rounding of the cheeks), "buffalo hump" (growth of supraclavicular and upper dorsal fat pads), increased axial fat, and a loss of fat on the extremities. As in the diabetic state there is a decreased formation of fat from carbohydrate. In general, the glucocorticoids tend to act antagonistically to insulin with respect to its actions on carbohydrate and fat metabolism.

Cortisol is the most active naturally occurring steroid in the human in terms of its effects on liver glycogen formation. On a molar basis, cortisone is about 65 percent as effective as cortisol, and corticosterone is about 35 percent as effective; aldosterone and deoxycorticosterone show less than 1 percent of the activity of cortisol. The effects of cortisol on protein, carbohydrate, and fat metabolism are illustrated in Figure 34-6.

The glucocorticoids cause a rapid involution of the thymus. They also suppress lymphoid tissue activity and act to induce lymphopenia and eosinopenia. The lymphocytolytic effect of the glucocorticoids interferes with the conversion of lymphocytes to antibody-producing cells and consequently interferes with the immune response. The glucocorticoids are thus effective immunosuppressants and are used therapeutically to counteract rejection of organ or tissue transplants in humans.

The glucocorticoids profoundly influence the inflammatory response. They decrease capillary permeability and depress the release of histamine and the migration of inflammatory cells. All of these effects reduce the inflammatory response of many tissues. When glucocorticoids are given in excess over a period of hours or days, local inflammatory reactions to irritating substances are greatly reduced or delayed, hypersensitivity reactions to antigens are suppressed, and healing of wounds is delayed. Therefore, the actions of glucocorticoids can be deleterious to the combating of infections and the healing of wounds, although alterations of these reactions may be of benefit in certain acute hypersensitivity states.

The adrenal steroids may also have many diverse effects. Those include muscular weakness as a result of muscle wasting; increased rates of secretion of gastric hydrochloric acid and pepsin, leading to the development of gastric and duodenal ulcers; hypertension and increased capillary resistance and fragility; increased excitability of brain tissue, with euphoria and restlessness; and stimulation of bone marrow. The adrenal steroids are required for normal thyroid, gonad, adrenal medullary, sympathetic nervous system, and pituitary function. It is generally accepted that cortisol institutes a "permissive" action on many functions that allows the effects of other hormones to be manifested. The effects of the glucocorticoids are mediated through cytoplasmic receptors. The hormone-receptor complex then regulates the level of a specific messenger ribonucleic acid (mRNA). The changes in the synthesis of proteins is stimu-

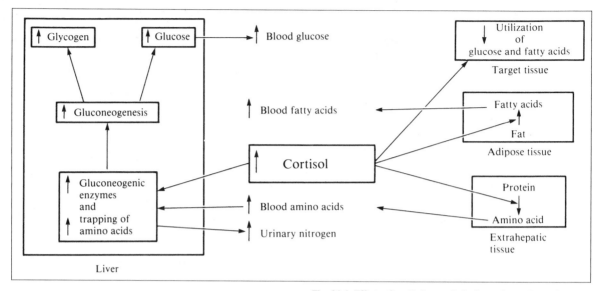

Fig. 34-6. Effects of cortisol on carbohydrate, fat, and protein metabolism.

lated by the mRNA. These then mediate the specific response of the target cell.

MINERALOCORTICOID EFFECTS. Another primary abnormality resulting from adrenalectomy is a disturbance in electrolyte metabolism. This alteration depends in part on the effect of renal handling of sodium, potassium, and water. In the absence of aldosterone, tubular reabsorption of sodium, chloride, and water decreases, and reabsorption of potassium increases. Therefore, adrenalectomy results in hyponatremia, hypochloremia, hyperkalemia, and extracellular dehydration. Each abnormality then contributes to hemoconcentration, acidosis, hypotension, decreased glomerular filtration rate, extrarenal uremia, and shock. The administration of aldosterone reverses these changes by virtue of its actions on the distal convoluted tubule, where it acts to increase the reabsorption of Na^+, along with Cl^- and water, to decrease the reabsorption of K^+, and to increase the urinary loss of H^+.

The effects of aldosterone are mediated through cytosolic receptors. The hormone-receptor complex acts on the nucleus to stimulate deoxyribonucleic acid (DNA)–directed RNA synthesis of a specific protein that enhances the transcellular movement of sodium from the filtrate into the cells of the distal convoluted tubule. A "sodium pump" then moves the sodium out of the cell on the serosal side into the extracellular fluid (ECF) and conserves sodium. The alterations induce an expansion of the ECF volume, a shift of

sodium to the intracellular compartment, and a depletion of potassium in both the extracellular and intracellular compartments. An extracellular metabolic alkalosis characterized by increased serum pH and carbon dioxide combining power, mild hypernatremia, and hypochloremia results (Fig. 34-7).

In addition to the renal influences, a deficiency or excess of adrenal steroid activity exerts a direct effect on the passage of sodium, potassium, and hydrogen ions across cell membranes. Excessive amounts of aldosterone may lead to an expansion of the ECF volume that is greater than the amount of exogenous water retained. Thus, intracellular fluid dehydration must occur. Adrenalectomy or hypoaldosteronism is accompanied by changes that lead to hyponatremia, hypochloremia, hyperkalemia, extracellular dehydration, and circulatory collapse. It should be pointed out that during continued administration of aldosterone to normal subjects, an "escape" from the intense sodium-retaining effect is observed within 5 to 10 days, even though steroid administration is continued. Aldosterone also decreases the sodium concentration and increases the potassium concentration in sweat and the secretions of the salivary and intestinal glands. Aldosterone is approximately 25 times more potent than deoxycorticosterone in its sodium-retaining effect, and about 800 times more potent than cortisol.

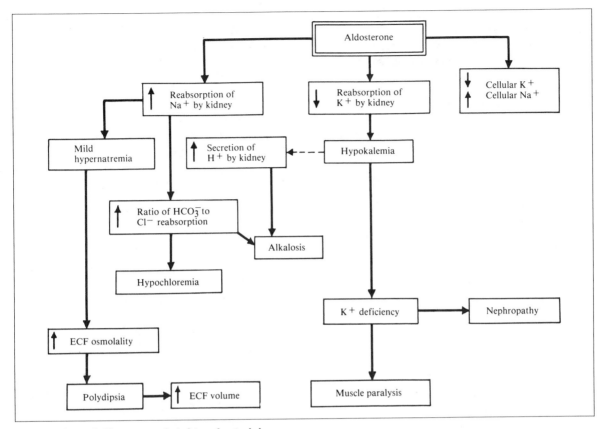

Fig. 34-7. Effects of aldosterone on electrolyte and water balance.

ROLE OF ADRENAL CORTEX IN STRESS. Following adrenalectomy or in adrenal insufficiency, there is little ability to tolerate changes in the internal and external environment, such as cold, heat, infections, trauma, or prolonged exercise. The resistance to environmental change in such subjects is very low, and they will succumb to stresses less severe than those tolerated by normal persons. In 1956, Selye showed that when an animal is subjected to a stressor, it reacts in a stereotyped manner. A *stressor* is any stimulus that causes an increase in the secretion rates of CRF, ACTH, and cortisol, leading to an increase in the concentration of cortisol in the plasma greater than that normally observed at the same hour of the day in undisturbed normal subjects during the normal activity of the sleep-wake cycle. The entire reaction to the stressor may be termed a *stress reaction.* If the same stimulus is given to an animal whose hypophysis or adrenals have been removed, its ability to survive the injury is minimal, but survival can be attained if

cortisol or ACTH is administered. Thus protection is afforded by glucocorticoids, and activation of the pituitary-adrenal axis must ensue. The adjustments that follow a stimulus must be useful to the subject in the attempt to maintain homeostasis. Many theories have been advanced to suggest how the adrenal steroids act to protect the subject, but none can account for the protection afforded. It should be emphasized that an increase in plasma cortisol concentration is essential for humans if they are to withstand severe stress.

REGULATION OF ADRENOCORTICAL FUNCTION. The removal of the anterior pituitary causes notable changes in adrenocortical function and morphology. Hypophysectomy results in atrophy of the adrenal cortex, a decrease in the secretion rate of cortisol to about 10 percent of that seen normally, and a decrease of about 50 percent in the rate of aldosterone secretion. The level of aldosterone secretion is maintained to such an extent that death does not usually ensue as a result of adrenocortical insufficiency if a very protected environment is maintained. Although signs of adrenocortical insufficiency are observed, the marked alterations in electrolyte metabolism normally seen following adrenalectomy are absent. Thus, in the absence of ACTH, aldosterone secretion rates can be maintained at levels that are compatible with life. However, ACTH is required for maximal aldosterone output by the adrenal cortex. Other factors that participate in the regulation of aldosterone secretion will be discussed later.

ACTH is a polypeptide containing 39 amino acids and has a molecular weight of about 4500. A 24–amino acid fragment of this hormone has biological activity equal to that of the parent compound, ACTH. ACTH is secreted by a specific basophilic cell of the anterior pituitary, the corticotrope. The corticotrope synthesizes two precursor polypeptides of ACTH, prepro-opiocortin and pro-opiocortin. Prepro-opiocortin is a straight-chain polypeptide containing 265 amino acid residues. It contains a short signal peptide, an N-terminal fragment, ACTH, and β-lipotropin (β-LPH). Pro-opiocortin is formed by removal of the signal peptide

and is not released into the plasma. Opiocortin is formed following the removal of the N-terminal fragment and contains within its sequence ACTH and β-LPH. ACTH is secreted and contains within it the amino acid sequences of α-melanocyte–stimulating hormone (α-MSH), amino acids 1 to 13, and a corticotropinlike intermediate lobe peptide (CLIP), residues 18 to 39. β-LPH has 91 residues and is also secreted into the plasma. It contains within its sequence γ-LPH (1–58), β-endorphin (61–91), and β-MSH (41–58). The powerful opiate met-enkephalin is also contained within the β-endorphin sequence (61–65). A schematic representation of prepro-opiocortin and its subunits is presented in Figure 34-8.

There is a sharp diurnal variation in plasma ACTH. Its concentration in plasma during the early morning (0600 hrs) is about 55 to 65 pg per milliliter in normal, undisturbed subjects, but similar subjects have a plasma ACTH concentration of about 25 to 35 pg per milliliter in the early evening (1800 hrs). The plasma cortisol levels are consistent with the plasma ACTH levels, and the concentration of each of these

Fig. 34-8. Prepro-opiocortin and its daughter polypeptides. MSH = melanocyte-stimulating hormone; LPH = lipotropin; CLIP = corticotropinlike intermediate lobe peptide.

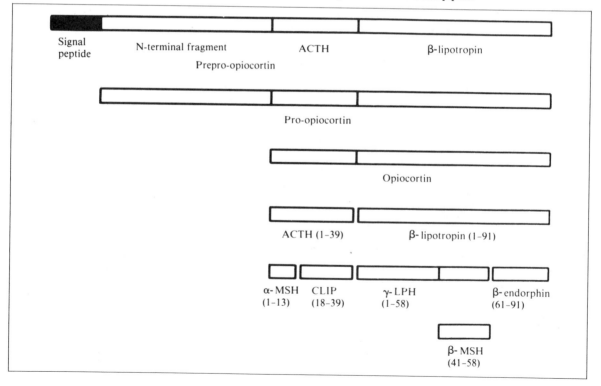

hormones in the plasma is related to the sleep-activity cycle. These changes are shown in Figure 34-4.

Available evidence indicates that ACTH stimulates the secretion of cortisol by affecting enzymatic steps early in the synthetic pathway by accelerating the conversion of cholesterol to Δ^5-pregnenolone, and ACTH appears to have little action on the synthesis of steroids on later steps. The early effects of ACTH on adrenal steroid synthesis are mediated through cyclic AMP. This activates phosphorylase, and glycogen is converted to glucose 1-phosphate and then to glucose 6-phosphate. Increased amounts of generated nicotinamide adenine dinucleotide phosphate then serve as an energy source for steroid synthesis. The more long-term effects of ACTH on protein synthesis and growth of the gland appear to be dependent on an increased synthesis of mRNA, increased total RNA content, an increased number of polysomes, and increased mitotic activity.

When ACTH is administered to a normal animal or one that has had a hypophysectomy, there is a marked increase in the rates of secretion of all of the adrenal steroids. ACTH also induces an increase in the size of cells in the zona fasciculata and zona reticularis and increases the mitotic activity of the cells. The hypertrophic and hyperplastic changes result in an increase in adrenal size. An increase in adrenal blood flow also follows ACTH administration. Further, the hormone acutely depletes the lipid granules of cortical cells and reduces both the cholesterol and ascorbic acid concentrations in the gland.

Changes in the secretion of ACTH by the adenohypophysis follow removal of the adrenal cortex or the administration of large amounts of cortisol. In the former case, an increase in ACTH secretion occurs; in the latter, ACTH secretion may be completely inhibited.

An intricate homeostatic mechanism regulates the normal secretion of cortisol by the adrenal cortex (Fig. 34-9). The rate of synthesis and secretion of ACTH is regulated directly by a hormone, corticotropin releasing factor (CRF), which is secreted by neurons located in the basal medial hypothalamus and is a polypeptide that has not been characterized. It is released into the hypophyseal portal circulation, through which it gains access to the corticotropes, where, through the mediation of cyclic AMP, it stimulates ACTH synthesis and secretion. The regulation of ACTH secretion thus involves factors that regulate the secretion of CRF and effect the response of the corticotropes to CRF. These factors include the neural influences on the CRF-secreting cells, the influence of cortisol on the CRF-secreting and ACTH-secreting elements, and other humoral agents that act on the hypothalamohypophyseal unit.

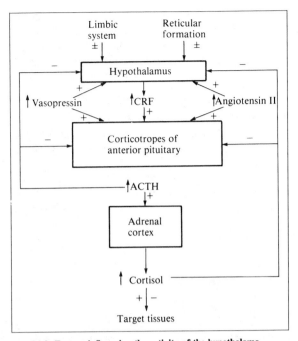

Fig. 34-9. Factors influencing the activity of the hypothalamo-hypophyseal-adrenocortical system. CRF = corticotropin-releasing factor; + = stimulation; − = inhibition.

Early studies showed that removal of the anterior pituitary from its vascular connection, the hypophyseal portal vessels, to the hypothalamus notably decreased ACTH secretion and abolished the normal stress response. Similarly, destruction of the basal medial hypothalamus abolished all stress responses. Stimulatory neural inputs to the basal medial hypothalamus and the CRF-secreting neurons are several and have been shown to arise in the amygdaloseptal complex and medial reticular formation. Lesions of the rostral midbrain and posterior diencephalon depress CRF release and cortisol secretion. The stimuli that activate the CRF-ACTH-cortisol stress response must in part act via the various stimulatory pathways. Inhibitory neural inputs into the neurons that secrete CRF arise in the hippocampus. Electrical stimulation of the posterior hypothalamus decreases CRF and ACTH secretion, and lesions of the posterior midbrain increase CRF and ACTH secretion. Stimulation of specific areas of the lateral reticular formation also inhibit CRF and cortisol secretion. Present evidence would indicate that the synapses that stimulate the CRF-secreting neurons are cholinergic in nature, whereas those that inhibit the CRF-secreting neurons are noradrenergic. The circadian rhythm of the CRF-ACTH-cortisol system is probably imposed via neural inputs into the anterior hypothalamus, because lesions in this area abolish the rhythm, but still permit the normal stress response of the system to occur. The location of the area that supplies the rhythmic input to the system is not known.

It has been known for many years that a reciprocal relationship exists between the adrenal cortex and the adenohypophysis. Early studies revealed that the systemic administration of cortisol induces a decrease in ACTH secretion and a resultant decrease in cortisol secretion and adrenal atrophy, and that pretreatment with high doses of cortisol prior to exposure to an external or internal stimulus abolishes the normal stress response (e.g., an increase in ACTH secretion). It has also been shown that adrenalectomy or the absence of cortisol results in the secretion of excessive amounts of ACTH. The absence of cortisol likewise leads to excessive β-MSH secretion and high blood levels of β-MSH. Normal blood cortisol levels inhibit β-MSH secretion. More recent studies indicate that the infusion of minute amounts of cortisol solutions or the implantation of cortisol pellets into the anterior pituitary acts to inhibit the release of ACTH under a variety of conditions. By means of such in vitro techniques as the monolayer culture of pituitary cells or isolated incubated glands, it has been shown that the ability of the corticotropes to secrete ACTH in response to CRF is inhibited by adding cortisol to the medium. Cortisol also acts on the hypothalamus to decrease CRF secretion; cortisol administration lowers the content of CRF in the hypothalamus, and the implantation of cortisol in the hypothalamus decreases CRF secretion. Cortisol also decreases CRF secretion by exerting an inhibitory action on the amygdala.

Two other humoral agents, vasopressin and angiotensin II, act on the hypothalamohypophyseal-adrenocortical system to increase the activity of the system. Each acts directly at the level of the hypothalamus and at the level of the corticotrope to stimulate CRF and ACTH respectively. Vasopressin appears to potentiate the action of CRF on the corticotropes. It should be pointed out that high systemic levels of both vasopressin and angiotensin II are required to induce their CRF-stimulating and ACTH-stimulating effects. Figure 34-7 illustrates the factors that regulate the activity of the hypothalamohypophyseal-adrenocortical system. Psychological factors such as anxiety also affect this system and increase its activity. Similarly, various sedatives, tranquilizers, and anesthetics on initial administration stimulate the stress response, but on prolonged administration they may suppress it. Each acts on the system at the level of the central nervous system.

CONTROL OF ALDOSTERONE SECRETION. The rate of aldosterone secretion is not completely independent on ACTH, because following hypophysectomy it is reduced to about one-half the normal rate. Also, varying degrees of atrophy of the zona glomerulosa follow hypophysectomy, and the degree of atrophy is directly related to the time interval between hypophysectomy and autopsy.

The factors that alter aldosterone output are many, but the alterations in sodium balance and in the volume of the body fluids stand out as especially important. Increased output of aldosterone, without a simultaneous rise in cortisol output, has been shown to follow restriction of dietary sodium, high potassium intake, water depletion, thoracic postcaval constriction, and aortic constriction. Hemorrhage, anxiety, surgical stress, and trauma also stimulate aldosterone secretion, but there is a concurrent elevation of cortisol secretion. This would indicate that hypothalamohypophyseal activation of ACTH release has occurred.

Davis first demonstrated that the renin-angiotensin system is an important factor in the regulation of aldosterone secretion by the zona glomerulosa. Hyperaldosteronism resulting from chronic thoracic caval constriction or occurring during chronic sodium depletion, acute blood loss, or acute aortic constriction is accompanied by a decrese in renal blood flow, an increase in the renin content of the kidney, and hypergranulation (with occasional hyperplasia) of the juxtaglomerular cells of the kidney. The juxtaglomerular apparatus has been described and its functional role discussed in Chapter 22. Nephrectomy in the animal without a hypophysis consistently causes a further marked reduction in the secretion of aldosterone by the zona glomerulosa. Present evidence would indicate that a decrease in renal perfusion pressure or in filtered sodium load acts on the renal juxtaglomerular complex, and an increase in renin production follows. The renin then acts on the circulating angiotensinogen to form the decapeptide angiotensin I. The latter, in the presence of chloride and a converting enzyme, is converted to the octapeptide angiotensin II, a potent hypertensive agent. Angiotensin II acts directly on the zona glomerulosa to increase aldosterone secretion. The aldosterone then acts on the nephron and increases sodium reabsorption and, along with it, water reabsorption. The extracellular fluid is expanded, renal perfusion pressure is elevated, and the stimulus for renin production is removed. Angiotensin II can stimulate aldosterone secretion in the hypophysectomized, nephrectomized animal without increasing cortisol secretion, and it stimulates the in vitro conversion of cholesterol to aldosterone.

Angiotensin II, vasopressin (antidiuretic hormone), and alpha-adrenergic agonists each directly inhibit renin se-

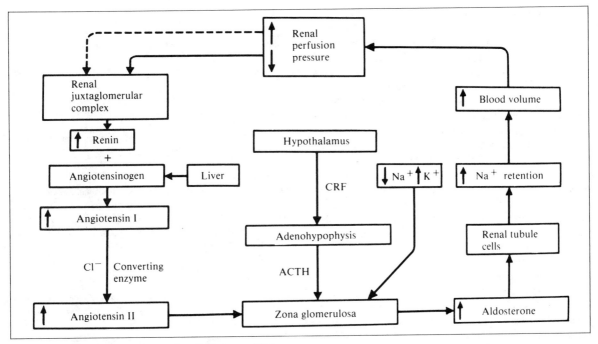

Fig. 34-10. Factors influencing the secretion of aldosterone. CRF = corticotropin-releasing factor; —— = stimulation; ---- = inhibition.

cretion, whereas beta-adrenergic agonists stimulate renin secretion. A receptor system in the macula densa detects changes in the rate of delivery of sodium to the distal tubule, and in the presence of low delivery rates of sodium the secretion of renin is stimulated. Conversely, when the sodium delivery rate to the distal nephron is high, the secretion of renin is inhibited.

Circulating Na^+ and K^+ exert an effect on aldosterone secretion. High K^+ and low Na^+ act directly on the zona glomerulosa to increase aldosterone secretion, because, in the absence of the adenohypophysis and/or kidneys, aldosterone secretion remains high in either circumstance. Also, using the isolated perfused adrenal gland, lowering the Na^+, or increasing the K^+ concentration of the Na^+ perfusate provokes a rise in aldosterone secretion. The factors that regulate aldosterone secretion are shown in Figure 34-10.

DISEASES OF THE ADRENAL CORTEX

Observation of the scheme of biosynthetic pathways of the adrenocortical hormones leads to the conclusion that many types of functional disorders may be produced as a result either of enzyme deficiency or excess or faulty pituitary function. Various forms of congenital adrenal hyperplasia or adrenogenital syndrome are frequently observed in the newborn. Congenital adrenal hyperplasia occurs as a result of a deficiency of a specific enzyme or enzymes in the biosynthetic pathways of steroids, so that a deficiency of steroids is produced beyond the deficiency (block), and an excessive synthesis and secretion of steroids is produced in front of the block. Most forms of congenital adrenal hyperplasia include a deficiency in glucocorticoid secretion and, as a result of the increased ACTH secretion, include adrenal stimulation with hypertrophy and, in the most common forms, excessive secretion of adrenal androgens during early fetal life and extending to postnatal life. In the most marked, but rarest, form, cholesterol cannot be converted to Δ^5-pregnenolone, and thus no adrenal steroids can be synthesized. In another form the enzyme 3β-hydroxy-dehydrogenase is lacking, and consequently no progesterone, 17α-hydroxyprogesterone, or Δ^4-androstenedione is synthesized. These two forms are not compatible with postnatal life. A condition in which there is a deficiency in 17α-hydroxylase is compatible with life, because the pathways leading to corticosterone and aldosterone are not affected.

The two most prevalent forms of congenital adrenal hyperplasia involve enzymatic defects in 21-hydroxylase and 11-hydroxylase. These two deficiencies do not permit biosynthesis along either the glucocorticoid or the mineralocorticoid pathway, and andrenal steroidogenesis is shunted to the androgen-producing pathway. Therefore, each of these two conditions is characterized by adrenal insufficiency with excessive androgen secretion and masculinization, whether present in the female or in the male.

Cushing's syndrome includes clinical situations that result from excessive and long-term action of the glucocorticoids. Several primary factors may lead to exposure to high plasma level of glucocorticoids: excessive secretion of ACTH by the anterior pituitary, a glucocorticoid-secreting tumor of the adrenal gland, and administration of excessive amounts of glucocorticoids. The last is very common because of the use of glucocorticoids as immunosuppressant agents and in the treatment of certain diseases.

Hyperaldosterone states have been described and are generally associated with hypertension. Primary aldosteronism (Conn's syndrome) results from adenomatous hyperplasia of the zona glomerulosa. Secondary hyperaldosteronism generally results from excessive stimulation of the zona glomerulosa by the renin-angiotensin system.

Conditions leading to adrenocortical insufficiency may arise in a primary dysfunction of the adrenal cortex, as in Addison's disease, in which, in its severest form, one observes changes similar to those seen following adrenalectomy. These include high plasma concentrations of both ACTH and MSH, because of the lack of the feedback effect of cortisol on the anterior pituitary. Adrenal insufficiency may also be the indirect result of a deficiency in ACTH secretion by the anterior pituitary. This is nearly always accompanied by concomitant decreases in the secretion of all anterior pituitary hormones.

BIBLIOGRAPHY

Claman, H. N. Glucocorticosteroids. I. Anti-inflammatory mechanisms. *Hosp. Pract.* 123–134, July 1983.

Claman, H. N. How corticosteroids work. *J. Allergy Clin. Immunol.* 55:145–151, 1975.

Cryer, P. E. Physiology and pathophysiology of the human sympathoadrenal neuroendocrine system. *N. Engl. J. Med.* 303:436–444, 1980.

Davis, J. O., and Freeman, R. H. Mechanisms regulating renin release. *Physiol. Rev.* 56:1–56, 1976.

Finkelstein, M., and Schaefer, J. M. Inborn errors of steroid biosynthesis. *Physiol. Rev.* 59:353–406, 1979.

Fuller, R. W. Control of epinephrine synthesis and secretion. *Fed. Proc.* 32:1772–1781, 1973.

Imura, H., and Nakai, Y. "Endorphins" in pituitary and other tissues. *Annu. Rev. Physiol.* 43:265–278, 1981.

Leung, K., and Munck, A. Peripheral actions of glucocorticoids. *Annu. Rev. Physiol.* 37:245–272, 1975.

Mulrow, P. J. The adrenal cortex. *Annu. Rev. Physiol.* 43:409–434, 1972.

Parrillo, J. E., and Fauci, A. S. Mechanisms of glucocorticoid action on immune processes. *Annu. Rev. Pharmacol. Toxicol.* 19:179–201, 1979.

Peach, M. J. Renin-angiotensin system: Biochemistry and mechanism of action. *Physiol. Rev.* 57:313–370, 1977.

Phifer, R. F., Orth, D. N., and Spicer, S. S. Specific demonstration of the human hypophysial adrenocortico-melanotropic (ACTH/MSH) cell. *J. Clin. Endocrinol. Metab.* 39:684–692, 1974.

Reid, I. A., Morris, B. J. and Ganong, W. F. The renin-angiotensin system. *Annu. Rev. Physiol.* 40:377–410, 1978.

Selye, H. *The Stress of Life.* New York: McGraw-Hill, 1956.

Seron-Ferre, M., and Jaffe, R. B. The fetal adrenal gland. *Annu. Rev. Physiol.* 43:141–162, 1981.

Swartz, S. L., and Dluhy, R. G. Corticosteroids: Clinical pharmacology and therapeutic use. *Drugs* 16:238–255, 1978.

Yasuda, N., Greer, M. A., and Alzawa, T. Corticotropin-releasing factor. *Endocr. Rev.* 3:123–140, 1982.

PHYSIOLOGY OF REPRODUCTION

35

Ward W. Moore

GENETIC BASIS OF SEX

The sex of an individual is the outcome of two distinct processes: sex determination and sex differentiation. Sex determination is regulated by genetic phenomena, and genes carry the information that lays the groundwork and dictates the development of the primordial or indifferent gonad. Sex differentiation is an ordered and sequential process and refers to the course of development of the gonads, genital ducts, external genitalia, and secondary sex characteristics. It results from two development processes: the direction of differentiation of the gonads toward ovaries or testes and the differentiation of the genital ducts and external genitalia toward femaleness or maleness. Genetic, or chromosomal, sex is established at conception and governs the development of gonadal sex. Gonadal sex then controls the development of somatic, or phenotypical, sex. Each mature ovum contains 22 chromosomes, termed *autosomes,* plus an X chromosome. Each mature spermatozoon also contains 22 autosomes and either an X chromosome or

a Y chromosome. It is the sex chromosomes, X and Y, that determine the sex of an individual. The male zygote possesses an XY pair of sex chromosomes, while the female zygote possesses an XX pair.

The first specific process of division (meiosis) that leads to the formation of the mature male germ cell involves a division of chromosomes, half of each pair to one daughter cell and half to another. Thus, the two daughter cells bear different sex chromosomes, one an X and the other a Y. In the female the corresponding division produces two cells, each of which carries an X chromosome. When the ovum and spermatozoon unite, the chances are equal for the formation of an XX zygote (female genotype) or an XY zygote (male genotype). The genotypical sex of the zygote is regulated by the components contributed by the male germ cell and is determined at conception. The primary determinant of differentiation of the gonads appears to be the presence or absence of the H-Y antigen, an antigen on the zygote cell surface. The control of the expression of the H-Y antigen involves interactions between regulatory sites on the Y chromosome, the X chromosome, and some autosomes. The H-Y antigen carries determinants that masculinize the gonad, and in the absence of this masculinizing signal the gonad will undergo ovarian differentiation. Environmental and hormonal factors do not markedly influence gonadal sex differentiation. Sex chromatin tests, are now available that can determine the true sex of individuals varying in age from very young embryos to very old adults.

In a few instances, abnormal chromosome combinations occur during the transfer of chromosomes from one cell to the other. In one type the fertilized ovum contains only one X chromosome and no Y chromosome. It will result in an XO genotype, yield a chromatin-negative test, and produce an individual who either has rudimentary gonads or no gonads. The condition is known as Turner's syndrome, or ovarian dysgenesis. Not infrequently, another type of abnormal chromosomal complement, XXY, occurs. The individual resulting from this combination has a positive chromatin test, male external genitalia, and abnormal seminiferous tubules with aspermatogenesis. The condition is referred to as Klinefelter's syndrome, or seminiferous tubule dysgenesis. Other abnormal XY genotypes include XXX, XYY, and types of mosaicism such as XO/XX and XO/XY.

SEXUAL DIFFERENTIATION AND DEVELOPMENT

The reproductive system passes through a period of early embryonal development during which it is difficult, both grossly and microscopically, to tell the sexes apart. This period is referred to as the indifferent stage. Nalbandov (1976) has stated the following:

"In this stage all anlagen for subsequent differentiation into complete male and female systems are present in rudimentary form; the blueprint, as it were, is finished. All the materials for the later elaboration of the fittings and furnishing of structures are present, but no attempt is made to arrange the internal furnishings permanently until final orders are received for the emphasis of either male or female aspects of the different structures."

A bilateral longitudinal ridge, the genital ridge, appears medial to the mesonephros during the fifth week of embryonic life in the human. The surface of this ridge is covered by germinal epithelium and contains the primordial germ cells that have migrated from yolk-sac endoderm. Specific histological changes occur in this structure, the *indifferent gonad*, during the seventh week, and they indicate that the gonad is committed to either the male or female sex. In the presence of H-Y antigen the gonad will develop into a testis,

and the cells of the germinal epithelium will become organized within the medulla of the gonad, form the *anlagen* of the seminiferous tubules, and subsequently join the cords of the mesonephros to form a continuous network of tubules. Mesenchymal cells in the medulla develop into Leydig cells, which become apparent at the end of the eighth week. In the absence of H-Y antigen the gonad will become an ovary, and no changes in the indifferent gonad will be apparent until well after the tenth week. At this time the cortex of the gonad accumulates a nest of cells, which are differentiated into ovarian follicles, each containing a primary oocyte.

Two hormones secreted by the fetal testes, testosterone and anti-müllerian hormone, are responsible for the formation of the male phenotype. The fetus possesses the basic duct systems for both sexes, the mesonephric duct system and the müllerian duct system. If the ducts are to differentiate in accordance with the gonadal differentiation, either the mesonephric ducts or the müllerian ducts must degenerate. Only after continued development does one system gain ascendancy and become definite, whereas the other regresses and becomes vestigial.

The embryonic Sertoli cells of the male gonad secrete a hormone, a glycoprotein, called müllerian-inhibiting substance or anti-müllerian hormone. This hormone promotes regression of the müllerian ducts. The müllerian ducts begin to regress in the male at about the tenth week of gestation, and, under the influence of testosterone, secreted by the fetal Leydig cells, the mesonephric system develops, and differentiation proceeds toward the male phenotype. Differentiation of the external genitalia toward the male phenotype is dependent on the presence of an enzyme, 5-alpha-reductase, in the target cells and the conversion of testosterone to dihydroxytestosterone (DHT). Thus, those organs derived from the urogenital sinus, tubercle, and swellings in the male require DHT for differentiation. In the female the mesonephric system begins to disappear late in the thirteenth week, and the müllerian system becomes dominant. In the absence of the two fetal testicular hormones, testosterone and AMH, or their receptors, differentiation of the reproductive tract and the external genitalia proceeds toward the female phenotype. The origins of various parts of the genital system in both sexes are given in Table 35-1.

Varying degrees of masculinization are seen in the newborn human female with congenital adrenal hyperplasia. This is caused by a genetic defect in which simple mendelian recessive genes combine and lead to a deficiency in an enzyme system concerned with adrenal steroid synthesis. The result is excess production of adrenal androgen. The earlier in embryonic life the defect is manifested, the more

Table 35-1. Homologies between the male and female reproductive systems in humans.

Male	Primordial structure	Female
GONAD		
Testes		Ovary
Rete testes		Rete ovarii*
GENITAL DUCTS		
Vas efferens	Mesonephric tubules	Epoophoron*
Paradidymis*		Paroophoron*
Epididymis	Mesonephric duct	Gartner's duct*
Vas deferens		
Ejaculatory duct		
Seminal vesicles		
	Müllerian duct	Hydatid*
		Oviduct
		Uterus
		Vagina (upper one-third)
EXTERNAL GENITALIA		
Urethra	Urogenital sinus	Urethra
Prostatic utricle*		Vagina (lower two-thirds)
Prostate		Urethral glands
Bulbourethral glands		Vestibular glands
Glans penis	Genital tubercle	Glans clitoris
Corpus penis		Corpus clitoris
Raphe penis	Urogenital folds	Labia minora
Scrotum	Labioscrotal swellings	Labia majora

*Rudimentary.

marked are the aberrant changes observed in the newborn female genital system. The newborn male with defects in adrenal 21-hydroxylase, or 11-hydroxylase, or both also has a high rate of adrenal androgen secretion and is precociously and markedly masculinized. The genesis of these disorders is discussed in the preceding chapter. Also, in the absence of appropriate receptors for testosterone or failure to convert it to DHT, the external female phenotype is present in XY males at birth.

Under normal circumstances the effect of the maternal hormones is insufficient to cause profound changes in the fetus. The maternal estrogens and progestins cause cervical enlargement, hypertrophy of the vaginal epithelium, and growth of the mammary gland of the fetus. However, certain synthetic estrogens and progestins with androgenic properties, when administered early in pregnancy, may produce marked structural alterations (masculinization) in the female fetus. The placenta is permeable to most steroid hormones, but at birth the infant normally shows little evidence of excessive stimulation by other maternal hormones.

It should be noted that none of the fetal endocrine glands studied has been found to be indispensable for fetal survival; removal of the hypophysis, thyroid, parathyroid, gonads, or adrenals has been accomplished in the fetuses of laboratory animals, and these fetuses survive. Homeostasis of the fetus is generally maintained by way of maternal homeostatic mechanisms, but fetal hormones do play an indispensable role in normal development.

ENDOCRINOLOGY OF THE MALE

The testes of the male have a dual function, one of gamete formation and the other of hormone production, and both are essential for normal reproductive function. Gametogenesis is dependent to a considerable extent on androgen production by the testes, and both are dependent on the secretion of the gonadotropins luteinizing hormone (LH) and follicle-stimulating hormone (FSH), by the anterior pituitary, (AP). In turn, the AP is dependent on normal hypothalamic function, the secretion of gonadotropin-releasing hormone (GnRH), and intact hypophyseal portal vessels. Each step is indispensable if the male is to reach full reproductive capacity, but neither is essential to life.

The typical histological picture of the testes is of seminiferous tubules separated by interstitial connective tissue. This pattern is observed from prenatal stages throughout life, with decided variations according to age. At birth, the seminiferous tubules have not yet formed lumina, are small, and contain only spermatogonia. Leydig cells are present and functional at birth, but then regress and are difficult to identify until the approach of puberty. At this time, maturation changes, such as thickening of the tunica propria, initiation of spermatogenesis, proliferation of the Leydig cells, and the secretion of testosterone occur once again. The changes are dependent on increased hypophyseal gonadotropin production. Thus, from about the second or third month of life, by which time maternal and placental hormonal influences have disappeared, until the eighth to the eleventh year, sexual development in the male is largely in abeyance, and individuals of both sexes are in state of physiological hypogonadism.

During the infantile period the gonads are potentially able to function in an adult fashion, for they respond to gonadotropins. The secondary sex organs respond either to androgens or to estrogens. Therefore, gross insensitivity of the target organs cannot be responsible for the lack of sexual maturation in childhood. The gonads of immature animals, when transplanted to mature animals, function as mature gonads; but the gonads of mature animals, when transplanted to immature animals, become quiescent and atrophy.

The adenohypophysis of the newborn, when transplanted into the sella turcica of a mature animal following hypophysectomy, stimulates and repairs gonadal function more rapidly than the pituitary of a mature animal when transplanted into the sella turcica of an infant animal. Thus, the fetal hypophysis contains gonadotropins, and the gland of the prepubertal individual is potentially capable of stimulating the gonads. These observations form the basis of the hypothesis that the secretion of gonadotropins at puberty is dependent on a neural or neurohumoral phenomenon that is inhibited during childhood. The hypothesis is strengthened by experimental observations that lesions in the amygdala, stria terminalis, or anterior hypothalamus result in precocious puberty. Similarly, many naturally occuring lesions in the central nervous system (CNS) induce the production of gonadotropins at an early age. These lesions have one characteristic in common, namely, involvement of the hypothalamus. The manner in which the CNS-induced inhibition of gonadotropic secretion is removed at puberty is not known.

REGULATION OF TESTICULAR FUNCTIONS. The transition from boyhood to manhood normally begins between the eighth and eleventh years, although it is usually not apparent on gross examination until the tenth to thirteenth year of age. Each stage that occurs with the transition is initiated by the action of gonadotropins, whose secretion is dependent on altered CNS activity and increased GnRH secretion.

As a result of the increase in FSH and LH secretion by the AP, the seminiferous tubules and Sertoli cells are stimulated, and the Leydig cells start to proliferate and undergo functional changes. They begin to synthesize androgens (masculinizing steroids), the primary and most potent of which is testosterone.

The testicular androgens are steroidal in nature and are 19-carbon compounds, in contrast to the 21-carbon glucocorticoids and mineralocorticoids and the 18-carbon estrogenic steroids. In the testes, as in the adrenal cortex, the hormones stem from cholesterol, but the primary synthetic pathway in the testes is via Δ^5-pregnenolone, progesterone, 17α-hydroxyprogesterone, and Δ^4-androstenedione is acted on at the $17-\alpha$ position by a reductase to yield

testosterone, the main testicular androgen. The secretory rate of testosterone is stimulated by LH. Other androgens that are secreted in very small amounts by the Leydig cells are Δ^4-androstenedione, dehydroepiandrosterone, and DHT. Testosterone in the plasma is bound to a globulin called testosterone-estradiol–binding globulin, which as the name implies, also binds the ovarian hormone estradiol. The affinity of testosterone-estradiol–binding globulin is much higher for testosterone than for estradiol, and about 96 to 98 percent circulates in the bound state and 2 to 4 percent in the free and diffusible state. The testosterone level in the plasma of the prepubertal boy is generally less that 0.5 ng per milliliter and increases with advancing puberty to an adult level that has a range of 5 to 12 ng per milliliter. Both serum LH and testosterone concentrations increase throughout puberty, the increases in testosterone following those of LH. The major site of inactivation of the testicular androgens is the liver, where they are converted to androsterone, epiandrosterone, and etiocholanolone (Fig. 35-1). These metabolites are conjugated with glucuronide or sulfate, are excreted as water-soluble salts in the urine, and are regarded as 17-ketosteroids. The urinary 17-ketosteroids excretion is constant and equal in both boys and girls up to the age of 8 or 9 years and is of adrenal origin. Of the total of approximately 14 to 16 mg of 17-ketosteroids excreted per day by the adult male, about 10 mg is of adrenal origin. This figure is derived from the secretion of only 10 mg per day by eunuchs. Estradiol has been isolated from the human testes. At puberty the testicular secretion of estradiol increases, and the urinary excretion of estrogens increases. They both decrease following bilateral orchiectomy and increase following gonadotropin administration.

The mechanism of the initiation and maintenance of testicular gametogenic and secretory activity is best illustrated in the male whose hypophysis has been removed. The seminiferous tubules and Leydig cells are atrophic in such an individual, gametogenesis is halted, and testosterone secretion is minimal.

The administration of LH to the hypophysectomized animal stimulates the Leydig cells to respond by secretion of testosterone. There is no demonstrable effect of FSH on the Leydig cells, but it causes a significant stimulation of the seminiferous tubule and Sertoli cells. Its actions on the Sertoli cells include the stimulation of the secretion of a hormone, inhibin, and the secretion of a protein, androgen-binding protein, which is released into the lumen of the seminiferous tubules. Follicle-stimulating hormone alone cannot support spermatogenesis. To maintain complete spermatogenesis, both FSH and LH or testosterone must be administered. LH is probably the primary adenohypoph-

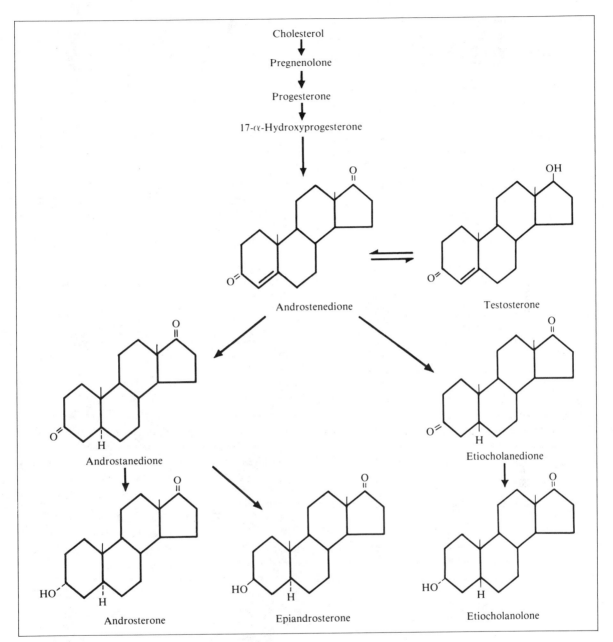

Fig. 35-1. Major biosynthetic pathway for the formation of the testicular androgen testosterone. The remaining steroids are metabolized and inactivated by the liver and excreted as 17-keto-steroids in the urine.

yseal hormone responsible for stimulation of the seminifer-ous tubule and Leydig cells, although FSH does stimulate the Sertoli cell directly. In the latter case, there is direct stimulation, while in the former an indirect action is mediated via the secretion of testosterone by the cells of Leydig. When LH is administered immediately after hypophysectomy, it has the capacity for stimulating the Leydig cells and advancing spermatogenesis and spermio-genesis. That FSH has a physiological role in spermatogen-esis is shown when administration is delayed in order to allow posthypophysectomy atrophy of the testes. In this instance, LH cannot act as a complete gonadotropin but must be supplemented with FSH in order to allow complete repair of testicular function.

The intravenous administration of testosterone in amounts adequate to maintain physiological levels in the plasma suppresses LH secretion but has no influence on FSH secretion; high plasma levels of testosterone are re-quired before FSH secretion is suppressed. Dihydroxytes-tosterone also depresses LH secretion, but it has no effect on FSH secretion. The hormone secreted by the Sertoli cells, inhibin, acts on the hypothalamopituitary complex and inhibits the secretion of FSH by the anterior pituitary. The administration of large amounts of testosterone to the intact male leads to the inhibition of gonadotropin secretion by the pituitary and results in testicular damage. The degree of testicular damage is directly related to the amount of testosterone administered. Castration of the male removes the inhibitory effects of endogenous testosterone and inhi-bin on the pituitary and results in marked increases in LH and FSH secretion respectively, and high concentrations of these hormones appear in the urine. The secretion of both FSH and LH is dependent on the release of the hormone GnRH from the hypothalamus. This hormone, a decapep-tide secreted by the hypothalamus, stimulates the synthesis and secretion of both FSH and LH; FSH and LH also act back, via a "short loop," to inhibit the secretion of GnRH. The relation among the hypothalamus, the anterior pitu-itary, and the testes is depicted in Figure 35-2.

EFFECTS OF THE ANDROGENS. The testicular steroids, with testosterone as the primary example, have been described as having androgenic effects, and these are associated with masculinization of the reproductive tract. They also induce growth effects on nonreproductive tissue, referred to as *anabolic* effects. Free testosterone may leave the plasma and enter cells by diffusion. It has several types of target cells, and it may be acted on and induces its effects in several ways. First, testosterone may bind directly with its receptor in the cytoplasm, and the testosterone-receptor

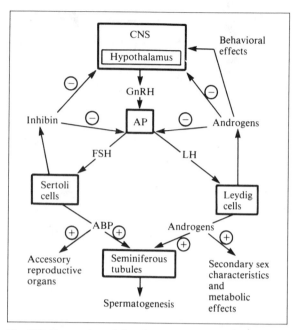

Fig. 35-2. Hormonal regulation of reproductive function in the male. ABP = androgen binding protein; AP = anterior pituitary; + = stimulation; − = inhibition.

complex then acts at the level of the nuclear chromatin to initiate testosterone-specific responses. Such responses may occur in skeletal muscle, the kidney, the pituitary, and the heart. Second, testosterone may be acted on in specific target cells by the enzyme 5-alpha-reductase, and be con-verted to DHT. The DHT then binds to its specific receptor in the cytoplasm; DHT-specific responses are then mediated following DHT-receptor complex interaction with nuclear chromatin. The production of DHT from testos-terone occurs primarily in the reproductive tract, external genitalia, and the skin of the adult. Third, testosterone may be transformed into estradiol under the influence of an aromatase in specific target cells. Estradiol then binds to its specific receptor in the cytosol, and the estradiol-receptor complex is transferred to the nucleus for initiation of estradiol-specific responses. Such conversions of testos-terone to estradiol occur in some parts of the brain. Fourth, testosterone may be acted on in specific target cells by a beta-reductase. The resulting steroids mediate their effects by binding to a beta steroid receptor in the cytosol and produce their effects in a manner similar to that described previously. The beta steroid effects of testosterone are most prominently induced in the bone marrow and in the liver.

Reproductive Tract and Secondary Sex Characteristics. The entire male reproductive tract, including the accessory glands and external genitalia, is dependent on testosterone, or DHT, or both for full morphological and functional development. Removal of these androgens results in marked reductions in the size and functions of the reproductive tract, and following castration of the male, respiratory activity of the accessory sex glands decreases markedly. Replacement therapy with testosterone reverses these effects, and its administration to an intact male induces hypertrophy and hyperplasia of all portions of the tract.

The androgens are necessary for normal spermatogenesis. They also initiate the recession of the male hairline, dictate the distribution and growth of hair on the face, body, and pubes, and stimulate hair growth in the axilla. In addition, they induce enlargement of the larynx and thickening of the vocal cords and thus lead to the deep voice of the adult male. The androgens, particularly testosterone, are necessary for both the quantitative and qualitative development of the secondary sex characteristics that distinguish the man from the boy.

Metabolism. The most notable effect of testosterone on metabolism is nitrogen retention. This is accomplished by increasing the rate of protein synthesis. Following testosterone administration the rate of urinary nitrogen excretion decreases, and a positive nitrogen balance results. In many species, testosterone has been shown to increase the rate of protein synthesis and to decrease the rate of protein breakdown. It should be noted that the male accessory sex apparatus accumulates nitrogen in advance of other tissues.

Following castration, some skeletal muscles cease to accumulate nitrogen, some accumulate it at a reduced rate, and others are unaffected. Androgens cause an increase in the working ability of muscles as a result of their stimulating action on the number, thickness, and tensile strength of muscle fibers, and they support glycogen, phosphocreatine and adenosine triphosphate synthesis. They are responsible for the greater muscle development in men as compared with women and regulate in part the distribution of body fat and the skeletal configuration of the adult male.

The androgens also stimulate the renal retention of sodium and potassium, and, together with water, but they are considerably less effective in this respect than aldosterone. The effects of testosterone on linear growth and bone metabolism are such that it stimulates growth of bone but limits the final size attained by long bone. It stimulates the deposition of protein matrix and causes the retention of calcium and phosphorus yet, at the same time, promotes closure of the epiphyses. Testosterone is a potent growth stimulator, but its growth-promoting effect is observed only in the presence of an otherwise normal hormonal environment.

Physiology of Spermatozoa. The gametogenic function of the testes is dependent on FSH secretion by the AP and testosterone secretion by the Leydig cells. Sertoli cells are intimately associated with the process of maturation of spermatozoa. Each Sertoli cell extends from the basement membrane of the seminiferous tubules to the lumen of the tubule and is joined to its neighbor cell by a very tight junction. These junctions provide a barrier, the *blood-testes barrier*, that limits the movement of molecules from the plasma to the lumen of the tubules, and vice versa. The Sertoli cells also secrete *androgen-binding protein* into the lumen of the seminiferous tubules. This protein binds avidly with testosterone and DHT and is transported to the epididymis, where it is degraded. It is presumed that androgen-binding protein facilitates the actions of these androgens in the epididymis. Its overall secretion rate is governed by FSH, and it may be maintained by testosterone if administered immediately following hypophysectomy in experimental animals. A small quantity of androgen-binding protein may also be secreted into the plasma by the Sertoli cell.

The seminiferous tubules constitute about 90 percent of the testicular mass. In adult testes the spermatogonia give rise to primary spermatocytes. Both of these cell types are diploid; i.e., the cells contain XY sex chromosomes. The primary spermatocytes undergo meiotic division and form haploid secondary spermatocytes, which in turn divide to form spermatids. After a series of transformations, called *spermiogenesis,* the spermatids yield sperm cells, or spermatozoa. The process is fairly slow, but it continues uninterrupted and may be maintained to extreme old age. It has been estimated that a man will produce 1×10^9 sperm for every ovum shed by a woman's ovaries.

The testes descend into the scrotum during late prenatal life. The main function of the scrotum in humans is to provide a testicular environment that is about 5° F lower than the core temperature. Failure of testicular descent, called *cryptorchism,* occurs commonly in humans and can be the result of androgen deficiency or anatomical obstruction. In either naturally occurring or experimental cryptorchism, spermatogenesis eventually stops completely. If spermatogonia are still present in the testes, spermatogenesis can be reinstituted on removal of the testes from the body cavity. However, after prolonged exposure to elevated temperature, testicular damage becomes irreversible. Temporary or permanent sterility in men may result from prolonged fever, even though the testes have descended into the scrotum. It is not known why high temperatures are injurious to sperm.

The manner in which sperm are transported through the seminiferous tubules and epididymis is unknown. Various mechanisms have been proposed. They include a continuous *vis a tergo* provided by the release of new sperm, conveyance in seminiferous tubule secretions, ciliary action by cells of the efferent ducts, and contraction of smooth muscles of the epididymal tubules. However, sperm within the seminiferous tubules and epididymis are nonmotile and do not become motile until they come into contact with the seminal plasma. That sperm may survive a long time in the epididymis can be demonstrated by blocking the ducts between the epididymis and the testes. By using this technique and studying the ejaculate, one can demonstrate that epididymal sperm remain viable up to 60 days. Fertile sperm may be present in the ejaculate for as long as six weeks after castration. On the other hand, sperm in the vas deferens rapidly lose their fertilizing capacity. Sperm are not found in the seminal vesicles, prostate, or any of the other accessory glands.

The seminal plasma consists of secretions of the testes and epididymis (less than 5 percent), seminal vesicles (about 30 percent), prostate (about 60 percent), and Cowper's glands (less than 5 percent). These secretions have two main functions: They furnish the sperm with metabolizable substrate and function as a suspending and activating medium.

A major constituent of the secretion of the seminal vesicles is fructose, which serves as the chief source of energy for the sperm. Major constituents of the prostatic contribution to seminal plasma are citric acid, calcium, acid phosphate, and proteolytic enzyme. The last is primarily responsible for the liquefaction of coagulated semen. The functional significance of acid phosphatase in seminal fluid is not known. This enzyme has clinical significance because it enters the bloodstream when malignant growth of the prostate occurs. The secretions of the prostate and seminal vesicles are not necessary for the production of viable sperm, because both testicular sperm and sperm taken from the epididymis are viable.

Erection is a vascular phenomenon that is dependent on the vascular structural pattern of the penis. It is the result of venous constriction and arterial dilation that allows blood to flow under high pressure into the erectile tissue. The erectile tissue is a spongelike system composed of vascular spaces that are relatively collapsed and contain little blood in the flaccid state. However, during erection these spaces become distended with blood, and the pressure approaches

that in the carotid artery. Erection may result from either physical or psychic stimuli; the motor pathways are via the sacral component of the craniosacral system, the *nervi erigentes,* and there is a center for reflex erection in the sacral cord. The subsidence of erection can result from stimulation of the sympathetic innervation of the penis.

Ejaculation actually refers to two distinct actions, emission and ejaculation, both of them reflex phenomena. The afferent arc of these reflexes originates in the sense organs of the glans penis, and the impulses are transmitted centrally through the internal pudendal nerves. The first action of ejaculation, which delivers sperm and seminal plasma into the urethra, is the result of contraction of smooth muscles of the genital tract and is called emission. This action includes peristaltic contractions of the testes, epididymis, and vas deferens that cause the expulsion of sperm into the internal urethra. It also includes the contractions of the seminal vesicles, prostate, and bulbourethral glands that result in the movement of their respective secretions into the urethra. Emission is evoked by stimulation of the hypogastric nerves. The second action, ejaculation, causes the expulsion of seminal fluid from the urethra to the exterior as a result of striated muscle contraction (bulbocavernosus muscle). This results from increased neural activity transmitted peripherally via the internal pudendal nerves.

The volume of ejaculate is from 2 to 6 ml in humans. The ejaculate contains more than 10^7 sperm per milliliter and has a slightly alkaline reaction (pH = 7.1–7.5) and a specific gravity of about 1.028. Coagulation of normal semen occurs promptly after ejaculation. However, it liquefies within several minutes, and the sperm attain full motility.

The life span of sperm, or the time during which they are able to fertilize ova, is relatively short (about 24–36 hr). However, sperm may be frozen to $-169°$ C, stored, and thawed without impairment of their fertilizing capacity. It has been claimed that sperm concentrations of less than 50×10^6 sperm per milliliter of ejaculate indicate sterility. Of this number introduced into the female reproductive tract, less than 10^4 sperm reach the oviducts, less than 10^2 reach the vicinity of the ovum, 10^1 may penetrate the zona pellucida of the ovum, but only 1 sperm enters the ovum and accomplishes fertilization. Therefore, the sine qua non for fertility in the male is the number of sperm per ejaculum. Other criteria, such as percentage of abnormal sperm (with respect to morphology) and volume of ejaculate, are also used in assessing semen quality. It should be pointed out that sperm are incapable of penetration of the zona pellucida that surrounds the ovum until they have been in the female reproductive tract for about 4 hours. During this time they undergo a process known as *capacitation,* which results in changes that permit the release of enzymes in the head of the sperm.

ENDOCRINOLOGY OF THE FEMALE

In addition to the production of gametes the basic function of the ovaries is to secrete hormones that regulate the activity of the female reproductive tract and determine the secondary sex characteristics. The gametogenic and endocrine aspects of ovarian function fluctuate rhythmically during the active reproductive life of the female. The periodic changes in the functional activities of the ovaries are determined by the interrelationships between the activity of the adenohypophysis and the ovaries. The adenohypophyseal gonadotropins FSH and LH are concerned with ovarian function as well as with testicular function. In addition, a third pituitary hormore, prolactin (PRL), is concerned with mammary gland function. Its role in regulating cyclic ovarian activity in humans is not known. The reproductive cycle of the female is a complex series of coordinated events involving the CNS, AP, ovaries, oviducts, uterus, cervix, and vagina and is referred to as the *menstrual cycle*. The female experiences a shorter reproductive life than does the male. The cyclic reproductive activity in the female commences at puberty (10 to 15 years of age) and is maintained for approximately 30 years. The cyclic activity gradually decreases, the ovaries involute, and reproductive cycles cease during the fifth decade of life. This transition is termed the *menopause* and is followed by hypogonadism for the remainder of the life span.

Reproductive functions and behavior are markedly influenced by the gonadal hormones. That these are not the only factors involved is emphasized by the adverse effects on the functions of the pituitary-gonadal axis of both female and male of the lack of certain nutritional factors and caloric intake, nongonadal endocrine disorders, changes in the external environment, and CNS and psychological factors.

OVARY. The size and microscopic appearance of the paired ovaries in the adult vary with the period of the reproductive cycle. The most important functional components of the mature ovary are the *follicles* and *corpora lutea*. Three stages of growth of the follicles can be noted. The immature stage (i.e., the primary follicle) is observed from the embryonic to the late life of the ovary. This structure makes up the bulk of the population of follicles. The primary follicle of the ovum is surrounded by several layers of granulosa cells and has no vitelline membrane. The secondary follicle is formed as a result of the development of the *zona pellucida,* which is a membrane surrounding the ovum. The tertiary follicle is characterized by the presence of an *antrum,* a fluid-filled space surrounding the ovum. The ovum, enclosed in granulosa cells, is bathed by liquid, the *liquor folliculi.* As the follicle grows, it moves toward the cortex of the ovary, and the antrum enlarges to produce the mature follicle. The mature follicle, called *graafian follicle,* extends throughout the thickness of the cortex and bulges as a blister on the free surface of the ovary. In addition, the mature ovary contains follicles that exhibit varying degrees of degeneration; these are the atretic follicles.

The *corpus luteum* is a temporary endocrine organ that forms following ovulation as a result of luteinization of the *granulosa* and *theca interna* cells of the ovarian follicle. Luteinization involves hypertrophy and hyperplasia of the granulosa and theca interna cells with the development of lipid inclusions within the cells. Immediately after ovulation, the cavity of the follicle becomes filled with blood and lymph. Gradually, the space occupied by the fluid is filled by the luteal cells, and a well-developed blood supply is formed within the mass of lutein cells. Seven to eight days after ovulation, regressive changes are seen, such as loss of lipid material and invasion of connective tissue. The span of functional activity of the corpus luteum is very constant and in humans lasts approximately 10 to 12 days, after which the gland becomes nonfunctional and referred to as the *corpus albicans*. During the ensuing weeks, further degenerative changes occur, and the organ is replaced by connective tissue.

OVARIAN HORMONES. The ovaries secrete three types of steroid hormones: estrogens, progestins, and androgens. The term *estrogen* refers to any substance that produces vaginal cornification, whereas the term *progestin* refers to any substance that produces secretory changes in the estrogen-primed uterus. The estrogens and progestins bind to specific cytoplasmic receptors in their target cells, and the hormone-receptor complex is transferred to the cell nucleus and interacts with acceptor sites on the chromatin. This interaction is associated with increased ribonucleic acid (RNA) polymerase and with increases in messenger RNA (mRNA) and protein synthesis, the latter two leading to steroid-specific responses. Cells of the theca interna of the ovarian follicle and thecal lutein cells of the corpus luteum are the source of ovarian estrogens. Progestins are produced by luteinizing granulosa cells. The estrogens and progestins stimulate the growth and development of the female reproductive tract and exert a variety of metabolic effects of other tissues.

The human ovary secretes the estrogens estradiol and estrone; a third compound with estrogenic activity, estriol, is a metabolic end product of estradiol and estrone. All the steroid hormones, regardless of their site of origin, are derived from cholesterol. The compound 17α-hydroxyprogesterone plays a pivotal role in the ovarian synthesis of estradiol and estrone. This becomes apparent if one refers to the pathways of adrenal and testicular steroid biosynthesis. The pathways of ovarian estrogen and progestin biosynthesis are depicted in Figure 35-3. The ovarian estrogens are secreted into the bloodstream, where they may bind with testosterone-estradiol–binding globulin, the same globulin that binds testosterone. It has less affinity for estradiol than for testosterone, and approximately 30 percent of the circulating estradiol remains in the "free" form and is in equilibrium with testosterone-estradiol–binding globulin. The liver inactivates estrogens. The important role of the liver in estrogen inactivation is further emphasized by the fact that hyperestrogenism accompanies certain types of hepatic disorders. It results from failure of hepatic removal of estrogens from the circulation. The estrogens undergo oxidation in the liver to form estriol, and the liver conjugates each of the estrogens with sulfate or glucuronide. Large amounts of conjugated estrogens are excreted in the bile, and small amounts appear in the urine.

Progesterone is secreted into the plasma by luteal tissue during its short life span. The circulating level of progesterone, like that of the estrogens, is very low. It is rapidly metabolized by the liver to *pregnanediol* and *pregnanolone*. Only a small amount (less than 10 percent) of administered progesterone can be recovered in the urine as pregnanediol or pregnanolone. Both appear in the urine as free compounds, but mostly they are conjugated with glucuronide. The excretion of progesterone metabolites will be considered further when the reproductive cycle, pregnancy, and lactation are discussed.

EFFECTS OF ESTROGENS. The estrogens have a striking influence on the cells of the female reproductive tract, and hardly a tissue in the body is unaffected by them. Their effects on the genital tract of the female are most dramatically observed when estrogens are administered to an animal following ovariectomy.

Vagina. The vaginal epithelium of either an immature animal or an animal following ovariectomy is very thin and shows marked atrophic changes. Estrogen administration brings about a sharp increase in mitotic activity and stratification of the vaginal epithelial cells. Cornification of the epithelial cells results from a loss of blood supply to the superficial cells, which occurs because of the rapid growth (increase in thickness) of the vaginal epithelium. Estrogens increase the glycogen and mucopolysaccharide content of the vaginal mucosa, decrease the pH (to about 4.5), increase the vascularity of the structure, and produce a slight degree of edema. The appearance of cornified cells in a vaginal smear or lavage is diagnostic of estrogen action.

Uterus. Estrogen administration causes proliferation of the uterine cervical mucosa and the secretion of alkaline mucus from the cervical glands. The estrogens also have a pronounced effect on the endometrium, inducing endometrial mitoses (hyperplasia), increasing the height of the lining epithelium (hypertrophy), increasing the blood supply and capillary permeability of uterine vessels, and increasing the uterine water and electrolyte content. They also enhance uterine anaerobic and aerobic glycolysis and increase the rate of synthesis of uterine RNA and deoxyribonucleic acid (DNA), protein, and glycogen. Thus, the uterus increases in weight and size on account of marked proliferative changes following estrogen administration. Contrarily, ovariectomy results in definite atrophic changes in all uterine tissue, particularly the endometrium and myometrium.

Following ovariectomy, an animal's smooth muscle is relatively inactive. Estrogen administration increases myometrial contractility and motility and the sensitivity of uterine smooth muscle to oxytocin. These changes result from the ability of estrogens to increase the content of the myometrial contractile substance, actomyosin, and a greater isometric tension can therefore be developed. Action potentials are rarely observed in uterine muscle taken from an animal that has had an ovariectomy, but the frequency increases following estrogen administration and brings about greater myometrial activity. It has also been shown that the membrane potential of uterine muscle is elevated by estrogen over that of the castrate. This effect of estrogen is dependent on the presence of Ca^{2+}. Estrogens appear to set the membrane potential at a critical level, so that the myometrial cells show an increased excitability (decreased threshold), increased spontaneous activity, and increased sensitivity (reactivity) to appropriate stimuli.

Estrogen administration also stimulates proliferation of the epithelium and smooth muscle of the uterine tubes or the oviducts; the effects here are similar to those on the body of the uterus. Estrogen administration can lead to failure of implantation of the ovum, largely because of its effects on the musculature of the oviduct; estrogen stimulates the smooth muscle of the oviduct to increase the peristaltic action in the direction from the uterus to the ovary. This may be of benefit to sperm transport, but if the activity is too great, it prevents the passage of the ovum down the fallopian tube to the uterus.

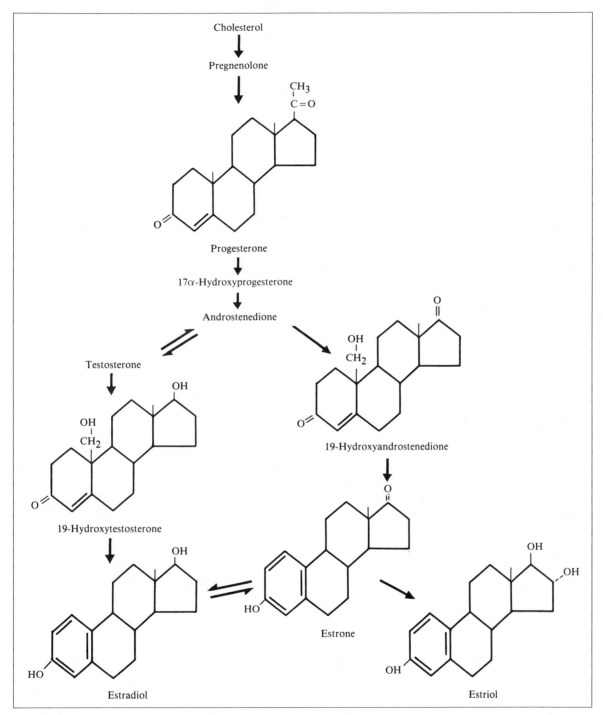

Fig. 35-3. Major pathways for the biosynthesis of ovarian estrogens
and progestins.

Metabolism. Estrogen administration results in renal retention of Na^+ and Cl^- and decreased urine volume, with subsequent increases in blood volume, extracellular fluid (ECF) volume, and body weight. The action of estrogen on renal Na^+ handling is independent of changes in renal hemodynamics and the secretions of the adrenal cortex. Estrogens also promote $1,25(OH)_2D_3$ formation in the kidney and increase thyroxine-binding globulin and transcortin production in the liver.

The estrogens also promote protein synthesis and affect skeletal growth by increasing osteoblast activity and the rate of deposition of calcium and protein in bone. Bone growth (length) is limited by estrogens because they promote closure of the epiphyseal plates, which are the growth centers of bone.

Secondary Sex Characteristics. The secondary sex characteristics of the female are dependent on the estrogens. Estrogens act on the mammary gland to induce growth and proliferation of stromal tissue. The tendency in the female for adipose tissue to concentrate in the buttocks, hips, thighs, and mammary gland, plus the presence of a uniform layer of subcutaneous fat over the body, leads to a rounded contour with varying convexities and results in a form that is distinctly less angular than that of the male. The female skeleton is smaller than that of the male, with marked differences in the pelvis, which is adapted in the female to aid in the carriage and delivery of young. The hair of the body is fine, and scalp hair grows faster and is more permanent. Ovariectomy in experimental animals abolishes mating responses and sexual behavior patterns. Libido in the human female is relatively unaffected by ovariectomy in the adult, but it may be decreased if ovariectomy is performed before puberty.

Each of the effects of estrogen on the reproductive tract and mammary gland may be considered indicative and diagnostic of puberty, because the estrogen secretion heralds the onset of ovarian endocrine activity. The primary point is that estrogens cause proliferative changes in the female reproductive system. The role of estrogen in cyclic reproductive behavior will be discussed later.

EFFECTS OF PROGESTINS. The effects of progesterone, per se, are never observed under physiological conditions and are not normally apparent until after the actions of estrogens have been operative. The effects of estrogens are generalized as being proliferative, whereas progesterone acts on proliferated tissue to differentiate it into a secretory type of tissue. Progesterone acts on the reproductive tract and mammary glands in such a manner as to prepare the tract for implantation of the fertilized ovum and to maintain gestation and lactation.

Reproductive Tract. Progesterone transforms the cornified vaginal epithelium to a mucified condition and decreases cervical secretions. The progestational response of the endometrium, which is a result of priming with estrogen followed by progesterone stimulation, is characterized by further endometrial growth, including thickening of the epithelium and accumulation of water. The secretory changes induced by progesterone include enlargement of endometrial stromal cells and the formation of tortuous glands with marked glycogen synthesis and deposition. Thus, the highly secretory endometrium provides the environment that is necessary for implantation of the blastocyst. The effect of progesterone on the myometrium assures that expulsion of the implanted blastocyst does not occur. The membrane potential of the uterine smooth muscle is further elevated, resulting in a decrease in its excitability. The hyperpolarized membrane of the progesterone-dominated smooth muscle cell of the uterus shows poor conduction, lowered spontaneous activity, and reduced sensitivity to specific stimuli. It follows that the progesterone-dominated uterus cannot participate in the organized contractions that result in expulsion of the fetus. Progesterone also results in decreased smooth muscle activity in the oviducts. It acts on the estrogen-primed mammary gland and stimulates lobuloalveolar growth, which is the hallmark of the mature mammary gland.

Body Temperature. Variations in the awakening basal body temperature occur and have been correlated with cyclic ovarian activity. The phase of the cycle during which estrogen is dominant is characterized by a low basal body temperature. At ovulation time, a slight drop in temperature occurs, but when progesterone secretions rise, the body temperature is elevated over that observed during the early stage of the cycle (Fig. 35-4). The basal body temperature remains elevated as long as progesterone secretion is maintained. The mechanism of this action is not understood. It should be obvious, because of the small temperature changes, that even minor infections that elevate the body temperature obscure the cyclic variations in basal body temperature.

Miscellaneous Effects. Progesterone, because of its key position in steroid biosynthetic pathways, might be expected to mimic the effects of a variety of steroids. It possesses adrenocorticoid activity in that it prolongs the survival of adre-

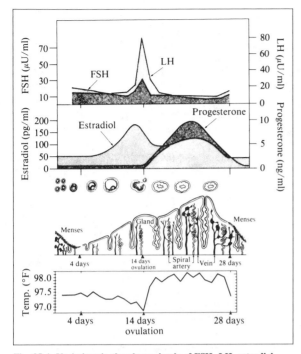

Fig. 35-4. Variations in the plasma levels of FSH, LH, estradiol, and progesterone throughout the human menstrual cycle. Changes in the ovarian elements, follicles, and corpus luteum, an endometrial morphology, and basal body temperature are also shown.

nalectomized animals. It has slight protein catabolic activity and androgenic activity, and aids in the maintenance of testicular weight in hypophysectomized animals. It also tends to induce a slight hyperventilation, which results in a lowering of the PCO_2 of arterial blood.

REPRODUCTIVE CYCLE

The cyclic variations in reproductive activity occur as the result of neural and hormonal interaction between the CNS, hypothalamus, the AP, ovaries, and reproductive tract. The normal cycle is characterized by maturation of ovarian follicles, ovulation, corpus luteum formation, and preparation of the reproductive tract for the implantation of a fertilized ovum. If conception and implantation do not occur, these processes are repeated. The hypothalamic hormone GnRH, the pituitary hormones FSH and LH, and the ovarian hormones estradiol and progesterone are all involved in inducing the changes in cyclic activity at all levels of the complex system. The relationships between these organs and hor-

mones during the menstrual cycle are depicted in Figure 35-4.

HORMONAL CHANGES AND THEIR REGULATION. When one considers the first day of the human menstrual cycle to be coincident with the onset of menstruation, it is clear that plasma gonadotropic hormone levels during the cycle are characterized by midcycle peaks of both FSH and LH immediately prior to ovulation. The plasma level of LH is relatively constant during the early part of the cycle, prior to the peak, and slightly higher than that observed during the latter half of the cycle. The plasma LH level tends to decline gradually after ovulation, but as menstruation approaches, a modest increase occurs in circulating LH. The plasma FSH pattern during the first half of the cycle shows a moderate early elevation followed by a slight decline until the midcycle peak. After the peak the FSH concentration gradually decreases, but it starts to increase a few days prior to menstruation. It has been observed that both FSH and LH plasma levels are not maintained in a steady manner throughout any given day of the cycle; in fact, the plasma concentrations of both are actually maintained by frequent surges of secretion by the pituitary that are much more vigorous during the midcycle peaks. These surges in gonadotropin secretion are in response to surges of GnRH secretion by the hypothalamus.

The concentrations of estradiol in the plasma follows a biphasic pattern, with low concentrations during the early and late phases of the cycle. The plasma estradiol level rises rapidly and peaks prior to the surge in FSH and LH secretion. The concentration declines during the time of ovulation and then shows a second gradual rise and fall prior to menstruation. The plasma progesterone concentration, in contrast to the plasma estradiol level, remains very low during the first half of the cycle. It then increases markedly after ovulation and corpus luteum formation, remains high for several days, and decreases to a low level prior to menstruation. The changes in plasma concentration of the four hormones are shown in Figure 35-4. Not included in the figure are the 17α-hydroxyprogesterone levels during the cycle. The plasma concentration of this steroid is very low during the first part of the cycle and rises at midcycle along with FSH and LH. It then follows the same general pattern observed with progesterone. The observation that the ovaries of animals whose hypophysis has been removed remain immature, but that gonads of such animals can be restored or maintained with pituitary extracts, particularly the gonadotropins, is taken as proof that the pituitary regulates ovarian function. The pituitary that is removed from its normal site and transplanted to some extrasellar site is incapable of maintaining normal cyclic gonadal function.

As its name indicates, FSH stimulates growth of the ovarian follicle, which includes hyperplasia of the granulosa cells, enlargement of the antrum, and a marked increase in ovarian weight. These changes occur when FSH is administered to immature rats after hypophysectomy. Since changes in the reproductive tract indicative of estrogen secretion do not occur, it is concluded that FSH alone is incapable of stimulating estrogen secretion. If after hypophysectomy the immature animal is treated with LH, the ovarian interstitial cells are stimulated, but no estrogen secretion results. However, the combination of FSH and LH notably increases ovarian weight, follicle growth, and estrogen secretion. The rate of follicular growth and estrogen secretion increases under the influence of FSH and LH. The surge of LH release at midcycle induces ovulation and causes luteinization of the granulosa and theca interna cells of the ovulated follicle. The secretion of progesterone and estradiol by the corpus luteum is stimulated by LH, and present studies indicate that the secretion of small amounts of LH is required for the maintenance of a functional corpus luteum in humans. Once formed, the human corpus luteum appears to have a preset life span of approximately 12 days.

Various studies indicate that the ovarian steroids estradiol and progesterone regulate gonadotropic hormone secretion by the AP by exerting both negative and positive feedback controls over both FSH and LH secretion. The type of feedback control exerted by the steroids appears to be dependent on their concentrations in plasma. Early investigations showed that removal of the ovaries results in marked increases in gonadotropic hormone secretion by the AP. Similarly, after the termination of reproductive function, i.e., at menopause in the human female, the plasma estradiol concentration is minimal and gonadotropin levels are high. In both instances, the administration of estradiol lowers plasma FSH and LH concentrations. Thus, the negative feedback effect of estradiol can account for the relatively low levels of FSH and LH observed throughout most of the cycle. It has been shown that single or multiple injections of estrogens to anovulatory women can induce the surge of LH secretion and induce ovulation. Progesterone blocks this estrogen-induced surge. On the basis of the evidence, it has been concluded that rising plasma estradiol levels immediately prior to midcycle exert a positive feedback effect and thus trigger massive bursts of GnRh and LH secretion.

As the functional activity of the corpus luteum diminishes, the negative feedback effect of estradiol on FSH and LH is diminished slightly, thereby permitting modest increases in FSH and LH secretion. A new hormonal cycle can then begin. In the event of pregnancy, however, the functional life of the corpus luteum is prolonged, and cyclic activity ceases throughout the duration of the pregnancy.

As a result of observing the effects of emotional states, malnutrition, light, and temperature on gonadal function, modulation of gonadal activity by the CNS has been obvious for many years. It is well known that several species of animals ovulate only after appropriate stimulation, e.g., coitus. These species have been referred to as the induced or reflex ovulators.

There can be little doubt that the functional integrity of the AP-ovary system is dependent on a control exercised by the nervous system. The nervous control over the AP secretion of FSH and LH is exerted via the secretion of GnRH from the hypothalamus. This hormone is a decapeptide and has been synthesized. The feedback effects of estradiol are directed at the hypothalamus and probably not at the pituitary. Similarly, the two gonadotropins, FSH and LH, appear to operate through feedback mechanisms at the hypothalamic level to inhibit gonadotropin release.

Available evidence indicates that the hypothalamus acts as a center that integrates brainstem and spinal reflexes into patterns of behavior observed in mating in the female. The activity of the hypothalamus may be modified by influences from the neocortex, rhinencephalon, and brainstem reticular formation.

CYCLIC CHANGES IN THE REPRODUCTIVE TRACT. The structural and functional changes in the reproductive tract are well known and are repeated without much variation through the years. Changes in uterine and vaginal function are directly regulated by the ovarian estrogens and progestins. The changes that occur during the preovulatory stage or early part of the cycle depend upon estrogen secretion, whereas the postovulatory stage is influenced by estrogen and progesterone. Thus, the early part of the cycle is characterized by follicle development and proliferation of the reproductive tract and has been termed the proliferative or follicular phase of the cycle. During this phase the vaginal epithelium undergoes rapid growth, and the vaginal smear is eventually transformed into one dominated by cornified cells. The uterus begins to accumulate fluid, becomes highly contractile, and undergoes proliferative changes under the influence of estrogen until the time of ovulation. Following ovulation and corpus luteum formation, reproductive tract function is altered by progesterone.

The phase of the cycle that follows ovulation is called the secretory or luteal phase. During this phase the vaginal epithelium is infiltrated with leukocytes, and the vaginal smear is characterized by the presence of leukocytes, mucin, and cornified cells. The vaginal epithelium becomes thin, and leukocytes are abundant, so that the vaginal smear contains nucleated epithelial cells and leukocytes. Immediately after ovulation the uterus diminishes in vascularity and contractility. The increasing titer of progesterone transforms the proliferated endometrium into an organ with highly coiled glands that have the ability to secrete large amounts of glycogen. Thus, the uterus is prepared for nidation. Following regression of the corpus luteum a new set of follicles is stimulated by the rising levels of FSH, and the cycle is repeated. Cyclic variations in the activity of the fallopian tubes, or oviducts, and in the uterine cervix are also observed, and these changes are dependent on the circulating estrogens and progesterone.

The reproductive cycle differs among species in several ways. In general, the nonprimate mammalian species exhibit a period of sexual activity during a particular phase of the cycle. This period coincides with the time of maximal estrogen secretion prior to ovulation. The period of sexual receptivity, estrus, or "heat," is characterized by behavioral changes designed to attract the male at the time when ova are most easily fertilized.

The reproductive cycle in the primates is called the menstrual cycle, and in the normal nonpregnant female the interval that extends from the onset of a period of uterine bleeding to the onset of the next period of bleeding describes its duration. The mean cycle length is 28 days, but the range is very large, from 20 to 35 days.

The menstrual cycle has been conveniently divided into three stages on the basis of the endometrial histological picture. The first stage, menstruation or menses, lasts about 4 to 6 days and is characterized by the occurrence of hemorrhage in the endometrial stroma. The endometrium degenerates, and together with interstitial blood, is sloughed into the uterine lumen and exits through the vagina as menstrual discharge. The second or proliferative stage is characterized by repair and proliferation of the endometrium and proceeds under the influence of estrogens during the next 8 to 10 days. The endometrium undergoes marked changes that provide the appropriate environment for the implantation of the fertilized egg during the third or secretory stage (lasting about 9–10 days). This final stage is remarkably constant in duration and dependent on ovulation and the secretion of progesterone by the corpus luteum. If a fertilized ovum is not available, or if implantation does not occur, the corpus luteum and the endometrium degenerate, and the cycle is repeated.

Menstruation should not be considered an actively induced process, because it occurs as a result of cessation of stimulation of the endometrium by the ovarian hormones. It is a result of withdrawal of the ovarian steroids, as shown by the following observations: (1) Ovariectomy, if performed during the last 2 or 3 weeks of the cycle, precipitates uterine bleeding 2 to 6 days postoperatively. (2) The administration of estrogens following ovariectomy results in a proliferative type of endometrium, and the withdrawal of estrogen treatment results in menstruation. (3) In the absence of ovaries, simultaneous administration of estrogen and progesterone causes the development of a secretory endometrium, and the withdrawal of both steroids results in menstruation within 2 to 3 days. (4) The administration of large amounts of estrogen prevents menstruation for long periods of time. Figure 35-4 depicts the changes in the endometrium throughout the menstrual cycle.

Transplantation of the endometrium to the anterior chambers of the eye has enabled investigators to observe directly the factors that are associated with menstruation. The observations have shown that alterations in blood flow through the endometrial vessels are responsible for menstruation. Prior to menstruation (2–6 days), the coiled arteries deep in the endometrium undergo slight constriction and offer increased resistance to blood flow, and because of the inadequate blood flow the endometrium regresses and becomes necrotic. Immediately prior to menstruation, the coiled arteries constrict further and reduce, to a greater degree, the flow to the endometrium. Then, after varying intervals, the arteries dilate and hemorrhage occurs. The arteries then constrict again, and hemorrhage ceases from the artery involved. It should be remembered that each artery bleeds only once each cycle; thus, each artery dilates, bleeds, then constricts, but not all coiled arteries do so simultaneously, and a small area may be sloughed and repaired before other areas are sloughed. Therefore, as a result of decreasing estrogen and progesterone secretion, the endometrium is deprived of essential materials, and it regresses. The spiral arteries undergo intermittent contraction and cause recurrent ischemia, so that small hemorrhages occur in the tissue, cellular necrosis occurs, and menstruation ensues.

CONTROL OF FERTILITY. Currently, much attention is being given to the study of the control of fertility and family planning. Modern methods of contraception include the use of oral contraceptives (the "pill") and intrauterine devices (IUD). The oral contraceptives in use today consist of synthetic estrogenic and progestational steroids. The combination of synthetic hormones is taken cyclically, one pill daily, for 21 consecutive days of each 28-day menstrual cycle. The steroids act on the CNS and hypophysis to prevent the midcycle surge of GnRH, FSH, and LH. The pill prevents ovulation and corpus luteum formation, and the ovaries tend to atrophy. Because of the direct stimulating action of the steroids, the endometrium does not undergo atrophy. On withdrawal of the steroids, the endometrium breaks down, menstruation ensues, and a new cycle can be induced. These substances also act on the endometrium and myometrium to create an environment that is hostile to fertilization and implantation. The pill has a high degree of effectiveness in preventing ovulation and thus pregnancy.

Since primitive times, intrauterine devices have been used in an attempt to control conception. Today, new designs and materials have made such devices effective. The IUD does not prevent ovulation but acts by creating an intrauterine environment that is hostile to fertilization of the ovum, or implantation of the blastocyst, or both. Well-designed and unbiased studies have shown that both the pill and the IUD are very effective contraceptive agents that produce only minimal side effects.

PREGNANCY

Pregnancy may occur only after a series of events that terminate in the oviducts, where zygote formation takes place as a result of the union of an ovum and a sperm. The ovum of the human may remain viable in the female reproductive tract as long as 36 hours. It has been estimated that in women of proved fertility, only one in four conceives when coitus with a fertile male is completed during her midcycle, or fertile period. Sperm deposited in the vagina must be transported up the reproductive tract to the oviducts. In many species, sperm appear in the oviducts within several minutes after having been deposited in the vagina during midcycle, or estrus. The rapid transport of sperm through the uterus and into the oviducts is the result of mechanical propulsion provided by contraction of the uterine musculature. On the basis of work in experimental animals, it has been suggested that the increase in uterine muscle activity is caused by the reflex release of oxytocin from the hypothalamoneurohypophyseal system, or by the presence of prostaglandins in the semen, or by both.

Fertilization occurs in the ampullary portion of the oviduct and is generally accomplished within 15 to 20 hours after ovulation. The muscular and ciliary activity of the oviduct propels the zygote toward the uterus, and by the fourth day after ovulation it is in a 12- to 16-cell stage and reaches the uterine cavity. Generally, by the seventh day the cell mass, or blastocyst, makes contact with and adheres to the uterus and penetrates and passes through the uterine epithelium, which soon covers it.

The mechanism that causes the corpus luteum to persist is the result of the action of hormone secreted by the embryonic trophoblast. This hormone, human chorionic gonadotropin (HCG), is luteotropic and is responsible for preventing the demise of the corpus luteum. Because of persistent luteal function the synthesis and secretion of estrogen and progesterone continue, the endometrium fails to undergo its usual menstrual regression, menstruation does not occur, and overt cyclic ovarian and uterine activity is absent throughout the pregnancy. Pregnancy can be established and maintained only if adequate amounts of progesterone are secreted prior to and throughout the course of that pregnancy.

The placenta serves as an organ of exchange between mother and fetus. It also serves as an endocrine organ, secreting at least four types of hormones: the two gonadotropins, HCG and human placental lactogen (HPL), estrogens, and progesterone. Human chorionic gonadotropin is a glycoprotein and is structurally similar to LH and mimics many of its effects; HPL, also called placental somatomammotropin, is structurally similar to prolactin and growth hormone. It has actions on its target cells similar to those of prolactin and growth hormone. In addition, other steroidal substances, such as cortisol and aldosterone, are secreted by the placenta. The placenta may also secrete TSH-like and ACTH-like polypeptides. The first to appear is HCG, which is generally detected in the plasma about 8 days after presumed ovulation. It maintains the functional activity of the corpus luteum during the early stages of pregnancy and reaches a peak circulating level of approximately 10 to 16 μg per milliliter at about 9 to 11 weeks. The plasma levels of HCG throughout pregnancy are shown in Figure 35-5. The presence of HCG serves as the basis for all of the pregnancy tests. Also depicted in Figure 35-5 are the plasma levels of HPL. This hormone has been detected in the trophoblast as early as the third week of gestation, can be detected in the plasma by the sixth week, and attains a level of 8 to 12 μg per milliliter prior to parturition. The rising HPL levels parallel the increase in placental weight throughout the gestation period.

The endocrine function of the ovary can be replaced by hormonal products of the placenta early in the second month of human pregnancy, as shown by the fact that bilateral ovariectomy may be performed during the second month of gestation, yet pregnancy continues to a successful termination at the end of the fortieth week. The levels of urinary estrogens and pregnanediol increase gradually throughout pregnancy and reach a maximum a few days before parturition. The estrogen level increases to about 150 to 175 ng per milliliter of plasma, and the plasma progester-

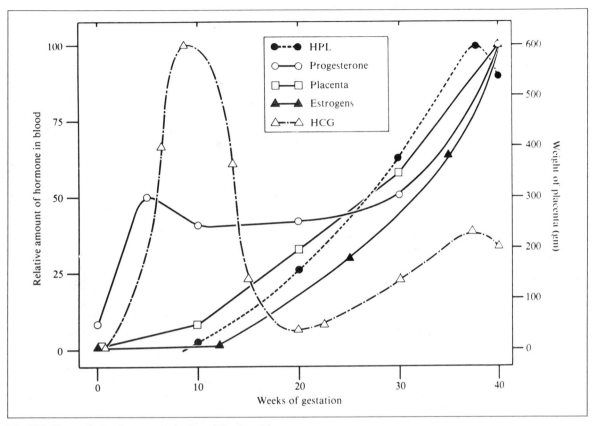

Fig. 35-5. Changes in the plasma concentrations of the placental gonadotropins HCG and HPL and estrogens and progesterone throughout normal human pregnancy. The placental weight is also depicted.

one concentration may increase up to 200 mg per milliliter prior to parturition (see Fig. 35-5).

The fetus and the placenta complement each other in the synthesis of certain steroid metabolites, for neither is capable of completing the entire biosynthetic pathway alone. Many steroid metabolites produced during pregnancy result from an interplay between the fetus, the placenta, and the mother. Fetal well-being can be assessed by measuring the metabolites that arise primarily from the fetus. One such metabolite is estriol. The fetal adrenal cortex produces large amounts of the 19-carbon steroids dehydroepiandrosterone sulfate and 16α-hydroxydehydroepiandrosterone sulfate. The latter is transported to the placenta via the placental arteries and converted to estriol. The mother contributes only small amounts of 18- and 19-carbon steroids to the placental pathway for the synthesis of estriol. This unit is termed the *fetal-placental unit*. Figure 35-6 illustrates the major pathway of estriol sythesis in the fetal-placental unit and the distribution of estriol in the fetus, amniotic fluid, placenta, and mother.

Thyroxine-binding globulin and transcortin levels also increase during pregnancy. The increases in the binding proteins reduce the amount of free circulating thyroxine (T_4) and cortisol and indirectly lead to increases in the rates of secretion of TSH and ACTH. However, euthyroid and eucortical states are maintained after a new equilibrium is established.

The uterus must serve several related functions during pregnancy. Initially, it must be prepared to receive the fertilized ovum and allow implantation. Once implantation has occurred, it must grow and adapt to the growing product of conception, and it must be prepared to assume delivery at term.

In the nonpregnant state the uterus is thick walled, weighs about 60 gm, is about 7 cm in length, and is contained within the pelvis. At the end of pregnancy it is thin

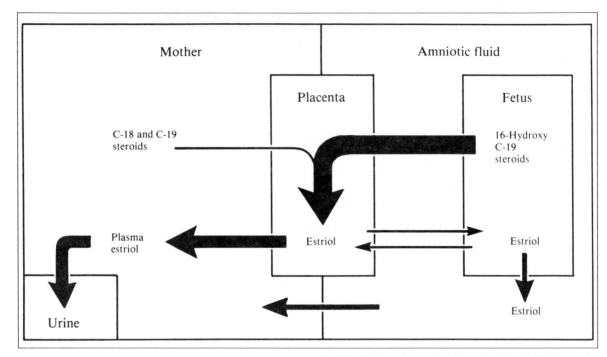

Fig. 35-6. Major pathway of estriol synthesis by the fetal-placental unit in the human. The width of the arrow indicates the relative importance of the pathway.

walled, weighs more than 1 kg, is about 35 cm long, extends up to and elevates the diaphragm, and may contain a total volume of up to 5 liters. Thus, the pregnant state imposes widespread changes on the maternal organism. Nearly all of them represent adaptive responses that permit the mother to deliver a normal infant without harm to either her or her child.

In order to supply adequate amounts of oxygen and other nutrients to the fetus, the blood flow through the gravid uterus must increase. Studies in humans indicate that uterine blood flow at term in a single pregnancy is about 500 ml per minute and in a twin pregnancy, over 1000 ml per minute. On the other hand, 24 to 48 hours after delivery the blood flow decreases to a rate close to that observed in the nonpregnant state. Vasomotor fibers to the uterus are thoracolumbar in origin; their stimulation results in constriction of the vessels they supply. However, the dominant control over the uterine vessels is exercised by hormones; for instance, the administration of estrogen induces uterine hyperemia prior to any metabolic changes in the uterine tissue. With increased blood flow to the pregnant uterus, various adaptive changes occur in the cardiovascular system, and the uterine blood flow is maintained so that oxygen is supplied in excess of the oxygen requirement of the tissue.

The mother undergoes numerous metabolic changes in response to the demands of the growing fetus and placenta. At the end of an average pregnancy the fetus weighs about 3.5 kg, the placenta weighs 500 to 600 gm, and there is about 1 liter of amniotic fluid. The total blood volume is increased about 1.5 liters in excess of that observed in the nonpregnant state. The increased retention of water, fetal growth, and mechanical factors place marked demands on the cardiovascular and urinary systems and nutritional status of the mother.

The cardiovascular system adapts by increasing cardiac output, heart rate, and pulse pressure. The cardiac output, however, is significantly reduced when the pregnant woman is in the supine position, because the venous return is reduced as a result of compression of the inferior vena cava and iliac veins by the uterus. An increased femoral venous pressure contributes to the dependent edema often seen during the late stages of pregnancy. An increase in cutaneous blood flow is observed, and this contributes to the loss of excess heat generated by the increased metabolic rate.

Both glomerular filtration rate, and renal plasma flow increase during pregnancy. Late in pregnancy, both of these variables, urine volume, and sodium excretion are markedly affected by posture, with significant decreases in each occuring with a change from the lateral recumbent position to the supine or upright position. The hemodynamic changes are related to the reduced venous return to the heart, while the changes in water and electrolyte excretion are probably related to the decreased glomerular filtration rate and reflex increases in antidiuretic hormone and aldosterone respectively. As the uterus enlarges during late pregnancy, ureteral compression may occur at the pelvic rim, and the urinary bladder is pushed forward and upward, tending to impair venous and lymphatic drainage.

The metabolic demands placed on the maternal system to supply needs for fetal, uterine, placental, and maternal growth are extensive. The metabolic rate and heat production increase about 25 percent as compared with the nonpregnant state. Most of the increase in oxygen consumption is required to meet the demands of the metabolic activity of the products of conception. On the average the total weight gain in pregnancy is in excess of 10 kg. The dietary requirement for protein and total calories for pregnancy increases about 10% over that in the nonpregnant state. The requirement for all vitamins increases 10 to 30 percent. The increased amounts of calcium, phosphorus, iodine, iron, and magnesium required for fetal growth and development must be supplied in the mother's diet. The oxygen demand is easily met by the respiratory system. The minute volume increases, and the hyperventilation tends to cause a respiratory alkalosis by lowering the plasma PCO_2. A slight reduction in the plasma bicarbonate compensates for the respiratory alkalosis. The functional residual capacity is reduced as a result of elevation of the diaphragm by the enlarging uterus. All other pulmonary function tests are within normal limits throughout the duration of pregnancy.

PARTURITION

The average duration of a normal human pregnancy is 10 lunar months, 280 days after the onset of the last menstrual flow, or approximately 265 days after ovulation and conception. The termination of a normal pregnancy, parturition or labor, depends on dilation of the cervix and alterations in the contractile activity of the uterine muscle from a state of quiescence to one of increased activity. Cervical dilation appears to result from mechanical pressure transmitted by uterine contractions. A variety of factors influence the activity of the contractile elements of the myometrial cells, and there is no general agreement as to the relative importance of each in initiating labor. However, once labor be-

gins, whatever the trigger, a series of events carries the delivery process to completion.

Some of the factors that influence myometrial activity are progesterone, estrogens, oxytocin, the prostaglandins, the autonomic innervation of the uterus, and mechanical factors. Several experimental observations have shown that progesterone plays a predominant role in the maintenance of pregnancy, and massive doses of progesterone may maintain pregnancy beyond term. The concept has been advanced that a progesterone block of uterine smooth muscle maintains pregnancy and that withdrawal of the block is responsible for the onset of parturition. This concept is based on the theory that the placenta gives up the bulk of its progesterone by direct diffusion to the myometrium. Under this circumstance, one would expect that the concentration gradient of progesterone would be greatest at implantation sites rather than at interplacental sites. The latter has been shown to be the case in experimental animals. The concept is further strengthened by observations in women who bear twins in a bicornuate uterus (the septum divides the uterus into two horns). The twins have separate placentas and can be born as much as several weeks apart. Thus, in the same woman, conditions at the same moment in pregnancy can be appropriate for the maintenance of pregnancy as well as for its termination, indicating that local effects of the various factors appear to predominate.

The estrogens increase the synthesis of high-energy phosphate and actomyosin in myometrial cells. They decrease the membrane potential of the contractile cells to the range in which the spontaneous discharge of action potentials occurs. The estrogens may also influence the myometrium by making it more sensitive to the actions of oxytocin and the prostaglandins. Further, they indirectly influence the uterus by increasing the release of oxytocin from the neurohypophysis or the release of prostaglandins from the uterus.

It is well known that the metabolic clearance of oxytocin decreases progressively throughout pregnancy, and that the sensitivity of the myometrium to a constant dose of oxytocin increases up to the thirty-sixth week of pregnancy. There is little doubt that the induction of labor can be accomplished by the administration of oxytocin after progesterone withdrawal and the ascendancy of an estrogen-dominated uterus. During parturition, some of the stimuli (in addition to the estrogens) that provoke the release of oxytocin arise primarily from mechanical stimulation of the uterus, cervix, and vagina.

Some prostaglandins are potent stimulators of myometrial activity and have been used clinically for the induction of labor and in therapeutic abortions. It has been shown that the uterine decidua is a source of prostaglandins in pregnancy. The concentrations of the prostaglandins in the uterus and in the maternal plasma rise prior to parturition.

Uterine activity in pregnancy can be increased through the sympathetic innervation of the uterus. The action of alpha-adrenergic (excitatory) activity is increased by high estrogen levels and by reduced progesterone levels, and the adrenergic fibers constitute the motor limb of a spinal reflex arc that is stimulated by cervical dilatation. Alpha-adrenergic activity also increases the sensitivity of the myometrium to oxytocin. The uterine innervation is not essential for normal delivery.

Mechanical stimuli influence the time of parturition. Increasing the volume of the uterus increases myometrial activity. Mechanical stimulation of the cervix and vagina augment uterine contractions, probably by stimulating oxytocin release. Disruption of the relationship between the uterine decidua and the placenta tends to stimulate prostaglandin release.

Labor is divided into three stages. Stage one begins with the first pain and ends when dilatation of the cervix is complete. Stage two is the stage of expulsion and ends with the birth of the baby. Stage three ends with delivery of the placenta. The factors that regulate the highly synchronized and integrated sequence of events constituting parturition or labor in the human are well known, but the precise functions and importance of each remain to be determined.

LACTATION

Lactation consists of mammogenesis (development of the mammary glands), lactogenesis (milk secretion), galactopoiesis (maintenance of lactation), and milk ejection. Mammary glands are modified sweat glands and are made up of 15 to 25 lobes. The immature gland comprises short ducts that radiate from the nipple to each lobe. Each duct has many side branches extending from it to supply the lobules. The estrogens are primarily concerned with growth of the duct system; following the administration of estrogen, the duct system becomes extensively arborized. Progesterone acts in conjunction with estrogens to stimulate duct and alveolar growth. Physiological doses of estrogen and progesterone produce full mammary growth in either the castrated male or the female. However, both the administration of estrogen plus progesterone to an animal following hypophysectomy and the administration of the pituitary hormones FSH and LH, which stimulate estrogen secretion by the ovary, fail to induce mammary development. The full mammary growth observed in late pregnancy is dependent on several hormones: estrogens, progesterone, prolactin, HPL, growth hormone, and cortisol. Thus, many hormones

are required to advance the mammary gland to full functional development. The hormonal environment at the end of pregnancy is such that the mammary gland is morphologically ready to begin lactogenesis.

Some synthesis and storage of milk may begin prior to parturition, but only at or shortly after parturition does a copious secretion and flow of milk occur. The integrity of the hypophysis is necessary for the initiation of milk secretion, because failure of lactation occurs following hypophysectomy, and the administration of extracts of the hypophysis provides a positive lactogenic stimulus. The increased secretion of milk at parturition is probably a result of diminished circulating titers of estrogen and progesterone that follow delivery of the placenta. The high levels of progesterone during pregnancy tend to inhibit lactogenesis, but the increasing levels of both prolactin and HPL tend to stimulate lactogenesis. Once the placenta is delivered, the check placed on prolactin release and lactogenesis by progesterone and estrogen is removed, and full lactogenesis becomes evident. Many hormones, such as growth hormone, thyroxine, cortisol, estrogen, and insulin, are required for milk secretion and its maintenance, but their exact roles remain obscure. Mechanical factors also play a prominent part in the maintenance of lactation, because as milk production begins, pressure in the glands rises and, if not relieved, reaches a point where milk secretion is retarded. As a result the mammary glands involute if and when nursing is terminated.

Galactopoiesis requires both adequate maternal nutrition and fluid intake and the permissive effects of numerous hormones. The maintenance of human lactation is influenced by many social and psychological factors. Two neuroendocrine reflexes ensure the secretion of prolactin and oxytocin and aid in the maintenance of lactation.

In the nonpregnant female the fasting plasma PRL level averages about 10 ng per milliliter and does not vary with the menstrual cycle. Essentially the same concentration is observed in the male. During pregnancy the plasma PRL increases, and postpartum levels at the onset of nursing are about three times the average for nonnursing females. Marked increases in plasma PRL are induced by suckling or by manual stimulation of the breast. The response to suckling is prompt (within 10 min), and plasma levels as high as 250 to 300 ng per milliliter are attained within 30 minutes of the onset of suckling. Spinal cord transection between the last thoracic and first lumbar vertebrae leads to denervation of the caudal nipples of the rat, but leaves the more cranial nipples innervated. If the cranial nipples are covered so as to force suckling from the denervated nipples, milk cannot be removed from these glands, and the glands involute. However, if suckling is permitted from the intact nipples, full lactation and milk removal occurs from the denervated nipples. Thus, suckling induces PRL secretion over a neural pathway.

Suckling also stimulates the secretion of oxytocin from the posterior pituitary, via a reflex. This reflex, the milk ejection, or letdown, reflex, is composed of sensory stimuli associated with suckling acting on receptors in the nipples of the mammary gland. Impulses are carried centrally, where they facilitate the release of oxytocin from the HNS (see Chap. 30). Oxytocin acts as the efferent limb of the reflex arc and is carried via the plasma to the mammary gland, where it causes contraction of myoepithelial cells that surround the alveoli of the mammary gland. The milk contained within the alveoli is forced out into the duct system and cisterns of the mammary gland, from which it can then be removed by suckling. This reflex may be easily conditioned in either experimental animals or humans. On the other hand, it may be inhibited either by emotional factors or by the administration of epinephrine. In vitro experiments have demonstrated that the effect of epinephrine is via its vasoconstricting action. The effect of emotional stress on the suckling reflex may be mediated via epinephrine release.

In summary: Lactation is the result of the action of many hormones and neural elements. The mammary gland develops through the action of estrogens and progestins, and full secretory function is attained through the action of PRL, STH, TSH, ACTH, and probably insulin. Full lactation and milk secretion are maintained only in animals with intact innervation of the mammary gland and the neurohypophysis, and active stimulation of the mammary gland through suckling triggers the milk letdown reflex by stimulating oxytocin secretion.

BIBLIOGRAPHY

Bardin, C. W., and Catteral, J. F. Testosterone: A major determinant of extragenital sexual dimorphism. *Science* 211:1285–1293, 1981.

Bardin, C. W., Musto, N., Gunsalus, G., Kotite, N., Cheng, S.-L., Larrea, F., and Becker, R. Extracellular androgen binding protein. *Annu. Rev. Physiol.* 43:189–198, 1981.

Boyar, R. M. Control of the onset of puberty. *Annu. Rev. Med.* 29:509–520, 1978.

Chan, L., and O'Malley, B. W. Mechanism of action of the sex steroid hormones. *N. Engl. J. Med.* 294:1322–1328, 1372–1381, 1430–1437, 1976.

Droegmueller, W., and Bressler, R. Effectiveness and risks of contraception. *Annu. Rev. Med.* 31:329–343, 1980.

Espey, L. L. Ovulation as an inflammatory reaction—a hypothesis. *Biol. Reprod.* 22:73–106, 1980.

Frantz, A. G. Prolactin, *N. Engl. J. Med.* 298:201–207, 1978.

Griffin, J. E., and Wilson, J. D. The syndromes of androgen resistance. *N. Engl. J. Med.* 302:198–209, 1980.

Hsueh, A. J. W., and Jones, P. B. C. Gonadotropin-releasing hormone. *Annu. Rev. Physiol.* 45:83–94, 1983.

Lee, P. A., Jaffe, R. B., and Midgley, A. R., Jr. Serum gonadotropin, testosterone, and prolactin concentration throughout puberty in boys. *J. Clin. Endocrinol. Metab.* 39:664–672, 1974.

Leong, D. A., Frawley, L. S., and Neill, J. D. The neuroendocrine control of prolactin secretion. *Annu. Rev. Physiol.* 45:109–127, 1983.

Lipsett, M. B. Physiology and pathology of the Leydig cell. *N. Engl. J. Med.* 303:682–688, 1980.

McCann, S. M. Luteinizing hormone releasing hormone. *N. Engl. J. Med.* 296:797–802, 1977.

Means, A. R., Dedman, J. R., Tash, T. S., Tindall, D. J., van Sickle, M., and Welsh, M. J. Regulation of the testis Sertoli cell by follicle stimulating hormone. *Annu. Rev. Physiol.* 42:59–70, 1980.

Nalbandov, A. V. *Reproductive Physiology* (3rd ed.). San Francisco: Freeman, 1976.

Noel, G. L., Suh, H. K., and Frantz, A. G. Prolactin release during nursing and breast stimulation in post-partum and nonpospartum subjects. *J. Clin. Endocrinol. Metab.* 38:413–423, 1974.

Richards, J. S. Maturation of ovarian follicles: Actions and interactions of pituitary and ovarian hormones on follicular cell differentiation. *Physiol. Rev.* 60:51–89, 1980.

Segal, S. J. The physiology of human reproduction. *Sci. Am.* 231:52–62, 1974.

Shiu, R. P. C., and Friesen, H. G. Mechanism of action of prolactin in the control of mammary gland function. *Annu. Rev. Physiol.* 42:83–96, 1980.

Simpson, E. R., and McDonald, P. C. Endocrine physiology of the placenta. *Annu. Rev. Physiol.* 43:163–188, 1981.

Thornburn, G. D., and Challis, J. R. G. Endocrine control of parturition. *Physiol. Rev.* 59:863–918, 1979.

Topper, Y. J., and Freeman, C. S. Multiple hormone interactions in the developmental biology of the mammary gland. *Physiol. Rev.* 60:1049–1108, 1980.

Wilson, J. D., George, F. W., and Griffin, J. E. The hormonal control of sexual development. *Science* 211:1278–1284, 1981.

Yen, S. S. C. Gonadotropin releasing hormone. *Annu. Rev. Med.* 26:403–417, 1975.